Atomic Masses of the Elements

Name	Symbol	Atomic Number	Atomic Mass	Name	Symbol	Atomic Number	Atomic Mass
Actinium	Ac	89	227	Mercury	Hg	80	200.6
Aluminum	Al	13	26.98	Molybdenum	Mo	42	95.94
Americium	Am	95	243	Neodymium	Nd	60	144.2
Antimony	Sb	51	121.8	Neon	Ne	10	20.18
Argon	Ar	18	39.95	Neptunium	Np	93	237
Arsenic	As	33	74.92	Nickel	Ni	28	58.69
Astatine	At	85	210	Niobium	Nb	41	92.91
Barium	Ba	56	137.3	Nitrogen	N	7	14.01
Berkelium	Bk	97	247	Nobelium	No	102	259
Beryllium	Be	4	9.012	Osmium	Os	76	190.2
Bismuth	Bi	83	209.0	Oxygen	O	8	16.00
Bohrium	Bh	107	262	Palladium	Pd	46	106.4
Boron	B	5	10.81	Phosphorus	P	15	30.97
Bromine	Br	35	79.90	Platinum	Pt	78	195.1
Cadmium	Cd	48	112.4	Plutonium	Pu	94	244
Calcium	Ca	20	40.08	Polonium	Po	84	209
Californium	Cf	98	251	Potassium	K	19	39.10
Carbon	C	6	12.01	Praseodymium	Pr	59	140.9
Cerium	Ce	58	140.1	Promethium	Pm	61	145
Cesium	Cs	55	132.9	Protactinium	Pa	91	231
Chlorine	Cl	17	35.45	Radium	Ra	88	226
Chromium	Cr	24	52.00	Radon	Rn	86	222
Cobalt	Co	27	58.93	Rhenium	Re	75	186.2
Copper	Cu	29	63.55	Rhodium	Rh	45	102.9
Curium	Cm	96	247	Rubidium	Rb	37	85.47
Dubnium	Db	105	262	Ruthenium	Ru	44	101.1
Dysprosium	Dy	66	162.5	Rutherfordium	Rf	104	261
Einsteinium	Es	99	252	Samarium	Sm	62	150.4
Erbium	Er	68	167.3	Scandium	Sc	21	44.96
Europium	Eu	63	152.0	Seaborgium	Sg	106	263
Fermium	Fm	100	257	Selenium	Se	34	78.96
Fluorine	F	9	19.00	Silicon	Si	14	28.09
Francium	Fr	87	223	Silver	Ag	47	107.9
Gadolinium	Gd	64	157.3	Sodium	Na	11	22.99
Gallium	Ga	31	69.72	Strontium	Sr	38	87.62
Germanium	Ge	32	72.59	Sulfur	S	16	32.06
Gold	Au	79	197.0	Tantalum	Ta	73	180.9
Hafnium	Hf	72	178.5	Technetium	Tc	43	98
Hassium	Hs	108	265	Tellurium	Te	52	127.6
Helium	He	2	4.003	Terbium	Tb	65	158.9
Holmium	Ho	67	164.9	Thallium	Tl	81	204.4
Hydrogen	H	1	1.008	Thorium	Th	90	232.0
Indium	In	49	114.8	Thulium	Tm	69	168.9
Iodine	I	53	126.9	Tin	Sn	50	118.7
Iridium	Ir	77	192.2	Titanium	Ti	22	47.88
Iron	Fe	26	55.85	Tungsten	W	74	183.9
Krypton	Kr	36	83.80	Uranium	U	92	238.0
Lanthanum	La	57	138.9	Vanadium	V	23	50.94
Lawrencium	Lr	103	260	Xenon	Xe	54	131.3
Lead	Pb	82	207.2	Ytterbium	Yb	70	173.0
Lithium	Li	3	6.941	Yttrium	Y	39	88.91
Lutetium	Lu	71	175.0	Zinc	Zn	30	65.38
Magnesium	Mg	12	24.31	Zirconium	Zr	40	91.22
Manganese	Mn	25	54.94	—	—	110	269
Meitnerium	Mt	109	266	—	—	111	272
Mendelevium	Md	101	258	—	—	112	277
				—	—	114	289

PLATINUM EDITION

General, Organic, and Biological

Chemistry

STRUCTURES OF LIFE

Karen C. Timberlake

PEARSON

Benjamin
Cummings

San Francisco Boston New York
Capetown Hong Kong London Madrid Mexico City
Montreal Munich Paris Singapore Sydney Tokyo Toronto

Publisher	Jim Smith
Managing Editor	Joan Marsh
Project Editor	Claudia Herman
Production Supervisor	Vivian McDougal
Production Editor	Jean Lake
Director of Marketing	Stacy Treco
Marketing Manager	Christy Lawrence
Market Development Manager	Susan Winslow
Art Coordinators	Betty Gee, Side By Side Studios
	Anthony J. Asaro
	Jean Lake
Composition	Pre-Press, Side By Side Studios
Copyeditor	Cathy Cobb
Proofreader	Martha Ghent
Art Direction, Text and Cover Designer, Page Layout	Mark Ong, Side By Side Studios
Illustrators	Laura Brown, Carin Caine, Wen Chao, J.B. Woolsey Associates, Pre-Press, Side By Side Studios
Photo Research	Anthony J. Asaro
Text Photography	John Bagley, Richard Tauber
Cover Photographs	John Bagley, Richard Tauber
Cover Illustration	Blakeley Kim

Library of Congress Cataloging-in-Publication Data

Timberlake, Karen.
   General, organic, and biological chemistry: structures of life/Karen C. Timberlake–
Platinum ed.
   p. cm.
   Includes index.
   ISBN 0-8053-8914-8
   1. Chemistry. I. Title.

QD33.2.T56 2003
540—dc21

2003040902

ISBN 0-8053-8914-8

4 5 6 7 8 9 10   QWV   07 06 05

www.aw.com/bc

Brief Contents

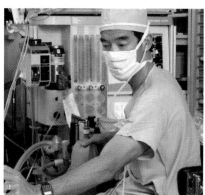

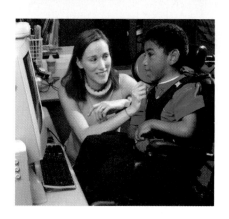

Contents

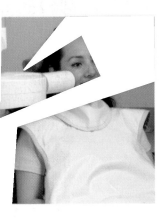

4   Compounds and Their Bonds   110

5   Energy and States of Matter   157

16 Carbohydrates 517

17 Carboxylic Acids and Esters 549

18 Lipids 575

Applications and Activities

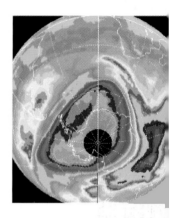

ENVIRONMENTAL NOTE

EXPLORE YOUR WORLD

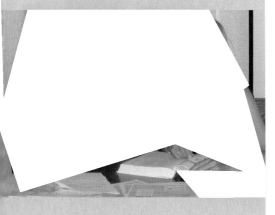

About the Author

Karen Timberlake is professor emeritus of chemistry at Los Angeles Valley College, where she taught chemistry for allied health and preparatory chemistry for 36 years. She received her bachelor's degree in chemistry from the University of Washington in 1962 and her master's degree in biochemistry from the University of California at Los Angeles in 1965. She has also taken graduate courses in science education during a sabbatical at the University of Northern Colorado.

Professor Timberlake has been writing chemistry textbooks for general, organic, and biological chemistry for 28 years. During that time, her name has become associated with the strategic use of pedagogical tools that promote student success in chemistry and the application of chemistry to real-life situations in health and medicine. More than one million students have learned chemistry using texts, laboratory manuals, and study guides written by Karen Timberlake. In addition to *General, Organic, and Biological Chemistry: Structures of Life,* Platinum Edition, she is also the author of *Chemistry: An Introduction to General, Organic, and Biological Chemistry* and the accompanying *Study Guide with Solutions for Selected Problems, Laboratory Manual,* and *Essentials Laboratory Manual.*

Professor Timberlake belongs to numerous science and education organizations including the American Chemical Society (ACS) and the National Science Teachers Association (NSTA). In 1987, she was the Western Regional Winner of Excellence in College Chemistry Teaching Award given by the Chemical Manufacturers Association. She has participated in education grants for science teaching including the Los Angeles Collaborative for Teaching Excellence (LACTE) and the Title III grant at her college. She often speaks at conferences and education meetings on using student-centered teaching methods in chemistry to promote the learning success of students.

Her husband, Bill, is also a chemistry professor and contributes much to Professor Timberlake's textbooks. When the Professors Timberlake are not writing textbooks, they relax by hiking, traveling to Mexico and Europe, trying new restaurants, and playing lots of tennis. The Timberlakes' son, John, and daughter-in-law, Cindy, live in Long Beach, California, where he is in curriculum sales. The Timberlakes' grandson, Daniel, who was born a year ago in September, does not know what he wants to be yet.

To the Student

I hope that this textbook helps you discover exciting new ideas and gives you a rewarding experience as you develop an understanding and appreciation of the role of chemistry in your life. If you would like to share your experience with chemistry, or have questions and comments about this text, I would appreciate hearing from you.

Karen C. Timberlake
Email: khemist@aol.com

Preface

Welcome to *General, Organic, and Biological Chemistry: Structures of Life,* Platinum Edition. My main objective in writing this text is to relate the structure and behavior of matter to its functions in health and life because all too often students find chemistry a series of facts to be memorized. This text is designed to be used in a one-year course that includes general, organic, and biological chemistry. It is written for students who are preparing for careers in health-related professions such as nursing, dietetics, respiratory therapy, or laboratory technology.

The Theme Emphasizes the Relationship Between Structures and Life

Throughout the text, the structures of inorganic, organic, and biochemical compounds and molecules are related to their function. The discussion of bonding and shapes of molecules in Chapter 4 provides a foundation for understanding the structure of organic and biochemical molecules. The topic of stereochemistry and chiral organic molecules, which begins in Chapter 15 with Aldehydes, is revisited as an important concept in the understanding of the structures of carbohydrates, amino acids, and chiral drugs. The structures of molecules are related to their physical and chemical properties such as solubility in water, density, and boiling point. The structural levels of proteins are related to their function, while chemical processes that denature proteins emphasize the importance of structure to activity. Throughout the text, the atomic structure of matter is highlighted by macro-to-micro illustrations that relates the atomic level to the macroscopic structures of real-life materials. In this way the chemical concepts and structures of molecules are continuously related to the behavior and function of biomolecules in the body. My goal in this text is to provide a learning environment that makes the study of chemistry an engaging and positive experience. It is also my goal to help every student become a critical thinker by instilling in them the scientific concepts that form the basis for making important decisions about issues concerning health and the environment.

Pedagogical Features

Students are often challenged by the study of chemistry and have difficulty finding the relevance of chemistry to their career paths. Many features in this text illustrate how chemistry is connected to students' lives and their interest in allied health careers. Chapter openers begin with interviews with health care professionals who discuss their type of work. Health Notes and Environmental Notes in every chapter continually relate the chemistry to current topics in health, medicine, and research, which support the student's interest in a health career.

A comprehensive learning program in each chapter provides many learning tools. Learning Goals in every section preview the concepts the student is to learn. The Sample Problems in every section model successful problem-solving techniques for the student. In this text I have redesigned the typical linear process of reading a whole chapter and then doing problems into a more cyclic process that combines each topic with immediate applications using Explore Your World activities, Health and Environmental Notes, and immediate problem solving. Questions and Problems now follow each section to encourage the student to apply newly learned material immediately to the problem-solving experience. Students are encouraged to be active learners by working problems and thinking about real-life issues as they progress through each chapter. Additional Problems found at the end of the chapter integrate the topics in the entire chapter to promote further study and critical thinking. Answers for Study Checks and odd-numbered problems located at the end of the text give immediate feedback to problem solving. Chapter Reviews give a brief overview of the important concepts in each section of the chapter. Key Terms and Summary of Reactions remind students of the new vocabulary and new reactions presented in each chapter.

Throughout the text, a strong art program visually connects the real-life world of materials familiar to students with their atomic-level structure. Students realize that all of the things they see every day have an atomic level of organization and structure that relates to their behavior and function.

Macro-to-micro illustrations from typical materials to their atomic level of organization continually relate structure to ordinary objects. Every figure throughout the text contains a question that requires the student to study that figure and relate the visual representation to the text. Molecular structures prevalent throughout the organic and biochemistry chapters aid in the visualization of the three-dimensional structures of organic and biochemical molecules. The plentiful use of three-dimensional structures takes the descriptions of molecules in the text to a visual level, which stimulates the imagination of the student. The use of color in diagrams, organic equations, and biochemical pathways clarifies and organizes the features of rather complex ideas.

Chapter Organization

In each textbook I write, I consider it essential to relate every chemical concept to real-life issues of health and environment. In this text I use the theme of Structures of Life to indicate the relationship between structures and function. The first part, Chapters 1 through 10, begin with measurement and the need to understand numerical structures of the metric system in the world of health care. In Chapters 2 and 3, I look at the structure of atoms, electron configuration, and the nucleus in radioactive atoms. I placed nuclear radiation early in the text because it extends the concepts of subatomic particles, atomic number, and mass to radioisotopes. In addition, students are introduced to the application of the atomic world to the field of nuclear medicine. Chapter 4 describes how the structures of atoms lead to ionic and covalent bonds as well as the shape and polarity of molecules. This introduction to the three-dimensional shape of molecules provides a basis for the shape of organic and biochemical compounds. In Chapter 5, I discuss energy, its measurement, and its relationship to nutritional energy values. I look at the structures of the three states of matter and the energy involved in changes to state.

In the next chapters, I look at the interaction of reactants in chemical reactions. Chapter 6 describes the reactants and products in combination, decomposition, and replacement reactions as well as oxidation and reduction reactions, which helps students organize and recognize the types of chemical reactions. I then discuss other features related to an understanding of reactions including energy in reactions, rates of reactions, and a qualitative discussion of chemical equilibrium and reversible reactions. Building on this basic understanding of equations, Chapter 7 discusses quantitative aspects of reactions including the mole, percent composition, mass calculations in reactions, percent yield, and equilibrium constants. Chapter 8 discusses gases and the gas

laws and Chapter 9 introduces the student to the important health care topics of solutions, electrolytes, solubility, concentrations, and osmosis and dialysis. Chapter 10 completes the general chemisty section with a discussion of acids and bases, conjugate acid–base pairs, pH, buffers, the role in health of these topics, and finally acid–base titration.

Although it is traditional to place the topics of organic and biochemistry in two separate sections, I have placed chapters on carbohydrates and lipids next to the organic chapters that discuss similar functional groups. Thus I begin the second part of the text with Chapter 11 with the shape of organic compounds, physical and chemical properties, and an overview of functional groups and constitutional isomers that form the structure of organic chemistry and form a basis for understanding the biomolecules of living systems. In Chapters 12 through 15, I describe the structures and nomenclature of saturated and unsaturated hydrocarbons, alcohols, thiols, ethers, aldehydes, and ketones. In each chapter, physical and chemical properties are related to the structure and functional groups in each family of organic compounds. In Chapter 15, I introduce the concept of chiral compounds and enantiomers, which I revisit in the chapters on carbohydrates and amino acids to show its importance in the nature of biomolecules and chiral drugs and their behavior in living systems. Chapter 16 integrates a biochemistry chapter, Carbohydrates, which contains the functional groups discussed in the previous chapters. To the student, the application of organic chemistry to biochemistry is often the most exciting part of their study of chemistry because of relationship to health and medicine. Chapter 17 discusses Carboxylic Acids and Esters, which is followed by Chapter 18 on Lipids, which contain the same functional groups in larger molecules such as triacylglycerols and glycerophospholipids. I discuss the role of these lipids and cholesterol in cell membranes as well as the lipids that function as bile salts and steroid hormones.

Chapters 19 and 20 continue with the physical and chemical properties of organic compounds, Amines and Amides, followed by related biomolecules, Amino Acids and Proteins. The importance of the structure of proteins from primary to quaternary is related to the shapes and activity of proteins. Chapter 21 relates the importance of the three-dimensional shape of proteins to their function as enzymes. The student learns that the shape of an enzyme is a factor in enzyme regulation and how end products might change the shape to increase or decrease the rate of an enzyme-catalyzed reaction. We also see that proteins change shape and lose function when subjected to pH changes and high temperatures. The important role of water-soluble vitamins as coenzymes is related to enzyme function. Chapter 22 describes the nucleic acids and their importance as biomolecules that store and direct information for cellular components, growth, and reproduction.

The final three chapters, Chapters 23, 24, and 25, discuss the metabolic pathways of biomolecules from the digestion of foodstuffs to the synthesis of ATP. Chapter 23 describes the stages of metabolism and the digestion of carbohydrates, our most important fuel. Then we see how the metabolic pathway of glycolysis degrades the glucose molecule to pyruvate. Under aerobic conditions, we learn that pyruvate is converted to acetyl CoA. We also look at how the storage molecule glycogen is replenished and how glucose is synthesized from noncarbohydrate sources. In Chapter 24, we look at the central role of the citric acid cycle in utilizing acetyl CoA to produce reduced coenzymes under aerobic conditions for the electron transport system and oxidative phosphorylation. Details on the structure and function of ATP synthase are included. Chapter 25 discusses the digestion of lipids and proteins and the metabolic pathways that convert fatty acids and amino acids to energy. We also learn how excess carbohydrates are converted to triacylglycerols in adipose tissue and how the intermediates of the citric acid cycle are converted to nonessential amino acids. Finally, we summarize the relationships between the catabolic and anabolic pathways in metabolism.

Because a chemistry course for allied health may be taught within different time frames, it may be difficult to cover all the chapters in this text. However, each chapter is now a complete package, which allows some chapters to be skipped or the order of presentation to be changed.

New for the Platinum Edition

This edition includes **Media Icons** in the margins to direct students to tutorials and case studies on The ChemistryPlace™ for GOB Chemistry Platinum. The ChemistryPlace™ Website is a robust resource providing animations of concepts in the book and case studies to contextualize chemical concepts in a real-world setting. Now the connections between the book and Website are clearly visible.

The **Paired Problems** within the chapter are highlighted with alternating colors to emphasize the connections between odd and even problems. It is even easier to find problems to work out in class or assign as homework.

The new **Walk-Through** in the front of the book will help instructors and students find the book's features easily.

In addition, a **Media Grid** in the front of the book correlates the media with the book.

For the first time, each new copy of the book includes a free copy of the *Understanding the Human Genome Project* booklet by Michael A. Palladino. Now you can learn and teach about this cutting-edge work and how it applies to the future of medicine and biotechnology.

On The ChemistryPlace™ Website, new features make it an even more valuable asset:

- Ten new **case studies** provide scenarios such as food irradiation and alcohol toxicity to show how chemical concepts apply to familiar situations.
- Over 2000 **Powerpoint® Slides**—complete with art, photos, and tables from the text—cover concepts from every section of the book. The Powerpoint® slides can be viewed online for a review of topics as well as downloaded and customized for classroom presentation.
- New **Flashcards** enable students to test themselves on the Key Terms in every chapter.
- In addition to **Chapter Quizzes**, a new set of **Cumulative Quizzes** give students the experience of testing on several chapters of material at one time.
- New **Tutorials** cover biochemistry topics in more detail than before.
- The **Research Navigator**™ makes finding topical news and journal articles only a click away in *The New York Times* Search by Subject™ Archive and the EBSCO ContentSelect™.

For instructors, the new **Image Presentation Library CD-ROM** features all the visuals from the text in a high-resolution format. Customizable Powerpoint® lecture outlines, featuring over 2000 slides, are also available on the CD-ROM. The **Instructor's Integration Guide** cross-references supplements such as the Instructor's Presentation Library CD-ROM, transparencies, lab manual, and media, showing where each can be utilized in relation to the text. The **Transparency Package,** with 300 full-color transparency acetates, now features 100 transparency masters with problems from the book for use in the classroom. In addition, **Blackboard, CourseCompass,** and **WebCT** Course Management Systems provide powerful course management capability. **WebAssign** is also available for homework assignments that are graded and recorded online.

Media Icons

Media Icons in the margins direct students to tutorials and case studies on The ChemistryPlace™ Website.

WEB TUTORIAL
The Shape of Molecules
CASE STUDY
Hazardous Materials

The ChemistryPlace™

New features include:

- Ten new case studies
- Flashcards
- New tutorials
- Cumulative quizzes
- PowerPoint® lecture outlines by the author for both presentation and student use
- The Research Navigator™, which makes topical chemistry news available for students

Other features include quizzes, study goals, tutorials, career focus, Web links, a glossary, an interactive periodic table, and a calculator. The online syllabus manager allows instructors to post class notes and assignments on the Web.

QUESTIONS AND PROBLEMS

Naming and Writing Ionic Formulas

4.35 Write names for the following ionic compounds:
   a. Al_2O_3 b. $CaCl_2$ c. Na_2O
   d. Mg_3N_2 e. KI

4.36 Write names for the following ionic compounds:
   a. $MgCl_2$ b. K_3P c. Li_2S
   d. LiBr   e. MgO

4.37 Why is a Roman numeral placed after the name of most transition metals ions?

4.38 The compound $CaCl_2$ is named calcium chloride; the compound $CuCl_2$ is named copper(II) chloride. Explain why a Roman numeral is used in one name but not the other.

4.39 Write the names of the following Group 4A and transition metal ions (include the Roman numeral when necessary):
   a. Fe^{2+} b. Cu^{2+} c. Zn^{2+} d. Pb^{4+}

Paired Problems

Paired Problems within the chapter are highlighted with alternating colors.

Media Grid

The **Media Grid** in the front of the book correlates the media with the book.

WebAssign

Selected problems are available as electronic homework through WebAssign.

Free Booklet

A free booklet, ***Understanding the Human Genome Project***, by Michael A. Palladino, is included with each copy of the book.

Michael A. Palladino

Understanding the Human Genome Project

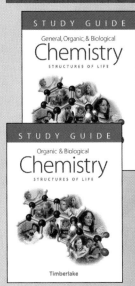
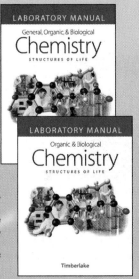

A Step-by-Step Approach

The author understands the learning challenges facing students in this course, so she breaks down chemical concepts and calculations into manageable pieces to walk students through the problem-solving process.

6.3 Balancing a Chemical Equation

We have seen that a chemical equation must be balanced. In many cases, we can use a method of trial and error to balance an equation. However, Table 6.3 gives a set of steps that can be useful.

To practice using the steps from Table 6.3, let us balance the reaction of the gas methane, CH_4, with oxygen to produce carbon dioxide and water. This is the principal reaction that occurs in the flame of a gas burner you use in the laboratory and in the flame of a gas stove.

Step 1 **Use the correct formulas.** As a first step, we write the equation using the correct formulas for the reactants and products.

$$CH_4 + O_2 \longrightarrow CO_2 + H_2O$$

Step 2 **Determine if the equation is balanced.** When we compare the atoms on the reactant side with the atoms on the product side, we see that there are more hydrogen atoms on the left side and more oxygen atoms on the right.

$$CH_4 + O_2 \longrightarrow CO_2 + H_2O$$

1 C	1 C
4 H	2 H Not balanced
2 O	3 O Not balanced

Step 3 **Balance the equation one element at a time.** First we balance the hydrogen atoms by placing a coefficient of 2 in front of the formula for water. Then we balance the oxygen atoms by placing a coefficient of 2 in front of the formula for oxygen. There are now four oxygen atoms and four hydrogen atoms in both the reactants and products.

$$CH_4 + 2O_2 \longrightarrow CO_2 + 2H_2O$$

Step 4 **Check to see if the equation is balanced.** Rechecking the balanced equation shows that the numbers of atoms of carbon, hydrogen, and oxygen are the same for both the reactants and the products. The equation is balanced using the lowest possible whole numbers as coefficients.

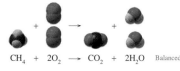

$$CH_4 + 2O_2 \longrightarrow CO_2 + 2H_2O \quad \text{Balanced}$$

Reactants	*Products*
1 C atom	1 C atom
4 H atoms	4 H atoms
4 O atoms	4 O atoms

Suppose you had written the equation as follows:

$$2CH_4 + 4O_2 \longrightarrow 2CO_2 + 4H_2O \quad \text{Incorrect}$$

Although there are equal numbers of atoms on both sides of the equation, this is not written correctly. All the coefficients must be divided by 2 to obtain the lowest possible whole numbers.

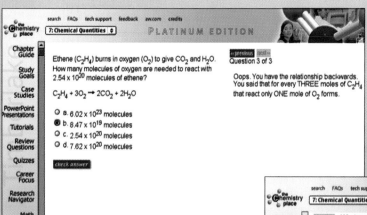

Review Questions

On The ChemistryPlace™ Website, **Review Questions** give students more interactive practice for each section of the book.

Quizzes

The ChemistryPlace™ Website includes two **Quizzes** per chapter, with questions unique to the site, for extra learning practice. Now students can test their knowledge of several chapters at once with new **Cumulative Quizzes** that draw connections between topics in groups of chapters. Hints and feedback are provided for all quizzes, and results can be e-mailed to instructors.

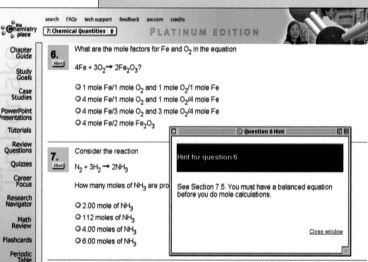

Sample Problems

Numerous **Sample Problems** appear throughout the text to immediately demonstrate the application of each new concept. The worked-out solution gives step-by-step explanations, provides a problem-solving model, and illustrates required calculations.

Questions and Problems

Questions and Problems at the end of each section encourage students to apply concepts and begin problem solving after each section. Answers to odd-numbered problems are given at the end of the chapter.

Additional Problems

Additional Problems at the end of each chapter combine concepts from that chapter and previous chapters. These problems are more complex mathematically and require more critical thinking; some involve real-world applications.

Student Friendly Approach

KEEPING STUDENTS ENGAGED IS THE ULTIMATE GOAL

Writing Style

Karen Timberlake is known for her accessible writing style, based on a carefully paced and simple development of chemical ideas, suited to the background of allied health students. She precisely defines terms and sets clear goals for each section. Her clear analogies help students to visualize and understand key chemical concepts.

Learning Goals

At the beginning of each section, a **Learning Goal** clearly identifies the key concept of the section, providing a road map for studying. All information contained in that section relates back to the Learning Goal.

LEARNING GOAL

State a chemical equation in words and determine the number of atoms in the reactants and products.

WEB TUTORIAL
Chemical Reactions and Equations

6.2 Chemical Equations

When you build a model airplane, prepare a new recipe, mix a medication, or clean a patient's teeth, you follow a set of directions. These directions tell you what materials to use and the products you will obtain. In chemistry, a **chemical equation** tells us the materials we need and the products that will form in a chemical reaction.

Writing a Chemical Equation

Suppose you work in a bicycle shop, assembling wheels and bodies into bicycles. You could represent this process by a simple equation:

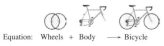

Equation: Wheels + Body ⟶ Bicycle

Art Program

The exciting new art program is not only beautifully rendered, but pedagogically effective as well.

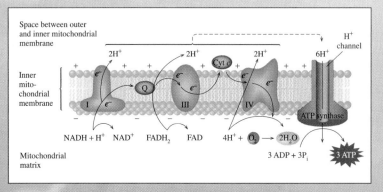

Macro-to-Micro Art

Photographs and drawings illustrate the atomic structure of recognizable objects, putting chemistry in context and connecting the atomic world to the macroscopic world.

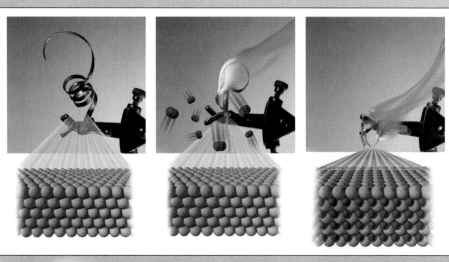

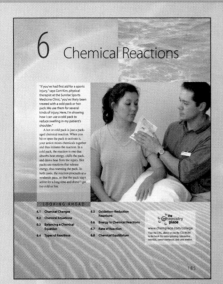

Chapter Opener

Each chapter begins with an interview with a professional in allied health or other relevant fields. These interviews illustrate how health professionals interact with chemistry in their careers.

Career Focus

Within the chapters are additional examples of professionals using chemistry.

Explore Your World

Hands-on activities that use everyday materials encourage students to actively explore selected chemistry topics.

EXPLORE YOUR WORLD

Fat Storage and Blubber

Obtain two medium-sized plastic baggies and a can of Crisco or other type of shortening used for cooking. You will also need a bucket or large container with water and ice cubes. Place several tablespoons of the shortening in one of the baggies. Place the second baggie inside and tape the outside edges of the bag. With your hand inside the inner baggie, move the shortening around to cover your hand. With one hand inside the double baggie, submerge both your hands in the container of ice water. Measure the time it takes for one hand to feel uncomfortably cold. Experiment with different amounts of shortening.

Questions

1. How effective is the bag with "blubber" in protecting your hand from the cold?
2. How does "blubber" help an animal survive starvation?
3. How would twice the amount of shortening affect your results?
4. Why would animals in warm climates, such as camels and migratory birds, need to store fat?
5. If you placed 300 g of shortening in the baggie, how many moles of ATP could it provide if used for energy production (assume it produces the same ATP as myristic acid)?

Environmental Note

Environmental Notes delve into high interest issues such as global warming, pheromones, acid rain, radon, food irradiation, and recycling of plastics.

ENVIRONMENTAL NOTE

Nuclear Power Plants

In a nuclear power plant, the quantity of uranium-235 is held below a critical mass, so it cannot sustain a chain reaction. The fission reactions are slowed by placing control rods, which absorb some of the fast-moving neutrons, among the uranium samples. In this way, less fission occurs, and there is a slower, controlled production of energy. The heat from the controlled fission is used to produce steam. The steam drives a generator, which produces electricity. Approximately 10% of the electrical energy produced in the United States is generated in nuclear power plants.

Although nuclear power plants help meet some of our energy needs, there are some problems. One of the most serious problems is the production of radioactive by-products that have very long half-lives. It is essential that these waste products be stored safely for a very long time in a place where they do not contaminate the environment. Early in 1990, the Environmental Protection Agency gave its approval for the storage of radioactive hazardous wastes in chambers 2150 ft underground. In 1998 the Waste Isolation Pilot Plant (WIPP) repository site in New Mexico was ready to receive plutonium waste from former U.S. bomb factories. Although authorities claim the caverns are safe, some

Containment shell

Control rods

Nuclear reactor

Hot water or molten sodium

Steam

Steam

Heat

Cool water

Steam generator

Fuel

Water way

Health Note

HEALTH NOTE

Hot Packs and Cold Packs

In a hospital, at a first-aid station, or at an athletic event, a *cold pack* may be used to reduce swelling from an injury, remove heat from inflammation, or decrease capillary size to lessen the effect of hemorrhaging. Inside the plastic container of a cold pack, there is a compartment containing solid ammonium nitrate (NH_4NO_3) that is separated from a compartment containing water. The pack is activated when it is hit or squeezed hard enough to break the walls between the compartments and cause the ammonium nitrate to mix with the water (shown as H_2O over the reaction arrow). In an endothermic process, one mole of NH_4NO_3 that dissolves absorbs 6.3 kcal of heat from the water. The temperature drops and the pack becomes cold and ready to use:

Endothermic Reaction in a Cold Pack

$$6.3 \text{ kcal} + NH_4NO_3(s) \xrightarrow{H_2O} NH_4NO_3(aq)$$

Hot packs are used to relax muscles, lessen aches and cramps, and increase circulation by expanding capillary size. Constructed in the same way as cold packs, a hot pack may contain the salt $CaCl_2$. The dissolving of the salt in water is exothermic and releases 18 kcal per mole of salt. The temperature rises and the pack becomes hot and ready to use:

Exothermic Reaction in a Hot Pack

$$CaCl_2(s) \xrightarrow{H_2O} CaCl_2(aq) + 18 \text{ kcal}$$

Health Notes in each chapter apply chemical concepts to relevant topics of health and medicine. These topics include weight loss and weight gain, artificial fats, anabolic steroids, alcohol, genetic diseases, viruses, and cancer.

Questions Paired with Figures

Questions paired with figures challenge students to think critically about photos and illustrations.

Figure 11.4 Propane, C_3H_8, is an organic compound, whereas sodium chloride, NaCl, is an inorganic compound.

Why is propane used as a fuel?

Acknowledgments

The preparation of this new text was a continuous effort of many people for a two-year period. As in my work on many editions of *Chemistry: An Introduction to General, Organic, and Biological Chemistry,* I am thankful for the support, encouragement, and dedication of many people who put in hours of tireless effort to produce a high quality book that provides an outstanding learning package. Once again the editorial team at Benjamin Cummings has done an excellent job. I appreciate the work of my Publisher, Jim Smith, who supported my vision of a new book with a new art program and learning package. My Project Editor, Claudia Herman, is like an angel who continually encouraged me on the development of this new text, while she skillfully coordinated reviews, art, website materials, interviews, and all the things it takes to make a book come together. Jean Lake, Production Editor, brilliantly coordinated all phases of the manuscript to the final pages of a beautiful book. Cathy Cobb, Copyeditor, precisely edited the manuscript to make sure the words were correct to help students learn chemistry. Betty Gee was absolutely invaluable in coordinating two photographers and four different artists as they created an art program of over seven hundred pieces.

I am especially proud of the new art program in this text, which lends beauty and understanding to chemistry. The photographs are the creative work of Mark Ong, Art Director, and photographers John Bagley and Richard Tauber, who took wonderfully vivid photographs that tell the story of the beauty of chemistry and its relationship to the structures of life. Photos of chemical compounds make students feel as though they are seeing a chemical in the laboratory. The macro-to-micro features that relate real-life photos to their atomic and molecular structure are a fantastic learning tool for students. There are new three-dimensional ball-and-stick models by J.B. Woolsey and Associates that provide students with the visual impressions and dimensions of organic and biochemical compounds. The entire art program is new and geared to the students' interest in allied health careers. Photos and interviews with health care professionals open each chapter and appear within chapters. I applaud Mark's creative eye as he art-directed and interviewed different professionals in the health care field. It was a big effort and is the keynote for every chapter. Thanks to the Production Coordinator, Tony Asaro, who gathered all the materials for many of the photos in the text, conducted many of the interviews, and agreed to be the "patient" in some of the photos.

It has been especially helpful to have accuracy reviewers for this new text. Without them, much would have gone unnoticed and uncorrected. Thanks to Professors Juliette Bryson and Yuri Zhorov, who read many galley and page proofs, for their valuable comments and corrections. Thanks to Martha Ghent for the hours of proofreading and to Lynn Carlson for her outstanding preparation of the Test Bank. Thanks to an enthusiastic and tireless Marketing Team, Christy Lawrence, Marketing Manager, Stacy Treco, Director of Marketing, and Chalon Bridges, Market Development Manager, whose interviews with professors provided valuable information for the direction of this book.

This text also reflects the contributions of many professors who took the time to review and edit the manuscript, provide outstanding comments, help, and suggestions. I am extremely grateful to an incredible group of peers for their careful assessment of all the new ideas for the text, suggested additions, corrections, changes, and deletions, and an incredible amount of feedback about the best direction for the text. In addition, I appreciate the time that health care professionals took to let us take photos and discuss their work with them. I admire and appreciate every one of you.

Platinum Edition Reviewers

John Hardee, Henderson State University
Melvin Merken, Worcester State College
Sandra Neuendorf, University of Wisconsin, Oshkosh
Elva Mae Nicholson, Eastern Michigan University
Kim Percell, Cape Fear Community College
Terri Pope, Cuyahoga Community College
Jamie Schneider, Winona State University
David Winters, Tidewater Community College, Virginia Beach

First Edition Reviewers

Peter Balanda, Ferris State University
Mark Benvenuto, University of Detroit Mercy
Tom Burkholder, Central Connecticut State University
G. Lynn Carlson, University of Wisconsin, Parkside
Ana Ciereszko, Miami-Dade Community College
Mark Chiu, Seton Hall University
Patricia Draves, University of Central Arkansas
Fabian Fang, California State University, Bakersfield
Don Glover, Bradley University
Sharon Kapica, County College of Morris
Robert Kolodny, Armstrong-Atlantic State
Patti Landers, University of Oklahoma
Richard Langley, Steven F. Austin State University
Tim Lubben, Northwestern College
Charlene McMahon, Carroll College
Melvin Merken, Worcester State College
Pamela Mork, Concordia College
Barbara Mowery, Thomas Nelson Community College
Shane Phillips, California State University, Stanislaus
Elizabeth Roberts-Kirchhoff, University of Detroit Mercy
Richard Sheardy, Seton Hall University
Kevin Siebenlist, Marquette University
Howard Silverstein, Georgia Perimeter College
Steve Socol, McHenry County College
John Sowa, Seton Hall University
Karen Wiechelman, University of Louisiana at Lafayette
Don Williams, Hope College
Suzanne Williams, Northern Michigan University

1 Measurements

"I use measurement in just about every part of my nursing practice," says registered nurse Vicki Miller. "When I receive a doctor's order for a medication, I have to verify that order. Then I draw a carefully measured volume from an IV or a vial to create that particular dose. Some dosage orders are specific to the size of the patient. I measure the patient's weight and calculate the dosage required for the weight of that patient."

Nurses use measurement each time they take a patient's temperature, height, weight, or blood pressure. Measurement is used to obtain the correct amounts for injections and medications and to determine the volumes of fluid intake and output. For each measurement, the amounts and units are recorded in the patient's records.

LOOKING AHEAD

the Chemistry place

www.chemplace.com/college

Visit the URL above or use the CD-ROM in the book for extra quizzing, interactive tutorials, career resources, and case studies.

You may wonder why this chemistry book begins with measurement. Think about your day; you probably made some measurements. Perhaps you checked your weight by stepping on a scale. If you did not feel well, you may have used a thermometer to determine your temperature. To make some soup, you added 2 cups of water to a package mix. If you stopped at the gas station, you watched the gas pump measure the number of gallons of gasoline you put in the car.

Chemistry and measurement are an important part of our everyday life. Levels of toxic materials in our air, soil, and water are discussed in our newspaper. We read about radon in our homes, holes in the ozone layer, trans fatty acids, global warming, and DNA analysis. Understanding chemistry and measurement helps us make proper choices about our world.

Measurement is an essential part of health careers such as nursing, dental hygiene, respiratory therapy, nutrition, and veterinary technology. The temperature, height, and weight of a patient are measured and recorded. Samples of blood and urine are collected and sent to a laboratory where glucose, pH, urea, and protein are measured by the lab technicians.

By studying measurement, you will develop skills for solving problems and learning the language of numbers in chemistry. If you intend to go into a health career, measurement and an understanding of measured values will be an important part of your evaluation of the health of a patient.

1.1 Units of Measurement

LEARNING GOAL

Write the names and abbreviations for the metric or SI units used in measurements of length, volume, and mass.

Suppose that during a recent medical examination the nurse recorded your mass (weight) as 70.0 kilograms (70.0 kg), your height as 1.78 meters (1.78 m), and your temperature as 37.0 degrees Celsius (37.0°C). The system of measurement used in these clinical evaluations is the *metric system*. Perhaps you would be more familiar with these measurements if they were stated in the U.S. system. Then your weight would be 154 pounds (154 lb), your height 5 feet 10 inches (5 ft 10 in.), and your temperature 98.6 degrees Fahrenheit (98.6°F).

The **metric system** is used by scientists and health professionals throughout the world. It is also the common measuring system in all but a few countries in the world. In 1960, a modification of the metric system called the *International System of Units,* Système International (**SI**), was adopted to provide additional uniformity. In this text, we will use metric units and introduce some of the SI units that are in use today.

Length

The **meter (m)** is used to measure length in both the metric system and SI. It is 39.4 inches (in.), which makes a meter slightly longer than a yard (yd). A smaller unit of length, the **centimeter (cm),** is about as wide as your little finger. For comparison, there are 2.54 cm in 1 in.

1 m   = 39.4 in.

2.54 cm = 1 in.

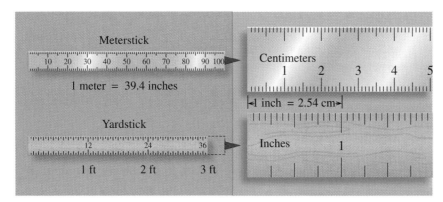

Figure 1.1 compares metric and U.S. units for length.

Volume

Volume is the amount of space occupied by a substance. The metric unit *liter* is commonly used to measure volume. A **liter (L)** is slightly larger than the quart (qt), which is used in the U.S. system. The **milliliter (mL)** is more convenient for measuring smaller volumes of fluids in hospitals and laboratories.

$$1 \text{ L} = 1.06 \text{ qt}$$
$$946 \text{ mL} = 1 \text{ qt}$$

A comparison of metric and U.S. units for volume appears in Figure 1.2.

Mass

The **mass** of an object is a measure of the quantity of material it contains. Everything has mass, including rocks, water, people, and dogs. In the metric system, the unit for mass is the **gram (g).** The SI unit of mass, the **kilogram (kg),** is used for larger masses such as body weight. It takes 2.20 lb to make 1 kg, and 454 g is needed to equal 1 pound.

$$1 \text{ kg} = 2.20 \text{ lb}$$
$$454 \text{ g} = 1 \text{ lb}$$

A summary of some metric and SI units is given in Table 1.1.

You may be more familiar with the term *weight* than with mass. Weight is a measure of the gravitational pull on an object. On Earth, an astronaut with a mass of 75.0 kg has a weight of 165 lb. On the moon where the gravitational pull is one-sixth that of Earth, the astronaut has a weight of 27.5 lb. However, the mass of the astronaut is the same as on

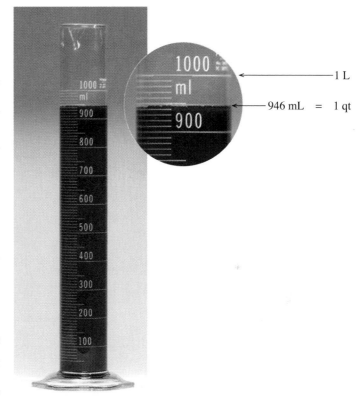

Figure 1.2 Volume is the space occupied by a substance. In the metric system, volume is based on the liter, which is slightly larger than a quart.

Q How many milliliters are in 1 quart?

Table 1.1	Units of Measurement	
Measurement	**Metric**	**SI**
Length	Meter (m)	Meter (m)
Volume	Liter (L)	Cubic meter (m³)
Mass	Gram (g)	Kilogram (kg)
Time	Second (s)	Second (s)
Temperature	Celsius (°C)	Kelvin (K)

Figure 1.3 On an electronic balance, mass is shown in grams as a digital readout.

Q How many grams are in 1 pound of candy?

Figure 1.4 A thermometer is used to determine the temperature of a substance.

Q What kinds of temperature readings have you made today?

Earth, 75.0 kg. Scientists measure mass rather than weight because mass does not vary.

In a chemistry laboratory, a balance is used to measure the mass of a substance as shown in Figure 1.3.

Temperature

You probably use a thermometer to see how hot something is, or how cold it is outside, or perhaps to determine if you have a fever. (See Figure 1.4.) The **temperature** of an object tells us how hot or cold that object is. In the metric system, temperature is measured on the Celsius temperature scale. On a **Celsius (°C)** thermometer, water has a freezing point of 0°C and a boiling point of 100°C. On a Fahrenheit (°F) thermometer, water has a freezing point of 32°F and a boiling point of 212°F. In the SI system, temperature is based on the **Kelvin (K) temperature scale,** where the lowest temperature possible is assigned as 0 K. We will discuss the relationships between these temperature scales in section 1.9.

SAMPLE PROBLEM 1.1

Units of Measurement

Complete the following table:

Type of Measurement	Metric Unit	Abbreviation
length	_____	_____
_____	liter	_____
_____	_____	g

Solution

length	meter	m
volume	liter	L
mass	gram	g

Study Check

What type of measurement uses the unit centimeter?

Answers to the Study Checks can be found at the end of this text in the Answer section. Checking your answers will let you know if you understand the material in this section.

QUESTIONS AND PROBLEMS

Units of Measurement

In every chapter, each dark-red, odd-numbered exercise in Questions and Problems is paired with the next even-numbered exercise. Answers for odd-numbered Questions and Problems are given at the end of this text. Complete solutions to odd-numbered Questions and Problems are in the *Study Guide.*

1.1 Compare the units you would use and the units that a student in France would use to measure the following:
   a. weight   **b.** height
   c. the amount of gasoline to fill the gas tank   **d.** temperature

1.2 Suppose that a friend tells you the following information. Why are the statements confusing and how would you make them clear?
a. I rode my bicycle for 15 today.   **b.** My dog weighs 45.
c. It is hot today. It is 40.   **d.** I lost 3.5 last week.

1.3 What are some common units used for measuring length in the metric system and the U.S. system?

1.4 What are some common units used for measuring mass in the metric system and the U.S. system?

1.5 What is the name of the unit and the type of measurement (mass, volume, length, or temperature) indicated for each of the following quantities?
a. 4.8 m   **b.** 325 g   **c.** 1.5 mL   **d.** 480 m   **e.** 28°C

1.6 What is the name of the unit and the type of measurement (mass, volume, length, or temperature) indicated for each of the following quantities?
a. 0.8 L   **b.** 3.6 m   **c.** 4 kg   **d.** 35 g   **e.** 373 K

1.2 Scientific Notation

LEARNING GOAL

Write a number in scientific notation.

In chemistry and science, we use numbers that are very, very small to measure things as tiny as the width of a human hair, which is 0.000 008 m, or things that are even smaller. Or perhaps we want to know the number of hairs on the average scalp, which is about 100 000 hairs. In both measurements, it is convenient to write the numbers in scientific notation. (In this section we have added spaces to help make the places easier to count.)

Width of a human hair	0.000 008 m	8×10^{-6} m
Hairs on a human scalp	100 000 hairs	1×10^5 hairs

Powers of 10

The number 1000 is the same as multiplying 10 three times. The three is written as a power (exponent) of 10.

$$1000 = 10 \times 10 \times 10 = 1 \times 10^3 \overset{\text{Power of 10}}{\longleftarrow}$$

Writing a Number in Scientific Notation

When a number is written in **scientific notation,** there are two parts: a coefficient and a power of 10. For example, the number 2400 is written as 2.4×10^3. The value 2.4 is the coefficient and 10^3 is the power of 10. The coefficient was determined by moving the decimal point to the left to give a number between 1 and 10. Counting the number of places (three) we moved the decimal gives the power of 10. For a large number, the power of 10 is always positive.

$$2\,4\,0\,0. \;=\; 2.4 \;\times\; 1\,0\,0\,0 \;=\; 2.4 \;\times\; 10^3$$

← 3 places

Number between 1 and 10   Power of 10

When a small number is written in scientific notation, the power of 10 will be negative. For example, 0.01 is the same as dividing by 10 two times, which is written as 10 with a power of -2.

$$0.01 = \frac{1}{10 \times 10} = \frac{1}{10^2} = 1 \times 10^{-2}$$

To write 0.0086 in scientific notation, we write two parts, 8.6×10^{-3}. The coefficient is 8.6 and 10^{-3} is the power of 10. For this small number, the decimal point was moved to the right three places, which gives 10^{-3} as the power of 10.

$$0.0086 = 8.6 \times \frac{1}{10 \times 10 \times 10} = 8.6 \times 10^{-3}$$

3 places ⟶   Number   Power
                                        between   of 10
                                        1 and 10

On many scientific calculators, a number can be converted into scientific notation using the appropriate keys. For example, the number 0.0052 can be entered followed by hitting the 2nd or 3rd function key and the SCI key. The scientific notation appears in the calculator window as a coefficient and the power of ten as a superscript.

0.0052 [2ⁿᵈ or 3ʳᵈ function key] [SCI] $= 5.2^{-3}$ or $5.2 - 03$

This calculator reading is written in scientific notation as 5.2×10^{-3}.

Converting Scientific Notation to a Standard Number

When a number in scientific notation has a positive power of 10, the standard number is written by moving the decimal point to the right for the same number of places as the power of 10. Placeholder zeros are used to give additional decimal places.

$$4.3 \times 10^3 = 4.3 \times 1000 = 4.300 = 4300$$

For a number in scientific notation with a negative power of 10, the standard number is written by moving the decimal point to the left for the same number of places. Placeholder zeros are added in front of the coefficient as needed.

$$4.3 \times 10^{-3} = 4.3 \times \frac{1}{1000} = 0004.3 = 0.0043$$

SAMPLE PROBLEM 1.2

Scientific Notation

1. Write the following measurements in scientific notation.
   a. 350 g   **b.** 0.000 16 L   **c.** 5 220 000 m
2. Write the following as standard numbers.
   a. 2.85×10^2 L   **b.** 7.2×10^{-3} m   **c.** 2.4×10^5 g

Solution

1. **a.** 3.5×10^2 g   **b.** 1.6×10^{-4} L   **c.** 5.22×10^6 m
2. **a.** 285 L   **b.** 0.0072 m   **c.** 240 000 g

Study Check

Write the following measurements in scientific notation.

   a. 425 000 m   **b.** 0.000 000 8 g

QUESTIONS AND PROBLEMS

Scientific Notation

1.7 Write the following measurements in scientific notation:
   a. 55 000 m   **b.** 480 g
   c. 0.000 005 cm   **d.** 0.000 14 s

1.8 Write the following measurements in scientific notation:
   a. 180 000 000 g   **b.** 0.000 06 m
   c. 750 000 g   **d.** 0.15 m

1.9 Which number in each pair is larger?
   a. 7.2×10^3 or 8.2×10^2 **b.** 4.5×10^{-4} or 3.2×10^{-2}
   c. 1×10^4 or 1×10^{-4} **d.** 0.000 52 or 6.8×10^{-2}

1.10 Which number in each pair is smaller?
   a. 4.9×10^{-3} or 5.5×10^{-9} **b.** 1250 or 3.4×10^2
   c. 0.000 000 4 or 5×10^{-8} **d.** 4×10^8 or 4×10^{-10}

1.11 Write the following as standard numbers:
   a. 1.2×10^4 **b.** 8.25×10^{-2} **c.** 4×10^6

1.12 Write the following as standard numbers:
   a. 3.6×10^{-5} **b.** 8.75×10^4 **c.** 3×10^{-2}

1.3 Measured and Exact Numbers

LEARNING GOAL

Identify a number as measured or exact.

Whenever you make a measurement, you use some type of measuring device. For example, you may use a meterstick to measure your height, a scale to check your weight, and a thermometer to take your temperature. **Measured numbers** are the numbers you obtain when you measure a quantity such as your height, weight, or temperature.

Measured Numbers

Suppose you are going to measure the lengths of the objects in Figure 1.5. You would select a ruler with a scale marked on it. By observing the lines on the scale, you determine the measurement for each object. Perhaps the divisions on the scale are marked as 1 cm. Another ruler might be marked in divisions of 0.1 cm. To report the length, you would first read the numerical value of the marked line. Finally, you would make an *estimation* of the distance between the smallest marked lines. This estimated value is the final digit in a measured number.

An estimation is made by visually breaking up the space between the smallest marked lines, usually into tenths. For example, in Figure 1.5a, the end of the object falls between the lines marked 4 cm and 5 cm. That means that the length is 4 cm plus an estimated digit. If you estimate that the end is halfway between 4 cm and 5 cm, you would report its length as 4.5 cm. However, someone else might report the length as 4.4 cm. The last digit in measured numbers will differ because people do not estimate in the same way. The ruler shown in Figure 1.5b is marked with lines at 0.1 cm. With this ruler, you can estimate to the 0.01-cm place. Perhaps you would report the length of the object as 4.55 cm, while someone else may report its length as 4.56 cm. Both results are acceptable.

(a)

(b)

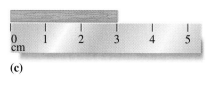

(c)

Figure 1.5 The lengths of the rectangular objects are measured as **(a)** 4.5 cm and **(b)** 4.55 cm.

Q What is the length of the object in (c)?

Temperature Scales on Thermometers

Look at some different thermometers such as an oral thermometer, a meat thermometer, a candy thermometer, an outdoor thermometer, or a car temperature gauge. Determine the temperature unit for each and the temperature range from low to high for each.

Questions

1. What is the value of the marked lines on the temperature scale for each thermometer?
2. What is the estimated place that would be reported in a measurement from the scale on each type of thermometer?

Because there are variations in estimation, there is always some *uncertainty* in every measurement. Sometimes a measurement ends right on a marked line. In order to indicate the uncertainty of the measurement, a zero is written as the estimated figure. The zero in the uncertainty place indicates that the measurement was not above or below the value of that line. For example, in Figure 1.5c, we would write the measurement as 3.0 cm to indicate that the length is not 2.9 cm or 3.1 cm.

Exact Numbers

Exact numbers are numbers we obtain by counting items or from a definition that compares two units in the same measuring system. Suppose a friend asks you to tell her the number of coats in your closet or the number of bicycles in your garage or the number of classes you are taking in school. To do this you must have counted the items. It was not necessary for you to use any type of measuring tool. Suppose an instructor asks you to state the number of seconds in one minute or the number of hours in one day. Again without using any measuring device, you know the answers. The relationships for time are defined as 60 seconds in one minute, and 24 hours in one day. For more examples of exact numbers, see Table 1.2.

Table 1.2   Examples of Some Exact Numbers

Counted Numbers	Defined Equalities	
	U.S. System	Metric System
Eight doughnuts	1 ft = 12 in.	1 L = 1000 mL
Two baseballs	1 qt = 4 cups	1 m = 100 cm
Five capsules	1 lb = 16 ounces	1 kg = 1000 g

SAMPLE PROBLEM 1.3

Types of Numbers

Identify each of the following as an exact or a measured number:

a. number of eggs in a 3-egg omelet
b. amount of coffee needed to make 10-cup pot of coffee
c. number of tea bags needed to make a pot of tea
d. amount of sugar needed to sweeten the coffee or tea

Solution

a. exact   **b.** measured   **c.** exact   **d.** measured

Study Check

Are the 2 pieces of toast or the 6 pancakes you ate for breakfast exact or measured numbers?

QUESTIONS AND PROBLEMS

Measured and Exact Numbers

1.13 What is the estimated digit in each of the following measured numbers?
      a. 8.6 m   **b.** 45.25 g   **c.** 25.0°C

1.14 What is the estimated digit in each of the following measured numbers?
   a. 125.04 g   **b.** 5.057 m   **c.** 525.8°C

1.15 There are 4 chairs at the dining table. Why is the number 4 here called an exact number?

1.16 There are 7 days in one week. Why is the number 7 here called an exact number?

1.17 Are the numbers in each of the following statements measured or exact?
   a. A patient weighs 155 lb.
   b. The basket holds 8 apples.
   c. In the metric system, 1 kg is equal to 1000 g.
   d. The distance from Denver, Colorado, to Houston, Texas, is 1720 km.

1.18 Are the numbers in each of the following statements measured or exact?
   a. There are 31 students in the laboratory.
   b. The oldest known flower lived 120 000 000 years ago.
   c. The largest gem ever found, an aquamarine, has a mass of 104 kg.
   d. A laboratory test shows a blood cholesterol level of 184 mg/dL.

1.4 Significant Figures

LEARNING GOAL

Determine the number of significant digits in measured numbers.

WEB TUTORIAL
Significant Figures

When we work in a medical environment or in a laboratory, it is important to make measurements that are accurate and that can be duplicated. A measurement is **accurate** when it is close to the real size or actual value of what we are measuring. That means we need to be careful when we determine a patient's temperature or take a blood pressure. It is also important to use a measuring instrument that is properly adjusted. Using a measuring device that is not in good working order will not give an accurate result even with careful measuring.

A measurement is **precise** if it can be repeated many times with results that are very close in value to each other. For example, suppose that four students measure your height as 152.3 cm, 152.4 cm, 152.1 cm, and 152.3 cm. We would say that the measurements were precise. If your actual height is 152.3 cm, each of the measurements was also accurate.

SAMPLE PROBLEM 1.4

Accuracy and Precision

The standard diameter for a large pizza at the Pizza Shack is 14 inches. However, the new pizza maker has prepared four pizzas that are each 10 inches in diameter.

a. Is the size of the pizzas accurate? Why?
b. Is the size of the pizzas precise? Why?

Solution

a. No. They are not accurate because the diameter of each pizza is not close to the standard diameter required for a large pizza.
b. Yes. They are precise because all of the pizzas are close in size to each other.

Study Check

What should be the size of the large pizzas to be both accurate and precise?

Counting Significant Figures

In measured numbers, the **significant figures** are all the reported numbers including the estimated digit. When we do calculations, we will need to count the significant figures in each of the measured numbers used in the calculations. All *nonzero* numbers are counted as significant figures. Zeros may or may not be significant depending on their position in a number. Table 1.3 gives the rules and examples of counting significant figures.

Table 1.3   Significant Figures in Measured Numbers

Rule	Measured Number	Number of Significant Figures
1. A number is a *significant figure* if it is		
a. a nonzero digit	4.5 g	2
	122.35 m	5
b. a zero between nonzero digits	205 m	3
	5.082 kg	4
c. a zero at the end of a decimal number	50. L	2
	25.0°C	3
	16.00 g	4
d. any digit in the coefficient of a number written in scientific notation	4.0×10^5 m	2
	5.70×10^{-3} g	3
2. A number is *not significant* if it is		
a. a zero at the beginning of a decimal number	0.0004 lb	1
	0.075 m	2
b. a zero used as a placeholder in a large number without a decimal point	850 000 m	2
	1 250 000 g	3

When one or more zeros in a large number are significant digits, writing the number in scientific notation can show them more clearly. For example, if the first zero in the measurement 500 m is significant, it is shown by writing 5.0×10^2 m. In this text, we will also place a decimal point after a zero if that zero is part of the measurement. For example, writing the measurement 250. g would indicate that the zero is included in the three significant figures. It could also be written as 2.50×10^2 g. Unless noted otherwise, we will assume that zeros at the end of large numbers are not significant.

Number	*Number of Significant Digits*	*Scientific Notation*
400 000 g	1	4×10^5 g
400 000 g	2	4.0×10^5 g
400 000 g	3	4.00×10^5 g

SAMPLE PROBLEM 1.5

Significant Figures

Identify each of the following numbers as measured or exact; give the number of significant figures in each of the measured numbers.

a. 42.2 g   **b.** 3 eggs   **c.** 0.0005 cm
d. 450 000 km   **e.** 9 planets

Solution

a. measured; three   **b.** exact   **c.** measured; one

d. measured; two   **e.** exact

Study Check

State the number of significant figures in each of the following measured numbers:

a. 0.00035 g   **b.** 2000 m   **c.** 2.0045 L

QUESTIONS AND PROBLEMS

Significant Figures

1.19 For each measurement, indicate if the zeros are significant figures or not.
   a. 0.0038 m   **b.** 5.04 cm   **c.** 800. L
   d. 3.0×10^{-3} kg   **e.** 85 000 m

1.20 For each measurement, indicate if the zeros are significant figures or not.
   a. 20.05 g   **b.** 5.00 m   **c.** 0.000 02 L
   d. 120 000 years   **e.** 8.05×10^2 L

1.21 How many significant figures are in each of the following measured quantities?
   a. 11.005 g   **b.** 0.000 32 m   **c.** 36 000 000 km
   d. 1.80×10^4 g   **e.** 0.8250 L   **f.** 30.0°C

1.22 How many significant figures are in each of the following measured quantities?
   a. 20.60 mL   **b.** 1036.48 g   **c.** 4.00 m
   d. 20.8°C   **e.** 60 800 000 g   **f.** 5.0×10^{-3} L

1.23 Write each of the following numbers in scientific notation with two significant figures:
   a. 5000 L   **b.** 30 000 g   **c.** 100 000 m   **d.** 0.000 25 cm

1.24 Write each of the following numbers in scientific notation with three significant figures:
   a. 5 100 000   **b.** 2600   **c.** 40 000   **d.** 0.00820

1.5 Significant Figures in Calculations

In science and medicine, we measure many things: the length of a bacterium, the volume of a medication, and the mass of cholesterol in a blood sample. The numbers from these measurements are often used in calculations. However, the calculation cannot change the precision of a measured number. Therefore, we report a calculated answer that reflects the precision of the original measurements.

Using a calculator will usually help you do calculations faster than you can without one. However, calculators cannot think for you. It is up to you to enter the numbers correctly, press the right function keys, and adjust the calculator display to give the appropriate answer.

Rounding Off

If we calculate the area of a carpet that measures 5.5 m by 3.5 m, we obtain the number 19.25 as the square meters in the area. However, all four digits cannot be

significant because an answer cannot be more precise than the measured numbers used. Because each of the measurement numbers has just two significant figures, the calculated result must be *rounded off* to give an answer that also has two significant figures: 19 m². A final answer *cannot be more precise than any measurement used in the calculation.*

For exact numbers we do not have to determine the number of significant figures in the final answer. There is no uncertainty in an exact number to affect the outcome of any calculation.

When you obtain a calculator result, determine the number of significant figures you should keep for the answer and round off the calculator results using the following rules.

WEB TUTORIAL
Significant Figures

Rules for Rounding Off

1. If the first digit to be dropped is *4 or less,* it and all following digits are simply dropped from the number.

	Three Significant Figures	*Two Significant Figures*
Example 1: 8.4234 rounds off to	8.42	8.4

2. If the first digit to be dropped is *5 or greater,* the last retained digit of the number is increased by 1.

Example 2: 14.780 rounds off to	14.8	15

SAMPLE PROBLEM 1.6

Rounding Off

Round off each of the following numbers to three significant figures:

a. 35.7823 m   **b.** 0.002627 L
c. 3826.8 g   **d.** 1.2836 kg

Solution

a. 35.8 m   **b.** 0.00263 L
c. 3830 g   **d.** 1.28 kg

Study Check

Round off each of the numbers in Sample Problem 1.6 to two significant figures.

Multiplication and Division

In multiplication and division, the final answer is written so it has the same number of digits as the measurement with the *fewest* significant figures (SFs).

Example 1

Multiply the following measured numbers: 24.65 × 0.67.

24.65	⊗	0.67	⊜	16.5155	⟶	17
Four SFs		Two SFs		Calculator display		Final answer, rounded to two SFs

The answer in the calculator display has more digits than the data allows. The measurement 0.67 has the least number of significant figures, two. Therefore, the calculator answer is rounded off to two significant figures.

Example 2

Solve the following:

$$\frac{2.85 \times 67.4}{4.39}$$

To do this problem on a calculator, enter the number and then press the operation key. In this case, we might press the keys in the following order:

2.85 ☒ 67.4 ÷ 4.39 =   43.756264 ⟶ 43.8

Three SFs   Three SFs   Three SFs   Calculator   Final answer, rounded
                                          display   to three SFs

All of the measurements in this problem have three significant figures. Therefore, the calculator result is rounded off to give an answer, 43.8, that has three significant figures.

Adding Significant Zeros

Sometimes, a calculator displays a small whole number. To give an answer with the correct number of significant figures, significant zeros are written after the calculator result. For example, suppose the calculator display is 4, but you used measurements that have three significant numbers. Placing two significant zeros after the 4 gives the correct answer, 4.00.

$$\frac{8.00}{2.00} = \qquad 4 \quad \longrightarrow \quad 4.00$$

Three SFs   Calculator display   Final answer, two zeros
                                       added to give three SFs

SAMPLE PROBLEM 1.7

Significant Figures in Measured Numbers

Perform the following calculations of measured numbers. Give the answers with the correct number of significant figures.

a. 56.8×0.37 **b.** $\dfrac{71.4}{11}$ **c.** $\dfrac{(2.075)(0.585)}{(8.42)(0.0045)}$ **d.** $\dfrac{25.0}{5.00}$

Solution

a. 21   **b.** 6.5   **c.** 32   **d.** 5.00 (Must add significant zeros)

Study Check

Solve:

a. 45.26×0.01088 **b.** $2.6 \div 324$ **c.** $\dfrac{4.0 \times 8.00}{16}$

Phlebotomist

As part of the medical team, phlebotomists collect and process blood for laboratory tests. They work directly with patients, calming them if necessary prior to the collection of blood. Phlebotomists are trained to collect blood in a safe manner and provide patient care if fainting occurs. Blood is drawn through venipuncture methods such as syringe, vacutainer, blood culture, and fingerstick. They also prepare patients for procedures such as glucose tolerance tests. In the preparation of specimens for analysis, a phlebotomist determines media, inoculation method, and reagents for culture setup.

Addition and Subtraction

In addition or subtraction, the answer is written so it has the same number of decimal places as the measurement having the fewest decimal places.

Example 3

Add:

	2.045	Three decimal places
+	34.1	One decimal place
	36.145	Calculator display
	36.1	Answer, rounded to one decimal place

Example 4

Subtract:

	255	Ones place
−	175.65	Two decimal places
	79.35	Calculator display
	79	Answer, rounded to ones place

When numbers are added or subtracted to give answers ending in zero, the zero does not appear after the decimal point in the calculator display. For example, 14.5 g − 2.5 g = 12.0 g. However, if you do the subtraction on your calculator, the display shows 12. To give the correct answer, a significant zero is written after the decimal point.

Example 5

	14.5 g	One decimal place
−	2.5 g	One decimal place
	12.	Calculator display
	12.0 g	Answer, zero written after the decimal point

SAMPLE PROBLEM 1.8

Significant Figures in Addition and Subtraction

Perform the following calculations and give the answer with the correct number of decimal places:

a. 27.8 cm + 0.235 cm
b. 104.45 mL + 0.838 mL + 46 mL
c. 153.247 g − 14.82 g

Solution

a. 28.0 cm   **b.** 151 mL   **c.** 138.43 g

Study Check

Solve:
a. 82.45 mg + 1.245 mg + 0.00056 mg   **b.** 4.259 L − 3.8 L

QUESTIONS AND PROBLEMS

Significant Figures in Calculations

1.25 Why do we usually need to round off calculations that use measured numbers?

1.26 Why do we sometimes add a zero to a number in a calculator display?

1.27 How is the number of significant figures determined for multiplication and/or division?

1.28 How is the number of decimal places determined for addition and/or subtraction?

1.29 Round off each of the following numbers to three significant figures.
a. 1.854 **b.** 184.2038 **c.** 0.004738265
d. 8807 **e.** 1.832149

1.30 Round off each of the numbers in problem 1.29 to two significant figures.

1.31 For the following problems, give answers with the correct number of significant figures:
a. 45.7×0.034 **b.** 0.00278×5
c. $\dfrac{34.56}{1.25}$ **d.** $\dfrac{(0.2465)(25)}{1.78}$

1.32 For the following problems, give answers with the correct number of significant figures:
a. 400×185 **b.** $\dfrac{2.40}{(4)(125)}$
c. $0.825 \times 3.6 \times 5.1$ **d.** $\dfrac{3.5 \times 0.261}{8.24 \times 20.0}$

1.33 For the following problems, give answers with the correct number of decimal places:
a. $45.48 \text{ cm} + 8.057 \text{ cm}$ **b.** $23.45 \text{ g} + 104.1 \text{ g} + 0.025 \text{ g}$
c. $145.675 \text{ mL} - 24.2 \text{ mL}$ **d.** $1.08 \text{ L} - 0.585 \text{ L}$

1.34 For the following problems, give answers with the correct number of decimal places:
a. $5.08 \text{ g} + 25.1 \text{ g}$ **b.** $85.66 \text{ cm} + 104.10 \text{ cm} + 0.025 \text{ cm}$
c. $24.568 \text{ mL} - 14.25 \text{ mL}$ **d.** $0.2654 \text{ L} - 0.2585 \text{ L}$

Table 1.4 Daily Values for Selected Nutrients

Nutrient	Amount Recommended
Protein	44 g
Vitamin C	60 mg
Vitamin B_{12}	6 μg
Calcium	1 g
Iron	18 mg
Iodine	150 μg
Sodium	2400 mg
Zinc	15 mg

1.6 SI and Metric Prefixes

The U.S. Food and Drug Administration has determined the daily values (DV) of nutrients for adults and children age 4 or older. Some of these recommended daily values are listed in Table 1.4.

The special feature of the metric system of units is that a **prefix** can be attached to any unit to increase or decrease its size by some factor of 10. For example, in the daily values, the prefixes *milli* and *micro* are used to make the smaller units, milligram (mg) and microgram (μg). Table 1.5 lists some of the metric prefixes, their symbols, and their decimal values.

The prefix *centi* is like cents in a dollar. One cent would be a centidollar or $\frac{1}{100}$ of a dollar. That also means that one dollar is the same as 100 cents. The prefix *deci* is like dimes in a dollar. One dime would be a decidollar or $\frac{1}{10}$ of a dollar. That also means that one dollar is the same as 10 dimes.

WEB TUTORIAL
The Metric System

Table 1.5   Metric and SI Prefixes

Prefix	Symbol	Meaning	Numerical Value	Scientific Notation
Prefixes That Increase the Size of the Unit				
giga	G	one billion	1 000 000 000	10^9
mega	M	one million	1 000 000	10^6
kilo	k	one thousand	1 000	10^3
Prefixes That Decrease the Size of the Unit				
deci	d	one-tenth	0.1	10^{-1}
centi	c	one-hundredth	0.01	10^{-2}
milli	m	one-thousandth	0.001	10^{-3}
micro	μ	one-millionth	0.000 001	10^{-6}
nano	n	one-billionth	0.000 000 001	10^{-9}

The relationship of a unit to its base unit can be expressed by replacing the prefix with its numerical value. For example, when the prefix *kilo* in *kilometer* is replaced with its value of 1000, we find that a kilometer is equal to 1000 meters. Other examples follow.

1 **kilo**meter (1 km) = **1000** meters (1000 m)

1 **kilo**liter (1 kL) = **1000** liters (1000 L)

1 **kilo**gram (1 kg) = **1000** grams (1000 g)

SAMPLE PROBLEM 1.9

Prefixes

Fill in the blanks with the correct numerical value:
a. kilogram   = _____ grams
b. millisecond   = _____ second
c. deciliter   = _____ liter

Solution

a. The numerical value of *kilo* is 1000; 1000 grams.
b. The numerical value of *milli* is 0.001; 0.001 second.
c. The numerical value of *deci* is 0.1; 0.1 liter.

Study Check

Write the correct prefix in the blanks:
a. 1 000 000 seconds   = _____ seconds
b. 0.01 meter   = _____ meter

Measuring Length

An ophthalmologist may measure the diameter of the retina of the eye in centimeters (cm), whereas a surgeon may need to know the length of a nerve in millimeters (mm). When the prefix *centi* is used with the unit *meter*, it indicates the unit *centimeter*, a length that is one-hundredth of a meter (0.01 m). A *millimeter* measures a length of 0.001 m. There are 1000 mm in a meter.

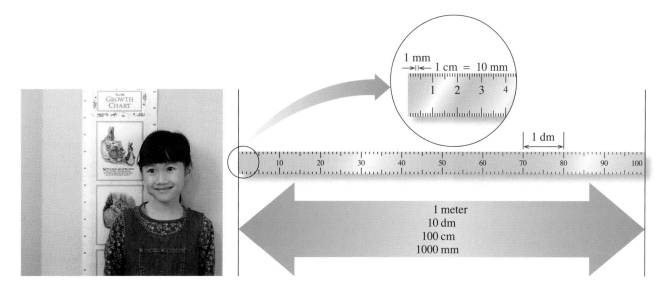

If we compare the lengths of a millimeter and a centimeter, we find that 1 mm is 0.1 cm; there are 10 mm in 1 cm. These comparisons are examples of **equalities**, which show the relationship between two units that measure the same quantity. For example, in the equality 1 m = 100 cm, each quantity describes the same length but in a different unit. Note that each quantity in the equality expression has both a number and a unit.

Figure 1.6 The metric length of one meter is the same length as 10 dm, 100 cm, and 1000 mm.

Q How many millimeters (mm) are in 1 centimeter (cm)?

An Equality

First Quantity   *Second Quantity*

1   m   =   100   cm

Number + unit   Number + unit

Some Length Equalities

1 m = 100 cm

1 m = 1000 mm

1 cm = 10 mm

Some metric units for length are compared in Figure 1.6.

Measuring Volume

Volumes of 1 L or smaller are common in the health sciences. When a liter is divided into 10 equal portions, each portion is a deciliter (dL). There are 10 dL in 1 L. Laboratory results for blood work are often reported in deciliters. Notice the values listed in Table 1.6 for some substances in the blood.

When a liter is divided into a thousand parts, each of the smaller volumes is called a milliliter. In a 1-L container of physiological saline, there are 1000 mL of solution (see Figure 1.7).

Some Volume Equalities

1 L = 10 dL

1 L = 1000 mL

1 dL = 100 mL

Table 1.6 Some Typical Laboratory Test Values

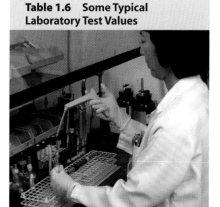

Substance in Blood	Typical Range
Albumin	3.5–5.0 g/dL
Ammonia	20–150 μg/dL
Calcium	8.5–10.5 mg/dL
Cholesterol	105–250 mg/dL
Iron (male)	80–160 μg/dL
Protein (total)	6.0–8.0 g/dL

Figure 1.7 A plastic intravenous fluid container contains 1000 mL.

Q How many liters of solution are in the intravenous fluid container?

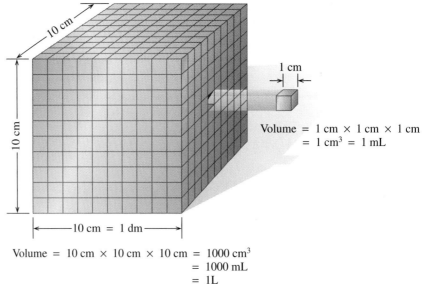

Volume $= 10 \text{ cm} \times 10 \text{ cm} \times 10 \text{ cm} = 1000 \text{ cm}^3$
$= 1000 \text{ mL}$
$= 1 \text{L}$

Figure 1.8 A cube measuring 10 cm on each side has a volume of 1000 cm³, or 1 L; a cube measuring 1 cm on each side has a volume of 1 cm³ (cc) or 1 mL.

Q What is the relationship between a milliliter (mL) and a cubic centimeter (cm³)?

The **cubic centimeter** (**cm³** or **cc**) is the volume of a cube whose dimensions are 1 cm on each side. A cubic centimeter has the same volume as a milliliter, and the units are often used interchangeably.

$$1 \text{ cm}^3 = 1 \text{ cc} = 1 \text{ mL}$$

When you see *1 cm,* you are reading about length; when you see *1 cc* or *1 cm³* or *1 mL,* you are reading about volume. A comparison of units of volume is illustrated in Figure 1.8.

Measuring Mass

When you get a physical examination, your mass is recorded in kilograms, whereas the results of your laboratory tests are reported in grams, milligrams (mg), or micrograms (µg). A kilogram is equal to 1000 g. One gram represents the same mass as 1000 mg, and one mg equals 1000 µg.

Some Mass Equalities

1 kg $= 1000$ g

1 g $\;= 1000$ mg

1 mg $= 1000$ µg

SAMPLE PROBLEM 1.10

Writing Metric Relationships

1. Identify the larger unit in each of the following pairs:
   a. centimeter or kilometer   **b.** L or dL   **c.** mg or µg

2. Complete the following list of metric equalities:

    a. 1 L = _____ dL   **b.** 1 km = _____ m

    c. 1 m = _____ cm   **d.** 1 cm³ = _____ mL

Solution

1. a. kilometer   **b.** L   **c.** mg

2. a. 10 dL   **b.** 1000 m   **c.** 100 cm   **d.** 1 mL

Study Check

Complete the following equalities:

a. 1 kg = _____ g   **b.** 1 mL = _____ L

QUESTIONS AND PROBLEMS

SI and Metric Prefixes

1.35 What are some advantages and disadvantages of the United States converting completely to the metric (SI) system?

1.36 What parts of the U.S. currency system are similar to the units in the metric system?

1.37 The speedometer in the margin is marked in both km/h and mi/h or mph. What is the meaning of each abbreviation?

1.38 In Canada, a highway sign gives a speed limit as 80 km/h. According to the speedometer in problem 1.37, would you be exceeding the speed limit of 55 mph if you were in the United States?

1.39 How does the prefix *kilo* affect the gram unit in *kilogram*?

1.40 How does the prefix *centi* affect the meter unit in *centimeter*?

1.41 Write the abbreviation for each of the following units:

    a. milligram   **b.** deciliter   **c.** kilometer

    d. kilogram   **e.** microliter

1.42 Write the complete name for each of the following units:

    a. cm   **b.** kg   **c.** dL   **d.** Gm   **e.** μg

1.43 Write the numerical values for each of the following prefixes:

    a. centi   **b.** kilo   **c.** milli   **d.** deci   **e.** mega

1.44 Write the complete name (prefix + unit) for each of the following numerical values:

    a. 0.10 g   **b.** 0.000 001 g   **c.** 1000 g

    d. $\frac{1}{100}$ g   **e.** 0.001 g

1.45 Complete the following metric relationships:

    a. 1 m = _____ cm   **b.** 1 km = _____ m

    c. 1 mm = _____ m   **d.** 1 L = _____ mL

1.46 Complete the following metric relationships:

    a. 1 kg = _____ g   **b.** 1 mL = _____ L

    c. 1 g = _____ kg   **d.** 1 g = _____ mg

1.47 For each of the following pairs, which is the larger unit?

    a. milligram or kilogram   **b.** milliliter or microliter

    c. cm or km   **d.** kL or dL

1.48 For each of the following pairs, which is the smaller unit?

    a. mg or g   **b.** centimeter or millimeter

    c. mm or μm   **d.** mL or dL

1.7 Problem Solving Using Conversion Factors

Many problems in chemistry and the health sciences require a change of units. You make changes in units every day. For example, suppose you spent 2.0 hours (hr) on your homework, and someone asked you how many minutes that was. You would answer 120 minutes (min). You knew how to change from hours to minutes because you knew an equality (1 hr = 60 min) that related the two units. To do the problem, the equality is written in the form of a fraction called a **conversion factor.** One of the quantities is the numerator, and the other is the denominator. Be sure to include the units when you write the conversion factors. Two factors are always possible from any equality.

Two Conversion Factors for the Equality 1 hr = 60 min

$$\frac{\text{Numerator}}{\text{Denominator}} \quad \begin{array}{c} \longrightarrow \\ \longrightarrow \end{array} \quad \frac{60 \text{ min}}{1 \text{ hr}} \quad \text{and} \quad \frac{1 \text{ hr}}{60 \text{ min}}$$

These factors are read as "60 minutes per 1 hour," and "1 hour per 60 minutes." The term *per* means "divide." Some common relationships are given in Table 1.7. It is important that the equality you select to construct a conversion factor is a true relationship.

Table 1.7   Some Common Equalities

Quantity	U.S.	Metric (SI)	Metric–U.S.
Length	1 ft = 12 in.	1 km = 1000 m	2.54 cm = 1 in.
	1 yard = 3 ft	1 m = 1000 mm	1 m = 39.4 in.
	1 mile = 5280 ft	1 cm = 10 mm	1 km = 0.621 mi
Volume	1 qt = 4 cups	1 L = 1000 mL	946 mL = 1 qt
	1 qt = 2 pints	1 dL = 100 mL	1 L = 1.06 qt
	1 gallon = 4 qts	1 mL = 1 cm^3	
Mass	1 lb = 16 oz	1 kg = 1000 g	1 kg = 2.20 lb
		1 g = 1000 mg	454 g = 1 lb
Time		1 hr = 60 min	
		1 min = 60 sec	

the Chemistry place

WEB TUTORIAL
The Metric System

Metric Conversion Factors

We can write metric conversion factors for the metric relationships we have studied. For example, from the equality for meters and centimeters, we can write the following factors:

Metric Equality	*Conversion Factors*
1 m = 100 cm	$\dfrac{100 \text{ cm}}{1 \text{ m}}$ and $\dfrac{1 \text{ m}}{100 \text{ cm}}$

Both are proper conversion factors for the relationship; one is just the inverse of the other. The usefulness of conversion factors is enhanced by the fact that we can turn a conversion factor over and use its inverse.

Figure 1.9 In the U.S., the contents of many packaged foods are listed in both U.S. and metric units.

Q **What are some advantages of using the metric system?**

Metric–U.S. System Conversion Factors

Suppose you need to convert from pounds, a unit in the U.S. system, to kilograms in the metric (or SI) system. A relationship you could use is

1 kg = 2.20 lb

The corresponding conversion factors would be

$$\frac{2.20 \text{ lb}}{1 \text{ kg}} \quad \text{and} \quad \frac{1 \text{ kg}}{2.20 \text{ lb}}$$

Figure 1.9 illustrates the contents of some packaged foods in both metric and U.S. units.

SAMPLE PROBLEM 1.11

Writing Conversion Factors from Equalities

Write conversion factors for the relationship for the following pairs of units:
a. Milligrams and grams
b. Minutes and hours
c. Quarts and milliliters

Solution

	Equality	*Conversion Factors*	
a.	1 g = 1000 mg	$\frac{1 \text{ g}}{1000 \text{ mg}}$ and	$\frac{1000 \text{ mg}}{1 \text{ g}}$
b.	1 hr = 60 min	$\frac{1 \text{ hr}}{60 \text{ min}}$ and	$\frac{60 \text{ min}}{1 \text{ hr}}$
c.	1 qt = 946 mL	$\frac{1 \text{ qt}}{946 \text{ mL}}$ and	$\frac{946 \text{ mL}}{1 \text{ qt}}$

Study Check

Write the equality and conversion factors for the relationship between inches and centimeters.

Problem Solving

The process of problem solving in chemistry requires you to convert a quantity given initially in one unit to the same quantity in different units. Using one or more of the conversion factors we have discussed accomplishes this conversion.

Quantity (Initial unit) $\times$ Conversion factor = Same quantity (New unit)

Suppose you have a patient with a weight of 165 lb. In order to give a medication, you need to know the patient's mass in kilograms.

Step 1　The quantity 165 lb is the given or stated quantity.

Given: 165 lb

Step 2　It is helpful to decide on a unit plan. For this problem our unit plan is to change from pounds to kilograms.

Unit Plan: lb $\longrightarrow$ kg

Given　　　　Unit for answer

Step 3　The relationship for pounds and kilograms can be written as conversion factors.

Relationship: 1 kg = 2.20 lb

Conversion Factors: $\dfrac{1 \text{ kg}}{2.20 \text{ lb}}$　and　$\dfrac{2.20 \text{ lb}}{1 \text{ kg}}$

Step 4　Now we can write the setup to solve the problem using our unit plan (step 2) and one of the conversion factors (step 3). First, write down the given quantity, 165 lb. Then multiply it by the conversion factor that has the unit lb in the denominator because that will cancel out the given unit in the numerator. The unit kg in the numerator (top number) gives the desired unit for the answer.

Problem Setup:

Unit for answer goes here

$$165 \text{ lb} \quad \times \quad \frac{1 \text{ kg}}{2.20 \text{ lb}} \quad = \quad 75.0 \text{ kg}$$

Given　　　　　Conversion factor　　　Answer
(initial unit)　　(cancels initial unit)　(desired unit)

Take a look at the way the units cancel. The unit that you want in the answer is the one that remains after all the other units have canceled out. This is a helpful way to check a problem.

$$\text{lb} \quad \times \quad \frac{1 \text{ kg}}{\text{lb}} \quad = \quad \text{kg} \quad \text{Answer unit}$$

Do the calculations on your calculator to give the following:

165　÷　2.20　=　75.0　(add zero)

The value of 75.0 combined with the desired unit, kilograms, gives the final answers of 75.0 kg. With few exceptions, answers to numerical problems must contain a number and a unit.

Suppose the problem were set up in the following way. If the units do not cancel, you know there is an error in the setup.

$$165 \text{ lb} \quad \times \quad \frac{2.20 \text{ lb}}{1 \text{ kg}} \quad = \quad \frac{363 \text{ lb}^2}{\text{kg}} \quad \textbf{Wrong units!}$$

SAMPLE PROBLEM 1.12

Problem Solving Using Metric Factors

The Daily Value (DV) for sodium is 2400 mg. How many grams of sodium is that?

Solution

Step 1   Given: 2400 mg

Step 2   Unit Plan: mg $\longrightarrow$ g   Unit for answer

Step 3   Metric Equality: 1 g = 1000 mg

Conversion Factors: $\dfrac{1 g}{1000 \text{ mg}}$ and $\dfrac{1000 \text{ mg}}{1 g}$

Step 4   Problem Setup:

Unit for answer
goes here

$$2400 \; \cancel{\text{mg}} \quad \times \quad \frac{1 \text{ g}}{1000 \; \cancel{\text{mg}}} \quad = \quad 2.4 \text{ g of sodium needed daily}$$

Given   Metric factor   Answer (in grams)

Study Check

A can containing 473 mL of frozen orange juice is mixed with 1415 mL of water. How many liters of orange juice were prepared?

Using Two or More Conversion Factors

In many problems you will need to use two or more steps in your unit plan. Then two or more conversion factors will be required. These can be constructed from the equalities you have learned or those stated in the problem. In setting up the problem, one factor follows the other. Each factor is arranged to cancel the preceding unit until you obtain the desired unit.

SAMPLE PROBLEM 1.13

Problem Solving Using Two Factors

A recipe for salsa requires 3.0 cups of tomato sauce. If only metric measures are available, how many milliliters of tomato sauce are needed? (There are 4 cups in 1 quart.)

Solution

You may not know a relationship between cups and milliliters. However, you do know how to change cups to quarts, and quarts to milliliters.

Step 1   Given: 3.0 cups

Step 2   Unit Plan: cups $\longrightarrow$ quarts $\longrightarrow$ milliliters   Unit for answer

Step 3   Relationships and Conversion Factors:

1. 1 qt = 4 cups $\dfrac{1 \text{ qt}}{4 \text{ cups}}$ and $\dfrac{4 \text{ cups}}{1 \text{ qt}}$

2. 1 qt = 946 mL $\dfrac{1 \text{ qt}}{946 \text{ mL}}$ and $\dfrac{946 \text{ mL}}{1 \text{ qt}}$

Step 4   Problem Setup: Use factor 1 to convert from cups to quarts:

$$3.0 \text{ cups} \times \frac{1 \text{ qt}}{4 \text{ cups}} = 0.75 \text{ qt}$$

Then use factor 2 to convert from quarts to milliliters:

$$0.75 \text{ qt} \times \frac{946 \text{ mL}}{1 \text{ qt}} = 710 \text{ mL}$$

In a continuous setup all the factors are used one after the other. The unit in the denominator of each new factor always cancels the unit in the previous numerator.

$$3.0 \text{ cups} \times \frac{1 \text{ qt}}{4 \text{ cups}} \times \frac{946 \text{ mL}}{1 \text{ qt}} = 710 \text{ mL}$$

Given   U.S.   Metric−U.S.   Answer
quantity   factor   factor   (in milliliters)

All the calculations are done in a sequence on the calculator. No intermediate values are written down, only the final answer.

3.0   ÷   4   ×   946   =   710
Two SFs   Exact   Three SFs   Two SFs

Study Check

One medium bran muffin contains 4.2 g of fiber. How many ounces (oz) of fiber are obtained by eating three medium bran muffins, if 1 lb = 16 oz? (*Hint:* Number of muffins ⟶ g of fiber ⟶ lb ⟶ oz)

Using a sequence of two or more conversion factors is a very efficient way to set up and solve problems, especially if you are using a calculator. Once you have the problem set up, the calculations can be done without writing out the intermediate values. This process is worth practicing until you understand unit cancellation and the mathematical calculations. A summary of steps in problem solving is given in Table 1.8.

Table 1.8   Steps in Problem Solving

Step 1   Identify the given quantity and unit.

Step 2   Write a unit plan to help you think about changing units from the given to the answer unit. Be sure you can supply a conversion factor for each change.

Step 3   Determine the equalities and corresponding conversion factors you will need to change from one unit to another.

Step 4   Set up the problem according to your unit plan. Arrange each conversion factor to cancel the preceding unit. Check that the units cancel to give the unit of the answer.

Step 5   Carry out the calculations and give a final answer with the correct number of significant figures and unit.

Clinical Calculations Using Conversion Factors

Conversion factors are also useful for calculating medications. For example, if an antibiotic is available in 5-mg tablets, the dosage can be written as a conversion factor, 5 mg/1 tablet. In many hospitals, the apothecary unit of grains (gr) is still in use; there are 65 mg in 1 gr. When you do a medication problem, you often start with a doctor's order that contains the quantity to give the patient. The medication dosage is used as a conversion factor.

SAMPLE PROBLEM 1.14

Clinical Calculations with Factors

Synthroid is used as a replacement or supplemental therapy for diminished thyroid function. A dosage of 0.200 mg is prescribed with tablets that contain 50 μg Synthroid. How many tablets are required to provide the prescribed medication?

Solution

Step 1   Given: 0.200 mg Synthroid

Step 2   Unit Plan: mg $\longrightarrow$ μg $\longrightarrow$ tablets   Unit for answer

Step 3   Relationships and Conversion Factors

1. 1 mg = 1000 μg $\dfrac{1 \text{ mg}}{1000 \ \mu g}$ and $\dfrac{1000 \ \mu g}{1 \text{ mg}}$

2. 1 tablet = 50 μg Synthroid $\dfrac{1 \text{ tablet}}{50 \ \mu g}$ and $\dfrac{50 \ \mu g}{1 \text{ tablet}}$

Step 4   Problem Setup:

$$0.200 \ \cancel{\text{mg}} \text{ Synthroid} \times \frac{1000 \ \cancel{\mu g}}{1 \ \cancel{\text{mg}}} \times \frac{1 \text{ tablet}}{50 \ \cancel{\mu g} \text{ Synthroid}} = 4 \text{ tablets}$$

Initial unit   Metric factor   Dosage factor   Answer

Study Check

An antibiotic dosage of 500 mg is ordered. If the antibiotic is supplied in liquid form as 250 mg in 5.0 mL, how many mL would be given?

SAMPLE PROBLEM 1.15

Converting Clinical Factors

Some important functions of calcium in the body are bone formation, muscle contraction, and blood coagulation. For an adult, the normal range for serum calcium is 8.7–10.2 mg/dL. A calcium deficiency (hypocalcemia) can cause muscle spasms, cramps, and convulsions. Determine whether a calcium level falls within the normal range if a 2.0-mL serum sample from a patient contains 180 μg of calcium.

Solution

The relationship of the lab test can be set up as two units. Then the units are converted to the desired units.

$$\frac{180 \ \cancel{\mu g} \text{ calcium}}{2.0 \ \cancel{\text{mL}}} \times \frac{1 \text{ mg}}{1000 \ \cancel{\mu g}} \times \frac{100 \ \cancel{\text{mL}}}{1 \text{ dL}} = \frac{9.0 \text{ mg}}{\text{dL}}$$

The lab value 9.0 mg/dL is within the normal range for calcium.

A patient has a serum calcium of 0.050 g/L. Does this patient have a normal calcium, hypocalcemia, or hypercalcemia (low or high calcium level in the blood)?

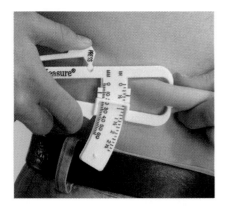

Figure 1.10 The thickness of the skin fold at the waist measured in millimeters (mm) is used to determine the amount of body fat.

Q **What is the percent body fat of an athlete with a body mass of 120 kg and 18 kg body fat?**

Using Percents as Conversion Factors

Sometimes percents are given in a problem. To work with a percent, it is convenient to write it as a conversion factor. To do this, we choose units from the problem to express the numerical relationship of the part to 100 parts of the whole. For example, an athlete might have 18% body fat by weight. (See Figure 1.10). We can use a choice of units for weight or mass to write the following conversion factors:

$$\frac{18 \text{ kg body fat}}{100 \text{ kg body mass}} \qquad \frac{18 \text{ lb body fat}}{100 \text{ lb body weight}}$$

If a question is asked about the number of kilograms of body fat in a marathon runner with a total body mass of 48 kg, we would state the units in the factor in kg body fat. Note that the units in the factor for the percent cancel the given.

$$48 \text{ kg body mass} \times \frac{18 \text{ kg body fat}}{100 \text{ kg body mass}} = 8.6 \text{ kg body fat}$$

Given   Percent factor

If the problem asks about the pounds of body fat in a dancer weighing 126 lb, we would state the factor in lb body fat.

$$126 \text{ lb body weight} \times \frac{18 \text{ lb body fat}}{100 \text{ lb body weight}} = 23 \text{ lb body fat}$$

SAMPLE PROBLEM 1.16

Using Percent Factors

A wine contains 13% alcohol by volume. How many milliliters of alcohol are contained in a glass containing 125 mL of the wine?

Solution

The factor for 13% alcohol can be written using mL units as:

$$\frac{13 \text{ mL alcohol}}{100 \text{ mL wine}}$$

The factor is then used with the given to set up the problem.

$$125 \text{ mL wine} \times \frac{13 \text{ mL alcohol}}{100 \text{ mL wine}} = 16 \text{ mL alcohol}$$

There are 85 students in a chemistry class. On the last test, 20% received grades of A. How many As were earned on the test?

QUESTIONS AND PROBLEMS

Problem Solving Using Conversion Factors

1.49 Why can two conversion factors be written for an equality such as
1 m = 100 cm?

1.50 How can you check that you have written the correct conversion factors for an equality?

1.51 What equality is expressed by the conversion factor $\dfrac{1000\ g}{1\ kg}$?

1.52 What equality is expressed by the conversion factor $\dfrac{1\ m}{100\ cm}$?

1.53 Write a numerical relationship and conversion factors for each of the following statements:
a. One yard is 3 feet.
b. One mile is 5280 feet.
c. One minute is 60 seconds.
d. A car goes 27 miles on 1 gallon of gas.

1.54 Write a numerical relationship and conversion factors for each of the following statements:
a. One gallon is 4 quarts.
b. At the store, oranges are $1.29 per lb.
c. There are 7 days in 1 week.
d. One dollar has four quarters.

1.55 Write the numerical relationship and conversion factors for the following pairs of units:
a. centimeters and meters   **b.** milligrams and grams
c. liters and milliliters   **d.** deciliters and milliliters

1.56 Write the numerical relationship and corresponding conversion factors for the following pairs of units:
a. centimeters and inches   **b.** pounds and kilograms
c. pounds and grams   **d.** quarts and milliliters

1.57 When you convert one unit to another, how do you know which unit to place in the denominator?

1.58 When you convert one unit to another, how do you know which unit to place in the numerator?

1.59 Use U.S. system conversion factors to solve the following problems:
a. How many yards are in 24 ft?
b. How many seconds are in 15 min?
c. You need 3.5 qt of oil for your car. How many gallons is that?

1.60 Use U.S. system conversion factors to solve the following problems:
a. A necklace contains 2.6 oz of gold. How many pounds is that?
b. You ran a total of 2.4 miles today. How far is that in feet? (1 mi = 5280 ft)
c. One game at the arcade requires one quarter. You have $3.50 in your pocket. How many games can you play?

1.61 Use metric conversion factors to solve the following problems:
a. The height of a student is 175 centimeters. How tall is the student in meters?
b. A cooler has a volume of 5500 mL. What is the capacity of the cooler in liters?
c. A hummingbird has a mass of 0.0055 kg. What is the mass of the hummingbird in grams?

1.62 Use metric conversion factors to solve the following problems:
   a. The Daily Value of phosphorus is 800 mg. How many grams of phosphorus are recommended?
   b. A glass of orange juice contains 0.85 dL of juice. How many milliliters of orange juice is that?
   c. A package of chocolate instant pudding contains 2840 mg of sodium. How many grams of sodium is that?

1.63 Solve the following problems using one or more conversion factors:
   a. A container holds 0.750 qt of liquid. How many milliliters of lemonade will it hold?
   b. What is the mass in kilograms of a person who weighs 165 lb?
   c. The femur, or thighbone, is the longest bone in the body. In a 6-ft-tall person, the femur might be 19.5 in. long. What is the length of that person's femur in millimeters?
   d. A person on a diet has been losing weight at the rate of 3.5 lb per week. If the person has been on the diet for 6 weeks, how many kilograms were lost?

1.64 Solve the following problems using one or more conversion factors:
   a. You need 4.0 ounces of a steroid ointment. If there are 16 oz in 1 lb, how many grams of ointment does the pharmacist need to prepare?
   b. During surgery, a patient receives 5.0 pints of plasma. How many milliliters of plasma were given? (1 quart = 2 pints)
   c. A piece of plastic tubing measuring 560 mm in length is needed for an intravenous setup. How many feet of tubing are required?
   d. Zippy the snail moves at the rate of 2.0 inches per hour. How many centimeters does Zippy travel in 4.0 hours?

1.65 Using conversion factors, solve the following clinical problems:
   a. You have used 250 L of distilled water for a dialysis patient. How many gallons of water is that?
   b. A patient needs 0.024 g of a sulfa drug. There are 8-mg tablets in stock. How many tablets should be given?
   c. The daily dose of ampicillin for the treatment of an ear infection is 115 mg/kg of body weight. What is the daily dose for a 34-lb child?

1.66 Using conversion factors, solve the following clinical problems:
   a. The physician has ordered 1.0 g of tetracycline to be given every 6 hours to a patient. If your stock on hand is 500-mg tablets, how many will you need for 1 day's treatment?
   b. An intramuscular medication is given at 5.00 mg/kg of body weight. If you give 425 mg of medication to a patient, what is the patient's weight in pounds?
   c. A physician has ordered 325 mg of atropine, intramuscularly. If atropine were available as 0.50 g/mL of solution, how many milliliters would you need to give?

1.67 a. Oxygen makes up 46.7% by mass of the earth's crust. How many grams of oxygen are present if a sample of the earth's crust has a mass of 325 g?
   b. Magnesium makes up 2.1% by mass of the earth's crust. How many grams of magnesium are present if a sample of the earth's crust has a mass of 1.25 g?

1.68 a. Water is 11.2% by mass hydrogen. How many kilograms of water would contain 5.0 g of hydrogen?
   b. Water is 88.8% by mass oxygen. How many grams of water would contain 2.25 kg of oxygen?

1.8 Density

Why do objects sink or float? In Figure 1.11, we see that a sample of lead sinks in water, but a cork floats. Why does this happen? We need to look at density. The density of lead is greater than the density of water, which makes the lead object sink. The cork floats because cork is less dense than water. Differences in density determine whether an object will sink or float.

The mass and volume of any object can be measured. However, the separate measurements do not tell us how tightly packed the substance might be. However, if we compare the mass of the object to its volume, we obtain a relationship called **density.**

$$\text{Density} = \frac{\text{mass of substance}}{\text{volume of substance}}$$

In the metric system, the densities of solids and liquids are usually expressed as grams per cubic centimeter (g/cm^3) or grams per milliliter (g/mL). The density of gases is usually stated as grams per liter (g/L). Table 1.9 gives the densities of some common substances.

Table 1.9   Densities of Some Common Substances

Solids (at 25°C)	Density (g/mL)	Liquids (at 25°C)	Density (g/mL)	Gases (at 0°C)	Density (g/L)
Cork	0.26	Gasoline	0.66	Hydrogen	0.090
Wood (maple)	0.75	Ethyl alcohol	0.79	Helium	0.179
Ice	0.92	Olive oil	0.92	Methane	0.714
Sugar	1.59	Water (at 4°C)	1.000	Neon	0.90
Bone	1.80	Plasma (blood)	1.03	Nitrogen	1.25
Aluminum	2.70	Urine	1.003–1.030	Air (dry)	1.29
Cement	3.00	Milk	1.04	Oxygen	1.43
Diamond	3.52	Mercury	13.6	Carbon dioxide	1.96
Silver	10.5				
Lead	11.3				
Gold	19.3				

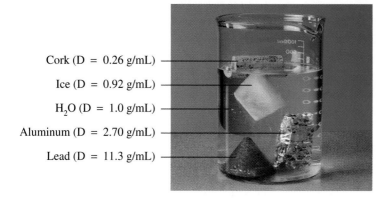

Cork (D = 0.26 g/mL)
Ice (D = 0.92 g/mL)
H₂O (D = 1.0 g/mL)
Aluminum (D = 2.70 g/mL)
Lead (D = 11.3 g/mL)

Figure 1.11 Objects that sink in water are more dense than water; objects float if they are less dense.

Q Why does a cork float and a piece of lead sink?

Mass of zinc object

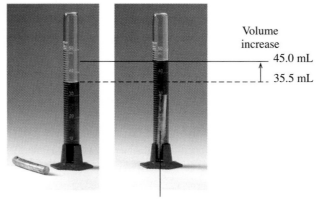

Volume increase
45.0 mL
35.5 mL

Submerged zinc object

Figure 1.12 The density of a solid can be determined by volume displacement because a submerged object displaces a volume of water equal to its own volume.

Q What is the density of the zinc object?

Density of Solids

The density of a solid is calculated from its mass and volume. When a solid is completely submerged, it displaces a volume of water that is *equal to its own volume*. In Figure 1.12, the water level rises from 35.5 mL to 45.0 mL. This means that 9.5 mL of water is displaced and that the volume of the object is 9.5 mL.

SAMPLE PROBLEM 1.17

Calculating Density

A 50.0-mL sample of buttermilk has a mass of 56.0 g. What is the density of the buttermilk?

Solution

To calculate density, substitute the mass (g) and volume (mL) of the buttermilk into the expression for density.

$$\text{Density} = \frac{\text{mass}}{\text{volume}} = \frac{56.0 \text{ g}}{50.0 \text{ mL}} = \frac{1.12 \text{ g}}{1 \text{ mL}} = 1.12 \text{ g/mL} \quad \begin{array}{l} \text{Two units in} \\ \text{this answer} \end{array}$$

Study Check

A copper sample has a mass of 44.65 g and a volume of 5.0 cm³. What is the density of copper, expressed as grams per cubic centimeter (g/cm³)?

SAMPLE PROBLEM 1.18

Using Volume Displacement to Calculate Density

A lead weight used in the belt of a scuba diver has a mass of 226 g. When the weight is carefully placed in a graduated cylinder containing 200.0 mL of water, the water level rises to 220.0 mL. What is the density of the lead weight (g/mL)?

Solution

Both the mass and volume of the lead weight are needed to calculate its density. Its mass, 226 g, is given. The volume of the lead weight is equal to the volume of water displaced, which is calculated as follows:

Volume displaced	220.0 mL
	−200.0 mL
	20.0 mL Answer to first decimal place

Volume of the lead weight = 20.0 mL

Three SFs

$$\text{Density of lead} = \frac{\text{mass of lead}}{\text{volume of lead}} = \frac{226\text{ g}}{20.0\text{ mL}} = 11.3\text{ g/mL}$$

Three SFs Three SFs in answer

In the expression for density, be sure to use the volume of water the object displaces and *not* the original volume of water.

Study Check

A total of 0.50 lb of glass marbles is added to 425 mL of water. The water level rises to a volume of 528 mL. What is the density (g/mL) of the glass marbles?

EXPLORE YOUR WORLD

Sink or Float?

1. Fill a large container or bucket with water. Place a can of diet and a can of nondiet soft drink in the water. What happens? Using information on the label, how might you account for your observations?
2. Design an experiment to determine the substance that is the most dense in each of the following:
   a. water and vegetable oil
   b. water and ice
   c. rubbing alcohol and ice
   d. vegetable oil, water, and ice

SAMPLE PROBLEM 1.19

Sink or Float?

From their densities, determine whether each of the following substances will sink or float when placed in seawater, which has a density of 1.025 g/mL.

Substance	***Density***
Gasoline	0.66 g/mL
Asphalt	1.2 g/mL
Cardboard	0.69 g/cm^3

Solution

The gasoline and cardboard will float because they are less dense than seawater. Asphalt will sink because it is more dense than water.

Study Check

Using Table 1.9, explain why helium-filled balloons float in air.

Problem Solving Using Density

Because density is a comparison of two units, it can be used as a conversion factor. For example, if the mass and the density of a sample are known, the volume of the sample can be calculated.

SAMPLE PROBLEM 1.20

Problem Solving Using Density

Olive oil has a density of 0.92 g/mL. How many mL of olive oil would you measure out if you needed 12 g of olive oil?

Solution

Step 1 **Given:** 12 g olive oil

Step 2 **Unit Plan:** g $\longrightarrow$ mL

Step 3   Relationships and Conversion Factors:

Density: 0.92 g = 1 mL $\dfrac{0.92 \text{ g}}{1 \text{ mL}}$ and $\dfrac{1 \text{ mL}}{0.92 \text{ g}}$

Step 4   Problem Setup:

$$12 \text{ g olive oil} \quad \times \quad \dfrac{1 \text{ mL olive oil}}{0.92 \text{ g olive oil}} \quad = \quad 13 \text{ mL olive oil}$$

Given   Density factor   Answer to two SFs

Study Check

If there is 0.25 mL of mercury in a thermometer, how many grams of mercury are there?

Specific Gravity

Specific gravity (sp gr) is a ratio between the density of a substance and the density of water. Specific gravity is calculated by dividing the density of a sample by the density of water. In this text, we will use the metric value of 1.00 g/mL for the density of water. A substance with a specific gravity of 1.00 has the same density as water. A substance with a specific gravity of 3.00 is three times as dense as water, whereas a substance with a specific gravity of 0.50 is just one-half as dense as water.

$$\text{Specific gravity} = \dfrac{\text{density of sample}}{\text{density of water}}$$

In the calculations for specific gravity, the units of density must match. Then all units cancel to leave only a number. *Specific gravity is one of the few unitless values you will encounter in chemistry.*

An instrument called a hydrometer is often used to measure the specific gravity of fluids such as battery fluid or a sample of urine. In Figure 1.13, a hydrometer is used to measure the specific gravity of a fluid.

SAMPLE PROBLEM 1.21

Specific Gravity

What is the specific gravity of coconut oil that has a density of 0.925 g/mL?

Solution

$$\text{sp gr oil} = \dfrac{\text{density of oil}}{\text{density of water}} = \dfrac{0.925 \text{ g/ml}}{1.00 \text{ g/ml}} = 0.925 \quad \text{No units}$$

Study Check

What is the specific gravity of ice if 35.0 g of ice has a volume of 38.2 mL?

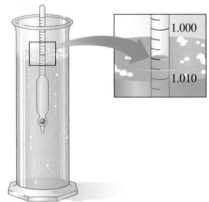

Figure 1.13 When the specific gravity of beer measures 1.010 or less with a hydrometer, the fermentation process is complete.

Q If the hydrometer reading is 1.006, what is the density of the liquid?

SAMPLE PROBLEM 1.22

Problem Solving with Specific Gravity

John took 2.0 teaspoons (tsp) of cough syrup (sp gr 1.20) for a persistent cough. If there are 5.0 mL in 1 tsp, what was the mass (in grams) of the cough syrup?

Solution

Step 1   Given: 2.0 tsp

Step 2   Unit Plan: tsp $\longrightarrow$ mL $\longrightarrow$ g   Unit for answer

Step 3   Relationships and Conversion Factors:

$$1 \text{ tsp} = 5.0 \text{ mL} \qquad \frac{1 \text{ tsp}}{5.0 \text{ mL}} \quad \text{and} \quad \frac{5.0 \text{ mL}}{1 \text{ tsp}}$$

For problem solving, it is convenient to convert the specific gravity value to density.

Density = (sp gr) g/mL = 1.20 g/mL

Relationship: 1 mL = 1.20 g

Conversion Factors: $\dfrac{1 \text{ mL}}{1.20 \text{ g}}$ and $\dfrac{1.20 \text{ g}}{1 \text{ mL}}$

Step 4   Problem Setup:

$$2.0 \text{ tsp} \times \frac{5.0 \text{ mL}}{1 \text{ tsp}} \times \frac{1.20 \text{ g}}{1 \text{ mL}} = 12 \text{ g of syrup}$$

Study Check

An ebony carving has a mass of 275 g. If ebony has a specific gravity of 1.33, what is the volume of the carving?

QUESTIONS AND PROBLEMS

Density

1.69 In an old trunk, you find a piece of metal that you think may be aluminum, silver, or lead. In lab you find it has a mass of 217 g and a volume of 19.2 cm^3. Using Table 1.9, what is the metal you have found?

1.70 Suppose you have two 100-mL graduated cylinders. In each cylinder there are 40.0 mL of water. You also have two cubes: one is lead and the other is aluminum. Each cube measures 2.0 cm on each side. After you carefully lower each into the water of its own cylinder, what will the new water level be in each of the cylinders?

1.71 What is the density (g/mL) of each of the following samples?
   a. A 20.0-mL sample of a salt solution that has a mass of 24.0 g.
   b. A solid object with a mass of 1.65 lb and a volume of 170 mL.
   c. A gem has a mass of 45.0 g. When the gem is placed in a graduated cylinder containing 20.0 mL of water, the water level rises to 34.5 mL.

1.72 What is the density of each of the following samples?
   a. A medication, if 3.00 mL has a mass of 3.85 g.
   b. The fluid in a car battery, if it has a volume of 125 mL and a mass of 155 g.
   c. A 5.00-mL urine sample from a patient suffering from symptoms resembling those of diabetes mellitus. The mass of the urine sample is 5.025 g.

1.73 Use the density value to solve the following problems:
   a. What is the mass, in grams, of 150 mL of a liquid with a density of 1.4 g/mL?
   b. What is the mass of a glucose solution that fills a 0.500-L intravenous bottle if the density of the glucose solution is 1.15 g/mL?

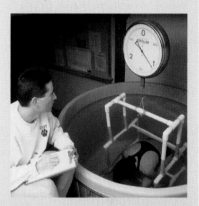

c. A sculptor has prepared a mold for casting a bronze figure. The figure has a volume of 225 mL. If bronze has a density of 7.8 g/mL, how many ounces of bronze are needed in the preparation of the bronze figure?

1.74 Use the density value to solve the following problems:

a. A graduated cylinder contains 18.0 mL of water. What is the new water level after 35.6 g of silver metal with a density of 10.5 g/mL is submerged in the water?

b. A thermometer containing 8.3 g of mercury has broken. If mercury has a density of 13.6 g/mL, what volume spilled?

c. A fish tank holds 35 gallons of water. Using the density of 1.0 g/mL for water, determine the number of pounds of water in the fish tank.

1.75 Solve the following specific gravity problems:

a. A urine sample has a density of 1.030 g/mL. What is the specific gravity of the sample?

b. A liquid has a volume of 40.0 mL and a mass of 45.0 g. What is the specific gravity of the liquid?

c. The specific gravity of a vegetable oil is 0.85. What is its density?

1.76 Solve the following specific gravity problems:

a. A 5% glucose solution has a specific gravity of 1.02. What is the mass of 500 mL of glucose solution?

b. A bottle containing 325 g of cleaning solution is used for carpets. If the cleaning solution has a specific gravity of 0.850, what volume of solution was used?

c. Butter has a specific gravity of 0.86. What is the mass, in grams, of 2.15 L of butter?

LEARNING GOAL

Given a temperature, calculate a corresponding temperature on another scale.

1.9 Temperature

When we take the **temperature** of a substance, we are measuring the intensity of heat. Temperature tells us the direction that heat will flow between two substances. Heat always flows from a substance with a higher temperature to a substance with a lower temperature until the temperatures of both are the same. When you drink hot coffee or touch a hot stove, heat flows to your mouth or hand, which is at a lower temperature. When you touch an ice cube, it feels cold because heat flows from your hand to the colder ice cube.

We use thermometers to determine how "hot" or "cold" a substance is. When a thermometer is placed in a substance, the heat flow causes the liquid in the thermometer—alcohol or mercury—to expand or contract.

Celsius and Fahrenheit Temperatures

Temperatures in science, and in most of the world, are measured and reported in *Celsius (°C)* units. In the United States, everyday temperatures are commonly reported in *Fahrenheit (°F)* units. A typical room temperature of 22°C would be the same as 72°F. A normal body temperature of 37.0°C is 98.6°F.

On the Celsius and Fahrenheit scales, the temperatures of melting ice and boiling water are used as reference points. On the Celsius scale, the freezing point of water is defined as 0°C, and the boiling point as 100°C. On a Fahrenheit scale,

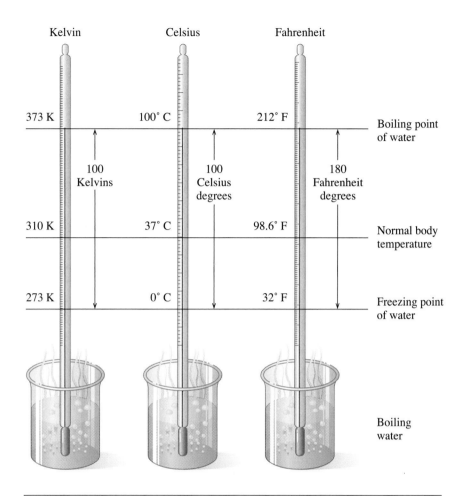

Figure 1.14 A comparison of the Fahrenheit, Celsius, and Kelvin temperature scales between the freezing and boiling points of water.

Q **What is the difference in the values for freezing on the Celsius and Fahrenheit temperature scales?**

water freezes at 32°F and boils at 212°F. On each scale, the temperature difference between freezing and boiling is divided into smaller units, or degrees. On the Celsius scale, there are 100 units between the freezing and boiling of water compared with 180 units on the Fahrenheit scale. That makes a Celsius unit almost twice the size of a Fahrenheit degree: 1°C = 1.8°F. (See Figure 1.14.)

$$180 \text{ Fahrenheit units} = 100 \text{ Celsius units}$$

$$\frac{180 \text{ Fahrenheit units}}{100 \text{ Celsius units}} \quad \text{or} \quad \frac{1.8 \text{ Fahrenheit units}}{1 \text{ Celsius unit}}$$

In a chemistry laboratory or a hospital, you will measure temperatures in Celsius units. If you do need to convert to Fahrenheit, multiply the Celsius temperature by the above conversion factor to change to Fahrenheit units, and add 32 degrees. The 32 you add adjusts the freezing point on the Celsius scale of 0°C to 32°F on the Fahrenheit scale.

$$°F = \frac{1.8°F}{1°C}(°C) + 32° \longrightarrow °F = 1.8 \, (°C) + 32°$$

Changes °C   Adjusts freezing
to °F   points

SAMPLE PROBLEM 1.23

Temperature Conversion

While traveling in China, you discover that your temperature is 38.2°C. What is your temperature measured with a Fahrenheit thermometer?

Solution

Using the temperature conversion equation, we write:

$$°F = 1.8(°C) + 32°$$

In the equation, *the values of 1.8 and 32 are exact numbers*. Therefore, the answer is reported to the same decimal place as the initial temperature.

$$\begin{aligned} °F &= 1.8(38.2) + 32 \\ &= 68.8 + 32 \quad \text{Exact} \\ &= 100.8°F \quad \text{Answer to first decimal place} \end{aligned}$$

Study Check

When making ice cream, you use rock salt to chill the mixture. If the temperature drops to −11°C, what is it in °F?

SAMPLE PROBLEM 1.24

Converting to Celsius Temperature

You are going to cook a turkey at 325°F. If you use an oven with Celsius settings, at what temperature should you set it?

Solution

To convert from Fahrenheit to Celsius, we rearrange the temperature equation.

$$\begin{aligned} °F &= 1.8\,(°C) + 32 \\ °F - 32 &= 1.8\,(°C) \\ \frac{°F - 32}{1.8} &= °C \end{aligned}$$

Entering the temperature of 325°F into the equation gives:

$$°C = \frac{(325 - 32)}{1.8} = \frac{\overset{\text{Three SFs}}{293}}{\underset{\text{Exact}}{1.8}} = 163°C$$

On your calculator, first subtract (325 − 32), and then divide by 1.8.

Study Check

Your patient has a temperature of 103.6°F. What is this temperature on a Celsius thermometer?

Kelvin Temperature Scale

Scientists tell us that the coldest temperature possible is $-273°C$ (more precisely, $-273.15°C$). On the Kelvin scale, this temperature, called *absolute zero,* has the value of 0 Kelvin (0 K). Units on the Kelvin scale are called kelvins (K); no degree symbol is used. Because there are no lower temperatures, the Kelvin scale has no negative numbers. Between the freezing and boiling points of water, there are 100 kelvins, which makes a kelvin equal in size to a Celsius unit.

$1 \text{ K} = 1°C$

To calculate a Kelvin temperature, add 273 to the Celsius temperature:

$\text{K} = °C + 273$

Table 1.10 gives a comparison of some temperatures on the three scales.

Table 1.10   A Comparison of Temperatures

Example	Fahrenheit (°F)	Celsius (°C)	Kelvin (K)
Sun	9937	5503	5776
A hot oven	450	232	505
A desert	120	49	322
A high fever	104	40	313
Room temperature	72	22	295
Water freezes	32	0	273
A northern winter	−76	−60	213
Helium boils	−452	−269	4
Absolute zero	−459	−273	0

SAMPLE PROBLEM 1.25

Converting Celsius Temperature to Kelvins

What is a normal body temperature of 37°C on the Kelvin scale?

Solution

To convert from Celsius temperature to a Kelvin temperature, we use the equation

$\text{K} = °C + 273$

$\text{K} = 37 + 273$

$ = 310. \text{ K}$

Study Check

You are cooking a pizza at 375°F. What is that temperature in kelvins? (*Hint:* Convert to Celsius temperature first.)

SAMPLE PROBLEM 1.26

Converting from Kelvin to Celsius Temperature

In a cryogenic laboratory, a technician freezes a sample using liquid nitrogen at 77 K. What is the temperature in °C?

HEALTH NOTE

Variation in Body Temperature

Normal body temperature is considered to be 37.0°C, although it varies throughout the day and from person to person. Oral temperatures of 36.1°C are common in the morning and climb to a high of 37.2°C between 6 P.M. and 10 P.M. Temperatures above 37.2°C for a person at rest are usually an indication of disease. Individuals involved in prolonged exercise may also experience elevated temperatures. Body temperatures of marathon runners can range from 39°C to 41°C as heat production during exercise exceeds the body's ability to lose heat.

Changes of more than 3.5°C from the normal body temperature begin to interfere with bodily functions. Temperatures above 41°C can lead to convulsions, particularly in children, which may cause permanent brain damage. Heatstroke occurs above 41.1°C. Sweat production stops and the skin becomes hot and dry. The pulse rate is elevated, and respiration becomes weak and rapid. The person can become lethargic and lapse into a coma. Damage to internal organs is a major concern, and treatment, which must be immediate, may include immersing the person in an ice-water bath.

In hypothermia, body temperature can drop as low as 28.5°C. The person may appear cold and pale and have an irregular heartbeat. Unconsciousness can occur if the body temperature drops below 26.7°C. Respiration becomes slow and shallow, and oxygenation of the tissues decreases. Treatment involves providing oxygen and increasing blood volume with glucose and saline fluids. Internal temperature may be restored by injecting warm fluids (37.0°C) into the peritoneal cavity.

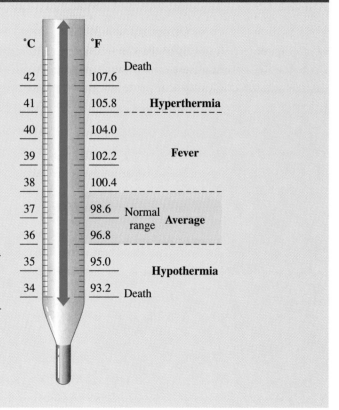

Solution

To find the Celsius temperature, we rearrange the Kelvin equation:

$$K = °C + 273$$

$$K - 273 = °C$$

Using our given temperature of 77 K, we calculate the Celsius temperature for the freezing of the sample.

$$°C = 77\ K - 273 = -196°C$$

Study Check

On the planet Mercury, the average night temperature is 13 K, and the average day temperature is 683 K. What are these temperatures in Celsius degrees?

QUESTIONS AND PROBLEMS

Measuring Temperature

1.77 Your friend who is visiting from France just took her temperature. When she reads 99.8, she becomes concerned that she is quite ill. How would you explain the temperature to your friend?

1.78 You have a friend who is using a recipe for flan from a Mexican cookbook. You notice that he set your oven temperature at 175°F. What would you advise him to do?

1.79 Solve the following temperature conversions:

    a. 37.0°C = _____ °F   **b.** 65.3°F = _____ °C

    c. −27°C = _____ K   **d.** 62°C = _____ K

    e. 114°F = _____ °C   **f.** 72°F = _____ K

1.80 Solve the following temperature conversions:

    a. 25°C = _____ °F

    b. 155°C = _____ °F

    c. −25°F = _____ °C

    d. 224 K = _____ °C

    e. 545 K = _____ °C

    f. 875 K = _____ °F

1.81 **a.** A patient with heat stroke has a temperature of 106°F. What does this read on a Celsius thermometer?

    b. Because high fevers can cause convulsions in children, the doctor wants to be called if the child's temperature goes over 40.0°C. Should the doctor be called if a child has a temperature of 103°F?

1.82 **a.** Hot compresses are being prepared for the patient in room 32B. The water is heated to 145°F. What is the temperature of the hot water in °C?

    b. During extreme hypothermia, a young woman's temperature dropped to 20.6°C. What was her temperature on the Fahrenheit scale?

Chapter Review

1.1 Units of Measurement

In science, physical quantities are described in units of the metric or International System (SI). Some important units are meter (m) for length, liter (L) for volume, gram (g) and kilogram (kg) for mass, and Celsius (°C) for temperature.

1.2 Scientific Notation

Large and small numbers can be written using scientific notation in which the decimal point is moved to give a coefficient between 1 and 10 and the number of decimal places moved shown as a power of 10. A large number will have a positive power of 10, while a small number will have a negative power of 10.

1.3 Measured and Exact Numbers

A measured number is any number derived from using a measuring device to determine a quantity. An exact number is obtained by counting items or from a definition; no measuring device is used.

1.4 Significant Figures

Significant figures are the numbers reported in a measurement including the last estimated digit. Zeros in front of a decimal number or at the end of a large number are not significant.

1.5 Significant Figures in Calculations

In multiplication or division, the measured number in the calculation that has the smallest number of significant figures determines the number of significant figures in an answer. In addition or subtraction, the final answer reflects the last place that all the measured numbers have in common.

1.6 SI and Metric Prefixes

Prefixes placed in front of a unit change the size of the unit by factors of 10. Prefixes such as *centi*, *milli*, and *micro* provide smaller units; prefixes such as *kilo* provide larger units. An equality relates two metric units that measure the same quantity of length, volume, or mass. Examples of metric equalities are 1 m = 100 cm; 1 L = 1000 mL; 1 kg = 1000 g.

1.7 Problem Solving Using Conversion Factors

Conversion factors are used to express a relationship in the form of a fraction. Two factors can be written for any relationship in the metric or U.S. system. Factors are useful when changing a quantity expressed in one unit to one expressed in another.

1.8 Density

The density of a substance is a ratio of its mass to its volume, usually g/mL or g/cm^3. The units of density can be used as a factor to convert between the mass and volume of a substance. Specific gravity (sp gr) compares the density of a substance to the density of water, 1.00 g/mL.

1.9 Temperature

1.9 Temperature

In science, temperature is measured in Celsius units, °C, or kelvins, K. In the United States, the Fahrenheit scale, °F, is still in use.

Key Terms

accuracy How close a measurement is to the true value.

Celsius (°C) temperature scale A temperature scale on which water has a freezing point of 0°C and a boiling point of 100°C.

centimeter (cm) A unit of length in the metric system; there are 2.54 cm in 1 in.

conversion factor A ratio in which the numerator and denominator are quantities from an equality or given relationship. For example, the conversion factors for the relationship 1 kg = 2.20 lb are written as the following:

$$\frac{2.20\ \text{lb}}{1\ \text{kg}} \quad \text{and} \quad \frac{1\ \text{kg}}{2.20\ \text{lb}}$$

cubic centimeter (cm³, cc) The volume of a cube that has 1-cm sides, equal to 1 mL.

density The relationship of the mass of an object to its volume expressed as grams per cubic centimeter (g/cm³), grams per milliliter (g/mL), or grams per liter (g/L).

equality A relationship between two units that measure the same quantity.

exact number A number obtained by counting or definition.

gram (g) The metric unit used in measurements of mass.

Kelvin (K) temperature scale A temperature scale on which the lowest possible temperature is 0 K, which makes the freezing point of water 273 K and the boiling point of water 373 K.

kilogram (kg) A metric mass of 1000 g, equal to 2.20 lb. The kilogram is the SI standard unit of mass.

liter (L) The metric unit for volume that is slightly larger than a quart.

mass A measure of the quantity of material in an object.

measured number A number obtained when a quantity is determined by using a measuring device.

meter (m) The metric unit for length that is slightly longer than a yard. The meter is the SI standard unit of length.

metric system A system of measurement used by scientists and in most countries of the world.

milliliter (mL) A metric unit of volume equal to one-thousandth of a L (0.001 L).

precision Ability to repeat a measurement and obtain values close together.

prefix The part of the name of a metric unit that precedes the base unit and specifies the size of the measurement. All prefixes are related on a decimal scale.

scientific notation A form of writing large and small numbers using a coefficient between 1 and 10, followed by a power of 10.

SI units An International System of units that modifies the metric system.

significant figures The numbers recorded in a measurement.

specific gravity (sp gr) A relationship between the density of a substance and the density of water:

$$\text{sp gr} = \frac{\text{density of sample}}{\text{density of water}}$$

temperature An indicator of the hotness or coldness of an object.

volume The amount of space occupied by a substance.

Additional Problems

This group of questions and problems is related to the topics in this chapter. However, they do not all follow the chapter order and they require you to combine concepts and skills from several sections. These problems will help you increase your critical thinking skills and help you prepare for your next exam.

1.83 During a workout at the gym, you set the treadmill at a pace of 55.0 meters per minute. How many minutes will you walk if you cover a distance of 7500 feet?

1.84 A fish company delivers 22 kg of salmon, 5.5 kg of crab, and 3.48 kg of oysters to your seafood restaurant.
   a. What is the total mass, in kilograms, of the seafood?
   b. What is the total number of pounds?

1.85 In France, grapes are 1.4 Euros per kilogram. What is the cost of grapes in dollars per pound, if the exchange rate is 1.1 Euros per dollar?

1.86 In Mexico, avocados are 125 pesos per kilogram. What is the cost in cents of an avocado that weighs 0.45 lb if the exchange rate is 7.5 pesos to the dollar?

1.87 Bill's recipe for onion soup calls for 4.0 lb of thinly sliced onions. If an onion has an average mass of 115 g, how many onions does Bill need?

1.88 The price of 1 pound (lb) of potatoes is $1.75. If all the potatoes sold today at the store bring in $1420, how many kilograms (kg) of potatoes did grocery shoppers buy?

1.89 The following nutrition information is listed on a box of crackers:
Serving size 0.50 oz (6 crackers)
Fat 4 g per serving Sodium 140 mg per serving
  a. If the box has a net weight (contents only) of 8.0 oz, about how many crackers are in the box?
  b. If you ate 10 crackers, how many ounces of fat are you consuming?
  c. How many grams of sodium are used to prepare 50 boxes of crackers?

1.90 A dialysis unit requires 75,000 mL of distilled water. How many gallons of water are needed? (1 gal = 4 qt)

1.91 To prevent bacterial infection, a doctor orders 4 tablets of amoxicillin per day for 10 days. If each tablet contains 250 mg of amoxicillin, how many ounces of the medication are given in 10 days?

1.92 Celeste's diet restricts her intake of protein to 24 g per day. If she eats an 8.0-oz burger that is 15.0% protein, has she exceeded her protein limit for the day? How many ounces of a burger would be allowed for Celeste?

1.93 What is a cholesterol level of 1.85 g/L in the standard units of mg/dL?

1.94 An object has a mass of 3.15 oz. When it is submerged in a graduated cylinder initially containing 325.2 mL of water, the water level rises to 442.5 mL. What is the density (g/mL) of the object?

1.95 The density of lead is 11.3 g/mL. The water level in a graduated cylinder initially at 215 mL rises to 285 mL after a piece of lead is submerged. What is the mass in grams of the lead?

1.96 A graduated cylinder contains 155 mL of water. A 15.0-g piece of iron (density = 7.86 g/cm^3) and a 20.0-g piece of lead (density = 11.3 g/cm^3) are added. What is the new water level in the cylinder?

1.97 How many cubic centimeters (cm^3) of olive oil have the same mass as 1.00 L of gasoline? (See Table 1.9.)

1.98 Ethyl alcohol has a specific gravity of 0.79. What is the volume, in quarts, of 1.50 kg of alcohol?

1.99 A typical adult body contains 55% water. If a person weighs 175 lb, how many kilograms of water do they have in their body?

1.100 A typical adult body contains 55% water. If a person has a mass of 65 kg, how many pounds of water do they have in their body?

1.101 **a.** Some athletes have as little as 3.0% body fat. If such a person has a body mass of 45 kg, how much body fat does that person have?
  b. In a process called liposuction, a doctor removes some adipose tissue from a person's body. If body fat has a density of 0.94 g/mL and 3.0 liters of fat are removed, how many pounds of fat were removed from the patient? (See the Health Note "Determination of Percentage of Body Fat.")

1.102 When a 50.0-kg person is immersed in a water tank for body fat testing, she has an underwater mass of 2.0 kg. This difference in body mass is equal to the mass of water displaced by the body. The density of water is 1.00 g/mL. (See the Health Note "Determination of Percentage of Body Fat.")
  a. What volume of water is displaced?
  b. What is the volume, in L, of the person?
  c. What is the body density, in g/mL, of that person?

1.103 A urinometer is a hydrometer used to measure the specific gravity of a sample of urine. (See the Health Note "Specific Gravity of Urine.")
  a. What is the density of a urine sample that has a specific gravity of 1.012?
  b. What is the mass of a 5.00-mL urine sample with a specific gravity of 1.022?

1.104 A normal range for specific gravity of urine is 1.003 to 1.030. A 10.0-mL sample of urine has a mass of 10.31 g. What is the specific gravity of the urine? Is the urine considered normal? Why or why not? (See the Health Note "Specific Gravity of Urine.")

2 Atoms and Elements

"Many of my patients have diabetes, ulcers, hypertension, and cardiovascular problems," says Sylvia Lau, registered dietitian. "If a patient has diabetes, I discuss foods that raise blood sugar such as fruit, milk, and starches. I talk about how dietary fat contributes to weight gain and complications from diabetes. For stroke patients, I suggest diets low in fat and cholesterol because high blood pressure increases the risk of another stroke."

If a lab test shows low levels of iron, zinc, iodine, magnesium, or calcium, a dietitian discusses foods that provide those essential elements. For instance, she may recommend more beef for an iron deficiency, whole grain for zinc, leafy green vegetables for magnesium, dairy products for calcium, and iodized table salt and seafood for iodine.

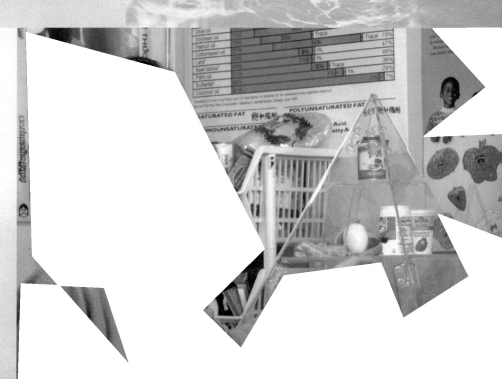

LOOKING AHEAD

the **C**hemistry place

www.chemplace.com/college

Visit the URL above or use the CD-ROM in the book for extra quizzing, interactive tutorials, career resources, and case studies.

Everything that has mass and occupies space is called *matter*. The primary substances of all matter are elements, of which there are now 114 different kinds. Of these, 88 occur in nature and are found in different combinations, providing the great number of substances that make up our world.

There are several elements you know. You use aluminum in the form of foil, or in pots and pans, or in the cans of soft drinks. Jewelry is made of the element gold, or silver, or perhaps platinum. The elements calcium and phosphorus form the structure of bones and teeth.

There are other elements that are important to our state of health, but they are present in small amounts. Iron is needed in our hemoglobin to carry oxygen through the bloodstream to our cells. Copper is needed in the formation of red blood cells, and zinc functions in the formation of collagen. Our thyroid depends on the proper level of iodine to produce the necessary amount of thyroid hormones.

Deficiencies in certain of these elements can lead to improper functioning of enzymes and abnormal growth and bone formation. Low levels of iron can lead to anemia, while low levels of iodine can cause hypothyroidism and goiter. Someone working in health care is aware of the physical symptoms that indicate a deficiency of an element. Lab tests assess whether elements such as iron, copper, zinc, or iodine are within normal ranges in a patient's blood serum. A dietitian may recommend beef for iron and zinc, whole grains and leafy green vegetables for magnesium, dairy products for calcium, and iodized table salt and seafood for iodine.

Some elements are used as medication. A condition called bipolar disorder involves severe mood swings. The element lithium is used to reduce the severity of the manic episodes. Because of side effects, the dosage of lithium must be monitored regularly by assessing blood levels of lithium.

Table 2.1 Some Elements and Their Names

Element	Source of Name
Uranium	The planet Uranus
Titanium	Titans (mythology)
Chlorine	*Chloros*, "greenish yellow" (Greek)
Iodine	*Ioeides*, "violet" (Greek)
Magnesium	Magnesia, a mineral
Californium	California
Curium	Marie and Pierre Curie

2.1 Elements and Symbols

Elements are primary substances from which all other things are built. They cannot be broken down into simpler substances. Many of the elements were named for planets, mythological figures, minerals, colors, geographic locations, and famous people. Some sources of the names of elements are listed in Table 2.1.

LEARNING GOAL

Given the name of an element, write its correct symbol; from the symbol, write the correct name.

Chemical Symbols

Chemical symbols are one- or two-letter abbreviations for the names of the elements. Only the first letter of an element's symbol is capitalized; a second letter, if there is one, is lowercase. That way, we know when a different element is indicated. If both letters are capitalized, it represents the symbols of two different elements. For example, the element cobalt has the symbol Co. However, the two capital letters CO specify two elements, carbon (C) and oxygen (O).

One-Letter Symbols		*Two-Letter Symbols*	
C	carbon	Co	cobalt
S	sulfur	Si	silicon
N	nitrogen	Ne	neon
I	iodine	Ni	nickel

Although most of the symbols use letters from the current names, some are derived from their ancient Latin or Greek names. For example, Na, the symbol for sodium, comes from the Latin word *natrium*. The symbol for iron, Fe, is derived from the Latin name *ferrum*. Table 2.2 lists the names and symbols of some common elements. Learning their names and symbols will greatly help your learning of chemistry. A complete list of all the elements and their symbols appears on the inside front cover of this text.

SAMPLE PROBLEM 2.1

Writing Chemical Symbols

What are the chemical symbols for the following elements?

a. carbon   **b.** nitrogen   **c.** chlorine   **d.** copper

Solution

a. C   **b.** N   **c.** Cl   **d.** Cu

Study Check

What are the chemical symbols for the elements silicon, sulfur, and silver?

EXPLORE YOUR WORLD

Origins of Element Names

Use a reference such as a dictionary, the Internet (look for an interactive periodic table), or the CRC *Handbook of Chemistry and Physics* (Chemical Rubber Publishing Company). Look up an element and determine the origin of its name. For example, polonium is named for Poland and uranium is named for the planet Uranus. Make a list of the element names and their origins.

Questions

1. What elements are named for a country, a region, a state, or a city?
2. What elements are named for people or figures in mythology?
3. What other sources do you find for the names of elements?

HEALTH NOTE

Latin Names for Elements in Clinical Usage

In medicine, the Latin name natrium may be used for sodium, an important electrolyte in body fluids and cells. An increase in serum sodium, hypernatremia, may occur when water is lost due to profuse sweating, severe diarrhea or vomiting, or when there is inadequate water intake. A decrease in sodium, hyponatremia, may occur when a person takes in a large amount of water or hypotonic fluid replacement solutions. Conditions that occur in cardiac failure, liver failure, and malnutrition can also cause hyponatremia.

The Latin name kalium may be used for potassium, the most common electrolyte inside the cells. Potassium regulates osmotic pressure, acid base/balance, nerve and muscle excitability, and the function of cellular enzymes. Serum potassium measures potassium outside the cells, which amounts to only 2 percent of total body potassium. An increase in serum potassium (hyperkalemia) may occur when cells are severely injured, in renal failure when potassium is not properly excreted, and in Addison's disease. A severe loss of potassium (hypokalemia) may occur during excessive vomiting, diarrhea, renal tubular defects, and glucose/insulin therapy.

Table 2.2   Names and Symbols of Some Common Elements

Aluminum

Carbon

Copper

Gold

Name[a]	Symbol
Aluminum	Al
Argon	Ar
Arsenic	As
Barium	Ba
Boron	B
Bromine	Br
Cadmium	Cd
Calcium	Ca
Carbon	C
Chlorine	Cl
Chromium	Cr
Cobalt	Co
Copper (*cuprum*)	Cu
Fluorine	F
Gold (*aurum*)	Au
Helium	He
Hydrogen	H
Iodine	I
Iron (*ferrum*)	Fe
Lead (*plumbum*)	Pb
Lithium	Li
Magnesium	Mg
Manganese	Mn
Mercury (*hydrargyrum*)	Hg
Neon	Ne
Nickel	Ni
Nitrogen	N
Oxygen	O
Phosphorus	P
Platinum	Pt
Potassium (*kalium*)	K
Radium	Ra
Silicon	Si
Silver (*argentum*)	Ag
Sodium (*natrium*)	Na
Strontium	Sr
Sulfur	S
Tin (*stannum*)	Sn
Titanium	Ti
Uranium	U
Zinc	Zn

Iron

Silver

Sulfur

[a]Names given in parentheses are ancient Latin or Greek words from which the symbols are derived.

HEALTH NOTE

Elements Essential to Health

Several elements are essential for the well-being and survival of the human body. The four elements oxygen, carbon, hydrogen, and nitrogen are the elements that make up carbohydrates, fats, proteins, and DNA. Most of the hydrogen and oxygen is found in water, which makes up 55–60% of our body mass. Some examples and the amounts present in a 60-kg person are listed in Table 2.3.

Table 2.3   Elements Essential to Health

Element	Symbol	Amount in a 60-kg Person	Where Found
Oxygen	O	39 kg	Water, carbohydrates, fats, proteins
Carbon	C	11 kg	Carbohydrates, fats, proteins
Hydrogen	H	6 kg	Water, carbohydrates, fats, proteins
Nitrogen	N	2 kg	Proteins, DNA, RNA
Calcium	Ca	1 kg	Bones, teeth
Phosphorus	P	0.6 kg	Bones, teeth, DNA, RNA
Potassium	K	0.2 kg	Inside cells (important in conduction of nerve impulses)
Sulfur	S	0.2 kg	Some amino acids
Sodium	Na	0.1 kg	Body fluids (important in nerve conduction and fluid balance)
Magnesium	Mg	0.1 kg	Inside cells (important in enzyme function)
Chlorine	Cl	0.1 kg	Outside of cells (major electrolyte)

Table 2.4   Some Physical Properties of Copper

Color	Reddish-orange
Odor	Odorless
Melting point	1083°C
Boiling point	2567°C
State at 25°C	Solid
Luster	Very shiny
Conduction of electricity	Excellent
Conduction of heat	Excellent

SAMPLE PROBLEM 2.2

Naming Chemical Elements

Give the name of the element that corresponds to each of the following chemical symbols:

a. Zn   **b.** K   **c.** H   **d.** Fe

Solution

a. zinc   **b.** potassium   **c.** hydrogen   **d.** iron

Study Check

What are the names of the elements with the chemical symbols Mg, Al, and F?

Physical Properties

Suppose a friend asks you to look into a mirror and describe what you see. You might tell her the color of your hair, eyes, and skin, or the length and texture of your hair. Perhaps you have freckles or dimples. These are descriptions of some of your physical properties. Every element has **physical properties** too, which are those characteristics that can be observed or measured without affecting the identity of an element. The physical properties of a substance include its shape, color, odor, taste, density, hardness, melting point, and boiling point. For example, copper has the physical properties listed in Table 2.4.

SAMPLE PROBLEM 2.3

Physical Properties

What physical property could you use to distinguish between copper and silver?

Solution

Copper has a red-orange color and silver has a gray color.

Study Check

What are some physical properties of the mercury in a thermometer?

QUESTIONS AND PROBLEMS

Elements and Symbols

Answers for odd-numbered Questions and Problems are given at the end of this text. Complete solutions to odd-numbered Questions and Problems are in the *Study Guide*.

2.1 Write the symbols for the following elements:
  a. copper **b.** silicon **c.** potassium **d.** nitrogen
  e. iron **f.** barium **g.** lead **h.** strontium

2.2 Write the symbols for the following elements:
  a. oxygen **b.** lithium **c.** sulfur **d.** aluminum
  e. hydrogen **f.** neon **g.** tin **h.** gold

2.3 Write the name of the element for each symbol.
  a. C **b.** Cl **c.** I **d.** Hg
  e. F **f.** Ar **g.** Zn **h.** Ni

2.4 Write the name of the element for each symbol.
  a. He **b.** P **c.** Na **d.** Mg
  e. Ca **f.** Br **g.** Cd **h.** Si

2.5 What elements are in the following substances?
  a. table salt, NaCl **b.** plaster casts, $CaSO_4$
  c. Demerol, $C_{15}H_{22}ClNO_2$ **d.** antacid, $CaCO_3$

2.6 What elements are in the following substances?
  a. water, H_2O **b.** baking soda, $NaHCO_3$
  c. lye, NaOH **d.** sugar, $C_{12}H_{22}O_{11}$

2.2 The Periodic Table

By the late 1800s, scientists began to notice that certain elements looked alike and behaved much the same way in reactions. A Russian chemist, Dmitri Mendeleev, suggested that the elements could be arranged in groups or families according to their similar properties. Today, we call this arrangement of elements the **periodic table.** (See Figure 2.1.)

Groups and Periods

The elements in the periodic table are divided into **representative elements** and **transition elements**. Each vertical column on the periodic table contains a **group**

LEARNING GOAL
Use the periodic table to identify the group and the period of an element and whether it is a metal or a nonmetal.

Periodic Table of Elements

Representative elements

Alkali metals ↓   Alkaline earth metals ↓

Halogens ↓   Noble gases ↓

1
Group 1A

18
Group 8A

Period number

Transition elements

13 Group 3A   14 Group 4A   15 Group 5A   16 Group 6A   17 Group 7A

Period	Group 1A	Group 2A	3 3B	4 4B	5 5B	6 6B	7 7B	8	9 8B	10	11 1B	12 2B	Group 3A	Group 4A	Group 5A	Group 6A	Group 7A	Group 8A
1	1 H 1.008																	2 He 4.003
2	3 Li 6.941	4 Be 9.012											5 B 10.81	6 C 12.01	7 N 14.01	8 O 16.00	9 F 19.00	10 Ne 20.18
3	11 Na 22.99	12 Mg 24.31											13 Al 26.98	14 Si 28.09	15 P 30.97	16 S 32.06	17 Cl 35.45	18 Ar 39.95
4	19 K 39.10	20 Ca 40.08	21 Sc 44.96	22 Ti 47.88	23 V 50.94	24 Cr 52.00	25 Mn 54.94	26 Fe 55.85	27 Co 58.93	28 Ni 58.69	29 Cu 63.55	30 Zn 65.38	31 Ga 69.72	32 Ge 72.59	33 As 74.92	34 Se 78.96	35 Br 79.90	36 Kr 83.80
5	37 Rb 85.47	38 Sr 87.62	39 Y 88.91	40 Zr 91.22	41 Nb 92.91	42 Mo 95.94	43 Tc 98	44 Ru 101.1	45 Rh 102.9	46 Pd 106.4	47 Ag 107.9	48 Cd 112.4	49 In 114.8	50 Sn 118.7	51 Sb 121.8	52 Te 127.6	53 I 126.9	54 Xe 131.3
6	55 Cs 132.9	56 Ba 137.3	57* La 138.9	72 Hf 178.5	73 Ta 180.9	74 W 183.9	75 Re 186.2	76 Os 190.2	77 Ir 192.2	78 Pt 195.1	79 Au 197.0	80 Hg 200.6	81 Tl 204.4	82 Pb 207.2	83 Bi 209.0	84 Po 209	85 At 210	86 Rn 222
7	87 Fr 223	88 Ra 226	89† Ac 227	104 Rf 261	105 Db 262	106 Sg 263	107 Bh 262	108 Hs 265	109 Mt 266	110 — 269	111 — 272	112 — 277	114 — 289		116 — 289			

*Lanthanides	58 Ce 140.1	59 Pr 140.9	60 Nd 144.2	61 Pm 145	62 Sm 150.4	63 Eu 152.0	64 Gd 157.3	65 Tb 158.9	66 Dy 162.5	67 Ho 164.9	68 Er 167.3	69 Tm 168.9	70 Yb 173.0	71 Lu 175.0
†Actinides	90 Th 232.0	91 Pa 231	92 U 238.0	93 Np 237	94 Pu 244	95 Am 243	96 Cm 247	97 Bk 247	98 Cf 251	99 Es 252	100 Fm 257	101 Md 258	102 No 259	103 Lr 260

☐ Metals   ☐ Metalloids   ☐ Nonmetals

Figure 2.1 Groups and periods in the periodic table.

Q **What is the symbol of the alkali metal in period 3?**

of elements, which have similar properties. One numbering system assigns the values of 1 (one) through 18 (eighteen) to the columns across the periodic table. In another system, the representative elements use the letter "A" with the numbers 1A–8A; the letter "B" is used for the transition elements. Columns 1 and 2 correspond to Groups 1A and 2A, and columns 13–18 are the same columns as Groups 3A to 8A. On some periodic tables, the values may be written with Roman numerals: IA to VIIIA.

A single horizontal row in the periodic table is called a **period.** Each row is counted from the top of the table down as Period 1, Period 2, and so on. The first period contains only the elements hydrogen (H) and helium (He). The second period, which is the second row of elements, contains lithium (Li), beryllium (Be), boron (B), carbon (C), nitrogen (N), oxygen (O), fluorine (F), and neon (Ne). The third period begins with sodium (Na) and ends at argon (Ar). There are seven periods in the periodic table.

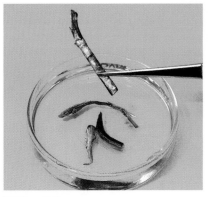

Lithium (Li)

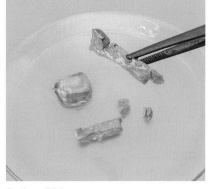

Sodium (Na)

Potassium (K)

Classification of Groups

Several groups in the periodic table have special names. Group 1A elements, lithium (Li), sodium (Na), potassium (K), rubidium (Rb), cesium (Cs), and francium (Fr), are a family of elements known as the **alkali metals** (see Figure 2.2). The elements within this group are soft, shiny metals that are good conductors of heat and electricity and have relatively low melting points. Alkali metals react vigorously with water and form white products when they combine with oxygen.

Group 2A elements, beryllium (Be), magnesium (Mg), calcium (Ca), strontium (Sr), barium (Ba), and radium (Ra), are called the **alkaline earth metals.** They are also shiny metals like those in Group 1A, but they are not as reactive.

The **halogens** are found on the right side of the periodic table in Group 7A. They include the elements fluorine (F), chlorine (Cl), bromine (Br), and iodine (I), as shown in Figure 2.3.

Group 8A contains the **noble gases,** helium (He), neon (Ne), argon (Ar), krypton (Kr), xenon (Xe), and radon (Rn). They are quite unreactive and are seldom found in combination with other elements.

Figure 2.2 Lithium (Li), sodium (Na), and potassium (K) are some alkali metals from Group 1A.

Q What physical properties do these alkali metals have in common?

Figure 2.3 Chlorine (Cl_2), bromine (Br_2), and iodine (I_2) are samples of halogens from Group 7A.

Q What elements are in the halogen group?

SAMPLE PROBLEM 2.4

Groups and Periods in the Periodic Table

State whether each set represents elements in a group, a period, or neither:

a. F, Cl, Br, I   **b.** Na, Al, P   **c.** K, Al, O

Solution

a. The elements F, Cl, Br, and I are part of a group of elements; they all appear in the Group 7A.
b. The elements Na, Al, and P all appear in the third row or third period in the periodic table.
c. Neither. The elements K, Al, and O are not part of the same group and they do not belong to the same period.

Study Check

a. What elements are found in Period 2?
b. What elements are found in Group 2A?

Stating Group and Period Number

Identify the period number and the group number for the following elements:

a. calcium   **b.** tin

Solution

The period is found by counting down the rows of the elements on the periodic table, and the group number is found at the top of the column that contains that element.

a. Calcium (Ca) is in Period 4 and Group 2A.

b. Tin (Sn) is in Period 5 and Group 4A.

Study Check

Give the symbols of the elements that are represented by the following period and group numbers:

a. Period 3, Group 5A   **b.** Period 6, Group 8A

Metals, Nonmetals, and Metalloids

Another feature of the periodic table is the heavy zigzag line that separates the elements into the *metals* and the *nonmetals.* The metals are those elements on the left of the line *except for hydrogen,* and the nonmetals are the elements on the right.

In general, most **metals** are shiny solids. They can be shaped into wires (ductile) or hammered into a flat sheet (malleable). Metals are good conductors of heat and electricity. They usually melt at higher temperatures than nonmetals. All of the metals are solids at room temperature, except for mercury (Hg), which is a liquid. Some typical metals are sodium (Na), magnesium (Mg), copper (Cu), gold (Au), silver (Ag), iron (Fe), and tin (Sn).

Nonmetals are not very shiny, malleable, or ductile, and they are often poor conductors of heat and electricity. They typically have low melting points and low densities. You may have heard of nonmetals such as hydrogen (H), carbon (C), nitrogen (N), oxygen (O), chlorine (Cl), and sulfur (S).

Table 2.5   Some Characteristics of a Metal, a Metalloid, and a Nonmetal

Silver (Ag)	Antimony (Sb)	Sulfur (S)	
Metal	Metalloid	Nonmetal	
Shiny	Blue-grey, shiny	Dull, yellow	
Extremely ductile	Brittle	Brittle	
Can be hammered into sheets (malleable)	Shatters when hammered	Shatters when hammered	
Good conductor of heat and electricity	Poor conductor of heat and electricity	Poor conductor, good insulator	
Used in coins, jewelry, tableware	Used to harden lead, color glass and plastics	Used in gunpowder, rubber, fungicides	
Density 10.5 g/mL	Density 6.7 g/mL	Density 2.1 g/mL	
Melting point 962°C	Melting point 630°C	Melting point 113°C	

HEALTH NOTE

Some Important Trace Elements in the Body

Some metals and nonmetals known as trace elements are essential to the proper functioning of the body. Although they are required in very small amounts, their absence can disrupt major biological processes and cause illness. The trace elements listed in Table 2.6 are present in the body combined with other elements.

Table 2.6   Some Important Trace Elements in the Body

Element	Adult DV[a]	Biological Function	Deficiency Symptoms	Dietary Sources
Iron (Fe)	10 mg (males) 18 mg (females)	Formation of hemoglobin; enzymes	Dry skin, spoon nails, decreased hemoglobin count, anemia	Liver and other organ meats, oysters, red or dark meat, green leafy vegetables, fortified breads and cereals, egg yolk
Copper (Cu)	2.0–5.0 mg	Necessary in many enzyme systems; growth; aids formation of red blood cells and collagen	Uncommon; anemia, decreased white cell count; bone demineralization	Nuts, organ meats, whole grains, shellfish, eggs, poultry, leafy green vegetables
Zinc (Zn)	15 mg	Amino acid metabolism; enzyme systems; energy production; collagen	Retarded growth and bone formation; skin inflammation; loss of taste and smell; poor healing	Oysters, crab, lamb, beef, organ meats, whole grains
Manganese (Mn)	2.5–5.0 mg	Necessary for some enzyme systems; collagen formation; central nervous system; fat and carbohydrate metabolism; blood clotting	Abnormal skeletal growth; impairment of central nervous system	Whole grains, wheat germ, legumes, pineapple, figs
Iodine (I)	150 μg	Necessary for activity of thyroid gland	Hypothyroidism; goiter; cretinism	Seafood, iodized salt
Fluorine (F)	1.5–4.0 mg	Necessary for solid tooth formation and retention of calcium in bones with aging	Dental cavities	Tea, fish, water in some areas, supplementary drops, toothpaste

[a]DV, daily value.

The elements located along the heavy line are called **metalloids** and include B, Si, Ge, As, Sb, and Te. Metalloids are elements that exhibit some properties that are typical of the metals and other properties that are characteristic of the nonmetals. For example, they are better conductors of heat and electricity than the nonmetals, but not as good as the metals. The metalloids are semiconductors because they can act as conductors or insulators. Table 2.5 compares some characteristics of silver, a metal, with those of antimony, a metalloid, and sulfur, a nonmetal.

SAMPLE PROBLEM 2.6

Classifying Elements as Metals and Nonmetals

Using a periodic table, classify the following elements as metals or nonmetals:

a. Na   **b.** C   **c.** Cl   **d.** Cu

Solution

a. Metal; sodium is located to the left of the heavy zigzag line that separates metals and nonmetals.

b. Nonmetal; carbon is on the right side of the zigzag line.

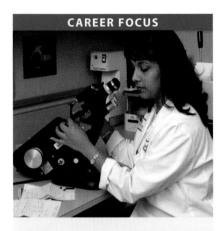

Optician

"When a patient brings in a prescription, I help select the proper lenses, put them into a frame, and fit them properly on the patient's face," says Suranda Lara, optician, Kaiser Hospital. "If a prescription requires a thinner and lighter-weight lens, we formulate that lens. So we have to understand the different materials used to make lenses. Sometimes patients come in with their own glasses that they want to convert to sunglasses. We remove the lenses and put them into a tint bath, which turns them into sunglasses."

Opticians fit and adjust eyewear for patients who have had their eyesight tested by an ophthalmologist or optometrist. Optics and mathematics are used to select materials for frames and lenses that are compatible with patients' facial measurements and lifestyles.

c. Nonmetal; chlorine is on the right side of the zigzag line.

d. Metal; copper is located to the left of the zigzag line.

Study Check

Which of the following elements, nickel or nitrogen, would be a gas and a poor conductor of heat?

QUESTIONS AND PROBLEMS

The Periodic Table

2.7 Identify the group or period number described by each of the following statements:
  a. contains the elements C, N, and O
  b. begins with helium
  c. the alkali metals
  d. ends with neon

2.8 Identify the group or period number described by each of the following statements:
  a. contains Na, K, and Rb
  b. the row that begins with Li
  c. the noble gases
  d. contains F, Cl, Br, and I

2.9 Classify the following as an alkali metal, alkaline earth metal, transition element, halogen, or noble gas:
  a. Ca   **b.** Fe   **c.** Xe   **d.** Na   **e.** Cl

2.10 Classify the following as an alkali metal, alkaline earth metal, transition element, halogen, or noble gas:
  a. Ne   **b.** Mg   **c.** Cu   **d.** Br   **e.** Ba

2.11 Give the symbol of the element described by the following:
  a. Group 4A, Period 2   **b.** a noble gas in Period 1
  c. an alkali metal in Period 3   **d.** Group 2A, Period 4
  e. Group 3A, Period 3

2.12 Give the symbol of the element described by the following:
  a. an alkaline earth metal in Period 2
  b. Group 5A, Period 3
  c. a noble gas in Period 4
  d. a halogen in Period 5
  e. Group 4A, Period 4

2.13 Is each of the following elements a metal or nonmetal?
  a. calcium   **b.** sulfur
  c. an element that is shiny   **d.** does not conduct heat
  e. located in Group 7A   **f.** phosphorus
  g. magnesium   **h.** silver

2.14 Is each of the following elements a metal or nonmetal?
  a. located in Group 2A   **b.** a good conductor of electricity
  c. chlorine   **d.** iron
  e. an element that is not shiny **f.** oxygen
  g. nitrogen   **h.** aluminum

2.3 The Atom

LEARNING GOAL

Describe the electrical charge and location in an atom for a proton, a neutron, and an electron.

the Chemistry place

WEB TUTORIAL
Atoms and Isotopes

All the elements listed on the periodic table are made up of small particles called atoms. An **atom** is the smallest particle of an element that retains the characteristics of that element. You have probably seen the element aluminum. Imagine that you are tearing a piece of aluminum foil into smaller and smaller pieces. Now imagine that you have a piece so small that you can no longer break it down further. Then you would have an atom of aluminum, the smallest particle of an element that still retains the characteristics of that element.

Billions of atoms are packed together to build you and everything around you. The paper in this book contains atoms of carbon, hydrogen, and oxygen. The ink on this paper, even the dot over the letter *i,* contains huge numbers of atoms. There are as many atoms in that dot as there are seconds in 10 billion years.

The concept of the atom is relatively recent. Although the Greek philosophers in 500 B.C. reasoned that everything must contain minute particles they called *atomos,* the idea of atoms did not become a scientific theory until 1808. Then John Dalton (1766–1844) developed an atomic theory that proposed that atoms were responsible for the combinations of elements found in compounds.

Atomic Theory

1. All matter is made up of tiny particles called atoms.
2. All atoms of a given element are similar to one another and different from atoms of other elements.
3. Atoms of two or more different elements combine to form compounds. A particular compound is always made up of the same kinds of atoms and always has the same number of each kind of atom.
4. A chemical reaction involves the rearrangement, separation, or combination of atoms. Atoms are never created or destroyed during a chemical reaction.

Atoms are the building blocks of everything we see around us; yet, we cannot see an atom or even a billion atoms with the naked eye. However, when billions and billions of atoms are packed together, the characteristics of each atom are added to those of the next until we can see the characteristics we associate with the element. For example, a small piece of the shiny, copper-colored element we call copper consists of many, many copper atoms. Through a special kind of microscope called a scanning tunneling microscope, we can now see images of individual atoms, such as the atoms of carbon in graphite shown in Figure 2.4.

Figure 2.4 Graphite, a form of carbon, magnified millions of times by a scanning tunneling microscope. This instrument generates an image of the atomic structure. The round yellow objects are atoms.

Q Why is a microscope with extremely high magnification needed to see these atoms?

Parts of an Atom

By the early part of the twentieth century, growing evidence indicated that atoms were not solid spheres, as Dalton had imagined. New experiments showed that atoms were composed of even smaller bits of matter called **subatomic particles.** The chemistry of an element depends upon the subatomic particles that are the building blocks of the atoms.

There are three subatomic particles of interest to us, the proton, neutron, and electron. Two of these carry electrical charges. The **proton** has a positive charge (+), and the **electron** carries a negative charge (−). The **neutron** has no electrical charge; it is neutral.

Like charges repel; they push away from each other. When you brush your hair on a dry day, electrical charges that are alike build up on the brush and in your hair;

Positive charges
repel

Negative charges
repel

Unlike charges
attract

Figure 2.5 Like charges repel and unlike charges attract.

Q Why are the electrons attracted to the protons in the nucleus of an atom?

as a result your hair flies away from the brush. Opposite or unlike charges attract. The crackle of clothes taken from the clothes dryer indicates the presence of electrical charges. The clinginess of the clothing is due to the attraction of opposite, unlike charges, as shown in Figure 2.5.

Mass

All of the subatomic particles are extremely small compared with the things you see around you. One proton has a mass of 1.7×10^{-24} g, and the neutron is about the same. The mass of the electron is even less. Because the mass of a subatomic particle is so minute, chemists use a small unit of mass called an **atomic mass unit (amu).** An amu is defined as one-twelfth of the mass of the carbon atom with 6 protons and 6 neutrons, a standard with which the mass of every other atom is compared. In biology, the atomic mass unit is called a dalton in honor of John Dalton. On the amu scale, the proton and neutron each have a mass of about 1 amu. Because the electron mass is so small it is usually ignored in atomic mass calculations. Table 2.7 summarizes some information about the subatomic particles in an atom.

Table 2.7   Particles in the Atom

Subatomic Particle	Symbol	Electrical Charge	Approximate Mass (amu)	Location in Atom
Proton	p or p^+	1+	1	Nucleus
Neutron	n or n^0	0	1	Nucleus
Electron	e^-	1−	0.0005 ($1/2000$)	Outside nucleus

Structure of the Atom

By the early 1900s, scientists knew that atoms contained subatomic particles, but they did not know how those particles were arranged. In 1911, an experiment led Ernest Rutherford to propose that most of the mass of the atom was contained in a small region at the center of the atom. In his gold-foil experiment, positively charged alpha particles were aimed at a very thin sheet of gold foil as seen in Figure 2.6. He thought that the heavy alpha particles would travel in straight paths and go right on through the gold foil. But much to his surprise, a few alpha particles changed direction as they passed through the gold foil. Some were deflected so much that they seemed to go back in the direction of the source of the particles. According to Rutherford, it was as though he had shot a cannonball at a piece of tissue paper, and it bounced back at him.

Nucleus of the Atom

From the gold-foil experiment, Rutherford concluded that the alpha particles would change direction only if they came close to some type of dense, positively charged region in the atom. This dense core of an atom called the **nucleus,** where the protons and neutrons are located, has a positive charge and contains most of the mass of an atom. Most of an atom is empty space, which is occupied only by fast-moving electrons. (See Figure 2.7.) If an atom were the size of a football stadium, the nucleus would be about the size of a golf ball placed in the center of the field.

Source of alpha particles
Beam of alpha particles
Gold foil
Fluorescent screen
Lead shield

(a)

Alpha particles
Nucleus
Atoms of gold foil

(b)

SAMPLE PROBLEM 2.7

Identifying Subatomic Particles

Complete the following table for subatomic particles:

Name	Symbol	Mass (amu)	Charge	Location in Atom
Electron	_____	_____	_____	_____
_____	_____	1	0	_____

Solution

Name	Symbol	Mass (amu)	Charge	Location in Atom
Electron	e^-	0.0005 ($\frac{1}{2000}$)	1−	Outside nucleus
Neutron	n^0	1	0	Nucleus

Study Check

Give the symbol, mass, electrical charge, and location of a proton in the atom.

QUESTIONS AND PROBLEMS

The Atom

2.15 Is a proton, neutron, or electron described by each of the following?
   a. has the smallest mass   **b.** carries a positive charge
   c. is found outside the nucleus   **d.** is electrically neutral

2.16 Is a proton, neutron, or electron described by each of the following?
   a. has a mass about the same as a proton's   **b.** is found in the nucleus
   c. is found in the larger part of the atom   **d.** carries a negative charge

2.17 What did Rutherford's gold-foil experiment tell us about the organization of the subatomic particles in an atom?

2.18 Why does the nucleus in every atom have a positive charge?

2.19 Which of the following particles have opposite charges?
   a. two protons   **b.** a proton and an electron
   c. two electrons   **d.** a proton and a neutron

2.20 Which of the following pairs of particles have the same charges?
   a. two protons   **b.** a proton and an electron
   c. two electrons   **d.** an electron and a neutron

2.21 On a dry day, your hair flies away when you brush it. How would you explain this?

2.22 Sometimes clothes removed from the dryer cling together. What kinds of charges are on the clothes?

Figure 2.6(a) Positive alpha particles are aimed at a piece of gold foil. **(b)** Alpha particles that come close to the atomic nuclei are deflected from their straight path.

Q **From this experiment, what did Rutherford conclude about the size of the nucleus in an atom?**

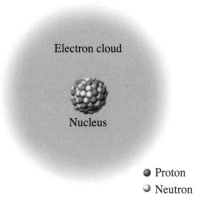

Electron cloud

Nucleus

● Proton
◔ Neutron

Figure 2.7 In an atom, the protons and neutrons are found in the nucleus; the electrons are located outside the nucleus.

Q **Why can we say that the atom is mostly empty space?**

WEB TUTORIAL
Atoms and Isotopes

2.4 Atomic Number and Mass Number

All of the atoms of the same element always have the same number of protons. This feature distinguishes atoms of one element from atoms of all the other elements. An **atomic number,** which is equal to the number of protons in the nucleus of an atom, is used to identify each element.

Atomic number = number of protons in an atom

On the inside front cover of this text is a list of all the elements, their chemical symbols, and their atomic numbers. Next to it is a periodic table, which gives all of the elements in order of increasing atomic number. In the periodic table, the atomic number is the whole number that appears above the symbol for each element. For example, a hydrogen atom, with atomic number 1, has 1 proton; a lithium atom, with atomic number 3, has 3 protons; an atom of carbon, with atomic number 6, has 6 protons; and gold, with atomic number 79, has 79 protons.

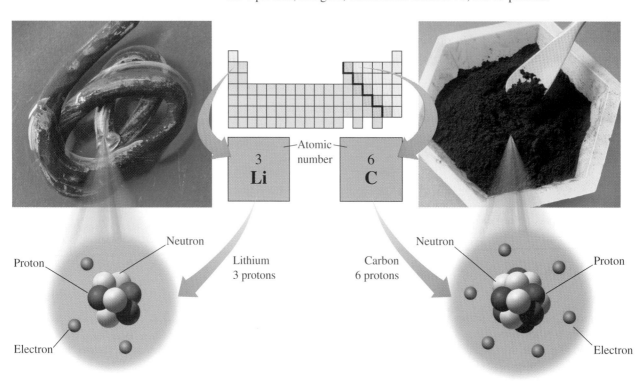

Atoms Are Neutral

An atom is electrically neutral. That means that the number of protons in an atom is equal to the number of electrons. This electrical balance gives an atom an overall charge of zero. Thus, in every atom, the atomic number also gives the number of electrons.

SAMPLE PROBLEM 2.8

Using Atomic Number to Find the Number of Protons and Electrons

Using the periodic table in Figure 2.1, state the atomic number, number of protons, and number of electrons for an atom of each of the following elements:

a. nitrogen   **b.** magnesium   **c.** bromine

Solution

a. atomic number 7; 7 protons and 7 electrons
b. atomic number 12; 12 protons and 12 electrons
c. atomic number 35; 35 protons and 35 electrons

Study Check

Consider an atom that has 26 electrons.
a. How many protons are in its nucleus?   **b.** What is its atomic number?
c. What is its name, and what is its symbol?

Mass Number

We now know that the protons and neutrons determine the mass of the nucleus. For any atom, the **mass number** is the sum of the number of protons and neutrons in the nucleus.

Mass number = number of protons + number of neutrons

For example, an atom of oxygen that contains 8 protons and 8 neutrons has a mass number of 16. An atom of iron that contains 26 protons and 30 neutrons has a mass number of 56. Table 2.8 illustrates the relationship between atomic number, mass number, and the number of protons, neutrons, and electrons in some atoms of different elements.

Table 2.8   Composition of Some Atoms of Different Elements

Element	Symbol	Atomic Number	Mass Number	Number of Protons	Number of Neutrons	Number of Electrons
Hydrogen	H	1	1	1	0	1
Nitrogen	N	7	14	7	7	7
Chlorine	Cl	17	37	17	20	17
Iron	Fe	26	56	26	30	26
Gold	Au	79	197	79	118	79

SAMPLE PROBLEM 2.9

Calculating Mass Number

Calculate the mass number of an atom by using the information given:
a. 5 protons and 6 neutrons
b. 18 protons and 22 neutrons
c. atomic number 48 and 64 neutrons

Solution

a. Mass number = 5 + 6 = 11
b. Mass number = 18 + 22 = 40
c. Mass number = 48 + 64 = 112

What is the mass number of a silver atom that has 60 neutrons?

SAMPLE PROBLEM 2.10

Calculating Numbers of Protons and Neutrons

For an atom of phosphorus that has a mass number of 31, determine the following:
a. the number of protons
b. the number of neutrons
c. the number of electrons

Solution

a   On the periodic table, the atomic number of phosphorus is 15. A phosphorus atom has 15 protons.

b.   The number of neutrons in this atom is found by subtracting the atomic number from the mass number. The number of neutrons is 16.

$$\text{Mass number} - \text{atomic number} = \text{number of neutrons}$$
$$31 \quad - \quad 15 \quad = \quad 16$$

c.   Because an atom is neutral, there is an electrical balance of protons and electrons. Because the number of electrons is equal to the number of protons, the phosphorus atom has 15 electrons.

Study Check

How many neutrons are in the nucleus of a bromine atom that has a mass number of 80?

QUESTIONS AND PROBLEMS

Atomic Number and Mass Number

2.23 Would you use atomic number, mass number, or both to obtain the following?
  a. number of protons in an atom
  b. number of neutrons in an atom
  c. number of particles in the nucleus
  d. number of electrons in a neutral atom

2.24 What do you know about the subatomic particles from the following?
  a. atomic number
  b. mass number
  c. mass number − atomic number
  d. mass number + atomic number

2.25 Write the names and symbols of the elements with the following atomic numbers:
  a. 3   **b.** 9   **c.** 20   **d.** 30
  e. 10   **f.** 14   **g.** 53   **h.** 8

2.26 Write the names and symbols of the elements with the following atomic numbers:
  a. 1   **b.** 11   **c.** 19   **d.** 26
  e. 35   **f.** 47   **g.** 15   **h.** 2

2.27 How many protons and electrons are there in a neutral atom of the following?
   a. magnesium   **b.** zinc   **c.** iodine   **d.** potassium

2.28 How many protons and electrons are there in a neutral atom of the following?
   a. carbon   **b.** fluorine   **c.** calcium   **d.** sulfur

2.29 Complete the following table for neutral atoms.

Name of Element	Symbol	Atomic Number	Mass Number	Number of Protons	Number of Neutrons	Number of Electrons
	Al		27			
		12			12	
Potassium					20	
				16	15	
			56			26

2.30 Complete the following table for neutral atoms.

Name of Element	Symbol	Atomic Number	Mass Number	Number of Protons	Number of Neutrons	Number of Electrons
	N		15			
Calcium			42			
				38	50	
		14			16	
		56	138			

2.5 Isotopes and Atomic Mass

We have seen that all atoms of the same element have the same number of protons and electrons. However, the atoms of any one element are not completely identical because they can have different numbers of neutrons. **Isotopes** are atoms of the same element that have different numbers of neutrons. For example, all atoms of the element magnesium (Mg) have 12 protons. However, some magnesium atoms have 12 neutrons, others have 13 neutrons, and still others have 14 neutrons. The differences in numbers of neutrons for these magnesium atoms cause their mass numbers to be different but not their chemical behavior. The three isotopes of magnesium have the same atomic number but different mass numbers.

To distinguish between the different isotopes of an element, we can write an **atomic symbol** that indicates the mass number and the atomic number of the atom.

WEB TUTORIAL
Atoms and Isotopes

Atomic Symbol for an Isotope of Magnesium

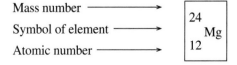

Mass number ⟶

Symbol of element ⟶ $^{24}_{12}\text{Mg}$

Atomic number ⟶

An isotope may be referred to by its name or symbol followed by the mass number, such as magnesium-24 or Mg-24. Magnesium has three naturally occurring isotopes, as shown in Table 2.9.

Table 2.9   Isotopes of Magnesium

Atomic symbol	$^{24}_{12}\text{Mg}$	$^{25}_{12}\text{Mg}$	$^{26}_{12}\text{Mg}$
Number of protons	12	12	12
Number of electrons	12	12	12
Mass number	**24**	**25**	**26**
% abundance	78.9%	10.0%	11.1%
Number of neutrons	**12**	**13**	**14**

Atomic structure of Mg

Isotopes of Mg

$^{24}_{12}\text{Mg}$ $^{25}_{12}\text{Mg}$ $^{26}_{12}\text{Mg}$

SAMPLE PROBLEM 2.11

Identifying Protons and Neutrons in Isotopes

State the number of protons and neutrons in the following isotopes of neon (Ne):
a. $^{20}_{10}\text{Ne}$ **b.** $^{21}_{10}\text{Ne}$ **c.** $^{22}_{10}\text{Ne}$

Solution

The atomic number of Ne is 10; each isotope has 10 protons. The number of neutrons in each isotope is found by subtracting the atomic number (10) from each mass number.
a. 10 protons; 10 neutrons (20 − 10)
b. 10 protons; 11 neutrons (21 − 10)
c. 10 protons; 12 neutrons (22 − 10)

Study Check

Write a symbol for the following isotopes:
a. a nitrogen atom with 8 neutrons
b. an atom with 20 protons and 22 neutrons
c. an atom with mass number 27 and 14 neutrons

Atomic Mass

In laboratory work, a scientist generally uses samples that contain many atoms of an element. Among those atoms are all of the various isotopes with their different masses. To obtain a convenient mass to work with, chemists use the mass of an "average atom" of each element. This average atom has an **atomic mass,** which is the weighted average mass of all of the naturally occurring isotopes of that element.

To calculate an atomic mass, the percent abundance of each isotope must be known. For example, in chlorine (Cl), 75.8% of the atoms have a mass number of 35; the other 24.2% have a mass number of 37. Using these values and the mass number of each isotope (which is quite close to each isotope's mass in amu), we can calculate the average atomic mass for the element.

Determination of the Average Atomic Mass of Chlorine

Isotope	Percent of Sample		Mass Number		Contribution to Average Atom
$^{35}_{17}Cl$	$\dfrac{75.8}{100}$	$\times$	35	$=$	26.5 amu
$^{37}_{17}Cl$	$\dfrac{24.2}{100}$	$\times$	37	$=$	8.95 amu
			Atomic mass Cl	$=$	35.45 amu

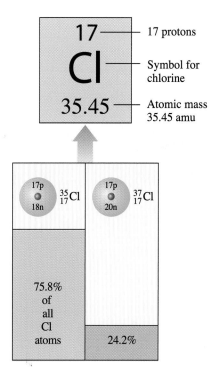

This calculation illustrates the way in which isotopes determine the overall atomic mass for an element. Most elements consist of several isotopes, and this is one reason atomic masses listed on the periodic table are seldom whole numbers.

Table 2.10 lists the isotopes of some selected elements and their average atomic masses. On the periodic table, the atomic mass is given below the symbol of each element.

Table 2.10   Isotopes and Average Atomic Mass of Some Elements

Element	Isotopes	Average Atomic Mass
Lithium	$^{6}_{3}Li$ $^{7}_{3}Li$	6.941 amu
Carbon	$^{12}_{6}C$ $^{13}_{6}C$ $^{14}_{6}C$	12.01 amu
Oxygen	$^{16}_{8}O$ $^{17}_{8}O$ $^{18}_{8}O$	16.00 amu
Sulfur	$^{32}_{16}S$ $^{33}_{16}S$ $^{34}_{16}S$ $^{36}_{16}S$	32.06 amu
Chlorine	$^{35}_{17}Cl$ $^{37}_{17}Cl$	35.45 amu
Copper	$^{63}_{29}Cu$ $^{65}_{29}Cu$	63.55 amu

Calculating Average Atomic Mass

Calculate the average atomic mass for the isotopes of magnesium illustrated in Table 2.9. The abundance of ^{24}Mg is 78.9%, ^{25}Mg is 10.0%, and ^{26}Mg is 11.1%.

Solution

Multiply the mass (its mass number) of each isotope by the abundance and total:

Abundance %   Mass

$\dfrac{78.9}{100}$	$\times$	24	$=$	18.9
$\dfrac{10.0}{100}$	$\times$	25	$=$	2.50
$\dfrac{11.1}{100}$	$\times$	26	$=$	2.89

Average atomic mass of Mg = 24.3

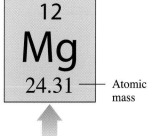

Study Check

There are two naturally occurring isotopes of boron, ^{10}B with an abundance of 19.9%, and ^{11}B with 80.1%. What is the average atomic mass of boron?

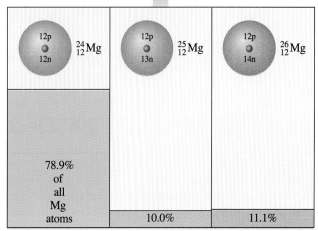

QUESTIONS AND PROBLEMS

Isotopes and Atomic Mass

2.31 What are the number of protons, neutrons, and electrons in the following isotopes?

a. $^{27}_{13}$Al   **b.** $^{52}_{24}$Cr   **c.** $^{34}_{16}$S   **d.** $^{56}_{26}$Fe

2.32 What are the number of protons, neutrons, and electrons in the following isotopes?

a. $^{2}_{1}$H   **b.** $^{14}_{7}$N   **c.** $^{26}_{14}$Si   **d.** $^{70}_{30}$Zn

2.33 Write the atomic symbols for isotopes with the following:

a. 15 protons and 16 neutrons

b. 35 protons and 45 neutrons

c. 13 electrons and 14 neutrons

d. a chlorine atom with 18 neutrons

2.34 Write the atomic symbols for isotopes with the following:

a. an oxygen atom with 10 neutrons

b. 4 protons and 5 neutrons

c. 26 electrons and 30 neutrons

d. a mass number of 24 and 13 neutrons

2.35 There are four isotopes of sulfur with mass numbers 32, 33, 34, and 36.

a. Write the atomic symbol for each of these atoms.

b. How are these isotopes alike?

c. How are they different?

d. Why is the atomic mass of sulfur listed on the periodic table not a whole number?

2.36 There are four isotopes of strontium with mass numbers 84, 86, 87, 88.

a. Write the atomic symbol for each of these atoms.

b. How are these isotopes alike?

c. How are they different?

d. Why is the atomic mass of strontium listed on the periodic table not a whole number?

2.37 What information do you need about an element and its isotopes in order to calculate its average atomic mass?

2.38 Copper consists of two isotopes, copper-63 and copper-65. If the atomic mass for copper on the periodic table is 63.55, are there more atoms of Cu-63 or Cu-65 in a sample of copper?

2.39 Two isotopes of gallium are naturally occurring with ^{69}Ga at 60.1% and ^{71}Ga at 39.9%. What is the atomic mass of gallium?

2.40 Strontium consists of four naturally occurring isotopes ^{84}Sr (0.56%), ^{86}Sr (9.86%), ^{87}Sr (7.00%), and ^{88}Sr (82.58%). What is the atomic mass of strontium?

LEARNING GOAL

Given the name or symbol of one of the first 18 elements in the periodic table, write the electron arrangement and use to explain the periodic law.

2.6 Electron Energy Levels

Electrons are constantly moving within the large space of an atom, which means they possess energy. However, they do not all have the same energy. We may think of the energy levels of an atom as similar to the rungs on a ladder. The lowest energy level would be the first rung of the ladder; the second energy level would be the second rung. As you climb up or down the ladder, you must step from one rung to the next. You cannot stop at a level between the rungs. In atoms, electrons exist only in the available energy levels. Generally, the energy levels closest to the nucleus contain electrons with the lowest energies, whereas energy levels farther away contain electrons with higher energies. Unlike the ladder, the lower energy levels are far apart compared to the higher energy levels that are closer together.

Electrons of similar energy are grouped in energy levels called **shells.** The maximum number of electrons allowed in each shell is given by the formula $2n^2$ where n is the number of the main energy level. As shown in Table 2.11, shell 1, the lowest energy level, can hold up to 2 electrons; shell 2 can hold up to 8 electrons, shell 3 can take 18 electrons, and shell 4 has room for 32 electrons. In the atoms of the elements known today, electrons can occupy energy levels as high as shell 7.

Electron Shell Arrangements for the First 18 Elements

The electron shell arrangement of an atom gives the number of electrons in each shell. The electron shell arrangements for the first 18 elements can be written by placing electrons in shells beginning with the lowest energy. The single electron of hydrogen and the 2 electrons of helium can be placed in shell 1.

As shown in Table 2.12, the elements of the second period (lithium, Li, to neon, Ne) have enough electrons to fill the first shell and part or all of the second shell. For example, lithium has 3 electrons. Two of those electrons complete shell 1. The remaining electron goes into the second shell. As we go across period 2, more and more electrons enter the second shell. For example, an atom of carbon, with a total of 6 electrons, fills shell 1 with 2 electrons, and 4 remaining electrons enter the second shell. The last element in Period 2 is neon. The 10 electrons in an atom of neon completely fill the first and second shells.

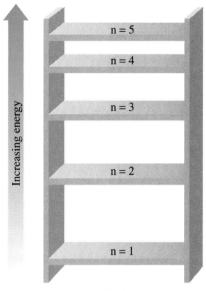

Nucleus

Table 2.11 Capacity of Some Electron Shells

Electron Shell	Maximum Number of Electrons
1	2
2	8
3	18
4	32

WEB TUTORIAL
Bohr's Shell Model

Table 2.12 Electron Shell Arrangements for the First 18 Elements

Element	Symbol	Atomic Number	Number of Electrons in Shell		
			1	2	3
Hydrogen	H	1	1		
Helium	He	2	2		
Lithium	Li	3	2	1	
Beryllium	Be	4	2	2	
Boron	B	5	2	3	
Carbon	C	6	2	4	
Nitrogen	N	7	2	5	
Oxygen	O	8	2	6	
Fluorine	F	9	2	7	
Neon	Ne	10	2	8	
Sodium	Na	11	2	8	1
Magnesium	Mg	12	2	8	2
Aluminum	Al	13	2	8	3
Silicon	Si	14	2	8	4
Phosphorus	P	15	2	8	5
Sulfur	S	16	2	8	6
Chlorine	Cl	17	2	8	7
Argon	Ar	18	2	8	8

In an atom of sodium, atomic number 11, the first and second electron shells are filled and the last electron enters the third shell. The rest of the elements in the third period continue to add to the third shell. For example, a sulfur atom with 16 electrons has 2 electrons in the first shell, 8 electrons in the second shell, and 6 electrons in the third shell. At the end of period 3, we find that argon has 8 electrons in the third shell. The filling of electron shells after argon is discussed in the next section.

SAMPLE PROBLEM 2.13

Writing Electron Shell Arrangements

Write the electron shell arrangement for each of the following:

a. oxygen   **b.** chlorine

Solution

a. Oxygen has an atomic number of 8. Therefore, there are 8 electrons in the electron arrangement:

$2e^-$ $6e^-$

b. An atom of chlorine has 17 protons and 17 electrons. The electrons are arranged as follows:

$2e^-$ $8e^-$ $7e^-$

Study Check

What element has the following electron shell arrangement?
$2e^-$ $8e^-$ $2e^-$

The Periodic Law

As we discussed earlier, the elements in any vertical column or group in the periodic table exhibit similar physical and chemical properties. There is also a regular pattern of change in properties from one group to the next across every period. This regular pattern of change in physical and chemical properties as atomic numbers increase is known as the **periodic law.**

We also find that there is a similarity in the electron arrangements. All the elements in each group have the same number of electrons in their outer shells. The outer shell is the shell with the highest energy that is occupied by one or more electrons. For example, the elements in Group 1A such as lithium, sodium, and potassium, all have one electron in their outer shell. We see this again in other groups such as Group 7A where all the elements including fluorine, chlorine, and bromine have seven electrons each in their outer shell. Thus, the similarity of properties among elements in a group can be attributed to having the same number of electrons in their outer shell. As we look across each period, there is a regular increase of one electron from group to group.

Group Number

The **group numbers** 1A–8A appear at the top of the periodic table. Each group number is equal to the number of electrons in the outer shell of the elements in that column. All elements in Group 1A have 1 electron in their outer shells, elements in Group 2A have 2 electrons in their outer shells, elements in Group 3A have 3 electrons in their outer shells, and so on. We are most interested in the number of electrons in the outer shells because these electrons have the greatest effect on the way atoms form compounds. Table 2.13 shows the electron arrangement by group for some representative elements.

Table 2.13   Electron Shell Arrangements, by Group, for Some Representative Elements

Group Number	Element	Number of Electrons in Shell			
		1	2	3	4
1A	Hydrogen	1			
	Lithium	2	1		
	Sodium	2	8	1	
	Potassium	2	8	8	1
2A	Beryllium	2	2		
	Magnesium	2	8	2	
	Calcium	2	8	8	2
3A	Boron	2	3		
	Aluminum	2	8	3	
	Gallium	2	8	18	3
4A	Carbon	2	4		
	Silicon	2	8	4	
	Germanium	2	8	18	4
5A	Nitrogen	2	5		
	Phosphorus	2	8	5	
	Arsenic	2	8	18	5
6A	Oxygen	2	6		
	Sulfur	2	8	6	
	Selenium	2	8	18	6
7A	Fluorine	2	7		
	Chlorine	2	8	7	
	Bromine	2	8	18	7
8A	Helium	2			
	Neon	2	8		
	Argon	2	8	8	
	Krypton	2	8	18	8

SAMPLE PROBLEM 2.14

Using Group Numbers

Using the periodic table, write the group number and the number of electrons in the outer electron level of the following elements:

a. sodium   **b.** sulfur   **c.** aluminum

Solution

a. Sodium is in Group 1A; sodium has 1 electron in the outer electron level.

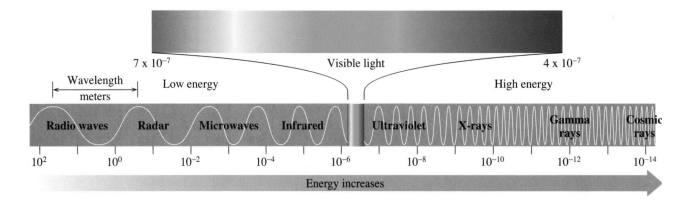

7 x 10⁻⁷   Visible light   4 x 10⁻⁷

Figure 2.8 The electromagnetic spectrum contains radiation with long and short wavelengths in the visible and invisible regions.

Q How does the energy of ultraviolet light compare to that of microwaves?

b. Sulfur is in Group 6A; sulfur has 6 electrons in the outer electron level.

c. Aluminum is in Group 3A; aluminum has 3 electrons in the outer electron level.

Study Check

What is the group number, total number of electrons, and name of an atom that has 5 electrons in the third electron level?

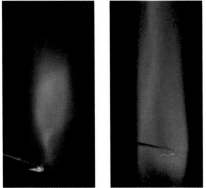

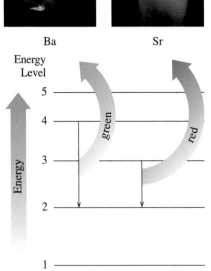

Figure 2.9 The heat in a flame test provides energy for electrons in atoms of strontium and barium to jump to higher energy levels. When the electrons fall to a lower energy level, energy is emitted in the visible range of red light and green light.

Q How does the energy of the red light emitted by strontium compare to that of the green light emitted by barium?

Electromagnetic Spectrum

When we listen to a radio, use a microwave oven, turn on a light, see the colors of a rainbow, or have an X ray, we are using various forms of electromagnetic radiation. Light and other electromagnetic radiation consist of energy particles called photons that move as a wave of energy. In a wave, the distance between the peaks is called the *wavelength*. While all the forms of electromagnetic radiation travel at the speed of light, 3.0×10^8 meters per second, they differ in energy and wavelength. High-energy radiation such as gamma radiation has short wavelengths; low-energy radiation such as radio waves has long wavelengths. (See Figure 2.8.)

When sunlight passes through a prism or crystal, the light separates in a color spectrum: red, orange, yellow, green, blue, indigo, and violet. These are the same colors we see in a rainbow that forms when sunlight passes through raindrops acting as prisms. In this visible spectrum, which is the only part of the electromagnetic spectrum that we can see, red light has the longest wavelength and violet light has the shortest wavelength.

Energy-Level Changes

Atoms can absorb energy from heat, light, or electricity. When some of the electrons in those atoms absorb a certain amount of energy, they jump to higher energy levels. When the electrons drop back to lower more stable energy levels, energy is emitted. (See Figure 2.9.) If the energy emitted is in the visible range, we see colors of the visible spectrum. Sodium streetlights and neon lights are examples of how electrons of atoms absorb energy and emit energy as yellow and red light.

In the neon signs, the energy emitted—as electrons drop to lower energy levels—produces colors in the visible range.

EXPLORE YOUR WORLD

Light Released from Candy

For this investigation, you will need some candies with wintergreen such as Lifesaver or other brands of wintergreen mints or Necco wafers. You also need pliers, a room that can be darkened, and some clear tape. Cover the end of the pliers with clear tape, break the candies in half, and darken the room. Wait 3–5 minutes so that your eyes can adjust to the darkness before you proceed. Place a piece of a candy in the pliers and crush it. Look for visible light to be emitted. If you have a friend or relative you can do this with, watch each other as you bite and chew a candy.

The crushing action introduces energy, which is absorbed by the electrons, causing them to jump to a higher energy level. When the electrons drop from a higher energy level back to a lower level, energy is released, but this time in the form of visible light energy that you can see.

Questions

1. How does the change in energy levels cause energy to be emitted as visible light?
2. What color(s) of light is emitted?
3. Are there differences in the brightness of the light?

ENVIRONMENTAL NOTE

Biological Reactions to UV Light

Our everyday life depends on sunlight, but exposure to sunlight can have damaging effects on living cells, and too much exposure can even cause their death. The light energy, especially ultraviolet (UV), excites electrons and may lead to unwanted chemical reactions. The list of damaging effects of sunlight includes sunburn; wrinkling; premature aging of the skin; changes in the DNA of the cells, which can lead to skin cancers and melanomas; inflammation of the eyes; and perhaps cataracts. Some drugs, like the acne medications Accutane and Retin-A, as well as antibiotics, diuretics, sulfonamides, and estrogens, make the skin extremely photosensitive and can cause undesirable changes in its reaction to sunlight. Using a sunscreen is now recommended by doctors to prevent the adverse effects of sun exposure.

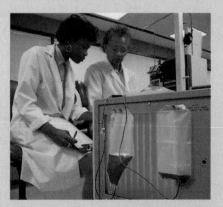

High-energy radiation is the most damaging biologically. Most of the radiation in this range is absorbed in the epidermis of the skin. The degree to which radiation is absorbed depends on the thickness of the epidermis, the hydration of the skin, the amount of coloring pigments and proteins of the skin, and the arrangement of the blood vessels. In light-skinned people, 85–90% of the radiation is absorbed by the epidermis, with the rest reaching the dermis layer. In dark-skinned people, 90–95% of the radiation is absorbed by the epidermis, with a smaller percentage reaching the dermis.

However, medicine does take advantage of the beneficial effect of sunlight. Phototherapy can be used to treat certain skin conditions including psoriasis, eczema, and dermatitis. In the treatment of psoriasis, for example, oral drugs are given to make the skin more photosensitive; exposure to UV radiation follows. Low-energy radiation is used to break down bilirubin in neonatal jaundice. Sunlight is also a factor in stimulating the immune system.

In cutaneous T-cell lymphoma, an abnormal increase in T cells causes painful ulceration of the skin. The skin is treated by photophoresis, in which the patient receives a photosensitive chemical, and then blood is removed from the body and exposed to ultraviolet light. The blood is returned to the patient, and the treated T cells stimulate the immune system to respond to the cancer cells.

QUESTIONS AND PROBLEMS

Electron Energy Levels

2.41 Electrons exist in specific energy levels. Explain.

2.42 In what order do electrons fill the energy levels 1–3 for the first 18 elements on the periodic table?

2.43 How many electrons are in shell 2 of the following elements?
    a. sodium   **b.** nitrogen   **c.** sulfur
    d. helium   **e.** chlorine

2.44 How many electrons are in shell 3 of the following elements?
    a. oxygen   **b.** sulfur   **c.** phosphorus
    d. argon   **e.** fluorine

2.45 Write the electron shell arrangement for each of the following elements:
    Example: sodium 2, 8, 1
    a. carbon   **b.** argon   **c.** sulfur
    d. silicon   **e.** an atom with 13 protons and 14 neutrons
    f. nitrogen

2.46 Write the electron shell arrangement for each of the following atoms:
    Example: sodium 2, 8, 1
    a. phosphorus   **b.** neon   **c.** oxygen
    d. an atom with atomic number 18   **e.** aluminum   **f.** silicon

2.47 Identify the elements that have the following electron shell arrangements:

Energy Level:	1	2	3
a.	$2e^-$	$1e^-$	
b.	$2e^-$	$8e^-$	$2e^-$
c.	$1e^-$		
d.	$2e^-$	$8e^-$	$7e^-$
e.	$2e^-$	$6e^-$	

2.48 Identify the elements that have the following electron shell arrangements:

Energy Level:	1	2	3
a.	$2e^-$	$5e^-$	
b.	$2e^-$	$8e^-$	$6e^-$
c.	$2e^-$	$4e^-$	
d.	$2e^-$	$8e^-$	$8e^-$
e.	$2e^-$	$8e^-$	$3e^-$

2.49 Answer the following with *gain* or *emit*:
    a. Electrons can jump to higher energy levels when they _____ a specific amount of energy.
    b. When electrons drop to lower energy levels, they _____ a certain amount of energy.

2.50 **a.** Comparing X rays, infrared, and radio waves as shown in Figure 2.8, which has the highest energy?
    b. Why do we protect ourselves from X rays but not radio waves?

2.51 The elements boron and aluminum are in the same group on the periodic table.
    a. Write the electron arrangements for B and Al.
    b. How many electrons are in the outer energy level of each atom?
    c. What is their group number?

2.52 The elements fluorine and chlorine are in the same group on the periodic table.
    a. Write the electron arrangements for F and Cl.
    b. How many electrons are in each of their outer energy levels?
    c. What is their group number?

2.53 What is the number of electrons in the outer energy level and the group number for each of the following elements?

Example: fluorine $7e^-$; Group 7A

a. magnesium   **b.** chlorine   **c.** oxygen   **d.** nitrogen
e. barium   **f.** bromine

2.54 What is the number of electrons in the outer energy level and the group number for each of the following elements?

Example: fluorine $7e^-$; Group 7A

a. lithium   **b.** silicon   **c.** neon   **d.** argon
e. tin   **f.** cesium

2.55 Why do Mg, Ca, and Sr have similar properties?

2.56 Name two elements that would exhibit physical and chemical behavior similar to chlorine.

2.7 Subshells and Orbitals

The electrons in the electron shells or main energy levels can be described in yet more detail. Within each shell, the electrons with identical energy are grouped as **subshells,** which are identified by the letters *s, p, d,* and *f.* The *s* subshell has the lowest energy, followed by the *p* subshell, then the *d* subshell, and finally the highest energy subshell, the *f* subshell.

Order of Increasing Energy for Subshells

$$s \longrightarrow p \longrightarrow d \longrightarrow f$$

Lowest
energy

Highest
energy

The number of subshells in each shell is equal to the energy level. As shown in Figure 2.10, shell 1 has only one subshell, 1*s.* Shell 2 has two subshells, 2*s* and 2*p.*

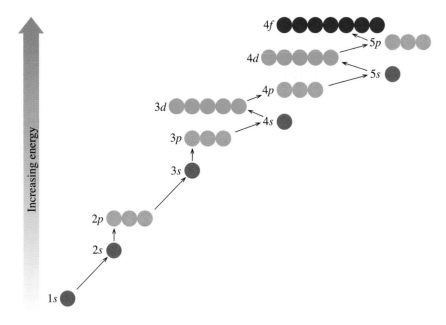

Figure 2.10 Energy levels of subshells and the order of electron filling up to the 4*f* subshell.

Q Why does the 3*d* subshell fill after the 4*s*?

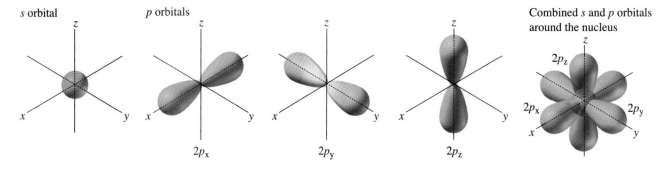

s orbital   p orbitals   Combined s and p orbitals around the nucleus

$2p_x$ $2p_y$ $2p_z$

Figure 2.11 The shapes of orbitals differ; s orbitals are spherical and p orbitals are dumbbell shaped. Each orbital holds a maximum of two electrons.

Q If there are three p orbitals in a p subshell, how many total electrons are in a p subshell?

WEB TUTORIAL
Bohr's Shell Model

Shell 3 has three subshells, 3s, 3p, and 3d. The fourth shell consists of four subshells, 4s, 4p, 4d, and 4f. In the third and fourth shells, the subshells are so close in energy that there is an overlap of subshells. The 4s subshell is lower in energy than the 3d subshell.

Orbitals

An **orbital** is described as a region in space around the nucleus in which an electron of a certain energy is most likely to be found. Each orbital can hold only one or two electrons.

A certain type of orbital makes up each type of subshell. An s orbital is spherical, with the nucleus at the center. (See Figure 2.11.) That means that the electrons of every s subshell are most likely to be contained in a spherical space. Each s orbital has a maximum of two electrons, and there is just one s orbital for every s subshell.

A p subshell consists of three p orbitals, which have shapes like dumbbells. A p orbital has two lobes that make it look like a "dumbbell." For a p subshell, three p orbitals are arranged in three different directions (x, y, z axes) around the nucleus. Because each p orbital can hold up to two electrons, the total set of p orbitals accounts for six electrons in a p subshell.

Number of Electrons in Subshells

Each type of subshell holds a specific number of electrons. An s subshell holds 1 or 2 electrons. A p subshell takes up to 6 electrons, a d subshell can hold up to 10 electrons, and an f subshell has a maximum capacity of 14 electrons. The number of electrons in the subshells adds up to give the total number of electrons in each electron shell or energy level, as shown in Table 2.14.

Table 2.14   Electron Capacity in Subshells for Energy Levels 1–4

Shell	Subshell	Orbitals	Maximum Number of Electrons	Shell Capacity
1	1s	1s	2	2
2	2s	2s	2	8
	2p	2p 2p 2p	6	
3	3s	3s	2	18
	3p	3p 3p 3p	6	
	3d	3d 3d 3d 3d 3d	10	
4	4s	4s	2	32
	4p	4p 4p 4p	6	
	4d	4d 4d 4d 4d 4d	10	
	4f	4f 4f 4f 4f 4f 4f 4f	14	

Subshells

Describe the third electron shell in terms of the following:
a. number and notation of subshells
b. number and type of orbitals in each subshell
c. maximum number of electrons in each subshell
d. maximum number of electrons in the third shell

Solution

a. The third shell has three subshells: 3*s*, 3*p*, and 3*d*.
b. There is one 3*s* orbital in the 3*s* subshell, three 3*p* orbitals in the 3*p* subshell, and five 3*d* orbitals in the 3*d* subshell.
c. The 3*s* subshell can accommodate 2 electrons, the 3*p* subshell can hold up to 6 electrons, and the 3*d* subshell will take up to 10 electrons.
d. The maximum number of electrons for the third shell is 18.

Study Check

State the number of electrons that would fill the following:
a. the second shell   **b.** the 3*p* subshell   **c.** a 5*d* orbital

Subshells and Orbitals

2.57 What would be the maximum number of electrons in the following?
   a. 2*p* orbital   **b.** 2*p* subshell
   c. shell 2   **d.** 3*s* orbital
2.58 What would be the maximum number of electrons in the following?
   a. 1*s* subshell   **b.** 3*s* subshell
   c. shell 3   **d.** 4*s* orbital

2.8 Electron Configurations

Write electron configurations using subshell notation.

The **electron configuration** describes the organization of electrons in the subshells of atoms. Electrons are arranged starting with the lowest energy subshell 1*s* and proceeding to each higher energy subshell in order. The number of electrons in each subshell is written as a superscript. The subshells of lower numbers represent the filled inner shells. The subshells with the highest number represent the outermost shell.

Subshell Blocks on the Periodic Table

The electron arrangement of the elements is related to their position in the periodic table. Different sections of the table called blocks correspond to the *s*, *p*, *d*, and *f* subshells as shown in Figure 2.12. The electron configuration can be written by following the subshell blocks across the periodic table starting with period 1.

Figure 2.12 Electron configuration follows the order of subshells on the periodic table.

Q **If neon is in the Group 8A, period 2, how many electrons are in the 1s, 2s, and 2p subshells of neon?**

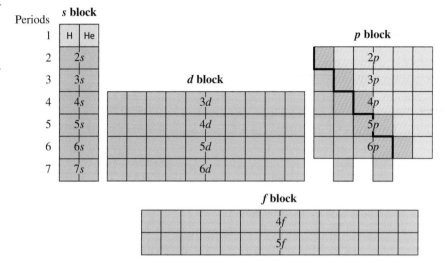

1. The *s* **block,** which is formed by Group 1A and Group 2A, shows the number of electrons in each of the *s* subshells. The period number gives the energy level beginning with 1*s*, 2*s*, and so on.
2. The *p* **block** begins with Group 3A and ends at Group 8A. The six elements in the *p* block on the periodic table correspond to the six electrons in a *p* subshell. The period number gives the particular energy level of each *p* subshell beginning with 2*p*.
3. The *d* **block** appears after atomic number 20 (calcium) with ten elements of the transition metals. The energy level of the *d* block is one less than the period number. For example, in period 4 the 4*s* subshell comes first. The next block is the 3*d* subshell, which holds 10 electrons. The next subshell is 4*p*.
4. The *f* **block** indicates the 14 electrons in the *f* subshell. Elements that have atomic numbers higher than 57 (La) have electrons in the 4*f* block. The energy level of each *f* block is two less than the corresponding period number.

Period 1 Hydrogen and Helium

We can begin to build the electron configuration for period 1, which contains hydrogen and helium. The first two electrons go into the 1*s* subshell, which is the subshell with the lowest energy.

First period

We write the electron configuration by listing each subshell that contains one or more electrons and indicating that number as a superscript.

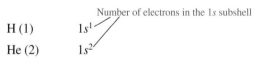

Number of electrons in the 1*s* subshell

H (1) $1s^1$
He (2) $1s^2$

Period 2 Lithium to Neon

Period 2 begins with lithium. The first two electrons fill the 1*s* subshell. The third electron goes into the next available subshell with the lowest energy, the 2*s*. Going across period 2, another electron completes the $2s$ for beryllium. The next six electrons fill the orbitals of the $2p$ subshell, which is next lowest in energy.

Second period

Li	Be
$2s^1$	$2s^2$

B	C	N	O	F	Ne
$2p^1$	$2p^2$	$2p^3$	$2p^4$	$2p^5$	$2p^6$

The complete electron configurations for the elements in period 2 always begin with the $1s$ subshell followed by the filling of the subshells in period 2. The arrangement of electrons in neon completes shells 1 and 2. Examples of electron configurations for some elements in Period 2 are shown below:

Li (3)	$1s^2\ 2s^1$		O (8)	$1s^2\ 2s^2\ 2p^4$
B (5)	$1s^2\ 2s^2\ 2p^1$		Ne (10)	$1s^2\ 2s^2\ 2p^6$

Period 3 Sodium to Argon

In period 3, electrons go into the $3s$ and $3p$ subshells, which are complete in argon. However, the $3d$ subshell is vacant.

Third period

Na	Mg
$3s^1$	$3s^2$

Al	Si	P	S	Cl	Ar
$3p^1$	$3p^2$	$3p^3$	$3p^4$	$3p^5$	$3p^6$

As electron configurations become longer, we may use an abbreviated or shorthand form called the **noble gas notation.** In this notation, the symbol [Ne] is used to show the completed subshells $1s^2 2s^2 2p^6$ of neon. Examples of the electron configurations and their noble gas notations for some elements in period 3 are shown below:

		Noble Gas Notation
Na (11)	$1s^2\ 2s^2\ 2p^6\ 3s^1$	[Ne] $3s^1$
Al (13)	$1s^2\ 2s^2\ 2p^6\ 3s^2\ 3p^1$	[Ne] $3s^2\ 3p^1$
S (16)	$1s^2\ 2s^2\ 2p^6\ 3s^2\ 3p^4$	[Ne] $3s^2\ 3p^4$
Ar (18)	$1s^2\ 2s^2\ 2p^6\ 3s^2\ 3p^6$	[Ne] $3s^2\ 3p^6$

Period 4 Potassium and Calcium

On the periodic table, the $4s$ block fills before the $3d$. This occurs because the $4s$ subshell has a lower energy level than a $3d$ subshell. The next two electrons fill the $4s$. In the noble gas notation, the filled subshells are the same as [Ar].

Fourth period

K	Ca
$4s^1$	$4s^2$

Examples of the electron configurations and their noble gas notations for potassium and calcium are shown below:

		Noble Gas Notation
K (19)	$1s^2\ 2s^2\ 2p^6\ 3s^2\ 3p^6\ 4s^1$	[Ar] $4s^1$
Ca (20)	$1s^2\ 2s^2\ 2p^6\ 3s^2\ 3p^6\ 4s^2$	[Ar] $4s^2$

Transition Metals

The electrons in the ten elements of the transition metal group then fill the $3d$ subshell.

Sc	Ti	V	Cr	Mn	Fe	Co	Ni	Cu	Zn
$3d^1$	$3d^2$	$3d^3$	$(4s^1)3d^5$	$3d^5$	$3d^6$	$3d^7$	$3d^8$	$(4s^1)3d^{10}$	$3d^{10}$

The configurations of scandium, iron, and zinc are shown below:

Sc (21)	$1s^2\ 2s^2\ 2p^6\ 3s^2\ 3p^6\ 4s^2\ 3d^1$	$[Ar]\ 4s^2\ 3d^1$
Fe (26)	$1s^2\ 2s^2\ 2p^6\ 3s^2\ 3p^6\ 4s^2\ 3d^6$	$[Ar]\ 4s^2\ 3d^6$
Zn (30)	$1s^2\ 2s^2\ 2p^6\ 3s^2\ 3p^6\ 4s^2\ 3d^{10}$	$[Ar]\ 4s^2\ 3d^{10}$

Some of the transition metals show some variation in the order of filling; for instance, some move an s electron to the d subshell. This occurs for elements such as chromium and copper because it gives them a half-filled (5 electrons) or filled (10 electrons) d subshell, which is especially stable. For example, we might expect copper (Cu) to have 9 electrons in the $3d$ subshell. However, it gains the greater stability of a full subshell of ten $3d$ electrons by moving an electron from the $4s$.

Noble Gas Notation

Cu (29)	$1s^2\ 2s^2\ 2p^6\ 3s^2\ 3p^6\ 4s^1\ 3d^{10}$	$[Ar]\ 4s^1\ 3d^{10}$

Gallium to Krypton

Starting with gallium, the next six electrons enter the $4p$ block on the periodic table.

Ga	Ge	As	Se	Br	Kr
$4p^1$	$4p^2$	$4p^3$	$4p^4$	$4p^5$	$4p^6$

Noble Gas Notation

Ga (31)	$1s^2\ 2s^2\ 2p^6\ 3s^2\ 3p^6\ 4s^2\ 3d^{10}\ 4p^1$	$[Ar]\ 4s^2\ 3d^{10}\ 4p^1$
As (33)	$1s^2\ 2s^2\ 2p^6\ 3s^2\ 3p^6\ 4s^2\ 3d^{10}\ 4p^3$	$[Ar]\ 4s^2\ 3d^{10}\ 4p^2$
Br (35)	$1s^2\ 2s^2\ 2p^6\ 3s^2\ 3p^6\ 4s^2\ 3d^{10}\ 4p^5$	$[Ar]\ 4s^2\ 3d^{10}\ 4p^5$
Kr (36)	$1s^2\ 2s^2\ 2p^6\ 3s^2\ 3p^6\ 4s^2\ 3d^{10}\ 4p^6$	$[Ar]\ 4s^2\ 3d^{10}\ 4p^6$

The complete electron configurations for elements 1–36 are shown in Table 2.15.

SAMPLE PROBLEM 2.16

Electron Configuration

Use the subshell blocks on the periodic table to write the electron configuration for each of the following:

1. sulfur, atomic number 16
2. bromine, atomic number 35.

Solution

1. Sulfur is in period 3, Group 6A. Period 1 will fill the $1s^2$ and period 2 will complete $2s^2$ and $2p^6$. In period 3, the $3s$ block is complete $3s^2$ and there are four electrons in the $3p$ block $3p^4$.

S $1s^2\,2s^2\,2p^6\,3s^2\,3p^4$

2. Bromine is located in Group 7A in period 4. Going across the periodic table starting with period 1, the subshell blocks are filled.

Period 1 $1s^2$ (filled)
Period 2 $2s^2$ and then $2p^6$ (filled)
Period 3 $3s^2$ and then $3p^6$ (filled)
Period 4 $4s^2$, then $3d^{10}$ (filled) and finally $4p^5$ (five electrons)

Br $1s^2\,2s^2\,2p^6\,3s^2\,3p^6\,4s^2\,3d^{10}\,4p^5$

Table 2.15   Electron Configurations for Elements 1–36

Element	Atomic Number	Electron Configuration
H	1	$1s^1$
He	2	$1s^2$
Li	3	$1s^2\,2s^1$
Be	4	$1s^2\,2s^2$
B	5	$1s^2\,2s^2\,2p^1$
C	6	$1s^2\,2s^2\,2p^2$
N	7	$1s^2\,2s^2\,2p^3$
O	8	$1s^2\,2s^2\,2p^4$
F	9	$1s^2\,2s^2\,2p^5$
Ne	10	$1s^2\,2s^2\,2p^6$
Na	11	$1s^2\,2s^2\,2p^6\,3s^1$
Mg	12	$1s^2\,2s^2\,2p^6\,3s^2$
Al	13	$1s^2\,2s^2\,2p^6\,3s^2\,3p^1$
Si	14	$1s^2\,2s^2\,2p^6\,3s^2\,3p^2$
P	15	$1s^2\,2s^2\,2p^6\,3s^2\,3p^3$
S	16	$1s^2\,2s^2\,2p^6\,3s^2\,3p^4$
Cl	17	$1s^2\,2s^2\,2p^6\,3s^2\,3p^5$
Ar	18	$1s^2\,2s^2\,2p^6\,3s^2\,3p^6$
K	19	$1s^2\,2s^2\,2p^6\,3s^2\,3p^6\,4s^1$
Ca	20	$1s^2\,2s^2\,2p^6\,3s^2\,3p^6\,4s^2$
Sc	21	$1s^2\,2s^2\,2p^6\,3s^2\,3p^6\,4s^2\,3d^1$
Ti	22	$1s^2\,2s^2\,2p^6\,3s^2\,3p^6\,4s^2\,3d^2$
V	23	$1s^2\,2s^2\,2p^6\,3s^2\,3p^6\,4s^2\,3d^3$
Cr	24	$1s^2\,2s^2\,2p^6\,3s^2\,3p^6\,4s^1\,3d^5$ (1/2 filled d subshell is stable)
Mn	25	$1s^2\,2s^2\,2p^6\,3s^2\,3p^6\,4s^2\,3d^5$
Fe	26	$1s^2\,2s^2\,2p^6\,3s^2\,3p^6\,4s^2\,3d^6$
Co	27	$1s^2\,2s^2\,2p^6\,3s^2\,3p^6\,4s^2\,3d^7$
Ni	28	$1s^2\,2s^2\,2p^6\,3s^2\,3p^6\,4s^2\,3d^8$
Cu	29	$1s^2\,2s^2\,2p^6\,3s^2\,3p^6\,4s^1\,3d^{10}$ (fills the d subshell for stability)
Zn	30	$1s^2\,2s^2\,2p^6\,3s^2\,3p^6\,4s^2\,3d^{10}$
Ga	31	$1s^2\,2s^2\,2p^6\,3s^2\,3p^6\,4s^2\,3d^{10}\,4p^1$
Ge	32	$1s^2\,2s^2\,2p^6\,3s^2\,3p^6\,4s^2\,3d^{10}\,4p^2$
As	33	$1s^2\,2s^2\,2p^6\,3s^2\,3p^6\,4s^2\,3d^{10}\,4p^3$
Se	34	$1s^2\,2s^2\,2p^6\,3s^2\,3p^6\,4s^2\,3d^{10}\,4p^4$
Br	35	$1s^2\,2s^2\,2p^6\,3s^2\,3p^6\,4s^2\,3d^{10}\,4p^5$
Kr	36	$1s^2\,2s^2\,2p^6\,3s^2\,3p^6\,4s^2\,3d^{10}\,4p^6$

What is the electron configuration for cobalt (Co)?

Group Numbers and the Outer Shell Electrons

The number of electrons in the s and p subshells corresponds to the Group A numbers. These electrons form the outer shell of electrons, which are important in the bonding patterns of atoms. The outermost electrons are shown for Periods 1 to 4 in Table 2.16. The 1–18 designations on the periodic table (see Figure 2.1) indicate the s, p, and d electrons in the highest energy levels. In the United States, the A-B designations are still in use. The letter "A" is used for the representative elements where the numbers 1A−8A refer to the s and p electrons in the outer shell. The letter "B" is used for the transition elements and includes the d electrons.

Table 2.16   Outer shell Electrons For Representative Elements in Periods 1–4

1A	2A	3A	4A	5A	6A	7A	8A
1 H $1s^1$							2 He $1s^2$
3 Li $2s^1$	4 Be $2s^2$	5 B $2s^22p^1$	6 C $2s^22p^2$	7 N $2s^22p^3$	8 O $2s^22p^4$	9 F $2s^22p^5$	10 Ne $2s^22p^6$
11 Na $3s^1$	12 Mg $3s^2$	13 Al $3s^23p^1$	14 Si $3s^23p^2$	15 P $3s^23p^3$	16 S $3s^23p^4$	17 Cl $3s^23p^5$	18 Ar $3s^23p^6$
19 K $4s^1$	20 Ca $4s^2$	31 Ga $4s^24p^1$	32 Ge $4s^24p^2$	33 As $4s^24p^3$	34 Se $4s^24p^4$	35 Br $4s^24p^5$	36 Kr $4s^24p^6$

QUESTIONS AND PROBLEMS

Electron Configurations

2.59 Write the electron configuration ($1s^2\ 2s^2$) for the following elements:
   a. magnesium   **b.** phosphorus   **c.** argon
   d. sulfur   **e.** chlorine   **f.** zinc
   g. strontium   **h.** iodine

2.60 Write the electron configuration ($1s^2\ 2s^2$) for the following elements:
   a. potassium   **b.** sodium   **c.** beryllium
   d. nitrogen   **e.** carbon   **f.** calcium
   g. manganese   **h.** cesium

2.61 Identify the element that has the following electron arrangement:
   a. $1s^1$ **b.** $1s^2\ 2s^2\ 2p^3$ **c.** $1s^2\ 2s^2\ 2p^6\ 3s^1$
   d. $1s^2\ 2s^2\ 2p^6$ **e.** [Ne] $3s^2\ 3p^2$ **f.** [Ar] $4s^2\ 3d^{10}\ 4p^5$

2.62 Identify the element that has the following electron arrangement:
   a. $1s^2\ 2s^2\ 2p^2$ **b.** $1s^2\ 2s^1$ **c.** $1s^2\ 2s^2\ 2p^6\ 3s^2\ 3p^4$
   d. $1s^2\ 2s^2\ 2p^6\ 3s^2\ 3p^6\ 4s^2$ **e.** [Ne] $3s^2\ 3p^4$ **f.** [Kr] $5s^2\ 4d^{10}\ 5p^2$

Chapter Review

2.1 Elements and Symbols

Elements are the primary substances of matter. Chemical symbols are one- or two-letter abbreviations of the names of the elements.

2.2 The Periodic Table

The periodic table is an arrangement of the elements by increasing atomic number. A vertical column on the periodic table containing elements with similar properties is called a group. A horizontal row is called a period. Elements in Group 1A are called the alkali metals; Group 2A, alkaline earth metals; Group 7A, the halogens; and Group 8A, the noble gases. On the periodic table, metals are located on the left of the heavy zigzag line, and nonmetals are to the right of the heavy zigzag line. Elements located on the heavy line are called metalloids.

2.3 The Atom

An atom is the smallest particle that retains the characteristics of an element. Atoms are composed of three subatomic particles. Protons have a positive charge ($+$), electrons carry a negative charge ($-$), and neutrons are electrically neutral. The protons and neutrons, each with a mass of about 1 amu, are found in the tiny, dense nucleus. Electrons with a much smaller mass are located outside the nucleus.

2.4 Atomic Number and Mass Number

The atomic number gives the number of protons in all the atoms of the same element. In a neutral atom, there is an equal number of protons and electrons. The mass number is the total number of protons and neutrons in an atom.

2.5 Isotopes and Atomic Mass

Atoms that have the same number of protons but different numbers of neutrons are called isotopes. The atomic mass of an element is the weighted average mass of all the isotopes in a naturally occurring sample of that element.

2.6 Electron Energy Levels

Every electron has a specific amount of energy. In an atom, the electrons of similar energy are grouped in specific energy levels or shells. The first shell nearest the nucleus can hold 2 electrons, the second shell can hold 8 electrons, the third shell will take up to 18 electrons. The electron arrangement is written by placing the number of electrons in that atom in order from the lowest energy levels and filling to higher levels. The similarity of behavior for the elements in a group is related to having the same number of electrons in their outermost shells. The group number for an element gives the number of electrons in its outermost shell.

2.7 Subshells and Orbitals

Within each energy level or shell, the electrons are grouped into subshells. Each subshell consists of orbitals, which represent the space where electrons of that energy are likely to be found. An s orbital has a spherical shape and there is one s orbital in an s subshell. A p orbital has two lobes like a dumbbell and there are three p orbitals in each p subshell. A d subshell consists of five d orbitals and an f subshell has seven f orbitals. Because there can be two electrons in any kind of orbital, there are 2 electrons possible in an s subshell, 6 electrons in a p subshell, 10 electrons in a d subshell, and 14 electrons in an f subshell.

2.8 Electron Configurations

The electron configuration gives the organization of the electrons occupying each subshell. For a particular element, electrons are placed in the subshells starting with the one with lowest energy and going in order to the next highest energy until all the electrons have been used. The subshells with the highest period numbers represent the outermost electrons or the outer shell of the atom.

Key Terms

alkali metals Elements of Group 1A except hydrogen; these are soft, shiny metals with one outer shell electron.

alkaline earth metals Group 2A elements, which have 2 electrons in their outer shells.

atom The smallest particle of an element that retains the characteristics of the element.

atomic mass The weighted average mass of all the naturally occurring isotopes of an element.

atomic mass unit (amu) A small mass unit used to describe the mass of very small particles such as atoms and subatomic particles; 1 amu is equal to one-twelfth the mass of a carbon-12 atom.

atomic number A number that is equal to the number of protons in an atom.

atomic symbol An abbreviation used to indicate the mass number and atomic number of an isotope.

chemical symbol An abbreviation that represents the name of an element.

electron A negatively charged subatomic particle having a very small mass that is usually ignored in calculations; its symbol is e^-.

electron configuration An organization of electrons within the atom given by increasing energy subshells.

element A primary substance that cannot be separated into any simpler substances.

group A vertical column in the periodic table that contains elements having similar physical and chemical properties.

group number A number that appears at the top of each vertical column (group) in the periodic table and indicates the number of electrons in the outermost shell.

halogen Group 7A elements of fluorine, chlorine, bromine, and iodine.

isotope An atom that differs only in mass number from another atom of the same element. Isotopes have the same atomic number (number of protons) but different numbers of neutrons.

mass number The total number of neutrons and protons in the nucleus of an atom.

metal An element that is shiny, malleable, and a good conductor of heat and electricity. The metals are located to the left of the zigzag line in the periodic table.

metalloid Elements with properties of both metals and nonmetals located along the heavy zigzag line on the periodic table.

neutron A neutral subatomic particle having a mass of 1 amu and found in the nucleus of an atom; its symbol is n or n^0.

noble gas An element in Group 8A of the periodic table, generally unreactive and seldom found in combination with other elements.

nonmetal An element with little or no luster that is a poor conductor of heat and electricity. The nonmetals are located to the right of the zigzag line in the periodic table.

nucleus The compact, very dense center of an atom, containing the protons and neutrons of the atom.

orbital The region around the nucleus where electrons of a certain energy are more likely to be found. The s orbitals are spherical; the p orbitals have two lobes like a dumbbell.

period A horizontal row of elements in the periodic table.

periodic law The repetition of similar chemical and physical properties with increasing atomic number due to the reappearance of the same number of electrons in the outermost shells of atoms of those elements.

periodic table An arrangement of elements by increasing atomic number such that elements having similar chemical behavior are grouped in vertical columns.

physical property A characteristic that can be observed or measured without affecting the identity of an element, including shape, color, odor, taste, density, hardness, melting point, and boiling point.

proton A positively charged subatomic particle having a mass of 1 amu and found in the nucleus of an atom; its symbol is p or p^+.

representative element An element found in Groups 1A through 8A of the periodic table.

shell An energy level containing electrons of similar energies.

subatomic particle A particle within an atom; protons, neutrons, and electrons are subatomic particles.

subshell A group of electrons within a shell that have the same energy.

transition element An element in the B section located between Groups 2A and 3A on the periodic table.

Additional Problems

2.63 Why is Co the symbol for cobalt, but not CO?

2.64 Which of the following is correct? Write the correct symbol if needed.
a. copper, Co   b. silicon, SI   c. iron, Fe
d. fluorine, Fl   e. potassium, P   f. sodium, Na
g. gold, Au   h. lead, PB

2.65 Give the symbol and name of the element found in the following group and period on the periodic table.
a. Group 2A, Period 3   b. Group 7A, Period 4
c. Group 3A, Period 3   d. Group 6A, Period 2

2.66 Give the group and period number for the following elements:
a. potassium   b. phosphorus
c. carbon   d. neon

2.67 The following statements are *false*. Reword the underlined word or phrase to make a *true* statement.
a. The proton is a _neutral_ particle.
b. The electrons are found _in the nucleus_.
c. The nucleus is the _largest_ part of the atom.
d. The _neutron_ has a negative charge.
e. Most of the mass of an atom is due to its _electrons_.

2.68 State the number of protons and neutrons in the following atoms:
a. ^2_1H b. $^{37}_{17}Cl$ c. $^{106}_{48}Cd$ d. $^{224}_{83}Bi$

2.69 The most abundant isotope of iron is Fe-56.
a. How many protons, neutrons, and electrons are in this isotope?
b. What is the symbol of another isotope of iron with 25 neutrons?

c. What is the symbol of a different atom with the same mass number of 51 and 27 neutrons?

2.70 Cadmium, atomic number 48, consists of eight naturally occurring isotopes. Do you expect any of the isotopes to have the atomic mass listed on the periodic table for cadmium? Explain.

2.71 Distinguish between
   a. atoms and isotopes
   b. atomic number and mass number

2.72 Distinguish between
   a. elemental symbol and atomic symbol
   b. atomic number and atomic mass

2.73 Consider the following atoms in which the chemical symbol of the element is represented by X.

$^{16}_{8}X$ $^{16}_{9}X$ $^{18}_{10}X$ $^{17}_{8}X$ $^{18}_{8}X$
   a. What atoms have the same number of protons?
   b. Which atoms are isotopes? Of what element?
   c. Which atoms have the same mass number?
   d. What atoms have the same number of neutrons?

2.74 Five isotopes of zinc are zinc-64, zinc-66, zinc-67, zinc-68, and zinc-70.
   a. Write the atomic symbols, including atomic number and mass number for each of these atoms.
   b. Give the number of protons, electrons, and neutrons for each of the zinc isotopes.

2.75 A sample of copper has two naturally occurring isotopes. The isotope Cu-63 makes up 69.2% of all the copper atoms, and Cu-65 makes up 30.8% of the sample. What average atomic mass would you calculate for copper?

2.76 Silicon has three isotopes that occur in nature. ^{28}Si has 92.2% abundance, ^{29}Si has 4.70% abundance, and ^{30}Si has 3.10% abundance. What is the atomic mass of silicon?

2.77 If the diameter of a sodium atom is 3.14×10^{-8} cm, how many sodium atoms would fit along a line exactly 1 inch long?

2.78 A lead atom has a mass of 3.4×10^{-22} g. How many lead atoms are in a cube of lead that has a volume of 2.00 cm³, if the density of lead is 11.3 g/cm³?

2.79 Write the names of two representative elements that have the following number of electrons in their outer energy levels:
   a. 2 electrons   **b.** 6 electrons
   c. 8 electrons   **d.** 3 electrons

2.80 Explain the difference between the following:
   a. atomic mass and average atomic mass
   b. atom and nucleus

2.81 Write the names of two elements that are in the following groups:
   a. halogens   **b.** noble gases
   c. alkali metals   **d.** alkaline earth metals

2.82 The following are trace elements that have been found to be crucial to the biochemical and physiological processes in the body. Indicate whether each is a metal or nonmetal.
   a. zinc   **b.** cobalt   **c.** manganese (Mn)
   d. iodine   **e.** copper   **f.** selenium (Se)

2.83 Indicate the maximum number of each of the following:
   a. subshells in shell 3
   b. orbitals in the $2p$ subshell
   c. electrons in $3d$ subshell
   d. electrons in a $3p$ orbital

2.84 Indicate the maximum number of each of the following:
   a. electrons in shell 2
   b. number of subshells in shell 4
   c. electrons in $2d$ orbital
   d. electrons in the $4d$ subshell

2.85 Indicate the element that meets the following conditions:
   a. has 3 electrons in shell 4
   b. has two $2p$ electrons
   c. begins to fill the $5s$ subshell
   d. has 2 electrons in the $4d$ subshell
   e. completes the $4p$ subshell

2.86 Indicate the element that meets the following conditions:
   a. has 5 electrons in shell 3
   b. has four $3p$ electrons
   c. begins to fill the $5p$ subshell
   d. has 1 electron in the $4p$ subshell
   e. completes the $7s$ subshell

2.87 Using the periodic table, write the electron configuration ($1s^2\ 2s^2$) for the following elements:
   a. arsenic   **b.** rubidium   **c.** barium
   d. silicon   **e.** bromine   **f.** tin

2.88 Identify the element that has the following electron arrangement:
   a. $1s^2\ 2s^2\ 2p^5$ **b.** [Ne]$3s^2\ 3p^1$
   c. $1s^2\ 2s^2\ 2p^6\ 3s^2\ 3p^6\ 4s^2\ 3d^6$
   d. $1s^2\ 2s^2\ 2p^6\ 3s^2\ 3p^6$ **e.** [Kr]$5s^2\ 4d^{10}$

3 Nuclear Radiation

"Everything we do in this department involves radioactive materials," says Julie Goudak, nuclear medicine technologist at Kaiser Hospital. "The radioisotopes are given in several ways. The patient may ingest an isotope, breathe it in, or receive it by an IV injection. We do many diagnostic tests, particularly of the heart function, to determine if a patient needs a cardiac CAT scan."

A nuclear medicine technologist administers isotopes that emit radiation to determine the level of function of an organ such as the thyroid or heart, to detect the presence and size of a tumor, or to treat disease. A radioisotope locates in a specific organ and its radiation is used by a computer to create an image of that organ. From this data, a physician can make a diagnosis and design a treatment program.

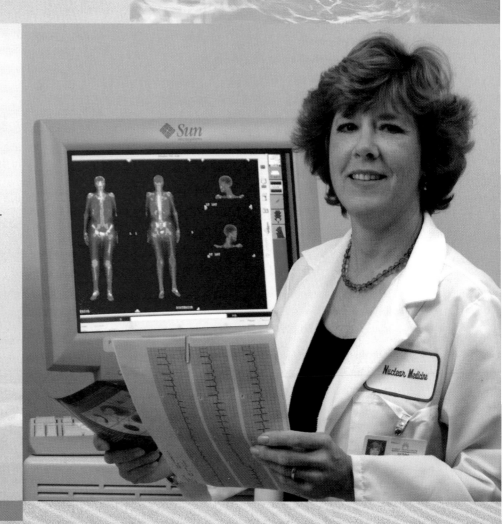

LOOKING AHEAD

the Chemistry place

www.chemplace.com/college

Visit the URL above or use the CD-ROM in the book for extra quizzing, interactive tutorials, career resources, and case studies.

A female patient, aged 50, complains of nervousness, irritability, increased perspiration, brittle hair, and muscle weakness. Her hands are shaky at times and her heart often beats rapidly. She has been experiencing weight loss. The doctor decides to test for hyperthyroidism. To get a detailed look at the thyroid, a thyroid scan is ordered. The patient is given a small amount of an iodine radioisotope, which will be taken up by the thyroid. The scan shows a higher than normal rate of uptake of the radioactive iodine indicating an overactive thyroid gland, a condition called hyperthyroidism. Treatment for hyperthyroidism includes the use of drugs to lower the level of thyroid hormone, the use of radioactive iodine to destroy thyroid cells, or surgical removal of part or the entire thyroid. In our case, the nuclear physician decides to use radioactive iodine. To begin treatment, the patient drinks a solution containing radioactive iodine. In the following few weeks, the cells that take up the radioactive iodine are destroyed by the radiation. Tests show that the patient's thyroid is smaller, and that the blood level of thyroid hormone is normal once again.

With the discovery of X rays in 1895 and the production of artificial radioactive substances in 1934, the field of nuclear medicine was established. In 1937, the first radioactive isotope was used to treat a patient with leukemia at the University of California at Berkeley. Major strides in the use of radioactivity in medicine occurred in 1946, when a radioactive iodine isotope was successfully used to diagnose thyroid function and to treat hyperthyroidism and thyroid cancer. In the 1970s and 1980s, a variety of radioactive substances were used to produce an image of an organ such as liver, spleen, thyroid gland, kidneys and the brain, and to detect heart disease. Today, procedures in nuclear medicine provide information about the function and structure of every organ in the body, which allows the nuclear physician to diagnose and treat diseases early.

3.1 Natural Radioactivity

LEARNING GOAL
Describe alpha, beta, and gamma radiation.

Most naturally occurring isotopes of elements up to atomic number 19 have stable nuclei. Elements with higher atomic numbers (20 to 83) consist of a mixture of isotopes, some of which may have unstable nuclei. When the nucleus of an isotope is unstable, it is **radioactive,** which means that it will spontaneously emit energy to become more stable. This energy, called **radiation,** may take the form of particles such as alpha (α) particles or beta (β) particles or pure energy such as gamma (γ) rays. Elements with atomic numbers of 84 and higher consist only of radioactive isotopes. So many protons and neutrons are crowded together in their nuclei that the strong repulsions between the protons makes those nuclei unstable.

In Chapter 2, we wrote symbols to distinguish the different isotopes of an element. Recall that an atom's mass number is equal to the sum of the protons and neutrons in the nucleus, and its atomic number is equal to the number of protons. In the symbol for an isotope, the mass number is written in the upper left corner of the symbol, and the atomic number is written in the lower left corner. For example, a radioactive isotope of iodine used in the diagnosis and treatment of thyroid conditions has a mass number of 131 and an atomic number of 53. We can write its symbol as shown on the next page.

Mass number (protons and neutrons)
Element
Atomic number (protons)

$$^{131}_{53}\text{I}$$

This isotope is also called iodine-131 or I-131. Radioactive isotopes are named by writing the mass number after the element's name or symbol. When necessary, we can obtain the atomic number from the periodic table. Table 3.1 compares some stable, nonradioactive isotopes with some radioactive isotopes.

Table 3.1   Stable and Radioactive Isotopes of Some Elements

Magnesium	Iodine	Uranium
Stable Isotopes		
$^{24}_{12}\text{Mg}$	$^{127}_{53}\text{I}$	None
Magnesium-24	Iodine-127	
Radioactive Isotopes		
$^{23}_{12}\text{Mg}$	$^{125}_{53}\text{I}$	$^{235}_{92}\text{U}$
Magnesium-23	Iodine-125	Uranium-235
$^{27}_{12}\text{Mg}$	$^{131}_{53}\text{I}$	$^{238}_{92}\text{U}$
Magnesium-27	Iodine-131	Uranium-238

Types of Radiation

We cannot feel, taste, or smell radiation, but high-energy radiation is capable of creating havoc within the cells of our bodies. Different forms of radiation are emitted from an unstable nucleus when a change takes place among its protons and neutrons. Energy is released, and a new, more stable nucleus is formed. One type of radiation consists of alpha particles. An **alpha particle** contains 2 protons and 2 neutrons, which gives it a mass number of 4 and an atomic number of 2. Because it has 2 protons, an alpha particle has a charge of 2+. That makes it identical to a helium nucleus. In equations, it is written as the Greek letter alpha (α) or as the symbol for helium.

$$\alpha \quad \text{or} \quad ^{4}_{2}\text{He}$$
Alpha particle

Another type of radiation occurs when a radioisotope emits **beta particles.** A beta particle, which is a high-energy electron, has a charge of 1− and a mass number of 0. It is represented by the Greek letter beta (β) or by the symbol for the electron (e^-).

$$\beta \quad \text{or} \quad ^{0}_{-1}e$$
Beta particle

Beta particles are produced from unstable nuclei when neutrons change into protons. The high-energy electrons do not exist in the nucleus until there is a transformation of neutrons within the nucleus according to the following nuclear equation.

$$^{1}_{0}\text{n} \longrightarrow ^{1}_{1}\text{H} + ^{0}_{-1}e$$

| Neutron in the nucleus | New proton remains in the nucleus | New electron formed and emitted as a beta particle |

Gamma rays are high-energy radiation, similar to X rays, released as an unstable nucleus undergoes a rearrangement of its particles to give a more stable, lower energy nucleus. A gamma ray is shown as the Greek letter gamma (γ). Because gamma rays are energy only, there is no mass or charge associated with their symbol.

γ
Gamma ray

Table 3.2 summarizes the types of radiation we will use in nuclear equations.

Table 3.2 Some Common Forms of Radiation

Type of Radiation	Symbol		Mass Number	Charge
Alpha particle	α,	^4_2He	4	2+
Beta particle	β,	$^{\;0}_{-1}e$	0	1−
Gamma ray, X ray	γ		0	0
Proton	^1_1H,	^1_1p	1	1+
Neutron	^1_0n		1	0

SAMPLE PROBLEM 3.1

Writing Symbols for Radiation Particles

Write the symbol for an alpha particle.

Solution

The alpha particle contains 2 protons and 2 neutrons. It has a mass number of 4 and an atomic number of 2.

α or ^4_2He

Study Check

What is the symbol used for beta radiation?

Radiation Protection

Because many cells in the body are sensitive to radiation, it is important that the radiologist, doctor, and nurse working with radioactive isotopes use proper radiation protection. Proper **shielding** is necessary to prevent exposure. Alpha particles are the heaviest of the radiation particles; they travel only a few centimeters in the air before they collide with air molecules, acquire electrons, and become helium atoms. A piece of paper, clothing, and our skin are protection against alpha particles. Lab coats and gloves will also provide sufficient shielding. However, if ingested or inhaled, alpha particles can bring about serious internal damage because their mass and high charge causes much ionization in a short distance.

Beta particles have a very small mass and move much faster and farther than alpha particles, traveling as much as several meters through air. They can pass through paper and penetrate as far as 4–5 mm into body tissue. External exposure to beta particles can burn the surface of the skin, but they are stopped before they

HEALTH NOTE

Biological Effects of Radiation

When high-energy radiation strikes molecules in its path, electrons may be knocked away. The result of this *ionizing radiation* is the formation of unstable ions or radicals. A free radical is a particle that has an unpaired electron. For example, when radiation passes through the human body, it may interact with water molecules, removing electrons and producing H_2O^+ ions or free radicals.

When ionizing radiation strikes the cells of the body, the unstable ions or free radicals that form can cause undesirable chemical reactions. The cells in the body most sensitive to radiation are the ones undergoing rapid division—those of the bone marrow, skin, reproductive organs, and intestinal lining, as well as all cells of growing children. Damaged cells may lose their ability to produce necessary materials. For example, if radiation damages cells of the bone marrow, red blood cells may no longer be produced. If sperm cells or ova or the cells of a fetus are damaged, birth defects may result. In contrast, cells of the nerves, muscles, liver, and adult bones are much less sensitive to radiation because they undergo little or no cellular division.

Cancer cells are another example of rapidly dividing cells. Because cancer cells are highly sensitive to radiation, large doses of radiation are used to destroy them. The surrounding normal tissue, dividing at a slower rate, shows a greater resistance to radiation and suffers less damage. In addition, normal tissue is able to repair itself more readily than cancerous tissue. However, this repair is not always complete. Long-term effects of ionizing radiation include a shortened life span, malignant tumors, leukemia, anemia, and genetic mutations.

can reach the internal organs. Heavy clothing such as lab coats and gloves are needed to protect the skin from beta particles. (See Figure 3.1.)

Gamma rays travel great distances through the air and pass through many materials, including body tissues. Only very dense shielding, such as lead or concrete, will stop them. Because they can penetrate so deeply, exposure to gamma rays can be extremely hazardous. Even the syringe used to give an injection of a gamma-emitting radioactive isotope is placed inside a special lead-glass cover.

When preparing radioactive materials, the radiologist wears special gloves and works behind leaded windows. Long tongs are used within the work area to pick up vials of radioactive material, keeping them away from the hands and body. (See Figure 3.2.) Table 3.3 summarizes the shielding materials required for the various types of radiation.

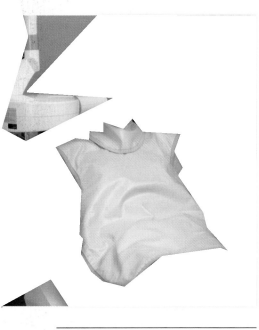

Figure 3.1 Shielding materials protect a person from alpha, beta, and gamma radiation.

Q **Why is lead or concrete needed for protection from gamma rays when paper and clothing are sufficient protection from alpha particles?**

Table 3.3 Properties of Ionizing Radiation and Shielding Required

Property	Alpha	Beta	Gamma
Characteristics	Helium nucleus	Electron	High-energy rays
Symbols	α, $^4_2$He	β, $^0_{-1}e$	γ, $^0_0\gamma$
Travel Distance in Air	2–4 cm	200–300 cm	500 m
Tissue Depth	0.05 mm	4–5 mm	50 cm or more
Shielding	Paper, clothing	Heavy clothing, lab coats, gloves,	Lead, thick concrete
Typical Source	Radium-226	Carbon-14	Technetium-99m

Try to keep the time you must spend in a radioactive area to a minimum. A certain amount of radiation is emitted every minute. Remaining in a radioactive area twice as long exposes a person to twice as much radiation.

Keep your distance! The greater the distance from the radioactive source, the lower the intensity of radiation received. If you double your distance from the radiation source, the intensity of radiation drops to $\left(\frac{1}{2}\right)^2$ or one-fourth of its previous value. This is one of the reasons dentists and x-ray technicians leave the room and stand behind a shield or lead-lined wall to take your X rays. They are exposed to radiation every day and must minimize the amount of radiation they receive.

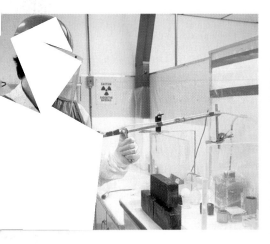

Figure 3.2 A person working with radioisototopes wears protective clothing and gloves and stands behind a lead shield.

Q **What types of radiation does the lead shield block?**

SAMPLE PROBLEM 3.2

Radiation Protection

How does the type of shielding for alpha radiation differ from that used for gamma radiation?

Solution

Alpha radiation is stopped by paper and clothing. However, lead or concrete is needed for protection from gamma radiation.

Study Check

Besides shielding, what other methods help reduce exposure to radiation?

QUESTIONS AND PROBLEMS

Natural Radioactivity

The answers for the odd-numbered Questions and Problems throughout this chapter are given at the end of this text. The complete solutions to the odd-numbered Questions and Problems are in the *Study Guide*.

3.1 **a.** How are an alpha particle and a helium nucleus similar? different?
   b. What symbols are used for alpha particles?

3.2 **a.** How are a beta particle and an electron similar? different?
   b. What symbols are used for beta particles?

3.3 Naturally occurring potassium consists of three isotopes, potassium-39, potassium-40, and radioactive potassium-41.
   a. Write the atomic symbols for each isotope.
   b. In what ways are the isotopes similar, and in what ways do they differ?

3.4 Naturally occurring iodine is iodine-127. Medically, radioactive isotopes of iodine-125 and iodine-130 are used.
   a. Write the atomic symbols for each isotope.
   b. In what ways are the isotopes similar, and in what ways do they differ?

3.5 Supply the missing information in the following table:

Medical Use	Atomic Symbol	Mass Number	Number of Protons	Number of Neutrons
Heart imaging	$^{201}_{81}\text{Tl}$			
Radiation therapy		60	27	
Abdominal scan			31	36
Hyperthyroidism	$^{131}_{53}\text{I}$			
Leukemia treatment		32		17

3.6 Supply the missing information in the following table:

Medical Use	Atomic Symbol	Mass Number	Number of Protons	Number of Neutrons
Cancer treatment	$^{60}_{27}\text{Co}$			
Brain scan		99	43	
Blood flow		141	58	
Bone scan		85		47
Lung function	$^{133}_{54}\text{Xe}$			

3.7 Write a symbol for the following:
   a. alpha particle   **b.** neutron   **c.** beta particle
   d. nitrogen-15   **e.** iodine-125

3.8 Write a symbol for the following:
   a. proton   **b.** gamma ray   **c.** electron
   d. barium-131   **e.** cobalt-60

3.9 Identify the symbol for X in each of the following:
   a. $^{0}_{-1}\text{X}$ **b.** $^{4}_{2}\text{X}$ **c.** $^{1}_{0}\text{X}$ **d.** $^{24}_{11}\text{X}$ **e.** $^{14}_{6}\text{X}$

3.10 Identify the symbol for X in each of the following:
   a. $^{1}_{1}\text{X}$ **b.** $^{32}_{15}\text{X}$ **c.** $^{0}_{0}\text{X}$ **d.** $^{59}_{26}\text{X}$ **e.** $^{85}_{38}\text{X}$

3.11 a. Why does beta radiation penetrate further in solid material than alpha radiation?

    b. How does ionizing radiation cause damage to cells of the body?

    c. Why does the x-ray technician leave the room when you receive an X ray?

    d. What is the purpose of wearing gloves when handling radioisotopes?

3.12 a. As a nurse in an oncology unit, you sometimes give an injection of a radioisotope. What are three ways you can minimize your exposure to radiation?

    b. Why are cancer cells more sensitive to radiation than nerve cells?

    c. What is the purpose of placing a lead apron on a patient who is receiving routine dental X rays?

    d. Why are the walls in a radiology office built of thick concrete blocks?

3.2 Nuclear Equations

When a nucleus spontaneously breaks down by emitting radiation, the process is called *radioactive decay*. It can be shown as a *nuclear equation* using the symbols for the original radioactive nucleus, the new nucleus, and the radiation emitted.

$$\text{Radioactive nucleus} \longrightarrow \text{new nucleus} + \text{radiation}\ (\alpha, \beta, \gamma)$$

A nuclear equation is balanced when the sum of the mass numbers and the sum of the atomic numbers of the particles and atoms on one side of the equation are equal to their counterparts on the other side.

Alpha Emitters

Alpha emitters are radioisotopes that decay by emitting alpha particles. For example, uranium-238 decays to thorium-234 by emitting an alpha particle.

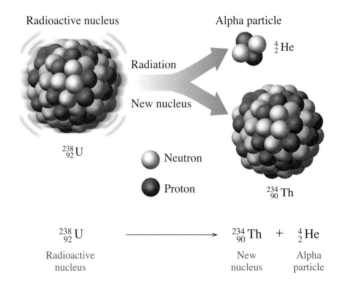

The alpha particle emitted contains 2 protons, which gives the new nucleus 2 fewer protons, or 90 protons. That means that the new nucleus has an atomic number of 90 and is therefore thorium (Th). Because the alpha particle has a mass number of 4, the mass number of the thorium isotope is 234, 4 less than that of the original uranium nucleus.

Completing a Nuclear Equation

In another example of radioactive decay, radium-226 emits an alpha particle to form a new isotope whose mass number, atomic number, and identity we must determine. We would first write an incomplete nuclear equation:

$$^{226}_{88}\text{Ra} \longrightarrow \boxed{?} \; + \; ^{4}_{2}\text{He}$$

In the equation, the mass number, 226, of the radium is equal to the combined mass numbers of the alpha particle and the new nucleus. We can calculate the mass number of the new nucleus by subtracting:

$226 - 4 = 222$ Mass number of new nucleus

The atomic number of radium (88) must equal the sum of the atomic numbers of the alpha particle and the new nucleus. Therefore, we obtain the atomic number of the new nucleus by the following calculation:

$88 - 2 = 86$ Atomic number of new nucleus

In the periodic table, the element that has an atomic number of 86 is radon (Rn). We complete the nuclear equation by writing the atomic symbol for radon-222:

$$^{226}_{88}\text{Ra} \longrightarrow \boxed{^{222}_{86}\text{Rn}} \; + \; ^{4}_{2}\text{He}$$

SAMPLE PROBLEM 3.3

Writing an Equation for Alpha Decay

Smoke detectors, now required in many homes and apartments, contain an alpha emitter such as americium-241. The alpha particles ionize air molecules, producing a constant stream of electrical current. However, when smoke particles enter the detector, they interfere with the formation of ions in the air, and the electric current is interrupted. This causes the alarm to sound and warns the occupants of the danger of fire. Complete the following nuclear equation for the decay of americium-241:

$$^{241}_{95}\text{Am} \longrightarrow \boxed{?} \; + \; ^{4}_{2}\text{He}$$

Solution

The mass number of the new nucleus is 237, obtained by subtracting the mass number of the alpha particle (4) from the mass number of americium (241).

Mass number of Am	−	Mass number of alpha particle	=	Mass number of new nucleus
241	−	4	=	237

The atomic number of the new nucleus is obtained by subtracting the atomic number of the alpha particle (2) from the atomic number of americium (95).

Atomic number of Am	−	Atomic number of alpha particle	=	Atomic number of new nucleus
95	−	2	=	93

Radon in Our Homes

The presence of radon has become a much publicized environmental and health issue because of radiation danger. Radioactive isotopes that produce radon, such as radium-226 and uranium-238, are naturally present in many types of rocks and soils. Radium-226 emits an alpha particle and is converted into radon gas, which diffuses out of the rocks and soil.

$$^{226}_{88}\text{Ra} \longrightarrow ^{222}_{86}\text{Rn} \; + \; ^{4}_{2}\text{He}$$

As uranium-238 decays, it also forms radium-226, which in turn produces radon. Uranium-238 has been found in particularly high levels in an area between Pennsylvania and New England.

Outdoors, radon gas poses little danger because it dissipates in the air. However, if the source of radon is under a house or building, the gas can enter the house through cracks in the foundation or other openings, where the radon can be inhaled by those living or working there. Inside the lungs, radon emits alpha particles to form polonium-218, which is known to cause cancer when present in the lungs.

$$^{222}_{86}\text{Rn} \longrightarrow ^{218}_{84}\text{Po} \; + \; ^{4}_{2}\text{He}$$

Some researchers have estimated that 10% of all lung cancer deaths in the United States are due to radon. The Environmental Protection Agency (EPA) recommends that the maximum level of radon not exceed 4 picocuries (pCi) per liter of air in a home. One (1) picocurie (pCi) is equal to 10^{-12} curies (Ci); curies are described in Section 3.4. In California, 1% of all the houses surveyed exceeded the EPA's recommended maximum radon level.

The element whose atomic number is 93 is neptunium, Np. The completed nuclear equation for the reaction is written as follows:

$$^{241}_{95}\text{Am} \longrightarrow ^{237}_{93}\text{Np} + ^{4}_{2}\text{He}$$

Study Check

Write a balanced nuclear equation for the alpha emitter polonium-214.

Beta Emitters

A beta emitter is a radioactive isotope that decays by emitting beta particles. To form a beta particle, the unstable nucleus converts a neutron into a proton. The newly formed proton adds to the number of protons already in the nucleus and increases the atomic number by 1. However, the mass number of the newly formed nucleus stays the same. For example, carbon-14 decays by emitting a beta particle and forming a nitrogen nucleus.

In the nuclear equation of a beta emitter, the mass number of the radioactive nucleus and the mass number of the new nucleus are the same, and the atomic number of the new nucleus increases by 1, indicating a change of one element into another. Below is the nuclear equation for the beta decay of carbon-14:

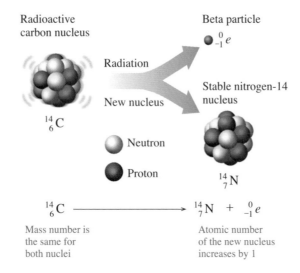

Radioactive carbon nucleus

Beta particle

$^{0}_{-1}e$

Radiation

Stable nitrogen-14 nucleus

New nucleus

$^{14}_{6}\text{C}$

○ Neutron

● Proton

$^{14}_{7}\text{N}$

$$^{14}_{6}\text{C} \longrightarrow ^{14}_{7}\text{N} + ^{0}_{-1}e$$

Mass number is the same for both nuclei

Atomic number of the new nucleus increases by 1

SAMPLE PROBLEM 3.4

Writing an Equation for Beta Decay

Cobalt-60, a radioactive isotope used in the treatment of cancer, decays by emitting a beta particle. Write the nuclear equation for its decay.

Solution

The radioactive cobalt isotope has a mass number of 60. Looking at the periodic table, we find that the atomic number of cobalt is 27. The products of the nuclear decay are a beta particle and a new nucleus:

$$^{60}_{27}\text{Co} \longrightarrow \boxed{?} + ^{0}_{-1}e$$

Because the mass number does not change in beta emission, we can assign a mass number of 60 to the new nucleus. The atomic number of the new nucleus will be 1 more than that of cobalt, or 28.

$$\text{Atomic number of Co} = \text{Atomic number of new nucleus} - 1$$
$$27 = 28 - 1$$

Because the element that has an atomic number of 28 is nickel (Ni), we can write the equation for the beta decay of the radioactive cobalt as

$$^{60}_{27}\text{Co} \longrightarrow {}^{60}_{28}\text{Ni} + {}^{0}_{-1}e$$

Study Check

Iodine-131, a beta emitter, is used to check thyroid function and to treat hyperthyroidism. Write its nuclear equation.

Gamma Emitters

There are very few pure gamma emitters, although gamma radiation accompanies most alpha and beta radiation. In radiology, one of the most commonly used gamma emitters is technetium (Tc). The excited state called metastable technetium is written as technetium-99m, Tc-99m, or ^{99m}Tc. By emitting energy in the form of gamma rays, the excited nucleus becomes more stable.

$$^{99m}_{43}\text{Tc} \longrightarrow {}^{99}_{43}\text{Tc} + \gamma$$

Figure 3.3 summarizes the changes in the nucleus for alpha, beta, and gamma radiation.

Beta Emitters in Medicine

The radioactive isotopes of several biologically important elements are beta emitters. When a radiologist wants to treat a malignancy within the body, a beta emitter may be used. The short range of penetration into the tissue by beta particles is advantageous for certain conditions. For example, some malignant tumors increase the fluid within the body tissues. A compound containing phosphorus-32, a beta emitter, is injected into the body cavity where the tumor is located. The beta particles travel only a few millimeters through the tissue, so only the malignancy and any tissue within that range are affected. The growth of the tumor is slowed or stopped, and the production of fluid decreases. Phosphorus-32 is also used to treat leukemia, polycythemia vera (excessive production of red blood cells), and lymphomas.

$$^{32}_{15}\text{P} \longrightarrow {}^{32}_{16}\text{S} + {}^{0}_{-1}e$$

Another beta emitter, iron-59, is used in blood tests to determine the level of iron in the blood, and the rate of production of red blood cells by the bone marrow.

$$^{59}_{26}\text{Fe} \longrightarrow {}^{59}_{27}\text{Co} + {}^{0}_{-1}e$$

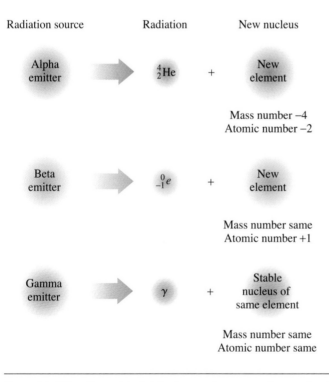

Radiation source	Radiation	New nucleus
Alpha emitter	$^{4}_{2}\text{He}$ +	New element
		Mass number −4 Atomic number −2
Beta emitter	$^{0}_{-1}e$ +	New element
		Mass number same Atomic number +1
Gamma emitter	γ +	Stable nucleus of same element
		Mass number same Atomic number same

Figure 3.3 When the nuclei of alpha, beta, and gamma emitters emit radiation, new, more stable nuclei are produced.

Q What changes occur in the number of protons and neutrons when an alpha emitter gives off radiation?

QUESTIONS AND PROBLEMS

Nuclear Equations

3.13 Write a balanced nuclear equation for the alpha decay of each of the following radioactive isotopes:

a. $^{208}_{84}\text{Po}$ **b.** $^{232}_{90}\text{Th}$ **c.** $^{251}_{102}\text{No}$ **d.** $^{220}_{86}\text{Rn}$

3.14 Write a balanced nuclear equation for the alpha decay of each of the following radioactive isotopes:

a. $^{243}_{96}\text{Cm}$ **b.** $^{252}_{99}\text{Es}$ **c.** $^{251}_{98}\text{Cf}$ **d.** $^{261}_{107}\text{Bh}$

3.15 Write a balanced nuclear equation for the beta decay of each of the following radioactive isotopes:

a. $^{25}_{11}\text{Na}$ **b.** $^{20}_{8}\text{O}$ **c.** $^{92}_{38}\text{Sr}$ **d.** $^{42}_{19}\text{K}$

3.16 Write a balanced nuclear equation for the beta decay of each of the following radioactive isotopes:

a. potassium-42 **b.** iron-59 **c.** iron-60 **d.** barium-141

3.17 Complete each of the following nuclear equations:

a. $^{28}_{13}\text{Al} \longrightarrow ? + ^{0}_{-1}e$ **b.** $? \longrightarrow ^{86}_{36}\text{Kr} + ^{1}_{0}\text{n}$

c. $^{66}_{29}\text{Cu} \longrightarrow ^{66}_{30}\text{Zn} + ?$ **d.** $? \longrightarrow ^{4}_{2}\text{He} + ^{234}_{90}\text{Th}$

3.18 Complete each of the following nuclear equations:

a. $^{11}_{6}\text{C} \longrightarrow ^{7}_{4}\text{Be} + ?$ **b.** $^{35}_{16}\text{S} \longrightarrow ? + ^{0}_{-1}e$

c. $? \longrightarrow ^{90}_{39}\text{Y} + ^{0}_{-1}e$ **d.** $^{210}_{83}\text{Bi} \longrightarrow ? + ^{4}_{2}\text{He}$

3.3 Producing Radioactive Isotopes

Today, more than 1500 radioisotopes are produced by converting stable, nonradioactive isotopes into radioactive ones. To do this, a stable atom is bombarded by fast-moving alpha particles, protons, or neutrons. When one of these particles is absorbed by a stable nucleus, the nucleus becomes unstable and the atom is now a radioactive isotope, or **radioisotope.** The process of changing one element into another is called **transmutation.**

When a nonradioactive isotope such as boron-10 is bombarded by an alpha particle, it is converted to nitrogen-13, a radioisotope. In this bombardment reaction, a neutron is emitted.

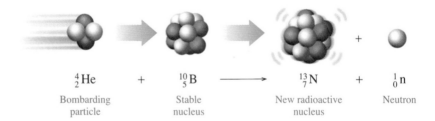

$^{4}_{2}\text{He}$ + $^{10}_{5}\text{B}$ $\longrightarrow$ $^{13}_{7}\text{N}$ + $^{1}_{0}\text{n}$

Bombarding particle   Stable nucleus   New radioactive nucleus   Neutron

All of the known elements that have atomic numbers greater than 92 have been produced by bombardment; none of these elements occurs naturally. Most have been produced in only small amounts and exist for such a short time that it is difficult to study their properties. An example is element 105, dubnium, which is produced when californium-249 is bombarded with nitrogen-15.

$$^{249}_{98}\text{Cf} + ^{15}_{7}\text{N} \longrightarrow ^{260}_{105}\text{Db} + 4\,^{1}_{0}\text{n}$$

Technetium-99m is a radioisotope used in nuclear medicine for several diagnostic procedures, including the detection of brain tumors and examinations of the liver and spleen. The source of technetium-99m is molybdenum-99, which is produced in a nuclear reactor by neutron bombardment of molybdenum-98.

$$^{98}_{42}\text{Mo} + ^{1}_{0}\text{n} \longrightarrow ^{99}_{42}\text{Mo}$$

Many radiology laboratories have a small generator containing the radioactive molybdenum-99, which decays to give the technetium-99m radioisotope.

$$^{99}_{42}\text{Mo} \longrightarrow ^{99\text{m}}_{43}\text{Tc} + ^{0}_{-1}e$$

The technetium-99m radioisotope has a half-life of 6 hours and decays by emitting gamma rays. Gamma emission is most desirable for diagnostic work because the gamma rays pass through the body to the detection equipment.

$$^{99\text{m}}_{43}\text{Tc} \longrightarrow ^{99}_{43}\text{Tc} + \gamma$$

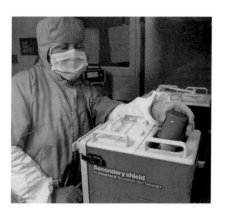

SAMPLE PROBLEM 3.5

Writing Equations for Transmutations

Gallium-67 is used in the treatment of lymphomas. It is produced by the bombardment of zinc-66 by a proton. Write the equation for this nuclear bombardment.

Solution

The reactants in this nuclear equation are zinc-66 and a proton. The radioactive isotope formed is gallium-67.

$$\underset{\text{Proton}}{^{66}_{30}\text{Zn} + ^{1}_{1}\text{H}} \longrightarrow \underset{\text{New radioisotope}}{^{67}_{31}\text{Ga}}$$

Study Check

Write the equation for the bombardment of aluminum-27 by an alpha particle to produce the radioactive isotope phosphorus-30 and one neutron.

SAMPLE PROBLEM 3.6

Completing a Nuclear Equation for a Bombardment Reaction

Write the missing symbol in the following bombardment reaction:

$$\underset{\text{Proton}}{^{58}_{28}\text{Ni} + ^{1}_{1}\text{H}} \longrightarrow \boxed{?} + ^{4}_{2}\text{He}$$

Solution

The sum of the mass numbers for nickel and hydrogen is 59. Therefore, the mass number of the new isotope must be 59 minus 4, or 55. The sum of the atomic numbers is 29. The atomic number of the new isotope is 29 minus 2, or 27. The element that has atomic number 27 is cobalt (Co).

$$\underset{\text{Proton}}{^{58}_{28}\text{Ni} + ^{1}_{1}\text{H}} \longrightarrow \underset{\text{New isotope}}{\boxed{^{55}_{27}\text{Co}}} + ^{4}_{2}\text{He}$$

Study Check

Complete the following bombardment equation:

$$\boxed{?} + {}^{4}_{2}\text{He} \longrightarrow {}^{17}_{8}\text{O} + {}^{1}_{1}\text{H}$$

QUESTIONS AND PROBLEMS

Producing Radioactive Isotopes

3.19 Complete each of the following bombardment reactions:

a. ${}^{9}_{4}\text{Be} + {}^{1}_{0}\text{n} \longrightarrow ?$

b. ${}^{32}_{16}\text{S} + ? \longrightarrow {}^{32}_{15}\text{P}$

c. $? + {}^{1}_{0}\text{n} \longrightarrow {}^{24}_{11}\text{Na} + {}^{4}_{2}\text{He}$

d. When Al-27 is bombarded by an alpha particle, it forms Si-30. What other particle is produced?

3.20 Complete each of the following bombardment reactions:

a. ${}^{40}_{18}\text{Ar} + ? \longrightarrow {}^{43}_{19}\text{K} + {}^{1}_{1}\text{H}$

b. ${}^{238}_{92}\text{U} + {}^{1}_{0}\text{n} \longrightarrow ?$

c. $? + {}^{1}_{0}\text{n} \longrightarrow {}^{14}_{6}\text{C} + {}^{1}_{1}\text{H}$

d. When an alpha particle bombards N-14, it forms a radioisotope and a proton. What is the radioisotope that is produced?

LEARNING GOAL

Describe the detection and measurement of radiation.

WEB TUTORIAL
Radiation and its Biological Effects

3.4 Radiation Detection and Measurement

One of the most common instruments for detecting beta and gamma radiation is the Geiger counter. It consists of a metal tube filled with a gas such as argon. When radiation enters a window on the end of the tube, it produces ions in the gas; the ions produce an electrical current. Each burst of current is amplified to give a click and a readout on a meter.

$$\text{Ar} + \text{radiation} \longrightarrow \text{Ar}^{+} + e^{-}$$

Radiation is measured in several different ways. We can measure the activity of a radioactive sample or determine the impact of radiation on biological tissue.

Activity

When a radiology laboratory obtains a radioisotope, the activity of the sample is measured in terms of the number of nuclear disintegrations per second. The **curie (Ci),** the original unit of activity, was defined as the number of disintegrations that occur in one second for 1 g of radium, which is equal to 3.7×10^{10} disintegrations per second. The curie was named for Marie Curie, a Polish scientist, who along with her husband Pierre discovered the radioactive elements radium and polonium. A newer unit of radiation activity is the **becquerel (Bq),** which is one disintegration per second.

Radiation Absorbed

After we measure the activity of a radioisotope, we often want to know how much radiation the tissues in the body absorb. The **rad (radiation absorbed dose)** is a unit that measures the amount of radiation absorbed by a gram of material such as body tissue. A newer unit for absorbed dose is the **gray (Gy),** which is equal to 100 rads.

Biological Damage (Equivalent Dose)

The **rem (radiation equivalent in humans)** measures the biological effects of different kinds of radiation. Alpha particles don't penetrate the skin. But if they should enter the body by some other route, they cause a lot of damage even though the particles travel a short distance in tissue. High-energy radiation such as beta particles and high-energy protons and neutrons that penetrate the skin and travel into tissue cause more damage. Gamma rays are damaging because they travel a long way through tissue and create a great deal of ionization.

To determine the **equivalent dose** or rem dose, the absorbed dose (rads) is multiplied by a factor that adjusts for biological damage caused by a particular form of radiation. For beta and gamma radiation, and X rays, the factor is 1, so the biological damage in rems is the same as the absorbed radiation (rads). For high-energy protons and neutrons, the factor is about 10, and for alpha particles it is 20.

Biological damage (rem) = absorbed dose (rad) × factor

Often the measurement for an equivalent dose will be in units of millirems **(mrem).** One rem is equal to 1000 mrem. A newer unit is the **sievert (Sv).** One sievert is equal to 100 rems. Table 3.4 summarizes the units used to measure radiation.

Table 3.4   Some Units of Radiation Measurement

Measurement	Common Unit	SI Unit	Relationship
Activity	curie (Ci) = 3.7×10^{10} disintegrations/s	becquerel (Bq) = 1 disintegration/s	1 Ci = 3.7×10^{10} Bq
Absorbed Dose	rad	gray (Gy)	1 Gy = 100 rad
Biological Damage	rem = rad × factor	1 sievert (Sv)	1 Sv = 100 rem

Measuring Activity

A treatment for leukemia uses phosphorus-32, a beta emitter, which has an activity of 2 millicuries (mCi). How many beta particles are emitted in 1s?

Solution

We can calculate the number of beta particles from the radioisotope's activity.

$$2\,\cancel{mCi} \times \frac{1\,\cancel{Ci}}{1000\,\cancel{mCi}} \times \frac{3.7 \times 10^{10}\,\beta \text{ particles}}{\cancel{s}\,\cancel{Ci}} \times 1\cancel{s}$$

$$= 7 \times 10^7\,\beta \text{ particles}$$

Study Check

An iodine-131 source has an activity of 0.25 Ci. How many radioactive atoms will disintegrate in 1.0 min?

Background Radiation

We are all exposed to low levels of radiation every day. Naturally occurring radioactive isotopes are part of the atoms of wood, brick, and concrete in our homes and the buildings where we work and go to school. This radioactivity, called background radiation, is present in the soil, in the food we eat, in the water we drink, and in the air we breathe. For example, one of the naturally occurring isotopes of potassium, potassium-40, is radioactive. It is found in the body because it is always present in any potassium-containing food. Other naturally occurring radioisotopes in air and food are carbon-14, radon-222, strontium-90, and iodine-131. Table 3.5 lists some common sources of radiation.

We are constantly exposed to radiation (cosmic rays) produced in space by the sun. At higher elevations, the amount of radiation from outer space is greater because there are fewer air molecules to absorb the radiation. People living at high altitudes or flying in an airplane receive more radiation from cosmic rays than those who live at sea level. For example, a person living in Denver receives about twice the cosmic radiation as a person living in Los Angeles.

A person living close to a nuclear power plant normally does not receive much additional radiation, perhaps 0.1 millirem (mrem) in 1 year. (One rem equals 1000 mrem.) However, in the accident at the Chernobyl nuclear power plant in 1986, it is estimated that people in a nearby town received as much as 1 rem/hr.

In addition to naturally occurring radiation from construction materials in our homes, we receive radiation from television. In the medical clinic, dental and chest X rays also add to our radiation exposure. The average person in the United States receives about 0.17 rem or 170 mrem of radiation annually.

Table 3.5 Average Annual Radiation Received by a Person in the United States

Source	Dose (mrem)
Natural	
The ground	15
Air, water, food	30
Cosmic rays	40
Wood, concrete, brick	50
Medical	
Chest X ray	50
Dental X ray	20
Upper gastrointestinal tract X ray	200
Other	
Television	2
Air travel	1
Radon	200[a]
Cigarette smoking	35

[a]Varies widely.

Radiation Sickness

The larger the dose of radiation received at one time, the greater the effect on the body. Exposure to radiation under 25 rem usually cannot be detected. Whole-body exposure of 100 rem produces a temporary decrease in the number of white blood cells. If the exposure to radiation is 100 rem or higher, the person suffers the symp-

Radiation and Food

Foodborne illnesses caused by pathogenic bacteria such as *Salmonella*, *Listeria*, and *Escherichia coli* have become a major health concern in the United States. The Centers for Disease Control and Prevention estimates that each year *E. coli* in contaminated foods infects 20,000 people in the United States, and that 500 people die. *E. coli* has been responsible for outbreaks of illness from contaminated ground beef, fruit juices, lettuce, and alfalfa sprouts.

The Food and Drug Administration (FDA) has approved the use of 0.3 kilogray (0.3 kGy) to 1 kGy of ionizing radiation produced by cobalt-60 or cesium-137 for the treatment of foods. The irradiation technology is much like that used to sterilize medical supplies. Cobalt pellets are placed in stainless steel tubes, which are arranged in racks. When food moves through the series of racks, the gamma rays pass through the food and kill the bacteria.

It is important for consumers to understand that when food is irradiated, it never comes in contact with the radioactive source. The gamma rays pass through the food to kill bacteria, but that does not make the food radioactive. The radiation kills bacteria because it stops their ability to divide and grow. We cook or heat food thoroughly for the same purpose. Radiation, as well as heat, has little effect on the food itself because its cells are no longer dividing or growing. Thus irradiated food is not harmed although a small amount of vitamin B_1 and C may be lost.

Currently, tomatoes, blueberries, strawberries, and mushrooms are being irradiated to allow them to be harvested when completely ripe and extend their shelf life (see Figure 3.4). The FDA has also approved the irradiation of pork, poultry, and beef in order to decrease potential infections and to extend shelf life. Currently, irradiated vegetable and meat products are available in

(a)

(b)

Figure 3.4 **(a)** The FDA requires this symbol to appear on irradiated retail foods. **(b)** After 2 weeks, the irradiated strawberries on the right show no spoilage. Mold is starting to grow on the nonirradiated ones on the left.

Q Why are irradiated foods used on space ships and in nursing homes?

retail markets in South Africa. Apollo 17 astronauts ate irradiated foods on the moon, and some U.S. hospitals and nursing homes now use irradiated poultry to reduce the possibility of infections among patients. The extended shelf life of irradiated food also makes it useful for campers and military personnel. Soon consumers concerned about food safety will have a choice of irradiated meats, fruits, and vegetables at the market.

toms of radiation sickness: nausea, vomiting, fatigue, and a reduction in white-cell count. A whole-body dosage greater than 300 rem can lower the white-cell count to zero. The victim suffers diarrhea, hair loss, and infection.

CASE STUDY
Food Irradiation

Lethal Dose

Exposure to radiation of about 500 rem is expected to cause death in 50% of the people receiving that dose. This amount of radiation is called the lethal dose for one-half the population, or the LD_{50}. The LD_{50} varies for different life-forms, as Table 3.6 shows. Radiation dosages of about 600 rem would be fatal to all humans within a few weeks.

Table 3.6 Lethal Doses of Radiation for Some Life-Forms	
Life-Form	LD_{50} (rem)
Insect	100,000
Bacterium	50,000
Rat	800
Human	500
Dog	300

Maximum Permissible Dose

Any person working with radiation, such as a radiologist, a radiation technician, or a nurse whose patient has received a radioisotope, must wear some type of detection apparatus to measure exposure to radiation. A standard for occupational exposure is the maximum permissible dose (MPD). If this dose is not exceeded, the probability of injury is minimized. The detection apparatus is usually a badge, ring, or pin containing a small piece of photographic film. The window that covers the film absorbs light and beta rays, but gamma rays penetrate the covering and expose the film. The darker the film, the more the person has been exposed to gamma radiation. These badges are checked periodically to prevent exposure to more than the maximum permissible dose. The maximum permissible dose for occupational exposure is 5 rem (5000 mrem) per year.

QUESTIONS AND PROBLEMS

Radiation Detection and Measurement

3.21 **a.** How does a Geiger counter detect radiation?
   b. What are the SI unit and the older unit that describe the activity of a radioactive sample?
   c. What are the SI unit and the older unit that describe the radiation dose absorbed by tissue?
   d. What is meant by the term kilogray?

3.22 **a.** What is background radiation?
   b. What are the SI unit and the older unit that describe the biological effect of radiation?
   c. What is meant by the terms mCi and mrem?
   d. Why is a factor used to determine the dose equivalent?

3.23 **a.** A sample of iodine-131 has an activity of 3.0 Ci. How many disintegrations occur in the iodine-131 sample in 20. s?
   b. The recommended dosage of iodine-131 is 4.20 μCi/kg of body weight. How many microcuries of iodine-131 are needed for a 70.0-kg patient with hyperthyroidism?

3.24 **a.** The dosage of technetium-99m for a lung scan is 20 μCi/kg of body weight. How many millicuries should be given to a 50.0-kg patient? (1 mCi = 1000 μCi)
   b. A patient receives 50 mrads in a chest X ray. What is that amount in grays? What would be the dose equivalent in mrems?
   c. Suppose a person absorbed 50 mrads of alpha radiation. What would be the dose equivalent in mrems? How does it compare with the mrems in part b?

3.25 Why would an airline pilot be exposed to more background radiation than the person who works at the ticket counter?

3.26 In radiation therapy, a patient receives high doses of radiation. What symptoms of radiation sickness might the patient exhibit?

LEARNING GOAL

Given a half-life, calculate the amount of radioisotope remaining after one or more half-lives.

3.5 Half-Life of a Radioisotope

The **half-life** of a radioisotope is the time it takes for one-half of a sample to decay. For example, iodine-131, a radioisotope used in diagnosis and treatment of thyroid disorders, has a half-life of 8 days. If we begin with a sample containing 20 grams of iodine-131, there would be 10 g remaining after 8 days. In another 8 days, we would have 5 g of iodine-131 left. Thus, in each half-life of 8 days, one-half of the radioactive sample has disintegrated.

$$20 \text{ g } ^{131}\text{I} \xrightarrow[10 \text{ g } ^{131}\text{I decay}]{8 \text{ days}} 10 \text{ g } ^{131}\text{I} \xrightarrow[5 \text{ g } ^{131}\text{I decay}]{8 \text{ days}} 5 \text{ g } ^{131}\text{I}$$

This information can be summarized as follows:

Time elapsed	0	8 days	16 days	24 days
Number of half-lives	0	1	2	3
Quantity of iodine-131	20. g	10. g	5.0 g	2.5 g

A **decay curve** is a diagram of the decay of a radioactive isotope. Figure 3.5 shows such a curve for the iodine-131 we have discussed.

Naturally occurring isotopes of the elements usually have long half-lives, as shown in Table 3.7. They disintegrate slowly and produce radiation over a long period of time, even hundreds or millions of years. In contrast, many of the radioisotopes used in nuclear medicine have much shorter half-lives. They disintegrate rapidly and produce almost all their radiation in a short period of time. For example, technetium-99m emits half of its radiation in the first 6 hr. This means that a small amount of the radioisotope given to a patient is essentially gone within 2 days. The decay products of technetium-99m are totally eliminated by the body.

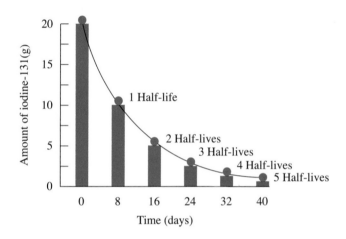

Figure 3.5 The decay curve for iodine-131 shows that one-half of the radioactive sample decays and one-half remains radioactive after each half-life of 8 days.

Q How many grams of the 20-g sample remain radioactive after 2 half-lives?

Table 3.7 Half-Lives of Some Radioisotopes

Element	Radioisotope	Half-Life	Type of Radiation
Naturally Occurring Radioisotopes			
Carbon	^{14}C	5730 yr	β
Potassium	^{40}K	1.3×10^9 yr	β, γ
Radium	^{226}Ra	1600 yr	α, γ
Uranium	^{238}U	4.5×10^9 yr	α, γ
Some Medical Radioisotopes			
Carbon	^{11}C	20 min	$\beta^{+\ a}$
Chromium	^{51}Cr	28 days	γ
Iodine	^{131}I	8 days	β, γ
Iodine	^{125}I	60 days	γ
Iron	^{59}Fe	46 days	β, γ
Phosphorus	^{32}P	14 days	β
Oxygen	^{15}O	2 min	$\beta^{+\ a}$
Potassium	^{42}K	12 hr	β, γ
Sodium	^{24}Na	15 hr	β, γ
Strontium	^{85}Sr	64 days	γ
Technetium	^{99m}Tc	6.0 hr	γ

^{a}Note: β^+ is a positron, which has the same mass as an electron but has a positive charge.

WEB TUTORIAL
Nuclear Chemistry

EXPLORE YOUR WORLD

Modeling Half-Lives

Model 1

Obtain a deck of cards. Place 32 cards faceup on the table. The cards facing up represent radioactive atoms. Using a clock with a second hand, begin timing. After 1 minute, turn half the cards facedown. They represent the number of atoms that emit radiation during the first half-life. How many radioactive atoms (faceup cards) are left? When the next minute passes, turn half of the remaining cards facedown. How many radioactive atoms (faceup cards) are left? For each of the next 3 minutes, continue to turn over half the radioactive atoms (faceup cards). Write your own definition of radioactive half-life.

Model 2

Obtain a piece of paper and a licorice stick or celery stalk. Draw a vertical and a horizontal axis on the paper. Label the vertical axis as radioactive atoms and the horizontal axis as minutes. Place the licorice stick or celery against the vertical axis and mark its height for zero minutes. In the next minute, cut the licorice stick or celery in two. (You can eat the half if you are hungry.) Place the shortened licorice stick or celery at 1 minute on the horizontal axis and mark its height. Every minute cut the licorice stick or celery in half again and mark the shorter height at the corresponding time. Keep reducing the length by half until you cannot divide the licorice or celery in half any more. Connect the points you made for each minute. What does the curve look like? How does this curve represent the concept of a half-life for a radioisotope?

SAMPLE PROBLEM 3.8

Calculating Quantity of Radioisotopes

Nitrogen-13, which has a half-life of 10 min, is used to image organs in the body. For the diagnostic procedure, the patient receives an injection of a compound containing the radioisotope. Originally, the nitrogen-13 has an activity of 40 microcuries (μCi). If the procedure requires 30 min, what is the remaining activity of the radioisotope?

Solution

$$\text{Number of half-lives} = 30 \cancel{\text{min}} \times \frac{1 \text{ half-life}}{10 \cancel{\text{min}}}$$

$$= 3$$

The activity of the radioisotope after 3 half-lives is

$$40 \ \mu\text{Ci} \xrightarrow{10 \text{ min}} 20 \ \mu\text{Ci} \xrightarrow{10 \text{ min}} 10 \ \mu\text{Ci} \xrightarrow{10 \text{ min}} 5 \ \mu\text{Ci}$$

Another way to calculate the activity of radioactive nitrogen-13 left in the sample is to construct a chart to show the number of half-lives, elapsed time, and the amount of radioactive isotope that is left in the sample.

Time elapsed	0	10 min	20 min	30 min
Number of half-lives elapsed	0	1	2	3
Activity of N-13 remaining	40 μCi	20 μCi	10 μCi	5 μCi

Study Check

Iron-59, used in the determination of bone-marrow function, has a half-life of 46 days. If the laboratory receives a sample of 8.0 g of iron-59, how many grams are still active after 184 days?

SAMPLE PROBLEM 3.9

Dating Using Half-Lives

In Los Angeles, the remains of ancient animals have been unearthed at La Brea tar pits. Suppose a bone sample from the tar pits is subjected to the carbon-14 dating

ENVIRONMENTAL NOTE

Dating Ancient Objects

A technique known as radiological dating is used by geologists, archaeologists, and historians as a way to determine the age of ancient objects. The age of an object derived from plants or animals (such as wood, fiber, natural pigments, bone, and cotton or woolen clothing) is determined by measuring the amount of carbon-14, a naturally occurring radioactive form of carbon. In 1960, Willard Libby received the Nobel Prize for the work he did developing carbon-14 dating techniques during the 1940s. Carbon-14 is produced in the upper atmosphere by the bombardment of $^{14}_{7}N$ by high-energy neutrons from cosmic rays.

$$^{1}_{0}n \quad + \quad ^{14}_{7}N \quad \longrightarrow \quad ^{14}_{6}C \quad + \quad ^{1}_{1}H$$

Neutron from cosmic rays   Nitrogen in atmosphere   Radioactive carbon-14   Proton

The carbon-14 reacts with oxygen to form radioactive carbon dioxide, $^{14}CO_2$. Because carbon dioxide is continuously absorbed by living plants during the process of photosynthesis, some carbon-14 will be taken into the plant. After the plant dies, no more carbon-14 is taken up, and the amount of carbon-14 contained in the plant steadily decreases as it undergoes radioactive β decay.

$$^{14}_{6}C \quad \longrightarrow \quad ^{14}_{7}N \quad + \quad ^{0}_{-1}e$$

Scientists use the half-life of carbon-14 (5730 years) to calculate the amount of time that has passed since the plant died, a process called **carbon dating.** The smaller the amount of carbon-14 remaining in the sample, the greater the number of half-lives that have passed. Thus, the approximate age of the sample can be determined. For example, a wooden beam found in an ancient Indian dwelling might have one-half of the carbon-14 found in living plants today. Thus, the dwelling was probably constructed about 5730 years ago, one half-life of carbon-14. Carbon-14 dat-

ing was used to determine that the Dead Sea Scrolls are about 2000 years old.

A radiological dating method used for determining the age of rocks is based on the radioisotope uranium-238, which decays through a series of reactions to lead-206. The uranium-238 isotope has a very long half-life, about 4×10^9 (4 billion) years. Measurements of the amounts of uranium-238 and lead-206 enable geologists to determine the age of rock samples. The older rocks will have a higher percentage of lead-206 because more of the uranium-238 has decayed. The age of rocks brought back from the moon by the *Apollo* missions was determined using uranium-238. They were found to be about 4×10^9 years old, approximately the same age calculated for Earth.

method. If the sample shows that two half-lives have passed, when did the animal live?

Solution

We can calculate the age of the bone sample by using the half-life of carbon-14 (5730 years).

$$2 \text{ half-lives} \times \frac{5730 \text{ years}}{1 \text{ half-life}} = 11{,}500 \text{ years}$$

We would estimate that the animal lived 11,500 years ago, or about 9000 B.C.

Study Check

Suppose that a piece of wood found in a tomb had $\frac{1}{8}$ (3 half-lives) of its original C-14 activity. About how many years ago was the wood part of a living tree?

QUESTIONS AND PROBLEMS

Half-Life of a Radioisotope

3.27 What is meant by the term half-life?

3.28 Why are radioisotopes with short half-lives used for diagnosis in nuclear medicine?

3.29 Technetium-99m is an ideal radioisotope for scanning organs because it has a half-life of 6.0 hr and is a pure gamma emitter. Suppose that 80.0 mg were prepared in the technetium generator this morning. How many milligrams would remain after the following intervals?
   a. one half-life   **b.** two half-lives
   c. 18 hr   **d.** 24 hr

3.30 A sample of sodium-24 with an activity of 12 mCi is used to study the rate of blood flow in the circulatory system. If sodium-24 has a half-life of 15 hr, what is the activity of the sodium after $2\frac{1}{2}$ days?

3.31 Strontium-85, used for bone scans, has a half-life of 64 days. How long will it take for the radiation level of strontium-85 to drop to one-fourth of its original level? To one-eighth?

3.32 Fluorine-18, which has a half-life of 110 min, is used in PET scans. If 100 mg of fluorine-18 is shipped at 8:00 A.M., how many milligrams of the radioisotope are still active if the sample arrives at the radiology laboratory at 1:30 P.M.?

3.6 Medical Applications Using Radioactivity

LEARNING GOAL

Describe the use of radioisotopes in medicine.

Suppose a radiologist wants to determine the condition of an organ in the body. How is this done? In some cases, the patient is given a radioisotope that is known to concentrate in that organ. In general, the cells in the body cannot differentiate between a nonradioactive atom and a radioactive one. All atoms of an element, including any radioactive isotopes, have the same electron arrangement and the same chemistry in the body. The difference is that radioactive atoms can be detected because they emit radiation and the nonradioactive atoms do not. Some radioisotopes used in nuclear medicine are listed in Table 3.8.

After a patient receives a radioisotope, the radiologist determines the level and location of radioactivity emitted by the radioisotope. An apparatus called a scanner is used to produce an image of the organ. The scanner moves slowly across the patient's body above the region where the organ containing the radioisotope is located. The gamma rays emitted from the radioisotope in the organ can be used to expose a photographic plate, producing a **scan** of the organ. On a scan, an area of decreased or increased radiation can indicate such conditions as a disease of the organ, a tumor, a blood clot, or edema.

A common method of determining thyroid function is the use of radioactive iodine uptake (RAIU). Taken orally, the radioisotope iodine-131 mixes with the iodine already present in the thyroid. Twenty-four hours later, the amount of iodine taken up by the thyroid is determined. A detection tube held up to the area of the thyroid gland detects the radiation coming from the iodine-131 that has located there. (See Figure 3.6.)

The iodine uptake is directly proportional to the activity of the thyroid. A patient with a hyperactive thyroid will have a higher than normal level of radioactive iodine, whereas a patient with a hypoactive thyroid will record low values.

If the patient has hyperthyroidism, treatment is begun to lower the activity of the thyroid. One treatment involves giving the patient a therapeutic dosage of radioactive iodine, which has a higher radiation count than the diagnostic dose. The

Table 3.8   Medical Applications of Radioisotopes

Isotope	Half-Life	Medical Application
Ce-141	32.5 days	Gastrointestinal tract diagnosis; measuring myocardial blood flow
Co-60	5.3 yr	External radiation therapy; sterilize surgical instruments and medicines
Ga-67	78 hr	Abdominal imaging; tumor detection
Ga-68	68 min	Detect pancreatic cancer
P-32	4.3 days	Treatment of leukemia, polycythemia vera (excess red blood cells), pancreatic cancer
I-123	13.1 hr	Imaging brain, thyroid, kidney, and heart; measuring cerebral blood flow; and detection of neurological disease
I-125	60 days	Treatment of brain cancer; osteoporosis detection
I-131	8 days	Imaging thyroid; treatment of Graves' disease, goiter, and hyperthyroidism; treatment of thyroid and prostate cancer
Sr-85	65 days	Detection of bone lesions; brain scans
Sr-89	50 days	Alleviation of bone cancer pain; treatment of prostate cancer
Tc-99m	6 hr	Imaging of skeleton and heart muscle, brain, liver, heart, lungs, bone, spleen, kidney, and thyroid; *most widely used radioisotope in nuclear medicine*
Tl-201	73 hr	Heart imaging; diagnosis and location of myocardial infarction

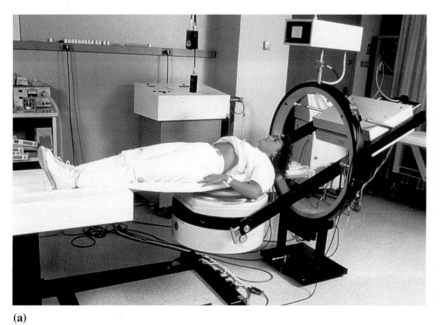

(a)

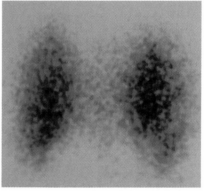

(b)

Figure 3.6 (a) A scanner is used to detect radiation from a radioisotope that has accumulated in an organ. **(b)** A scan of the thyroid shows the accumulation of radioactive iodine-131 in the thyroid.

Q What type of radiation would move through body tissues to create a scan?

radioactive iodine goes to the thyroid where its radiation destroys some of the thyroid cells. The thyroid produces less thyroid hormone, bringing the hyperthyroid condition under control.

Positron Emission Tomography (PET)

Radioisotopes that emit a positron are used in an imaging method called positron emission tomography (PET). A positron is a particle emitted from the nucleus that has the same mass as an electron but has a positive charge.

Radiation Doses in Diagnostic and Therapeutic Procedures

We can compare the levels of radiation exposure commonly used during diagnostic and therapeutic procedures in nuclear medicine. In diagnostic procedures, the radiologist minimizes radiation damage by using only enough radioisotope to evaluate the condition of an organ or tissue. The doses used in **radiation therapy** are much greater than those used for diagnostic procedures. For example, a therapeutic dose would be used to destroy the cells in a malignant tumor. Although there will be some damage to surrounding tissue, the healthy cells are more resistant to radiation and can repair themselves. (See Table 3.9.)

Table 3.9   Radiation Doses Used for Diagnostic and Therapeutic Procedures

Organ/Condition	Dose (rem)
Diagnostic	
Liver	0.3
Thyroid	50.0
Lung	2.0
Therapeutic	
Lymphoma	4500
Skin cancer	5000–6000
Lung cancer	6000
Brain tumor	6000–7000

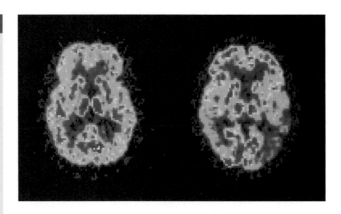

Figure 3.7 These PET scans of the brain show a normal brain on the left and a brain affected by Alzheimer's disease on the right.

Q When positrons collide with electrons, what type of radiation is produced that gives an image of an organ?

$$\beta^+ \quad \text{or} \quad {}^{0}_{+1}e$$
Positron

Carbon-11 is an example of a radioisotope that emits a positron when it decays.

$${}^{11}_{6}C \longrightarrow {}^{0}_{+1}e + {}^{11}_{5}B$$
Positron

A positron is produced when a proton changes to a neutron.

$${}^{1}_{1}H \longrightarrow {}^{0}_{+1}e + {}^{1}_{0}n$$

The new nucleus retains the same mass number but has a lower atomic number. The positron exists for only a moment before it collides with an electron producing gamma rays, which can be detected.

$${}^{0}_{+1}e + {}^{0}_{-1}e \longrightarrow 2\gamma$$
Positron   Electron   Gamma rays
              in an atom   produced

Medically, positron emitters such as carbon-11, oxygen-15, and nitrogen-13 are used to diagnose conditions involving blood flow, metabolism, and particularly brain function. Glucose labeled with carbon-11 is used to detect damage in the brain from epilepsy, stroke, and Parkinson's disease. The gamma rays from the positrons emitted by carbon-11 are detected by computerized equipment to create a three-dimensional image of the organ. (See Figure 3.7.)

SAMPLE PROBLEM 3.10

Medical Application of Radioactivity

In hyperthyroidism, the thyroid produces an overabundance of thyroid hormone. Why would a radioisotope such as iodine-131 with an activity of 10 μCi be used to image the thyroid?

Solution

The thyroid takes up the iodine-131 because the thyroid uses most of the iodine in the body. An activity level of 10 μCi is high enough to determine the level of thyroid function, but low enough to keep cell damage to a minimum.

Study Check

The evaluation of the thyroid image determined that the patient had an enlarged thyroid. A few weeks later, that same patient was given 100 mCi of iodine-131 intravenously. Why was the activity of the iodine-131 sample used for treatment so much greater than the activity of the iodine-131 used in the diagnostic procedure?

QUESTIONS AND PROBLEMS

Medical Applications Using Radioactivity

3.33 Bone and bony structures contain calcium and phosphorus.
  a. Why would the radioisotopes of calcium-47 and phosphorus-32 be used in the diagnosis and treatment of bone diseases?
  b. During nuclear tests, scientists were concerned that strontium-85, a radioactive product, would be harmful to the growth of bone in children. Explain.

3.34 a. Technetium-99m emits only gamma radiation. Why would this type of radiation be used in diagnostic imaging rather than an isotope that also emits beta or alpha radiation?
  b. A patient with polycythemia vera (excess production of red blood cells) receives radioactive phosphorus-32. Why would this treatment reduce the production of red blood cells in the bone marrow of the patient?

HEALTH NOTE

Other Imaging Methods

Computed Tomography (CT)

Another imaging method used to detect changes within the body is computed tomography (CT). A computer monitors the degree of absorption of 30,000 x-ray beams directed at the brain at successive layers. The differences in absorption based upon the densities of the tissues and fluids in the brain provide a series of images of the brain. This technique is successful in the identification of brain hemorrhages, tumors, and atrophy. (See Figure 3.8.)

Magnetic Resonance Imaging (MRI)

Magnetic resonance imaging (MRI) is a powerful imaging technique that does not involve x-ray radiation. It is the least invasive imaging method available. MRI is based on the absorption of energy when the protons in hydrogen atoms are excited by a strong magnetic field. Hydrogen atoms make up 63% of all the atoms in the body. In the nucleus, the spin of the single proton acts like a tiny magnet. With no external field, the spins of the protons have random orientations. However, when placed within a large magnet, the spins of the protons align with the magnetic field. A magnet aligned with the field has a lower energy than one that is aligned against the field. As the MRI scan proceeds, radio-frequency pulses of energy are applied. When a nucleus absorbs certain energy, its proton "flips" and becomes aligned against the field. Because hydro-

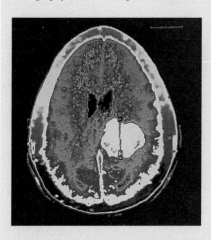

Figure 3.8 A CT scan shows a brain tumor (yellow area) in the center of the right side of the brain.

Q What is the type of radiation used to give a CT scan?

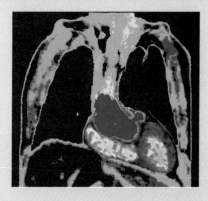

Figure 3.9 An MRI scan of the heart and lungs, with the left ventricle shown in red.

Q What is the source of energy in MRI?

gen atoms in the body are in different chemical environments, frequencies of different energies are absorbed. The energies absorbed are calculated and converted to color images of the body. MRI is particularly useful to image soft tissues because soft tissues contain large amounts of water. (See Figure 3.9.)

3.35 In a diagnostic test for leukemia, a patient receives 4.0 mL of a solution containing selenium-75. If the activity of the selenium-75 is 45 μCi/mL, what is the dose received by the patient?

3.36 A vial contains radioactive iodine-131 with an activity of 2.0 mCi per milliliter. If the thyroid test requires 3.0 mCi in an "atomic cocktail," how many milliliters are used to prepare the iodine-131 solution?

3.7 Nuclear Fission and Fusion

In the 1930s, scientists bombarding uranium-235 with neutrons discovered that the U-235 nucleus splits into two medium-weight nuclei and produces a great amount of energy. This was the discovery of a new kind of nuclear reaction called nuclear **fission**. The energy generated by nuclear fission, splitting the atom, was called atomic energy. When uranium-235 absorbs a neutron, it breaks apart to form two smaller nuclei, several neutrons, and a great amount of energy. A typical equation for nuclear fission is

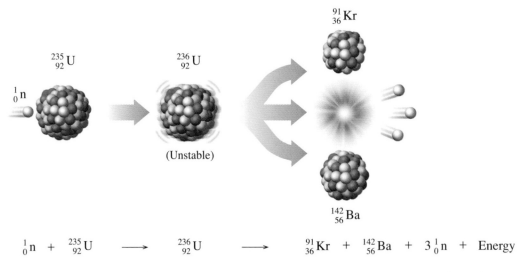

$$ {}^{1}_{0}n + {}^{235}_{92}U \longrightarrow {}^{236}_{92}U \longrightarrow {}^{91}_{36}Kr + {}^{142}_{56}Ba + 3\,{}^{1}_{0}n + \text{Energy} $$

If we could weigh these products with great accuracy, we would find that their total mass is slightly less than the mass of the starting materials. The missing mass has been converted into energy, consistent with the famous equation derived by Albert Einstein:

$$ E = mc^2 $$

E is the energy released, m is the mass lost, and c is the speed of light, 3×10^8 m/s. Even though the mass loss is very small, when it is multiplied by the speed of light squared the result is a large value for the energy released. The fission of 1 g of uranium-235 produces about as much energy as the burning of 3 tons of coal.

Chain Reaction

Fission begins when a neutron collides with the nucleus of a uranium atom. The resulting nucleus is unstable and splits into smaller nuclei. This fission process also releases several neutrons and large amounts of gamma radiation and energy. The

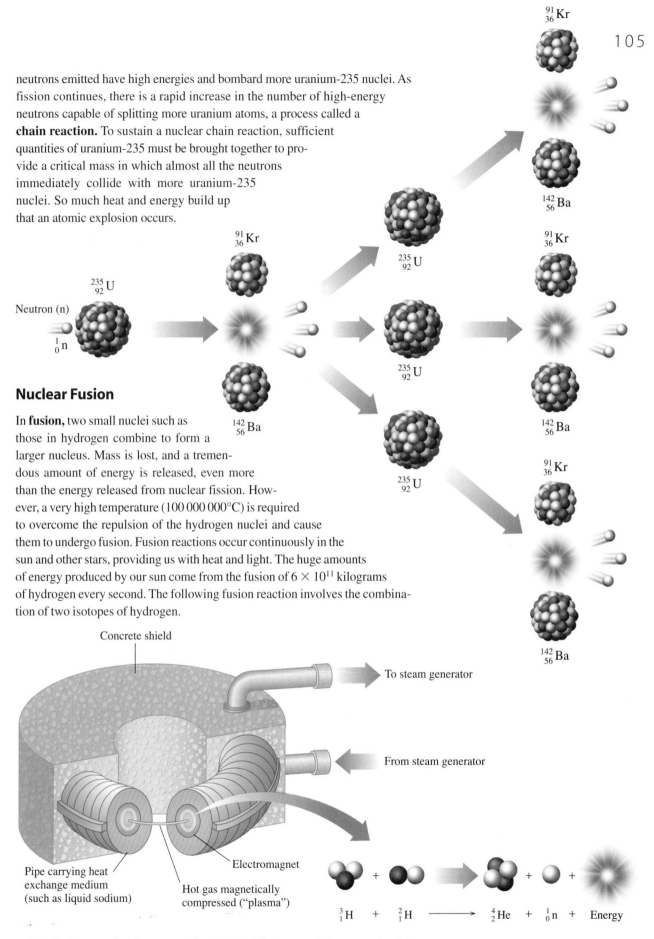

neutrons emitted have high energies and bombard more uranium-235 nuclei. As fission continues, there is a rapid increase in the number of high-energy neutrons capable of splitting more uranium atoms, a process called a **chain reaction.** To sustain a nuclear chain reaction, sufficient quantities of uranium-235 must be brought together to provide a critical mass in which almost all the neutrons immediately collide with more uranium-235 nuclei. So much heat and energy build up that an atomic explosion occurs.

Nuclear Fusion

In **fusion,** two small nuclei such as those in hydrogen combine to form a larger nucleus. Mass is lost, and a tremendous amount of energy is released, even more than the energy released from nuclear fission. However, a very high temperature (100 000 000°C) is required to overcome the repulsion of the hydrogen nuclei and cause them to undergo fusion. Fusion reactions occur continuously in the sun and other stars, providing us with heat and light. The huge amounts of energy produced by our sun come from the fusion of 6×10^{11} kilograms of hydrogen every second. The following fusion reaction involves the combination of two isotopes of hydrogen.

The fusion reaction has tremendous potential as a possible source for future energy needs. One of the advantages of fusion as an energy source is that hydrogen

Nuclear Power Plants

In a nuclear power plant, the quantity of uranium-235 is held below a critical mass, so it cannot sustain a chain reaction. The fission reactions are slowed by placing control rods, which absorb some of the fast-moving neutrons, among the uranium samples. In this way, less fission occurs, and there is a slower, controlled production of energy. The heat from the controlled fission is used to produce steam. The steam drives a generator, which produces electricity. Approximately 10% of the electrical energy produced in the United States is generated in nuclear power plants.

Although nuclear power plants help meet some of our energy needs, there are some problems. One of the most serious problems is the production of radioactive by-products that have very long half-lives. It is essential that these waste products be stored safely for a very long time in a place where they do not contaminate the environment. Early in 1990, the Environmental Protection Agency gave its approval for the storage of radioactive hazardous wastes in chambers 2150 ft underground. In 1998 the Waste Isolation Pilot Plant (WIPP) repository site in New Mexico was ready to receive plutonium waste from former U.S. bomb factories. Although authorities claim the caverns are safe, some people are concerned with the safe transport of the radioactive waste by trucks on the highways.

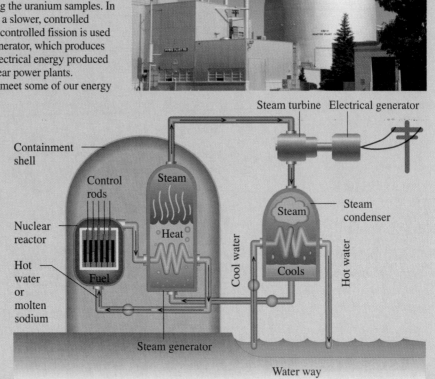

is plentiful in the oceans. Although scientists expect some radioactive waste from fusion reactors, the amount is expected to be much less than that from fission, and the waste products should have shorter half-lives. However, fusion is still in the experimental stage because the extremely high temperatures needed have been difficult to reach and even more difficult to maintain. Research groups around the world are attempting to develop the technology needed to make the fusion reaction a reality in our lifetime.

SAMPLE PROBLEM 3.11

Identifying Fission and Fusion

Classify the following as pertaining to nuclear fission, nuclear fusion, or both:

a. Small nuclei combine to form larger nuclei.
b. Large amounts of energy are released.
c. Very high temperatures are needed for reaction.

Solution

a. fusion   **b.** both fusion and fission   **c.** fusion

Study Check

Would the following reaction be an example of a fission or fusion reaction?

$$_1^2H + {}_1^1H \longrightarrow {}_2^3He$$

QUESTIONS AND PROBLEMS

Nuclear Fission and Fusion

3.37 What is nuclear fission?

3.38 How does a chain reaction occur in nuclear fission?

3.39 Complete the following fission reaction:

$$^{235}_{92}U + {}^{1}_{0}n \longrightarrow {}^{131}_{50}Sn + ? + 2\,{}^{1}_{0}n + energy$$

3.40 In another fission reaction, U-235 bombarded with a neutron produces Sr-94, another small nucleus, and 3 neutrons. Write the complete equation for the fission reaction.

3.41 Indicate whether each of the following are characteristic of the fission or fusion process or both:
    a. Neutrons bombard a nucleus.
    b. The nuclear process occurring in the sun.
    c. A large nucleus splits into smaller nuclei.
    d. Small nuclei combine to form larger nuclei.

3.42 Indicate whether each of the following are characteristic of the fission or fusion process or both:
    a. Very high temperatures are required to initiate the reaction.
    b. Less radioactive waste is produced.
    c. Hydrogen nuclei are the reactants.
    d. Large amounts of energy are released when the nuclear reaction occurs.

Chapter Review

3.1 Natural Radioactivity

Radioactive isotopes have unstable nuclei that break down (decay), spontaneously emitting alpha (α), beta (β), and gamma (γ) radiation. Because radiation can damage the cells in the body, proper protection must be used: shielding, limiting the time of exposure, and distance.

3.2 Nuclear Equations

A balanced equation is used to represent the changes that take place in the nuclei of the reactants and products. The new isotopes and the type of radiation emitted can be determined from the symbols that show the mass numbers and atomic numbers of the isotopes in the nuclear reaction.

3.3 Producing Radioactive Isotopes

A radioisotope is produced artificially when a nonradioactive isotope is bombarded by a small particle such as a proton, or an alpha or beta particle. Many radioactive isotopes used in nuclear medicine are produced in this way.

3.4 Radiation Detection and Measurement

In a Geiger counter, radiation ionizes gas in a metal tube, which produces an electrical current. The curie (Ci) measures the number of nuclear transformations of a radioactive sample. Activity is also measured in becquerel (Bq) units. The amount of radiation absorbed by a substance is mea-

sured in rads or the gray (Gy). The rem and the sievert (Sv) are units used to determine the biological damage from the different types of radiation.

3.5 Half-Life of a Radioisotope

Every radioisotope has its own rate of emitting radiation. The time it takes for one-half of a radioactive sample to decay is called its half-life. For many medical radioisotopes, such as Tc-99m and I-131, half-lives are short. For other isotopes, usually naturally occurring ones such as C-14, Ra-226, and U-238, half-lives are extremely long.

3.6 Medical Applications Using Radioactivity

In nuclear medicine, radioisotopes are given that go to specific sites in the body. By detecting the radiation they emit, an evaluation can be made about the location and extent of an injury, disease, tumor, or the level of function of a particular organ. Higher levels of radiation are used to treat or destroy tumors.

3.7 Nuclear Fission and Fusion

In fission, a large nucleus breaks apart into smaller pieces, releasing one or more types of radiation and a great amount of energy. In fusion, small nuclei combine to form a larger nucleus while great amounts of energy are released.

Key Terms

alpha particle A nuclear particle identical to a helium ($_2^4$He, or α) nucleus (2 protons and 2 neutrons).

becquerel (Bq) A unit of activity of a radioactive sample equal to one disintegration per second.

beta particle A particle identical to an electron ($_{-1}^{\;\;0}e$, or β) that forms in the nucleus when a neutron changes to a proton and an electron.

carbon dating A technique used to date ancient specimens that contain carbon. The age is determined by the amount of active C-14 that remains in the sample.

chain reaction A fission reaction that will continue once it has been initiated by a high-energy neutron bombarding a heavy nucleus such as U-235.

curie (Ci) A unit of radiation equal to 3.7×10^{10} disintegrations/s.

decay curve A diagram of the decay of a radioactive element.

equivalent dose The measure of biological damage from an absorbed dose that has been adjusted for the type of radiation.

fission A process in which large nuclei are split into smaller pieces, releasing large amounts of energy.

fusion A reaction in which large amounts of energy are released when small nuclei combine to form larger nuclei.

gamma ray High-energy radiation (γ) emitted to make a nucleus more stable.

gray (Gy) A unit of absorbed dose equal to 100 rads.

half-life The length of time it takes for one-half of a radioactive sample to decay.

rad (radiation absorbed dose) A measure of an amount of radiation absorbed by the body.

radiation Energy or particles released by radioactive atoms.

radiation therapy The use of high doses of radiation to destroy harmful tissues in the body.

radioactive The process by which an unstable nucleus breaks down with the release of high-energy radiation.

radioisotope A radioactive atom of an element.

rem (radiation equivalent in humans) A measure of the biological damage caused by the various kinds of radiation (rad $\times$ radiation biological factor).

scan The image of a site in the body created by the detection of radiation from radioactive isotopes that have accumulated in that site.

shielding Materials used to provide protection from radioactive sources.

sievert (Sv) A unit of biological damage (equivalent dose) equal to 100 rems.

transmutation The formation of a radioactive nucleus by bombarding a stable nucleus with fast-moving particles.

Additional Problems

3.43 Carbon-12 is a nonradioactive isotope of carbon, and carbon-14 is a radioactive isotope. What is similar and different about the two carbon isotopes?

3.44 Give the number of protons, neutrons, and electrons in atoms of the following isotopes:
a. boron-10   **b.** zinc-72
c. iron-59   **d.** gold-198

3.45 Describe alpha, beta, and gamma radiation in terms of the following:
a. type of radiation   **b.** symbols
c. depth of tissue penetration
d. type of shielding needed for protection

3.46 When you have dental X rays at the dentist's office, the technician places a heavy lead apron over you, and then leaves the room to take your X rays. What is the purpose of these activities?

3.47 Write the balanced nuclear equations for each of the following emitters:
a. thorium-225 (α)   **b.** bismuth-210 (α)
c. cesium-137 (β)   **d.** tin-126 (β)

3.48 Write the balanced nuclear equations for each of the following emitters:
a. potassium-40 (β)   **b.** sulfur-35 (β)
c. platinum-190 (α)   **d.** radium-210 (α)

3.49 Complete each of the following nuclear equations:

a. $_7^{14}\text{N} + _2^4\text{He} \longrightarrow \boxed{?} + _1^1\text{H}$

b. $_{13}^{27}\text{Al} + _2^4\text{He} \longrightarrow _{14}^{30}\text{Si} + \boxed{?}$

c. $_{92}^{235}\text{U} + _0^1\text{n} \longrightarrow _{38}^{90}\text{Sr} + 3_0^1\text{n} + \boxed{?}$

3.50 Complete each of the following nuclear equations:

a. $^{59}_{27}\text{Co} + \boxed{?} \longrightarrow ^{56}_{25}\text{Mn} + ^{4}_{2}\text{He}$

b. $\boxed{?} \longrightarrow ^{14}_{7}\text{N} + ^{0}_{-1}e$

c. $^{76}_{36}\text{Kr} + ^{0}_{-1}e \longrightarrow \boxed{?}$

3.51 Write the atomic symbols and a complete nuclear equation for the following:
  a. When two oxygen-16 atoms collide, one of the products is an alpha particle. What is the other product?
  b. When californium-249 is bombarded by oxygen-18, a new element, seaborgium-263, and four neutrons are produced.
  c. Radon-222 emits an alpha particle, and the product emits another alpha particle. Write the two nuclear reactions.

3.52 Write the symbols and complete the following nuclear equations:
  a. When polonium-210 emits an alpha particle, the product is lead-206.
  b. Bismuth-211 decays by emitting an alpha particle. The product is another radioisotope that emits a beta particle. Write equations for the two nuclear changes.

3.53 If the amount of radioactive phosphorus-32 in a sample decreases from 1.2 g to 0.30 g in 28 days, what is the half-life of phosphorus-32?

3.54 If the amount of radioactive iodine-123 in a sample decreases from 0.4 g to 0.1 g in 26.2 hours, what is the half-life of iodine-123?

3.55 Iodine-131, a beta emitter, has a half-life of 8.0 days.
  a. Write the nuclear equation for the beta-decay of iodine-131.
  b. How many grams of a 12.0-g sample of iodine-131 would remain after 40 days?
  c. How many days have passed if 48 g of iodine-131 decayed to 3.0 grams of iodine-131?

3.56 Cesium-137, a beta emitter, has a half-life of 30 years.
  a. Write the nuclear equation for cesium-137.
  b. How many grams of a 16 g sample of cesium-137 would remain after 90 years?
  c. How many years will be needed for 28 grams of cesium-137 to decay to 3.5 g of cesium-137?

3.57 A nurse was accidentally exposed to potassium-42 while doing some brain scans for possible tumors. The error was not discovered until 36 hours later when the activity of the potassium-42 sample was 2.0 μCi. If potassium-42 has a half-life of 12 hr, what was the activity of the sample at the time the nurse was exposed?

3.58 A wooden object from the site of an ancient temple has a carbon-14 activity of 10 counts per minute compared with a reference piece of wood cut today that has an activity of 40 counts per minute. If the half-life for carbon-14 is 5730 years, what is the age of the ancient wood object?

3.59 A 120-mg sample of technetium-99m is used for a diagnostic test. If technetium-99m has a half-life of 6.0 hr, how much of the technetium-99m sample remains 24 hr after the test?

3.60 The half-life of oxygen-15 is 124 seconds. If a sample of oxygen-15 has an activity of 4000 becquerels, how many minutes will elapse before it reaches an activity of 500 becquerels?

3.61 Why does a radiation technician wear a film badge?

3.62 When can exposure to radiation be detected? At about what exposure level will a person begin to suffer symptoms of radiation sickness?

3.63 What is the purpose of irradiating meats, fruits, and vegetables?

3.64 The irradiation of foods was approved in the United States in the 1980s.
  a. Why have we not seen many irradiated products in our markets?
  b. Would you buy foods that have been irradiated? Why or why not?

3.65 What is the difference between fission and fusion?

3.66 **a.** What are the products in the fission of uranium-235 that make possible a nuclear chain reaction?
  b. What is the purpose of placing control rods among uranium samples in a nuclear reactor?

3.67 Where does fusion occur naturally?

3.68 Why are scientists continuing to try to build a fusion reaction even though very high temperatures have been difficult to reach and maintain?

4 Compounds and Their Bonds

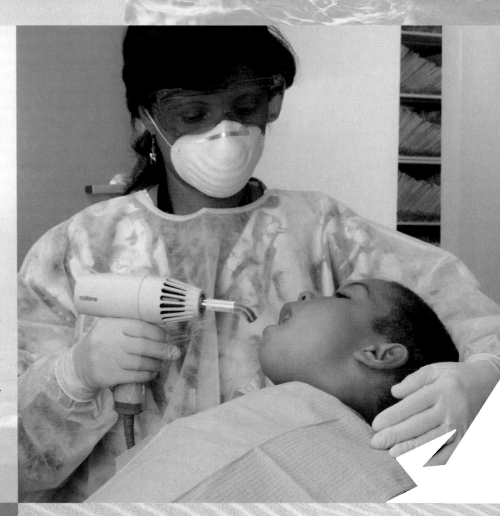

"One way to prevent cavities in children is to apply a thin, plastic coating called a sealant to their teeth," says Dr. Pam Alston, a dentist in private practice. "We look for teeth with deep grooves and pits that trap food. We clean the teeth and apply an etching agent, which helps the sealant bind to the teeth. Then we apply the liquid sealant, which fills in the grooves and pits, and use ultraviolet light to solidify the coating."

The use of fluoride compounds, such as SnF_2 in toothpaste and NaF in water, and mouth rinses have greatly reduced tooth decay. The fluoride ion replaces the hydroxide ion to form $Ca_{10}(PO_4)_6F_2$, which strengthens the enamel and makes it less susceptible to decay. Other compounds used in dentistry are the anesthetic known as laughing gas, N_2O, and Novocaine, $C_{13}H_{20}N_2O_2$.

LOOKING AHEAD

the
Chemistry
place

www.chemplace.com/college

Visit the URL above or use the CD-ROM in the book for extra quizzing, interactive tutorials, career resources, and case studies.

Almost everything around you is in the form of a **compound,** which is a substance containing two or more of the elements we discussed in Chapter 2. Your food contains compounds known as carbohydrates, proteins, and lipids. Perhaps you have used the compound called salt, which is composed of sodium and chlorine as NaCl. When you prepare your cereal, you may have added some sugar, which is composed of carbon, hydrogen, and oxygen as $C_{12}H_{22}O_{11}$. In the hospital, patients are given a saline solution that contains NaCl or a solution that contains glucose, $C_6H_{12}O_6$. Anesthetics and medications are also compounds. The anesthetic Halothane is $C_2HBrClF_3$. The antibiotic Amoxicillin is $C_{16}H_{19}N_3O_5S$. The antidepressant Prozac is $C_{17}H_{18}F_3NO$.

A drug may be developed from a natural source or it may be synthesized in a chemistry laboratory. Digitoxin, $C_{41}H_{64}O_{13}$, a cardiac tonic, comes from the dried leaves of the purple foxglove plant. For many years, insulin was obtained from animal sources although it is now produced by recombinant DNA methods. A natural form of penicillin is extracted from cultures of the *Penicillium* mold. In the laboratory, it is converted to *penicillin* G, which is used to synthesize the variety of penicillins used today.

A medication is a substance that is administered to diagnose, treat, or prevent a disease. Some recent news reports indicate that medication errors and adverse drug events are responsible for many deaths in the United States. A medication may have several names including its generic name, its chemical name, and a variety of brand names given to it by the manufacturers. When different pharmaceutical companies manufacture the same drug, it will have different brand names. For example, Halothane is a brand name for a common anesthetic. Other brand names for the same compound are Fluothane and Rhodialothan. Its chemical name is 2-bromo-2-chloro-1,1,1-trifluoroethane. While the drugs vary in their intensity and action, they must be prepared with uniform strengths so that a patient will receive consistent drug dosages. In the United States, the Federal Food, Drug, and Cosmetic Act required the development of official drug standards. Information for those drugs listed in the *United States Pharmacopoeia (USP)* includes their source, characteristics, tests for purity, storage methods, and normal dosages.

4.1 Valence Electrons

LEARNING GOAL

Using the periodic table, write the electron-dot structure for a representative element.

For the atoms of the representative elements in the periodic table, the electrons in the outer shell determine their chemical properties. These influential electrons, called **valence electrons,** are located in the valence shell, which is the outermost energy level of an atom. All of the elements in Group 1A have 1 valence electron. The elements in Group 2A have 2 valence electrons; similarly, the elements in Group 3A have 3 valence electrons, elements in Group 5A have 5 valence electrons, and elements in Group 7A have 7 valence electrons. If we compare the electron arrangement and the group number of a representative element, we find that the number of valence electrons in an atom is equal to the group number. Some examples of the relationship between electron arrangement, group number, and number of valence electrons are shown in Table 4.1.

Table 4.1   Electron Configuration and Valence Electrons

Element	Group Number	Electron Configuration	Number of Valence Electrons
Li	1A	$1s^2\,\mathbf{2s^1}$	1
Ca	2A	$1s^2\,2s^2\,2p^6\,3s^2\,3p^6\,\mathbf{4s^2}$	2
Al	3A	$1s^2\,2s^2\,2p^6\,\mathbf{3s^2\,3p^1}$	3
C	4A	$1s^2\,\mathbf{2s^2\,2p^2}$	4
N	5A	$1s^2\,\mathbf{2s^2\,2p^3}$	5
S	6A	$1s^2\,2s^2\,2p^6\,\mathbf{3s^2\,3p^4}$	6
Cl	7A	$1s^2\,2s^2\,2p^6\,\mathbf{3s^2\,3p^5}$	7
Ne	8A	$1s^2\,\mathbf{2s^2\,2p^6}$	8

SAMPLE PROBLEM 4.1

Counting Valence Electrons

Using the electron arrangement, state the number of valence electrons in atoms of each of the following elements:

a. O   **b.** Na

Solution

a. The element oxygen has 6 valence electrons, as shown in its electron configuration.

$$1s^2\,2s^2\,2p^4$$

b. The element sodium has 1 valence electron, because there is 1 electron in its highest energy level, shell 3.

$$1s^2\,2s^2\,2p^6\,3s^1$$

Study Check

How many electrons are in the outer shell of sulfur?

Atoms of magnesium

$\overset{\displaystyle \cdot}{\text{M}}\text{g}\cdot$

Dot structure

$1s^2\,2s^2\,2p^6\,\boxed{3s^2}$

Electron configuration of magnesium

Electron-Dot Structures

An **electron-dot structure** is a convenient way to represent the valence shell. Valence electrons are shown as dots placed on the sides, top, or bottom of the symbol for the element. It does not matter on which of the four sides you place the dots. However, 1 to 4 valence electrons are arranged as single dots. When there are more than 4 electrons, the electrons begin to pair up. Any of the following would be an acceptable electron-dot structure for magnesium, which has 2 valence electrons:

Possible Electron-Dot Structures for the 2 Valence Electrons in Magnesium

$\overset{\displaystyle \cdot}{\text{M}}\text{g}\cdot$ $\overset{\displaystyle \cdot}{\text{M}}\text{g}$ $\cdot\overset{\displaystyle \cdot}{\text{M}}\text{g}$ $\cdot\text{M}\underset{\displaystyle \cdot}{\text{g}}\cdot$ $\text{M}\underset{\displaystyle \cdot}{\text{g}}\cdot$ $\cdot\text{M}\underset{\displaystyle \cdot}{\text{g}}$

Electron-dot structures for the first 20 elements are given in Table 4.2. Note that although helium has just 2 valence electrons, it appears in Group 8A with the rest of the noble gases because shell 1, its outermost shell, is filled.

Table 4.2   Electron-Dot Structures for the First Twenty Elements

	Group Number							
	1A	**2A**	**3A**	**4A**	**5A**	**6A**	**7A**	**8A**
Number of Valence Electrons	$1e^-$	$2e^-$	$3e^-$	$4e^-$	$5e^-$	$6e^-$	$7e^-$	$8e^-$
	H·							He:
	Li·	Be·	·B·	·C·	·N·	·O:	·F:	:Ne:
	Na·	Mg·	·Al·	·Si·	·P·	·S:	·Cl:	:Ar:
	K·	Ca·						

SAMPLE PROBLEM 4.2

Writing Electron-Dot Structures

Write the electron-dot structure for each of the following elements:

a. chlorine   **b.** aluminum   **c.** argon

Solution

a. Because the group number for chlorine is 7A, we can state that chlorine has 7 valence electrons.

·Cl:

b. Aluminum, in Group 3A, has 3 valence electrons.

·Al·

c. Argon, in Group 8A, has 8 valence electrons.

:Ar:

Study Check

What is the electron-dot structure for the alkaline-earth metal in Period 4?

QUESTIONS AND PROBLEMS

The answers for the odd-numbered Questions and Problems throughout this chapter are given at the end of this text. The complete solutions to odd-numbered Questions and Problems are in the *Study Guide*.

Valence Electrons

4.1   Where are the valence electrons of an atom?

4.2   Why is the group number for a representative element the same as the number of valence electrons?

4.3   Write the electron configuration and the number of valence electrons in an atom of each of the following elements:
  a. N   **b.** O   **c.** Ar   **d.** K   **e.** S

4.4   Write the electron configuration and the number of valence electrons in an atom of each of the following elements:
  a. Na   **b.** Al   **c.** Cl   **d.** Mg   **e.** P

4.5   Write the group number and electron-dot structure for each element:
  a. sulfur   **b.** nitrogen   **c.** calcium
  d. sodium   **e.** potassium

4.6   Write the group number and electron-dot structure for each element:
  a. carbon   **b.** oxygen   **c.** fluorine
  d. lithium   **e.** chlorine

4.7   Using the symbol M for a metal atom, draw the electron-dot structure for an atom of a metal in the following groups:
  a. Group 1A   **b.** Group 2A

4.8   Using the symbol Nm for a nonmetal atom, draw the electron-dot structure for an atom of a nonmetal in the following groups:
  a. Group 5A   **b.** Group 7A

4.9   The alkali metals are in the same family on the periodic table. What is their group number and how many valence electrons does each have?

4.10 The halogens are in the same family on the periodic table. What is their group number and how many valence electrons does each have?

4.2 Octet Rule and Ions

Atoms of most of the elements on the periodic table combine to form compounds except for the atoms of the noble gases. One explanation is that noble gases have a particularly stable **octet** of 8 valence electrons except for helium, which is stable with 2 electrons filling its first shell.

Compounds result from the formation of chemical bonds between two or more different elements. Ionic bonds occur when the atoms of the elements lose and gain valence electrons. Covalent bonds occur when atoms share valence electrons.

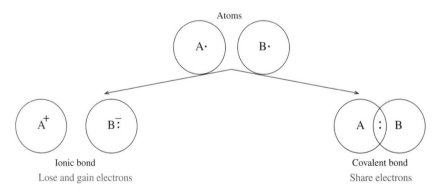

In both ionic bonds and in covalent bonds, the atoms achieve the electron configuration of a noble gas. Known as the **octet rule,** this tendency for atoms to acquire a noble gas arrangement provides a key to our understanding of the ways in which atoms bond and form compounds.

Ionization Energy

The **ionization energy** of an atom is the energy that is needed to remove an electron from the valence shell of the atom.

$$Na + \text{ionization energy} \longrightarrow Na^+ + e^-$$

Figure 4.1 illustrates the relative ionization energies for the representative elements. Typically the metals of Groups 1A, 2A, and 3A, on the left side of the periodic

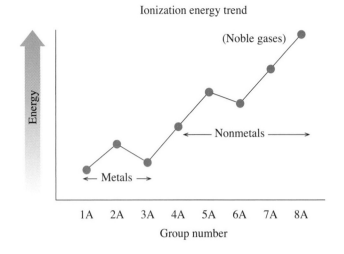

Figure 4.1 Trends in ionization energies for groups of representative elements on the periodic table.

Q Why does a nonmetal in Group 7A have a higher ionization energy than a metal in Group 1A?

table have small ionization energies while the nonmetals in Groups 5A, 6A, 7A, and the noble gases (8A) have high values. These differences in the ionization energies mean that when metals react, they lose valence electrons more easily than the nonmetals do. To attain a stable octet, metals lose electrons and the nonmetals tend to gain electrons.

SAMPLE PROBLEM 4.3

Octet Rule

Write the electron-dot structures for neon (Ne), sodium (Na), and fluorine (F). Which atoms have octets and which do not?

Solution

$$:\overset{..}{\underset{..}{Ne}}:\qquad Na\cdot\qquad \cdot\overset{..}{\underset{..}{F}}:$$

Neon, a noble gas, has an octet of 8 valence electrons. Sodium, however, has just 1 valence electron, and fluorine has 7. Sodium and fluorine would be expected to lose, gain, or share electrons to give each atom an octet.

Study Check

Which of the following elements have octets?
a. carbon   **b.** calcium   **c.** krypton (Kr)

Positive Ions

When metals react, they lose their valence electrons in order to attain a noble gas arrangement. By losing electrons, they form **ions,** which are atoms or groups of atoms that have a different number of electrons than protons. For example, when a sodium atom loses its outer electron, the next shell of electrons has a noble gas arrangement like neon.

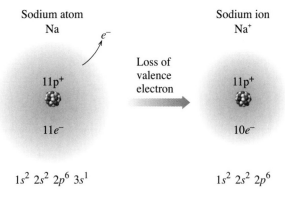

By losing an electron, sodium now has 10 electrons instead of 11. Because there are still 11 protons in its nucleus, the atom is no longer neutral. It has become a sodium ion and has an electrical charge, called an **ionic charge,** of 1+. In the symbol for the ion, the ionic charge is written in the upper right-hand corner, as in Na^+.

Metals in ionic compounds have lost their valence electrons to form positively charged ions. Positive ions are also called **cations** (pronounced *cat'-ions*). Magne-

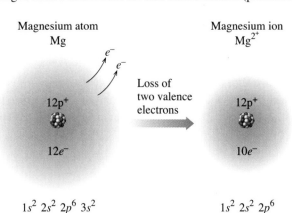

sium, a metal in Group 2A, attains a noble gas arrangement like neon by losing 2 valence electrons to form a positive ion with a 2+ ionic charge. A metal ion is named by its element name.

Negative Ions

A nonmetal forms a negative ion when it gains one or more electrons to fill its valence shell. For example, when an atom of chlorine with 7 valence electrons obtains one more electron, it completes the noble gas arrangement of argon. The resulting particle is a chloride ion having a negative ionic charge (Cl^-). Ions with negative ionic charges are called **anions** (pronounced *an'-ions*). Nonmetal ions are named by using *ide* at the end of their element names.

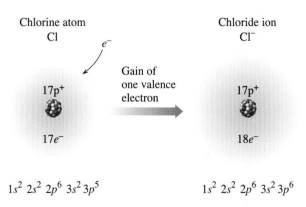

SAMPLE PROBLEM 4.4

Calculating Ionic Charge

Write the symbol and the ionic charge for each of the following ions:
a. a nitride ion that has 7 protons and 10 electrons
b. a calcium ion that has 20 protons and 18 electrons
c. a lithium ion that has 3 protons and 2 electrons

Solution

a. $7p^+$ and $10e^- = N^{3-}$ **b.** $20p^+$ and $18e^- = Ca^{2+}$
c. $3p^+$ and $2e^- = Li^+$

Study Check

How many protons and electrons are in a bromide ion, Br^-?

Ionic Charges from Group Numbers

Group numbers can be used to determine the ionic charges for most ions of the representative elements. We have seen that metals lose electrons to form positive ions. The elements in Groups 1A, 2A, and 3A lose the same number of electrons (1, 2, and 3, respectively). Group 1A metals form ions with 1+ charges, Group 2A metals form ions with 2+ charges, and Group 3A metals form ions with 3+ charges.

The nonmetals from Groups 5A, 6A, and 7A form negative ions. Group 5A nonmetals form ions with 3− charges, Group 6A nonmetals form ions with 2− charges, and Group 7A nonmetals form ions with 1− charges. The elements of Group 4A do not typically form ions. Table 4.3 lists the ionic charges for some common ions of representative elements.

Table 4.3   Ionic Charges for Representative Elements

Group Number	Number of Valence Electrons	Electron Change to Give an Octet	Ionic Charge	Examples
Metals				
1A	1	Lose 1	1+	Li^+, Na^+, K^+
2A	2	Lose 2	2+	Mg^{2+}, Ca^{2+}
3A	3	Lose 3	3+	Al^{3+}
Nonmetals				
5A	5	Gain 3	3−	N^{3-}, P^{3-}
6A	6	Gain 2	2−	O^{2-}, S^{2-}
7A	7	Gain 1	1−	F^-, Cl^-, Br^-, I^-

SAMPLE PROBLEM 4.5

Writing Ions

Consider the elements aluminum and oxygen.
a. Identify each as a metal or a nonmetal.

b. State the number of valence electrons for each.

c. State the number of electrons that must be lost or gained for each to acquire an octet.

d. Write the symbol of each resulting ion, including its ionic charge.

Solution

	Aluminum	*Oxygen*
a.	metal	nonmetal
b.	3 valence electrons	6 valence electrons
c.	loses $3e^-$	gains $2e^-$
d.	Al^{3+}	O^{2-}

Study Check

What are the symbols for the ions formed by potassium and sulfur?

HEALTH NOTE

Some Important Ions in the Body

A number of ions in body fluids have important physiological and metabolic functions. Some of them are listed in Table 4.4. Ringer's solution is used to maintain or correct electrolyte balance in body fluids.

Table 4.4   Ions in the Body

Ion	Occurrence	Function	Source	Result of Too Little	Result of Too Much
Na^+	Principal cation outside the cell	Regulation and control of body fluids	Salt, cheese, pickles	Hyponatremia, anxiety, diarrhea, circulatory failure, decrease in body fluid	Hypernatremia, little urine, thirst, edema
K^+	Principal cation inside the cell	Regulation of body fluids and cellular functions	Bananas, orange juice, milk, prunes, potatoes	Hypokalemia (hypopotassemia), lethargy, muscle weakness, failure of neurological impulses	Hyperkalemia (hyperpotassemia), irritability, nausea, little urine, cardiac arrest
Ca^{2+}	Cation outside the cell; 90% of calcium in the body in bone as $Ca_3(PO_4)_2$ or $CaCO_3$	Major cation of bone; muscle smoothant	Milk, yogurt, cheese, greens, and spinach	Hypocalcemia, tingling fingertips, muscle cramps, tetany	Hypercalcemia, relaxed muscles, kidney stones, deep bone pain
Mg^{2+}	Cation outside the cell; 70% of magnesium in the body in bone structure	Essential for certain enzymes, muscles, and nerve control	Widely distributed (part of chlorophyll of all green plants), nuts, whole grains	Disorientation, hypertension, tremors, slow pulse	Drowsiness
Cl^-	Principal anion outside the cell	Gastric juice, regulation of body fluids	Salt	Same as for Na^+	Same as for Na^+

QUESTIONS AND PROBLEMS

Octet Rule and Ions

4.11 a. How does the octet rule explain the formation of a sodium ion?
b. What noble gas has the same electronic arrangement as the sodium ion?
c. Why do you think Group 1A and Group 2A elements are found in many compounds, but not Group 8A elements?

4.12 a. How does the octet rule explain the formation of a chloride ion?
b. What noble gas has the same electronic arrangement as the chloride ion?
c. Why do you think Group 7A elements are found in many compounds, but not Group 8A elements?

4.13 Write the electron-dot arrangement for each of the following elements. Indicate which elements have a complete valence shell and which do not.
a. neon **b.** oxygen **c.** lithium **d.** argon

4.14 Write the electron-dot arrangement for each of the following elements. Indicate which elements have a complete valence shell and which do not.
a. magnesium **b.** nitrogen **c.** helium **d.** potassium

4.15 State the number of electrons that must be lost by atoms of each of the following elements to acquire a noble gas electron arrangement:
a. Li **b.** Mg **c.** Al **d.** Cs **e.** Ba

4.16 State the number of electrons that must be gained by atoms of each of the following elements to acquire a noble gas electron arrangement:
a. Cl **b.** O **c.** N **d.** I **e.** P

4.17 What noble gas has the same electronic arrangement as each of the following ions?
a. Na^+ **b.** Mg^{2+} **c.** K^+ **d.** O^{2-} **e.** F^-

4.18 What noble gas has the same electronic arrangement as each of the following ions?
a. Li^+ **b.** Sr^{2+} **c.** S^{2-} **d.** Al^{3+} **e.** Br^-

4.19 State the number of electrons lost or gained when the following elements form ions:
a. Mg **b.** P **c.** Group 7A **d.** Na **e.** Al

4.20 State the number of electrons lost or gained when the following elements form ions:
a. O **b.** Group 2A **c.** F **d.** Li **e.** N

4.21 Write the symbols of the ions with the following number of protons and electrons:
a. 3 protons, 2 electrons **b.** 9 protons, 10 electrons
c. 12 protons, 10 electrons **d.** 26 protons, 23 electrons
e. 30 protons, 28 electrons

4.22 How many protons and electrons are in the following ions?
a. O^{2-} **b.** K^+ **c.** Br^- **d.** S^{2-} **e.** Sr^{2+}

4.23 Write the symbol for the ion of each of the following:
a. chloride **b.** potassium **c.** oxide **d.** aluminum

4.24 Write the symbol for the ion of each of the following:
a. fluoride **b.** calcium **c.** sodium **d.** lithium

4.25 What is the name of each of the following ions?
a. K^+ **b.** S^{2-} **c.** Ca^{2+} **d.** N^{3-}

4.26 What is the name of each of the following ions?
a. Mg^{2+} **b.** Ba^{2+} **c.** I^- **d.** Cl^-

LEARNING GOAL

Using charge balance, write the correct formula for an ionic compound.

4.3 Ionic Compounds

Ionic compounds consist of positive and negative ions. The ions are held together by strong electrical attractions between the opposite charges, called **ionic bonds.**

The physical and chemical properties of an ionic compound such as NaCl are very different from those of the original elements. For example, the original elements of NaCl were sodium, a soft, shiny metal, and chlorine, a yellow-green poisonous gas. Yet, as positive and negative ions, they form table salt, NaCl, a white, crystalline substance that is common in our diet. Scientists diagram the solid form of NaCl as alternating Na^+ and Cl^- ions packed together in a lattice structure, as shown in Figure 4.2.

(a)

Figure 4.2 **(a)** The elements sodium and chlorine react to form the ionic compound sodium chloride, the compound that makes up table salt. **(b)** Crystals of NaCl under magnification. **(c)** A diagram of the arrangements of Na^+ and Cl^- packed together in a NaCl crystal.

Q What is the type of bonding between Na^+ and Cl^- ions in salt?

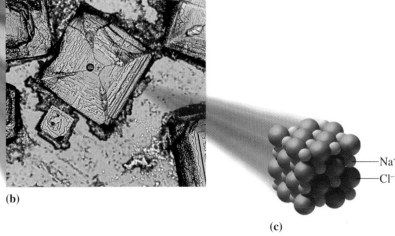

(b)

(c)

Charge Balance in Ionic Compounds

The **formula** of an ionic compound indicates the number and kinds of ions that make up the ionic compound. The sum of the ionic charges in the formula is always zero. For example, the NaCl formula indicates that there is one sodium ion, Na^+, for every chloride ion, Cl^-, in the compound. Note that the ionic charges of the ions do not appear in the formula of the ionic compound.

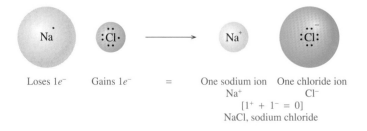

Subscripts in Formulas

The formula of any ionic compound has a zero overall charge. That means that the total amount of positive charge is equal to the total amount of negative charge. Consider a compound of magnesium and chlorine. To achieve an octet, a Mg atom loses its 2 valence electrons to form Mg^{2+}. Each Cl atom gains 1 electron to form Cl^-, which has a complete valence shell. In this example, two Cl^- ions are needed to balance the positive charge of Mg^{2+}. This gives the formula, $MgCl_2$, magnesium chloride, in which a subscript of 2 shows that two Cl^- were needed for charge balance.

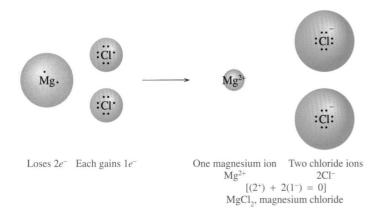

Loses $2e^-$ Each gains $1e^-$

One magnesium ion   Two chloride ions
Mg^{2+} $2Cl^-$
$[(2^+) + 2(1^-) = 0]$
$MgCl_2$, magnesium chloride

SAMPLE PROBLEM 4.6

Diagramming an Ionic Compound

Diagram the formation of the ionic compound aluminum fluoride, AlF_3.

Solution

In their electron-dot structures, aluminum has 3 valence electrons and fluorine has 7. The aluminum loses its 3 valence electrons, and each fluorine atom gains an electron, to give ions with noble gas arrangements in the ionic compound AlF_3.

Loses $3e^-$ Gains $3e^-$ Al^{3+} Three F^-
$[(3^+) + 3(1^-) = 0]$

Study Check

Diagram the formation of lithium sulfide, Li_2S.

Writing Ionic Formulas from Ionic Charges

The subscripts in the formula of an ionic compound represent the number of positive and negative ions that give an overall charge of zero. Thus, we can now write a formula directly from the ionic charges of the positive and negative ions. Suppose we wish to write the formula of the ionic compound containing Na^+ and S^{2-} ions. To balance the ionic charge of the S^{2-} ion, we will need to place two Na^+ ions in the formula. This gives the formula Na_2S, which has an overall charge of zero.

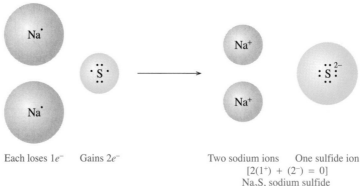

Each loses $1e^-$ Gains $2e^-$ Two sodium ions   One sulfide ion
                                           $[2(1^+) + (2^-) = 0]$
                                           Na_2S, sodium sulfide

---

SAMPLE PROBLEM 4.7

Writing Formulas from Ionic Charges

Use ionic charge balance to write the formula for the ionic compound containing Na^+ and N^{3-}.

Solution

Determine the number of each ion needed for charge balance. The charge for nitrogen $(3-)$ is balanced by three Na^+ ions. Writing the positive ion first gives the formula Na_3N.

Study Check

Use ionic charges to determine the formula of the compound that would form when calcium and chlorine react.

QUESTIONS AND PROBLEMS

Ionic Compounds

4.27 Which of the following pairs of elements are likely to form an ionic compound?
   a. lithium and chlorine   **b.** oxygen and chlorine
   c. potassium and oxygen   **d.** sodium and neon
   e. sodium and magnesium

4.28 Which of the following pairs of elements are likely to form an ionic compound?
   a. helium and oxygen   **b.** magnesium and chlorine
   c. chlorine and bromine   **d.** potassium and sulfur
   e. sodium and potassium

4.29 Using electron-dot structures, diagram the formation of the following ionic compounds:
   a. KCl   **b.** $CaCl_2$ **c.** Na_3N

4.30 Using electron-dot structures, diagram the formation of the following ionic compounds:
   a. MgS   **b.** $AlCl_3$ **c.** Li_2O

4.31 Write the correct ionic formula for compounds formed between the following ions:
   a. Na^+ and O^{2-} **b.** Al^{3+} and Br^- **c.** Ba^{2+} and O^{2-}
   d. Mg^{2+} and Cl^- **e.** Al^{3+} and S^{2-}

4.32 Write the correct ionic formula for compounds formed between the following ions:
  a. Al^{3+} and Cl^- **b.** Ca^{2+} and S^{2-} **c.** Li^+ and S^{2-}
  d. K^+ and N^{3-} **e.** K^+ and I^-

4.33 Write the correct formula for ionic compounds formed by the following metals and nonmetals:
  a. sodium and sulfur **b.** potassium and nitrogen
  c. aluminum and iodine **d.** lithium and oxygen

4.34 Write the correct formula for ionic compounds formed by the following metals and nonmetals:
  a. calcium and chlorine **b.** barium and bromine
  c. sodium and phosphorus **d.** magnesium and oxygen

4.4 Naming and Writing Ionic Formulas

LEARNING GOAL

Given the formula of an ionic compound, write the correct name.

The name of a metal ion is the same as its elemental name. The name of a nonmetal ion is obtained by replacing the end of its elemental name with *ide*. Table 4.5 lists the names of some important metal and nonmetal ions.

Table 4.5 Formulas and Names of Some Common Ions

Group Number	Formula of Ion	Name of Ion	Group Number	Formula of Ion	Name of Ion
	Metals			**Nonmetals**	
1A	Li^+	Lithium	5A	N^{3-}	Nitride
	Na^+	Sodium		P^{3-}	Phosphide
	K^+	Potassium	6A	O^{2-}	Oxide
2A	Mg^{2+}	Magnesium		S^{2-}	Sulfide
	Ca^{2+}	Calcium	7A	F^-	Fluoride
	Ba^{2+}	Barium		Cl^-	Chloride
3A	Al^{3+}	Aluminum		Br^-	Bromide
				I^-	Iodide

SAMPLE PROBLEM 4.8

Naming Ions

Write the name of each of the following ions:

a. Al^{3+} **b.** S^{2-}

Solution

a. Aluminum ion; a metal ion uses the elemental name.
b. Sulfide ion; a nonmetal changes the ending of the name of the element to *ide*.

Study Check

Write the name of each of the following ions:

a. Ca^{2+} **b.** N^{3-} **c.** Br^-

Physical Therapy

"Physical therapists need to understand how the body, muscles, and joints function in order to recognize when something is not working and which area needs strengthening," says Vincent Leddy, physical therapist. "Chemistry is important to understanding the body's physiology and how chemical changes in the body affect movement. I went into physical therapy because I enjoy teaching movement to children. I am guiding Maggie into her chair but allowing her to move as much as she can by herself. I help her by giving gentle pressure and reassurance that she'll be safe in the transition. I work on getting Maggie to use her body, and the occupational therapist works on Maggie's fine motor skills and how she uses her hand on the switch or picks up objects. By using both physical and occupational therapy, we enhance the child's performance."

Naming Ionic Compounds Containing Two Elements

In the name of an ionic compound made up of two elements, the metal ion is named followed by the name of the nonmetal ion. Subscripts are never mentioned; they are understood as a result of the charge balance of the ions in the compound.

Compound	Metal Ion	Nonmetal Ion	Name
NaF	Na^+	F^-	
	Sodium	Fluoride	Sodium fluoride
$MgBr_2$	Mg^{2+}	Br^-	
	Magnesium	Bromide	Magnesium bromide
Al_2O_3	Al^{3+}	O^{2-}	
	Aluminum	Oxide	Aluminum oxide

SAMPLE PROBLEM 4.9

Naming Ionic Compounds

Write the name of each of the following ionic compounds:

a. Na_2O **b.** Mg_3N_2

Solution

Compound	Ions and Names		Name of Compound
a. Na_2O	Na^+	O^{2-}	
	Sodium	Oxide	Sodium oxide
b. Mg_3N_2	Mg^{2+}	N^{3-}	
	Magnesium	Nitride	Magnesium nitride

Study Check

Name the compound $CaCl_2$.

Metals That Form Two Positive Ions

Like all metals, the transition metals and the representative metals in Group 4A also form positive ions. However, these metals typically form two or more kinds of positive ions because they lose their outer s electrons first followed by the loss of one or more d electrons. For example, in some ionic compounds, iron is in the Fe^{2+} form, but in different compounds, it takes the Fe^{3+} form. Copper also forms two different ions: Cu^+ is present in some compounds and Cu^{2+} in others. When a metal can form two or more ions, it is not possible to predict the ionic charge from the group number. We say that it has a variable valence.

When two different ions are possible for a metal, a naming system is needed to identify the particular cation in a compound. To do this, a Roman numeral that matches the ionic charge is placed in parentheses after the elemental name of the metal. For the cations of iron, Fe^{2+} is named iron(II), and Fe^{3+} is named iron(III). An older system names Fe^{2+} as the *ferrous ion* and Fe^{3+} as the *ferric ion*. Table 4.6 lists the ions of some common metals that produce two or more ions.

Figure 4.3 shows some common ions and their location on the periodic table. Typically, the transition metals form two or more positive ions. However, zinc, cad-

Table 4.6   Some Metals That Form Two Positive Ions

Element	Possible Ions	Systematic Name	Older Name
Iron	Fe^{2+}	Iron(II)	Ferrous
	Fe^{3+}	Iron(III)	Ferric
Copper	Cu^+	Copper(I)	Cuprous
	Cu^{2+}	Copper(II)	Cupric
Gold	Au^+	Gold(I)	Aurous
	Au^{3+}	Gold(III)	Auric
Tin	Sn^{2+}	Tin(II)	Stannous
	Sn^{4+}	Tin(IV)	Stannic
Lead	Pb^{2+}	Lead(II)	Plumbous
	Pb^{4+}	Lead(IV)	Plumbic

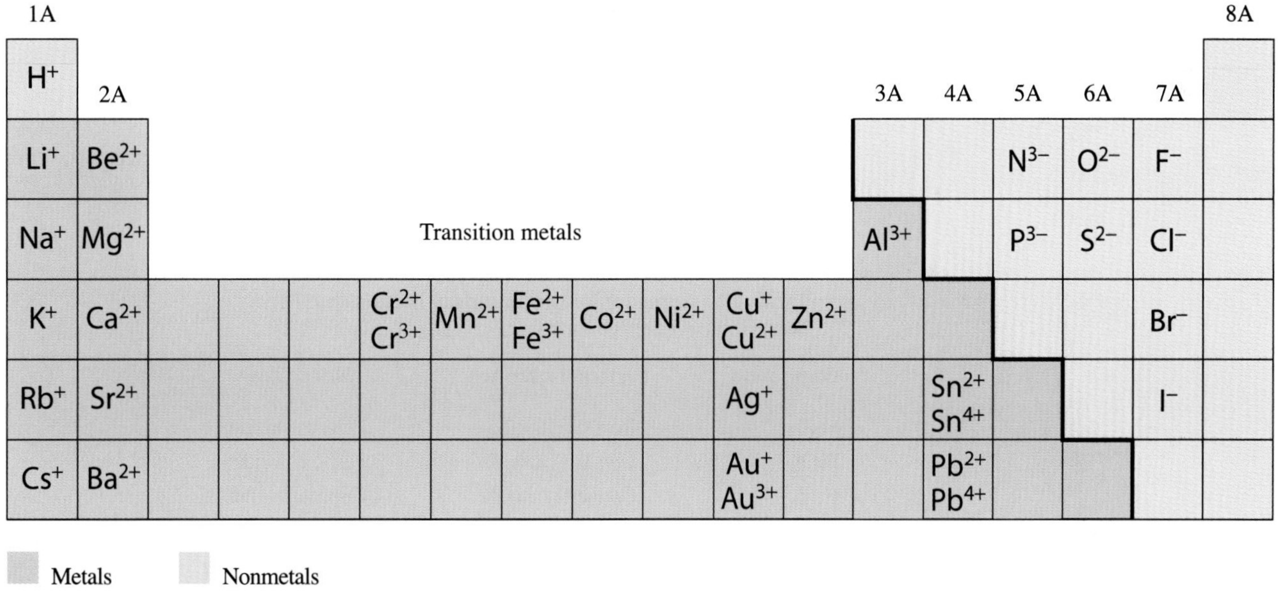

Metals   Nonmetals

mium, and silver form only one ion. The ionic charges of silver, cadmium, and zinc are fixed like Group 1A and 2A metals, so their elemental names are sufficient when naming their ionic compounds.

Figure 4.3 On the periodic table, positive ions are produced by metals in Groups 1A, 2A, and 3A and transition metals. Negative ions are produced by Groups 5A, 6A, and 7A.

Q What are the typical ions produced by calcium, copper, and oxygen?

SAMPLE PROBLEM 4.10

Writing Simple Ions

Write the symbol of the following ions:

a. copper(II)   **b.** iron(III)   **c.** zinc

Solution

a. Cu^{2+} **b.** Fe^{3+} **c.** Zn^{2+}

Study Check

What is the name of the Cu^+ ion?

Naming Compounds That Include Transition Metals

When a metal forms more than one type of positive ion, the ionic charge of that metal in a particular formula must be included as a Roman numeral in parentheses when the metal is named. The selection of the correct Roman numeral depends upon the calculation of the ionic charge of the metal in the formula of a specific ionic compound. For example, we know that in the formula $CuCl_2$ the positive charge of the copper ion must balance the negative charge of two chloride ions. Because we know that chloride ions each have a $1-$ charge, there must be a total negative charge of $2-$. Balancing the $2-$ by the positive charge gives a charge of $2+$ for Cu (a Cu^{2+} ion):

$CuCl_2$

Cu charge	+	Cl^- charge	= 0
(?)	+	$2(1-)$	= 0
$(2+)$	+	$2-$	= 0

Because copper forms ions with two different charges, Cu^+ and Cu^{2+}, we need to use the Roman numeral system and place (II) after *copper* when naming the compound:

Copper(II) chloride

Table 4.7 lists names of some ionic compounds in which the metals form two types of positive ion.

A summary of how ionic compounds are named follows.

Table 4.7   Some Ionic Compounds of Metals That Form Two Kinds of Positive Ions

Compound	Systematic Name	Older Name
$FeCl_2$	Iron(II) chloride	Ferrous chloride
$FeCl_3$	Iron(III) chloride	Ferric chloride
Cu_2S	Copper(I) sulfide	Cuprous sulfide
$CuCl_2$	Copper(II) chloride	Cupric chloride
$SnCl_2$	Tin(II) chloride	Stannous chloride
$PbBr_4$	Lead(IV) bromide	Plumbic bromide

Rules for Naming Ionic Compounds

Number of Ions Possible for Metal	Name of Metal Ion	Name of Nonmetal Ion
One positive ion: Groups 1A, 2A, 3A, Zn, Ag, Cd	Use element name	Element name ends in *ide*
More than one positive ion: Transition metals and Group 4A metals	Use element name with a Roman numeral in parentheses that is equal to the positive charge	Element name ends in *ide*

SAMPLE PROBLEM 4.11

Naming Ionic Compounds

Write the name for each of the following ionic compounds:

a. FeO   **b.** AlF_3 **c.** Cu_2S

Solution

a. Because iron is a transition metal that forms more than one ion, it is necessary to determine the ionic charge of Fe in FeO:

Fe charge $+$ O^{2-} charge $= 0$

? $+ (2-)$ $= 0$

$(2+)$ $+ (2-)$ $= 0$

The charge on Fe is therefore $2+$. So we place the Roman numeral II in parentheses after *iron:*

Iron(II) oxide

b. Because aluminum is in Group 3A and forms only Al^{3+}, the name of the metal is sufficient. A Roman numeral is not needed.

AlF_3 Aluminum fluoride

c. As a transition metal, copper forms more than one positive ion. Balancing the charge in Cu_2S, we determine the ionic charge of the copper ion:

2(Cu charge) $+$ S^{2-} charge $= 0$

$2(?)$ $+ (2-)$ $= 0$

$2(1+)$ $+ (2-)$ $= 0$

The charge on Cu is therefore $1+$. So we place the Roman numeral I in parentheses after copper:

Copper(I) sulfide

Although the total positive charge is $2+$, it is divided between two copper ions; therefore, each copper ion in the formula is Cu^+.

Study Check

Write the name of the compound whose formula is $AuCl_3$.

Writing Formulas from the Name of an Ionic Compound

The formula of an ionic compound can be written from its name because the first term in the name describes the metal ion and the second term specifies the nonmetal ion. Subscripts are added as needed to balance the charge. The steps for writing a formula from the name of an ionic compound follow.

Steps for Writing a Formula from the Name of an Ionic Compound

Step 1   Write the positive ion from the first term in the name and the negative ion from the second term.

Step 2   Balance the ionic charges to give an overall charge of zero.

Step 3   Write the formula, placing the symbol for the metal first and the symbol for the nonmetal second. Use appropriate subscripts for charge balance.

SAMPLE PROBLEM 4.12

Writing Formulas for Ionic Compounds

Write the correct formula for iron(III) chloride.

Solution

Step 1   Write the names of the metal ion and the nonmetal ion. Because the Roman numeral (III) indicates the ionic charge of the particular iron ion, we can write Fe^{3+} for the positive ion. The chloride ion has an ionic charge of $1-$.

Iron(III)   chloride

Fe^{3+} Cl^-

Step 2   Balance the charges.

Ions needed to balance charge: Fe^{3+} Cl^-

Cl^-

Cl^-

Total charge: $(3+) + 3(1-) = 0$

Step 3   Write the formula of the compound, metal first, using subscripts when needed.

$FeCl_3$

Study Check

What is the correct formula of the ionic compound aluminum sulfide?

QUESTIONS AND PROBLEMS

Naming and Writing Ionic Formulas

4.35 Write names for the following ionic compounds:
   a. Al_2O_3 **b.** $CaCl_2$ **c.** Na_2O
   d. Mg_3N_2 **e.** KI

4.36 Write names for the following ionic compounds:
   a. $MgCl_2$ **b.** K_3P **c.** Li_2S
   d. LiBr   **e.** MgO

4.37 Why is a Roman numeral placed after the name of most transition metals ions?

4.38 The compound $CaCl_2$ is named calcium chloride; the compound $CuCl_2$ is named copper(II) chloride. Explain why a Roman numeral is used in one name but not the other.

4.39 Write the names of the following Group 4A and transition metal ions (include the Roman numeral when necessary):
   a. Fe^{2+} **b.** Cu^{2+} **c.** Zn^{2+} **d.** Pb^{4+}

4.40 Write the names of the following Group 4A and transition metal ions (include the Roman numeral when necessary):
   a. Ag^+ **b.** Cu^+ **c.** Fe^{3+} **d.** Sn^{2+}

4.41 Write names for the following ionic compounds:
   a. $SnCl_2$ **b.** FeO   **c.** Cu_2S **d.** CuS

4.42 Write names for the following ionic compounds:
   a. Ag_3P **b.** PbS   **c.** SnO_2 **d.** $AuCl_3$

4.43 Indicate the charge on the metal ion in each of the following formulas:
   a. $AuCl_3$ **b.** Fe_2O_3 **c.** PbI_4 **d.** $SnCl_2$

4.44 Indicate the charge on the metal ion in each of the following formulas:
   a. $FeCl_2$ **b.** CuO **c.** Fe_2S_3 **d.** AlP

4.45 Write formulas for the following ionic compounds:
   a. magnesium chloride   **b.** sodium sulfide
   c. copper(I) oxide   **d.** zinc phosphide
   e. gold(III) nitride

4.46 Write formulas for the following ionic compounds:
   a. iron(III) oxide   **b.** barium fluoride
   c. tin(IV) chloride   **d.** silver sulfide
   e. copper(II) chloride

4.5 Covalent Bonds

When atoms of two nonmetals combine, covalent compounds result. The atoms, which are similar or even identical, are held together by sharing electrons. This type of bond is called a **covalent bond.**

The simplest covalent molecule is hydrogen, H_2. A hydrogen atom attains a noble gas arrangement when its first shell is filled with 2 electrons. To accomplish this, two hydrogen atoms share their single electrons and form a covalent bond:

$$H\cdot \; + \; \cdot H \longrightarrow H\!:\!H \longrightarrow H\!-\!H \; = \; H_2$$

| Electrons to share | A shared pair of electrons | A covalent bond | A hydrogen molecule |

The covalent bond representing a shared pair of electrons can also be written as a line between the two atoms. Atoms that are bonded through covalent bonds produce a **molecule.**

In most covalent molecules, each of the atoms has an octet of 8 electrons. For example, a fluorine molecule, F_2, consists of two fluorine atoms. Each atom needs one more valence electron for a noble gas arrangement. An octet for each is achieved when the fluorine atoms share a pair of electrons in their outer shells.

$$:\!\ddot{F}\!\cdot \; + \; \cdot\!\ddot{F}\!: \longrightarrow :\!\ddot{F}\!:\!\ddot{F}\!: \qquad :\!\ddot{F}\!-\!\ddot{F}\!: \; = \; F_2$$

| Electrons to share | A shared pair of electrons | A covalent bond | A fluorine molecule |

Hydrogen (H_2) and fluorine (F_2) are examples of nonmetal elements whose natural state is diatomic; that is, they contain two atoms. The elements that exist as diatomic molecules are listed in Table 4.8.

Sharing Electrons Between Atoms of Different Elements

The number of electrons that an atom shares and the number of covalent bonds it forms are usually equal to the number of electrons needed to acquire a noble gas arrangement. For example, carbon has 4 valence electrons. Because carbon needs to acquire 4 more electrons for an octet, it forms 4 covalent bonds by sharing its 4 valence electrons. Table 4.9 relates the group numbers of some nonmetals to the number of covalent bonds the elements typically form.

Table 4.8   Elements That Exist as Diatomic, Covalent Molecules

Element	Diatomic Molecule	Name
H	H_2	Hydrogen
N	N_2	Nitrogen
O	O_2	Oxygen
F	F_2	Fluorine
Cl	Cl_2	Chlorine
Br	Br_2	Bromine
I	I_2	Iodine

Table 4.9   Bonding Patterns of Typical Nonmetals in Covalent Compounds

1A	3A	4A	5A	6A	7A
H 1 bond					
	B 3 bonds	C 4 bonds	N 3 bonds	O 2 bonds	F 1 bond
		Si 4 bonds	P 3 (or 5) bonds	S 2 (4 or 6) bonds	Cl 1 (3 or 5) bonds
					Br 1 (3 or 5) bonds
					I 1 (3, 5, or 7) bonds

Methane, a component of natural gas, is a compound made of carbon and hydrogen. To attain an octet, each carbon shares 4 electrons, and each hydrogen shares 1 electron. In this molecule, a carbon atom forms four covalent bonds with four hydrogen atoms. The electron-dot structure for the molecule is written with the carbon atom in the center and the hydrogen atoms on the sides.

While the octet rule is useful, there are exceptions. We have already seen that a hydrogen (H_2) molecule requires just two electrons or a single bond to achieve the stability of the nearest noble gas, helium. In BF_3, the nonmetal B shares only 3 valence electrons to give a total of six valence electrons or 3 bonds. The nonmetals typically form octets. However, atoms such as P, S, Cl, Br, and I can share more of their valence electrons and expand the octets to stable valence shells of 10, 12, or even 14 electrons. In PCl_3, the P atom has an octet, but in PCl_5 the P atom has 10 valence electrons or 5 bonds. In H_2S, the S atom has an octet, but in SF_6, there are 12 valence electrons or 6 bonds to the sulfur atom.

Table 4.10 gives the electron-dot structures for several covalent molecules.

SAMPLE PROBLEM 4.13

Writing Electron-Dot Structures for Covalent Compounds

Draw an electron-dot structure for H_2S.

Solution

Sulfur, which has 6 valence electrons, shares 2 electrons to form two covalent bonds. Two atoms of hydrogen, each having 1 valence electron, will form two covalent bonds, one for each atom:

$$H:\overset{..}{\underset{..}{S}}:$$
$$H$$

Electron-dot structure for H_2S

Study Check

Write the electron-dot structure for NH_3.

Table 4.10 Electron-Dot Structures for Some Covalent Compounds		
CH_4	NH_3	H_2O

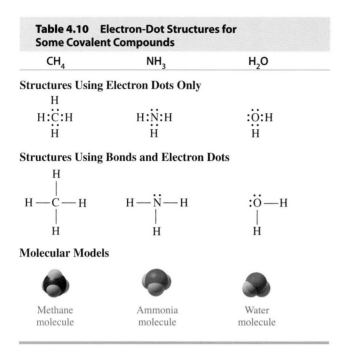

Multiple Covalent Bonds

Up to now, we have looked at covalent bonding in molecules having only single bonds. In many covalent compounds, atoms share two or three pairs of electrons to complete their electron octets. A **double bond** is the sharing of two pairs of electrons, and in a **triple bond,** three pairs of electrons are shared. Atoms of carbon, oxygen, nitrogen, and sulfur are most likely to form multiple bonds. Atoms of hydrogen and the halogens do not form double or triple bonds. For example, in HCN, the bond for H must be a single bond, while the bond between C and N is a triple bond.

$$H—C≡N\!:$$

Double and triple bonds are formed when single covalent bonds fail to complete the octets of all the atoms in the molecule. For example, in the electron-dot structure for the covalent compound N_2, an octet is achieved when each nitrogen atom shares 3 electrons. Thus, three covalent bonds, or a triple bond, will form.

We use the steps in the following example to write an electron-dot structure for SO_2.

Steps for Writing Electron-Dot Structures

Step 1 Determine the central atom of the structure, which is usually the single atom in the formula such as an atom of B, C, N, Si, P, or S. Place the other atoms around it. For more complex structures, a skeleton structure may be given.

For SO$_2$, the central atom is S, which is bonded to the two oxygen atoms.

O   S   O

Step 2   Find the total number (sum) of valence electrons by adding the number of valence electrons for each atom in the compound.

The total valence electrons for SO$_2$ = S   1 × 6e^- = 6e^-

O   2 × 6e^- = 12e^-

18e^-

Step 3   Place a pair of electrons (single bond) to attach each of the atoms to the central atom. Determine the total number of bonding electrons used.

O:S:O   4e^- used to connect atoms

Step 4   Subtract the number of bonding electrons used in step 3 from the total number of valence electrons. Place the remaining electrons as nonbonding pairs around the atoms and try to give each atom an octet. (Remember, hydrogen needs only 2 electrons, a single bond.)

18e^- − 4e^- (step 3) = 14e^- remaining

:Ö: S̈   :Ö: 18e^-

Not an octet

Step 5   If octets for all the atoms were not completed in step 4, move a nonbonding pair of electrons from an adjacent atom to the atom that needs an octet to form a multiple bond.

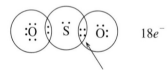

:Ö̈   S̈   :Ö: 18e^-

Double bond gives sulfur an octet

:Ö̈ — S̈ = Ö:

SAMPLE PROBLEM 4.14

Drawing Electron-Dot Structures Having Multiple Bonds

Draw the electron-dot structure for the compound CO$_2$. Carbon is the central atom.

Step 1   Arrange the atoms for CO$_2$ with C as the central atom: O   C   O

Step 2   The 4 valence electrons of carbon and the 6 electrons of two oxygen atoms gives a total of 16 valence electrons.

Total valence electrons for CO$_2$ = C   1 × 4e^- = 4e^-

O   2 × 6e^- = 12e^-

16e^-

Step 3   Use two pairs of electrons ($4e^-$) to attach each of the oxygen atoms to the central carbon atom.

O$:$C$:$O $4e^-$ used to connect atoms

Step 4   Place the remaining 12 valence electrons around the atoms:

$16e^- - 4e^- = 12e^-$

Need to share
↓

:Ö   : C̈   : Ö:

Only 6e$^-$ ↑   Only 6e$^-$
Need to share

Step 5   Octets for the oxygen atoms are not complete. Move two nonbonding pairs of electrons on the carbon atom between the carbon atom and the oxygen atoms to complete the octets for both oxygens.

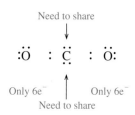 $16e^-$:Ö=C=Ö: CO_2

Octets   Molecule of
                                carbon dioxide

Study Check

Determine the number of valence electrons for the compound HCN and arrange them in the electron-dot structure.

QUESTIONS AND PROBLEMS

Covalent Bonds

4.47 Write the electron-dot structure for each of the following covalent molecules:
   a. Br_2 **b.** H_2 **c.** HF   **d.** OF_2

4.48 Write the electron-dot structure for each of the following covalent molecules:
   a. NCl_3 **b.** CCl_4 **c.** Cl_2 **d.** SiF_4

4.49 The following covalent compounds have double or triple bonds. Write their electron-dot structures. (The skeleton structure is indicated in parentheses.)
   a. C_2H_2 (H C C H)

   b. CH_2O $\left(\begin{array}{c} O \\ H\,C\,H \end{array} \right)$

   c. SO_3 $\left(\begin{array}{c} O \\ O\,S\,O \end{array} \right)$

4.50 The following covalent compounds have double or triple bonds. Write their electron-dot structures. (The skeleton structure is indicated in parentheses.)

   a. HCN   **b.** H_2CCH_2 $\left(\begin{array}{c} H\,H \\ H\,C\,C\,H \end{array} \right)$ **c.** HNO_2 (H O N O)

4.6 Naming and Writing Formulas of Covalent Compounds

Unlike the names of ionic compounds, the names of covalent compounds need prefixes because several different compounds can be formed from the same two nonmetals. For example, carbon and oxygen can form two different compounds, carbon monoxide, CO, and carbon dioxide, CO_2. Nitrogen and oxygen also form several different covalent molecules. We could not distinguish between them by using the name *nitrogen oxide*. Therefore, prefixes are used with molecular compounds composed of two nonmetals.

In the name of a covalent compound, the first nonmetal in the formula is named by its elemental name; the second nonmetal is named by its elemental name with the ending changed to *ide*. Subscripts indicating two or more atoms of an element are expressed as prefixes placed in front of each name. Table 4.11 lists some prefixes used in naming covalent compounds.

Table 4.11   Prefixes Used in Naming Covalent Compounds

Number of Atoms	Prefix
1	Mono
2	Di
3	Tri
4	Tetra
5	Penta
6	Hexa
7	Hepta
8	Octa

Some Covalent Compounds Formed by Nitrogen and Oxygen

NO	Nitrogen oxide
N_2O	Dinitrogen oxide
N_2O_3	Dinitrogen trioxide
N_2O_4	Dinitrogen tetroxide
N_2O_5	Dinitrogen pentoxide

In the name of a covalent compound, the prefix *mono* is usually omitted. When the vowels *o* and *o* or *a* and *o* appear together, the first vowel is omitted. Table 4.12 lists the formulas, names, and commercial uses of some other covalent compounds.

the Chemistry place

WEB TUTORIAL
Covalent Bonds

Table 4.12   Some Common Covalent Compounds

Formula	Name	Commercial Uses
CS_2	Carbon disulfide	Manufacture of rayon
CO_2	Carbon dioxide	Carbonation of beverages, fire extinguishers, propellant in aerosols, dry ice
SiO_2	Silicon dioxide	Manufacture of glass
NCl_3	Nitrogen trichloride	Bleaching of flour in some countries (prohibited in U.S.)
SO_2	Sulfur dioxide	Preserving fruits, vegetables; disinfectant in breweries; bleaching textiles
SO_3	Sulfur trioxide	Manufacture of explosives
SF_6	Sulfur hexafluoride	Electrical circuits
ClO_2	Chlorine dioxide	Bleaching pulp (for making paper), flour, leather
ClF_3	Chlorine trifluoride	Rocket propellant

SAMPLE PROBLEM 4.15

Naming Covalent Compounds

Name each of the following covalent compounds:

a. NCl_3 **b.** P_4O_6

Solution

a. In NCl_3, the first nonmetal is nitrogen. The second nonmetal is chlorine, which is named *chloride*. The subscript indicating three chlorine atoms is shown as the prefix *tri* in front of the name of the second element, trichloride.

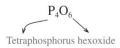

NCl_3
Nitrogen trichloride

b. In P_4O_6, the first nonmetal is phosphorus. Because there are four P atoms, it is named *tetraphosphorus*. The second nonmetal, consisting of six oxygen atoms, is named *hexoxide*. (When the vowels *a* and *o* appear together, as would happen in *hexa* plus *oxide*, the ending of the prefix is dropped.) Putting the names for the two nonmetals together, we have

P_4O_6
Tetraphosphorus hexoxide

Study Check

Scientists have recently combined some of the noble gases with other elements. One of the compounds produced was XeF_6. How would you name this compound?

SAMPLE PROBLEM 4.16

Writing Formulas from Names of Covalent Compounds

Write the formulas of the following covalent compounds:

a. sulfur dichloride   **b.** dinitrogen pentoxide

Solution

a. The first nonmetal is sulfur. Since there is no prefix given, we know that there is one atom of sulfur. The second nonmetal in the formula is chlorine. The prefix *di* indicates that there are two atoms of chlorine, which means that the subscript 2 appears with chlorine in the formula.

Sulfur dichloride

SCl_2

b. The first nonmetal in the name is nitrogen. The prefix *di* means that there are two atoms of nitrogen and therefore the subscript 2 is needed. The second nonmetal is oxygen. The prefix *pent(a)* means "five"—there are five atoms of oxygen so the subscript 5 is needed.

Dinitrogen pentoxide

N_2O_5

Study Check

What is the formula of iodine pentafluoride?

QUESTIONS AND PROBLEMS

Covalent Compounds

4.51 Name the following covalent compounds:
    a. PBr_3 **b.** CBr_4 **c.** SiO_2
    d. HF **e.** NI_3

4.52 Name the following covalent compounds:
    a. CS_2 **b.** P_2O_5 **c.** Cl_2O
    d. PCl_3 **e.** N_2O_4

4.53 Name the following covalent compounds:
    a. N_2O_3 **b.** NCl_3 **c.** $SiBr_4$
    d. PCl_5 **e.** SO_3

4.54 Name the following covalent compounds:
    a. SiF_4 **b.** IBr_3 **c.** CO_2
    d. SO_2 **e.** N_2O

4.55 Write the formulas of the following covalent compounds:
    a. carbon tetrachloride   **b.** carbon monoxide
    c. phosphorus trichloride   **d.** dinitrogen tetroxide

4.56 Write the formulas of the following covalent compounds:
    a. sulfur dioxide   **b.** silicon tetrachloride
    c. iodine pentafluoride   **d.** dinitrogen oxide

4.57 Write the formulas of the following covalent compounds:
    a. oxygen difluoride   **b.** boron trifluoride
    c. dinitrogen trioxide   **d.** sulfur hexafluoride

4.58 Write the formulas of the following covalent compounds:
    a. sulfur dibromide   **b.** carbon disulfide
    c. tetraphosphorus hexoxide   **d.** dinitrogen pentoxide

---

WEB TUTORIAL
Bonds and Bond Polarity

4.7 Bond Polarity

In covalent bonds, as we have seen, electrons may be shared between atoms that are identical, as in Cl_2, or between atoms that are different, as in HCl. When electrons are shared between identical nonmetal atoms, they are shared equally. However, when they are shared between atoms of different nonmetals, they are shared unequally because one of the atoms in the covalent bond has a stronger attraction for the pair of electrons than the other. The ability of an atom to attract the shared electrons is called its electronegativity, and it is indicated by a number called an **electronegativity value.** The element with the greater electronegativity value pulls the shared electrons closer to its nucleus.

In the table of electronegativity values shown in Figure 4.4, fluorine, the most electronegative element, has the highest value of 4.0. By contrast, the metals in Groups 1A and 2A have low electronegativity values. The electronegativity values increase going from left to right across each period of the periodic table, and upward through a group. The nonmetals, which have high electronegativity values, have a greater attraction for electrons than do the metals.

We use the electronegativity values to predict the type of bonding between atoms. The greatest difference in electronegativity values is 3.3 between fluorine (4.0) and cesium (0.7). A large difference in values indicates that one atom attracts the electrons and the other atom gives them up to form an ionic bond. Using a midpoint of about 1.7, we say that a difference of 1.7 or greater results in an ionic bond.

Electronegativity increases →

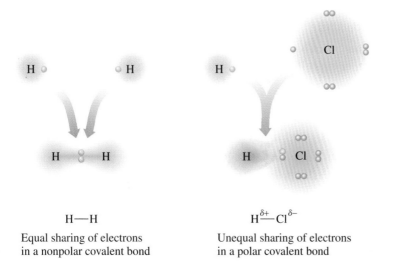

	1A	2A												3A	4A	5A	6A	7A	8A
													H 2.1						
	Li 1.0	Be 1.5												B 2.0	C 2.5	N 3.0	O 3.5	F 4.0	
	Na 0.9	Mg 1.2												Al 1.5	Si 1.8	P 2.1	S 2.5	Cl 3.0	
	K 0.8	Ca 1.0												Ga 1.6	Ge 1.8	As 2.0	Se 2.4	Br 2.8	
	Rb 0.8	Sr 1.0												In 1.7	Sn 1.8	Sb 1.9	Te 2.1	I 2.5	
	Cs 0.7	Ba 0.9												Tl 1.8	Pb 1.9	Bi 1.9	Po 2.0	At 2.1	

Electronegativity decreases ↓

Figure 4.4 Electronegativity values of representative elements indicate the ability of atoms to attract electrons.

Q What element on the periodic chart has the strongest attraction for electrons?

H—H

Equal sharing of electrons
in a nonpolar covalent bond

$H^{\delta+}$—$Cl^{\delta-}$

Unequal sharing of electrons
in a polar covalent bond

Figure 4.5 In the nonpolar covalent bond of H_2, electrons are shared equally. In the polar covalent bond of HCl, electrons are shared unequally.

Q H_2 has a nonpolar covalent bond, but HCl has a polar covalent bond. Explain.

When the difference is less than 1.7, the atoms both attract electrons and form covalent bonds. When two atoms have the same or similar electronegativity values, they essentially share electrons equally. The **polarity** of a bond is related to the difference in electronegativity. Bonds with electronegativity differences of less than 0.4 are defined as **nonpolar covalent bonds.** When the electronegativity difference lies between 0.4 and 1.7, electrons are pulled closer to the atom with the greater electronegativity value. The electrons are shared unequally, which is called a **polar covalent bond.** For example, in HCl the shared electron pair is pulled closer to the more electronegative chlorine atom, as shown in Figure 4.5. Table 4.13 gives some examples of how electronegativity differences are used to predict bond types.

In a polar covalent bond, the more electronegative atom may be shown with a partial negative charge, indicated by the lowercase Greek letter delta with a nega-

WEB TUTORIAL
Polar Attractions

Table 4.13   Predicting Bond Type from Electronegativity Differences

Molecule		Type of Electron Sharing	Electronegativity Difference[a]	Bond Type
H_2	H—H	Shared equally	$2.1 - 2.1 = 0$	Nonpolar covalent
Cl_2	Cl—Cl	Shared equally	$3.0 - 3.0 = 0$	Nonpolar covalent
HI	$\overset{\delta+ \quad \delta-}{\text{H—I}}$	Shared unequally	$2.5 - 2.1 = 0.4$	Polar covalent
HCl	$\overset{\delta+ \quad \delta-}{\text{H—Cl}}$	Shared unequally	$3.0 - 2.1 = 0.9$	Polar covalent
NaCl	$Na^+ Cl^-$	Electron transfer	$3.0 - 0.9 = 2.1$	Ionic
MgO	$Mg^{2+} O^{2-}$	Electron transfer	$3.5 - 1.2 = 2.3$	Ionic

[a]Values are taken from Figure 4.4.

tive sign, δ^-. The atom with the lower electronegativity is farther from the electron pair and is shown with a partial positive charge, indicated by the Greek letter delta with a positive sign, δ^+.

SAMPLE PROBLEM 4.17

Identifying Bond Type

Indicate whether bonds between the following atoms would be nonpolar covalent, polar covalent, or ionic. If polar covalent, indicate the partial charges with δ^+ and δ^- notation.

a. carbon and chlorine   **b.** fluorine and fluorine

Solution

a. The bond between the two nonmentals C and Cl has an electronegativity difference of 0.5. The bond is polar covalent. Because the chlorine atom has the greater electronegativity (3.0), the Cl is designated as partially negative, δ^-, in the polar bond.

$$\overset{\delta+ \quad \delta-}{\text{C—Cl}}$$

b. The bond F—F occurs between identical atoms. It is a nonpolar covalent bond, as indicated by an electronegativity difference of zero (0).

Study Check

Determine if the bonds between the following would be nonpolar covalent, polar covalent, or ionic. If polar covalent, show the partial charges with δ^+ and δ^-.

a. hydrogen and hydrogen   **b.** hydrogen and oxygen

A Review of Bonding

As we have seen, the types of bonds range from the nonpolar covalent bonds, in which electrons are shared equally, all the way to the ionic bond, in which electrons are completely transferred from one atom to another. This variation in bonding is continuous; there is no definite point where one type of bond stops and the next starts. However, for purposes of discussion, we can state some general rules (sum-

Table 4.14   Electronegativity Difference and Types of Bonds

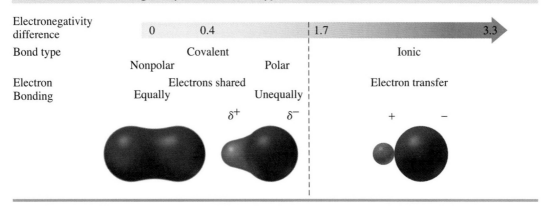

marized in Table 4.14) that may be used to predict the type of bond formed between two elements.

Rules for Predicting Bond Type

1. A bond between two atoms of the same element or between atoms having the same electronegativity value is a nonpolar covalent bond.
2. A covalent bond between atoms of two different nonmetals having different electronegativities is a polar covalent bond.
3. A bond between a metal and a nonmetal is usually ionic.

SAMPLE PROBLEM 4.18

Predicting Bond Polarity

Predict the type of bond between each of the following pairs of elements:

a. Ca and Cl   **b.** P and S   **c.** Br and Br

Solution

a. A bond between calcium, a metal, and chlorine, a nonmetal, would be ionic.
b. A bond between phosphorus and sulfur, two nonmetals with different eletronegativity values, is polar covalent.
c. A bond between atoms of the same element would be nonpolar covalent.

Study Check

Predict the type of bond for the following:

a. S—O   **b.** Na—O   **c.** O—O

QUESTIONS AND PROBLEMS

Bond Polarity

4.59 Identify the bonding between the following pairs of elements as ionic, polar covalent, or nonpolar covalent. If no bond forms, write *none*.
a. Cl and Cl   **b.** Ne and O   **c.** F and F
d. Mg and F   **e.** S and Cl

4.60 Identify the bonding between the following pairs of elements as ionic, polar covalent, or nonpolar covalent. If no bond forms, write *none*.
  a. Na and Cl   **b.** O and F   **c.** K and S
  d. H and H   **e.** N and H

4.61 Predict the type of bonding (ionic, polar covalent, or nonpolar covalent) in each of the following:
  a. SO_2 **b.** Na_2O **c.** PCl_3 **d.** Cl_2 **e.** NH_3

4.62 Predict the type of bonding (ionic, polar covalent, or nonpolar covalent) in each of the following:
  a. $CaCl_2$ **b.** H_2 **c.** OF_2 **d.** N_2O **e.** CO

4.63 Identify the atom that has the higher electronegativity in each of the following pairs:
  a. Na and F   **b.** N and F   **c.** S and Cl
  d. K and Br   **e.** Cl and Br

4.64 Identify the atom that has the higher electronegativity in each of the following pairs:
  a. Li and Na   **b.** Si and Cl   **c.** S and O
  d. I and F   **e.** Ca and Br

4.65 Place the symbols for partially positive (δ^+) and partially negative (δ^-) above the appropriate atoms in the following polar covalent bonds:
  a. H—F   **b.** C—Cl   **c.** N—O   **d.** N—F

4.66 Place the symbols for partially positive (δ^+) and partially negative (δ^-) above the appropriate atoms in the following polar covalent bonds:
  a. O—H   **b.** O—S   **c.** P—Cl   **d.** S—Cl

LEARNING GOAL

Write a formula of a compound containing a polyatomic ion.

4.8 Polyatomic Ions

In a **polyatomic ion,** a group of atoms has acquired an electrical charge. Most polyatomic ions consist of a nonmetal such as phosphorus, sulfur, carbon, or nitrogen covalently bonded to oxygen atoms. These oxygen-containing polyatomic ions have an ionic charge of $1-$, $2-$, or $3-$ because 1, 2, or 3 electrons were added to the atoms in the group to complete their octets. Only one of the common polyatomic ions, NH_4^+, is positively charged. Some models of polyatomic ions are shown in Figure 4.6.

Naming Polyatomic Ions

The names of the most common polyatomic ions end in *ate*. The *ite* ending is used for the names of related ions that have one less oxygen atom. Recognizing these endings will help you identify polyatomic ions in the name of a compound. The hydroxide ion (OH^-) and cyanide ion (CN^-) are exceptions to this naming pattern. There is no easy way to learn polyatomic ions. You will need to memorize the number of oxygen atoms and the charge associated with each ion, as shown in Table 4.15. By memorizing the formulas and the names of the ions shown in the boxes, you can derive the related ions. For example, the sulfate ion is SO_4^{2-}. We write the formula of the sulfite ion, which has one less oxygen atom, as SO_3^{2-}. The formula of hydrogen carbonate can be written by placing a hydrogen cation (H^+) in front of the formula for carbonate (CO_3^{2-}) and decreasing the charge from $2-$ to $1-$ to give HCO_3^-.

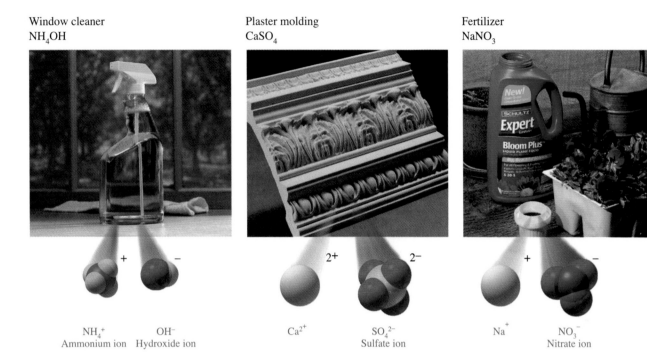

Window cleaner
NH_4OH

Plaster molding
$CaSO_4$

Fertilizer
$NaNO_3$

NH_4^+ OH^-
Ammonium ion   Hydroxide ion

Ca^{2+} SO_4^{2-}
Sulfate ion

Na^+ NO_3^-
Nitrate ion

Table 4.15   Names and Formulas of Some Common Polyatomic Ions

Nonmetal	Formula of Ion[a]	Name of Ion
Hydrogen	OH^-	Hydroxide
Nitrogen	NH_4^+	Ammonium
	$\boxed{NO_3^-}$	**Nitrate**
	NO_2^-	Nitrite
Chlorine	$\boxed{ClO_3^-}$	**Chlorate**
	ClO_2^-	Chlorite
Carbon	$\boxed{CO_3^{2-}}$	**Carbonate**
	HCO_3^-	Hydrogen carbonate (or bicarbonate)
	CN^-	Cyanide
Sulfur	$\boxed{SO_4^{2-}}$	**Sulfate**
	HSO_4^-	Hydrogen sulfate (or bisulfate)
	SO_3^{2-}	Sulfite
	HSO_3^-	Hydrogen sulfite (or bisulfite)
Phosphorus	$\boxed{PO_4^{3-}}$	**Phosphate**
	HPO_4^{2-}	Hydrogen phosphate
	$H_2PO_4^-$	Dihydrogen phosphate
	PO_3^{3-}	Phosphite

[a]Boxed formulas are the most common polyatomic ion for that element.

Figure 4.6 Many products contain poly-atomic ions, which are groups of atoms that carry an ionic charge.

**Q Why does the sulfate ion have a 2−
charge?**

Writing Formulas for Compounds Containing Polyatomic Ions

No polyatomic ion exists by itself. Like any ion, a polyatomic ion must be associated with ions of opposite charge. The bonding between polyatomic ions and other ions is one of electrical attraction. For example, the compound sodium sulfate consists of sodium ions (Na^+) and sulfate ions (SO_4^{2-}) held together by ionic bonds.

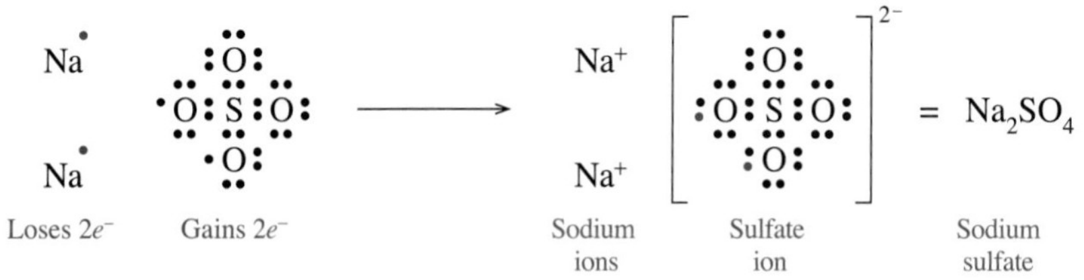

To write correct formulas for compounds containing polyatomic ions we follow the same rules of charge balance that we used for writing the formulas of simple ionic compounds. The total negative and positive charges must equal zero. For example, consider the formula for a compound containing calcium ions and carbonate ions. The ions are written as

$$Ca^{2+} \qquad CO_3{}^{2-}$$

Calcium ion   Carbonate ion

Ionic charge: $(2+) + (2-) = 0$

Because one ion of each balances the charge, the formula is written as

$CaCO_3$
Calcium carbonate

When more than one polyatomic ion is needed for charge balance, parentheses are used to enclose the formula of the ion. A subscript is written outside the closing parenthesis to indicate the number of polyatomic ions. Consider the formula for magnesium nitrate. The ions in this compound are the magnesium ion and the nitrate ion, a polyatomic ion.

$$Mg^{2+} \qquad NO_3{}^-$$

Magnesium ion   Nitrate ion

To balance the positive charge of 2+, two nitrate ions are needed. The formula, including the parentheses around the nitrate ion, is as follows:

$$NO_3{}^-$$
$$Mg^{2+}$$
$$NO_3{}^-$$
$$(2+) + 2(1-) = 0$$

Magnesium nitrate

$$Mg(NO_3)_2$$

Parentheses enclose the formula of the nitrate ions

Subscript outside the parentheses indicates the use of two nitrate ions

SAMPLE PROBLEM 4.19

Writing Formulas Having Polyatomic Ions

Write the formula for a compound containing aluminum ions and bicarbonate ions.

Solution

The positive ion in the compound is the aluminum ion, Al^{3+}, and the negative ion is the bicarbonate ion, $HCO_3{}^-$, which is a polyatomic ion.

$$Al^{3+} \qquad HCO_3{}^-$$

Aluminum ion   Bicarbonate ion

Three $HCO_3{}^-$ ions are required to balance the Al^{3+} charge.

Polyatomic Ions in Bone and Teeth

Bone structure consists of two parts: a solid mineral material and a second phase made up primarily of collagen proteins. The mineral substance is a compound called hydroxyapatite, a solid formed from calcium ions, phosphate ions, and hydroxide ions. This material is deposited in the web of collagen to form a very durable bone material.

$Ca_{10}(PO_4)_6(OH)_2$
Calcium hydroxyapatite

In most individuals, bone material is continuously being absorbed and re-formed. After age 40, more bone material may be lost than formed, a condition called osteoporosis. Bone mass reduction occurs at a faster rate in women than in men and at different rates in different parts of the skeleton. The reduction in bone mass can be as much as 50% over a period of 30 to 40

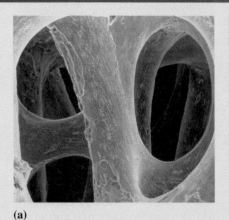

(a)

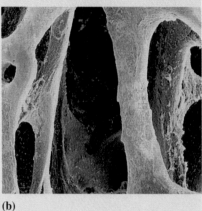

(b)

years. Scanning electron micrographs (SEMs) show **(a)** normal bone and **(b)** bone in osteoporosis due to calcium loss. It is recommended that persons over 35, especially women, include a daily calcium supplement in their diet.

Al^{3+} HCO_3^-
           HCO_3^-
           HCO_3^-
$(3+) + 3 \times (1-) = 0$

The formula for the compound is written by enclosing the formula of the bicarbonate ion, HCO_3^-, in parentheses and writing the subscript 3 just outside the last parenthesis:

$Al(HCO_3)_3$

Study Check

Write the formula for a compound containing ammonium ions and phosphate ions.

Naming Compounds Containing Polyatomic Ions

When naming ionic compounds containing polyatomic ions, we write the positive ion, usually a metal, first, and then we write the name of the polyatomic ion. It is important that you learn to recognize the polyatomic ion in the formula and name it correctly. As with other ionic compounds, no prefixes are used.

Na_2SO_4 $FePO_4$ $Al_2(CO_3)_3$

$Na_2\boxed{SO_4}$ $Fe\boxed{PO_4}$ $Al_2(\boxed{CO_3})_3$

Sodium sulfate   Iron(III) phosphate   Aluminum carbonate

Table 4.16 lists the formulas and names of some ionic compounds that include polyatomic ions and also gives their uses in medicine and industry.

SAMPLE PROBLEMS 4.20

Naming Compounds Containing Polyatomic Ions

Name the following ionic compounds:

a. $CaSO_4$ **b.** $Cu(NO_2)_2$

Names of Compounds Listed on Products

Calcium carbonate is the active ingredient in some antacids. One of the inert ingredients is silicon dioxide. Are these compounds ionic or covalent compounds? What are their formulas? Look at the labels of other food and/or health products. See if you can recognize some of the names. What are some of the ionic compounds? What compounds have covalent bonds? Try to write some of their formulas. You may use a reference book such as the *Handbook of Chemistry and Physics* or the *Merck Index* to find or check some of the formulas.

Table 4.16   Some Compounds That Contain Polyatomic Ions

Formula	Name	Use
$BaSO_4$	Barium sulfate	Radiopaque medium
$CaCO_3$	Calcium carbonate	Antacid, calcium supplement
$Ca_3(PO_4)_2$	Calcium phosphate	Calcium replenisher
$CaSO_3$	Calcium sulfite	Preservative in cider and fruit juices
$CaSO_4$	Calcium sulfate	Plaster casts
$AgNO_3$	Silver nitrate	Topical anti-infective
$NaHCO_3$	Sodium bicarbonate	Antacid
$Zn_3(PO_4)_2$	Zinc phosphate	Dental cements
$FePO_4$	Iron(III) phosphate	Food and bread enrichment
K_2CO_3	Potassium carbonate	Alkalizer, diuretic
$Al_2(SO_4)_3$	Aluminum sulfate	Antiperspirant, anti-infective
$AlPO_4$	Aluminum phosphate	Antacid
$MgSO_4$	Magnesium sulfate	Cathartic, Epsom salts

Solution

Compound Formula	Ions Present	Name of Compound
a. $CaSO_4$	Ca^{2+}, calcium SO_4^{2-}, sulfate	Calcium sulfate
b. $Cu(NO_2)_2$	Cu^{2+}, copper(II) NO_2^-, nitrite <small>One less oxygen than nitrate, NO_3^-</small>	Copper(II) nitrite

Study Check

What is the name of $Ca_3(PO_4)_2$?

Summary of Naming Compounds

Throughout this chapter we have examined strategies for naming ionic and covalent compounds. Now we can summarize the rules, as illustrated in Figure 4.7. In general, compounds having two elements are named by stating the first element, followed by the second element with an *ide* ending. If the first element is a metal, the compound is ionic; if the first element is a nonmetal, the compound is covalent. For ionic compounds, it is necessary to determine whether the metal can form more than one type of positive ion; if so, a Roman numeral following the name of the metal indicates the particular ionic charge. In naming covalent compounds having two elements, prefixes are necessary to indicate the number of atoms of each nonmetal as shown in that particular formula. Ionic compounds having three or more elements include some type of polyatomic ion. They are named by ionic rules, but have an *ate* or *ite* ending when the polyatomic ion has a negative charge. Table 4.17 summarizes some naming rules for elements, ionic compounds, and covalent compounds.

SAMPLE PROBLEM 4.21

Naming Ionic and Covalent Compounds

Name the following compounds:

a. Na_3P **b.** $CuSO_4$ **c.** SO_3

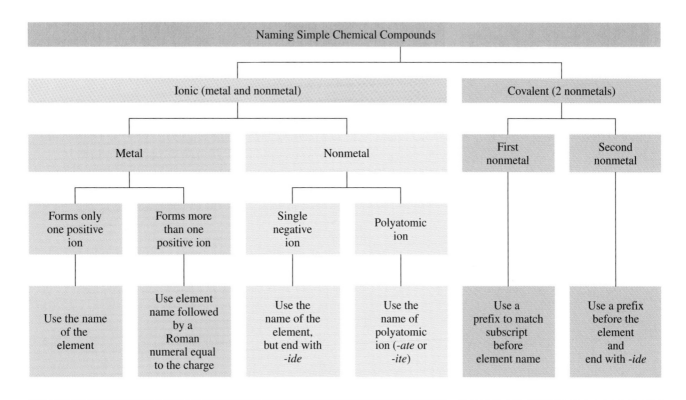

Figure 4.7 A flowchart of how ionic and covalent compounds are named.

Q Why are the names of some metal ions followed by a Roman numeral in the name of a compound?

Table 4.17   Rules for Naming Elements, Ionic Compounds, and Covalent Compounds

Type	Formula Feature	Naming Procedure
Element	Symbol of element; may be diatomic. *Examples:* Ca Cl$_2$	Use name of element. *Examples:* Calcium Chlorine
Ionic compound (two elements)	Symbol of metal followed by symbol of nonmetal; subscripts used for charge balance. *Examples:* Na$_2$O Fe$_2$S$_3$	Use element name for metal; Roman numeral required if two or more positive ions possible. For nonmetal use element name with *ide* ending. *Examples:* Sodium oxide Iron(III) sulfide
Ionic compound (more than two elements)	Usually symbol of metal followed by a polyatomic ion composed of nonmetals; parentheses may enclose polyatomic ion for charge balance. *Examples:* Mg(NO$_3$)$_2$ CuSO$_4$ (NH$_4$)$_2$CO$_3$	Use element name for metal, with Roman numeral if needed, followed by name of polyatomic ion. *Examples:* Magnesium nitrate Copper(II) sulfate Ammonium carbonate
Covalent compound (two elements)	Symbols of two nonmetals; subscripts show number of atoms in a molecule. *Examples:* N$_2$O$_3$ CCl$_4$	Place prefixes before each element name if there are two or more atoms in the formula; add *ide* ending to second element. *Examples:* Dinitrogen trioxide Carbon tetrachloride

Solution

a. Na_3P is an ionic compound. Na is a metal that forms a single ion, Na^+, which is named sodium. P^{3-}, a single negative ion, is named phosphide. The compound Na_3P is sodium phosphide.

b. $CuSO_4$ is an ionic compound. Cu is a transition metal that forms more than one ion. Cu^{2+} is the positive ion that balances the charge on SO_4^{2-}. Roman numeral II specifies the charge in the name copper(II). SO_4^{2-} is a polyatomic ion named sulfate. The compound $CuSO_4$ is named copper(II) sulfate.

c. SO_3 is covalent because it consists of two nonmetals. The first element S is named sulfur (no prefix is needed). The subscript three calls for the prefix *tri* and the element oxygen becomes oxide. The compound SO_3 is named sulfur trioxide.

Study Check

What is the name of $Fe(NO_3)_2$?

QUESTIONS AND PROBLEMS

Polyatomic Ions

4.67 Write the formulas including the charge for the following polyatomic ions:
   a. bicarbonate   b. ammonium
   c. phosphate   d. hydrogen sulfate

4.68 Write the formulas including the charge for the following polyatomic ions:
   a. nitrite   b. sulfite   c. hydroxide   d. phosphite

4.69 Name the following polyatomic ions:
   a. SO_4^{2-} b. CO_3^{2-} c. PO_4^{3-} d. NO_3^-

4.70 Name the following polyatomic ions:
   a. OH^- b. HSO_3^- c. CN^- d. NO_2^-

4.71 Complete the following table with the formula of the compound:

	OH^-	NO_2^-	CO_3^{2-}	HSO_4^-	PO_4^{3-}
Li^+					
Cu^{2+}					
Ba^{2+}					

4.72 Complete the following table with the formula of the compound:

	OH^-	NO_3^-	HCO_3^-	SO_3^{2-}	PO_4^{3-}
NH_4^+					
Al^{3+}					
Pb^{4+}					

4.73 Write the formula for the polyatomic ion in each of the following and name each compound:
   a. Na_2CO_3 b. NH_4Cl c. Li_3PO_4
   d. $Cu(NO_2)_2$ e. $FeSO_3$

4.74 Write the formula for the polyatomic ion in each of the following and name each compound:
   a. KOH b. $NaNO_3$ c. $CuCO_3$ d. $NaHCO_3$ e. $BaSO_4$

4.75 Write the correct formula for the following compounds:
    a. barium hydroxide   **b.** sodium sulfate   **c.** iron(II) nitrate
    d. zinc phosphate   **e.** iron(III) carbonate

4.76 Write the correct formula for the following compounds:
    a. aluminum chlorate   **b.** ammonium oxide
    c. magnesium bicarbonate   **d.** sodium nitrite
    e. copper(I) sulfate

4.77 Name the compounds that are found in the following sources:
    a. $Al_2(SO_4)_3$ antiperspirant
    b. $CaCO_3$ antacid
    c. N_2O "laughing gas," inhaled anesthetic
    d. Na_3PO_4 cathartic
    e. $(NH_4)_2SO_4$ fertilizer
    f. Fe_2O_3 pigment

4.78 Name the compounds that are found in the following sources:
    a. N_2 Earth's atmosphere
    b. $Mg_3(PO_4)_2$ antacid
    c. $FeSO_4$ iron supplement in vitamins
    d. $MgSO_4$ Epsom salts
    e. Cu_2O fungicide
    f. SnF_2 dental caries prophylactic

4.9 Shapes of Molecules

Now that we have counted valence electrons and written electron-dot structures, we can look at the three-dimensional shapes of molecules. The shape is important in our understanding of how molecules interact with enzymes or certain antibiotics or produce our sense of taste and smell.

To determine shape, we use the **valence-shell electron-pair repulsion (VSEPR) theory,** which looks at the geometry of the electron groups around a central atom. In the VSEPR theory, we arrange the electron groups as far apart as possible to reduce the repulsion between the electrons. To determine the shape of a molecule, we first write its electron-dot structure. Then we determine its shape from the number of atoms and the number of nonbonding (lone) pairs of electrons around the central atom. A bonded atom or a lone pair is an electron group.

Shapes of Molecules with No Lone Pairs Around the Central Atom

In $BeCl_2$, there are two chlorine atoms bonded to the central beryllium atom. According to the VSEPR theory, these electron groups are opposite each other at an angle of 180° to give a **linear** shape.

Linear
180°

$$:\overset{..}{\underset{..}{Cl}}:Be:\overset{..}{\underset{..}{Cl}}: \qquad :\overset{..}{\underset{..}{Cl}}{-}Be{-}\overset{..}{\underset{..}{Cl}}:$$

Another example of a linear molecule is CO_2, where each bonded atom counts as one electron group.

$$:\overset{..}{\underset{..}{O}}::C::\overset{..}{\underset{..}{O}}: \qquad :\overset{..}{\underset{..}{O}}{=}C{=}\overset{..}{\underset{..}{O}}:$$

LEARNING GOAL

Predict the three-dimensional structure of a molecule.

WEB TUTORIAL
The Shape of Molecules

In BF$_3$, three fluorine atoms are bonded to the central atom. These electron groups are farthest apart when they are arranged at angles of 120°. This shape is called **trigonal planar**.

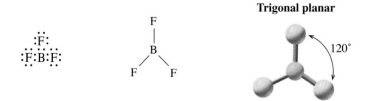

Trigonal planar

In a molecule of CH$_4$, four hydrogen atoms are bonded to the central carbon atom. The four bonds have a maximum separation when they are at an angle of about 109°, which forms a **tetrahedral** shape.

Tetrahedral

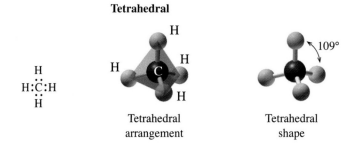

Tetrahedral
arrangement

Tetrahedral
shape

Shapes of Molecules with Lone Pairs Around the Central Atom

In many molecules, there are one or more lone (unshared) pairs of electrons around the central atom. We predict the shape of the molecule from the number of bonded atoms after we determine the geometry of all the electron groups. For example, in NH$_3$, the four electron pairs around the central nitrogen atom have a tetrahedral arrangement. However, the shape of the molecule is based on the three hydrogen atoms, which is called a **pyramidal** shape.

Pyramidal

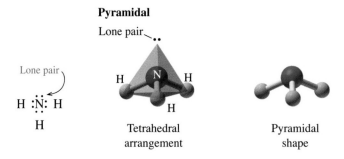

Tetrahedral
arrangement

Pyramidal
shape

In H$_2$O, the four electron groups have the geometry of a tetrahedron at angles of 109°. However, the shape determined by the two hydrogen atoms is called **bent**.

Bent

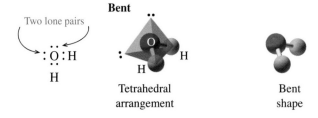

Tetrahedral
arrangement

Bent
shape

Rules for Using VSEPR Theory to Predict Shape

1. Write the electron dot structure.
2. Count all the electron groups around the central atom.
3. Use VSEPR to determine the arrangement of the electron groups.
4. Count the number of atoms bonded to the central atom.
5. Use the bonded atoms to identify the shape of the molecule. If all the electron groups are atoms, then the shape is the same as the electron group arrangement. Molecules with one or more lone pairs use special names. Some examples of shapes of molecules are shown in Table 4.18.

Table 4.18   Examples of Shapes of Molecules

Molecule	Electron-Dot Structure	Bonded Atoms	Molecular Shape (angle)	
Two (2) electron groups around the central atom				
$BeCl_2$	:C̈l:Be:C̈l:	2	linear (180°)	
CO_2	:Ö::C::Ö:	2	linear (180°)	
Three (3) electron groups				
BF_3	:F̈:B:F̈: with :F̈: above	3	trigonal planar (120°)	
SO_2	:Ö:S with :Ö: below	2	bent (120°)	
Four (4) electron groups around the central atom				
CH_4	H:C̈:H with H above and H below	4	tetrahedral (109°)	
NH_3	H:N̈:H with H below	3	pyramidal (109°)	
H_2O	:Ö:H with H below	2	bent (109°)	

Shapes of Molecules

Use VSEPR theory to predict the shape of the following molecules or ions:

a. PH_3 **b.** $SiCl_4$ **c.** NO_3^-

Solution

a. The electron-dot structure for PH_3 indicates four electron groups around the central P atom with three bonded atoms and one nonbonding pair.

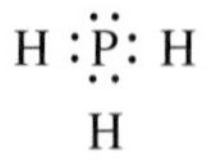

$$H : \ddot{P} : H$$
$$H$$

The molecule would have a pyramidal shape.

b. The electron-dot structure for $SiCl_4$ has four bonded atoms.

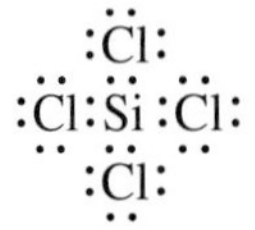

$$:\ddot{Cl}:$$
$$:\ddot{Cl}:Si:\ddot{Cl}:$$
$$:\ddot{Cl}:$$

With four bonded atoms and no lone pairs, the $SiCl_4$ molecule would have a tetrahedral shape.

c. In the polyatomic ion, NO_3^-, the total number of valence electrons is 24 (N = 5, 3 O = 18, and a negative charge = 1). Note that the -1 charge adds one more electron to the total valence electrons.

$$:\ddot{O}:N::\ddot{O}:^-$$
$$:\ddot{O}:$$

There are two single bonds and a double bond for a total of three electron groups. With no unshared electron pairs on N, the shape would be trigonal planar.

Study Check

Predict the shape of the polyatomic ion NH_4^+.

Shapes of Molecules

4.79 What is the shape of a molecule with each of the following:
    a. Two bonded atoms and no lone pairs
    b. Three bonded atoms and one lone pair

4.80 What is the shape of a molecule with each of the following:
    a. Four bonded atoms
    b. Two bonded atoms and two lone pairs

4.81 In the molecule PCl_3, the four electron groups around the phosphorus atom are arranged in a tetrahedral geometry. However, the shape of the molecule is called pyramidal. Why does the shape of the molecule have a different name from the name of the electron group geometry?

4.82 In the molecule H_2S, the four electron groups around the sulfur atom are arranged in a tetrahedral geometry. However, the shape of the molecule is called bent. Why does the shape of the molecule have a different name from the name of the electron group geometry?

4.83 Compare the electron-dot structures of BH_3 and NH_3. Why do these molecules have different shapes?

4.84 Compare the electron-dot structures CH_4 and H_2O. Why do these molecules have the same angles but different names for their shapes?

4.85 Use the VSEPR theory to predict the shape of each molecule:

   a. Br_2 **b.** OF_2

   c. HCN   **d.** CCl_4

HEALTH NOTE

The Chemical Sense of Taste and Smell

In the body, the way molecules such as sugar, alcohol, or antibiotics act is strongly influenced by shape and polarity. Slight differences in the structure and shape of molecules can cause your senses to detect very different tastes. Along the tongue there are taste buds that contain gustatory (taste) receptors. These receptors known as chemoreceptors interact with molecules of certain sizes and shapes. When we eat, the chemicals in food dissolve in the saliva of the mouth and when the solutions of chemicals come in contact with the receptors in the taste buds, molecules bind with taste receptors that fit their shape. Although there are many kinds of chemicals in our foods, we can sense only four primary tastes: sour, salty, bitter, and sweet. The tip of the tongue is most sensitive to sweet and salty chemicals, while the middle of the tongue responds to sour substances and the back of the tongue responds to bitter tastes. For example, sugar molecules bind to receptors in the front part of the tongue, which stimulates the release of a neurotransmitter. A signal moves along the cranial nerves to the cerebral cortex of the brain, which interprets it as a sweet taste.

The sense of smell (olfaction) is similar. Molecules in the gaseous phase dissolve in the mucous lining along the upper part of the nasal cavity. There the molecules interact with the olfactory receptor neurons that produce a sense of smell. Some scientists propose that molecules bind with seven differently shaped receptors to give seven primary odors: camphor, musk, floral, peppermint, ether, pungent, and putrid.

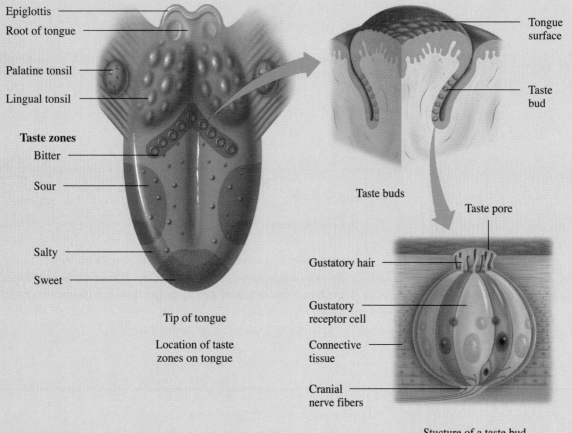

Epiglottis
Root of tongue
Palatine tonsil
Lingual tonsil
Taste zones
Bitter
Sour
Salty
Sweet
Tip of tongue

Location of taste zones on tongue

Tongue surface
Taste bud

Taste buds

Taste pore
Gustatory hair
Gustatory receptor cell
Connective tissue
Cranial nerve fibers

Stucture of a taste bud

4.86 Use the VSEPR theory to predict the shape of each molecule:
    a. CF_4 **b.** NCl_3
    c. SCl_2 **d.** CS_2

4.87 Write the electron-dot structure and predict the shape for the polyatomic ions:
    a. CO_3^{2-} **b.** SO_4^{2-}

4.88 Write the electron-dot structure and predict the shape for the polyatomic ions:
    a. NO_2^- **b.** PO_4^{3-}

LEARNING GOAL

Classify a molecule as polar or nonpolar.

WEB TUTORIAL
The Shape of Molecules

4.10 Polar and Nonpolar Molecules

We have seen that covalent bonds in molecules can be polar or nonpolar. Molecules can also be polar or nonpolar depending on their shape. Diatomic molecules such as H_2 or Cl_2 are nonpolar because they contain one nonpolar covalent bond.

H—H   Cl—Cl
     Nonpolar

Molecules with two or more polar bonds can also be **nonpolar** if the polar bonds have a symmetrical arrangement in the molecule. In a polar bond, the atom with the greater electronegativity is more negative and the other atom is more positive. This separation of charge called a **dipole** is represented by the symbols δ^+ and δ^- or as an arrow with a plus sign at one end ($\longrightarrow$). When the polar bonds or dipoles in a molecule cancel each other, the molecule is nonpolar. For example, CO_2 and CCl_4 contain polar bonds. However, the symmetry of the polar bonds cancels the dipoles, which makes CO_2 and CCl_4 molecules nonpolar.

Examples of Nonpolar Molecules with Polar Bonds

δ^- δ^+ δ^-
O=C=O

$\longleftrightarrow + \longrightarrow$

Net dipole = 0

The four individual bond moments add up to zero (they cancel)

Net dipole = 0

In a **polar molecule,** one end of the molecule is more negatively charged than another end. Polarity in a molecule occurs when the dipoles do not cancel each other. This depends on the type of atoms, the electron pairs around the central atom, and the shape of the molecule. For example, the HCl molecule is polar because electrons are shared unequally in a polar covalent bond.

H:Cl:

δ^+ δ^-
H $\longleftarrow$ Cl
Positive end   Negative end

In polar molecules with three or more atoms, the shape of the molecule determines whether the dipoles cancel or not. Often there are lone pairs around the central atom. In H_2O, the dipoles do not cancel, which makes the molecule positive at one end and negative at the other end. This gives the molecule a dipole.

:Ö:H
  H

δ^- :Ö: δ^-

δ^+ H   H δ^+ **Net dipole**

Negative end

Positive end

In the molecule NH_3, there are three dipoles, but they do not cancel.

Net dipole

SAMPLE PROBLEM 4.23

Polarity of Molecules

Determine whether each of the following molecules is polar or nonpolar:

a. CBr_4 **b.** OF_2

Solution

a. The electron-dot structure for CBr_4 has four bonded atoms.

:Br:
:Br:C:Br:
:Br:

The molecule would have a tetrahedral shape. With four identical atoms bonded to the central and no lone pairs, CBr_4 would be nonpolar.

b. The electron dot structure for OF_2 shows four electron groups with two bonded atoms.

:Ö:F:
:F:

The molecule would have a bent shape. With two lone pairs, the molecule is polar.

Study Check

Would PCl_3 be a polar or nonpolar molecule?

QUESTIONS AND PROBLEMS

Polar and Nonpolar Molecules

4.89 The molecule Cl_2 is nonpolar, but HCl is polar. Explain.

4.90 The molecules CH_4 and CH_3Cl both contain four bonds. Why is CH_4 nonpolar whereas CH_3Cl is polar?

4.91 Write the symbols δ^+ and δ^- over the atoms that are in polar bonds.
  a. Cl — Cl   **b.** C — Cl   **c.** N — O
  d. H — O   **e.** P — F   **f.** S — H

4.92 Write the symbols δ^+ and δ^- over the atoms that are in polar bonds.
  a. F — Cl   **b.** P — H   **c.** C — O
  d. S — O   **e.** N — F   **f.** O — Cl

4.93 Identify the following molecules as polar or nonpolar:

   a. CS_2 **b.** NF_3

   c. CBr_4 **d.** SO_3

4.94 Identify the following molecules as polar or nonpolar:

   a. H_2S **b.** PBr_3

   c. $SiCl_4$ **d.** SO_2

Chapter Review

4.1 Valence Electrons

The electrons located in the outermost energy level of an atom play a major role in determining its chemical properties. An electron-dot structure represents the valence electrons in an atom.

4.2 Ions and the Octet Rule

The nonreactivity of the noble gases is associated with the stable arrangement of 8 electrons, an octet, in their valence shells; helium needs 2 electrons for stability. Atoms of elements other than the noble gases achieve stability by losing, gaining, or sharing their valence electrons with other atoms in the formation of compounds. Metals of the representative elements form an octet by losing their valence electrons to form positively charged cations: Group 1A, 1+, Group 2A, 2+, and Group 3A, 3+. When they react with metals, nonmetals gain electrons to form octets in their valence shells. As anions, they have negative charges: Group 5A, 3−, Group 6A, 2−, Group 7A, 1−.

4.3 Ionic Compounds

The total positive and negative ionic charge must balance in the formula of an ionic compound. Charge balance in a formula is achieved by using subscripts after each symbol so that the overall charge is zero.

4.4 Naming and Writing Ionic Formulas

In naming ionic compounds, the positive ion is given first, followed by the name of the negative ion. Ionic compounds containing two elements end with *ide*. When the metal can form more than one positive ion, its ionic charge is determined from the total negative charge in the formula. Typically transition metals form cations with two or more ionic charges. The charge is given as a Roman numeral in the name, such as iron(II) and iron(III) for the cations of iron with 2+ and 3+ ionic charges.

4.5 Covalent Bonds

In a covalent bond, electrons are shared by atoms of nonmetals. By sharing, each of the atoms achieves a noble gas arrangement. In a nonpolar covalent bond, electrons are shared equally by atoms. In a polar covalent bond, electrons are unequally shared because they are attracted by the more electronegative atom. In some covalent compounds, double or triple bonds are needed to provide an octet.

4.6 Naming and Writing Formulas of Covalent Compounds

Two nonmetals can form two or more different covalent compounds. In their names, prefixes are used to indicate the subscript in the formula. The ending of the second nonmetal is changed to *ide*.

4.7 Bond Polarity

Electronegativity values indicate the ability of atoms to attract electrons. In general, the electronegativity of a metal is low; a nonmetal is high. In a nonpolar covalent bond, the difference in electronegativities is up to 0.4; in a polar covalent bond, the difference can be up to 1.7. Greater than 1.7, the bond is considered to be ionic.

4.8 Polyatomic Ions

A group of nonmetal atoms that carries an electrical charge, for example, the carbonate ion has the formula CO_3^{2-}. Most polyatomic ions have names that end with *ate* or *ite*.

4.9 Shapes of Molecules

The shape of a molecule is determined from the electron-dot structure and the number of bonded atoms and lone pairs. The electron arrangement of two electron groups around a central atom is linear; three electron groups trigonal planar; and four is tetrahedral. When all the electron groups are bonded to atoms, the shape has the same name as the electron arrangement. A central atom with two bonded atoms and one or two lone pairs has a bent shape. A central atom with three bonded atoms and one lone pair has a pyramidal shape.

4.10 Polar and Nonpolar Molecules

A polar covalent bond is a covalent bond in which one end has a more electronegative atom than at the other end, or a dipole. Nonpolar molecules contain nonpolar covalent bonds or have an arrangement of bonded atoms that causes the dipoles to cancel out. In polar molecules, the dipoles do not cancel because there are nonidentical bonded atoms or lone pairs.

Key Terms

anion A negatively charged ion such as Cl^-, O^{2-}, or SO_4^{2-}.

bent The shape of a molecule with two bonded atoms and one lone pair or two lone pairs.

cation A positively charged ion such as Na^+, Mg^{2+}, or Al^{3+}.

compound A combination of atoms in which noble gas arrangements are attained through electron transfer or electron sharing.

covalent bond A sharing of valence electrons by atoms.

dipole The separation of positive and negative charge in a polar bond indicated by an arrow that is drawn from the more positive atom to the more negative atom.

double bond A sharing of two pairs of electrons by two atoms.

electron-dot structure The representation of an atom that shows each valence electron as a dot above, below, or beside the symbol of the element.

electronegativity value A number that indicates the relative ability of an element to attract electrons.

formula The group of symbols that represent the atoms or ions in a compound.

ion An atom or group of atoms having an electrical charge because of a loss or gain of electrons.

ionic bond The attraction between oppositely charged ions that results from the transfer of electrons.

ionic charge The difference between the number of protons (positive) and the number of electrons (negative), written in the upper right corner of the symbol for the element or polyatomic ion.

ionic compound A compound of positive and negative ions held together by ionic bonds.

ionization energy The amount of energy required to remove an electron from the valence shell of an atom.

linear The shape of a molecule that has two bonded atoms and no lone pair.

molecule The smallest unit of two or more atoms held together by covalent bonds.

nonpolar covalent bond A covalent bond in which the electrons are shared equally.

nonpolar molecule A molecule that has only nonpolar bonds or in which bond dipoles cancel.

octet 8 valence electrons.

octet rule Elements in Groups 1–7A react with other elements by forming ionic or covalent bonds to produce a noble gas arrangement, usually 8 electrons in the outer shell.

polar covalent bond A covalent bond in which the electrons are shared unequally.

polarity A measure of the unequal sharing of electrons, indicated by the difference in electronegativity values.

polar molecule A molecule in which bond dipoles do not cancel.

polyatomic ion A group of covalently bonded nonmetal atoms that has an overall electrical charge.

pyramidal The shape of a molecule that has three bonded atoms and one lone pair.

tetrahedral The shape of a molecule with four bonded atoms.

trigonal planar The shape of a molecule with three bonded atoms and no lone pair.

triple bond A sharing of three pairs of electrons by two atoms.

valence electrons The electrons present in the outermost shell of an atom, which are largely responsible for the chemical behavior of the element.

valence-shell electron-pair repulsion (VSEPR) theory A theory that predicts the shape of a molecule by moving the electron pairs on a central atom as far apart as possible to minimize the repulsion of the negative regions.

Additional Problems

4.95 What are the elements in Period 3 represented by the following?

    a. $\cdot \ddot{X} \cdot$ **b.** $X\cdot$ **c.** $\cdot \dot{X} \cdot$ **d.** $\cdot \dot{\ddot{X}} \cdot$

4.96 Write the electron configuration of the following:

    a. Ar   **b.** Na^+ **c.** S^{2-}

    d. Cl^- **e.** Ca^{2+}

4.97 Write the electron configuration of the following:

    a. N^{3-} **b.** Ne   **c.** F^-

    d. Al^{3+} **e.** Li^+

4.98 What noble gas has the same arrangement as the following ions?

    a. Al^{3+} **b.** Cl^- **c.** Ca^{2+}

    d. Na^+ **e.** N^{3-}

4.99 Consider an ion with the symbol X^{2+} formed from a representative element.

    a. What is the group number of the element?

    b. What is the electron-dot structure of the element?

    c. If X is in Period 2, what is the element?

4.100 Consider the following electron-dot structures for representative elements X and Y:

X·　　·Ÿ·

a. What are the group numbers of X and Y?
b. Will a compound of X and Y be ionic or covalent?
c. What ions would be formed by X and Y?
d. What would be the formula of a compound of X and Y?
e. What would be the formula of a compound of X and chlorine?
f. What would be the formula of a compound of Y and chlorine?

4.101 One of the ions of tin is tin(IV).
a. What is the symbol for this ion?
b. How many protons and electrons are in the ion?
c. What is the formula of tin(IV) oxide?
d. What is the formula of tin(IV) phosphate?

4.102 As shown in the Health Note "Polyatomic Ions in Bone and Teeth," the mineral portion of bone is composed of calcium hydroxyapatite, $Ca_{10}(PO_4)_6(OH)_2$. Name the ions in the formula.

4.103 Classify the following compounds as ionic or covalent.
a. Li_2O　　**b.** N_2O　　**c.** CF_4
d. Cl_2O　　**e.** MgF_2　　**f.** CO
g. $CaCl_2$　　**h.** K_3N

4.104 Hydrogen and helium are both very light gases. Why is helium gas, but not hydrogen gas, used to fill balloons and blimps?

4.105 Name the following compounds:
a. $FeCl_3$　　**b.** Cl_2O_7　　**c.** Br_2
d. $Ca_3(PO_4)_2$　　**e.** PCl_3　　**f.** $Al_2(CO_3)_3$
g. $PbCl_4$　　**h.** $MgCO_3$　　**i.** NO_2
j. $SnSO_4$　　**k.** $Ba(NO_3)_2$　　**l.** CuS

4.106 Write the formulas for the following compounds:
a. tin(II) carbonate _____
b. lithium phosphide _____
c. silicon tetrachloride _____
d. iron(III) sulfide _____
e. carbon dioxide _____
f. calcium bromide _____
g. sodium carbonate _____
h. nitrogen dioxide _____
i. aluminum nitrate _____
j. copper(I) nitride _____
k. potassium phosphate _____
l. lead(IV) sulfide _____

4.107 Predict the shape and polarity of each of the following molecules:
a. A central atom with three identical bonded atoms and no lone pair.
b. A central atom with two bonded atoms and one lone pair.
c. A central atom with two identical atoms and no lone pairs.

4.108 Predict the shape and polarity of each of the following molecules:
a. A central atom with four identical bonded atoms and no lone pairs.
b. A central atom with three identical bonded atoms and one lone pair.
c. A central atom with four bonded atoms that are not identical and no lone pair.

4.109 Predict the shape and polarity of each of the following molecules:
a. H_2S　　**b.** NF_3　　**c.** BF_3

4.110 Predict the shape and polarity of each of the following molecules:
a. F_2　　**b.** CF_4　　**c.** HF

4.111 Predict the shape of each of the following polyatomic ions:
a. NO_3^-　　**b.** PO_4^{3-}　　**c.** SO_4^{2-}

4.112 Predict the shape of each of the following polyatomic ions:
a. CO_3^{2-}　　**b.** SO_3^{2-}　　**c.** NO_2^-

5 Energy and States of Matter

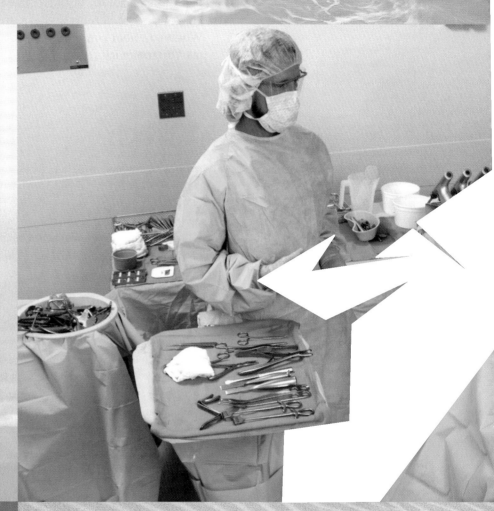

"As a surgical technologist, I assist the doctors during surgeries," says Christopher Ayars, surgical technologist, Kaiser Hospital. "I am there to help during general or orthopedic surgery by passing instruments, holding retractors, and maintaining the sterile field. Our equipment for surgery is sterilized by steam that is heated to 270°F, which is the same as 130°C."

Surgical technologists assist with surgical procedures by preparing and maintaining surgical equipment, instruments, and supplies; providing patient care in an operating room setting; preparing and maintaining a sterile field; and ensuring that there are no breaks in aseptic technique. Instruments, which have been sterilized, are wrapped and sent to surgery where they are checked again before they are opened.

LOOKING AHEAD

the
Chemistry
place

www.chemplace.com/college

Visit the URL above or use the CD-ROM in the book for extra quizzing, interactive tutorials, career resources, and case studies.

Energy makes it possible to do work. You use energy to do the work of running, playing tennis, studying, and breathing. You obtain that energy from the foods you eat. Energy takes many forms, but you are probably most familiar with heat. In your home, you use heat energy to cook food, heat water, warm the room, and dry your hair.

Matter is everything around us that has mass and occupies space. There are three states of matter: solid, liquid, or gas. For example, the water in an ice cube, an ice rink, or an iceberg is solid. Water from a faucet or in a pool is a liquid. Water becomes a gas when it evaporates from wet clothes or boils in a pan on the stove.

When you heat water in a tea kettle, the water boils and changes to a gas. You make ice cubes by using a freezer to remove heat from the water in an ice cube tray.

LEARNING GOAL

Describe some forms of energy.

5.1 Energy

When you are running, walking, dancing, or thinking, you are using energy to do **work.** In fact, **energy** is defined as the ability to do work. Suppose you are climbing a steep hill. Perhaps you become too tired to go on. We could say that you do not have sufficient energy to do any more work. Now suppose you sit down and have lunch. In a while you will have obtained some energy from the food, and you will be able to do more work and complete the climb.

Potential and Kinetic Energy

All energy can be classified as potential energy or kinetic energy. **Potential energy** is stored energy, whereas **kinetic energy** is the energy of motion. Any object that is moving has kinetic energy. A boulder resting on top of a mountain has potential energy because of its location. If the boulder rolls down the mountain, the potential energy becomes kinetic energy. Water stored in a reservoir has potential energy. When the water goes over the dam, the potential energy is converted to kinetic energy. Even the food you eat has potential energy stored in its chemical bonds. When you digest food, you convert its potential energy to kinetic energy to do biological work.

SAMPLE PROBLEM 5.1

Forms of Energy

Identify the energy in each of the following as potential or kinetic:

a. gasoline   **b.** skating   **c.** a candy bar

Solution

a. potential energy (stored)   **b.** kinetic energy (motion)
c. potential energy (stored)

Study Check

Would the energy in a stretched rubber band be kinetic or potential?

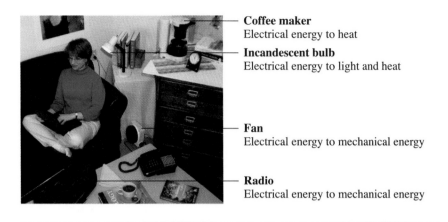

Coffee maker
Electrical energy to heat

Incandescent bulb
Electrical energy to light and heat

Fan
Electrical energy to mechanical energy

Radio
Electrical energy to mechanical energy

Figure 5.1 Examples of electrical energy converted into light, heat, and mechanical energy.

Q **Is heat a form of kinetic or potential energy? Why?**

Forms of Energy

Energy from the sun is stored as chemical energy during the formation of trees, plants, and fuels. When wood is burned, chemical energy is converted to heat and light. **Heat** is associated with the motion of particles in a substance. A frozen pizza feels cold because the particles in the pizza are moving very slowly. As heat is added, the motions of the particles increase, and the pizza becomes warm. Eventually the particles have enough energy to make the pizza hot and ready to eat.

In an electric power plant, energy from the burning of fossil fuels such as natural gas or coal is used to produce steam, which turns the turbines and generates electrical energy. In your home, electrical energy is converted to light energy when you switch on a light bulb, to mechanical energy when you use a mixer or a washing machine, and to heat when you use a hair dryer or a toaster (see Figure 5.1). In your body, chemical energy from the food you eat is converted to mechanical energy when you use muscles to pedal a bicycle, run a marathon, or mow the lawn.

QUESTIONS AND PROBLEMS

Energy

The answers for the odd-numbered Questions and Problems throughout this chapter are given at the end of this text. The complete solutions to odd-numbered Questions and Problems are in the *Study Guide*.

5.1 Discuss the changes in the potential and kinetic energy of a roller-coaster ride as the roller-coaster climbs up a ramp and goes down the other side.

5.2 Discuss the changes in the potential and kinetic energy of a ski jumper taking the elevator to the top of the jump and going down the ramp.

5.3 Indicate whether each statement describes potential or kinetic energy:
   a. water at the top of a waterfall **b.** kicking a ball
   c. the energy in a lump of coal **d.** a skier at the top of a hill

5.4 Indicate whether each statement describes potential or kinetic energy:
   a. the energy in your food **b.** a tightly wound spring
   c. an earthquake **d.** a car speeding down the freeway

5.5 What are the forms of energy involved in the following examples?
   Example: Striking a match converts chemical energy into heat and light.
   a. using a hair dryer **b.** using an electric fan
   c. burning gasoline in a car engine
   d. sunlight falling on a solar water heater

Global Warming

The amount of carbon dioxide (CO_2) gas in our atmosphere is on the increase as we burn more gasoline, coal, and natural gas. The algae in the oceans and the plants and trees in the forests normally absorb carbon dioxide, but they cannot keep up with the continued increase. The cutting of trees in the rain forests (deforestation) reduces the amount of carbon dioxide removed from the atmosphere. Many of the trees are also burned as land is cleared. It has been estimated that deforestation may account for 15–30% of the carbon dioxide that remains in the atmosphere each year.

The carbon dioxide in the atmosphere acts like the glass in a greenhouse. When sunlight warms the earth's surface, some of the heat is trapped by carbon dioxide. As CO_2 levels rise, more heat is trapped. It is not yet clear how severe the effects of global warming might be. Some scientists estimate that by around the year 2030, the atmospheric level of carbon dioxide could double and cause the temperature of Earth's atmosphere to rise by 2–5°C. If that should happen, it would have a profound impact on Earth's climate. For example, an increase in the melting of snow and ice could raise the ocean levels by as much as 2 m, which is enough to flood many cities located on the ocean shorelines.

Worldwide efforts are being made to reduce fossil fuel use and to slow or stop deforestation. It will require cooperation throughout the world to avoid the bleak future that some scientists predict should global warming continue unchecked.

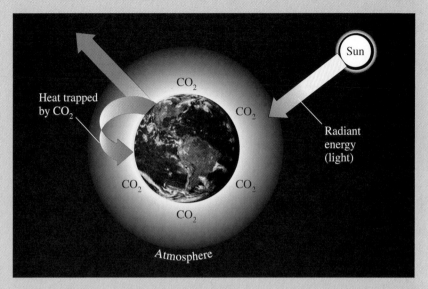

5.6 What are the forms of energy involved in the following examples?
a. turning on a light switch   **b.** cooking food on a gas stove
c. using a microwave oven   **d.** burning a log in the fireplace

Given the mass of a sample, and the temperature change, calculate the heat lost or gained.

5.2 Measuring Heat

In chemistry and the health sciences, heat is commonly measured in calories, from the Latin *caloric* meaning "heat." A **calorie (cal)** is the amount of heat needed to raise the temperature of exactly 1 g of water by exactly 1°C. A **kilocalorie (kcal)** is used for larger amounts of heat. The **joule (J),** pronounced "jewel," is the SI unit of energy. One calorie is the same amount of energy as 4.18 joules. One kilojoule is 1000 joules.

> 1 kcal = 1000 calories
>
> 1 cal = 4.18 J
>
> 1 kJ   = 1000 J

Specific Heat

All substances absorb energy, but some need only a small amount of heat to become hot and others need more. The amount of heat that raises the temperature of 1 g of a substance by 1°C is known as its **specific heat.**

$$\text{Specific heat} = \frac{\text{Amount of heat (cal or J)}}{1 \text{ g} \times 1°\text{C}}$$

On a hot day at the beach, the ocean water is cool to the touch, the air is warm; but the sand is so hot, it can burn your feet. Sand has a lower specific heat than water. Therefore, sand reaches a higher temperature than water. Because water has a high specific heat, one of its main functions in the body is to absorb large amounts of heat without causing fluctuations in body temperature.

When you cook, you may use aluminum or copper pans. If you compare specific heats, you find that the amount of heat that raises 1 g of water 1°C increases the temperature of 1 g of aluminum by about 5°C, and 1 g of copper by 10°C. Materials such as aluminum and copper can heat quickly to cook food. Table 5.1 lists the specific heats of a variety of materials.

Calculations Using Specific Heat

The heat gained or lost by a substance is calculated by multiplying its mass, temperature change, and specific heat. The temperature change often noted as ΔT is the difference in the temperatures of the sample before and after heat is added or removed. (The Greek letter Delta, symbol Δ, means "change in"; read ΔT as "delta T.")

$$\text{Heat gain or loss} = (\text{mass}) \times \left(\begin{smallmatrix}\text{temperature}\\\text{change}(\Delta T)\end{smallmatrix}\right) \times (\text{specific heat})$$

$$\text{Heat (calories)} = \text{g} \times °\text{C} \times \frac{\text{cal}}{\text{g} °\text{C}}$$

$$\text{Heat (joules)} = \text{g} \times °\text{C} \times \frac{\text{J}}{\text{g} °\text{C}}$$

Table 5.1   Specific Heats of Some Substances

Substance	Specific Heat (cal/g °C)	(J/g °C)
Water	1.00	4.18
Ethanol	0.58	2.4
Aluminum	0.22	0.92
Sand	0.19	0.79
Iron	0.11	0.46
Copper	0.093	0.39
Silver	0.057	0.24
Gold	0.031	0.13

SAMPLE PROBLEM 5.2

Calculating Heat

How many calories must be added to warm 45 g of aluminum from 12°C to 176°C?

Solution

The temperature change (ΔT) is the difference between the temperatures.

$\Delta T = 176°\text{C} - 12°\text{C} = 164°\text{C}$

The heat is calculated by multiplying the mass of aluminum, the temperature change, and the specific heat of aluminum (0.22 cal/g °C).

$$45 \text{ g} \times 164°\text{C} \times \frac{0.22 \text{ cal}}{\text{g} °\text{C}} = 1600 \text{ cal}$$

Mass ΔT Specific heat   Heat energy

Study Check

How many calories are needed to heat 30.0 g of copper from 75°C to 125°C if the specific heat of copper is 0.093 cal/g °C?

EXPLORE YOUR WORLD

Loss and Gain of Heat

Obtain three matching cups or glasses. Place 2 cups of cold tap water in the first one, and 2 cups of hot tap water in the second one. If you have a candy thermometer, determine the temperature of each. Otherwise, use your finger to quickly test each water sample. Pour about one-half of each water sample into a third cup or glass. Compare the temperature of the mixed water sample with the initial two water samples.

Questions

1. How does the temperature of the mixed water sample compare to the original hot water sample and cold water sample?
2. What happened to some of the heat of the hot water in the mixed sample?
3. Why did the temperature of the cold sample change in the mixed sample?

EXPLORE YOUR WORLD

Specific Heat

Obtain two identical containers, such as bowls or pie tins, and a thermometer. Place water in one container and an equal amount of vegetable oil in the other. Set the containers in the sun. Record the temperature of each sample initially and every 10 minutes after that for a total of 60 minutes.

Questions

1. Which sample had the higher temperature after 60 minutes?
2. Using time and temperature, how would you graph this experiment?
3. Which substance has the smaller specific heat?

SAMPLE PROBLEM 5.3

Calculating Heat Loss in Joules

A 225-g sample of hot tea cools from 95.0°C to 23.0°C. What is the heat loss in joules assuming that tea has the same specific heat as water (4.18 J/g °C)?

Solution

$$\Delta T = 95.0°\text{C} - 23.0°\text{C} = 72.0°\text{C}$$

The number of joules lost during cooling is calculated as follows:

$$\underset{\text{Mass}}{225\,\cancel{\text{g}}} \times \underset{\Delta T}{72.0°\cancel{\text{C}}} \times \underset{\text{Specific heat}}{\frac{4.18\text{ J}}{\cancel{\text{g}}\,°\cancel{\text{C}}}} = \underset{\text{Heat released}}{67{,}700\text{ J}} \quad \text{or} \quad 6.77 \times 10^4\text{ J}$$

Study Check

How much heat in joules is released when 15 g of gold cools from 215°C to 35°C? (See Table 5.1.)

SAMPLE PROBLEM 5.4

Calculating Mass of a Sample

Ethanol has a specific heat of 0.58 cal/g °C. When 650 cal of heat are added to a sample of ethanol, its temperature rises from 18°C to 32°C. What is the mass (g) of the ethanol sample?

Solution

The equation for the calculation of heat is rearranged to solve for mass.

$$\text{Mass (g)} = \frac{\text{Amount of heat}}{\text{Specific heat} \times \Delta T}$$

$$\text{Mass (g)} = \frac{650\text{ cal}}{0.58\text{ cal/g °C} \times 14°\text{C}} = 80.\text{ g}$$

Study Check

When 2.1 kcal are added to a piece of iron, its temperature rises from 15° to 122°C. What is the mass (g) of the iron sample?

QUESTIONS AND PROBLEMS

Measuring Heat Energy

5.7 If the same amount of heat is supplied to samples of 10.0 g each of aluminum, iron, and copper all at 15°C, which sample would reach the highest temperature?

5.8 Substances A and B are the same mass and at the same initial temperature. When they are heated, the final temperature of A is 55°C higher than the temperature of B. What does this tell you about the specific heats of A and B?

5.9 Convert each of the following:
   a. 3500 cal to kcal   **b.** 28 cal to joules
   c. 425 J to cal   **d.** 4.5 kJ to cal

5.10 Convert each of the following:
   a. 8.1 kcal to cal   **b.** 325 J to kJ
   c. 2550 cal to kJ   **d.** 2.50 kcal to J

5.11 What is the amount of heat required in each of the following?
   a. calories to heat 25 g of water from 15°C to 25°C
   b. joules to heat 150 g of water from 0°C to 75°C
   c. joules to heat 10.0 g of copper from 25°C to 275°C
   d. kilojoules to heat 150 g of water in a kettle from 15°C to 77°C
   e. joules to heat 10.0 g of silver from 15°C to 237°C

5.12 What is the amount of heat involved in each of the following?
   a. calories given off when 85 g of water cools from 45°C to 25°C
   b. joules given off when 0.50 kg of water cools from 85°C to 65°C
   c. kilocalories given off when 250 g of sand cools from 155°C to 42°C
   d. joules given off when 25 g of water cools from 86°C to 61°C
   e. kilocalories absorbed when 5.0 kg of water increases from 22°C to 28°C

5.13 Calculate the unknown quantity in each of the following examples:
   a. mass of gold when 450 cal raises the temperature from 15°C to 27°C
   b. mass of water when 2.5 kcal raises the temperature from 8°C to 36°C
   c. specific heat of 25 g of a metal if 320 cal of heat raises its temperature from 15°C to 117°C
   d. specific heat of 36 g of a metal if 1.5 kcal of heat raises its temperature from 4°C to 221°C

5.14 Calculate the unknown quantity in each of the following examples:
   a. mass of silver when 180 cal raises the temperature from 14°C to 26°C
   b. mass of water when 3.2 kcal raises the temperature from 5°C to 45°C
   c. specific heat of 28 g of a metal if 260 cal of heat raises its temperature from 3°C to 215°C
   d. specific heat of 45 g of a metal that requires 2.1 kcal of heat to raise its temperature from 2°C to 186°C

5.3 Energy and Nutrition

When you are watching your food intake, the Calories you are counting are actually kilocalories. In the field of nutrition, it is common to use the **Calorie, Cal** (with an uppercase C) to mean 1000 cal, or 1 kcal. Nutritional values may also be given in kilojoules (kJ).

Nutritional Calories

1 Cal = 1 kcal

1 kcal = 1000 cal

1 kcal = 4.18 kJ

The number of Calories in a food is determined by using an apparatus called a **calorimeter,** shown in Figure 5.2. A sample is placed in a steel container within the calorimeter and water is added to fill a surrounding chamber. When the food burns (combustion), the heat emitted causes an increase in the temperature of the water.

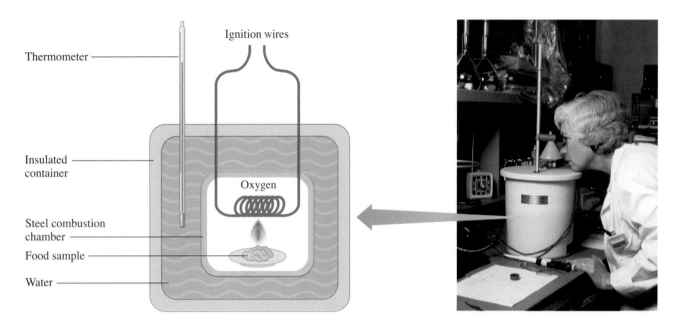

Figure 5.2 The temperature change is observed during the combustion of a food sample in a calorimeter. The energy released by the combustion of a food sample is determined by measuring the mass of the food sample, the mass of water in the calorimeter, and the change in temperature of the water.

Q **How is the increase in water temperature used to determine the number of Calories in a food sample?**

By calculating the kilocalories (or Calories) absorbed by the water, we can determine the caloric content of the food as shown in Sample Problem 5.5.

SAMPLE PROBLEM 5.5

Calculating Caloric Values

A 2.3-g sample of butter, a fat, is placed in a calorimeter containing 1900 g of water at an initial temperature of 17°C. After combustion of the butter, the water has a temperature of 28°C. What is the caloric value (kcal/g) of butter?

Solution

The heat absorbed by the water is calculated as follows:

$$1900 \; \cancel{g} \times 11°\cancel{C} \times \frac{1.00 \; cal}{\cancel{g} \; °\cancel{C}} = 21,000 \; cal$$

Mass $\times$ ΔT $\times$ Specific heat
of water

$$21,000 \; \cancel{cal} \times \frac{1 \; kcal}{1000 \; \cancel{cal}} = 21 \; kcal$$

Because the fat provided the 21 kcal of heat, the caloric value of butter is calculated as follows:

$$\frac{21 \; kcal}{2.3 \; g \; fat} = 9.1 \; kcal/g \; fat$$

Study Check

A 4.5 g sample of glucose, a carbohydrate, is placed in a calorimeter. The water in the container has a mass of 1500 g and an initial temperature of 15°C. After all the glucose is burned, the water temperature is 27°C. What is the caloric value for glucose?

Caloric Food Values

The **caloric values** are the kilocalories per gram of the three types of food: carbohydrates, fats, and proteins. These values are listed in Table 5.2.

Table 5.2 Caloric Values for the Three Food Types

Food Type	Carbohydrate	Fat (lipid)	Protein
Caloric value	$\dfrac{4\ \text{kcal}}{1\ \text{g}}$	$\dfrac{9\ \text{kcal}}{1\ \text{g}}$	$\dfrac{4\ \text{kcal}}{1\ \text{g}}$

If the composition of a food is known in terms of the mass of each food type, the caloric content (the total number of Calories) can be calculated.

$$\text{Kilocalories} \;=\; \underset{\substack{\text{Mass of carbohydrate,}\\ \text{fat, or protein}}}{\cancel{g}} \;\times\; \underset{\text{Caloric value}}{\dfrac{\text{kcal}}{\cancel{g}}}$$

The caloric content of a packaged food is listed in the nutritional information on the package, usually in terms of the number of Calories in a serving. The general composition and caloric content of some foods are given in Table 5.3.

Table 5.3 General Composition and Caloric Content of Some Foods

Food	Protein (g)	Fat (g)	Carbohydrate (g)	Calories (kcal[a])
Banana, 1 medium	1	trace	26	110
Beans, red kidney, 1 cup	15	1	42	240
Beef, lean, 3 oz	22	5	trace	130
Carrots, raw, 1 cup	1	trace	11	50
Chicken, no skin, 3 oz	20	3	0	110
Egg, 1 large	6	6	trace	80
Milk, 4% fat, 1 cup	9	9	12	165
Milk, nonfat, 1 cup	9	trace	12	85
Oil, olive, 1 tbs	0	14	0	130
Potato, baked	3	trace	23	105
Salmon, 3 oz	17	5	0	115
Steak, 3 oz	20	27	0	320
Yogurt, lowfat, 1 cup	8	4	13	120

[a]Values in kcal are rounded to nearest five.

SAMPLE PROBLEM 5.6

Caloric Content for a Food

How many kilocalories are in a piece of chocolate cake that contains 35 g of carbohydrate, 10 g of fat, and 5 g of protein? Report the answer to the nearest tens place.

Solution

Using the caloric values for carbohydrate, fat, and protein (Table 5.2), we can calculate the total number of kcal:

Nutrition Facts

Serving Size 14 crackers (31g)
Servings Per Container About 7

Amount Per Serving

Calories 120   Calories from Fat 35

% Daily Value*

Total Fat 4g	**6%**
Saturated Fat 0.5g	**3%**
Polyunsaturated Fat 0.5%	
Monounsaturated Fat 1.5g	
Cholesterol 0mg	**0%**
Sodium 310mg	**13%**
Total Carbohydrate 19g	**6%**
Dietary Fiber Less than 1g	**4%**
Sugars 2g	
Proteins 2g	

Vitamin A 0%	•	Vitamin C 0%
Calcium 4%	•	Iron 6%

*Percent Daily Values are based on a 2,000 calorie diet. Your daily values may be higher or lower depending on your calorie needs.

	Calories:	2,000	2,500
Total Fat	Less than	65g	80g
Sat Fat	Less than	20g	25g
Cholesterol	Less than	300mg	300mg
Sodium	Less than	2,400mg	2,400mg
Total Carbohydrate		300g	375g
Dietary Fiber		25g	30g

Calories per gram:
Fat 9 • Carbohydrate 4 • Protein 4

CASE STUDY
Calories from Hidden Sugar

Carbohydrate: $35 \text{ g} \times \dfrac{4 \text{ kcal}}{1 \text{ g}} = 140 \text{ kcal}$

Fat: $10 \text{ g} \times \dfrac{9 \text{ kcal}}{1 \text{ g}} = 90 \text{ kcal}$

Protein: $5 \text{ g} \times \dfrac{4 \text{ kcal}}{1 \text{ g}} = \underline{20 \text{ kcal}}$

Total caloric content: 250 kcal

Study Check

A 1-oz (28 g) serving of oat-bran hot cereal with half a cup of whole milk contains 22 g of carbohydrate, 7 g of fat, and 10 g of protein. If you eat two servings of the oat bran for breakfast, how many kilocalories will you obtain?

HEALTH NOTE

Losing and Gaining Weight

The number of kilocalories needed in your daily diet depends on your age, sex, and physical activity. Some general levels of energy needs are given in Table 5.4.

When food intake exceeds energy output, a person's body weight increases. Food intake is usually regulated by the hunger center in the hypothalamus, located in the brain. The regulation of food intake is normally proportional to the nutrient stores in the body. If these nutrient stores are low, you feel hungry; if they are high, you do not feel like eating.

Weight reduction occurs when food intake is less than energy output. Many diet products contain cellulose, which has no nutritive value but provides bulk and makes you feel full. Some diet drugs depress the hunger center and must be used with caution because they excite the nervous system and can elevate blood pressure. Because muscular exercise is an important way to expend energy, an increase in daily exercise aids weight loss. Table 5.5 lists some activities and the amount of energy they require.

Table 5.4   Typical Energy Requirements

Age (yr)	Weight (lb)	Mass (kg)	Energy (kcal)
Young Adult			
Female (13–19)	115	52	2400
Male (13–19)	130	59	2980
Adult			
Female	121	55	2200
Male	143	65	3000

Table 5.5   Energy Expended by a 70-kg (155-lb) Person

Activity	Energy Expended (kcal/hr)
Sleeping	60
Sitting	100
Walking	200
Swimming	500
Running	550

QUESTIONS AND PROBLEMS

Energy and Nutrition

5.15 Use the following data for foods burned in a calorimeter to calculate the number of Calories (kilocalories) for each food:

Food	Mass (g) of Water in a Calorimeter	Temperature (°C) Initial	Temperature (°C) Final
a. celery (1 stalk)	500.	25	35
b. waffle (7-in. diameter)	5000.	20.	62
c. popcorn (1 cup, no oil)	1000.	25	50.

5.16 A 0.50-g sample of octane, a component in gasoline, is burned in a calorimeter. If the water in the calorimeter has a mass of 1200 g and an initial temperature of 22°C, what is the amount of heat in kilojoules per gram produced by the octane if the final temperature is 26°C?

5.17 Using the caloric values for foods, determine each of the following:
  a. The total Calories for 1 cup of orange juice that has 26 g of carbohydrate, no fat, and 2 g of protein.
  b. The grams of carbohydrate in 1 apple if the apple has no fat and no protein and provides 72 kcal of energy.
  c. The number of Calories in 1 tablespoon of vegetable oil, which contains 14 g of fat and no carbohydrate or protein.
  d. How many Calories are in 1 breakfast roll if it has 30. g of carbohydrates, 15 g of fat, and 5 g of protein?

5.18 Using the caloric values for foods, determine each of the following:
  a. The total Calories for 2 tablespoons of crunchy peanut butter that contains 6 g of carbohydrate, 16 g of fat, and 7 g of protein.
  b. The grams of protein in 1 cup of soup that has 110 Cal with 7 g of fat and 9 g of carbohydrate.
  c. How many grams of sugar (carbohydrate) are in 1 can of cola if there are 140 Cal and no fat and no protein?
  d. How many grams of fat are in 1 avocado if there are 405 Calories, 13 g of carbohydrate, and 5 g of protein?

5.19 One cup of clam chowder contains 9 g of protein, 12 g of fat, and 16 g of carbohydrate. How many kilocalories are in the clam chowder? In Europe, the energy value on the label would be given as kJ. How many kJ should be on the label?

5.20 A high-protein diet contains 70. g of carbohydrate, 150 g of protein, and 5.0 g of fat. How many kilocalories does this diet provide? How many kilojoules is that?

5.4 States of Matter

Matter is anything that occupies space and has mass. This book, the food you eat, the water you drink, your cat or dog, and the air you breathe are just a few examples of matter. On Earth, matter exists in one of three *physical states:* solid, liquid, or gas. All matter is made up of tiny particles. In a **solid,** very strong attractive forces hold the particles close together. They are arranged in such a rigid pattern that they can only vibrate slowly in their fixed positions. This gives a solid a definite shape and volume. For many solids, this rigid structure produces a crystal as seen in quartz and amethyst (see Figure 5.3).

In a **liquid,** the particles have enough energy to move freely in random directions. They are still close to each other with strong attractions to maintain a definite volume, but there is no rigid structure. Thus, when oil, water, or vinegar is poured

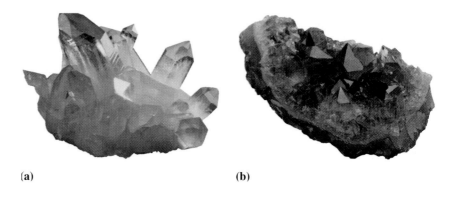

(a)   (b)

Figure 5.3 The solid states of **(a)** quartz and **(b)** amethyst, a purple form of quartz.

Q Why do these crystals have a regular shape?

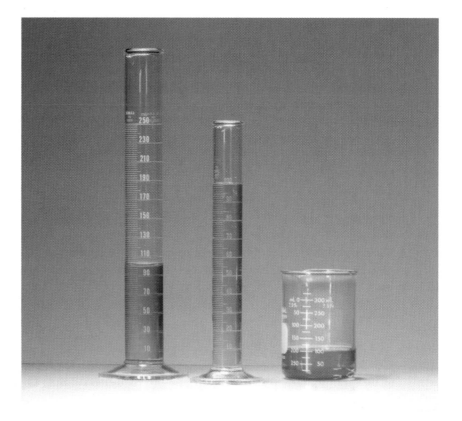

Figure 5.4 A liquid with a volume of 100 mL takes the shape of its container.

Q Why does a liquid have a definite volume, but not a definite shape?

from one container to another, the liquid maintains its own volume but takes the shape of the new container (see Figure 5.4).

The air you breathe is made of gases, mostly nitrogen and oxygen. In a **gas,** the molecules move at high speeds creating great distances between molecules. This behavior allows gases to fill their container. Gases have no definite shape or volume of their own; they take the shape and volume of their container, as shown in Figure 5.5. Table 5.6 compares some of the properties of the three states of matter.

Table 5.6   Some Properties of Solids, Liquids, and Gases

Property	Solid	Liquid	Gas
	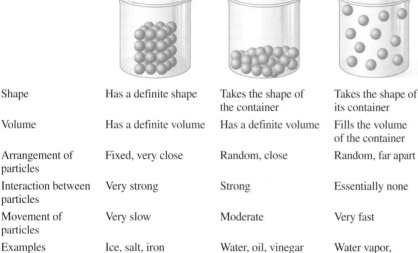		
Shape	Has a definite shape	Takes the shape of the container	Takes the shape of its container
Volume	Has a definite volume	Has a definite volume	Fills the volume of the container
Arrangement of particles	Fixed, very close	Random, close	Random, far apart
Interaction between particles	Very strong	Strong	Essentially none
Movement of particles	Very slow	Moderate	Very fast
Examples	Ice, salt, iron	Water, oil, vinegar	Water vapor, helium, air

Figure 5.5 A gas takes the shape and volume of its container.

Q Why does a gas fill the volume of a container?

EXPLORE YOUR WORLD

States of Matter

1. Place an object such as an eraser or marble in a glass. Transfer the object to another container and observe the shape and volume of the object in the new container.
2. Remove the object and place one cup of water in the container. Transfer all the water to another container. Measure the volume of the water again.
3. Open a bottle of household ammonia or perfume and stand a few feet away. Measure the time it takes for you to detect its odor.

Questions

1. What can you say about the interaction between and arrangement of the particles in the solid object?

2. a. What happens to the shape of the water when you place it in a different container?
   b. Did the volume of the water change when you poured it into a different container?
   c. What can you say about the arrangement and interaction between the particles in a liquid?
3. a. Could you smell the ammonia (or perfume) right away? After a few minutes?
   b. How did the ammonia (or perfume) molecules get from the bottle to your nose?
   c. What can you say about the interaction and relationship between gas particles?
   d. Summarize the properties of solids, liquids, and gases.

States of Matter

Identify each of the following as a solid, a liquid, or a gas:

a. oxygen in the air
b. a substance with a fixed arrangement of particles
c. safflower oil
d. a substance that has great distances between its particles

Solution

a. gas   **b.** solid   **c.** liquid   **d.** gas

Study Check

What state of water has a definite volume but takes the shape of its container?

QUESTIONS AND PROBLEMS

States of Matter

5.21 Indicate whether each of the following describes a gas, a liquid, or a solid:
   a. This substance has no definite volume or shape.
   b. The particles in this substance do not interact strongly with each other.
   c. The particles of this substance are held in a definite structure.

5.22 Indicate whether each of the following describes a gas, a liquid, or a solid:
   a. The substance has a definite volume but takes the shape of the container.
   b. The particles of this substance are very far apart.
   c. This substance occupies the entire volume of the container.

5.5 Attractive Forces Between Particles

In gases, the interactions are minimal, which allows gas molecules to move far apart from each other. In solids and liquids, there are sufficient interactions between the particles to hold them close together, although some solids have low melting points while others have very high melting points. Such differences in properties are explained by looking at the various kinds of attractive forces between particles.

In ionic compounds, the positive and negative ions are held together by ionic bonds. These are strong attractive forces that require large amounts of energy to pull the ions apart and melt the ionic solid. As a result, ionic solids have high melting points. For example, the ionic solid NaCl melts at 801°C. In compounds held together by covalent bonds, there are other types of attractive forces that exist in the solid. These attractive forces, which are weaker than ionic bonds, include dipole–dipole attractions, hydrogen bonding, and dispersion forces.

In Chapter 4, we learned that some covalent molecules have dipoles. For polar molecules, attractive forces called **dipole–dipole attractions** occur between the positive end of one molecule and the negative end of another. For example, in a sample of HCl, the positive hydrogen end of one dipole attracts the negative chlorine atom in another molecule.

$$\begin{array}{cc} \overset{\delta^+}{H}-\overset{\delta^-}{Cl}\cdots & \overset{\delta^+}{H}-\overset{\delta^-}{Cl} \end{array}$$

In certain polar molecules, strong dipoles occur when a hydrogen atom is attached to an atom of fluorine, oxygen, or nitrogen, all of which have high electronegativity values. In a special type of dipole–dipole attraction called a **hydrogen bond,** a strong attractive force occurs between the partially positive hydrogen atom and the strongly electronegative atoms of F, O, or N. Hydrogen bonds are a major factor in the formation and structure of biological molecules such as proteins and DNA.

Examples of Hydrogen Bonding

$$\begin{array}{ccc} \overset{\delta^-}{-N}-\overset{\delta^+}{H} & \cdots & \overset{\delta^-}{N}-\overset{\delta^+}{H} \\ | & & | \end{array}$$

$$\begin{array}{ccc} \overset{\delta^-}{O}-\overset{\delta^+}{H} & \cdots & \overset{\delta^-}{O}-\overset{\delta^+}{H} \\ | & & | \end{array}$$

$$\begin{array}{ccc} \overset{\delta^-}{F}-\overset{\delta^+}{H} & \cdots & \overset{\delta^-}{F}-\overset{\delta^+}{H} \end{array}$$

Hydrogen bonds are the strongest types of attractive forces between polar molecules. Polar compounds have higher melting and boiling points than nonpolar compounds with a similar mass. For example, the nonpolar compound propane C_3H_8 (molar mass 44) has a melting point of $-188°C$ and a boiling point of $-42°C$. In contrast, the polar compound ethanol C_2H_5OH (molar mass 46) has a melting point of $-114°C$ and a boiling point of $79°C$. The polar O—H provides hydrogen bonding between the ethanol molecules. As a result, ethanol requires more energy and higher temperatures to liquefy and vaporize its molecules. A comparison of the melting points and types of attractive forces of some substances is shown in Table 5.7.

Nonpolar compounds do form solids, but at very low temperatures. Very weak attractions called **dispersion forces** occur when temporary dipoles form within nonpolar molecules. Usually we think that the electrons in a nonpolar molecule are distributed symmetrically. However, at any moment, the motion of the electrons may cause more electrons to be present at one end of the molecule, which forms a temporary dipole. Although these dispersion forces are the weakest of the dipole interactions, they make it possible for nonpolar molecules to form liquids and solids. As the size of the nonpolar compound increases, there are more electrons that can produce more temporary dipoles. Larger nonpolar molecules have higher melting and boiling points. The various types of attractions between particles in solids and liquids are summarized in Table 5.8.

Table 5.7   Melting Points and Attractive Forces of Selected Substances

Substance	Melting Point	Attractive Force
MgF_2	1248°C	Ionic
Na_2S	920°C	
KF	860°C	
NaCl	801°C	
H_2O	0°C	Hydrogen bonds
NH_3	−78°C	(H and F, O, or N)
CH_3OH	−98°C	
CH_3SH	−123°C	Dipole-dipole
CH_3F	−142°C	(polar molecules)
CH_4	−183°C	Dispersion
C_3H_8	−188°C	(nonpolar molecules)
F_2	−220°C	

Table 5.8 Comparison of Bonding Forces

Type of Force		Typical Bond Strength	Example
Ionic Compounds			
Ionic bonds	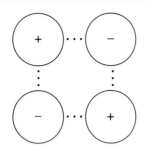	150–3000 kcal/mole	$Na^+ \cdots Cl^-$ $Cl^- \cdots Na^+$
Molecular Compounds			
Hydrogen bond (X = F, O, or N)	$\overset{\delta^-}{X}\!-\!\overset{\delta^+}{H} \cdots \overset{\delta^-}{X}\!-\!\overset{\delta^+}{H}$	5–10 kcal/mole	$\overset{\delta^-}{O}\!-\!\overset{\delta^+}{H} \cdots \overset{\delta^-}{O}\!-\!\overset{\delta^+}{H}$ with $\overset{\delta^+}{H}$ groups
Dipole–dipole	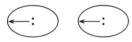	0.1–1 kcal/mole	$\overset{\delta^+}{Br}\!-\!\overset{\delta^-}{Cl} \cdots \overset{\delta^+}{Br}\!-\!\overset{\delta^-}{Cl}$
Dispersion forces (temporary shift of electrons)		0.01 kcal/mole	$F\!-\!F \cdots F\!-\!F$ $\overset{\delta^-}{F}\!-\!\overset{\delta^+}{F} \cdots \overset{\delta^-}{F}\!-\!\overset{\delta^+}{F}$

SAMPLE PROBLEM 5.8

Attractive Forces Between Particles

Indicate the major type of molecular interaction

a. dipole–dipole   **b.** hydrogen bonding   **c.** dispersion forces

expected of each of the following:

1. H—F   **2.** F—F   **3.** PCl_3

Solution

1. (b) H—F is a polar molecule that interacts with other H—F molecules by hydrogen bonding.
2. (c) Because F—F is nonpolar, only dispersion forces provide attractive forces.
3. (a) The polarity of the PCl_3 molecules provides dipole–dipole interactions.

Study Check

Why is the boiling point of H_2S lower than that of water H_2O?

QUESTIONS AND PROBLEMS

Attractive Forces Between Particles

5.23 Identify the major type of interactive force that occurs in the following substances:
    a. BrF   **b.** KCl   **c.** CCl_4
    d. HF   **e.** Cl_2

5.24 Identify the major type of interactive force that occurs in the following substances:
    a. HCl   **b.** MgF_2 **c.** PBr_3
    d. Br_2 **e.** NH_3

5.25 Identify the substance that would have the higher boiling point in each pair and explain your choice.
    a. HF or HBr   **b.** NaF or HF
    c. $MgBr_2$ or PBr_3 **d.** CH_4 or C_4H_{10}

5.26 Identify the substance that would have the higher boiling point in each pair and explain your choice.
    a. $MgCl_2$ or PCl_3 **b.** H_2O or H_2Se
    c. NH_3 or PH_3 **d.** F_2 or HF

5.6 Melting and Freezing

Matter undergoes a **change of state** when it is converted from one state to another state. When an ice cube melts in a drink, a change of state has occurred.

When heat is added to a solid, the particles in the rigid structure move faster. At a temperature called the **melting point (mp),** the particles in the solid gain sufficient energy to overcome the energy of the attractive forces that hold them together. The particles in the solid separate and move about in random patterns. The substance is **melting,** changing from a solid to a liquid.

If the temperature is lowered, the reverse process takes place. Kinetic energy is lost, the particles slow down, and attractive forces pull the particles

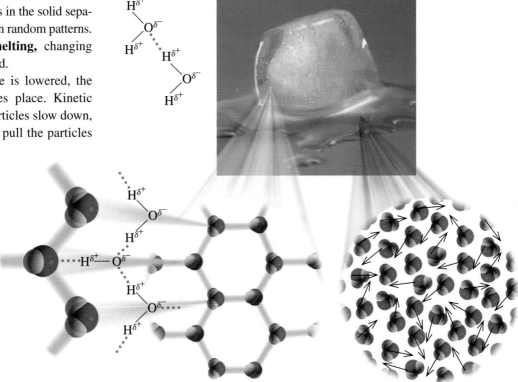

Solid   Melting   Liquid
         Freezing

close together. The substance is **freezing.** A liquid changes to a solid at the **freezing point (fp),** which is the same temperature as the melting point. Every substance has its own freezing (melting) point: water freezes at 0°C; ice melts at 0°C; gold freezes (melts) at 1064°C; nitrogen freezes (melts) at −210°C.

Suppose we have a glass containing ice and water. At warm temperatures, the ice melts, forming more liquid. At cold temperatures, heat is lost and the liquid freezes. However, at the melting (freezing) point of 0°C, ice melts at the same rate that water freezes. The reversible processes of melting and freezing are in balance. We say that the ice–liquid system has reached dynamic equilibrium.

Heat of Fusion

During melting, energy called the **heat of fusion** is needed to separate the particles of a solid. For example, 80. calories of heat are needed to melt exactly 1 g of ice.

Heat of Fusion for Water

$$\frac{80.\ \text{cal}}{1\ \text{g ice}}$$

The heat of fusion (80. cal/g) is also the heat that must be removed to freeze 1 g of water. Water is sometimes sprayed in fruit orchards during very cold weather. If the air temperature drops to 0°C, the water begins to freeze. Heat is released in this change of state, which warms the air and protects the fruit.

To determine the heat needed to melt a sample of ice, multiply the mass of the ice by its heat of fusion. There is no temperature change in the calculation because temperature remains constant as long as the ice is melting.

Calculating Heat to Melt (or Freeze) Water

Heat = mass × heat of fusion
cal   = g   × 80. cal/g

Heat of Fusion

Ice cubes at 0°C with a mass of 26 g are added to your soft drink.

a. How much heat (calories) will be absorbed to completely melt the ice at 0°C?
b. What happens to the temperature of your drink? Why?

Solution

a. The heat in calories required to melt the ice is calculated as follows:

$$26\ \cancel{g} \times \frac{80.\ \text{cal}}{1\ \cancel{g}} = 2100\ \text{cal}$$

Mass   Heat of fusion

b. The drink will be colder because heat from the drink is providing the energy to melt the ice.

Study Check

In a freezer, 120 g of water at 0°C is placed in an ice cube tray. How many kilojoules of heat must be removed to form ice cubes?

CAREER FOCUS

Histologist

"While a patient is in surgery for skin cancer, some tissue around the cancer is sent to us," says Mary Ann Pipe, histology technician. "Using the Mohs surgical technique, we place the tissue block on a glass slide, chill it to –30°C in a machine called a cryostat, and freeze it for longer in another machine called a heat extractor. From this frozen block of tissue, we cut extremely thin slices—one-thousandth of an inch—from different depths. We prepare three separate slides from skin at three different depths up to the surface of the skin. We stain the cells pink and blue by placing the slides in hemotoxin, and then in eosin. The slices are a tissue map that the doctor can easily read to determine if the margins around the skin cancer are clear or whether more tissue must be removed."

Sublimation

In a process called **sublimation,** the particles on the surface of a solid absorb enough heat to change directly to a gas with no temperature change. For example, dry ice, which is solid carbon dioxide, sublimes at $-78°C$. It is called "dry" because it does not form a liquid. You may have noticed that mothballs seem to disappear at room temperature. They have sublimed, a process we can detect by the odor of their vapor. In very cold areas, snow does not melt but sublimes directly to vapor. In a frost-free refrigerator, ice on the walls of the freezer sublimes when warm air is circulated through the compartment during the defrost cycle.

Freeze-dried foods prepared by sublimation are convenient for long storage and for camping and hiking. A food that has been frozen is placed in a vacuum chamber where it dries as the ice sublimes. The dried food retains all of its nutritional value and needs only water to be edible. A food that is freeze-dried does not need refrigeration because bacteria cannot grow without moisture. The reverse process of sublimation is called **deposition.**

QUESTIONS AND PROBLEMS

Melting and Freezing

5.27 Which of the following is a change of state?
   a. An iceberg is solid water.
   b. Water on the street freezes during a cold wintry night.
   c. A wood log is chopped to make kindling.
   d. Butter is melted to put on popcorn.

5.28 Which of the following is a change of state?
   a. Apple chips are made by freeze-drying apple slices.
   b. Dough is used to make different pasta shapes.
   c. Solid nitrogen melts at 63 K.
   d. Salt is dissolved in water to make a gargle for the throat.

5.29 Identify each of the following changes of state as melting, freezing, or sublimation.
   a. The solid structure of a substance breaks down as liquid forms.
   b. Heat is added to silver at its melting point of 962°C.
   c. Coffee is freeze-dried.
   d. Water on the street turns to ice during a cold wintry night.

5.30 Identify each of the following changes of state as melting, freezing, or sublimation.
   a. Dry ice in an ice-cream cart disappears.
   b. Snow on the ground turns to liquid water.
   c. Heat is removed from 125 g of liquid water at 0°C.
   d. On a cold wintry day, rain turns to snow.

5.31 Calculate the heat needed at 0°C to make each of the following changes of state. Indicate whether heat was absorbed or released.
   a. calories to melt 65 g of ice
   b. joules to melt 17.0 g of ice
   c. kilocalories to freeze 225 g of water
   d. kilojoules to freeze 50.0 g of water

5.32 Calculate the heat needed at 0°C to make each of the following changes of state. Indicate whether heat was absorbed or released.
   a. calories to freeze 35 g of water
   b. joules to freeze 250 g of water
   c. kilocalories to melt 140 g of ice
   d. kilojoules to melt 5.0 kg of ice

Describe the change of state between gas and liquid and calculate the energy involved.

5.7 Boiling and Condensation

Water in a mud hole disappears, unwrapped food dries out, and clothes hung on a line dry. **Evaporation** takes place when fast-moving water molecules have enough energy to escape from the surface. (See Figure 5.6a.) The loss of the "hot" water molecules removes heat and leaves the liquid cooler.

As water vapor particles cool, they lose kinetic energy and slow down. In **condensation,** the water molecules form liquid again as attractive forces pull them together, a process that is the reverse of evaporation. You may have noticed that condensation forms on the mirror when you take a shower. Because heat is lost as a liquid condenses, it is a warming process. That is why, when a rain storm is approaching, we notice a warming of the air as gaseous water molecules condense to rain.

Liquid–Gas Equilibrium

With enough time, water in an open container will all evaporate. However, if a tight fitting cover is placed on the container, the water level will go down just a small amount. At first, the water molecules evaporate from the surface. Then, some of the vapor molecules begin to condense and return to liquid. Eventually, the number of evaporating molecules is equal to the number condensing. The reverse processes of evaporation and condensation have equalized. As a result of the liquid–gas equilibrium, the water level does not go any lower.

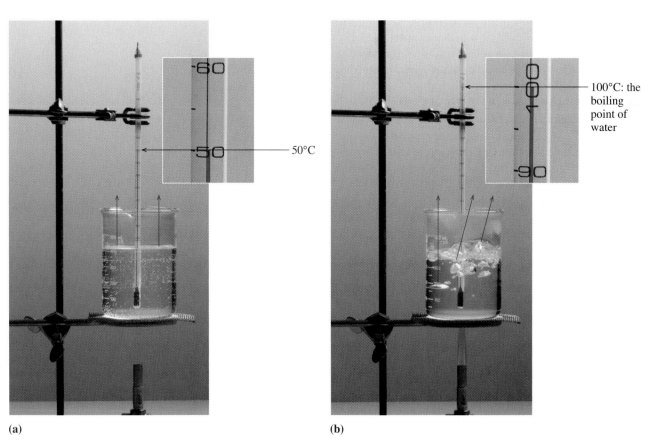

(a)   (b)

Figure 5.6 (a) Evaporation occurs at the surface of a liquid. **(b)** Boiling occurs as bubbles of gas form throughout the liquid.

Q Why does water evaporate faster at 80°C than at 20°C?

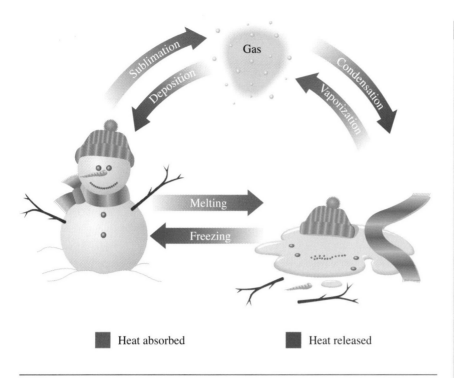

Figure 5.7 A summary of the changes of state.

Q Is heat added or released when liquid water freezes?

Boiling

As the heat added to a liquid increases, more and more particles evaporate. At the **boiling point (bp),** the particles throughout the liquid have the energy needed to change into a gas. The **boiling** of the liquid occurs as gas bubbles form throughout the liquid, then rise to the surface and escape. (See Figure 5.6b.) All the changes of state are illustrated in Figure 5.7.

SAMPLE PROBLEM 5.10

Identifying States and Changes of States

Give the state or change of state described in the following:

a. particles on the surface of a liquid escaping to form vapor
b. a liquid changing to a solid
c. gas bubbles forming throughout a liquid

Solution

a. evaporation   **b.** freezing   **c.** boiling

Study Check

What is the phase change that occurs in condensation?

Heat of Vaporization

The energy needed to vaporize exactly 1 g of liquid to gas is called the **heat of vaporization.** For water, 540 calories, a great amount of heat, is needed to convert 1 g of water to vapor.

HEALTH NOTE

Steam Burns

Hot water at 100°C will cause burns and damage to the skin. However, getting steam on the skin is even more dangerous. Let us consider 25 g of hot water at 100°C. If this water falls on a person's skin, the temperature of the water will drop to body temperature, 37°C. The heat released during cooling burns the skin. The amount of heat can be calculated from the temperature change of 63°C:

$$25 \text{ g} \times 63°C \times \frac{1.00 \text{ cal}}{\text{g} °C} = 1{,}600 \text{ cal heat}$$

For comparison, we can calculate the amount of heat released when 25 g of steam at 100°C hits the skin. First, the steam condenses to water (liquid) at 100°C:

$$25 \text{ g} \times \frac{540 \text{ cal}}{1 \text{ g}} = 14{,}000 \text{ cal heat released}$$

Now the temperature of the 25 g of water drops from 100°C to 37°C, releasing still more heat—1600 cal, as we saw earlier. Now we can calculate the total amount of heat released from the condensation and cooling of the steam as follows:

Condensation (100°C)   = 14,000 cal
Cooling (100°C to 37°C)   =   1,600 cal
Heat released   = 16,000 cal (rounded)

The amount of heat released from steam is 10 times greater than the heat from the same amount of hot water.

Heat of Vaporization for Water

$$\frac{540 \text{ cal}}{1 \text{ g water}}$$

When 1 g of water condenses, the heat of vaporization, 540 calories, is the amount of energy that must be removed. Therefore, 540 cal/g is also the *heat of condensation* of water. To calculate the amount of heat needed to vaporize (or condense) a sample of water, multiply the mass of the sample by the heat of vaporization. As before, no temperature change occurs for a change of state.

Calculating Heat to Vaporize (or Condense) Water

Heat = mass × heat of vaporization
cal   = g   × 540 cal/g

SAMPLE PROBLEM 5.11

Calculating Heat for Vaporization

In a sauna, 150 g of water is converted to steam at 100°C. How many kilocalories of heat are needed?

Solution

The heat in calories to vaporize 150 g of water at 100°C is calculated by multiplying its mass by the heat of vaporization of water. To convert calories to kilocalories, we use the equality: 1 kcal = 1000 cal

$$150 \text{ g} \times \frac{540 \text{ cal}}{1 \text{ g}} \times \frac{1 \text{ kcal}}{1000 \text{ cal}} = 81 \text{ kcal}$$

Study Check

When steam from a pan of boiling water reaches a cool window, it condenses. How much heat in kilocalories (kcal) is released when 25 g of steam condenses at 100°C?

QUESTIONS AND PROBLEMS

Boiling and Condensation

5.33 Identify each of the following changes of state as evaporation, boiling, or condensation:
   a. The water vapor in the clouds changes to rain.
   b. Wet clothes dry on a clothesline.
   c. Lava flows into the ocean and steam forms.
   d. After a hot shower, your bathroom mirror is covered with water.

5.34 Identify each of the following changes of state as evaporation, boiling, or condensation:
   a. At 100°C, the water in a pan changes to steam.
   b. On a cool morning, the windows in your car fog up.
   c. A shallow pond dries up in the summer.
   d. Your teakettle whistles when the water is ready for tea.

5.35 **a.** How does perspiration during heavy exercise cool the body?
   b. Why do clothes dry more quickly on a hot summer day than on a cold winter day?
   c. Why do wet clothes stay wet in a plastic bag?

5.36 **a.** When a sports injury occurs during a game, a spray such as ethyl chloride may be used to numb an area of the skin. Explain how a substance like ethyl chloride that evaporates quickly can numb the skin.
   b. Why does water in a wide, flat, shallow dish evaporate more quickly than the same amount of water in a tall, narrow glass?
   c. Which will dry out faster: a sandwich on a plate or a sandwich in plastic wrap?

5.37 Calculate the heat change at 100°C in each of the following problems. Indicate whether heat was absorbed or released.
   a. calories to vaporize 10.0 g of water
   b. joules to vaporize 50.0 g of water
   c. kilocalories to condense 8.0 kg of steam
   d. kilojoules to condense 170 g of steam

5.38 Calculate the heat change at 100°C in each of the following problems. Indicate whether heat was absorbed or released.
   a. calories to condense 10.0 g of steam
   b. joules to condense 75 g of steam
   c. kilocalories to vaporize 44 g of water
   d. kilojoules to vaporize 5.0 kg of water

5.8 Heating and Cooling Curves

LEARNING GOAL

Draw a heating or cooling curve for a substance and calculate the heat involved.

Diagrams for Changes of State

When heat is added to a solid, it begins to melt, forming a liquid. As more heat is added, the liquid warms until it boils and forms a gas. Each of these changes of state can be illustrated visually as a diagram called a **heating curve.** (See Figure 5.8.) In a heating curve, the temperature is shown on the vertical axis. The addition of heat is shown on the horizontal axis. Each of the segments represents a warming step or a change of state.

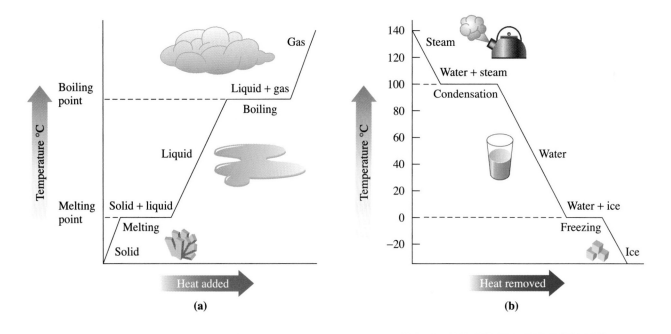

Fig. 5.8 **(a)** A heating curve diagrams the temperature increases and changes in state as heat is added. **(b)** A cooling curve for water.

Q **What does the plateau at 100°C represent on the cooling curve for water?**

Steps on a Heating Curve

The first diagonal line indicates a warming of a solid as heat is added. When the melting temperature is reached, solid begins to change to a liquid. There is no change in temperature as the solid changes to liquid. This process is shown as a flat line, or plateau, at the melting point on the heating curve. During the melting process, all of the heat is used to break apart the particles of the solid while temperature remains constant.

After all the particles are in the liquid state, the temperature begins to rise again as shown by the next diagonal line. As more heat is added, the liquid warms and the particles move about more vigorously until the boiling point is reached. Now the liquid begins to boil, indicated as another flat region on the diagram at its boiling point. Again there is no temperature change as long as liquid is changing into gas. After all the liquid has changed to gas, additional heat will cause the temperature to rise again.

Steps on a Cooling Curve

When heat is removed from a substance such as water vapor, the gas cools and condenses. A **cooling curve** is a diagram of this cooling process. (See Figure 5.8.) On the cooling curve, the temperature is plotted on the vertical axis and the removal of heat on the horizontal axis. If a gas is cooled, heat is lost and the temperature drops. At the condensation point (same as the boiling point), the gas begins to condense, forming liquid. This process is indicated by a flat line (plateau) on the cooling curve at the condensation point.

After all of the gas has changed into liquid, the particles within the liquid cool as indicated by the downward sloping line that shows the temperature decrease. At the freezing point, the particles in the liquid slow so much that solid begins to form. A second flat line at the freezing point indicates the change of state from liquid to solid (freezing). Once all of the substance is frozen, more heat can be lost, which lowers the solid's temperature below its freezing point.

Combining Energy Calculations

Up to now, we have calculated one step in a heating or cooling curve. However, many problems require a combination of steps that include a temperature change as well as a change of state. The heat is calculated for each step separately and then added together to find the total energy as seen in Sample Problem 5.12.

SAMPLE PROBLEM 5.12

Combined Heat Calculations

Calculate the calories needed to convert 15.0 g of water at 25°C to steam at 100°C. You will need to use the specific heat of liquid water, 1.00 cal/g °C, and the heat of vaporization for water, 540 cal/g.

Solution

In the heating curve for water from 25°C to steam at 100°C, we find that two steps are needed for the calculation. One step is the heat needed to warm the liquid from 25°C to 100°C, and the second step is the heat for the vaporization of the water to gas.

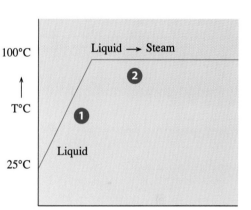

Step 1 Heat to warm the water (liquid) from 25°C to 100°C:

$$15.0 \text{ g} \times 75°C \times \frac{1.00 \text{ cal}}{\text{g} °C} = 1100 \text{ cal}$$

Mass ΔT Specific heat of water

Step 2 Heat to vaporize the water at 100°C:

$$15.0 \text{ g} \times \frac{540 \text{ cal}}{1 \text{ g}} = 8100 \text{ cal}$$

Mass   Heat of vaporization of water

We calculate the total heat by adding step 1 and step 2:

1. Heating water   1100 cal
2. Changing liquid to steam   8100 cal
   Total heat needed: 9200 cal

Study Check

How many calories are released when 25.0 g of steam condenses at 100°C, cools to 0°C, and freezes? (*Hint:* The solution will require three energy steps.)

QUESTIONS AND PROBLEMS

Heating and Cooling Curves

5.39 Draw a heating curve for a sample of ice that is heated from −20°C to 140°C. Indicate the segment of the graph that corresponds to each of the following:
   a. solid   **b.** melting point   **c.** liquid
   d. boiling point   **e.** gas

5.40 Draw a cooling curve for a sample of steam that cools from 110°C to −10°C. Indicate the segment of the graph that corresponds to each of the following:

   a. solid   **b.** freezing point   **c.** liquid

   d. condensation point (boiling point)   **e.** gas

5.41 Using the values for the heat of fusion, specific heat of water, and/or heat of vaporization, calculate the amount of heat energy in each of the following:

   a. calories needed to warm 20.0 g of water at 15°C to 72°C (one step)

   b. calories need to melt 50.0 g of ice at 0°C and to warm the liquid to 65°C (two steps)

   c. kilojoules given off when 15 g of steam condenses at 100°C and the liquid cools to 0°C (two steps)

   d. kilocalories needed to melt 24 g of ice at 0°C, to warm the liquid to 100°C, and to vaporize it at 100°C (three steps)

5.42 Using heat of fusion, specific heat of water, and/or heat of vaporization, calculate the amount of heat energy in each of the following:

   a. calories to condense 125 g of steam at 100°C and to cool the liquid to 15°C (two steps)

   b. joules to melt a 525-g ice cube at 0°C and to warm the liquid to 15°C (two steps)

   c. kilocalories to condense 85 g of steam at 100°C, cool the liquid and freeze it at 0°C (three steps)

   d. calories to warm 55 mL of water (density = 1.0g/mL) from 10.°C to 100°C and vaporize it at 100°C (two steps)

5.43 A bag containing 250 g of ice at 0°C was placed on a burn on a patient's hand. When the bag was removed, the ice had melted, and the liquid water had come to a temperature of 24°C. How many kilocalories of heat were absorbed by the ice and cool water? How many kilojoules is that?

5.44 A 115-g sample of steam at 100°C is emitted from a volcano. It condenses, cools, and falls as snow at 0°C. How many kilocalories of heat are released? How many kilojoules is that?

Chapter Review

5.1 Energy

Energy is the ability to do work. Potential energy is stored energy; kinetic energy is the energy of motion.

5.2 Measuring Heat

Specific heat is the amount of energy required to raise the temperature of exactly 1 g of a substance by exactly 1°C. The heat lost or gained by a substance is determined by multiplying its mass (g), the temperature change (ΔT), and its specific heat (cal/g °C).

5.3 Energy and Nutrition

The nutritional Calorie is the same amount of energy as 1 kcal or 1000 calories. The caloric content of a food is the sum of calories from carbohydrate, fat, and protein.

5.4 States of Matter

Matter is anything that has mass and occupies space. The three states of matter are solid, liquid, and gas.

5.5 Attractive Forces Between Particles

In ionic solids, oppositely charged ions are held in a rigid lattice structure by ionic bonds. Attractive forces called dipole–dipole attractions and hydrogen bonds hold the solid and liquid states of polar covalent compounds together. Nonpolar compounds form solids and liquids by temporary dipoles called dispersion forces.

5.6 Melting and Freezing

A substance undergoes a physical change when its shape, size, or state changes, but its identity does not change. Melting occurs when the particles in a solid absorb enough energy to break apart and form a liquid. The amount of energy required to convert exactly 1 g of solid to liquid is called the heat of fusion. For water, 80. cal are needed to melt 1 g of ice, or must be removed to freeze 1 g of water. Sublimation is a special process whereby a solid changes directly to a gas.

5.7 Boiling and Condensation

Evaporation occurs when particles in a liquid state absorb enough energy to break apart and form gaseous particles. Boiling is the vaporization of liquid at its boiling point. The heat of vaporization is the amount of heat needed to convert exactly 1 g of liquid to vapor. For water, 540 cal are needed to vaporize 1 g of water, or must be removed to condense 1 g of steam.

5.8 Heating and Cooling Curves

A heating or cooling curve illustrates the changes in temperature and state as heat is added to or removed from a substance. Plateaus on the graph indicate changes of state. The total heat absorbed or removed from a substance undergoing temperature changes and changes of state is the sum of energy calculations for change(s) of state and change(s) in temperature.

Key Terms

boiling The formation of bubbles of gas throughout a liquid.

boiling point (bp) The temperature at which a substance exists as a liquid and gas; liquid changes to gas (boils), and gas changes to liquid (condenses).

caloric value The kilocalories obtained per gram of the food types: carbohydrate, fat, and protein.

calorie (cal) The amount of heat energy that raises the temperature of exactly 1 g of water exactly 1°C.

Calorie (Cal) The dietary unit of energy, equal to 1000 cal, or 1 kcal.

calorimeter An instrument used to measure the amount of heat released by burning a sample.

change of state The transformation of one state of matter to another; for example, solid to liquid, liquid to solid, liquid to gas.

condensation The change of state of a gas to a liquid.

cooling curve A diagram that illustrates temperature changes and changes of states for a substance as heat is removed.

deposition The change of a gas directly to a solid; the reverse of sublimation.

dipole–dipole attractions Attractive forces between oppositely charged ends of polar molecules.

dispersion forces Weak dipole bonding that results from a momentary polarization of nonpolar molecules in a substance.

energy The ability to do work.

evaporation The formation of a gas (vapor) by the escape of high-energy molecules from the surface of a liquid.

freezing A change of state from liquid to solid.

freezing point (fp) The temperature at which the solid and liquid forms of a substance are in equilibrium; a liquid changes to a solid (freezes), a solid changes to a liquid (melts).

gas A state of matter characterized by no definite shape or volume. Particles in a gas move rapidly.

heat The energy associated with the motion of particles in a substance.

heat of fusion The energy required to melt exactly 1 g of a substance. For water, 80. cal are needed to melt 1 g of ice; 80. cal are released when 1 g of water freezes.

heat of vaporization The energy required to vaporize 1 g of a substance. For water, 540 calories are needed to vaporize exactly 1 g of liquid: 1 g of steam gives off 540 cal when it condenses.

heating curve A diagram that shows the temperature changes and changes of state of a substance as it is heated.

hydrogen bond The attraction between a partially positive H and a strongly electronegative atom of F, O, or N.

joule (J) The SI unit of heat energy; 4.18 J = 1 cal.

kinetic energy A type of energy that is required for actively doing work; energy of motion.

kilocalorie (kcal) An amount of heat energy equal to 1000 calories.

liquid A state of matter that takes the shape of its container but has a definite volume.

matter Anything that has mass and occupies space.

melting The conversion of a solid to a liquid.

melting point (mp) The temperature at which a solid becomes a liquid (melts). It is the same temperature as the freezing point.

potential energy An inactive type of energy that is stored for use in the future.

solid A state of matter that has its own shape and volume.

specific heat A quantity of heat that changes the temperature of exactly 1 g of a substance by exactly 1°C.

sublimation The change of state in which a solid is transformed directly to a gas without forming a liquid.

work An activity that requires energy.

Additional Problems

5.45 On a hot day, the beach sand gets hot, but the water stays cool. Explain.

5.46 Why do drops of liquid water form on a glass of iced tea?

5.47 When it rains or snows, the air temperature seems warmer. Explain.

5.48 **a.** Water is sprayed on the ground of an orchard when temperatures are near freezing to keep the fruit from freezing. Explain.
  b. How many kilocalories of heat are released if 5.0 kg of water at 15°C is sprayed on the ground and cools and freezes at 0°C?

5.49 If you used the 2000 Calories you expend in energy in one day to heat 50,000 g of water at 20°C, what would be its new temperature?

5.50 A 0.50-g sample of vegetable oil is placed in a calorimeter. When the sample is burned, 18.9 kJ are given off. What is the caloric value (kcal/g) of the oil?

5.51 A typical diet in the United States provides 15% of the calories from protein, 45% from carbohydrates, and the remainder from fats. Calculate the grams of protein, carbohydrate, and fat to be included each day in diets having the following caloric requirements:
  a. 1200 kcal   **b.** 1900 kcal   **c.** 2600 kcal

5.52 You have just eaten a quarter-pound cheeseburger, French fries, and chocolate shake. How many hours will a 70.0-kg person need to run to "burn off" the kilocalories in this meal? See Table 5.5 (assume you are a 70-kg person).

Item	Protein (g)	Fat (g)	Carbohydrate (g)
Cheeseburger	31	29	34
French fries	3	11	26
Chocolate shake	11	9	60.

5.53 If you want to lose 1 pound of "fat," which is 15% water, how many kilocalories do you need to expend? How many kilojoules is that?

5.54 Calculate the Cal (kcal) in 1 cup of whole milk: 12 g of carbohydrate, 9 g of fat, and 9 g of protein.

5.55 A hot-water bottle contains 725 g of water at 65°C. If the water cools to body temperature (37°C), how many kilojoules of heat could be transferred to sore muscles?

5.56 When 1.0 g of gasoline burns, 11,500 calories of energy are given off. If the density of gasoline is 0.74

g/mL, how many kilocalories of energy are obtained from 1.5 gallons of gasoline?

5.57 Select the substance in each group that would have the highest boiling point.
  a. NF_3 HF F_2 **b.** H_2O SO_2 O_2
  c. HCl   KCl OCl_2 **d.** He CH_4 NH_3

5.58 Describe the type of compound that could have each of the following types of attractive forces:
  a. dipole–dipole   **b.** hydrogen bonds
  c. dispersion forces

5.59 The compounds C_2H_6 and C_8H_{18} are both nonpolar. The melting point of C_2H_6 (ethane) is $-172°C$ and the melting point of C_8H_{18} (octane) is $-57°C$. What accounts for this difference in melting points?

5.60 Why would NH_3 have a higher boiling point ($-33°C$) than PH_3 ($-88°C$)?

5.61 Indicate the major type of attractive force
  (1) ionic   (2) dipole–dipole
  (3) hydrogen bond   (4) dispersion forces
  that occurs between particles of the following substances:
  a. NH_3 **b.** HF   **c.** CH_4
  d. $CHCl_3$ **e.** H_2O **f.** LiCl

5.62 An ice cube tray holds 325 g of water. If the water initially has a temperature of 25°C, how many kilojoules of heat must be removed to cool and freeze the water at 0°C?

5.63 A 3.0-kg block of lead is taken from a furnace at 300.°C and placed on a large block of ice at 0°C. The specific heat of lead is 0.028 cal/g °C. If all the heat given up by the lead is used to melt ice, how much ice is melted after the temperature of the lead drops to 0°C?

5.64 The melting point of benzene is 5.5°C and its boiling point is 80.1°C. Sketch a heating curve for benzene from 0°C to 100°C.
  a. What is the state of benzene at 15°C?
  b. What happens on the curve at 5.5°C?
  c. What is the state of benzene at 63°C?
  d. What is the state of benzene at 98°C?
  e. At what temperature will both liquid and gas be present?

5.65 A sample of 50.0 g of ice at 0°C is added to 50.0 g of water (liquid) at 0°C. How many kcal are needed to melt the ice, warm all the liquid water to 100°C, and change it to steam at 100°C?

6 Chemical Reactions

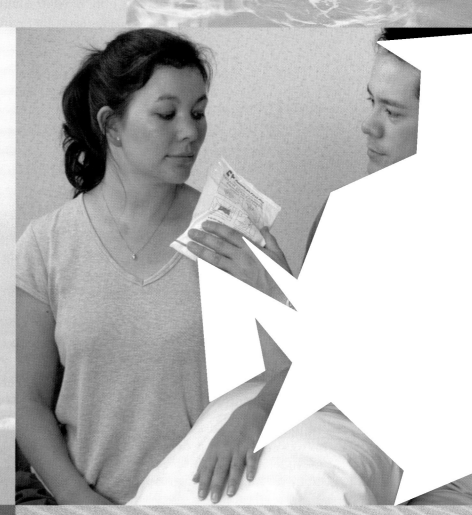

"If you've had first aid for a sports injury," says Cort Kim, physical therapist at the Sunrise Sports Medicine Clinic, "you've likely been treated with a cold pack or hot pack. We use them for several kinds of injury. Here, I'm showing how I can use a cold pack to reduce swelling in my patient's shoulder."

A hot or cold pack is just a packaged chemical reaction. When you hit or open the pack to activate it, your action mixes chemicals together and thus initiates the reaction. In a cold pack, the reaction is one that absorbs heat energy, chills the pack, and draws heat from the injury. Hot packs use reactions that release energy, thus warming the pack. In both cases, the reaction proceeds at a moderate pace, so that the pack stays active for a long time and doesn't get too cold or hot.

LOOKING AHEAD

the
Chemistry place

www.chemplace.com/college

Visit the URL above or use the CD-ROM in the book for extra quizzing, interactive tutorials, career resources, and case studies.

Chemical reactions occur everywhere. In the engines of our cars, gasoline burns to provide energy to move the car, play the radio, and run the air conditioner. Chemical reactions in our bodies provide energy to move our muscles. In nature, chemical reactions occur in the leaves of plants to convert carbon dioxide and water into carbohydrates.

Chemists use chemical equations to describe the reactants and products of chemical reactions. Most chemical reactions involve a loss or gain of energy. The rate of a reaction is measured by the speed at which reactants form products. When you increase the cooking temperature, food cooks faster. Generally, the rate of reaction slows down when the temperature is lowered.

In a reversible reaction, the products change back into the reactants. We will see how chemical equilibrium is reached when the rates of the forward and reverse reactions are the same.

6.1 Chemical Changes

In a **physical change,** the appearance of a substance is altered, but not its composition. When liquid water becomes a gas, or freezes to a solid, it is still water (see Figure 6.1). If we smash a rock or tear a piece of paper, only the size of the material changes. The smaller pieces are still rock or paper because there was no change in the composition of the substances.

In a **chemical change,** the reacting substances change into new substances that have different compositions and different properties. New properties may involve a change in color, a change in temperature, or the formation of bubbles or a solid. For instance, when silver tarnishes, the bright silver metal (Ag) reacts with sulfur (S) to become the dull, blackish substance we call tarnish (Ag_2S) (see Figure 6.1). Table 6.1 gives some examples of some typical physical and chemical changes.

A **chemical reaction** always involves chemical change because atoms of the reacting substances form new combinations with new properties. A chemical reaction takes place when rust forms on a nail or piece of iron: the iron (Fe) is reacting with oxygen (O_2) to produce the new substance, rust (Fe_2O_3). Perhaps you have placed an antacid tablet in a glass of water and noticed it fizzing and bubbling as sodium bicarbonate ($NaHCO_3$) reacts to form carbon dioxide (CO_2) gas (see Figure 6.2).

Table 6.1   Comparison of Some Chemical and Physical Changes

Chemical Changes	Physical Changes
Rusting nail	Melting ice
Bleaching a stain	Boiling water
Burning a log	Sawing a log in half
Tarnishing silver	Tearing paper
Fermenting grapes	Breaking a glass
Souring of milk	Pouring milk

WEB TUTORIAL
What is Chemistry?

SAMPLE PROBLEM 6.1

Classifying Chemical and Physical Change

Classify each of the following changes as physical or chemical:

a. water freezing into an icicle   **b.** burning a match
c. breaking a chocolate bar   **d.** digesting a chocolate bar

Solution

a. Physical. Freezing water involves only a change from liquid water to ice. No change has occurred in the substance water.

b. Chemical. Burning a match causes the formation of heat, light, and ash that were not present before striking the match.

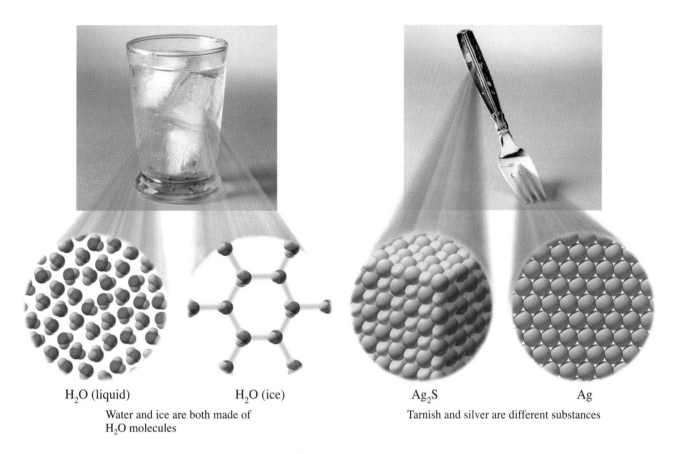

H₂O (liquid)   H₂O (ice)   Ag₂S   Ag

Water and ice are both made of   Tarnish and silver are different substances
H₂O molecules

Figure 6.1 A chemical change produces new substances; a physical change does not.

Q Why is melting ice considered a physical change?

Figure 6.2 Examples of chemical reactions involve chemical change: iron (Fe) reacts with oxygen (O_2) to form rust (Fe_2O_3) and an antacid ($NaHCO_3$) tablet in water forms bubbles of carbon dioxide (CO_2).

Q What is the evidence for chemical change in these chemical reactions?

c. Physical. Breaking a chocolate bar does not affect its composition.

d. Chemical. The digestion of the chocolate bar converts its components into new substances.

Study Check

Classify the following changes as physical or chemical:

a. chopping a carrot   **b.** developing a Polaroid picture
c. inflating a balloon

QUESTIONS AND PROBLEMS

Chemical Changes

6.1 Classify each of the following changes as chemical or physical:
   a. chewing a gum drop
   b. ignition of fuel in the space shuttle
   c. drying clothes
   d. neutralizing stomach acid with an antacid tablet
   e. formation of snowflakes
   f. an exploding dynamite stick

6.2 Classify each of the following changes as chemical or physical:
   a. fogging the mirror during a shower
   b. tarnishing of a silver bracelet
   c. breaking a bone
   d. mending a broken bone
   e. burning paper
   f. picking up the leaves in the yard

LEARNING GOAL

State a chemical equation in words and determine the number of atoms in the reactants and products.

the Chemistry place

WEB TUTORIAL
Chemical Reactions and Equations

6.2 Chemical Equations

When you build a model airplane, prepare a new recipe, mix a medication, or clean a patient's teeth, you follow a set of directions. These directions tell you what materials to use and the products you will obtain. In chemistry, a **chemical equation** tells us the materials we need and the products that will form in a chemical reaction.

Writing a Chemical Equation

Suppose you work in a bicycle shop, assembling wheels and bodies into bicycles. You could represent this process by a simple equation:

Equation: Wheels + Body ⟶ Bicycle

When you burn charcoal in a grill, the carbon in the charcoal combines with oxygen to form carbon dioxide. We can represent this reaction by a chemical equation that is much like the one for the bicycle:

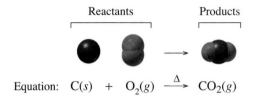

Reactants   Products

Equation: $C(s)$ + $O_2(g)$ $\xrightarrow{\Delta}$ $CO_2(g)$

In an equation, the formulas of the **reactants** are written on the left of the arrow and the formulas of the **products** on the right. When there are two or more formulas on the same side, they are separated by plus ($+$) signs. The delta sign (Δ) indicates that heat was used to start the reaction.

Sometimes, as in the charcoal case, the formulas in an equation may include letters, in parentheses, that give the physical state of the substances as solid (s), liquid (l), or gas (g). If a substance is dissolved in water, it is an aqueous (aq) solution. Table 6.2 summarizes some of the symbols used in equations.

When a reaction takes place, the bonds between the atoms of the reactants are broken and new bonds are formed to give the products. Atoms cannot be gained, lost, or changed into other types of atoms during a reaction. Therefore, a reaction must be written as a **balanced equation,** which shows the same number of atoms for each element on both sides of the arrow. Let's see if the equation we wrote above for burning carbon is balanced:

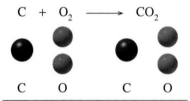

C + O_2 $\longrightarrow$ CO_2

C	O	C	O

Reactant atoms   =   Product atoms

The answer is yes: this equation is *balanced* because there is one carbon atom and two oxygen atoms on each side of the arrow.

Now consider the reaction in which hydrogen reacts with oxygen to form water. First we write the formulas of the reactants and products:

H_2 + O_2 $\longrightarrow$ H_2O

Is the equation balanced? To find out, we add up the atoms of each element on either side of the arrow. No; the equation is *not balanced*. There are two oxygen atoms to the left of the arrow, but only one to the right. Thus, the atoms on the left side do not match the atoms on the right side. To balance this equation, we place whole numbers called **coefficients** in front of the formulas. First we write a coefficient of 2 in front of the H_2O formula. Now the product has four hydrogen atoms. That means we must also write a coefficient of 2 in front of the formula H_2 in the reactants.

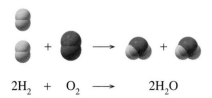

$2H_2$ + O_2 $\longrightarrow$ $2H_2O$

Table 6.2 Some Symbols Used in Writing Equations	
Symbol	**Meaning**
$+$	Separates two or more formulas
$\longrightarrow$	Reacts to form products
Δ	The reactants are heated
(s)	Solid
(l)	Liquid
(g)	Gas
(aq)	Aqueous

Now the number of hydrogen atoms and oxygen atoms are the same in the reactants as in the products. The equation is *balanced*.

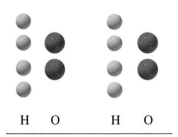

H   O   H   O

Reactant atoms = Product atoms

SAMPLE PROBLEM 6.2

Chemical Equations

Hydrogen and nitrogen react to form ammonia, NH_3.

$$3H_2 + N_2 \longrightarrow 2NH_3$$
Ammonia

a. What are the coefficients in the equation?

b. How many atoms of each element are in the reactants and products of the equation?

Solution

a. The coefficients are three (3) in front of H_2; one (1), which is understood, in front of N_2; and two (2) in front of NH_3.

b. In the reactants there are 6 hydrogen atoms and 2 nitrogen atoms. There are also 6 hydrogen atoms and 2 nitrogen atoms in the product.

Study Check

When ethane (C_2H_6) burns in oxygen, the products are carbon dioxide and water. The balanced equation is as follows.

$$2C_2H_6 + 7O_2 \longrightarrow 4CO_2 + 6H_2O$$

State the total number of atoms of each element on each side of the equation.

QUESTIONS AND PROBLEMS

Chemical Equations

6.3 State the number of atoms of each element on the reactant and on the product sides of the following equations:

    a. $2NO + O_2 \longrightarrow 2NO_2$

    b. $5C + 2SO_2 \longrightarrow CS_2 + 4CO$

    c. $2C_2H_2 + 5O_2 \longrightarrow 4CO_2 + 2H_2O$

    d. $N_2H_4 + 2H_2O_2 \longrightarrow N_2 + 4H_2O$

6.4 State the number of atoms of each element on the reactant and on the product sides of the following equations:

a. $CH_4 + 2O_2 \longrightarrow CO_2 + 2H_2O$
b. $4P + 5O_2 \longrightarrow P_4O_{10}$
c. $4NH_3 + 6NO \longrightarrow 5N_2 + 6H_2O$
d. $6CO_2 + 6H_2O \longrightarrow C_6H_{12}O_6 + 6O_2$
  Glucose

6.5 Determine whether each of the following equations is balanced or not balanced:

a. $S + O_2 \longrightarrow SO_3$ **b.** $2Al + 3Cl_2 \longrightarrow 2AlCl_3$
c. $H_2 + O_2 \longrightarrow H_2O$ **d.** $C_3H_8 + 5O_2 \longrightarrow 3CO_2 + 4H_2O$

6.6 Determine whether each of the following equations are balanced or not balanced:

a. $PCl_3 + Cl_2 \longrightarrow PCl_5$ **b.** $CO + 2H_2 \longrightarrow CH_3OH$
c. $2KClO_3 \longrightarrow 2KCl + O_2$ **d.** $Mg + N_2 \longrightarrow Mg_3N_2$

6.7 All of the following are balanced equations. State the number of atoms of each element in the reactants and in the products.

a. $2Na + Cl_2 \longrightarrow 2NaCl$ **b.** $PCl_3 + 3H_2 \longrightarrow PH_3 + 3HCl$
c. $P_4O_{10} + 6H_2O \longrightarrow 4H_3PO_4$

6.8 All of the following are balanced equations. State the number of atoms of each element in the reactants and in the products.

a. $2N_2 + 3O_2 \longrightarrow 2N_2O_3$
b. $Al_2O_3 + 6HCl \longrightarrow 2AlCl_3 + 3H_2O$
c. $C_5H_{12} + 8O_2 \longrightarrow 5CO_2 + 6H_2O$

6.3 Balancing a Chemical Equation

We have seen that a chemical equation must be balanced. In many cases, we can use a method of trial and error to balance an equation. However, Table 6.3 gives a set of steps that can be useful.

Table 6.3 Steps for Balancing Chemical Equations

Step 1   **Use the correct formulas.** Check the formulas of the reactants and products.

Step 2   **Determine if the equation is balanced.** Count the number of atoms of each element on the reactant side and the product side. Count a polyatomic ion as a group if its formula is unchanged.

Step 3   **Balance the equation one element at a time.** Pick an element, and balance the equation for that element by placing coefficients in front of the formulas to equalize the number of atoms of that element on both sides of the arrow. A typical starting place is with a formula that has subscripts. Never try to balance an equation by changing the subscripts of a formula—that would make the formula incorrect. Balance O_2 and other diatomic elements last.

Step 4   **Check to see if the equation is balanced.** Sometimes balancing one element will put another element out of balance. If that happens, repeat steps 3 and 4 until the equation is balanced. Make sure the coefficients you use are the smallest possible whole numbers. If you can divide all the coefficients by 2 or some other number, do so to make them as small as possible.

To practice using the steps from Table 6.3, let us balance the reaction of the gas methane, CH_4, with oxygen to produce carbon dioxide and water. This is the principal reaction that occurs in the flame of a gas burner you use in the laboratory and in the flame of a gas stove.

LEARNING GOAL

Write a balanced chemical equation from the formulas of the reactants and products for a reaction.

the Chemistry place

WEB TUTORIAL
Chemical Reactions and Equations

Step 1   **Use the correct formulas.** As a first step, we write the equation using the correct formulas for the reactants and products.

$$CH_4 + O_2 \longrightarrow CO_2 + H_2O$$

Step 2   **Determine if the equation is balanced.** When we compare the atoms on the reactant side with the atoms on the product side, we see that there are more hydrogen atoms on the left side and more oxygen atoms on the right.

$$CH_4 + O_2 \longrightarrow CO_2 + H_2O$$

1 C   1 C

4 H   2 H Not balanced

2 O   3 O Not balanced

Step 3   **Balance the equation one element at a time.** First we balance the hydrogen atoms by placing a coefficient of 2 in front of the formula for water. Then we balance the oxygen atoms by placing a coefficient of 2 in front of the formula for oxygen. There are now four oxygen atoms and four hydrogen atoms in both the reactants and products.

$$CH_4 + 2O_2 \longrightarrow CO_2 + 2H_2O$$

Step 4   **Check to see if the equation is balanced.** Rechecking the balanced equation shows that the numbers of atoms of carbon, hydrogen, and oxygen are the same for both the reactants and the products. The equation is balanced using the lowest possible whole numbers as coefficients.

$$CH_4 + 2O_2 \longrightarrow CO_2 + 2H_2O \quad \text{Balanced}$$

Reactants	*Products*
1 C atom	1 C atom
4 H atoms	4 H atoms
4 O atoms	4 O atoms

Suppose you had written the equation as follows:

$$2CH_4 + 4O_2 \longrightarrow 2CO_2 + 4H_2O \quad \text{Incorrect}$$

Although there are equal numbers of atoms on both sides of the equation, this is not written correctly. All the coefficients must be divided by 2 to obtain the lowest possible whole numbers.

SAMPLE PROBLEM 6.3

Balancing Equations

Balance the following equations:

a. $Al + Cl_2 \longrightarrow AlCl_3$

b. $Na_3PO_4 + MgCl_2 \longrightarrow Mg_3(PO_4)_2 + NaCl$

Solution

a. Step 1 The correct formulas are written in the equation.

$$Al + Cl_2 \longrightarrow AlCl_3$$

Step 2 When we compare the number of atoms on the reactant and product sides, we find that the chlorine atoms are not balanced.

Reactants   ***Product***

$$Al + Cl_2 \longrightarrow AlCl_3$$

1 Al   1 Al

2 Cl   3 Cl   Not balanced

Step 3 In the formulas with chlorine, there is an odd–even relationship. A 2 placed in front of $AlCl_3$ will give an even number of 6 Cl atoms. Any number we use to multiply the Cl_2 will double the number of Cl atoms. To balance the 6 Cl in the product, we place a 3 in front of Cl_2. Finally, we must rebalance the 2 Al atoms that are now in the product by placing a 2 in front of the Al on the left side.

Step 4 When we recheck the balance of atoms, we see that the equation is balanced.

Reactants   ***Product***

$$2Al + 3Cl_2 \longrightarrow 2AlCl_3 \quad \text{Balanced}$$

~~1 Al~~ 2 Al = ~~1 Al~~ 2 Al

~~2 Cl~~ 6 Cl = ~~3 Cl~~ 6 Cl

b. Step 1 In the equation, the correct formulas are written.

$$Na_3PO_4 + MgCl_2 \longrightarrow Mg_3(PO_4)_2 + NaCl$$

Step 2 When we compare the number of ions on the reactant and product sides, we find that the equation is not balanced. In this equation, it is more convenient to balance the phosphate polyatomic ion as a group instead of its individual atoms.

Reactants   ***Products***

$$Na_3PO_4 + MgCl_2 \longrightarrow Mg_3(PO_4)_2 + NaCl$$

3 Na^+ 1 Na^+ Not balanced

1 PO_4^{3-} 2 PO_4^{3-} Not balanced

1 Mg^{2+} 3 Mg^{2+} Not balanced

2 Cl^- 1 Cl^- Not balanced

Step 3 We begin with the formula of $Mg_3(PO_4)_2$ because there are several subscripts. A 3 in front of $MgCl_2$ will balance magnesium and a 2 in front of Na_3PO_4 will balance the phosphate ion. Looking again at each of the ions in the reactants and products, we see that the sodium and chloride ions are not yet equal. A 6 is needed in front of the NaCl to complete the balance.

	Reactants		*Products*	

$$2Na_3PO_4 + 3MgCl_2 \longrightarrow Mg_3(PO_4)_2 + 6NaCl \quad \text{Balanced}$$

$6\,Na^+$	$=$	$6\,Na^+$
$2\,PO_4^{3-}$	$=$	$2\,PO_4^{3-}$
$3\,Mg^{2+}$	$=$	$3\,Mg^{2+}$
$6\,Cl^-$	$=$	$6\,Cl^-$

Study Check

Balance the following equations:

a. $Fe + O_2 \longrightarrow Fe_3O_4$
b. $C_3H_8 + O_2 \longrightarrow CO_2 + H_2O$

QUESTIONS AND PROBLEMS

Balancing a Chemical Equation

6.9 Balance the following equations:
   a. $N_2 + O_2 \longrightarrow NO$
   b. $HgO \longrightarrow Hg + O_2$
   c. $Fe + O_2 \longrightarrow Fe_2O_3$
   d. $Na + Cl_2 \longrightarrow NaCl$
   e. $Cu_2O + O_2 \longrightarrow CuO$

6.10 Balance the following equations:
   a. $Al + Cl_2 \longrightarrow AlCl_3$
   b. $P_4 + O_2 \longrightarrow P_4O_{10}$
   c. $C_4H_8 + O_2 \longrightarrow CO_2 + H_2O$
   d. $Sb_2S_3 + HCl \longrightarrow SbCl_3 + H_2S$
   e. $Fe_2O_3 + C \longrightarrow Fe + CO$

6.11 Balance the following equations:
   a. $Mg + AgNO_3 \longrightarrow Mg(NO_3)_2 + Ag$
   b. $CuCO_3 \longrightarrow CuO + CO_2$
   c. $Al + CuSO_4 \longrightarrow Cu + Al_2(SO_4)_3$
   d. $Pb(NO_3)_2 + NaCl \longrightarrow PbCl_2 + NaNO_3$
   e. $Al + HCl \longrightarrow AlCl_3 + H_2$

6.12 Balance the following equations:
   a. $Zn + H_2SO_4 \longrightarrow ZnSO_4 + H_2$
   b. $Al + H_2SO_4 \longrightarrow Al_2(SO_4)_3 + H_2$
   c. $K_2SO_4 + BaCl_2 \longrightarrow BaSO_4 + KCl$
   d. $CaCO_3 \longrightarrow CaO + CO_2$
   e. $Al_2(SO_4)_3 + KOH \longrightarrow Al(OH)_3 + K_2SO_4$

LEARNING GOAL

Identify a reaction as a combination, decomposition, replacement, or combustion reaction.

6.4 Types of Reactions

A great number of reactions occur in nature, in biological systems, and in the laboratory. However, there are some general patterns among all reactions that help us classify reactions. Some reactions may fit into two reaction types.

Figure 6.3 This combination reaction is

$$2\,Mg(s)\ +\ O_2(g)\ \xrightarrow{\Delta}\ 2MgO(s)$$

Q What happens to the reactants in a combination reaction?

Combination Reactions

In a **combination reaction,** two or more elements or simple compounds bond to form one product. For example, sulfur and oxygen combine to form the product sulfur dioxide.

Two or more   combine   a single
reactants   to yield   product

[A] + [B]   ⟶   [A B]

Combination reaction

⚪ + 🔴 ⟶ 🔴⚪

$$S(s)\ +\ O_2(g)\ \longrightarrow\ SO_2(g)$$

🧪 **the Chemistry place**

WEB TUTORIAL
Chemical Reactions and Equations

In Figure 6.3, the elements magnesium and oxygen combine to form a single product, magnesium oxide.

$$2Mg(s)\ +\ O_2(g)\ \longrightarrow\ 2MgO(s)$$

In other examples of combination reactions, elements or simple compounds combine to form a single product.

$$N_2(g)\ +\ 3H_2(g)\ \longrightarrow\ 2NH_3(g)\quad\text{Ammonia}$$
$$Cu(s)\ +\ S(s)\ \longrightarrow\ CuS(s)$$
$$MgO(s)\ +\ CO_2(g)\ \longrightarrow\ MgCO_3(s)$$

Decomposition Reactions

In a **decomposition reaction,** a reactant splits into two or more simpler products. For example, when mercury(II) oxide is heated, the products are mercury and oxygen.

A   splits   two or more
reactant   into   products

[A B]   ⟶   [A] + [B]

$$2HgO(s) \longrightarrow 2Hg(l) + O_2(g)$$

In another example of a decomposition reaction, calcium carbonate breaks apart into simpler compounds of calcium oxide and carbon dioxide.

$$CaCO_3(s) \longrightarrow CaO(s) + CO_2(g)$$

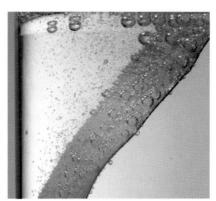

Figure 6.4 The equation for this single replacement reaction is

$$Zn(s) + 2HCl(aq) \longrightarrow ZnCl_2(aq) + H_2(g)$$

Q What changes in the formulas of the reactants identify this reaction as a single replacement?

Replacement Reactions

In replacement reactions, elements in compounds are replaced by other elements. In a **single replacement reaction,** a reacting element switches place with an element in the reacting compound.

In the single replacement reaction shown in Figure 6.4, zinc replaces hydrogen in hydrochloric acid, $HCl(aq)$.

$$Zn(s) + 2HCl(aq) \longrightarrow ZnCl_2(aq) + H_2(g)$$

In the following single replacement reaction, the halogen (Group 7A) chlorine replaces bromine in the compound potassium bromide.

$$Cl_2(g) + 2KBr(s) \longrightarrow 2KCl(s) + Br_2(l)$$

In a **double replacement reaction,** the positive ions in the reacting compounds switch places.

Single replacement

One element   replaces   another element in the compound

  +   ⟶

Double replacement

replace
Two elements   each other   in compounds

 + ⟶

For example, in the reaction shown in Figure 6.5, barium ions change places with sodium ions in the reactants to form sodium chloride and a white solid precipitate of barium sulfate.

$$BaCl_2(aq) + Na_2SO_4(aq) \longrightarrow BaSO_4(s) + 2NaCl(aq)$$

When sodium hydroxide and hydrochloric acid (HCl) react, sodium and hydrogen ions switch places, forming sodium chloride and water.

$$NaOH(aq) + HCl(aq) \longrightarrow NaCl(aq) + HOH(l)$$

Figure 6.5 The equation for this double replacement reaction is

$$Na_2SO_4(aq) + BaCl_2(aq) \longrightarrow BaSO_4(s) + 2NaCl(aq)$$

Q How do the changes in the formulas of the reactants identify this equation as a double replacement reaction?

Combustion Reactions

The burning of a log in a fireplace and the burning of fuel in the engine of a car are examples of combustion reactions. In a **combustion reaction,** fuel and oxygen are required and often the reaction produces carbon dioxide, water, and heat. For example, methane gas (CH_4) reacts with oxygen to produce carbon dioxide and water. The heat produced by this combustion reaction cooks our food and heats our homes.

$$CH_4(g) + 2O_2(g) \longrightarrow CO_2(g) + 2H_2O + \text{heat}$$

Combustion reactions also occur in the cells of the body in order to metabolize food, which provides building materials and energy for the activities we want to do. We absorb oxygen (O_2) from the air to burn glucose ($C_6H_{12}O_6$) from our food, and eventually our cells produce CO_2, H_2O, and energy.

$$C_6H_{12}O_6 + 6O_2 \longrightarrow 6CO_2 + 6H_2O + \text{energy}$$

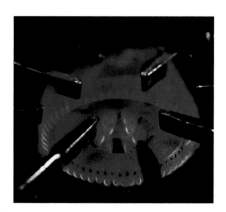

When a combustion reaction adds oxygen to form a single product, it can also be categorized as a combination reaction. For example, in the following reactions, sulfur and aluminum are burned in oxygen, producing oxide products.

$$S + O_2 \longrightarrow SO_2$$
$$2Al + 3O_2 \longrightarrow 2Al_2O_3$$

Table 6.4 summarizes the reaction types and gives examples.

Table 6.4   Summary of Reaction Types

Reaction Type	Example
Combination	
$A + B \longrightarrow AB$	$Ca(s) + Cl_2(g) \longrightarrow CaCl_2(s)$
Decomposition	
$AB \longrightarrow A + B$	$Fe_2S_3(s) \longrightarrow 2Fe(s) + 3S(s)$
Single Replacement	
$A + BC \longrightarrow AC + B$	$Cu(s) + 2AgNO_3(aq) \longrightarrow Cu(NO_3)_2(aq) + 2Ag(s)$
Double Replacement	
$AB + CD \longrightarrow AD + CB$	$BaCl_2(aq) + K_2SO_4(aq) \longrightarrow BaSO_4(s) + 2KCl(aq)$
Combustion	
$AH_4 + 2O_2 \longrightarrow AO_2 + 2H_2O + \text{heat}$	$CH_4(g) + 2O_2(g) \longrightarrow CO_2(g) + 2H_2O(g) + \text{heat}$
$B + O_2 \longrightarrow BO_2$	$S(s) + O_2(g) \longrightarrow SO_2(g)$

SAMPLE PROBLEM 6.4

Classification of Reactions

Classify the following reactions as combination, decomposition, single replacement, double replacement, or combustion.

a. $2Fe_2O_3 + 3C \longrightarrow 3CO_2 + 4Fe$
b. $2Na + Cl_2 \longrightarrow 2NaCl$
c. $Fe_2S_3 \longrightarrow 2Fe + 3S$
d. $C_2H_4 + 3O_2 \longrightarrow 2CO_2 + 2H_2O$
e. $BaCl_2 + K_2SO_4 \longrightarrow BaSO_4 + 2KCl$

Solution

a. single replacement   **b.** combination   **c.** decomposition
d. combustion   **e.** double replacement

Study Check

Nitrogen monoxide and oxygen react to form nitrogen dioxide. Write the balanced equation and identify the reaction type(s).

QUESTIONS AND PROBLEMS

Types of Reactions

6.13 a. Why is the following reaction called a decomposition reaction?

$$2Al_2O_3 \longrightarrow 4Al + 3O_2$$

b. Why is the following reaction called a single replacement reaction?

$$Br_2 + BaI_2 \longrightarrow BaBr_2 + I_2$$

HEALTH NOTE

Smog and Health Concerns

There are two types of smog. One, photochemical smog, requires sunlight to initiate reactions that produce pollutants such as nitrogen oxides and ozone. The other type of smog, industrial or London smog, occurs in areas where coal containing sulfur is burned and the unwanted product sulfur dioxide is emitted.

Photochemical smog is most prevalent in cities where people are dependent on cars for transportation. On a typical day in Los Angeles, for example, nitrogen monoxide (NO) emissions from car exhausts increase as traffic increases on the roads. The nitrogen monoxide is formed when N_2 and O_2 react at high temperatures in car and truck engines.

$$N_2 + O_2 \xrightarrow{\text{Heat}} 2NO$$

Then, NO reacts with oxygen in the air to produce NO_2, a reddish brown gas that is irritating to the eyes and damaging to the respiratory tract.

$$2NO + O_2 \longrightarrow 2NO_2$$

When NO_2 is exposed to sunlight, it is converted into NO and oxygen atoms.

$$NO_2 \xrightarrow{\text{Sunlight}} NO + O$$
$$\text{Oxygen atoms}$$

Oxygen atoms are so reactive that they combine with oxygen molecules in the atmosphere, forming ozone.

$$O + O_2 \longrightarrow O_3$$
$$\text{Ozone}$$

In the upper atmosphere (the stratosphere), ozone is beneficial because it protects us from harmful ultraviolet radiation that comes from the sun. However, in the lower atmosphere, ozone irritates the eyes and respiratory tract, where it causes coughing, decreased lung function, and fatigue. It also causes deterioration of fabrics, cracks rubber, and damages trees and crops.

Industrial smog is prevalent in areas where coal with a high sulfur content is burned to produce electricity. During combustion, the sulfur is converted to sulfur dioxide:

$$S + O_2 \longrightarrow SO_2$$

The SO_2 is damaging to plants, suppresses growth, and it is corrosive to metals such as steel. SO_2 is also damaging to humans and can cause lung impairment and respiratory difficulties. The SO_2 in the air reacts with more oxygen to form SO_3. SO_3 combines with water in the air to form sulfuric acid, which makes the rain acidic.

$$2SO_2 + O_2 \longrightarrow 2SO_3$$
$$SO_3 + H_2O \longrightarrow H_2SO_4$$
$$\text{Sulfuric acid}$$

The presence of sulfuric acid in rivers and lakes causes an increase in the acidity of the water, reducing the ability of animals and plants to survive.

6.14 **a.** Why is the following reaction called a combination reaction?

$$H_2 + Br_2 \longrightarrow 2HBr$$

b. Why is the following reaction called a double replacement reaction?

$$AgNO_3 + NaCl \longrightarrow AgCl + NaNO_3$$

6.15 Classify each of the following reactions as a combination, decomposition, single replacement, double replacement, or combustion reaction:
a. $4Fe + 3O_2 \longrightarrow 2Fe_2O_3$
b. $Mg + 2AgNO_3 \longrightarrow Mg(NO_3)_2 + 2Ag$
c. $CuCO_3 \longrightarrow CuO + CO_2$
d. $NaOH + HCl \longrightarrow NaCl + H_2O$
e. $C_4H_8 + 6O_2 \longrightarrow 4CO_2 + 4H_2O$
f. $ZnCO_3 \longrightarrow CO_2 + ZnO$
g. $Al_2(SO_4)_3 + 6KOH \longrightarrow 2Al(OH)_3 + 3K_2SO_4$
h. $Pb + O_2 \longrightarrow PbO_2$

6.16 Classify each of the following reactions as a combination, decomposition, single replacement, double replacement, or combustion reaction:
a. $CuO + 2HCl \longrightarrow CuCl_2 + H_2O$
b. $2Al + 3Br_2 \longrightarrow 2AlBr_3$
c. $2C_2H_2 + 5O_2 \longrightarrow 4CO_2 + 2H_2O$
d. $Pb(NO_3)_2 + 2NaCl \longrightarrow PbCl_2 + 2NaNO_3$
e. $2Mg + O_2 \longrightarrow 2MgO$
f. $Fe_2O_3 + 3C \longrightarrow 2Fe + 3CO$
g. $C_6H_{12}O_6 \longrightarrow 2C_2H_6O + 2CO_2$
h. $BaCl_2(aq) + K_2CO_3(aq) \longrightarrow BaCO_3(s) + 2KCl(aq)$

6.17 Try your hand at predicting the products that would result from the following types of reactions. Balance each equation you write:
a. combination: $Mg + Cl_2 \longrightarrow$
b. decomposition: $HBr \longrightarrow$
c. single replacement: $Mg + Zn(NO_3)_2 \longrightarrow$
d. double replacement: $K_2S + Pb(NO_3)_2 \longrightarrow$
e. combustion: $C_2H_6 + O_2 \longrightarrow$

6.18 Try your hand at predicting the products that would result from the reactions of the following. Balance each equation you write:
a. combination: $Ca + S \longrightarrow$
b. decomposition: $PbO_2 \longrightarrow$
c. single replacement: $KI + Cl_2 \longrightarrow$
d. double replacement: $CuCl_2 + Na_2S \longrightarrow$
e. combustion: $C_2H_4 + O_2 \longrightarrow$

6.5 Oxidation–Reduction Reactions

Oxidation–reduction reactions are defined as a loss and a gain of electrons among the reactants. In an oxidation reaction, one element or ion loses electrons. Simultaneously, a reduction reaction occurs as another element or ion gains electrons. You may like to use the shorthand phrase "LEO the lion goes GER" to remind yourself that a loss of electrons is oxidation (LEO), and a gain of electrons is a reduction (GER).

LEARNING GOAL

Define the terms oxidation and reduction.

When we looked at the formation of ionic compounds, we also saw that metals lost electrons and nonmetals gained electrons. Now we can say that metals are oxidized and nonmetals are reduced.

Consider the elements and ions in the reaction where calcium loses electrons to form calcium ion and where sulfur gains electrons to form sulfide ion. The elemental forms of calcium and sulfur have no ionic charge; they are neutral.

$$Ca + S \longrightarrow Ca^{2+} + S^{2-} = CaS$$

If we write the formation of each ion of the compound, we find that there is an oxidation and a reduction. In every oxidation–reduction reaction, the number of electrons exchanged among the reactants must be equal.

$$Ca \longrightarrow Ca^{2+} + 2e^- \quad \text{Oxidation (LEO)}$$
$$S + 2e^- \longrightarrow S^{2-} \quad \text{Reduction (GER)}$$

The loss of electrons is an **oxidation.** We say that calcium was oxidized. At the same time, the sulfur atom gained electrons. This gain of electrons is called **reduction.** We say that the sulfur was reduced. The two parts must occur simultaneously and with the same total number of electrons. We can write the two parts as follows.

$$\overset{\displaystyle \text{Oxidation}}{Ca + S \quad \longrightarrow \quad \underset{\displaystyle \text{Reduction}}{CaS}}$$

If we look at a single replacement reaction, we also find that it has two parts: an oxidation and a reduction. Consider the reaction of zinc and copper(II) sulfate. (See Figure 6.6.)

$$Zn + CuSO_4 \longrightarrow ZnSO_4 + Cu$$

We can rewrite the equation to show the ions of the compounds.

$$Zn + Cu^{2+} + SO_4^{2-} \longrightarrow Zn^{2+} + SO_4^{2-} + Cu$$

Now we can show that zinc loses 2 electrons to form Zn^{2+}, and Cu^{2+} gains 2 electrons to form Cu.

$$Zn \longrightarrow Zn^{2+} + 2e^- \quad \text{Oxidation}$$
$$Cu^{2+} + 2e^- \longrightarrow Cu \quad \text{Reduction}$$

In this single replacement reaction, zinc was oxidized and copper(II) ion was reduced.

Figure 6.6 In this single replacement reaction Zn (s) is oxidized to Zn^{2+} and Cu^{2+} is reduced to Cu(s):

$$Zn(s) + CuSO_4(aq) \longrightarrow$$
$$Cu(s) + ZnSO_4(aq)$$

Q In the oxidation, does Zn(s) lose or gain electrons?

SAMPLE PROBLEM 6.5

Oxidation–Reduction Reactions

In photographic film, the following decomposition reaction occurs in the presence of light. What is oxidized and what is reduced?

$$2AgBr \xrightarrow{\text{Light}} 2Ag + Br_2$$

Solution

In the AgBr compound, silver ion has an ionic charge of 1+ and the bromide ion has an ionic charge of 1−. We can write the reaction as follows.

$$2Ag^+ + 2Br^- \longrightarrow 2Ag + Br_2$$

We can determine the oxidation reaction and the reduction reaction by writing the changes for each ion separately.

$$2Br^- \longrightarrow Br_2 + 2e^- \quad \text{Oxidation}$$
$$2Ag^+ + 2e^- \longrightarrow 2Ag \quad \text{Reduction}$$

In the reaction, bromide ion is oxidized, and silver ion is reduced.

Study Check

In the following combination reaction, what is oxidized and what is reduced?

$$2Li + F_2 \longrightarrow 2LiF$$

Gain and Loss of Oxygen

An early definition described oxidation as a reaction that added oxygen to an element or compound. When you see rust forming on a car or a nail, the iron is being oxidized. Aluminum and sulfur are also oxidized when they add oxygen. In each reaction, oxygen is reduced to O^{2-}.

$$4Fe + 3O_2 \longrightarrow 2Fe_2O_3 \quad \text{Oxidation: add O}$$
$$4Al + 3O_2 \longrightarrow 2Al_2O_3 \quad \text{Oxidation: add O}$$
$$S + O_2 \longrightarrow SO_2 \quad \text{Oxidation: add O}$$

Loss and Gain of Hydrogen in Biological Systems

In organic and biochemical systems, the oxidation of carbon compounds often involves the loss of hydrogen atoms. Therefore, in a reduction reaction, there is a gain of hydrogen atoms. For example, methyl alcohol (CH_3OH), a poisonous substance, is metabolized in the body by the following reactions.

$$CH_3OH \longrightarrow H_2CO + 2H \quad \text{Oxidation: loss of H atoms}$$
Methyl alcohol   Formaldehyde

The formaldehyde can be oxidized further, this time by the addition of oxygen, to produce formic acid.

$$2H_2CO + O_2 \longrightarrow 2H_2CO_2 \quad \text{Oxidation: add O}$$
Formaldehyde   Formic acid

EXPLORE YOUR WORLD

Oxidation of Fruits and Vegetables

Freshly cut surfaces of fruits and vegetables discolor when exposed to oxygen in the air. Cut three slices of a fruit or vegetable such as apple, potato, avocado, or banana. Leave one piece on the kitchen counter (uncovered). Wrap one piece in plastic wrap and leave on the kitchen counter. Dip one piece in lemon juice (vitamin C, also called ascorbic acid, an antioxidant) and leave uncovered.

Questions

1. What changes take place in each sample after 1–2 hours?
2. Why would wrapping fruits and vegetables slow the rate of discoloration?
3. If lemon juice contains vitamin C (an antioxidant), why would dipping a fruit or vegetable in lemon juice affect the oxidation reaction on the surface of the fruit or vegetable?
4. Other kinds of antioxidants are vitamin E, citric acid, and BHT. Look for these antioxidants on the labels of cereals, potato chips, and other foods in your kitchen. Why are antioxidants added to food products that will be stored on our kitchen shelves?

Finally, formic acid is oxidized to carbon dioxide and water.

$$2H_2CO_2 + O_2 \longrightarrow 2CO_2 + 2H_2O \quad \text{Oxidation: add O}$$

Formic acid

The intermediate products of the oxidation of methyl alcohol are quite toxic, causing blindness and possibly death as they interfere with key reactions in the cells of the body.

In summary, we find that the particular definition of oxidation and reduction we use depends on the process that occurs in the reaction. All of these definitions are summarized in Table 6.5. Oxidation always involves a loss of electrons, but it may also be seen as an addition of oxygen, or the loss of hydrogen atoms. A reduction always involves a gain of electrons, and may also be seen as the loss of oxygen, or the gain of hydrogen.

Table 6.5   Characteristics of Oxidation and Reduction

Oxidation	
Always Involves	May Involve
Loss of electrons	Addition of oxygen
	Loss of hydrogen

Reduction	
Always Involves	May Involve
Gain of electrons	Loss of oxygen
	Gain of hydrogen

SAMPLE PROBLEM 6.6

Identification of Oxidation–Reduction Reactions

Determine if each of the following is an oxidation–reduction reaction.

a. $Zn + Cl_2 \longrightarrow ZnCl_2$
b. $C_2H_4 + 3O_2 \longrightarrow 2CO_2 + 2H_2O$
c. $BaCl_2 + Na_2CO_3 \longrightarrow BaCO_3 + 2NaCl$

Solution

a. Yes. Zinc loses electrons; it is oxidized. $Zn \longrightarrow Zn^{2+} + 2e^-$
   Chlorine gains electrons; it is reduced. $Cl_2 + 2e^- \longrightarrow 2Cl^-$
b. Yes. The addition of oxygen indicates that an oxidation–reduction reaction has occurred.
c. No. There are no changes in ionic charges between reactants and products. For example, barium is Ba^{2+} in $BaCl_2$ and in $BaCO_3$. The sodium is Na^+ and chlorine is Cl^- in all the compounds. There is no transfer of electrons in this double replacement reaction.

Study Check

Indicate what is oxidized and what is reduced in the following equation:

$$2Fe + 3Br_2 \longrightarrow 2FeBr_3$$

QUESTIONS AND PROBLEMS

Oxidation–Reduction Reactions

6.19 Indicate whether each of the following describes an oxidation or a reduction:
   a. loss of hydrogen atoms   b. gain of electrons
   c. addition of oxygen atoms   d. always accompanies oxidation

6.20 Indicate whether each of the following describes an oxidation or a reduction:
   a. loss of electrons   b. gain of hydrogen atoms
   c. loss of oxygen atoms   d. always accompanies reduction

6.21 In the following reactions, identify what is oxidized, and what is reduced.
   a. $Zn + Cl_2 \longrightarrow ZnCl_2$ b. $Cl_2 + 2NaBr \longrightarrow 2NaCl + Br_2$
   c. $2PbO \longrightarrow 2Pb + O_2$ d. $2Fe^{3+} + Sn^{2+} \longrightarrow 2Fe^{2+} + Sn^{4+}$

6.22 In the following reactions, identify what is oxidized, and what is reduced.
   a. $2Li + F_2 \longrightarrow 2LiF$ b. $Cl_2 + 2KI \longrightarrow 2KCl + I_2$

c. $Zn + Cu^{2+} \longrightarrow Zn^{2+} + Cu$

d. $Fe + CuSO_4 \longrightarrow FeSO_4 + Cu$

6.23 In the mitochondria of human cells, energy for the production of ATP is provided by the oxidation and reduction reactions of the iron ions in the cytochromes of the electron transport chain. Identify each of the following reactions as oxidation or reduction.

a. $Fe^{3+} + e^- \longrightarrow Fe^{2+}$ **b.** $Fe^{2+} \longrightarrow Fe^{3+} + e^-$

6.24 Chlorine (Cl_2) is a strong germicide used to disinfect drinking water and to kill microbes in swimming pools. If the product is Cl^-, was the elemental chlorine oxidized or reduced?

6.25 When linoleic acid, an unsaturated fatty acid, reacts with hydrogen, it forms a saturated fatty acid.

$$C_{18}H_{32}O_2 + 2H_2 \longrightarrow C_{18}H_{36}O_2$$

Is linoleic acid oxidized or reduced in the hydrogenation reaction?

6.26 In one of the reactions in the citric acid cycle, which provides energy for ATP synthesis, succinic acid is converted to fumaric acid.

$$\underset{\text{Succinic acid}}{C_4H_6O_4} \longrightarrow \underset{\text{Fumaric acid}}{C_4H_4O_4} + 2H$$

The reaction is accompanied by a coenzyme, flavin adenine dinucleotide (FAD).

$$FAD + 2H \longrightarrow FADH_2$$

a. Is succinic acid oxidized or reduced?

b. Is FAD oxidized or reduced?

c. Why would the two reactions occur together?

6.6 Energy in Chemical Reactions

For a chemical reaction to take place, the molecules of the reactants must come in contact with each other. If there are no collisions between the molecules (or atoms), no reaction is possible. When the molecules collide, bonds between atoms are broken and new bonds can form. The energy needed to break apart those bonds is called the **activation energy.** If the energy of a collision is less than the activation energy, the molecules bounce apart without reacting.

The concept of activation energy is analogous to climbing over a hill. To reach a destination on the other side, we must expend energy to climb to the top of the hill. Once we are at the top, we can easily roll down the other side. The energy needed to get us from our starting point to the top of the hill would be our activation energy.

Exothermic and Endothermic Reactions

As the bonds of the products form, energy is released into the reaction system. However, that energy may be greater or less than the activation energy of the reaction. The **heat of reaction** is the energy difference between the reactants and the products. In **exothermic reactions,** the energy of the products is lower than that of the reactants and heat is given off. In **endothermic reactions,** the energy of the products is higher than the energy of the reactants and heat must be absorbed for products to form.

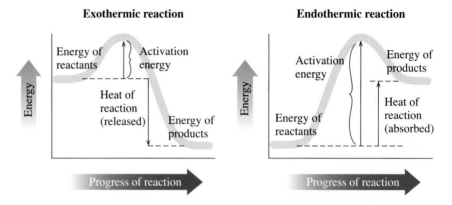

For example, the combustion of methane (CH_4) is an exothermic reaction. Heat (213 kcal per mole of CH_4 burned) is released along with the products CO_2 and H_2O.

Exothermic Reaction

$$CH_4 + 2O_2 \longrightarrow CO_2 + 2H_2O + 213 \text{ kcal}$$
Methane   Heat released

The formation of hydrogen iodide is an endothermic reaction. In the overall energy change, heat is absorbed when the hydrogen and iodine undergo reaction.

Endothermic Reaction

$$H_2 + I_2 + 12 \text{ kcal of heat} \longrightarrow 2HI$$
Heat absorbed

Reaction	Energy Change	Heat in the Equation
Exothermic	Heat released	Product side
Endothermic	Heat absorbed	Reactant side

SAMPLE PROBLEM 6.7

Exothermic and Endothermic Reactions

In the reaction of one mole of solid carbon with oxygen gas, the energy of the carbon dioxide product is 94 kcal lower than the energy of the reactants.

a. Is the reaction exothermic or endothermic?
b. What is the equation for the reaction?

Solution

a. exothermic   **b.** $C + O_2 \longrightarrow CO_2 + 94 \text{ kcal}$

Study Check

When two moles of ethanol (C_2H_5OH) react with oxygen (O_2), the products are carbon dioxide, water, and 326 kcal of heat. Write a balanced equation including the heat of reaction.

HEALTH NOTE

Hot Packs and Cold Packs

In a hospital, at a first-aid station, or at an athletic event, a *cold pack* may be used to reduce swelling from an injury, remove heat from inflammation, or decrease capillary size to lessen the effect of hemorrhaging. Inside the plastic container of a cold pack, there is a compartment containing solid ammonium nitrate (NH_4NO_3) that is separated from a compartment containing water. The pack is activated when it is hit or squeezed hard enough to break the walls between the compartments and cause the ammonium nitrate to mix with the water (shown as H_2O over the reaction arrow). In an endothermic process, each gram of NH_4NO_3 that dissolves absorbs 79 cal of heat from the water. The temperature drops and the pack becomes cold and ready to use:

Endothermic Reaction in a Cold Pack

$$6.3 \text{ kcal } + NH_4NO_3(s) \xrightarrow{H_2O} NH_4NO_3(aq)$$

Hot packs are used to relax muscles, lessen aches and cramps, and increase circulation by expanding capillary size. Constructed in the same way as cold packs, a hot pack may contain the salt $CaCl_2$. The dissolving of the salt in water is exothermic and releases 160 cal per gram of salt. The temperature rises and the pack becomes hot and ready to use:

Exothermic Reaction in a Hot Pack

$$CaCl_2(s) \xrightarrow{H_2O} CaCl_2(aq) + 18 \text{ kcal}$$

QUESTIONS AND PROBLEMS

Energy in Chemical Reactions

6.27 a. Why do chemical reactions require an energy of activation?
  b. What is the function of a catalyst?
  c. In an exothermic reaction, is the energy of the products higher or lower?
  d. Draw an energy diagram for an exothermic reaction.

6.28 a. What is measured by the heat of reaction?
  b. How does the heat of reaction differ in exothermic and endothermic reactions?
  c. In an endothermic reaction, is the energy of the products higher or lower than the reactants?
  d. Draw an energy diagram for an endothermic reaction.

6.29 Classify the following as exothermic or endothermic reactions:
  a. 125 kcal are released.
  b. In the energy diagram, the energy level of the products is higher than the reactants.
  c. The metabolism of glucose in the body provides energy.

6.30 Classify the following as exothermic or endothermic reactions:
  a. In the energy diagram, the energy level of the products is lower than the reactants.
  b. In the body, the synthesis of proteins requires energy.
  c. 30 kcal are absorbed.

6.31 Classify the following as exothermic or endothermic reactions:
  a. gas burning in a Bunsen burner:

$$CH_4(g) + 2O_2(g) \longrightarrow CO_2(g) + 2H_2O(g) + 213 \text{ kcal}$$

  b. dehydrating limestone:

$$Ca(OH)_2 + 15.6 \text{ kcal} \longrightarrow CaO + H_2O$$

  c. formation of aluminum oxide and iron from aluminum and iron(III) oxide:

$$2Al + Fe_2O_3 \longrightarrow Al_2O_3 + 2Fe + 204 \text{ kcal}$$

EXPLORE YOUR WORLD

Hotter or Colder?

Place 1 cup of water in a glass and test the temperature with your finger. You can use a candy thermometer if you have one. Add a tablespoon of baking soda and stir.

Questions

1. What happened to the temperature of water when you added baking soda?
2. Is the reaction exothermic or endothermic?
3. Would the heat for the reaction be written on the reactant or the product side of an equation for the reaction?
4. Predict what will happen if you add another tablespoon of baking soda. Test your prediction.

6.32 Classify the following as exothermic or endothermic reactions:
a. the combustion of propane:

$$C_3H_8 + 5O_2 \longrightarrow 3CO_2 + 4H_2O + 531 \text{ kcal}$$

b. the formation of "table" salt:

$$2Na + Cl_2 \longrightarrow 2NaCl + 196 \text{ kcal}$$

c. decomposition of phosphorus pentachloride:

$$PCl_5 + 16 \text{ kcal} \longrightarrow PCl_3 + Cl_2$$

LEARNING GOAL

Describe the factors that affect the rate of a reaction.

6.7 Rate of Reaction

The **rate** (or speed) **of reaction** is measured by the amount of reactant used up, or the amount of product formed, in a certain period of time. Reactions with low activation energies go faster than reactions with high activation energies. The rate of a reaction can be affected by the temperature, the amounts of reactants in the container, and catalysts.

At higher temperatures, the increase in kinetic energy makes the reactants move faster and collide more often, and provides more collisions with the required energy of activation. Reactions almost always go faster at higher temperatures. For every 10°C increase in temperature, most reaction rates approximately double. If we want food to cook faster, we raise the temperature. When body temperature rises, there is an increase in the pulse rate, rate of breathing, and metabolic rate. On the other hand, we slow down a reaction by lowering the temperature. We refrigerate perishable foods to make them last longer. In some cardiac surgeries, body temperature is lowered to 28°C so the heart can be stopped when less oxygen is required by the brain. This is also the reason that some people have survived submersion in icy lakes for long periods of time.

The rate of a reaction also increases when reactants are added. Then there are more collisions between the reactants and the reaction goes faster. For example, a patient having difficulty breathing may be given a breathing mixture with a higher oxygen content than the atmosphere. The increase in the number of oxygen molecules in the lungs increases the rate at which oxygen combines with hemoglobin. The increased rate of oxygenation of the blood means that the patient can breathe more easily.

$$Hb + O_2 \longrightarrow HbO_2$$
Hemoglobin   Oxygen   Oxyhemoglobin

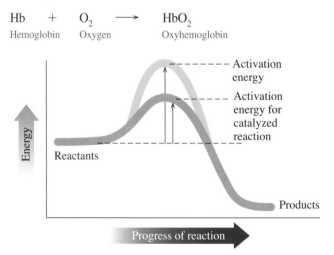

Another way to speed up a reaction is to lower the energy of activation. This can be done by adding a **catalyst,** which speeds up the reaction but is not itself changed or used up. Earlier, we discussed the energy required to climb a hill. If instead we found a tunnel through the hill, we won't need as much energy to get to the other side. A catalyst acts by providing an alternate pathway with a lower energy requirement. That makes it possible for more collisions to form product successfully. Catalysts have found many uses in industry. In the production of margarine, the reaction of hydrogen with vegetable oils is normally very slow. However, when finely divided platinum is present, the reaction occurs rapidly. In the body, biocatalysts called enzymes make most metabolic reactions go at the rates necessary for proper cellular activity.

A summary of the factors affecting reaction rate is given in Table 6.6.

Table 6.6   Factors That Increase Reaction Rate

Factor	Reason
More reactants	More collisions
Higher temperature	More collisions, more collisions with energy of activation
Adding a catalyst	Lowers energy of activation

SAMPLE PROBLEM 6.8

Factors That Affect the Rate of Reaction

Indicate whether the following changes will increase, decrease, or have no effect upon the rate of reaction:

a. increase in temperature
b. decrease in the number of reactants
c. adding a catalyst

Solution

a. increase   **b.** decrease   **c.** increase

Study Check

How does the lowering of temperature affect the rate of reaction?

EXPLORE YOUR WORLD

How Fast Can It Go?

Obtain three matching glasses. Half fill one with hot water, one with room temperature water, and one with cold (ice) water. Add a few drops of food coloring or ink to the water in each glass. Do not disturb.

Questions

1. In which glass does the color spread most rapidly?
2. In which water sample is the spread of color the slowest?
3. How does the temperature of each water sample affect the rate at which the color spreads throughout the water?

QUESTIONS AND PROBLEMS

Rate of Reaction

6.33 a. What is meant by the rate of a reaction?
 b. Why does bread grow mold more quickly at room temperature than in the refrigerator?

6.34 a. How does a catalyst affect the activation energy?
 b. Why is pure oxygen used in respiratory distress?

6.35 How would each of the following change the rate of the reaction shown here?

$$2SO_2 + O_2 \longrightarrow 2SO_3$$

a. adding SO_2 **b.** raising the temperature
c. adding a catalyst   **d.** removing some SO_2

6.36 How would each of the following change the rate of the reaction shown here?

$$2NO(g) + 2H_2(g) \longrightarrow N_2(g) + 2H_2O(g)$$

a. adding more NO   **b.** lowering the temperature
c. removing some H_2 **d.** adding a catalyst

the Chemistry place

WEB TUTORIAL
Equilibrium

6.8 Chemical Equilibrium

In many reactions, the products can interact and change back into the reactants. When a reaction occurs in both the forward and reverse directions, it is called a **reversible reaction.** For example, the reactants SO_2 and O_2 react to produce SO_3. Initially, the reaction of the reactants will go in the forward direction. As the product accumulates, more collisions occur between the SO_3 molecules to form the reactants again. This is known as the reverse reaction.

Forward reaction: $2SO_2(g) + O_2(g) \longrightarrow 2SO_3(g)$

Reverse reaction: $2SO_3(g) \longrightarrow 2SO_2(g) + O_2(g)$

The forward and reverse reactions are usually shown in a single equation by using a double arrow. A reversible reaction is two reactions that occur in opposite directions at the same time.

$$2SO_2(g) + O_2(g) \rightleftharpoons 2SO_3(g)$$

Chemical Equilibrium

Eventually the rate of the reverse reaction becomes equal to the rate of the forward reaction. Then the reaction has reached **chemical equilibrium.** There is no further change in the amounts of the reactants and the products. The same equilibrium mixture is obtained whether the reaction starts with the reactants or with just the products as shown in Figure 6.7. However, this does not mean that the amount of reactant becomes equal to the amount of product although that might happen in some reversible reactions. When SO_2 and O_2 are placed in a container and equilibrium is reached, there is mostly SO_3 in the container with only small amounts of SO_2 and O_2 (see Figure 6.8). This indicates that the forward reaction produces a large amount of product before the reverse reaction re-forms the reactants. In this case, we say that equilibrium favors the formation of products. Such a reaction is considered to be nearly complete.

In other reversible reactions, equilibrium is reached when only small amounts of products have formed. For example, when the reaction of $COCl_2$ reaches equilibrium, there is mostly $COCl_2$ and just small amounts of the products CO and Cl_2. The equilibrium favors the formation of the reactants as shown in Figure 6.9.

$$COCl_2(g) \rightleftharpoons CO(g) + Cl_2(g)$$

Figure 6.7 One sample initially contains SO_2 and O_2, while another sample contains only SO_3. At equilibrium, both have the same proportions of reactants and product.

Q Why is the same equilibrium mixture obtained from reactants as from products?

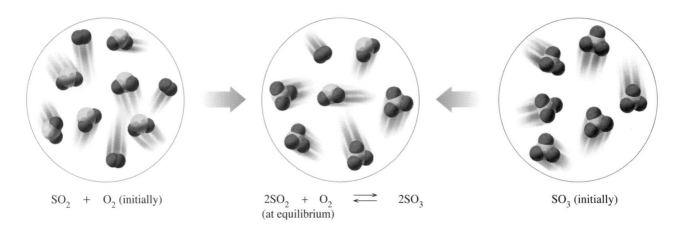

$SO_2 + O_2$ (initially) $2SO_2 + O_2 \rightleftharpoons 2SO_3$
(at equilibrium) SO_3 (initially)

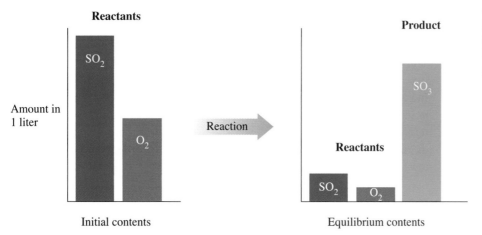

Figure 6.8 In the reaction of SO_2 and O_2, the equilibrium favors the formation of product SO_3.

Q If the closed container initially contains SO_3 molecules, how would the graph of contents look at equilibrium?

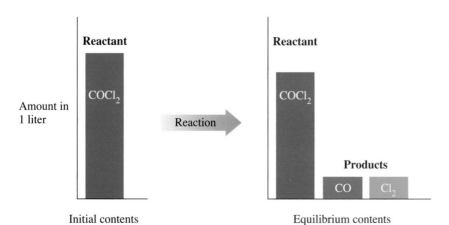

Figure 6.9 At equilibrium, the reaction $COCl_2 \rightleftharpoons CO + Cl_2$ favors the reactant because the reaction mixture at equilibrium contains mostly $COCl_2$.

Q Starting with only $COCl_2$ in a closed container, how do the forward and reverse reactions change as equilibrium is reached?

Effect of Concentration on Equilibrium

We can change the rate of the forward or reverse reaction by changing the ratio of products to reactants. When we alter a reaction that is at equilibrium, we say that there is a stress on the equilibrium. Then, according to **LeChâtelier's principle,** the rates of forward and reverse reactions will change to relieve that stress.

Consider the reaction of nitrogen and oxygen to form nitrogen monoxide. At equilibrium, the rates of the forward and reverse reactions are equal.

$$N_2(g) + O_2(g) \rightleftharpoons 2NO(g)$$

If O_2 is added to the system, it places a stress on the system. To remove this stress, the rate of the forward reaction increases in order to form more product. On the other hand, if some O_2 is removed from the equilibrium system, the rate of the reverse reaction increases to replenish O_2.

Stress: Add O_2
Shift: $\longrightarrow$
$$N_2(g) + O_2(g) \rightleftharpoons 2NO(g)$$
Stress: Remove O_2
Shift: $\longleftarrow$

If some NO is removed, the forward reaction will try to replace the NO. The reaction shifts toward the products.

Suppose that some NO is added to the equilibrium system. With more NO molecules, the reverse reaction speeds up. The reaction shifts toward the reactants.

Modeling an Equilibrium System

Obtain four or five tennis balls. Sit on the floor with another person sitting several feet away. Person 1 takes all the tennis balls. Begin the forward reaction by rolling the tennis balls, one by one, from person 1 to person 2. When person 2 receives two of the tennis balls, he or she can begin to roll the tennis balls back (reverse reaction).

Questions

1. What is the only direction possible at the beginning of this reaction (forward or reverse)?
2. When does your reaction reach equilibrium?
3. What happens to the equilibrium if you give two additional tennis balls to the reactant person?
4. What happens to the equilibrium if you take one or two of the product tennis balls out?

Stress: Remove NO
Shift: $\longrightarrow$

$$N_2(g) + O_2(g) \rightleftharpoons 2NO(g)$$

Stress: Add NO
Shift: $\longleftarrow$

In general, we can think of adding or removing one of the substances in a reaction as having the effect of *pushing* the reaction towards the products, or *pulling* the reaction back to the reactants. Adding more reactant *pushes* the forward reaction in the direction of the products. Adding more product *pushes* the reverse reaction back toward the reactants. If some product is removed, the reaction is *pulled* in the direction of the products. If a reactant is removed, the reaction is *pulled* in the direction of the reactants.

Effect of Temperature on Equilibrium

When the temperature of an equilibrium system is increased, the reaction that is favored is the one that removes the heat. For example, when heat is added to an exothermic reaction, the equilibrium favors the reactants because the reverse reaction requires heat. If heat is removed (temperature decreases) from an exothermic reaction, the forward reaction is favored because it produces heat.

Add heat

$$2SO_2(g) + O_2(g) \rightleftharpoons 2SO_3(g) + heat \qquad \text{Exothermic}$$

Remove heat

When heat is added to an endothermic reaction, the equilibrium favors the forward reaction because it absorbs heat. If heat is removed, the reverse reaction is favored because it produces heat.

Add heat

$$N_2(g) + O_2(g) + heat \rightleftharpoons 2NO(g) \qquad \text{Endothermic}$$

Remove heat

Table 6.7 summarizes the changes in equilibrium when changes are made in the amounts of products or reactants or temperature.

Table 6.7   Effects of Changes on Equilibrium

Factor	Change (stress)	Reaction Favored to Remove Stress
Concentration	Add more reactant	Forward
	Remove reactant	Reverse
	Add product	Reverse
	Remove product	Forward
Temperature	Raise T of endothermic reaction	Forward
	Lower T of endothermic reaction	Reverse
	Raise T of exothermic reaction	Reverse
	Lower T of exothermic reaction	Forward

SAMPLE PROBLEM 6.9

Factors Affecting Equilibrium

How do each of the following affect the equilibrium of the reaction shown here?

$$N_2(g) + 3H_2(g) \rightleftharpoons 2NH_3(g) + heat$$

a. add N_2
b. lower the temperature
c. add NH_3

Solution

a. The addition of N_2 pushes the reaction toward products; the equilibrium shifts toward the products.
b. A lower temperature removes heat, which favors the exothermic direction; the equilibrium shifts toward the products.
c. The addition of NH_3 favors the reverse reaction toward reactants; the equilibrium shifts toward the reactants.

Study Check

If some H_2 is removed from the reaction in the problem, how will the equilibrium be affected?

QUESTIONS AND PROBLEMS

Chemical Equilibrium

6.37 What is meant by the term "reversible reaction"?

6.38 When does a reversible reaction reach equilibrium?

6.39 Hydrogen chloride can be made by reacting hydrogen gas and chlorine gas.

$$H_2(g) + Cl_2(g) \rightleftharpoons 2HCl(g) + 44 kcal$$

What effect does each of the following changes have on equilibrium?
a. add H_2 **b.** add heat   **c.** remove HCl   **d.** remove Cl_2

6.40 Ammonia is produced by reacting nitrogen gas and hydrogen gas.

$$N_2(g) + 3H_2(g) \rightleftharpoons 2NH_3(g) + 22 kcal$$

What is the effect of each of the following changes on equilibrium?
a. remove N_2 **b.** remove heat
c. add NH_3 **d.** add H_2

6.41 In the lower atmosphere, oxygen is converted to ozone (O_3) by the energy provided from lightning.

$$3O_2(g) + heat \rightleftharpoons 2O_3(g)$$

What is the effect of each of the following changes on equilibrium?
a. add O_2 **b.** add O_3 **c.** add heat   **d.** remove heat

6.42 When heated, carbon reacts with water to produce carbon monoxide and hydrogen.

$$C(s) + H_2O(g) + heat \rightleftharpoons CO(g) + H_2(g)$$

What is the effect of each of the following changes on equilibrium?
a. add heat   **b.** remove heat
c. remove CO   **d.** add H_2O

Oxygen, Hemoglobin, and Carbon Monoxide Poisoning

When we take in air, the O_2 moves through our bronchial passages and diffuses through the air sacs in the lungs into the bloodstream. In the blood, a large molecule called hemoglobin (Hb) picks up the O_2 in a reversible reaction. At the air sacs in the lungs, the high concentration of O_2 favors the forward reaction to produce HbO_2.

$$Hb + O_2 \rightleftharpoons HbO_2$$

When a person is placed in a hyperbaric chamber where O_2 is at a higher concentration, the forward reaction favors the formation of more oxygenated hemoglobin. The increased amount of HbO_2 carries more O_2 to the tissues of the body, and helps to fight infections and cancer. In the body tissues, O_2 concentration is low, so the reverse reaction is favored and the HbO_2 releases the O_2.

Carbon monoxide (CO) is produced in an incomplete combustion reaction where ventilation is not adequate, such as in a furnace that isn't working properly. CO is attracted to Hb 140 times more strongly than O_2 is, and it attaches to the same binding site on the Hb as the O_2. When a molecule of HbCO forms, the Hb has no sites available to pick up O_2. In carbon monoxide poisoning, high concentrations of CO push the reaction toward the formation of HbCO in the blood, which can lead to dizziness and even death.

$$Hb + CO \rightleftharpoons HbCO$$

One way to treat carbon monoxide poisoning is to place a person in an atmosphere of pure oxygen. (Air is only 21% oxygen.) A higher concentration of O_2 will push the following reaction toward the products, which include the release of CO from the hemoglobin and a return to HbO_2.

$$HbCO + O_2 \rightleftharpoons HbO_2 + CO$$

the Chemistry place

CASE STUDY
Poison in the Home:
Carbon Monoxide

HEALTH NOTE

Homeostasis: Regulation of Body Temperature

In a physiological system of equilibrium called homeostasis, changes in our environment are balanced by changes in our bodies. It is crucial to our survival that we balance heat gain with heat loss. If we do not lose enough heat, our body temperature rises. At high temperatures, the body can no longer regulate our metabolic reactions. If we lose too much heat, body temperature drops. At low temperatures, essential functions proceed too slowly.

The skin plays an important role in the maintenance of body temperature. When the outside temperature rises, receptors in the skin send signals to the brain. The temperature-regulating part of the brain stimulates the sweat glands to produce perspiration. As perspiration evaporates from the skin, heat is removed and the body temperature is lowered.

In cold temperatures, epinephrine is released, causing an increase in metabolic rate, which increases the production of heat. Receptors on the skin signal the brain to contract the blood vessels. Less blood flows through the skin, and heat is conserved. The production of perspiration stops to lessen the heat lost by evaporation.

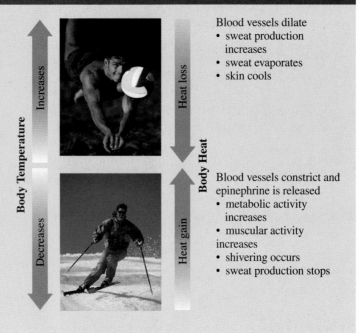

Blood vessels dilate
- sweat production increases
- sweat evaporates
- skin cools

Blood vessels constrict and epinephrine is released
- metabolic activity increases
- muscular activity increases
- shivering occurs
- sweat production stops

6.43 When you open a bottle of a soft drink, the drink eventually goes flat. How does LeChâtelier's principle explain this?

$$H_2CO_3(aq) \rightleftharpoons CO_2(g) + H_2O(l)$$

6.44 When you exercise, energy is produced by increasing the rate of the following reaction involving glucose. Why do you breathe faster when you exercise?

$$C_6H_{12}O_6(aq) + 6O_2(g) \rightleftharpoons 6CO_2(g) + 6H_2O(l) + energy$$

Glucose

Chapter Review

6.1 Chemical Changes

A chemical change occurs when the atoms of the initial substances rearrange to form new substances. When new substances form, a chemical reaction has taken place.

6.2 Chemical Equations

A chemical equation shows the formulas of the substances that react on the left of a reaction arrow, and the products that form on the right side of the reaction arrow.

6.3 Balancing a Chemical Equation

An equation is balanced by writing the smallest whole numbers (coefficients) in front of formulas to equalize the atoms of each element in the reactants and the products.

6.4 Types of Reactions

Many chemical reactions can be organized by reaction type: combination, decomposition, single replacement, double replacement, and/or combustion.

6.5 Oxidation–Reduction Reactions

When electrons are transferred in a reaction, it is an oxidation–reduction reaction. One reactant loses electrons, and another reactant gains electrons. Overall, the number of electrons lost and gained is equal.

6.6 Energy in Chemical Reactions

In a reaction, the reacting particles must collide with energy equal to or greater than the energy of activation. The heat of reaction is the energy difference between the initial energy of the reactants and the final energy of the products. In exothermic reactions, heat is released. In endothermic reactions, heat is absorbed.

6.7 Rate of Reaction

The rate of a reaction, which is the speed at which the reactants are converted to products, can be increased by adding more reactants, raising the temperature, or adding a catalyst.

6.8 Chemical Equilibrium

Chemical equilibrium occurs in a reversible reaction when the rate of the forward reaction is equal to the rate of the reverse reaction. The addition of reactants or removal of products favors the forward reaction. Removal of reactants or addition of products favors the reverse reaction. Adding or removing heat shifts the equilibrium of exothermic and endothermic reactions.

Key Terms

activation energy The energy needed upon collision to break apart the bonds of the reacting molecules.

balanced equation The final form of a chemical reaction that shows the same number of atoms of each element in the reactants and products.

catalyst A substance that increases the rate of reaction by lowering the activation energy.

chemical change The formation of a new substance with a different composition and properties than the initial substance.

chemical equation A shorthand way to represent a chemical reaction using chemical formulas to indicate the reactants and products.

chemical equilibrium The point at which the forward and reverse reactions take place at the same rate so that there is no further change in amounts of reactants and products.

chemical reaction The process by which a chemical change takes place.

coefficients Whole numbers placed in front of the formulas in an equation to balance the number of atoms or moles of atoms of each element in an equation.

combination reaction A reaction in which reactants combine to form a single product.

combustion reaction A reaction in which an element or a compound reacts with oxygen to form oxide products.

decomposition reaction A reaction in which a single reactant splits into two or more simpler substances.

double replacement reaction A reaction in which parts of two different reactants exchange places.

endothermic reaction A reaction that requires heat; the energy of the products is higher than the energy of the reactants.

exothermic reaction A reaction that releases heat; the energy of the products is lower than the energy of the reactants.

heat of reaction The energy released or absorbed during a chemical reaction equal to the energy difference between the reactants and products.

Le Châtelier's principle When a stress is placed on a system at equilibrium, the reaction shifts to relieve that stress.

oxidation The loss of electrons by a substance. Biological oxidation may involve the addition of oxygen, or the loss of hydrogen.

oxidation–reduction reaction A reaction in which the oxidation of one reactant is always accompanied by the reduction of another reactant.

physical change A change in which the physical properties change but the chemical composition does not change.

products The substances formed as a result of a chemical reaction.

rate of reaction The speed at which reactants are used to form product(s).

reactants The initial substances that undergo change in a chemical reaction.

reduction The gain of electrons by a substance. Biological reduction may involve the loss of oxygen or the gain of hydrogen.

reversible reaction A reaction in which a forward reaction occurs from reactants to products, and a reverse reaction occurs from products back to reactants.

single replacement reaction An element replaces a different element in a compound.

Additional Problems

6.45 Balance each of the following equations and identify the type of reaction.

    a. $Al + Cl_2 \longrightarrow AlCl_3$

    b. $Fe + H_2SO_4 \longrightarrow Fe_2(SO_4)_3 + H_2$

    c. $AgNO_3 + H_2S \longrightarrow Ag_2S + HNO_3$

    d. $Cl_2 + KI \longrightarrow KCl + I_2$

6.46 Balance each of the following equations and identify the type of reaction.

    a. $NI_3 \longrightarrow N_2 + I_2$

    b. $Mg + N_2 \longrightarrow Mg_3N_2$

    c. $Na_2SO_4 + Al(OH)_3 \longrightarrow Al_2(SO_4)_3 + NaOH$

    d. $Fe + HCl \longrightarrow FeCl_3 + H_2$

6.47 Complete and balance each of the following combustion reactions.

    a. $C_4H_8 + O_2 \longrightarrow$

    b. $C_5H_{12} + O_2 \longrightarrow$

    c. $C_6H_{12}O_6 + O_2 \longrightarrow$

6.48 Complete and balance each of the following combustion reactions.

    a. $C_2H_6 + O_2 \longrightarrow$

    b. $C_6H_6 + O_2 \longrightarrow$

    c. $C_3H_8O_3 + O_2 \longrightarrow$

6.49 Identify each of the following as an oxidation or a reduction reaction.

    a. $Zn^{2+} + 2e^- \longrightarrow Zn$

    b. $Al \longrightarrow Al^{3+} + 3e^-$

    c. $Pb \longrightarrow Pb^{2+} + 2e^-$

    d. $Cl_2 + 2e^- \longrightarrow 2Cl^-$

6.50 Write a balanced chemical equation for each of the following oxidation–reduction reactions.

    a. Sulfur reacts with molecular chlorine to form sulfur dichloride.

    b. Molecular chlorine and sodium bromide react to form molecular bromine and sodium chloride.

    c. Aluminum metal and iron(III) oxide react to produce aluminum oxide and elemental iron.

    d. Copper(II) oxide reacts with elemental C to form elemental copper and carbon dioxide.

6.51 In problem 6.50, show how electrons are lost and gained. Identify which element is oxidized and which element is reduced in each reaction.

6.52 When a gas barbecue is turned on, there is no reaction. Why is a match or lighter required to start burning the fuel?

6.53 In an endothermic reaction, nitrogen gas reacts with oxygen gas to produce nitrogen dioxide gas. Write the balanced equation for this reaction at equilibrium. Determine the effects on the equilibrium of each of the following:

    a. add heat   **b.** remove N_2

    c. add O_2 **d.** remove NO_2

    e. remove heat

6.54 The reaction between iron and oxygen that forms Fe_2O_3 also produces a lot of heat. However, the reaction is very slow. Would this indicate that the forward reaction has a high or low energy of activation? Explain.

6.55 At high altitudes, there is a lower concentration of O_2 in the air and less O_2 to combine with hemoglobin (Hb).

$$Hb + O_2 \rightleftharpoons HbO_2$$

People who live at high altitudes produce more hemoglobin molecules. How will an increase in Hb molecules affect the $Hb + O_2$ equilibrium?

6.56 Gasohol is a fuel containing ethyl alcohol (C_2H_6O) that burns in oxygen to give carbon dioxide, water, and heat. Write the balanced equation for the reaction. What is the effect of the following on the rate of the reaction?

    a. add oxygen   **b.** add a catalyst

7 Chemical Quantities

"In our food science laboratory I develop a variety of food products, from cake donuts to energy beverages," says Anne Cristofano, senior food technologist at Mattson & Company. "When I started the donut project, I researched the ingredients, then weighed them out in the lab. I added water to make a batter and cooked the donuts in a fryer. The batter and the oil temperature make a big difference. If I don't get the right taste or texture, I adjust the ingredients, such as sugar and flour, or adjust the temperature."

A food technologist studies the physical and chemical properties of food and develops scientific ways to process and preserve it for extended shelf life. The food products are tested for texture, color, and flavor. The results of these tests help improve the quality and safety of food.

LOOKING AHEAD

the Chemistry place

www.chemplace.com/college

Visit the URL above or use the CD-ROM in the book for extra quizzing, interactive tutorials, career resources, and case studies.

When we know the chemical equation for a reaction, we can determine the amount of product that can be produced. We do much the same thing at home when we use a recipe to make cookies or add the right quantity of water to make soup. At the automotive repair shop, a mechanic adjusts the carburetor or fuel injection system of an engine to allow for the correct amounts of fuel and oxygen so the engine will run properly. In the hospital, a respiratory therapist evaluates the level of CO_2 and O_2 in the blood. A certain amount of oxygen must reach the tissues for efficient metabolic reactions. If the oxygenation of the blood is low, the therapist will oxygenate the patient and recheck the blood levels.

In this chapter, we will describe the quantity of a compound as a mole and calculate the mass related to the formula of a compound. From a balanced equation, we can determine the mass and number of moles of a reactant and calculate the amount of product. Knowing how to determine the quantitative results of a chemical reaction is important to both chemists and medical personnel such as pharmacists and respiratory therapists.

LEARNING GOAL

Use Avogadro's number to determine the number of particles in a given number of moles.

WEB TUTORIAL
Stoichiometry

7.1 The Mole

At the store, you buy eggs by the dozen. In an office, pencils are ordered by the gross and paper by the ream. In a restaurant, soda is ordered by the case. In each of these examples, the terms *dozen, gross, ream,* and *case* count the number of items present. For example, when you buy a dozen eggs, you know you will get 12 eggs in the carton.

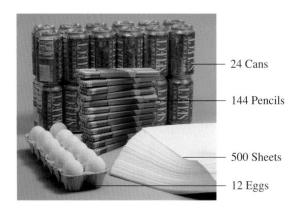

24 Cans

144 Pencils

500 Sheets

12 Eggs

In chemistry, particles such as atoms, molecules, and ions are counted by the **mole,** a unit that contains 6.02×10^{23} items. This very large number, called **Avogadro's number** after Amedeo Avogadro, an Italian physicist, looks like this when written with 3 significant figures:

Avogadro's Number

$602\ 000\ 000\ 000\ 000\ 000\ 000\ 000 = 6.02 \times 10^{23}$

One mole of any element always has Avogadro's number of atoms. For example, 1 mole of carbon contains 6.02×10^{23} carbon atoms; 1 mole of aluminum contains 6.02×10^{23} aluminum atoms; 1 mole of sulfur contains 6.02×10^{23} sulfur atoms.

1 mole of an element $= 6.02 \times 10^{23}$ atoms of that element

Table 7.1   Number of Particles in One-Mole Samples

Substance	Number and Type of Particles
1 mole carbon	6.02×10^{23} carbon atoms
1 mole aluminum	6.02×10^{23} aluminum atoms
1 mole sulfur	6.02×10^{23} sulfur atoms
1 mole water (H_2O)	6.02×10^{23} H_2O molecules
1 mole NaCl	6.02×10^{23} NaCl formula units
1 mole sucrose ($C_{12}H_{22}O_{11}$)	6.02×10^{23} sucrose molecules
1 mole vitamin C ($C_6H_8O_6$)	6.02×10^{23} vitamin C molecules

One mole of a compound contains Avogadro's number of molecules or formula units. Molecules are the particles of covalent compounds; **formula units** are the group of ions given by the formula of an ionic compound. For example, 1 mole of CO_2, a covalent compound, contains 6.02×10^{23} molecules of CO_2. One mole of NaCl, an ionic compound, contains 6.02×10^{23} formula units of NaCl (Na^+, Cl^-). Table 7.1 gives examples of the number of particles in some 1-mole quantities.

Avogadro's number tells us that one mole of any element or compound contains 6.02×10^{23} of the particular type of particles that make up a substance. We can use Avogadro's number as a conversion factor to convert between the moles of a substance and number of particles it contains.

$$\frac{6.02 \times 10^{23} \text{ particles}}{1 \text{ mole}} \quad \text{and} \quad \frac{1 \text{ mole}}{6.02 \times 10^{23} \text{ particles}}$$

SAMPLE PROBLEM 7.1

Calculating the Number of Particles in a Mole

a. How many atoms of aluminum are in 2.0 moles of Al?

b. How many moles of H_2O are in 2.6×10^{24} molecules of H_2O?

Solution

a. Avogadro's number is used to convert from moles to atoms of aluminum.

$$2.0 \text{ moles Al} \times \frac{6.02 \times 10^{23} \text{ atoms of Al}}{1 \text{ mole Al}} = 1.2 \times 10^{24} \text{ atoms of Al}$$

b. Avogadro's number is used to convert from molecules to moles of water.

$$2.6 \times 10^{24} \text{ molecules of } H_2O \times \frac{1 \text{ mole } H_2O}{6.02 \times 10^{23} \text{ molecules of } H_2O} = 4.3 \text{ moles } H_2O$$

Study Check

How many molecules of CO_2 are contained in 0.50 moles of CO_2?

Moles of Elements in a Formula

We have seen that the subscripts in a chemical formula of a compound indicate the number of atoms of each type of element. For example, in a molecule of aspirin, chemical formula $C_9H_8O_4$, there are 9 carbon atoms, 8 hydrogen atoms, and 4 oxygen atoms. The subscripts also state the number of moles of each element in one mole of aspirin: 9 moles of carbon atoms, 8 moles of hydrogen atoms, and 4 moles of oxygen atoms.

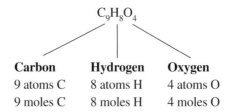

Carbon	**Hydrogen**	**Oxygen**
9 atoms C	8 atoms H	4 atoms O
9 moles C	8 moles H	4 moles O

SAMPLE PROBLEM 7.2

Moles of Elements in Compounds

a. How many moles of atoms of each element are present in 1 mole of caffeine, $C_8H_{10}N_4O_2$?

b. How many moles of chloride ions are in 2 moles of $AlCl_3$?

Solution

a. The subscripts in the formula of caffeine $C_8H_{10}N_4O_2$ are interpreted as moles: 8 moles C atoms, 10 moles H atoms, 4 moles N atoms, and 2 moles O atoms.

b. Any subscript in a formula can be written as a conversion factor for the moles of ions in the formula of an ionic compound.

$$2 \text{ moles } \cancel{AlCl_3} \times \frac{3 \text{ moles of } Cl^- \text{ ions}}{1 \text{ mole } \cancel{AlCl_3}} = 6 \text{ moles } Cl^- \text{ ions}$$

Study Check

Vitamin C (ascorbic acid) has the formula $C_6H_8O_6$. How many moles of each element are in 1 mole of vitamin C?

QUESTIONS AND PROBLEMS

The Mole

7.1 What is a mole?

7.2 What is Avogadro's number?

7.3 a. How many carbon atoms are in a mole of carbon?
    b. How many iron atoms are in 2.0 moles of iron?
    c. How many CO_2 molecules are in 0.10 mole of CO_2?

7.4 a. How many sulfur atoms are in 2.5 moles of sulfur?
    b. How many silver atoms are in 0.020 mole of silver?
    c. How many H_2O molecules are in 8.0 moles of H_2O?

7.5 Consider the formula for quinine $C_{20}H_{24}N_2O_2$.
    a. How many moles of hydrogen are in 1.0 mole of quinine?
    b. How many atoms of carbon are in 5.0 moles of quinine?
    c. How many atoms of nitrogen are in 0.020 mole of quinine?

7.6 Consider the formula for $Al_2(SO_4)_3$, which is used in antiperspirants.
    a. How many moles of sulfur are present in 3.0 moles of $Al_2(SO_4)_3$?
    b. How many aluminum ions are present in 0.40 mole $Al_2(SO_4)_3$?
    c. How many sulfate ions (SO_4^{2-}) are present in 1.5 moles $Al_2(SO_4)_3$?

7.7 a. How many atoms of C are in 2.5 moles of C?
    b. How many atoms of Cl are in 2.0 moles of PCl_3?

c. How many moles of SO_3 contain 5.0×10^{24} atoms of O?

7.8 a. How many atoms of Au are in 1.5 moles of Au?

  b. How many atoms of Br are in 0.30 moles of NBr_3?

  c. How many moles of C_2H_6O contain 5.0×10^{24} atoms of H?

7.2 Molar Mass

LEARNING GOAL

Given the chemical formula of a substance, calculate its molar mass.

WEB TUTORIAL
Stoichiometry

A single atom or molecule is much too small to weigh, even on the most accurate balance. In fact, it takes a huge number of atoms or molecules to make a piece of a substance that you can see. An amount of water that contains Avogadro's number of water molecules is only a few sips of water for you to drink. However, in the laboratory, we can use a balance to weigh out Avogadro's number of particles or 1 mole of a substance.

For any element, this quantity called its **molar mass** is the number of grams equal to the atomic mass of that element. We are counting out 6.02×10^{23} atoms of an element when we weigh out the number of grams equal to its molar mass. For example, if I need 1 mole of carbon (C) atoms, I would find the atomic mass of 12.01 for carbon on the periodic table. To obtain one mole of carbon atoms, I would weigh out 12.01 g of carbon. Thus, we can use the periodic table to determine the molar mass of any element because molar mass is the atomic mass in grams. One mole of sulfur has a molar mass of 32.06 grams and one mole of silver atoms has a molar mass of 107.9 grams.

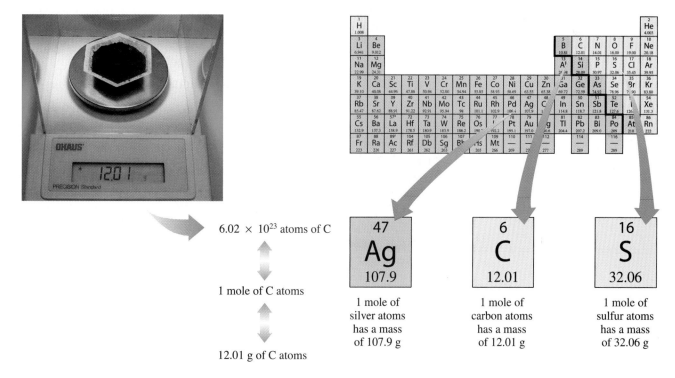

6.02 × 10²³ atoms of C

1 mole of C atoms

12.01 g of C atoms

| 47 |
| Ag |
| 107.9 |

1 mole of silver atoms has a mass of 107.9 g

| 6 |
| C |
| 12.01 |

1 mole of carbon atoms has a mass of 12.01 g

| 16 |
| S |
| 32.06 |

1 mole of sulfur atoms has a mass of 32.06 g

Molar Mass of Compounds

To determine the molar mass of a compound, multiply the molar mass of each element by its subscript in the formula, and add the results. For example, the molar

1 - Mole Quantities

| S | Fe | NaCl | $K_2Cr_2O_7$ | $C_{12}H_{22}O_{11}$ |

Figure 7.1   One-mole samples: sulfur, S (32.1 g), iron, Fe (55.9 g), salt, NaCl (58.5 g), potassium dichromate $K_2Cr_2O_7$ (294.2 g), and sugar, sucrose, $C_{12}H_{22}O_{11}$ (342.3 g).

Q How is the molar mass for $K_2Cr_2O_7$ obtained?

mass of sulfur trioxide, SO_3, is obtained by adding the molar masses of 1 mole of sulfur and 3 moles of oxygen.

Number of Moles in the Formula		*Molar Mass*		*Total Mass for Each*
1 mole S	×	$\dfrac{32.1 \text{ g S}}{1 \text{ mole S}}$	=	32.1 g S
3 moles O	×	$\dfrac{16.0 \text{ g O}}{1 \text{ mole O}}$	=	48.0 g O

Molar mass of SO_3 =   80.1 g

Figure 7.1 shows some 1-mole quantities of substances. Table 7.2 lists the molar mass for several 1-mole samples.

Table 7.2   The Molar Mass and Number of Particles in One-Mole Quantities

Substance	Molar Mass	Number of Particles in One Mole
1 mole carbon (C)	12.0 g	6.02×10^{23} C atoms
1 mole sodium (Na)	23.0 g	6.02×10^{23} Na atoms
1 mole iron (Fe)	55.9 g	6.02×10^{23} Fe atoms
1 mole NaF (preventative for dental caries)	42.0 g	6.02×10^{23} NaF formula units
1 mole $CaCO_3$ (antacid)	100.1 g	6.02×10^{23} $CaCO_3$ formula units
1 mole $C_6H_{12}O_6$ (glucose)	180.0 g	6.02×10^{23} glucose molecules
1 mole $C_8H_{10}N_4O_2$ (caffeine)	194.0 g	6.02×10^{23} caffeine molecules

SAMPLE PROBLEM 7.3

Calculating Molar Mass of Compounds

Find the molar mass of each compound.

a. $CaCO_3$, a compound used as an antacid and to replenish calcium in the diet

b. $Mg(NO_3)_2$ used in fireworks

Solution

a.

Number of Moles in the Formula		*Molar Mass*		*Total Mass for Each Element*
1 $\cancel{\text{mole Ca}}$	×	$\frac{40.1 \text{ g Ca}}{1 \cancel{\text{mole Ca}}}$	=	40.1 g Ca
1 $\cancel{\text{mole C}}$	×	$\frac{12.0 \text{ g C}}{1 \cancel{\text{mole C}}}$	=	12.0 g C
3 $\cancel{\text{moles O}}$	×	$\frac{16.0 \text{ g O}}{1 \cancel{\text{mole O}}}$	=	48.0 g O
		Molar mass of $CaCO_3$	=	100.1 g

b.

Number of Moles in the Formula		*Molar Mass*		*Total Mass for Each Element*
1 $\cancel{\text{mole Mg}}$	×	$\frac{24.3 \text{ g Mg}}{1 \cancel{\text{mole Mg}}}$	=	24.3 g Mg
2 $\cancel{\text{moles N}}$	×	$\frac{14.0 \text{ g N}}{1 \cancel{\text{mole N}}}$	=	28.0 g N
6 $\cancel{\text{moles O}}$	×	$\frac{16.0 \text{ g O}}{1 \cancel{\text{mole O}}}$	=	96.0 g O
		Molar mass of $Mg(NO_3)_2$	=	148.3 g

Study Check

Calculate the molar mass of salicylic acid, $C_7H_6O_3$.

EXPLORE YOUR WORLD

Calculating Moles in the Kitchen

The labels on food products list the components in grams and milligrams. It is possible to convert those values to moles, which is also interesting. Read the labels of some products in the kitchen and convert the amounts given in grams or milligrams to moles using molar mass. Here are some examples.

Questions

1. How many moles of NaCl are in a box of salt containing 746 g NaCl?
2. How many moles of sugar are contained in a 5-lb bag of sugar if sugar has the formula $C_{12}H_{22}O_{11}$?
3. A serving of cereal contains 90 mg potassium. If there are 11 servings of cereal in the box, how many moles of K^+ are present in the cereal in the box?

QUESTIONS AND PROBLEMS

Molar Mass

7.9 Calculate the molar mass (to 0.1 g) for each of the following compounds:

a. NaCl (table salt) **b.** Fe_2O_3 (rust)
c. Li_2CO_3 (antidepressant) **d.** $Al_2(SO_4)_3$ (antiperspirant)
e. $Mg(OH)_2$ (antacid) **f.** $C_{16}H_{19}N_3O_5S$ (Amoxicillin, an antibiotic)

7.10 Calculate the molar mass (to 0.1 g) for each of the following compounds:

a. $FeSO_4$ (iron supplement) **b.** Al_2O_3 (absorbent and abrasive)
c. $C_7H_5NO_3S$ (saccharin) **d.** C_3H_8O (rubbing alcohol)
e. $(NH_4)_2CO_3$ (baking powder) **f.** $Zn(C_2H_3O_2)_2$ (zinc dietary supplement)

7.3 Calculations Using Molar Mass

LEARNING GOAL

Given the number of moles of a substance, calculate the mass in grams; given the mass, calculate the number of moles.

The molar mass of an element or a compound is one of the most useful conversion factors in chemistry. Molar mass is used to change from moles of a substance to grams, or from grams to moles. To do these calculations, we use the molar mass as a conversion factor. For example, 1 mole of magnesium has a mass of 24.3 g. To express molar mass as an equality, we can write

WEB TUTORIAL
Stoichiometry

1 mole Mg = 24.3 g Mg

From this equality, two conversion factors can be written.

$$\frac{24.3 \text{ g Mg}}{1 \text{ mole Mg}} \quad \text{and} \quad \frac{1 \text{ mole Mg}}{24.3 \text{ g Mg}}$$

Conversion factors are written for compounds in the same way. For example, the molar mass of the compound H_2O is 18.0 g.

1 mole H_2O = 18.0 g H_2O

The conversion factors from the molar mass of H_2O are written as

$$\frac{18.0 \text{ g } H_2O}{1 \text{ mole } H_2O} \quad \text{and} \quad \frac{1 \text{ mole } H_2O}{18.0 \text{ g } H_2O}$$

We can now change from moles to grams, or grams to moles, using the conversion factors derived from the molar mass. (Remember, you must determine the molar mass of the substance first.)

SAMPLE PROBLEM 7.4

Converting Moles of an Element to Grams

Silver metal is used in the manufacture of tableware, mirrors, jewelry, and dental alloys. If the design for a piece of jewelry requires 0.750 mole of silver, how many grams of silver are needed?

Solution

Using the atomic mass, we write the molar mass of silver as an equality.

1 mole Ag = 107.9 g

From the equality, we can write conversion factors that relate grams and moles.

$$\frac{107.9 \text{ g Ag}}{1 \text{ mole Ag}} \quad \text{and} \quad \frac{1 \text{ mole Ag}}{107.9 \text{ g Ag}}$$

Unit plan: 0.750 mole Ag $\longrightarrow$ g Ag

The problem is set up by multiplying the given (0.750 mole Ag) by the conversion factor that cancels "mole Ag." For this problem, we use the first conversion factor.

$$0.750 \text{ mole Ag} \times \frac{107.9 \text{ g Ag}}{1 \text{ mole Ag}} = 80.9 \text{ g Ag}$$

Given   Molar mass   Answer (rounded to
          conversion factor   three significant figures)

Study Check

Calculate the number of grams of gold (Au) present in 0.124 mole of gold.

SAMPLE PROBLEM 7.5

Converting Moles of a Compound to Grams

Salicylic acid has a formula of $C_7H_6O_3$. How many grams of salicylic acid are in 2.25 moles of $C_7H_6O_3$?

Solution

The molar mass of salicylic acid is 138.0 g. The molar mass gives the equality

$$1 \text{ mole } C_7H_6O_3 = 138.0 \text{ g } C_7H_6O_3$$

From the molar mass, we can write two conversion factors:

$$\frac{138.0 \text{ g salicylic acid}}{1 \text{ mole salicylic acid}} \quad \text{and} \quad \frac{1 \text{ mole salicylic acid}}{138.0 \text{ g salicylic acid}}$$

Unit plan: 2.25 moles $C_7H_6O_3 \longrightarrow$ g $C_7H_6O_3$

In the problem setup, the given, 2.25 moles of salicylic acid, is multiplied by the first factor to cancel the unit "mole salicylic acid."

$$2.25 \cancel{\text{ moles salicylic acid}} \times \frac{138.0 \text{ g salicylic acid}}{1 \cancel{\text{ mole salicylic acid}}} = 311 \text{ g salicylic acid}$$

Given   Molar mass conversion factor   Answer (rounded to 3 SFs)

Study Check

Calculate the mass (grams) of 0.65 mole of ammonium carbonate, $(NH_4)_2CO_3$, a substance found in baking powders.

SAMPLE PROBLEM 7.6

Converting Mass of a Compound to Moles

Iodized salt provides sufficient iodine in a person's diet to prevent goiter, a condition that occurs in the thyroid when iodine levels are too low. Iodized salt is prepared by adding 0.010% by mass potassium iodide (KI) to table salt. If 0.0737 g of KI are added to each box of salt, how many moles of KI are required in the preparation of 250 boxes of salt?

Solution

The molar mass of KI is

$$1 \text{ mole KI} = 166.0 \text{ g KI}$$

Written in the form of conversion factors, this molar mass gives

$$\frac{166.0 \text{ g KI}}{1 \text{ mole KI}} \quad \text{and} \quad \frac{1 \text{ mole KI}}{166.0 \text{ g KI}}$$

Unit plan: boxes of salt $\longrightarrow$ g KI $\longrightarrow$ mole KI

The problem is set up with the given, 250 boxes of salt, multiplied by a factor of 0.0737 g KI per box. The final conversion factor cancels "g KI."

$$250 \cancel{\text{ boxes of salt}} \times \frac{0.0737 \cancel{\text{ g KI}}}{1 \cancel{\text{ box of salt}}} \times \frac{1 \text{ mole KI}}{166.0 \cancel{\text{ g KI}}} = 0.111 \text{ mole KI}$$

Given (exact)   Conversion factor   Molar mass   Answer (rounded to 3 SFs)
                                            conversion factor

Study Check

One gelcap of an antacid contains 311 mg $CaCO_3$ and 232 mg $MgCO_3$. In a recommended dosage of two gelcaps, how many moles each of $CaCO_3$ and $MgCO_3$ are present?

QUESTIONS AND PROBLEMS

Calculations Using Molar Mass

7.11 Calculate the number of grams in each of the following:

    a. 2.00 moles Na   **b.** 2.80 moles Ca

    c. 0.125 mole Sn

7.12 Calculate the number of grams in each of the following:

    a. 1.50 moles of K   **b.** 2.5 moles of C

    c. 0.25 mole P

7.13 Calculate the number of grams in each of the following:

    a. 0.500 mole NaCl   **b.** 1.75 moles Na_2O

    c. 0.225 mole H_2O

7.14 Calculate the number of grams in each of the following:

    a. 2.0 moles $MgCl_2$ **b.** 3.5 moles C_3H_8

    c. 5.00 moles C_2H_6O

7.15 How many grams are in 0.150 mole of each of the following?

    a. K   **b.** Cl_2 **c.** Na_2CO_3

7.16 How many grams are in 2.25 moles of each of the following?

    a. N_2 **b.** NaBr   **c.** C_6H_{14}

7.17 a. The compound $MgSO_4$ is called Epsom salts. How many grams will you need to prepare a bath containing 5.00 moles of Epsom salts?

    b. In a bottle of soda, there is 0.25 mole CO_2. How many grams of CO_2 are in the bottle?

7.18 a. Cyclopropane, C_3H_6, is an anesthetic given by inhalation. How many grams are in 0.25 mole of cyclopropane?

    b. The sedative Demerol hydrochloride has the formula $C_{15}H_{22}ClNO_2$. How many grams are in 0.025 mole of Demerol hydrochloride?

7.19 How many moles are contained in each of the following?

    a. 50.0 g Ag   **b.** 0.200 g C

    c. 15.0 g NH_3 **d.** 75.0 g SO_2

7.20 How many moles are contained in each of the following?

    a. 25.0 g Ca   **b.** 5.00 g S

    c. 40.0 g H_2O **d.** 100.0 g O_2

7.21 How many moles are in 25.0 g of each of the following compounds?

    a. CO_2 **b.** $Al(OH)_3$ **c.** $MgCl_2$

7.22 How many moles are in 4.00 g of each of the following compounds?

    a. NH_3 **b.** $Ca(NO_3)_2$ **c.** SO_3

7.23 A can of Drāno contains 480 g of NaOH. How many moles of NaOH are in the can of Drāno?

7.24 A gold nugget weighs 35.0 g. How many moles of gold are in the nugget?

7.25 How many moles of S are in each of the following quantities?

    a. 25 g of S   **b.** 125 g of SO_2 **c.** 275 g of Al_2S_3

7.26 How many moles of C are in each of the following quantities?

    a. 75 g of C   **b.** 0.25 mole of C_2H_6 **c.** 5.0×10^{24} molecules of CO_2

7.27 How many atoms of N are in each of the following quantities?

    a. 40.0 g of N   **b.** 1.5 moles of N_2O_4

    c. 2.0 moles of N_2

7.28 How many atoms of Ag are in each of the following quantities?

    a. 5.0 g of Ag   **b.** 0.40 moles of Ag_2S

    c. 0.75 g of AgCl

7.4 Percent Composition and Empirical Formulas

In our discussion of formulas in Chapter 4, we determined that the atoms of the elements in a compound are combined in a definite proportion. Thus a mole of any compound contains a definite proportion by mass of its elements. Using the molar mass of a compound, we can now determine the **percent composition** of a compound, which is the percent by mass of each element present. For example, we can use the molar mass of dinitrogen oxide, "laughing gas," N_2O, to calculate its percent composition.

WEB TUTORIAL
Stoichiometry

Step 1 Determine the molar mass of N_2O.

$$2 \text{ mole N} \quad \times \quad \frac{14.0 \text{ g N}}{1 \text{ mole N}} \quad = \quad 28.0 \text{ g N}$$

$$1 \text{ mole O} \quad \times \quad \frac{16.0 \text{ g O}}{1 \text{ mole O}} \quad = \quad 16.0 \text{ g O}$$

$$1 \text{ mole } N_2O \quad = \quad 44.0 \text{ g } N_2O$$

Step 2 Calculate the percent by mass of each element in N_2O.

$$\% \text{ N:} \quad \frac{28.0 \text{ g N}}{44.0 \text{ g } N_2O} \quad \times \quad 100 \quad = 63.6\% \text{ N}$$

$$\% \text{ O:} \quad \frac{16.0 \text{ g O}}{44.0 \text{ g } N_2O} \quad \times \quad 100 \quad = 36.4\% \text{ O}$$

Step 3 Check that the total percent is equal to 100.0%.

$$63.6\% + 36.4\% = 100.0\%$$

SAMPLE PROBLEM 7.7

Calculating Percent Composition

Calculate the percent composition of magnesium chloride, $MgCl_2$.

Solution

Step 1 Determine the molar mass of $MgCl_2$.

$$1 \text{ mole Mg} \quad \times \quad \frac{24.3 \text{ g Mg}}{1 \text{ mole Mg}} \quad = \quad 24.3 \text{ g Mg}$$

$$2 \text{ moles Cl} \quad \times \quad \frac{35.5 \text{ g}}{1 \text{ mole Cl}} \quad = \quad 71.0 \text{ g Cl}$$

$$1 \text{ mole } MgCl_2 \quad = \quad 95.3 \text{ g } MgCl_2$$

Step 2 Calculate the percent composition.

$$\% \text{ Mg:} \quad \frac{24.3 \text{ g Mg}}{95.3 \text{ g } MgCl_2} \quad \times \quad 100 \quad = \quad 25.5\% \text{ Mg}$$

$$\% \text{ Cl:} \quad \frac{71.0 \text{ g Cl}}{95.3 \text{ g } MgCl_2} \quad \times \quad 100 \quad = \quad 74.5\% \text{ Cl}$$

Step 3 Check that the total percent is equal to 100.0%.

$$25.5\% + 74.5\% = 100.0\%$$

Calculate the percent composition of potassium carbonate K_2CO_3.

Percent Composition of Common Substances

The percent composition can be calculated for products you may have in your medicine cabinet or kitchen. Practice the calculation of percent composition for some of the following substances.

1. salt $NaCl$
2. aspirin $C_9H_8O_4$
3. Penicillin $C_{16}H_{18}N_2O_5S$
4. sugar $C_{12}H_{22}O_{11}$
5. ibuprofen $C_{13}H_{18}O_2$
6. vanillin $C_8H_8O_3$

Empirical and Molecular Formulas

Up to now, the formulas you have seen written have been **molecular formulas,** which are the actual or true formulas of compounds. If we write a formula that represents the lowest whole-number ratio of the atoms in a compound, it is called the simplest or **empirical formula**. For example, the compound benzene, with molecular formula C_6H_6, has the empirical formula CH. Some molecular formulas and their empirical formulas are shown in Table 7.3.

Table 7.3   Examples of Molecular and Empirical Formulas

Name	Molecular (actual)	Empirical (simplest)
acetylene	C_2H_2	CH
benzene	C_6H_6	CH
ammonia	NH_3	NH_3
hydrazine	N_2H_4	NH_2
ribose	$C_5H_{10}O_5$	CH_2O
glucose	$C_6H_{12}O_6$	CH_2O

Finding Empirical Formulas

The empirical formula of a compound can be determined from the number of grams of the elements in the compound. Then the moles of each element can be calculated to find the simplest formula.

Step 1   Calculate the moles of each element.

Step 2   Divide each of the mole amounts by the smallest number of moles.

Step 3   Obtain a set of whole numbers and write as subscripts for the empirical formula.

SAMPLE PROBLEM 7.8

Calculating an Empirical Formula

A laboratory experiment indicates that a compound contains 6.88 g of iron (Fe) and 13.1 g of chlorine (Cl). Determine its empirical formula.

Solution

Step 1   Calculate the number of moles of each element.

$$6.88 \text{ g Fe} \times \frac{1 \text{ mole Fe}}{55.9 \text{ g Fe}} = 0.123 \text{ mole Fe}$$

$$13.1 \text{ g Cl} \times \frac{1 \text{ mole Cl}}{35.5 \text{ g Cl}} = 0.369 \text{ mole Cl}$$

Step 2   Divide each of the number of moles by the smallest mole value and round to whole numbers. In this case, the 0.123 mole Fe is the smallest mole value.

Chemistry of Fertilizers

Every year in the spring, homeowners and farmers add fertilizers to the soil to produce greener lawns and larger crops. Plants require several nutrients, but the major ones are nitrogen, phosphorus, and potassium. Nitrogen promotes green growth, phosphorus promotes strong root development for strong plants and abundant flowers, and potassium helps plants defend against diseases and weather extremes. The numbers on a package of fertilizer give the percentages each of N, P, and K by mass. For example, the numbers 30-3-4 indicate 30% nitrogen, 3% phosphorus, and 4% potassium.

Nitrogen (N) 30%
    Phosphorus (P) 3%
        Potassium (K) 4%

The major nutrient, nitrogen, is present in huge quantities as N_2 in the atmosphere, but plants cannot utilize nitrogen in this form. Bacteria in the soil convert atmospheric N_2 to usable forms by nitrogen fixation. To provide additional nitrogen to plants, several types of nitrogen-containing chemicals including ammonia, nitrates, and ammonium compounds are added to soil. The nitrates are absorbed directly, but ammonia and ammonium salts are first converted to nitrates by the soil bacteria.

Commercially, ammonia is prepared by a reaction of hydrogen and nitrogen. The ammonia gas is liquefied, placed in tanks, transported to farms, and applied directly to the soil. Ammonia is also reacted with acids to form ammonium salts.

$$3H_2(g) + N_2(g) \longrightarrow 2NH_3(g)$$
$$NH_3(aq) + HNO_3(aq) \longrightarrow NH_4NO_3(aq)$$
$$2NH_3(aq) + H_2SO_4(aq) \longrightarrow (NH_4)_2SO_4(aq)$$
$$2NH_3(aq) + H_3PO_4(aq) \longrightarrow (NH_4)_2HPO_4(aq)$$

The percent nitrogen in a fertilizer depends on the nitrogen compound. The percent nitrogen by mass in each type is calculated using percent composition.

Type of nitrogen fertilizer	*Percent nitrogen by mass*		
NH_3	$\dfrac{14.0 \text{ g N}}{17.0 \text{ g NH}_3}$	$\times$ 100%	= 82.3%
NH_4NO_3	$\dfrac{28.0 \text{ g N}}{80.0 \text{ g NH}_4NO_3}$	$\times$ 100%	= 35.0%
$(NH_4)_2 SO_4$	$\dfrac{28.0 \text{ g N}}{132.1 \text{ g } (NH_4)_2SO_4}$	$\times$ 100%	= 21.2%
$(NH_4)_2 HPO_4$	$\dfrac{28.0 \text{ g N}}{132.0 \text{ g } (NH_4)_2HPO_4}$	$\times$ 100%	= 21.2%

The choice of a fertilizer depends on its use and convenience. A fertilizer can be prepared as crystals or a powder, in a liquid solution, or as a gas such as ammonia. The ammonia and ammonium fertilizers are water soluble and quick-acting. Other forms may be made slow release by enclosing water-soluble ammonium salts in a thin plastic coating. The most commonly used fertilizer is NH_4NO_3 because it is easy to apply and has a high percent of N by mass.

$$\frac{0.123 \text{ mole Fe}}{0.123} = 1.00 \text{ mole Fe (rounds to 1)}$$

$$\frac{0.369 \text{ mole Cl}}{0.123} = 3.00 \text{ mole Cl (rounds to 3)}$$

Step 3 Write the whole numbers (1 and 3) as subscripts to obtain the empirical formula $FeCl_3$.

Study Check

A sample of a compound contains 3.24 g Na, 2.26 g S, and 4.51 g O. What is its empirical formula?

SAMPLE PROBLEM 7.9

Calculating Empirical Formula from Percent Composition

Calculate the empirical formula of a compound that has 74% Na and 26% O.

Clinical Laboratory Scientist

At a blood bank, blood is collected, typed, and tested for antibodies to determine appropriate transfusion therapy and compatibility of organ transplants. "I am placing reagents in the tubes and adding blood to the reagents to determine the blood type as A, B, AB, or O and Rh positive or negative," says Cecile Bush, Senior Clinical Laboratory Scientist. "I'm also setting up an antibody screen to see if the patient has any type of antibodies. It is very important that we make sure that the blood types of a blood or organ donor and recipient are compatible. If improperly matched, a recipient's immune system will produce antibodies that cause clotting in the blood, which blocks circulation in the capillaries of the recipient."

Solution

Step 1 Calculate the moles of each element. We must first change the percentages of the elements to grams by choosing a sample size of 100 g. Then the numerical values remain the same, but we have grams instead of the % signs. In 100 g of this compound, there are 74 g of Na and 26 g of O. Then the number of moles for each element is calculated.

$$74 \text{ g Na} \times \frac{1 \text{ mole Na}}{23.0 \text{ g Na}} = 3.2 \text{ mole Na}$$

$$26 \text{ g O} \times \frac{1 \text{ mole O}}{16.0 \text{ g O}} = 1.6 \text{ mole O}$$

Step 2 Divide each of the mole values by the smallest one.

$$\frac{3.2 \text{ mole Na}}{1.6} = 2.0 \text{ mole Na (rounds to 2)}$$

$$\frac{1.6 \text{ mole O}}{1.6} = 1.0 \text{ mole O (rounds to 1)}$$

Step 3 Use the whole number values to write the subscripts of the empirical formula, which is Na_2O.

Study Check

Determine the empirical formula of a compound that has a percent composition of K 44.9%, S 18.4%, and O 36.7%.

SAMPLE PROBLEM 7.10

Calculating an Empirical Formula

Calculate the empirical formula of a compound that has a percent composition 36% Al and 64% S.

Solution

Step 1 Calculate the moles of each element. First, we convert the percent composition of each element to the number of grams in a 100 g sample. In 100 g of this compound, there would be 36 g Al and 64 g S.

$$36 \text{ g Al} \times \frac{1 \text{ mole Al}}{27.0 \text{ g Al}} = 1.3 \text{ mole Al}$$

$$64 \text{ g S} \times \frac{1 \text{ mole S}}{32.1 \text{ g S}} = 2.0 \text{ mole S}$$

Step 2 Divide each of the mole amounts by the smallest one.

$$\frac{1.3 \text{ mole Al}}{1.3} = 1.0 \text{ mole Al}$$

$$\frac{2.0 \text{ mole S}}{1.3} = 1.5 \text{ moles S}$$

not 2.0

Step 3 At this point, we do not have a set of whole numbers. Whole numbers are obtained by multiplying all the values by a number (2 in this problem) that converts the fraction (1.5) to a whole number.

1.0 mole Al $\times$ 2   =   2 moles Al

1.5 moles S $\times$ 2   =   3 moles S

empirical formula Al_2S_3

Study Check

Determine the empirical formula of a compound that has a percent composition of Fe 70.0% and O 30.0%.

QUESTIONS AND PROBLEMS

Percent Composition and Empirical Formulas

7.29 Calculate the percent composition by mass of the following compounds:
   a. MgF_2 **b.** $Ca(OH)_2$ **c.** $C_4H_8O_4$

7.30 Calculate the percent composition by mass of sulfur in the following compounds:
   a. Na_2SO_4 **b.** Al_2S_3 **c.** SO_3

7.31 Calculate the empirical formula for each of the following compounds:
   a. 3.2 g N and 1.8 g O   **b.** 8.0 g C and 2.0 g H
   c. 1.6 g H, 22 g N, and 76 g O

7.32 Calculate the empirical formula for each of the following compounds:
   a. 2.9 g Ag and 0.43 g S   **b.** 22 g Na and 7.7 g O
   c. 19 g Na, 0.83 g H, 27 g S, and 53 g O

7.33 Calculate the empirical formula of each compound from the percent composition:
   a. 71% K, and 29% S   **b.** 55% Ga and 45% F
   c. 31% B and 69% O

7.34 Calculate the empirical formula of each compound from the percent composition:
   a. 56% Ca, 44% S   **b.** 78% Ba, 22% F
   c. 76% Zn, 24% P

7.5 Mole Relationships in Chemical Equations

In Chapter 6, we saw that equations are balanced in terms of the numbers of atoms in a chemical reaction. However, when we do experiments in the laboratory, prepare medications in the hospital, or follow a recipe for pancakes, we work with samples measured in grams that contain billions of atoms and molecules.

For any balanced equation the total amount of matter in the reactants must be equal to that of the products. Let us look at a pancake recipe and see what this means. If we weighed all of the ingredients, we would find that their total mass was equal to the mass of the pancakes. After those substances undergo a chemical reaction, their mass becomes part of the mass of the pancakes.

Reaction:

$$1 \text{ cup} \quad + \quad 3 \text{ eggs} \quad + \quad 2 \text{ cups} \qquad \longrightarrow \qquad 16 \text{ pancakes}$$
$$\text{milk} \qquad\qquad\qquad \text{pancake}$$
$$\text{mix}$$

Mass of the reactants   =   Mass of the products

Conservation of Mass

When tarnish forms, silver reacts with sulfur to form silver sulfide.

$$2Ag + S \longrightarrow Ag_2S$$

In this reaction, we see that the number of silver atoms that react is two times the number of sulfur atoms. If we had 200 silver atoms, we would need 100 sulfur atoms for a complete reaction. Now that we have learned the mole concept, we can interpret the equation in terms of moles. This equation tells us that 2 moles of silver is needed for every 1 mole of sulfur that reacts. Using the molar mass of each substance, we can also determine the mass in grams of the reactants and products. Let us look at all the ways we can now interpret a chemical equation:

$$2Ag + S \longrightarrow Ag_2S$$

2Ag   +   S   ⟶   Ag₂S

Mass of reactants   =   Mass of products

Number of atoms:

2 Ag atoms	+ 1 S atom	⟶ 1 Ag₂S formula unit (ions)
200 Ag atoms	+ 100 S atoms	⟶ 100 Ag₂S formula units
$2(6.02 \times 10^{23})$	$+ \ 6.02 \times 10^{23}$	⟶ 6.02×10^{23}
Ag atoms	S atoms	Ag₂S formula units

Number of moles:

2 moles Ag   + 1 mole S   ⟶ 1 mole Ag₂S

Number of grams:

2(107.9 g) Ag   + 32.1 g S   ⟶ 247.9 g Ag₂S

247.9 g reactants   ⟶ 247.9 g product

Conservation of Mass

Calculate the total mass of reactants and the total mass of products in the following equation:

$$CH_4 + 2O_2 \longrightarrow CO_2 + 2H_2O$$

Solution

Multiplying the number of moles (coefficients) in the equation by their molar masses gives the mass of the reactants and products.

	Reactants			*Products*	
	CH_4 +	$2O_2$	$\longrightarrow$	CO_2 +	$2H_2O$
Moles:	1 mole CH_4	2 mole O_2		1 mole CO_2	2 mole H_2O
Molar mass:	$\times$ 16.0 g/mole	$\times$ 32.0 g/mole		$\times$ 44.0 g/mole	$\times$ 18.0 g/mole
Mass:	16.0 g +	64.0 g	$\longrightarrow$	44.0 g +	36.0 g
Total mass:		80.0 g reactants	=	80.0 g products	

Study Check

For the following equation, compare the total mass of the reactants and products.

$$4NH_3 + 3O_2 \longrightarrow 2N_2 + 6H_2O$$

Mole–Mole Relationships in an Equation

When iron reacts with sulfur, the product is iron(III) sulfide.

$$2Fe(s) + 3S(s) \longrightarrow Fe_2S_3(s)$$

Iron (Fe)	+	Sulfur (S)		Iron(III) sulfide (Fe_2S_3)
$2Fe\ (s)$	+	$3S\ (s)$	$\longrightarrow$	$Fe_2S_3\ (s)$

Because the equation is balanced, we know the proportions of iron and sulfur in the reaction. For this reaction, we see that 2 moles of iron reacts with 3 moles of sulfur to form 1 mole of iron(III) sulfide. From the proportions given by the coefficients, we can write **mole–mole conversion factors** between reactants and product:

Fe and S: $\dfrac{2 \text{ moles Fe}}{3 \text{ moles S}}$ and $\dfrac{3 \text{ moles S}}{2 \text{ moles Fe}}$

Fe and Fe_2S_3: $\dfrac{2\ \text{moles Fe}}{1\ \text{mole Fe}_2S_3}$ and $\dfrac{1\ \text{mole Fe}_2S_3}{2\ \text{moles Fe}}$

S and Fe_2S_3: $\dfrac{3\ \text{moles S}}{1\ \text{mole Fe}_2S_3}$ and $\dfrac{1\ \text{mole Fe}_2S_3}{3\ \text{moles S}}$

SAMPLE PROBLEM 7.12

Writing Mole–Mole Conversion Factors

Consider the following balanced equation:

$$4Na + O_2 \longrightarrow 2Na_2O$$

Write the mole–mole conversion factors for
a. Na and O_2 **b.** Na and Na_2O

Solution

a. $\dfrac{4\ \text{moles Na}}{1\ \text{mole O}_2}$ and $\dfrac{1\ \text{mole O}_2}{4\ \text{moles Na}}$

b. $\dfrac{4\ \text{moles Na}}{2\ \text{moles Na}_2O}$ and $\dfrac{2\ \text{moles Na}_2O}{4\ \text{moles Na}}$

Study Check

Using the equation in Sample Problem 7.12, write the mole–mole conversion factors for O_2 and Na_2O.

WEB TUTORIAL
Stoichiometry

Using Mole–Mole Factors in Calculations

Whenever you prepare a recipe, adjust a carburetor for the proper mixture of fuel and air, or prepare medicines in a pharmaceutical laboratory, you need to know the proper amounts of reactants to use and how much of the product will form. Earlier we wrote all the possible conversion factors that can be obtained from the balanced equation: $2Fe + 3S \longrightarrow Fe_2S_3$. Now we will show how mole–mole factors are used in chemical calculations.

SAMPLE PROBLEM 7.13

Calculating Moles of a Reactant

In the reaction of iron and sulfur, how many moles of sulfur are needed to react with 6.0 moles of iron?

$$2Fe + 3S \longrightarrow Fe_2S_3$$

Solution

In this problem, we need to find the relationship between the moles of Fe and the moles of S.

$$\text{mole Fe} \longrightarrow \text{mole S}$$

Using the coefficients of Fe and S in the balanced equation, two mole–mole factors can be written:

$\dfrac{2\ \text{moles Fe}}{3\ \text{moles S}}$ and $\dfrac{3\ \text{moles S}}{2\ \text{moles Fe}}$

To solve the problem, we write the given quantity of 6.0 moles of Fe and convert to moles of S by using the mole–mole factor.

$$6.0 \text{ moles Fe} \times \frac{3 \text{ moles S}}{2 \text{ moles Fe}} = 9.0 \text{ moles of S}$$

Given   Mole–mole   Answer
               factor

Study Check

Using the equation in Sample Problem 7.13, calculate the number of moles of iron needed to completely react with 2.7 moles of sulfur.

SAMPLE PROBLEM 7.14

Finding the Moles of Product

In a combustion reaction, propane reacts with oxygen. How many moles of CO_2 can be produced when 6.0 moles of propane reacts?

$$C_3H_8 + 5O_2 \longrightarrow 3CO_2 + 4H_2O$$

Solution

We are given the moles of C_3H_8 and need to find the moles of CO_2.

$$\text{mole } C_3H_8 \longrightarrow \text{mole } CO_2$$

The coefficients in the equation tell us that 1 mole of C_3H_8 will produce 3 moles of CO_2, which can be written as two mole–mole factors as follows.

$$\frac{1 \text{ mole } C_3H_8}{3 \text{ moles } CO_2} \quad \text{and} \quad \frac{3 \text{ moles } CO_2}{1 \text{ mole } C_3H_8}$$

Now we can write the given of 6.0 moles C_3H_8 and multiply by the mole–mole factor that gives the moles of CO_2.

$$6.0 \text{ moles } C_3H_8 \times \frac{3 \text{ moles } CO_2}{1 \text{ mole } C_3H_8} = 18 \text{ moles } CO_2$$

Given   Mole–mole factor   Answer
               (coefficients)

Study Check

Using the equation in Sample Problem 7.14, calculate how many moles of oxygen must react to produce 2.0 moles of water.

QUESTIONS AND PROBLEMS

Mole Relationships in Chemical Equations

7.35 Give an interpretation of the following equations in terms of (1) number of particles and (2) number of moles:
   a. $2SO_2 + O_2 \longrightarrow 2SO_3$ **b.** $4P + 5O_2 \longrightarrow 2P_2O_5$

7.36 Give an interpretation of the following equations in terms of (1) number of particles and (2) number of moles:
   a. $2Al + 3Cl_2 \longrightarrow 2AlCl_3$ **b.** $4HCl + O_2 \longrightarrow 2Cl_2 + 2H_2O$

7.37 Calculate the total masses of the reactants and the products in each of the equations of problem 7.35.

7.38 Calculate the total masses of the reactants and the products in each of the balanced equations of problem 7.36.

7.39 Write all of the mole factors for the equations listed in problem 7.35.

7.40 Write all of the mole factors for the equations listed in problem 7.36.

7.41 The reaction of hydrogen with oxygen produces water.

$$2H_2(g) + O_2(g) \longrightarrow 2H_2O(g)$$

a. How many moles of O_2 are required to react with 2.0 moles of H_2?

b. If you have 5.0 moles of O_2, how many moles of H_2 are needed for the reaction?

c. How many moles of H_2O form when 2.5 moles of O_2 reacts?

7.42 Ammonia is produced by the reaction of hydrogen and nitrogen.

$$N_2(g) + 3H_2(g) \longrightarrow 2NH_3(g)$$
Ammonia

a. How many moles of H_2 are needed to react with 1.0 mole of N_2?

b. How many moles of N_2 reacted if 0.60 mole of NH_3 is produced?

c. How many moles of NH_3 are produced when 1.4 moles of H_2 reacts?

7.43 Carbon disulfide and carbon monoxide are produced when carbon is heated with sulfur dioxide.

$$5C + 2SO_2 \longrightarrow CS_2 + 4CO$$

a. How many moles of C are needed to react with 0.500 mole of SO_2?

b. How many moles of CO are produced when 1.2 moles of C reacts?

c. How many moles of SO_2 are required to produce 0.50 mole CS_2?

d. How many moles of CS_2 are produced when 2.5 moles of C reacts?

7.44 In the acetylene torch, acetylene gas (C_2H_2) burns in oxygen to produce carbon dioxide and water.

$$2C_2H_2(g) + 5O_2(g) \longrightarrow 4CO_2(g) + 2H_2O(g)$$

a. How many moles of O_2 are needed to react with 2.00 moles of C_2H_2?

b. How many moles of CO_2 are produced when 3.5 moles of C_2H_2 reacts?

c. How many moles of C_2H_2 are required to produce 0.50 mole H_2O?

d. How many moles of CO_2 are produced from 0.100 mole O_2?

LEARNING GOAL

Given the quantity in grams of a substance in a reaction, calculate the quantity of another substance in the reaction.

WEB TUTORIAL
Stoichiometry

7.6 Mass Calculations for Reactions

Often, when you perform a chemistry experiment, you need to use a laboratory balance to weigh out a certain mass of reactant. Once you know its mass you can convert grams to moles. Then, using mole–mole factors from the balanced equation, you can determine the moles and grams of other substances that react or are produced in the reaction.

SAMPLE PROBLEM 7.15

Calculating the Mass of a Product

In the formation of smog, nitrogen reacts with oxygen to produce nitrogen oxide. Calculate the grams of NO produced when 1.50 moles of O_2 reacts.

$$N_2(g) + O_2(g) \longrightarrow 2NO(g)$$

Solution

From the moles of O_2, you need to determine the mass, in grams, of NO.

mole O_2 $\longrightarrow$ g NO

From the equation we know that 1 mole of N_2 reacts with 1 mole of O_2 to form 2 moles of NO. First, we can change the given, 1.50 moles O_2, to moles of NO using the mole–mole factors derived from their coefficients in the balanced equation.

$$1.50 \text{ moles } O_2 \times \frac{2 \text{ moles NO}}{1 \text{ mole } O_2} = 3.00 \text{ moles NO}$$

Now the moles of NO can be converted to grams of NO using the molar mass.

$$3.00 \text{ moles NO} \times \frac{30.0 \text{ g NO}}{1 \text{ mole NO}} = 90.0 \text{ g NO}$$

These two steps can be combined to give a sequence that calculates the grams of NO from the moles of O_2.

mole O_2 $\longrightarrow$ mole NO $\longrightarrow$ g NO

The combined solution shows both conversion factors in the setup.

$$\underset{\text{Given}}{1.50 \text{ mole } O_2} \times \underset{\text{Mole factor}}{\frac{2 \text{ moles NO}}{1 \text{ mole } O_2}} \times \underset{\text{Molar mass NO}}{\frac{30.0 \text{ g NO}}{1 \text{ mole NO}}} = 90.0 \text{ g NO}$$

Study Check

Using the equation in Sample Problem 7.15, calculate how many grams of NO can be produced when 0.25 mole of N_2 reacts.

SAMPLE PROBLEM 7.16

Calculating the Mass of a Product

In a combustion reaction, acetylene (C_2H_2) burns with oxygen.

$$2C_2H_2 + 5O_2 \longrightarrow 4CO_2 + 2H_2O$$

How many grams of carbon dioxide are produced when 22.0 g of C_2H_2 are burned?

Solution

The question asks us to convert grams of C_2H_2 to grams of CO_2 produced.

g C_2H_2 $\longrightarrow$ g CO_2

First we convert g C_2H_2 to moles of C_2H_2. Then we change moles of C_2H_2 to moles of CO_2 by using their mole relationship in the balanced equation. Finally, the moles of CO_2 is changed to g CO_2.

g C_2H_2 $\longrightarrow$ mole C_2H_2 $\longrightarrow$ mole CO_2 $\longrightarrow$ g CO_2

$$\underset{\text{Given}}{22.0 \text{ g } C_2H_2} \times \underset{\text{Molar mass } C_2H_2}{\frac{1 \text{ mole } C_2H_2}{26.0 \text{ g } C_2H_2}} \times \underset{\text{Mole factor}}{\frac{4 \text{ moles } CO_2}{2 \text{ moles } C_2H_2}} \times \underset{\text{Molar mass } CO_2}{\frac{44.0 \text{ g } CO_2}{1 \text{ mole } CO_2}} = 74.5 \text{ g } CO_2$$

Study Check

Using the equation in Sample Problem 7.16, calculate how many grams of CO_2 can be produced when 25.0 g of O_2 reacts.

Summary of Calculations in Chemical Reactions

In all of these calculations using equations, we have used the molar masses of the substances and their mole factors from the coefficients in the balanced equation. We use the following steps to set up and solve problems that involve calculations and equations.

> **Steps in Finding the Moles or Masses of Substances in a Chemical Reaction**

Step 1 If necessary, convert the mass of substance A to moles using its molar mass.

grams of A $\longrightarrow$ moles of A

Step 2 Convert the moles of A to moles of the desired substance B using the mole factor derived from the coefficients of the balanced equation.

moles of A $\longrightarrow$ moles of B

Step 3 If needed, convert the moles of B to grams using its molar mass.

moles of B $\longrightarrow$ grams of B

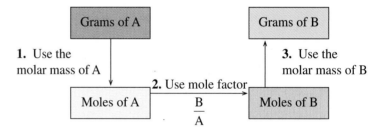

Think of the preceding diagram as a map. Locate the initial and final units for a problem and follow the steps to obtain your answer.

SAMPLE PROBLEM 7.17

Calculating the Mass of a Reactant

Propane gas, C_3H_8, a fuel for camp stoves and some specially equipped automobiles, reacts with oxygen to produce carbon dioxide and water. How many grams of O_2 are required to react with 22.0 g of C_3H_8?

$$C_3H_8(g) + 5O_2(g) \longrightarrow 3CO_2(g) + 4H_2O(g)$$

Solution

The equation indicates that 1 mole of C_3H_8 reacts with 5 moles of O_2 to produce 3 moles of CO_2 and 4 moles of H_2O. The 22.0 g of C_3H_8 are first changed to moles of C_3H_8. Then the moles of C_3H_8 are changed to moles of O_2, which can be converted to grams of O_2.

$$\text{g } C_3H_8 \longrightarrow \text{mole } C_3H_8 \longrightarrow \text{mole } O_2 \longrightarrow \text{g } O_2$$

$$22.0 \text{ g } \cancel{C_3H_8} \times \frac{1 \text{ mole } \cancel{C_3H_8}}{44.0 \text{ g } \cancel{C_3H_8}} \times \frac{5 \text{ moles } \cancel{O_2}}{1 \text{ mole } \cancel{C_3H_8}} \times \frac{32.0 \text{ g } O_2}{1 \text{ mole } \cancel{O_2}} = 80.0 \text{ g } O_2$$

Given   Molar mass C_3H_8 Mole factor   Molar mass O_2

Study Check

Using the equation in Sample Problem 7.17, calculate how many grams of C_3H_8 are needed to produce 15.0 g of H_2O.

QUESTIONS AND PROBLEMS

Mass Calculations for Reactions

7.45 Sodium reacts with oxygen to produce sodium oxide.

$$4Na(s) + O_2(g) \longrightarrow 2Na_2O(s)$$

a. How many grams of Na_2O are produced when 2.5 moles of Na react?
b. If you have 18.0 g of Na, how many grams of O_2 are required for reaction?
c. How many grams of O_2 are needed in a reaction that produces 75.0 g of Na_2O?

7.46 Nitrogen gas reacts with hydrogen gas to produce ammonia by the following equation:

$$N_2(g) + 3H_2(g) \longrightarrow 2NH_3(g)$$

a. If you have 1.8 moles of H_2, how many grams of NH_3 can be produced?
b. How many grams of H_2 are needed to react with 2.80 g of N_2?
c. How many grams of NH_3 can be produced from 12 g of H_2?

7.47 Ammonia and oxygen react to form nitrogen and water.

$$4NH_3 + 3O_2 \longrightarrow 2N_2 + 6H_2O$$
Ammonia

a. How many grams of O_2 are needed to react with 8.0 moles of NH_3?
b. How many grams of N_2 can be produced when 6.50 g of O_2 reacts?
c. How many grams of water are formed from the reaction of 34 g of NH_3?

7.48 Iron(III) oxide reacts with carbon to give iron and carbon monoxide.

$$Fe_2O_3 + 3C \longrightarrow 2Fe + 3CO$$

a. How many grams of C are required to react with 2.50 moles of Fe_2O_3?
b. How many grams of CO are produced when 36.0 g of C reacts?
c. How many grams of Fe can be produced when 6.00 g of Fe_2O_3 reacts?

7.7 Percent Yield

LEARNING GOAL

Given the actual quantity of product, determine the percent yield for the reaction.

Up to this point, we have solved calculations with equations as though the amount of product were the maximum amount possible or 100%. However, most chemical reactions do not produce 100% of the possible product. There are side reactions that give other substances. Some of the product may be lost as we transfer products from one container to another. The equilibrium for the reaction can also prevent the complete conversion of reactants to products.

Suppose we are running a chemical reaction in the laboratory. We first weigh out the reactants and calculate how much product we can expect. The amount of product we obtain by calculation is the **theoretical yield** for the reaction. After the reaction ends, the amount of product we collect is the **actual yield.** We can explain this using an analogy of exam scores from class. If an exam has 100 points, the best score theoretically is 100 points, which could be obtained if a student answers all the questions correctly. If a student earns 85 points, the actual score is 85.

If we know the actual yield and the theoretical yield of a product obtained in the laboratory, we can use the two values to express the **percent yield** of a reaction. For example, the student who earned 85 points out of 100 points possible received a percent yield of 85% on the exam.

$$\text{Percent yield} = \frac{\text{actual yield}}{\text{theoretical yield}} \times 100\%$$

The actual yield will be less than the expected theoretical yield, which means that the percent yield must be less than 100%. To calculate percent yield, we use the following steps:

Step 1 Use the balanced equation to calculate the theoretical yield of product from the given reactant.

Step 2 Calculate the percent yield by setting up the actual yield (given) divided by the theoretical yield.

SAMPLE PROBLEM 7.18

Calculating Percent Yield

Nitrogen reacts with hydrogen to produce ammonia.

$$N_2(g) + 3H_2(g) \longrightarrow 2NH_3(g)$$

What is the percent yield if the reaction of 35.5 g of N_2 produces 26.0 g of NH_3?

Solution

Step 1 The theoretical yield is calculated from the unit plan:

$$g\ N_2 \longrightarrow \text{moles } N_2 \longrightarrow \text{moles } NH_3 \longrightarrow g\ NH_3$$

$$35.5\ \cancel{g\ N_2} \times \frac{1\ \cancel{\text{mole } N_2}}{28.0\ \cancel{g\ N_2}} \times \frac{2\ \cancel{\text{moles } NH_3}}{1\ \cancel{\text{mole } N_2}} \times \frac{17.0\ g\ NH_3}{1\ \cancel{\text{mole } NH_3}} = 43.1\ g\ NH_3$$

Step 2 The ratio is set up to calculate the percent yield.

$$\text{Percent yield} = \frac{26.0\ g\ NH_3\ \text{(actual yield)}}{43.1\ g\ NH_3\ \text{(theoretical yield)}} \times 100\% = 60.3\%$$

This percent yield means that 60.3% of the possible amount of NH_3 was produced by the reaction.

Study Check

If 5.00 g H_2 reacts according to the equation in Sample Problem 7.18, what is the percent yield if 22.0 g of NH_3 are produced?

QUESTIONS AND PROBLEMS

Percent Yield

7.49 Carbon disulfide is produced by the reaction of carbon and sulfur dioxide

$$5C(s) + 2SO_2(g) \longrightarrow CS_2(g) + 4CO(g)$$

a. What is the percent yield for the reaction if 40.0 g of carbon produces 36.0 g of carbon disulfide?

b. What is the percent yield for the reaction if 32.0 g of sulfur dioxide produces 12.0 g of carbon disulfide?

7.50 Iron(III) oxide reacts with carbon monoxide to produce iron and carbon dioxide.

$$Fe_2O_3(s) + 3CO(g) \longrightarrow 2Fe(s) + 3CO_2(g)$$

a. What is the percent yield for the reaction if 65.0 g of iron(III) oxide produces 15.0 g of iron?

b. What is the percent yield for the reaction if 75.0 g carbon monoxide produces 15.0 g of carbon dioxide?

7.51 Aluminum reacts with oxygen to produce aluminum oxide.

$$4Al(s) + 3O_2(g) \longrightarrow 2Al_2O_3(s)$$

The reaction of 50.0 g aluminum and sufficient oxygen has a 75.0% yield. How many grams of aluminum oxide are produced?

7.52 Propane (C_3H_8) burns in oxygen to produce carbon dioxide and water.

$$C_3H_8(g) + 5O_2(g) \longrightarrow 3CO_2(g) + 4H_2O(g)$$

Calculate the mass of CO_2 that can be produced if the reaction of 45.0 g of propane and sufficient oxygen has a 60.0% yield.

7.8 Equilibrium Constants

LEARNING GOAL

Calculate the equilibrium constant K_{eq} for a reversible reaction at equilibrium given the moles of reactants and products in one liter of reaction mixture.

In Chapter 6, we learned that reversible reactions reach equilibrium when the forward and reverse reactions take place at the same rate. We might think of a ski resort as an analogy. Early in the morning, most people are riding up the ski lift. Eventually, the number of skiers riding on the ski lift becomes equal to the skiers going down the slopes. If the number of people riding the ski lift becomes equal to the number of people skiing down the slopes, we would say that equilibrium is reached. There is no further change in the number of skiers on the hill at any given time.

At equilibrium, the amounts of reactants and products in a container become constant. That makes it possible to express the relationship of reactants and products in any system at equilibrium as a ratio of the products to the reactants known as an **equilibrium constant**, K_{eq}.

WEB TUTORIAL
Equilibrium

$$K_{eq} = \frac{\text{products}}{\text{reactants}}$$

For the ski-lift analogy, this could be written as

$$K_{eq} = \frac{\text{Skiers going down the slope}}{\text{Skiers going up the ski lift}}$$

Consider the general reaction

$$aA + bB \rightleftharpoons cC + dD$$

In this equation, the capital letters A and B are the reactants and C and D are the products. The small letters are their coefficients in the balanced equation. In the equilibrium constant, the products are multiplied in the numerator and divided by the reactants in the denominator.

$$K_{eq} = \frac{[C]^c[D]^d}{[A]^a[B]^b}$$

The brackets around each substance indicate the concentration, which is the number of moles of that substance in one liter of the reaction mixture. The coefficients for each substance are placed as superscripts or exponents.

SAMPLE PROBLEM 7.19

Write the K_{eq}

Write the expression for the equilibrium constant for the following reactions:

a. $H_2(g) + I_2(g) \rightleftharpoons 2HI(g)$

b. $3H_2(g) + N_2(g) \rightleftharpoons 2NH_3(g)$

Solution

a. The product HI is placed in a bracket and the coefficient of 2 is used as an exponent in the numerator of the K_{eq} expression. The reactants H_2 and I_2 are placed in brackets in the denominator.

$$K_{eq} = \frac{[HI]^2}{[H_2][I_2]}$$

b. The product NH_3 is placed in a bracket and the coefficient of 2 is used as an exponent in the numerator of the K_{eq} expression. The reactants N_2 and H_2 are placed in brackets in the denominator and the coefficient of 3 is used as the exponent for the H_2. The coefficient of 1 for N_2 is understood.

$$K_{eq} = \frac{[NH_3]^2}{[N_2][H_2]^3}$$

Study Check

Write the balanced chemical equation that would give the following equilibrium constant expression.

$$K_{eq} = \frac{[NO_2]^2}{[NO]^2[O_2]}$$

When the K_{eq} is greater than 1 (>1), the equilibrium favors the products. Consider the following reaction, which has a large K_{eq}.

$$2CO(g) + O_2(g) \rightleftharpoons 2CO_2(g)$$

$$K_{eq} = \frac{[CO_2]^2}{[CO]^2[O_2]} = 2 \times 10^{11} \quad \frac{\text{Mostly product}}{\text{Little reactant}} \quad \text{equilibrium favors products}$$

A K_{eq} that is less than 1 (<1) favors the reactants. Consider the following reaction, which has a small K_{eq}.

$$N_2(g) + O_2(g) \rightleftharpoons 2NO(g)$$

$$K_{eq} = \frac{[NO]^2}{[N_2][O_2]} = 2 \times 10^{-9} \quad \frac{\text{Little product}}{\text{Mostly reactant}} \quad \text{equilibrium favors reactants}$$

SAMPLE PROBLEM 7.20

Equilibrium Position

For the following reactions and their equilibrium constants, indicate whether the equilibrium favors the reactants or products:

a. $2H_2(g) + O_2(g) \rightleftharpoons 2H_2O(g)$ $\quad K_{eq} = 3 \times 10^{81}$

b. $PCl_5(g) \rightleftharpoons PCl_3(g) + Cl_2(g)$ $\quad K_{eq} = 2 \times 10^{-2}$

HEALTH NOTE

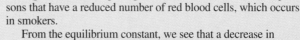

Oxygen-Hemoglobin Equilibrium and Hypoxia

In Chapter 6, we looked at the equilibrium between hemoglobin (Hb), oxygen, and oxyhemoglobin.

$$Hb + O_2 \rightleftharpoons HbO_2$$

When the O_2 level is high in the alveoli of the lung, the reaction favors the product HbO_2. In the tissues where O_2 concentration is low, the reverse reaction releases the oxygen from the hemoglobin. The equilibrium constant is written

$$K_{eq} = \frac{[HbO_2]}{[Hb][O_2]}$$

At normal atmospheric pressure, oxygen diffuses into the blood because the partial pressure of oxygen in the alveoli is higher than that in the blood. At altitudes above 8,000 ft, the decrease in the amount of oxygen in the air results in a significant reduction of oxygen to the blood and body tissues. At an altitude of 18,000 feet, a person will obtain 29% less oxygen. When oxygen levels are lowered, a person may experience hypoxia, which has symptoms that include increased respiratory rate, headache, decreased mental acuteness, fatigue, decreased physical coordination, nausea, vomiting, and cyanosis. A similar problem occurs in persons with a history

of lung disease that impairs gas diffusion in the alveoli or in persons that have a reduced number of red blood cells, which occurs in smokers.

From the equilibrium constant, we see that a decrease in oxygen will shift the equilibrium to the reactants. Such a shift depletes the concentration of HbO_2 and causes the hypoxia condition.

$$Hb + O_2 \longleftarrow HbO_2$$

Immediate treatment of altitude sickness includes hydration, rest, and if necessary, descending to a lower altitude. The adaptation to lowered oxygen levels requires about 10 days. During this time the bone marrow increases red blood cell production providing more red blood cells and hemoglobin. A person living at a high altitude can have 50% more red blood cells than someone at sea level. This increase in hemoglobin causes a shift in the equilibrium back to HbO_2 product. Eventually the higher concentration of HbO_2 will provide more oxygen to the tissues and the symptoms of hypoxia will lessen.

$$Hb + O_2 \longrightarrow HbO_2$$

For some who climb high mountains, it is important to stop and acclimatize for several days at increasing altitudes. At very high altitudes, it may be necessary to use an oxygen tank.

Solution:

a. A K_{eq} greater than 1 favors the products.
b. A K_{eq} less than 1 favors the reactants.

Study Check

A reaction has a K_{eq} of about 1. What can you say about the position of equilibrium for the reactants and products?

QUESTIONS AND PROBLEMS

Equilibrium Constants

7.53 Write the expression for the equilibrium constant K_{eq} for each of the following reactions:

    a. $CH_4(g) + 2H_2S(g) \rightleftharpoons CS_2(g) + 4H_2(g)$
    b. $2NO(g) \rightleftharpoons N_2(g) + O_2(g)$
    c. $2SO_3(g) + CO_2(g) \rightleftharpoons CS_2(g) + 4O_2(g)$

7.54 Write the expression for the equilibrium constant K_{eq} for each of the following reactions:

    a. $2HBr(g) \rightleftharpoons H_2(g) + Br_2(g)$
    b. $CO(g) + 2H_2(g) \rightleftharpoons CH_3OH(g)$
    c. $CH_4(g) + 2H_2S(g) \rightleftharpoons CS_2(g) + 4H_2(g)$

7.55 Use the value of the K_{eq} to determine if the equilibrium favors the reactants or the products for each of the following reactions

a. $H_2(g) + F_2(g) \rightleftharpoons 2HF(g)$ $K_{eq} = 1 \times 10^{13}$

b. $2H_2O(g) \rightleftharpoons 2H_2(g) + O_2(g)$ $K_{eq} = 3 \times 10^{-82}$

7.56 Use the value of the K_{eq} to determine if the equilibrium favors the reactants or the products for each of the following reactions

a. $N_2(g) + O_2(g) \rightleftharpoons 2NO(g)$ $K_{eq} = 2 \times 10^{-9}$

b. $2SO_2(g) + O_2(g) \rightleftharpoons 2SO_3(g)$ $K_{eq} = 8 \times 10^2$

Chapter Review

7.1 The Mole

One mole of an element contains 6.02×10^{23} atoms; a mole of a compound contains 6.02×10^{23} molecules or formula units.

7.2 Molar Mass

The molar mass (g/mole) of any substance is the mass in grams equal numerically to its atomic mass, or the sum of the atomic masses, which have been multiplied by their subscripts in a formula.

7.3 Calculations Using Molar Mass

The molar mass is the mass in grams of 1 mole of an element or a compound. It becomes an important conversion factor when it is used to change a quantity in grams to moles, or to change a given number of moles to grams.

7.4 Percent Composition and Empirical Formulas

The percent composition of a compound is obtained by dividing the mass in grams of each element in a compound by the molar mass of that compound. The empirical formula can be calculated by determining the mole ratio from the grams of the elements present in a sample of a compound or the percent composition.

7.5 Mole Relationships in Chemical Equations

In a balanced equation, the total mass of the reactants is equal to the total mass of the products. The coefficients in

an equation describing the relationship between the moles of any two components are used to write mole–mole factors. When the number of moles for one substance is known, a mole–mole factor is used to find the moles of a different substance in the reaction.

7.6 Mass Calculations for Reactions

In calculations using equations, the molar masses of the substances and their mole factors are used to change the number of grams of one substance to the corresponding grams of a different substance.

7.7 Percent Yield in a Chemical Reaction

The percent yield of a reaction indicates the percent of product actually produced during a reaction. The percent yield is calculated by dividing the actual yield in grams of a product by the theoretical yield in grams.

7.8 Equilibrium Constants

An equilibrium constant K_{eq} is the ratio of the amount of products to the amount of reactants in 1.0 L of a reaction mixture of a reversible reaction. The components are raised to powers with exponents that are equal to the coefficients in the chemical equation. A large value of K_{eq} indicates that the equilibrium favors the products and has gone nearly to completion whereas value of K_{eq} that is less than one favors the reactants.

Key Terms

actual yield The actual amount of product produced by a reaction.

Avogadro's number The number of items in a mole, equal to 6.02×10^{23}.

empirical formula The simplest or smallest whole-number ratio of the atoms in a formula.

equilibrium constant The ratio of the amount of products to the amount of reactants in 1.0 L of a reaction mixture of a reversible reaction with each component raised to an exponent equal to the coefficient of that compound in the chemical equation.

formula unit The group of ions represented by the formula of an ionic compound.

molar mass The mass in grams of 1 mole of an element equal numerically to its atomic mass. The molar mass of a compound is equal to the sum of the masses of the elements in the formula.

mole A group of atoms, molecules, or formula units that contains 6.02×10^{23} of these items.

molecular formula The actual formula that gives the number of atoms of each type of element in the compound.

mole–mole conversion factor A conversion factor that relates the number of moles of two compounds in an equation derived from their coefficients.

percent composition The percent by mass of the elements in a formula.

percent yield The ratio of the actual yield of a reaction to the theoretical yield possible for the reaction.

theoretical yield The maximum amount of product that a reaction can produce from a given amount of reactant.

Additional Problems

7.57 Calculate the percent composition for each of the following compounds:

a. K_2CrO_4 **b.** $Al(HCO_3)_3$ **c.** $C_6H_{12}O_6$

7.58 Determine the percent phosphorus in each of the following compounds:

a. P_4O_{10} **b.** Mg_3P_2 **c.** $Ca_3(PO_4)_2$

7.59 Calculate the empirical formula of each compound that contains

a. 2.20 g S and 7.81 g F

b. 6.35 g Ag, 0.825 g N, and 2.83 g O

c. 89.2 g Au and 10.9 g O

7.60 Calculate the empirical formula of each compound from the percent composition.

a. 61% Sn and 39% F

b. 25.9% N and 74.1% O

c. 22.1% Al, 25.4% P, and 52.5% O

7.61 A typical human body contains 60.0% water by mass. If a person weighs 70.0 kg, how many moles of water are present in that person's body?

7.62 A blood level of 400.0 mg of alcohol per 100.0 mL of blood can cause coma and possibly be fatal. How many moles of alcohol (C_2H_6O) are present in 10.0 mL of blood?

7.63 At a winery, glucose ($C_6H_{12}O_6$) in grapes undergoes fermentation to produce ethanol (C_2H_6O) and carbon dioxide.

$$C_6H_{12}O_6 \longrightarrow 2C_2H_6O + 2CO_2$$
Glucose   Ethanol

a. How many moles of glucose are required to form 124 g of ethanol?

b. How many grams of ethanol would be formed from the reaction of 0.240 kg of glucose?

7.64 Gasohol is a fuel containing ethanol (C_2H_6O) that burns in oxygen (O_2) to give carbon dioxide and water.

a. State the reactants and products for this reaction in the form of a balanced equation.

b. How many moles of O_2 are needed to completely react with 4.0 moles of C_2H_6O?

c. If a car produces 88 g of CO_2, how many grams of O_2 are used up in the reaction?

d. If you add 125 g of C_2H_6O to your gas, how many grams of CO_2 and H_2O can be produced?

7.65 Balance the following equation:

$$NH_3 + F_2 \longrightarrow N_2F_4 + HF$$

a. How many moles of each reactant are needed to produce 4.00 moles of HF?

b. How many grams of F_2 are required to react with 1.50 moles of NH_3?

c. How many grams of N_2F_4 can be produced when 3.40 g of NH_3 reacts?

7.66 When 50.0 g of iron(III) oxide reacts with carbon monoxide, 32.8 g of iron is produced. What is the percent yield of the reaction?

$$Fe_2O_3(s) + 3CO(g) \longrightarrow 2Fe(s) + 3CO_2(g)$$

7.67 Acetylene gas C_2H_2 burns in oxygen to produce carbon dioxide and water. If 62.0 g of CO_2 are produced when 22.5 g of C_2H_2 react with sufficient oxygen, what is the percent yield for the reaction?

7.68 A reaction of nitrogen and sufficient hydrogen produced 30.0 g of ammonia, which is a 65.0% yield for the reaction. How many grams of nitrogen reacted?

$$N_2(g) + 3H_2(g) \longrightarrow 2NH_3(g)$$

8 Gases

"When oxygen levels in the blood are low, the cells in the body don't get enough oxygen," says Sunanda Tripathi, registered nurse, Santa Clara Valley Medical Center. "We use a nasal cannula to give supplemental oxygen to a patient. At a flow rate of 2 liters per minute, a patient breathes in a gaseous mixture that is about 28% oxygen compared to 21% in ambient air."

When a patient has a breathing disorder, the flow and volume of oxygen into and out of the lungs is measured. A ventilator may be used if a patient has difficulty breathing. When pressure is increased, the lungs expand. When the pressure of the incoming gas is reduced, the lung volume contracts to expel carbon dioxide. These relationships—known as gas laws—are an important part of ventilation and breathing.

LOOKING AHEAD

the
Chemistry
place

www.chemplace.com/college

Visit the URL above or use the CD-ROM in the book for extra quizzing, interactive tutorials, career resources, and case studies.

We all live at the bottom of a sea of gases called the atmosphere. The most important of these gases is oxygen, which constitutes about 21% of the atmosphere. Without oxygen, life on this planet would be impossible because oxygen is vital to all life processes of plants and animals. Ozone (O_3), formed in the upper atmosphere by the interaction of oxygen with ultraviolet light, absorbs some of the harmful radiation before it can strike the earth's surface. The other gases in the atmosphere include nitrogen (78% of the atmosphere), argon, carbon dioxide (CO_2), and water vapor. Carbon dioxide gas, a product of combustion and metabolism, is used by plants in photosynthesis, which produces the oxygen that is essential for humans and animals.

The atmosphere has become a dumping ground for other gases, such as methane, chlorofluorohydrocarbons (CFCs), sulfur dioxide, and nitrogen oxides. The chemical reactions of these gases with sunlight and oxygen in the air are contributing to air pollution, ozone depletion, global warming, and acid rain. Such chemical changes can seriously affect our health and the way all of us live. A knowledge of gases and some of the laws that govern gas behavior can help us understand the nature of matter and allow us to make decisions concerning important environmental and health issues.

8.1 Gases and Kinetic Theory

LEARNING GOAL

Describe the kinetic theory of gases.

WEB TUTORIAL
Properties of Gases

The behavior of gases is quite different from that of liquids and solids. Gas particles are far apart, whereas particles of both liquids and solids are held close together as strong attractive forces become more important at lower temperatures. This means that a gas has no definite shape or volume and will completely fill any container. Because there are great distances between its particles, a gas is less dense than a solid or liquid and can be compressed. A model for the behavior of a gas, called the **kinetic molecular theory of gases,** helps us understand gas behavior.

Kinetic Molecular Theory of Gases

1. **A gas consists of small particles (atoms or molecules) that move randomly with rapid velocities.** Gas molecules moving in all directions at high speeds cause a gas to fill the entire volume of a container.
2. **The attractive forces between the particles of a gas can be neglected.** Gas particles move far apart and fill a container of any size and shape.
3. **The actual volume occupied by gas molecules is extremely small compared to the volume that the gas occupies.** The volume of the container is considered equal to the volume of the gas. Most of the volume of a gas is empty space, which allows gases to be easily compressed.
4. **The average kinetic energy of gas molecules is proportional to the Kelvin temperature.** Gas particles move faster as the temperature increases. At higher temperatures, gas particles hit the walls of the container with more force, which produces higher pressures.
5. **Gas particles are in constant motion, moving rapidly in straight paths.** When gas particles collide, they rebound and travel in new directions. When they collide with the walls of the container, they exert gas pressure. An increase in the number or force of collisions against the walls of the container causes an increase in the pressure of the gas.

The kinetic theory helps explain some of the characteristics of gases. For example, we can quickly smell perfume from a bottle that is opened on the other side of a room, because its particles move rapidly in all directions. They move faster at higher temperatures, and more slowly at lower temperatures. Sometimes tires and gas-filled containers explode when temperatures are too high. From the kinetic theory, we know that gas particles move faster when heated, hit the walls of a container with more force, and cause a buildup of pressure inside a container.

SAMPLE PROBLEM 8.1

Kinetic Theory of Gases

How would you explain the following characteristics of a gas using the kinetic theory of gases?

a. The particles of gas are not attracted to each other.
b. Gas particles move faster when the temperature is increased.

Solution

a. The distance between the particles of a gas is so great that the particles are not attracted to each other.
b. Temperature increases the kinetic energy of the gas particles, which makes them move faster.

Study Check

Why can you smell food that is cooking in the kitchen when you are in a different part of the house?

When we talk about a gas, we describe it in terms of four properties: pressure, volume, temperature, and the amount of gas.

Pressure (*P*)

Gas particles are extremely small. However, when billions and billions of gas particles hit against the walls of a container, they exert a force known as the pressure of the gas. The more molecules that strike the wall, the greater the pressure. If we heat the container to make the molecules move faster, they smash into the walls of the container more often and with increased force, which increases the pressure. The gas molecules of oxygen and nitrogen in the air around us are exerting pressure on us all the time. We call the pressure exerted by the air atmospheric pressure. As you go to higher altitudes the atmospheric pressure is less because there are fewer molecules of oxygen and nitrogen in the air. The most common units used for gas measurement are the atmosphere (atm) and millimeters of mercury (mm Hg). On the TV weather report, you may hear or see the atmospheric pressure given in inches of mercury, or kilopascals in countries other than the United States.

Volume (*V*)

The volume of gas equals the size of the container in which the gas is placed. When you inflate a tire or a basketball, you are adding more gas particles, which increases the number of particles hitting the walls of the tire or basketball, and its volume increases. Sometimes, on a cool morning, a tire looks flat. The volume of the tire has decreased because a lower temperature decreases the speed of the molecules, which reduces the force of their impacts on the walls of the tire. The most common units for volume measurement are liters (L) and milliliters (mL).

Temperature (*T*)

The temperature of a gas is related to the kinetic energy of its particles. For example, if we have a gas at 200 K in a rigid container and heat it to a temperature of 400 K, the gas particles will have twice the kinetic energy that they did at 200 K. That also means that the gas at 400 K exerts twice the pressure of the gas at 200 K. Although you measure gas temperature using a Celsius thermometer, all comparisons of gas behavior and all calculations related to temperature must use the Kelvin temperature. No one has quite created the conditions for absolute zero (K), but we predict that the particles will have zero kinetic energy, and the gas will exert zero pressure at this temperature.

Amount of Gas (*n*)

When you add air to a bicycle tire, you are increasing the amount of gas, which results in a higher pressure in the tire. Usually we measure the amount of gas by its mass (grams). In gas law calculations, we need to change the grams of gas to moles.

A summary of the four properties of a gas are given in Table 8.1.

Table 8.1 Properties That Describe a Gas

Property	Description	Unit(s) of Measurement
Pressure (*P*)	The force exerted by gas against the walls of the container	atmosphere (atm); mm Hg; torr
Volume (*V*)	The space occupied by the gas	liter (L); milliliter (mL)
Temperature (*T*)	Determines the kinetic energy and rate of motion of the gas particles	Celsius (°C); Kelvin (K) *required in calculations*
Amount (*n*)	The quantity of gas present in a container	grams (g); moles (*n*) *required in calculations*

SAMPLE PROBLEM 8.2

Properties of Gases

Identify the property of a gas that is described by each of the following:

a. increases the kinetic energy of gas particles
b. the force of the gas particles hitting the walls of the container
c. the space that is occupied by a gas

Solution

a. temperature **b.** pressure **c.** volume

Study Check

When helium is added to a balloon, the number of grams of gas increases. What property of a gas is described?

EXPLORE YOUR WORLD

Forming a Gas

Obtain baking soda and a jar or a plastic bottle. You will also need an elastic glove that fits over the mouth of the jar or a balloon that fits snugly over the top of the plastic bottle. Place a cup of vinegar in the jar or bottle. Sprinkle some baking soda into the fingertips of the glove or into the balloon. Carefully fit the glove or balloon over the top of the jar or bottle. Slowly lift the fingers of the glove or the balloon so that the baking soda falls into the vinegar. Watch what happens. Squeeze the glove or balloon.

Question

1. Describe the properties of gas that you observe as the reaction takes place between vinegar and baking soda.
2. How do you know that a gas was formed?

QUESTIONS AND PROBLEMS

Properties of Gases

8.1 Use the kinetic theory of gases to explain each of the following:
a. Gases move faster at higher temperatures.

b. Gases can be compressed much more than liquids or solids.

c. Gases have low densities.

8.2 Use the kinetic molecular theory of gases to explain each of the following:

a. A container of non-stick cooking spray explodes when thrown into a fire.

b. The air in a hot-air balloon is heated to make the balloon rise.

8.3 Identify the property of a gas that is measured in each of the following:

a. 350 K

b. space occupied by a gas

c. 2.00 g of O_2

d. force of gas particles striking the walls of the container

8.4 Identify the property of a gas that is measured in each of the following measurements:

a. determines the kinetic energy of the gas particles

b. 1.0 atm   **c.** 10.0 L   **d.** 0.50 mole of He

WEB TUTORIAL
Properties of Gases

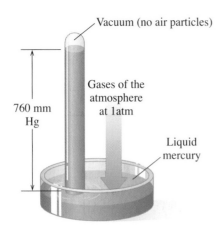

Figure 8.1 A barometer: The pressure exerted by the gases in the atmosphere is equal to the downward pressure of a mercury column in a closed glass tube. The height of the mercury column measured in mm Hg is called atmospheric pressure.

Q Why does the height of the mercury column change from day to day?

8.2 Gas Pressure

When water boils in a pan covered with a lid, the collisions of the molecules of steam (gas) lift up the lid. The gas molecules exert **pressure,** which is defined as a force acting on a certain area.

$$\text{Pressure } (P) = \frac{\text{force}}{\text{area}}$$

The air that covers the surface of the earth, the atmosphere, contains vast numbers of gas particles. Because the air particles have mass, they are pulled toward the earth by gravity, where they exert an **atmospheric pressure.**

The atmospheric pressure can be measured using a **barometer,** as shown in Figure 8.1. A long glass tube is closed on one end and filled with mercury, then it is inverted and its open end is placed in a dish of mercury.

Eventually, the mercury in the tube reaches a level where its weight exerts a downward pressure that is equal to the atmospheric pressure. At a pressure of exactly 1 atmosphere (atm), the mercury column would be exactly 760 mm high. We say that the atmospheric pressure is 760 mm Hg (millimeters of mercury), which is the same as a standard **atmosphere (atm).** Atmospheric pressure changes with variations in weather and altitude. On a hot, sunny day, the air is more dense, which causes more pressure on the mercury surface. The mercury column rises indicating a higher atmospheric pressure. On a rainy day, the atmosphere exerts less pressure, which causes the mercury column to go lower. In the weather report, this type of weather is called a low-pressure system. Atmosphere pressure is greatest at sea level. Above sea level, the density of the gases in the air decreases, which causes lower atmospheric pressures.

Atmospheric Pressure and Altitude

At sea level, where there are more air particles, the atmospheric pressure is about 1 atm. In Los Angeles, atmospheric pressure can fluctuate between 730 and 760 mm

Measuring Blood Pressure

The measurement of your blood pressure is one of the important measurements a doctor or nurse makes during a physical examination. Acting like a pump, the heart contracts to create the pressure that pushes blood through the circulatory system. During contraction, the blood pressure is called systolic and is at its highest. When the heart muscles relax, the blood pressure is called diastolic and falls. Normal range for systolic pressure is 100–120 mm Hg, and for diastolic pressure, 60–80 mm Hg, usually expressed as a ratio such as 100/80. These values are somewhat higher in older people. When blood pressures are elevated, such as 140/90, there is a greater risk of stroke, heart attack, or kidney damage. Low blood pressure prevents the brain from receiving adequate oxygen, causing dizziness and fainting.

The blood pressures are measured by a sphygmomanometer, an instrument consisting of a stethoscope and an inflatable cuff connected to a tube of mercury called a manometer. After the cuff is wrapped around the upper arm, it is pumped up with air until it cuts off the flow of blood through the arm. With the stethoscope over the artery, the air is slowly released from the cuff. When the pressure equals the systolic pressure, blood starts to flow again, and the noise it makes is heard through the stethoscope. As air continues to be released, the cuff deflates until no sound in the

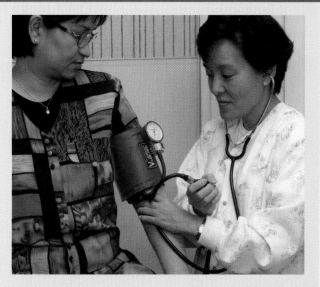

artery is heard. That second pressure reading is noted as the diastolic pressure, the pressure when the heart is not contracting.

Hg, depending on the weather. At higher altitudes, there are fewer air particles and the atmospheric pressure is less than 1 atm. In Denver (1.6 km above sea level), a typical atmospheric pressure might be 630 mm Hg, and if you were climbing Mount Everest (9.3 km above sea level) it might get as low as 270 mm Hg.

Divers must be concerned about increasing pressures on their ears and lungs when they dive below the surface of the ocean. Because water is denser than air, the pressure on a diver increases rapidly as the diver descends. At a depth of 33 ft below the surface of the ocean, an additional 1 atmosphere of pressure is exerted by the water on a diver, for a total of 2 atm. At 100 ft down, there is a total pressure of 4 atm on a diver. The air tanks a diver carries continuously adjust the pressure of the breathing mixture to match the increase of pressure.

One atmosphere of pressure may also be expressed as 760 **torr,** a pressure unit named to honor Evangelista Torricelli, the inventor of the barometer. Because they are equal, units of torr and mm Hg are used interchangeably.

1 atm = 760 mm Hg = 760 torr

1 mm Hg = 1 torr

In SI units, pressure is measured in pascals (Pa); 1 atm is equal to 101,325 Pa. Because a pascal is a very small unit, it is likely that pressures would be reported in kilopascals.

1 atm = 1.01×10^5 Pa = 101 kPa

If you have a barometer in your home, it probably gives pressure in inches of mercury. One atmosphere is also equal to the pressure of a column of mercury that is 29.9 inches high. Weather reports in the United States are given in inches of mercury. A lowering of pressure often indicates rain or snow, whereas an increase in pressure (referred to as a high-pressure system) usually brings dry and sunny weather.

The American equivalent of 1 atm is 14.7 pounds per square inch (lb/in.²). When you use a pressure gauge to check the air pressure in the tires of a car it may read 30–35 psi. Table 8.2 summarizes the various units used in the measurement of pressure.

Table 8.2   Units for Measuring Pressure

Unit	Abbreviation	Unit Equivalent to 1 atm
Atmosphere	atm	1 atm (exact)
Millimeters of Hg	mm Hg	760 mm Hg
Torr	torr	760 torr
Inches of Hg	in. Hg	29.9 in. Hg
Pounds per square inch	lb/in.² (psi)	14.7 lb/in.²
Pascal	Pa	101,325 Pa

SAMPLE PROBLEM 8.3

Units of Pressure

A sample of neon gas has a pressure of 0.50 atm. Give the pressure of the neon in

a. millimeters of Hg   **b.** inches of Hg

Solution

a. The equality 1 atm $=$ 760 mm Hg can be written as conversion factors:

$$\frac{760 \text{ mm Hg}}{1 \text{ atm}} \quad \text{or} \quad \frac{1 \text{ atm}}{760 \text{ mm Hg}}$$

Using the appropriate conversion factor, the problem is set up as

$$0.50 \text{ atm} \times \frac{760 \text{ mm Hg}}{1 \text{ atm}} = 380 \text{ mm Hg}$$

b. One atm is equal to 29.9 in. Hg. Using this equality as a conversion factor in the problem setup, we obtain

$$0.50 \text{ atm} \times \frac{29.9 \text{ in. Hg}}{1 \text{ atm}} = 15 \text{ in. Hg}$$

Study Check

What is the pressure in atmospheres for a gas that has a pressure of 655 torr?

QUESTIONS AND PROBLEMS

Gas Pressure

8.5 What units are used to measure the pressure of a gas?

8.6 Which of the following statement(s) describes the pressure of a gas?
   a. the force of the gas particles on the walls of the container
   b. the number of gas particles in a container
   c. the volume of the container
   d. 3.00 atm
   e. 750 torr

8.7 An oxygen tank contains oxygen (O_2) at a pressure of 2.00 atm. What is the pressure in the tank in terms of the following units?

    a. torr   **b.** lb/in.^{2} **c.** mm Hg

8.8 On a climb up Mt. Whitney, the atmospheric pressure is 467 mm Hg. What is the pressure in terms of the following units?

    a. atm   **b.** torr   **c.** in. Hg

8.3 Pressure and Volume (Boyle's Law)

Imagine that you can see air particles hitting the walls inside a bicycle tire pump. What happens to the pressure inside the pump as we push down on the handle? As the air is compressed, the air particles are crowded together. In the smaller volume, more collisions occur, and the air pressure increases.

When a change in one property (in this case, volume) causes a change in another property (in this case, pressure), those properties are related. Furthermore, when one change causes a change in the opposite direction, such as a decrease in volume causing an increase in pressure, the properties have an **inverse relationship.** The relationship between the pressure and volume of a gas is known as **Boyle's law.** The law states that the volume (V) of a sample of gas changes inversely with the pressure (P) of the gas as long as there has been no change in the temperature (T) or amount of gas (n), as illustrated in Figure 8.2.

If we change the volume or pressure of a gas sample without any change occurring in the temperature or in the amount of the gas, the new pressure and volume will give the same product as the initial pressure and volume. Because the PV product has the same value under both conditions, we can set the initial and final PV values equal to each other.

Boyle's Law

$$P_1V_1 = P_2V_2 \quad \text{No change in number of moles and temperature}$$

Use the pressure–volume relationship (Boyle's law) to determine the new pressure or volume of a certain amount of gas at a constant temperature.

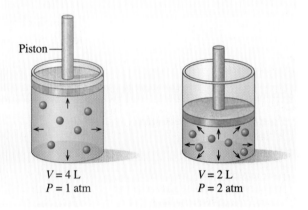

Piston —

$V = 4\,L$
$P = 1$ atm

$V = 2\,L$
$P = 2$ atm

Figure 8.2 Boyle's law: As volume decreases, gas molecules become more crowded, which causes the pressure to increase. Pressure and volume are inversely related.

Q If the volume of a gas increases, what will happen to its pressure?

the **C**hemistry place

WEB TUTORIAL
Properties of Gases

SAMPLE PROBLEM 8.4

Calculating Pressure When Volume Changes

A sample of hydrogen gas (H_2) has a volume of 5.0 L and a pressure of 1.0 atm. What is the new pressure if the volume is decreased to 2.0 L at constant temperature?

Solution

In this problem, we want to know the final pressure (P_2) for the change in volume. In calculations with gas laws, it is helpful to organize the data in a table.

Conditions 1	*Conditions 2*
$V_1 = 5.0\,L$	$V_2 = 2.0\,L$
$P_1 = 1.0$ atm	$P_2 = ?$

For a *PV* relationship, we use Boyle's law and solve for P_2 by dividing both sides by V_2.

$$P_1V_1 = P_2V_2$$

$$\frac{P_1V_1}{V_2} = \frac{P_2\cancel{V_2}}{\cancel{V_2}}$$

$$P_2 = \frac{P_1V_1}{V_2}$$

From the table, we see that the volume has decreased. Because pressure and volume are inversely related, the pressure must increase. When we substitute in the values, we see the ratio of the volumes (volume factor) is greater than 1, which increases the pressure.

$$P_2 = \frac{1.0 \text{ atm} \times 5.0\,\cancel{L}}{2.0\,\cancel{L}} = 2.5 \text{ atm}$$

<small>Volume factor increases pressure</small>

Note that the units of volume must cancel and the final pressure is in atmospheres.

Study Check

A sample of helium gas has a volume of 312 mL at 648 torr. If the volume expands to 825 mL at constant temperature, what is the new pressure in torr?

HEALTH NOTE

Pressure–Volume Relationship in Breathing

The importance of Boyle's law becomes more apparent when you consider the mechanics of breathing. Our lungs are elastic, balloon-like structures contained within an airtight chamber called the thoracic cavity. The diaphragm, a muscle, forms the flexible floor of the cavity.

Inspiration

The process of taking a breath of air begins when the diaphragm flattens, and the rib cage expands, causing an increase in the volume of the thoracic cavity. The elasticity of the lungs allows them to expand when the thoracic cavity expands. According to Boyle's law, the pressure inside the lungs will decrease when their volume increases. This causes the pressure inside the lungs to fall below the pressure of the atmosphere. This difference in pressures produces a **pressure gradient** between the lungs and the atmosphere. In a pressure gradient, molecules flow from an area of greater pressure to an area of lower pressure. Thus, we inhale as air flows into the lungs (*inspiration*), until the pressure

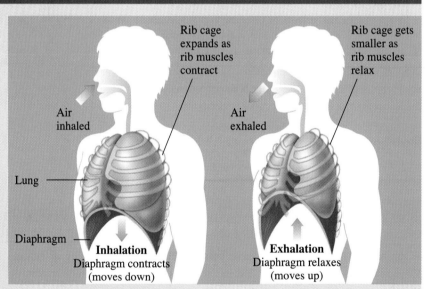

within the lungs becomes equal to the pressure of the atmosphere.

Expiration

Expiration, or the exhalation phase of breathing, occurs when the diaphragm relaxes and moves back up into the thoracic cavity to its resting position. This reduces the volume of the thoracic cavity,

which squeezes the lungs and decreases their volume. Now the pressure in the lungs is greater than the pressure of the atmosphere, so air flows out of the lungs. Thus, breathing is a process in which pressure gradients are continuously created between the lungs and the environment as a result of the changes in the volume and pressure.

EXPLORE YOUR WORLD

Medicine Dropper "Diver"

You can observe the effect of pressure change on the ability of a medicine dropper to sink or float in water. Obtain a plastic bottle with a cap and a medicine dropper. A 2-liter soda bottle or a 1-liter water bottle works well. The dropper needs to fit through the top of the bottle. Take up water in the medicine dropper until it is one-half to three-quarters full. The rest of the space is air (gas). Place the medicine dropper in a glass of water. It should just float at the surface. If not, adjust the water level.

Fill the plastic bottle with water up to about 1 to $1\frac{1}{2}$ inches from the top. Put the dropper with water in the bottle. Tightly cap the bottle. When you squeeze the sides of the bottle, the dropper should sink. If you cannot make the dropper sink, change the

amount of water you take into the dropper, or the amount of water in the bottle. You may have to adjust several times.

Questions

1. What does squeezing the sides of the bottle do to the pressure of the gas in the bottle and in the dropper?
2. What happens to the volume of gas in the dropper when you squeeze the bottle?
3. How does this cause the dropper to sink?
4. Why does the dropper rise when you release the sides of the bottle?
5. What is the relationship between the pressure and volume of a gas?

SAMPLE PROBLEM 8.5

Calculating Volume When Pressure Changes

In the hospital respiratory unit, the gauge on a 12-L tank of compressed oxygen reads 4500 mm Hg. How many liters of oxygen at constant temperature can be delivered from the tank at a pressure of 750 mm Hg?

Solution

Placing our information in a table gives the following:

Conditions 1	Conditions 2
P_1 = 4500 mm Hg	P_2 = 750 mm Hg
V_1 = 12 L	V_2 = ?

Using Boyle's law, we solve for V_2. According to Boyle's law, a decrease in the pressure will cause an increase in the volume. When we substitute in the values, the ratio of pressures (pressure factor) is greater than 1, which increases the volume.

$$V_2 = V_1 \times \frac{P_1}{P_2}$$

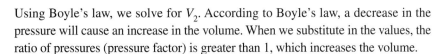

$$V_2 = 12\ \text{L} \times \underbrace{\frac{4500\ \cancel{\text{mm Hg}}}{750\ \cancel{\text{mm Hg}}}}_{\substack{\text{Pressure factor} \\ \text{increases volume}}} = 72\ \text{L}$$

Study Check

A sample of methane gas (CH_4) has a volume of 125 mL at 0.600 atm pressure and 25°C. How many milliliters will it occupy at a pressure of 1.50 atm and 25°C?

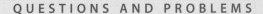

QUESTIONS AND PROBLEMS

Pressure and Volume (Boyle's Law)

8.9 Why do scuba divers need to exhale air when they ascend to the surface of the water?

8.10 Why does a sealed bag of chips expand when you take it to a higher altitude?

8.11 What happens to the volume of your lungs during expiration when you get rid of some CO_2?

8.12 How do respirators (or CPR) help a person obtain oxygen and expel carbon dioxide?

8.13 The air in a cylinder with a piston has a volume of 220 mL and a pressure of 650 mm Hg.

a. If a change results in a higher pressure inside the cylinder, does cylinder A or B represent the final volume? Explain your choice.

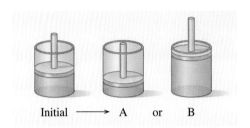

Initial ⟶ A   or   B

b. If the pressure inside the cylinder increases to 1.2 atm, what is the final volume of the cylinder? Complete the following data table:

Property	Initial	Final
Pressure (P)		
Volume (V)		

8.14 A balloon is filled with helium gas. When the following changes are made at constant temperature, which of these diagrams (A, B, or C) shows the new volume of the balloon?

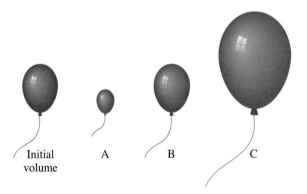

Initial volume   A   B   C

a. The balloon floats to a higher altitude where the outside pressure is lower.

b. The balloon is taken inside the house, but the atmospheric pressure remains the same.

c. The balloon is put in a hyperbaric chamber in which the pressure is increased.

8.15 A gas with a volume of 4.0 L is contained in a closed container. Indicate the changes in its pressure when the volume undergoes the following changes at constant temperature.

a. The volume is compressed to 2 L.

b. The volume is allowed to expand to 12 L.

c. The volume is compressed to 0.40 L.

8.16 A gas at a pressure of 2.0 atm is contained in a closed container. Indicate the changes in its volume when the pressure undergoes the following changes at constant temperature.

a. The pressure increases to 6.0 atm.

b. The pressure drops to 1.0 atm.

c. The pressure drops to 0.40 atm.

8.17 A 10.0-L balloon contains He gas at a pressure of 655 mm Hg. What is the new pressure of the He gas at each of the following volumes if there is no change in temperature?

    a. 20.0 L   **b.** 2.50 L   **c.** 1500 mL   **d.** 120 mL

8.18 The air in a 5.00-L tank has a pressure of 1.20 atm. What is the new pressure of the air when the air is placed in tanks that have the following volumes, if there is no change in temperature?

    a. 1.00 L   **b.** 2500 mL   **c.** 750 mL   **d.** 8.0 L

8.19 A sample of nitrogen (N_2) gas has a volume of 4.5 L at a pressure of 760 mm Hg. What is the new pressure if the gas sample is compressed to a volume of 2.0 L if there is no change in temperature?

8.20 An emergency tank of oxygen holds 20.0 L of oxygen (O_2) at a pressure of 15.0 atm and 22°C. When the gas is released at 22°C, it provides 300.0 L of oxygen. What is the pressure at which the oxygen is released?

8.21 A sample of nitrogen (N_2) has a volume of 50.0 L at a pressure of 760. mm Hg. What is the volume of the gas at each of the following pressures if there is no change in temperature?

    a. 1500 mm Hg   **b.** 2.0 atm   **c.** 0.500 atm   **d.** 850 torr

8.22 A sample of methane (CH_4) has a volume of 25 mL at a pressure of 0.80 atm. What is the volume of the gas at each of the following pressures if there is no change in temperature?

    a. 0.40 atm   **b.** 2.00 atm   **c.** 2500 mm Hg   **d.** 80.0 torr

8.23 Cyclopropane, C_3H_6, is a general anesthetic. A 5.0-L sample has a pressure of 5.0 atm. What is the volume of the anesthetic given to a patient at a pressure of 1.0 atm?

8.24 The volume of air in a person's lungs is 615 mL at a pressure of 760 mm Hg. Inhalation occurs as the pressure in the lungs drops to 752 mm Hg. To what volume did the lungs expand?

8.25 Use the words *inspiration* and *expiration* to describe the part of the breathing cycle that occurs as a result of each of the following:

a. The diaphragm contracts (flattens out).

b. The volume of the lungs decreases.

c. The pressure within the lungs is less than the atmosphere.

8.26 Use the words *inspiration* and *expiration* to describe the part of the breathing cycle that occurs as a result of each of the following:

a. The diaphragm relaxes, moving up into the thoracic cavity.

b. The volume of the lungs expands.

c. The pressure within the lungs is greater than the atmosphere.

8.4 Temperature and Volume (Charles' Law)

When preparing a hot-air balloon for a flight, the air in the balloon is heated with a small propane heater. As the air warms, its volume increases. The resulting decrease in density allows the balloon to rise.

To study the effect of changing temperature on the volume of a gas, we must not change the pressure or the amount of the gas. Suppose we increase the Kelvin tem-

perature of a gas sample. The kinetic theory shows that the activity (kinetic energy) of the gas will also increase. To keep pressure constant, the volume of the container must increase. (See Figure 8.3.) By contrast, if the temperature of the gas is lowered, the volume of the container must be reduced to maintain the same pressure.

Suppose that you are going to take a ride in a hot-air balloon. The captain turns on a propane burner to heat the air inside the balloon. As the temperature rises, the air particles move faster and spread out, causing the volume of the balloon to increase. Eventually, the air in the balloon becomes less dense than the air outside, and the balloon and its passengers rise. In fact, it was in 1787 that Jacques Charles, a balloonist as well as a physicist, proposed that the volume of a gas is related to the temperature. This became **Charles' law,** which states that the volume (V) of a gas is directly related to the temperature (K) when there is no change in the pressure (P) or amount (n) of gas. A **direct relationship** is one in which the related properties increase or decrease together. For two conditions, initial and final, we can write Charles' law as follows.

Charles' Law

$$\frac{V_1}{T_1} = \frac{V_2}{T_2} \qquad \text{No change in number of moles and pressure}$$

Remember that all temperatures used in gas law calculations must be converted to their corresponding Kelvin (K) temperatures.

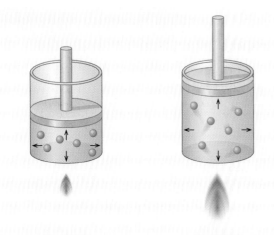

$T = 200\ \text{K}$
$V = 1\ \text{L}$

$T = 400\ \text{K}$
$V = 2\ \text{L}$

Figure 8.3 Charles's law: The Kelvin temperature of a gas is directly related to the volume of the gas when there is no change in the pressure. When the temperature increases making the molecules move faster, the volume must increase to maintain constant pressure.

Q If the temperature of a gas decreases at constant pressure, how will the volume change?

WEB TUTORIAL
Properties of Gases

Calculating Volume When Temperature Changes

A sample of neon gas has a volume of 5.0 L and a temperature of 17°C. Find the new volume of the gas after the temperature has been increased to 47°C at constant pressure.

Solution

When the temperatures are given in degrees Celsius, they must be changed to Kelvins.

$$T_1 = 17°\text{C} + 273 = 290\ \text{K}$$
$$T_2 = 47°\text{C} + 273 = 320\ \text{K}$$

Conditions 1	Conditions 2
$T_1 = 290\ \text{K}$	$T_2 = 320\ \text{K}$
$V_1 = 5.0\ \text{L}$	$V_2 = ?$

In this problem, we want to know the final volume (V_2) when the temperature increases. Using Charles' law, we solve for V_2 by multiplying both sides by T_2.

$$\frac{V_1}{T_1} = \frac{V_2}{T_2}$$

$$V_2 = \frac{V_1 T_2}{T_1}$$

From the table, we see that the temperature has increased. Because temperature is directly related to volume, the volume must increase. When we substitute in the values, we see that the ratio of the temperatures (temperature factor) is greater than 1, which increases the volume.

$$V_2 = 5.0 \text{ L} \times \frac{320 \text{ } \boldsymbol{K}}{290 \text{ } \boldsymbol{K}} = 5.5 \text{ L}$$

Temperature factor
increases volume

Study Check

A mountain climber inhales 480 mL of air at a temperature of $-8°C$. What volume will the air occupy in the lungs if the climber's body temperature is 37°C?

QUESTIONS AND PROBLEMS

Temperature and Volume (Charles' Law)

8.27 Select the diagram that shows the new volume of a balloon when the following changes are made at constant pressure.

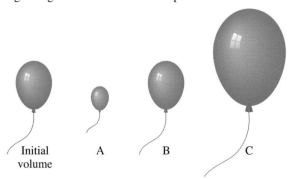

Initial volume   A   B   C

    a. The temperature is changed from 150 K to 300 K.
    b. The balloon is placed in a freezer.
    c. The balloon is first warmed, and then returned to its starting temperature.

8.28 Indicate whether the final volume of gas in each of the following is the same, larger, or smaller than the initial volume:
    a. A volume of 500 mL of air on a cold winter day at 5°C is breathed into the lungs, where body temperature is 37°C.
    b. The heater used to heat 1400 L of air in a hot-air balloon is turned off.
    c. A balloon filled with helium at the amusement park is left in a car on a hot day.

8.29 What change in volume occurs when gases for hot-air balloons are heated prior to their ascent?

8.30 On a cold, wintry morning, the tires on a car appear flat. How has their air volume changed overnight?

8.31 A balloon contains 2500 mL of helium gas at 75°C. What is the new volume of the gas when the temperature changes to the following, if n and P are not changed?
    a. 55°C   **b.** 680 K   **c.** $-25°C$ **d.** 240 K

8.32 A gas has a volume of 4.00 L at 0°C. What final temperature in degrees Celsius is needed to cause the volume of the gas to change to the following, if n and P are not changed?
    a. 10.0 L   **b.** 1200 mL   **c.** 2.50 L   **d.** 50.0 mL

8.5 Temperature and Pressure (Gay-Lussac's Law)

If we could watch the molecules of a gas as the temperature rises, we would notice that they move faster and hit the sides of the container more often and with greater force. If we keep the volume of the container the same, we would observe an increase in the pressure. A temperature–pressure relationship, also known as **Gay-Lussac's law,** states that the pressure of a gas is directly related to its Kelvin temperature. This means that an increase in temperature increases the pressure of a gas, and a decrease in temperature decreases the pressure of the gas, provided the volume and number of moles of the gas remain the same. (See Figure 8.4.) The ratio of pressure (P) to temperature (T) is the same under all conditions as long as volume (V) and amount of gas (n) do not change.

Gay-Lussac's Law

$$\frac{P_1}{T_1} = \frac{P_2}{T_2}$$ No change in number of moles and volume

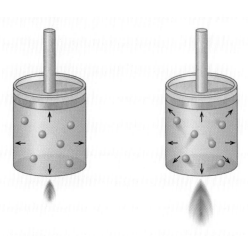

$T = 200$ K
$P = 1$ atm

$T = 400$ K
$P = 2$ atm

Figure 8.4 Gay-Lussac's law: The pressure of a gas is directly related to the temperature of the gas. When the Kelvin temperature of a gas is doubled, the pressure is doubled at constant volume.

Q How does a decrease in the temperature of a gas affect its pressure at constant volume?

WEB TUTORIAL
Properties of Gases

SAMPLE PROBLEM 8.7

Calculating Pressure When Temperature Changes

Aerosol containers can be dangerous if they are heated because they can explode. Suppose a container of hair spray with a pressure of 4.0 atm at a room temperature of 25°C is thrown into a fire. If the temperature of the gas inside the aerosol can reaches 402°C, what will be its pressure? The aerosol container may explode if the pressure inside exceeds 8.0 atm. Would you expect it to explode?

Solution

We must first change the temperatures to kelvins.

$$T_1 = 25°C + 273 = 298 \text{ K}$$
$$T_2 = 402°C + 273 = 675 \text{ K}$$

Conditions 1	Conditions 2
$P_1 = 4.0$ atm	$P_2 = ?$
$T_1 = 298$ K	$T_2 = 675$ K

Using Gay-Lussac's law, we can solve for P_2.

$$\frac{P_1}{T_1} = \frac{P_2}{T_2}$$

$$P_2 = \frac{P_1 T_2}{T_1}$$

From the table, we see that the temperature has increased. Because pressure and temperature are directly related, the pressure must increase. When we substitute in the values, we see the ratio of the temperatures (temperature factor) is greater than 1, which increases pressure.

$$P_2 = 4.0 \text{ atm} \times \frac{675 \text{ K}}{298 \text{ K}} = 9.1 \text{ atm}$$

Temperature factor
increases pressure

Because the calculated pressure exceeds 8.0 atm, we expect the can to explode.

Study Check

In a storage area where the temperature has reached 55°C, the pressure of oxygen gas in a 15.0-L steel cylinder is 965 torr. To what temperature would the gas have to be cooled to reduce the pressure to 850. torr?

Vapor Pressure and Boiling Point

In Chapter 5, we learned that liquid molecules with sufficient kinetic energy can break away from the surface of the liquid as they become gas particles or vapor. In an open container, all the liquid will eventually evaporate. In a closed container, the vapor accumulates and creates pressure called **vapor pressure.** Each liquid exerts its own vapor pressure at a given temperature. As temperature increases, more vapor forms, and vapor pressure increases. Table 8.3 lists the vapor pressure of water at various temperatures.

A liquid reaches its boiling point when its vapor pressure becomes equal to the external pressure. As boiling occurs, bubbles of the gas form within the liquid and quickly rise to the surface. For example, at an atmospheric pressure of 760 mm Hg, water will boil at 100°C, the temperature at which its vapor pressure reaches 760 mm Hg.

At higher altitudes, atmospheric pressures are lower and the boiling point of water is lower than 100°C. Earlier, we saw that the typical atmospheric pressure in Denver is 630 mm Hg. This means that water in Denver needs a vapor pressure of 630 mm Hg to boil. Because water has a vapor pressure of 630 mm Hg at 95°C, water boils at 95°C in Denver. Water boils at even lower temperatures at still lower atmospheric pressures, as Table 8.4 shows.

Table 8.3   Vapor Pressure of Water

Temperature (C°)	Vapor Pressure (mm Hg)
0	5
10	9
20	18
30	32
37	47[a]
40	55
50	93
60	149
70	234
80	355
90	528
100	760

[a]At body temperature.

Table 8.4   Altitude, Atmospheric Pressure, and the Boiling Point of Water

Location	Altitude (km)	Atmospheric Pressure (mm Hg)	Boiling Point (°C)
Sea level	0	760	100
Los Angeles	0.09	752	99
Las Vegas	0.70	700	98
Denver	1.60	630	95
Mount Whitney	4.00	467	87
Mount Everest	9.30	270	70

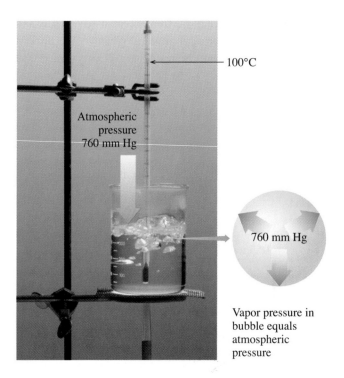

100°C

Atmospheric
pressure
760 mm Hg

760 mm Hg

Vapor pressure in
bubble equals
atmospheric
pressure

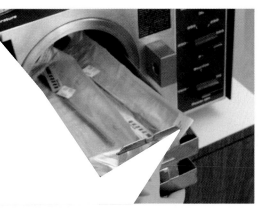

People who live at high altitudes often use pressure cookers to obtain higher temperatures when preparing food. When the external pressure is greater than 1 atm, a temperature higher than 100°C is needed to boil water. Laboratories and hospitals use devices called autoclaves to sterilize laboratory and surgical equipment. An autoclave, like a pressure cooker, is a closed container that increases the total pressure above the liquid so it will boil at higher temperatures. Table 8.5 shows how the boiling point of water increases as pressure increases.

QUESTIONS AND PROBLEMS

Temperature and Pressure (Gay-Lussac's Law)

8.33 Why do aerosol cans explode if heated?

8.34 How can the tires on a car look flat on a cold, winter morning, but have a blowout when the car is driven on hot pavement in the desert?

8.35 In a fire at an oxygen-tank factory, some of the tanks filled with oxygen explode. Explain.

8.36 In Gay-Lussac's law, what properties of gases are considered as directly related and which properties are held constant?

8.37 Solve for the new pressure when each of the following temperature changes occurs, with n and V constant:
  a. A gas sample has a pressure of 1200 torr at 155°C. What is the final pressure of the gas after the temperature has dropped to 0°C?
  b. An aerosol can has a pressure of 1.40 atm at 12°C. What is the final pressure in the aerosol can if it is used in a room where the temperature is 35°C?

8.38 Solve for the new temperature in degrees Celsius when pressure is changed.
  a. A 10.0-L container of helium gas has a pressure of 250 torr at 0°C. To what Celsius temperature does the sample need to be heated to obtain a pressure of 1500 torr?

Table 8.5 Pressure and the Boiling Point of Water

Pressure (mm Hg)	Boiling Point (°C)
760	100
800	100.4
1075	110
1520 (2 atm)	120
2026	130
7600 (10 atm)	180

b. A 500.0-mL sample of air at 40.°C and 740 mm Hg is cooled to give a pressure of 680 mm Hg.

8.39 Match the terms *vapor pressure, atmospheric pressure,* and *boiling point* to the following descriptions.
   a. the temperature at which bubbles of vapor appear within the liquid
   b. the pressure exerted by a gas above the surface of its liquid
   c. the pressure exerted on the earth by the particles in the air
   d. the temperature at which the vapor pressure of a liquid becomes equal to the external pressure

8.40 In which pair(s) would boiling occur?

Atmospheric Pressure	Vapor Pressure
a. 760 mm Hg	700 mm Hg
b. 480 torr	480 mm Hg
c. 1.2 atm	912 mm Hg
d. 1020 mm Hg	760 mm Hg
e. 740 torr	1.0 atm

8.41 Give an explanation for the following observations:
   a. Water boils at 87°C on the top of Mt. Whitney.
   b. Food cooks more quickly in a pressure cooker than in an open pan.

8.42 Give an explanation for the following observations:
   a. A wet towel dries faster on a windy day.
   b. Water used to sterilize surgical equipment is heated to 120°C at 2.0 atm in an autoclave.

8.6 The Combined Gas Law

All pressure–volume–temperature relationships for gases that we have studied may be combined into a single relationship called the **combined gas law.** This expression is useful for studying the effect of changes in two of these variables on the third as long as the amount of gas (number of moles) remains constant.

Combined Gas Law

$$\frac{P_1V_1}{T_1} = \frac{P_2V_2}{T_2} \quad \text{No change in moles of gas}$$

By remembering the combined gas law, we can derive any of the gas laws by omitting those properties that do not change. Table 8.6 summarizes the pressure–volume–temperature relationships of gases.

LEARNING GOAL

Use the combined gas law to find the new pressure, volume, or temperature of a gas when changes in two of these properties are given.

WEB TUTORIAL
Properties of Gases

SAMPLE PROBLEM 8.8

Using the Combined Gas Law

A 25.0-mL bubble is released from a diver's air tank at a pressure of 4.00 atm and a temperature of 11°C. What is the volume of the bubble when it reaches the ocean surface, where the pressure is 1.00 atm and the temperature is 18°C?

Solution

We must first change the temperature to kelvins.

$$T_1 = 11°C + 273 = 284 \text{ K}$$
$$T_2 = 18°C + 273 = 291 \text{ K}$$

Conditions 1	Conditions 2
P_1 = 4.00 atm	P_2 = 1.00 atm
V_1 = 25.0 mL	V_2 = ?
T_1 = 284 K	T_2 = 291 K

Because the pressure and temperature are both changing, we must use the combined gas law to solve for V_2.

$$\frac{P_1 V_1}{T_1} = \frac{P_2 V_2}{T_2}$$

$$V_2 = \frac{V_1 P_1 T_2}{P_2 T_1}$$

From the data table, we determine that the pressure decrease and the temperature increase will both increase the volume.

$$V_2 = 25.0 \text{ mL} \times \frac{4.00 \text{ atm}}{1.00 \text{ atm}} \times \frac{291 \text{ K}}{284 \text{ K}} = 102 \text{ mL}$$

$$= \quad \times \text{ Pressure} \quad \times \text{ Temperature}$$

factor   factor
increases   increases
volume   volume

Study Check

A weather balloon is filled with 15.0 L of helium at a temperature of 25°C and a pressure of 685 mm Hg. What is the pressure of the helium in the balloon in the upper atmosphere when the temperature is −35°C and the volume becomes 34.0 L?

Table 8.6   Summary of Gas Laws

Combined Gas Law	Properties Held Constant	Relationship	
$\dfrac{P_1 V_1}{T_1} = \dfrac{P_2 V_2}{T_2}$	T, n	$P_1 V_1 = P_2 V_2$	Boyle's Law
$\dfrac{P_1 V_1}{T_1} = \dfrac{P_2 V_2}{T_2}$	P, n	$\dfrac{V_1}{T_1} = \dfrac{V_2}{T_2}$	Charles' Law
$\dfrac{P_1 V_1}{T_1} = \dfrac{P_2 V_2}{T_2}$	V, n	$\dfrac{P_1}{T_1} = \dfrac{P_2}{T_2}$	Gay-Lussac's Law

QUESTIONS AND PROBLEMS

The Combined Gas Law

8.43 Write the expression for the combined gas law. What gas laws are combined to make the combined gas law?

8.44 Rearrange the variables in the combined gas law to give an expression for the following:

a. V_2 **b.** P_2

8.45 A sample of helium gas has a volume of 6.50 L at a pressure of 845 mm Hg and a temperature of 25°C. What is the pressure of the gas in atm when the volume and temperature of the gas sample are changed to the following?

a. 1850 mL and 325 K
b. 2.25 L and 12°C
c. 12.8 L and 47°C

8.46 A sample of argon gas has a volume of 735 mL at a pressure of 1.20 atm and a temperature of 112°C. What is the volume of the gas in milliliters when the pressure and temperature of the gas sample are changed to the following?

a. 658 mm Hg and 281 K
b. 0.55 atm and 75°C
c. 15.4 atm and −15°C

8.47 A 100.0-mL bubble of hot gases at 225°C and 1.80 atm escapes from an active volcano. What is the new volume of the bubble outside the volcano where the temperature is −25°C and the pressure is 0.80 atm?

8.48 A scuba diver 40 ft below the ocean surface inhales 50.0 mL of compressed air in a scuba tank at a pressure of 3.00 atm and a temperature of 8°C. What is the pressure of air in the lungs if the gas expands to 150.0 mL at a body temperature of 37°C?

WEB TUTORIAL
Properties of Gases

8.7 Volume and Moles (Avogadro's Law)

Describe the relationship between the amount of a gas and its volume and use this relationship in calculations.

In our study of the gas laws, we have looked at changes in properties for a specified amount (*n*) of gas. Now we will consider how the properties of a gas change when there is a change in number of moles or grams. For example, when you blow up a balloon, its volume increases because you add more air molecules. If a basketball gets a hole in it, and some of the air leaks out, its volume decreases. In 1811, Amedeo Avogadro stated that the volume of a gas is directly related to the number of moles of a gas when temperature and pressure are not changed. We refer to this statement as **Avogadro's law.** If the number of moles of a gas are doubled, then the volume will double as long as we do not change the pressure or the temperature. (See Figure 8.5.) For two conditions, we can write

Avogadro's Law

$$\frac{V_1}{n_1} = \frac{V_2}{n_2}$$ No change in pressure or temperature.

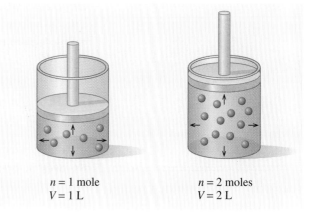

$n = 1$ mole
$V = 1$ L

$n = 2$ moles
$V = 2$ L

SAMPLE PROBLEM 8.9

Calculating Volume for a Change in Moles

A balloon with a volume of 220 mL is filled with 2.0 moles of helium. To what volume will the balloon expand if 3.0 moles of helium are added, to give a total of 5.0 moles of helium (the pressure and temperature do not change)?

Figure 8.5 Avogadro's law: The volume of a gas is directly related to the number of moles of the gas. If the number of moles is doubled, the volume must double at constant temperature and pressure.

Q If a balloon has a leak, what happens to its volume?

Solution

A data table for our given information can be set up as follows:

Conditions 1	Conditions 2
$V_1 = 220$ mL	$V_2 = ?$
$n_1 = 2.0$ moles	$n_2 = 5.0$ moles

Using Avogadro's law, we can solve for V_2.

$$\frac{V_1}{n_1} = \frac{V_2}{n_2}$$

$$V_2 = \frac{V_1 n_2}{n_1}$$

From the table, we see that the number of moles has increased. Because the number of moles and volume are directly related, the volume must increase at constant pressure and temperature. When we substitute in the values, we see the ratio of the moles (mole factor) is greater than 1, which increases volume.

$$V_2 = 220 \text{ mL} \times \frac{5.0 \text{ moles}}{2.0 \text{ moles}} = 550 \text{ mL}$$

New   Initial   Mole factor
volume volume that increases volume

Study Check

At a certain temperature and pressure, 8.00 g of oxygen has a volume of 5.00 L. What is the volume after 4.00 g of oxygen is added to the balloon?

STP and Molar Volume

Using Avogadro's law, we can say that any two gases will have equal volumes if they contain the same number of moles of gas at the same temperature and pressure. To help us make comparisons between different gases, arbitrary conditions called standard temperature (273 K) and pressure (1 atm) (abbreviated **STP**) were selected:

STP Conditions
Standard temperature is 0°C (273 K)

Standard pressure is 1 atm (760 mm Hg)

At STP, 1 mole of any gas occupies a volume of 22.4 L. This value is known as the *molar volume* of a gas. (See Figure 8.6.) Suppose we have three containers, one filled with 1 mole of oxygen gas (O_2), another filled with 1 mole of nitrogen gas (N_2), and one filled with 1 mole of helium gas (He). When the gases are at STP conditions (1 atm and 273 K), each has a volume of 22.4 L. Thus, the volume of 1 mole of any gas at STP is 22.4 L, the **molar volume** of a gas.

Molar Volume
The volume of 1 mole of gas at STP = 22.4 L

As long as a gas is at STP conditions (0°C and 1 atm), its molar volume can be used as a conversion factor to convert between the number of moles of gas and its volume.

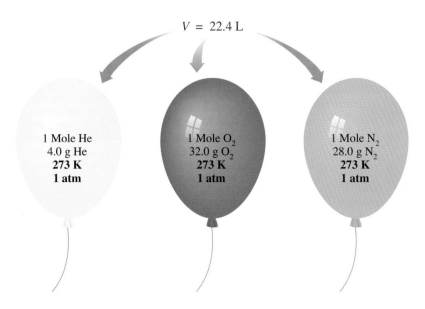

$V = 22.4 \, L$

1 Mole He
4.0 g He
273 K
1 atm

1 Mole O_2
32.0 g O_2
273 K
1 atm

1 Mole N_2
28.0 g N_2
273 K
1 atm

Figure 8.6 Avogadro's law indicates that one mole of any gas at STP has a molar volume of 22.4 L.

Q What volume of gas is occupied by 16.0 g of methane gas, CH_4, at STP?

Molar Volume Conversion Factors

$$\frac{1 \text{ mole gas (STP)}}{22.4 \text{ L}} \quad \text{and} \quad \frac{22.4 \text{ L}}{1 \text{ mole gas (STP)}}$$

SAMPLE PROBLEM 8.10

Calculations Using Molar Volume

What is the volume of 64.0 g of O_2 gas at STP?

Solution

The mass of the gas is changed to moles by its molar mass, which is 32.0 g/mole. Because the gas is at STP, we can use the molar volume to convert from moles to liters of gas.

$$64.0 \text{ g } O_2 \times \underbrace{\frac{1 \text{ mole } O_2}{32.0 \text{ g } O_2}}_{\substack{\text{Molar mass} \\ \text{of } O_2}} \times \underbrace{\frac{22.4 \text{ L } O_2}{1 \text{ mole } O_2}}_{\substack{\text{Molar volume} \\ \text{of a gas}}} = 44.8 \text{ L } O_2$$

Study Check

How many grams of nitrogen (N_2) gas are in 5.6 L of the gas at STP?

QUESTIONS AND PROBLEMS

Volume and Moles (Avogadro's Law)

8.49 What happens to the volume of a bicycle tire or a basketball when you use an air pump to add air?

8.50 Sometimes when you blow up a balloon and release it, it flies around the room. What is happening to the air that was in the balloon and its volume?

8.51 A sample containing 1.50 moles of neon gas has a volume of 8.00 L. What is the new volume of the gas in liters when the following changes occur in the quantity of the gas at constant pressure and temperature?
a. A leak allows one-half of the neon atoms to escape.

  b. A sample of 25.0 g of neon is added to the neon gas already in the container.

  c. A sample of 3.50 moles of O_2 is added to the neon gas already in the container.

8.52 A sample containing 4.80 g of O_2 gas has a volume of 15.0 L. Pressure and temperature remain constant.

  a. What is the new volume if 0.50 mole of O_2 gas is added?

  b. Oxygen is released until the volume is 10.0 L. How many moles of O_2 are removed?

  c. What is the volume after 4.00 g of He is added to the O_2 gas already in the container?

8.53 Use the molar volume of a gas to solve the following at STP:

  a. the number of moles of O_2 in 44.8 L of O_2 gas

  b. the number of moles of CO_2 in 4.00 L of CO_2 gas

  c. the volume (L) of 6.40 g of O_2

  d. the volume (mL) occupied by 50.0 g of neon

8.54 Use molar volume to solve the following problems at STP:

  a. the volume (L) occupied by 2.5 moles of N_2

  b. the volume (mL) occupied by 0.420 mole of He

  c. the number of grams of neon contained in 11.2 L of Ne gas

  d. the number of moles of H_2 in 1600 mL of H_2 gas

8.8 The Ideal Gas Law

The four properties used in the measurement of a gas—pressure (P), volume (V), temperature (T), and amount of a gas (n)—can be combined to give a single expression called the **ideal gas law,** which is written as follows:

Ideal Gas Law

$$PV = nRT$$

Rearranging the ideal gas law shows that the four gas properties equal a constant, R.

$$\frac{PV}{nT} = R$$

To calculate the value of R, we substitute the STP conditions for molar volume into the expression, which is 1 mole of any gas occupies 22.4 L at STP (273 K and 1 atm).

$$R = \frac{(1.00\ \text{atm})(22.4\ \text{L})}{(1.00\ \text{mole})(273\ \text{K})} = \frac{0.0821\ \text{L} \cdot \text{atm}}{\text{mole} \cdot \text{K}}$$

The value for R, the **universal gas constant,** is 0.0821 L · atm per mole · K. If we use 760 mm Hg for the pressure we obtain another useful value for R of 62.4 L · mm Hg per mole · K.

$$R = \frac{(760\ \text{mm Hg})(22.4\ \text{L})}{(1.00\ \text{mole})(273\ \text{K})} = \frac{62.4\ \text{L} \cdot \text{mm Hg}}{\text{mole} \cdot \text{K}}$$

The ideal gas law is a useful expression when you are given the measurements for any three of the four properties of a gas.

In working problems using the ideal gas law, the units of each variable must match the units in the R you select.

Property	Unit
Pressure (P)	atm or mm Hg
Volume (V)	L
Amount (n)	moles
Temperature (T)	K

SAMPLE PROBLEM 8.11

Calculations Using the Ideal Gas Law

Nitrous oxide, N_2O, is an anesthetic known as "laughing gas." What is the pressure in atmosphere of 0.35 mole of N_2O at 22°C contained in a 5.0 L tank?

Solution

Organizing the measured quantities indicates that we know three of the four properties of the gas.

P = ?

V = 5.0 L

n = 0.35 mole

T = 295 K (22 + 273)

To calculate the pressure of N_2O gas, we first solve the ideal gas law for P:

$$PV = nRT$$

$$P = \frac{nRT}{V}$$

Now, we can place the given quantities into the expression. All of the units cancel except for the pressure in atmospheres.

$$P = \frac{(0.35 \ \text{mole})(0.0821 \ L \cdot \text{atm})(295 \ K)}{(5.0 \ L)(\text{mole} \cdot K)} = 1.7 \ \text{atm}$$

Study Check

Chlorine gas, Cl_2, is used to purify the water in swimming pools. How many grams of chlorine are in a 7.0 L tank if the gas has a pressure of 760 mm Hg and a temperature of 24°C? (*Hint:* Solve for the number of moles of Cl_2 using the ideal gas equation.)

QUESTIONS AND PROBLEMS

Ideal Gas Law

8.55 Calculate the pressure, in atmospheres, of 2.00 moles of helium gas in a 10.0 L container at 27°C.

8.56 What is the volume in liters of 4.0 moles of methane gas, CH_4, at 18°C and 1.40 atm?

8.57 An oxygen gas container has a volume of 20.0 L. How many grams of oxygen are in the container if the gas has a pressure of 845 mm Hg at 22°C?

8.58 A 10.0 g sample of krypton has a temperature of 25°C at 575 mm Hg. What is the volume, in milliliters, of the krypton gas?

8.59 A 25.0 g sample of nitrogen, N_2, has a volume of 50.0 L and a pressure of 630. mm Hg. What is the temperature of the gas?

8.60 A 0.226 g sample of carbon dioxide, CO_2, has a volume of 525 mL and a pressure of 455 mm Hg. What is the temperature of the gas?

LEARNING GOAL

Use partial pressures to calculate the total pressure of a mixture of gases.

8.9 Partial Pressures (Dalton's Law)

Many gas samples are a mixture of gases. For example, the air you breathe is a mixture of mostly oxygen and nitrogen gases. For gas mixtures, we use the gas laws we have studied because particles of all gases behave in the same way. Therefore, the total pressure of the gases in a mixture is a result of the collisions of the gas particles regardless of what type of gas they are.

In a gas mixture, each gas exerts its **partial pressure,** which is the pressure it would exert if it were the only gas in the container. **Dalton's law** states that the total pressure of a gas mixture is the sum of the partial pressures of the gases in the mixture.

$$P_{total} = P_1 + P_2 + P_3 + \cdots$$

Total pressure   =   Sum of the partial pressures
of a gas mixture   of the gases in the mixture

Suppose we have two separate tanks, one filled with helium at 2.0 atm and the other filled with argon at 4.0 atm. When the gases are combined in a single tank with the same volume and temperature, the number of gas molecules, not the type of gas, determines the pressure in a container. There the pressure of the gas mixture would be 6.0 atm, which is the sum of their individual or partial pressures.

$$P_{total} = P_{He} + P_{Ar}$$
$$= 2.0 \text{ atm} + 4.0 \text{ atm}$$
$$= 6.0 \text{ atm}$$

$P_{He} = 2.0 \text{ atm}$ $P_{Ar} = 4.0 \text{ atm}$

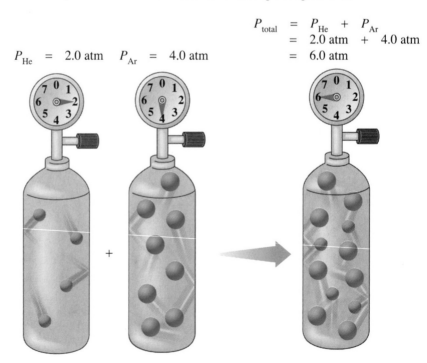

Calculating the Total Pressure of a Gas Mixture

A 10-L gas tank contains propane (C_3H_8) gas at a pressure of 300. torr. Another 10-L gas tank contains methane (CH_4) gas at a pressure of 500. torr. In preparing a gas fuel mixture, the gases from both tanks are combined in a 10-L container at the same temperature. What is the pressure of the gas mixture?

Solution

Using Dalton's law of partial pressure, we find that the total pressure of the gas mixture is the sum of the partial pressures of the gases in the mixture.

$$P_{total} = P_{propane} + P_{methane}$$
$$= 300.\ torr + 500.\ torr$$
$$= 800.\ torr$$

Therefore, when both propane and methane are placed in the same container, the total pressure of the mixture is 800. torr.

Study Check

A gas mixture consists of helium with a partial pressure of 315 mm Hg, nitrogen with a partial pressure of 204 mm Hg, and argon with a partial pressure of 422 mm Hg. What is the total pressure in atmospheres?

Air Is a Gas Mixture

The air you breathe is a mixture of gases. What we call the atmospheric pressure is actually the sum of the partial pressures of the gases in the air. Table 8.7 lists partial pressures for the gases in air on a typical day.

Table 8.7   Typical Composition of Air

Gas	Partial Pressure (mm Hg)	Percentage (%)
Nitrogen, N_2	594.0	78
Oxygen, O_2	160.0	21
Carbon dioxide, CO_2	0.3	} 1
Water vapor, H_2O	5.7	
Total air	760.0	100

Partial Pressure of a Gas in a Mixture

A mixture of oxygen and helium is prepared for a scuba diver who is going to descend 200 ft below the ocean surface. At that depth, the diver breathes a gas mixture that has a total pressure of 7.0 atm. If the partial pressure of the oxygen in the tank at that depth is 1.5 atm, what is the partial pressure of the helium?

Solution

From Dalton's law of partial pressures, we know that the total pressure is equal to the sum of the partial pressures:

Oxygen in the Air

For this activity, you will need some candles of different heights, a shallow bowl, a glass or jar tall enough to fit over the tallest candle, and a watch with a second hand. Using some melted wax from a burning candle, stick one candle to the bottom of a shallow bowl. Light the candle and place a jar over the candle. Observe the candle flame. Measure the time that the candle continues to burn.

Stick another candle that has a different height to the bottom of the bowl. You can try three candles if you wish. Light all the candles and carefully lower the jar over all the candles. The opening of the jar should be sitting on the bottom of the bowl. Measure the time that the candles continue to burn. Watch the order in which they are extinguished.

Remove the jar and place some water around the candles. Light the candles again, and place the jar over them. Observe the water level inside the jar as the candles go out.

Questions

1. What happens to the flame when you cover the candle with a jar? Explain.
2. How does the time for one candle's flame to go out compare to the time for two or three candles? Explain.
3. In what order were the candles extinguished? Explain.
4. What happens to the water level inside the jar after the candle flames go out?
5. Approximately how far up the jar did the water rise?
6. If you express the change in water level in the jar as a percentage, and the atmospheric pressure is 760 mm Hg, what would you calculate for the partial pressure of oxygen in the air?

$$P_{total} = P_{O_2} + P_{He}$$

To solve for the partial pressure of helium (P_{He}), we rearrange the expression to give the following:

$$P_{He} = P_{total} - P_{O_2}$$
$$P_{He} = 7.0 \text{ atm} - 1.5 \text{ atm}$$
$$= 5.5 \text{ atm}$$

Thus, in the gas mixture that the diver breathes, the partial pressure of the helium is 5.5 atm.

Study Check

An anesthetic consists of a mixture of cyclopropane gas, C_3H_6, and oxygen gas, O_2. If the mixture has a total pressure of 825 torr, and the partial pressure of the cyclopropane is 73 torr, what is the partial pressure of the oxygen in the anesthetic?

Blood Gases

Our cells continuously use oxygen and produce carbon dioxide. Both gases move in and out of the lungs through the membranes of the alveoli, the tiny air sacs at the ends of the airways in the lungs. An exchange of gases occurs in which oxygen from the air diffuses into the lungs and into the blood, while carbon dioxide produced in the cells is carried to the lungs to be exhaled. In Table 8.8, partial pressures are given for the gases in air that we inhale (inspired air), air in the alveoli, and the air that we exhale (expired air). The partial pressure of water vapor increases within the lungs because the vapor pressure of water is 47 mm Hg at body temperature.

At sea level, oxygen normally has a partial pressure of 100 mm Hg in the alveoli of the lungs. Because the partial pressure of oxygen in venous blood is 40 mm Hg, oxygen diffuses from the alveoli into the bloodstream. The oxygen combines with hemoglobin, which carries it to the tissues of the body where the partial pressure of oxygen can be very low, less than 30 mm Hg. Oxygen diffuses from the blood where the partial pressure of O_2 is high into the tissues where O_2 pressure is low.

As oxygen is used in the cells of the body during metabolic processes, carbon dioxide is produced, so the partial pressure of CO_2 may be as high as 50 mm Hg or more. Carbon dioxide diffuses from the tissues into the bloodstream and is carried

Table 8.8 Partial Pressures of Gases During Breathing

Gas	Partial Pressure (mm Hg)		
	Inspired Air	Alveolar Air	Expired Air
Nitrogen, N_2	594.0	573	569
Oxygen, O_2	160.0	100	116
Carbon dioxide, CO_2	0.3	40	28
Water vapor, H_2O	5.7	47	47
Total	760.0	760	760

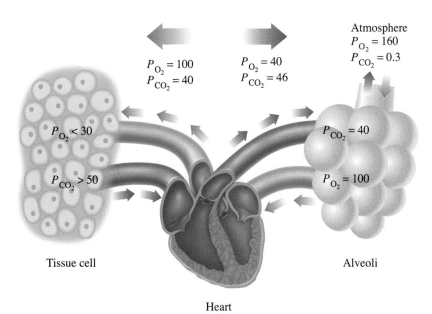

Figure 8.7 During gas exchange, differences in partial pressures cause oxygen and carbon dioxide to diffuse across alveoli and tissue membranes.

Q Why does oxygen diffuse from the alveoli into the blood and from the blood into the tissues of the body?

CASE STUDY
Scuba Diving
and Blood Gases

to the lungs. There it diffuses out of the blood, where CO_2 has a partial pressure of 46 mm Hg, into the alveoli, where the CO_2 is at 40 mm Hg and is exhaled. (See Figure 8.7.) Table 8.9 gives the partial pressures of blood gases in the tissues and in oxygenated and deoxygenated blood.

Table 8.9   Partial Pressures of Oxygen and Carbon Dioxide in Blood and Tissues

Gas	Partial Pressure (mm Hg)		
	Oxygenated Blood	Deoxygenated Blood	Tissues
O_2	100	40	30 or less
CO_2	40	46	50 or greater

QUESTIONS AND PROBLEMS

Partial Pressures (Dalton's Law)

8.61 A typical air sample in the lungs contains oxygen at 100 mm Hg, nitrogen at 573 mm Hg, carbon dioxide at 40 mm Hg, and water vapor at 47 mm Hg. Why are these pressures called partial pressures?

8.62 Suppose a mixture contains helium and oxygen gases. If the partial pressure of helium is the same as the partial pressure of oxygen, what do you know about the number of helium atoms compared to the number of oxygen molecules? Explain.

8.63 In a gas mixture, the partial pressures are nitrogen 425 torr, oxygen 115 torr, and helium 225 torr. What is the total pressure (torr) exerted by the gas mixture?

8.64 In a gas mixture, the partial pressures are argon 415 mm Hg, neon 75 mm Hg, and nitrogen 125 mm Hg. What is the total pressure (atm) exerted by the gas mixture?

8.65 A gas mixture containing oxygen, nitrogen, and helium exerts a total pressure of 925 torr. If the partial pressures are oxygen 425 torr and helium 75 torr, what is the partial pressure (torr) of the nitrogen in the mixture?

HEALTH NOTE

Hyperbaric Chambers

A burn patient may undergo treatment for burns and infections in a hyperbaric chamber, a device in which pressures can be obtained that are two to three times greater than atmospheric pressure. A greater oxygen pressure increases the level of dissolved oxygen in the blood and tissues, where it fights bacterial infections. High levels of oxygen are toxic to many strains of bacteria. The hyperbaric chamber may also be used during surgery to help counteract carbon monoxide (CO) poisoning, and to treat some cancers.

The blood is normally capable of dissolving up to 95% of the oxygen. Thus, if the partial pressure of the oxygen is 2280 mm Hg (3 atm), 95% of that or 2160 mm Hg of oxygen can dissolve in the blood where it saturates the tissues. In the case of carbon monoxide poisoning, this oxy-

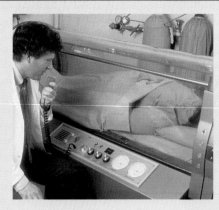

gen can replace the carbon monoxide that has attached to the hemoglobin.

A patient undergoing treatment in a hyperbaric chamber must also undergo decompression (reduction of pressure) at a rate that slowly reduces the concentration of dissolved oxygen in the blood. If

decompression is too rapid, the oxygen dissolved in the blood may form gas bubbles in the circulatory system.

If divers do not decompress slowly, they suffer a similar condition called the bends. While below the surface of the ocean, divers breathe air at higher pressures. At such high pressures, nitrogen gas will dissolve in their blood. If they ascend to the surface too quickly, the dissolved nitrogen forms bubbles in the blood that can produce life-threatening blood clots. The gas bubbles can also appear in the joints and tissues of the body and be quite painful. A diver suffering from the bends is placed immediately in a decompression chamber where pressure is first increased and then slowly decreased. The dissolved nitrogen can then diffuse through the lungs until atmospheric pressure is reached.

8.66 A gas mixture containing oxygen, nitrogen, and neon exerts a total pressure of 1.20 atm. If helium added to the mixture increases the pressure to 1.50 atm, what is the partial pressure (atm) of the helium?

8.67 In certain lung ailments such as emphysema, there is a decrease in the ability of oxygen to diffuse into the blood.
   a. How would the partial pressure of oxygen in the blood change?
   b. Why does a person with severe emphysema sometimes use a portable oxygen tank?

8.68 An accident to the head can affect the ability of a person to ventilate (breathe in and out), and so can certain drugs.
   a. What would happen to the partial pressures of oxygen and carbon dioxide in the blood if a person cannot properly ventilate?
   b. When a person with hypoventilation is placed on a ventilator, an air mixture is delivered at pressures that are alternately above the air pressure in the person's lung, and then below. How will this move oxygen gas into the lungs, and carbon dioxide out?

Chapter Review

8.1 Gases and Kinetic Theory

In a gas, particles are so far apart and moving so fast that their attractions are unimportant. A gas is described by the physical properties of pressure (P), volume (V), temperature (T), and amount in moles(n).

8.2 Gas Pressure

A gas exerts pressure, the force of the gas particles striking the surface of a container. Gas pressure is measured in units of torr, mm Hg, atm, and pascal.

8.3 Pressure and Volume (Boyle's Law)

The volume (V) of a gas changes inversely with the pressure (P) of the gas if there is no change in the amount and temperature: $P_1V_1 = P_2V_2$. This means that the pressure increases if volume decreases; pressure decreases if volume increases.

8.4 Temperature and Volume (Charles' Law)

The volume (V) of a gas is directly related to its Kelvin temperature (T) when there is no change in the amount and pressure of the gas:

$$\frac{V_1}{T_1} = \frac{V_2}{T_2}$$

Therefore, if temperature increases, the volume of the gas increases; if temperature decreases, volume decreases.

8.5 Temperature and Pressure (Gay-Lussac's Law)

The pressure (P) of a gas is directly related to its Kelvin temperature (T).

$$\frac{P_1}{T_1} = \frac{P_2}{T_2}$$

This means that an increase in temperature (T) increases the pressure of a gas, or a decrease in temperature decreases the pressure, as long as the amount and volume stay constant. Vapor pressure is the pressure of the gas that forms when a liquid evaporates. At the boiling point of a liquid, the vapor pressure equals the external pressure.

8.6 The Combined Gas Law

Gas laws combine into a relationship of pressure (P), volume (V), and temperature (T).

$$\frac{P_1 V_1}{T_1} = \frac{P_2 V_2}{T_2}$$

This expression is used to determine the effect of changes in two of the variables on the third.

8.7 Volume and Moles (Avogadro's Law)

The volume (V) of a gas is directly related to the number of moles (n) of the gas when the pressure and temperature of the gas do not change.

$$\frac{V_1}{n_1} = \frac{V_2}{n_2}$$

If the moles of gas are increased, the volume must increase; or if the moles of gas are decreased, the volume decreases. At standard temperature (273 K) and pressure (1 atm) abbreviated STP, 1 mole of any gas has a volume of 22.4 L.

8.8 The Ideal Gas Law

The ideal gas law gives the relationship of all the quantities P, V, n, and T that describe and measure a gas. $PV = nRT$. Any of the four variables can be calculated if the other three are known.

8.9 Partial Pressures (Dalton's Law)

In a mixture of two or more gases, the total pressure is the sum of the partial pressures of the individual gases.

$$P_{total} = P_1 + P_2 + P_3 + \cdots$$

The partial pressure of a gas in a mixture is the pressure it would exert if it were the only gas in the container.

Key Terms

atmosphere (atm) The pressure exerted by a column of mercury 760 mm high.

atmospheric pressure The pressure exerted by the atmosphere.

Avogadro's law A gas law that states that the volume of gas is directly related to the number of moles of gas in the sample when pressure and temperature do not change.

barometer An instrument used to measure atmospheric pressure.

Boyle's law A gas law stating that the pressure of a gas is inversely related to the volume when temperature and moles of the gas do not change; that is, if volume decreases, pressure increases.

Charles' law A gas law stating that the volume of a gas changes directly with a change in Kelvin temperature when pressure and moles of the gas do not change.

combined gas law A relationship that combines several gas laws relating pressure, volume, and temperature.

$$\frac{P_1 V_1}{T_1} = \frac{P_2 V_2}{T_2}$$

Dalton's law A gas law stating that the total pressure exerted by a mixture of gases in a container is the sum of the partial pressures that each gas would exert alone.

direct relationship A relationship in which two properties increase or decrease together.

Gay-Lussac's law A gas law stating that the pressure of a gas changes directly with a change in temperature when the number of moles of a gas and its volume are held constant.

ideal gas law A law that combines the four measured properties of a gas in the equation $PV = nRT$.

inverse relationship A relationship in which two properties change in opposite directions.

kinetic molecular theory of gases A model used to explain the behavior of gases.

molar volume A volume of 22.4 L occupied by 1 mole of a gas at STP conditions of 0°C (273 K) and 1 atm.

partial pressure The pressure exerted by a single gas in a gas mixture.

pressure The force exerted by gas particles that hit the walls of a container.

pressure gradient A difference in pressure causing gas particles to move from the area of high pressure to the area of low pressure.

STP Standard conditions of 0°C (273 K) temperature and 1 atm pressure used for the comparison of gases.

torr A unit of pressure equal to 1 mm Hg; 760 torr =1 atm.

universal gas constant _R_ A numerical value that relates the quantities _P, V, n,_ and _T_ in the ideal gas law, _PV = nRT._

vapor pressure The pressure exerted by the particles of vapor above a liquid.

Additional Problems

8.69 When a gas is heated at constant volume, the pressure increases. Use ideas from the kinetic molecular theory to explain.

8.70 When you drink through a straw, you breathe in to reduce the pressure in your mouth. What pressure is exerted on the top of the liquid you are drinking? Why does the liquid move up the straw?

8.71 At a restaurant, a customer chokes on a piece of food. You put your arms around the person's waist and use your fists to push up on the person's abdomen, an action called the Heimlich maneuver.
  a. How would this action change the volume of the chest and lungs?
  b. Why does it cause the person to expel the food item from the airway?

8.72 An airplane is pressurized to 650 mm Hg, which is the atmospheric pressure at a ski resort at 13,000 ft altitude.
  a. If air is 21% oxygen, what is the partial pressure of oxygen on the plane?
  b. If the partial pressure of oxygen drops below 100 mm Hg, passengers become drowsy. If this happens, oxygen masks are released. What is the total cabin pressure at which oxygen masks are dropped?

8.73 A fire extinguisher has a pressure of 150 lb/in.^{2} at 25°C. What is the pressure in atmospheres if the fire extinguisher is used at a temperature of 75°C?

8.74 A weather balloon has a volume of 750 L when filled with helium at 8°C at a pressure of 380 torr. What is the new volume of the balloon, where the pressure is 0.20 atm and the temperature is −45°C?

8.75 A sample of hydrogen (H_2) gas at 127°C has a pressure of 2.00 atm. At what temperature (°C) will the pressure of the H_2 decrease to 0.25 atm?

8.76 Nitrogen (N_2) is prepared and a sample of 250 mL is collected over water at 30°C and a total pressure of 745 mm Hg.

  a. Using the vapor pressure of water in Table 8.3, what is the partial pressure of the nitrogen?
  b. What is the volume of the nitrogen at STP?

8.77 A gas mixture with a total pressure of 2400 torr is used by a scuba diver. If the mixture contains 2.0 moles of helium and 6.0 moles of oxygen, what is the partial pressure of each gas in the sample?

8.78 What is the total pressure in mm Hg of a gas mixture containing argon gas at 0.25 atm, helium gas at 350 mm Hg, and nitrogen gas at 360 torr?

8.79 A gas mixture contains oxygen and argon at partial pressures of 0.60 atm and 425 mm Hg. If nitrogen gas added to the sample increases the total pressure to 1250 torr, what is the partial pressure in torr of the nitrogen added?

8.80 A gas mixture contains helium and oxygen at partial pressures of 255 torr and 0.450 atm. What is the total pressure in mm Hg of the mixture after it is placed in a container one-half the volume of the original container?

8.81 What is the density (g/L) of oxygen gas at STP?

8.82 At a party, a helium-filled balloon floats close to some hot lights and bursts. What might be a reason for this?

8.83 Compare the partial pressures of the following respiratory gases:
  a. oxygen in the lungs and in the blood coming to the alveoli
  b. oxygen in arterial blood and venous blood
  c. carbon dioxide in the tissues and arterial blood
  d. carbon dioxide in the venous blood and the lungs

8.84 For each of the comparisons in problem 8.83, describe the direction of diffusion for each gas.

8.85 When heated, calcium carbonate decomposes to give calcium oxide and carbon dioxide gas.

$$CaCO_3(s) \longrightarrow CaO(s) + CO_2(g)$$

If 2.00 moles of $CaCO_3$ react, how many liters of CO_2 gas are produced at STP?

8.86 In the reaction

$$2H_2(g) + O_2(g) \longrightarrow 2H_2O(g)$$

a. If 16.00 g of O_2 react at 0.800 atm and 127°C, what volume of oxygen was used?
b. How many moles of water are produced in part a?
c. If the water is produced at 760 mm Hg and 115°C, what volume of water was collected?

8.87 Magnesium reacts with oxygen to form magnesium oxide. How many liters of oxygen gas at STP are needed to completely react 8.0 g of magnesium?

$$2Mg(s) + O_2(g) \longrightarrow 2MgO(s)$$

8.88 In the formation of smog, nitrogen and oxygen gas react to form nitrogen dioxide. How many grams of NO_2 will be produced when 2.0 L of nitrogen at 840 mm Hg and 24°C are completely reacted?

$$N_2(g) + 2O_2(g) \longrightarrow 2NO_2(g)$$

8.89 How many molecules of CO_2 are in 35.0 L of CO_2 (g) at 1.2 atm and 5°C?

8.90 A steel cylinder with a volume of 15.0 L is filled with 50.0 g nitrogen gas at 25°C. What is the pressure of the N_2 gas in the cylinder?

8.91 A 2.00 L container is filled with methane gas CH_4 at a pressure of 2500 mm Hg and a temperature of 18°C. How many grams of methane are in the container?

8.92 A container is filled with 4.0×10^{22} O_2 molecules at 5°C and 845 mm Hg. What is the volume in mL of the container?

8.93 Your space ship has docked at a space station above Mars. The temperature inside the space station is a carefully controlled 24°C at a pressure of 745 mm Hg. A balloon with a volume of 425 mL drifts into the airlock where the temperature is −95°C and the pressure is 0.115 atm. What is the new volume of the balloon? Assume that the balloon is very elastic.

8.94 What is the molar mass of a gas if 1.15 g of the gas has a volume of 225 mL at STP? (*Hint:* Find moles of gas first.)

8.95 A sample of gas with a mass of 1.62 g occupies a volume of 941 mL at a pressure of 748 torr and a temperature of 20.0°C. What is the molar mass of the gas? (*Hint:* Find moles of gas first.)

8.96 At STP, 762 mL of a gas has a mass of 1.02 g. If the gas has an empirical formula of CH_3, what is the molecular formula of the compound?

8.97 How many liters of H_2 gas can be produced at STP from 25.0 g of Zn?

$$Zn(s) + 2HCl(aq) \longrightarrow ZnCl_2(aq) + H_2(g)$$

8.98 Hydrogen gas can be produced in the laboratory through the reaction of magnesium metal with hydrochloric acid.

$$Mg(s) + 2HCl \longrightarrow MgCl_2(aq) + H_2(g)$$

What is the volume, in liters, of H_2 gas produced at 24°C and 835 mm Hg, from the reaction of 12.0 g of Mg?

8.99 Nitrogen dioxide reacts with water to produce oxygen and ammonia.

$$4NO_2(g) + 6H_2O(g) \longrightarrow 7O_2(g) + 4NH_3(g)$$

a. How many liters of O_2 at STP are produced when 2.5×10^{23} molecules of NO_2 react?
b. A 5.00 L sample of H_2O (g) reacts at a temperature of 375°C and a pressure of 725 mm Hg. How many grams of NH_3 can be produced?

8.100 A 1.00 g sample of dry ice (CO_2) is placed in a container that has a volume of 4.6 L and a temperature of 24.0°C. Calculate the pressure in mm Hg inside the container after all the dry ice is converted to gas.

$$CO_2(s) \longrightarrow CO_2(g)$$

8.101 Aluminum oxide can be formed from its elements.

$$4Al(s) + 3O_2(g) \longrightarrow 2Al_2O_3(s)$$

What volume of oxygen is needed at STP to completely react 5.4 g of aluminum?

9 Solutions

"There is a lot of chemistry going on in the body, including how drugs interact," says Josephine Firenze, registered nurse, Kaiser Hospital.

Normally, the body maintains a homeostasis of fluids and electrolytes. Conditions that alter the composition of body fluids can lead to convulsions, coma, or death. To treat the disease process and to establish homeostasis, a patient may be given intravenous fluid therapy. Solutions that are compatible with body fluids such as a 5% glucose or a 0.9% saline are used. An infusion pump delivers a desired number of milliliters per hour to the patient. During IV therapy, a patient is checked for fluid overload indicated by edema, swelling, and a greater fluid input than output.

LOOKING AHEAD

the Chemistry place

www.chemplace.com/college

Visit the URL above or use the CD-ROM in the book for extra quizzing, interactive tutorials, career resources, and case studies.

Solutions are everywhere around us. Each consists of a mixture of at least one substance dissolved in another. The air we breathe is a solution of oxygen and nitrogen gases. Carbon dioxide gas dissolved in water makes carbonated drinks. When we make solutions of coffee or tea, we use hot water to dissolve substances from coffee beans or tea leaves. The ocean is also a solution, consisting of many salts such as sodium chloride dissolved in water. In a hospital, the antiseptic tincture of iodine is a solution of iodine dissolved in alcohol.

Our body fluids contain water and dissolved substances such as glucose and urea and electrolytes such as K^+, Na^+, Cl^-, Mg^{2+}, HCO_3^-, and HPO_4^{2-}. Proper amounts of each of these dissolved substances and water must be maintained in the body fluids. Small changes in electrolyte levels can seriously disrupt cellular process and endanger our health. Therefore, the measurement of their concentrations is a valuable diagnostic tool.

In the processes of osmosis and dialysis, water, essential nutrients, and waste products enter and leave the cells of the body. In osmosis, water flows in and out of the cells of the body. In dialysis, small particles in solution as well as water diffuse through semipermeable membranes. The kidneys regulate the amount of water and electrolytes that are excreted. If there is an excess of water, the kidneys will form dilute urine. If the electrolytes are too concentrated, the kidneys will remove the excess, which produces concentrated urine.

9.1 Properties of Water

LEARNING GOAL

Describe hydrogen bonding and surface tension in water.

Water is one of the most common substances in nature. In the H_2O molecule, an oxygen atom shares electrons with two hydrogen atoms. Because the oxygen atom is much more electronegative, the O—H bonds are polar. In each of the polar bonds, the oxygen atom has a partial negative (δ^-) charge, and the hydrogen atom has a partial positive (δ^+) charge. Because of the arrangement of the polar bonds, water is a *polar substance*.

WEB TUTORIAL
Hydrogen Bonding

Hydrogen Bonds in Water

Hydrogen bonds occur between molecules where hydrogen is attracted to the strongly electronegative atoms of O, N, or F in other molecules. A hydrogen bond is an attraction between the oxygen atom of one water molecule and a hydrogen atom in another water molecule. In the diagram, hydrogen bonds are shown as dots between the water molecules. Although hydrogen bonds are much weaker than covalent or ionic bonds, there are many of them linking molecules together. As a result hydrogen bonding plays an important role in the properties of water and biological compounds such as proteins and DNA.

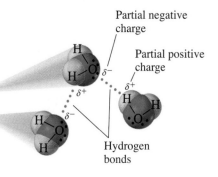

Partial negative charge

Partial positive charge

Hydrogen bonds

HEALTH NOTE

Water in the Body

The average adult contains about 60% water by weight, and the average infant about 75%. About 60% of the body's water is contained within the cells as intracellular fluids; the other 40% makes up extracellular fluids, which include the interstitial fluid in tissue and the plasma in the blood. These external fluids carry nutrients and waste materials between the cells and the circulatory system.

Every day you lose between 1500 and 3000 mL of water from the kidneys as urine, from the skin as perspiration, from the lungs as you exhale, and from the gastrointestinal tract. Serious dehydration can occur in an adult if there is a 10% net loss in total body fluid, and a 20% loss of fluid can be fatal. An infant suffers severe dehydration with a 5–10% loss in body fluid.

Water loss is continually replaced by the liquids and foods in the diet and from metabolic processes that produce water in the cells of the body. Table 9.1 lists the % by mass of water contained in some foods.

24 Hours

Water gain		Water loss	
		Urine	1500 mL
Liquid	1000 mL	Perspiration	300 mL
Food	1200 mL	Breath	600 mL
Metabolism	300 mL	Feces	100 mL
Total	2500 mL	Total	2500 mL

Table 9.1   Percentage of Water in Some Foods

Food	Water (% by mass)	Food	Water (% by mass)
Vegetables		**Meats/Fish**	
Carrot	88	Chicken, cooked	71
Celery	94	Hamburger, broiled	60
Cucumber	96	Salmon	71
Tomato	94	**Grains**	
Fruits		Cake	34
Apple	85	French bread	31
Banana	76	Noodles, cooked	70
Cantaloupe	91	**Milk Products**	
Grapefruit	89	Cottage cheese	78
Orange	86	Milk, whole	87
Strawberry	90	Yogurt	88
Watermelon	93		

Surface Tension

Imagine that you have filled a glass up to its rim with water. Carefully adding a few more drops of water or dropping in some pennies does not cause the water to overflow. Instead the water seems to adhere to itself, forming a dome that rises above the rim of the glass. This effect is the result of the polarity of water. Throughout the liquid in the glass, water molecules are attracted in all directions by surrounding water molecules. However, the water molecules on the surface are pulled like a skin toward the rest of the water in the glass. As a result, the water molecules on the surface become more tightly packed, a feature called **surface tension.** Because of surface tension, a needle floats on the top of water, certain water bugs can travel across the surface of a pond or lake, and drops of water are spherical.

When compounds called surfactants are added to water, they disrupt the hydrogen bonding between the water molecules. As a result, the surface tension is decreased and the water spreads out rather than forming drops. Soap, detergents, shampoos, and fabric softeners are examples of surfactants we use every day.

The lungs produce a surfactant that allows efficient exchange of oxygen and carbon dioxide between the alveoli and the capillaries. When premature babies lack an

EXPLORE YOUR WORLD

Surface Tension

Obtain a glass or jar that is smooth around the top. Fill it with water as full as you can without running over. Obtain several pennies. Predict the number of pennies you can add before water runs down the side. Holding a penny vertically, carefully slide it into the glass of water. Keep adding pennies until you see water run down the outside of the glass. Observe the surface of the water. Repeat the test with containers of different shapes and sizes.

Questions

1. How did your prediction compare with the number of pennies you were able to add before the water ran over?
2. What happened to the surface of the water? Explain.
3. How does the number of pennies added compare for containers of different shapes or sizes?

EXPLORE YOUR WORLD

Surfactants

Place some water in a small bowl. Carefully place a sewing needle on the surface. Push the needle below the surface of the water. Pick up the needle and float it on the water surface again. Now add a few drops of a liquid detergent or touch the water surface with the tip of a bar of soap.

Using a new water sample in the bowl, sprinkle pepper on the surface of the water. Observe. Add a few drops of liquid detergent or touch the surface with the tip of a bar of soap. Soaps and detergents contain surfactants, which decrease the surface tension.

Questions

1. Why does a needle float on the surface of the water?
2. What makes the needle sink when soap is added?
3. How did the detergent or soap change the appearance of the pepper on the surface of the water?

adequate amount of this surfactant, they have difficulty breathing, which is a condition known as respiratory distress syndrome (RDS). A surfactant may be placed in the lungs and the oxygen increased until the baby produces a sufficient supply of its own surfactant.

SAMPLE PROBLEM 9.1

Hydrogen Bonding in Water

Why does hydrogen bonding occur between water molecules?

Solution

Water molecules are polar because they contain polar bonds between hydrogen atoms and highly electronegative oxygen atoms. Hydrogen bonds form when the negatively charged oxygen atom is attracted to the positively charged hydrogen atoms of other water molecules.

Study Check

Why do the molecules at the surface of liquid water behave differently than the water molecules within the liquid?

QUESTIONS AND PROBLEMS

Properties of Water

9.1 **a.** Draw a water molecule and describe its polarity.
 b. Show how hydrogen bonding occurs between two water molecules.

9.2 **a.** How does the surface of the water in a pond support the weight of small insects?
 b. How does a surfactant affect the surface tension of water?

Define solute and solvent; describe the formation of a solution.

Salt

Solute: The substance present in lesser amount

Water

Solvent: The substance present in greater amount

9.2 Solutions

A **solution** is a mixture in which one substance called the **solute** is uniformly dispersed in another substance called the **solvent.** Because the solute and the solvent do not react with each other, they can be mixed in varying proportions. A little salt dissolved in water tastes slightly salty. When more salt dissolves, the water tastes very salty. Usually, the solute (in this case, salt) is the substance present in the smaller amount, whereas the solvent (in this case, water) is present in the larger amount. In a solution, the particles of the solute are evenly dispersed among the molecules of the solvent. (See Figure 9.1.)

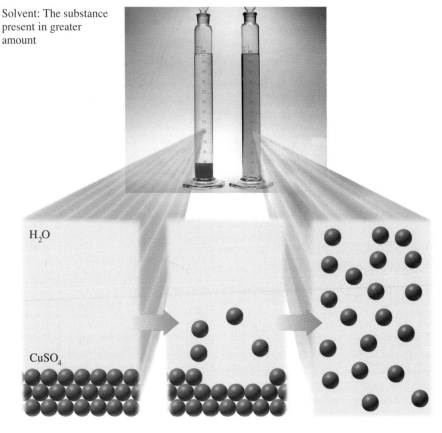

H_2O

$CuSO_4$

Figure 9.1 A solution of copper (II) sulfate ($CuSO_4$) forms as particles of solute dissolve, move away from the crystals, and become evenly dispersed among the solvent (water) molecules.

Q What does the uniform blue color indicate about the $CuSO_4$ solution?

Identifying a Solute and a Solvent

Identify the solute and the solvent in each of the following solutions:

a. 15 g of sugar dissolved in 100 g of water
b. 75 mL of water mixed with 25 mL of isopropyl alcohol (rubbing alcohol)

Solution

a. Sugar, the smaller quantity, is the solute; water is the solvent.

b. Isopropyl alcohol with the smaller volume is the solute. Water is the solvent.

Study Check

A tincture of iodine is prepared with 0.10 g of I_2 and 10.0 mL of ethyl alcohol. What is the solute, and what is the solvent?

Types of Solutes and Solvents

Solutes and solvents may be solids, liquids, or gases. The solution that forms has the same physical state as the solvent. When sugar is dissolved in a glass of water, a liquid sugar solution forms. Sugar is the solute, and water is the solvent. Soda water and soft drinks are prepared by dissolving CO_2 gas in water. The CO_2 gas is the solute, and water is the solvent. Table 9.2 lists some solutes and solvents and their solutions.

Table 9.2   Some Examples of Solutions

Type	Example	Solute	Solvent
Gas Solutions			
Gas in a gas	Air	Oxygen (gas)	Nitrogen (gas)
Liquid Solutions			
Gas in a liquid	Soda water	Carbon dioxide (gas)	Water (liquid)
	Household ammonia	Ammonia (gas)	Water (liquid)
Liquid in a liquid	Vinegar	Acetic acid (liquid)	Water (liquid)
Solid in a liquid	Seawater	Sodium chloride (solid)	Water (liquid)
	Tincture of iodine	Iodine (solid)	Alcohol (liquid)
Solid Solutions			
Liquid in a solid	Dental amalgam	Mercury (liquid)	Silver (solid)
Solid in a solid	Brass	Zinc (solid)	Copper (solid)
	Steel	Carbon (solid)	Iron (solid)

Like Dissolves Like

Gases form solutions easily because their particles are moving so rapidly that they are far apart and attractions to the other gas particles are not important. When solids or liquids form solutions, there must be an attraction between the solute particles and the solvent particles. Then the particles of the solute and solvent will mix together. If there is no attraction between a solute and a solvent, their particles do not mix and no solution forms.

A salt such as NaCl will form a solution with water because the Na^+ and Cl^- ions in the salt are attracted to the positive and negative parts of the individual water molecules. You can dissolve table sugar, $C_{12}H_{22}O_{11}$, in water because the molecule has many polar —OH groups that attract water molecules.

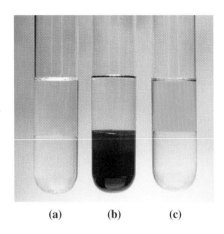

(a)   (b)   (c)

Figure 9.2 Like dissolves like. (**a**) The test tubes contain an upper layer of water (polar) and a lower layer of CH_2Cl_2 (nonpolar). (**b**) The nonpolar solute I_2 dissolves in the nonpolar layer. (**c**) The ionic solute $Ni(NO_3)_2$ dissolves in the water.

Q Which layer would dissolve polar molecules of sugar?

EXPLORE YOUR WORLD

Like Dissolves Like

Mix together small amounts of the following substances:
a. oil and water
b. water and vinegar
c. salt and water
d. sugar and water
e. salt and oil

Questions

1. Which of the mixtures formed a solution? Which did not?
2. Why do some mixtures form solutions, but others do not?

However, compounds containing nonpolar molecules such as iodine (I_2), oil, or grease do not dissolve in water because water is polar. Nonpolar solutes require nonpolar solvents for a solution to form. The expression "like dissolves like" is a way of saying that the polarities of a solute and a solvent must be similar in order to form a solution. Figure 9.2 illustrates the formation of some polar and nonpolar solutions.

SAMPLE PROBLEM 9.3

Polar and Nonpolar Solutes

Indicate whether each of the following substances will dissolve in water. Explain.

a. KCl   **b.** octane, C_8H_{18}, a compound in gasoline
c. ethanol, C_2H_6O, a substance in mouthwash

Solution

a. Yes. KCl is an ionic compound.
b. No. C_8H_{18} is a nonpolar substance.
c. Yes. C_2H_6O is a polar substance.

Study Check

Will oil, a nonpolar substance, dissolve in hexane, a nonpolar solvent?

Dissolving a Solute

An ionic compound such as sodium chloride, NaCl, is held together by ionic bonds between positive Na^+ ions and negative Cl^- ions. It dissolves in water because water is a polar solvent. When solid NaCl crystals are placed in water, the process of dissolution begins as the ions on the surface of the crystal come in contact with water molecules. (See Figure 9.3.) The negatively charged oxygen atom at one end of a water molecule attracts the positive Na^+ ions. The positively charged hydrogen atoms at the other end of a water molecule attract the negative Cl^- ions. The attractive forces of several water molecules provide the energy to break the ionic bonds between the Na^+ and Cl^- ions in the NaCl crystal. As the water molecules pull the ions into solution, a new surface of the NaCl crystal is exposed to the solvent. During a process called **hydration,** the dissolved Na^+ and Cl^- ions are surrounded by water molecules, which diminishes their attraction to other ions and helps keep them in solution.

Hydrates

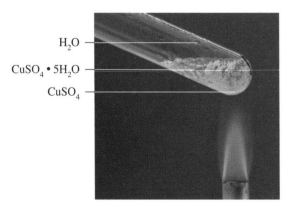

H_2O
$CuSO_4 \cdot 5H_2O$
$CuSO_4$

In solid ionic compounds called **hydrates,** there are water molecules that are part of the crystal structure. These water molecules are called the **water of hydration.** When heated strongly, the bonds to the water molecules are broken, and the water evaporates. For example, in the blue crystals of $CuSO_4 \cdot 5H_2O$, the salt $CuSO_4$ is bonded to five water molecules indicated by a large dot. During heating, water is lost and the blue hydrate turns into the white anhydrate $CuSO_4$. Adding a few drops of water to the anhydrate reverses the reaction.

$$CuSO_4 \cdot 5H_2O(s) \underset{- \text{ heat}}{\overset{+ \text{ heat}}{\rightleftarrows}} CuSO_4(s) + 5\,H_2O(g)$$

copper(II) sulfate   copper(II) sulfate   water of hydration
pentahydrate (blue)   anhydrate (white)

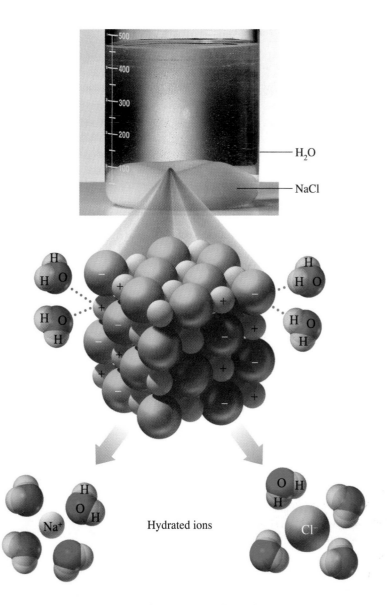

H₂O

NaCl

H
H O

H
H O

H
H O

H
H O

H
O
H
Na⁺

O H
H

Cl⁻

Hydrated ions

Figure 9.3 Ions on the surface of a crystal of NaCl dissolve in water as they are attracted to the polar water molecules that pull the ions into solution and surround them.

Q What helps keep the Na⁺ and Cl⁻ ions in solution?

Some anhydrates are used as drying agents and can be found in small packets inside containers. They keep the air dry by bonding water molecules to form their hydrates. The formulas of some hydrates are listed in Table 9.3.

$$CaCl_2 \ + \ 2H_2O \ \longrightarrow \ CaCl_2 \bullet 2H_2O$$
  anhydrate   hydrate

Table 9.3   Some Hydrates, Their Formulas, and Uses

Name	Formula	Uses
Magnesium sulfate heptahydrate	$MgSO_4 \bullet 7H_2O$	Epsom salts, laxative
Sodium sulfate decahydrate	$Na_2SO_4 \bullet 10H_2O$	Dyes and textile printing
Calcium sulfate hemihydrate	$CaSO_4 \bullet \frac{1}{2}H_2O$	Plaster of Paris, casts
Calcium chloride dihydrate	$CaCl_2 \bullet 2H_2O$	Deicer, fire extinguishers, and refrigerants
Iron(II) sulfate heptahydrate	$FeSO_4 \bullet 7H_2O$	Iron supplement, fertilizer
Aluminum chloride hexahydrate	$AlCl_3 \bullet 6H_2O$	Antiperspirant
Sodium tetraborate decahydrate	$Na_2B_4O_7 \bullet 10H_2O$	Borax

SAMPLE PROBLEM 9.4

Percent Water in a Hydrate

Calculate the percent by mass of water in the iron supplement iron(II) sulfate heptahydrate.

Solution

Iron(II) sulfate heptahydrate has the formula $FeSO_4 \cdot 7H_2O$, which has a molar mass of 278 g. The mass of water in the formula is 126 g.

$$\text{Percent } H_2O \text{ (by mass)} = \frac{g\ H_2O}{g\ \text{hydrate}} \times 100\%$$

$$= \frac{126\ g}{278\ g} \times 100\%$$

$$= 45.3\%$$

Study Check

What is the percent water by mass in aluminum chloride hexahydrate?

QUESTIONS AND PROBLEMS

Solutions

9.3 Identify the solute and the solvent in each solution composed of the following:
   a. 10.0 g of NaCl and 100.0 g of H_2O
   b. 50.0 mL of ethanol, $C_2H_5OH(l)$, and 10.0 mL of H_2O
   c. 0.20 L of O_2 and 0.80 L of N_2

9.4 Identify the solute and the solvent in each solution composed of the following:
   a. 50.0 g of silver and 4.0 g of mercury
   b. 100.0 mL of water and 5.0 g of sugar
   c. 1.0 g of I_2 and 50.0 mL of alcohol

9.5 Water is a polar solvent; CCl_4 is a nonpolar solvent. In which solvent are each of the following more likely to be soluble?
   a. KCl, ionic
   b. I_2, nonpolar
   c. sugar, polar
   d. gasoline, nonpolar

9.6 Water is a polar solvent; hexane is a nonpolar solvent. In which solvent are each of the following more likely to be soluble?
   a. vegetable oil, nonpolar
   b. benzene, nonpolar
   c. $LiNO_3$, ionic
   d. Na_2SO_4, ionic

9.7 Describe the formation of an aqueous KI solution.

9.8 Describe the formation of an aqueous LiBr solution.

9.9 Write the formula of sodium sulfate decahydrate and calculate its percent water by mass.

9.10 Write the formula of magnesium sulfate heptahydrate and calculate its percent water by mass.

9.3 Electrolytes and Nonelectrolytes

LEARNING GOAL
Describe the solution process for electrolytes and nonelectrolytes in water.

Solutions containing ions are conductors of electricity. The ions come from substances called **electrolytes.** To test for electrolytes, we can use an apparatus that consists of a battery and a pair of electrodes connected by wires to a light bulb. The light bulb glows when current flows through it, and that only happens when there are ions moving through the solution.

In water, a **strong electrolyte** exists only as ions in solution. When there are many ions in solution, the light bulb glows brightly. However, in a solution containing **weak electrolytes,** there are only a few ions. With a small number of ions in solution, the light bulb glows dimly. When the electrodes are placed in pure water or a sugar solution, the light bulb does not glow at all. Covalent substances that dissolve in water and do not conduct electricity are **nonelectrolytes.**

Strong Electrolytes

Sodium chloride (NaCl), a strong electrolyte, consists of sodium ions and chloride ions held together by strong ionic bonds. In water, the ions are attracted to water molecules and separate from the solid, a process called dissociation. A solution forms that looks like water but tastes salty because it contains dissolved sodium ions and chloride ions. We write an equation for the dissociation of NaCl in water to show that solid NaCl separates into hydrated ions of $Na^+(aq)$ and $Cl^-(aq)$:

$$NaCl(s) \xrightarrow[\substack{100\% \\ \text{dissociation}}]{H_2O} \underset{\text{Ions in solution}}{Na^+(aq) + Cl^-(aq)}$$

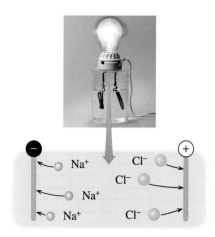

Strong electrolyte

Other soluble ionic compounds dissolve in a similar way. When we write an ionic equation, we balance the electrical charges of the ions. The total positive charge must equal the total negative charge. For example, magnesium chloride, $MgCl_2$, dissolves in water to give one magnesium ion for every two chloride ions in solution.

$$MgCl_2(s) \xrightarrow{H_2O} Mg^{2+}(aq) + 2Cl^-(aq)$$
$$\underset{\substack{\text{Charge balance for ions} \\ \text{in solution}}}{}$$

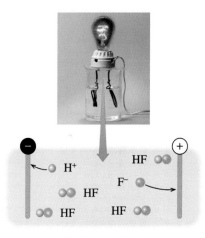

Weak electrolyte

Weak Electrolytes

A weak electrolyte is typically a polar covalent compound that dissolves in water mostly as nondissociated molecules. However, a few of the molecules ionize to produce a small number of ions in solution. For example, most HF molecules that dissolve in water remain as molecules, although a few HF molecules do ionize, producing H^+ and F^- ions. Thus, in water, weak electrolytes are only slightly ionized, which means that most of the dissolved substance is in the molecular form. There are really two reactions taking place in this solution. One is the ionization of HF molecules written as

$$HF \xrightarrow{\text{Ionization}} H^+ + F^-$$

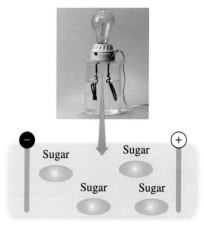

Nonelectrolyte

As the number of H^+ and F^- build up in solution, some of the ions recombine to give HF molecules.

$$HF \xleftarrow{\quad\text{Recombination}\quad} H^+ + F^-$$

Eventually the reverse reaction goes as fast as the forward reaction. Then the concentrations of HF, H^+, and F^- no longer change. An equilibrium exists between the HF molecules and its ions.

$$HF \underset{\text{A few ionize}}{\rightleftharpoons} H^+ + F^-$$

Nonelectrolytes

Nonelectrolytes are polar covalent compounds that dissolve in water as molecules. They produce no ions and their solutions do not conduct electricity. For example, sucrose (sugar) is a nonelectrolyte that dissolves in water to produce a solution of sugar molecules.

$$\underset{\text{Sucrose}}{C_{12}H_{22}O_{11}(s)} \xrightarrow{\quad H_2O \quad} \underset{\text{Solution of sucrose molecules}}{C_{12}H_{22}O_{11}(aq)}$$

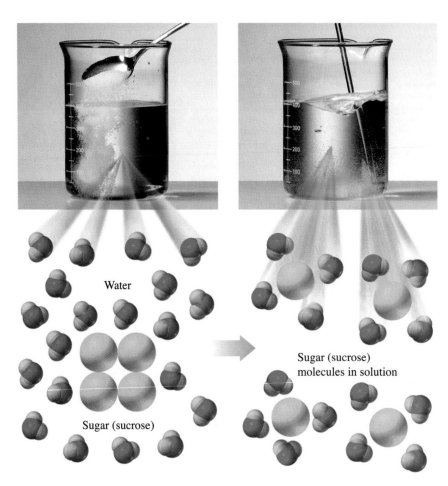

Water

Sugar (sucrose)

Sugar (sucrose) molecules in solution

In an aqueous solution:

A strong electrolyte is completely (100%) ionized.

A weak electrolyte is slightly ionized.

A nonelectrolyte forms no ions, only molecules.

SAMPLE PROBLEM 9.5

Solutions of Electrolytes and Nonelectrolytes

Indicate whether solutions of each of the following contain ions, molecules, or both:

a. Na_2SO_4, a strong electrolyte
b. CH_3OH, a nonelectrolyte

Solution

a. A solution of Na_2SO_4 contains the ions of the salt, Na^+ and SO_4^{2-}.
b. A nonelectrolyte such as CH_3OH dissolves in water as molecules.

Study Check

Boric acid, H_3BO_3, is a weak electrolyte. Would you expect a boric acid solution to contain ions, molecules, or both?

QUESTIONS AND PROBLEMS

Electrolytes and Nonelectrolytes

9.11 KF is a strong electrolyte, and HF is a weak electrolyte. How are they different?

9.12 NaOH is a strong electrolyte, and CH_3OH is a nonelectrolyte. How are they different?

9.13 The following soluble salts are strong electrolytes. Write a balanced equation for their dissociation in water:
 a. KCl   **b.** $CaCl_2$ **c.** K_3PO_4 **d.** $Fe(NO_3)_3$

9.14 The following salts are strong electrolytes. Write a balanced equation for their dissociation in water:
 a. LiBr   **b.** $NaNO_3$ **c.** $FeCl_3$ **d.** $Mg(NO_3)_2$

9.15 Indicate whether aqueous solutions of the following will contain ions only, molecules only, or molecules and some ions:
 a. acetic acid ($HC_2H_3O_2$), found in vinegar, a weak electrolyte
 b. NaBr, a salt
 c. fructose ($C_6H_{12}O_6$), a nonelectrolyte

9.16 Indicate whether aqueous solutions of the following will contain ions only, molecules only, or molecules and some ions:
 a. Na_2SO_4, a salt
 b. ethanol, C_2H_5OH, a nonelectrolyte
 c. HCN, hydrocyanic acid, a weak electrolyte

9.17 Indicate the type of electrolyte represented in the following equations:

 a. $K_2SO_4 \xrightarrow{H_2O} 2K^+ + SO_4^{2-}$

 b. $NH_4OH \underset{}{\overset{H_2O}{\rightleftharpoons}} NH_4^+ + OH^-$

 c. $C_6H_{12}O_6(s) \xrightarrow{H_2O} C_6H_{12}O_6(aq)$

9.18 Indicate the type of electrolyte represented in the following equations:

a. $CH_3OH \xrightarrow{H_2O} CH_3OH(aq)$

b. $MgCl_2 \xrightarrow{H_2O} Mg^{2+} + 2Cl^-$

c. $HClO \underset{}{\overset{H_2O}{\rightleftharpoons}} H^+ + ClO^-$

LEARNING GOAL

Calculate the number of equivalents for a given quantity of an electrolyte.

9.4 Equivalents

We have seen that electrolytes dissolve in solution as ions. When several electrolytes are dissolved in solution, many different ions are present. Body fluids typically contain a mixture of several electrolytes, such as Na^+, Cl^-, K^+, and Ca^{2+}. When we have different ions in solution, we measure each individual ion in terms of an **equivalent (Eq),** which is the amount of that ion equal to 1 mole of positive or negative electrical charge. For example, 1 mole of Na^+ ions and 1 mole of Cl^- ions are each 1 equivalent because they each contain 1 mole of charge. This is the same for ions such as K^+, H^+, OH^-, and HCO_3^-. For an ion with a charge of $2+$ or $2-$, there are 2 equivalents for each mole. Some examples of ions and equivalents are shown in Table 9.4.

Table 9.4   Equivalents of Electrolytes

Ion	Electrical Charge	Number of Equivalents in 1 mole
Na^+	$1+$	1 Eq
Ca^{2+}	$2+$	2 Eq
Fe^{3+}	$3+$	3 Eq
Cl^-	$1-$	1 Eq
SO_4^{2-}	$2-$	2 Eq

We can use the number of equivalents in 1 mole as a conversion factor. For example, Ca^{2+} contains 2 equivalents in 1 mole of ions, which can be written as two possible conversion factors.

$$\frac{2 \text{ Eq } Ca^{2+}}{1 \text{ mole } Ca^{2+}} \quad \text{and} \quad \frac{1 \text{ mole } Ca^{2+}}{2 \text{ Eq } Ca^{2+}}$$

SAMPLE PROBLEM 9.6

Calculating Equivalents

How many equivalents are in 6.0 g of Ca^{2+}?

Solution

The conversion factor of molar mass converts the grams to moles. Because there are 2 equivalents in 1 mole of calcium ion, we calculate the number of equivalents in our sample.

$$6.0 \text{ g Ca}^{2+} \times \frac{1 \text{ mole Ca}^{2+}}{40.1 \text{ g Ca}^{2+}} \times \frac{2 \text{ Eq Ca}^{2+}}{1 \text{ mole Ca}^{2+}} = 0.30 \text{ Eq Ca}^{2+}$$

Study Check

How many grams of Cl^- are present in 2.0 Eq of Cl^-?

SAMPLE PROBLEM 9.7

Electrolyte Concentration

In body fluids, concentrations of electrolytes are often expressed as milliequivalent (mEq) per liter. A typical concentration for Na^+ in the blood is 138 mEq/L. How many grams of sodium ion are in 1.00 L of blood?

Solution

Using the volume and the electrolyte concentration (in mEq/L) we can find the number of equivalents in 1.00 L of blood.

$$1.00 \text{ L} \times \frac{138 \text{ mEq}}{1 \text{ L}} \times \frac{1 \text{ Eq}}{1000 \text{ mEq}} = 0.138 \text{ Eq Na}^+$$

We can then convert equivalents to moles (for Na^+ there is 1 Eq per mole), and moles to grams using molar mass.

$$0.138 \text{ Eq Na}^+ \times \frac{1 \text{ mole Na}^+}{1 \text{ Eq Na}^+} \times \frac{23.0 \text{ g Na}^+}{1 \text{ mole Na}^+} = 3.17 \text{ g Na}^+$$

Study Check

A Ringer's solution for intravenous fluid replacement contains 155 mEq Cl^- per liter of solution. If a patient receives 1250 mL of Ringer's solution, how many grams of chloride were given?

QUESTIONS AND PROBLEMS

Equivalents

9.19 Indicate the number of equivalents in each of the following:
    a. 1 mole K^+ **b.** 2 moles OH^-
    c. 1 mole Ca^{2+} **d.** 3 moles CO_3^{2-}

9.20 Indicate the number of equivalents in each of the following:
    a. 1 mole Mg^{2+} **b.** 0.5 mole H^+
    c. 4 moles Cl^- **d.** 2 moles Fe^{3+}

9.21 Calculate the number of equivalents in each of the following:
    a. 25.0 g Cl^- **b.** 15.0 g Fe^{3+}
    c. 4.0 g Ca^{2+} **d.** 1.0 g H^+

9.22 Calculate the number of equivalents in each of the following:
    a. 10.0 g Na^+ **b.** 8.0 g OH^-
    c. 20.0 g Mg^{2+} **d.** 4.0 g Al^{3+}

9.23 If blood plasma typically contains 3.0 mEq/L of Mg^{2+}, how many grams of Mg^{2+} would be in 5.0 L of plasma?

9.24 If blood plasma typically contains 110 mEq/L of Cl^-, how many grams of Cl^- would be in 5.0 L of plasma?

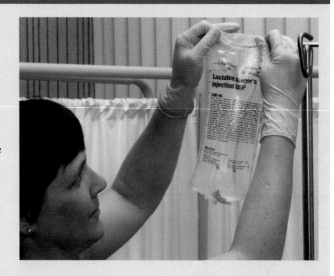

HEALTH NOTE

Electrolytes in Body Fluids

The concentrations of electrolytes present in body fluids and in intravenous fluids given to a patient are often expressed in milliequivalents per liter (mEq/L) of solution.

$$1 \text{ Eq} = 1000 \text{ mEq}$$

Table 9.5 gives the concentrations of some typical electrolytes in blood plasma. There is a charge balance because the total number of positive charges is equal to the total number of negative charges. The use of a specific intravenous solution depends on the nutritional, electrolyte, and fluid needs of the individual patient. Examples of various types of solutions are given in Table 9.6.

Table 9.5   Some Typical Concentrations of Electrolytes in Blood Plasma

Electrolyte	Concentration (mEq/L)
Cations (Positive Ions)	
Na^+	138
K^+	5
Mg^{2+}	3
Ca^{2+}	4
Total	150
Anions (Negative Ions)	
Cl^-	110
HCO_3^-	30
HPO_4^{2-}	4
Proteins	6
Total	150

Table 9.6   Electrolyte Concentrations in Intravenous Replacement Solutions

Solution	Electrolytes (mEq/L)	Use
Sodium chloride (0.9%)	Na^+ 154, Cl^- 154	Replacement of fluid loss
Potassium chloride with 5% dextrose	K^+ 40, Cl^- 40	Treatment of malnutrition (low potassium levels)
Ringer's solution	Na^+ 147, K^+ 4, Ca^{2+} 4, Cl^- 155	Replacement of fluids and electrolytes lost through dehydration
Maintenance solution with 5% dextrose	Na^+ 40, K^+ 35, Cl^- 40, lactate$^-$ 20, HPO_4^{2-} 15	Maintenance of fluid and electrolyte levels
Replacement solution (extracellular)	Na^+ 140, K^+ 10, Ca^{2+} 5, Mg^{2+} 3, Cl^- 103, acetate$^-$ 47, citrate^{3-} 8	Replacement of electrolytes in extracellular fluids

9.25 A physiological saline solution contains 154 mEq/L each of Na^+ and Cl^-. How many grams each of Na^+ and Cl^- are in 1.0 L of the saline solution?

9.26 A solution to replace potassium loss contains 40. mEq/L each of K^+ and Cl^-. How many grams each of K^+ and Cl^- are in 1.5 L of the solution?

9.27 A solution contains 40. mEq/L of Cl^- and 15 mEq/L of HPO_4^{2-}. If Na^+ is the only cation in the solution, what is the Na^+ concentration in milliequivalents per liter?

9.28 A sample of Ringer's solution contains the following concentrations (mEq/L) of cations: Na^+ 147, K^+ 4, and Ca^{2+} 4. If Cl^- is the only anion in the solution, what is the Cl^- concentration in milliequivalents per liter?

LEARNING GOAL

Define solubility; distinguish between an unsaturated and a saturated solution. Identify an insoluble salt.

9.5 Solubility

The term **solubility** is used to describe the amount of a solute that can dissolve in a given amount of solvent. Many factors such as the type of solute, the type of solvent, and temperature affect a solute's solubility. Solubility, usually expressed in grams of solute in 100 grams of solvent, is the maximum amount of solute that can be dissolved at a certain temperature. If a solute readily dissolves when added to the solvent, the solution does not contain the maximum amount of solute. We call the solution an **unsaturated solution.** When a solution contains all the solute that can dissolve, it is a **saturated solution.** If we try to add more solute, undissolved solute will remain on the bottom of the container. For example, 36 g of NaCl can dissolve in 100 grams of water at 20°C. If we add 15 g of NaCl to 100 g of water at 20°C, it all dissolves; the solution is unsaturated. The NaCl solution becomes saturated when we dissolve a total of 36 g of NaCl.

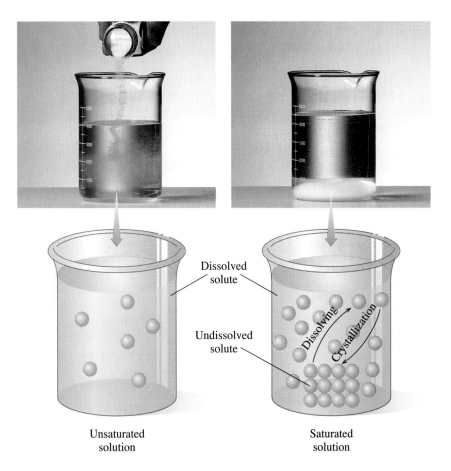

Dissolved solute

Undissolved solute

Dissolving

Crystallization

Unsaturated solution

Saturated solution

When a solution becomes saturated, equilibrium exists between the forward reaction that dissolves the solute and the reverse reaction of crystallization. Once equilibrium is reached, there is no further change in the amount of solid solute. The rate at which solute particles are removed from the surface of the solid is equal to the rate at which particles in solution crystallize out as solid.

$$\text{Solid solute} \underset{\text{crystallizes}}{\overset{\text{dissolves}}{\rightleftharpoons}} \text{saturated solution}$$

SAMPLE PROBLEM 9.8

Saturated Solutions

At 20°C, the solubility of KCl is 34 g/100 g of water. In the laboratory, a student mixes 45 g of KCl with 100 g of water at a temperature of 20°C.

a. How much of the KCl will dissolve?
b. Is the solution saturated?
c. What is the mass in grams of any solid KCl on the bottom of the container?

Solution

a. A total of 34 g of the KCl will dissolve because that is its solubility and, therefore, the maximum amount of KCl in solution at 20°C.

b. Yes, the solution is saturated.

c. The mass of the solid KCl is 11 g.

Study Check

At 50°C, the solubility of $NaNO_3$ is 114 g/100 g of water. How many grams of $NaNO_3$ are needed to make a saturated $NaNO_3$ solution with 50 g of water at 50°C?

Effect of Temperature

The solubility of most solids in water increases as temperature increases. For example, if you add sugar to ice tea, a layer of undissolved sugar forms on the bottom of the glass when saturation is reached. However, if the tea is heated, more sugar dissolves because the solubility of sugar is greater in hot water.

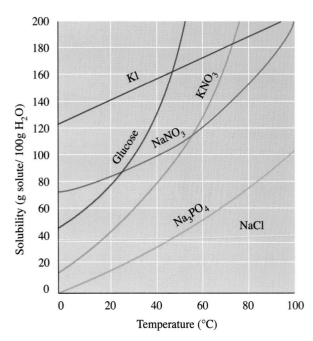

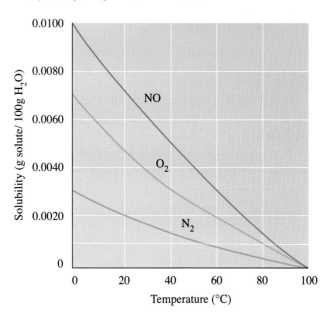

In contrast, gases are less soluble in water at higher temperature. As the temperature increases, increasing numbers of gas molecules escape from the solution. At high temperatures, bottles containing carbonated solutions may burst as more gas molecules leave the solution and increase the gas pressure inside the bottle. Biologists have found that increased temperatures in rivers and lakes cause the amount of dissolved oxygen to decrease until the warm water can no longer support a biological community. Electric generating plants are required to have their own ponds to use with their cooling towers to lessen the threat of thermal pollution.

Henry's Law

Henry's law states that the solubility of gas in a liquid is directly related to the pressure of that gas above the liquid. As a gas is compressed at higher pressures, there are more gas molecules available to enter and dissolve in the liquid. A can of soda is carbonated by using CO_2 gas at high pressure to increase the solubility of the CO_2 in the beverage. When you open the can at atmospheric pressure, the pressure on the CO_2 drops, which decreases the solubility of CO_2. As a result, bubbles of CO_2 rapidly escape from the solution. The burst of bubbles is even more noticeable when you open a warm can of soda.

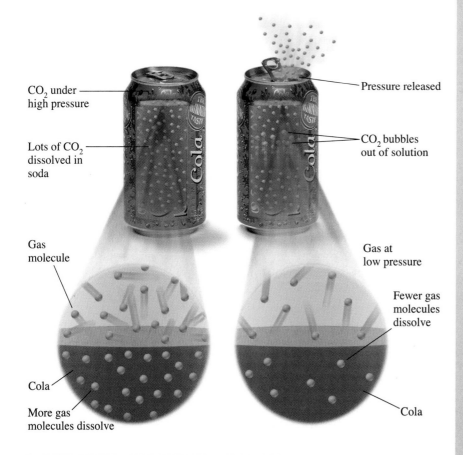

CO_2 under high pressure

Lots of CO_2 dissolved in soda

Gas molecule

Cola

More gas molecules dissolve

Pressure released

CO_2 bubbles out of solution

Gas at low pressure

Fewer gas molecules dissolve

Cola

SAMPLE PROBLEM 9.9

Factors Affecting Solubility

Indicate whether the solubility of the solute will increase or decrease in each of the following situations:

a. using 80°C water instead of 25°C water to dissolve sugar
b. effect on dissolved O_2 in a lake as it warms

Solution

a. An increase in the temperature increases the solubility of the sugar.

b. An increase in the temperature decreases the solubility of O_2 gas.

Study Check

At 10°C, the solubility of KNO_3 is 20 g/100 g H_2O. Would the value of 5 g/100 g H_2O or 65 g/100 g H_2O be the more likely solubility at 40°C? Explain.

WEB TUTORIAL
Solubility

Soluble and Insoluble Salts

Up to now, we have considered ionic compounds that dissolve in water; they are **soluble salts.** However, some ionic compounds do not separate into ions in water. They are **insoluble salts** that remain as solids even in contact with water.

Salts that are always soluble in water usually contain at least one of the following ions: Li^+, Na^+, K^+, NH_4^+, or NO_3^-. Most salts containing Cl^- are soluble, but $AgCl$, $PbCl_2$, or Hg_2Cl_2 are not; they are insoluble chloride salts. Similarly, most salts containing SO_4^{2-} are soluble, but a few are insoluble as shown in Table 9.7. Most other salts are insoluble and do not dissolve in water. (See Figure 9.4.) In an insoluble salt, attractions between its positive and negative ions are too strong for the polar water molecules to break. We use solubility rules to predict whether a salt (a solid ionic compound) would be expected to dissolve in water. Table 9.8 illustrates the use of these rules.

Figure 9.4 Mixing certain aqueous solutions produces insoluble salts.

Q What ions make each of these salts insoluble in water?

CdS

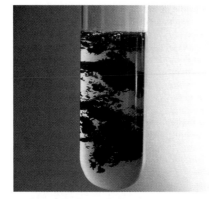
FeS

PbCrO₄

Ni(OH)₂

Table 9.7   Solubility Rules for Ionic Solids in Water

Soluble if salt contains		Insoluble if salt contains
NH_4^+ Li^+, Na^+, K^+ Nitrates NO_3^-, acetates $C_2H_3O_2^-$	but not with	Carbonates CO_3^{2-}, Sulfides S^{2-} Phosphates PO_4^{3-}, Hydroxides, OH^-
Chlorides Cl^-, Bromides Br^-, Iodides I^-	but not with	Ag^+, Pb^{2+}, or Hg_2^{2+}
Sulfates SO_4^{2-}	but not with	Ba^{2+}, Pb^{2+}, Ca^{2+}, Sr^{2+}

Table 9.8   Using Solubility Rules

Ionic Compound	Solubility in Water	Reasoning
K_2S	Soluble	Contains K^+
$Ca(NO_3)_2$	Soluble	Contains NO_3^-
$PbCl_2$	Insoluble	Is an insoluble chloride
$NaOH$	Soluble	Contains Na^+
$AlPO_4$	Insoluble	Contains no soluble ions

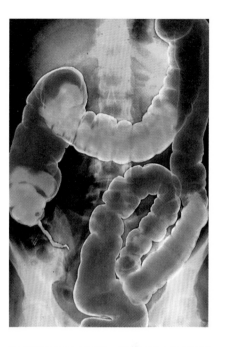

Figure 9.5 A barium sulfate enhanced X ray of the abdomen shows the large intestine.

In medicine, the insoluble salt $BaSO_4$ is used as an opaque substance to enhance X rays of the gastrointestinal tract. $BaSO_4$ is so insoluble that it does not dissolve in gastric fluids. (See Figure 9.5.) Other barium salts cannot be used, because they dissolve in water releasing Ba^{2+}, which is poisonous.

Q Why is BaSO₄ an insoluble substance?

Formation of a Solid

Sometimes, when we mix two solutions, a solid called a *precipitate* forms and drops to the bottom of the container. A solid forms when ions of an insoluble salt come in contact with one another. For example, when a solution of $AgNO_3$ (Ag^+ and NO_3^-) is mixed with a solution of $NaCl$ (Na^+ and Cl^-), the white insoluble salt $AgCl$ is produced. We can write the reaction as a double replacement equation.

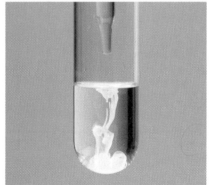

$$AgNO_3(aq) + NaCl(aq) \longrightarrow AgCl(s) + NaNO_3(aq)$$

In a *net ionic equation*, we write only the ions that form the insoluble salt $AgCl$, because Na^+ and NO_3^- ions remain in solution.

$$Ag^+ + Cl^- \longrightarrow AgCl(s)$$

SAMPLE PROBLEM 9.10

Soluble and Insoluble Salts

1. Predict whether each of the following salts is soluble in water:
   a. Na_3PO_4 **b.** $CaCO_3$
2. Solutions of $BaCl_2$ and K_2SO_4 are mixed and a white solid forms.
   a. Write the equation (double replacement) for the reaction.
   b. What is the white solid that forms?
   c. Write the net ionic equation.

Solution

1. a. Soluble. All salts with Na^+ are soluble.
   b. Insoluble. Most salts with CO_3^{2-} are insoluble.
2. a. $BaCl_2(aq) + K_2SO_4(aq) \longrightarrow BaSO_4(s) + 2KCl(aq)$
   b. $BaSO_4$ is the white solid.
   c. $Ba^{2+} + SO_4^{2-} \longrightarrow BaSO_4(s)$

Study Check

Would you expect the following salts to be soluble in water? Why?

a. $PbCl_2$ **b.** K_3PO_4 **c.** $FeCO_3$

QUESTIONS AND PROBLEMS

Solubility

9.29 State whether each of the following refers to a saturated or unsaturated solution:
   a. A crystal added to a solution does not change in size.
   b. A sugar cube completely dissolves when added to a cup of coffee.

9.30 State whether each of the following refers to a saturated or unsaturated solution:
   a. A spoonful of salt added to boiling water dissolves.
   b. A layer of sugar forms on the bottom of a glass of tea as ice is added.

Use this table for problems 9.31–9.34:

Substance	Solubility (g/100 g H_2O)	
	20°C	50°
KCl	34.0	42.6
$NaNO_3$	88.0	114.0
$C_{12}H_{22}O_{11}$ (sugar)	203.9	260.4

9.31 Using the above table, determine whether each of the following solutions will be saturated or unsaturated at 20°C:
   a. Adding 25.0 g of KCl to 100 g of H_2O
   b. Adding 11.0 g of $NaNO_3$ to 25 g of H_2O
   c. Adding 400.0 g of sugar to 125 g of H_2O

9.32 Using the above table, determine whether each of the following solutions will be saturated or unsaturated at 50°C:
   a. Adding 25.0 g of KCl to 50 g of H_2O
   b. Adding 150.0 g of $NaNO_3$ to 75 g of H_2O
   c. Adding 80.0 g of sugar to 25 g of H_2O

HEALTH NOTE

Gout and Kidney Stones: A Problem of Saturation in Body Fluids

The conditions of gout and kidney stones involve compounds in the body that exceed their solubility levels and form solid products. Gout affects adults, primarily men, over the age of 40. Attacks of gout may occur when the concentration of uric acid in blood plasma exceeds its solubility, which is 7 mg/100 mL of plasma at 37°C. Insoluble deposits of needle-like crystals of uric acid can form in the cartilage, tendons, and soft tissues where they cause painful gout attacks. They may also form in the tissues of the kidneys, where they can cause renal damage. High levels of uric acid in the body can be caused by an increase in uric acid production, failure of the kidneys to remove uric acid, or by a diet with an overabundance of foods containing purines, which are metabolized to uric acid in the body. Foods in the diet that contribute to high levels of uric acid include certain meats, sardines, mushrooms, asparagus, and beans. Drinking alcoholic beverages may also significantly increase uric acid levels and bring about gout attacks.

Treatment for gout involves diet changes and drugs. Depending on the levels of uric acid, a medication, such as probenecid, can be used to help the kidneys eliminate uric acid, or allopurinol, which blocks the production of uric acid by the body.

Kidney stones are solid materials that form in the urinary tract. Most kidney stones are composed of calcium phosphate and calcium oxalate, although they can be solid uric acid. The excessive ingestion of minerals and insufficient water intake can cause the concentration of mineral salts to exceed the solubility of the mineral salts and lead to the formation of kidney stones. When a kidney stone passes through the urinary tract, it causes considerable pain and discomfort, necessitating the use of painkillers and surgery. Sometimes ultrasound is used to break up kidney stones. Persons prone to kidney stones are advised to drink six to eight glasses of water every day to prevent saturation levels of minerals in the urine.

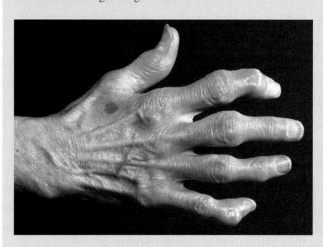

9.33 A solution containing 80.0 g of KCl in 200 g of H$_2$O at 50°C is cooled to 20°C.
   a. How many grams of KCl remain in solution at 20°C?
   b. How many grams of solid KCl came out of solution after cooling?

9.34 A solution containing 80.0 g of NaNO$_3$ in 75 g of H$_2$O at 50°C is cooled to 20°C.
   a. How many grams of NaNO$_3$ remain in solution at 20°C?
   b. How many grams of solid NaNO$_3$ came out of solution after cooling?

9.35 Explain the following observations:
   a. More sugar dissolves in hot tea than in ice tea.
   b. Champagne in a warm room goes flat.
   c. A warm can of soda has more spray when opened than a cold one.

9.36 Explain the following observations:
   a. An open can of soda loses its "fizz" quicker at room temperature than in the refrigerator.
   b. Chlorine gas in tap water escapes as the sample warms to room temperature.
   c. Less sugar dissolves in iced coffee than in hot coffee.

9.37 Predict whether each of the following ionic compounds is soluble in water:
   a. LiCl   **b.** AgCl   **c.** BaCO$_3$ **d.** K$_2$O   **e.** Fe(NO$_3$)$_3$

CASE STUDY
Kidney Stones and Saturated Solutions

9.38 Predict whether each of the following ionic compounds is soluble in water:
  a. PbS   **b.** NaI   **c.** Na_2S
  d. Ag_2O **e.** $CaSO_4$

9.39 Determine whether a solid forms when solutions containing the following salts are mixed. If so, write the equation (double replacement) for the reaction, and identify the insoluble salt.
  a. KCl and Na_2S **b.** $AgNO_3$ and K_2S
  c. $CaCl_2$ and Na_2SO_4

9.40 Determine whether a solid forms when solutions containing the following salts are mixed. If so, write the equation (double replacement) for the reaction, and identify the insoluble salt.
  a. Na_3PO_4 and $AgNO_3$ **b.** K_2SO_4 and Na_2CO_3
  c. $Pb(NO_3)_2$ and Na_2CO_3

LEARNING GOAL

Calculate the percent concentration of a solute in a solution; use percent concentration to calculate the amount of solute or solution.

9.6 Percent Concentration

The amount of solute dissolved in a certain amount of solution is called the **concentration** of the solution. Although there are many ways to express a concentration, they all specify a certain amount of solute in a given amount of solution.

$$\frac{\text{Concentration}}{\text{of a solution}} = \frac{\text{amount of solute}}{\text{amount of solution}}$$

Mass Percent

The **mass percent** (% m/m) concentration of a solution is the percent by mass of solute in a certain mass of solution. This is also known as weight percent (% wt/wt). In the laboratory, both the solute and the solution are weighed on a balance.

$$\text{mass \%} = \frac{\text{mass of solute (g)}}{\text{mass of solution (g)}} \times 100\%$$

Suppose we prepared a solution by mixing 8.0 g of KCl (solute) with 42.0 g of water (solvent). Together the mass of the solute and mass of solvent give the mass of the solution (8.0 g + 42.0 g = 50.0 g). The mass % is calculated by substituting in the values into the mass percent expression.

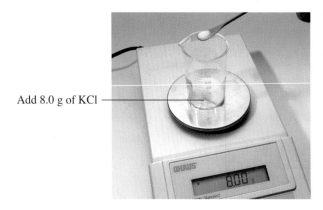

Add 8.0 g of KCl

8.00 g

Add water until the solution weighs 50.0 g

50.0 g

50.00 g

$$\frac{8.0 \text{ g of KCl}}{50.0 \text{ g of solution}} \times 100\% = 16\% \text{ (m/m)}$$

8.0 g KCl + 42.0 g H$_2$O
(solute + solvent)

SAMPLE PROBLEM 9.11

Mass Percent Concentration

What is the mass percent of a solution prepared by dissolving 30.0 g NaOH in 120.0 g of H$_2$O?

Solution

We find the mass of the solution by adding the mass of the solute and solvent.

30.0 g NaOH + 120.0 g H$_2$O = 150.0 g of solution

$$\text{mass } \% = \frac{30.0 \text{ g NaOH}}{150.0 \text{g}} \times 100 = 20.0\% \text{ (m/m)}$$

Study Check

What is the mass percent of NaCl in a solution made by dissolving 2.0 g of NaCl in 56.0 g of H$_2$O?

Volume Percent

Because the volumes of liquids or gases can be easily measured, the concentrations of their solutions are often expressed as **volume percent** (% v/v). The units of volume used in the ratio must be the same, for example, both in milliliters or both in liters.

$$\text{volume } \% = \frac{\text{volume (mL) solute}}{\text{volume (mL) solution}} \times 100\%$$

We interpret a volume/volume percent as the volume of solute in 100 mL of solution. In the wine industry, a label that reads 12% (v/v) means 12 mL of alcohol in 100 mL of wine.

Mass/Volume Percent

A **mass/volume percent** (% m/v), or weight/volume percent (% w/v), is calculated by dividing the grams of the solute by the volume (mL) of solution and multiplying by 100. Widely used in hospitals and pharmacies, the preparation of intravenous solutions and medicines involves the mass/volume percent.

$$\text{mass/volume } \% = \frac{\text{grams of solute}}{\text{milliliters of solution}} \times 100\%$$

The mass/volume percent (% m/v) indicates the grams of a substance that are contained in 100 mL of a solution. For example, a 5% (m/v) glucose solution contains 5 g of glucose in 100 mL of solution. The volume of solution represents the combination of the volume of the glucose and H$_2$O.

Calculating Percent Concentration

A student prepared a solution by dissolving 5.0 g of KI in enough water to give a final volume of 250.0 mL. What is the mass/volume percent of the KI solution?

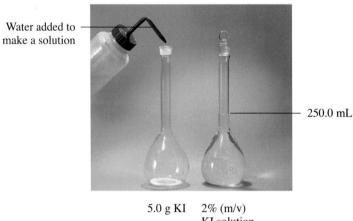

Water added to make a solution

250.0 mL

5.0 g KI   2% (m/v) KI solution

Solution

$$\text{mass/volume \%} = \frac{\overset{\text{Mass of solute}}{5.0 \text{ g KI}}}{\underset{\text{Volume of solution}}{250.0 \text{ mL solution}}} \times 100\% = 2.0\% \text{ (m/v)KI}$$

Study Check

What is the mass/volume percent, % (m/v), of Br_2 in a solution prepared by dissolving 12 g of bromine (Br_2) in enough carbon tetrachloride to make 250 mL of solution?

Percent Concentrations as Conversion Factors

In the preparation of solutions, we often need to calculate the amount of solute or solution. Then the percent concentration is useful as a conversion factor. Some examples of percent concentrations, their meanings, and possible conversion factors are given in Table 9.9.

Table 9.9   Conversion Factors from Percent Concentrations

Percent Concentration	Meaning	Conversion Factors		
10% (m/m) KCl	There are 10 g of KCl in 100 g of solution.	$\dfrac{10 \text{ g KCl}}{100 \text{ g solution}}$	and	$\dfrac{100 \text{ g solution}}{10 \text{ g KCl}}$
5% (m/v) glucose	There are 5 g of glucose in 100 mL of solution.	$\dfrac{5 \text{ g glucose}}{100 \text{ mL solution}}$	and	$\dfrac{100 \text{ mL solution}}{5 \text{ g glucose}}$
12% (v/v) ethanol	There are 12 mL of ethanol in 100 mL of solution.	$\dfrac{12 \text{ mL ethanol}}{100 \text{ mL solution}}$	and	$\dfrac{100 \text{ mL solution}}{12 \text{ mL ethanol}}$

SAMPLE PROBLEM 9.13

Using Mass/Volume Percent

A laboratory technician must prepare 2.00 L of a 5% (m/v) glucose solution. How many grams of glucose must be measured out?

Solution

Using the 5% (m/v) concentration as a conversion factor, we calculate the grams of glucose to weigh out for the solution as

$$2.00 \cancel{L} \times \frac{1000 \cancel{mL}}{1 \cancel{L}} \times \frac{5 \text{ g glucose}}{100 \cancel{mL}} = 100 \text{ g glucose}$$

Given   Metric factor   Percent factor

Study Check

How many grams of NaCl are needed to prepare 3.0 L of a 1% (m/v) NaCl solution?

QUESTIONS AND PROBLEMS

Percent Concentration

9.41 What is the difference between a 5% (m/m) glucose solution and a 5% (m/v) glucose solution?

9.42 What is the difference between a 10% (v/v) methyl alcohol (CH_3OH) solution and a 10% (m/m) methyl alcohol solution?

9.43 Calculate the mass percent, % (m/m), for the solute in each of the following solutions:
  a. 25 g of KCl and 125 g of H_2O
  b. 8.0 g of $CaCl_2$ in 80.0 g of solution
  c. 12 g of sugar in 225 g of tea (solution)

9.44 Calculate the mass percent, % (m/m), for the solute in each of the following solutions:
  a. 75 g of NaOH in 325 g of solution
  b. 2.0 g of KOH in 20.0 g of H_2O
  c. 48.5 g of Na_2CO_3 in 250.0 g of solution

9.45 Calculate the mass/volume percent, % (m/v) for the solute in each of the following solutions:
  a. 75 g of Na_2SO_4 in 250 mL of solution
  b. 0.50 g of KI in 15.0 mL of solution
  c. 39 g of sucrose in 355 mL of a carbonated drink

9.46 Calculate the mass/volume percent, % (m/v), for the solute in each of the following solutions:
  a. 2.50 g KCl in 50.0 mL of solution
  b. 7.5 g of casein in 120 mL of low-fat milk
  c. 0.78 g LiBr in 24 mL of solution

9.47 Calculate the amount of solute (g or mL) needed to prepare the following solutions:
  a. 50.0 mL of a 5.0% (m/v) KCl solution
  b. 1250 mL of a 4.0% (m/v) NH_4Cl solution
  c. 250 mL of a 10.0% (v/v) acetic acid solution

9.48 Calculate the amount of solute (g or mL) needed to prepare the following solutions:

a. 150 mL of a 40.0% (m/v) $LiNO_3$ solution

b. 450 mL of a 2.0% (m/v) KCl solution

c. 225 mL of a 15% (v/v) isopropyl alcohol solution

9.49 A mouthwash contains 22.5% alcohol by volume. If the bottle of mouthwash contains 355 mL, what is the volume in milliliters of the alcohol?

9.50 A bottle of champagne is 11% alcohol by volume. If there are 750 mL of champagne in the bottle, how many milliliters of alcohol are present?

9.51 A patient receives 100 mL of 20% (m/v) mannitol solution every hour.

a. How many grams of mannitol are given in 1 hour?

b. How many grams of mannitol does the patient receive in 15 hours?

9.52 A patient receives 250 mL of a 4.0% (m/v) amino acid solution twice a day.

a. How many grams of amino acids are in 250 mL of solution?

b. How many grams of amino acids does the patient receive in 1 day?

9.53 Calculate the amount of solution (g or mL) that contains each of the following amounts of solute.

a. 5.0 g of $LiNO_3$ from a 25% (m/v) $LiNO_3$ solution

b. 40.0 g of KOH from a 10.0% (m/v) KOH solution

c. 2.0 mL of acetic acid from a 10.0% (v/v) acetic acid solution

9.54 Calculate the amount of solution (g or mL) that contains each of the following amounts of solute.

a. 7.50 g of NaCl from a 2.0% (m/v) NaCl solution

b. 4.0 g of NaOH from a 25% (m/v) NaOH solution

c. 20.0 g of KBr from an 8.0% (m/v) KBr solution

9.55 A patient needs 100 g of glucose in the next 12 hours. How many liters of a 5% (m/v) glucose solution must be given?

9.56 A patient received 2.0 g of NaCl in 8 hours. How many milliliters of a 0.90% (m/v) NaCl (saline) solution were delivered?

9.7 Molarity

When the solutes of solutions take part in reactions, chemists are interested in the number of reacting particles. For this purpose, chemists use **molarity (M),** a concentration that states the number of moles of solute in exactly 1 liter of solution. The molarity of a solution can be calculated knowing the moles of solute and the volume of solution.

$$\text{Molarity (M)} = \frac{\text{moles of solute}}{\text{liters of solution}}$$

For example, if 1.0 mole of NaCl were dissolved in enough water to prepare 1.0 L of solution, the resulting NaCl solution has a molarity of 1.0 M. The abbreviation M indicates the units of moles per liter (moles/L).

$$M = \frac{\text{moles of solute}}{\text{liters of solution}} = \frac{1.0 \text{ mole NaCl}}{1.0 \text{ L of solution}} = \frac{1.0 \text{ mole NaCl}}{1 \text{ L}} = 1.0 \text{ M NaCl}$$

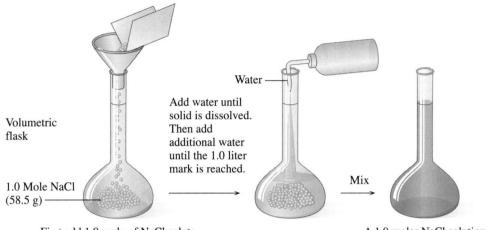

Volumetric flask

Add water until solid is dissolved. Then add additional water until the 1.0 liter mark is reached.

Water

1.0 Mole NaCl (58.5 g)

First add 1.0 mole of NaCl solute

Mix

A 1.0 molar NaCl solution

SAMPLE PROBLEM 9.14

Calculating Molarity

What is the molarity (M) of 60.0 g of NaOH in 0.250 L of solution?

Solution

Because molarity requires moles of solute, we convert grams of NaOH to moles of NaOH using the molar mass NaOH (40.0).

$$60.0 \ \underset{\text{Grams of NaOH}}{\cancel{\text{g NaOH}}} \times \underset{\text{Molar mass}}{\frac{1 \ \text{mole NaOH}}{40.0 \ \cancel{\text{g NaOH}}}} = 1.50 \ \text{moles NaOH}$$

$$M = \frac{\overset{\text{Moles of solute}}{1.50 \ \text{moles NaOH}}}{\underset{\text{Liters of solution}}{0.250 \ \text{L}}} = 6.00 \ \text{M solution}$$

Study Check

What is the molarity of a solution that contains 75 g of KNO_3 dissolved in 350 mL of solution?

Molarity as a Conversion Factor

When we need to calculate the moles of solute or the volume of solution, the molarity is used as a conversion factor. Examples of conversion factors from molarity are given in Table 9.10.

SAMPLE PROBLEM 9.15

Molarity as a Conversion Factor

How many moles of NaCl are present in 4.0 L of a 2.0 M NaCl solution?

Solution

Using the molarity factor that cancels volume (L), we can calculate the moles of NaCl.

$$4.0 \, \cancel{L} \times \frac{2.0 \text{ moles NaCl}}{1 \, \cancel{L}} = 8.0 \text{ moles NaCl}$$

Given volume   Molarity as a conversion factor

Study Check

How many moles of HCl are present in 750 mL of a 6.0 M HCl solution?

Table 9.10   Some Examples of Molar Solutions

Molarity	Meaning	Conversion Factors		
6.0 M HCl	6.0 moles of HCl in 1 liter of solution	$\dfrac{6.0 \text{ moles HCl}}{1 \text{ L}}$	and	$\dfrac{1 \text{ L}}{6.0 \text{ moles HCl}}$
0.20 M NaOH	0.20 mole of NaOH in 1 liter of solution	$\dfrac{0.20 \text{ mole NaOH}}{1 \text{ L}}$	and	$\dfrac{1 \text{ L}}{0.20 \text{ mole NaOH}}$

To prepare a solution, we must convert the number of moles of solute needed into grams. Using the volume and the molarity of the solution with the molar mass of the solute, we can calculate the number of grams of solute necessary. This type of calculation is illustrated in Sample Problem 9.16.

SAMPLE PROBLEM 9.16

Using Molarity

How many grams of KCl would you need to weigh out to prepare 0.250 L of a 2.00 M KCl solution?

Solution

The number of moles of KCl is determined by using the given volume of solution and the molarity as a conversion factor.

$$0.250 \, \cancel{\text{L solution}} \times \frac{2.00 \text{ moles KCl}}{1 \, \cancel{\text{L solution}}} = 0.500 \text{ mole KCl}$$

Given volume   Molarity factor

Knowing the moles of solute needed, the number of grams of KCl is calculated using the molar mass of KCl (74.6 g/mol).

$$0.500 \, \cancel{\text{mole KCl}} \times \frac{74.6 \text{ g KCl}}{1 \, \cancel{\text{mole KCl}}} = 37.3 \text{ g KCl}$$

Moles KCl   Molar mass KCl

Study Check

How many grams of $NaHCO_3$ are in 325 mL of a 4.50 M $NaHCO_3$ solution?

The volume of a solution can be calculated if the number of moles of solute and the molarity of the solution are given.

SAMPLE PROBLEM 9.17

Calculating Volume of a Molar Solution

What volume, in liters, of 1.5 M HCl solution is needed to provide 6.0 moles of HCl?

Solution

Using the molarity as a conversion factor, we set up the following calculation:

$$6.0 \; \text{moles HCl} \times \frac{1 \; \text{L solution}}{1.5 \; \text{moles HCl}} = 4.0 \; \text{L HCl}$$

Given   Molarity factor

Study Check

How many milliliters of 2.0 M NaOH solution will provide 20.0 g of NaOH? (*Hint:* Convert grams of NaOH to moles of NaOH first.)

QUESTIONS AND PROBLEMS

Molarity

9.57 Calculate the molarity (M) of the following solutions:
   a. 2.0 moles of glucose in 4.0 L of solution
   b. 0.10 mole of NaCl in 40.0 mL of solution
   c. 4.0 g of KOH in 2.0 L of solution

9.58 Calculate the molarity (M) of the following solutions:
   a. 0.50 mole glucose in 0.200 L of solution
   b. 0.35 mole LiBr in 350 mL of solution
   c. 36.5 g of HCl in 1.0 L of solution

9.59 Calculate the moles of solute needed to prepare each of the following:
   a. 1.0 L of a 3.0 M NaCl solution
   b. 0.40 L of a 1.0 M KBr solution
   c. 250 mL of a 4.0 M NaCl solution

9.60 Calculate the moles of solute needed to prepare each of the following:
   a. 5.0 L of a 2.0 M $CaCl_2$ solution
   b. 4.0 L of a 0.10 M NaOH solution
   c. 25 mL of a 1.0 M KBr solution

9.61 Calculate the grams of solute needed to prepare each of the following solutions:
   a. 2.0 L of a 1.5 M NaOH solution
   b. 4.0 L of a 0.20 M KCl solution
   c. 25 mL of a 1.0 M NaCl solution

9.62 Calculate the grams of solute needed to prepare each of the following solutions:
   a. 2.0 L of a 6.0 M NaOH solution
   b. 5.0 L of a 0.10 M $CaCl_2$ solution
   c. 1500 mL of a 2.0 M NaCl solution

9.63 What volume in liters provides the following amounts of solute?
   a. 3.0 moles of NaOH from a 2.0 M NaOH solution
   b. 15 moles of NaCl from a 1.5 M NaCl solution
   c. 16 g of NaOH from a 6.0 M NaOH solution

9.64 What volume in liters provides the following amounts of solute?

a. 0.100 mole of KCl from a 4.0 M KCl solution
b. 125 g of NaOH from a 1.0 M NaOH solution
c. 5.0 moles of HCl from a 6.0 M HCl solution

9.8 Colloids and Suspensions

From its properties, identify a mixture as a solution, a colloid, or a suspension.

EXPLORE YOUR WORLD

Tyndall Effect

If you have a small flashlight, you can observe the Tyndall effect in various solutions. Obtain solutions such as skim milk, soft drinks, vinegar, tea, salt water, and cleaning solutions. Shine a beam of light through each solution. Observe how the beam of light appears in the solution.

Questions

1. Which of the solutions contain colloids?
2. Why don't some solutions scatter light?

The solute particles in a solution play an important role in determining the properties of that solution. In most of the solutions discussed so far, the solute is dissolved as small particles that are uniformly dispersed throughout the solvent to give a homogeneous solution. When you observe a solution, such as salt water, you cannot visually distinguish the solute from the solvent. The solution appears transparent even when a light shines through it because the particles are too small to scatter the light. The particles are so small that they go through filters and through semipermeable membranes. A **semipermeable membrane** allows solvent molecules such as water and very small solute particles to pass through, but not large solute molecules.

Colloids

The particles in colloidal dispersions, or **colloids,** are much larger than solute particles in a solution. Colloidal particles are large molecules, such as proteins, or groups of molecules or ions. Colloids are homogeneous mixtures that do not separate or settle out. Colloidal particles are small enough to pass through filters but too large to pass through semipermeable membranes.

When a beam of light shines through a colloid, it produces the **Tyndall effect,** in which the large solute particles scatter the light and make the light beam visible. The Tyndall effect causes a light beam to be visible in fog because the water droplets in the air scatter the light. When there is smoke or smog in the air, the blues and greens of sunlight are scattered more, so that we observe the oranges and reds of a beautiful sunset.

Types of Colloids

Colloids are classified by the type of solute and the dispersing medium. There are four types of colloids: aerosols, foams, emulsions, and sols. *Aerosols* consist of liquid or solid particles dispersed in a gas. *Foams* are dispersions of gases in liquids or solids. An *emulsion* is a liquid dispersed in another liquid or a solid. For example, when milk is not homogenized, the suspended particles of cream separate. The process of homogenization breaks the cream particles into colloids that remain dispersed within the milk solution. In a *sol,* particles of a solid are dispersed in a liquid or solid to give a more rigid solution. Table 9.11 lists several examples of colloids.

Suspensions

Suspensions are heterogeneous, nonuniform mixtures that are very different from solutions or colloids. The particles of a suspension are so large that they can often be seen with the naked eye. They are trapped by filters and semipermeable membranes.

The weight of the suspended solute particles causes them to settle out soon after mixing. If you stir muddy water, it mixes but then quickly separates as the suspended particles settle to the bottom and leave clear liquid at the top. You can find suspensions among the medications in a hospital or in your medicine cabinet.

Table 9.11 Colloids

Examples	Type	Substance Dispersed	Dispersing Medium
Fog, clouds, sprays	Aerosol	Liquid	Gas
Dust, smoke	Aerosol	Solid	Gas
Shaving cream, whipped cream, soapsuds	Foam	Gas	Liquid
Styrofoam, marshmallows	Foam	Gas	Solid
Mayonnaise, butter, homogenized milk, hand lotions	Emulsion	Liquid	Liquid
Cheese, butter	Emulsion	Liquid	Solid
Blood plasma, paints (latex), gelatin	Sol	Solid	Liquid
Cement, pearls	Sol	Solid	Solid

Table 9.12 Comparison of Solutions, Colloids, and Suspensions

Type of Mixture	Type of Particle	Effect of Light	Settling	Separation
Solution	Small particles such as atoms, ions, or small molecules	Transparent	Particles do not settle	Particles cannot be separated by filters or semipermeable membranes
Colloid	Larger molecules or groups of molecules or ions	Tyndall effect occurs	Particles do not settle	Particles can be separated by semipermeable membranes but not by filters
Suspension	Very large particles that may be visible	Opaque (not transparent)	Particles settle rapidly	Particles can be separated by filters

HEALTH NOTE

Colloids and Solutions in the Body

Colloids in the body are isolated by semipermeable membranes. For example, the intestinal lining allows solution particles to pass into the blood and lymph circulatory systems. However, the colloids from foods are too large to pass through the membrane, and they remain in the intestinal tract. Digestion breaks down large colloidal particles, such as starch and protein, into smaller particles, such as glucose and amino acids, that can pass through the intestinal membrane and enter the circulatory system. Certain foods, such as bran, a fiber, cannot be broken down by human digestive processes, and they move through the intestine intact.

Because large proteins, such as enzymes, are colloids, they remain inside cells. However, many of the substances that must be obtained by cells, such as oxygen, amino acids, electrolytes, glucose, and minerals, can pass through cellular membranes. Waste products, such as urea and carbon dioxide, pass out of the cell to be excreted.

These include Kaopectate, calamine lotion, antacid mixtures, and liquid penicillin. It is important to "shake well before using" to suspend all the particles before giving a medication that is a suspension.

Water-treatment plants make use of the properties of suspensions to purify water. When coagulants such as aluminum sulfate or ferric sulfate are added to untreated water, they react with impurities to form large suspended particles called floc. In the water-treatment plant, a system of filters traps the suspended particles but clean water passes through.

Figure 9.6 illustrates some properties of solutions, colloids, and suspensions, and Table 9.12 compares the different types of solutions.

SAMPLE PROBLEM 9.18

Classifying Types of Mixtures

Classify each of the following as a solution, colloid, or suspension:

a. a homogeneous mixture in which a beam of light is visible
b. a mixture that settles rapidly upon standing

Figure 9.6 Properties of different types of solutions: (**a**) suspensions settle out; (**b**) suspensions are separated by a filter; (**c**) solution particles go through a semipermeable membrane, but colloids and suspensions do not.

Q **A filter paper can be used to separate suspension particles from a solution, but a semipermeable membrane is needed to separate colloids from a solution. Explain.**

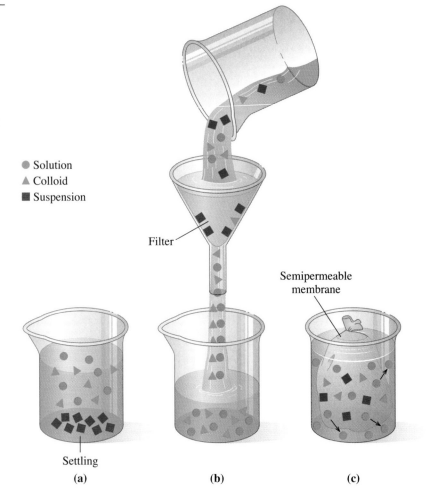

● Solution
▲ Colloid
■ Suspension

Filter

Semipermeable membrane

Settling

(a)   (b)   (c)

c. a mixture whose solute particles pass through both filters and membranes

Solution

a. colloid   **b.** suspension   **c.** solution

Study Check

Enzymes are large protein molecules that catalyze chemical reactions inside the cells of the body. If they cannot pass through the cell membrane, are they solutions or colloids?

QUESTIONS AND PROBLEMS

Colloids and Suspensions

9.65 Identify the following as characteristic of a solution, colloid, or suspension:
   a. a transparent mixture that cannot be separated by a semipermeable membrane
   b. a mixture that appears cloudy when a beam of light is passed through it
   c. a mixture that settles out upon standing
9.66 Identify the following as characteristic of a solution, colloid, or suspension:
   a. Particles of this mixture remain inside a semipermeable membrane, but pass through filters.

b. A beam of light is not visible in this solution.

c. The particles of solute in this solution are very large and visible.

9.67 What is an emulsion?

9.68 What is an aerosol?

9.9 Osmosis and Dialysis

The movement of water into and out of the cells of plants as well as our own bodies is an important biological process. In a process called **osmosis,** the solvent water moves through a semipermeable membrane from a solution that has a lower concentration of solute into a solution where the solute concentration is higher. This movement, or osmosis, of water happens in a direction that equalizes (or attempts to equalize) the concentrations on both sides of the membrane. For example, when water is separated from sucrose (sugar) solution by a semipermeable membrane, water molecules, but not the larger sucrose molecules, move through the membrane to dilute the sucrose solution. As more water flows across the membrane, the level of the sugar solution rises as the level of water on the other side decreases.

Semipermeable membrane

More water molecules flow into the sucrose solution where the concentration of water is lower.

At some point, the levels of the liquids on each side of the semipermeable membrane stop changing. Because the liquid level of the sugar solution is higher than the solvent side, the weight of the higher level of solution exerts a downward pressure called osmotic pressure. **Osmotic pressure** is the pressure that prevents the flow of additional water into the more concentrated solution. The water molecules flow back and forth between the two compartments as the system comes to equilibrium, but the levels remain fixed. The osmotic pressure of a solution depends on the number of solute particles in the solution. Pure water has no osmotic pressure. The greater the number of particles dissolved in a solution, the higher its osmotic pressure.

If a pressure greater than the osmotic pressure is applied to a solution, osmosis is reversed and solvent flows out of the solution and the level in the solvent compartment increases. This process, known as reverse osmosis, is used in desalinization plants in some parts of the world to produce drinking water from sea (salt) water.

SAMPLE PROBLEM 9.19

Osmotic Pressure

A 2% sucrose solution and an 8% sucrose solution are separated by a semipermeable membrane.

a. Which sucrose solution exerts the greater osmotic pressure?
b. In what direction does water flow initially?
c. Which solution will have the higher level of liquid at equilibrium?

Solution

a. The 8% sucrose solution has the higher solute concentration, more solute particles, and the greater osmotic pressure.
b. Initially, water will flow out of the 2% solution into the more concentrated 8% solution.
c. The level of the 8% solution will be higher.

Study Check

If a 10% glucose solution is separated from a 5% glucose solution by a semipermeable membrane, which solution will decrease in volume?

Isotonic Solutions

Because the cell membranes in biological systems are semipermeable, osmosis is an ongoing process. The solutes in body solutions such as blood, tissue fluids, lymph, and plasma all exert osmotic pressure. Most intravenous solutions are **isotonic solutions,** which exert the same osmotic pressure as body fluids. *Iso* means "equal to," and *tonic* refers to the osmotic pressure of the solution in the cell. In the hospital, isotonic solutions or **physiological solutions** include 0.90% (m/v) NaCl solution and 5% (m/v) glucose solution. Although they do not contain the same particles as the body fluids, they exert the same osmotic pressure.

Hypotonic and Hypertonic Solutions

A red blood cell placed in an isotonic solution retains its normal volume because there is an equal flow of water into and out of the cell. (See Figure 9.7a.) However, if a red blood cell is placed in a solution that is not isotonic, the differences in osmotic pressure inside and outside the cell can drastically alter the volume of the cell. When a red blood cell is placed in pure water, a **hypotonic solution** (*hypo* means "lower than"), water flows into the cell by osmosis. (See Figure 9.7b.) The increase in fluid causes the cell to swell, and possibly burst, a process called **hemolysis.** A similar process occurs when you place dehydrated food, such as

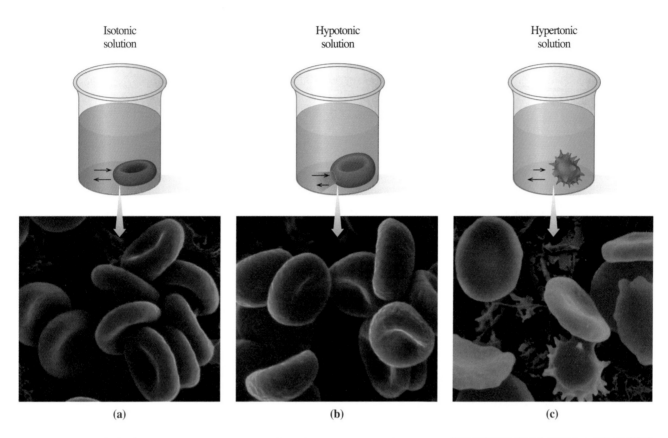

| Isotonic solution | Hypotonic solution | Hypertonic solution |

(a)   **(b)**   **(c)**

raisins or dried fruit, in water. The water enters the cells and the food becomes plump and smooth.

If a red blood cell is placed in a **hypertonic solution,** which has a higher solute concentration (*hyper* means "greater than"), water leaves the cell by osmosis. Suppose a red blood cell is placed in a 10% (m/v) NaCl solution. Because the osmotic pressure in the red blood cell is equal to that of a 0.90% (m/v) NaCl solution, the 10% (m/v) NaCl solution has a much greater osmotic pressure. As water is lost, the cell shrinks, a process called **crenation.** (See Figure 9.7c.) A similar process occurs when making pickles, which uses a hypertonic salt solution that causes the cucumbers to shrivel as they lose water.

Figure 9.7 (a) In an isotonic solution, a red blood cell retains its normal volume. (b) Hemolysis: In a hypotonic solution, water flows into a red blood cell, causing it to swell and burst. (c) Crenation: In a hypertonic solution, water leaves the red blood cell, causing it to shrink.

Q **What happens to a red blood cell placed in a 4% NaCl solution?**

Osmolarity

Every molecule and every ion of solute that dissolves in a solution contributes to its osmotic pressure. A solution that has more dissolved particles has a higher osmotic pressure than a solution with fewer dissolved particles. We have seen that molarity describes the moles of substance in solution. However, when ionic compounds dissolve in water, one mole of the substance forms two or more moles of ions. Because osmosis depends on the number of particles in solution, we use a concentration term called **osmolarity** to indicate the number of moles of solute particles in one liter of solution.

When one mole of NaCl dissolves in one liter of water, it produces one mole of Na^+ ions and one mole of Cl^- ions, which is two moles of particles. This would be a 1 M NaCl solution or a 2 Osm (osmolar) solution.

$$1 \text{ mole NaCl} \longrightarrow 1 \text{ mole } Na^+ \text{ and } 1 \text{ mole } Cl^- = 2 \text{ moles ions}$$

We can use a solution of K_2SO_4 as an example. When K_2SO_4 dissolves, two moles of K^+ ions and one mole of SO_4^{2-} ions are produced.

Everyday Osmosis

1. Place a few pieces of dry fruit such as raisins, prunes, or banana chips in water. Observe them after 1 hour or more. Look at them again the next day.
2. Place some grapes in a concentrated salt-water solution. Observe them after 1 hour or more. Look at them again the next day.
3. Place one potato slice in water and another slice in a concentrated salt-water solution. After 1–2 hours, observe the shapes and size of the slices. Look at them again the next day.

Questions

1. How did the shape of the dried fruit change after being in water? Explain.
2. How did the appearance of the grapes change after being in a concentrated salt solution? Explain.
3. How does the appearance of the potato slice that was placed in water compare to the appearance of the potato slice placed in salt water? Explain.
4. At the grocery store, why are sprinklers used to spray water on fresh produce such as lettuce, carrots, and cucumbers?

$$K_2SO_4(s) \xrightarrow{H_2O} 2K^+(aq) + SO_4^{2-}(aq)$$

$$\underset{1 \text{ mole}}{} \qquad \underset{3 \text{ moles of ions} = 3 \text{ osmoles}}{}$$

Using the total number of moles of ions as a factor, we can convert the molarity of a solution to its osmolarity. For example, a 0.10 M K_2SO_4 solution would be 0.30 Osm K_2SO_4.

$$0.10 \text{ M } K_2SO_4 \times \frac{3 \text{ osmoles}}{1 \text{ mole } K_2SO_4} = 0.30 \text{ Osm } K_2SO_4$$

The physiological solutions 0.9% (m/v) NaCl and 5% (m/v) glucose are isotonic solutions because they have the same number of particles and therefore the same osmolarities as body fluids. Normal serum has an osmolarity of 0.30 Osm. Let's see how the osmolarity of a 0.9% (m/v) NaCl compares.

$$\frac{0.9 \text{ g NaCl}}{100 \text{ mL}} \times \frac{1 \text{ mole NaCl}}{58.5 \text{ g NaCl}} \times \frac{2 \text{ osmoles}}{1 \text{ mole NaCl}} \times \frac{1000 \text{ mL}}{1 \text{ L}} = 0.3 \text{ Osm NaCl}$$

We can do a similar calculation to show that a 5% (m/v) glucose solution also has the same number of particles and therefore the same osmotic pressure as body fluids. Because glucose dissolves as molecules, one mole of glucose produces one mole of molecules.

$$\frac{5 \text{ g glucose}}{100 \text{ mL}} \times \frac{1 \text{ mole glucose}}{180 \text{ g glucose}} \times \frac{1 \text{ osmole}}{1 \text{ mole glucose}} \times \frac{1000 \text{ mL}}{1 \text{ L}} = 0.3 \text{ Osm}$$

Therefore, the osmolarity of a 0.9% (m/v) NaCl and the osmolarity of a 5% (m/v) glucose are the same as the osmolarities of serum and will exert the same osmotic pressures.

SAMPLE PROBLEM 9.20

Isotonic, Hypotonic, and Hypertonic Solutions

Describe each of the following solutions as isotonic, hypotonic, or hypertonic. Indicate whether a red blood cell placed in each solution will undergo hemolysis, crenation, or no change.

a. a 5% (m/v) glucose solution   **b.** a 0.2% (m/v) NaCl solution

Solution

a. A 5% (m/v) glucose solution is isotonic. A red blood cell will not undergo any change.
b. A 0.2% (m/v) NaCl solution is hypotonic. A red blood cell will undergo hemolysis.

Study Check

What is the effect of a 10% (m/v) glucose solution on a red blood cell?

Dialysis

Dialysis is a process that is similar to osmosis. In dialysis, a semipermeable membrane, called a dialyzing membrane, permits small solute molecules and ions as well as solvent water molecules to pass through, but it retains large particles, such as colloids. Dialysis is a way to separate solution particles from colloids.

Suppose we fill a cellophane bag with a solution containing NaCl, glucose, starch, and protein and place it in pure water. Cellophane is a dialyzing membrane, and the sodium ions, chloride ions, and glucose molecules will pass through it into the surrounding water. However, starch and protein remain inside because they are colloids. Water molecules will flow by osmosis into the cellophane bag. Eventually, the concentrations of sodium ions, chloride ions, and glucose molecules inside and outside the dialysis bag become equal. To remove more NaCl or glucose, the cellophane bag must be placed in a fresh sample of pure water.

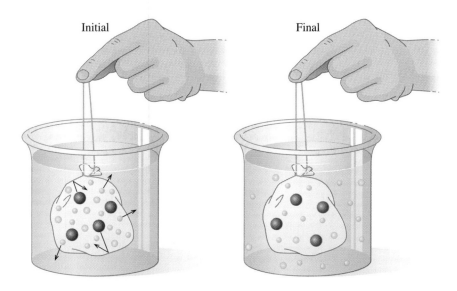

   ○ Solution particles such as Na⁺, Cl⁻, glucose
   ● Colloidal particles such as protein, starch

QUESTIONS AND PROBLEMS

Osmosis and Dialysis

9.69 a. How do plants obtain water from the ground?
b. How does a pickle get all shriveled up?

9.70 a. Why should you not drink large quantities of seawater?
b. Why does salt preserve foods?

9.71 A 10% (m/v) starch solution is separated from pure water by an osmotic membrane.
a. Which solution has the higher osmotic pressure?
b. In which direction will water flow initially?
c. In which compartment will the volume level rise?

9.72 Two solutions, a 0.1% (m/v) albumin solution and a 2% (m/v) albumin solution, are separated by a semipermeable membrane. (Albumins are colloidal proteins.)
a. Which compartment has the higher osmotic pressure?
b. In which direction will water flow initially?
c. In which compartment will the volume level rise?

9.73 Indicate the compartment (A or B) that will increase in volume for each of the following pairs of solutions separated by semipermeable membranes:

HEALTH NOTE

Dialysis by the Kidneys and the Artificial Kidney

The fluids of the body undergo dialysis by the membranes of the kidneys, which remove waste materials, excess salts, and water. In an adult, each kidney contains about 2 million nephrons. At the top of each nephron, there is a network of arterial capillaries called the glomerulus.

As blood flows into the glomerulus, small particles, such as amino acids, glucose, urea, water, and certain ions, will move through the capillary membranes into the nephron. As this solution moves through the nephron, substances still of value to the body (such as amino acids, glucose, certain ions, and 99% of the water) are reabsorbed. The major waste product, urea, is excreted in the urine.

Hemodialysis

If the kidneys fail to dialyze waste products, increased levels of urea can become life-threatening in a relatively short time. A person with kidney failure must use an artificial kidney, which cleanses the blood by **hemodialysis.**

A typical artificial kidney machine contains a large tank filled with about 100 L of water containing selected electrolytes. In the center of this dialyzing bath (dialysate), there is a dialyzing coil or membrane made of cellulose tubing. As the patient's blood flows through the dialyzing coil, the highly concentrated waste products dialyze out of the blood. No blood is lost because the membrane is not permeable to large particles such as red blood cells.

Dialysis patients do not produce much urine. As a result, they retain large amounts of water between dialysis treatments, which produces a strain on the heart. The intake of fluids for a dialysis patient may be restricted to as little as a few teaspoons of water a day. In the dialysis procedure, the pressure of the blood is increased as it circulates through the dialyzing coil so water can be squeezed out of the blood. For some dialysis patients, 2–10 L of water may be removed during one treatment. Dialysis patients have from two to three treatments a week, each treatment requiring about 5–7 hr. Some of the newer treatments require less time. For many patients, dialysis is done at home with a home dialysis unit.

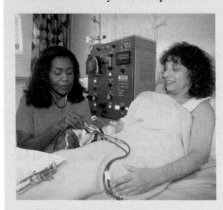

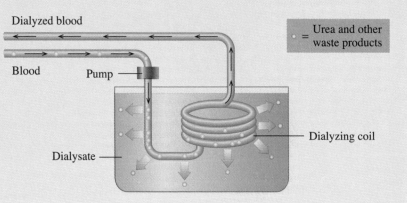

A	B
a. 5% (m/v) glucose	10% (m/v) glucose
b. 4% (m/v) albumin	8% (m/v) albumin
c. 0.1% (m/v) NaCl	10% (m/v) NaCl

9.74 Indicate the compartment (A or B) that will increase in volume for each of the following pairs of solutions separated by semipermeable membranes:

A	B
a. 20% (m/v) glucose	10% (m/v) glucose
b. 10% (m/v) albumin	2% (m/v) albumin
c. 0.5% (m/v) NaCl	5% (m/v) NaCl

9.75 Are the following solutions isotonic, hypotonic, or hypertonic compared with a red blood cell?
  a. distilled H_2O **b.** 1% (m/v) glucose
  c. 0.90% (m/v) NaCl   **d.** 5% (m/v) glucose

9.76 Will a red blood cell undergo crenation, hemolysis, or no change in each of the following solutions?
  a. 1% (m/v) glucose   **b.** 2% (m/v) NaCl
  c. 5% (m/v) NaCl   **d.** 0.1% (m/v) NaCl

9.77 Each of the following mixtures is placed in a dialyzing bag and immersed in distilled water. Which substances will be found outside the bag in the distilled water?
  a. NaCl solution
  b. starch (colloid) and alanine (amino acid solution)

  c. NaCl solution and starch (colloid)
  d. urea solution

9.78 Each of the following mixtures is placed in a dialyzing bag and immersed in distilled water. Which substances will be found outside the bag in the distilled water?
  a. KCl solution and glucose solutions
  b. an albumin solution (colloid)
  c. an albumin solution (colloid), KCl solution, and glucose solution
  d. urea solution and NaCl solution

9.79 What is the osmolarity of the following solutions?
  a. 0.20 M NaCl   **b.** 0.50 M $CaCl_2$ **c.** 0.10 M K_3PO_4

9.80 Which solution has the greatest osmotic pressure: 0.1 M NaCl, 0.1 M $CaCl_2$, or a 0.1 M glucose?

Chapter Review

9.1 Properties of Water

The polar O—H bond leads to hydrogen bonding between water molecules. Surface tension is a result of the hydrogen bonds in water.

9.2 Solutions

A solution forms when a solute dissolves in a solvent. The expression "like dissolves like" means that a polar or ionic solute dissolves in a polar solvent and a nonpolar solute requires a nonpolar solvent.

9.3 Electrolytes and Nonelectrolytes

An ionic solute dissolves in water, a polar solvent, because the polar water molecules attract and pull the ions into solution where they become hydrated. Substances that release ions are called electrolytes because the solution will conduct an electrical current. Strong electrolytes are completely ionized, whereas weak electrolytes are only partially ionized. Nonelectrolytes are substances that dissolve in water to produce molecules and cannot conduct electrical currents.

9.4 Equivalents

An equivalent is the amount of a electrolyte that carries one mole of positive or negative charge. One mole of Na^+ is 1 equivalent. One mole of Ca^{2+} has 2 equivalents. In fluid replacement solutions, the concentrations of electrolytes are expressed as mEq/L of solution.

9.5 Solubility

A solution that contains the maximum amount of dissolved solute is a saturated solution. The solubility of a solute is the maximum amount of a solute that can dissolve in 100 g of solvent. A solution containing less than the maximum amount of dissolved solute is unsaturated. An increase in temperature increases the solubility of most solids in water, but decreases the solubility of gases in water. Salts that are soluble in water usually contain Li^+, Na^+, K^+, NH_4^+, NO_3^-, or acetate $C_2H_3O_2^-$.

9.6 Percent Concentration

The concentration of a solution is the amount of solute dissolved in a certain amount of solution. Mass percent expresses the ratio of the mass of solute to the mass of solution multiplied by 100. Percent concentration is also expressed as volume/volume and mass/volume ratios. In calculations of grams or milliliters of solute or solution, the percent concentration is used as a conversion factor.

9.7 Molarity

Molarity is the moles of solute per liter of solution. Units of molarity, moles/liter, are used in conversion factors to solve for moles of solute or volume of solution.

9.8 Colloids and Suspensions

Colloids contain particles that show the Tyndall effect, do not settle out, and pass through filters but not through semipermeable membranes. Suspensions have very large particles that settle out of solution.

9.9 Osmosis and Dialysis

In osmosis, solvent (water) passes through a semipermeable membrane from a solution of a lower solute concentration to a solution of a higher concentration. Isotonic solutions have osmotic pressures equal to that of body fluids. A red blood cell maintains its volume in an isotonic solution, but swells and may burst (hemolysis) in a hypotonic solution, and shrinks (crenates) in a hypertonic solution. In dialysis, water and small solute particles pass through a dialyzing membrane, while larger particles are retained.

Key Terms

colloid A mixture having particles that are moderately large. Colloids scatter light (the Tyndall effect) and pass through filters but cannot pass through semipermeable membranes.

concentration A measure of the amount of solute that is dissolved in a specified amount of solution.

crenation The shriveling of a cell due to water leaving the cell when the cell is placed in a hypertonic solution.

dialysis A process in which water and small solute particles pass through a semipermeable membrane.

electrolyte A substance that produces ions when dissolved in water; its solution conducts electricity.

equivalent (Eq) The amount of a positive or negative ion that supplies 1 mole of electrical charge.

hemodialysis A mechanical cleansing of the blood by an artificial kidney using the principle of dialysis.

hemolysis A swelling and bursting of red blood cells in a hypotonic solution due to an increase in fluid volume.

Henry's law The solubility of a gas in a liquid is directly related to the pressure of that gas above the liquid.

hydrate A solid ionic compound that contains water molecules as part of the crystal structure.

hydration The process of surrounding dissolved ions by water molecules.

hydrogen bond The attraction between a partially positive hydrogen atom in one molecule and a highly electronegative atom such as oxygen in another molecule.

hypertonic solution A solution that has a higher osmotic pressure than the red blood cells of the body.

hypotonic solution A solution that has a lower osmotic pressure than the red blood cells of the body.

insoluble salt An ionic compound that does not dissolve in water.

isotonic solution A solution that has the same osmotic pressure as that of the red blood cells of the body.

mass percent The grams of solute in exactly 100 grams of solution.

mass/volume percent The grams of solute in exactly 100 mL of solution.

molarity (M) The number of moles of solute in exactly 1 L of solution.

nonelectrolyte A substance that dissolves in water as molecules; its solution will not conduct an electrical current.

osmolarity A concentration term that gives the moles of solute particles in one liter of solution.

osmosis The flow of a solvent, usually water, through a semipermeable membrane into a solution of higher solute concentration.

osmotic pressure The pressure that prevents the flow of water into the more concentrated solution.

physiological solution A solution that exerts the same osmotic pressure as normal body fluids.

saturated solution A solution containing the maximum amount of solute that can dissolve at a given temperature. Any additional solute will remain undissolved in the container.

semipermeable membrane A membrane that permits the passage of certain substances while blocking or retaining others.

solubility The maximum amount of solute that can dissolve in exactly 100 g of solvent, usually water, at a given temperature.

soluble salt An ionic compound that dissolves in water.

solute The component in a solution that changes state upon dissolving; if no change in state occurs, it is the component present in the smaller quantity.

solution A homogeneous mixture in which the solute is made up of small particles (ions or molecules) that can pass through filters and semipermeable membranes.

solvent The substance in which the solute dissolves; usually the component present in greatest amount.

strong electrolyte A compound that ionizes completely when it dissolves in water. Its solution is a good conductor of electricity.

surface tension A characteristic of water in which the hydrogen bonding between water molecules pulls its surface molecules together so tightly that they form a "skin."

suspension A mixture in which the solute particles are large enough and heavy enough to settle out and be retained by both filters and semipermeable membranes.

Tyndall effect The opaque path of a light beam through a colloid as light is scattered by the colloidal particles.

unsaturated solution A solution that contains less solute than can be dissolved.

volume percent A percent concentration that relates the volume of the solute to the volume of the solution.

water of hydration The amount of water contained in the crystal structure of a hydrate.

weak electrolyte A substance that produces only a few ions along with many molecules when it dissolves in water. Its solution is a weak conductor of electricity.

Additional Problems

9.81 Why does iodine dissolve in hexane, but not in water?

9.82 How does temperature and pressure affect the solubility of solids and gases in water?

9.83 If NaCl has a solubility of 36.0 g at 20°C, how many grams of water are needed to prepare a saturated solution containing 80.0 g NaCl?

9.84 If the solid NaCl in a saturated solution of NaCl continues to dissolve, why is there no change in the concentration of the NaCl solution?

9.85 Why would a solution made by mixing solutions of $NaNO_3$ and KCl be clear, while a combination of KCl and $Pb(NO_3)_2$ solution produces a solid?

9.86 Indicate whether each of the following is soluble in water.
  a. KCl   **b.** $MgSO_4$ **c.** PbS
  d. $AgNO_3$ **e.** $Ca(OH)_2$

9.87 Calculate the mass percent of a solution containing 15.5 g of Na_2SO_4 and 75.5 g of H_2O.

9.88 How many grams of K_2CO_3 must be added to 750 mL of solution to prepare a 3.5% (m/v) K_2CO_3 solution?

9.89 A patient receives all her nutrition from fluids given through the vena cava. Every 12 hours, 750 mL of a solution that is 4% (m/v) amino acids (protein) and 25% (m/v) glucose (carbohydrate) is given along with 500 mL of a 10% (m/v) lipid (fat).
  a. In 1 day, how many grams of amino acids, glucose, and lipid are given to the patient?
  b. How many kilocalories does she obtain in 1 day?

9.90 An 80-proof brandy is 40.0% (v/v) ethyl alcohol. The "proof" is twice the percent concentration of alcohol in the beverage. How many milliliters of alcohol are present in 750 mL of brandy?

9.91 How many milliliters of a 12% (v/v) propyl alcohol solution would you take to obtain 4.5 mL of propyl alcohol?

9.92 How many liters of a 5.0% (m/v) glucose solution would you take to obtain 75 g of glucose?

9.93 If you were in the laboratory, how would you prepare 250 mL of a 2.0 M KCl solution?

9.94 What is the molarity of a solution containing 15.6 g of KCl in 274 mL of solution?

9.95 A solution is prepared with 70.0 g of HNO_3 and 130.0 g of H_2O. It has a density of 1.21 g/mL.
  a. What is the mass percent of the HNO_3 solution?
  b. What is the total volume of the solution?
  c. What is the mass/volume percent?
  d. What is its molarity (M)?

9.96 What is the molarity of a 15% (m/v) NaOH solution?

9.97 How many grams of solute are in each of the following solutions?
  a. 2.5 L of 3.0 M $Al(NO_3)_3$
  b. 75 mL of 0.50 M $C_6H_{12}O_6$

9.98 Why would a dialysis unit (artificial kidney) use isotonic concentrations of NaCl, KCl, $NaHCO_3$, and glucose in the dialysate?

9.99 Why would solutions with high salt content be used to prepare dried flowers?

9.100 A patient on dialysis has a high level of urea, a high level of sodium, and a low level of potassium in the blood. Why is the dialyzing solution prepared with a high level of potassium but no sodium or urea?

9.101 Why is it dangerous to drink seawater even if you are stranded on a desert island?

9.102 In some countries, pure water is obtained from seawater by reverse osmosis. What do you think such a process involves?

9.103 Calculate the osmolarity of each of the following solutions:
  a. 0.20 M NaCl   **b.** 0.050 M K_2CO_3
  c. 0.2 M $CaCl_2$ **d.** 0.1 M Na_2SO_4

9.104 In 1 liter of Ringer's solution there are 0.147 mole of NaCl, 0.004 mole KCl, and 0.004 mole of $CaCl_2$. What is the osmolarity of Ringer's solution? Why is it used as a physiological solution?

9.105 A semipermeable membrane separates two compartments, A and B. Indicate in each case the side where the level will rise.

	A	B
a.	1% (m/v) starch	6% (m/v) starch
b.	0.1 M glucose	0.1 M $CaCl_2$
c.	1% KCl	1% (m/v) KCl
d.	5% (m/v) glucose	0.9% (m/v) NaCl
e.	10% (m/v) sucrose	2% (m/v) sucrose

9.106 What happens to red blood cells placed in the following solutions?

a. 0.05% (m/v) NaCl   **b.** 7% (m/v) glucose

c. 0.90% (m/v) NaCl   **d.** 2% (m/v) glucose

9.107 Given the equation

$$Pb(NO_3)_2\ (aq) + 2KCl(aq) \longrightarrow PbCl_2\ (s) + 2KNO_3\ (aq)$$

a. How many grams of $PbCl_2$ will be formed from 50.0 mL of 1.5 M KCl?

b. How many milliliters of 2.0 M $Pb(NO_3)_2$ will react with 50.0 mL of 1.5 M KCl?

9.108 In the reaction

$$Mg(s) + 2HCl\ (aq) \longrightarrow MgCl_2(aq) + H_2(g)$$

a. How many milliliters of 6.0 M HCl are required to react with 15.0 g of magnesium?

b. How many liters of hydrogen at STP will be formed when 0.50 L of 2.0 M HCl reacts with magnesium?

c. What is the molarity of an HCl solution, if 8.5 L of H_2 gas are produced at 735 mm Hg and 25°C when 250.0 mL of HCl react with magnesium?

10 Acids and Bases

"In a stat lab, we are sent blood samples of patients in emergency situations," says Audrey Trautwein, clinical laboratory technician, Stat Lab, Santa Clara Valley Medical Center. "We may need to assess the status of a trauma patient in ER or a patient who is in surgery. For example, an acidic blood pH diminishes cardiac function, and affects the actions of certain drugs. In a stat situation, it is critical that we obtain our results fast. This is done using a blood gas analyzer. As I put a blood sample into the analyzer, a small probe draws out a measured volume, which is tested simultaneously for pH, P_{O_2} and P_{CO_2} as well as electrolytes, glucose, and hemoglobin. In about one minute we have our test results, which are sent to the doctor's computer."

LOOKING AHEAD

the Chemistry place

www.chemplace.com/college

Visit the URL above or use the CD-ROM in the book for extra quizzing, interactive tutorials, career resources, and case studies.

Lemons, grapefruits, and vinegar taste sour because they contain acids. We have acid in our stomach that helps us digest food; we produce lactic acid in our muscles when we exercise. Acid from bacteria turns milk sour to make cottage cheese or yogurt. Bases are solutions that neutralize acids. Sometimes we take antacids such as milk of magnesia to offset the effects of too much stomach acid.

The pH of a solution describes its acidity. The pH of body fluids, including blood and urine, is regulated primarily by the lungs and the kidneys. Major changes in the pH of the body fluids can severely affect biological activities within the cells. Buffers are present to prevent large fluctuations in pH.

10.1 Acids and Bases

The term *acid* comes from the Latin word *acidus,* which means "sour." We are familiar with the sour tastes of vinegar and lemons and other common acids in foods.

In the nineteenth century, Arrhenius was the first to describe **acids** as substances that produce hydrogen ions (H^+) when they dissolve in water. For example, hydrogen chloride ionizes in water to give hydrogen ions, H^+, and chloride ions, Cl^-. The hydrogen ions, H^+, give acids a sour taste, change blue litmus to red, and corrode some metals.

$$HCl(g) \xrightarrow{\text{H}_2\text{O}} H^+(aq) + Cl^-(aq)$$

Polar covalent   Ionization
compound   in water

Naming Acids

When an acid dissolves in water to produce hydrogen ion and a simple nonmetal anion, the prefix *hydro-* is used before the name of the nonmetal and its *–ide* ending is changed to *–ic acid.* For example, hydrogen chloride (HCl) dissolves in water to form HCl (*aq*), which is named hydrochloric acid.

Table 10.1   Naming Common Acids

Acid	Name of Acid	Anion	Name of Anion
HCl	**hydro**chloric **acid**	Cl^-	chlor**ide**
HBr	**hydro**brom**ic acid**	Br^-	brom**ide**
HNO_3	nitr**ic acid**	NO_3^-	nit**rate**
HNO_2	nit**rous acid**	NO_2^-	nit**rite**
H_2SO_4	sulfur**ic acid**	SO_4^{2-}	sulf**ate**
H_2SO_3	sulfur**ous acid**	SO_3^{2-}	sulf**ite**
H_2CO_3	carbon**ic acid**	CO_3^{2-}	carbon**ate**
H_3PO_4	phosphor**ic acid**	PO_4^{3-}	phosph**ate**
$HClO_3$	chlor**ic acid**	ClO_3^-	chlor**ate**
$HClO_2$	chlor**ous acid**	ClO_2^-	chlor**ite**
CH_3COOH	acet**ic acid**	CH_3COO^-	acet**ate**

When an acid contains a polyatomic ion, the name of the acid comes from the name of the polyatomic ion. The *–ate* in the name is replaced with *–ic acid*. For example, HNO_3, which contains the nitrate ion (NO_3^-), is named nitric acid. If the acid contains a polyatomic ion with an *–ite* ending, its name ends with *–ous acid*. You can determine the *–ous acid* formula by using one oxygen less than the polyatomic ion of the *–ic acid*. The names of some common acids and their anions are listed in Table 10.1.

Bases

You may be familiar with some bases such as antacids, drain openers, and oven cleaners. According to the Arrhenius theory, **bases** are ionic compounds that dissociate into a metal ion and hydroxide ions (OH^-) when they dissolve in water. For example, sodium hydroxide is an Arrhenius base that dissociates in water to give sodium ions, Na^+, and hydroxide ions, OH^-.

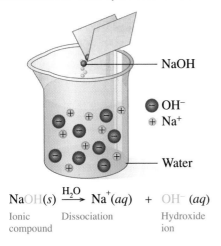

NaOH

● OH^-
⊕ Na^+

Water

$$NaOH(s) \xrightarrow{H_2O} Na^+(aq) + OH^-(aq)$$

Ionic compound   Dissociation   Hydroxide ion

Most Arrhenius bases are from Groups 1A and 2A, such as NaOH, KOH, LiOH, and $Ca(OH)_2$. There are other bases such as $Al(OH)_3$ and $Fe(OH)_3$, but they are fairly insoluble. The hydroxide ions (OH^-) give Arrhenius bases common characteristics such as a bitter taste and soapy feel.

Naming Bases

Typical Arrhenius bases are named as hydroxides.

Bases	Name
NaOH	sodium **hydroxide**
KOH	potassium **hydroxide**
$Ca(OH)_2$	calcium **hydroxide**
$Al(OH)_3$	aluminum **hydroxide**

SAMPLE PROBLEM 10.1

Names of Acids and Bases

Name the following as acids or bases.

a. H_3PO_4 **b.** NaOH   **c.** HNO_2 **d.** HBr

Solution

a. phosphoric acid   **b.** sodium hydroxide
c. nitrous acid   **d.** hydrobromic acid

Learning Check

Give the name for: H_2SO_4 and KOH.

SAMPLE PROBLEM 10.2

Dissociation of an Arrhenius Base

Write an equation for the dissociation of LiOH(s) in water.

Solution

LiOH dissociates in water to give a basic solution of lithium ions (Li^+) and hydroxide (OH^-).

$$LiOH(s) \xrightarrow{H_2O} Li^+(aq) + OH^-(aq)$$

Study Check

Calcium hydroxide is used in some antacids to counteract excess stomach acid. Write an equation for dissociation of calcium hydroxide.

Brønsted–Lowry Acids and Bases

Early in the twentieth century, Brønsted and Lowry expanded the definition of acids and bases. A **Brønsted–Lowry acid** donates a proton (hydrogen ion, H^+) to another substance, and a **Brønsted–Lowry base** accepts a proton.

A Brønsted–Lowry acid is a proton (H^+) donor.

A Brønsted–Lowry base is a proton (H^+) acceptor.

A proton (H^+) does not actually exist in water. Its attraction to polar water molecules is so strong that the proton bonds to the water molecule and forms a **hydronium ion, H_3O^+**.

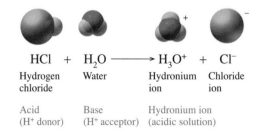

We can now write the formation of a hydrochloric acid solution as a transfer of a proton from hydrogen chloride to water. By accepting a proton in the reaction, water is acting as a base according to the Brønsted–Lowry concept.

In another reaction ammonia (NH_3) reacts with water. Because the nitrogen atom of NH_3 has a stronger attraction for a proton, water acts as an acid by donating a proton.

$$NH_3 + H_2O \rightleftarrows NH_4^+ + OH^-$$

Ammonia	Water		
Base (H^+ acceptor)	Acid (H^+ donor)	Ammonium ion	Hydroxide ion (basic solution)

Table 10.2 compares some characteristics of acids and bases.

Table 10.2 Some Characteristics of Acids and Bases

Characteristic	Acids	Bases
Arrhenius	H^+	OH^-
Behavior in water	H^+ donor	H^+ acceptor
Electrolytes	Yes	Yes
Taste	Sour	Bitter, chalky
Feel	May sting	Soapy, slippery
Litmus	Red	Blue
Phenolphthalein	Colorless	Red
Neutralization	Neutralizes bases	Neutralizes acids

SAMPLE PROBLEM 10.3

Acids and Bases

In each of the following equations, identify the reactant that is an acid (H^+ donor) and the reactant that is a base (H^+ acceptor).

a. $HBr + H_2O \longrightarrow H_3O^+ + Br^-$ **b.** $H_2O + CN^- \rightleftarrows HCN + OH^-$

Solution

a. HBr, acid; H_2O, base **b.** H_2O, acid; CN^-, base

Study Check

When HNO_3 reacts with water, water acts as a base (H^+ acceptor). Write the equation for the reaction.

EXPLORE YOUR WORLD

Acids and Bases in Products at Home

A variety of acids and bases are in many of the foods we eat. Citric acid and ascorbic acid (vitamin C) give citrus fruits their sour taste. Vinegar is a 5% acetic acid solution. Many skin-care products now include acids from fruits. Antacids contain bases such as aluminum hydroxide and calcium hydroxide as well as calcium carbonate, which is a Brønsted–Lowry base because the carbonate ion accepts protons.

Read the labels of several products you use in your home, such as soft drinks, antacids, cereals, skin-care products, vitamins, vinegar, and juices. Look for the names of acids and bases, such as citric acid, acetic acid, lactic acid, folic acid, aluminum hydroxide, sodium hydroxide, and magnesium hydroxide. Make a list of the various products and the acids and bases each contains. Write out as many of the formulas of the acids and bases as you can. To find the formulas of some acids and bases, use a chemistry reference book such as the *Merck Index*. You could also do a search on the Internet.

QUESTIONS AND PROBLEMS

Acids and Bases

10.1 Indicate whether each of the following statements is characteristic of an acid or a base:

a. has a sour taste
b. neutralizes bases
c. produces H_3O^+ ions in water
d. is named potassium hydroxide
e. is a proton (H^+) acceptor

10.2 Indicate whether each of the following statements is characteristic of an acid or a base:

a. neutralizes acids   b. produces OH^- in water
c. is a proton (H^+) donor   d. has a soapy feel
e. turns litmus red

10.3 Name each of the following as an acid or base:
a. HCl   b. $Ca(OH)_2$ c. H_2CO_3
d. HNO_3 e. H_2SO_3

10.4 Name each of the following as an acid or base:
a. $Al(OH)_3$ b. HBr   c. H_2SO_4
d. KOH   e. HNO_2

10.5 Write formulas for the following acids and bases:
a. magnesium hydroxide   b. hydrofluoric acid
c. phosphoric acid   d. lithium hydroxide
e. copper(II) hydroxide

10.6 Write formulas for the following acids and bases:
a. barium hydroxide   b. hydriodic acid   c. nitric acid
d. iron(III) hydroxide   e. sodium hydroxide

10.7 In each of the following equations, identify the acid (proton donor) and base (proton acceptor) for the reactants:
a. $HI + H_2O \longrightarrow H_3O^+ + I^-$
b. $F^- + H_2O \rightleftharpoons HF + OH^-$

10.8 In each of the following equations, identify the acid (proton donor) and base (proton acceptor) for the reactants:
a. $CO_3^{2-} + H_2O \rightleftharpoons HCO_3^- + OH^-$
b. $H_2SO_4 + H_2O \longrightarrow H_3O^+ + HSO_4^-$

LEARNING GOAL

Identify conjugate acid–base pairs for Brønsted–Lowry acids and bases.

WEB TUTORIAL
Nature of Acids and Bases

10.2 Conjugate Acid–Base Pairs

According to the Brønsted–Lowry theory, the reaction between an acid and base involves proton transfer. When an acid (HA) donates H^+ to a base (B), the products are A^- and BH^+. Because they are also an acid and a base, a reverse reaction can occur in which the acid BH^+ donates H^+ to the base (A^-). When a pair of molecules or ions are related by the loss or gain of one H^+, it is called a **conjugate acid–base pair**. Because protons are transferred in both a forward and reverse reaction, each acid–base reaction contains two conjugate acid–base pairs. In this general reaction, the acid HA has a conjugate base A^- and the base B has a conjugate acid BH^+.

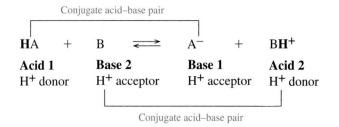

Now we can identify the conjugate acid–base pairs in a reaction such as hydrofluoric acid and water. Because the reaction is reversible, the conjugate acid H_3O^+ can transfer a proton to the conjugate base F^- and re-form the acid HF. Using the relationship of loss and gain of one H^+, we identify the conjugate acid–base pairs as HF and F^- along with H_3O^+ and H_2O.

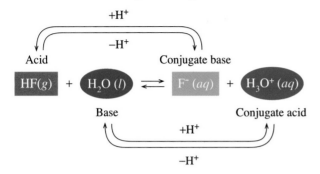

In another proton-transfer reaction, ammonia, NH_3, accepts H^+ from H_2O to form the conjugate acid NH_4^+ and conjugate base OH^-. Each of these conjugate acid–base pairs, NH_3 and NH_4^+ as well as H_2O and OH^-, are related by the loss and gain of one H^+. Table 10.3 gives some more examples of conjugate acid–base pairs.

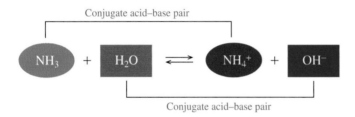

SAMPLE PROBLEM 10.4

Conjugate Acid–Base Pairs

Write the formula of the conjugate base of each of the following Brønsted–Lowry acids.

a. $HClO_3$ **b.** H_2CO_3 **c.** $H_2PO_4^-$

Solution

The conjugate base forms when the acid donates a proton.

a. ClO_3^- is the conjugate base of $HClO_3$
b. HCO_3^- is the conjugate base of H_2CO_3
c. HPO_4^{2-} is the conjugate base of $H_2PO_4^-$

Study Check

Write the conjugate acid of each of the following Brønsted–Lowry bases.

a. HS^- **b.** Cl^- **c.** NO_2^-

SAMPLE PROBLEM 10.5

Identifying Conjugate Acid–Base Pairs

Write the Brønsted–Lowry equation for the reaction of acid HBr and the base NH_3. Identify the conjugate acid–base pairs.

Solution

In the reaction, HBr donates H^+ to NH_3. The resulting Br^- is the conjugate base and NH_4^+ is the conjugate acid.

$$HBr + NH_3 \longrightarrow Br^- + NH_4^+$$

The conjugate acid–base pairs are HBr and Br^- along with NH_3 and NH_4^+.

Study Check

In the following reaction, identify the conjugate acid–base pairs.

$$HCN + SO_4^{2-} \rightleftharpoons CN^- + HSO_4^-$$

QUESTIONS AND PROBLEMS

Conjugate Acid–Base Pairs

10.9 Write the formula and name of the conjugate base for each of the following acids:
  a. HF
  b. H_2O
  c. H_2CO_3
  d. HSO_4^-

10.10 Write the formula and name of the conjugate base for each of the following acids:
  a. HCO_3^-
  b. H_3O^+
  d. HPO_4^{2-}
  d. HNO_2

10.11 Write the formula and name of the conjugate acid for each of the following bases:
  a. CO_3^{2-}
  b. H_2O
  c. $H_2PO_4^-$
  d. Br^-

10.12 Write the formula and name of the conjugate acid for each of the following bases:
  a. SO_4^{2-}
  b. CN^-
  c. OH^-
  d. ClO_2^-, chlorite ion

10.13 Identify the acid and base on the left side of the following equations and identify their conjugate species on the right side.
  a. $H_2CO_3(aq) + H_2O(l) \rightleftharpoons H_3O^+(aq) + HCO_3^-(aq)$
  b. $NH_4^+(aq) + H_2O(l) \rightleftharpoons H_3O^+(aq) + NH_3(aq)$
  c. $HCN(aq) + NO_2^-(aq) \rightleftharpoons CN^-(aq) + HNO_2(aq)$

10.14 Identify the acid and base on the left side of following equations and identify their conjugate species on the right side.
  a. $H_3PO_4(aq) + H_2O(l) \rightleftharpoons H_3O^+(aq) + H_2PO_4^-(aq)$
  b. $CO_3^{2-}(aq) + H_2O(l) \rightleftharpoons OH^-(aq) + HCO_3^-(aq)$
  c. $H_3PO_4(aq) + NH_3(aq) \rightleftharpoons NH_4^+(aq) + H_2PO_4^-(aq)$

10.15 When ammonium chloride dissolves in water, the ammonium ion NH_4^+ donates a proton to water. Write a balanced equation for the reaction.

10.16 When sodium carbonate dissolves in water, the carbonate ion CO_3^{2-} acts as a base. Write a balanced equation for the reaction.

10.3 Strengths of Acids and Bases

Acids are typically classified according to their ability to donate protons. Strong acids give up protons easily, whereas in a weak acid most of the molecules keep their protons and just a few give them up. Bases are classified in terms of their ability to accept protons. Strong bases have a strong attraction for protons, whereas weak bases have little attraction for protons.

Strong and Weak Acids

The six **strong acids** listed in Table 10.3 are such good proton donors that they dissociate almost completely ($\approx 100\%$) to give H_3O^+ ions.

$$HCl + H_2O \longrightarrow H_3O^+ + Cl^-$$

Most other acids are **weak acids** that dissociate only slightly and produce small amounts of H_3O^+ ions. Even at high concentrations, weak acids produce low concentrations of hydronium ions. (See Figure 10.1.) The acidic products you drink or use at home are weak acids. In carbonated soft drinks, CO_2 is dissolved in water to form carbonic acid, H_2CO_3. Citric acid is a weak acid found in fruits and fruit juices such as lemons, oranges, and grapefruit. The vinegar we put on a salad is a 5% acetic acid solution. In the equation for acetic acid in water, a longer arrow indicates the direction of dissociation.

$$\underset{\text{Acetic acid}}{CH_3COOH} + H_2O \underset{\longrightarrow}{\xleftarrow{\hspace{1em}}} H_3O^+ + \underset{\text{Acetate ion}}{CH_3COO^-}$$

Table 10.3   Some Conjugate Acid–Base Pairs

Acid		Conjugate Base	
Strong acids			
Perchloric acid	$HClO_4$	Perchlorate ion	ClO_4^-
Sulfuric acid	H_2SO_4	Hydrogen sulfate ion	HSO_4^-
Hydroiodic acid	HI	Iodide ion	I^-
Hydrobromic acid	HBr	Bromide ion	Br^-
Hydrochloric acid	HCl	Chloride ion	Cl^-
Nitric acid	HNO_3	Nitrate ion	NO_3^-
Weak acids			
Hydronium ion	H_3O^+	Water	H_2O
Hydrogen sulfate ion	HSO_4^-	Sulfate ion	SO_4^{2-}
Nitrous acid	HNO_2	Nitrite ion	NO_2^-
Phosphoric acid	H_3PO_4	Dihydrogen phosphate ion	$H_2PO_4^-$
Acetic acid	CH_3COOH	Acetate ion	CH_3COO^-
Hydrofluoric acid	HF	Fluoride ion	F^-
Extremely weak acids			
Carbonic acid	H_2CO_3	Bicarbonate ion	HCO_3^-
Hydrosulfuric acid	H_2S	Hydrogen sulfide ion	HS^-
Ammonium ion	NH_4^+	Ammonia	NH_3
Hydrocyanic acid	HCN	Cyanide ion	CN^-
Bicarbonate ion	HCO_3^-	Carbonate ion	CO_3^{2-}
Hydrogen sulfide ion	HS^-	Sulfide ion	S^{2-}
Water	H_2O	Hydroxide ion	OH^-

Increasing acid strength

Increasing base strength

Figure 10.1 A strong acid such as HCl is completely dissociated (≈100%), whereas a weak acid such as CH₃COOH contains mostly molecules and a few ions.

Q **What is the difference between a strong acid and a weak acid?**

WEB TUTORIAL
Nature of Acids and Bases

Strong and Weak Bases

The Arrhenius bases such as LiOH, KOH, NaOH, and Ca(OH)$_2$ are **strong bases** that dissociate completely (100%). Because these strong bases are ionic compounds, they dissociate in water to give an aqueous solution of a metal ion and hydroxide ion.

$$KOH(s) \xrightarrow{\text{H}_2\text{O}} K^+(aq) + OH^-(aq)$$

Strong bases, such as NaOH (also known as lye), are used in household products to remove grease in ovens and to clean drains. Because high concentrations of hydroxide ions cause severe damage to the skin and eyes, the use of such products in the home or in the chemistry laboratory should be carefully supervised. If you spill an acid or a base on your skin or get some in your eyes, be sure to flood the area immediately with water.

The **weak bases** are poor acceptors of protons. A typical weak base, ammonia, NH$_3$, is found in window cleaners. In water, some ammonia molecules accept protons to form ammonium hydroxide.

$$NH_3(g) + H_2O(l) \rightleftharpoons NH_4^+(aq) + OH^-(aq)$$

Direction of Equilibrium

There is a relationship between the components in each conjugate acid–base pair. The strong acids that donate protons easily have weak conjugate bases that do not readily accept protons. As the strength of the acid decreases, the strength of its conjugate base increases. Weak acids have strong conjugate bases.

In any acid–base reaction, there are two acids and two bases. However, one acid is stronger than the other acid and one base is stronger than the other base. By comparing their relative strengths, we can determine the direction of the reaction. For example, the strong acid H_2SO_4 gives up protons to water. The hydronium ion H_3O^+ produced is a weaker acid than H_2SO_4 and the conjugate base HSO_4^- is a weaker base than water. For reactions involving proton transfer, the prevalent direction is toward the weaker acid and base in the products, which can be shown by a long arrow.

$$H_2SO_4 \; + \; H_2O \; \rightleftharpoons \; H_3O^+ \; + \; HSO_4^- \qquad \text{Favors products}$$

| Stronger acid | Stronger base | Weaker acid | Weaker base |

Let's look at another reaction in which water donates a proton to carbonate CO_3^{2-} to form HCO_3^- and OH^-. From Table 10.3, we see that HCO_3^- is a stronger acid than H_2O. We also see that OH^- is a stronger base than CO_3^{2-}. The equilibrium favors the weaker acid and base reactants as shown by the long arrow for the reverse reaction.

$$CO_3^{2-} \; + \; H_2O \; \rightleftharpoons \; HCO_3^- \; + \; OH^-$$

| Weaker base | Weaker acid | Stronger acid | Stronger base |

SAMPLE PROBLEM 10.6

Strengths of Acids and Bases

For the following questions select from one of the following:

$$HCO_3^- \qquad HSO_4^- \qquad HNO_2$$

a. Which is the strongest acid?
b. Which acid has the strongest conjugate base?

Solution

As derived from the information in Table 10.3:

a. The strongest acid in this group is HSO_4^-.
b. The weakest acid in this group, HCO_3^-, has the strongest conjugate base CO_3^{2-}.

Study Check

Which is the strongest base: F^- or NH_3?

SAMPLE PROBLEM 10.7

Direction of Reaction

Does the direction of the following reaction favor the reactants on the left or the products on the right?

$$HF(aq) \; + \; H_2O(l) \; \rightleftharpoons \; H_3O^+(aq) \; + \; F^-(aq)$$

Solution

From Table 10.3, we see that HF is a weaker acid than H_3O^+ and H_2O is a weaker base than F^-. The reaction favors the reverse direction and therefore the reactants on the left side.

$$HF(aq) \; + \; H_2O(l) \; \rightleftharpoons \; H_3O^+(aq) \; + \; F^-(aq)$$

| Weaker acid | Weaker base | Stronger acid | Stronger base |

Study Check

Does the reaction of nitric acid and water favor the reactants or the products?

QUESTIONS AND PROBLEMS

Strengths of Acids and Bases

10.17 What is meant by the phrase "A strong acid has a weak conjugate base"?

10.18 What is meant by the phrase "A weak acid has a strong conjugate base"?

10.19 Identify the stronger acid in each pair:
   a. HBr or HNO_2 **b.** H_3PO_4 or HSO_4^-
   c. HCN or H_2CO_3

10.20 Identify the stronger acid in each pair:
   a. NH_4^+ or H_3O^+ **b.** H_2SO_4 or HCN
   c. H_2O or H_2CO_3

10.21 Identify the weaker acid in each pair:
   a. HCl or HSO_4^- **b.** HNO_2 or HF
   c. HCO_3^- or NH_4^+

10.22 Identify the weaker acid in each pair:
   a. HNO_3 or HCO_3^- **b.** HSO_4^- or H_2O
   c. H_2SO_4 or H_2CO_3

10.23 Predict whether the equilibrium for each of the following reactions favors the reactants or the products.
   a. $H_2CO_3(aq) + H_2O(l) \rightleftharpoons H_3O^+(aq) + HCO_3^-(aq)$
   b. $NH_4^+(aq) + H_2O(l) \rightleftharpoons H_3O^+(aq) + NH_3(aq)$
   c. $HCl(aq) + NH_3(aq) \rightleftharpoons Cl^-(aq) + NH_4^+(aq)$

10.24 Predict whether the equilibrium for each of the following reactions favors the reactants or the products.
   a. $H_3PO_4(aq) + H_2O(l) \rightleftharpoons H_3O^+(aq) + H_2PO_4^-(aq)$
   b. $CO_3^{2-}(aq) + H_2O(l) \rightleftharpoons OH^-(aq) + HCO_3^-(aq)$
   c. $HS^-(aq) + H_2O(l) \rightleftharpoons H_3O^+(aq) + S^{2-}(aq)$

10.25 Write an equation for the acid–base reaction between ammonium ion and sulfate ion. Why does the equilibrium favor the reactants?

10.26 Write an equation for the acid–base reaction between nitrous acid and sulfate ion. Why does the equilibrium favor the reactants?

LEARNING GOAL

Write the equilibrium expression for a weak acid or weak base.

10.4 Dissociation Constants

We have seen that reactions of weak acids in water reach equilibrium. If HA is a weak acid, the concentration of H_3O^+ and A^- will be small, which means that the equilibrium will favor the reactants. (See Figure 10.2.)

$$HA + H_2O \rightleftharpoons H_3O^+ + A^-$$

Acid Dissociation Constants

Because the reaction of a weak acid in water reaches equilibrium, we can write its equilibrium expression. Recall that in an equilibrium expression the molar concentrations of the products are divided by the molar concentration of the reactants.

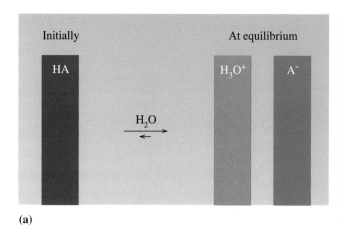

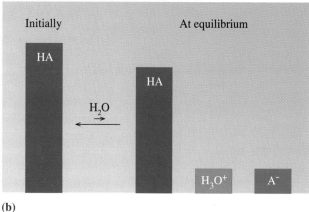

(a)

(b)

$$K_{eq} = \frac{[H_3O^+][A^-]}{[HA][H_2O]}$$

In this standard equilibrium expression, the concentration of water is included. However, water is also the solvent, which means that its concentration will remain essentially the same. Therefore, the water concentration is combined with the equilibrium constant to give a new constant called the **acid dissociation constant, K_a**.

$$K_{eq}[H_2O] = K_a = \frac{[H_3O^+][A^-]}{[HA]} \quad \text{Acid dissociation constant}$$

Let's consider a sample of carbonic acid that dissociates in water to form bicarbonate ion and hydronium ion.

$$H_2CO_3(aq) + H_2O(l) \rightleftharpoons HCO_3^-(aq) + H_3O^+(aq)$$

The K_a expression for carbonic acid is

$$K_a = \frac{[H_3O^+][HCO_3^-]}{[H_2CO_3]} = 4.5 \times 10^{-7}$$

The K_a measured for carbonic acid at 25°C is quite small, which confirms that the reaction of carbonic acid in water favors the reactants. Usually the concentration units are omitted in the values given for equilibrium constants.

A similar equation and equilibrium expression can be written for a weak base such as ammonia in water.

$$NH_3(aq) + H_2O(l) \rightleftharpoons NH_4^+(aq) + OH^-(aq)$$

We can write a **base dissociation constant, K_b**.

$$K_b = \frac{[NH_4^+][OH^-]}{[NH_3]} = 1.8 \times 10^{-5}$$

The K_b for ammonia is small because the reaction favors the reactants. We can conclude that weak acids and weak bases will have small K_a and K_b values because their reactions favor the reactants. The smaller the K_a or K_b value, the weaker the acid or base. On the other hand, strong acids and bases, which are essentially 100% dissociated, would have large K_a and K_b values, although these values are not usually measured. Table 10.4 gives some K_a and K_b values for selected weak acids and bases.

We have described strong and weak acids in several ways. Table 10.5 summarizes the characteristics of acids in terms of strengths and equilibrium position.

Fig 10.2 At equilibrium, (a) a strong acid has a high concentration of H_3O^+ and A^-, and (b) a weak acid has a high concentration of HA and low concentration of H_3O^+ and A^-.

Q How does the height of the H_3O^+ and A^- in the bar diagram change for a very weak acid?

Table 10.4 K_a and K_b Values for Selected Weak Acids and Bases

Acids

Phosphoric acid	H_3PO_4	7.5×10^{-3}
Hydrofluoric acid	HF	7.2×10^{-4}
Nitrous acid	HNO_2	4.5×10^{-4}
Formic acid	HCOOH	1.8×10^{-4}
Acetic acid	CH_3COOH	1.8×10^{-5}
Carbonic acid	H_2CO_3	4.3×10^{-7}
Dihydrogen phosphate	$H_2PO_4^-$	6.2×10^{-8}
Hydrocyanic acid	HCN	4.9×10^{-10}
Hydrogen phosphate	HPO_4^{2-}	2.2×10^{-13}

Bases

Carbonate	CO_3^{2-}	2.2×10^{-4}
Ammonia	NH_3	1.8×10^{-5}

Table 10.5   Characteristics of Acids

Characteristic	Strong Acids	Weak Acids
Equilibrium position	Toward ionized products	Toward reactants
K_a	Large	Small
$[H_3O^+]$ and $[A^-]$	$\approx 100\%$ of [HA]	Small percent of [HA]
Conjugate bases	Weak	Strong

SAMPLE PROBLEM 10.8

Acid Dissociation Constants

Write the expression for the acid dissociation constant for nitrous acid.

Solution

The equation for the dissociation of nitrous acid is written.

$$HNO_2 + H_2O \rightleftharpoons H_3O^+ + NO_2^-$$

The acid dissociation constant is written as the concentrations of the products divided by the concentration of the weak acid.

$$K_a = \frac{[H_3O^+][NO_2^-]}{[HNO_2]} = 4.5 \times 10^{-4}$$

Study Check

Is nitrous acid a weaker or stronger acid than carbonic acid? Why?

QUESTIONS AND PROBLEMS

Dissociation Constants

10.27 Consider the following acids and their dissociation constants:

$H_2SO_3(aq) + H_2O(l) \rightleftharpoons H_3O^+(aq) + HSO_3^-(aq)\ K_a = 1.2 \times 10^{-2}$
$HS^-(aq) + H_2O(l) \rightleftharpoons H_3O^+(aq) + S^{2-}(aq)\ K_a = 1.3 \times 10^{-19}$
   a. Which is the stronger acid, H_2SO_3 or HS^-?
   b. What is the conjugate base of H_2SO_3?
   c. Which acid has the weaker conjugate base?
   d. Which acid has the stronger conjugate base?
   e. Which acid produces more ions?

10.28 Consider the following acids and their dissociation constants:
   $HPO_4^{2-}(aq) + H_2O(l) \rightleftharpoons H_3O^+(aq) + PO_4^{3-}(aq)$
   $K_a = 2.2 \times 10^{-13}$
   $HCOOH(aq) + H_2O(l) \rightleftharpoons H_3O^+(aq) + HCOO^-(aq)$
   $K_a = 1.8 \times 10^{-4}$
   a. Which is the weaker acid, HPO_4^{2-} or HCOOH?
   b. What is the conjugate base of HPO_4^{2-}?
   c. Which acid has the weaker conjugate base?
   d. Which acid has the stronger conjugate base?
   e. Which acid produces more ions?

10.29 Phosphoric acid ionizes to form dihydrogen phosphate and hydronium ion. Phosphoric acid has a K_a of 7.5×10^{-3}. Write the equation and equilibrium constant for the dissociation of the acid.

10.30 Aniline, $C_6H_5NH_2$, a weak base with a K_b of 4.0×10^{-10} has a conjugate acid $C_6H_5NH_3^+$. Write the equation and equilibrium constant for the dissociation of the base.

10.5 Ionization of Water

We have seen that in some acid–base reactions water acts as an acid and in other reactions as a base. Does that mean water can be both an acid and a base? Yes, this is exactly what happens with water molecules in pure water. Let's see how this happens.

WEB TUTORIAL
The pH Scale

Ionization of Water

In water, one water molecule donates a proton to another to produce H_3O^+ and OH^-, which means that water can behave as both an acid and a base. Let's take a look at the conjugate acid–base pairs in water.

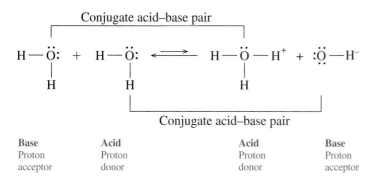

Because there is both a forward and reverse reaction, the ionization of water reaches equilibrium.

$H_2O(l) + H_2O(l) \rightleftharpoons H_3O^+(aq) + OH^-(aq)$

Its equilibrium expression is written

$$K_{eq} = \frac{[H_3O^+][OH^-]}{[H_2O][H_2O]}$$

As before, the concentration of water as the solvent remains essentially constant. We simplify this expression by combining the concentration of water with the equilibrium constant, which gives a new constant known as the **ion-product constant for water (K_w).** Because the amount of dissociation is small, the concentration of water stays essentially constant and is included with K_{eq} to give K_w.

$$K_{eq}[H_2O][H_2O] = K_w = [H_3O^+][OH^-]$$

In pure water, the transfer of a proton between two water molecules produces equal numbers of H_3O^+ and OH^-. Experiments have determined that, in pure water, the concentrations of H_3O^+ and OH^- at 25°C are each 1.0×10^{-7} M. Square brackets around the symbols indicate their concentrations in moles per liter (M).

Pure water $[H_3O^+] = [OH^-] = 1.0 \times 10^{-7}$ M

When we place these concentrations into the ion-product constant, the K_w value is 1.0×10^{-14}. As before, the concentration units are omitted in the K_w value.

$$K_w = [H_3O^+] \times [OH^-]$$
$$= (1.0 \times 10^{-7} \text{ M})(1.0 \times 10^{-7} \text{ M}) = 1.0 \times 10^{-14}$$

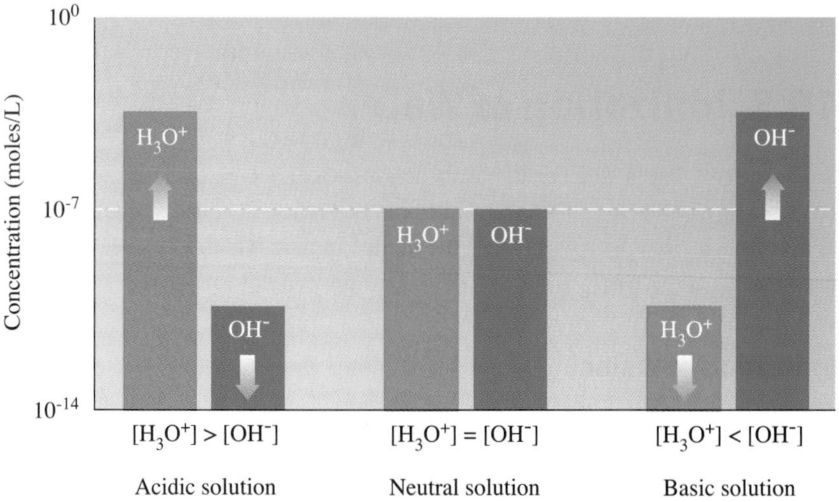

Figure 10.3 In a neutral solution, $[H_3O^+]$ and $[OH^-]$ are equal. In acidic solutions, the $[H_3O^+]$ is greater than the $[OH^-]$. In basic solutions, the $[OH^-]$ is greater than the $[H_3O^+]$.

Q **Is a solution that has a $[H_3O^+]$ of 1.0 × 10^{-3} M acidic, basic, or neutral?**

Table 10.6   Examples of $[H_3O^+]$ and $[OH^-]$ in Neutral, Acidic, and Basic Solutions

Type of Solution	$[H_3O^+]$	$[OH^-]$	K_w
Neutral	1.0×10^{-7} M	1.0×10^{-7} M	1.0×10^{-14}
Acidic	1.0×10^{-2} M	1.0×10^{-12} M	1.0×10^{-14}
Acidic	2.5×10^{-5} M	4.0×10^{-10} M	1.0×10^{-14}
Basic	1.0×10^{-8} M	1.0×10^{-6} M	1.0×10^{-14}
Basic	5.0×10^{-11} M	2.0×10^{-4} M	1.0×10^{-14}

The K_w value of 1.0×10^{-14} is important because it applies to pure water as well as any aqueous solution. We have seen that pure water has equal $[H_3O^+]$ and $[OH^-]$. Suppose that acid is added, which increases the $[H_3O^+]$. To maintain the K_w, there is a shift in equilibrium as H_3O^+ combines with OH^-, which decreases the $[OH^-]$. If base is added, $[OH^-]$ increases. Again, the K_w remains constant as some OH^- combines with H_3O^+, which decreases the $[H_3O^+]$. (See Figure 10.3.) Thus, all aqueous solutions have H_3O^+ and OH^-. An increase in the concentration of one of the ions will cause an equilibrium shift that decreases the other ion. Examples are shown in Table 10.6.

When $[H_3O^+]$ and $[OH^-]$ are equal in a solution, it is **neutral.** In an *acidic* solution, the $[H_3O^+]$ is greater than the $[OH^-]$. For example, if the $[H_3O^+]$ is 1.0×10^{-4} M, the $[OH^-]$ would be 1.0×10^{-10} M.

$$[OH^-] = \frac{K_w}{[H_3O^+]} = \frac{1.0 \times 10^{-14}}{1.0 \times 10^{-4}} = 1.0 \times 10^{-10} \text{ M}$$

In a *basic* solution, the $[OH^-]$ is greater than the $[H_3O^+]$. For example, if the $[OH^-]$ is 1.0×10^{-6} M, the $[H_3O^+]$ must be 1.0×10^{-8} M.

$$[H_3O^+] = \frac{K_w}{[OH^-]} = \frac{1.0 \times 10^{-14}}{1.0 \times 10^{-6}} = 1.0 \times 10^{-8} \text{ M}$$

In other words, the product of the ions $[H_3O^+][OH^-]$ is always equal to 1.0×10^{-14} at 25°C. Therefore, we can use the K_w to calculate the $[H_3O^+]$ if the $[OH^-]$ is known, or the $[OH^-]$ if the $[H_3O^+]$ is known.

$$K_w = [H_3O^+] \times [OH^-]$$

$$[OH^-] = \frac{K_w}{[H_3O^+]} \qquad [H_3O^+] = \frac{K_w}{[OH^-]}$$

SAMPLE PROBLEM 10.9

Calculating $[H_3O^+]$ and $[OH^-]$ in Solution

A vinegar solution has a $[H_3O^+]$ of 2.0×10^{-3} M at 25°C.

a. What is the $[OH^-]$?
b. Is the solution acidic, basic, or neutral?

Solution

In an aqueous solution (25°C)

$$K_w = [H_3O^+] \times [OH^-] = 1.0 \times 10^{-14}$$

Rearranging the K_w for $[OH^-]$ gives

$$[OH^-] = \frac{K_w}{[H_3O^+]} = \frac{1.0 \times 10^{-14}}{2.0 \times 10^{-3}} = 5.0 \times 10^{-12} \text{ M}$$

b. Because the $[H_3O^+] = 2.0 \times 10^{-3}$ M is greater than 1.0×10^{-7} M, it is an acidic solution.

Study Check

What is the $[H_3O^+]$ of an ammonia cleaning solution with an $[OH^-] = 4.0 \times 10^{-4}$ M? Is the solution acidic, basic, or neutral?

QUESTIONS AND PROBLEMS

Ionization of Water

10.31 Why are the concentrations of H_3O^+ and OH^- equal in pure water?

10.32 What is the meaning and value of K_w at 25°C?

10.33 In an acidic solution, how does the concentration of H_3O^+ compare to the concentration of OH^-?

10.34 If a base is added to pure water, why does the $[H_3O^+]$ decrease?

10.35 Indicate whether the following are acidic, basic, or neutral solutions:
    a. $[H_3O^+] = 2.0 \times 10^{-5}$ M   **b.** $[H_3O^+] = 1.4 \times 10^{-9}$ M
    c. $[OH^-] = 8.0 \times 10^{-3}$ M   **d.** $[OH^-] = 3.5 \times 10^{-10}$ M

10.36 Indicate whether the following are acidic, basic, or neutral solutions:
    a. $[H_3O^+] = 6.0 \times 10^{-12}$ M   **b.** $[H_3O^+] = 1.4 \times 10^{-4}$ M
    c. $[OH^-] = 5.0 \times 10^{-12}$ M   **d.** $[OH^-] = 4.5 \times 10^{-2}$ M

10.37 Calculate the $[OH^-]$ of each aqueous solution with the following $[H_3O^+]$:
    a. coffee, 1.0×10^{-5} M   **b.** soap, 1.0×10^{-8} M
    c. cleanser, 5.0×10^{-10} M   **d.** lemon juice, 2.5×10^{-2} M

10.38 Calculate the $[OH^-]$ of each aqueous solution with the following $[H_3O^+]$:
    a. NaOH, 1.0×10^{-12} M   **b.** aspirin, 6.0×10^{-4} M
    c. milk of magnesia, 1.0×10^{-9} M   **d.** stomach acid, 5.2×10^{-2} M

10.39 Calculate the $[OH^-]$ of each aqueous solution with the following $[H_3O^+]$:
    a. vinegar, 1.0×10^{-3} M   **b.** urine, 5.0×10^{-6} M
    c. ammonia, 1.8×10^{-12} M   **d.** NaOH, 4.0×10^{-13} M

10.40 Calculate the $[OH^-]$ of each aqueous solution with the following $[H_3O^+]$:
    a. baking soda, 1.0×10^{-8} M   **b.** orange juice, 2.0×10^{-4} M
    c. milk, 5.0×10^{-7} M   **d.** bleach, 4.8×10^{-12} M

LEARNING GOAL

Calculate pH from $[H_3O^+]$; given the pH, calculate $[H_3O^+]$ and $[OH^-]$ of a solution.

the Chemistry place

WEB TUTORIAL
The pH Scale

10.6 The pH Scale

Many kinds of careers such as respiratory therapy, wine and beer making, medicine, agriculture, spa cleaning, and soap manufacturing require personnel to measure the $[H_3O^+]$ and $[OH^-]$ of solutions. The proper levels of acidity are necessary for soil to support plant growth and prevent algae in swimming pool water. Measuring the acidity levels of blood and urine checks the function of the kidneys.

pH

We have seen that the $[H_3O^+]$ can range by many factors of ten from 1.0 M to 1.0×10^{-14} M. Such values are rather inconvenient to work with, especially if the acidity of many samples is to be measured. Therefore a simpler way of describing acidity is used: the pH scale. On the pH scale, a number between 0 and 14 represents the H_3O^+ concentration. A pH value less than 7 corresponds to an acidic solution; a pH value greater than 7 indicates a basic solution. (See Figure 10.4.)

Acidic solution	pH < 7	$[H_3O^+] > 1.0 \times 10^{-7}$ M
Neutral solution	pH = 7	$[H_3O^+] = 1.0 \times 10^{-7}$ M
Basic solution	pH > 7	$[H_3O^+] < 1.0 \times 10^{-7}$ M

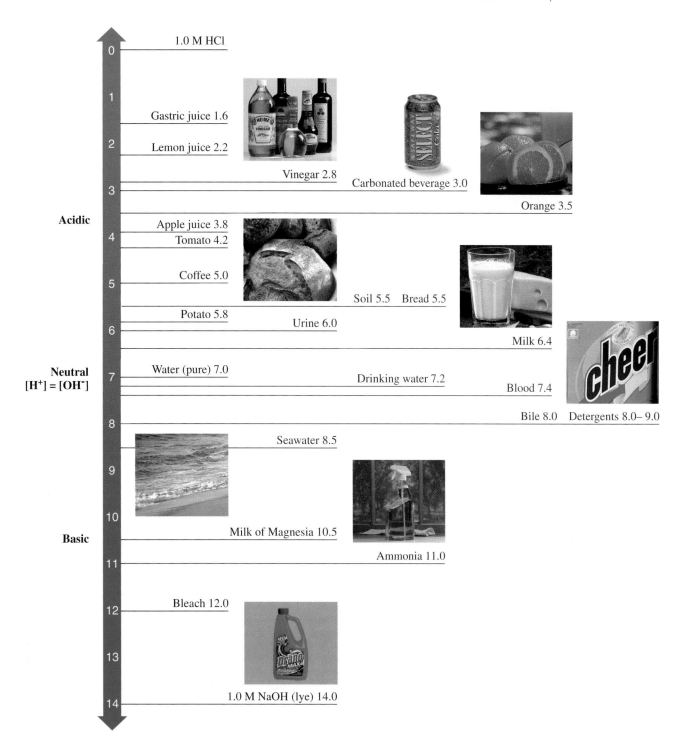

Figure 10.4 On the pH scale, values below 7 are acidic, a value of 7 is neutral, and values above 7 are basic.

Q Is apple juice an acidic, basic, or a neutral solution?

In the laboratory, a pH meter is commonly used to determine the pH of a solution. There are also indicators and pH papers that turn specific colors when placed in solutions of different pH values. The pH is found by comparing the colors to a color chart. (See Figure 10.5.)

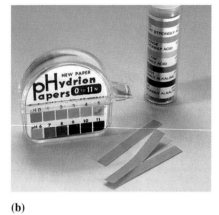

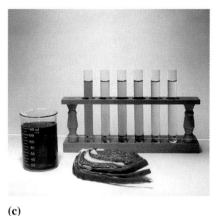

(a)   (b)   (c)

Fig 10.5 The pH of a solution can be determined using **(a)** a pH meter, **(b)** pH paper, and **(c)** indicators that turn different colors corresponding to different pH values.

Q **If a pH meter reads 4.00, is the solution acidic, basic, or neutral?**

Calculating the pH of Solutions

The pH scale is a log scale that corresponds to the hydrogen-ion concentrations of aqueous solutions. Mathematically, **pH** is the negative logarithm (base 10) of the H_3O^+ concentration.

$$pH = -\log[H_3O^+]$$

Essentially, the negative powers of ten in the concentrations are converted to positive numbers. For example, a lemon juice solution with $[H_3O^+] = 1.0 \times 10^{-2}$ M has a pH of 2.00. This can be calculated using the pH equation. For whole numbers in the $[H_3O^+]$, remember to add the correct number of significant zeros to the resulting pH obtained on your calculator.

$$pH = -\log [1.0 \times 10^{-2}]$$
$$pH = -(-2.00)$$
$$= 2.00$$

Let's look at how we determine the number of significant figures in the pH. For a logarithm, the number of decimal places in the pH value is equal to the number of significant figures in the $[H_3O^+]$. The number in the one's place is the power of ten.

$$[H_3O^+] = \mathbf{1.0} \times 10^{-3} \qquad\qquad pH = 3.\mathbf{00}$$

2 significant figures   2 decimal places

Because pH is a log scale, a change of 1 pH unit corresponds to a ten-fold change in $[H_3O^+]$. It is important to note that the pH decreases as the $[H_3O^+]$ increases. For example, a solution with a pH of 2.00 has 10 times more $[H_3O^+]$ than a solution with a pH of 3.00.

SAMPLE PROBLEM 10.10

Calculating pH

Determine the pH for the following solutions:

a. $[H_3O^+] = 1.0 \times 10^{-5}$ M   **b.** $[H_3O^+] = 2.5 \times 10^{-8}$ M

Solution

a. $pH = -\log [1.0 \times 10^{-5}] = 5.00$

b. Enter the 2.5×10^{-8} into your calculator and take the log. Change the sign $(+/-$ key) to obtain the pH and round off to give two decimal places.

$$pH = -\log [2.5 \times 10^{-8}] = 7.60$$

Study Check

What is the pH of bleach with $[OH^-]$ of 2.4×10^{-3} M?

SAMPLE PROBLEM 10.11

Calculating pH

Determine the pH of the following solutions.

a. $[H_3O^+] = 1.0 \times 10^{-5}$ M   **b.** $[H_3O^+] = 6.0 \times 10^{-8}$ M
c. $[OH^-] = 1.0 \times 10^{-3}$ M

Solution

a. Because the coefficient in the $[H_3O^+]$ is 1.0, the pH of the solution is the power of 10 without the negative sign, followed by two zeros.

$$[H_3O^+] = 1.0 \times 10^{-5} \qquad pH = 5.00$$

b. Enter the concentration into your calculator, press the *log* key, and change signs as follows:

6 [EE or EXP] 8 [+/−] [log] [+/−] =   7.22184875

On some calculators, a different sequence is used to enter data:

[+/−] [log] 6 [EE or EXP] [+/−] 8

In this problem, the coefficient 6.0 has two significant figures. Therefore, the calculator answer is rounded to give two decimal places and the pH value is reported as 7.22.

c. Because $[OH^-]$ is given, we have to calculate the $[H_3O^+]$ using the ion product of water.

$$[H_3O^+][OH^-] = 1.0 \times 10^{-14}$$

$$[H_3O^+] = \frac{1.0 \times 10^{-14}}{1.0 \times 10^{-3}}$$

$$= 1.0 \times 10^{-11}$$

$$pH = 11.00$$

Study Check

What is the pH of a bleach whose $[OH^-]$ is 1.0×10^{-4} M? (*Hint:* Calculate the $[H_3O^+]$ first.)

Sometimes, we need to determine the $[H_3O^+]$ from the pH. For whole number pH values, we can use the negative pH value as the power of 10 in the H_3O^+ concentration.

$$[H_3O^+] = 1 \times 10^{-pH}$$

EXPLORE YOUR WORLD

Using Vegetables and Flowers as pH Indicators

Many flowers and vegetables with strong color, especially reds and purples, contain compounds that change color with changes in pH. Some examples are red cabbage, cranberry juice, and cranberry drinks.

Materials Needed

Red cabbage, water, and a saucepan; or cranberry juice or drinks

Several glasses or small glass containers and some tape and a pen or pencil to mark the containers

Several colorless household solutions such as vinegar, lemon juice, other fruit juices, baking soda, antacids, aspirin, window cleaners, soaps, shampoos, and detergents

Procedure

1. Obtain a bottle of cranberry juice or cranberry drink, or use a red cabbage to prepare the red cabbage pH indicator, as follows: Tear up several red cabbage leaves and place them in a saucepan and cover with water. Boil for about 5 minutes. Cool and collect the purple solution.

2. Place small amounts of each household solution into separate, clear glass containers and mark what each one is. If the sample is a solid or a thick liquid, add a small amount of water. Add some cranberry juice or some red cabbage indicator until you obtain a color.
3. Observe the colors of the various samples. The colors that indicate acidic solutions are the pink and orange colors (pH 1–4) and the pink to lavender colors (5–6). A neutral solution has about the same purple color as the indicator. Bases will give blue to green color (pH 8–11) or a yellow color (pH 12–13).
4. Arrange your samples by color and pH. Classify each of the solutions as acidic (1–6), neutral (7), or basic (8–13).
5. Try to make an indicator using other colorful fruits or flowers.

Questions

1. Which products that tested acidic listed an acid on their labels?
2. Which products that tested basic listed a base on their labels?
3. How many products were neutral?
4. Which flowers or vegetables behaved as indicators?

HEALTH NOTE

Stomach Acid, HCl

When a person sees, smells, thinks about, or tastes food, the gastric glands in the stomach begin to secrete an HCl solution that is strongly acidic. In a single day, a person may secrete as much as 2000 mL of gastric juice.

The HCl in the gastric juice activates a digestive enzyme called pepsin that breaks down proteins in food entering the stomach. The secretion of HCl continues until the stomach has a pH of about 2, which is the optimum pH for activating the digestive enzymes without ulcerating the stomach lining. Normally, large quantities of viscous mucus are secreted within the stomach to protect its lining from acid and enzyme damage.

SAMPLE PROBLEM 10.12

Calculating $[H_3O^+]$ from pH

Determine $[H_3O^+]$ for solutions having the following pH values:

a. pH = 3.0   **b.** pH = 12.0

Solution

For pH values that are whole numbers, the $[H_3O^+]$ can be written 1.0×10^{-pH}.

a. $[H_3O^+] = 1 \times 10^{-3}$ M   **b.** $[H_3O^+] = 1 \times 10^{-12}$ M

Study Check

What is the $[H_3O^+]$ and $[OH^-]$ of beer if it has a pH of 5.0?

---

pOH

The pOH scale is similar to the pH scale except that pOH corresponds to the $[OH^-]$ of an aqueous solution.

$$pOH = -\log[OH^-]$$

Solutions with high $[OH^-]$ values have low pOH. The sum of the pOH and pH of any solution equals the pK_w.

$$K_w = 1.0 \times 10^{-14}$$
$$pK_w = -\log[K_w] = 14.00$$
$$= (-\log[H_3O^+]) + (-\log[OH^-])$$
$$= pH + pOH = 14.00$$

A comparison of $[H_3O^+]$, $[OH^-]$, and their corresponding pH and pOH values is given in Table 10.7.

Table 10.7 A Comparison of $[H_3O^+]$, $[OH^-]$, and Corresponding pH Values at 25°C

$[H_3O^+]$	pH	$[OH^-]$	pOH	
10^0	0	10^{-14}	14	
10^{-1}	1	10^{-13}	13	
10^{-2}	2	10^{-12}	12	
10^{-3}	3	10^{-11}	11	Acidic
10^{-4}	4	10^{-10}	10	
10^{-5}	5	10^{-9}	9	
10^{-6}	6	10^{-8}	8	
10^{-7}	7	10^{-7}	7	Neutral
10^{-8}	8	10^{-6}	6	
10^{-9}	9	10^{-5}	5	
10^{-10}	10	10^{-4}	4	
10^{-11}	11	10^{-3}	3	Basic
10^{-12}	12	10^{-2}	2	
10^{-13}	13	10^{-1}	1	
10^{-14}	14	10^0	0	

SAMPLE PROBLEM 10.13

Calculating pOH and pH

a. What is the pOH and pH of seawater if the $[OH^-]$ is 1.0×10^{-6} M?
b. What is the pOH and pH of a sample of wine if $[H_3O^+]$ is 1.5×10^{-3} M?

Solution

a. $pOH = -\log[OH^-] = -\log [1.0 \times 10^{-6}] = 6.00$
From the pK_w we know that $pH + pOH = 14.00$. Solving for pH gives
$pH = 14.00 - pOH = 14.00 - 6.00 = 8.00$
b. $pH = -\log[H_3O^+] = -\log [1.5 \times 10^{-3}] = 2.82$
From the pK_w we know that $pH + pOH = 14.00$. Solving for pOH gives
$pOH = 14.00 - pH = 14.00 - 2.82 = 11.18$

Study Check

What is the pOH and pH of a solution of milk of magnesia with $[OH^-]$ of 5.0×10^{-4} M?

QUESTIONS AND PROBLEMS

The pH Scale

10.41 Why does a neutral solution have a pH of 7.00?
10.42 If you know the $[OH^-]$, how can you determine the pH of a solution?
10.43 State whether each of the following solutions is acidic, basic, or neutral:
    a. blood, pH 7.38   **b.** vinegar, pH 2.8
    c. drain cleaner, pOH 2.8   **d.** coffee, pH 5.52

Acid Rain

Rain typically has a pH of 6.2. It is slightly acidic because carbon dioxide in the air combines with water to form carbonic acid. However, in many parts of the world, rain has become considerably more acidic, with pH values as low as 3 being reported. One cause of acid rain is the sulfur dioxide (SO_2) gas produced when coal that contains sulfur is burned.

In the air, the SO_2 gas reacts with oxygen to produce SO_3, which then combines with water to form sulfuric acid, H_2SO_4, a strong acid.

$$S + O_2 \longrightarrow SO_2$$
$$2SO_2 + O_2 \longrightarrow 2SO_3$$
$$SO_3 + H_2O \longrightarrow H_2SO_4$$

In parts of the United States, acid rain has made lakes so acidic they are no longer able to support fish and plant life. Limestone ($CaCO_3$) is sometimes added to these lakes to neutralize the acid. In Eastern Europe, acid rain has brought about an environmental disaster. Nearly 40% of the forests in Poland have been severely damaged, and some parts of the land are so acidic that crops will not grow. Throughout Europe and the United States, monuments made of marble (a form of $CaCO_3$) are deteriorating as acid rain dissolves the marble.

$$2H^1 + CaCO_3 \longrightarrow Ca^{2+} + H_2O + CO_2$$

1935   1994

Marble statue in Washington Square Park

Efforts to slow or stop the damaging effects of acid rain include the reduction of sulfur emissions. This will require installation of expensive equipment in coal-burning plants to absorb more of the SO_2 gases before they are emitted. In some outdated plants, this may be impossible, and they will need to be closed. It is a difficult problem for engineers and scientists, but one that must be solved.

10.44 State whether each of the following solutions is acidic, basic, or neutral:
    a. soda, pH 3.22   **b.** shampoo, pOH 8.3
    c. laundry detergent, pOH 4.56   **d.** rain, pH 5.8

10.45 A solution with a pH of 3 is 10 times more acidic than a solution with pH 4. Explain.

10.46 A solution with a pH of 10 is 100 times more basic than a solution with pH 8. Explain.

10.47 Calculate the pH of each solution given the following $[H_3O^+]$ or $[OH^-]$ values.
    a. $[H_3O^+] = 1.0 \times 10^{-4}$ M   **b.** $[H_3O^+] = 3.0 \times 10^{-9}$ M
    c. $[OH^-] = 1.0 \times 10^{-5}$ M   **d.** $[OH^-] = 2.5 \times 10^{-11}$ M

10.48 Calculate the pH of each solution given the following $[H_3O^+]$ or $[OH^-]$ values.
    a. $[H_3O^+] = 1.0 \times 10^{-8}$ M   **b.** $[H_3O^+] = 5.0 \times 10^{-6}$ M
    c. $[OH^-] = 4.0 \times 10^{-2}$ M   **d.** $[OH^-] = 8.0 \times 10^{-3}$ M

10.49 Complete the following table:

$[H_3O^+]$	$[OH^-]$	pH	pOH	Acidic, Basic, or Neutral?
	1.0×10^{-6} M			
		3.00		
2.8×10^{-5} M				
			2.00	

10.50 Complete the following table:

$[H_3O^+]$	$[OH^-]$	pH	pOH	Acidic, Basic, or Neutral?
		10.00		
				Neutral
			5.00	
6.4×10^{-12} M				

10.7 Reactions of Acids and Bases

LEARNING GOAL
Write balanced equations for reactions of acids and bases.

Typical reactions of acids and bases include the reactions of acids with metals, bases, and carbonate or bicarbonate ions. For example, when you drop an antacid tablet in water, the bicarbonate ion and citric acid in the tablet react to produce carbon dioxide bubbles, a salt, and water.

Acids and Metals

Acids react with certain metals known as *active metals* to produce hydrogen gas (H_2) and the salt of that metal. Active metals include potassium, sodium, calcium, magnesium, aluminum, zinc, iron, and tin. In these single replacement reactions, the metal loses electrons and the metal ion replaces the hydrogen in the acid.

$$\underset{\text{Metal}}{Mg(s)} + \underset{\text{Acid}}{2HCl(aq)} \longrightarrow \underset{\text{Salt}}{MgCl_2(aq)} + \underset{\text{Hydrogen}}{H_2(g)}$$

$$\underset{\text{Metal}}{Zn(s)} + \underset{\text{Acid}}{2HCl(aq)} \longrightarrow \underset{\text{Salt}}{ZnCl_2(aq)} + \underset{\text{Hydrogen}}{H_2(g)}$$

Acids and Carbonates and Bicarbonates

When strong acids are added to a carbonate or bicarbonate, the reaction produces bubbles of carbon dioxide gas, a salt, and water. In the reaction, H^+ is transferred to the carbonate to give carbonic acid, H_2CO_3, which breaks down rapidly to CO_2 and H_2O. The net ionic equation is written by omitting the metal ions and chloride ions that are not reacting.

$$HCl(aq) + NaHCO_3(aq) \longrightarrow CO_2(g) + H_2O + NaCl(aq)$$
$$H^+(aq) + HCO_3^-(aq) \longrightarrow CO_2(g) + H_2O$$

$$2HCl(aq) + Na_2CO_3(aq) \longrightarrow CO_2(g) + H_2O + 2NaCl(aq)$$
$$2H^+(aq) + CO_3^{2-}(aq) \longrightarrow CO_2(g) + H_2O$$

Acids and Hydroxides

Neutralization is a reaction between an acid and a base to form a salt and water. For example, the neutralization reaction HCl and NaOH is written

$$\underset{\text{Acid}}{HCl} + \underset{\text{Base}}{NaOH} \longrightarrow \underset{\text{Salt}}{NaCl} + \underset{\text{Water}}{H_2O}$$

In the neutralization, the H^+ reacts with OH^- to form water, leaving Na^+ and Cl^- ions in solution. By omitting the unreacted ions from the overall equation, the ionic equation for *neutralization* is

$$H^+ + Cl^- + Na^+ + OH^- \longrightarrow Na^+ + Cl^- + H_2O$$

$$H^+ \quad + \quad OH^- \longrightarrow H_2O$$

Equations for neutralization reactions are balanced so that the H^+ from the acid is balanced by the OH^-. For example, when aluminum hydroxide in antacids neutralizes stomach acid the products are a salt, aluminum chloride, and water.

$$3HCl + Al(OH)_3 \longrightarrow AlCl_3 + 3H_2O$$

SAMPLE PROBLEM 10.14

Reactions of Acids

Write a balanced equation for the reaction of HNO_3 with each of the following:

a. Al   **b.** K_2CO_3 **c.** $Mg(OH)_2$

Solution

a. $6HNO_3 + 2Al \longrightarrow 2Al(NO_3)_3 + 3H_2(g)$
b. $2HNO_3 + K_2CO_3 \longrightarrow CO_2(g) + H_2O + 2KNO_3$
c. $2HNO_3 + Mg(OH)_2 \longrightarrow Mg(NO_3)_2 + 2H_2O$

Study Check

Write the equation for the reaction between H_2SO_4 and $NaHCO_3$.

QUESTIONS AND PROBLEMS

Reactions of Acids and Bases

10.51 Complete and balance the equations for the following reactions:
  a. $ZnCO_3 + HCl \longrightarrow$
  b. $Zn + HCl \longrightarrow$
  c. $HCl + NaHCO_3 \longrightarrow$
  d. $HNO_3 + Mg(OH)_2 \longrightarrow$

10.52 Complete and balance the equations for the following reactions:
  a. $KHCO_3 + HCl \longrightarrow$
  b. $Ca + HNO_3 \longrightarrow$
  c. $H_2SO_4 + Al(OH)_3 \longrightarrow$
  d. $Na_2CO_3 + H_2SO_4 \longrightarrow$

10.53 Balance each of the following neutralization reactions:
  a. $HCl + Mg(OH)_2 \longrightarrow MgCl_2 + H_2O$
  b. $H_3PO_4 + LiOH \longrightarrow Li_3PO_4 + H_2O$

10.54 Balance each of the following neutralization reactions:
  a. $HNO_3 + Ba(OH)_2 \longrightarrow Ba(NO_3)_2 + H_2O$
  b. $H_2SO_4 + Al(OH)_3 \longrightarrow Al_2(SO_4)_3 + H_2O$

10.55 Write a balanced equation for the neutralization of each of the following:
  a. H_2SO_4 and $NaOH$
  b. HCl and $Fe(OH)_3$
  c. H_2CO_3 and $Mg(OH)_2$

10.56 Write a balanced equation for the neutralization of each of the following:
  a. H_3PO_4 and $NaOH$
  b. HI and $LiOH$
  c. HNO_3 and $Ca(OH)_2$

Everyday Neutralization Reactions

Baking soda, $NaHCO_3$, can also be used as an antacid. When $NaHCO_3$ reacts with acetic acid in vinegar, the products are $CO_2(g)$, H_2O, and CH_3COONa (salt).

$$NaHCO_3 + CH_3COOH \longrightarrow$$
$$CO_2(g) + H_2O + CH_3COONa$$

For this exploration, you will need baking soda, a glass, vinegar, and a pH indicator such as red cabbage juice (if available) or cranberry juice.

Place 1 tablespoon of baking soda in the glass. Add some pH indicator to give a color. Add one teaspoon of vinegar and observe the reaction. Continue to add more vinegar, one teaspoon at a time, and observe the reaction and color changes.

Try the neutralization using an antacid instead of baking soda. From the package label, identify the base(s) in the antacid product.

Questions

1. How does the pH change as you add more acid to the baking soda (base) or antacid (base)?
2. How do you know the base (baking soda or antacid) is being neutralized?
3. When do you know that the base has been neutralized?
4. What are the formulas of some bases that are typically used to prepare antacids?
5. What is the purpose of a base in an antacid?

Antacids

Antacids are substances used to neutralize excess stomach acid (HCl). Some antacids are mixtures of aluminum hydroxide and magnesium hydroxide. These hydroxides are not very soluble in water, so the levels of available OH^- are not damaging to the intestinal tract. However, aluminum hydroxide has the side effects of producing constipation and binding phosphate in the intestinal tract, which may cause weakness and loss of appetite. Magnesium hydroxide has a laxative effect. These side effects are less likely when a combination of the antacids is used.

$$Al(OH)_3 + 3HCl \longrightarrow AlCl_3 + 3H_2O$$
$$Mg(OH)_2 + 2HCl \longrightarrow MgCl_2 + 2H_2O$$

Some antacids use calcium carbonate to neutralize excess stomach acid. About 10% of the calcium is absorbed into the bloodstream, where it elevates the levels of serum calcium. Calcium carbonate is not recommended for patients who have peptic ulcers or a tendency to form kidney stones.

$$CaCO_3 + 2HCl \longrightarrow H_2O + CO_2(g) + CaCl_2$$

Still other antacids contain sodium bicarbonate. This type of antacid has a tendency to increase blood pH and elevate sodium levels in the body fluids. It also is not recommended in the treatment of peptic ulcers.

$$NaHCO_3 + HCl \longrightarrow NaCl + CO_2(g) + H_2O$$

The neutralizing substances in some antacid preparations are given in Table 10.8.

Table 10.8   Basic Compounds in Some Antacids

Antacid	Base(s)
Amphojel	$Al(OH)_3$
Milk of magnesia	$Mg(OH)_2$
Mylanta, Maalox, Di-Gel, Gelusil, Riopan	$Mg(OH)_2$, $Al(OH)_3$
Bisodol	$CaCO_3$, $Mg(OH)_2$
Titralac, Tums, Pepto-Bismol	$CaCO_3$
Alka-Seltzer	$NaHCO_3$, $KHCO_3$

10.8 Acidity of Salt Solutions

A **salt** is another term used for ionic compounds. When a salt dissolves in water, it separates into cations and anions. If one of the ions has an attraction for water, the pH changes to make an acidic or basic solution.

Salts That Form Neutral Solutions

Earlier we saw that when strong acids such as HCl and HNO_3 dissolve in water, they form weak conjugate bases Cl^- and NO_3^- along with H_3O^+. Therefore if a salt such as NaCl or $NaNO_3$ dissolves in water, the ions Cl^- and NO_3^- do not attract H^+ from water. The metal ions in the salt such as Na^+, which are from a strong base, do not produce H^+. Therefore, a solution of a salt containing the ions of a strong base and a strong acid will be neutral. For example, salts such as NaCl, KCl, $NaNO_3$, KNO_3, and KBr form solutions with pH of 7.

Salts That Form Basic Solutions

Some salts contain ions that are conjugate bases of weak acids. Suppose the salt NaF is dissolved in water.

$$NaF(s) \xrightarrow{\text{H}_2\text{O}} Na^+(aq) + F^-(aq)$$

The anion F^- is the conjugate base of HF, a weak acid. Therefore, F^- will act as a strong base by removing a proton from water. The hydroxide ion produced makes the salt solution basic. The metal ion Na^+ has no effect on the pH of the solution.

$$F^-(aq) + H_2O(l) \rightleftharpoons HF(aq) + OH^-(aq)$$

Therefore, salts containing ions of a weak acid and strong base produce basic solutions. Salts such as NaCN, KNO_2, and Na_2SO_4 will produce solutions with a pH greater than 7.

Salts That Form Acidic Solutions

Suppose the salt NH_4Cl is dissolved in water.

$$NH_4Cl(s) \xrightarrow{\text{H}_2\text{O}} NH_4^+(aq) + Cl^-(aq)$$

As a weak acid, the cation NH_4^+ donates a proton to water producing H_3O^+. The anion Cl^- has no effect on pH.

$$NH_4^+(aq) + H_2O(l) \rightleftharpoons H_3O^+ + NH_3(aq)$$

Therefore a salt containing the ions of a weak base and a strong acid will produce an acidic aqueous solution. Salts such as NH_4Br, NH_4NO_3, and $NaHSO_4$ form solutions with pH values less than 7.

Sometimes a salt may contain the ions of a weak acid and a weak base. For example, when NH_4F dissolves in water, the ions NH_4^+ and F^- will both affect the pH of the solution. Then, it is the stronger conjugate acid or base that determines whether the solution is acidic or basic. The salt solution would be neutral only if their conjugate strengths were equal.

Table 10.9   Properties of Some Common Ions in Water

	Species Producing Ion	pH Effect	Cations	Anions
Neutral	Strong acid; strong base	None	Li^+, Na^+, K^+ Mg^{2+}, Ca^{2+}, Ba^{2+}	Cl^-, Br^-, I^- NO_3^-, ClO_4^-
Basic	Weak acid; strong base	>7	None	CH_3COO^-, F^-, CO_3^{2-}, S^{2-}, PO_4^{3-}, SO_4^{2-}, CN^-, NO_2^-, HCO_3^-, HS^-
Acidic	Strong acid; weak base	<7	NH_4^+	HSO_4^-

Predicting the Acidity of Salt Solutions

Predict whether solutions of the following salts would be acidic, basic, or neutral:

a. KCN   **b.** NH_4Br **c.** $NaNO_3$

Solution

a. This is a salt of a weak acid and a strong base. The cation K^+ has no effect on pH. The anion CN^- is the conjugate base of the weak acid HCN. The CN^- will attract protons from water making a basic solution.

b. This is a salt of a strong acid and a weak base. The Br^- will not react with water, but the NH_4^+, which is a weak acid, will donate a proton to water to make an acidic solution.

c. $NaNO_3$ is a salt of a strong acid and a strong base. The solution will remain neutral.

Study Check

Will a Na_3PO_4 solution be acidic, basic, or neutral?

QUESTIONS AND PROBLEMS

Acidity of Salt Solutions

10.57 Why does a salt containing a cation from a strong base and an anion from a weak acid form a basic solution?

10.58 Why does a salt containing a cation from a weak base and an anion from a strong acid form an acidic solution?

10.59 Predict whether each of the following salts will form an acidic, basic, or neutral solution. For acidic and basic solutions, write an equation for the reaction that takes place.
a. $MgCl_2$ **b.** NH_4NO_3
c. Na_2CO_3 **d.** K_2S

10.60 Predict whether each of the following salts will form an acidic, basic, or neutral solution. For acidic and basic solutions, write an equation for the reaction that takes place.
a. Na_2SO_4 **b.** KBr
c. $BaCl_2$ **d.** NH_4I

10.9 Buffers

When a small amount of acid or base is added to pure water, the pH changes drastically. However, if a solution is buffered, there is little change in pH. A **buffer solution** resists a change in pH when small amounts of acid or base are added. For example, blood is buffered to maintain a pH of about 7.4. If the pH of the blood goes even slightly above or below this value, changes in our uptake of oxygen and our metabolic process can be drastic enough to cause death. Even though we are

LEARNING GOAL

Describe the role of buffers in maintaining the pH of a solution.

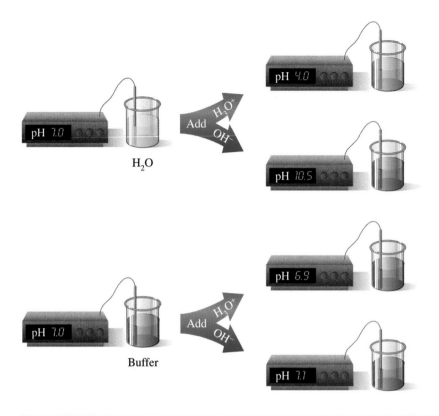

Figure 10.6 Adding an acid or a base to water changes the pH drastically, but a buffer resists pH change when small amounts of acid or base are added.

Q Why does the pH change several pH units when acid is added to water, but not when acid is added to a buffer?

WEB TUTORIAL
pH and Buffers

constantly obtaining acids and bases from foods and biological processes, the buffers in the body so effectively absorb those compounds that blood pH remains essentially unchanged. (See Figure 10.6.)

In a buffer, an acid must be present to react with any OH^- that is added, and a base must be available to react with any added H_3O^+. However, that acid and base must not be able to neutralize each other. Therefore, a combination of an acid–base conjugate pair is used in buffers. Most buffer solutions consist of nearly equal concentrations of a weak acid and a salt containing its conjugate base. (See Figure 10.6.) Buffers may also contain a weak base and the salt of the weak base, which contains its conjugate acid.

For example, a typical buffer contains acetic acid (CH_3COOH) and a salt such as sodium acetate (CH_3COONa). As a weak acid, acetic acid dissociates slightly in water to form H_3O^+ and a very small amount of CH_3COO^-. The presence of the salt provides a much larger concentration of acetate ion (CH_3COO^-), which is necessary for its buffering capability.

$$CH_3COOH(aq) + H_2O(l) \rightleftharpoons H_3O^+(aq) + CH_3COO^-(aq)$$
Large amount   Large amount

Let's see how this buffer solution maintains the H_3O^+ concentration. When a small amount of acid is added, it will combine with the acetate ion (anion) as the equilibrium shifts to the reactant acetic acid. There will be a small decrease in the $[CH_3COO^-]$ and a small increase in $[CH_3COOH]$, but the $[H_3O^+]$ will not change very much.

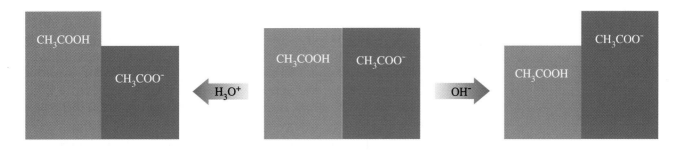

Fig 10.7 The buffer described here consists of about equal concentrations of acetic acid (CH₃COOH) and its conjugate base of acetate ion (CH₃COO⁻). Adding H₃O⁺ to the buffer uses up some CH₃COO⁻, whereas adding OH⁻ neutralizes some CH₃COOH. The pH of the solution is maintained as long as the added amounts of acid or base are small compared to the concentrations of the buffer components.

Q How does this acetic acid/acetate ion buffer maintain pH?

$$CH_3COOH(aq) \ + \ H_2O(l) \ \longleftarrow \ H_3O^+(aq) \ + \ CH_3COO^-(aq)$$

If a small amount of base is added to this buffer solution, it is neutralized by the acetic acid and water is produced. The concentration of [CH₃COOH] decreases slightly and the [CH₃COO⁻] increases slightly, but again the [H₃O⁺] does not change very much. (See Figure 10.7.)

$$CH_3COOH(aq) \ + \ OH^-(aq) \ \longrightarrow \ H_2O(aq) \ + \ CH_3COO^-(aq)$$

SAMPLE PROBLEM 10.16

Identifying Buffer Solutions

Indicate whether each of the following would make a buffer solution:

a. HCl, a strong acid, and NaCl
b. H₃PO₄, a weak acid
c. HF, a weak acid, and NaF

Solution

a. No. A solution of a strong acid and its salt is completely ionized.
b. No. A weak acid is not sufficient for a buffer; the salt of the weak acid is also needed.
c. Yes. This mixture contains a weak acid and its salt.

Study Check

Will a mixture of NaCl and Na₂CO₃ make a buffer solution? Explain.

Calculating the pH of a Buffer

By rearranging the K_a expression to give [H₃O⁺] we can obtain the ratio of the acetic acid/acetate buffer.

$$K_a = \frac{[H_3O^+][CH_3COO^-]}{[CH_3COOH]}$$

$$[H_3O^+] = K_a \times \frac{[CH_3COOH]}{[CH_3COO^-]}$$

Because K_a is a constant, the $[H_3O^+]$ is determined by the $[CH_3COOH]/$ $[CH_3COO^-]$ ratio. As long as the addition of small amounts of either acid or base changes the ratio of $[CH_3COOH]/[CH_3COO^-]$ only slightly, the changes in $[H_3O^+]$ will be small and the pH will be maintained. It is important to note that the amount of acid or base that is added must be small compared to the supply of the buffer components CH_3COOH and CH_3COO^-. However, if a large amount of acid or base is added, the buffering capacity of the system may be exceeded. Buffers are also prepared from other conjugate acid–base pairs such as $H_2PO_4^-/HPO_4^{2-}$, $HPO_4^{2-}/$ PO_4^{3-}, HCO_3^-/CO_3^{2-}, NH_4^+/NH_3.

SAMPLE PROBLEM 10.17

pH of a Buffer

The K_a for acetic acid, CH_3COOH, is 1.8×10^{-5}. What is the pH of a buffer prepared with 1.0 M CH_3COOH and 1.0 M CH_3COO^-?

Solution

CASE STUDY
Hyperventilation
and Blood pH

$$CH_3COOH + H_2O \rightleftharpoons H_3O^+ + CH_3COO^-$$

$$K_a = \frac{[H_3O^+][CH_3COO^-]}{[CH_3COOH]}$$

$$[H_3O^+] = K_a \times \frac{[CH_3COOH]}{[CH_3COO^-]}$$

Substituting these values in the expression for $[H_3O^+]$ gives

$$[H_3O^+] = 1.8 \times 10^{-5} \times \frac{[1.0\ M]}{[1.0\ M]}$$

$$[H_3O^+] = 1.8 \times 10^{-5}$$

Using the concentration of $[H_3O^+]$ in the pH expression gives the pH of the buffer.

$$pH = -\log[1.8 \times 10^{-5}] = 4.74$$

Study Check

One of the conjugate acid–base pairs that buffers the blood is $H_2PO_4^-/HPO_4^{2-}$ with a K_a of 6.2×10^{-8}. What is the pH of a buffer that is 0.50 M $H_2PO_4^-$ and 0.50 M HPO_4^{2-}?

QUESTIONS AND PROBLEMS

Buffers

10.61 Which of the following represent a buffer system? Explain.
    a. NaOH and NaCl   **b.** H_2CO_3 and $NaHCO_3$
    c. HF and KF   **d.** KCl and NaCl

10.62 Which of the following represent a buffer system? Explain.
    a. H_3PO_4 **b.** $NaNO_3$
    c. CH_3COOH and CH_3COONa **d.** HCl and NaOH

10.63 Consider the buffer system of hydrofluoric acid, HF, and its salt, NaF.

$$HF \rightleftharpoons H^+ + F^-$$

HEALTH NOTE

Buffers in the Blood

The arterial blood has a normal pH of 7.35–7.45. If changes in H_3O^+ lower the pH below 6.8 or raise it above 8.0, cells cannot function properly and death may result. In our cells, CO_2 is continually produced as an end product of cellular metabolism. Some CO_2 is carried to the lungs for elimination, and the rest dissolves in body fluids such as plasma and saliva, forming carbonic acid. As a weak acid, carbonic acid dissociates to give bicarbonate and H_3O^+. More of the anion HCO_3^- is supplied by the kidneys to give an important buffer system in the body fluid: the H_2CO_3/HCO_3^- buffer.

$$CO_2 + H_2O \underset{}{\overset{H_2O}{\rightleftharpoons}} H_2CO_3 \rightleftharpoons H_3O^+ + HCO_3^-$$

Excess H_3O^+ entering the body fluids reacts with the HCO_3^- and excess OH^- reacts with the carbonic acid.

$$H_2CO_3 + H_2O \longleftarrow H_3O^+ + HCO_3^- \quad \text{Equilibrium shifts left}$$
$$H_2CO_3 + OH^- \longrightarrow H_2O + HCO_3^- \quad \text{Equilibrium shifts right}$$

For the carbonic acid, we can write the equilibrium expression as

$$K_a = \frac{[H_3O^+][HCO_3^-]}{[H_2CO_3]}$$

To maintain the normal blood pH (7.35–7.45), the ratio of H_2CO_3/HCO_3^- needs to be about 1 to 10, which is obtained by typical concentrations in the blood of 0.0024 M H_2CO_3 and 0.024 M HCO_3^-.

$$[H_3O^+] = K_a \times \frac{[H_2CO_3]}{[HCO_3^-]}$$

$$= 4.3 \times 10^{-7} \times \frac{[0.0024]}{[0.024]} = 4.3 \times 10^{-7} \times 0.10 = 4.3 \times 10^{-8}$$

$$pH = -\log(4.3 \times 10^{-8}) = 7.37$$

In the body, the concentration of carbonic acid is closely associated with the partial pressure of CO_2. Table 10.10 lists the normal values for arterial blood. If the CO_2 level rises, producing more H_2CO_3, the equilibrium produces more H_3O^+, which lowers the pH. This condition is called **acidosis**. Difficulty with ventilation or gas diffusion can lead to respiratory acidosis, which can happen in emphysema or when the medulla of the brain is affected by an accident or depressive drugs.

A lowering of the CO_2 level leads to a high blood pH, a condition called **alkalosis**. Excitement, trauma, or a high temperature may cause a person to hyperventilate, which expels large amounts of CO_2. As the partial pressure of CO_2 in the blood falls below normal, the equilibrium shifts from H_2CO_3 to CO_2 and H_2O. This shift decreases the $[H_3O^+]$ and raises the pH. Table 10.11 lists some of the conditions that lead to changes in the blood pH and some possible treatments. The kidneys also regulate H_3O^+ and HCO_3^- components, but more slowly than the adjustment made by the lungs through ventilation.

Table 10.10   Normal Values for Blood Buffer in Arterial Blood

P_{CO_2}	40 mm Hg
H_2CO_3	2.4 mmoles/L of plasma
HCO_3^-	24 mmoles/L of plasma
pH	7.35–7.45

Table 10.11   Acidosis and Alkalosis: Symptoms, Causes, and Treatments

Respiratory Acidosis: CO_2 ↑ pH ↓

Symptoms:	Failure to ventilate, suppression of breathing, disorientation, weakness, coma
Causes:	Lung disease blocking gas diffusion (e.g., emphysema, pneumonia, bronchitis, and asthma); depression of respiratory center by drugs, cardiopulmonary arrest, stroke, poliomyelitis, or nervous system disorders
Treatment:	Correction of disorder, infusion of bicarbonate

Respiratory Alkalosis: CO_2 ↓ pH ↑

Symptoms:	Increased rate and depth of breathing, numbness, light-headedness, tetany
Causes:	Hyperventilation due to anxiety, hysteria, fever, exercise; reaction to drugs such as salicylate, quinine, and antihistamines; conditions causing hypoxia (e.g., pneumonia, pulmonary edema, and heart disease)
Treatment:	Elimination of anxiety-producing state, rebreathing into a paper bag

Metabolic Acidosis: H^+ ↑ pH ↓

Symptoms:	Increased ventilation, fatigue, confusion
Causes:	Renal disease, including hepatitis and cirrhosis; increased acid production in diabetes mellitus, hyperthyroidism, alcoholism, and starvation; loss of alkali in diarrhea; acid retention in renal failure
Treatment:	Sodium bicarbonate given orally, dialysis for renal failure, insulin treatment for diabetic ketosis

Metabolic Alkalosis: H^+ ↓ pH ↑

Symptoms:	Depressed breathing, apathy, confusion
Causes:	Vomiting, diseases of the adrenal glands, ingestion of excess alkali
Treatment:	Infusion of saline solution, treatment of underlying diseases

a. What is the purpose of the buffer system?

b. Why is a salt of the acid needed?

c. How does the buffer react when some H^+ is added?

d. How does the buffer react when some OH^- is added?

10.64 Consider the buffer system of nitrous acid, HNO_2, and its salt, $NaNO_2$.

$$HNO_2 \rightleftharpoons H^+ + NO_2^-$$

a. What is the purpose of a buffer system?

b. What is the purpose of $NaNO_2$ in the buffer?

c. How does the buffer react when some H^+ is added?

d. How does the buffer react when some OH^- is added?

10.65 Nitrous acid has a K_a of 4.5×10^{-4}. What is the pH of a buffer solution containing 0.10 M HNO_2 and 0.10 M NO_2^-?

10.66 Acetic acid has a K_a of 1.8×10^{-5}. What is the pH of a buffer solution containing 0.15 M CH_3COOH (acetic acid) and 0.15 M CH_3COO^-?

10.67 Compare the pH of a HF buffer that contains 0.10 M HF and 0.10 M NaF with another HF buffer that contains 0.060 M HF and 0.120 M NaF. See Table 10.4.

10.68 Compare the pH of a H_2CO_3 buffer that contains 0.10 M H_2CO_3 and 0.10 M $NaHCO_3$ with another H_2CO_3 buffer that contains 0.15 M H_2CO_3 and 0.050 M $NaHCO_3$. See Table 10.4.

10.10 Dilutions

LEARNING GOAL

Describe the preparation of a dilute solution of an acid or base given the molarity or percent concentration of the solution.

In chemistry, we often need to prepare a dilute solution from a more concentrated solution. In a process called **dilution,** we add water to a solution to make a larger volume. For example, you might prepare some orange juice by adding three cans of water to the original can of concentrated orange juice.

When more water is added, the solution volume increases, causing a decrease in the concentration. However, *the amount of solute does not change.* Therefore the initial amount of solute before dilution is equal to the final amount of solute in the diluted solution. (See Figure 10.8.) By diluting the solution, we decrease the concentration.

Amount of solute in the concentrated solution = amount of solute in the diluted solution

The amount of the solute depends on the concentration, C, and the volume, V.

$$C_1V_1 = C_2V_2$$

We have seen that the concentration of a solution can be expressed in % or molarity, M.

$$\%_1V_1 = \%_2V_2$$

or

$$M_1V_1 = M_2V_2$$

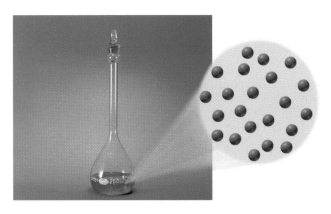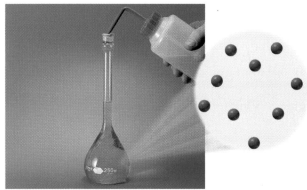

Figure10.8 When water is added to a concentrated solution, there is no change in the number of particles, but the solute particles can spread out as the volume of the diluted solution increases.

Q What is the concentration of the diluted solution after an equal volume of water is added to a sample of 6 M HCl?

Diluting a Base

What is the new % (m/v) concentration when water is added to 75 mL of 20.% (m/v) KOH to make 500. mL of a diluted KOH solution?

Solution

Solving for the new concentration in the expression

$$\%_1V_1 = \%_2V_2$$

gives

$$\%_2 = \frac{\%_1V_1}{V_2}$$

$$\%_2 = \frac{20.\% \times 75 \text{ mL}}{500. \text{ mL}} = 3.0\% \text{ (m/v) KOH (diluted)}$$

Study Check

You need to prepare 200 mL of 2.0% (m/v) NaOH solution from a 10.0% (m/v) NaOH solution. What volume of 10.0% NaOH is needed?

Diluting an Acid

What volume (mL) of a 0.20 M HCl solution can be prepared by diluting 50.0 mL of a 1.0 M HCl solution?

Solution

Solving for the new volume in the expression

$$M_1V_1 = M_2V_2$$

gives

$$V_2 = \frac{M_1 V_1}{M_2}$$

$$V_2 = \frac{(1.0 \text{ M})(50.0 \text{ mL})}{0.20 \text{ M}}$$

$$= 250 \text{ mL}$$

Study Check

What volume (mL) of 8.0 M HCl must be placed in a volumetric flask to prepare 1.0 L of 0.50 M HCl?

QUESTIONS AND PROBLEMS

Diluting Acids and Bases

10.69 To make tomato soup, you add one can of water to the condensed soup. Why is this a dilution?

10.70 A can of frozen lemonade calls for the addition of three cans of water to make a pitcher of the beverage. Why is this a dilution?

10.71 Calculate the final concentration of each of the following:
  a. Water is added to 25.0 mL of a 6.0 M HCl solution until the volume is exactly 200 mL.
  b. A 10.0-mL sample of 25% (m/v) KOH solution is diluted with water to give a final volume of exactly 500 mL.

10.72 Calculate the final concentration of each of the following:
  a. Water is added to 5.0 mL of a 3.5 M H_2SO_4 solution until the volume is exactly 250 mL.
  b. A 5.0-mL sample of 50.0% (m/v) acetic acid (CH_3COOH) solution is diluted with water to give a final volume of exactly 100 mL.

10.73 Determine the volume (mL) of acid needed to prepare the following:
  a. 250 mL of a 0.200 M HNO_3 solution using a 4.00 M HNO_3 solution.
  b. 750 mL of a 0.100 M H_3PO_4 solution using a 6.0 M H_3PO_4 solution.

10.74 Determine the volume of base (mL) needed to prepare the following:
  a. 10.0 mL of a 0.050 M $Ba(OH)_2$ solution using a 0.20 M $Ba(OH)_2$ solution.
  b. 500 mL of a 0.20% NaOH solution using a 10.0% NaOH solution.

10.75 You need to dilute 25.0 mL of a 3.0 M HCl solution to 0.15 M HCl. How many mL of water should you add?

10.76 If you are diluting 40.0 mL of a 2.5 M HBr solution to 0.50 M HBr, how many mL of water will you need to add?

LEARNING GOAL

Calculate the molarity or volume of an acid or base from titration information.

10.11 Acid–Base Titration

Suppose we need to find the molarity of an HCl solution of unknown concentration. We can do this by a laboratory procedure called **titration** in which we neutralize an acid sample with a known amount of base. In our titration, we first place a

Figure 10.9 The titration of an acid. A known volume of an acid is placed in a flask with an indicator and titrated with a measured volume of NaOH to the neutralization point.

Q What data is needed to determine the molarity of the acid in the flask?

measured volume of the acid in a flask and add a few drops of an **indicator** such as phenolphthalein. In an acidic solution, phenolpthalein is colorless. Then we fill a buret with a NaOH solution of known molarity and carefully add NaOH to the acid in the flask, as shown in Figure 10.9.

In the titration, we neutralize the acid by adding a volume of base that contains a matching number of moles of OH^-. We know that neutralization has taken place when the phenolphthalein in the solution changes from colorless to pink. This is called the neutralization endpoint. From the volume added and molarity of the NaOH, we can calculate the number of moles of NaOH and then the concentration of the acid.

SAMPLE PROBLEM 10.20

Titration of an Acid

A 10.0-mL sample of HCl is titrated with 1.0 M NaOH. The equation for the reaction is

$$NaOH + HCl \longrightarrow NaCl + H_2O$$

If 20.0 mL of base were required to neutralize the HCl, what is the molarity of the acid?

Solution

Step 1 Finding the Reacting Moles of Base

From the volume and the molarity of the NaOH, calculate the moles of NaOH used in the titration.

$$20.0 \text{ mL} \times \frac{1 \text{ L}}{1000 \text{ mL}} \times \frac{1.0 \text{ mole NaOH}}{1 \text{ L}} = 0.020 \text{ mole NaOH}$$
Molarity of NaOH

Step 2 Finding the Reacting Moles of Acid

In the neutralization equation, 1 mole of NaOH reacts with 1 mole of HCl. The moles of HCl are calculated as follows:

$$0.020 \text{ mole NaOH} \times \frac{1 \text{ mole HCl}}{1 \text{ mole NaOH}} = 0.020 \text{ mole HCl}$$

Step 3   Calculating the Molarity of the Acid

To calculate the molarity of the acid, place the moles of acid and its initial volume in the expression for molarity.

$$\text{Molarity (M) of acid} = \frac{\text{moles of acid}}{\text{volume of acid}} = \frac{0.020 \text{ mole HCl}}{0.0100 \text{ L}} = 2.0 \text{ M HCl}$$

10.0-mL HCl sample
expressed in liters

Study Check

In a titration, a 25.0-mL HCl sample required 50.0 mL of a 0.100 M NaOH solution for neutralization. What is the molarity of the HCl sample?

SAMPLE PROBLEM 10.21

Titration of an Acid

A 5.00-mL sample of H_2SO_4 requires 40.0 mL of 2.0 M NaOH for complete neutralization. What is the molarity of the acid? The equation for the neutralization is

$$H_2SO_4 + 2NaOH \longrightarrow Na_2SO_4 + 2H_2O$$

Solution

Using the steps for titration calculations we find the following:

Step 1   From the volume and the molarity of the NaOH, calculate the moles of NaOH.

$$40.0 \text{ mL} \times \frac{1 \text{ L}}{1000 \text{ mL}} \times \frac{2.0 \text{ moles NaOH}}{1 \text{ L}} = 0.080 \text{ mole NaOH}$$

Molarity of NaOH

Step 2   From the neutralization equation, we know that 2 moles of NaOH react with 1 mole of H_2SO_4. The moles of H_2SO_4 in the sample are calculated as follows:

$$0.080 \text{ mole NaOH} \times \frac{1 \text{ mole H}_2\text{SO}_4}{2 \text{ moles NaOH}} = 0.040 \text{ mole H}_2\text{SO}_4$$

Step 3   To calculate the molarity of the acid, place the number of moles of acid and its initial volume in the expression for molarity.

$$\text{Molarity (M) of acid} = \frac{\text{moles of acid}}{\text{volume of acid}} = \frac{0.040 \text{ mole H}_2\text{SO}_4}{0.0050 \text{ L}} = 8.0 \text{ M H}_2\text{SO}_4$$

5.0-mL H_2SO_4 sample
expressed in liters

Study Check

What is the volume (mL) of 0.100 M NaOH needed to completely neutralize 5.00 mL of 0.200 M H_3PO_4? (*Hint:* Write the balanced equation for neutralization.)

QUESTIONS AND PROBLEMS

Acid–Base Titration

10.77 If you need to determine the concentration of a solution of formic acid, HCOOH, how would you proceed?

10.78 If you need to determine the concentration of a sample of acetic acid, CH₃COOH, how would you proceed?

10.79 What is the molarity of the acid in each of the following titrations?

a. a 5.0-mL sample of HCl solution that is titrated with 22.0 mL of 2.0 M NaOH solution

$$HCl + NaOH \longrightarrow NaCl + H_2O$$

b. a 10.0-mL sample of sulfuric acid solution that is titrated with 15.0 mL of 1.0 M NaOH solution

$$H_2SO_4 + 2NaOH \longrightarrow Na_2SO_4 + 2H_2O$$

c. a 10.0-mL sample of phosphoric acid solution that is titrated with 16.0 mL of 1.0 M NaOH solution

$$H_3PO_4 + 3NaOH \longrightarrow Na_3PO_4 + 3H_2O$$

10.80 What is the molarity of the acid in each of the following titrations?

a. a 20.0-mL sample of HCl solution that is titrated with 8.0 mL of 6.0 M NaOH solution

$$HCl + NaOH \longrightarrow NaCl + H_2O$$

b. a 10.0-mL sample of sulfuric acid solution that is titrated with 25.0 mL of 0.50 M NaOH solution

$$H_2SO_4 + 2NaOH \longrightarrow Na_2SO_4 + 2H_2O$$

c. a 30.0-mL sample of phosphoric acid solution that is titrated with 18.0 mL of 3.0 M NaOH solution

$$H_3PO_4 + 3NaOH \longrightarrow Na_3PO_4 + 3H_2O$$

Chapter Review

10.1 Acids and Bases

According to the Arrhenius theory, an acid produces H^+, and a base produces OH^- in aqueous solutions. According to the Brønsted–Lowry theory, acids are proton (H^+) donors, and bases are proton acceptors.

10.2 Conjugate Acid–Base Pairs

Two conjugate acid–base pairs are present in an acid–base reaction. Each acid–base pair is related by the loss or gain of one H^+. For example, when the acid HF donates a H^+, the F^- it forms is its conjugate base because F^- is capable of accepting a H^+. The other acid–base pair would be H_2O and H_3O^+.

10.3 Strengths of Acids and Bases

In strong acids, all the H^+ in the acid is donated to H_2O; in a weak acid, only a small percentage of acid molecules produce H_3O^+. Strong bases are hydroxides of Groups 1A and 2A that dissociate completely in water. An important weak base is ammonia, NH_3.

10.4 Dissociation Constants

In water, weak acids and weak bases produce only a few ions when equilibrium is reached. The reaction for a weak acid can be written as $HA + H_2O \rightleftharpoons H_3O^+ + A^-$. The acid dissociation product is written as $K_a = \dfrac{[H_3O^+][A^-]}{[HA]}$.

For a weak base, $B + H_2O \rightleftarrows BH^+ + OH^-$, the base dissociation constant K_b is written as $K_b = \dfrac{[BH^+][OH^-]}{[B]}$.

10.5 Ionization of Water

In pure water, a few molecules transfer protons to other water molecules, producing small, but equal, amounts of $[H_3O^+]$ and $[OH^-]$ such that each has a concentration of 1×10^{-7} mole/L. The ion product, K_w, $[H_3O^+][OH^-] = 1 \times 10^{-14}$, applies to all aqueous solutions. In acidic solutions, the $[H_3O^+]$ is greater than the $[OH^-]$. In basic solutions, the $[OH^-]$ is greater than the $[H_3O^+]$.

10.6 The pH Scale

The pH scale is a range of numbers from 0 to 14 related to the $[H_3O^+]$ of the solution. A neutral solution has a pH of 7. In acidic solutions, the pH is below 7, and in basic solutions the pH is above 7. Mathematically, pH is the negative logarithm of the hydronium ion concentration ($-\log[H_3O^+]$).

10.7 Reactions of Acids and Bases

When an acid reacts with a metal, hydrogen gas and a salt are produced. The reaction of an acid with a carbonate or bicarbonate produces carbon dioxide, a salt, and water. In neutralization, an acid reacts with a base to produce a salt and water.

10.8 Acidity of Salt Solutions

A salt of a weak acid contains an anion that removes protons from water and makes the solution basic. A salt of a weak base contains an ion that donates a proton to water producing an acidic solution. Salts of strong acids and bases produce neutral solutions because they contain ions that do not affect the pH.

10.9 Buffers

A buffer solution resists changes in pH when small amounts of acid or a base are added. A buffer contains either a weak acid and its salt or a weak base and its salt. The weak acid picks up added OH^-, and the anion of the salt picks up added H^+. Buffers are important in maintaining the pH of the blood.

10.10 Dilutions

When water is added to a solution, the volume increases. The solute is now distributed throughout a larger volume, which dilutes the solution and decreases the concentration.

10.11 Acid–Base Titration

In a laboratory procedure called a titration, an acid sample is neutralized with a known amount of a base. From the volume and molarity of the base, the concentration of the acid is calculated.

Key Terms

acid A substance that dissolves in water and produces hydrogen ions (H^+), according to the Arrhenius theory. All acids are proton donors, according to the Brønsted–Lowry theory.

acid dissociation constant (K_a) The product of the ions from the dissociation of a weak acid divided by the concentration of the weak acid.

acidosis A physiological condition in which the blood pH is lower than 7.35.

alkalosis A physiological condition in which the blood pH is higher than 7.45.

base A substance that dissolves in water and produces hydroxide ions (OH^-) according to the Arrhenius theory. All bases are proton acceptors, according to the Brønsted–Lowry theory.

base dissociation constant (K_b) The product of the ions from the dissociation of a weak base divided by the concentration of the weak base.

Brønsted–Lowry acids and bases An acid is a proton donor, and a base is a proton acceptor.

buffer solution A mixture of a weak acid or a weak base and its salt that resists changes in pH when small amounts of an acid or base are added.

conjugate acid–base pair An acid and base that differ by one H^+. When an acid donates a proton, the product is its conjugate base, which is capable of accepting a proton in the reverse reaction.

dilution A process by which water (solvent) is added to a solution to increase the volume and decrease (dilute) the concentration of the solute.

hydronium ion, H_3O^+ The H_3O^+ ion formed by the attraction of a proton (H^+) to an H_2O molecule.

indicator A substance added to a titration sample that changes color when the acid or base is neutralized.

ion-product constant of water, K_w The product of $[H_3O^+]$ and $[OH^-]$ in solution; $K_w = [H_3O^+][OH^-]$.

neutral A solution with equal concentrations of $[H_3O^+]$ and $[OH^-]$.

neutralization A reaction between an acid and a base to form a salt and water.

pH A measure of the $[H_3O^+]$ in a solution; pH = $-\log[H_3O^+]$.

pOH A measure of the $[OH^-]$ in a solution; pOH = $-\log[OH^-]$.

salt An ionic compound that contains a metal ion and a nonmetal or polyatomic ion.

strong acid An acid that completely ionizes in water.

strong base A base that completely ionizes in water.

titration The addition of base to an acid sample to determine the concentration of the acid.

weak acid An acid that ionizes only slightly in solution.

weak base A base that ionizes only slightly in solution.

Additional Problems

10.81 Name each of the following:
  a. H_2SO_4 **b.** KOH   **c.** $Ca(OH)_2$
  d. HCl   **e.** HNO_2

10.82 Are the following examples of body fluids acidic, basic, or neutral?
  a. saliva, pH 6.8   **b.** urine, pH 5.9
  c. pancreatic juice, pH 8.0   **d.** bile, pH 8.4
  e. blood, pH 7.45

10.83 What are some similarities and differences between strong and weak acids?

10.84 What are some ingredients found in antacids? What do they do?

10.85 One ingredient in some antacids is $Mg(OH)_2$.
  a. If the base is not very soluble in water, why is it considered a strong base?
  b. What is the neutralization reaction of $Mg(OH)_2$ with stomach acid, HCl?

10.86 Acetic acid, which is the acid in vinegar, is a weak acid. Why?

10.87 Using Table 10.3, determine which is the stronger acid in each of the following pairs:
  a. HF or HCN   **b.** H_3O^+ or NH_4^+
  c. HNO_2 or CH_3COOH **d.** H_2O or HCO_3^-

10.88 Using Table 10.3, determine which is the stronger base in each of the following pairs:
  a. H_2O or Cl^- **b.** OH^- or NH_3
  c. SO_4^{2-} or NO_2^- **d.** CO_3^{2-} or H_2O

10.89 Determine the pH and pOH for the following solutions:
  a. $[H_3O^+] = 2.0 \times 10^{-8}$ M
  b. $[H_3O^+] = 5.0 \times 10^{-2}$ M
  c. $[OH^-] = 3.5 \times 10^{-4}$ M
  d. $[OH^-] = 0.005$ M

10.90 Are the solutions in problem 10.89 acidic, basic, or neutral?

10.91 What are the $[H_3O^+]$ and $[OH^-]$ for a solution with the following pH values?
  a. 3.0   **b.** 6.0   **c.** 8.0   **d.** 11.0

10.92 Solution A has a pH of 4.0 and solution B has a pH of 6.0.
  a. Which solution is more acidic?
  b. What is the $[H_3O^+]$ in each?
  c. What is the $[OH^-]$ in each?

10.93 What is the value of $[OH^-]$ in a solution that contains 0.20 g of NaOH in 0.25 L of solution?

10.94 What is the value of $[H_3O^+]$ in a solution that contains 1.5 g of HNO_3 in 0.50 L of solution?

10.95 What is the pH and pOH of a solution prepared by dissolving 2.5 g HCl in water to make 425 mL of solution?

10.96 What is the pH and pOH of a solution prepared by dissolving 1.0 g $Ca(OH)_2$ in water to make 875 mL of solution?

10.97 Will solutions of the following salts be acidic, basic, or neutral?
  a. KF   **b.** NaCN
  c. NH_4NO_3 **d.** NaBr

10.98 Will solutions of the following salts be acidic, basic, or neutral?
  a. K_2SO_4 **b.** KNO_2
  c. $MgCl_2$ **d.** NH_4Cl

10.99 A buffer is made by dissolving H_3PO_4 and NaH_2PO_4 in water.
  a. Write an equation that shows how this buffer neutralizes small amounts of acids.
  b. Write an equation that shows how this buffer neutralizes small amounts of base.
  c. Calculate the pH of this buffer if it is 0.10 M H_3PO_4 and 0.10 M $H_2PO_4^{2-}$; the K_a for H_3PO_4 is 7.5×10^{-3}.

10.100 A buffer is made by dissolving CH_3COOH and CH_3COONa in water.

    a. Write an equation that shows how this buffer neutralizes small amounts of acid.

    b. Write an equation that shows how this buffer neutralizes small amounts of base.

    c. Calculate the pH of this buffer if it is 0.10 M CH_3COOH and 0.10 M CH_3COO^-; the K_a for CH_3COOH is 1.8×10^{-5}.

10.101 A person with untreated diabetes has an increase in acidic compounds in the blood. When admitted, this person has an increase in pulse and respiratory rates.

    a. Why does the body respond by hyperventilating?

    b. What treatment might be used to correct the condition?

10.102 In the blood plasma, pH is maintained by the carbonic acid–bicarbonate buffer system.

    a. How is pH maintained when acid is added to the buffer system?

    b. How is pH maintained when base is added to the buffer system?

10.103 Calculate the volume (mL) of a 0.50 M NaOH solution that will completely neutralize the following:

    a. 25.0 mL of a 0.10 M HCl solution

    b. 10.0 mL of a 2.0 M H_2SO_4 solution

10.104 How many milliliters of 1.0 M NaOH solution are needed to completely neutralize 25.0 mL of 2.0 M H_2SO_4 solution?

10.105 Calculate the volume (mL) of a 2.0 M HCl solution that will completely neutralize the following:

    a. 35.0 mL of a 0.15 M $Ba(OH)_2$ solution

    b. 10.0 mL of a 2.0 M KOH solution

10.106 A 10.0 mL sample of vinegar, which is an aqueous solution of acetic acid, CH_3COOH, requires 16.5 mL of 0.500 M NaOH to reach the endpoint in a titration. What is the molarity of the acetic acid solution?

11 Introduction to Organic Chemistry

"The purpose of our research was to create a way to make Taxol," says Paul Wender, Francis W. Bergstrom Professor of organic chemistry and head of the Wender research group at Stanford University. "Taxol is a chemotherapy drug originally derived from the bark of the Pacific yew tree. However, removing the bark from yew trees destroys them, so we need a renewable resource. We worked out a synthesis that began with turpentine, which is both renewable and inexpensive. Initially, Taxol was used with patients who did not respond to chemotherapy. The first person to be treated was a woman who was diagnosed with terminal ovarian cancer and given three to six months to live. After a few treatments with Taxol, she was declared 98% disease–free. A drug like Taxol can save many lives, which is one reason that a study of organic chemistry is so important."

LOOKING AHEAD

the **Chemistry** place

www.chemplace.com/college

Visit the URL above or use the CD-ROM in the book for extra quizzing, interactive tutorials, career resources, and case studies.

Organic chemistry is the chemistry of carbon compounds. The element carbon has a special role in chemistry because it bonds with other carbon atoms to give a vast array of molecules. The variety of molecules is so great that we find organic compounds in many common products we use such as gasoline, medicine, shampoos, plastic bottles, and perfumes. The food we eat is composed of different organic compounds that supply us with fuel for energy and the carbon atoms needed to build and repair the cells of our bodies. Large organic molecules make up the proteins in hair and skin, the lipids in cell membranes and adipose tissue, and the DNA in cell nuclei.

Although many organic compounds occur in nature, chemists have synthesized even more. The cotton, wool, or silk in your clothes contain naturally occurring organic compounds, whereas materials such as polyester, nylon, or plastic have been synthesized through organic reactions. Sometimes it is convenient to synthesize a molecule in the lab even though that molecule is also found in nature. For example, vitamin C synthesized in a laboratory has the same structure as the vitamin C in oranges or lemons. In these next chapters, you will learn about the structures and reactions of organic molecules, which will provide a foundation for understanding the more complex molecules of biochemistry.

LEARNING GOAL

Identify the number of bonds for carbon and other atoms in organic molecules.

WEB TUTORIAL
Introduction to Organic Molecules
Organic Molecules and Isomers

11.1 Organic Compounds

In Chapter 4, we learned that a carbon atom has four valence electrons. By sharing those valence electrons to achieve an octet, a carbon atom forms four covalent bonds. In all organic molecules, all carbon atoms always have four covalent bonds. This is important to remember when you write the structures of organic compounds.

$$
\cdot \overset{\displaystyle \cdot}{\underset{\displaystyle \cdot}{C}} \cdot \ + \ 4H\cdot \longrightarrow \quad H : \overset{\displaystyle H}{\underset{\displaystyle H}{\overset{\displaystyle ..}{\underset{\displaystyle ..}{C}}}} : H \quad = \quad H-\overset{\displaystyle H}{\underset{\displaystyle H}{C}}-H
$$

Methane

In organic molecules with two or more carbon atoms, one or more of the valence electrons is used to form covalent bonds between carbon atoms.

$$
H : \overset{H}{\underset{H}{C}} : \overset{H}{\underset{H}{C}} : H \quad = \quad H-\overset{H}{\underset{H}{C}}-\overset{H}{\underset{H}{C}}-H \qquad \text{carbon–carbon single bond}
$$

Ethane

$$
\overset{H}{\underset{H}{C}} :: \overset{H}{\underset{H}{C}} \quad = \quad \overset{H}{}\diagdown C = C \diagup \overset{H}{} \qquad \text{carbon–carbon double bond}
$$

Ethene (ethylene)

In organic compounds, carbon atoms are most likely to bond with hydrogen, oxygen, nitrogen, sulfur, and halogens such as chlorine. Hydrogen with one valence electron forms a single covalent bond. An octet is achieved by nitrogen forming three covalent bonds and oxygen and sulfur each forming two covalent bonds. The halogens with seven valence electrons form one covalent bond to complete octets. Table 11.1 lists the number of covalent bonds most often formed by elements found in organic compounds.

Table 11.1 Covalent Bonds for Elements in Organic Compounds

Element	Group	Covalent Bonds	Structure of Atoms
H	1A	1	H—
C	4A	4	—C—
N	5A	3	—N—
O, S	6A	2	—Ö— —S̈—
F, Cl, Br, I	7A	1	—Ẍ: (X = F, Cl, Br, I)

SAMPLE PROBLEM 11.1

Carbon Bonds

Complete the following structures by adding the correct number of hydrogen atoms:

a. C—C—C   **b.** C—O—C   **c.** C—C—C (with Cl above third C)

Solution

We need to add hydrogen atoms to give each carbon atom four covalent bonds, oxygen two bonds, and chlorine one bond.

a.

      H H H
      | | |
  H—C—C—C—H
      | | |
      H H H

b.

      H   H
      | |
  H—C—O—C—H
      | |
      H   H

c.

      H H Cl
      | | |
  H—C—C—C—H
      | | |
      H H H

Study Check

Complete the following structure by adding hydrogen atoms.
C—C—N

QUESTIONS AND PROBLEMS

Organic Compounds

11.1 Add hydrogen atoms to the following to give a correct structure:
   a. C—C—C   **b.** C—C—O
   c. C=C—C—N

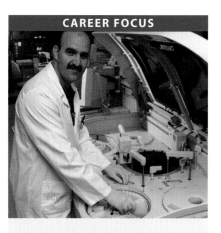

11.2 Add hydrogen atoms to the following to give a correct structure:

a. $C—C—\overset{\displaystyle O}{\overset{\|}{C}}—C$ **b.** $C—C—N$ **c.** $C—\overset{\displaystyle C}{\overset{|}{C}}—C{=}C$

11.3 Determine whether each of the following formulas is correct or incorrect:

a. $H—\overset{\displaystyle Cl}{\overset{|}{C}}—H$

b. $H—\overset{\displaystyle H}{\underset{\displaystyle H}{\overset{|}{\underset{|}{C}}}}—H—\overset{\displaystyle H}{\underset{\displaystyle H}{\overset{|}{\underset{|}{C}}}}—H$

c. $H—C{\equiv}C—H$

d. $H—\overset{\displaystyle H}{\underset{\displaystyle H}{\overset{|}{\underset{|}{C}}}}—\overset{\displaystyle H}{\overset{|}{N}}—H$

11.4 Determine whether each of the following formulas is correct or incorrect:

a. $H—\overset{\displaystyle Br}{\underset{\displaystyle F}{\overset{|}{\underset{|}{C}}}}—Cl$

b. $H—\overset{\displaystyle O}{\underset{\displaystyle H}{\overset{\|}{\underset{|}{C}}}}—O—H$

c. $H—\overset{\displaystyle H}{\overset{|}{C}}—\overset{\displaystyle H}{\underset{\displaystyle H}{\overset{|}{\underset{|}{C}}}}—O—H$

d. $H—\overset{\displaystyle H}{\underset{\displaystyle H}{\overset{|}{\underset{|}{C}}}}—\overset{\displaystyle H}{\overset{|}{N}}—\overset{\displaystyle H}{\underset{\displaystyle H}{\overset{|}{\underset{|}{C}}}}—H$

LEARNING GOAL

Discuss the reason that a carbon atom bonded to four other atoms has a tetrahedral shape.

the **Chemistry** place

WEB TUTORIAL
Introduction to Organic Molecules

Organic Molecules and Isomers

11.2 The Tetrahedral Structure of Carbon

In most organic molecules, a carbon atom is bonded to four other atoms. The VSEPR theory (Chapter 4) predicts that when four bonds are arranged as far apart as possible, they have a tetrahedral shape. For example, in the simplest organic molecule methane, CH_4, the bonds to hydrogen are directed to the corners of a tetrahedron with bond angles of 109.5°. The three-dimensional structure of methane can be illustrated as a ball-and-stick model or a space-filling model.

Chemists often use a shorthand notation to represent the three-dimensional structure. Wedge-shaped bonds show bonds that come toward the reader, while dashed lines depict bonds that go back, away from the reader. (See Figure 11.1.) However, we also write the two-dimensional structure for convenience with the understanding that the bonds to the carbon atoms are in a tetrahedral shape.

Tetrahedral Shapes in Carbon Chains

When there are two or more carbon atoms in a molecule, each carbon retains the tetrahedral shape when it is bonded to four other atoms. For example, the ball-and-stick model of ethane C_2H_6 is based on two tetrahedra attached to each other. (See Figure 11.2.) In this structure, straight lines depict the atoms in the plane of the page, wedges indicate bonds coming forward, and dashed lines show bonds going back of the plane. All the bond angles are close to 109.5°.

SAMPLE PROBLEM 11.2

Structure of Carbon Compounds

In a propane molecule with 3 carbon atoms, what is the geometry around each carbon atom?

$H—\overset{\displaystyle H}{\underset{\displaystyle H}{\overset{|}{\underset{|}{C}}}}—\overset{\displaystyle H}{\underset{\displaystyle H}{\overset{|}{\underset{|}{C}}}}—\overset{\displaystyle H}{\underset{\displaystyle H}{\overset{|}{\underset{|}{C}}}}—H$

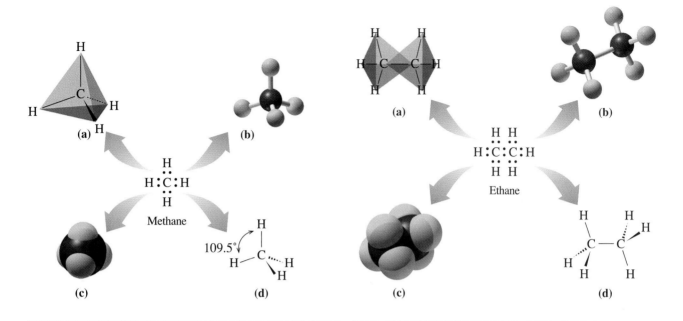

Figure 11.1 Three-dimensional representations of methane, CH_4: **(a)** tetrahedron, **(b)** ball-and-stick model, **(c)** space-filling model, **(d)** wedges and dashes.

Q Why does methane have a tetrahedral shape and not a flat shape?

Figure 11.2 Three-dimensional representations of ethane, C_2H_6: **(a)** tetrahedral shape of each carbon, **(b)** ball-and-stick model, **(c)** space-filling model, **(d)** wedges and dashes.

Q How is the tetrahedral shape maintained in a molecule with two carbon atoms?

Solution

Each carbon atom is bonded to four other atoms, which gives a tetrahedral shape to each carbon in the molecule.

Study Check

Why does a carbon atom bonded to four bromine atoms have a tetrahedral shape?

QUESTIONS AND PROBLEMS

The Tetrahedral Structure of Carbon

11.5 Why is the structure of the CH_4 molecule three-dimensional rather than two-dimensional?

11.6 In the abbreviated three-dimensional structure of CH_3—CH_3, what do the wedges and dashes represent?

11.3 Polarity of Organic Molecules

A covalent bond between two carbon atoms is nonpolar because the electrons are shared equally. However, when a carbon atom bonds to a different atom, the bond is polar covalent. Most of the elements found in organic compounds are more electronegative than carbon, with the exception of hydrogen.

LEARNING GOAL

Predict whether an organic molecule is polar or nonpolar.

Solubility and Polarity of Hydrocarbons

Fill a small bowl about half full with water. Add a few drops of some mineral oil or a small amount of petroleum jelly and mix. Mineral oil, a laxative and lubricant, is a mixture of liquid hydrocarbons, and petroleum jelly or Vaseline is a colloidal mixture of solid and liquid hydrocarbons.

Questions

1. What do your observations tell you about the solubility of hydrocarbons in water?
2. If water is polar, would you classify mineral oil or petroleum jelly as polar or nonpolar substances?
3. From your observations, are these hydrocarbons more or less dense than water? Explain.

Element	F	O	N	Cl	Br	C	H
Electronegativity value	4.0	3.5	3.0	3.0	2.8	2.5	2.1

As the carbon bonds to atoms with higher electronegativity values, the polarity of the bonds increases.

$$H_3C—CH_3 \qquad H_3C—Br \qquad H_3C—NH_2 \qquad H_3C—OH$$

Nonpolar ⟶ Increasing polarity of covalent bonds

Polarity of Organic Molecules

The polarity of a molecule is determined by the polarity of its bonds and the structure of the molecule. A bond between identical atoms is nonpolar, which makes molecules such as H_2 and Cl_2 nonpolar. A polar bond forms between atoms of different electronegativity values, which makes a molecule such as HCl polar. However, when molecules consist of several polar bonds, the arrangement of the bonds determines whether it is a polar or nonpolar molecule. If a molecule contains a symmetrical arrangement of polar bonds so that the dipoles cancel out, the molecule is nonpolar. For example, CCl_4 is nonpolar because it has a tetrahedral structure that cancels out the C—Cl dipoles. However, if there are different atoms bonded to carbon, the dipoles are not likely to cancel each other, which makes the molecule polar. For example, the CH_3Cl molecule is polar because the C—Cl bond is much more polar than the C—H bonds. (See Figure 11.3.)

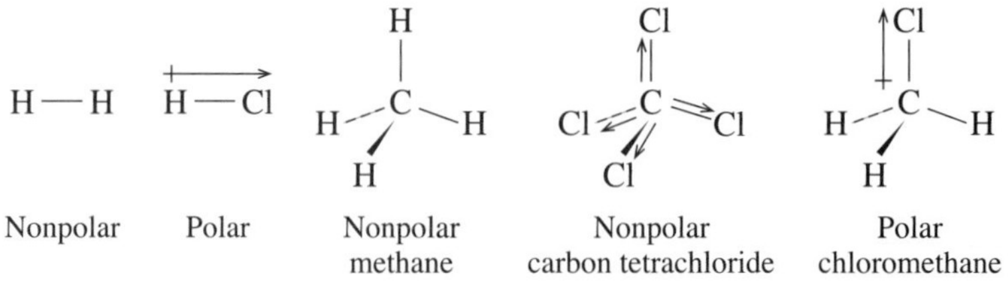

Figure 11.3 Effects of polarity of bonds on the polarity of molecules.

Q Why are the CH_4 and CCl_4 molecules nonpolar?

H—H   H—Cl
Nonpolar   Polar

Nonpolar methane

Nonpolar carbon tetrachloride

Polar chloromethane

Polarity of Organic Molecules

Predict whether each of the following molecules will be polar or nonpolar.
a. F_2 **b.** CF_4 **c.** CCl_3F

Solution

a. A bond between two identical atoms such as F—F is nonpolar, which makes the F_2 molecule nonpolar.
b. The CF_4 molecule is nonpolar because the polarities of C—F bonds cancel out in the tetrahedral shape.
c. The CCl_3F molecule is polar because the C—F bond is more polar than the C—Cl bonds.

Study Check

Why is CH_3F a polar molecule?

Polarity of Organic Molecules

11.7 Predict whether each of the following molecules will be polar or nonpolar:
    a. Br_2 **b.** CBr_4 **c.** $CHBr_3$ **d.** CH_3Br

11.8 Predict whether each of the following molecules will be polar or nonpolar:
    a. HI   **b.** CH_3OH **c.** CH_2BrF **d.** CCl_4

11.4 Properties of Organic Compounds

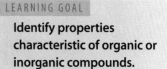

Identify properties characteristic of organic or inorganic compounds.

The polarities of molecules have a strong influence on the physical and chemical behavior of organic compounds. **Organic compounds** containing carbon and hydrogen have covalent bonds and form nonpolar molecules. As a result, the attractions between molecules are weak, which accounts for the low melting and boiling points of carbon compounds. Most organic compounds are not soluble in water. For example, vegetable oil, which is a mixture of organic compounds, does not dissolve in water, but floats on top. Small organic compounds with oxygen or nitrogen atoms are somewhat soluble because the electronegative atom forms hydrogen bonds with water.

Many of the organic compounds undergo combustion and burn vigorously in air. In contrast, many of the inorganic compounds, which we studied earlier, are ionic, which leads to high melting and boiling points. Inorganic compounds that are ionic or polar covalent are usually soluble in water and many produce ions and conduct electrical currents. Most inorganic substances do not burn in air. Table 11.2 contrasts some of the properties associated with organic and inorganic compounds such as propane, C_3H_8, and sodium chloride, NaCl. (See Figure 11.4.)

Table 11.2   Some Properties of Organic and Inorganic Compounds

Property	Organic	Example: C_3H_8	Inorganic	Example: NaCl
Bonding	Covalent	Covalent	Many are ionic, some covalent	Ionic
Polarity of bonds	Nonpolar, unless a more electro-negative atom is present	Nonpolar	Most are ionic or polar covalent, a few are non-polar covalent	Ionic
Melting point	Low	$-188°C$	High	$801°C$
Boiling point	Low	$-42°C$	High	$1413°C$
Flammability	High	Burns in air	Low	Does not burn
Solubility in water	Not soluble, unless a polar group is present	No	Most are soluble, unless nonpolar	Yes
Particles in solution	Molecules	Molecules	Many produce ions in water	Na^+Cl^-

Figure 11.4 Propane, C_3H_8, is an organic compound, whereas sodium chloride, NaCl, is an inorganic compound.

Q Why is propane used as a fuel?

SAMPLE PROBLEM 11.4

Properties of Organic Compounds

Indicate whether the following properties are characteristic of organic or inorganic compounds:

a. Not soluble in water   **b.** High melting point   **c.** Burns in air

Solution

a. Many organic compounds are not soluble in water

b. Inorganic compounds are most likely to have high melting points.

c. Organic compounds are most likely to be flammable.

Study Check

Octane floats on water. What type of compound is octane?

QUESTIONS AND PROBLEMS

Properties of Organic Molecules

11.9   Identify the following as formulas of organic or inorganic compounds:
   a. KCl   **b.** C_4H_{10} **c.** CH_3CH_2OH
   d. H_2SO_4 **e.** $CaCl_2$ **f.** CH_3CH_2Cl

11.10 Identify the following as formulas of organic or inorganic compounds:
   a. $C_6H_{12}O_6$ **b.** Na_2CO_3 **c.** I_2
   d. C_2H_5Cl **e.** $C_{10}H_{22}$ **f.** CH_4

11.11 Identify the following properties as most typical of organic or inorganic compounds:

a. soluble in water   b. low boiling point
c. burns in air   d. solid at room temperature

11.12 Identify the following properties as most typical of organic or inorganic compounds:

a. high melting point   b. gas at room temperature
c. covalent bonds   d. produces ions in water

11.5 Functional Groups

LEARNING GOAL

Classify organic molecules according to their functional groups.

WEB TUTORIAL
Functional Groups

Organic compounds number in the millions and more are synthesized every day. It might seem that the task of learning organic chemistry would be overwhelming. However, within this vast number of compounds, there are characteristic structural features called **functional groups,** which are a certain group of atoms that react in a predictable way. Compounds with the same functional group undergo similar chemical reactions. The identification of functional groups allows us to classify organic compounds according to their structure and to name compounds within each family. For now we will focus on recognizing the patterns of atoms that make up the functional groups.

Alkanes, Alkenes, and Alkynes

The **hydrocarbons,** which contain only carbon and hydrogen, are the simplest of the organic compounds. The **alkanes** contain only carbon–carbon single bonds. They are also called **saturated hydrocarbons** because they cannot add any more hydrogen atoms to the structure. The alkanes are not very reactive compared to compounds with functional groups, but they serve as a basic structure for the rest of the organic molecules.

Hydrocarbons with double or triple bonds are called **unsaturated hydrocarbons,** because they can add atoms of hydrogen, oxygen, or a halogen. The **alkenes** contain a functional group that is a double bond between two adjacent carbon atoms; **alkynes** contain a triple bond. The alkenes and the alkynes are much more reactive than the single-bonded alkanes.

As we proceed, we will write the structures in an abbreviated form called **condensed structural formulas.** In a condensed structural formula, the hydrogen atoms attached to each carbon are written adjacent to the symbol C for carbon. Thus, a CH_3— is the abbreviation for a carbon attached to three hydrogen atoms, whereas —CH_2— shows a carbon attached to two hydrogen atoms. The functional group is usually written separately so it can be easily recognized.

	An alkane	An alkene	An alkyne
Structure	H—C—C—H (with H's)	H—C=C—H (with H's)	H—C≡C—H
Functional group	—C—C—	—C=C—	—C≡C—
Condensed structural formulas	CH_3—CH_3	CH_2=CH_2	HC≡CH

An alcohol

An ether

Alcohols and Ethers

The characteristic functional group in **alcohols** is the **hydroxyl (—OH) group** bonded to a carbon atom. In **ethers,** the characteristic structural feature is an oxygen atom bonded to two carbon atoms. The oxygen atom also has two unshared pairs of electrons, but they will not be shown in the structural formulas.

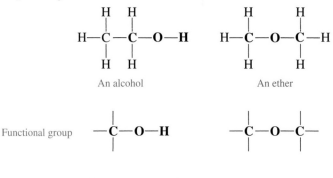

An alcohol   An ether

Functional group

Condensed structural formulas $CH_3—CH_2—OH$ $CH_3—O—CH_3$

An aldehyde

A ketone

Aldehydes and Ketones

The aldehydes and ketones contain a **carbonyl group** (C=O), which is a carbon with a double bond to oxygen. In an **aldehyde,** the carbon atom of the carbonyl group is bonded to another carbon and one hydrogen atom. Only the simplest aldehyde, CH_2O, has a carbonyl group attached to two hydrogen atoms. In a **ketone,** the carbonyl group is bonded to two other carbon atoms.

$CH_3—\overset{\overset{\textbf{O}}{\|}}{\textbf{C}}—\textbf{H}$ $H_3C—\overset{\overset{\textbf{O}}{\|}}{\textbf{C}}—CH_3$

An aldehyde   A ketone

Functional group

SAMPLE PROBLEM 11.5

Classifying Organic Compounds

Classify the following organic compounds according to their functional groups:

a. $CH_3—CH_2—CH_2—OH$ **b.** $CH_3—CH=CH—CH_3$

c. $CH_3—CH_2—\overset{\overset{O}{\|}}{C}—CH_2—CH_3$

Solution

a. alcohol   **b.** alkene   **c.** ketone

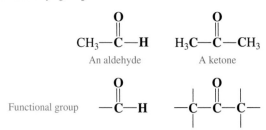

Study Check

Why is $CH_3—CH_2—O—CH_3$ an ether, but $CH_3—\overset{\overset{OH}{|}}{CH}—CH_3$ is an alcohol?

Carboxylic Acids and Esters

In **carboxylic acids,** the functional group is the *carboxyl group,* which is a combination of the *carbo*nyl and hydro*xyl* groups.

$$CH_3-\overset{\overset{\displaystyle O}{\|}}{C}-O-H \quad \text{or} \quad CH_3COOH \quad \text{or} \quad CH_3CO_2H$$

A carboxylic acid

A carboxylic acid

$$\text{Functional group} \quad -\overset{\overset{\displaystyle O}{\|}}{C}-O-H \quad \text{or} \quad -COOH \quad \text{or} \quad -CO_2H$$

An **ester** is similar to a carboxylic acid, except the oxygen of the carboxyl group is attached to a carbon and not to hydrogen.

$$CH_3-\overset{\overset{\displaystyle O}{\|}}{C}-O-CH_3 \quad \text{or} \quad CH_3COOCH_3 \quad \text{or} \quad CH_3CO_2CH_3$$

An ester

An ester

$$\text{Functional group} \quad -\overset{\overset{\displaystyle O}{\|}}{C}-O-\overset{\displaystyle |}{\underset{\displaystyle |}{C}}- \quad \text{or} \quad -COOC- \quad \text{or} \quad -CO_2C-$$

Amines

In **amines,** the central atom is a nitrogen atom. Amines are derivatives of ammonia, NH_3, in which carbon atoms replace one, two, or three of the hydrogen atoms. Amines are classified as primary, secondary, or tertiary amines according to the number of carbon groups bonded to the nitrogen atoms. A nitrogen atom has an unshared pair of electrons, but the lone pair will not be shown in the structural formulas.

A primary amine

$$NH_3 \qquad CH_3-NH_2 \qquad CH_3-\underset{\displaystyle CH_3}{\overset{\displaystyle |}{NH}} \qquad CH_3-\underset{\displaystyle CH_3}{\overset{\displaystyle |}{N}}-CH_3$$

| Ammonia | A primary amine (1°) | A secondary amine (2°) | A tertiary amine (3°) |

A list of the common functional groups in organic compounds is shown in Table 11.3.

A secondary amine

SAMPLE PROBLEM 11.6

Identifying Functional Groups

Classify the following organic compounds according to their functional groups:

a. $CH_3-CH_2-NH-CH_3$ **b.** $CH_3CO_2CH_2CH_3$

c. $CH_3-CH_2-\overset{\overset{\displaystyle O}{\|}}{C}-OH$

Solution

a. amine **b.** ester **c.** carboxylic acid

Study Check

How does a primary amine differ from a secondary amine?

Table 11.3   Classification of Organic Compounds

Class	Example	Functional Group	Characteristic
Alkane	H_3C—CH_3	—C—C—	Carbon–carbon single bond
Alkene	H_2C=CH_2	C=C	Carbon–carbon double bond
Alkyne	HC≡CH	—C≡C—	Carbon–carbon triple bond
Aromatic			Six-atom ring
Haloalkane	CH_3—Cl	—F,—Cl,—Br,—I	One or more halogen atoms
Alcohol	CH_3—CH_2—OH	—OH	Hydroxyl group (—OH)
Ether	H_3C—O—CH_3	—O—	Oxygen atom bonded to two carbons
Thiol	CH_3—SH	—SH	A —SH group bonded to carbon
Aldehyde	CH_3—C(=O)—H	—C(=O)—H	Carbonyl group (carbon-oxygen double bond) with —H
Ketone	H_3C—C(=O)—CH_3	—C(=O)—	Carbonyl group (carbon–oxygen double bond) between carbon atoms
Carboxylic acid	CH_3—C(=O)—O—H	—C(=O)—O—H	Carboxyl group (carbon oxygen double bond and —OH)
Ester	CH_3—C(=O)—O—CH_3	—C(=O)—O—	Carboxyl group with —H replaced by a carbon
Amine	CH_3—NH_2	—N—	Nitrogen atom with one or more carbon groups
Amide	CH_3—C(=O)—NH_2	—C(=O)—N—	Carbonyl group bonded to nitrogen

Functional Groups in Familiar Compounds

The flavors and odors of foods and our uses for fuels and many household products can be attributed to the functional groups of organic compounds. As we discuss these familiar products, look for the functional groups we have described.

Hydrocarbons, particularly alkanes, are the compounds that make up natural gas, motor oils, gasoline, and other fuels. The coal and crude oil that provides these hydrocarbons was formed 200 million years ago from the pressures on buried organic matter deep in the earth. Because the hydrocarbons are flammable we use compounds such as methane, propane, and butane as energy sources

$$CH_4 \qquad CH_3-CH_2-CH_3$$
Methane   Propane

$$CH_3-CH_2-CH_2-CH_3$$
Butane

When high temperatures and pressures are used to join small alkenes, extremely long carbon chains called polymers are formed. You may recognize the names of some polymers such as Styrofoam, Teflon, polyethylene, and polystyrene, which are used to make plastic bottles, photographic film, plastic utensils, outdoor clothing, and plastic parts for the body. The polymer polyvinyl chloride (PVC) is used in plastic pipes, garden hoses, and shower curtains. Even in long polymers, every carbon atom that is bonded to four other atoms has a tetrahedral shape.

$$Many \quad H-\overset{\overset{\displaystyle H}{|}}{C}=\overset{\overset{\displaystyle H}{|}}{C}-Cl \longrightarrow$$

Polyvinyl chloride (PVC)

Perhaps you wrap food in a plastic wrap to keep it fresh. This polymer contains many chlorine atoms that help the plastic adhere to the containers.

Polyvinylidene chloride (Saran)

Ethyl alcohol is the alcohol found in alcoholic beverages. Isopropyl alcohol is another alcohol commonly used to disinfect skin before giving injections and to treat cuts.

$$CH_3-CH_2-OH \qquad CH_3-\overset{\overset{\displaystyle OH}{|}}{CH}-CH_3$$
Ethyl alcohol   Isopropyl alcohol

Ketones and aldehydes are in many items we use or eat each day. Acetone or dimethyl ketone is produced in great amounts commercially. Acetone is used as an organic solvent because it dissolves a wide variety of organic substances. You may be familiar with acetone as fingernail polish remover. Ketones and

aldehydes used in the food industry are found in flavorings such as vanilla, cinnamon, and spearmint. When we buy a small bottle of liquid flavorings, the aldehyde or ketone is dissolved in alcohol because the compounds are not very soluble in water. The aldehyde butyraldehyde adds a "buttery" taste to foods and margarine.

$$CH_3-CH_2-CH_2-\overset{\overset{\displaystyle O}{\|}}{C}-H$$
Butyraldehyde "butter" flavoring

The sour tastes of vinegar and fruit juices and the pain from ant stings are all due to carboxylic acids. Acetic acid is the carboxylic acid that makes up vinegar. Aspirin also contains a carboxylic acid group. Esters found in fruits produce the pleasant aromas and tastes of bananas, oranges, pears, and pineapples. Esters are also used as solvents in many household cleaners, polishes, and glues.

$$CH_3-\overset{\overset{\displaystyle O}{\|}}{C}-OH$$
(acetic acid in vinegar)

$$CH_3-\overset{\overset{\displaystyle O}{\|}}{C}-O-CH_2-CH_2-CH_3$$
Propyl acetate (pears)

$$CH_3-NH_2$$

$$CH_3-\overset{\overset{\displaystyle O}{\|}}{C}-O-CH_2-CH_2-CH_2-CH_2-CH_3$$
Pentyl acetate (bananas)

One of the characteristics of fish is their odor, which is due to amines. Amines produced when proteins decay have a particularly pungent and offensive odor.

$$H_2N-CH_2-CH_2-CH_2-CH_2-NH_2$$
Putrescine

$$H_2N-CH_2-CH_2-CH_2-CH_2-CH_2-NH_2$$
Cadaverine

Alkaloids are biologically active amines synthesized by plants to ward off insects and animals. Some typical alkaloids include caffeine, nicotine, histamine, and the decongestant epinephrine. Many are painkillers and hallucinogens such as morphine, LSD, marijuana, and cocaine. Certain parts of our neurons have receptor sites that respond to the various alkaloids. By modifying the structures of certain alkaloids to eliminate side effects, chemists have synthesized painkillers and drugs such as Novocain, codeine, and Valium.

QUESTIONS AND PROBLEMS

Functional Groups

11.13 Identify the class of compounds that contains each of the following functional groups.
  a. hydroxyl group attached to a carbon chain
  b. carbon–carbon double bond
  c. carbonyl group attached to two hydrogen atoms
  d. carboxyl group in which the hydrogen is replaced by a carbon atom

11.14 Identify the class of compounds that contains each of the following functional groups:
  a. a nitrogen atom attached to one or more carbon atoms
  b. carboxyl group
  c. oxygen atom bonded to two carbon atoms
  d. a carbonyl group between two carbon atoms

11.15 Classify the following molecules according to their functional groups. The possibilities are alcohol, ether, ketone, carboxylic acid, or amine:

a. CH_3—CH_2—O—CH_2—CH_3

b.
$$
\begin{array}{c}
\overset{\displaystyle OH}{|} \\
CH_3-CH-CH_3
\end{array}
$$

c.
$$
\begin{array}{c}
\overset{\displaystyle O}{\parallel} \\
CH_3-C-CH_2-CH_3
\end{array}
$$

d. CH_3—CH_2—CH_2—$COOH$

e. CH_3—CH_2—NH_2

11.16 Classify the following molecules according to their functional groups. The possibilities are alkene, aldehyde, carboxylic acid, ester, or amine:

a. $CH_3CH_2COOCH_2CH_3$

b.
$$
\begin{array}{c}
\overset{\displaystyle CH_3}{|} \\
CH_3-N-CH_3
\end{array}
$$

c.
$$
\begin{array}{c}
\overset{\displaystyle O}{\parallel} \\
CH_3-CH_2-CH_2-C-H
\end{array}
$$

d. CH_3—CH_2—CH_2—$COOH$

e. CH_3—CH=CH—CH_3

LEARNING GOAL

Write condensed structural formulas for constitutional isomers.

WEB TUTORIAL
Isomers: Diversity
in Molecules

11.6 Constitutional Isomers

As we begin to look at the structure of organic compounds, we will find that several compounds can have the same molecular formula. This property called **isomerism** occurs when compounds have identical molecular formulas, but a different arrangement of atoms. There are several kinds of isomers in organic chemistry, but for now we will look at the **constitutional isomers** (or structural isomers) in which atoms of identical molecular formulas are connected in a different order.

With the exception of some small molecules such as CH_4 and C_2H_6, most organic molecules have isomers and that number increases as the number of carbon atoms increases. For example, there are 3 constitutional isomers with a molecular formula C_5H_{12}, 75 with a molecular formula of $C_{10}H_{22}$, and 4,327 with a molecular formula of $C_{15}H_{32}$. We are not going to write all of their structures, but it does demonstrate another reason for the large number of organic compounds.

We can illustrate this idea by looking at the constitutional isomers with molecular formula C_2H_6O. There are two different compounds with two different functional groups that have this formula. One isomer, ethyl alcohol, is a liquid at room temperature and boils at 78°C. The other isomer, dimethyl ether, is a gas at room temperature with a boiling point of −24°C. In the alcohol, the oxygen atom is connected to the end of the carbon atoms, but in the ether, it is between the carbon atoms. (See Figure 11.5.) This difference in the way the carbon and oxygen atoms connect strongly influences the physical and chemical behavior. Ethyl alcohol reacts vigorously with sodium metal producing hydrogen gas. Dimethyl ether does not react with sodium. We conclude that the difference in the structural formulas of these two constitutional isomers accounts for their markedly different physical and chemical properties.

Ethyl alcohol

$$
\begin{array}{cc}
\text{H} \quad \text{H} & \text{H} \qquad \text{H} \\
| \quad\; | & |\qquad\quad | \\
\text{H—C—C—O—H} & \text{H—C—O—C—H} \\
| \quad\; | & |\qquad\quad | \\
\text{H} \quad \text{H} & \text{H} \qquad \text{H}
\end{array}
$$

Ethyl alcohol   Dimethyl ether

Dimethyl ether

Functional group	hydroxyl group (—OH)	oxygen atom (—O—)
Molecular formula	C_2H_6O	C_2H_6O
Condensed formula	CH_3—CH_2—OH	CH_3—O—CH_3
Molar mass	46	46
Room temperature	liquid	gas
Melting point	−117°C	−138°C
Boiling point	78°C	−25°C
Reaction with sodium metal	vigorous; produces H_2	none

Figure 11.5 The constitutional isomers of ethyl alcohol and dimethyl ether can be seen in their ball-and-stick models.

Q **What makes ethyl alcohol and dimethyl ether constitutional isomers?**

SAMPLE PROBLEM 11.7

Constitutional isomers

Indicate whether the following pair of molecules represents constitutional isomers or not.

$$
\begin{array}{cc}
\quad\; \text{O} & \quad\;\; \text{O} \\
\quad\; || & \quad\;\; || \\
\text{CH}_3\text{—CH}_2\text{—C—H} & \text{CH}_3\text{—C—CH}_3
\end{array}
$$

Aldehyde (C_3H_6O)

Solution

These two compounds are constitutional isomers because they have identical molecular formulas C_3H_6O, but the oxygen atom forms a double bond to a different carbon atom in the chain.

$$
\begin{array}{cc}
\quad\; \text{O} & \quad\;\; \text{O} \\
\quad\; || & \quad\;\; || \\
\text{CH}_3\text{—CH}_2\text{—C—H} & \text{CH}_3\text{—C—CH}_3
\end{array}
$$

Aldehyde (C_3H_6O)   Ketone (C_3H_6O)

Ketone (C_3H_6O)

Study Check

Would you expect the two compounds in Sample Problem 11.7 to have the same boiling point? Explain.

HEALTH NOTE

Organic Compounds in Living Organisms

Our investigation into the structure and variety of organic compounds will be the foundation for understanding of the structures and functions of large molecules of life. You may already be familiar with the major biomolecules of carbohydrates, lipids, proteins, and nucleic acids such as the DNA in the nuclei of our cells. Each of these compounds consists of many carbon atoms bonded to hydrogen, but also to oxygen and nitrogen. We might say that the four major elements of life are C, H, O, and N. The tetrahedral geometry for carbon that we described for methane is also true in structures of long chains of carbon atoms.

In plants, photosynthesis combines CO_2 and water to form monosaccharides such as glucose and polysaccharides such as starch and cellulose. Glucose has a six-carbon chain with an aldehyde group and several hydroxyl groups. A variation in the direction of a single —— OH group in the three-dimensional structures gives another monosaccharide galactose. Thus structure plays an important role in the understanding of carbohydrates. When glucose molecules join by ether links, long polymers of complex carbohydrates called polysaccharides are formed, which we know as starch and cellulose. A difference in the direction of the bond between the glucose molecules allows us to digest the starches in our cereals and bread, but not the cellulose in their packaging.

Lipids containing carbon, hydrogen, and oxygen make up a variety of components in our body but you are probably most familiar with body fat and steroids. Body fat is composed of fatty acids, which are carboxylic acids with long carbon chains, bonded to glycerol by ester bonds. Because most lipids are insoluble,

Nucleic acids, DNA in cells

Carbohydrates for energy

Lipids, body fats, cell walls, hormones

Protein in skin, hair, muscles, hemoglobin, antibodies, enzymes

they play a major role in the structure of cell walls, which separate cell contents from aqueous fluids outside the cells. Steroids, a group of lipids that contain four carbon rings, include cholesterol, hormones, neurotransmitters, and vitamins. By modifying structures of steroids, chemists have developed drugs that mimic certain physiological behavior in the body. For example, changes in the structure of progesterone produced a compound that is now used in birth control pills.

Proteins are composed of carbon, hydrogen, oxygen, and nitrogen. These large polymers consist of small molecules called amino acids that have both amine and carboxylic acid functional groups. The order of amino acids in a protein determines the structure and function of that protein in the body. There are proteins such as hemoglobin that transport oxygen to our cells, contract our muscles, form antibodies, make up our hair, skin, and

enzymes that catalyze biological reactions. In an enzyme, a part of the protein structure binds and reacts only with compounds of a certain shape. Substances that have structures similar to that of the proper reactant can attach to the enzyme and inhibit enzyme action. In enzyme regulation, an end product binds to the enzyme in a way that alters protein structure and decreases enzyme activity.

Nucleic acids are long chains of nucleotides, which contain a carbohydrate called ribose, a nitrogen base, and a phosphate group. Our nucleic acids carry the messages for the production of all the proteins produced in cells. Each time a cell divides the DNA (deoxyribonucleic acid) is duplicated, which places the same instructions for protein synthesis in each new cell. In genetic diseases, mutations alter the nucleic acids causing a change in the proteins, which affect enzyme structure and function in the cell.

SAMPLE PROBLEM 11.8

Writing Constitutional Isomers

Write condensed structural formulas of two alcohols that are constitutional isomers of C_3H_8O.

Solution

Each formula contains three carbon atoms and the hydroxyl group (—OH). We can write two different alcohols by connecting the hydroxyl group to different carbon atoms.

$$\begin{array}{c} \text{OH} \\ | \\ \text{C—C—C} \end{array} \qquad \text{C—C—C—OH}$$

We complete the structural formula of the constitutional isomers by adding the correct number of hydrogen atoms.

$$\underset{\displaystyle CH_3-\overset{\textstyle OH}{\overset{|}{CH}}-CH_3}{} \qquad CH_3-CH_2-CH_2-OH$$

Study Check

Draw a condensed structural formula of two amines that are constitutional isomers with the formula C_2H_7N.

QUESTIONS AND PROBLEMS

Constitutional Isomers

11.17 Indicate whether each of the following pairs represents constitutional isomers, identical compounds, or different compounds that are not constitutional isomers:

a. $CH_3-CH_2-CH_2-\overset{\overset{\textstyle O}{\|}}{C}-H$ and $CH_3-CH_2-\overset{\overset{\textstyle O}{\|}}{C}-CH_3$

b. CH_3-CH_2-OH and $HO-CH_2-CH_3$

c. $CH_3-CH_2-CH_2-CH_3$ and $CH_3-\overset{\overset{\textstyle CH_3}{|}}{CH}-CH_3$

d. $CH_3-CH_2-CH_2-OH$ and $CH_3-\overset{\overset{\textstyle OH}{|}}{CH}-CH_3$

e. CH_3CH_2COOH and CH_3COOCH_3 f. $CH_3-\overset{\overset{\textstyle OH}{|}}{CH}-CH_3$ and $CH_3-\overset{\overset{\textstyle O}{\|}}{C}-CH_3$

11.18 Indicate whether each of the following pairs of molecules represents constitutional isomers, identical compounds, or different compounds that are not constitutional isomers:

a. $CH_3-CH_2-\overset{\overset{\textstyle CH_3}{|}}{CH}-CH_2-CH_3$ and $CH_3-\overset{\overset{\textstyle CH_3}{|}}{CH}-\overset{\overset{\textstyle CH_3}{|}}{CH}-CH_3$

b. $CH_3-O-CH_2-CH_3$ and $CH_3-CH_2-O-CH_3$ c. $CH_3-CH_2-\overset{\overset{\textstyle O}{\|}}{C}-H$ and $CH_2=CH-CH_2-OH$

d. $CH_3-CH_2-NH_2$ and $CH_3-\overset{\overset{\textstyle H}{|}}{N}-CH_3$ e. $CH_3-\overset{\overset{\textstyle O}{\|}}{C}-O-CH_3$ and $CH_3-CH_2-\overset{\overset{\textstyle O}{\|}}{C}-O-H$

f. $CH_3-CH_2-\overset{\overset{\textstyle O}{\|}}{C}-H$ and $CH_3-CH_2-\overset{\overset{\textstyle O}{\|}}{C}-O-H$

11.19 Draw condensed structural formulas of a carboxylic acid and two esters that have the molecular formula $C_3H_6O_2$.

11.20 Draw condensed structural formulas for two alcohols and two ethers with a molecular formula $C_4H_{10}O$.

Chapter Review

11.1 Organic Compounds

In organic compounds, carbon atoms share four valence electrons to form four covalent bonds. In organic molecules with two or more carbon atoms, additional valence electrons are used to form carbon–carbon covalent bonds. Although carbon is most often bonded to hydrogen, it can also bond to oxygen, nitrogen, sulfur, or a halogen such as chlorine or bromine.

11.2 The Tetrahedral Structure of Carbon

The VSEPR theory predicts that the bonds in a carbon atom bonded to four atoms are arranged as far apart as possible to give a tetrahedral shape. In the simplest organic molecule, methane, CH_4, the four bonds that bond hydrogen to the carbon atom are directed out to the corners of a tetrahedron with bond angles of $109.5°$.

11.3 Polarity of Organic Molecules

Polar bonds result when carbon is bonded to atoms with higher electronegativity values such as oxygen, nitrogen, or one of the halogens. If polar bonds are arranged symmetrically, as in a tetrahedron, their dipoles cancel, which makes the molecule nonpolar. A molecule of CCl_4 is nonpolar even though it consists of polar $C-Cl$ bonds. However, if one of the bonds is more polar ($C-Cl$) than the others as in CH_3Cl, the dipoles do not cancel, which makes the molecule polar.

11.4 Properties of Organic Compounds

Most organic compounds have covalent bonds and form nonpolar molecules. Often they have low melting points and low boiling points, are not very soluble in water, produce molecules in solutions, and burn vigorously in air. In contrast, many inorganic compounds are ionic or contain polar covalent bonds and form polar molecules. Many have high melting and boiling points, are usually soluble in water, produce ions in water, and do not burn in air.

11.5 Functional Groups

An organic molecule contains a characteristic group of atoms called a functional group that determines the molecule's family name and chemical reactivity. Functional groups are used to classify organic compounds, act as reactive sites in the molecule, and provide a system of naming for organic compounds. Some common functional groups include the hydroxyl group ($-OH$) in alcohols, the carbonyl group ($C=O$) in aldehydes and ketones, and a nitrogen atom ($-NH_2$) in amines.

11.6 Constitutional Isomers

Constitutional isomers (or structural isomers) are compounds that have the same molecular formula, but differ in the order that their atoms are connected. They have the same numbers of atoms, the same molar mass, but differ in their physical and chemical properties such as solubility, boiling points, and reactivity.

Key Terms

alcohols A class of organic compounds that contains the hydroxyl ($-OH$) group bonded to a carbon atom.

aldehydes A class of organic compounds that contains a carbonyl group ($C=O$) bonded to at least one hydrogen atom.

alkanes Hydrocarbons containing only single bonds between carbon atoms.

alkenes Hydrocarbons that contain carbon–carbon double bonds.

alkynes Hydrocarbons that contain carbon–carbon triple bonds.

amines A class of organic compounds that contains a nitrogen atom bonded to carbon atoms.

carbonyl group A functional group that contains a double bond between a carbon atom and an oxygen atom.

carboxylic acids A class of organic compounds that contains the functional group ($-COOH$).

condensed structural formulas An abbreviated structure in which the hydrogen atoms attached to each carbon atom are written adjacent to the symbol for carbon.

constitutional isomers Isomers having the same molecular formula but a different order of bonding in the structure.

esters A class of organic compounds that contains a $-COO-$ group with an oxygen atom bonded to carbon.

ethers A class of organic compounds that contains an oxygen atom bonded to two carbon atoms.

functional group A group of atoms that determine the physical and chemical properties and naming of a class of organic compounds.

hydrocarbons Organic compounds consisting of only carbon and hydrogen.

hydroxyl group The group of atoms ($-OH$) characteristic of alcohols.

isomerism The property of organic compounds in which identical molecular formulas have different arrangements of atoms.

ketones A class of organic compounds in which a carbonyl group is bonded to two carbon atoms.

organic compounds Compounds made of carbon that typically have covalent bonds, nonpolar molecules, low

melting and boiling points, are insoluble in water, and flammable.

saturated hydrocarbon A compound of carbon and hydrogen that contains only single carbon–carbon (C—C) bonds.

unsaturated hydrocarbon A compound of carbon and hydrogen that contains at least one double carbon–carbon (C=C) or triple (C≡C) bond.

Additional Problems

11.21 Compare organic and inorganic compounds in terms of:
- **a.** types of bonds
- **b.** solubility in water
- **c.** melting points
- **d.** flammability

11.22 Identify each of the following compounds as organic or inorganic:
- **a.** Na_2SO_4
- **b.** $CH_2=CH_2$
- **c.** Cr_2O_3
- **d.** $C_{12}H_{22}O_{11}$

11.23 Match the following physical and chemical properties with the compounds butane, C_4H_{10}, or potassium chloride, KCl.
- **a.** melts at $-138°C$
- **b.** burns vigorously in air
- **c.** melts at $770°C$
- **d.** produces ions in water
- **e.** is a gas at room temperature

11.24 Match the following physical and chemical properties with the compounds cyclohexane, C_6H_{12}, or calcium nitrate, $Ca(NO_3)_2$.
- **a.** contains only covalent bonds
- **b.** melts above $500°C$
- **c.** insoluble in water
- **d.** liquid at room temperature
- **e.** produces ions in water

11.25 Complete the following structures by adding the correct number of hydrogen atoms:
- **a.** C—C
- **b.** C—C—O
- **c.** C—C—$\overset{\overset{\displaystyle O}{\|}}{C}$
- **d.** C—O—C—C

11.26 Identify the errors in the following formulas:
- **a.** CH_4F
- **b.** $CH_3—CH_2$
- **c.** $CH_3≡CH_3$
- **d.** $CH_3CH=\overset{\overset{\displaystyle O}{\|}}{C}—H$

11.27 Distinguish between the following:
- **a.** a hydroxyl group and carbonyl group

- **b.** an alcohol and an ether
- **c.** a carboxylic acid and an ester

11.28 Classify the following as alkanes, alkenes, alkynes, alcohols, ethers, aldehydes, ketones, carboxylic acids, esters, or amines.
- **a.** CH_3NH_2
- **b.** $CH_3—\overset{\overset{\displaystyle O}{\|}}{C}—CH_3$
- **c.** $CH_3—\overset{\overset{\displaystyle OH}{|}}{CH}—CH_3$
- **d.** $CH_3—CH_2—CH_3$
- **e.** $CH_3—C≡CH$
- **f.** $CH_3—O—CH_2—CH_3$

11.29 Identify the pairs of compounds that are constitutional isomers, identical compounds, or different compounds that are not constitutional isomers:
- **a.** $CH_3—\overset{\overset{\displaystyle Cl}{|}}{CH}—Cl$ and $Cl—CH_2—CH_2—Cl$
- **b.** $CH_3CH_2\overset{\overset{\displaystyle O}{\|}}{CH}$ and $CH_3CH_2\overset{\overset{\displaystyle O}{\|}}{COH}$
- **c.** $CH_3—O—CH_2CH_2CH_3$ and $CH_3CH_2—O—CH_2CH_3$
- **d.** $CH_3—O—CH_2CH_3$ and $CH_3—\overset{\overset{\displaystyle O}{\|}}{C}—CH_3$
- **e.** $CH_3—\overset{\overset{\displaystyle O}{\|}}{C}—O—CH_3$ and $CH_3—CH_2—\overset{\overset{\displaystyle O}{\|}}{C}—OH$
- **f.** $CH_3—\overset{\overset{\displaystyle O}{\|}}{C}—CH_2—CH_3$ and $CH_3—CH_2—\overset{\overset{\displaystyle O}{\|}}{C}—CH_3$
- **g.** $CH_3—CH_2—CH_2—CH_3$ and $CH_3—\overset{\overset{\displaystyle CH_3}{|}}{CH}—CH_2—CH_3$

11.30 Draw condensed structural formulas for the following constitutional isomers with a molecular formula C_3H_9N:
  a. two primary amines
  b. a secondary amine
  c. a tertiary amine

11.31 Predict whether each of the following molecules will be polar or nonpolar:
  a. CH_3F **b.** $CHCl_3$ **c.** CF_4
  d. CH_3—CH_2—Cl **e.** CH_3OH **f.** CH_2F_2

11.32 Write structural formulas of two esters and a carboxylic acid that are constitutional isomers with molecular formula $C_3H_6O_2$.

11.33 Match each of the following terms with the corresponding description:

hydrocarbon, alcohol, ether, aldehyde, ketone, carboxylic acid, ester, amine, saturated hydrocarbon, unsaturated hydrocarbon, functional group, tetrahedral, constitutional isomers

  a. An organic compound that contains a hydroxyl group bonded to a carbon.
  b. A hydrocarbon that contains one or more carbon–carbon double bonds.
  c. An organic compound in which the carbon of a carbonyl group is bonded to a hydrogen.

  d. A hydrocarbon that contains only carbon–carbon single bonds.
  e. An organic compound in which the carbon of a carbonyl group is bonded to a hydroxyl group.
  f. An organic compound that contains a nitrogen atom bonded to one or more carbon atoms.
  g. The three-dimensional shape of a carbon bonded to four hydrogen atoms.
  h. Organic compounds with identical molecular formulas that differ in the order the atoms are connected.
  i. An organic compound in which the hydrogen atom of a carboxyl group is replaced by a carbon atom.
  j. Organic compounds that contain only carbon and hydrogen atoms.
  k. An organic compound that contains an oxygen atom bonded to two carbon atoms.
  l. A hydrocarbon that contains a carbon–carbon triple bond.
  m. A charactistic group of atoms that make compounds behave and react in a particular way.
  n. An organic compound in which the carbonyl group is bonded to two carbon atoms.

12 Alkanes

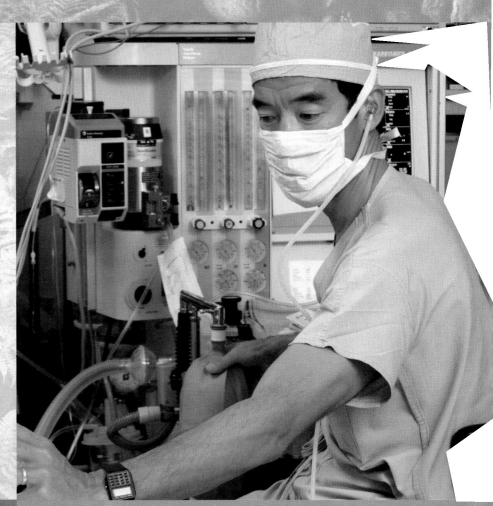

"During surgery, I work with the surgeon to provide a safe level of anesthetics that renders the patient free from pain," says Mark Noguchi, nurse anesthetist (CRNA), Kaiser Hospital. "We do spinal and epidural blocks as well as general anesthetics, which means the patient is totally asleep. We use a variety of pharmaceutical agents including halothane ($C_2HBrClF_3$), and bupivacain ($C_{18}H_{28}N_2O$), as well as muscle relaxants such as midazolam ($C_{18}H_{13}ClFN_3$) to achieve the results we want for the surgical situation. We also assess the patient's overall hemodynamic status. If blood is lost, we replace components such as plasma, platelets, and coagulation factors. We also monitor the heart rate and run EKGs to determine cardiac function."

LOOKING AHEAD

the Chemistry place

www.chemplace.com/college

Visit the URL above or use the CD-ROM in the book for extra quizzing, interactive tutorials, career resources, and case studies.

At the beginning of the nineteenth century, scientists classified chemical compounds as inorganic and organic. An inorganic compound was a substance that was composed of minerals, and an organic compound was a substance that came from an organism, thus the use of the word "organic." It was thought that some type of "vital force," which could only be found in living cells, was required to synthesize an organic compound. This perception was shown to be incorrect in 1828, when the German chemist Friedrick Wöhler synthesized urea, a product of protein metabolism, by heating an inorganic compound ammonium cyanate.

$$NH_4CNO \xrightarrow{\text{Heat}} H_2N-\overset{\overset{\displaystyle O}{\|}}{C}-NH_2$$

Ammonium cyanate (inorganic)   Urea, organic

We now define organic chemistry as the study of hydrocarbons and their derivatives. As we go into the twenty-first century, we have identified around ten million organic compounds. When carbon atoms bond, they form chains of carbon atoms that can be hundreds or thousands of atoms long. There are also many carbon compounds with identical molecular formulas, but different structures because they have different arrangements of atoms. In this chapter, we look at alkanes and the ways their atoms are arranged in carbon chains, chains with carbon branches, and rings of carbon atoms.

12.1 Alkanes

In Chapter 11, we looked at several families of organic compounds. Among these are the hydrocarbons, compounds that contain only carbon and hydrogen. The **alkanes** are hydrocarbons that contain only carbon–carbon single bonds. The general formula for an alkane is C_nH_{2n+2}, where n is the number of carbon atoms. The number of hydrogen atoms is twice the number of carbon atoms plus two more. Some examples follow:

General Formula (C_nH_{2n+2}) *for Some Alkanes*	*C atoms*	*H atoms*
CH_4	1	$4 = (2 \times 1) + 2$
C_2H_6	2	$6 = (2 \times 2) + 2$
C_4H_{10}	4	$10 = (2 \times 4) + 2$
$C_{10}H_{22}$	10	$22 = (2 \times 10) + 2$

When relatively few organic compounds were known, names were assigned in a random fashion. Today the **IUPAC system** (International Union of Pure and Applied Chemistry) determines the protocol for naming organic compounds. The names for the first four continuous-chain saturated alkanes are methane, ethane, propane, and butane. Then the IUPAC system uses Greek prefixes such as *pent (5)*, *hex (6)*, and *hept (7)* to indicate the number of carbon atoms in the chain. The suffix *ane* indicates that the compounds are in the alkane family.

Table 12.1 IUPAC Names for the First Ten Continuous-Chain Alkanes

Number of Carbon Atoms	Prefix	Name	Molecular Formula	Condensed Structural Formula
1	Meth	Methane	CH_4	CH_4
2	Eth	Ethane	C_2H_6	CH_3—CH_3
3	Prop	Propane	C_3H_8	CH_3—CH_2—CH_3
4	But	Butane	C_4H_{10}	CH_3—CH_2—CH_2—CH_3
5	Pent	Pentane	C_5H_{12}	CH_3—CH_2—CH_2—CH_2—CH_3
6	Hex	Hexane	C_6H_{14}	CH_3—CH_2—CH_2—CH_2—CH_2—CH_3
7	Hept	Heptane	C_7H_{16}	CH_3—CH_2—CH_2—CH_2—CH_2—CH_2—CH_3
8	Oct	Octane	C_8H_{18}	CH_3—CH_2—CH_2—CH_2—CH_2—CH_2—CH_2—CH_3
9	Non	Nonane	C_9H_{20}	CH_3—CH_2—CH_2—CH_2—CH_2—CH_2—CH_2—CH_2—CH_3
10	Dec	Decane	$C_{10}H_{22}$	CH_3—CH_2—CH_2—CH_2—CH_2—CH_2—CH_2—CH_2—CH_2—CH_3

It is important to learn the prefixes in Table 12.1 because they indicate the same number of carbon atoms in the names of other kinds of organic compounds.

Structural Formulas

The **structural formulas** for an alkane show the order of bonding for the carbon atoms in the chain. In the **expanded structural formula**, all the individual atoms and their covalent bonds are drawn. In a **condensed structural formula**, each carbon atom is grouped with its bonded hydrogen atoms. A subscript indicates the number of hydrogen atoms bonded to each carbon atom. The C—H bonds are understood but not written individually.

In **continuous alkanes**, the carbon atoms are connected in a row. However, in chains of three carbon atoms or more, the carbon atoms do not actually lie in a straight line. The tetrahedral shape of carbon (Chapter 11) arranges the carbon bonds in a zigzag pattern, which is seen in the ball-and-stick model of hexane. (See Figure 12.1.) An abbreviated structure called the **line-bond formula** shows only the bonds from carbon to carbon. The ends of the lines and the corners where the lines meet are understood to be carbon atoms attached to the proper number of hydrogen atoms to give four bonds. In Table 12.2, we see how these structural formulas are written to represent the first three alkanes methane, ethane, and propane. Note that the molecular formula gives the total number of carbon and hydrogen atoms, but does not indicate the arrangement of atoms in the molecule.

Figure 12.1 A ball-and-stick model of hexane.

Q Why do the carbon atoms in hexane appear to be arranged in a zigzag chain?

Conformation of Alkanes

Another structural property of alkanes is the rotation of atoms around each carbon–carbon single bond. In an alkane, the groups attached to each carbon are

Table 12.2   Writing Structural Formulas for Some Alkanes

Alkane	Methane	Ethane	Propane
Molecular formula	CH_4	C_2H_6	C_3H_8
Structural formulas			

Expanded

H | H—C—H | H	H H | | H—C—C—H | | H H	H H H | | | H—C—C—C—H | | | H H H

Condensed

CH_4	CH_3—CH_3	CH_3—CH_2—CH_3 or CH_2 between CH_3 and CH_3

Line-bond

	—	⋀

not in fixed positions, but rotate freely about the bond connecting the two carbon atoms. This motion is analogous to the independent rotation of the wheels of a toy car. The different arrangements that occur during the rotation about a single bond are called **conformations.**

Suppose we could look down the center C—C bond in butane, C_4H_{10}, as it rotates. Sometimes the CH_3 groups line up in front of each other, and at other times they are opposite each other. (See Figure 12.2.) As the CH_3 groups turn around the single bond, many conformations are possible, some of which are shown in Figure 12.3. Although conformations constantly change during the rotation, the conformation in which the CH_3 groups are farthest apart is the most stable. The conformations of butane are depicted by a variety of two-dimensional structural formulas in Table 12.3. It is important to recognize that all of these structural formulas represent the same continuous chain of four carbon atoms.

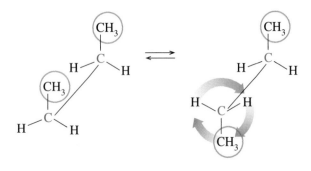

Figure 12.2 As the CH_3 groups rotate freely around a carbon–carbon single bond in butane, they have different conformations.

Q Why are the CH_3 groups in butane close to each other in one conformation and far apart from each other in another conformation?

SAMPLE PROBLEM 12.1

Drawing Expanded, Condensed, and Line-Bond Structural Formulas

Draw expanded, condensed, and line-bond structural formulas for pentane.

Solution

The expanded structural formula for pentane, C_5H_{12}, is written with 5 carbon atoms bonded in a continuous chain connected to hydrogen atoms. In the condensed structural formula each carbon atom is grouped with its attached hydrogens. The line-bond formula shows only the bonds connecting the carbon atoms.

H H H H H | | | | | H—C—C—C—C—C—H | | | | | H H H H H	CH_3—CH_2—CH_2—CH_2—CH_3	⋁⋁
Expanded formula	Condensed	Line-bond

Table 12.3   Some Structural Formulas and Conformations for Butane C$_4$H$_{10}$

Expanded structural formula

$$\begin{array}{ccccccc} & H & & H & & H & & H \\ & | & & | & & | & & | \\ H-\!\!\!\!\!& C & \!\!\!-\!\!\!& C & \!\!\!-\!\!\!& C & \!\!\!-\!\!\!& C & \!\!\!-\!\!\!H \\ & | & & | & & | & & | \\ & H & & H & & H & & H \end{array}$$

Condensed structural formulas

CH$_3$—CH$_2$—CH$_2$—CH$_3$

$$\begin{array}{cc} CH_2\!\!-\!\!CH_2 \\ | \quad\quad | \\ CH_3 \quad CH_3 \end{array}$$

$$\begin{array}{c} CH_3 \\ | \\ CH_2\!\!-\!\!CH_2 \\ | \\ CH_3 \end{array}$$

$$\begin{array}{c} CH_3 \\ | \\ CH_2 \\ | \\ CH_2 \\ | \\ CH_3 \end{array}$$

$$\begin{array}{c} CH_3\!\!-\!\!CH_2 \\ \quad\quad | \\ \quad CH_2\!\!-\!\!CH_3 \end{array}$$

$$\begin{array}{c} CH_3 \\ | \\ CH_2\!\!-\!\!CH_2\!\!-\!\!CH_3 \end{array}$$

$$\begin{array}{c} CH_3 \qquad\quad CH_2 \\ \diagdown \qquad\qquad \diagdown \\ CH_2 \qquad\quad CH_3 \end{array}$$

Line-bond formula

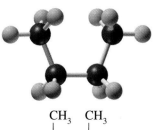

$$\begin{array}{cc} CH_3 & CH_3 \\ | & | \\ CH_2\!\!-\!\!\!& CH_2 \end{array}$$

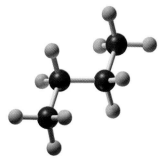

$$\begin{array}{c} CH_3 \\ | \\ CH_2\!\!-\!\!CH_2 \\ | \\ CH_3 \end{array}$$

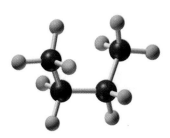

$$\begin{array}{c} CH_3 \\ | \\ CH_3\!\!-\!\!CH_2\!\!-\!\!CH_2 \end{array}$$

Figure 12.3 In the possible conformations of butane, the CH$_3$ groups may be close to each other or far apart, which is the most stable arrangement.

Q Why is it possible for a butane molecule to have many conformations?

Study Check

Draw the condensed structural formula for the following line-bond formula:

QUESTIONS AND PROBLEMS

Alkanes

12.1 Use the general formula for alkanes to determine the following:
  a. the molecular formula of an alkane with 7 carbon atoms
  b. the number of hydrogen atoms bonded to 5 carbon atoms
  c. the number of carbon atoms that are bonded to 10 hydrogen atoms

12.2 Use the general formula for alkanes to determine the following:
  a. the molecular formula of an alkane with 9 carbon atoms
  b. the number of hydrogen atoms bonded to 8 carbon atoms
  c. the number of carbon atoms that are bonded to 26 hydrogen atoms

12.3 Write the stated type of structural formula for a continuous chain of each of the following alkanes:
  a. an expanded structural formula for propane
  b. the condensed structural formula for hexane
  c. a line-bond formula for hexane

12.4 Write the stated type of structural formula for a continuous chain of each of the following alkanes:
  a. an expanded structural formula for butane
  b. the condensed structural formula for octane
  c. a line-bond formula for decane

12.5 Give the IUPAC name for each of the following alkanes.

a. CH_3
$|$
$CH_2—CH_2—CH_2$
$|$
CH_3

b. $\wedge\wedge\wedge$

c. $CH_2—CH_3$
$|$
CH_2
$|$
$CH_3—CH_2—CH_2$

12.6 Give the IUPAC name for each of the following alkanes.

a. CH_4

b. $\vee\wedge\vee\wedge\vee$

c. CH_3
$|$
CH_2
$|$
CH_3

Write the IUPAC names for alkanes.

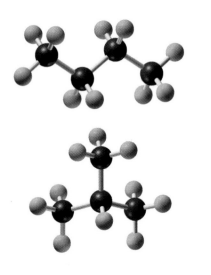

Figure 12.4 The constitutional isomers of C_4H_{10} have the same number and type of atoms, which are bonded in a different order.

Q **What makes these molecules constitutional isomers?**

12.2 IUPAC Naming System for Alkanes

When two molecules have the same molecular formula but different arrangements of atoms, they are called **constitutional isomers** (Chapter 11). For example, there are two different condensed structural formulas for alkanes with the molecular formula C_4H_{10}. (See Figure 12.4.) Although they contain the same type and number of atoms, the atoms are connected in a different order. In one molecule, the carbon atoms are in a continuous carbon chain. In the other, a carbon side group called a **branch** or **substituent** replaces a hydrogen atom on a carbon chain. An alkane with one or more branches is called a **branched-chain alkane.** Constitutional isomers have different properties such as boiling point, melting point, and density, as shown in Table 12.4.

SAMPLE PROBLEM 12.2

Constitutional Isomers

Identify each pair of structural formulas as constitutional isomers or conformations of the same molecule.

a. CH_3 CH_3
$|$ $|$
$CH_2—CH_2$ and $CH_2—CH_2—CH_3$
$|$
CH_3

b. CH_3
$|$
$CH_3—CH—CH_2—CH_2—CH_3$ and $CH_3—CH—CH—CH_3$ with CH_3 CH_3 above

Solution

a. The structural formulas represent conformations of the same molecule because each has four atoms in a continuous chain.

b. These are constitutional isomers because the molecular formula C_6H_{14} is identi-

Table 12.4   Some Properties of Constitutional Isomers of C$_4$H$_{10}$

Condensed formula	CH$_3$—CH$_2$—CH$_2$—CH$_3$	
Melting point	–138°C	–159°C
Boiling point	–1°C	–12°C
Density	0.58 g/mL	0.55 g/mL

cal but the atoms are bonded in a different order. One has one CH$_3$ side group attached to a five-carbon chain, and the other has two CH$_3$ side groups attached to a four-carbon chain.

Study Check

The following compound has the same molecular formula as the compounds in part b of the preceding Sample Problem. Is the following structural formula a constitutional isomer or identical to one of the molecules in part b?

$$CH_3—CH_2—\overset{\overset{\displaystyle CH_3}{|}}{CH}—CH_2—CH_3$$

Classifying Carbon Atoms in Hydrocarbons

In a hydrocarbon, each carbon atom is classified according to the number of carbon atoms connected to it. A **primary** (1°) **carbon** is bonded to one other carbon atom. A **secondary** (2°) **carbon** has two carbon atoms attached to it. A **tertiary** (3°) **carbon** is bonded to three other carbon atoms.

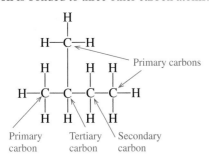

Alkyl Groups in Branched Alkanes

In the IUPAC names for branched-chain alkanes, the substituents are named as alkyl groups. An **alkyl group** is an alkane that is missing one hydrogen atom. The alkyl group is named by replacing the *ane* ending of the corresponding alkane name with *yl*. Alkyl groups do not exist on their own: they must be attached to something such as a carbon chain. Some of the common alkyl groups are illustrated in Table 12.5.

Some of the alkyl groups use common names such as isopropyl and isobutyl. The classification of the carbon in the alkyl group attached to a carbon chain is used in the names secondary-butyl (*sec*-butyl) and tertiary-butyl (*tert*-butyl).

the
Chemistry
place

WEB TUTORIAL
Introduction to
Organic Molecules

Table 12.5   Names and Formulas of Some Common Alkyl Groups

Alkane	Name of Alkane	Corresponding Alkyl Group	Name of Alkyl Group
One carbon			
CH_4	Methane	CH_3—	Methyl
Two carbons			
CH_3—CH_3	Ethane	CH_3—CH_2—	Ethyl
Three carbons			
CH_3—CH_2—CH_3	Propane	CH_3—CH_2—CH_2—	Propyl
		CH_3—$\overset{\mid}{CH}$—CH_3	Isopropyl
Four carbons			
CH_3—CH_2—CH_2—CH_3	Butane	CH_3—CH_2—CH_2—CH_2—	Butyl
		CH_3—$\overset{\mid}{CH}$—CH_2—CH_3	*sec*-Butyl (secondary butyl)
CH_3—$\overset{\overset{\displaystyle CH_3}{\mid}}{CH}$—$CH_3$	Isobutane	CH_3—$\overset{\overset{\displaystyle CH_3}{\mid}}{CH}$—$CH_2$—	Isobutyl
		CH_3—$\overset{\overset{\displaystyle CH_3}{\mid}}{\underset{\mid}{C}}$—$CH_3$	*tert*-Butyl or *t*-butyl (tertiary butyl)

Classification of Carbon in Alkyl Groups

Primary (1°) Carbon	Secondary (2°) Carbon	Tertiary (3°) Carbon
CH_3—CH_2—CH_2—$\overset{\overset{\displaystyle H}{\mid}}{\underset{\underset{\displaystyle H}{\mid}}{C}}$—	CH_3—CH_2—$\overset{\overset{\displaystyle CH_3}{\mid}}{\underset{\underset{\displaystyle H}{\mid}}{C}}$—	H_3C—$\overset{\overset{\displaystyle CH_3}{\mid}}{\underset{\underset{\displaystyle CH_3}{\mid}}{C}}$—
Butyl group	*sec*-Butyl group	*tert*-Butyl group

Naming Branched-Chain Alkanes

In the IUPAC system of naming, the longest carbon chain is the main chain, which is numbered to give the location of substituents attached to it. Some common names are still used for compounds with one to four carbon atoms, but compounds with five or more carbon atoms follow the IUPAC rules. Let's take a look at the way the IUPAC system is used to name the following alkane.

$$CH_3—\overset{\overset{\displaystyle CH_3}{\mid}}{CH}—CH_2—CH_2—CH_3$$

Rule 1   Find the longest continuous chain of carbon atoms and name it as the main chain. In this example, the longest chain has five carbon atoms, which is *pentane*.

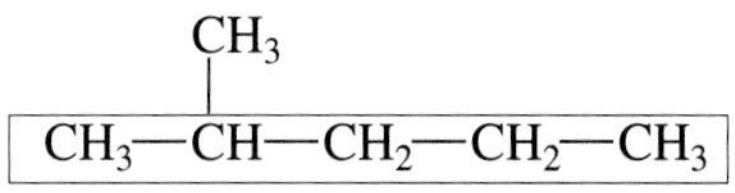

$$CH_3-CH-CH_2-CH_2-CH_3 \qquad \text{Pentane}$$
$$\quad\quad\;\; | $$
$$\quad\quad CH_3$$

Rule 2   **Number the carbon in the main chain starting from the end nearest a substitutent.** Once you start numbering the main chain, continue in that same direction.

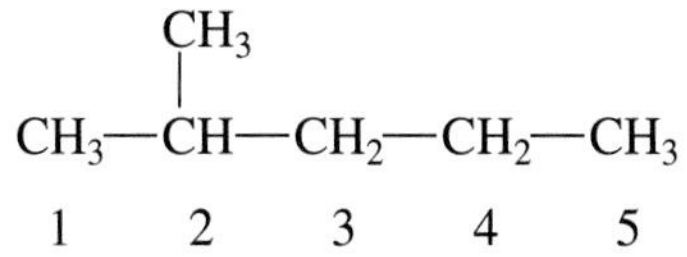

$$CH_3-CH-CH_2-CH_2-CH_3 \qquad \text{Pentane}$$
$$1 \quad\; 2 \quad\;\; 3 \quad\;\; 4 \quad\;\; 5$$

Rule 3   **Give the location and name of each alkyl group in front of the name of the main chain.** Use a prefix (di-, tri-, tetra- and so on) to indicate a group that appears more than once. In the name, hyphens separate numbers from words and commas separate numbers. When multiple substituents allow numbering from both ends of the main chain, use the direction that gives the lowest series of numbers.

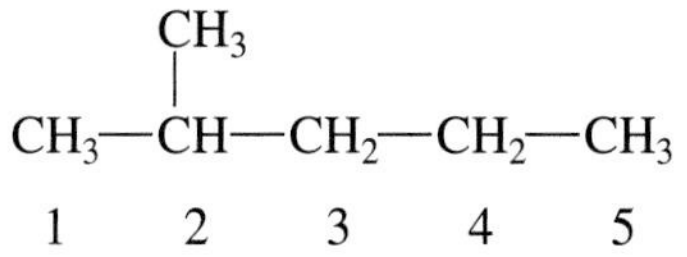

$$CH_3-CH-CH_2-CH_2-CH_3 \qquad \text{2-Methylpentane}$$
$$1 \quad\; 2 \quad\;\; 3 \quad\;\; 4 \quad\;\; 5$$

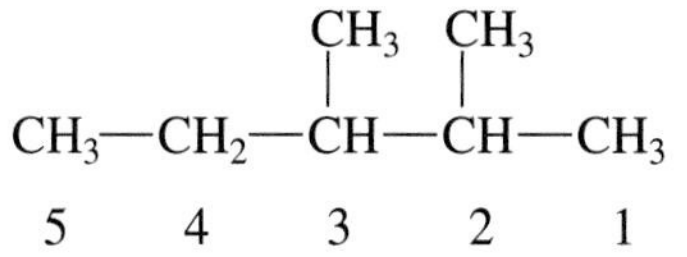

2,3-**Di**methylpentane
not 3,4-dimethylpentane

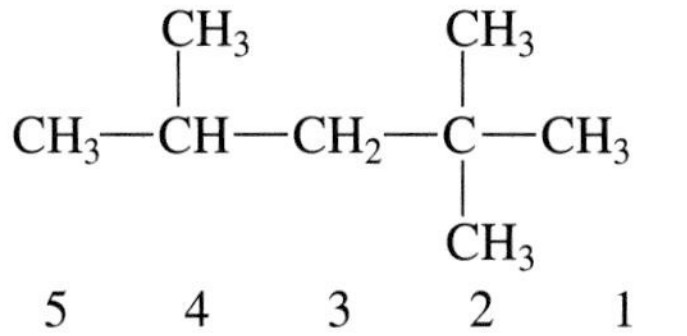

2,2,4-**Tri**methylpentane
not 2,4,4-trimethylpentane

Rule 4   **List the substituents in alphabetical order.** The prefixes for repeated substituents are *not* used in deciding alphabetical order.

Examples of Alphabetical Order of Alkyl Substituents

Ethyl, methyl, propyl   Triethyl, dimethyl   Ethyl, isopropyl, dimethyl

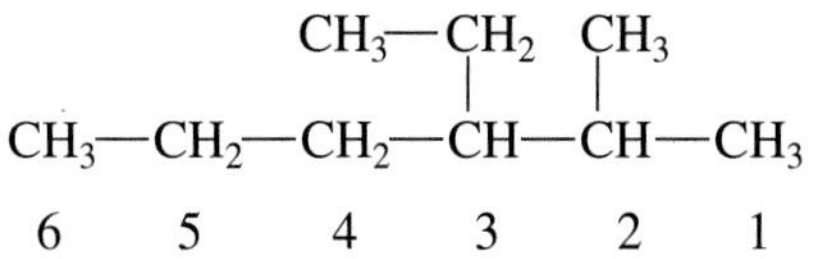

3-Ethyl-2-methylhexane
not 2-methyl-3-ethylhexane
not 4-ethyl-5-methylhexane

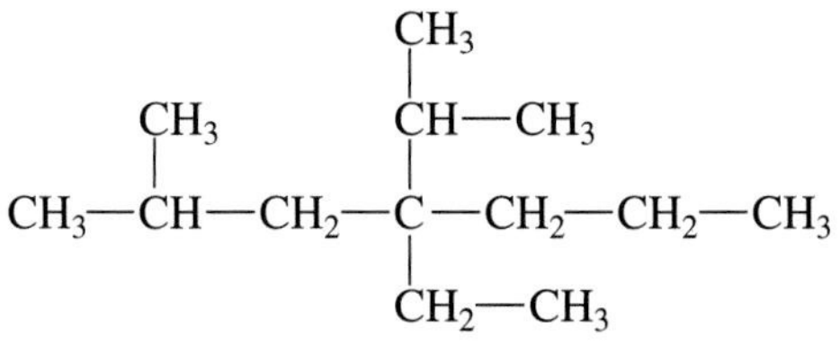

4-Ethyl-4-isopropyl-2-methylheptane
not 2-methyl-4-ethyl-4-isopropylheptane

Be sure that you always find the longest continuous carbon chain. It may not be the most obvious horizontal one. Name the following alkane.

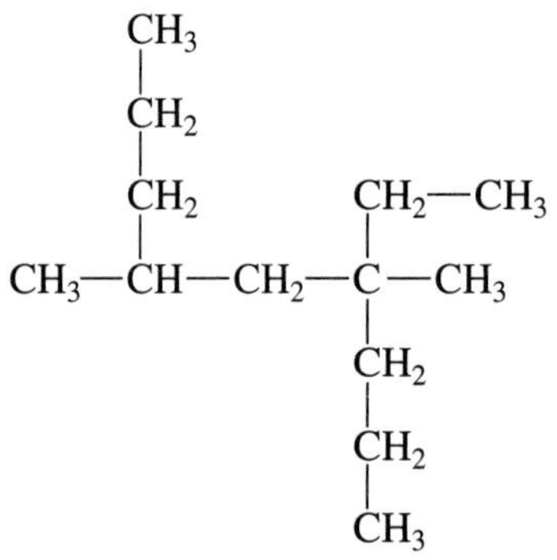

CH3
|
CH2
|
CH2   CH2—CH3
| |
CH3—CH—CH2—C—CH3
            |
            CH2
            |
            CH2
            |
            CH3

How many carbon atoms are in the main chain? If you said five, look again. There is a chain of nine carbon atoms that goes around some corners!

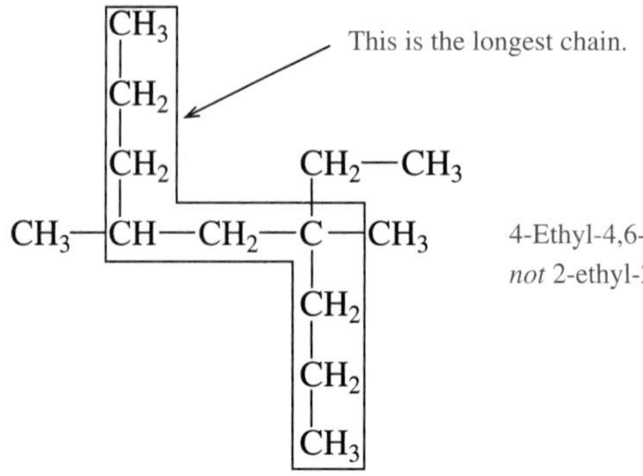

CH3   This is the longest chain.
|
CH2
|
CH2   CH2—CH3
| |
CH3—CH—CH2—C—CH3   4-Ethyl-4,6-dimethylnonane
            | *not* 2-ethyl-2,4-dipropylpentane
            CH2
            |
            CH2
            |
            CH3

SAMPLE PROBLEM 12.3

Writing IUPAC names

Give the IUPAC name for each of the following alkanes:

         CH3   CH3   CH3   CH3 CH3
          | | | | |
a. 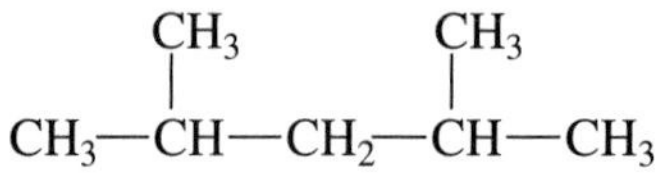 CH3—CH—CH2—CH—CH3   **b.** 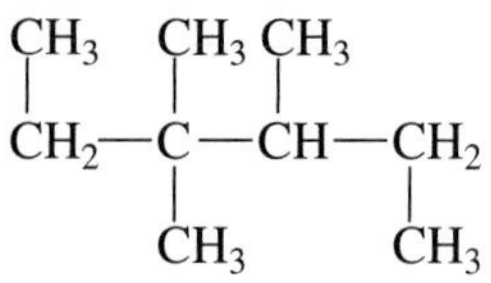 CH2—C—CH—CH2
                                               | |
                                              CH3   CH3

                  CH3
                   |
       CH3—CH2 CH—CH3   CH3
               | | |
c. CH3—CH2—CH—CH—CH2—CH—CH3

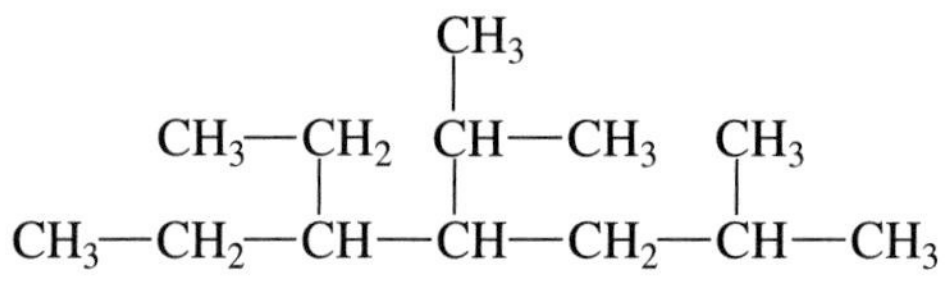

Solution

a. The main chain of five carbon atoms is *pentane.* The two methyl groups named *dimethyl* are on carbons 2 and 4. The IUPAC name is 2,4-dimethylpentane.

b. The longest chain of six carbons is *hexane.* There are three methyl groups indicated by the prefix *tri,* two on carbon 3 and one on carbon 4. The IUPAC name is 3,3,4-trimethylhexane.

c. The main chain with seven carbon atoms is *heptane.* Counting from *right to left,* there is a methyl group on carbon 2, an isopropyl on carbon 4, and an ethyl group on carbon 5. Placing the alkyl group names in alphabetical order gives the IUPAC name 5-ethyl-4-isopropyl-2-methylheptane.

Study Check

Give the IUPAC name of the following compound:

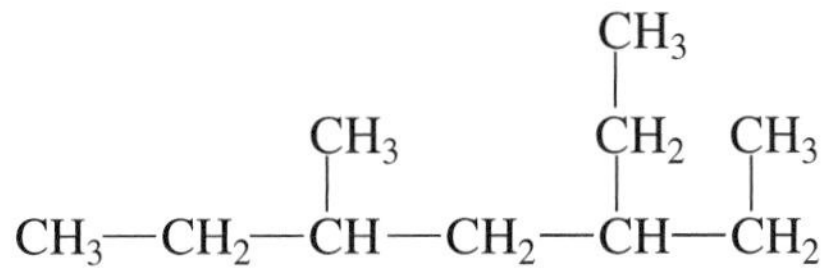

$$CH_3-CH_2-\underset{\underset{\displaystyle CH_3}{|}}{CH}-CH_2-\underset{\underset{\displaystyle CH_2}{|}\;\underset{\displaystyle CH_3}{}}{CH}-CH_2$$

with a CH₃ branch at top

QUESTIONS AND PROBLEMS

IUPAC Naming System for Alkanes

12.7 Indicate whether each of the following pairs of structural formulas represent constitutional isomers or different conformations.

a. 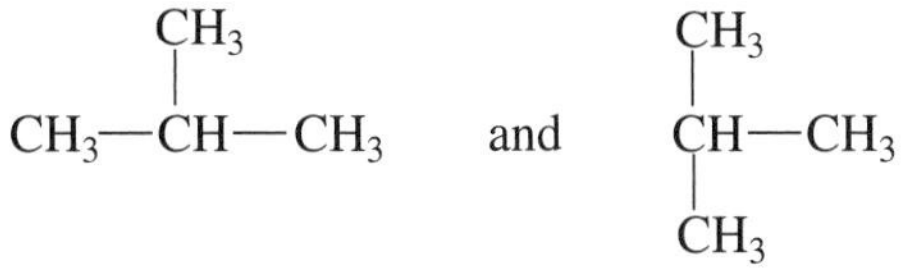 $CH_3-\underset{\underset{\displaystyle CH_3}{|}}{CH}-CH_3$ and $\underset{\underset{\displaystyle CH_3}{|}}{\overset{\overset{\displaystyle CH_3}{|}}{CH}}-CH_3$

b. 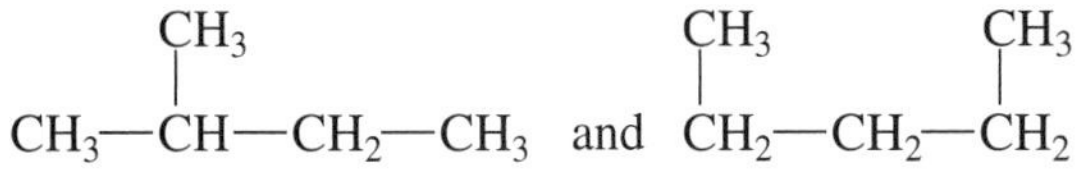 $CH_3-\underset{\underset{\displaystyle CH_3}{|}}{CH}-CH_2-CH_3$ and $\underset{\underset{\displaystyle CH_2}{|}}{\overset{\overset{\displaystyle CH_3}{|}}{CH_2}}-CH_2-\overset{\overset{\displaystyle CH_3}{|}}{}$

c. 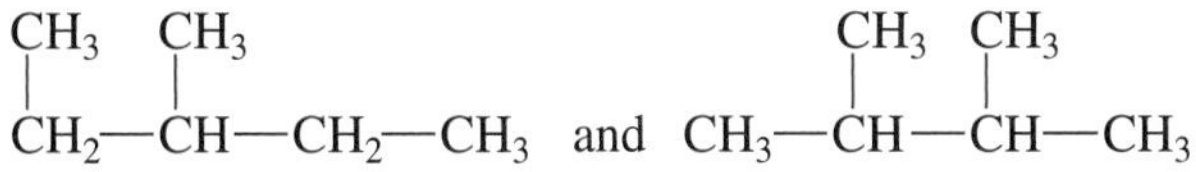 $\underset{\underset{\displaystyle CH_2}{|}}{\overset{\overset{\displaystyle CH_3}{|}}{CH_2}}-\overset{\overset{\displaystyle CH_3}{|}}{CH}-CH_2-CH_3$ and $CH_3-\overset{\overset{\displaystyle CH_3}{|}}{CH}-\overset{\overset{\displaystyle CH_3}{|}}{CH}-CH_3$

12.8 Indicate whether each of the following pairs of structural formulas represent constitutional isomers or different conformations.

a. 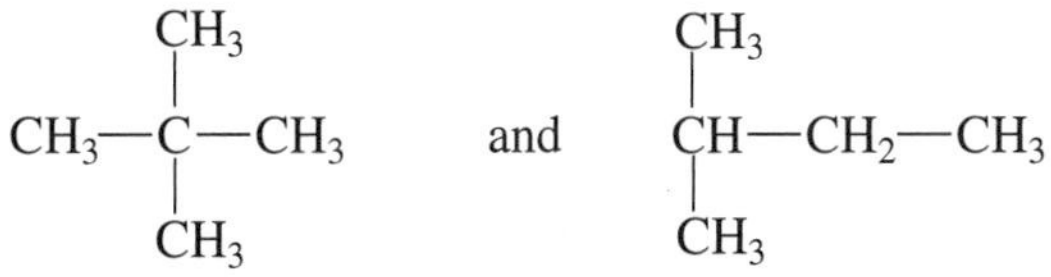 $CH_3-\underset{\underset{\displaystyle CH_3}{|}}{\overset{\overset{\displaystyle CH_3}{|}}{C}}-CH_3$ and $\underset{\underset{\displaystyle CH_3}{|}}{\overset{\overset{\displaystyle CH_3}{|}}{CH}}-CH_2-CH_3$

b. $CH_3-\overset{\overset{\displaystyle CH_3}{|}}{CH}-\overset{\overset{\displaystyle CH_3}{|}}{CH}-\overset{\overset{\displaystyle CH_3}{|}}{CH_2}$ and $CH_3-\overset{\overset{\displaystyle CH_3}{|}}{CH}-CH_2-\overset{\overset{\displaystyle CH_3}{|}}{CH}-CH_3$

c. $CH_3-\overset{\overset{\displaystyle CH_3}{|}}{CH}-CH_2-CH_3$ and $CH_3-CH_2-\overset{\overset{\displaystyle CH_3}{|}}{CH}-CH_3$

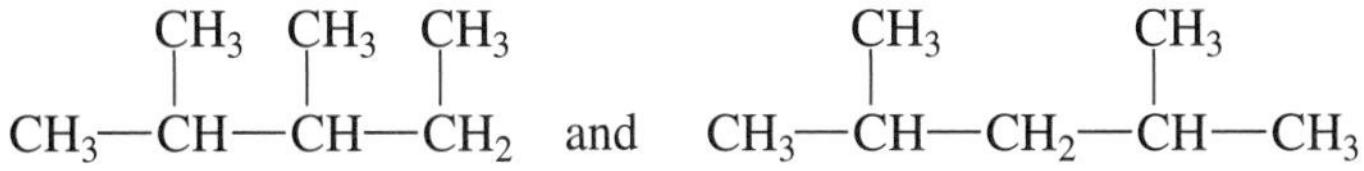

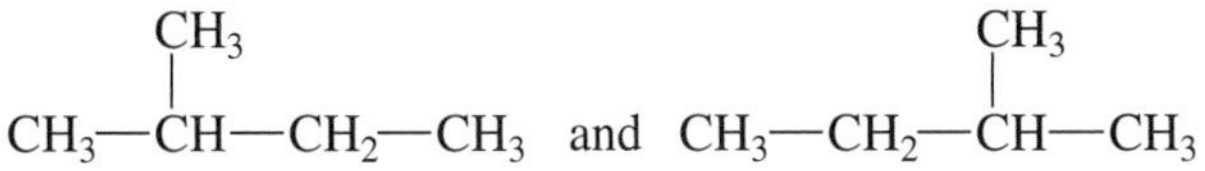

12.9 Name each of the following alkyl groups:

a. $CH_3-CH_2-CH_2-$ b. 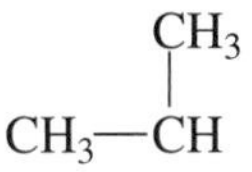 $CH_3-\overset{\overset{\displaystyle CH_3}{|}}{CH}$ c. $CH_3-CH_2-CH_2-CH_2-$ d. CH_3-

12.10 Name each of the following alkyl groups:

a. CH_3-CH_2- b. 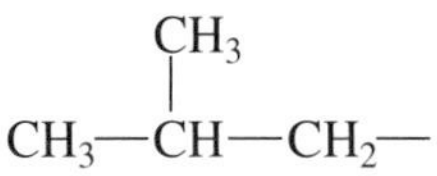 $CH_3-\overset{\overset{\displaystyle CH_3}{|}}{CH}-CH_2-$ c. 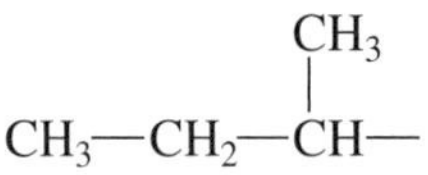 $CH_3-CH_2-\overset{\overset{\displaystyle CH_3}{|}}{CH}-$ d. 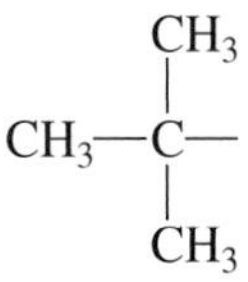 $CH_3-\underset{\underset{\displaystyle CH_3}{|}}{\overset{\overset{\displaystyle CH_3}{|}}{C}}-$

12.11 Give the IUPAC name for each of the following alkanes:

a. 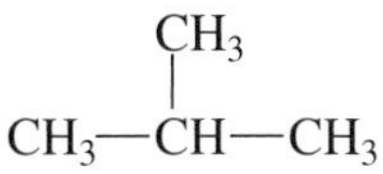 $CH_3-\overset{\overset{\displaystyle CH_3}{|}}{CH}-CH_3$ b. 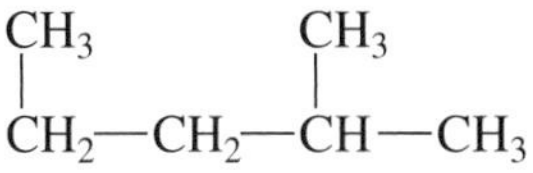 $\overset{\overset{\displaystyle CH_3}{|}}{CH_2}-CH_2-\overset{\overset{\displaystyle CH_3}{|}}{CH}-CH_3$ c. 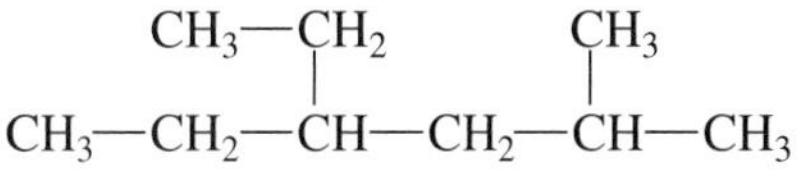 $CH_3-CH_2-\overset{\overset{\displaystyle CH_3-CH_2}{|}}{CH}-CH_2-\overset{\overset{\displaystyle CH_3}{|}}{CH}-CH_3$

$$CH_3-CH-CH_3$$
d. $CH_3-CH-CH-CH_2-CH_2-CH_3$
$$\qquad\qquad CH_3$$

$$\qquad\qquad\qquad\qquad CH_3$$
$$\qquad CH_3 \qquad CH-CH_3$$
e. $CH_3-CH_2-CH-CH_2-CH-CH_2-CH_2-CH_3$

12.12 Give the IUPAC name for each of the following alkanes:

$$\quad CH_3 \; CH_3$$
a. $CH_3-CH-CH-CH_3$

$$\qquad CH_3-CH_2\,CH_3$$
b. $CH_3-CH_2-C-CH-CH_2-CH_3$
$$\qquad\qquad\qquad CH_3$$

$$\qquad CH_3$$
$$\qquad CH_2$$
$$\qquad CH_2 \; CH_3 \qquad\; CH_3$$
c. $CH_3-CH_2-CH_2-CH-CH-CH_2-CH-CH_3$

$$\quad CH_3 \qquad CH_2-CH_3$$
d. $CH_2-CH-CH-CH_2-CH_2-CH_3$
$$\qquad\qquad\; CH_3$$

$$\qquad\qquad\qquad CH_3$$
$$\qquad\quad CH_3 \; CH-CH_3$$
e. $CH_3-CH_2-CH_2-CH-CH-CH_2-CH-CH_2-CH_3$
$$\qquad\qquad\qquad\qquad CH_2-CH_3$$

12.13 Give the IUPAC name for each of the following line-bond formulas:

a. **b.** **c.**

12.14 Give the IUPAC name for each of the following line-bond formulas:

a. **b.** **c.**

12.3 Drawing Structural Formulas

Suppose you are asked to draw the condensed structural formula of 2,3-dimethyl-hexane. The IUPAC name gives all the information needed to draw the structure. The alkane name at the end gives the number of carbon atoms in the main chain. The first part of the IUPAC name gives the substituents and where they are attached. We can break down the name in the following way.

2,3-Dimethylhexane				
2,3-	di	methyl	hex	ane
Substituents on carbons 2 and 3	Two identical groups	CH_3- alkyl group	6 Carbon atoms in main chain	Single C—C bonds

To draw the condensed structural formula, we proceed as follows:

Step 1   Draw the main chain of carbon atoms.

C—C—C—C—C—C   2,3-Dimethyl**hexane**

Step 2   Draw the substituents on the carbon atoms indicated by the location numbers.

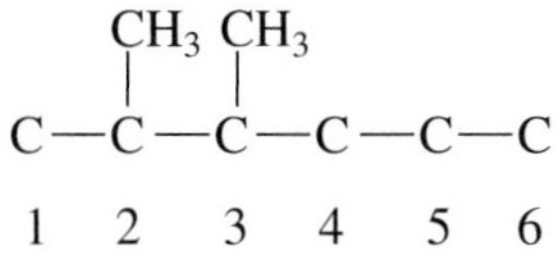

$$CH_3 \quad CH_3$$
$$\underset{1 \quad 2 \quad 3 \quad 4 \quad 5 \quad 6}{C-C-C-C-C-C}$$ **2,3-Dimethyl**hexane

Step 3   Add the correct number of hydrogen atoms to give four bonds to each carbon atom.

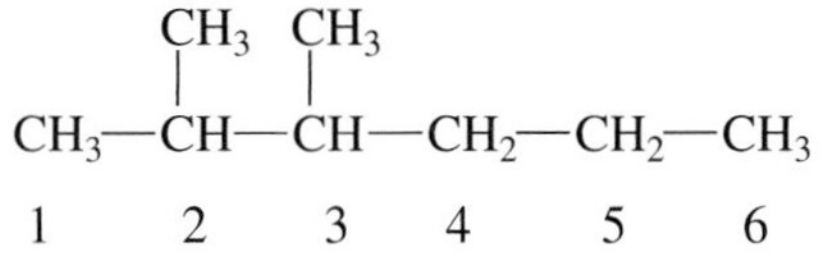

$$CH_3 \quad CH_3$$
$$\underset{1 \qquad 2 \qquad 3 \qquad 4 \qquad 5 \qquad 6}{CH_3-CH-CH-CH_2-CH_2-CH_3}$$ 2,3-Dimethylhexane

---

SAMPLE PROBLEM 12.4

Drawing Structural Formulas from IUPAC Names

Draw the structural formulas for each of the following alkanes:

a. 3-ethyl-5-methylheptane

b. 4-*tert*-butyloctane

Solution

a. Using the alkane name, heptane, at the end of the IUPAC name, draw a chain of seven carbon atoms and number it.

$$\underset{1 \quad 2 \quad 3 \quad 4 \quad 5 \quad 6 \quad 7}{C-C-C-C-C-C-C}$$

The beginning of the name indicates an ethyl group on carbon 3 and a methyl group on carbon 5.

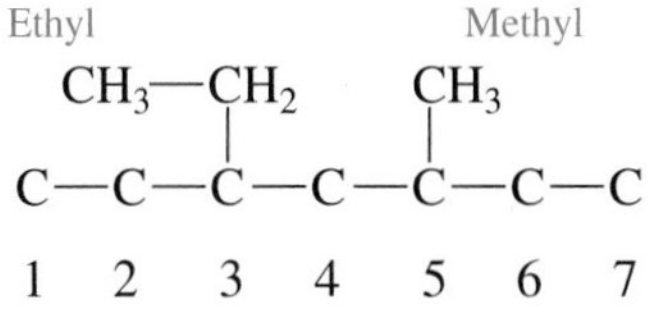

Ethyl   Methyl
$$CH_3-CH_2 \qquad CH_3$$
$$\underset{1 \quad 2 \quad 3 \quad 4 \quad 5 \quad 6 \quad 7}{C-C-C-C-C-C-C}$$

Complete the formula by adding the correct number of hydrogen atoms to each carbon.

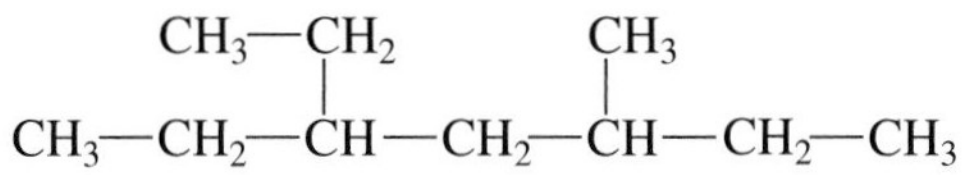

$$CH_3-CH_2 \qquad\qquad CH_3$$
$$CH_3-CH_2-CH-CH_2-CH-CH_2-CH_3$$

3-Ethyl-5-methylheptane

b. Draw a main chain of eight carbon atoms with a *tert*-butyl group on carbon 4.

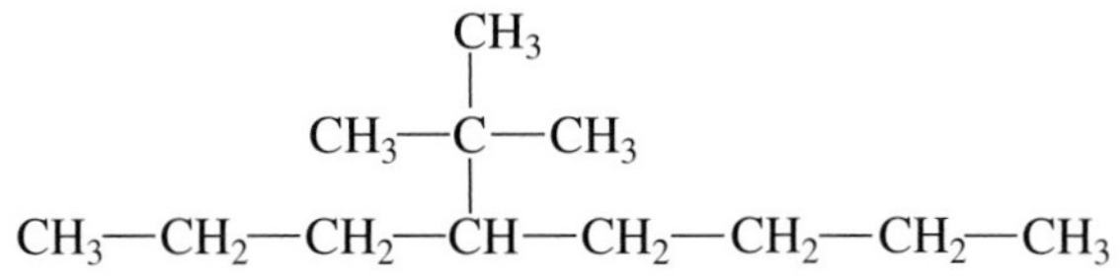

$$CH_3$$
$$CH_3-C-CH_3$$
$$CH_3-CH_2-CH_2-CH-CH_2-CH_2-CH_2-CH_3$$

4-*tert*-Butyloctane

What is the structural formula for 3-ethyl-2,4-dimethylpentane?

Drawing Constitutional Isomers

A common problem in organic chemistry is how to draw the constitutional isomers of a particular molecular formula such as C_6H_{14}. Writing the continuous chain of six carbon atoms is obvious, but drawing the other isomers can get more difficult. Using this example, we can proceed to draw the condensed structural formulas and the line-bond formulas as follows.

A System for Writing Constitutional Isomers of an Alkane

Step 1   Draw the longest continuous (unbranched) chain.

In this example, we draw the structure formula of a carbon chain with six carbon atoms.

$$CH_3-CH_2-CH_2-CH_2-CH_2-CH_3 \qquad \diagup\diagdown\diagup\diagdown\diagup\diagdown \qquad \text{Hexane}$$

Step 2   Remove one carbon from the chain and attach it as a methyl group to the shorter main chain in as many different locations as possible.

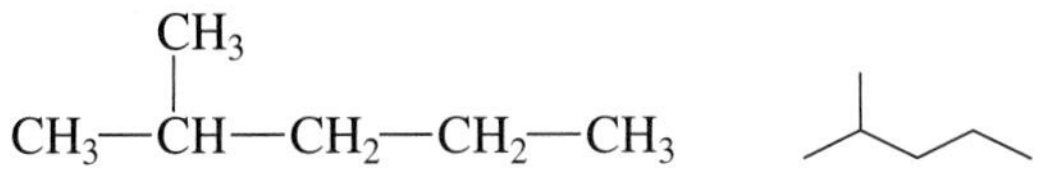

$$\begin{array}{c} CH_3 \\ | \\ CH_3-CH-CH_2-CH_2-CH_3 \end{array} \qquad \qquad \text{2-Methylpentane}$$

$$\begin{array}{c} CH_3 \\ | \\ CH_3-CH_2-CH-CH_2-CH_3 \end{array} \qquad \qquad \text{3-Methylpentane}$$

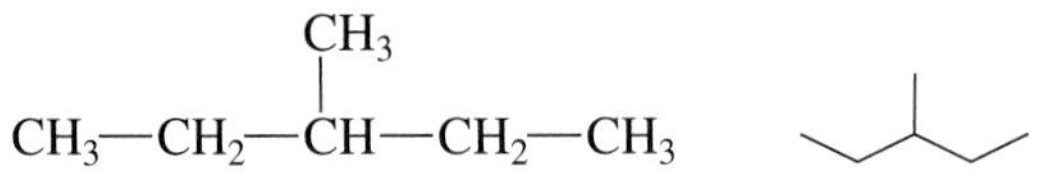

Make sure that you do not repeat one of the isomers. Each different isomer has a different name. For example, moving the methyl group to the next carbon atom in pentane repeats the structure and name 2-methylpentane.

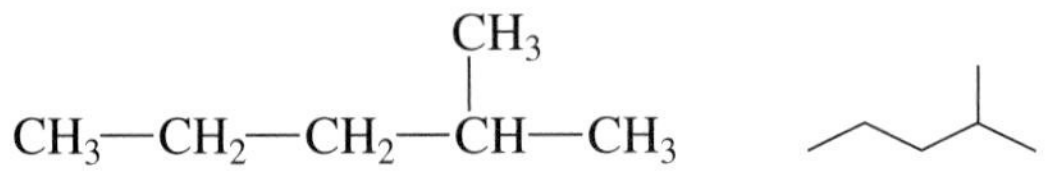

$$\begin{array}{c} CH_3 \\ | \\ CH_3-CH_2-CH_2-CH-CH_3 \end{array} \qquad \qquad \text{2-Methylpentane}$$

Step 3   Remove another carbon atom from the main chain and attach as another alkyl group.

As we proceed, the substitution increases for the middle carbon atoms in the shortened main chain

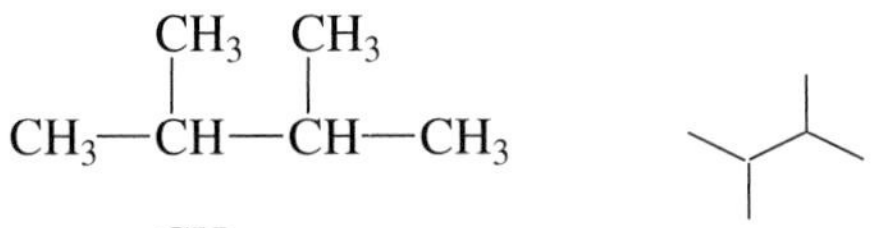

$$\begin{array}{c} CH_3 \quad CH_3 \\ | \qquad | \\ CH_3-CH-CH-CH_3 \end{array} \qquad \qquad \text{2,3-Dimethylbutane}$$

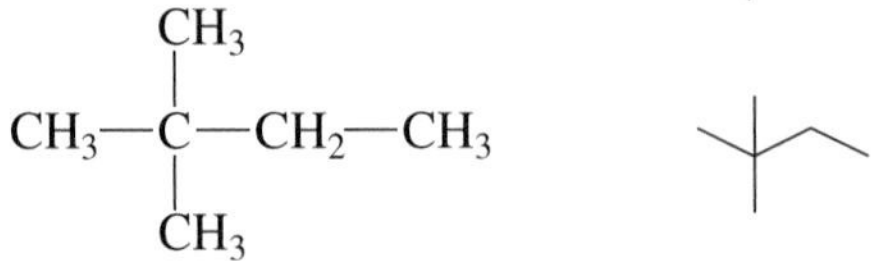

$$\begin{array}{c} CH_3 \\ | \\ CH_3-C-CH_2-CH_3 \\ | \\ CH_3 \end{array} \qquad \qquad \text{2,2-Dimethylbutane}$$

The chain cannot be shortened again because one of the above structures would be repeated. Therefore, we have drawn five constitutional isomers for C_6H_{14}.

SAMPLE PROBLEM 12.5

Drawing Constitutional Isomers

There are three constitutional isomers with molecular formula C_5H_{12}. Draw and name each of their condensed structural formulas.

Solution

One isomer is a main chain of five carbon atoms.

$$CH_3—CH_2—CH_2—CH_2—CH_3 \quad \text{Pentane}$$

Shorten the main chain by one carbon and attach a methyl group.

$$\begin{array}{c} CH_3 \\ | \\ CH_3—CH—CH_2—CH_3 \end{array} \quad \text{2-Methylbutane}$$

Shorten the main chain again and attach two methyl groups.

$$\begin{array}{c} CH_3 \\ | \\ CH_3—C—CH_3 \\ | \\ CH_3 \end{array} \quad \text{2,2-Dimethylpropane}$$

Study Check

What are the names of four constitutional isomers of dimethylpentane?

QUESTIONS AND PROBLEMS

Drawing Structural Formulas

12.15 Draw a condensed structural formula for each of the following alkanes:
   a. 2-methylbutane **b.** 3,3-dimethylpentane
   c. 2,3,5-trimethylhexane **d.** 3-ethyl-2,5-dimethyloctane
   e. 4-isopropyl-2-methylheptane **f.** 4-propylnonane

12.16 Draw a condensed structural formula for each of the following alkanes:
   a. 3-ethylpentane
   b. 3-ethyl-2-methylpentane
   c. 4-propylheptane
   d. 2,2,3,5-tetramethylhexane
   e. 4-ethyl-2,2-dimethyloctane
   f. 3-ethyl-4-isobutyl-2,3-dimethyldecane

12.17 Give the IUPAC names for the constitutional isomers of each of the following:
   a. methylheptane
   b. C_7H_{16} with a main chain of 5 carbon atoms

12.18 Give the IUPAC names for the constitutional isomers of each of the following:
   a. methyloctane **b.** dimethylhexane

12.4 Haloalkanes

A **haloalkane** is an alkane in which halogen atoms replace one or more hydrogen atoms in a carbon chain. In the IUPAC system, halogen atoms are named as substituents: fluorine is *fluoro*, chlorine is *chloro*, bromine is *bromo*, and iodine is *iodo*. The halo-substituents are numbered and arranged alphabetically just as we did with the alkyl groups. Simple haloalkanes are commonly named as alkyl halides; the carbon group is named as an alkyl group followed by the halide name.

Methylene chloride

Chloroform

Carbon tetrachloride

$$CH_3—Cl \qquad CH_3—CH_2—Br \qquad \overset{\displaystyle F}{\underset{}{CH_3—\overset{|}{C}H—CH_3}}$$

IUPAC:	chloromethane	bromoethane	2-fluoropropane
Common:	methyl chloride	ethyl bromide	isopropyl fluoride

$$Cl—CH_2—CH_2—Br \qquad CH_3—CH_2—\overset{\displaystyle Cl}{\overset{|}{C}H}—\overset{\displaystyle CH_3}{\overset{|}{C}H}—CH_3$$

IUPAC: 1-bromo-2-chloroethane   3-chloro-2-methylpentane

Methane compounds with two or more halogens have another set of common names that do not indicate their structures.

$$Cl—CH_2—Cl \qquad Cl—\overset{\displaystyle Cl}{\overset{|}{C}H}—Cl \qquad Cl—\overset{\displaystyle Cl}{\underset{\displaystyle Cl}{\overset{|}{\underset{|}{C}}}}—Cl$$

CH_2Cl_2 $CHCl_3$ CCl_4

IUPAC: dichloromethane   trichloromethane   tetrachloromethane

Common: methylene chloride   chloroform   carbon tetrachloride

SAMPLE PROBLEM 12.6

Naming Haloalkanes

Freon 11 and Freon 12 are compounds known as chlorofluorocarbons (CFCs) and are widely used as refrigerants and aerosol propellants. What are their IUPAC names?

HEALTH NOTE

Common Uses of Halogenated Alkanes

Some common uses of halogenated alkanes include solvents and anesthetics. For many years, carbon tetrachloride was widely used by dry cleaners and in-home spot removers to take oils and grease out of clothes. However this use was discontinued when it was found to be toxic to the liver, where it can cause cancer. Today, dry cleaners use other halogenated compounds such as methylene chloride, 1,1,1-trichloroethane, and 1,1,2-trichloro-1,2,2-trifluoroethane.

$$F—\overset{\displaystyle F}{\underset{\displaystyle F}{\overset{|}{\underset{|}{C}}}}—\overset{\displaystyle Cl}{\underset{\displaystyle H}{\overset{|}{\underset{|}{C}}}}—Br \qquad \text{Halothane (Fluothane)}$$

For minor surgeries, a local anesthetic such as chloroethane (ethyl chloride) $CH_3—CH_2—Cl$ is applied to an area of the skin where it evaporates quickly, which cools the skin causing a loss of sensation.

CH_2Cl_2 $Cl_3C—CH_3$ $FCl_2C—CClF_2$

methylene   1,1,1-trichloroethane   1,1,2-trichloro-
chloride   1,2,2-trifluoroethane

General anesthetics are compounds that are inhaled or injected to cause a loss of sensation so that surgery or other procedures can be done without causing pain to the patient. As nonpolar compounds they are soluble in the nonpolar nerve membranes, where they decrease the ability of the nerve cells to conduct the sensation of pain. Chloroform $CHCl_3$ was once used as an anesthetic, but it is toxic and may be carcinogenic. One of the most widely used general anesthetics is halothane, also called Fluothane. It has a pleasant odor, is nonexplosive, has few side effects, undergoes few reactions within the body, and is eliminated quickly.

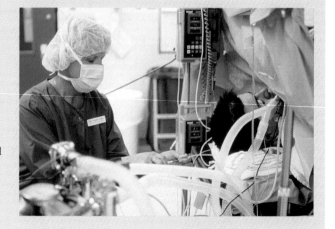

Cl
|
Cl—C—F
|
Cl

Freon 11

Cl
|
F—C—F
|
Cl

Freon 12

Solution

Freon 11, trichlorofluoromethane; Freon 12, dichlorodifluoromethane.

Study Check

Halothane or Fluothane is the commercial name of a haloalkane widely used as an anesthetic. What is its IUPAC name?

ENVIRONMENTAL NOTE

CFCs and Ozone Depletion

The compounds called chlorofluorocarbons (CFCs) were used as propellants for hairs sprays, paints, and as refrigerants in home and car air conditioners. Two widely used CFCs, Freon 11 (CCl_3F) and Freon 12 (CCl_2F_2), were developed in the 1920s as nontoxic refrigerants, which were safer than the sulfur dioxide and ammonia used at the time.

Cl
|
Cl—C—F
|
Cl

Freon 11

Cl
|
F—C—F
|
Cl

Freon 12

In the stratosphere, a layer of ozone (O_3) absorbs the ultraviolet (UV) radiation of the sun and acts as a protective shield for plants and animals on earth. Ozone is produced in the stratosphere when oxygen reacts with ultraviolet light and breaks into oxygen atoms that quickly combine with oxygen molecules to form ozone.

$$O_2 \xrightarrow{\text{UV}} O + O$$
$$O_2 + O \longrightarrow O_3$$

In the 1970s scientists became concerned that CFCs entering the atmosphere were accelerating the depletion of ozone and threatening the stability of the ozone layer. CFCs decompose in the upper atmosphere in the presence of UV light to produce highly reactive chlorine atoms.

$$CCl_3F \xrightarrow{\text{UV light}} CCl_2F + Cl$$

The reactive chlorine atoms catalyze the breakdown of ozone molecules.

$$Cl + O_3 \longrightarrow ClO + O_2$$
$$ClO + O_3 \longrightarrow Cl + 2O_2$$

It has been estimated that one chlorine atom can destroy as many as 100,000 ozone molecules. Normally, there is a balance between the ozone and oxygen in the atmosphere but the rapid destruction of ozone has upset that equilibrium.

Reports of polar ozone depletion over Antarctica in March 1985 prompted scientists to call for a freeze on the production of CFCs. In some areas as much as 50% of the ozone had been depleted, and at certain times of the year an ozone hole appears.

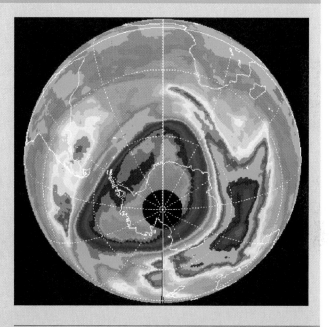

In the color image, the pink areas have the lowest levels of ozone.

There is evidence of thinning in the ozone layer over the Arctic as well but to a somewhat lesser degree due to warmer temperatures. It is interesting that in the lower atmosphere, ozone is an automobile and industrial pollutant, but in the stratosphere ozone is a life-protecting compound.

Today the use of CFCs is being phased out. However, it is expected that ozone levels will remain low for several decades due to the stability of CFCs. Chemical companies are developing substitutes to CFCs that are not as damaging to the ozone. Replacement compounds such as hydrochlorofluorocarbons (HCFCs) contain chlorine atoms, but these compounds break down in the lower atmosphere reducing the amount of chlorine that reaches the stratosphere. Hydrofluorocarbons (HFCs), which contain no chlorine, are being considered as another replacement for CFCs. However, the potential effects of fluorine compounds on ozone destruction must be determined.

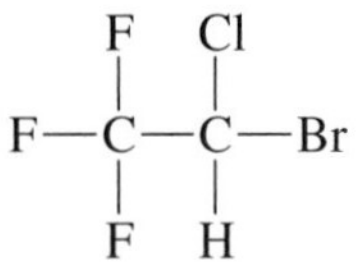

$$F-\underset{\underset{\displaystyle F}{|}}{\overset{\overset{\displaystyle F}{|}}{C}}-\underset{\underset{\displaystyle H}{|}}{\overset{\overset{\displaystyle Cl}{|}}{C}}-Br \qquad \text{Halothane}$$

QUESTIONS AND PROBLEMS

Haloalkanes

12.19 Give the IUPAC and common names for each of the following compounds:

a. CH_3-CH_2-Br **b.** $CH_3-CH_2-CH_2-F$

c. 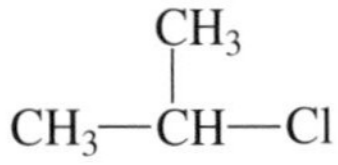 $CH_3-\underset{\underset{\displaystyle}{}}{\overset{\overset{\displaystyle CH_3}{|}}{C}H}-Cl$ **d.** $CHCl_3$

12.20 Give the IUPAC and common names for each of the following compounds:

a. 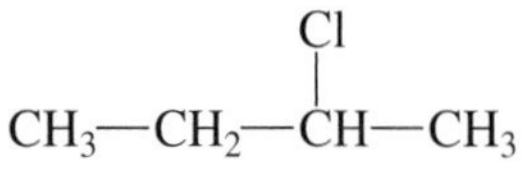 $CH_3-CH_2-\underset{\underset{\displaystyle}{}}{\overset{\overset{\displaystyle Cl}{|}}{C}H}-CH_3$ **b.** CCl_4

c. 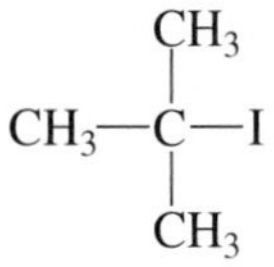 $CH_3-\underset{\underset{\displaystyle CH_3}{|}}{\overset{\overset{\displaystyle CH_3}{|}}{C}}-I$ **d.** CH_3F

12.21 Give the IUPAC name for each of the following compounds:

a. 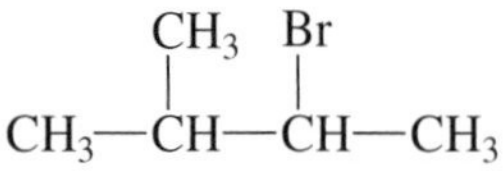 $CH_3-\underset{\underset{\displaystyle}{}}{\overset{\overset{\displaystyle CH_3}{|}}{C}H}-\underset{\underset{\displaystyle}{}}{\overset{\overset{\displaystyle Br}{|}}{C}H}-CH_3$ **b.**

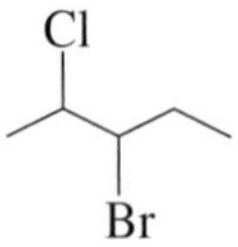

c. 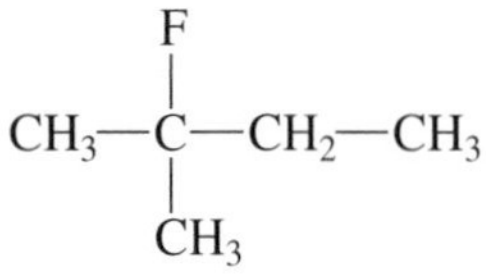 $CH_3-\underset{\underset{\displaystyle CH_3}{|}}{\overset{\overset{\displaystyle F}{|}}{C}}-CH_2-CH_3$

12.22 Give the IUPAC name for each of the following compounds:

a. 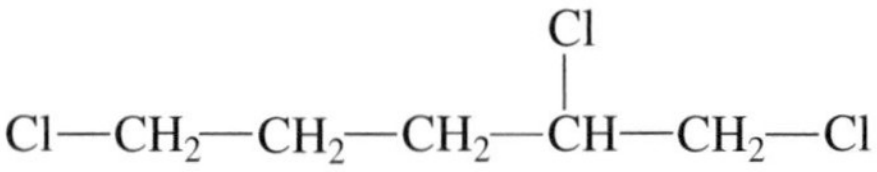 $Cl-CH_2-CH_2-CH_2-\underset{\underset{\displaystyle}{}}{\overset{\overset{\displaystyle Cl}{|}}{C}H}-CH_2-Cl$ **b.**

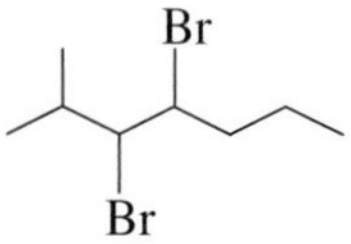

c. 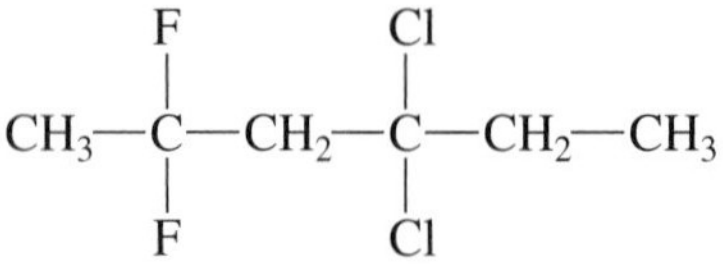 $CH_3-\underset{\underset{\displaystyle F}{|}}{\overset{\overset{\displaystyle F}{|}}{C}}-CH_2-\underset{\underset{\displaystyle Cl}{|}}{\overset{\overset{\displaystyle Cl}{|}}{C}}-CH_2-CH_3$

12.23 Write the condensed structural formula for each of the following compounds:

a. 2-chloropropane
b. 2-bromo-3-chlorobutane
c. methyl bromide
d. butyl chloride
e. 1,1-dibromo-2-chloro-4-fluoropentane
f. tetrabromomethane

12.24 Write the condensed structural formula for each of the following compounds:

a. 1,1,2,2-tetrabromopropane b. 2,2-dibromo-3,4-dichloropentane

c. isopropyl bromide   d. methylene chloride

e. 2,3-dichloro-2-methybutane f. dibromodichloromethane

12.25 The topical anesthetics, methyl chloride and ethyl chloride, sprayed on the skin lower the temperature of the skin and numb nerve endings. What are the structural formulas of the anesthetics?

12.26 The isomers of $C_2H_4Cl_2$ are used as industrial solvents. Write the condensed structural formulas of two isomers and give their IUPAC names.

12.27 Write four isomers of C_4H_9Cl and give their IUPAC names.

12.28 Write four isomers of $C_3H_6F_2$ and give their IUPAC names.

12.5 Cycloalkanes

LEARNING GOAL

Give the IUPAC name for a cycloalkane; draw the structural formulas from the IUPAC name.

Up to now, we have looked at carbon atoms in continuous or branched chains. However, hydrocarbons can also form cyclic structures called **cycloalkanes.** Cycloalkanes, which have a general formula of C_nH_{2n}, have two fewer hydrogen atoms than the corresponding alkanes. Thus, the simplest cycloalkane, cyclopropane, C_3H_6, has a ring of three carbon atoms bonded to six hydrogen atoms. There are several ways to draw the structural formula of cyclopropane. The line-bond formula, which omits the hydrogen atoms and looks like a simple geometric figure, is a convenient way to show cyclic structures. Each corner of the triangle represents a carbon atom with four bonds to other carbon and hydrogen atoms.

Cyclopropane C_3H_6

The ball-and-stick models and their structural formulas for several cycloalkanes are shown in Table 12.6.

IUPAC Names for Cycloalkanes

The names of cycloalkanes are similar to the names of the continuous-chain alkanes except that the prefix *cyclo* appears in front of the name of the alkane. When one substituent is attached to a carbon atom in the ring, the name of the substituent is placed in front of the cycloalkane name. No number is needed for a single alkyl group or halogen atom because the carbon atoms in the cycloalkane are equivalent. However, if two or more groups are attached, the ring is numbered to show the location of each group. The numbering starts by assigning carbon 1 to the substituent that gives the lowest numbers to the other substituents. Therefore, we may count clockwise or counterclockwise around a cycloalkane to give the lowest combination of numbers to the substituents.

methylcyclopentane

1,3-dimethylcyclopentane
not 2,4-dimethylcyclohexane

1-chloro-3-methylcyclohexane
not 1-methyl-3-chlorocyclohexane

2,4-dichloro-1-methylcyclohexane
not 1,3-dichloro-4-methylcyclohexane

Table 12.6   Formulas of Some Common Cycloalkanes

Condensed Structural Formula	Geometric Formula	Name	
CH_2 $H_2C{-}CH_2$	△	Cyclopropane	
$H_2C{-}CH_2$ $H_2C{-}CH_2$	□	Cyclobutane	
CH_2 $H_2C \quad CH_2$ $H_2C{-}CH_2$	⬠	Cyclopentane	
CH_2 $H_2C \quad CH_2$ $H_2C \quad CH_2$ CH_2	⬡	Cyclohexane	

SAMPLE PROBLEM 12.7

Naming Cycloalkanes

Give the IUPAC name for each of the following cycloalkanes:

a. 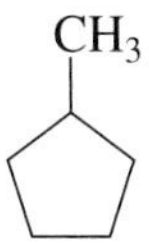 CH_3

b. 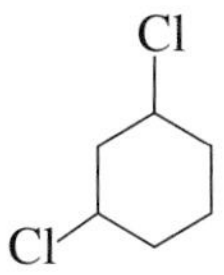   Cl

c. 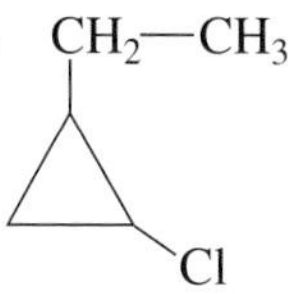 CH_2—CH_3

Cl

Cl

Cl

Solution

a. The pentagon indicates a main chain that is a ring with five carbon atoms, cyclopentane. The single substituent is a methyl group. The compound is named methylcyclopentane.

b. The ring of six carbon atoms is numbered to give the lowest numbers starting with carbon 1. It is named 1,3-dichlorocyclohexane.

c. The groups are numbered starting from carbon 1 with the substituent that is first alphabetically. It is named 1-chloro-2-ethylcyclopropane.

Study Check

What is the IUPAC name of the following compound?

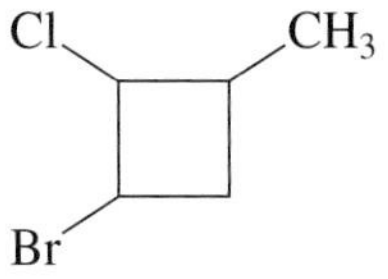

Cl CH_3

Br

SAMPLE PROBLEM 12.8

Writing Structural Formulas for Cycloalkanes

Write the structural formula for 1-chloro-4-isopropyl cyclohexane.

Solution

The name cyclohexane tells us that the substituents are bonded to a ring of six carbon atoms. The first part indicates a chlorine atom on carbon 1 and an isopropyl group on carbon 4.

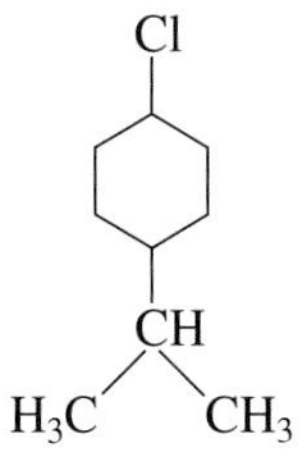

Cl

CH

H_3C CH_3

Study Check

What is the structural formula for 1-bromo-3-ethylcyclopentane?

Constitutional Isomers of Cycloalkanes

Constitutional isomers are possible for cycloalkanes. For example, the molecular formula C_5H_{10} can be written with rings containing three, four, or five carbon atoms.

Constitutional Isomers of C_5H_{10}

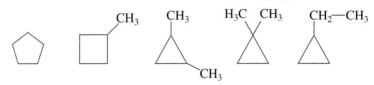

Cis–Trans Isomers

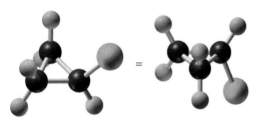

We saw that alkanes have rotation around the carbon–carbon single bonds. However, in cycloalkanes, there is no rotation of the carbon atoms in the ring, which gives the carbon ring two distinct sides. This structural characteristic produces stereoisomers called **cis–trans isomers** that differ only in the orientation of atoms in space. For example, chlorocyclopropane has just one isomer. Writing the chlorine atom on the "up" side of the ring is the same as writing it on the "down" side. By turning over the molecule, the chlorine atom will point down instead of up.

However, there are two stereoisomers of 1,2-dichlorocyclopropane. In the cis isomer, the two chlorine atoms are on the same side; both can be written "up" or both can be written "down." In the trans isomer, the chlorine atoms are on opposite sides of the carbon ring; if one is written "up," the other is "down." There is no way to change from one isomer into the other without breaking the bonds in the cyclic carbon structure. (See Figure 12.5.)

Chlorine atoms are
on the same side

cis-1,2-dichlorocyclopropane

Chlorine atoms are
on opposite sides

trans-1,2-dichlorocyclopropane

Figure 12.5 The cis–trans isomers of 1,2-dichlorocyclopropane are called *cis*-1,2-dichlorocyclopropane and *trans*-1,2-dichloropropane.

Q What prevents the molecules from changing between the cis- and trans isomers?

SAMPLE PROBLEM 12.9

Cis–Trans Isomers

Name each of the following as cis or trans isomers.

a.

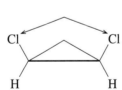

b.

c.

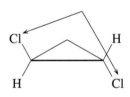

Solution

a. Two methyl groups are on opposite sides of cyclobutane, which gives the name *trans*-1,3-dimethylcyclobutane.

b. Two bromine atoms groups are on the same side of cyclopentane; *cis*-1,3-dibromocyclopentane.

c. Two methyl groups are on opposite sides of a cyclohexane; *trans*-1,2-dimethylcyclohexane.

Study Check

Draw the structural formula of *cis*-1,2-dimethylcyclopropane.

QUESTIONS AND PROBLEMS

Cycloalkanes

12.29 Use the general formula for cycloalkanes to determine the following:
    a. The molecular formula of a cycloalkane with 5 carbon atoms.
    b. The number of hydrogen atoms in a cycloalkane with 4 carbon atoms.
    c. The number of carbon atoms in a cycloalkane bonded to 12 hydrogen atoms.

12.30 Use the general formula for cycloalkanes to determine the following:
    a. The molecular formula of a cycloalkane with 6 carbon atoms.
    b. The number of hydrogen atoms in a cycloalkane with 5 carbon atoms.
    c. The number of carbon atoms in a cycloalkane bonded to 16 hydrogen atoms.

12.31 Give the IUPAC name for each of the following cycloalkanes:

a. **b.** **c.**

d. **e.** **f.**

12.32 Give the IUPAC name for each of the following cycloalkanes.

a. **b.** **c.**

d. **e.** **f.**

12.33 Draw the structural formulas for each of the following cycloalkanes.
    a. methylcyclopentane
    b. isopropylcyclohexane

c. 1,3-dimethylcyclobutane

d. 1-bromo-2,3-dimethylcyclopentane

12.34 Draw the structural formula for each of the following cycloalkanes.

    a. bromocyclopropane

    b. ethylcyclohexane

    c. 1,2-dichlorocyclobutane

    d. 1,3-dibromocyclopentane

12.35 Give the names of three cycloalkanes that are constitutional isomers of cyclopentane.

12.36 Give the names of four cycloalkanes that are constitutional isomers of cyclohexane.

12.37 Identify whether each of the following structures is a cis or trans isomer. If so, give its IUPAC name using the cis or trans prefix.

a. 　**b.** 　**c.** 　**d.**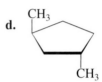

12.38 Identify whether each of the following structures is a cis or trans isomer. If so, give its IUPAC name using the cis or trans prefix.

a. 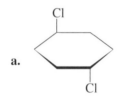　**b.** 　**c.** 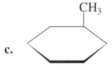　**d.**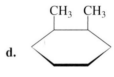

12.39 Draw structural formulas for each of the following compounds.

    a. *cis*-1,2-dimethylcyclopentane

    b. *trans*-1,3-dichlorocyclohexane

    c. *cis*-1-bromo-2-methylcyclobutane

12.40 Draw structural formulas for each of the following compounds.

    a. *trans*-1,2-dibromocyclopropane

    b. *cis*-1,4-dichlorocyclohexane

    c. *trans*-1-chloro-2-ethylcyclopentane

LEARNING GOAL

Identify the physical properties of alkanes and cycloalkanes.

12.6 Physical Properties of Alkanes and Cycloalkanes

Useful alkanes include the gasoline and diesel fuels that power our cars and the heating oils that heat our homes. You may have used a mixture of hydrocarbons such as mineral oil or petrolatum as a laxative or to soften your skin. The differences in uses of many of the alkanes and cycloalkanes result from their physical properties including solubility, density, and boiling point.

Solubility and Density

Alkanes and cycloalkanes are nonpolar, which makes them insoluble in water. However, they are soluble in nonpolar solvents such as other alkanes or cycloalkanes. Alkanes have densities from 0.65 g/mL to about 0.70 g/mL, which is less dense than the density of water (1.0 g/mL). If there is an oil spill in the ocean, the alkane mixtures in the crude oil remain on the surface and spread over a large area. In the *Exxon Valdez* oil spill in 1989, 40 million liters of oil covered over 25,000 square kilometers of water in Prince William Sound, Alaska. (See Figure 12.6.) If the crude oil reaches the beaches and inlets, there can be considerable damage to beaches, shellfish, fish, birds, and wildlife habitats. Even today, there is still oil on the surface, or just beneath the surface, in some areas of Prince William Sound. Cleanup includes both mechanical and chemical methods. In one method, a nonpolar compound that is "oil-attracting" is used to pick up oil, which is then scraped off into recovery tanks.

Melting and Boiling Points

Alkanes have the lowest melting and boiling points of all the organic compounds. The attractions between nonpolar alkanes in the solid and liquid states result from dispersion forces. In longer carbon chains, the greater number of electrons produces more attractions between molecules, which results in higher melting and boiling points.

CH_4 $CH_3 — CH_3$ $CH_3 — CH_2 — CH_3$
methane, bp −164°C   ethane, bp −89°C   propane, bp −42°C

The boiling points of branched alkanes are generally lower than continuous alkanes with the same number of carbon atoms. The branched chain alkanes tend to be more compact, which reduces the amount of contact between the molecules. Cycloalkanes have higher boiling points than the continuous-chain alkanes. Because rotation of carbon bonds is restricted, cycloalkanes maintain a rigid structure. Those rigid structures are like a set of dishes that can be stacked closely together with many points of contact and therefore attractions to each other. We can compare the boiling points of alkanes and cycloalkanes with five carbon atoms as shown in Table 12.7.

Some Uses of Alkanes

The first four alkanes—methane, ethane, propane, and butane—are gases at room temperature and are widely used as heating fuels. Tanks of liquid propane and butane are used to provide fuels for heating homes and cooking on the barbecues. As chlorofluorocarbons (CFCs) are phased out, butane has replaced Freon as a propellant in aerosol containers.

Alkanes having 5–8 carbon atoms (pentane, hexane, heptane, and octane) are liquids at room temperature. They are highly volatile, which makes them useful in fuels such as gasoline. Liquid alkanes with 9–17 carbon atoms have higher boiling points and are found in kerosene, diesel, and jet fuels. Motor oil is a mixture of high-molecular-weight liquid hydrocarbons and is used to lubricate the internal components of engines. Mineral oil is a mixture of liquid hydrocarbons and is used as a laxative and a lubricant. Alkanes with 18 or more carbon atoms are waxy solids at room temperature. The high-molecular-weight alkanes, known as paraffins, are used in waxy coatings of fruits and vegetables to retain moisture, inhibit mold growth, and enhance appearance. (See Figure 12.7.) Petrolatum, or Vaseline, is a mixture of liquid hydrocarbons, which have low boiling points, that are encapsu-

Figure 12.6 In oil spills, large quantities of oil spread over the water.

Q What physical properties cause oil to remain on the surface of water?

CASE STUDY
Hazardous Materials

Figure 12.7 The solid alkanes that make up waxy coatings on fruits and vegetables help retain moisture, inhibit mold, and enhance appearance.

Q Why does the waxy coating help the fruits and vegetables retain moisture?

Table 12.7   Comparison of Boiling Points of Alkanes and Cycloalkanes with Five Carbons

Formula	Name	Boiling Point (°C)
Cycloalkanes		
⬠	cyclopentane	49
▢—CH$_3$	methylcyclobutane	36.3
Continuous chain		
CH$_3$—CH$_2$—CH$_2$—CH$_2$—CH$_3$	pentane	36
Branched chains		
CH$_3$—CH(CH$_3$)—CH$_2$—CH$_3$	2-methylbutane	28
CH$_3$—C(CH$_3$)(CH$_3$)—CH$_3$	dimethylpropane	10

lated in solid hydrocarbons. It is used in ointments and cosmetics and as a lubricant and a solvent.

Crude Oil

Crude oil or petroleum contains a wide variety of hydrocarbons. At an oil refinery, the components in crude oil are separated by fractional distillation, a process that removes groups or fractions of hydrocarbons by continually heating the mixture to higher temperatures. (See Table 12.8.) Fractions containing alkanes with longer carbon chains require higher temperatures before they reach their boiling temperature and form gases. The gases are removed and passed through a distillation column where they cool and condense back to liquids. (See Figure 12.8.) The major use of crude oil is to obtain gasoline, but a barrel of crude oil is only about 35%

Table 12.8   Typical Alkane Mixtures Obtained by Distillation of Crude Oil

Distillation Temperatures (°C)	Number of Carbon Atoms	Product
Below 30	1–4	Natural gas
30–200	5–12	Gasoline
200–250	12–16	Kerosene, jet fuel
250–350	15–18	Diesel fuel, heating oil
350–450	18–25	Lubricating oil
Nonvolatile residue	Over 25	Asphalt, tar

Figure 12.8 At an oil refinery, crude oil is separated into hydrocarbon fractions by separating groups with different boiling point ranges.

Q How does fractional distillation separate crude oil into different compounds?

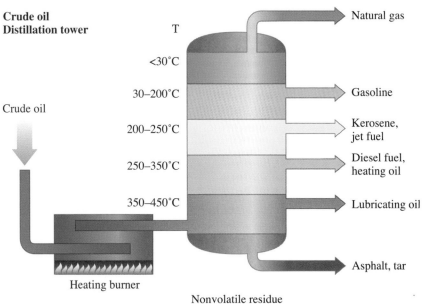

gasoline. To increase the production of gasoline, heating oils are broken down by "cracking" to give the lower weight alkanes.

SAMPLE PROBLEM 12.10

Properties of Alkanes and Cycloalkanes

Indicate the compound in each pair that has the higher boiling point:

a. propane or pentane

b. 2-methylbutane or cyclopentane

Solution

a. Pentane, with the longer continuous chain, has the higher boiling point in the pair.

b. Cyclopentane. The rigid structure of a cycloalkane provides more contact between molecules than do the branched-chain molecules with the same number of carbons.

Study Check

Which of the following compounds would have the highest boiling point: 2-methylpentane, hexane, or cyclohexane?

QUESTIONS AND PROBLEMS

Physical Properties of Alkanes

12.41 Heptane, C_7H_{16}, has a density of 0.68 g/mL, and boils at 98°C.
   a. What is the structural formula of heptane?
   b. Is it a solid, liquid, or gas at room temperature?
   c. Is it soluble in water?
   d. Will it float or sink in water?
   e. Would you expect the branched isomers of heptane to have higher or lower boiling points?

12.42 Nonane, C_9H_{20}, has a density of 0.79 g/mL.
   a. What is the structural formula of nonane?
   b. Is it a solid, liquid, or gas at room temperature?
   c. Is it soluble in water?
   d. Will it float or sink in water?
   e. If heptane has a boiling point of 98°C, would you expect nonane to have a boiling point of 75°C or 151°C?

12.43 In each of the following pairs of hydrocarbons, which one would you expect to have the higher boiling point?
   a. pentane or heptane   **b.** propane or cyclopropane
   c. hexane or 2-methylpentane

12.44 In each of the following pairs of hydrocarbons, which one would you expect to have the higher boiling point?
   a. propane or butane   **b.** hexane or cyclohexane
   c. 2,2-dimethylpentane or heptane

LEARNING GOAL

Write balanced equations for the combustion and halogenation of alkanes and cycloalkanes.

12.7 Chemical Properties of Alkanes and Cycloalkanes

Carbon–carbon single bonds are difficult to break, which makes alkanes the least reactive family of organic compounds. However, alkanes burn readily in oxygen and will undergo substitution reactions with halogens.

Combustion

The growth of civilization was strongly influenced by the discovery of fire. Heat from burning wood was used to make pottery and glass, extract metals, make weapons, and forge tools. An alkane undergoes **combustion** when it reacts with oxygen to produce carbon dioxide, water, and energy.

$$\text{Alkane} + O_2 \longrightarrow CO_2 + H_2O + \text{energy}$$

Methane is the gas we use to cook our foods and heat our homes. Propane is the gas used in portable heaters and gas barbecues. (See Figure 12.9.) Gasoline, a mixture of liquid hydrocarbons, is the fuel that powers our cars, lawn mowers, and snow blowers. As alkanes, they all undergo combustion. The equations for the combustion of methane (CH_4) and propane (C_3H_8) follow:

$$CH_4 + 2O_2 \longrightarrow CO_2 + 2H_2O + \text{energy}$$
$$C_3H_8 + 5O_2 \longrightarrow 3CO_2 + 4H_2O + \text{energy}$$

In the cells of our bodies, energy is produced by the combustion of glucose. Although a series of reactions is involved, we can write the overall combustion of glucose in our cells as follows:

$$C_6H_{12}O_6 + 6O_2 \xrightarrow{\text{enzymes}} 6CO_2 + 6H_2O + \text{energy}$$

SAMPLE PROBLEM 12.11

Combustion

Write a balanced equation for the complete combustion of butane.

Solution

The balanced equation for the complete combustion of butane can be written

$$2C_4H_{10} + 13O_2 \longrightarrow 8CO_2 + 10H_2O$$

Study Check

Write a balanced equation for the complete combustion of the following:

$$CH_3-\overset{\overset{\displaystyle CH_3}{|}}{CH}-CH_2-CH_3$$

Figure 12.9 The propane fuel in the tank undergoes combustion, which provides energy.

Q What is the balanced equation for the combustion of propane?

HEALTH NOTE

Incomplete Combustion

You may already know that it is dangerous to burn natural gas, oil, or wood in a closed room where ventilation and fresh air are not adequate. A gas heater, fireplace, or wood stove must have proper ventilation. If the supply of oxygen is limited, incomplete combustion produces carbon monoxide. The incomplete combustion of methane in natural gas is written as:

$$2CH_4(g) + 3O_2(g) \longrightarrow 2CO(g) + 4H_2O(g) + \text{heat}$$

Limited oxygen supply — Carbon monoxide

Carbon monoxide (CO) is a colorless, odorless, poisonous gas. When inhaled, CO passes into the bloodstream, where it attaches to hemoglobin (as you saw in the Health Note on Oxygen, Hemoglobin and Carbon Monoxide Poisoning in Chapter 6). When CO binds to the hemoglobin, it reduces the amount of oxygen (O_2) reaching the organs and cells. As a result, a healthy person can experience a reduction in exercise capability, visual perception, and manual dexterity.

When the amount of hemoglobin bound to CO (COHb) is 10% or less, a person may experience shortness of breath, mild headache, and drowsiness, which are symptoms that may be mistaken for the flu. Heavy smokers can have as high as 9% COHb in their blood. When as much as 30% of the hemoglobin is bound to CO, a person may experience more severe symptoms including dizziness, mental confusion, severe headache, and nausea. If 50% or more of the hemoglobin is bound to CO, a person could become unconscious and die if not treated immediately with oxygen.

Halogenation of Alkanes (Substitution)

Alkanes react with halogen to produce a mixture of halogenated compounds. In a **halogenation reaction,** atoms of a halogen bond to a carbon atom. This type of reaction is also called a **substitution** reaction because halogen atoms replace one or more hydrogen atoms in an alkane. When halogenation uses chlorine Cl_2 (Cl—Cl), it is also called chlorination and with bromine Br_2 (Br—Br), it is called bromination. When a mixture of an alkane such as methane and chlorine are heated or exposed to light, halogenation can take place.

$$\text{Alkane} + \text{halogen} \xrightarrow{\text{light or heat}} \text{haloalkane and hydrogen halide}$$

The first halogenation of methane occurs when one chlorine atom replaces a hydrogen atom as follows:

methane · chlorine → chloromethane · hydrogen chloride
alkane · *halogen* → haloalkane · *hydrogen halide*

In the presence of heat or light and chlorine, the reaction continues. Chlorine replaces hydrogen atoms in chloromethane, which gives a mixture of halogenated products.

+ HCl   + HCl   + HCl

When halogenation occurs with larger alkanes, the halogen atoms may substitute for any of the hydrogen atoms. For example, the first substitution of propane (monohalogenation) produces both 1-bromopropane and 2-bromopropane. Because secondary hydrogens are more reactive than primary hydrogens, there is more 2-bromopropane produced.

When cycloalkanes undergo monohalogenation, a single product results.

SAMPLE PROBLEM 12.12

Halogenation of Alkanes

Write the structural formula of the monohalogenated product formed in the following reactions:

a. $CH_3CH_3 + Br_2 \xrightarrow{\text{light or heat}}$

b. 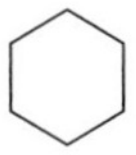 $+ Cl_2 \xrightarrow{\text{light or heat}}$

Solution

a. CH_3CH_2Br **b.**

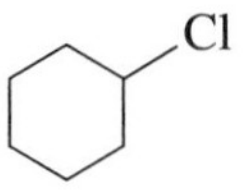

Study Check

Write the structural formulas of the organic products that form when two bromine atoms replace two hydrogen atoms in the halogenation of ethane.

QUESTIONS AND PROBLEMS

Chemical Properties of Alkanes and Cycloalkanes

12.45 Write a balanced equation for the complete combustion of each of the following compounds:
    a. ethane, C_2H_6 **b.** cyclopropane, C_3H_6
    c. octane   **d.** cyclohexane

12.46 Write a balanced equation for the complete combustion of each of the following compounds:
    a. hexane, C_6H_{14} **b.** cyclopentane, C_5H_{10}
    c. nonane   **d.** 2-methylbutane

12.47 Write the condensed structural formulas of the monohalogenated products from the chlorination of each of the following:
    a. ethane   **b.** cyclopentane
    c. 2-methylpropane

12.48 Write the condensed structural formulas of the monohalogenated products from the bromination of each of the following:
    a. butane   **b.** pentane
    c. cyclobutane

Chapter Review

12.1 Alkanes

Alkanes are hydrocarbons with the general formula C_nH_{2n+2} that have only C—C single bonds. In the expanded structural formula, a separate line is drawn for every bonded atom. A condensed structural formula depicts groups composed of each carbon atom and its attached hydrogen atoms. In a line-bond formula, only the carbon–carbon bonds are drawn. Conformational isomers exist because of free rotation of groups around the C—C single bonds.

12.2 IUPAC Naming System for Alkanes

The IUPAC system is used to name organic compounds in a systematic manner. The IUPAC name indicates the number of carbon atoms and family of the compounds. For example, in the name *pentane*, the prefix *pent* indicates a main chain of five carbon atoms and the ending *ane* indicates an alkane.

12.3 Drawing Structural Formulas

For a continuous alkane, the carbon atoms are connected in a chain and bonded to hydrogen atoms. Substituents such as alkyl groups and/or halogen atoms can replace hydrogen atoms on the main chain. Constitutional isomers have the same molecular formula but differ in the order of bonding. Conformational isomers, which are the different orientations of groups in a molecule, result as groups rotate freely about C—C bonds.

12.4 Haloalkanes

A haloalkane contains one or more F, Cl, Br, or I atoms. In the IUPAC system, halogen atoms are named as fluoro, chloro, bromo, or iodo substituents attached to the main chain. In the common name, alkyl halides, the name of the alkyl group precedes the halide, for example methyl chloride.

12.5 Cycloalkanes

In cycloalkanes, the carbon atoms form a ring or cyclic structure with the general formula C_nH_{2n}. The name is written by placing the prefix *cyclo* before the alkane name with the same number of carbon atoms. Substituents are named as alkyl and halo groups and numbered if two or more are bonded to the ring. Cyclic structures with two attached groups may have cis and trans isomers, which differ in the orientation of atoms in space. The two substituents are on the same side of the ring in the cis isomer, but on opposite sides in the trans isomer.

12.6 Physical Properties of Alkanes and Cycloalkanes

As nonpolar molecules, alkanes are not soluble in water. They are usually less dense than water. With only weak attractions, they have low melting and boiling points. For alkanes of similar mass, cycloalkanes have higher boiling points and branched alkanes have lower boiling points than the continuous, nonbranched chain alkanes.

12.7 Chemical Properties of Alkanes and Cycloalkanes

Although the C—C bonds in alkanes resist most reactions, alkanes undergo combustion and halogenation. In combustion or burning, alkanes react with oxygen to produce carbon dioxide and water. In the halogenation of alkanes, halogen atoms replace one or more hydrogen atoms. A source of energy such as UV radiation or heat is required for this substitution reaction.

Summary of Naming

Type	Example	Characteristic	Structure
Alkane	Propane	single C—C, C—H bonds	CH_3—CH_2—CH_3
	Methylpropane		CH_3—$\overset{\overset{CH_3}{\|}}{CH}$—$CH_3$
Haloalkane	1-Chloropropane	halogen atom	CH_3—CH_2—CH_2—Cl
Cycloalkane	Cyclobutane	carbon ring	

Summary of Reactions

Combustion

Alkane + O_2 $\longrightarrow$ CO_2 + H_2O + energy

Halogenation (substitution by halogen)

Alkane + halogen $\xrightarrow{\text{Light or heat}}$ haloalkane + hydrogen halide

CH_4 + Cl_2 $\xrightarrow{\text{Light or heat}}$ CH_3Cl + HCl

+ Cl_2 $\xrightarrow{\text{Light or heat}}$ + HCl

Key Terms

alkane A hydrocarbon that has only carbon–carbon single bonds.

alkyl group An alkane minus one hydrogen atom that bonds to a main chain. Alkyl groups are named like the alkanes except a *yl* ending replaces *ane*.

branch A carbon group or halogen bonded to the main carbon chain.

branched-chain alkane A hydrocarbon containing a substituent bonded to the main chain.

cis–trans isomer In cycloalkanes, two substituents bonded to the ring can differ in the orientation in space. In cis isomers, the groups are on the same side of the ring but on opposite sides in the trans isomer.

combustion A chemical reaction in which an alkane reacts with oxygen to produce CO_2, H_2O, and energy.

condensed structural formula A structural formula that shows the arrangement of the carbon atoms in a molecule but groups each carbon atom with its bonded hydrogen atoms (CH_3, CH_2, or CH).

conformations The different orientations of groups in a molecule resulting from the free rotation of groups about C—C bonds.

constitutional isomers Molecules that have the same molecular formula but different arrangement of atoms.

continuous alkane An alkane in which the carbon atoms are connected in a row, one after the other.

cycloalkane An alkane that is a ring or cyclic structure.

expanded structural formula A type of structural formula that shows the arrangement of the atoms by drawing each bond in the hydrocarbon as C—H or C—C.

haloalkane A type of alkane that contains one or more halogen atoms.

halogenation reaction A substitution reaction that occurs in the presence of light or heat in which halogen atoms replace hydrogen atoms in an alkane.

IUPAC system An organization known as International Union of Pure and Applied Chemistry that determines the system for naming organic compounds.

line-bond formula A type of structural formula that shows only the bonds from carbon to carbon.

primary carbon A carbon bonded to one other carbon atom.

secondary carbon A carbon bonded to two other carbon atoms.

structural formula A type of formula that shows the order of bonding for the carbon atoms in the molecule.

substituent Groups of atoms such as an alkyl group or a halogen bonded to the main chain or ring of carbon atoms.

substitution A reaction such as halogenation in which halogen atoms replace one or more hydrogen atoms in a hydrocarbon.

tertiary carbon A carbon bonded to three other carbon atoms.

Additional Problems

12.49 Identify whether each of the following pairs of structural formulas represent constitutional isomers, the same molecule, or different molecules.

a.

b.

c.

d.

12.50 Identify whether each of the following pairs of structural formulas represent constitutional isomers, the same molecule, or different molecules.

a.

b.

c.

d.

12.51 Write the name of each of the following alkyl groups:

a. CH_3—

b. CH_3—CH_2—CH_2—

c. 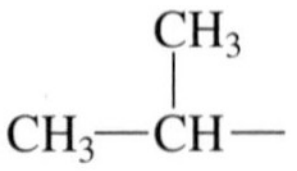
$$CH_3-\overset{\overset{\displaystyle CH_3}{|}}{CH}-$$

12.52 Write the name of each of the following alkyl groups:

a. CH_3—CH_2—

b. 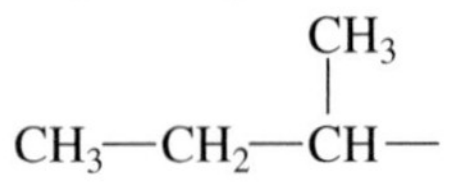
$$CH_3-CH_2-\overset{\overset{\displaystyle CH_3}{|}}{CH}-$$

c.
$$CH_3-\overset{\overset{\displaystyle CH_3}{|}}{\underset{\underset{\displaystyle CH_3}{|}}{C}}-$$

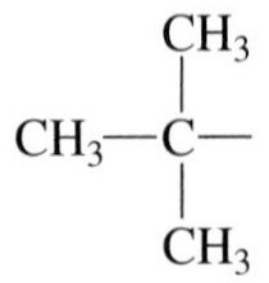

12.53 Give the IUPAC names for each of the following molecules:

a. 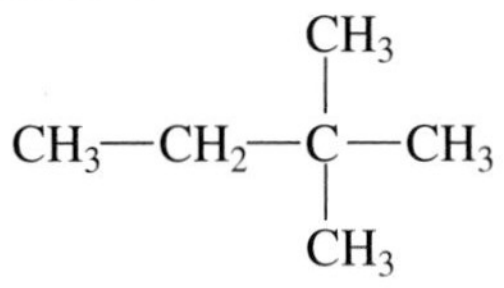
$$CH_3-CH_2-\overset{\overset{\displaystyle CH_3}{|}}{\underset{\underset{\displaystyle CH_3}{|}}{C}}-CH_3$$

b. CH_3—CH_2—Cl

c. 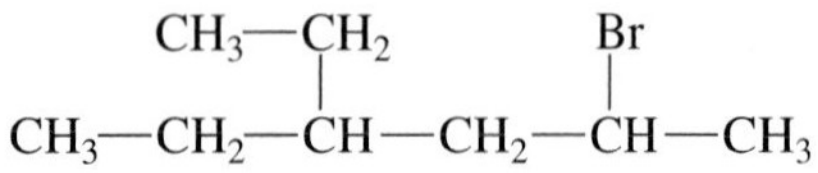
$$CH_3-CH_2-\overset{\overset{\displaystyle CH_3-CH_2}{|}}{CH}-CH_2-\overset{\overset{\displaystyle Br}{|}}{CH}-CH_3$$

d. 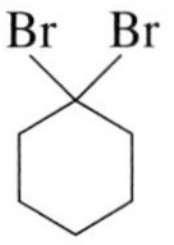 Br   Br

12.54 Give the IUPAC names for each of the following molecules:

a.

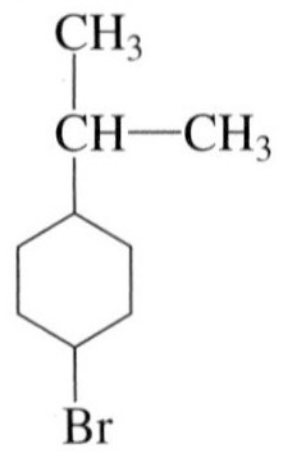

b.

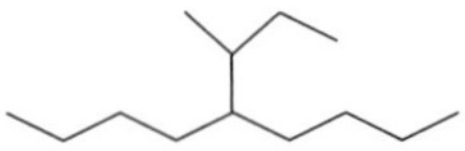

c. 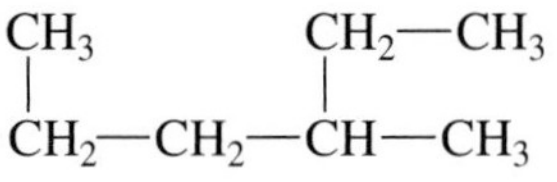
$$CH_3 \qquad CH_2-CH_3$$
$$CH_2-CH_2-\overset{|}{CH}-CH_3$$

d. 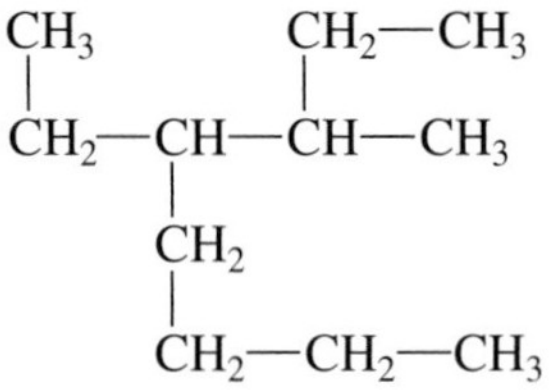
$$CH_3 \qquad CH_2-CH_3$$
$$CH_2-\overset{|}{CH}-\overset{|}{CH}-CH_3$$
$$\overset{|}{CH_2}$$
$$\overset{|}{CH_2}-CH_2-CH_3$$

12.55 Write the condensed structural formula for each of the following molecules:

a. 3-ethylhexane

b. 2,3-dimethylpentane

c. 1,3-dichloro-3-methylheptane

d. bromocyclobutane

e. 1-bromo-4-isopropylcyclohexane

f. *sec*-butyl chloride

12.56 Write the condensed structural formula for each of the following molecules:

a. ethylcyclopropane

b. 1,3-dimethylcyclohexane

c. isopropylcyclopentane

d. 1,1-dimethylcyclopentane

e. 2-bromo-1,1-dichlorocyclopentane

f. *tert*-butyl chloride

12.57 Draw the line-bond structural formula for each of the following molecules:

a. pentane

b. 2,3-dimethylhexane

c. 2-bromo-4-isopropylheptane

12.58 Draw the line-bond structural formula for each of the following molecules:

a. butane

b. 3-ethyl-2-methylhexane

c. 3,4,5-trimethyloctane

12.59 Write structural formulas and IUPAC names for two constitutional isomers of each of the following:

a. hexane

b. 3-ethylpentane

c. 2,2-dimethylbutane

d. 1,1-dibromocyclohexane   (cyclohexanes only)

12.60 Write structural formulas and IUPAC names for two constitutional isomers of each of the following:

a. pentane

b. 1,1-dibromocyclobutane (cyclobutanes only)

c. 2,2-dichlorobutane

d. 1,2-dimethylcyclopentane (cyclopentanes only)

12.61 The following names are incorrect. Write the correct IUPAC name.
  a. 3-methylbutane
  b. 3,4-dimethylpentane
  c. 2,4-dibromocyclohexane
  d. 1,4-dimethylbutane

12.62 The following names are incorrect. Write the correct IUPAC name.
  a. 1-ethylbutane
  b. 4,5-dimethylpentane
  c. 2,3-dichlorocyclopentane
  d. 2-isopropylbutane

12.63 Draw at least six constitutional isomers of $C_5H_{11}Cl$.

12.64 Draw at least five constitutional isomers of $C_4H_6Br_2$.

12.65 Draw the structural formulas for each of the following:
  a. *sec*-butyl bromide
  b. *cis*-1-bromo-2-chlorocyclopropane
  c. *trans*-1,3-dimethylcyclopentane
  d. *cis*-1,2-dibromocyclobutane

12.66 Draw the structural formulas for each of the following:
  a. *cis*-1,2-dimethylcyclobutane
  b. *trans*-1-bromo-2-chlorocyclopentane
  c. isopropyl fluoride
  d. *trans*-1-chloro-2-methylcyclohexane

12.67 In an automobile engine, "knocking" occurs when the combustion of gasoline occurs too rapidly. The octane number of gasoline represents the ability of a gasoline mixture to reduce knocking. A sample of gasoline is compared with heptane, rated 0 because it reacts with severe knocking, and 2,2,4-trimethylpentane (isooctane), which has a rating of 100 because of its low knocking. Write the condensed structural formula, molecular formula, and equation for the complete combustion of isooctane.

12.68 Draw the structures of the following halogenated compounds, which are used as refrigerants.
  a. Freon 14, tetrafluoromethane
  b. Freon 114, 1,2-dichloro-1,1,2,2-tetrafluoroethane
  c. Freon C318, octafluorocyclobutane

12.69 Identify the compound in each pair that has the higher boiling point.
  a. pentane or heptane
  b. pentane or cyclopentane

  c. hexane or 2-methylpentane
  d. cyclobutane or cyclohexane

12.70 Identify the compound in each pair that has the higher boiling point.
  a. butane or octane
  b. cyclopentane or pentane
  c. 2,2-dimetylbutane or cyclohexane
  d. propane or pentane

12.71 Write a balanced equation for the complete combustion of each of the following:
  a. propane
  b. C_5H_{12}
  c. cyclobutane
  d. octane

12.72 Write a balanced equation for the complete combustion of each of the following:
  a. hexane
  b. methylcyclohexane
  c. cyclopentane
  d. 2-methylpropane

12.73 Draw the structural formulas of the monohalogenated organic products that result from the chlorination of each of the following:
  a. ethane
  b. propane
  c. cyclopentane

12.74 Draw the structural formulas of the monohalogenated organic products that result from the bromination of each of the following:
  a. pentane
  b. cyclohexane
  c. 2-methylbutane

12.75 The density of pentane, a component of gasoline, is 0.63 g/mL. The heat of combustion for pentane is 845 kcal per mole (3536 kJ/mole).
  a. Write an equation for the complete combustion of pentane.
  b. What is the molar mass?
  c. How much heat is produced when 1 gallon of pentane is burned (1 gallon = 3.78 liters)?
  d. How many liters of CO_2 at STP are produced from the complete combustion of 1 gallon of pentane?

13 Unsaturated Hydrocarbons

"When we have a hazardous materials spill, the first thing we do is isolate it," says Don Dornell, assistant fire chief, Burlingame Fire Station. "Then our technicians and a county chemist identify the product from its flammability and solubility in water so we can use the proper materials to clean up the spill. We use different methods for alcohol, which mixes with water, than for gasoline, which floats. Because hydrocarbons are volatile, we use foam to cover them and trap the vapors. At oil refineries, we will use foams, but many times we squirt water on the tanks to cool the contents below their boiling points. By knowing the boiling point of the product and its density and vapor density, we know if it floats or sinks in water and where its vapors will go."

LOOKING AHEAD

13.1 Alkenes and Alkynes

13.2 Naming Alkenes and Alkynes

13.3 Cis–Trans Isomers

13.4 Addition Reactions

13.5 Polymerization

13.6 Aromatic Compounds

13.7 Properties of Aromatic Compounds

the Chemistry place

www.chemplace.com/college

Visit the URL above or use the CD-ROM in the book for extra quizzing, interactive tutorials, career resources, and case studies.

In Chapter 12, we looked at **saturated hydrocarbons**, which contain all single bonds between their carbon atoms. Now we will investigate unsaturated hydrocarbons, carbon compounds containing one or more carbon–carbon double bonds or triple bonds. When we cook with vegetable oils such as corn oil, safflower oil, or olive oil, we are using unsaturated lipids that have double bonds in their long carbon chains. Saturated fats from animal sources also have long chains of carbon atoms, but they are connected only by the single bonds of alkanes. If we compare the two types of fats, we find considerable differences in their physical and chemical properties. Vegetable oils are liquid at room temperature whereas animal fats are solid. Because double bonds are reactive, unsaturated fats can be oxidized by oxygen in the air, especially at warm temperatures, forming products that have rancid, unpleasant odors. The saturated fats are more resistant to reactions.

13.1 Alkenes and Alkynes

The **unsaturated hydrocarbons** have fewer hydrogen atoms than alkanes. **Alkenes** contain at least one double bond between carbons. If a straight-chain alkene contains just one double bond between carbons, it has the general formula C_nH_{2n}. Alkenes are useful compounds in industry and have important functions in plants and animals. Ethylene is an important industrial compound used to make a

plastic called polyethylene. The unsaturated sites are very reactive, which distinguishes unsaturated compounds from the alkanes. **Alkynes** contain at least one triple bond between carbons. If a straight-chain alkyne contains just one triple bond between carbons, it has the general formula C_nH_{2n-2}. (See Figure 13.1.) Only a small number of alkynes are found in nature. However, ethyne, commonly called acetylene, is used in welding where it burns at a very high temperature.

LEARNING GOAL

Identify structural formulas as alkenes, cycloalkenes, and alkynes.

Ethene

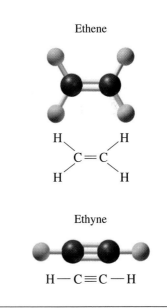

Ethyne

Figure 13.1 Ball-and-stick models of ethene and ethyne show the functional groups of double or triple bonds.

Q Why are these compounds called unsaturated hydrocarbons?

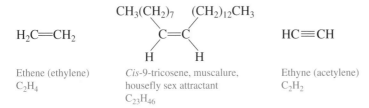

$H_2C=CH_2$

Ethene (ethylene)
C_2H_4

$CH_3(CH_2)_7$ $(CH_2)_{12}CH_3$
            $C=C$
      H H

Cis-9-tricosene, musculare, housefly sex attractant
$C_{23}H_{46}$

$HC\equiv CH$

Ethyne (acetylene)
C_2H_2

SAMPLE PROBLEM 13.1

Identifying Unsaturated Compounds

Classify each of the following structural formulas as an alkane, alkene, or alkyne. Write their molecular formulas.

a. $CH_3-C\equiv C-CH_3$ **b.** $CH_3-CH_2-CH_3$

c. $CH_3-CH_2-\overset{\overset{\displaystyle CH_3}{\displaystyle |}}{C}=CH-CH_2-CH_3$

Solution

The structural formula with a double bond is an alkene, and the one with a triple bond is an alkyne.

a. alkyne: C_4H_6 **b.** alkane; C_3H_8 **c.** alkene; C_7H_{14}

Study Check

Is the following structure a cycloalkane or cycloalkene?

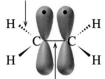

A σ bond between carbon and hydrogen

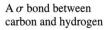

A σ bond between two carbons

Overlap of p orbitals forms a π bond

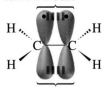

Figure 13.2 The π bond is formed by the overlapping of two p orbitals on adjacent carbon atoms.

Q What are the two types of bonds in a double bond?

Figure 13.3 In the flat ethene molecule, the bond angles are 120°. In ethyne, the bond angles are 180°.

Q What accounts for the different bond angles in double and triple bonds?

Alkene Structure

The simplest alkene C_2H_4 is called ethene, but it is more likely to be called by its common name ethylene. There are two CH_2 groups connected by a double bond. In the electron dot structure, the double bond is represented by two sets of electrons. We already know that carbon has four electrons and needs four bonds. In a double bond, each carbon atom is attached to three other atoms (one carbon and two hydrogens). The $C-C$ and two $C-H$ bonds each share a pair of electrons to form strong bonds called sigma (σ) bonds. To form a second bond between the carbon atoms, the single electrons in the p orbitals overlap to produce a weaker bond known as a pi (π) bond. Thus, a double bond is a combination of one sigma and one pi bond. (See Figure13.2.)

According to VSEPR theory (Chapter 4), three groups bonded to each carbon in the double bond are planar and arranged at angles of 120°. (See Figure 13.3.) The pi (π) bond is weaker than a sigma bond, and because it is easier to break, it is the key to the chemical properties of alkenes.

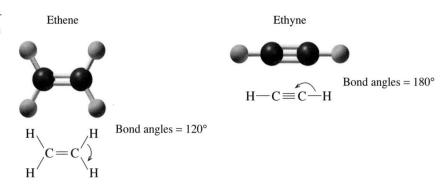

Ethene

Ethyne

Bond angles = 180°

$H-C\equiv C-H$

$\overset{H}{\underset{H}{\diagdown}}C=C\overset{H}{\underset{H}{\diagup}}$ Bond angles = 120°

Alkyne Structure

The simplest alkyne is called ethyne, but it is commonly known as acetylene. In ethyne there are two CH groups connected by a triple bond. We already know that carbon has four electrons and needs four bonds. In a triple bond, each carbon atom is attached to two other atoms (one carbon and one hydrogen). The $C-C$ and $C-H$ bonds each share a pair of electrons to form strong bonds called sigma (σ) bonds. We have seen how a pi (π) bond forms in alkenes when the p orbitals overlap. In a triple bond, two sets of p orbitals overlap forming two pi (π) bonds. Thus, a triple bond is a combination of one sigma and two pi bonds. (See Figure 13.4.)

According to VSEPR theory, two groups bonded to each carbon in the triple bond are linear and arranged at angles of 180°.

SAMPLE PROBLEM 13.2

Structure of Alkenes and Alkanes

What do double and triple bonds have in common and how do they differ?

Solution

In both types of bonds, a sigma bond connects two carbon atoms. The additional bonds are pi bonds, which are formed by overlapping one or two sets of p orbitals. A double bond has one sigma bond and one pi bond, whereas a triple bond has one sigma bond and two pi bonds.

Study Check

Why are the atoms arranged at angles of 120° in double bonds, but at 180° in triple bonds?

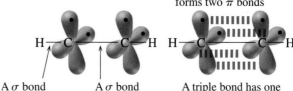

Two p orbitals

A σ bond between carbon and hydrogen

A σ bond between two carbons

Overlap of two p orbitals forms two π bonds

A triple bond has one σ bond and two π bonds

Figure 13.4 In ethyne, C_2H_2, the overlap of two p orbitals forms two π bonds required for the triple bond.

Q How is the requirement of four bonds for a carbon atom maintained in ethyne?

QUESTIONS AND PROBLEMS

Alkenes and Alkynes

13.1 Indicate whether each of the following statements describes a feature of an alkane, alkene, or alkyne:
   a. Has only sigma (σ) bonds in the molecule.
   b. The molecule contains one sigma (σ) and two pi (π) carbon–carbon bonds.
   c. The groups on a carbon atom are arranged at 120°.

13.2 Indicate whether each of the following statements describes a feature of an alkane, alkene, or alkyne:
   a. The molecule contains one sigma (σ) and one pi (π) carbon–carbon bond.
   b. The groups on the carbon atoms are arranged at 109°.
   c. The molecule is linear.

13.3 Identify the following as alkanes, alkenes, cycloalkenes, or alkynes:

   a. C_6H_{14}
   b. H—C—C=C—H (with H, H, H above and H below on first carbon)
   c. CH_3—CH_2—$C\equiv C$—H
   d. (structure)
   e. (structure with CH_3)

13.4 Identify the following as alkanes, alkenes, cycloalkenes, or alkynes:

   a. (with CH_3)
   b. (structure)
   c. CH_3—C=C—CH_3 (with CH_3 above and CH_3 below)
   d. (structure with $C\equiv CH$)
   e. (structure)

13.5 Write the condensed structural formulas of an alkene and cycloalkane with a molecular formula of C_3H_6.

13.6 Write the condensed structural formulas of two alkenes and a cycloalkane with a molecular formula of C_4H_8.

WEB TUTORIAL
Geometric Isomers

13.2 Naming Alkenes and Alkynes

The IUPAC names for alkenes and alkynes are similar to those of alkanes. The simplest alkene, ethene, often named by its common name, ethylene, is an important plant hormone involved in promoting the ripening of fruit. Commercially grown fruit, such as avocados, bananas, and tomatoes, are often picked before they are ripe. Before the fruit is brought to market, it is exposed to ethylene to accelerate the ripening process. Ethylene also accelerates the breakdown of cellulose in plants, which causes flowers to wilt and leaves to fall from trees.

The IUPAC name of the simplest alkyne is ethyne, although acetylene, its common name, is often used. See Table 13.1 for a comparison of the naming for alkanes, alkenes, and alkynes.

Table 13.1   Comparison of Names for Alkanes, Alkenes, and Alkynes

Alkane	Alkene	Alkyne
H_3C—CH_3	H_2C=CH_2	HC≡CH
Ethane	Ethene (ethylene)	Ethyne (acetylene)
CH_3—CH_2—CH_3	CH_3—CH=CH_2	CH_3—C≡CH
Propane	Propene	Propyne

IUPAC Rules for Naming Alkenes and Alkynes

For alkenes and alkynes, the longest carbon chain that contains the unsaturated site is the main chain, which is numbered to show the location of the unsaturated site.

1. **Name the longest carbon chain that contains the double or triple bond.** Replace the corresponding alkane ending with –*ene* for an alkene and –*yne* for an alkyne.

2. **Number the main chain from the end nearest the double or triple bond.** Place the lowest number of the carbon atom in the double or triple bond in front of the name of the main chain.

$$CH_3-CH_2-CH=CH_2 \qquad CH_3-CH=CH-CH_3 \qquad CH_3-C\equiv C-CH_3$$

$$\;\;4\quad\;\;\;3\quad\;\;2\quad\;\;1 \qquad\quad 1\quad\;\;\;2\quad\;\;\;3\quad\;\;4 \qquad\quad 1\quad\;\;2\quad\;\;3\quad\;\;4$$

1-butene   2-butene   2-butyne

3. **Place the number and names of substituents in front of the alkene or alkyne name.**

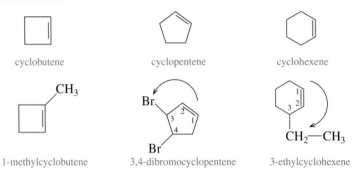

$$CH_2\!\!=\!\!CH\!-\!CH_2\!-\!\overset{\displaystyle CH_3}{\underset{\displaystyle |}{CH}}\!-\!CH_3 \qquad \overset{\displaystyle H_3C\;\;CH_3}{\underset{\displaystyle |\;\;\;\;\;|}{CH_3\!-\!C\!\!=\!\!C\!-\!CH_3}} \qquad \overset{\displaystyle CH_3}{\underset{\displaystyle |}{CH_3\!-\!CH\!-\!C\!\!\equiv\!\!CH}}$$

4-methyl-1-pentene   2,3-dimethyl-2-butene   3-methyl-1-butyne

4. **Cyclic alkenes are named as cycloalkenes with numbers for substituents.** With a substituent on the ring, the double bond is understood to be carbon 1 and 2 and the ring is numbered in the direction to give the lowest number to a substituent.

cyclobutene   cyclopentene   cyclohexene

1-methylcyclobutene   3,4-dibromocyclopentene   3-ethylcyclohexene

5. **Name a compound with two double bonds as a diene.** Use numbers to give the location of each double bond.

$$CH_2\!\!=\!\!CH\!-\!CH\!\!=\!\!CH_2 \qquad \overset{\displaystyle CH_3}{\underset{\displaystyle |}{CH_3\!-\!C\!\!=\!\!CH\!-\!CH\!\!=\!\!CH_2}}$$

1,3-butadiene   4-methyl-1,3-pentadiene

SAMPLE PROBLEM 13.3

Naming Alkenes and Alkynes

Write the IUPAC name for each of the following unsaturated compounds:

a. $CH_3CH_2CH\!\!=\!\!CHCH_3$ **b.** (cyclohexene ring)   **c.** $\overset{\displaystyle Cl}{\underset{\displaystyle |}{CH_3C\!\!=\!\!CHCH_3}}$ **d.** $HC\!\!\equiv\!\!CCH_2CH_2CH_3$ **e.** (methylcyclohexene ring) CH_3

Solutions

a. 2-pentene   **b.** cyclohexene   **c.** 2-chloro-2-butene
d. 1-pentyne   **e.** 1-methylcyclohexene

Study Check

Draw the structural formulas for each of the following:
a. 2-pentyne   **b.** 3-chlorocyclohexene

QUESTIONS AND PROBLEMS

Naming Alkenes and Alkynes

13.7 Compare the structural formulas of the following:
   a. propene and propyne   **b.** cyclohexane and cyclohexene
13.8 Compare the structural formulas of the following:
   a. 1-butyne and 2-butyne
   b. 1-methylcyclohexene and 3-methylcyclohexene

ENVIRONMENTAL NOTE

Fragrant Alkenes

The odors you associate with lemons, oranges, roses, and lavender are due to volatile compounds that are synthesized by the plants. Often it is unsaturated compounds that are responsible for the pleasant flavors and fragrances of many fruits and flowers. They were some of the first kinds of compounds to be extracted from natural plant material. In ancient times, they were highly valued in their pure forms. Limonene and myrcene give the characteristic odors and flavors to lemons and oranges and bay leaves, respectively. Geraniol and citronellal give roses and lemon grass their distinct aromas. In the food and perfume industries, these compounds are extracted or synthesized and used as perfumes and flavorings.

$$CH_3-\overset{\overset{\displaystyle CH_3}{|}}{C}=CH-CH_2-CH_2-\overset{\overset{\displaystyle CH_3}{|}}{CH}-CH_2-CH_2OH$$

Geraniol, roses

$$CH_3-\overset{\overset{\displaystyle CH_3}{|}}{C}=CH-CH_2-CH_2-\overset{\overset{\displaystyle CH_2}{||}}{C}-CH=CH_2$$

Myrcene, bay leaves

$$CH_3-\overset{\overset{\displaystyle CH_3}{|}}{C}=CH-CH_2-CH_2-\overset{\overset{\displaystyle CH_3}{|}}{C}=CH-CHO$$

Citronellal, lemon grass

Limonene, lemons, and oranges

13.9 Give the IUPAC name for each of the following:

a. $CH_2{=}CH_2$ **b.** $CH_3-\overset{\overset{\displaystyle CH_3}{|}}{C}=CH_2$ **c.** $CH_3-\overset{\overset{\displaystyle Br}{|}}{CH}-C{\equiv}C-CH_3$

d.    **e.** 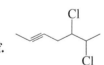   **f.**

13.10 Give the IUPAC name for each of the following:

a. $CH_2{=}CHCH_2CH_2CH_2CH_3$ **b.** $CH_3C{\equiv}CCH_2CH_2\overset{\overset{\displaystyle CH_3}{|}}{CH}CH_3$ **c.**    **d.**

e. $CH_3\overset{\overset{\displaystyle Cl}{|}}{CH}CH_2\overset{\overset{\displaystyle Cl}{|}}{CH}CH_2CH{=}CH_2$ **f.**

13.11 Draw the structural formula for each of the following compounds:
    a. propene   **b.** 1-pentene   **c.** 2-methyl-1-butene
    d. 3-methylcyclohexene   **e.** 2-chloro-3-hexyne

13.12 Draw the structural formula for each of the following compounds:
    a. 1-methylcyclopentene   **b.** 3-methyl-1-butyne
    c. 3,4-dimethyl-1-pentene   **d.** 4-ethyl-1-methylcyclohexene
    e. 1,2-dichlorocyclopentene

13.3 Cis–Trans Isomers

Several constitutional isomers can be written for alkenes similar to what we did with the structural formulas of alkanes, which gives the possibilities of a large number of constitutional isomers. Let's take a look at the structures for some constitutional isomers, including a cycloalkane that can be written for a molecular formula C_4H_8.

Constitutional Isomers of C_4H_8

H_2C=CH—CH_2—CH_3 CH_3—CH=CH—CH_3 H_2C=$\overset{\overset{\displaystyle CH_3}{|}}{C}$—$CH_3$ [ ]

1-butene   2-butene   2-methyl-1-propene   cyclobutane

Cis–Trans Isomers for Alkenes

In Chapter 12, we saw that cycloalkanes can have cis and trans isomers due to a lack of rotation of the carbon atoms in the ring. Alkenes can also have cis–trans isomers because there is no rotation around the carbons in the rigid double bond. As a result, any groups connected to the double bond remain fixed on one side or the other. In a **cis isomer,** two groups are on the same side of the double bond. In the **trans isomer,** the groups are on opposite sides of the double bond. For example, cis–trans isomers can be written for 2-butene. (See Figure 13.5.) In general, trans isomers are more stable than their cis counterparts because the large groups attached to the double bond are further apart. As with any pair of isomers, the cis–trans isomers of 2-butene are different compounds with different physical and chemical properties as shown in the following:

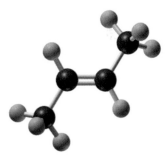

cis-2-butene

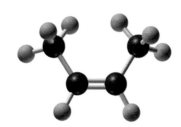

trans-2-butene

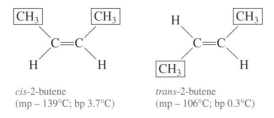

cis-2-butene
(mp – 139°C; bp 3.7°C)

trans-2-butene
(mp – 106°C; bp 0.3°C)

Not every alkene shows cis–trans isomerism. If one of the carbons in the double bond is attached to identical groups, the molecule does not have cis–trans isomers. This is the case of 1-butene and 2-methyl-1-propene, constitutional isomers of 2-butene. If the hydrogen atoms are interchanged on carbon 1, the same structure results. Alkynes do not have cis–trans isomers because the carbons in the triple bond are each attached to only one group.

Identical atoms ← $\overset{H}{\underset{H}{}} C$=$C \overset{CH_2—CH_3}{\underset{H}{}}$ $\overset{H}{\underset{H}{}} C$=$C \overset{CH_3}{\underset{CH_3}{}}$ → Identical groups

1-butene   2-methyl-1-propene

As long as the groups attached to the double bond are different, an alkene will show cis–trans isomers. Another example of cis–trans isomers is the following:

Figure 13.5 Ball-and-stick models of the cis and trans isomers of 2-butene.

Q What feature in 2-butene accounts for the different cis and trans isomers?

WEB TUTORIAL
Geometric Isomers

Same side

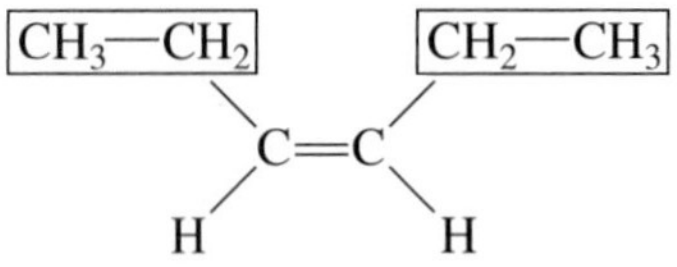

cis-3-hexene

Opposite side

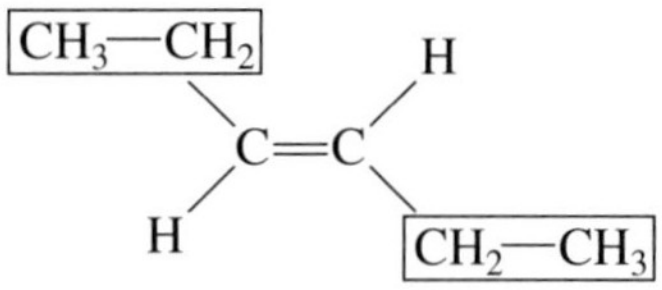

trans-3-hexene

SAMPLE PROBLEM 13.4

Writing and Naming Cis–Trans Isomers

Determine if each of the following shows cis–trans isomers. If so, write the names.

a. Br—CH=CH—Cl **b.** 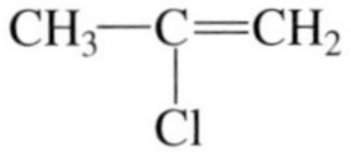

Solution

Draw the double bond and attach the separate groups or atoms to each carbon using the following model:

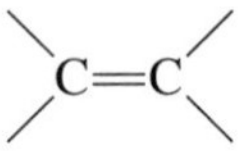

a. In a molecule of Br—CH=CH—Cl, the first carbon is bonded to a bromine atom and a hydrogen atom; the second carbon has a hydrogen atom and a chlorine atom. They can be drawn as follows:

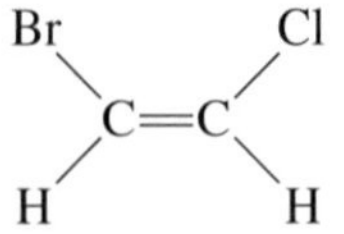

cis-1-bromo-2-chloroethene

In this drawing, the halogen atoms are on the same side and the hydrogen atoms are on the same side. It is the cis isomer.

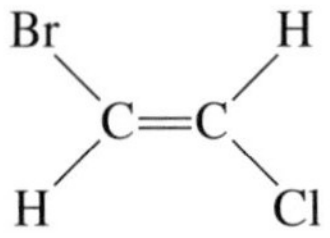

trans-1-bromo-2-chloroethene

In the trans isomer, the halogen atoms are drawn on the opposite sides of the double bond.

b. 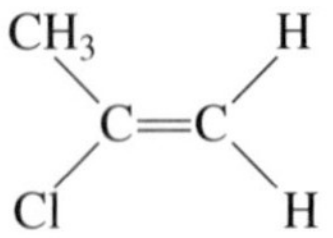

2-chloro-1-propene

In this drawing, there are two hydrogen atoms attached to one of the carbons in the double bond. This compound does not have cis–trans isomers.

Study Check

Draw the structural formula for *cis*-3-heptene.

Modeling Cis–Trans Isomers

Because cis–trans isomerism is not easy to imagine, here are some things you can do to understand the difference in rotation around a single bond compared to a double bond and how it affects groups that are attached to the carbon atoms in the double bond.

Using Your Hands and Fingers as Single and Double Bonds

Put the fingertips of your index fingers together. This is a model of a single bond. Consider the index fingers as a pair of carbon atoms and think of your thumbs and other fingers as other parts of a carbon chain. While your index fingers are touching, twist your hands and change the position of the thumbs relative to each other. Notice how the relationship of your other fingers changes.

Now place the tips of your index fingers and middle fingers together in a model of a double bond. As you did before, twist your hands to move the thumbs apart. What happens? Can you change the location of your thumbs relative to each other without breaking the double bond? The difficulty of moving your hands with two fingers touching represents the lack of rotation about a double bond. You have made a model of a cis isomer when both thumbs are on the same side. If you turn one hand

Cis-hands (*cis*-thumbs/fingers)

Trans-hands (*trans*-thumbs/fingers)

over so one thumb points down and one points up, you have made a model of a trans isomer.

Using Toothpicks and Gumdrops to Form Single and Double Bonds

Use toothpicks for bonds and gumdrops for atoms. Place a toothpick between two black gumdrops (carbon atoms). To each carbon atom attach three tooth-picks with a yellow gumdrop on the other

Trans-gumdrop isomer

end. Rotate one of the gumdrop carbon atoms to show the movement of the attached atoms.

Remove a toothpick and yellow gumdrop from each carbon atom. Place a second toothpick between the carbon atoms. Each carbon atom should now be attached to two noncarbon atoms. Make those two atoms different colors, such as yellow and red. First place the red gumdrops on the same side of the double bond, which makes a model of a cis isomer. Try to twist the double toothpicks. Can you do it? If you turn the whole molecule upside down, are the red gumdrops still on the same side?

Now switch one of the red gumdrops with the yellow gumdrop on the same carbon. The red gumdrops should be opposite or trans to each other. Twist the double bond again. Can you move the red gumdrops to the same side of the double bond? It should not happen without breaking the double bond. How does this illustrate two isomers of a double bond molecule when the atoms in the double bond are attached to different groups of atoms? If you attach two yellow or two red gumdrops to the same carbon atom, can you make cis–trans isomers? Why or why not?

Pheromones in Insect Communication

Insects and many other organisms emit minute quantities of chemicals called pheromones. Insects use pheromones to send messages to individuals of the same species. Some pheromones warn of danger, others call for defense, mark a trail, or attract the opposite sex. In the last 40 years, the structures of many pheromones have been chemically determined. One of the most studied is bombykol, the sex pheromone produced by the female of the silkworm moth species. The bombykol molecule is a 16 carbon chain with one cis double bond, one trans double bond, and an alcohol group. A few molecules of synthetic bombykol will attract male silkworm moths from distances of over one kilometer. The effectiveness of many of these pheromones depends on the cis or trans configuration of the double bonds in the molecules. A certain species will respond to one isomer but not the other.

Scientists are interested in synthesizing pheromones for use as nontoxic alternatives to pesticides. When used in a trap, bombykol can be used to isolate male silkworm moths. When a synthetic pheromone is released in several areas of a field or crop, the males cannot locate the females, which disrupts the reproductive cycle. This technique has been successful with controlling the oriental fruit moth, the grapevine moth, and the pink bollworm.

Bombykol, sex attractant for the silkworm moth

Cis–Trans Isomers for Night Vision

The retinas of the eyes consist of two types of cells, rods and cones. The rods on the edge of the retina allow us to see in dim light, and the cones, in the center, produce our vision in bright light. In the rods, there is a substance called rhodopsin that absorbs light. Rhodopsin is composed of *cis*-11-retinal, an unsaturated compound, attached to a protein. When rhodopsin absorbs light, the *cis*-11-retinal isomer is converted to its trans isomer, which changes its shape. The trans form no longer fits the protein and it separates from the protein. The change from the cis to trans

isomer and the separation from the protein generates an electrical signal that the brain converts into an image.

An enzyme (isomerase) converts the trans isomer back to the *cis*-11-retinal isomer and the rhodopsin re-forms. If there is a deficiency of rhodopsin in the rods of the retina, night blindness may occur. One common cause is a lack of vitamin A in the diet. In our diet, we obtain vitamin A from plant pigments containing β-carotene, which is found in foods such as carrots, squash, and spinach. In the small intestine, the β-carotene is converted to vitamin A, which can be converted to *cis*-11-retinal or stored in the liver for future use. Without a sufficient quantity of retinal, not enough rhodopsin is produced to enable us to see adequately in dim light.

Cis-trans isomers of retinal

11-*cis*-retinal → (Light) → 11-*trans*-retinal

QUESTIONS AND PROBLEMS

Cis-Trans Isomers

13.13 What is the difference between constitutional isomers and cis-trans isomers?

13.14 How does *cis*-2-butene differ from *trans*-2-butene?

13.15 Draw structural formulas for the constitutional isomers of molecular formula C_3H_5Cl.

13.16 Draw structural formulas for the constitutional isomers of molecular formula C_4H_6.

13.17 Which of the following can be written as cis–trans isomers?

a. $CH_2{=}CHCH_3$ b. $CH_3CH_2CH{=}CHCH_3$ c.

13.18 Which of the following do not have cis–trans isomers?

a.

b. $CH_3CH_2CH_2CH{=}CH_2$ c. $CH_2{=}CHCH_2CHCH_3$ (with CH_3 branch)

13.19 Write the IUPAC name of each of the following using cis or trans prefixes:

a.

b.

c.

13.20 Write the IUPAC name of each of the following using cis or trans prefixes:

a.
$$CH_3\text{—}C{=}C\text{—}CH_2CH_3$$ (with H, H below)

b.
$$CH_3\text{—}C{=}C\text{—}H, CH_2CH_2CH_2CH_3$$ (with H below)

c.
$$CH_3CH_2CH_2\text{—}C{=}C\text{—}H, CH_2CH_3$$ (with H below)

13.21 Draw the structural formula for each of the following:
 a. *trans*-2-butene   **b.** *cis*-2-pentene   **c.** *trans*-3-heptene

13.22 Draw the structural formula for each of the following:
 a. *cis*-3-hexene   **b.** *trans*-2-pentene   **c.** *cis*-4-octene

13.4 Addition Reactions

LEARNING GOAL

Write the structural formulas and names for the organic products of addition reactions of alkenes and alkynes.

For alkenes and alkynes, the most characteristic reaction is the **addition** of atoms or groups of atoms to the carbons of the double or triple bond. Addition occurs because the weak pi bonds in double and triple bonds are easily broken, which provides electrons for new single bonds. Many different kinds of reactants can be added to double and triple bonds to form more stable products. Some addition reactions require a catalyst but others do not. The general equation for the addition of a reactant A—B to an alkene can be written as follows:

$$\text{C=C} + A\text{—}B \xrightarrow{\text{Addition}} \overset{A\ B}{-C-C-}$$
Alkene

The addition reactions have different names that depend on the type of reactant we add to the alkene, as Table 13.2 shows.

Table 13.2 Reactants and Addition Reactions

Reactant Added	Name of Addition Reaction
H_2	Hydrogenation
Cl_2, Br_2	Halogenation
HCl, HBr, HI	Hydrohalogenation
HOH	Hydration

Hydrogenation

In a reaction called **hydrogenation**, atoms of hydrogen add to the carbons in a double or triple bond to form alkanes. A catalyst such as platinum (Pt), nickel (Ni), or palladium (Pd) is added to speed up the reaction. The general equation for hydrogenation can be written as follows:

$$\text{C=C} + H\text{—}H \xrightarrow{\text{Catalyst}} \overset{H\ H}{-C-C-}$$
Double bond (unsaturated)   Single bond (saturated)

Electrons from the pi bond are used to form these single bonds

Some examples of the hydrogenation of alkenes and alkynes follow:

$$CH_3\text{—}CH{=}CH\text{—}CH_3 + H\text{—}H \xrightarrow{Pt} CH_3\text{—}CH\text{—}CH\text{—}CH_3$$
2-Butene   Butane

$$\text{Cyclohexene} + H\text{—}H \xrightarrow{Ni} \text{Cyclohexane}$$

EXPLORE YOUR WORLD

Unsaturation in Fats and Oils

Read the labels on some containers of vegetable oils, margarine, peanut butter, and shortenings.

Questions

1. What terms on the label tell you that the compounds contain double bonds?
2. A label on a bottle of canola oil lists saturated, polyunsaturated, and monounsaturated fats. What do these terms tell you about the type of bonding in the fats?
3. A peanut butter label states that it contains partially hydrogenated vegetable oils or completely hydrogenated vegetable oils. What does this tell you about the type of reaction that took place in preparing the peanut butter?

The hydrogenation of alkynes requires two molecules of hydrogen to form the alkane product.

$$CH_3-C\equiv C-CH_3 + 2\,H-H \xrightarrow{Pt} CH_3-\overset{\overset{\displaystyle H}{|}}{\underset{\underset{\displaystyle H}{|}}{C}}-\overset{\overset{\displaystyle H}{|}}{\underset{\underset{\displaystyle H}{|}}{C}}-CH_3$$

2-Butyne   Butane

SAMPLE PROBLEM 13.5

Writing Equations for Hydrogenation

Write the structural formula for the product of the following hydrogenation reactions:

a. $CH_3-CH=CH_2 + H_2 \xrightarrow{Pt}$ **b.** ⬠ $+ H_2 \xrightarrow{Pt}$ **c.** $HC\equiv CH + 2H_2 \xrightarrow{Ni}$

Solution

In an addition reaction, hydrogen adds to the double or triple bond to give an alkane.

a. $CH_3-CH_2-CH_3$ **b.** ⬠   **c.** H_3C-CH_3

Study Check

Draw the structural formula of the product of the hydrogenation of 2-methyl-1-butene using a platinum catalyst.

HEALTH NOTE

Hydrogenation of Unsaturated Fats

Vegetable oils such as corn oil or safflower oil are unsaturated fats composed of fatty acids that contain double bonds. The process of hydrogenation is used commercially to convert the double bonds in the unsaturated fats in vegetable oils to saturated fats such as margarine, which are more solid. Adjusting the amount of added hydrogen produces partially hydrogenated fats such as soft margarine, solid margarine in sticks, and shortenings, which are used in cooking. For example, oleic acid is a typical unsaturated fatty acid in olive oil and has a cis-double bond at carbon 9. When oleic acid is hydrogenated, it is converted to stearic acid, a saturated fatty acid.

$$CH_3(CH_2)_7(CH_2)_7\overset{\overset{\displaystyle O}{\|}}{C}OH$$
$$C=C$$
$$HH$$

$+\ H_2 \xrightarrow{Pt} CH_3(CH_2)_7-CH_2-CH_2-(CH_2)_7\overset{\overset{\displaystyle O}{\|}}{C}OH$

Oleic acid (The cis isomer
is found in olive oil and other
unsaturated fats)

Stearic acid (found in
saturated fats)

Halogenation

In the **halogenation** reactions of alkenes or alkynes, halogen atoms such as chlorine or bromine are added to the double or triple bonds. The reaction occurs readily, without the use of any catalyst, and adds halogen atoms to yield a di- or tetrahaloalkane product. In the general equation for halogenation, the symbol X—X or X_2 is used for Cl_2 or Br_2.

$$\underset{}{\diagdown}\mathrm{C}{=}\mathrm{C}\underset{}{\diagup} + \text{ X—X} \longrightarrow \overset{\overset{X}{|}}{-\mathrm{C}}\overset{\overset{X}{|}}{-\mathrm{C}}-$$

Here are some examples of adding Cl_2 or Br_2 to alkenes:

$$\underset{\text{Ethene}}{\mathrm{CH_2}{=}\mathrm{CH_2}} + \mathrm{Cl{-}Cl} \longrightarrow \underset{\text{1,2-Dichloroethane}}{\overset{\overset{Cl}{|}}{\mathrm{CH_2}}{-}\overset{\overset{Cl}{|}}{\mathrm{CH_2}}}$$

$$\underset{\text{Cyclohexene}}{\bigcirc\!\!=} + \mathrm{Br{-}Br} \longrightarrow \underset{\text{1,2-Dibromocyclohexane}}{}$$

Cyclohexene   +   Br—Br   ⟶   1,2-Dibromocyclohexane

$$\underset{\text{Propyne}}{\mathrm{CH_3{-}C}{\equiv}\mathrm{CH}} + 2\mathrm{Cl{-}Cl} \longrightarrow \underset{\text{1,1,2,2-Tetrachloropropane}}{\mathrm{CH_3}{-}\overset{\overset{Cl}{|}}{\underset{\underset{Cl}{|}}{\mathrm{C}}}{-}\overset{\overset{Cl}{|}}{\underset{\underset{Cl}{|}}{\mathrm{CH}}}}$$

The addition reaction of bromine is sometimes used to test for the presence of double and triple bonds, as shown in Figure 13.6.

SAMPLE PROBLEM 13.6

Writing Products of Halogenation

Write the condensed structural formula of the product of the following reaction:

$$\underset{}{\mathrm{CH_3{-}\overset{\overset{CH_3}{|}}{C}{=}CH_2}} + \mathrm{Br_2} \longrightarrow$$

Solution

The addition of bromine to an alkene places a bromine atom on each of the carbon atoms of the double bond.

$$\mathrm{CH_3{-}\overset{\overset{CH_3}{|}}{\underset{\underset{Br}{|}}{C}}{-}\underset{\underset{Br}{|}}{CH_2}}$$

Study Check

What is the name of the product formed when chlorine is added to 1-butene?

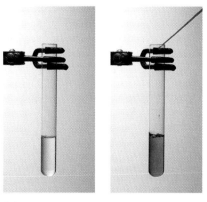

(a)

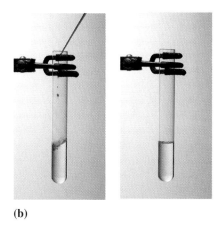

(b)

Figure 13.6 (a) When bromine is added to an alkane in the first test tube, the red color of bromine remains because the alkane does not react or reacts slowly. **(b)** When bromine is added to an alkene in the second test tube, the red color immediately disappears as bromine atoms add to the double bond.

Q Will the red color disappear when bromine is added to cyclohexane or cyclohexene?

Hydrohalogenation

In the reaction called **hydrohalogenation,** a hydrogen halide (HCl, HBr, or HI) adds to an alkene to yield a haloalkane. The hydrogen atom bonds to one carbon of the double bond, and the halogen atom adds to the other carbon. The general reaction, in which HX represents HCl, HBr, or HI, can be written as follows:

Alkene   Haloalkane (alkyl halide)

Two examples of hydrohalogenation follow:

$$CH_2{=}CH_2 + HCl \longrightarrow \overset{\overset{\displaystyle H}{|}}{CH_2}{-}\overset{\overset{\displaystyle Cl}{|}}{CH_2}$$

Ethene (ethylene)   Chloroethane (ethyl chloride)

$$CH_3{-}CH{=}CH{-}CH_3 + HBr \longrightarrow CH_3{-}\overset{\overset{\displaystyle H}{|}}{CH}{-}\overset{\overset{\displaystyle Br}{|}}{CH}{-}CH_3$$

2-Butene   2-Bromobutane

Steps in Addition Reactions of Alkenes

We have seen that in the addition reaction of an alkene, two groups add to the carbons in the double bond to give a saturated compound. To understand how the addition of H—X or H—OH takes place, we need to look at how the electrons from the pi bond are used to form new bonds. We can consider the steps involved when HBr adds to ethene. Initially, a proton from the HBr reacts with one of the carbons in the double bond. This makes the other carbon into a **carbocation** (a carbon cation), which has only three bonds and a positive charge. The carbocation reacts quickly with the bromide ion Br⁻.

Single product

Markovnikov's Rule

When HBr adds to a symmetrical alkene, a single product results. However, HBr can also add to a double bond with alkyl substituents in unsymmetrical alkenes.

The most stable carbocation that forms is the one with the most alkyl groups. Therefore, in the initial step, the proton adds to the carbon that is less substituted, which is also the carbon in the double that has the greater number of protons.

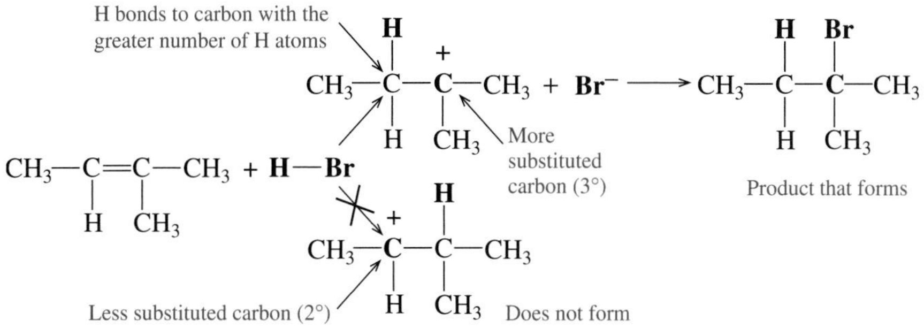

In 1870, Markovnikov, a Russian chemist, observed that the hydrohalogenation addition products of alkenes were limited to the more substituted halide product. His observation now called **Markovnikov's rule** states that when HX adds a double bond, the proton is bonded to the carbon atom that has the greater number of protons. Today, we know that this occurs because the proton adds to the position that produces the most stable carbocation, which also leads to the most substituted halide product.

SAMPLE PROBLEM 13.7

Addition to Alkenes

Predict the organic product for each of the following reactions:

a. $CH_3—CH\!=\!CH—CH_3 + HBr \longrightarrow$

b. 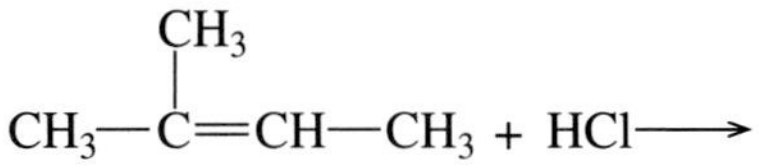
$$CH_3—\overset{\displaystyle CH_3}{\underset{}{C}}\!=\!CH—CH_3 + HCl \longrightarrow$$

Solution

a. This is a symmetrical alkene. Only one product forms when the H^+ and Br^- add to the carbons of the double bond

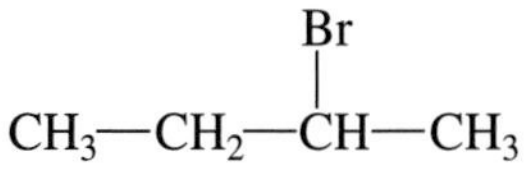

$$CH_3—CH_2—\overset{\displaystyle Br}{\underset{}{CH}}—CH_3$$

b. In the double bond of this unsymmetrical alkene, carbon 3 has the greater number of hydrogen atoms. Using Markovnikov's rule, the H from HCl adds to carbon 3 and the Cl adds to carbon 2. The product is the most substituted halide.

$$CH_3—\overset{\displaystyle CH_3}{\underset{\displaystyle Cl}{C}}—CH_2—CH_3$$

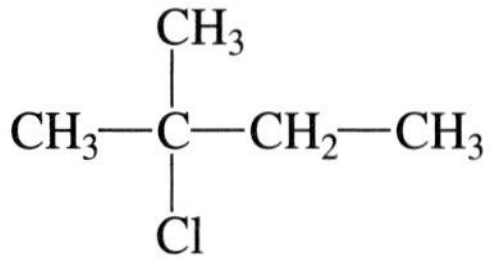

Study Check

Draw the structural formula of the organic product obtained when HBr adds to 1-methylcyclopentene.

Writing a Synthesis by Using an Alkene

What alkene would you start with to prepare the following compound?

$$\underset{\text{CH}_3-\overset{\displaystyle\overset{\text{Br}}{|}}{\text{CH}}-\text{CH}_2-\text{CH}_3}{}$$

Solution

The starting alkene could be 1-butene or 2-butene.

Study Check

Write an equation using an alkene and HCl to prepare 2-chloropentane.

Adding Water

Alkenes react with water (HOH) when the reaction is catalyzed by a strong acid such as H_2SO_4. In this reaction called **hydration,** H— attaches to one of the carbon atoms in the double bond, and —OH group to the other carbon. Hydration is used to prepare alcohols, which have the functional group —OH. In the general equation the acid is represented by H^+.

$$\overset{\diagdown}{\underset{\diagup}{}}\text{C}{=}\text{C}\overset{\diagup}{\underset{\diagdown}{}} + \text{H}{-}\text{OH} \xrightarrow{\text{H}^+} \overset{\text{H}\ \ \text{OH}}{\underset{|\ \ \ |}{-\text{C}-\text{C}-}}$$

Alkene   Alcohol

$$\text{CH}_2{=}\text{CH}_2 + \text{H}{-}\text{OH} \xrightarrow{\text{H}^+} \underset{\text{CH}_2-\text{CH}_2}{\overset{\text{H}\quad\ \ \text{OH}}{|\qquad\ |}} \longleftarrow \text{Functional group of alcohols}$$

Ethene   Ethanol (ethyl alcohol)

The addition of water to a double bond in which the carbon atoms are attached to different groups follows Markovnikov's rule.

$$\text{CH}_3{-}\text{CH}{=}\text{CH}_2 + \text{H}{-}\text{OH} \xrightarrow{\text{H}^+} \overset{\text{OH}\ \ \text{H}}{\underset{\text{CH}_3-\text{CH}-\text{CH}_2}{|\qquad\ |}} \ not\ \ \text{CH}_3{-}\text{CH}_2{-}\text{CH}_2{-}\text{OH}$$

Propene   2-Propanol

Writing Products of Hydration

Write the structural formulas for the products that form in the following hydration reactions:

a. $\text{CH}_3\text{CH}_2\text{CH}_2{-}\text{CH}{=}\text{CH}_2 + \text{HOH} \xrightarrow{\text{H}^+}$

b. $\boxed{\phantom{\text{||}}}\!\!| + \text{HOH} \xrightarrow{\text{H}^+}$

Solution

a. Water adds H— and —OH to the double bond. We use Markovnikov's rule to add the H— to the CH_2 in the double bond, and the —OH to the CH.

$$CH_3CH_2CH_2{-}\overset{\overset{\displaystyle OH}{\big\downarrow}}{C}H{=}\overset{\overset{\displaystyle H}{\big\downarrow}}{C}H_2 \xrightarrow{H^+} CH_3CH_2CH_2{-}\overset{\overset{\displaystyle OH}{\big|}}{C}H{-}CH_3$$

b. In cyclobutene, the H— adds to one side of the double bond, and the —OH adds to the other side.

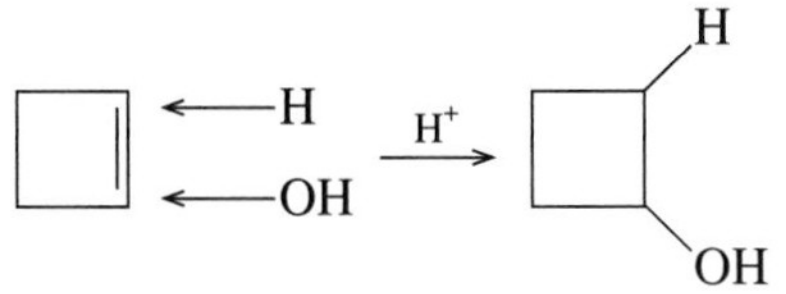

Study Check

Draw the structural formula for the alcohol obtained by the hydration of 2-methyl-2-butene.

QUESTIONS AND PROBLEMS

Addition Reactions

13.23 Give the condensed structural formulas and names of the products in each of the following reactions:

a. $CH_3{-}CH_2{-}CH_2{-}CH{=}CH_2 + H_2 \xrightarrow{Pt}$

b. $CH_2{=}\overset{\overset{\displaystyle CH_3}{\big|}}{C}{-}CH_2{-}CH_3 + Cl_2 \longrightarrow$

c.  $+ Br_2 \longrightarrow$

d. cyclopentene $+ H_2 \xrightarrow{Pt}$

e. 2-methyl-2-butene $+ Cl_2 \longrightarrow$

f. 2-pentyne $+ 2H_2 \xrightarrow{Pd}$

13.24 Give the condensed structural formulas and names of the products in each of the following reactions:

a. $CH_3{-}CH_2{-}CH{=}CH_2 + Br_2 \longrightarrow$

b. cyclohexene $+ H_2 \xrightarrow{Pt}$ **c.** *cis*-2-butene $+ H_2 \xrightarrow{Pt}$

d. $CH_3{-}\overset{\overset{\displaystyle CH_3}{\big|}}{C}{=}CH{-}CH_2{-}CH_3 + Cl_2 \longrightarrow$

e. 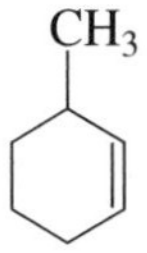 $+ Br_2 \longrightarrow$

f. $CH_3{-}\overset{\overset{\displaystyle CH_3}{\big|}}{C}H{-}C{\equiv}CH + 2Cl_2 \longrightarrow$

13.25 Give the condensed structural formulas of the products in each of the following reactions using Markovnikov's rule when necessary:

a. $CH_3{-}CH{=}CH{-}CH_3 + HBr \longrightarrow$

b. cyclopentene $+ HOH \xrightarrow{H^+}$ **c.** $CH_2{=}CH{-}CH_2{-}CH_3 + HCl \longrightarrow$

d. $CH_3{-}\overset{\overset{\displaystyle CH_3}{\big|}}{\underset{\underset{\displaystyle CH_3}{\big|}}{C}}{=}C{-}CH_3 + HI \longrightarrow$ **e.** $CH_3CH_2{-}\overset{\overset{\displaystyle CH_3}{\big|}}{C}{=}CHCH_3 + HBr$

f.
$$\underset{\text{CH}_3}{\bigcirc} + \text{ HOH } \xrightarrow{\text{H}^+}$$

13.26 Give the condensed structural formulas of the products in each of the following reactions using Markovnikov's rule when necessary:

a.
$$\text{CH}_3\underset{\overset{|}{\text{CH}_3}}{\text{C}}=\text{CH}-\text{CH}_3 + \text{HCl} \longrightarrow$$

b. $\text{CH}_3\text{CH}_2-\text{CH}=\text{CH}-\text{CH}_2\text{CH}_3 + \text{HOH} \xrightarrow{\text{H}^+}$

c.
$$\text{CH}_3\underset{\overset{|}{\text{CH}_3}}{\text{C}}=\text{CH}_2 + \text{HBr} \longrightarrow$$

d. 4-methylcyclopentene + HOH $\xrightarrow{\text{H}^+}$

e. $\bigcirc + \text{HBr} \longrightarrow$

f. $\text{CH}_3-\text{C}\equiv\text{C}-\text{CH}_3 + 2\text{HCl} \longrightarrow$

13.27 Write an equation including any catalysts for the following reactions:
  a. hydrogenation of 2-methylpropene
  b. addition of hydrogen chloride to cyclopentene
  c. addition of bromine to 2-pentene
  d. hydration of propene
  e. addition of chlorine to 2-butyne

13.28 Write an equation including any catalysts for the following reactions:
  a. hydration of 1-methylcyclobutene
  b. hydrogenation of 3-hexene
  c. addition of hydrogen bromide to 2-methyl-2-butene
  d. addition of chlorine to 2,3-dimethyl-2-pentene
  e. addition of HCl to 1-methylcyclopentene

LEARNING GOAL

Draw structural formulas of monomers that form a polymer or a three-monomer section of a polymer.

WEB TUTORIAL
Polymers

13.5 Polymerization

Polymers have been a part of life since prehistoric times. Cellulose in wood and starches obtained from vegetables are carbohydrate polymers made from many thousands of glucose molecules. Silk and wool are polymers made from amino acids. DNA, the molecule that carries genetic information, is a polymer of nucleotides.

Polymers are large molecules that consist of small repeating units called **monomers**. In the past hundred years, the plastics industry has made synthetic polymers that are in many of the materials we use every day, such as carpeting, plastic wrap, nonstick pans, plastic cups, and rain gear. In medicine, synthetic polymers are used to replace diseased or damaged body parts such as hip joints, teeth, heart valves, and blood vessels. (See Figure 13.7.)

Addition Polymers

Many of the synthetic polymers are made by addition reactions of monomers that are small alkenes. The conditions for many polymerization reactions require high temperatures and very high pressure (over 1000 atm). In an addition reaction, the polymer grows as monomers are added to the end of the chain. Polyethylene, a polymer made from ethylene $\text{CH}_2=\text{CH}_2$, is used in plastic bottles, film, and plastic dinnerware. (See Figure 13.8.) In the polymerization, a series of addition reac-

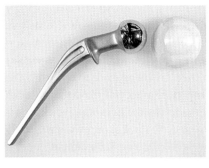

Figure 13.7 Synthetic polymers are used to replace diseased veins and arteries. A broken hip is repaired using a metal piece that fits into an artificial plastic cup socket.

Q Why are the substances in these plastic devices called polymers?

Polyethylene

Vinyl chloride

Polypropylene

Polytetrafluoroethylene (Teflon)

Polydichloroethylene (Saran)

Polystyrene

Figure 13.8 Synthetic polymers provide a wide variety of items that we use every day.

Q What are some alkenes used to make the polymers in these plastic items?

tions joins one monomer to the next until a long carbon chain forms that contains as many as 1000 monomers.

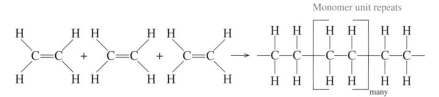

Ethene (ethylene) monomers

Polyethylene section

Table 13.3 lists several alkene monomers that are used to produce common synthetic polymers and Figure 13.8 shows examples of each. The alkane-like nature of these plastic synthetic polymers makes them unreactive. Thus, they do not decompose easily (they are nonbiodegradable) and have become contributors to pollution. Efforts are being made to make them more degradable. It is becoming increasingly important to recycle plastic material, rather than add to our growing landfills. You can identify the type of polymer used to manufacture a plastic item by looking for the recycle symbol (arrows in a triangle) found on the label or on the bottom of the plastic container. For example, either the number 5 or the letters PP inside the triangle is a code for a polypropylene plastic.

Table 13.3   Some Alkenes and Their Polymers

Monomer	Polymer Section	Common Uses
$CH_2{=}CH_2$ Ethene (ethylene)	Polyethylene	Plastic bottles, film, insulation materials
$CH_2{=}CH$ with Cl Chloroethene (vinyl chloride)	Polyvinyl chloride (PVC)	Plastic pipes and tubing, garden hoses, garbage bags
$CH_2{=}CH$ with CH_3 Propene (propylene)	Polypropylene	Ski and hiking clothing, carpets, artificial joints
$F{-}C{=}C{-}F$ with F, F Tetrafluoroethene	Polytetrafluoroethylene (Teflon)	Nonstick coatings
$CH_2{=}C{-}Cl$ with Cl 1,1-Dichloroethene	Polydichloroethylene (Saran)	Plastic film and wrap
$H_2C{=}CH$ with phenyl Phenylethene (styrene)	$-CH_2-CH-CH_2-CH-CH_2-CH-$ Polystyrene	Plastic coffee cups and cartons, insulation

1	2	3	4	5	6
PETE	HDPE	PV	LDPE	PP	PS
Polyethylene terephthalate	High-density polyethylene	Polyvinyl chloride	Low-density polyethylene	Polypropylene	Polystyrene

SAMPLE PROBLEM 13.10

Polymers

What are the starting monomers for the following polymers?

a. polypropylene

b. Saran

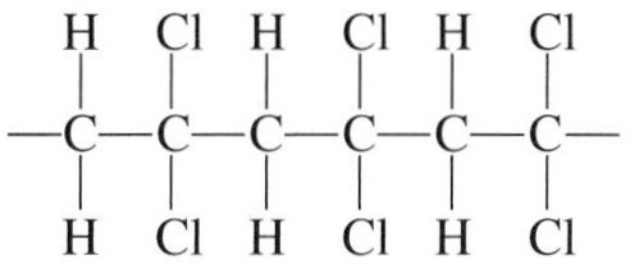

$$\begin{array}{cccccc} H & Cl & H & Cl & H & Cl \\ | & | & | & | & | & | \\ -C-&C-&C-&C-&C-&C- \\ | & | & | & | & | & | \\ H & Cl & H & Cl & H & Cl \end{array}$$

Solution

a. propene (propylene) $CH_2\!=\!\overset{\displaystyle CH_3}{\overset{\displaystyle |}{CH}}$

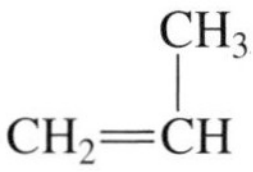

b. 1,1-dichloroethene, $CH_2\!=\!\overset{\displaystyle Cl}{\overset{\displaystyle |}{C}}\!-\!Cl$

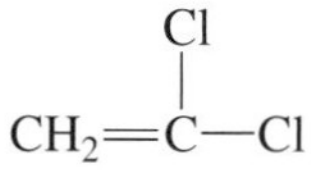

Study Check

What is the monomer for PVC?

QUESTIONS AND PROBLEMS

Addition Polymers of Alkenes

13.29 What is a polymer?

13.30 What is a monomer?

13.31 Write an equation that represents the formation of a part of the Teflon polymer from three of the monomer units.

13.32 Write an equation that represents the formation of a part of the polystyrene polymer from three of the monomer units.

13.33 A plastic called polyvinylidene difluoride, PVDF, is made from monomers of 1,1-difluoroethene. Write the structure of the polymer formed from the addition of three monomers of the 1,1-difluoroethene.

13.34 An alkene called acrylonitrile is the monomer used to form the polymer used in the fabric material called Orlon. Write an equation that represents the formation of a part of the polyacrylonitrile polymer from three of the monomer units.

Acrylonitrile $\quad CH_2\!=\!\overset{\displaystyle CN}{\overset{\displaystyle |}{CH}}$

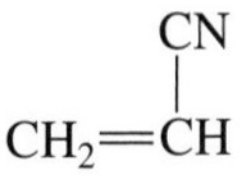

13.6 Aromatic Compounds

In 1825, Michael Faraday isolated a hydrocarbon called benzene, which had the molecular formula C_6H_6. Because many compounds containing benzene had fragrant odors, the family of benzene compounds became known as **aromatic compounds**. A molecule of **benzene** consists of a ring of six carbon atoms with one hydrogen atom attached to each carbon. Each carbon atom uses 3 valence electrons to bond to the hydrogen atom and two adjacent carbons. That leaves 1 valence electron to share in a double bond with an adjacent carbon. In 1865, August Kekulé dreamed that the carbon atoms were arranged in a flat ring with alternating single and double bonds between the carbon atoms. This idea led to two ways of writing the benzene structure, as follows:

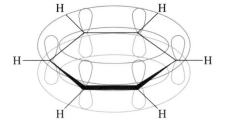

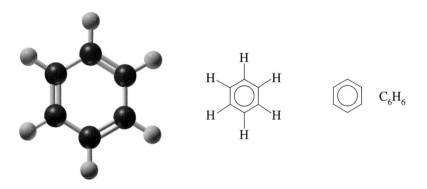

Structures for benzene

However, there is only one structure of benzene. Today we know that all the bonds in benzene are identical. In benzene, each carbon bonds to two other carbons and one hydrogen by sigma bonds. That leaves six *p* orbitals that overlap to form a continuous ring of orbitals above and below the carbon atoms. In this circle of *p* orbitals, the pi electrons are shared equally, a unique feature that makes aromatic compounds especially stable. The benzene structure is represented as a hexagon with a circle in the center.

Naming Aromatic Compounds

Aromatic compounds that contain a benzene ring with a single substituent are usually named as benzene derivatives. However, many of these compounds have been important in chemistry for many years and still use their common names. Some common names such as toluene are now accepted as IUPAC names.

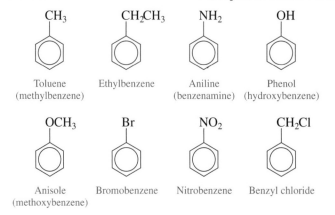

HEALTH NOTE

Aromatic Compounds in Health and Medicine

Aromatic compounds are common in nature and in medicine. Toluene is used as a starting reactant to make drugs, dyes, and explosives such as TNT (trinitrotoluene). The benzene ring is found in some amino acids (the building blocks of proteins), in pain relievers such as aspirin, acetaminophen, and ibuprofen, and flavorings such as vanillin.

Toluene   TNT (2,4,6-trinitrotoluene)   Aspirin   Vanillin   Ibuprofen

Acetaminophen   Phenylalanine, an amino acid   Methyl orange, textile dye

When a benzene ring is a substituent C_6H_5—, it is named as a phenyl group. A benzyl group is a benzene ring and a CH_2— group.

phenyl group   benzyl group

benzyl chloride   3-phenyl-1-butene

When there are two substituents on benzene, the ring is numbered to give the lowest numbers to the substituents. However, common names use the prefixes **ortho, meta,** and **para** to show the substituent arrangement. The prefix *ortho (o)* indicates a 1,2 arrangement, *meta (m)* is a 1,3 arrangement, and *para (p)* is used for 1,4 arrangements.

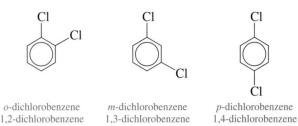

o-dichlorobenzene
1,2-dichlorobenzene

m-dichlorobenzene
1,3-dichlorobenzene

p-dichlorobenzene
1,4-dichlorobenzene

Common names are used for many disubstituted benzenes such as the constitutional isomers of dimethylbenzene.

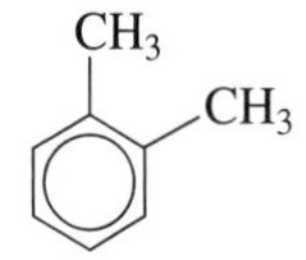

o-xylene
1,2-dimethylbenzene

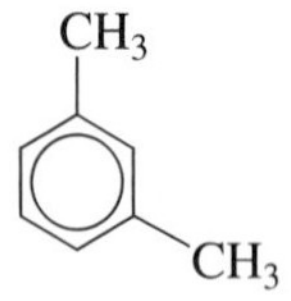

m-xylene
1,3-dimethylbenzene

p-xylene
1,4-dimethylbenzene

When there are three or more substituents on the benzene ring, numbers are used to show their arrangement. The substituents are numbered to give the lowest numbers and named alphabetically.

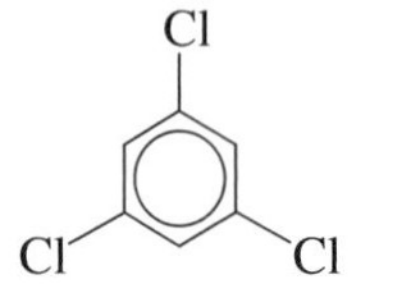

1,3,5-trichlorobenzene 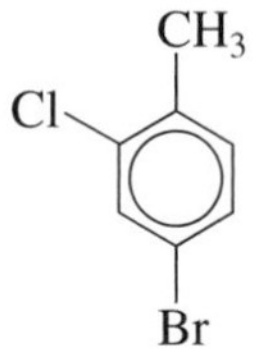   4-bromo-2-chlorotoluene 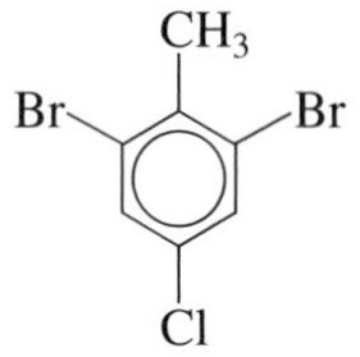   2,6-dibromo-4-chlorotoluene

SAMPLE PROBLEM 13.11

Naming Aromatic Compounds

Give IUPAC and common names for each of the following aromatic compounds:

a.    **b.** 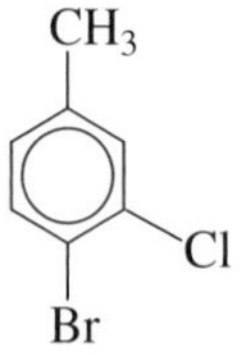   **c.**

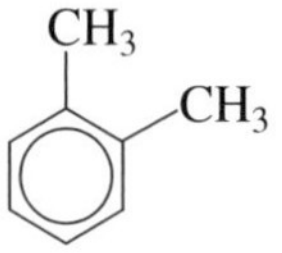

Solution

a. chlorobenzene; phenyl chloride
b. 4-bromo-3-chlorotoluene
c. 1,2-dimethylbenzene; o-xylene

Study Check

Name the following compound.

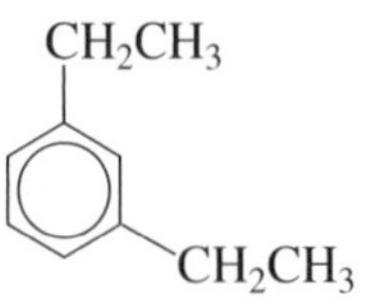

QUESTIONS AND PROBLEMS

Aromatic Compounds

13.35 Cyclohexane and benzene each have six carbon atoms. How are they different?

13.36 In the Health Note "Aromatic Compounds in Health and Medicine," what part of each molecule is the aromatic portion?

HEALTH NOTE

Polycyclic Aromatic Hydrocarbons (PAHs)

Large aromatic compounds known as polycyclic aromatic hydrocarbons are formed by fusing together two or more benzene rings edge-to-edge. In a fused ring compound, neighboring benzene rings share two or more carbon atoms. Naphthalene with two benzene rings is well known for its use in mothballs. Anthracene with three rings is used in the manufacture of dyes.

Napthalene

Anthracene

Phenanthrene

When a polycyclic compound contains phenanthrene, it may act as a carcinogen, a substance known to cause cancer. For example, some aromatic compounds in cigarette smoke cause cancer, as seen in the lung tissue of a heavy smoker. Benz[a]pyrene, a prod-

uct of combustion, has been identified in coal tar, tobacco smoke, barbecued meats, and automobile exhaust.

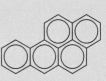
Benz[a]pyrene

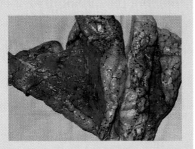

Compounds containing five or more fused benzene rings such as benz[a]pyrene are potent carcinogens. The molecules interact with the DNA in the cells, causing abnormal cell growth and cancer. Increased exposure to carcinogens increases the chance of DNA alterations in the cells.

13.37 Give the IUPAC and any common names for each of the following:

a.

b.

c.

d.

e.

f. CH₃—CH—CH₃

13.38 Give the IUPAC and any common names for each of the following:

a.

b.

c.

d.

e.

f.

13.39 Draw the structural formulas for each of the following compounds:
   a. methylbenzene
   b. *m*-dichlorobenzene
   c. 1-ethyl-4-methylbenzene
   d. *p*-chlorotoluene

13.40 Draw the structural formulas for each of the following compounds:
   a. benzene
   b. *o*-chloromethylbenzene
   c. propylbenzene
   d. 1,2,4-trichlorobenzene

13.7 Properties of Aromatic Compounds

The symmetrical structure of benzene allows the cyclic structures to stack close together, which contributes to higher melting points and boiling points of benzene and its derivatives. For example, hexane melts at $-95°C$, while benzene melts at $6°C$. Among the disubstituted benzene compounds, the para consitutional isomers are more symmetrical and have higher melting points than the ortho and meta isomers: o-xylene melts at $-26°C$ and m-xylene melts at $-48°C$, while p-xylene melts at $13°C$.

Aromatic compounds are less dense than water, although they are somewhat denser than other hydrocarbons. Halogenated benzene compounds are denser than water. Aromatic hydrocarbons are insoluble in water and are used as solvents for other organic compounds. Only those containing strongly polar functional groups such as —OH or —COOH will be more soluble. Benzene and other aromatic compounds are resistant to reactions that break up the aromatic system, although they are flammable, as are other hydrocarbon compounds.

Chemical Properties

The most important type of reaction for benzene and aromatic compounds is **substitution,** in which an atom or group of atoms replaces a hydrogen atom on a benzene ring. A substitution reaction, rather than addition, retains the stability of the aromatic bonding system. Substitution reactions of benzene include halogenation, nitration, and sulfonation.

Halogenation

In the chlorination or bromination of benzene, a chlorine or bromine atom replaces a hydrogen atom on the benzene ring. A catalyst such as $FeCl_3$ is required for chlorination; $FeBr_3$ is a catalyst in bromination.

Chlorobenzene

When toluene (methylbenzene) undergoes halogenation, a mixture of constitutional isomers is obtained as products. However, the presence of a methyl group in toluene has the effect of producing mostly ortho and para isomers. In most substitution reactions of toluene, the meta isomer is produced in very low amounts.

Toluene o-Chlorotoluene m-Chlorotoluene p-Chlorotoluene
                                    (very little)

Nitration

When benzene is heated with nitric acid, nitrobenzene is produced. Sulfuric acid (H_2SO_4) is required as a catalyst for the nitration.

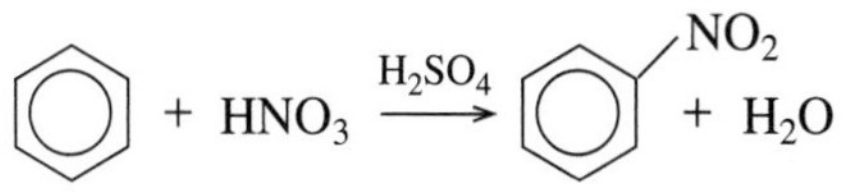

Nitrobenzene

Sulfonation

When benzene reacts with a mixture of $SO_3 + H_2SO_4$, known as "fuming sulfuric acid," the product is benzenesulfonic acid.

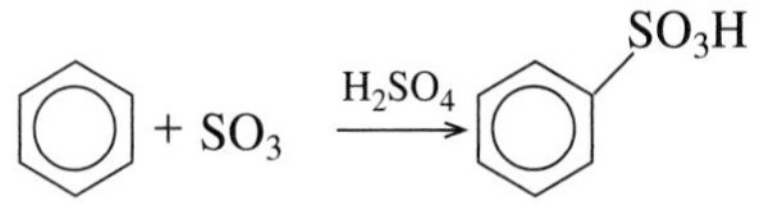

Benzenesulfonic acid

The sulfonation of aromatic compounds is one way to produce sulfa drugs.

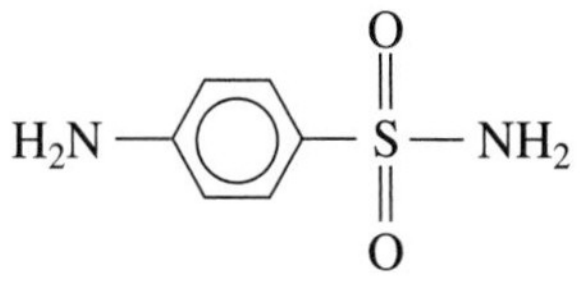

Sulfanilamide, a sulfa drug

SAMPLE PROBLEM 13.12

Reactions of Benzene

Write the structure of the organic product when benzene reacts with the following:
a. Br_2 and $FeBr_3$ **b.** HNO_3 and H_2SO_4

Solution

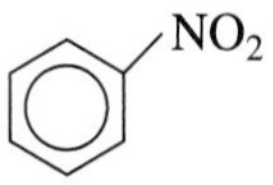

Study Check

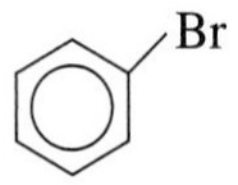

A chemist needs to synthesize chlorobenzene. If benzene is available in the lab, how could she prepare this compound?

QUESTIONS AND PROBLEMS

Reactions of Aromatic Compounds

13.41 Alkenes undergo addition reactions, but benzene does not. How does benzene react and why?

13.42 When toluene is reacted with chlorine in light, the product is CH_2Cl
How would you explain this result?

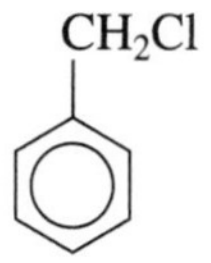

13.43 Draw the structures of the organic product(s), if any, for the following reactants:

a. Benzene + Cl_2 $\xrightarrow{\text{no catalyst}}$

b. Benzene + Cl_2 $\xrightarrow{\text{FeCl}_3}$

c. Benzene + HNO_3 $\xrightarrow{\text{H}_2\text{SO}_4}$

13.44 Draw the structures of the organic product(s), if any, for the following reactants:

a. Toluene + Br_2 $\xrightarrow{\text{FeBr}_3}$

b. Benzene + SO_3 $\xrightarrow{\text{H}_2\text{SO}_4}$

c. Benzene + HNO_3 $\xrightarrow{\text{H}_2\text{SO}_4}$

Chapter Review

13.1 Alkenes and Alkynes

Alkenes are unsaturated hydrocarbons that contain carbon–carbon double bonds ($C = C$) made of one sigma and one pi bond. Alkynes contain a triple bond ($C \equiv C$) composed of one sigma and two pi bonds.

13.2 Naming Alkenes and Alkynes

The IUPAC names of alkenes end with *ene,* while alkyne names end with *yne*. The main chain is numbered from the end nearest the double or triple bond. In a cycloalkene, the double bond is carbon 1 and 2, and the ring is numbered to give the lowest numbers to any substituents, which are named alphabetically.

13.3 Cis–Trans Isomers

Isomers of alkenes occur when the carbon atoms in the double bond are connected to different atoms or groups. In the cis isomer the attached groups are on the same side of the double bond, whereas in the trans isomer they are connected on the opposite sides of the double bond.

13.4 Addition Reactions

The addition of small molecules to the double bond is a characteristic reaction of alkenes. Hydrogenation adds hydrogen atoms to the double bond of an alkene to yield an alkane. Halogenation adds bromine or chlorine atoms to produce dihaloalkanes. Hydrogen halides and water can also add to a double bond. When there are a different number of groups attached to the carbons in the double bond, the

H from the reactant (HX or H—OH) adds to the carbon with the greater number of hydrogen atoms.

13.5 Polymerization

Polymers are long-chain molecules that consist of many repeating units of smaller carbon molecules called monomers. In nature, cellulose and starch are polymers of glucose and proteins are polymers of amino acids. Many materials that we use every day are synthetic polymers, including carpeting, plastic wrap, nonstick pans, and nylon. These synthetic materials are often made by addition reactions in which a catalyst links the carbon atoms from various kinds of alkene molecules.

13.6 Aromatic Compounds

Most aromatic compounds contain benzene, a cyclic structure containing six CH units. The structure of benzene is represented as a hexagon with a circle in the center. Many aromatic compounds use the parent name benzene, although common names such as toluene are retained. The benzene ring is numbered and the branches are listed in alphabetical order. For two branches, the positions are often shown by the prefixes *ortho* (1,2-), *meta* (1,3-), and *para* (1,4-).

13.7 Properties of Aromatic Compounds

Aromatic compounds undergo substitution reactions such as halogenation, nitration, and sulfonation. They do not undergo addition reactions, which would disrupt their stable aromatic bonding system.

Summary of Naming

Type	Example	Characteristic	Structure
Alkene	Propene (propylene)	double bond	$CH_3—CH=CH_2$
Cycloalkene	Cyclopropene	double bond in a carbon ring	(triangle)
Alkyne	Propyne	triple bond	$CH_3C\equiv CH$
Aromatic	Benzene	Aromatic ring of six carbons	(benzene ring)
	Methylbenzene or toluene		(toluene ring, CH_3)
	1,4-dichlorobenzene or *para*-dichlorobenzene		(ring, Cl top and bottom)

Summary of Reactions

Hydrogenation

Alkene + H_2 $\xrightarrow{Pt}$ alkane

$CH_2=CH—CH_3 + H_2 \xrightarrow{Pt} CH_3—CH_2—CH_3$

Alkyne + $2H_2$ $\xrightarrow{Pt}$ Alkane

$CH_3—C\equiv CH + 2H_2 \xrightarrow{Pt} CH_3—CH_2—CH_3$

Halogenation

Alkene + Cl_2 (or Br_2) $\longrightarrow$ dihaloalkane

$CH_2=CH—CH_3 + Cl_2 \longrightarrow \overset{Cl}{CH_2}—\overset{Cl}{CH}—CH_3$

Hydrohalogenation

Alkene + HX $\longrightarrow$ haloalkane

$CH_2=CH—CH_3 + HCl \longrightarrow CH_3—\overset{Cl}{CH}—CH_3$
Markovnikov's rule

Hydration of Alkenes

Alkene + H—OH $\xrightarrow{H^+}$ alcohol

$CH_2=CH—CH_3 + H—OH \xrightarrow{H^+} CH_3—\overset{OH}{CH}—CH_3$
Markovnikov's rule

Substitution Reactions of Benzene

(benzene) $\xrightarrow[FeCl_3]{Cl_2}$ (ring-Cl) Halogenation

$\xrightarrow[H_2SO_4]{HNO_3}$ (ring-NO_2) Nitration

$\xrightarrow[H_2SO_4]{SO_3}$ (ring-SO_3H) Sulfonation

Key Terms

addition A reaction in which atoms or groups of atoms bond to a double bond. Addition reactions include the addition of hydrogen (hydrogenation), halogens (halogenation), hydrogen halides (hydrohalogenation), or water (hydration).

alkene An unsaturated hydrocarbon containing a carbon–carbon double bond.

alkyne An unsaturated hydrocarbon containing a carbon–carbon triple bond.

aromatic compounds Compounds that contain the ring structure of benzene.

benzene A ring of six carbon atoms each of which is attached to a hydrogen atom, C_6H_6.

carbocation A carbon cation that has only three bonds and a positive charge and is formed during the addition reactions of hydration and hydrohalogenation.

cis isomer A geometric isomer in which similar groups are connected on the same side of the double bond.

cycloalkene A cyclic hydrocarbon that contains a double bond in the ring.

halogenation The addition of Cl_2 or Br_2 to an alkene to form halogen-containing compounds.

hydration An addition reaction in which the components of water, H— and —OH, bond to the carbon–carbon double bond to form an alcohol.

hydrogenation The addition of hydrogen (H_2) to the double bond of alkenes to yield alkanes.

hydrohalogenation The addition of a hydrogen halide such as HCl or HBr to a double bond.

Markovnikov's rule When adding HX or HOH to alkenes with different numbers of groups attached to the dou-

ble bonds, the H adds to the carbon that has the greater number of hydrogen atoms.

meta A method of naming that indicates two substituents at carbons 1 and 3 of benzene.

monomer The small organic molecule that is repeated many times in a polymer.

nitration The addition of a nitro group (—NO_2) to benzene.

ortho A method of naming that indicates two substituents at carbons 1 and 2 of a benzene ring.

para A method of naming that indicates two substituents at carbons 1 and 4 of a benzene ring.

polymer A very large molecule that is composed of many small, repeating structural units that are identical.

saturated hydrocarbons A compound of carbon and hydrogen in which the carbon chain consists of only single carbon–carbon bonds (H_3C—CH_3).

substitution The reactions of benzene and other aromatic compounds in which an atom or group of atoms replaces a hydrogen on a benzene ring .

sulfonation The reaction of benzene with SO_3 and H_2SO_4 to give benzenesulfonic acid.

trans isomer A geometric isomer in which similar groups are connected to opposite sides of the double bond in an alkene.

unsaturated hydrocarbons A compound of carbon and hydrogen in which the carbon chain contains at least one double (alkene) or triple carbon–carbon bond (alkyne). An unsaturated compound is capable of an addition reaction with hydrogen, which converts the double or triple bonds to single carbon–carbon bonds.

Additional Problems

13.45 Compare the formulas and bonding in propane, cyclopropane, propene, and propyne.

13.46 Compare the formulas and bonding in butane, cyclobutane, cyclobutene, and 2-butyne.

13.47 Give the IUPAC name for each of the following compounds:

a.

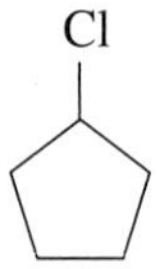

Cl

b.   Cl CH_3

$CH_3CHCH_2CHCH_3$

c. CH_3

CH_2=$CCH_2CH_2CH_3$

d. CH_3CH_2C≡CCH_3

e.   Cl

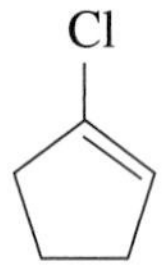

f. CH_3 H

C=C

H CH_2CH_3

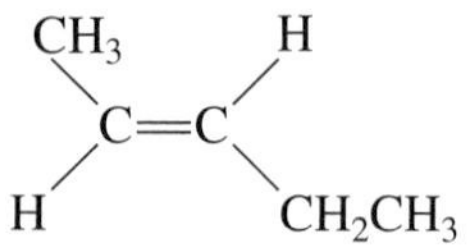

g.   Cl

Cl

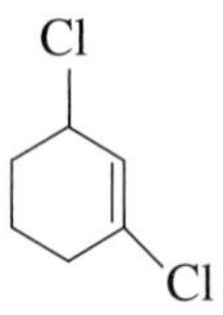

13.48 Write the condensed structures of each of the following compounds:

a. 1,2-dibromocyclopentane
b. 2-pentyne
c. *cis*-2-heptene
d. 3,3-dichloro-2-methylpentane
e. *trans*-3-hexene
f. 2-bromo-3-chlorocyclohexane
g. 2,3-dichloro-1-butene

13.49 Indicate if the following pairs of structures represent constitutional isomers, cis–trans isomers, or identical compounds.

a.

and

b.

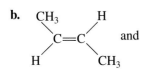

and

c. $CH_2\!=\!CH$ and $CH_3CH_2CH_2CH\!=\!CH_2$

d. $CH_3CHCH_2CHCH_3$ and $CH_3CH_2CHCH_2CH_2CH_3$

13.50 Draw the condensed structures and give the names for all the isomers of C_4H_8 including cyclic and cis–trans isomers.

13.51 Methylcyclopentane is formed by four different alkenes that react with hydrogen (H_2) in the presence of a Ni catalyst. Draw the condensed structures of each of these alkenes.

13.52 What role do cis–trans isomers play in night vision?

13.53 Write the cis and trans isomers for each of the following:
a. 2-pentene   **b.** 3-hexene
c. 2-butene   **d.** 2-hexene

13.54 Write the structures of the products, if any, for the following.

a. $CH_3CH\!=\!CHCH_3$ + H_2 $\xrightarrow{Ni}$

b. ⬠ + H_2 $\xrightarrow{Ni}$

c. $CH_3CH\!=\!CHCH_3$ + HBr $\longrightarrow$

d. ⬠ + HBr $\longrightarrow$

e. $CH_3CH\!=\!CHCH_3$ + $Cl_2\longrightarrow$

f. $CH_3CH_2CH_3$ + H_2 $\xrightarrow{Ni}$

g. $CH_3CH\!=\!CHCH_3$ + HOH $\xrightarrow{H^+}$

h. ⬡ + HOH $\xrightarrow{H^+}$

i. $CH_3C\!=\!CCH_3$ + HCl $\longrightarrow$ (with CH_3 groups)

13.55 What is the condensed structural formula of the organic compound needed to prepare each of the following products?

a. ? + H_2 $\xrightarrow{Ni}$ ⬡

b. ? + Br_2 $\longrightarrow$ $CH_3CHCHCH_2CH_3$ (Br Br)

c. ? + HCl $\longrightarrow$ CH_3CHCH_3 (Cl)

d. ? + HOH $\xrightarrow{H^+}$ (cyclopentane with OH)

13.56 Write the combustion reaction for the acetylene used in a welder's torch.

13.57 Copolymers contain more than one type of monomer. One copolymer used in medicine is made of alternating units of styrene and acrylonitrile. Write a section of the copolymer that would have these alternating units. (For structure of the styrene, see Table 13.3.)

$H_2C\!=\!CH$ (CN)
acrylonitrile

13.58 Lucite or Plexiglas is a polymer of methylmethacrylate. Write the part of the polymer that is made from the addition of three of these monomers.

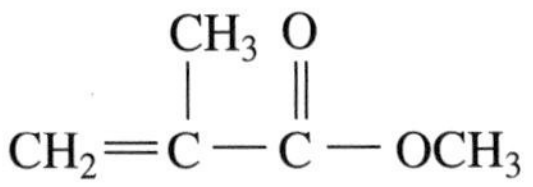

$$CH_2 = \underset{\underset{CH_3}{|}}{C} - \underset{\underset{O}{\|}}{C} - OCH_3$$

Methylmethacrylate

13.59 Name each of the following aromatic compounds:

a.   CH$_3$

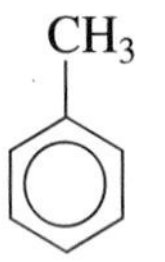

b.   CH$_3$ / Cl

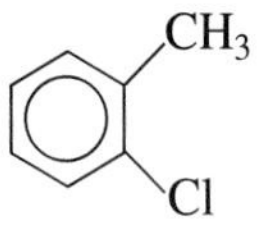

c.   CH$_3$ / CH$_2$CH$_3$

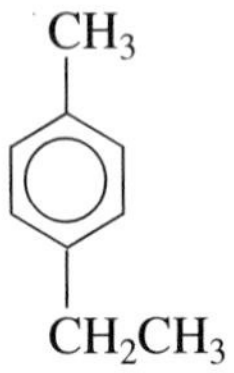

d.   CH$_2$CH$_3$ / CH$_2$CH$_3$

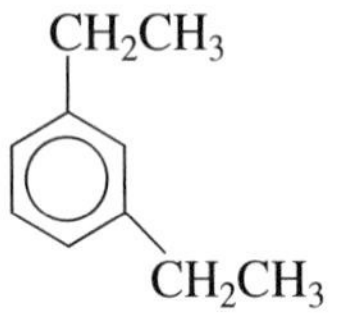

13.60 Write the structural formulas for each of the following:
   a. ethylbenzene
   b. *m*-dichlorobenzene
   c. 1,2,4-trimethylbenzene
   d. *p*-dimethylbenzene

13.61 Name the organic product(s) produced, if any, in each of the following reactions.

   a. benzene and Cl$_2$ $\xrightarrow{\text{FeCl}_3}$

   b. toluene and Br$_2$ $\xrightarrow{\text{FeBr}_3}$

   c. benzene and SO$_3$ $\xrightarrow{\text{H}_2\text{SO}_4}$

   d. benzene and Br$_2$ $\xrightarrow{\text{light}}$

13.62 What reactants and catalysts are needed to synthesize the following products?
   a. nitrobenzene
   b. benzenesulfonic acid
   c. bromobenzene

14 Alcohols, Phenols, Ethers, and Thiols

"We use mass spectrometry to analyze and confirm the presence of drugs," says Valli Vairavan, clinical lab technologist—Mass Spectrometry, Santa Clara Valley Medical Center. "A mass spectrometer separates and identifies compounds including drugs by mass. When we screen a urine sample, we look for metabolites, which are the products of drugs that have metabolized in the body. If the presence of one or more drugs such as heroin and cocaine is indicated, we confirm it by using mass spectrometry."

Drugs or their metabolites are detected in urine 24–48 hours after use. Cocaine metabolizes to benzoylecgonine and hydroxycocaine, morphine to morphine-3-glucuronide, and heroin to acetylmorphine. Amphetamines and methamphetamines are detected unchanged.

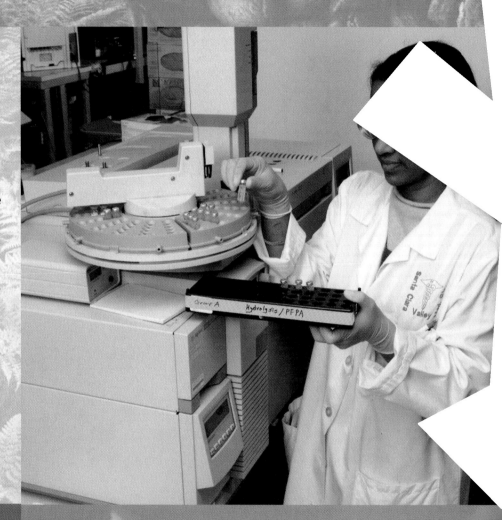

LOOKING AHEAD

the Chemistry place

www.chemplace.com/college

Visit the URL above or use the CD-ROM in the book for extra quizzing, interactive tutorials, career resources, and case studies.

In this chapter, we will look at organic compounds that contain single bonds to oxygen atoms or sulfur atoms. Alcohols, which contain the hydroxyl group (—OH) are commonly found in nature and used in industry and at home. For centuries, grains, vegetables, and fruits have been fermented to produce the ethanol present in alcoholic beverages. The hydroxyl group is important in biomolecules such as sugars and starches as well as in steroids such as cholesterol and estradiol. Menthol is a cyclic alcohol with a minty odor and flavor that is used in cough drops, shaving creams, and ointments. The ethers are compounds that contain an oxygen atom connected to two carbon atoms (—O—). Ethers are important solvents in chemistry and medical laboratories. Beginning in 1842, diethyl ether was used for about 100 years as a general anesthesia. Today less flammable and more easily tolerated anesthetics are used. Thiols, which contain an —SH group, give the strong odors we associate with garlic and onions.

Classify alcohols as primary, secondary, or tertiary.

14.1 Structure and Classification of Alcohols

In an **alcohol,** a **hydroxyl group (—OH)** replaces a hydrogen atom in an alkane. In a **phenol,** the hydroxyl group is attached to an aromatic ring. Both types of compounds have bent structures similar to water with an alkyl or aromatic group replacing one hydrogen atom (see Figure 14.1).

Figure 14.1 A hydrogen atom in water is replaced by an alkyl group in methanol and by an aromatic ring in phenol.

Q Why are the structures of alcohols and phenols similar to water?

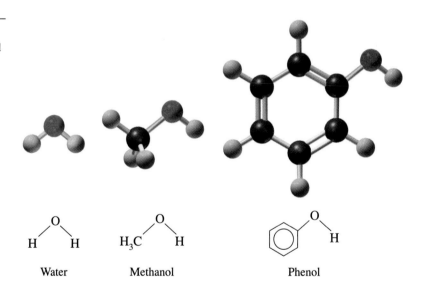

Water   Methanol   Phenol

WEB TUTORIAL
Alcohols and Thiols

Classification of Alcohols

When we studied alkanes in Chapter 12, we saw that carbon atoms can be classified as primary, secondary, or tertiary. Alcohols are classified in a similar way with an emphasis on the type of carbon atom bonded to the hydroxyl (—OH) group. In a **primary (1°) alcohol,** there is one alkyl group attached to the carbon bonded to the —OH group. Methanol is considered a primary alcohol. In a **secondary (2°) alcohol,** there are two alkyl groups attached. When there are three alkyl groups, the alcohol is classified as a **tertiary (3°) alcohol.**

Primary (1°) Secondary Tertiary
Alcohol (2°) Alcohol (3°) Alcohol

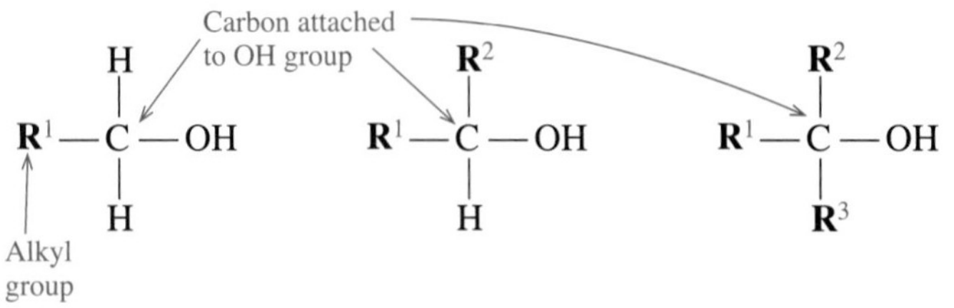

Carbon attached to OH group

$$R^1—\overset{\overset{\displaystyle H}{|}}{\underset{\underset{\displaystyle H}{|}}{C}}—OH \qquad R^1—\overset{\overset{\displaystyle R^2}{|}}{\underset{\underset{\displaystyle H}{|}}{C}}—OH \qquad R^1—\overset{\overset{\displaystyle R^2}{|}}{\underset{\underset{\displaystyle R^3}{|}}{C}}—OH$$

Alkyl group

Examples:

$$CH_3—CH_2—OH \qquad CH_3—\overset{\overset{\displaystyle CH_3}{|}}{CH}—OH \qquad CH_3—\overset{\overset{\displaystyle CH_3}{|}}{\underset{\underset{\displaystyle CH_3}{|}}{C}}—OH$$

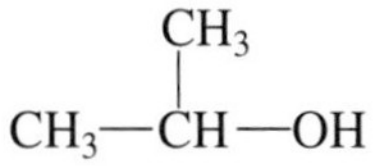

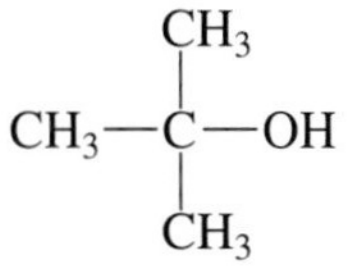

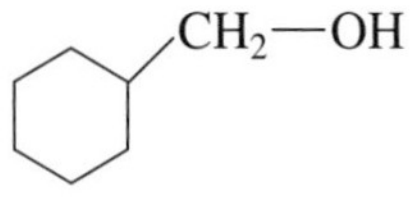

(One alkyl group) 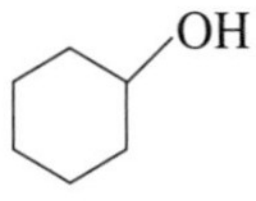   (Two alkyl groups) 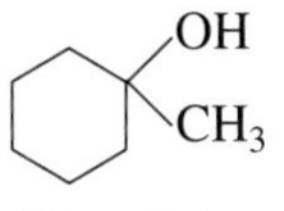   (Three alkyl groups)

SAMPLE PROBLEM 14.1

Classifying Alcohols

Classify each of the following alcohols as primary, secondary, or tertiary.

a. $CH_3—CH_2—CH_2—OH$

b. 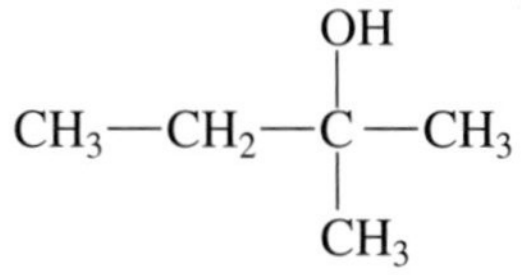 $CH_3—CH_2—\overset{\overset{\displaystyle OH}{|}}{\underset{\underset{\displaystyle CH_3}{|}}{C}}—CH_3$

c. OH (cyclopropane with OH)

Solution

a. One alkyl group attached to the carbon atom bonded to the —OH makes this a primary alcohol.

b. Three alkyl groups attached to the carbon atom bonded to the —OH makes this a tertiary alcohol.

c. Two alkyl groups attached to the carbon atom bonded to the —OH makes this a secondary alcohol.

Study Check

Classify the following as primary, secondary, or tertiary:

H_3C OH (cyclopentane ring)

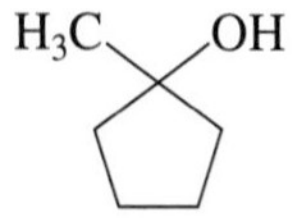

Structure and Classification of Alcohols

14.1 Classify each of the following as a primary, secondary, or tertiary alcohol:

a. CH_3—$\overset{\overset{\displaystyle CH_3}{|}}{CH}$—$CH_2$—$CH_2$—OH

b. CH_3—CH_2—CH_2—CH_2—OH

c. CH_3—$\overset{\overset{\displaystyle OH}{|}}{\underset{\underset{\displaystyle CH_3}{|}}{C}}$—$CH_2$—$CH_3$

d.

e.

14.2 Classify each of the following as a primary, secondary, or tertiary alcohol:

a.

b. CH_3—$\overset{\overset{\displaystyle CH_3}{|}}{CH}$—$CH_2$—OH

c. CH_2—OH (attached to benzene ring)

d. CH_3—CH_2—CH_2—$\overset{\overset{\displaystyle CH_3}{|}}{\underset{\underset{\displaystyle CH_3}{|}}{C}}$—OH

e.

14.2 Naming Alcohols, Phenols, and Thiols

LEARNING GOAL

Give IUPAC and common names for alcohols, phenols, and thiols; draw their condensed structural formulas.

the Chemistry place

WEB TUTORIAL
Alcohols and Thiols

The rules for IUPAC names of alcohols are similar to those we used to name other families of organic compounds. The alcohol family is indicated in the name by an *–ol* ending, which is numbered to show the location of the hydroxyl group on the main chain.

Step 1 Name the longest carbon chain containing the —OH group. Replace the *–e* in the alkane name by *–ol*. The common names of simple alcohols give the alkyl group followed by *alcohol*.

Step 2 Number the main chain starting at the end closest to the —OH group. Alcohols with two —OH groups are named *diols;* three —OH groups, *triols.*

CH_3OH
Methanol
(methyl alcohol)

CH_3OH_2OH
Ethanol
(ethyl alcohol)

$CH_3CH_2CH_2OH$
1-Propanol
(propyl alcohol)

$CH_3\overset{\overset{\displaystyle OH}{|}}{C}HCH_3$
2-Propanol
(isopropyl alcohol)

HO—CH_2—CH_2—OH
1,2-Ethanediol
(ethylene glycol)

$HOCH_2\overset{\overset{\displaystyle OH}{|}}{C}HCH_2OH$
1,2,3-Propanetriol
(glycerol)

Step 3 Name and number other substituents relative to the —OH group. The —OH takes precedence over double and triple bonds.

$$\underset{\underset{\text{4-Bromo-2-pentanol}}{}}{\overset{\overset{\text{Br}\quad\text{OH}}{|\qquad|}}{CH_3CHCH_2CHCH_3}}$$

$$\underset{\underset{\underset{\text{2-Chloro-3-methyl-3- hexanol}}{}}{\overset{|}{CH_3}}}{\overset{\overset{\text{Cl}\quad\text{OH}}{|\quad|}}{CH_3CHCCH_2CH_2CH_3}}$$

$$\underset{\text{2-Butene-1-ol}}{CH_3\!-\!CH\!=\!CH\!-\!CH_2\!-\!OH}$$

EXPLORE YOUR WORLD

Alcohols in Household Products
Read the labels on household products such as mouthwashes, cold remedies, alcoholic beverages, rubbing alcohol, and flavoring extracts. Look for names of alcohols such as ethyl alcohol, isopropyl alcohol, thymol, and menthol.

Questions
1. What part of the name tells you that it is an alcohol?
2. What alcohol is usually meant by the term "alcohol"?
3. What is the percent of alcohol in the products?
4. Write out the structures of the alcohols you find listed on the labels. You may need to use a reference book for some structures.

Step 4 Name a cyclic alcohol as a *cycloalkanol*. For other substituents, the ring is numbered with the —OH group on carbon 1.

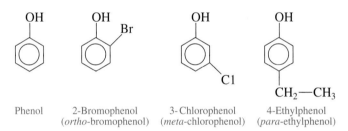

Cyclohexanol   2-Methylcyclopentanol   3-Chloro-5-methylcyclohexanol

Naming Phenols

The term *phenol* is the IUPAC name for a benzene ring bonded to a hydroxyl group (—OH) and is used in the name of the family of organic compounds derived from phenol. When there is a second substituent, the benzene ring is numbered starting from the carbon 1, which is bonded to the —OH group. The terms *ortho, meta,* and *para* are used for the common names of simple phenols.

Phenol   2-Bromophenol   3- Chlorophenol   4-Ethylphenol
            (*ortho*-bromophenol)   (*meta*-chlorophenol)   (*para*-ethylphenol)

Certain disubstituted phenols have common names based on historical uses. The methylphenols are commonly named as *cresols,* while benzenediols have a variety of common names.

3-Methylphenol   1,2-Benzenediol   1,3-Benzenediol   1,4-Benzenediol
(*meta*-cresol)   (catechol)   (resorcinol)   (hydroquinone)

SAMPLE PROBLEM 14.2

Naming Alcohols and Phenols

Give the IUPAC name for each of the following:

a. OH

b. $\underset{CH_3CHCH_2CHCH_3}{\overset{\overset{\text{CH}_3\quad\text{OH}}{|\qquad|}}{}}$

c. Br
OH

d. OH

CH₃

Solution

a. The ⎯OH is attached to a cyclopentane ring. Changing the *-e* to *-ol* gives the IUPAC name of *cyclopentanol*.

b. The parent chain is pentane; the alcohol is named *pentanol*. The carbon chain is numbered to give the position of the ⎯OH group on carbon 2 and the methyl group on carbon 4. The compound is named *4-methyl-2-pentanol*.

c. The compound is a *phenol* because the ⎯OH is attached to a benzene ring. The ring is numbered with carbon 1 attached to the ⎯OH in the direction that gives the bromine the lower number. The compound is named *2-bromophenol* or *o-bromophenol*.

d. The ⎯OH takes priority in naming this compound as 4-methylphenol. The common name is based on an isomer of cresol, which is *para*-cresol.

Study Check

Give the IUPAC name for the following:

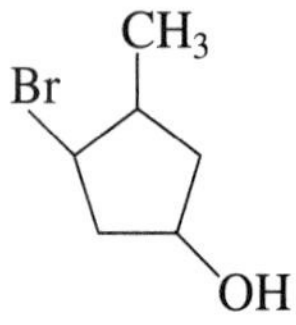

SAMPLE PROBLEM 14.3

Writing Condensed Structural Formulas of Alcohols and Phenols

a. The alcohol 3,5,5-trimethyl-1-hexanol is used as a plasticizer, a substance added to plastics to keep them pliable. Draw its condensed structural formula.

b. Thymol, used as an antiseptic in mouthwashes, has an IUPAC name of 5-methyl-2-isopropylphenol. Draw its condensed structural formula.

Solution

a. The parent chain of six carbon atoms has an ⎯OH group on carbon 1. Methyl groups are bonded to carbon 3 and to carbon 5.

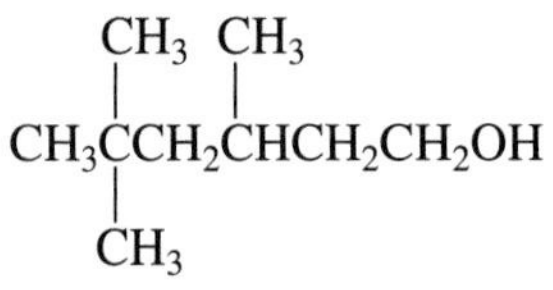

b. The phenol part of the name tells us to draw a benzene ring with an ⎯OH group on carbon 1. Then the alkyl groups are attached, methyl to carbon 5 and isopropyl to carbon 2.

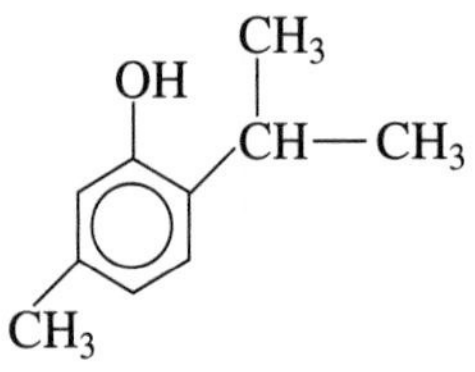

Study Check

Draw the condensed structural formula of 5-chloro-4-methyl-3-heptanol.

Thiols

Thiols are a family of sulfur-containing organic compounds that have a **sulfhydryl** (—SH) **group.** They have structures similar to alcohols except that a —SH group takes the place of an —OH group.

Naming Thiols

In the IUPAC system, thiols are named by adding *thiol* to the alkane name of the longest carbon chain bonded to the —SH group. As we did for alcohols, the location of the —SH group is indicated by numbering the main chain from the closest end.

$$CH_3—OH \qquad CH_3—SH \qquad \overset{\overset{\displaystyle SH}{|}}{CH_3—CH—CH_2—CH_3}$$

Methanol   Methanethiol   2-Butanethiol

An important property of thiols is a strong, usually disagreeable, odor. Methanethiol is the characteristic odor of oysters and cheddar cheese. (See Figure 14.2.) To help us detect natural gas (methane) leaks, a small amount of ethanethiol

Figure 14.2 Thiols are sulfur-containing compounds with an —SH group.

Q Why do thiols have structures similar to alcohols?

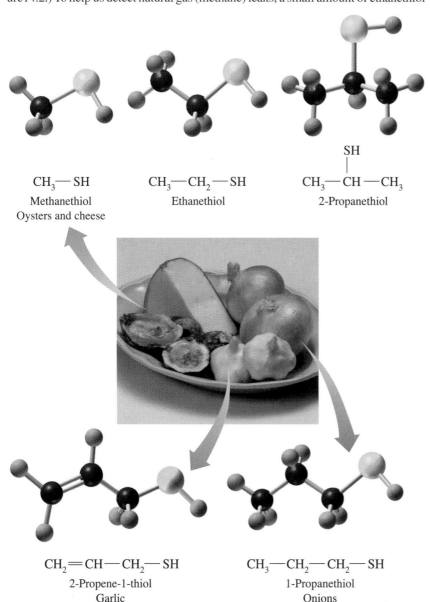

$CH_3—SH$
Methanethiol
Oysters and cheese

$CH_3—CH_2—SH$
Ethanethiol

$$\overset{\overset{\displaystyle SH}{|}}{CH_3—CH—CH_3}$$
2-Propanethiol

$CH_2\!=\!CH—CH_2—SH$
2-Propene-1-thiol
Garlic

$CH_3—CH_2—CH_2—SH$
1-Propanethiol
Onions

is added to the gas supply. There are thiols in the spray emitted when a skunk senses danger. The odor of onions is due to 1-propanethiol, which is also a lachrymator, a substance that makes eyes tear. Garlic contains thiols such as 2-propene-1-thiol. We can break this name down as follows.

2-	prop	ene	-1-	thiol
Carbon 2 has C=C	3 carbons in chain	Alkene	On carbon 1	—SH group

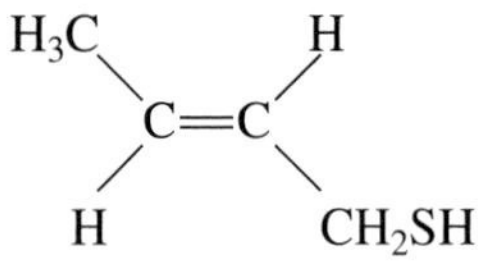

H_3C ⟍   ⟋H
    $C=C$
 H ⟋   ⟍ CH_2SH

$CH_3CH_2CH_2$—SH $CH_2=CH$—CH_2—SH

trans-2-Butene-1-thiol
(in skunk spray)

1-Propanethiol
(in onions)

2-Propene-1-thiol
(in garlic)

SAMPLE PROBLEM 14.4

Thiols

Draw the condensed structural formula of the following:

a. 1-butanethiol   **b.** cyclohexanethiol

Solution

a. This compound has a — SH group on the first carbon of a butane chain.

$CH_3CH_2CH_2CH_2$— SH

b. This compound has a — SH group on cyclohexane.

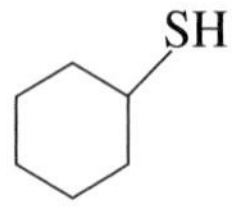

Study Check

What is the condensed structural formula of ethanethiol?

QUESTIONS AND PROBLEMS

Naming Alcohols, Phenols, and Thiols

14.3 Give the IUPAC name for each of the following alcohols:

a. CH_3CH_2OH

b. 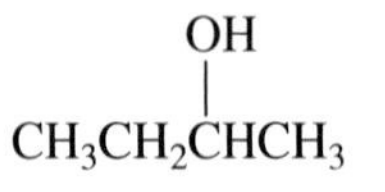
$\qquad$ OH
$\qquad$ |
$CH_3CH_2CHCH_3$

c. 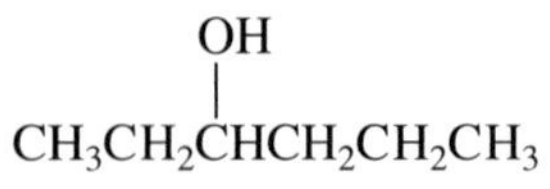
$\qquad\qquad$ OH
$\qquad\qquad$ |
$CH_3CH_2CHCH_2CH_2CH_3$

d. 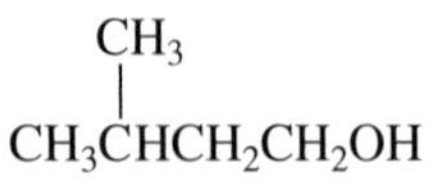
$\qquad$ CH_3
$\qquad$ |
$CH_3CHCH_2CH_2OH$

e. 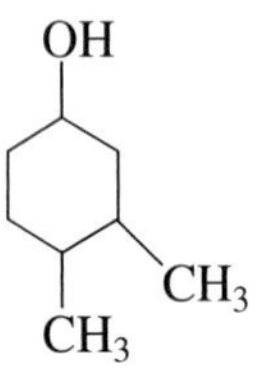
OH
CH_3
CH_3

f. 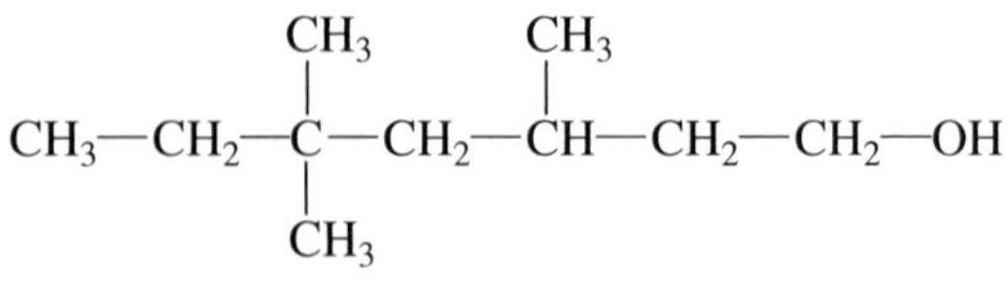
$\qquad\qquad\qquad$ CH_3 $\qquad$ CH_3
$\qquad\qquad\qquad$ | $\qquad\qquad$ |
CH_3—CH_2—C—CH_2—CH—CH_2—CH_2—OH
$\qquad\qquad\qquad$ |
$\qquad\qquad\qquad$ CH_3

14.4 Give the IUPAC name for each of the following alcohols:

a.
CH₂CH₃
—OH

b.
$$CH_3-\overset{\overset{\displaystyle Cl}{|}}{CH}-\overset{\overset{\displaystyle CH_3}{|}}{CH}-CH_2-OH$$

c. $CH_3CH_2\overset{\overset{\displaystyle CH_3}{|}}{C}H\overset{\overset{\displaystyle }{}}{C}HCH_2OH$
|
CH₃

d. $Cl-\overset{\overset{\displaystyle Cl}{|}}{C}H-CH_2-CH_2-\overset{\overset{\displaystyle OH}{|}}{C}H-CH_3$

e.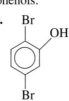
CH₃
OH
CH₃

f.
OH
|
CH₂
|
CH₂
|
CH₃—CH₂—CH—CH₂—CH₃

14.5 Write the condensed structural formula of each of the following alcohols:
  a. 1-propanol **b.** methyl alcohol **c.** 3-pentanol
  d. 2-methyl-2-butanol **e.** cyclohexanol **f.** 1,4-butanediol

14.6 Write the condensed structural formula of each of the following alcohols:
  a. ethyl alcohol **b.** 3-methyl-1-butanol
  c. 2,4-dichlorocyclohexanol **d.** propyl alcohol
  e. 1,3-cyclopentanediol **f.** 2,2,4-trimethyl-3-hexanol

14.7 Name each of the following phenols:

a. OH [benzene ring]

b. OH [benzene ring] Br

c. Cl [benzene ring] Cl OH

d. OH [benzene ring] Br

14.8 Name each of the following phenols:

a. OH [benzene ring] CH₂CH₃

b. Br OH [benzene ring] Br

c. OH [benzene ring] Cl

d. OH [benzene ring] Cl

14.9 Write the condensed structural formula of each of the following phenols:
  a. *m*-bromophenol **b.** *p*-chlorophenol **c.** 2,5-dichlorophenol
  d. *o*-phenylphenol **e.** 4-ethylphenol

14.10 Write the condensed structural formula of each of the following phenols:
  a. *o*-ethylphenol **b.** 2,4-dichlorophenol
  c. 2,4-dimethylphenol **d.** 2-ethyl-5-methylphenol
  e. *m*-phenylphenol

14.11 Give the IUPAC name for each of the following thiols:

a. CH_3-SH b. $CH_3-\overset{\overset{\displaystyle SH}{|}}{C}H-CH_3$ c. $CH_3-\overset{\overset{\displaystyle CH_3}{|}}{C}H-\overset{\overset{\displaystyle CH_3}{|}}{C}H-CH_2-SH$ d. [cyclobutane]SH

14.12 Give the IUPAC name for each of the following thiols:

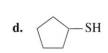

a. $CH_3CH_2CH_2-SH$ b. $CH_3-CH_2-CH_2-\overset{\overset{\displaystyle SH}{|}}{C}H-CH_3$ c. $CH_3-\overset{\overset{\displaystyle CH_3}{|}}{\underset{\underset{\displaystyle CH_3}{|}}{C}}-CH_2-SH$ d. [cyclopentane]—SH

14.3 Some Important Alcohols and Phenols

Methanol (methyl alcohol), the simplest alcohol, is found in many solvents and paint removers. It is sometimes called *wood alcohol* because methanol is produced when wood is heated to high temperatures in the absence of air. If ingested, methanol is oxidized to formaldehyde, which can cause headaches, blindness, and death. Methanol is used to make plastics, medicines, and fuels. In car racing, it is used as a fuel because it is less flammable and has a higher octane rating than gasoline. Today, methanol is synthesized by reacting carbon monoxide with hydrogen gas at high temperatures and pressures.

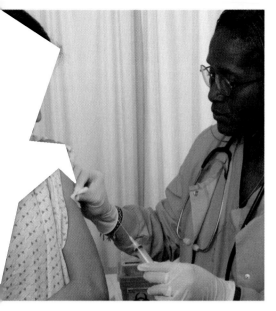

$$CO \ + \ 2H_2 \xrightarrow[\text{300–400°C, 200 atm}]{\text{ZnO/Cr}_2\text{O}_3 \text{ catalyst,}} CH_3{-}OH$$

Ethanol (ethyl alcohol) has been known since prehistoric times as an intoxicating product formed by the fermentation of grains and starches.

$$C_6H_{12}O_6 \xrightarrow{\text{fermentation}} 2CH_3{-}CH_2{-}OH \ + \ 2CO_2$$

Today, ethanol for commercial uses is produced from ethene and water reacted at high temperatures and pressures. It is used as a solvent for perfumes, varnishes, and some medicines, such as tincture of iodine. "Gasohol" is a mixture of ethanol and gasoline used as a fuel.

$$H_2C{=}CH_2 \ + \ H_2O \xrightarrow{\text{300°C, 200 atm, catalyst}} CH_3{-}CH_2{-}OH$$

2-Propanol (isopropyl alcohol), commonly referred to as *rubbing alcohol,* is used as an astringent because it evaporates rapidly and cools the skin, reducing the size of blood vessels near the surface. An antiseptic solution of isopropyl alcohol is used to clean the skin before an injection or taking a blood sample. It is also used to sterilize equipment because it destroys bacteria by coagulating protein. The toxicity of 2-propanol is similar to that of methanol.

1,2-Ethanediol (ethylene glycol) is used as antifreeze in heating and cooling systems. It is also a solvent for paints, inks, and plastics, and is used in the production of synthetic fibers such as Dacron. If ingested, it is extremely toxic. In the body, it is oxidized to oxalic acid, which forms insoluble salts in the kidneys that cause renal damage, convulsions, and death. Because its sweet taste is attractive to pets and children, ethylene glycol solutions must be carefully stored.

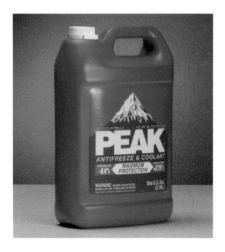

$$\underset{\text{1,2-Ethanediol (ethylene glycol)}}{HO{-}CH_2{-}CH_2{-}OH} \xrightarrow{[O]} \underset{\text{Oxalic acid}}{HO{-}\overset{\overset{\text{O}}{\|}}{C}{-}\overset{\overset{\text{O}}{\|}}{C}{-}OH}$$

1,2,3-Propanetriol (glycerol or glycerin), a trihydroxy alcohol, is a viscous liquid obtained from oils and fats during the production of soaps. The presence of several polar —OH groups makes it strongly attracted to water, a feature that makes glycerin useful as a skin softener in products such as skin lotions, cosmetics, shaving creams, and liquid soaps. It is also used as antifreeze and in fluid shock absorbers. A mixture of glycerin and a strong oxidizer such as potassium chlorate is extremely explosive and is used in the manufacture of dynamite.

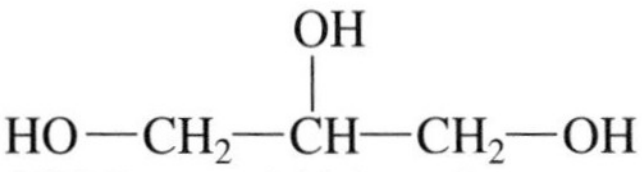

$$OH$$
$$HO—CH_2—CH—CH_2—OH$$
1,2,3-Propanetriol (glycerol)

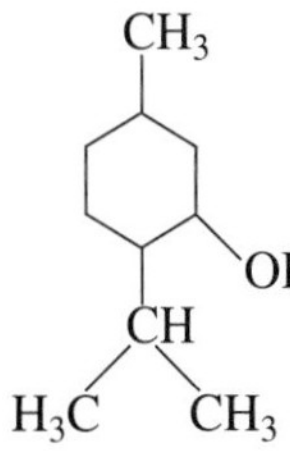

Menthol is a cyclohexanol with a peppermint taste and odor that is used in candy, throat lozenges, and nasal inhalers. It causes the mucous membranes to increase their secretions and soothes the respiratory tract.

Menthol

Derivatives of Phenol

Phenol and its derivatives are used as antiseptics in throat lozenges and mouthwashes. In household disinfectant sprays such as Lysol, the active ingredient is *ortho*-phenylphenol. Poison ivy contains a substance called urushiol that causes itching and blistering of the skin.

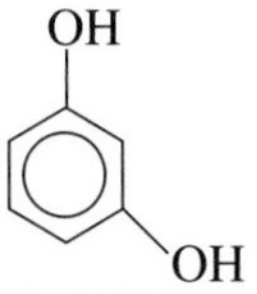

Resorcinol, antiseptic

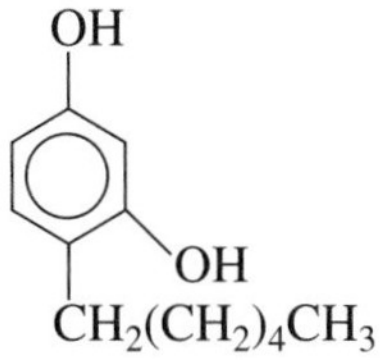

4-Hexylresorcinol, antiseptic
$$CH_2(CH_2)_4CH_3$$

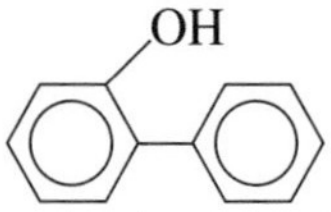

ortho-phenylphenol

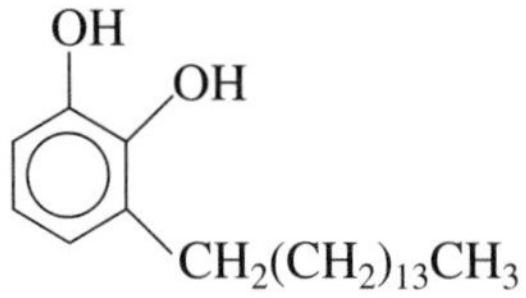

Urushiol
$$CH_2(CH_2)_{13}CH_3$$

Several of the essential oils of plants, which produce the odor or flavor of the plant, are derivatives of phenol. Eugenol is found in cloves, vanillin in vanilla bean, isoeugenol in nutmeg, and thymol in thyme and mint. Thymol has a pleasant, minty taste and is used in mouthwashes and by dentists to disinfect a cavity before adding a filling compound. (See Figure 14.3.)

Foods such as cereals and oils spoil when they react with the oxygen in the air. One way to prolong their shelf life is to add an antioxidant such as BHA or BHT. The phenol portion of these additives reacts readily with oxygen and keeps the food from spoiling.

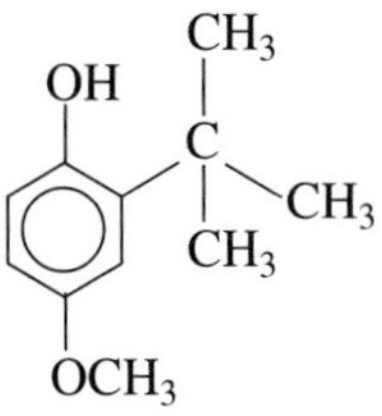

BHA (butylated hydroxyanisole)   BHT (butylated hydroxytoluene)

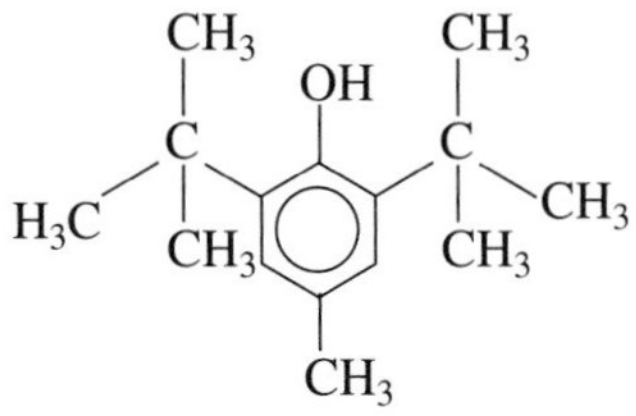

SAMPLE PROBLEM 14.5

Some Uses of Alcohols and Phenols

Identify the alcohol or phenol that best matches each of the following descriptions:
a. used to clean the skin prior to giving an injection
b. used in antifreeze
c. provides the odor and taste of cloves

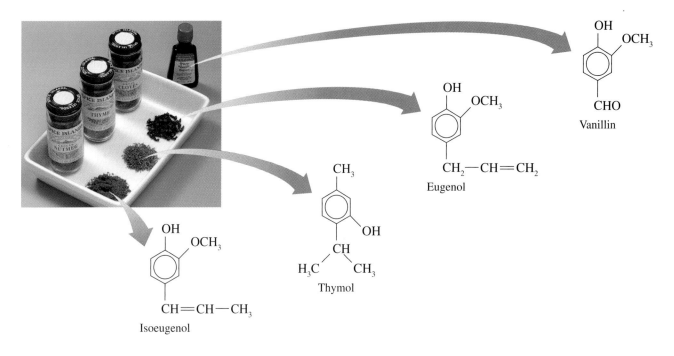

Figure 14.3 Derivatives of phenol are active ingredients found in the essential oils of cloves, vanilla, nutmeg, and mint.

Q If phenol is a structural feature common to these spices, what accounts for their different odors and tastes?

Solution

a. 2-propanol; isopropyl alcohol

b. 1,2-ethanediol; ethylene glycol

c. eugenol

Study Check

Is methanol or ethanol found in alcoholic beverages?

QUESTIONS AND PROBLEMS

Some Important Alcohols and Phenols

14.13 Identify the alcohol or phenol that best matches each of the following descriptions:
   a. found in alcoholic beverages
   b. gives a peppermint taste to candy and throat lozenges
   c. used as household disinfect in Lysol

14.14 Identify the alcohol or phenol that best matches each of the following descriptions:
   a. if ingested, it forms insoluble salts that cause renal damage
   b. added to foods to prevent reactions with oxygen that cause spoilage
   c. used to soften skin in skin lotions and shaving creams

LEARNING GOAL

Give the IUPAC and common names of ethers; draw the condensed structural formula.

14.4 Ethers

An **ether** contains an oxygen atom that is attached by single bonds to two carbon groups that are alkyls or aromatic rings. Ethers have a bent structure like water and alcohols except both hydrogen atoms are replaced by alkyl groups. (See Figure 14.4.)

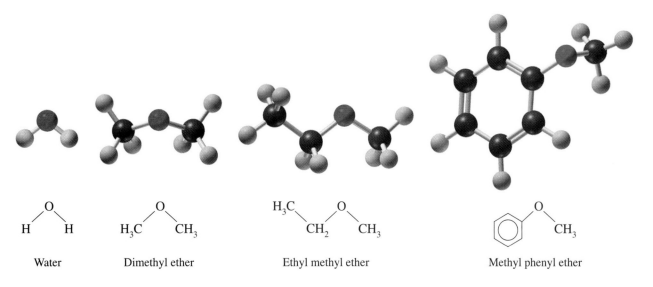

Water   Dimethyl ether   Ethyl methyl ether   Methyl phenyl ether

Naming Ethers

Most ethers are named by their common names. The IUPAC names are used only when the ether is more complex.

Step 1   For a simple ether, use the common name. Write the name of each alkyl or aryl (aromatic) group attached to the oxygen atom in alphabetical order followed by the word *ether.*

Step 2   For complex ethers, use the IUPAC name. Use the alkane name of the larger alkyl group as the main chain. Name the oxygen and smaller alkyl group as a substituent called an **alkoxy** (**alk**yl + **oxy**gen) **group.**

Step 3   Number the main chain beginning at the end nearest the alkoxy group and give the location of the alkoxy group on the main chain.

Alkoxy group

$\overbrace{CH_3-O}-\boxed{CH_2-CH_2-CH_3}$ main chain = propane
                         1   2   3

IUPAC name: 1-methoxypropane
Common name: methyl propyl ether

Some examples of naming ethers follow:

CH_3-O-CH_3 $CH_3CH_2-O-CH_2CH_3$ $CH_3-\overset{\overset{\displaystyle OCH_3}{|}}{CH}-CH_2-CH_3$

Methoxymethane   Ethoxyethane   2-Methoxybutane
(dimethyl ether)   (diethyl ether)

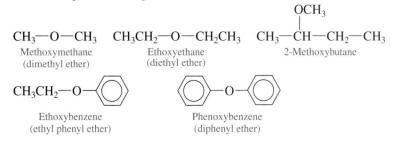

Ethoxybenzene   Phenoxybenzene
(ethyl phenyl ether)   (diphenyl ether)

Figure 14.4 The structures of ethers are similar to that of water.

Q What part of the structure of water is found in ethers?

SAMPLE PROBLEM 14.6

Ethers

Assign a common name and the IUPAC name to the following ethers:

a. $CH_3CH_2-O-CH_2CH_2CH_3$

b.

           OCH_3

Solution

a. The groups attached to the oxygen are an ethyl group and a propyl group. The common name is *ethyl propyl ether.* Naming the shorter alkyl group and the oxygen as ethoxy gives the IUPAC name of *1-ethoxypropane.*

b. The groups attached to the oxygen are a methyl group and a cyclobutyl group. The common name is *cyclobutyl methyl ether.* In the IUPAC name, the CH_3O— is named methoxy and the cyclobutane is the longer carbon chain, which gives the name *methoxycyclobutane.*

Study Check

What is the common name of ethoxybenzene?

Constitutional Isomers of Alcohols and Ethers

The general formula $C_nH_{2n+2}O$ represents the molecular formula of both alcohols and ethers formed from alkanes. For example, we can write structural formulas for constitutional isomers with the molecular formula C_2H_6O as follows:

$$CH_3—CH_2—OH \qquad\qquad CH_3—O—CH_3$$
Ethyl alcohol   Dimethyl ether

SAMPLE PROBLEM 14.7

Constitutional Isomers

Draw the structural formulas and give the common names of two alcohols and one ether with a molecular formula C_3H_8O.

Solution

To draw the structural formulas for alcohols, the hydroxyl group is bonded to two different atoms in a chain of three carbon atoms. For the ether, two alkyl groups are bonded to an oxygen atom.

$$CH_3—CH_2—CH_2—OH \qquad CH_3—\overset{\overset{\displaystyle OH}{\displaystyle |}}{CH}—CH_3 \qquad CH_3—CH_2—O—CH_3$$
Propyl alcohol   Isopropyl alcohol   Ethyl methyl ether

Study Check

Write the IUPAC names of the unbranched isomers of $C_4H_{10}O$.

Cyclic Ethers

Cyclic ethers contain an oxygen atom in a carbon ring. They are **heterocyclic compounds** because there is a ring with one or more atoms that are not carbon. The cyclic ethers are usually given common names. The five-atom rings with an oxygen atom use common names derived from the aromatic ring **furan.** The four-atom cyclic ethers are not common. The rings are numbered from the oxygen atom as 1.

HEALTH NOTE

Ethers as Anesthetics

Anesthesia is the loss of all sensation and consciousness. A general anesthetic is a substance that blocks signals to the awareness centers in the brain so the person has a loss of memory, a loss of feeling pain, and an artificial sleep. The term *ether* has been associated with anesthesia because diethyl ether was the most widely used anesthetic for more than a hundred years. Although it is easy to administer, ether is very volatile and highly flammable. A small spark in the operating room could cause an explosion. Since the 1950s, anesthetics such as Forane (isoflurane), Ēthrane (enflu-

rane), and Penthrane (methoxyflurane) have been developed that are not as flammable and do not cause nausea. Most of these anesthetics retain the ether group, but the addition of many halogen atoms reduces the volatility and flammability of the ethers. More recently, they have been replaced by halothane (1-bromo-1-chloro-2,2,2-trifluoroethane), discussed in Chapter 12, because of the side effects of the ether-type inhalation anesthetics.

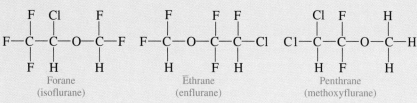

Forane (isoflurane)   Ēthrane (enflurane)   Penthrane (methoxyflurane)

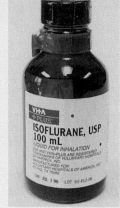

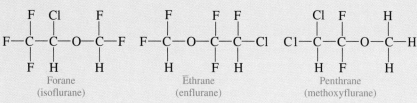

Ethylene oxide   Furan   3-Methylfuran   Tetrahydrofuran (THF)

A cyclic ether with six atoms is named *pyran,* which is the common name for an unsaturated ether ring of six atoms.

Pyran   Tetrahydropyran (THP)   4-Methylpyran

Cyclic ethers containing two oxygen atoms in a ring of six atoms are called *dioxanes*. The oxygen atoms are numbered because they can take different positions in the ring.

1,4-Dioxane   1,3-Dioxane

Dioxin is a term used for a group of highly toxic compounds composed of dioxanes bonded to aromatic rings. One of the most toxic is 2,3,7,8-tetrachlorodibenzodioxin (TCDD), now considered carcinogenic (cancer causing) because its structure interferes with DNA. Dioxin is formed during forest fires and as a by-product of many industrial processes involving chlorine such as chemical and pesticide manufacturing and pulp and paper bleaching. The herbicide Agent Orange used in Vietnam was contaminated by highly toxic dioxin, which formed during the synthesis of Agent Orange.

2,4,5-Trichlorophenoxyacetic acid (2,4,5-T; Agent Orange)

2,3,7,8-Tetrachlorodibenzodioxin (TCDD, "dioxin")

Cyclic Ethers

Identify the following as a cyclic alcohol, ether, or cyclic ether.

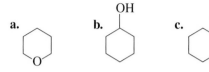

Solution

a. A cyclic ether has an oxygen atom in the ring.

b. A cyclic alcohol has a hydroxyl group bonded to a cyclic alkane.

c. An ether has an oxygen atom with single bonds to two alkyl or aryl groups.

Study Check

What is the difference between furan and pyran?

QUESTIONS AND PROBLEMS

Ethers

14.15 Give the IUPAC name and a common name for each of the following ethers:

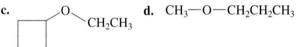

14.16 Give the IUPAC name and a common name for each of the following ethers:

a. $CH_3CH_2-O-CH_2CH_2CH_3$ **b.**   **c.**   **d.** CH_3-O-CH_3

14.17 Write the condensed structural formula for each of the following ethers:

   a. ethyl propyl ether   **b.** ethyl cyclopropyl ether
   c. methoxycyclopentane   **d.** 1-ethoxy-2-methylbutane
   e. 2,3-dimethoxypentane

14.18 Write the condensed structural formula for each of the following ethers:

   a. diethyl ether   **b.** diphenyl ether
   c. ethoxycyclohexane   **d.** 2-methoxy-2,3-dimethylbutane
   e. 1,2-dimethoxybenzene

14.19 Indicate whether each of the following pairs represent constitutional isomers, the same compound, or different compounds:

   a. 2-pentanol and 2-methoxybutane
   b. 2-butanol and cyclobutanol
   c. ethyl propyl ether and 2-methyl-1-butanol

14.20 Indicate whether each of the following pairs represent constitutional isomers, the same compound, or different compounds:

   a. 2-methoxybutane and 3-methyl-2-butanol
   b. 1-hexanol and dipropyl ether
   c. *tert*-butyl alcohol and diethyl ether

14.21 Give the name for each of the following cyclic ethers:

a. b. c.

14.22 Give the name for each of the following cyclic ethers:

a. b. c.

14.5 Physical Properties of Alcohols, Phenols, and Ethers

In Chapters 11 and 12, we learned that hydrocarbons, which are composed of only carbon and hydrogen, are nonpolar. In this chapter, we looked at compounds containing the element oxygen, which is strongly electronegative. As a result the oxygen and hydrogen atoms in the hydroxyl group —OH form hydrogen bonds. Although ethers also contain an oxygen atom, there is no hydrogen atom attached. Thus, ethers do not hydrogen bond with each other, but they do hydrogen bond with water.

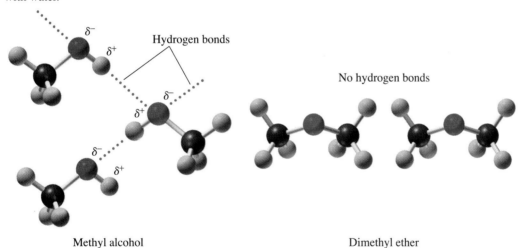

Methyl alcohol   Dimethyl ether

Boiling Points

Higher temperatures are required to provide the energy needed to break the hydrogen bonds between alcohol molecules. Thus, alcohols have higher boiling points than alkanes and ethers of similar mass. The boiling points of ethers are close to those of alkanes of similar mass because hydrogen bonding does not occur. The boiling points of some typical alkanes, alcohols, and ethers are listed in Table 14.1.

Solubility in Water

The oxygen atom in alcohols and ethers influences their solubility in water. In alcohols, the polar —OH group can hydrogen bond with water, which makes alcohols with one to four carbon atoms very soluble in water.

Water

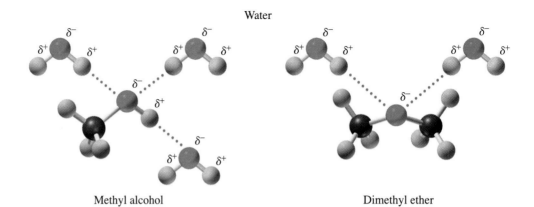

Methyl alcohol   Dimethyl ether

When there are more carbon atoms in the alkyl portion of the alcohol, the effect of the —OH group is diminished. The alkane portion does not participate in hydrogen bonding with water. Thus alcohols with five or more carbon atoms are not very soluble in water.

Nonpolar carbon chain ⟶ $\boxed{CH_3CH_2}$—OH
Soluble in water

$\boxed{CH_3—CH_2—CH_2—CH_2—CH_2—CH_2—CH_2—CH_2}$—OH
Insoluble in water

Ethers are more soluble in water than alkanes of similar mass because the oxygen can hydrogen bond with water, but ethers are not as soluble as alcohols. Ethers such as diethyl ether are very useful as solvents for hydrocarbons. However, ether vapors are highly flammable and react with oxygen to form explosive compounds. The utmost care must be taken when working with ethers.

Table 14.1 also compares the solubility of some alkanes, alcohols, and ethers by mass.

Table 14.1   Solubility and Boiling Points of Some Typical Alkanes, Alcohols, and Ethers

Compound	Structural Formula	Molar Mass (g/mole)	Boiling Point (°C)	Solubility in Water
Methane	CH_4	16	−161	No
Ethane	CH_3CH_3	30	−88	No
Methanol	CH_3—OH	32	65	Yes
Propane	$CH_3CH_2CH_3$	44	−42	No
Dimethyl ether	CH_3—O—CH_3	46	−23	Yes
Ethanol	CH_3CH_2—OH	46	78	Yes
Butane	$CH_3CH_2CH_2CH_3$	58	0	No
Ethyl methyl ether	CH_3—O—CH_2CH_3	60	8	Yes
1-Propanol	$CH_3CH_2CH_2$—OH	60	97	Yes
2-Propanol	$CH_3\overset{\displaystyle OH}{\underset{\displaystyle |}{C}}HCH_3$	60	83	Yes
Pentane	$CH_3CH_2CH_2CH_2CH_3$	72	36	No
Diethyl ether	CH_3CH_2—O—CH_2CH_3	74	35	Slightly
1-Butanol	$CH_3CH_2CH_2CH_2$—OH	74	118	Yes
2-Butanol	$CH_3\overset{\displaystyle OH}{\underset{\displaystyle |}{C}}HCH_2CH_3$	74	100	Yes

Solubility of Phenols

Phenol is soluble in water because the hydroxyl group ionizes slightly as a weak acid. In fact, an early name for phenol was *carbolic acid.* A concentrated solution of phenol is very corrosive and highly irritating to the skin; it can cause severe burns and ingestion can be fatal. Dilute solutions of phenol were previously used in hospitals as antiseptics, but they have generally been replaced.

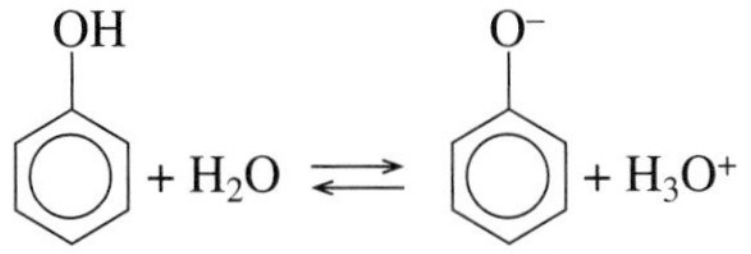

Phenol   Phenoxide ion

SAMPLE PROBLEM 14.9

Physical Properties of Alcohols, Ethers, and Phenols

Predict which compounds in each pair will be more soluble in water.
a. propane or ethanol   **b.** 1-propanol or 1-hexanol

Solution

a. Ethanol is more soluble because it can form hydrogen bonds with water.

b. The 1-propanol is more soluble because it has a shorter carbon chain.

Study Check

Dimethyl ether and ethanol both have molar masses of 46. However, ethanol has a much higher boiling point than dimethyl ether. How would you explain this difference in boiling points?

QUESTIONS AND PROBLEMS

Physical Properties of Alcohols, Phenols, and Ethers

14.23 Predict the compound with the higher boiling point in the following pairs:
   a. ethane or methanol   **b.** diethyl ether or 1-butanol
   c. 1-butanol or pentane

14.24 Glycerol (1,2,3-propanetriol) has a boiling point of 290°C. 1-Pentanol, which has about the same molar mass as glycerol, boils at 138°C. Why is the boiling point of glycerol so much higher?

14.25 Are each of the following soluble in water? Explain.
   a. CH_3CH_2OH **b.** CH_3-O-CH_3
   c. $CH_3CH_2CH_2CH_2CH_2CH_2-OH$
   d. $CH_3CH_2CH_3$ **e.**   OH

14.26 Give an explanation for the following observations:
   a. Ethanol is soluble in water, but propane is not.
   b. Dimethyl ether is soluble in water, but pentane is not.
   c. 1-Propanol is soluble in water, but 1-hexanol is not.

14.6 Reactions of Alcohols

In Chapter 12, we learned that hydrocarbons undergo combustion in the presence of oxygen. Alcohols burn with oxygen too. For example, in a restaurant, a dessert may be prepared by pouring a liquor on fruit or ice cream and lighting it. (See Figure 14.5.) The combustion of the ethanol in the liquor proceeds as follows:

$$CH_3{-}CH_2{-}OH + 3O_2 \longrightarrow 2CO_2 + 3H_2O + \text{energy}$$

Dehydration of Alcohols to Form Alkenes

Earlier we saw that alkenes can add water to yield alcohols. In a reverse reaction alcohols lose a water molecule when they are heated with an acid catalyst such as H_2SO_4. During the **dehydration** of an alcohol, a H— and —OH are removed from *adjacent carbon atoms of the same alcohol* to produce a water molecule. A double bond forms between the same two carbon atoms to produce an alkene product.

Figure 14.5 A flaming dessert is prepared using a liquor that undergoes combustion.

Q What is the equation for the combustion of the ethanol in the liquor?

Examples

The dehydration of a secondary alcohol can result in the formation of either of two products. **Saytzeff's rule** states that the major product is the one that results when the hydrogen (H—) is removed from the carbon atom with the smallest number of hydrogen atoms. This occurs because a hydrogen atom on a secondary carbon atom is easier to remove than a hydrogen from a primary carbon atom.

Adjacent carbon with the smallest number of H atoms

$$CH_3{-}CH{-}CH{-}CH_3 \xrightarrow[\text{Heat}]{H^+}$$

2-Butanol

$$CH_3{-}CH{=}CH{-}CH_3 + H_2O$$
2-Butene (major product: 90%)

$$CH_3{-}CH_2{-}CH{=}CH_2 + H_2O$$
1-Butene (minor product: 10%)

SAMPLE PROBLEM 14.10

Dehydration of Alcohols

Draw the condensed structural formulas for the alkenes produced by the dehydration of the following alcohols:

a.

$$\begin{array}{c} \text{OH} \\ | \\ \text{CH}_3\text{—CH}_2\text{—CH—CH}_2\text{—CH}_3 \end{array} \xrightarrow[\text{Heat}]{\text{H}^+}$$

b.

  $\xrightarrow[\text{Heat}]{\text{H}^+}$

Solution

a. $\text{CH}_3\text{—CH}_2\text{—CH}{=}\text{CH—CH}_3 + \text{H}_2\text{O}$

b. The —OH of this alcohol is removed along with an H from an adjacent carbon. Remember that the hydrogens are not drawn in this type of geometric figure.

Study Check

What is the name of the alkene produced by the dehydration of cyclopentanol?

SAMPLE PROBLEM 14.11

Predicting Reactants

Draw the structural formula of the alcohol that is needed to produce each of the following products.

a.

b. $\begin{array}{c} \text{CH}_3\text{—C}{=}\text{CH—CH}_3 \\ | \\ \text{CH}_3 \end{array}$

Solution

a. OH

b. $\begin{array}{c} \text{OH} \\ | \\ \text{CH}_3\text{—C—CH}_2\text{—CH}_3 \\ | \\ \text{CH}_3 \end{array}$ or $\begin{array}{c} \text{OH} \\ | \\ \text{CH}_3\text{—CH—CH—CH}_3 \\ | \\ \text{CH}_3 \end{array}$

Study Check

What is the name of an alcohol that forms 2-methylpropene?

Formation of Ethers

Ethers form when the dehydration of alcohols occurs at lower temperatures in the presence of an acid catalyst. Then the components of water are removed from two molecules: an H— from one alcohol and the —OH from another. When the remaining portions of the two alcohols join, an ether is produced.

$$\text{R—O}\boxed{\text{H + HO}}\text{—R} \xrightarrow[\text{Heat}]{\text{H}^+} \text{R—O—R} + \text{H}_2\text{O}$$

Example

$$\underset{\text{Methanol}}{\text{CH}_3\text{—OH}} + \underset{\text{Methanol}}{\text{HO—CH}_3} \xrightarrow[\text{Heat}]{\text{H}^+} \underset{\text{Dimethyl ether}}{\text{CH}_3\text{—O—CH}_3} + \text{H}_2\text{O}$$

Ether Formation

Write the structural formula and name of the ether produced by the following reaction:

$$2CH_3CH_2OH \xrightarrow[\text{Heat}]{H^+}$$

Solution

The structural formula of the ether product is written by attaching the alkyl groups of the reacting alcohols to an oxygen atom.

$$CH_3CH_2 - O - CH_2CH_3 \qquad \text{Diethyl ether}$$

Study Check

What alcohol is needed to form the ether named dicyclohexyl ether?

The reactions of alcohols have a central role in organic chemistry because alcohols can be converted to many of the other functional groups.

Oxidation of Alcohols

In Chapter 6, we learned that an oxidation was a loss of electrons and a reduction was a gain of electrons. However, in organic chemistry, it is convenient to think of **oxidation** as a loss of hydrogen atoms or the addition of oxygen. By counting the number of bonds between carbon and oxygen, we can decide if an oxidation reaction occurred. An increase in the number of carbon–oxygen bonds is the result of a loss of electrons. In a reduction, hydrogen is added, which gives a compound with fewer bonds between carbon and oxygen. (See Figure 14.6.)

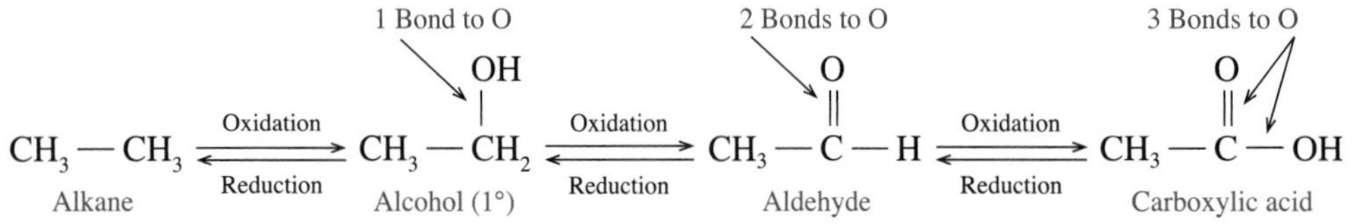

Figure 14.6 A compound is oxidized when the number of bonds increases between carbon and oxygen.

Q Is an aldehyde more or less oxidized than the corresponding primary alcohol?

Oxidation of Primary Alcohols

The oxidation of a primary alcohol produces an aldehyde, which contains a double bond between carbon and oxygen. The oxidation occurs by removing two hydrogen atoms, one from the —OH group and another from the carbon that is bonded to the —OH. In the laboratory, the oxidation requires some type of oxidizing agent such as O_2, $KMnO_4$, H_2CrO_4, or $K_2Cr_2O_7$. To indicate the presence of an oxidizing agent, reactions are often written with the symbol [O].

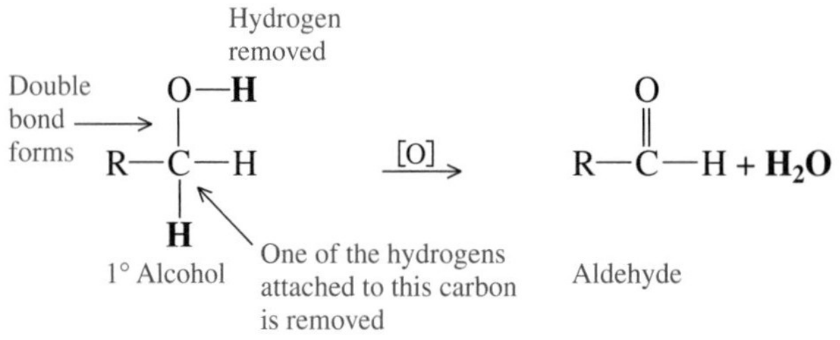

Examples

OH

H—C—H   [O]→ $\overset{\displaystyle O}{\overset{\|}{H-C-H}}$ + **H₂O**

     H

Methanol
(methyl alcohol)

Methanal
(formaldehyde)

OH

CH_3-CH_2 [O]→ $\overset{\displaystyle O}{\overset{\|}{CH_3-C-H}}$ + H_2O

Ethanol
(ethyl alcohol)

Ethanal
(acetaldehyde)

Aldehydes oxidize further, this time by the addition of another oxygen to form a carboxylic acid. This step occurs so readily that it is often difficult to isolate the aldehyde product during oxidation. We will learn more about carboxylic acids in Chapter 17.

$$\overset{\displaystyle O}{\overset{\|}{R-C-H}} \quad \overset{[O]}{\longrightarrow} \quad \overset{\displaystyle O}{\overset{\|}{R-C-OH}}$$

Aldehyde   Carboxylic acid

Example

$$\overset{\displaystyle O}{\overset{\|}{H-C-H}} \quad \overset{[O]}{\longrightarrow} \quad \overset{\displaystyle O}{\overset{\|}{H-C-OH}}$$

Methanal
(formaldehyde)

Methanoic acid
(formic acid)

Oxidation of Methanol

Methanol, "wood alcohol," is a highly toxic alcohol present in products such as windshield washer fluid, Sterno, and paint strippers. Methanol is rapidly absorbed in the gastrointestinal tract. In the liver, it is metabolized to formaldehyde and then formic acid, a substance that causes nausea, severe abdominal pain, and blurred vision. Blindness can occur because the intermediate products destroy the retina of the eye. As little as 4 mL of methanol can produce blindness. The formic acid, which is not readily eliminated from the body, lowers blood pH so severely that just 30 mL of methanol can lead to coma and death.

$$CH_3OH \quad \overset{\text{Liver enzymes}}{\underset{[O]}{\longrightarrow}} \quad \overset{\displaystyle O}{\overset{\|}{H-C-H}} \quad \longrightarrow \quad \overset{\displaystyle O}{\overset{\|}{H-C-OH}}$$

Methanol
(methyl alcohol)

Methanal
(formaldehyde)

Methanoic acid
(formic acid)

The treatment for methanol poisoning involves giving sodium bicarbonate to neutralize the formic acid in the blood. In some cases, ethanol is given intravenously to the patient. The enzymes in the liver pick up ethanol molecules to oxidize instead of methanol molecules, which gives more time for the methanol to be eliminated via the lungs without the formation of its dangerous oxidation products.

Oxidation of Secondary Alcohols

The oxidation of secondary alcohols is similar to that of primary alcohols except the products are ketones. One hydrogen is removed from the —OH and another from the carbon bonded to the —OH group. The result is a ketone that has the carbon–oxygen double bond attached to alkyl groups on both sides.

Two hydrogen atoms
removed

Double
bond →
forms

$$\underset{\substack{| \\ \text{H} \\ \text{2° Alcohol}}}{\overset{\text{O—H}}{\underset{|}{\text{R—C—R}}}} \xrightarrow{\text{[O]}} \underset{\text{Ketone}}{\overset{\text{O} \atop ||}{\text{R—C—R}}} + \textbf{H}_2\textbf{O}$$

Examples

$$\underset{\substack{2\text{-Propanol} \\ (\text{isopropyl alcohol})}}{\overset{\text{OH}}{\underset{\substack{| \\ \text{H}}}{\text{CH}_3\text{—C—CH}_3}}} \xrightarrow{\text{[O]}} \underset{\substack{\text{Propanone} \\ (\text{dimethyl ketone: acetone})}}{\overset{\text{O} \atop ||}{\text{CH}_3\text{—C—CH}_3}} + \textbf{H}_2\textbf{O}$$

Cyclohexanol $\xrightarrow{\text{[O]}}$ Cyclohexanone $=\text{O} + \text{H}_2\text{O}$

During vigorous exercise, lactic acid accumulates in the muscles and causes fatigue. When the activity level is decreased, oxygen enters the muscles and oxidizes the secondary —OH group in lactic acid to a ketone group in pyruvic acid. Pyruvic acid is metabolized further until it is completely oxidized to CO_2 and H_2O. The muscles in highly trained athletes are capable of taking up greater quantities of oxygen so that vigorous exercise can be maintained for longer periods of time.

Secondary alcohol

Keto group

$$\underset{\text{Lactic acid}}{\overset{\text{OH　O}}{\underset{\substack{| \quad ||}}{\text{CH}_3\text{—CH—COH}}}} \xrightarrow[\text{dehyrogenase}]{\text{Lactic acid}} \underset{\text{Pyruvic acid}}{\overset{\text{O　O}}{\underset{\substack{|| \quad ||}}{\text{CH}_3\text{—C—COH}}}}$$

Tertiary alcohols do not oxidize readily because there are no hydrogen atoms on the alcohol carbon. Because C—C bonds are usually too strong to oxidize, tertiary alcohols resist oxidation.

No
double
bond
forms

No hydrogen on
this carbon

$$\underset{\substack{| \\ \text{R} \\ \text{3° Alcohol}}}{\overset{\text{O—H}}{\underset{|}{\text{R—C—R}}}} \xrightarrow{\text{[O]}} \text{No oxidation product readily formed}$$

SAMPLE PROBLEM 14.13

Oxidation of Alcohols

Draw the structural formulas of the organic products that form when the following alcohols undergo oxidation.

a. CH$_3$CH$_2$CHCH$_3$ $\xrightarrow{\text{[O]}}$ $\overset{\text{OH}}{|}$ **b.** CH$_3$CH$_2$CH$_2$OH $\xrightarrow{\text{[O]}}$

Solution

a. Because the reactant is a secondary alcohol, oxidation produces a ketone and water.

CH$_3$CH$_2$$\overset{\displaystyle O}{\overset{\|}{C}}CH_3$

b. This primary alcohol can oxidize to an aldehyde and further oxidize to a carboxylic acid.

CH$_3$CH$_2$$\overset{\displaystyle O}{\overset{\|}{C}}$—H   CH$_3CH_2$$\overset{\displaystyle O}{\overset{\|}{C}}$—OH

Study Check

When propene is hydrated in the presence of an acid cataylst, the product forms a ketone when it is oxidized. Explain using equations.

CASE STUDY
Alcohol Toxicity

HEALTH NOTE

Oxidation of Alcohol in the Body

Ethanol is the most commonly abused drug in the United States. When ingested in small amounts, ethanol may produce a feeling of euphoria in the body although it is a depressant. In the liver, enzymes such as alcohol dehydrogenases oxidize ethanol to acetaldehyde, a substance that impairs mental and physical coordination. If the blood alcohol concentration exceeds 0.4%, coma or death may occur. Table 14.2 gives some of the typical behaviors exhibited at various levels of blood alcohol.

CH$_3$CH$_2$OH $\xrightarrow{\text{[O]}}$ CH$_3$$\overset{\displaystyle O}{\overset{\|}{C}}$H $\xrightarrow{\text{[O]}}$ 2CO$_2$ + H$_2$O

Ethanol   Ethanal
(ethyl alcohol)   (acetaldehyde)

The acetaldehyde produced from ethanol in the liver is further oxidized to acetic acid, which is converted to carbon dioxide and water in the citric acid (Krebs) cycle. Thus, the enzymes in the

Table 14.2 Typical Behaviors Exhibited by a 150-lb Person Consuming Alcohol

Number of Beers (12 oz) or Glasses of Wine (5 oz)	Blood Alcohol Level (w/v %)	Typical Behavior
1	0.025	Slightly dizzy, talkative
2	0.05	Euphoria, loud talking, and laughing
4	0.10	Loss of inhibition, loss of coordination, drowsiness, legally drunk in most states
8	0.20	Intoxicated, quick to anger, exaggerated emotions
12	0.30	Unconscious
16–20	0.40–0.50	Coma and death

liver can eventually break down ethanol, but the aldehyde and carboxylic acid intermediates can cause considerable damage while they are present within the cells of the liver.

A person weighing 150 lb requires about one hour to completely metabolize 10 ounces of beer. However, the rate of metabolism of ethanol varies between nondrinkers and drinkers. Typically, nondrinkers and social drinkers can metabolize 12–15 mg of ethanol/dL of blood in one hour, but an alcoholic can metabolize as much as 30 mg of ethanol/dL in one hour. Some effects of alcohol metabolism include an increase in liver lipids (fatty liver), an increase in serum triglycerides, gastritis, pancreatitis, ketoacidosis, alcoholic hepatitis, and psychological disturbances.

When the Breathalyzer test is used for suspected drunk drivers, the driver exhales a volume of breath into a solution containing the orange Cr^{6+} ion. If there is ethyl alcohol present in the exhaled air, the alcohol is oxidized, and the Cr^{6+} is reduced to give a green solution of Cr^{3+}.

CH$_3$CH$_2$OH + Cr^{6+} $\xrightarrow{\text{[O]}}$ CH$_3$$\overset{\displaystyle O}{\overset{\|}{C}}$OH + Cr^{3+}
Ethanol   Orange   Acetic acid   Green

Sometimes alcoholics are treated with a drug called Antabuse (disulfiram), which prevents the oxidation of acetaldehyde to acetic acid. As a result, acetaldehyde accumulates in the blood, which causes nausea, profuse sweating, headache, dizziness, vomiting, and respiratory difficulties. Because of these unpleasant side effects, the patient is less likely to use alcohol.

Oxidation of Thiols

Thiols also undergo oxidation by a loss of hydrogen atoms from the —SH groups. The oxidized product is called a **disulfide.**

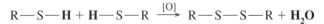

Example

$$CH_3—S—H + H—S—CH_3 \xrightarrow{[O]} CH_3—S—S—CH_3 + H_2O$$
Methanethiol   Dimethyl disulfide

Much of the protein in the hair is cross-linked by disulfide bonds, which occur mostly between the thiol groups of the amino acid cysteine.

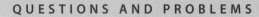

When a person is given a "perm," a reducing substance is used to break the disulfide bonds. While the hair is still wrapped around the curlers, an oxidizing substance is applied that causes new disulfide bonds to form between different parts of the protein hair strands, which gives the hair a new shape.

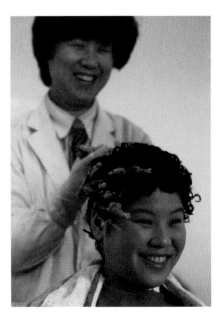

QUESTIONS AND PROBLEMS

Reactions of Alcohols

14.27 Draw the condensed structural formula of the alkene produced by each of the following dehydration reactions:

a. $CH_3—CH_2—CH_2—CH_2—OH \xrightarrow[\text{Heat}]{H^+}$

b.

c.

d.
$$CH_3—CH_2—CH_2—\overset{\overset{\displaystyle OH}{|}}{CH}—CH_3 \xrightarrow[\text{Heat}]{H^+}$$

14.28 Draw the condensed structural formula of the alkene produced by each of the following dehydration reactions:

a.
$$CH_3—\overset{\overset{\displaystyle CH_3}{|}}{CH}—CH_2—OH \xrightarrow[\text{Heat}]{H^+}$$

b.
$$CH_3—\overset{\overset{\displaystyle OH}{|}}{CH}—\overset{\overset{\displaystyle CH_3}{|}}{CH}—CH_2CH_3 \xrightarrow[\text{Heat}]{H^+}$$

c.   **d.**

14.29 Write the ether product from the reaction of each of the following:

a. $2CH_3OH \xrightarrow[\text{Heat}]{H^+}$

b. $2CH_3CH_2CH_2OH \xrightarrow[\text{Heat}]{H^+}$

14.30 Write the ether product from the reaction of each of the following:

a. $2CH_3CH_2OH \xrightarrow[\text{Heat}]{H^+}$

b.

$$\underset{\overset{|}{2CH_3CHCH_2OH}}{CH_3} \xrightarrow[\text{Heat}]{H^+}$$

14.31 What alcohol(s) could be used to produce each of the following compounds?

a. $CH_2{=}CH_2$

b. $CH_3{-}O{-}CH_2CH_3$

c.

14.32 What alcohol(s) could be used to produce each of the following compounds?

a. $CH_3CH_2{-}O{-}CH_2CH_3$

b.

$$\underset{\overset{|}{CH_3{-}C{=}CHCH_3}}{CH_3}$$

c.

14.33 Draw the condensed structural formula of the organic product when each of the following alcohols is oxidized [O] (if no reaction, write *none*):

a. $CH_3CH_2CH_2CH_2CH_2OH$

b. $\underset{\overset{|}{CH_3CH_2CHCH_3}}{OH}$

c. (cyclohexanol with OH)

d. $\underset{\overset{|}{OH}}{CH_3}{-}\underset{}{CH}{-}CH_2{-}\underset{\overset{|}{CH_3}}{CH}{-}CH_3$

e. $\underset{\overset{|}{CH_3}}{CH_3}{-}CH{-}CH_2{-}CH_2{-}OH$

14.34 Draw the condensed structural formula of the organic product when each of the following alcohols is oxidized [O] (if no reaction, write *none*):

a. (cyclobutane)$-CH_2OH$

b. $\underset{\overset{|}{CH_3{-}CH{-}CH_2{-}CH{-}OH}}{CH_3}$ with CH_3 below second CH

$$CH_3{-}\underset{\overset{|}{CH}}{CH}{-}CH_2{-}\underset{\overset{|}{CH_3}}{CH}{-}OH$$
with top CH_3

c. $\underset{\overset{|}{CH_3}}{CH_3{-}CH_2{-}\underset{\overset{|}{OH}}{C}{-}CH_3}$

d. $\underset{\overset{|}{CH_3CHCHCH_2CH_3}}{OHOH}$

e. (cyclobutane)$-OH$

14.35 Draw the condensed structural formula of the alcohol needed to give each of the following oxidation products:

a. $H{-}\overset{\overset{\displaystyle O}{\|}}{C}{-}H$

b. (cyclopentanone)

c. $CH_3{-}\overset{\overset{\displaystyle O}{\|}}{C}{-}CH_2{-}CH_3$

d.

$$\underset{\text{(benzaldehyde)}}{C_6H_5-\overset{\displaystyle O}{\overset{\|}{C}}-H}$$

e.

(3-methylcyclohexanone)

14.36 Draw the condensed structural formula of the alcohol needed to give each of the following oxidation products:

a. $CH_3-\overset{\displaystyle O}{\overset{\|}{C}}-H$

b. $CH_3-\overset{\displaystyle O}{\overset{\|}{C}}-\overset{\displaystyle CH_3}{\overset{|}{CH}}-CH_3$

c.

d. $CH_3-CH_2-\overset{\displaystyle O}{\overset{\|}{C}}-H$

e. $CH_3-\overset{\displaystyle CH_3}{\overset{|}{CH}}-CH_2-\overset{\displaystyle O}{\overset{\|}{C}}-H$

Chapter Review

14.1 Structure and Classification of Alcohols

The functional group of an alcohol is the hydroxyl group —OH bonded to a carbon chain. In a phenol, the hydroxyl group is bonded to an aromatic ring. Alcohols are classified according to the number of alkyl or aromatic groups bonded to the carbon that holds the —OH group. In a primary (1°) alcohol one alkyl group is attached to the hydroxyl carbon. In a secondary (2°) alcohol two alkyl groups are attached, and in a tertiary (3°) alcohol there are three alkyl groups bonded to the hydroxyl carbon. In thiols, the functional group is —SH, which is analogous to the —OH group of alcohols.

14.2 Naming Alcohols, Phenols, and Thiols

In the IUPAC system, the names of alcohols have *ol* endings, and the location of the —OH group is given by numbering the carbon chain. Simple alcohols are generally named by their common names with the alkyl name preceding the term *alcohol*. A cyclic alcohol is named as a cycloalkanol. An aromatic alcohol is named as a phenol.

14.3 Some Important Alcohols and Phenols

Methanol, the simplest alcohol, is used as a solvent and as a fuel in race cars. It is toxic if ingested. Ethanol is found in alcoholic beverages and used as a solvent for perfumes and varnishes. Isopropyl alcohol or rubbing alcohol is used as an astringent and to sterilize medical equipment. Menthol is a cyclohexanol with a peppermint taste that is used in candy

and throat lozenges. Phenol is used in antiseptics and disinfectants. Many plants produce oils containing derivatives of phenol such as eugenol in cloves and thymol in thyme.

14.4 Ethers

In an ether, an oxygen atom is connected by single bonds to two alkyl or aromatic groups, R—O—R. In the common names of ethers, the alkyl groups are listed alphabetically followed by the name *ether*. In the IUPAC name, the smaller alkyl group with the oxygen is named as an *alkoxy group* and is attached to the longer alkane chain, which is numbered to give the location of the alkoxy group.

14.5 Physical Properties of Alcohols, Phenols, and Ethers

The —OH group allows alcohols to hydrogen bond, which causes alcohols to have higher boiling points than alkanes and ethers of similar mass. Short-chain alcohols and ethers can hydrogen bond with water, which makes them soluble in water.

14.6 Reactions of Alcohols

At high temperatures, alcohols dehydrate in the presence of an acid to yield alkenes. At lower temperatures, two molecules of alcohol lose H and OH to produce an ether. Primary alcohols are oxidized to aldehydes, which can oxidize further to carboxylic acids. Secondary alcohols are oxidized to ketones. Tertiary alcohols do not oxidize.

Summary of Naming

Structure	Family	IUPAC Name	Common Name
CH_3—OH	Alcohol	Methanol	Methyl alcohol
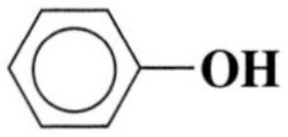 ⬡—OH	Phenol	Phenol	Phenol
CH_3—SH	Thiol	Methanethiol	
CH_3—O—CH_3	Ether	Methoxymethane	Dimethyl ether
 (furan ring with O)	Cyclic ether	Furan	
CH_3—S—S—CH_3	Disulfide	Dimethyldisulfide	

Summary of Reactions

Combustion of Alcohols

$$CH_3-CH_2-OH + 3O_2 \longrightarrow 2CO_2 + 3H_2O$$

Ethanol   Oxygen   Carbon dioxide Water

Intramolecular Dehydration of Alcohols to Form Alkenes

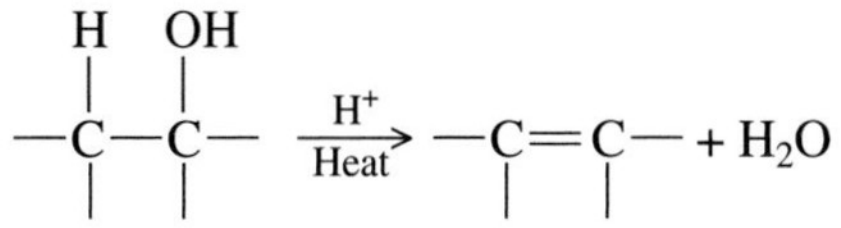

$$CH_3-CH_2-CH_2-OH \xrightarrow[\text{Heat}]{H^+}$$

1-Propanol

$$CH_3-CH=CH_2 + H_2O$$

Propene

Intermolecular Dehydration of Alcohols to Form Ethers

$$R-OH + HO-R \xrightarrow[\text{Heat}]{H^+} R-O-R + H_2O$$

Two alcohols   Ether

$$CH_3-OH + HO-CH_3 \xrightarrow[\text{Heat}]{H^+}$$

Methanol

$$CH_3-O-CH_3 + H_2O$$

Dimethyl ether

Oxidation of Primary Alcohols to Form Aldehydes

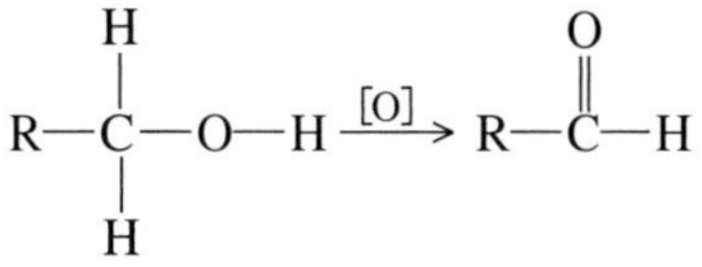

Primary alcohol   Aldehyde

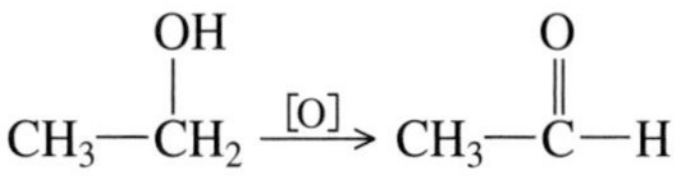

$$CH_3-CH_2-OH \xrightarrow{[O]} CH_3-\overset{O}{\overset{\|}{C}}-H$$

Ethanol   Acetaldehyde

Oxidation of Secondary Alcohols to Form Ketones

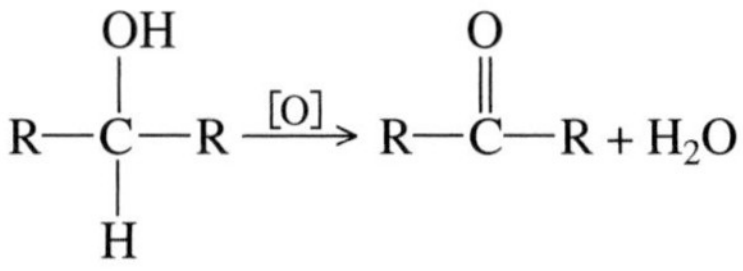

Secondary alcohol   Ketone

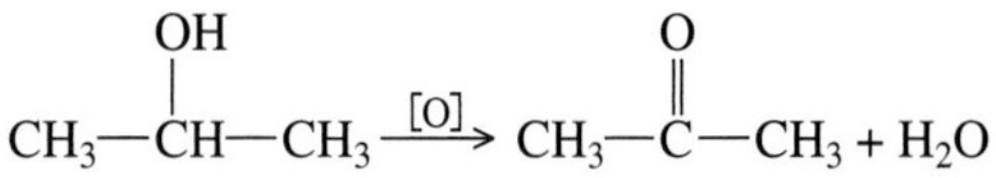

$$CH_3-\overset{OH}{\underset{}{CH}}-CH_3 \xrightarrow{[O]} CH_3-\overset{O}{\overset{\|}{C}}-CH_3 + H_2O$$

2-Propanol   Propanone

Oxidation of Thiols to Form Disulfides

$$R-S-H + H-S-R \xrightarrow{[O]} R-S-S-R + H_2O$$

Two thiols   Disulfide

$$CH_3-S-H + H-S-CH_3 \xrightarrow{[O]} CH_3-S-S-CH_3 + H_2O$$

Methanethiol   Dimethyl disulfide

Key Terms

alcohols   Organic compounds that contain the hydroxyl (—OH) functional group, attached to a carbon chain, R—OH.

alkoxy group   A group that contains oxygen bonded to an alkyl group.

cyclic ethers   Compounds that contain an oxygen atom in a carbon ring.

dehydration   A reaction that removes water from an alcohol in the presence of an acid to form alkenes at high temperature, or ethers at lower temperatures.

disulfides   Compounds formed from thiols, disulfides contain the —S—S— functional group.

ether   An organic compound, R—O—R, in which an oxygen atom is bonded to two alkyl or two aromatic groups, or a mix of the two.

furan   An unsaturated cyclic ether that is a five-atom ring with an oxygen atom.

heterocyclic compounds   Compounds composed of a ring that contains one or more atoms other than carbon.

hydroxyl group   The —OH functional group.

oxidation   The loss of two hydrogen atoms from a reactant to give a more oxidized compound, e.g., primary alcohols oxidize to aldehydes, secondary alcohols oxidize to ketones. An oxidation can also be the addition of an oxygen atom as in the oxidation of aldehydes to carboxylic acids.

phenol   An organic compound that has an —OH group attached to a benzene ring.

primary (1°) alcohol   An alcohol that has one alkyl group bonded to the alcohol carbon atom, R—CH$_2$OH.

Saytzeff's rule   In the dehydration of an alcohol, hydrogen is removed from the carbon that already has the smallest number of hydrogen atoms to form an alkene.

secondary (2°) alcohol   An alcohol that has two alkyl groups bonded to the carbon atom with the —OH group.

sulfhydryl group   The —SH functional group.

tertiary (3°) alcohol   An alcohol that has three alkyl groups bonded to the carbon atom with the —OH.

thiols   Organic compounds that contain a thiol group (—SH) in place of the —OH group of an alcohol.

Additional Problems

14.37 Classify each of the following as primary, secondary, or tertiary alcohols:

a.

b.

c. CH_3—CH—CH_2—OH
         |
         CH_3

d. CH_3—C—CH_2—CH—CH_3
         | |
         CH_3 OH
         |
         CH_3

e.   HO—CH_2—CH_2—CH_3

f.

14.38 Classify each of the following as primary, secondary, or tertiary alcohols:

a.

b.

c. CH_3—CH—CH_2—CH_3
         |
         CH_2—OH

d. CH_3—C—CH_2—CH—CH_3
         | |
         OH CH_3
         |
         CH_3

e. CH_3—CH_2—CH_2—CH_2—OH

f.

14.39 Identify each of the following as an alcohol, a phenol, an ether, a cyclic ether, or a thiol:

a. [structure: cyclohexane ring with OH, Cl, and CH₃ substituents]

b. [structure: benzene ring with O—CH₃]

c. CH₃—CH—CH₃ with SH group

d. CH₃—C—CH₂—CH—CH₃ with OH, CH₃, and CH₃ groups

e. CH₃—CH₂—CH₂—O—CH₃

f. [structure: furan ring with CH₃]

g. CH₃—CH—CH₂—CH—CH₃ with Br and OH groups

h. [structure: benzene ring with OH and CH₃]

14.40 Identify each of the following as an alcohol, a phenol, an ether, a cyclic ether, or a thiol:

a. [structure: benzene ring with OH and Cl]

b. CH₃—CH₂—CH₂—SH

c. [structure: cyclopentane ring]—O—CH₂—CH₃

d. CH₃—C—CH₂—CH—CH₃ with OH, CH₃, and CH₃ groups

e. CH₃—CH₂—CH—CH₂—CH₃ with O—CH₃ group

f. [structure: 1,4-dioxane ring with two O]

g. 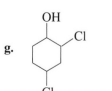 [structure: cyclohexane ring with OH, Cl, and Cl]

h. [structure: benzene ring with OH, CH₃, CH₃]

14.41 Give the IUPAC and common names (if any) for each of the compounds in problem 14.39.

14.42 Give the IUPAC and common names (if any) for each of the compounds in problem 14.40.

14.43 Draw the condensed structural formula of each of the following compounds:
  a. 3-methylcyclopentanol   **b.** *p*-chlorophenol
  c. 2-methyl-3-pentanol   **d.** phenyl ethyl ether
  e. 3-pentanethiol   **f.** *ortho*-cresol
  g. 2,4-dibromophenol

14.44 Draw the condensed structural formula of each of the following compounds:
  a. 3-methoxypentane   **b.** *meta*-chlorophenol
  c. 2,3-pentanediol   **d.** methyl propyl ether
  e. methanethiol   **f.** 3-methyl-2-butanol
  g. 3,4-dichlorocyclohexanol

14.45 Identify the alcohol described by each of the following:
  a. a syrupy liquid used in skin lotions
  b. used in antifreeze
  c. produced by fermentation of grains and sugars

14.46 Identify the alcohol described by each of the following:
  a. produced by the reaction of carbon monoxide and hydrogen at high temperatures
  b. used to clean the skin before an injection
  c. the "alcohol" in alcoholic drinks

14.47 Draw the structural formulas of all the alcohols with a molecular formula $C_4H_{10}O$.

14.48 Draw the structural formulas of all the ethers with a molecular formula $C_5H_{12}O$.

14.49 Which compound in each pair would you expect to have the higher boiling point?
  a. butane or 1-propanol
  b. 1-propanol or methyl ethyl ether
  c. ethanol or 1-butanol

14.50 Which compound in each pair would you expect to have the higher boiling point?
  a. propane or ethyl alcohol
  b. 2-propanol or 2-pentanol
  c. diethyl ether or 1-butanol

14.51 Explain why each of the following compounds would be soluble or insoluble in water.
  a. 2-propanol
  b. dimethyl ether
  c. 1-hexanol

14.52 Explain why each of the following compounds would be soluble or insoluble in water.
a. glycerol
b. butane
c. 1,3-hexanediol

14.53 Draw the condensed structural formula for the major product of each of the following reactions:

a. $CH_3-CH_2-CH_2-OH \xrightarrow{H^+, \text{ heat}}$

b. $CH_3-CH_2-CH_2-OH \xrightarrow{[O]}$

c. $CH_3-CH_2-\overset{\overset{\displaystyle OH}{|}}{CH}-CH_3 \xrightarrow{H^+, \text{ heat}}$

d. $CH_3-CH_2-\overset{\overset{\displaystyle OH}{|}}{CH}-CH_3 \xrightarrow{[O]}$

e. $2CH_3-CH_2-CH_2-OH \xrightarrow{H^+}$

f.
cyclohexanol with OH $\xrightarrow{H^+, \text{ heat}}$

g.
cyclohexanol with OH $\xrightarrow{[O]}$

14.54 Draw the condensed structural formula for the major product of each of the following reactions:

a. $CH_3-\overset{\overset{\displaystyle CH_3}{|}}{CH}-CH_2-OH \xrightarrow{H^+, \text{ heat}}$

b. $CH_3-\overset{\overset{\displaystyle CH_3}{|}}{CH}-\overset{\overset{\displaystyle OH}{|}}{CH}-CH_3 \xrightarrow{H^+, \text{ heat}}$

c. $CH_3-\overset{\overset{\displaystyle CH_3}{|}}{CH}-\overset{\overset{\displaystyle OH}{|}}{CH}-CH_3 \xrightarrow{[O]}$

d.
cyclopentane ring with OH and CH$_3$ $\xrightarrow{H^+, \text{ heat}}$

e.
cyclopentane ring with OH and CH$_3$ $\xrightarrow{[O]}$

f. $2CH_3-CH_2-OH \xrightarrow{H^+, \text{ heat}}$

g. $CH_3-CH_2-CH_2-\overset{\overset{\displaystyle OH}{|}}{CH}-CH_3 \xrightarrow{H^+, \text{ heat}}$

14.55 Sometimes several steps are needed to prepare a compound. Using a combination of the reactions we have studied, indicate how you might prepare the following from the starting substance given. For example, 2-propanol could be prepared from 1-propanol by first dehydrating the alcohol to give propene and then hydrating it again to give 2-propanol according to Markovnikov's rule.

$$CH_3-CH_2-CH_2-OH \xrightarrow{H^+, \text{ heat}} CH_3-CH=CH_2 + H_2O \xrightarrow{H^+}$$

$$CH_3-\overset{\overset{\displaystyle OH}{|}}{CH}-CH_3$$
2-Propanol

a. Prepare 2-chloropropane from 1-propanol.
b. Prepare 2-methylpropane from 2-methyl-2-propanol.

c. Prepare $CH_3-\overset{\overset{\displaystyle O}{\|}}{C}-CH_3$ from 1-propanol.

14.56 As in problem 14.55, indicate how you might prepare the following from the starting substance given:
a. Prepare 1-pentene from 1-pentanol.
b. Prepare chlorocyclohexane from cyclohexanol.
c. Prepare 1,2-dibromobutane from 1-butanol.

14.57 Identify the functional groups in the following molecule:

Testosterone

14.58 Identify the functional groups in the following molecule:

Tetrahydrocannabinol (THC)

14.59 Hexylresorcinol, an antiseptic ingredient used in mouthwashes and throat lozenges, has the IUPAC name of 4-hexyl-1,3-benzenediol. Draw its condensed structural formula.

14.60 Menthol, which has a minty flavor, is used in throat sprays and lozenges. Thymol is used as a topical antiseptic to destroy mold. Give each of their IUPAC names. What is similar and what is different about their structures?

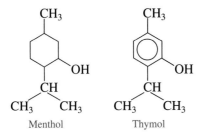

Menthol   Thymol

14.61 Write the condensed structural formulas for each of the following naturally occurring compounds:

a. 2,5-dichlorophenol, a defense pheromone of a grasshopper.

b. Skunk scent, a mixture of 3-methyl-1-butanethiol and *trans*-2-butene-1-thiol.

c. Pentachlorophenol, which is a wood preservative.

14.62 Dimethyl ether and ethyl alcohol both have the molecular formula C_2H_6O. One has a boiling point of $-24°C$, and the other, $79°C$. Draw the condensed structural formulas of each compound. Decide which boiling point goes with which compound and explain. Check the boiling points in a chemistry handbook.

15 Aldehydes, Ketones, and Chiral Molecules

"Dentures replace natural teeth that are extracted due to cavities, bad gums, or trauma," says Dr. Irene Hilton, dentist, La Clinica De La Raza. "I make an impression of teeth using alginate, which is a polysaccharide extracted from seaweed. I mix the compound with water and place the gel-like material in the patient's mouth, where it becomes a hard, cement-like substance. I fill this mold with gypsum ($CaSO_4$) and water, which form a solid to which I add teeth made of plastic or porcelain. When I get a good match to the patient's own teeth, I prepare a preliminary wax denture. This is placed in the patient's mouth to check the bite and adjust the position of the replacement teeth. Then a permanent denture is made using a hard plastic polymer (methyl methacrylate)."

LOOKING AHEAD

the Chemistry place

www.chemplace.com/college

Visit the URL above or use the CD-ROM in the book for extra quizzing, interactive tutorials, career resources, and case studies.

Many of the odors and flavors that you associate with solvents, flavorings, paint removers, and perfumes are from organic molecules called aldehydes and ketones. In biology, you may have seen specimens preserved in a solution of formaldehyde. You may have noticed the odor of acetone if you used paint or polish remover. Foods and perfumes with the odors and flavors of vanilla, almond, and cinnamon contain aldehydes that occur in nature.

The feature responsible for this wide variety of flavors and odors is a carbon–oxygen double bond called a **carbonyl group** (C=O). In this chapter we will study two families with carbonyl groups: aldehydes and ketones. They are prevalent in nature and play an important role in biochemical pathways. In later chapters, we will see how the carbonyl group influences the structures of carbohydrates, proteins, and nucleic acids. Aldehydes and ketones are also important compounds in industry, providing the solvents and reactants that make up many common materials we use in our lives.

15.1 Structure and Bonding

LEARNING GOAL

Identify compounds with the carbonyl group as aldehydes and ketones.

In an **aldehyde,** the carbon of the carbonyl group is bonded to at least one hydrogen atom. That carbon may also be bonded to another hydrogen, a carbon of an alkyl group, or an aromatic ring. (See Figure 15.1.) In a **ketone,** the carbonyl group is bonded to two alkyl groups or aromatic rings.

Writing Aldehyde and Ketone Structures

There are several ways to write the structural formulas of aldehydes and ketones. In the condensed structural formula, the aldehyde group may be drawn as separate atoms or it may be written as —CHO, with the double bond understood. An aldehyde would not be written as —COH, which looks like a hydroxy group. The keto

WEB TUTORIAL
Aldehydes and Ketones

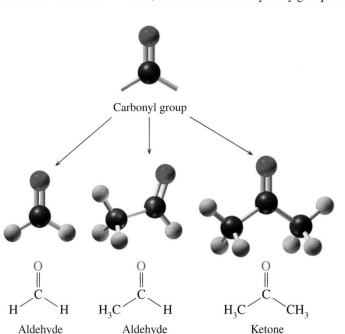

Carbonyl group

Aldehyde   Aldehyde   Ketone

Figure 15.1 The carbonyl group in aldehydes and ketones.

Q If aldehydes and ketones both contain a carbonyl group, how can you differentiate between compounds from each family?

group (C=O) is sometimes written as CO. For convenience, the line-bond formulas are also used with the functional group atoms written separately. Constitutional isomers, which have the same molecular formula, can be written for aldehydes and ketones as follows:

Formulas for Constitutional Isomers of C_3H_6O

Aldehyde

$$CH_3-CH_2-\overset{\overset{\displaystyle O}{\|}}{C}-H \;=\; CH_3-CH_2-CHO \;=$$

Ketone

$$CH_3-\overset{\overset{\displaystyle O}{\|}}{C}-CH_3 \;=\; CH_3-CO-CH_3 \;=$$

SAMPLE PROBLEM 15.1

Identifying Aldehydes and Ketones

Identify each of the following compounds as an aldehyde or ketone:

a. $CH_3-\overset{\overset{\displaystyle CH_3}{|}}{\underset{\underset{\displaystyle CH_3}{|}}{C}}-CH_2-\overset{\overset{\displaystyle O}{\|}}{C}-H$ b.

c.   d.

Solution

a. aldehyde   **b.** ketone   **c.** aldehyde   **d.** ketone

---

Study Check

Draw the condensed structural formula of a ketone that has a carbonyl group bonded to two ethyl groups.

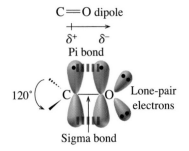

C=O dipole

$\delta^+ \quad \delta^-$

Pi bond

120°   C   O   Lone-pair electrons

Sigma bond

Figure 15.2 The sigma and pi bond in the carbon–oxygen double bond of the carbonyl group.

Q Why does the carbonyl bond have a dipole?

Structure of the Carbonyl Group

The carbonyl group consists of a carbon–oxygen double bond. Sigma bonds at angles of 120° bond the carbon to oxygen and two other atoms. The second bond is a pi bond that forms when a *p* orbital in the oxygen atom overlaps with a *p* orbital in carbon. (See Figure 15.2.) We saw the sigma and pi bonds in Chapter 13 when we described the double bonds between two carbon atoms in alkenes. The double bond in the carbonyl group is similar to that of alkenes, except the carbonyl group has a dipole. The oxygen atom with two lone pairs of electrons is also much more electronegative than the carbon atom. Therefore, the carbonyl group has a strong dipole with a partial negative charge (δ^-) on the oxygen, and a partial positive charge (δ^+) on the carbon. The polarity of the carbonyl group strongly influences the physical and chemical properties of aldehydes and ketones.

SAMPLE PROBLEM 15.2

Carbonyl Group

Using the dipole symbol, $+\!\!\!-\!\!\!\longrightarrow$, illustrate the polarity of the carbonyl group in each of the following compounds:

a. CH₃—C(=O)—CH₂—CH₃

b. CH₃—CH₂—C(=O)—H

Solution

a. CH₃—C(=O)—CH₂—CH₃

b. CH₃—CH₂—C(=O)—H

Study Check

Why does the carbonyl group have a dipole?

QUESTIONS AND PROBLEMS

Structure and Bonding

15.1 Identify the following compounds as aldehydes or ketones:

a. CH₃—CH₂—C(=O)—CH₃

b.

c.

d.

15.2 Identify the following compounds as aldehydes or ketones:

a.

b. CH₃—CH(CH₃)—C(=O)—H

c.

d.

15.3 Indicate if each of the following pairs of structural formulas represent (1) constitutional isomers, (2) the same compound, or (3) different compounds.

a. CH₃—C(=O)—CH₃ and CH₃—CH₂—C(=O)—H

b. 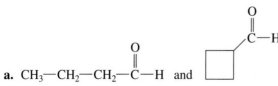 and

c. CH₃—C(=O)—CH₂—CH₃ and CH₃—CH₂—C(=O)—CH₃

15.4 Indicate if each of the following pairs of structural formulas represent (1) constitutional isomers, (2) the same compound, or (3) different compounds.

a. CH₃—CH₂—CH₂—C(=O)—H and

b. and

c. 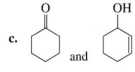 and

WEB TUTORIAL
Aldehydes and Ketones

15.2 Naming Aldehydes and Ketones

Aldehydes and ketones are some of the most important functional groups in organic chemistry. Because they have played a major role in organic chemistry for more than a century, the common names for unbranched ketones are still in use. In the common names the alkyl groups bonded to the carbonyl group are named as substituents and listed alphabetically followed by *ketone*. The name acetone, which is propanone, has been retained by the IUPAC system.

IUPAC Names for Ketones

In the IUPAC system, the name of a ketone is obtained by replacing the *-e* in the corresponding alkane name with *-one*.

Step 1   Name the longest carbon chain containing the carbonyl group by replacing the *-e* in the corresponding alkane name by *-one*.

Step 2   Number the main chain starting from the end nearest the carbonyl group. Place the number of the carbonyl carbon in front of the ketone name. (Propanone and butanone do not require numbers.)

$$CH_3-\overset{\overset{\displaystyle O}{\|}}{C}-CH_3 \qquad CH_3-CH_2-\overset{\overset{\displaystyle O}{\|}}{C}-CH_3 \qquad CH_3-CH_2-\overset{\overset{\displaystyle O}{\|}}{C}-CH_2-CH_3$$

Propanone   Butanone   3-Pentanone
(dimethyl ketone; acetone)   (ethyl methyl ketone)   (diethyl ketone)

Step 3   Name and number any substituents on the carbon chain.

$$CH_3-\overset{\overset{\displaystyle O}{\|}}{C}-\overset{\overset{\displaystyle CH_3}{|}}{CH}-CH_3 \qquad CH_3-\overset{\overset{\displaystyle Br}{|}}{CH}-\overset{\overset{\displaystyle O}{\|}}{C}-CH_2-CH_3$$

3-Methylbutanone   2-Bromo-3-pentanone   4-Chloro-5-methyl-2-hexanone

Step 4   For cyclic ketones, the prefix *cyclo* is used in front of the ketone name. Any substituents are located by numbering the ring starting with the carbonyl carbon as carbon 1.

Cyclopentanone   3-Methylcyclohexanone   2,3-Dichlorocyclopentanone

SAMPLE PROBLEM 15.3

Names of Ketones

Give the IUPAC name for the following ketone:

$$CH_3-\overset{\overset{\displaystyle CH_3}{|}}{CH}-CH_2-\overset{\overset{\displaystyle O}{\|}}{C}-CH_3$$

Solution

The longest chain is five carbon atoms. Counting from the right, the carbonyl group is on carbon 2 and a methyl group is on carbon 4. The IUPAC name is *4-methyl-2-pentanone*.

Study Check

What is the common name of 3-hexanone?

Naming Aldehydes

In the IUPAC names of aldehydes, the *-e* of the alkane name is replaced with *-al*.

Step 1 Name the longest carbon chain containing the carbonyl group by replacing the *-e* in the corresponding alkane name by *-al*. No number is needed for the aldehyde group because it always appears at the end of the chain. The IUPAC system names the aldehyde of benzene as benzaldehyde.

Benzaldehyde

The first four unbranched aldehydes are often referred to by their common names, which end in *aldehyde*. (See Figure 15.3.) The roots of these common names are derived from Latin or Greek words that indicate the source of the corresponding carboxylic acid. We will study carboxylic acids in the next chapter.

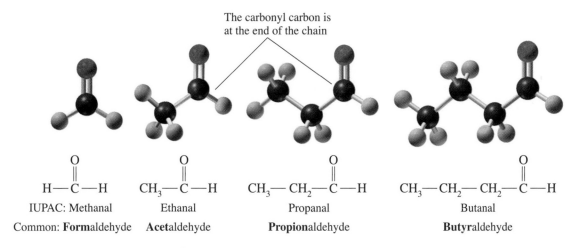

The carbonyl carbon is at the end of the chain

H—C—H	CH₃—C—H	CH₃—CH₂—C—H	CH₃—CH₂—CH₂—C—H
IUPAC: Methanal	Ethanal	Propanal	Butanal
Common: **Form**aldehyde	**Acet**aldehyde	**Propion**aldehyde	**Butyr**aldehyde

Step 2 Name and number any substituents on the carbon chain by counting the carbonyl carbon as carbon 1. An aldehyde group takes precedence in numbering over a ketone or an alcohol. When present with an aldehyde functional group, a ketone group is numbered and named oxo. When present with an aldehyde, an alcohol is named hydroxy.

Figure 15.3 In the structures of aldehydes, the carbonyl group is always the end carbon.

Q Why is the carbon in the carbonyl group in aldehydes always at the end of the chain?

2-Hydroxypropanal   3-Oxobutanal   3-Bromobenzaldehyde

SAMPLE PROBLEM 15.4

Naming Aldehydes

Give the IUPAC names for the following aldehydes:

a. $CH_3CH_2CH_2CH_2$—C—H   b. Cl—⟨ ⟩—CH   c. CH_3—CH—CH_2—C—H

Solution

a. pentanal
b. 4-chlorobenzaldehyde
c. The longest unbranched chain has four atoms with a methyl group on the third carbon. The IUPAC name is *3-methylbutanal.*

Study Check

What are the IUPAC and common names of the aldehyde with three carbon atoms?

QUESTIONS AND PROBLEMS

Naming Aldehydes and Ketones

15.5 Give the IUPAC name for each of the following compounds:

a. CH_3—CH_2—C—H   b. CH_3—CH_2—C—CH—CH_3 c. CH_3—CH—CH_2—C—H

d.   e.   f.

15.6 Give the IUPAC name for each of the following compounds:

a. CH_3—CH_2—CH_2—C—H   b. CH_3—CH—CH_2—C—CH_3 c.

d.   e.   f. CH_3—C—CH_2—CH—C—H

15.7 Give a common name for each of the following compounds:

a. CH_3—$\overset{\overset{\displaystyle O}{\|}}{C}$—H   **b.** CH_3—$\overset{\overset{\displaystyle O}{\|}}{C}$—$CH_2$—$CH_2$—$CH_3$ **c.** H—$\overset{\overset{\displaystyle O}{\|}}{C}$—H

15.8 Give the common name for each of the following compounds:

a. CH_3—$\overset{\overset{\displaystyle O}{\|}}{C}$—$CH_2$—$CH_3$ **b.** CH_3—CH_2—$\overset{\overset{\displaystyle O}{\|}}{C}$—$CH_2$—$CH_3$ **c.** CH_3—CH_2—$\overset{\overset{\displaystyle O}{\|}}{C}$—H

15.9 Write the condensed structural formula for each of the following compounds:
  a. acetaldehyde   **b.** 4-hydroxy-2-pentanone
  c. 2,3-dibromobutanal   **d.** methyl butyl ketone
  e. 3-methylpentanal   **f.** 3-oxopentanal

15.10 Write the condensed structural formula for each of the following compounds:
  a. propionaldehyde   **b.** butanal
  c. 4-bromo-3-hydroxyhexanal   **d.** 4-bromobutanone
  e. 2-oxobutanal   **f.** acetone

15.11 The IUPAC name of anisaldehyde used in perfumes is 4-methoxybenzalde-hyde. What is the structural formula of anisaldehyde?

15.12 The IUPAC name of vanillin, a naturally occurring compound in vanilla beans, is 4-hydroxy-3-methoxybenzaldehyde. What is the structural formula of vanillin?

15.3 Some Important Aldehydes and Ketones

Formaldehyde, the simplest aldehyde, is a colorless gas with a pungent odor. Industrially, it is a reactant in the synthesis of polymers used to make fabrics, insulation materials, carpeting, pressed wood products such as plywood, and plastics for kitchen counters. An aqueous solution called formalin, which contains 40% formaldehyde, is used as a germicide and to preserve biological specimens. Exposure to formaldehyde fumes can irritate eyes, nose, upper respiratory tract, and cause skin rashes, headaches, dizziness, and general fatigue.

Acetone, or propanone (dimethyl ketone), which is the simplest ketone, is a colorless liquid with a mild odor that has wide use as a solvent in cleaning fluids, paint and nail-polish removers, and rubber cement. (See Figure 15.4.) It is extremely flammable and care must be taken when using acetone. In the body, acetone may

WEB TUTORIAL
Aldehydes and Ketones

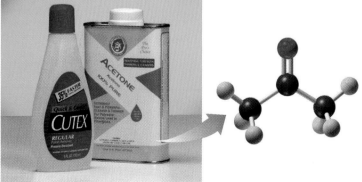

Figure 15.4 Acetone is used as a solvent in paint and nail-polish removers.

Q What is the IUPAC name of acetone?

be produced in uncontrolled diabetes, fasting, and high-protein diets when large amounts of fats are metabolized for energy.

Benzaldehyde (almond) Vanillin (vanilla) Cinnamaldehyde (cinnamon)

Several naturally occurring aromatic aldehydes are used to flavor food and as fragrances in perfumes. Benzaldehyde is found in almonds, vanillin in vanilla beans, and cinnamaldehyde in cinnamon.

SAMPLE PROBLEM 15.5

Glyceraldehyde is an intermediate in carbohydrate metabolism. What are the functional groups in glyceraldehyde?

Glyceraldehyde

ENVIRONMENTAL NOTE

Vanilla

Vanilla has been used as a flavoring for over a thousand years. After drinking a beverage made from powdered vanilla and cocoa beans with Emperor Montezuma in Mexico, Cortez took vanilla back to Europe where it became popular for flavoring and for scenting perfumes and tobacco. Thomas Jefferson introduced vanilla to the United States in the late 1700s. Today much of the vanilla we use in the world is grown in Mexico, Madagascar, Réunion, Seychelles, Tahiti, Ceylon, Java, the Philippines, and Africa.

The vanilla plant is a member of the orchid family. There are many species of *Vanilla*, but *Vanilla planifolia* (or *V. fragrans*) is considered to produce the best flavor. The vanilla plant grows like a vine,

and can grow to 100 feet in length. Its flowers are hand-pollinated to produce a green fruit that is picked in 8 or 9 months. It is sun-dried to form a long, dark brown pod, which is called "vanilla bean" because it looks like a string bean.

The flavor and fragrance of the vanilla bean comes from the black seeds found inside the dried bean; the seeds and pod are used to flavor desserts such as custards and ice cream. The extract of vanilla is made by chopping up vanilla beans and mixing them with a 35% alcohol–water mixture. The liquid, which contains the aldehyde vanillin, is drained from the bean residue and used for flavoring.

Vanillin

Solution

Glyceraldehyde has aldehyde and alcohol functional groups.

$$
\begin{array}{c}
\overset{\displaystyle O}{\underset{\displaystyle \|}{}} \\
\text{C—H} \qquad \longleftarrow \text{ Aldehyde} \\
\text{H—C—OH} \qquad \longleftarrow \text{ Alcohol} \\
\text{CH}_2\text{—OH}
\end{array}
$$

Study Check

The compound dihydroxyacetone (DHA) used in "sunless" tanning lotions darkens the skin without sun. What is its structural formula?

The flavor of butter or margarine is from butanedione; muscone is used to make musk perfumes; and oil of spearmint contains carvone.

$$
\text{CH}_3-\overset{\displaystyle O}{\underset{\displaystyle \|}{C}}-\overset{\displaystyle O}{\underset{\displaystyle \|}{C}}-\text{CH}_3
$$

Butanedione
(butter flavor)

Muscone
(musk)

Carvone
(spearmint oil)

Important hormones of the steroid family, such as cortisone, testosterone, and progesterone, contain one or more carbonyl groups.

Cortisone
(protein metabolism;
inflammation reducer)

Testosterone
(male sex hormone)

Progesterone
(female sex hormone)

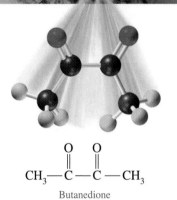

$$
\text{CH}_3-\overset{\displaystyle O}{\underset{\displaystyle \|}{C}}-\overset{\displaystyle O}{\underset{\displaystyle \|}{C}}-\text{CH}_3
$$

Butanedione

QUESTIONS AND PROBLEMS

Some Important Aldehydes and Ketones

15.13 Identify the aldehyde or ketone with each of the following uses:
 a. almond flavoring **b.** fingernail-polish remover
 c. making plywood

15.14 Identify the aldehyde or ketone with each of the following uses:
 a. butter flavoring **b.** produced in uncontrolled diabetes
 c. preserve tissues

LEARNING GOAL

Compare the boiling points and solubility of aldehydes and ketones to those of alkanes and alcohols.

15.4 Physical Properties

At room temperature, formaldehyde (bp −21°C) and acetaldehyde (bp 21°C) are gases. Aldehydes containing from 3 to 10 carbon atoms are liquids. The polar carbonyl group with a partially negative oxygen atom and a partially positive carbon atom has an influence on the boiling points and the solubility of aldehydes and ketones in water.

Boiling Points

The polar carbonyl group gives aldehydes and ketones higher boiling points than alkanes and ethers of similar mass. The increase in boiling points is due to dipole–dipole interactions.

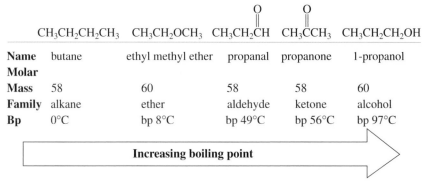

However, because there is no hydrogen on the oxygen atom, aldehydes and ketones cannot form hydrogen bonds with each other. Thus they have boiling points that are lower than alcohols.

	$CH_3CH_2CH_2CH_3$	$CH_3CH_2OCH_3$	$CH_3CH_2\overset{\overset{O}{\parallel}}{C}H$	$CH_3\overset{\overset{O}{\parallel}}{C}CH_3$	$CH_3CH_2CH_2OH$
Name	butane	ethyl methyl ether	propanal	propanone	1-propanol
Molar Mass	58	60	58	58	60
Family	alkane	ether	aldehyde	ketone	alcohol
Bp	0°C	bp 8°C	bp 49°C	bp 56°C	bp 97°C

Increasing boiling point →

Solubility of Aldehydes and Ketones in Water

Although aldehydes and ketones do not hydrogen bond with each other, the electronegative oxygen atom does hydrogen bond with water molecules. Carbonyl compounds with one to four carbons are very soluble in water. However, those with five carbon atoms or more are not very soluble because the alkyl portions diminish the effect of the polar carbonyl group. (See Figure 15.5.)

Table 15.1 compares the boiling points of some carbonyl compounds, as well as their solubilities in water.

SAMPLE PROBLEM 15.6

Boiling Point and Solubility

Would you expect ethanol CH_3CH_2OH to have a higher or lower boiling point than ethanal CH_3CHO? Explain.

Solution

Ethanol would have a higher boiling point because its molecules can hydrogen bond with each other, while molecules of ethanal cannot.

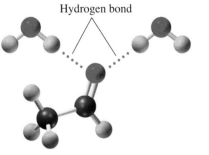

Hydrogen bond

Acetaldehyde

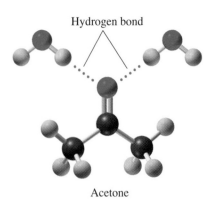

Hydrogen bond

Acetone

Figure 15.5 Hydrogen bonding of acetaldehyde and acetone with water.

Q Would you expect propanal to be soluble in water?

Study Check

If acetone molecules cannot hydrogen bond with each other, why is acetone soluble in water?

Physical Properties

15.15 Which compound in each of the following pairs would have the higher boiling point? Explain.

$$\text{O}$$
$$\|$$
a. $CH_3\!-\!CH_2\!-\!CH_3$ or $CH_3\!-\!CH_2\!-\!C\!-\!H$

b. propanal or pentanal   **c.** butanal or 1-butanol

15.16 Which compound in each of the following pairs would have the higher boiling point? Explain.

a. [OH structure] or [O structure]

b. butane or butanone   **c.** propanone or pentanone

15.17 Which compound in each of the following pairs would be more soluble in water? Explain.

$$\text{O} \qquad\qquad \text{O}\ \ \text{O}$$
$$\| \qquad\qquad \|\ \ \|$$
a. $CH_3\!-\!C\!-\!CH_2\!-\!CH_3$ or $CH_3\!-\!C\!-\!C\!-\!CH_3$

b. dimethyl ether or acetaldehyde **c.** acetone or 2-pentanone

15.18 Which compound in each of the following pairs would be more soluble in water? Explain.

a. $CH_3\!-\!CH_2\!-\!CH_3$ or $CH_3\!-\!CH_2\!-\!CHO$

b. propanone or 3-hexanone   **c.** propane or propanone

15.19 Would you expect an aldehyde with a formula of $C_8H_{16}O$ to be soluble in water? Explain.

15.20 Would you expect an aldehyde with a formula of C_3H_6O to be soluble in water? Explain.

Table 15.1   Comparison of Physical Properties of Some Selected Compounds

Compound	Boiling Point (°C)	Solubility in Water
Methanal (formaldehyde)	−21	Very soluble
Ethanal (acetaldehyde)	21	Very soluble
Propanal (propionaldehyde)	49	Soluble
Propanone (acetone)	56	Soluble
Butanal (butyraldehyde)	75	Soluble
Butanone	80	Soluble
Pentanal	103	Slightly soluble
2-Pentanone	102	Slightly soluble
3-Pentanone	102	Slightly soluble
Hexanal	129	Not soluble
2-Hexanone	127	Not soluble
3-Hexanone	124	Not soluble
Acetophenone	202	Not soluble

LEARNING GOAL
Identify chiral and achiral
carbon atoms in an organic
molecule.

15.5 Chiral Molecules

In the preceding chapters, we have looked at some types of isomers. Let's review those now. Molecules can be constitutional (also called structural) isomers when they have the same molecular formula, but different bonding arrangements. In the last chapter, we saw that alcohols and ethers can be constitutional isomers. In this chapter, we wrote constitutional isomers for aldehydes and ketones.

Constitutional Isomers

C_2H_6O CH_3—CH_2—OH CH_3—O—CH_3
                  Ethanol   Dimethyl ether

$$C_3H_6O \qquad CH_3-CH_2-\overset{\overset{O}{\|}}{C}-H \qquad CH_3-\overset{\overset{O}{\|}}{C}-CH_3$$
                  Propanal   Propanone

Another group of isomers called **stereoisomers** have identical molecular formulas too, but they are not constitutional isomers. In stereoisomers the atoms are bonded in the same sequence, but differ in the way they are arranged in space. Both the cis-trans isomers of cycloalkanes (Chapter 12), and the cis-trans isomers of alkenes (Chapter 13) are stereoisomers.

In this chapter, we take at look at stereoisomers called enantiomers and diastereoisomers. Enantiomers are stereoisomers that are mirror images, but cannot be superimposed on each other. Diastereoisomers are stereoisomers that are not mirror images. Many of the cis-trans stereoisomers such as *cis*- and *trans*-dimethyl-cyclopropane and *cis*- and *trans*-2-butene shown below are diastereoisomers, because the structures are not mirror images. We can summarize the types of isomers as follows:

Isomers
Compounds with the same molecular formula, but different structures

Constitutional Isomers
Isomers that differ in the order of bonding of atoms

CH_3—CH_2—OH

CH_3—O—CH_3

Stereoisomers
Isomers that have atoms bonded in the same order, but with different arrangements in space

Enantiomers
Mirror images of stereoisomers that cannot be superimposed

$$\begin{array}{ccc} \text{CHO} & & \text{CHO} \\ | & & | \\ \text{H}-\text{C}-\text{OH} & & \text{HO}-\text{C}-\text{H} \\ | & & | \\ \text{CH}_3 & & \text{CH}_3 \end{array}$$

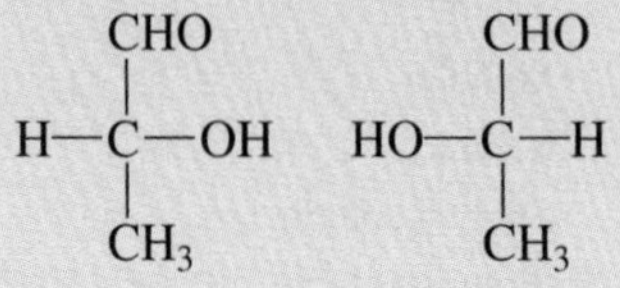

Diastereoisomers
Stereoisomers that are not mirror images

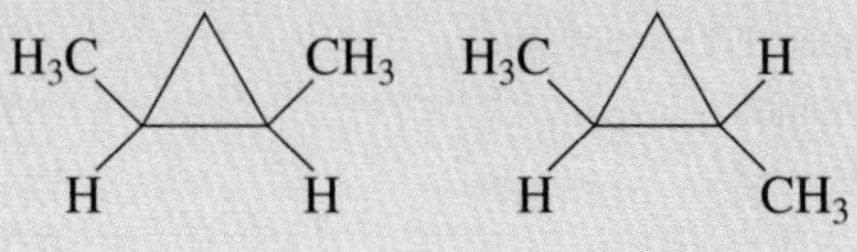

Left hand

Mirror image of right hand

Right hand

Figure 15.6 A pair of hands are chiral because they have mirror images that cannot be superimposed on each other.

Q Why are your shoes chiral objects?

Chirality

When stereoisomers have mirror images that are different, they are said to have "handedness." If you look at the palms of your hands, your thumbs are on opposite sides. If you turn your palms toward each other, you have mirror images. (See Figure 15.6.) The left hand is the mirror image of the right hand. Your hands are not superimposable. Objects such as hands that have nonsuperimposable mirror images are **chiral.** Left and right shoes are chiral; left- and right-handed golf clubs are chiral. When one mirror image can be superimposed on the other, the object is **achiral.** The object has no "handedness."

Molecules in nature also have mirror images, and often the "left-handed" stereoisomer has a different biological effect than does the "right-handed" one. For some compounds, one isomer has a certain odor, and the mirror image has a completely different odor. For example, one enantiomer of limonene smells like oranges, while its mirror image has the odor of lemons. When we think of how difficult it is to put a left-hand glove on our right hand, or a right shoe on our left foot, or use left-handed scissors if we are right handed, we begin to realize that certain properties of mirror images can be very different. (See Figure 15.7.)

Limonene

$$
\begin{array}{c}
CH_3 \\
| \\
\text{(ring structure)} \\
| \\
C \\
CH_3 \quad CH_2
\end{array}
$$

SAMPLE PROBLEM 15.7

Chiral Objects

Classify each of the following objects as chiral or achiral:

a. left ear   **b.** clear drinking glass   **c.** a glove

Solution

a. Chiral; the left ear cannot be superimposed on the right ear.

b. Achiral; mirror images of a clear drinking glass can be superimposed on each other.

c. Chiral; there is a right glove for the right hand and a left glove for the left hand.

Study Check

Would a bowling pin be chiral or achiral?

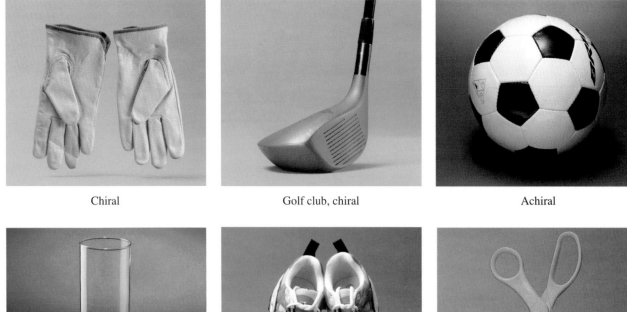

Chiral

Golf club, chiral

Achiral

Achiral

Chiral

Right-handed scissors, chiral

Figure 15.7 Everyday objects can be chiral or achiral.

Q **Why are some of the above objects chiral and others are achiral?**

Chiral Carbon Atoms

A carbon compound is chiral if it has at least one carbon atom bonded to four different atoms or groups. This type of carbon atom is called a **chiral carbon** because there are two different ways that it can bond to four atoms or groups of atoms. The resulting structures are mirror images of each other. Let's look at the mirror images of a carbon bonded to four different atoms. (See Figure 15.8.) If we line up the hydrogen and iodine atoms in the mirror images, the bromine and chorine atoms appear on opposite sides. No matter how we turn the models, we cannot align all four atoms at the same time. When stereoisomers cannot be superimposed, they are called **enantiomers.** If two or more atoms are the same, the atoms can be aligned (superimposed) and the mirror images represent the same structure. (See Figure 15.9.)

Figure 15.8 **(a)** The enantiomers of a chiral molecule are mirror images. **(b)** The enantiomers of a chiral molecule cannot be superimposed on each other.

Q **Why is the carbon atom in this compound a chiral carbon?**

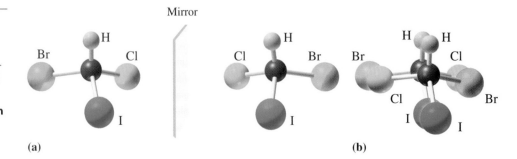

Mirror

(a)   **(b)**

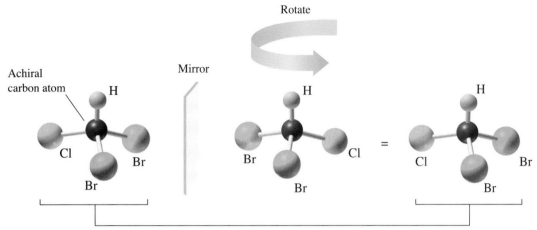

Figure 15.9 The mirror images of an achiral compound can be superimposed on each other.

Q **Why can the mirror images of the compound be superimposed?**

Chiral Carbons

Indicate whether the carbon in red is chiral or not chiral.

a. Cl—C—CH$_3$ with Cl above and H below

b. CH$_3$—C—CH$_2$—CH$_3$ with OH above and H below

c. CH$_3$—C—CH$_2$—CH$_3$ with O (double bond) above

d. CH$_3$—CH$_2$—C—C—H with CH$_3$ and O above and H below

Solution

a. Not chiral. Two of the substituents on the carbon are the same (Cl). A chiral carbon must be bonded to four different groups or atoms.

b. Chiral. Carbon 2 is bonded to four different groups: one OH, one CH$_3$, one CH$_2$—CH$_3$, and one H.

c. Not chiral. Carbon 2 is bonded to only three groups, not four.

d. Chiral. Carbon 2 is bonded to four different groups: one H, one CH$_3$, one CH$_2$—CH$_3$, and one CHO.

Study Check

Circle the two chiral carbons in the structural formula of the carbohydrate erythrose.

HO—CH$_2$—CH—CH—C—H with OH, OH, O substituents

Erythrose

Drawing Fischer Projections

Emil Fischer devised a simplified system for drawing stereoisomers that shows the arrangements of the atoms. Fischer received the Nobel Prize in 1902 for his contributions to carbohydrate and protein chemistry. Using his method called a **Fischer projection,** the bonds to a chiral atom are drawn as intersecting lines with the chiral carbon being at the center where the lines cross. The horizontal lines represent the bonds that come forward in the three-dimensional structure, and the vertical lines represent the bonds that point away.

By convention the carbon chain in the Fischer projection is written vertically with the most highly oxidized carbon at the top. For glyceraldehyde, the carbonyl group, which is the most highly oxidized group in the molecule, is written at the top. The letter L is assigned to the left-handed stereoisomer, which has the —OH group on the left of the chiral carbon. The letter D is assigned to the right-handed structure where the —OH is on the right of the chiral carbon. Let's look at how glyceraldehyde, the simplest sugar, is converted from a three-dimensional view to a Fischer projection. (See Figure 15.10.)

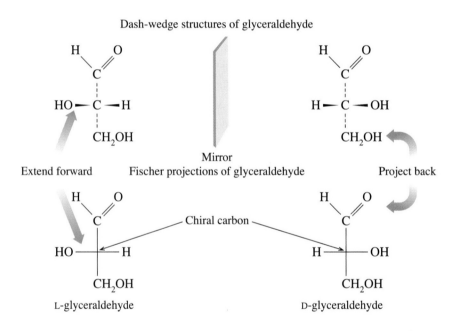

Dash-wedge structures of glyceraldehyde

Figure 15.10 In a Fischer projection, the chiral carbon atom is at the center with horizontal lines for bonds that extend toward the viewer and vertical lines for bonds that point away.

Q Why does glyceraldehyde have only one chiral carbon atom?

Fischer projections can also be written for larger compounds that have two or more chiral carbons. For example, the mirror images below are different. Each chiral atom is bonded to four different groups. To draw the mirror image of a Fischer projection, the positions of the substituents on the horizontal lines are reversed, while the groups on the vertical line are left unchanged.

We can also draw the mirror image of the carbohydrate erythrose, which has two chiral carbons.

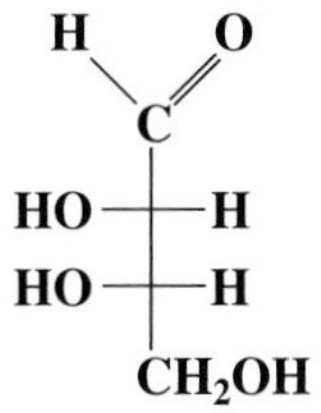

$$\begin{array}{cc}
\text{L-Erythrose} & \text{D-Erythrose}
\end{array}$$

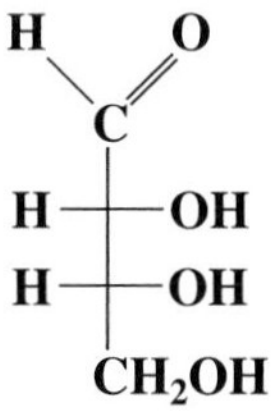

Fischer projections:

L-Erythrose:
```
    H    O
      C
HO──┼──H
HO──┼──H
    CH₂OH
```

D-Erythrose:
```
    H    O
      C
 H──┼──OH
 H──┼──OH
    CH₂OH
```

SAMPLE PROBLEM 15.9

Fischer Projections

Determine if each Fischer projection is a chiral compound. If so, identify it as the D or L isomer and draw the mirror image.

a. 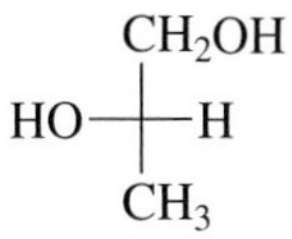
```
    CH₂OH
HO──┼──H
    CH₃
```

b. 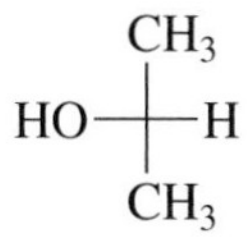
```
    CH₃
HO──┼──H
    CH₃
```

c. 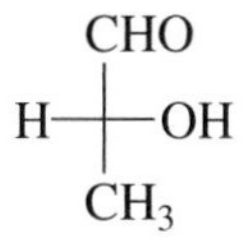
```
    CHO
 H──┼──OH
    CH₃
```

Solution

a. This chiral compound with four different substituents is the L isomer. The mirror image is written by reversing the H and OH on the horizontal lines.

```
    CH₂OH
 H──┼──OH
    CH₃
```

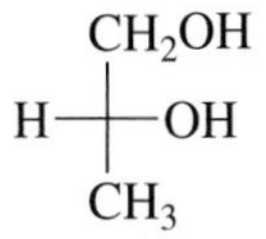

b. The compound is achiral because it has two identical groups (CH_3).

c. This chiral compound with four different substituents is the D isomer. The mirror image is written by reversing the H and OH on the horizontal lines.

```
    CHO
HO──┼──H
    CH₃
```

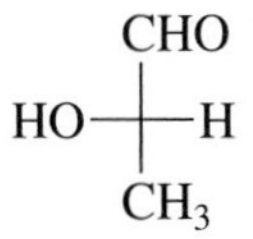

Study Check

Draw the Fischer projections for the D and L stereoisomers of 2-hydroxypropanal.

QUESTIONS AND PROBLEMS

Chiral Molecules

15.21 Identify each of the following structures as chiral or achiral. If chiral, indicate the chiral carbon.

a. 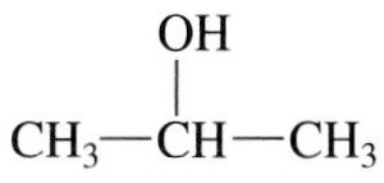
```
       OH
       │
CH₃──CH──CH₃
```

b. 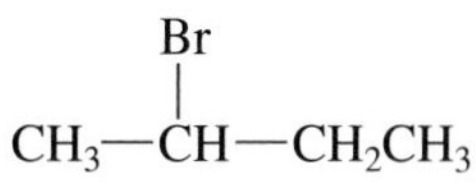
```
       Br
       │
CH₃──CH──CH₂CH₃
```

c. 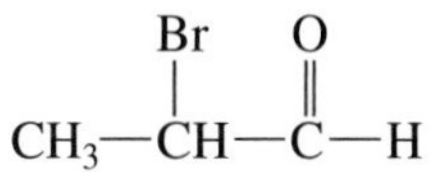
```
       Br  O
       │   ‖
CH₃──CH──C──H
```

d. 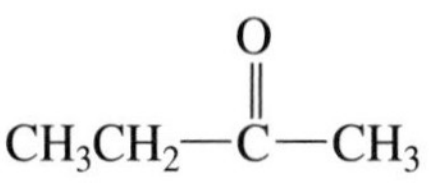
```
            O
            ‖
CH₃CH₂──C──CH₃
```

HEALTH NOTE

Enantiomers in Biological Systems

Most stereoisomers that are active in biological systems consist of only one enantiomer. Rarely are both enantiomers of biological molecules active. This happens because the enzymes and cell surface receptors on which metabolic reactions take place also have "handedness." Thus, only one of the enantiomers of a reactant or a drug interacts with its enzymes or receptors; the other is inactive. At the target site, the chiral receptor fits the arrangement of the substituents in only one enantiomer. Its mirror image, which is inactive, does not fit properly. This is similar to the idea that your right hand will only fit into a right-handed glove. (See Figure 15.11.)

A substance called carvone exists as two enantiomers. One enantiomer gives the odor of spearmint oil, while the other enantiomer produces the flavor of caraway seeds. Such differences in odor are due to the chiral receptor sites in the nose that fit the shape of one enantiomer, but not the other. Thus, our senses of smell and also taste are sensitive to the chirality of molecules.

Enantiomers of Carvone

From spearmint oil   From caraway seeds

In the brain, one stereoisomer of LSD causes hallucinations because it affects the production of serotonin, a chemical that is important in sensory perception. However, its enantiomer LSD produces little effect in the brain. The behavior of nicotine and epinephrine (adrenaline) also depends upon one of their enantiomers. One enantiomer of nicotine is more toxic than the other. In epinephrine (adrenaline), one enantiomer is responsible for the constriction of blood vessels.

Chiral carbon

Nicotine   Adrenalin (epinephrine)

A substance used to treat Parkinson's disease is L-dopa, which is converted to dopamine in the brain, where it raises the serotonin level. However, the D-dopa enantiomer has no biological effect.

L-Dopa, anti-Parkinsonian drug   D-Dopa has no biological effect

For many drugs, only one of the enantiomers is biologically active. However, for many years, drugs have been produced that were mixtures of their enantiomers. Today, drug researchers are using *chiral technology* to produce the active enantiomers of chiral drugs. Chiral catalysts are being designed that direct the formation of just one enantiomer rather than both. The benefits of producing only the active enantiomer include using a lower dose, enhancing activity, reducing interactions with other drugs, and eliminating possible harmful side effects from the nonactive enantiomer. Several active enantiomers are now being produced such as L-dopa and the active enantiomer of the popular analgesic ibuprofen used in Advil, Motrin, and Nuprin.

Ibuprofen

Figure 15.11 **(a)** The substituents on the biologically active enantiomer bind to all the sites on a chiral receptor; **(b)** its enantiomer does not bind properly and is not active biologically.

Q Why don't all the substituents of the mirror image of the active enantiomer fit into a chiral receptor site?

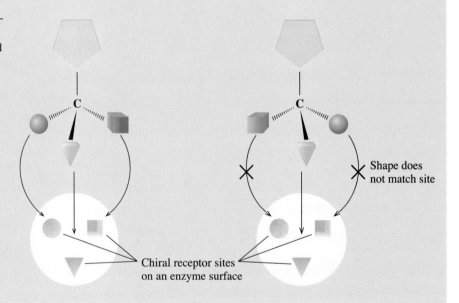

Shape does not match site

Chiral receptor sites on an enzyme surface

15.22 Identify each of the following structures as chiral or achiral. If chiral, indicate the chiral carbon.

a.
$$CH_3-\underset{\underset{CH_3}{|}}{\overset{\overset{Cl}{|}}{C}}-CH_2-\underset{\overset{Cl}{|}}{CH}-CH_3$$

b.
$$CH_3-\underset{\overset{Br}{|}}{C}=CH-CH_3$$

c.
$$CH_3-\underset{\underset{CH_3}{|}}{\overset{\overset{OH}{|}}{C}}-\underset{\overset{OH}{|}}{CH}-CH_3$$

d.
$$Br-CH_2-\underset{\overset{Cl}{|}}{CH}-CH_3$$

15.23 Identify the chiral carbon in each of the following naturally occurring compounds.

a. citronellol; one enantiomer has the geranium odor.

$$CH_3-\underset{\overset{CH_3}{|}}{C}=CH-CH_2-CH_2-\underset{\overset{CH_3}{|}}{CH}-CH_2-CH_2-OH$$

b. alanine, amino acid

$$H_2N-\underset{\overset{CH_3}{|}}{CH}-\overset{\overset{O}{\|}}{C}-OH$$

15.24 Identify the chiral carbon in each of the following naturally occurring compounds.

a. amphetamine (Benzedrine), stimulant, treatment of hyperactivity

$$\text{⬡}-CH_2-\underset{\overset{CH_3}{|}}{CH}-NH_2$$

b. Norepinephrine, increases blood pressure and nerve transmission

$$\underset{HO}{\overset{HO}{}}\text{⬡}\ \underset{\overset{OH}{|}}{CH}-CH_2-NH_2$$

15.25 Draw Fischer projections for each of the following dash-wedge structures.

a.
H
C
HO CH₃ Br

b.
CH₃
C
Cl OH Br

c.
CHO
C
HO H CH₂CH₃

15.26 Draw Fischer projections for each of the following dash-wedge structures.

a.
Br
C
HO Br CH₂OH

b.
CH₃
C
H OH CH₂OH

c.
CHO
C
HO H CH₂OH

15.27 Indicate whether each pair of Fischer projections represent enantiomers or identical structures.

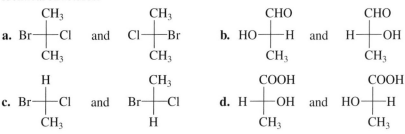

15.28 Indicate whether each pair of Fischer projections represent enantiomers or identical structures.

a.
$$\begin{array}{c} CH_2OH \\ Br{-}\!\!\!\!-\!\!\!\!-Cl \\ CH_3 \end{array}$$ and $$\begin{array}{c} CH_2OH \\ Cl{-}\!\!\!\!-\!\!\!\!-Br \\ CH_3 \end{array}$$ b. $$\begin{array}{c} CHO \\ H{-}\!\!\!\!-\!\!\!\!-H \\ CH_3 \end{array}$$ and $$\begin{array}{c} CHO \\ H{-}\!\!\!\!-\!\!\!\!-H \\ CH_3 \end{array}$$

c. $$\begin{array}{c} CH_3 \\ H{-}\!\!\!\!-\!\!\!\!-OH \\ CH_2CH_3 \end{array}$$ and $$\begin{array}{c} CH_3 \\ HO{-}\!\!\!\!-\!\!\!\!-H \\ CH_2CH_3 \end{array}$$ d. $$\begin{array}{c} COOH \\ H{-}\!\!\!\!-\!\!\!\!-NH_2 \\ CH_3 \end{array}$$ and $$\begin{array}{c} COOH \\ H_2N{-}\!\!\!\!-\!\!\!\!-H \\ CH_3 \end{array}$$

LEARNING GOAL

Draw the structural formulas of reactants and products for the oxidation or reduction of aldehydes and ketones.

15.6 Oxidation and Reduction

In Chapter 14, we saw that aldehydes produced by the oxidation of primary alcohols oxidize readily to carboxylic acids. In fact they oxidize so easily that even the aldehydes exposed to the air in the laboratory quickly form carboxylic acids. In contrast, ketones produced by the oxidation of secondary alcohols do not undergo further oxidation. Let's review examples of the oxidation reactions of primary and secondary alcohols that form aldehydes and ketones.

$$CH_3{-}CH_2{-}OH \xrightarrow{\text{Oxidation}} CH_3{-}\overset{\overset{\displaystyle O}{\|}}{C}{-}H \xrightarrow{\substack{\text{Further} \\ \text{oxidation}}} CH_3{-}\overset{\overset{\displaystyle O}{\|}}{C}{-}OH$$

Ethanol (1°)   Acetaldehyde   Acetic acid

$$CH_3{-}\overset{\overset{\displaystyle OH}{|}}{C}H{-}CH_3 \xrightarrow{\text{Oxidation}} CH_3{-}\overset{\overset{\displaystyle O}{\|}}{C}{-}CH_3 \xrightarrow{\substack{\text{Further} \\ \text{oxidation}}} \text{no reaction}$$

2-Propanol (2°)   Propanone

Tollens' Test

The ease of oxidation of aldehydes allows certain mild oxidizing agents to oxidize the aldehyde functional group without oxidizing other functional groups such as alcohols or ethers. In the laboratory, **Tollens' test** may be used to distinguish between an aldehyde and ketone. Tollens' reagent, a solution of Ag^+ ($AgNO_3$) and ammonia, oxidizes aldehydes, but not ketones. The silver ion is reduced to metallic silver, which forms a layer called a "silver mirror" on the inside of the container. Commercially, a similar process is used to make mirrors by applying a mixture of $AgNO_3$ and ammonia on glass with a spray gun. (See Figure 15.12.)

$$CH_3{-}\overset{\overset{\displaystyle O}{\|}}{C}{-}H + 2Ag^+ \xrightarrow{[O]} 2Ag(s) + CH_3{-}\overset{\overset{\displaystyle O}{\|}}{C}{-}OH$$

Acetaldehyde   Tollens reagent   Silver mirror   Acetic acid

Another test, called **Benedict's test,** gives a positive test with compounds that have an aldehyde functional group and an adjacent hydroxyl group. When Benedict's reagent containing Cu^{2+} ($CuSO_4$) ions is added to this type of aldehyde, a brick-red solid of Cu_2O forms. (See Figure 15.13.) The test is negative with simple aldehydes and ketones.

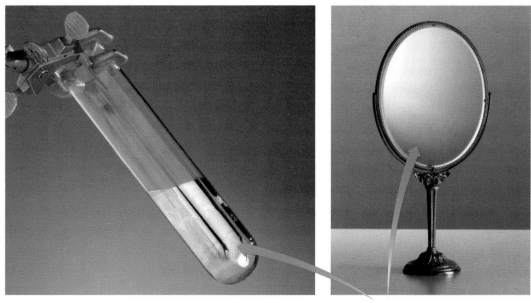

Figure 15.12 In Tollens' test, a silver mirror forms when the oxidation of an aldehyde reduces silver ion to metallic silver. The silvery surface of a mirror is formed in a similar way.

Q **What is the product of the oxidation of an aldehyde?**

$$Ag^+ + 1e^- \longrightarrow Ag(s)$$

$$\underset{\text{2-Hydroxypropanal}}{CH_3-\overset{\displaystyle OH}{\underset{|}{CH}}-\overset{\displaystyle O}{\underset{\|}{C}}-H} + \underset{\substack{\text{Benedict's}\\\text{reagent}}}{Cu^{2+}} \longrightarrow \underset{\text{Brick-red solid}}{Cu_2O(s)} + \underset{\text{2-Hydroxypropanoic acid}}{CH_3-\overset{\displaystyle OH}{\underset{|}{CH}}-\overset{\displaystyle O}{\underset{\|}{C}}-OH}$$

Because many sugars such as glucose contain this type of aldehyde grouping, Benedict's reagent can be used to determine the presence of glucose in blood or urine.

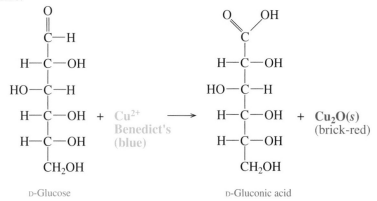

D-Glucose + Cu²⁺ Benedict's (blue) ⟶ D-Gluconic acid + Cu₂O(s) (brick-red)

Figure 15.13 The blue Cu^{2+} in Benedict's solution forms a brick-red solid of Cu_2O in a positive test for many sugars and aldehydes with adjacent hydroxyl groups.

Q **Which test tube contains an aldehyde with an adjacent hydroxyl group?**

Cu^{2+} $Cu_2O(s)$

SAMPLE PROBLEM 15.10

Alcohol Oxidation

Draw the condensed structural formula of the alcohol needed to give each of the following oxidation products:

a. $CH_3-\overset{\displaystyle O}{\underset{\|}{C}}-CH_2-CH_3$ **b.** $CH_3-\overset{\displaystyle CH_3}{\underset{|}{CH}}-\overset{\displaystyle O}{\underset{\|}{C}}-H$ **c.** $CH_3-\overset{\displaystyle O}{\underset{\|}{C}}-OH$

Solution

a. A secondary alcohol oxidizes to a ketone.

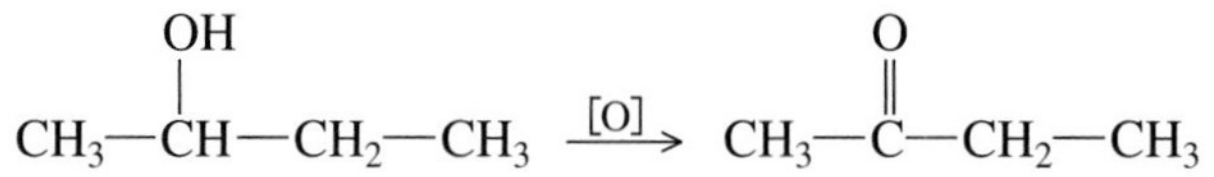

$$CH_3-\overset{\overset{\displaystyle OH}{|}}{CH}-CH_2-CH_3 \xrightarrow{[O]} CH_3-\overset{\overset{\displaystyle O}{\|}}{C}-CH_2-CH_3$$

b. A primary alcohol oxidizes to an aldehyde with a mild oxidizing agent.

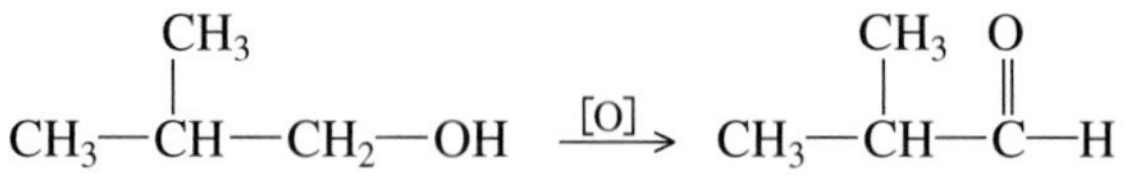

$$CH_3-\overset{\overset{\displaystyle CH_3}{|}}{CH}-CH_2-OH \xrightarrow{[O]} CH_3-\overset{\overset{\displaystyle CH_3}{|}}{CH}-\overset{\overset{\displaystyle O}{\|}}{C}-H$$

c. A primary alcohol oxidizes to an aldehyde, which oxidizes further to a carboxylic acid.

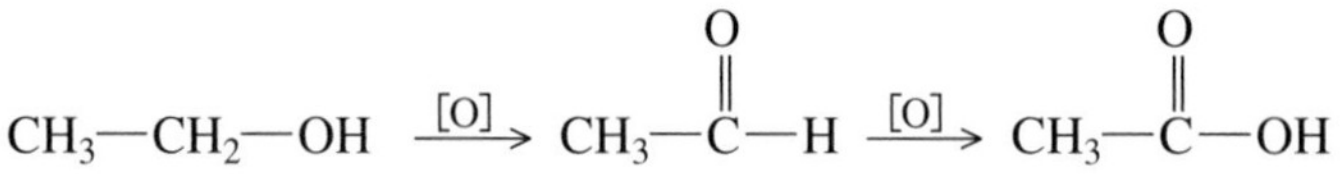

$$CH_3-CH_2-OH \xrightarrow{[O]} CH_3-\overset{\overset{\displaystyle O}{\|}}{C}-H \xrightarrow{[O]} CH_3-\overset{\overset{\displaystyle O}{\|}}{C}-OH$$

Study Check

What is the IUPAC name of the alcohol that oxidized to cyclohexanone?

SAMPLE PROBLEM 15.11

Tollens' Test

Draw the condensed structural formula of the product of oxidation, if any, when Tollens' reagent is added to each of the following compounds:

a. propanal   **b.** propanone   **c.** 2-methylbutanal

Solution

Tollens' reagent will oxidize aldehydes, but not ketones.

a. 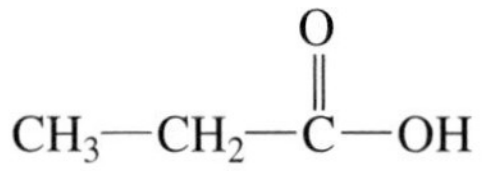 $CH_3-CH_2-\overset{\overset{\displaystyle O}{\|}}{C}-OH$ **b.** no reaction   **c.** 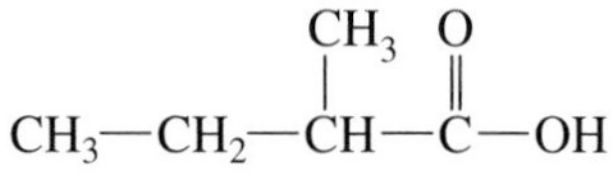 $CH_3-CH_2-\overset{\overset{\displaystyle CH_3}{|}}{CH}-\overset{\overset{\displaystyle O}{\|}}{C}-OH$

Study Check

Why does a silver mirror form when Tollens' reagent is added to a test tube containing benzaldehyde?

Reduction of Aldehydes and Ketones

Aldehydes and ketones are reduced by sodium borohydride ($NaBH_4$) or hydrogen (H_2). **Reduction** decreases the number of carbon–oxygen bonds by the addition of hydrogen or the loss of oxygen. Aldehydes are reduced to primary alcohols, and ketones to secondary alcohols. A catalyst such as nickel, platinum, or palladium is used with hydrogenation.

Aldehydes Reduce to Primary Alcohols

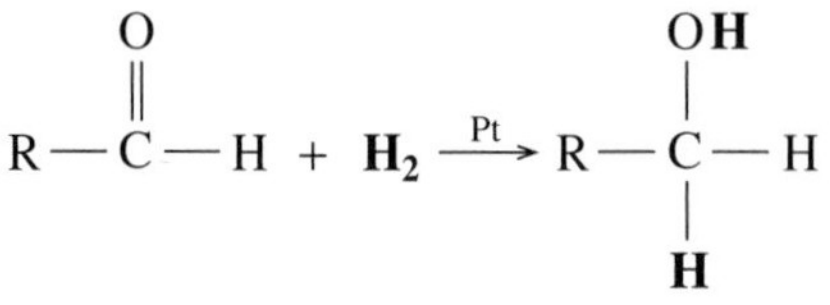

$$R-\overset{\overset{\displaystyle O}{\|}}{C}-H + H_2 \xrightarrow{Pt} R-\overset{\overset{\displaystyle O\textbf{H}}{|}}{\underset{\underset{\displaystyle \textbf{H}}{|}}{C}}-H$$

Aldehyde   1° alcohol

Ketones Reduce to Secondary Alcohols

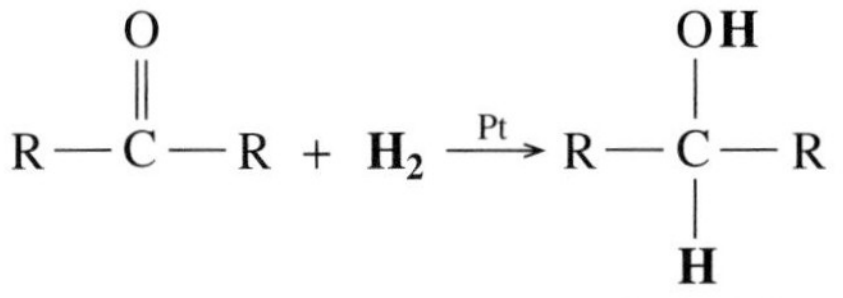

	Ketone		2° alcohol

Examples

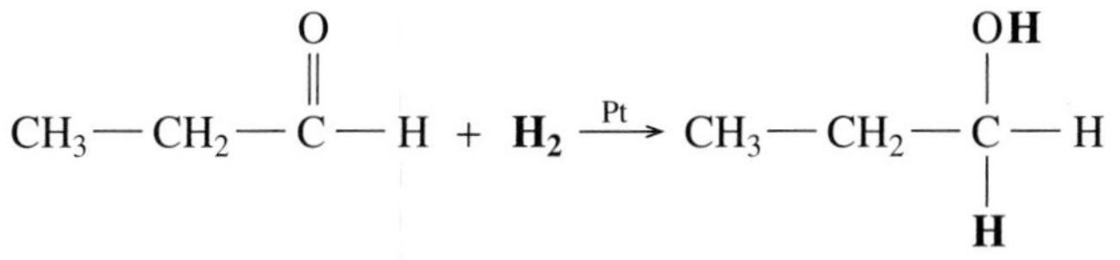

Propionaldehyde   1-Propanol (1° alcohol)

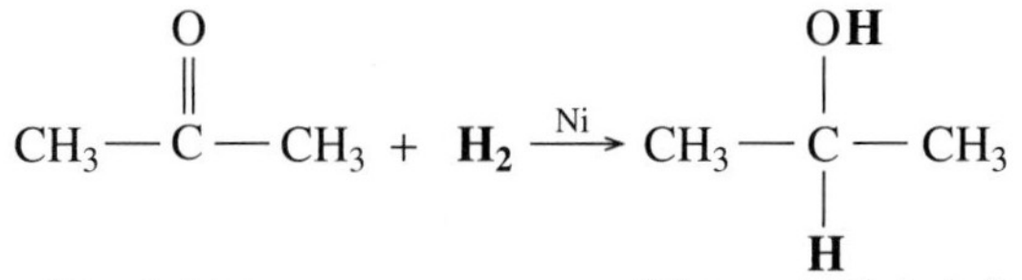

Dimethyl ketone   2-Propanol (2° alcohol)

SAMPLE PROBLEM 15.12

Reduction of Carbonyl Groups

Write an equation for the reduction of cyclopentanone in the presence of a nickel catalyst.

Solution

The reacting molecule is a cyclic ketone that has five carbon atoms. Hydrogen atoms will add to the carbon and oxygen in the carbonyl group to form the corresponding secondary alcohol.

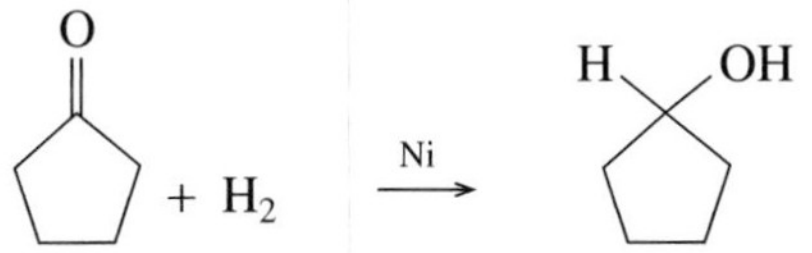

Cyclopentanone   Cyclopentanol

Study Check

What is the name of the product obtained from the hydrogenation of propionaldehyde?

QUESTIONS AND PROBLEMS

Oxidation and Reduction

15.29 Draw the condensed structural formula of the alcohol needed to give each of the following oxidation products:
   a. formaldehyde   **b.** cyclopentanone   **c.** 2-butanone
   d. benzaldehyde   **e.** 3-methylcyclohexanone

15.30 Draw the condensed structural formula of the alcohol needed to give each of the following oxidation products:
   a. acetaldehyde   **b.** 3-methylbutanone   **c.** cyclohexanone
   d. propionaldehyde   **e.** 3-methylbutanal

15.31 Draw the condensed structural formula of the organic product when each of the following alcohols is oxidized [O] (if no reaction, write *none*):

a. $CH_3CH_2CH_2CH_2CH_2OH$

b.
$$CH_3CH_2\overset{\overset{\displaystyle OH}{|}}{C}HCH_3$$

c.

d. $CH_3\!-\!\overset{\overset{\displaystyle OH}{|}}{C}H\!-\!CH_2\!-\!\overset{\overset{\displaystyle CH_3}{|}}{C}H\!-\!CH_3$

e. $CH_3\!-\!\overset{\overset{\displaystyle CH_3}{|}}{C}H\!-\!CH_2\!-\!CH_2\!-\!OH$

15.32 Draw the condensed structural formula of the organic product when each of the following alcohols is oxidized [O] (if no reaction, write *none*):

a.
CH_2OH

b.
$$CH_3\!-\!\overset{\overset{\displaystyle CH_3}{|}}{C}H\!-\!CH_2\!-\!\underset{\underset{\displaystyle CH_3}{|}}{C}H\!-\!OH$$

c. $CH_3\!-\!CH_2\!-\!\overset{\overset{\displaystyle OH}{|}}{\underset{\underset{\displaystyle CH_3}{|}}{C}}\!-\!CH_3$

d. $CH_3\overset{\overset{\displaystyle OH}{|}}{C}H\underset{\underset{\displaystyle OH}{|}}{C}HCH_2CH_3$

e.
OH

15.33 Give the condensed structural formula of the organic product formed when each of the following is reduced by hydrogen in the presence of a nickel catalyst:
  a. butyraldehyde
  b. acetone
  c. 3-bromohexanal
  d. 2-methyl-3-pentanone

15.34 Give the condensed structural formula of the organic product formed when each of the following is reduced by hydrogen in the presence of a nickel catalyst:
  a. ethyl propyl ketone
  b. formaldehyde
  c. 3-chlorocyclopentanone
  d. 2-pentanone

LEARNING GOAL

Write the products of the addition of alcohols to aldehydes and ketones.

15.7 Addition Reactions

One of the most common reactions of aldehydes and ketones is the addition of polar molecules to the carbonyl group. The carbonyl group is reactive because of the polarity of the $C\!=\!O$ double bond. In addition reactions, the negative part of the adding molecule bonds with the partially positively charged carbonyl carbon. The positive part, usually a proton, combines with the partially negatively charged carbonyl oxygen. This type of addition to the carbonyl group can be illustrated as follows:

Carbonyl group   Adding molecule
of aldehyde or
ketone

In general, aldehydes are more reactive than ketones because the carbonyl carbon is more positive in aldehydes. Also, the presence of two alkyl groups makes it more difficult for a molecule to form bonds with the carbon in the carbonyl group.

Addition of Water

The components of water add to aldehydes and ketones to give carbonyl hydrates in the presence of acid or base. The negative —OH group bonds with the carbonyl carbon, while the —H bonds to the negative oxygen. In water, the simplest aldehyde, formaldehyde, forms its hydrate called formalin, which is used to preserve tissues. Other aldehydes form hydrates in water as well, but not with as high a percentage as formaldehyde. The carbonyl group in ketones also reacts with water, but their hydrates are not very stable.

Chloral, which is an aldehyde with chlorine atoms, forms a hydrate known as chloral hydrate, the substance in "knock out" drops.

Acetal Formation

Similar to the addition of water to form hydrates, aldehydes and ketones react with alcohols in the presence of an acid catalyst to form **acetals.** (Ketal is an older term previously used for acetals from ketones.) In the acetal product, the two —OR groups are added to the carbonyl carbon and a molecule of water is eliminated. Recall that the symbol R represents an alkyl group. The general reaction is written as follows:

Examples of acetal formation

$$CH_3-\overset{\overset{\displaystyle O}{\|}}{C}-H \; + \; 2CH_3-OH \; \underset{\longleftarrow}{\overset{H^+}{\longrightarrow}} \; CH_3-\overset{\overset{\displaystyle OCH_3}{|}}{\underset{\underset{\displaystyle OCH_3}{|}}{C}}-H \; + \; H_2O$$

Acetaldehyde   Methyl alcohol   Acetaldehyde dimethyl acetal

$$CH_3-\overset{\overset{\displaystyle O}{\|}}{C}-CH_3 \; + \; 2CH_3CH_2-OH \; \underset{\longleftarrow}{\overset{H^+}{\longrightarrow}} \; CH_3-\overset{\overset{\displaystyle OCH_2CH_3}{|}}{\underset{\underset{\displaystyle OCH_2CH_3}{|}}{C}}-CH_3 \; + \; H_2O$$

Propanone   Ethyl alcohol   Propanone diethyl acetal

Cyclohexanone   + $2CH_3OH$ $\underset{\longleftarrow}{\overset{H^+}{\longrightarrow}}$ Cyclohexanone dimethyl acetal   + H_2O

Hemiacetal Intermediate

In the process of forming acetals, an intermediate called a **hemiacetal** forms when one of the two alcohol molecules adds to the carbonyl carbon. The term *hemi* indicates that the hemiacetal is halfway to an acetal. Most of the hemiacetal intermediates are unstable and difficult to isolate from the reaction mixture. In the next step, the second alcohol is added to produce the more stable acetal. Acetals are stable and can be isolated from the reaction mixture.

Examples of forming a hemiacetal intermediate

$$R-\overset{\overset{\displaystyle O}{\|}}{C}-R \; + \; R-OH \; \underset{\longleftarrow}{\overset{H^+}{\longrightarrow}} \; \left\{ R-\overset{\overset{\displaystyle OR}{|}}{\underset{\underset{\displaystyle OH}{|}}{C}}-H \right\} \; + \; R-OH \; \underset{\longleftarrow}{\overset{H^+}{\longrightarrow}} \; R-\overset{\overset{\displaystyle OR}{|}}{\underset{\underset{\displaystyle OR}{|}}{C}}-H \; + \; H_2O$$

Aldehyde   Alcohol   Hemiacetal intermediate   Acetal

$$CH_3-\overset{\overset{\displaystyle O}{\|}}{C}-H \; + \; CH_3OH \; \underset{\longleftarrow}{\overset{H^+}{\longrightarrow}} \; \left\{ CH_3-\overset{\overset{\displaystyle OCH_3}{|}}{\underset{\underset{\displaystyle OH}{|}}{C}}-H \right\} \; + \; CH_3OH \; \underset{\longleftarrow}{\overset{H^+}{\longrightarrow}} \; CH_3-\overset{\overset{\displaystyle OCH_3}{|}}{\underset{\underset{\displaystyle OCH_3}{|}}{C}}-H \; + \; H_2O$$

Acetaldehyde   Methyl alcohol   Hemiacetal intermediate   Acetaldehyde dimethyl acetal

Cyclohexanone   + CH_3OH $\underset{\longleftarrow}{\overset{H^+}{\longrightarrow}}$ $\left\{ \text{HO} \diagdown \diagup \text{OCH}_3 \text{ Hemiacetal intermediate} \right\}$ + CH_3OH $\underset{\longleftarrow}{\overset{H^+}{\longrightarrow}}$ Cyclohexanone dimethyl acetal   + H_2O

Both the step to the hemiacetal and the step to the acetal are reversible. The forward reaction to form the acetal is favored by removing water from the reaction mixture. The reverse reaction, which is the hydrolysis of an acetal, is favored by adding water to drive the equilibrium back to the ketone or aldehyde.

Identifying Addition Products

Identify each of the following structural formulas as a hemiacetal or acetal. Write the structural formulas for the carbonyl compounds and alcohols that are the reactants.

$$
\textbf{a.} \quad CH_3-\underset{\underset{H}{|}}{\overset{\overset{OCH_3}{|}}{C}}-OH
\qquad\qquad
\textbf{b.} \quad CH_3-\underset{\underset{O-CH_2CH_3}{|}}{\overset{\overset{O-CH_2CH_3}{|}}{C}}-CH_3
$$

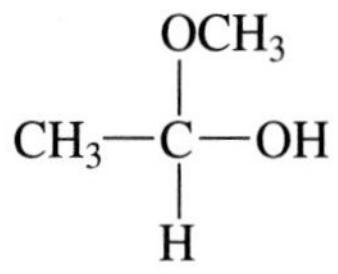

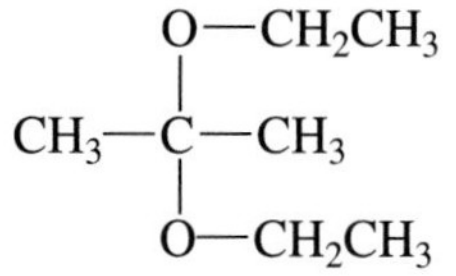

Solution

a. Hemiacetal. The reactants are CH_3—CHO and CH_3—OH.
b. Acetal. The reactants are CH_3—CO—CH_3 and CH_3CH_2—OH.

Study Check

Identify the compound below as a hemiacetal or an acetal.

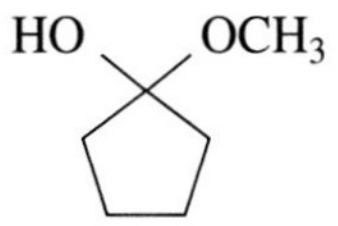

Acetals

Write the structural formula of the hemiacetal and acetal products when methanol adds to propionaldehyde.

Solution

To form the hemiacetal, the hydrogen from the alcohol adds to the oxygen of the carbonyl group to form a new hydroxyl group and the remaining part of the alcohol adds to the carbon atom in the carbonyl group. The acetal forms when a second molecule of methanol is added to the carbonyl carbon atom.

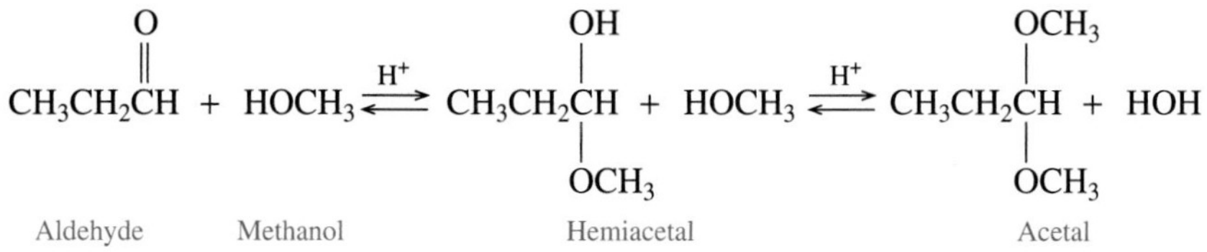

$$
\underset{\text{Aldehyde}}{CH_3CH_2\overset{\overset{O}{\|}}{C}H}
+ \underset{\text{Methanol}}{HOCH_3}
\overset{H^+}{\rightleftharpoons}
\underset{\text{Hemiacetal}}{CH_3CH_2\underset{\underset{OCH_3}{|}}{\overset{\overset{OH}{|}}{C}H}}
+ HOCH_3
\overset{H^+}{\rightleftharpoons}
\underset{\text{Acetal}}{CH_3CH_2\underset{\underset{OCH_3}{|}}{\overset{\overset{OCH_3}{|}}{C}H}}
+ HOH
$$

Study Check

What is the structural formula of the acetal produced when methanol adds to propanone?

Cyclic Hemiacetals

One very important type of hemiacetal that can be isolated is a cyclic hemiacetal that forms when the carbonyl group and the —OH group are in the *same* molecule.

Open chain   Cyclic hemiacetal

The five- and six-atom cyclic hemiacetals and acetals are more stable than their open-chain structures. For example, glucose, a simple sugar, forms a hemiacetal when the hydroxyl group on carbon 5 bonds with the carbonyl group. The hemiacetal of glucose is so stable that almost all the glucose (99%) exists as the hemiacetal in aqueous solution. We will discuss carbohydrates and their structures in Chapter 16.

Glucose   Formation of cyclic hemiacetal   New chiral carbon

An alcohol can add to the cyclic hemiacetal to form a cyclic acetal. This reaction is also very important in carbohydrate chemistry. It is the linkage used by glucose molecules to bond to other glucose molecules to form long chains.

Cyclic hemiacetal   Cyclic acetal

QUESTIONS AND PROBLEMS

Addition Reactions

15.35 Write the structural formula of the organic product formed by the addition of water to each of the following:
   a. acetaldehyde   **b.** formaldehyde

15.36 Write the structural formula of the organic product formed by the addition of water to each of the following:
   a. propanal   **b.** propanone

15.37 Indicate whether each of the following structural formulas is a hemiacetal, acetal, or neither.

a. $CH_3-CH_2-O-CH_2-OH$ **b.** $CH_3-CH_2-CH_2-\overset{\displaystyle OCH_3}{\underset{\displaystyle OH}{CH}}$ **c.** $CH_3-\overset{\displaystyle O-CH_2CH_3}{\underset{\displaystyle O-CH_2CH_3}{C}}-CH_2-CH_3$

d.   **e.**

15.38 Indicate whether each of the following structural formulas is a hemiacetal, acetal, or neither.

a. $CH_3{-}CH_2{-}O{-}CH_2{-}CH_3$ **b.** $HO{-}CH_2{-}CH_2{-}O{-}CH_2{-}CH_2{-}O{-}CH_3$

c. 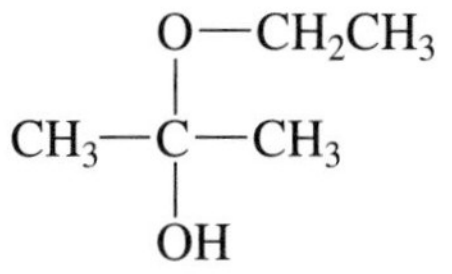
O—CH2CH3
CH3—C—CH3
OH

d. 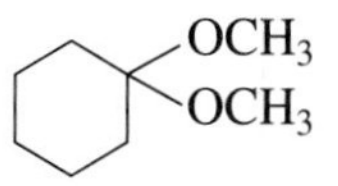
OCH3
OCH3

e. 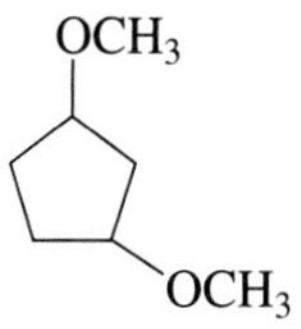
OCH3
OCH3

15.39 Draw the structural formula of the hemiacetal formed by adding methanol to each of the following compounds.

a. ethanal **b.** propanone
c. cyclopentanone **d.** butanal

15.40 Draw the structural formula of the hemiacetal formed by adding ethanol to each of the following compounds.

a. propanal **b.** 2-butanone
c. cyclohexanone **d.** formaldehyde

15.41 Draw the structural formulas of the acetal formed by adding a second methanol to the compounds in problem 15.39.

15.42 Draw the structural formulas of the acetal formed by adding a second ethanol to the compounds in problem 15.40.

Chapter Review

15.1 Structure and Bonding

Aldehydes and ketones contain a carbonyl group (C═O), which consists of a double bond between a carbon and an oxygen atom. Similar to the double bond in alkenes, the second bond is a pi bond, which forms when *p* orbitals overlap. However, in contrast to the C═C double bond, the C═O is strongly polar. In aldehydes, the carbonyl group appears at the end of carbon chains. In ketones, the carbonyl group occurs between two alkyl groups.

15.2 Naming Aldehydes and Ketones

In the IUPAC system, the *e* in the corresponding alkane is replaced with *al* for aldehydes, and *one* for ketones. For ketones with more than four carbon atoms in the main chain, the carbonyl group is numbered to show its location. Many of the simple aldehydes and ketones use common names.

15.3 Some Important Aldehydes and Ketones

Formaldehyde is the simplest aldehyde used in solution as formalin to preserve tissues as well as in the manufacture of many commercial products. Acetone is used as a solvent in paint and nail-polish removers. Many aldehydes and ketones are found in biological systems, flavorings, and drugs.

15.4 Physical Properties

Because they contain a polar carbonyl group, aldehydes and ketones have higher boiling points than alkanes and ethers. However, their boiling points are lower than alcohols because aldehydes and ketones cannot hydrogen bond with each other. Aldehydes and ketones can hydrogen bond with water molecules, which makes carbonyl compounds with one to four carbon atoms soluble in water.

15.5 Chiral Molecules

Chiral molecules are molecules with mirror images that cannot be superimposed on each other. These types of stereoisomers are called enantiomers. A chiral molecule must have at least one chiral carbon, which is a carbon bonded to four different atoms or groups of atoms. The Fischer projection is a simplified way to draw the arrangements of atoms by placing the chiral carbons at the center of crossed lines. The names of the mirror images are labeled D or L to differentiate between the enantiomers.

15.6 Oxidation and Reduction

Aldehydes are easily oxidized to carboxylic acids, but ketones do not oxidize further. Aldehydes, but not ketones, react with Tollens' reagent to give silver mirrors. In Benedict's test, aldehydes with adjacent hydroxyl groups reduce blue Cu^{2+} to give a brick-red Cu_2O solid. The reduction of aldehydes with hydrogen produces primary alcohols, while ketones are reduced to secondary alcohols.

15.7 Addition Reactions

Water and alcohols can add to the carbonyl group of aldehydes and ketones. The addition of one alcohol forms a hemiacetal, while the addition of two alcohols forms an acetal. Hemiacetals are not usually stable, except for cyclic hemiacetals, which are the most common form of simple sugars such as glucose.

Summary of Naming

Structure	Family	IUPAC Name	Common Name
	Aldehyde	Methanal	Formaldehyde
	Ketone	Propanone	Acetone; Dimethyl ketone

Summary of Reactions

Oxidation of Aldehydes to Carboxylic Acids

Reduction of Aldehydes to Primary Alcohols

Reduction of Ketones to Secondary Alcohols

Addition of Water to Aldehydes

Addition of Alcohols to Form Hemiacetals and Acetals

From aldehydes

Aldehyde   Alcohol   Hemiacetal   Acetal

Example

Formaldehyde   Methanol   Hemiacetal   Acetal

From ketones

Ketone   Alcohol   Hemiacetal   Acetal

Example

Acetone   Methanol   Hemiacetal   Acetal

Key Terms

acetal The product of the addition of two alcohols to an aldehyde or ketone.

achiral Molecules with mirror images that are superimposable.

aldehyde An organic compound with a carbonyl functional group and at least one hydrogen.

$$R-\overset{\overset{\displaystyle O}{\|}}{C}-H \ = \ R-CHO$$

Benedict's test A test for aldehydes with adjacent hydroxyl groups in which Cu^{2+} ($CuSO_4$) ions in Benedict's reagent are reduced to a brick-red solid of Cu_2O.

carbonyl group A functional group that contains a carbon–oxygen double bond ($C=O$).

chiral Objects or molecules that have mirror images that cannot be superimposed on each other.

chiral carbon A carbon atom that is bonded to four different atoms or groups of atoms.

enantiomers Stereoisomers that are mirror images that cannot be superimposed on each other.

Fischer projection A system for drawing stereoisomers that shows horizontal lines for bonds coming forward, and vertical lines for bonds going back with the chiral atom at the center.

hemiacetal The product of the addition of one alcohol to the double bond of the carbonyl group in aldehydes and ketones.

ketone An organic compound in which the carbonyl functional group is bonded to two alkyl groups.

$$R-\overset{\overset{\displaystyle O}{\|}}{C}-R \ = \ R-CO-R$$

reduction A decrease in the number of carbon–oxygen bonds by the addition of hydrogen to a carbonyl bond. Aldehydes are reduced to primary alcohols; ketones to secondary alcohols.

stereoisomers Isomers that have atoms bonded in the same order, but with different arrangements in space.

Tollens' test A test for aldehydes in which Ag^+ in Tollens' reagent is reduced to metallic silver, which forms a "silver mirror" on the walls of the container.

Additional Problems

15.43 Describe the bonds between the carbon and oxygen in the carbonyl group.

15.44 Why does the C=O double bond have a dipole, while the C=C does not?

15.45 Write the constitutional isomers for the carbonyl compounds of C_4H_8O.

15.46 Write the constitutional isomers for the carbonyl compounds of $C_5H_{10}O$.

15.47 Give the IUPAC and common names (if any) for each of the following compounds:

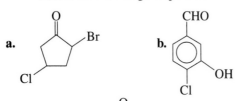

c. Cl—CH₂—CH₂—C(=O)—H

d. CH₃—CH₂—C(=O)—CH₂—CH(OH)—CH₃

e. CH₃—CH(Cl)—C(=O)—CH₂—CH₃   **f.**

15.48 Give the IUPAC and common names (if any) for each of the following compounds:

a. CH₃—CH₂—C(=O)—CH₃   **b.**

c.

d. CH₃—CH(CH₃)—CH(OH)—CH₂—C(=O)—H

e.   **f.** CH₃—C(=O)—CH₂—C(=O)—H

15.49 Draw the condensed structural formulas of each of the following:
a. 3-methylcyclopentanone
b. 4-chlorobenzaldehyde
c. 3-chloropropionaldehyde
d. ethyl methyl ketone
e. 3-methylhexanal
f. 2-heptanone, an alarm pheromone of bees

15.50 Draw the condensed structural formulas of each of the following:
a. propionaldehyde
b. 2-chlorobutanal
c. 2-methylcyclohexanone
d. 3,5-dimethylhexanal
e. 3-bromocyclopentanone
f. *trans*-2-hexenal, an alarm pheromone of an ant

15.51 Which of the following compounds are soluble in water?
a. CH₃—CH₂—CH₂—CH₃
b. CH₃—CH₂—C(=O)—H   c. CH₃—C(=O)—CH₃
d. CH₃—CH₂—CH₂—OH
e. CH₃—CH₂—C(=O)—CH₂—CH₂—CH₃

15.52 Which of the following compounds are soluble in water?
a. CH₃—CH₂—C(=O)—CH₃   b. H—C(=O)—H
c. CH₃—C(=O)—H   d. CH₃—CH₂—CH₃
e. CH₃—CH₂—CH(CH₃)—CH₂—CH₂—C(=O)—H

15.53 In each of the following pairs of compounds, select the compound with the higher boiling point:
a. CH₃—CH₂—OH or CH₃—C(=O)—H
b. CH₃—CH₂—CH₂—CH₃ or CH₃—CH₂—C(=O)—H
c. CH₃—CH₂—CH₂—OH or CH₃—C(=O)—CH₃

15.54 In each of the following pairs of compounds, select the compound with the higher boiling point:

a. $CH_3-\overset{\overset{\displaystyle O}{\|}}{C}-H$ or $CH_3-CH_2-CH_2-CH_2-\overset{\overset{\displaystyle O}{\|}}{C}-H$

b. $CH_3-CH_2-CH_2-CH_3$ or $CH_3-\overset{\overset{\displaystyle O}{\|}}{C}-CH_3$

c. $CH_3-CH_2-\overset{\overset{\displaystyle O}{\|}}{C}-H$ or $CH_3-\overset{\overset{\displaystyle OH}{|}}{CH}-CH_3$

15.55 Identify the chiral carbons, if any, in each of the following compounds.

a. $H-\overset{\overset{\displaystyle Cl}{|}}{\underset{\underset{\displaystyle Cl}{|}}{C}}-\overset{\overset{\displaystyle Cl}{|}}{\underset{\underset{\displaystyle H}{|}}{C}}-O-H$ b. $CH_3-\overset{\overset{\displaystyle H}{|}}{C}=\overset{\overset{\displaystyle CH_3}{|}}{C}-CH_3$

c. $HO-CH_2-\overset{\overset{\displaystyle OH}{|}}{CH}-CH_2-OH$

d. $CH_3-\overset{\overset{\displaystyle NH_2}{|}}{CH}-\overset{\overset{\displaystyle O}{\|}}{C}-H$

e. $CH_3-CH_2-\overset{\overset{\displaystyle Br}{|}}{CH}-CH_2-CH_2-CH_3$ f. (cyclohexanol with OH)

15.56 Identify the chiral carbons, if any, in each of the following compounds.

a. $CH_3-\overset{\overset{\displaystyle OCH_3}{|}}{CH}-CH_3$ b. $CH_3-\overset{\overset{\displaystyle OH}{|}}{CH}-\overset{\overset{\displaystyle O}{\|}}{C}-CH_3$

c. $CH_3-\overset{\overset{\displaystyle OH}{|}}{\underset{\underset{\displaystyle OH}{|}}{C}}-CH_3$ d. $CH_3-\overset{\overset{\displaystyle CH_3}{|}}{CH}-\overset{\overset{\displaystyle O}{\|}}{C}-CH_3$

e. $CH_3-\overset{\overset{\displaystyle Br}{|}}{\underset{\underset{\displaystyle OH}{|}}{C}}-CH_2-CH_3$ f. (cyclohexane with Cl at positions 1 and 4)

15.57 Identify each of the following pairs of Fischer projections as enantiomers or identical compounds:

a. $H-\overset{\overset{\displaystyle CH_2OH}{|}}{\underset{\underset{\displaystyle CH_2OH}{|}}{-}}-OH$ and $HO-\overset{\overset{\displaystyle CH_2OH}{|}}{\underset{\underset{\displaystyle CH_2OH}{|}}{-}}-H$

b. $H-\overset{\overset{\displaystyle CHO}{|}}{\underset{\underset{\displaystyle CH_2OH}{|}}{-}}-OH$ and $HO-\overset{\overset{\displaystyle CHO}{|}}{\underset{\underset{\displaystyle CH_2OH}{|}}{-}}-H$

c. $H-\overset{\overset{\displaystyle CH_3}{|}}{\underset{\underset{\displaystyle CH_2OH}{|}}{-}}-Cl$ and $H-\overset{\overset{\displaystyle CH_2OH}{|}}{\underset{\underset{\displaystyle CH_3}{|}}{-}}-Cl$

d. $H-\overset{\overset{\displaystyle OH}{|}}{\underset{\underset{\displaystyle CH_3}{|}}{-}}-OH$ and $HO-\overset{\overset{\displaystyle OH}{|}}{\underset{\underset{\displaystyle CH_3}{|}}{-}}-H$

15.58 Identify each of the following pairs of Fischer projections as enantiomers or identical compounds:

a. $H-\overset{\overset{\displaystyle CH_2CH_3}{|}}{\underset{\underset{\displaystyle CH_2OH}{|}}{-}}-Cl$ and $Cl-\overset{\overset{\displaystyle CH_2CH_3}{|}}{\underset{\underset{\displaystyle CH_2OH}{|}}{-}}-H$

b. $H-\overset{\overset{\displaystyle CH_2OH}{|}}{\underset{\underset{\displaystyle CH_2OH}{|}}{-}}-OH$ and $HO-\overset{\overset{\displaystyle CH_2OH}{|}}{\underset{\underset{\displaystyle CH_2OH}{|}}{-}}-H$

c. $H-\overset{\overset{\displaystyle CH_3}{|}}{\underset{\underset{\displaystyle CH_2OH}{|}}{-}}-Cl$ and $H-\overset{\overset{\displaystyle CH_3}{|}}{\underset{\underset{\displaystyle CH_2OH}{|}}{-}}-Cl$

d. $H-\overset{\overset{\displaystyle CHO}{|}}{\underset{\underset{\displaystyle CH_2OH}{|}}{-}}-OH$ and $HO-\overset{\overset{\displaystyle CHO}{|}}{\underset{\underset{\displaystyle CH_2OH}{|}}{-}}-H$

15.59 Draw the structural formula of the organic product when each of the following is oxidized:

a. $CH_3-CH_2-CH_2-OH$

b. $CH_3-\overset{\overset{\displaystyle OH}{|}}{CH}-CH_2-CH_2-CH_3$

c. $CH_3-CH_2-CH_2-\overset{\overset{\displaystyle O}{\|}}{C}-H$

d. (cyclohexanol with OH)

15.60 Draw the structural formula of the organic product when each of the following is oxidized:

a. 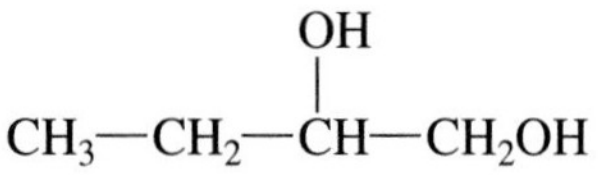 CH$_3$—CH$_2$—CH(OH)—CH$_2$OH

b. 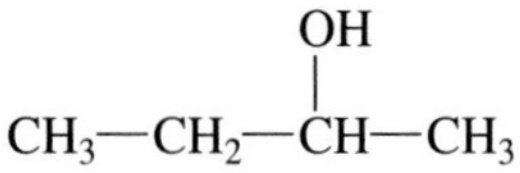 CH$_3$—CH$_2$—CH(OH)—CH$_3$

c. 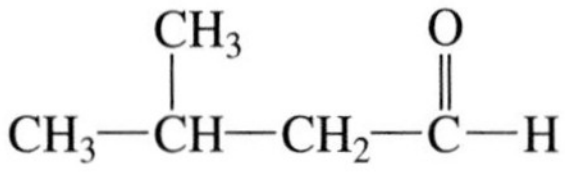 CH$_3$—CH(CH$_3$)—CH$_2$—C(=O)—H

d. 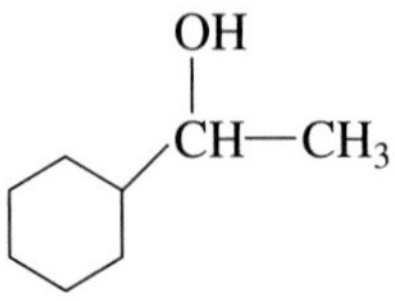 (cyclohexane ring)—CH(OH)—CH$_3$

15.61 Draw the structural formula of the organic product when hydrogen and a nickel catalyst reduce each of the following:

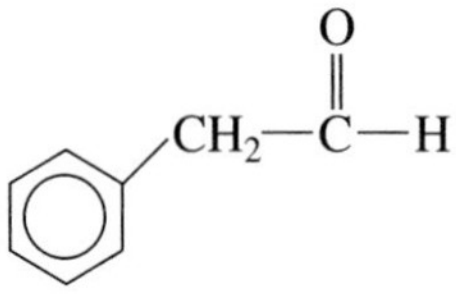

a. 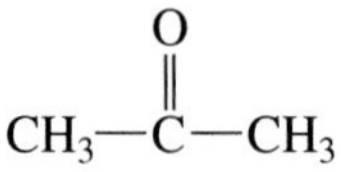 CH$_3$—C(=O)—CH$_3$

b. (benzene ring)—CH$_2$—C(=O)—H

c. CH$_3$—CH(CH$_3$)—CH$_2$—C(=O)—CH$_3$

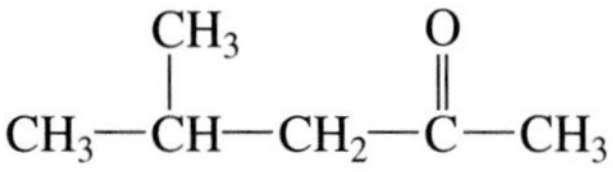

15.62 Draw the structural formula of the organic product when hydrogen and a nickel catalyst reduce each of the following:

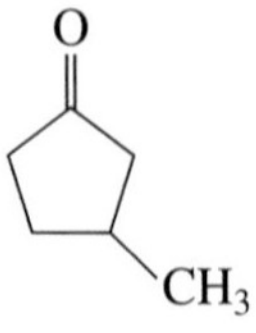

a. 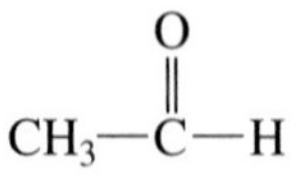 CH$_3$—C(=O)—H

b. (cyclopentanone with CH$_3$ substituent)

c. H—C(=O)—H

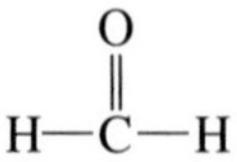

15.63 Using reactions such as dehydration, hydrogenation, oxidation, reduction, and hydration, indicate how you might prepare the following from the starting substance given:
a. propene to propanone
b. butanal to 1,2-dibromobutane
c. butanal to butanone

15.64 Using reactions such as dehydration, hydrogenation, oxidation, reduction, and hydration, indicate how you might prepare the following from the starting substance given:
a. pentanal to 1-pentene
b. 1-butanol to butanone
c. cyclohexene to cyclohexanone

15.65 Identify the following as hemiacetals or acetals. Give the names of the carbonyl compounds and alcohols used in their synthesis.

a. 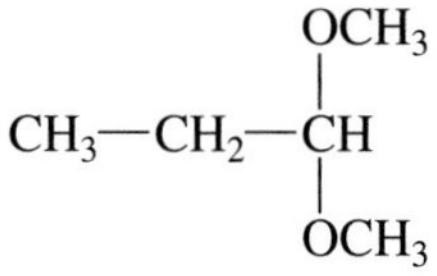 CH$_3$—CH$_2$—CH(OCH$_3$)(OCH$_3$)

b. 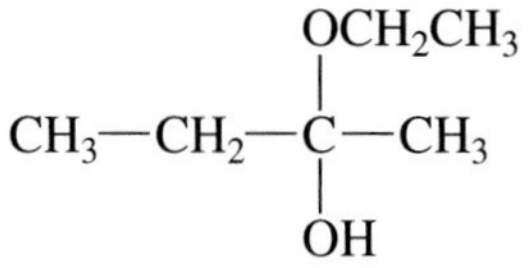 CH$_3$—CH$_2$—C(OCH$_2$CH$_3$)(OH)—CH$_3$

c. CH$_3$CH$_2$O—C(OCH$_2$CH$_3$)—(cyclohexane ring)

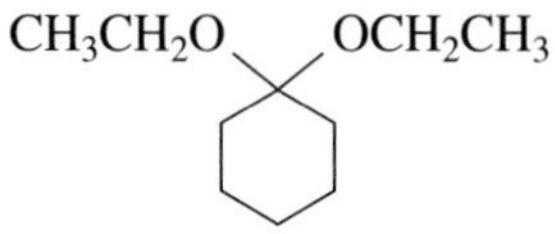

15.66 Identify the following as hemiacetals or acetals, and give the names of the carbonyl compounds and alcohols used in their synthesis.

a. 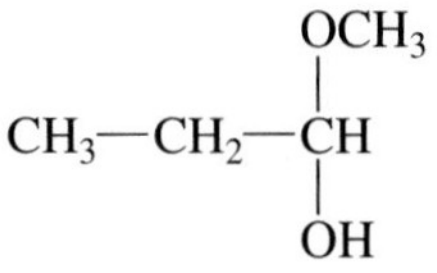 CH$_3$—CH$_2$—CH(OCH$_3$)(OH)

b. 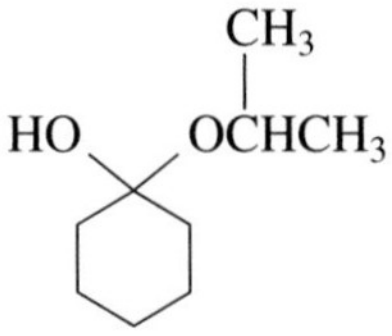 HO—C(OCHCH$_3$ with CH$_3$)—(cyclohexane ring)

c. CH$_3$—CH(OCH$_2$CH$_2$CH$_3$)(OCH$_2$CH$_2$CH$_3$)

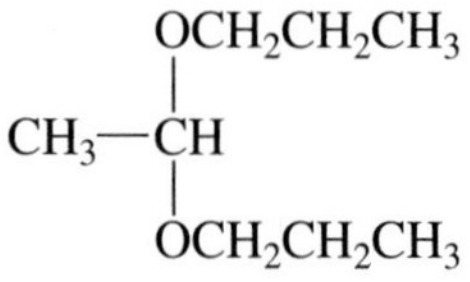

16 Carbohydrates

"We use a refractometer to measure sugar content in a small sample of juices from the grapes in different areas of the vineyard," says Leslie Bucher, laboratory director at Bouchaine Winery. "We also measure the alcohol content during fermentation and run tests for sulfur, pH, and total acid."

As grapes ripen, there is an increase in the sugars, which are the monosaccharides fructose and glucose. The sugar content is affected by soil conditions and the amount of sun and water. When the grapes are ripe and sugar content is at a desirable level, they are harvested. During fermentation, enzymes from yeast convert about half the sugar to ethanol, and half to carbon dioxide. Grapes harvested with 22.5% sugar will ferment to give a wine with 12.5–13.5% alcohol content.

LOOKING AHEAD

the Chemistry place

www.chemplace.com/college

Visit the URL above or use the CD-ROM in the book for extra quizzing, interactive tutorials, career resources, and case studies.

Of all the organic compounds in nature, carbohydrates are the most abundant. In plants, energy from the sun converts carbon dioxide and water into the carbohydrate glucose. Many of the glucose molecules are made into long-chain polymers of starch that store energy or into cellulose to build the structural framework of the plant. About 65% of the foods in our diet consist of carbohydrates. Each day you enjoy carbohydrates such as bread, pasta, potatoes, and rice. During digestion and cellular metabolism, the starches are broken down into glucose, which is oxidized further in our cells to provide our bodies with energy and to provide the cells with carbon atoms for building molecules of protein, lipids, and nucleic acids. Cellulose has other important uses, too. The wood in our furniture, the pages in this book, and the cotton in our clothing are made of cellulose.

Classify carbohydrates as monosaccharides, disaccharides, and polysaccharides.

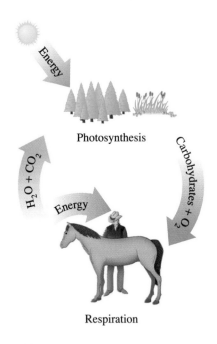

Figure 16.1 During photosynthesis, energy from the sun combines CO_2 and H_2O to form glucose $C_6H_{12}O_6$ and O_2. During respiration in the body, carbohydrates are oxidized to CO_2 and H_2O, while energy is produced.

Q What are the reactants and products of respiration?

16.1 Types of Carbohydrates

Carbohydrates such as table sugar, lactose in milk, and cellulose are all made of carbon, hydrogen, and oxygen. A carbohydrate is also called a **saccharide**, a word that comes from the Latin term *saccharum,* "sugar." Simple sugars, which have formulas of $C_n(H_2O)_n$, were once thought to be hydrates of carbon, thus the name *carbohydrate*. In a series of reactions called photosynthesis, energy from the sun is used to combine the carbon atoms from carbon dioxide (CO_2) and the hydrogen and oxygen atoms of water into the carbohydrate glucose.

$$6CO_2 + 6H_2O + \text{energy} \underset{\text{Respiration}}{\overset{\text{Photosynthesis}}{\rightleftharpoons}} \underset{\text{Glucose}}{C_6H_{12}O_6} + 6O_2$$

In our body tissues, glucose is oxidized in a series of metabolic reactions known as respiration, which releases chemical energy to do work in the cells. Carbon dioxide and water are produced and returned to the atmosphere. The combination of photosynthesis and respiration is called the carbon cycle, in which energy from the sun is stored in plants by photosynthesis and made available to us when the carbohydrates in our diets are metabolized. (See Figure 16.1.)

Types of Carbohydrates

The simplest carbohydrates are the **monosaccharides.** A monosaccharide cannot be split or hydrolyzed into smaller carbohydrates. One of the most common carbohydrates, glucose, $C_6H_{12}O_6$, is a monosaccharide. **Disaccharides** consist of two monosaccharide units joined together. A disaccharide can be split into two monosaccharide units. For example, ordinary table sugar, sucrose, $C_{12}H_{22}O_{11}$, is a disaccharide that can be hydrolyzed in the presence of an acid or an enzyme to give one molecule of glucose and one molecule of another monosaccharide, fructose.

$$\underset{\text{Sucrose}}{C_{12}H_{22}O_{11}} + H_2O \xrightarrow{\text{H}^+ \text{ or enzyme}} \underset{\text{Glucose}}{C_6H_{12}O_6} + \underset{\text{Fructose}}{C_6H_{12}O_6}$$

Polysaccharides are carbohydrates that are naturally occurring polymers containing many monosaccharide units. In the presence of an acid or an enzyme, a polysaccharide can be completely hydrolyzed to yield many molecules of monosaccharide.

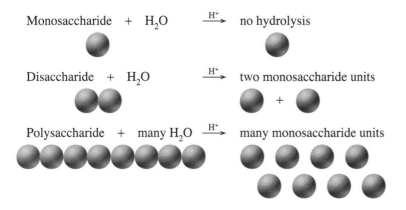

Monosaccharide + H₂O $\xrightarrow{H^+}$ no hydrolysis

Disaccharide + H₂O $\xrightarrow{H^+}$ two monosaccharide units

Polysaccharide + many H₂O $\xrightarrow{H^+}$ many monosaccharide units

---

SAMPLE PROBLEM 16.1

Types of Carbohydrates

Classify the following carbohydrates as mono-, di-, or polysaccharides:

a. When lactose, milk sugar, is hydrolyzed, two monosaccharide units are produced.

b. Cellulose, a carbohydrate in cotton, yields thousands of monosaccharide units when completely hydrolyzed.

Solution

a. A disaccharide contains two monosaccharide units.

b. A polysaccharide contains many monosaccharide units.

Study Check

Fructose found in fruits does not undergo hydrolysis. What type of carbohydrate is it?

QUESTIONS AND PROBLEMS

Classification of Carbohydrates

16.1 What reactants are needed for photosynthesis and respiration?

16.2 What is the relationship between photosynthesis and respiration?

16.3 What is a monosaccharide? A disaccharide?

16.4 What is a polysaccharide?

16.2 Classification of Monosaccharides

Monosaccharides are simple sugars that have an unbranched chain of three to eight carbon atoms, one of them in a carbonyl group and the rest attached to hydroxyl groups. There are two types of monosaccharide structures. In an **aldose,** the carbonyl group is on the first carbon (—CHO); a **ketose** contains the carbonyl group on the second carbon atom as a ketone (C═O).

WEB TUTORIAL
Carbohydrates

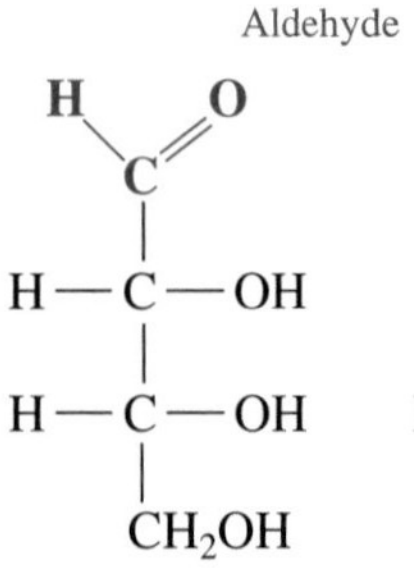

Aldehyde

Erythrose, a
polyhydroxy
aldehyde

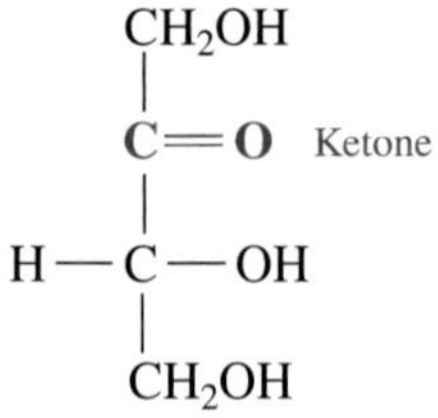

Erythrulose, a
polyhydroxy
ketone

A monosaccharide with three carbon atoms is a *triose*, one with four carbon atoms is a *tetrose*, a *pentose* has five carbons, and a *hexose* contains six carbons. We can use both classification systems to indicate the type of carbonyl group and the number of carbon atoms. An aldopentose is a five-carbon monosaccharide that is an aldehyde; a ketohexose would be a six-carbon monosaccharide that is a ketone. Some examples:

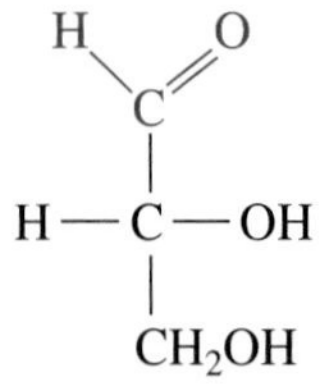

Glyceraldehyde
(aldotriose)

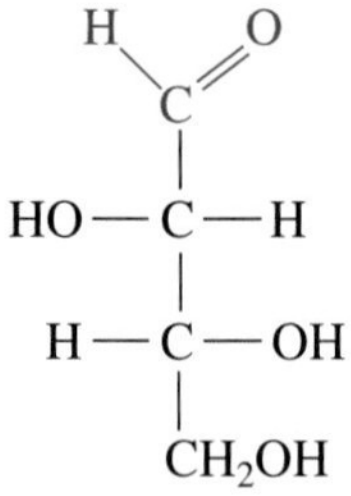

Threose
(aldotetrose)

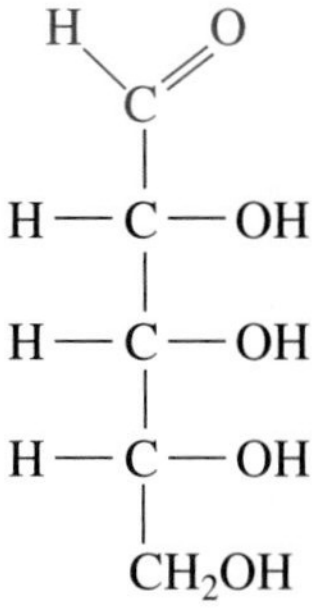

Ribose
(aldopentose)

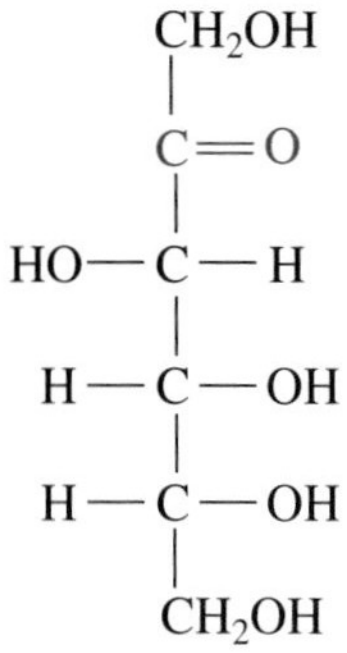

Fructose
(ketohexose)

SAMPLE PROBLEM 16.2

Monosaccharides

Classify each of the following monosaccharides to indicate their carbonyl group and number of carbon atoms:

a.

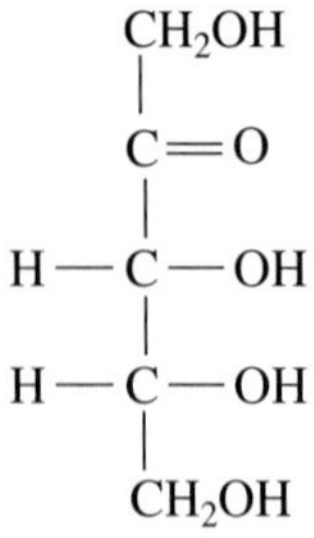

Ribulose

b.

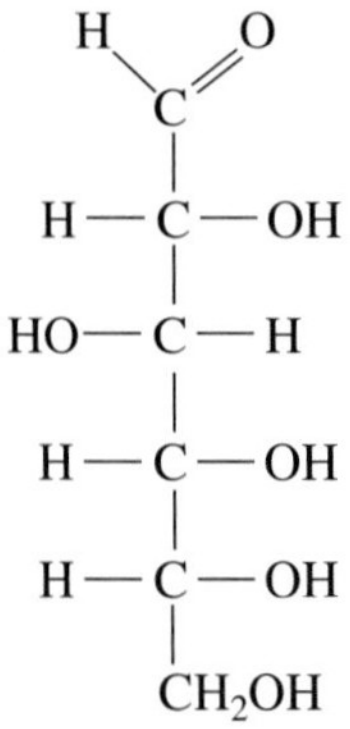

Glucose

Solution

a. The structural formula has a ketone group; ribulose is a ketose. Because there are five carbon atoms, it is a pentose. Combining these classifications makes it a ketopentose.

b. The structural formula has an aldehyde group; glucose is an aldose. Because there are six carbon atoms, it is an aldohexose.

Study Check

The simplest ketose is a triose named dihydroxyacetone. Draw its structural formula.

QUESTIONS AND PROBLEMS

Classification of Monosaccharides

16.5 What functional groups are found in all monosaccharides?

16.6 What is the difference between an aldose and a ketose?

16.7 What are the functional groups and number of carbons in a ketopentose?

16.8 What are the functional groups and number of carbons in an aldohexose?

16.9 Classify each of the following monosaccharides as an aldose or ketose.

a.
```
        CH₂OH
         |
         C=O
         |
   HO — C — H
         |
    H — C — OH
         |
    H — C — OH
         |
        CH₂OH
```
Fructose

b.
```
        CHO
         |
    H — C — OH
         |
    H — C — OH
         |
    H — C — OH
         |
        CH₂OH
```
Ribose

c.
```
        CH₂OH
         |
         C=O
         |
        CH₂OH
```
Dihydroxyacetone

d.
```
        CHO
         |
    H — C — OH
         |
   HO — C — H
         |
    H — C — OH
         |
        CH₂OH
```
Xylose

e.
```
        CHO
         |
    H — C — OH
         |
   HO — C — H
         |
   HO — C — H
         |
    H — C — OH
         |
        CH₂OH
```
Galactose

16.10 Classify each of the monosaccharides in problem 16.9 according to the number of carbon atoms in the chain.

16.3 D and L Notations from Fischer Projections

WEB TUTORIAL
Forms of Carbohydrates

In Chapter 15 we learned that compounds with chiral carbons can exist in forms that are mirror images of each other and cannot be superimposed. The monosaccharides are chiral because they contain one or more chiral carbons in their carbon chains. Thus a monosaccharide can exist as either of two molecular forms that are mirror images of each other. The Fischer projections that represent enantiomers are also used for sugars, particularly monosaccharides.

Let's take a look again at the Fischer projection for the simplest sugar, glyceraldehyde. By convention the carbon chain is written vertically with the aldehyde group (most oxidized carbon) at the top and the —CH₂OH group at the bottom. The center carbon is the chiral carbon to which four different groups are attached. The letter L is assigned to the stereoisomer if the —OH group is on the left of the chiral carbon. In D-glyceraldehyde, the —OH is on the right.

CHO
HO———H
CH₂OH
L-Glyceraldehyde

CHO
H———OH
CH₂OH
D-Glyceraldehyde

In an aldotetrose, there are two chiral carbon atoms, carbon 2 and 3. Then the bottom chiral atom, which is the chiral atom farthest from the carbonyl group, is used to assign the D or L configuration. The very bottom carbon atom in the Fischer projection of a carbohydrate — CH₂OH, is not chiral because it does not have four different groups bonded to it.

CHO
HO———H
HO———H
CH₂OH
L-Erythrose

CHO
H———OH
H———OH
CH₂OH
D-Erythrose

CHO
H———OH
HO———H
CH₂OH
L-Threose

CHO
HO———H
H———OH
CH₂OH
D-Threose

Most of the carbohydrates we will study have carbon chains with five or six carbon atoms, which means that they have several chiral carbons. However, we still use the chiral carbon furthest from the carbonyl group to determine the D or L isomer. Most of the naturally occurring sugars are the D isomers. The following are the isomers of ribose, which is a five-carbon monosaccharide, and glucose, a six-carbon monosaccharide.

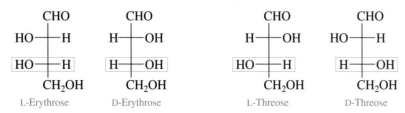

CHO
HO———H
HO———H
HO———H
CH₂OH
L-Ribose

CHO
H———OH
H———OH
H———OH
CH₂OH
D-Ribose

CHO
HO———H
H———OH
HO———H
HO———H
CH₂OH
L-Glucose

CHO
H———OH
HO———H
H———OH
H———OH
CH₂OH
D-Glucose

SAMPLE PROBLEM 16.3

Identifying D and L Isomers of Sugars

Is the following structure the D or L enantiomer of ribose?

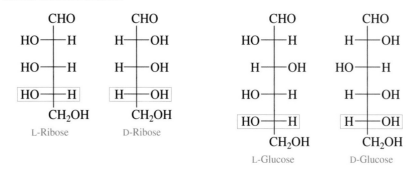

Solution

In ribose, carbon 4 is the chiral atom furthest from the carbonyl group. Because the hydroxyl group on carbon 4 is on the left, this enantiomer is L-ribose.

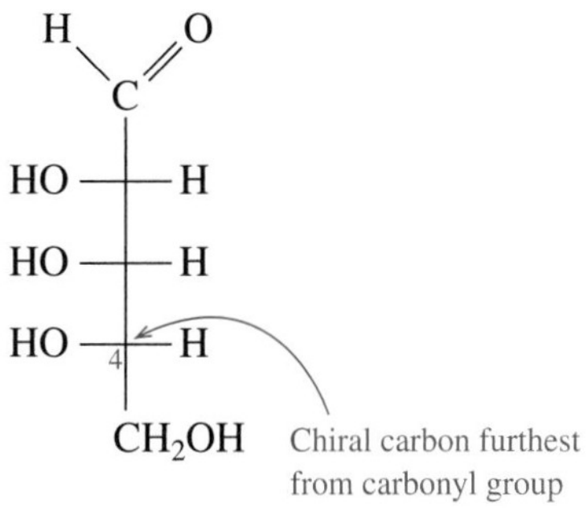

Chiral carbon furthest from carbonyl group

Study Check

Draw the Fischer projection for D-ribose.

QUESTIONS AND PROBLEMS

D and L Notations from Fischer Projections

16.11 What is a Fischer projection?

16.12 Write the Fischer projection formula for D-glyceraldehyde and L-glyceraldehyde.

16.13 State whether each of the following sugars is the D or L isomer:

a. Threose

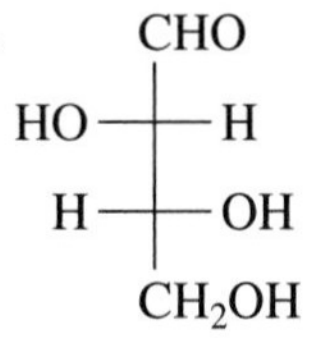

b. Xylulose

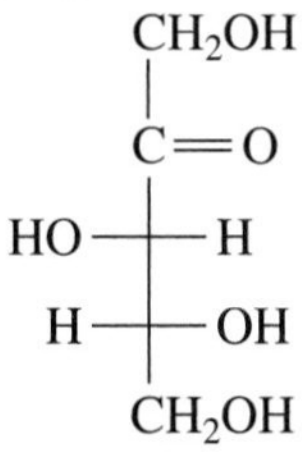

c. Mannose

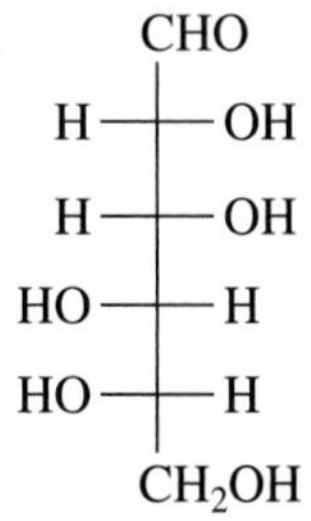

d. Allose

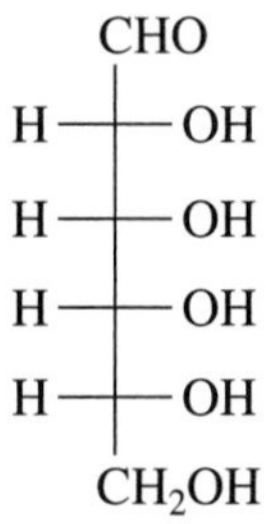

16.14 State whether each of the following sugars is the D or L isomer:

a. Ribulose

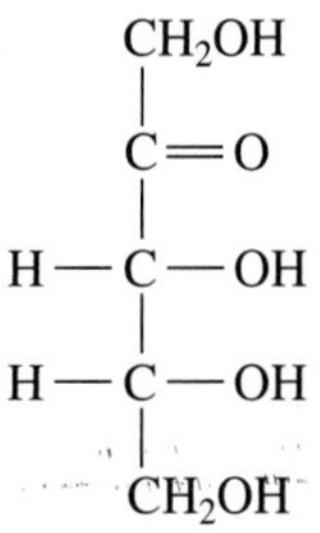

b. Sorbose

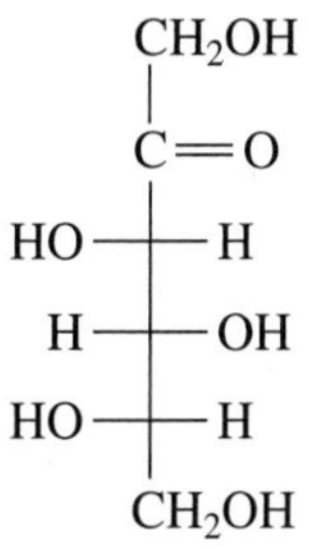

c. Glucose

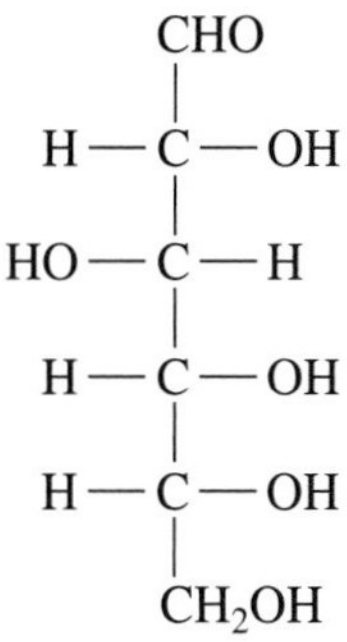

d. Ribose

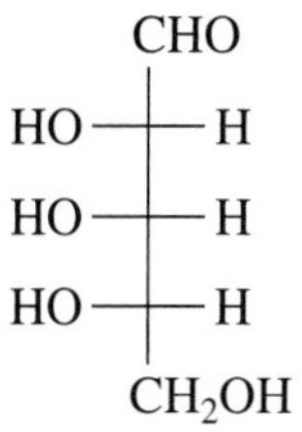

16.15   Write the enantiomers for a–d in problem 16.13.

16.16   Write the enantiomers for a–d in problem 16.14.

LEARNING GOAL

Draw the open-chain
structures for D-glucose,
D-galactose, and D-fructose.

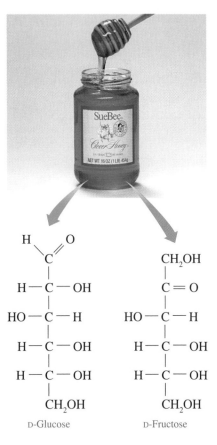

Figure 16.2 The sweet taste of honey is
due to the monosaccharides of glucose and
fructose.

**Q What are some differences in the
structures of glucose and fructose?**

WEB TUTORIAL
Forms of Carbohydrates

16.4 Structures of Some Important Monosaccharides

The hexoses glucose, galactose, and fructose are important monosaccharides. Although structural formulas for both the D and L isomers can be written for these hexoses, the D isomers are the more common form of carbohydrates found in nature. (See Figure 16.2.) We can write their open-chain structures as follows:

D-Glucose

The most common hexose is D-glucose, which has the molecular formula $C_6H_{12}O_6$. Also known as dextrose, **glucose** is found in fruits, vegetables, corn syrup, and honey. It is a building block of the disaccharides sucrose, lactose, and maltose and is the monosaccharide unit in polysaccharides such as starch, cellulose, and glycogen.

In the body, glucose normally occurs at a concentration of 70–90 mg/dL (1 dL = 100 mL) of blood. However, the amount of glucose depends on the time that has passed since eating. In the first hour after a meal, the level of glucose rises to about 130 mg/dL of blood, and then decreases over the next 2–3 hours as it is used in the tissues. Some glucose is converted to glycogen and stored in the liver and muscle. When the amount of glucose exceeds what is needed for energy or glycogen, the excess glucose is converted to fat, which can be stored in unlimited amounts.

Glycogen (liver and muscle)

Fat ← Excess — Glucose → Metabolism

Excess ↓

Urine $CO_2 + H_2O + energy$

Hyperglycemia and Hypoglycemia

A doctor may order a glucose tolerance test to evaluate the body's ability to return to normal glucose concentration in response to the ingestion of a specified amount of glucose. The patient fasts for 12 hours and then drinks a solution containing glucose. A blood sample is taken immediately, followed by more blood samples each half-hour for 2 hours, and then every hour for a total of 5 hours. If the blood glucose exceeds 140 mg/dL in plasma and remains high, hyperglycemia may be indicated. The term *glyc* or *gluco* refers to "sugar." The prefix *hyper* means above or over, and *hypo* is below or under. Thus the blood sugar level in *hyperglycemia* is above normal and below normal in *hypoglycemia*.

An example of a disease that can cause hyperglycemia is diabetes mellitus, which occurs when the pancreas is unable to produce sufficient quantities of insulin. As a result, glucose levels in the body fluids can rise as high as 350 mg/dL plasma. Symptoms of diabetes in people under the age of 40 include thirst, excessive urination, increased appetite, and weight loss. In older persons, diabetes is sometimes a consequence of excessive weight gain.

When a person is hypoglycemic, the blood glucose level rises and then decreases rapidly to levels as low as 40 mg/dL plasma. In some cases, hypoglycemia is caused by overproduction of insulin by the pancreas. Low blood glucose can cause dizziness, general weakness, and muscle tremors. A diet may be prescribed that consists of several small meals high in protein and low in carbohydrate. Some hypoglycemic patients are finding success with diets that include more complex carbohydrates rather than simple sugars.

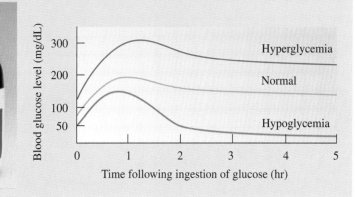

D-Galactose

Galactose is an aldohexose that does not occur in the free form in nature. It is obtained as a hydrolysis product of the disaccharide lactose, a sugar found in milk and milk products. Galactose is important in the cellular membranes of the brain and nervous system. The only difference in the structures of D-glucose and D-galactose is the arrangement of the —OH group on carbon 4.

In a condition called *galactosemia,* an enzyme needed to convert galactose to glucose is missing. The accumulation of galactose in the blood and tissues can lead to cataracts, mental retardation, and cirrhosis. The treatment for galactosemia is the removal of all galactose-containing foods, mainly milk and milk products, from the diet. If this is done for an infant immediately after birth, the damaging effects of galactose accumulation can be avoided.

D-**Fructose**

In contrast to glucose and galactose, **fructose** is a ketohexose. The structure of fructose differs from glucose at carbons 1 and 2 by the location of the carbonyl group.

D-Glucose   D-Fructose

Fructose is the sweetest of the carbohydrates, twice as sweet as sucrose (table sugar). This makes fructose popular with dieters because less fructose, and therefore fewer calories, are needed to provide a pleasant taste. After fructose enters the bloodstream, it is converted to its isomer, glucose. Fructose is found in fruit juices and honey; it is also called levulose and fruit sugar. Fructose is also obtained as one of the hydrolysis products of sucrose, the disaccharide known as table sugar.

SAMPLE PROBLEM 16.4

Monosaccharides

Ribulose has the following open-chain structure.

a. Identify the above compound D- or L-ribulose.
b. Write the open-chain structural formula of its enantiomer.

Solution

a. D-ribulose
b. The enantiomer, L-ribulose, has the following structural formula:

Study Check

What type of carbohydrate is ribulose?

QUESTIONS AND PROBLEMS

Structures of Some Important Monosaccharides

16.17 Draw the open-chain structure of D-glucose and L-glucose.

16.18 Draw the open-chain structure of D-fructose and L-fructose.

16.19 How does the open-chain structure of D-galactose differ from D-glucose?

16.20 How does the open-chain structure of D-fructose differ from D-glucose?

16.21 Identify a monosaccharide that fits each of the following descriptions:
  a. also called blood sugar
  b. not metabolized in galactosemia
  c. also called fruit sugar

16.22 Identify a monosaccharide that fits each of the following descriptions:
  a. high blood levels in diabetes
  b. obtained as a hydrolysis product of lactose
  c. the sweetest of the monosaccharides

16.5 Cyclic Structures of Monosaccharides

In Chapter 15, we saw that an aldehyde group reacts with one alcohol molecule to form a hemiacetal. For example, acetaldehyde reacts with methanol to form the following hemiacetal. In the product, the carbonyl carbon is bonded by an ether link to the alkyl group and to a new —OH group.

the Chemistry place

WEB TUTORIAL
Forms of Carbohydrates

$$CH_3-\overset{\overset{\displaystyle O}{\|}}{C}-H + CH_3-OH \rightleftharpoons CH_3-\overset{\overset{\displaystyle O-CH_3}{|}}{\underset{\underset{\displaystyle OH}{|}}{C}}-H$$

This same reaction occurs when a carbonyl group and —OH group are in the *same* molecule. The product, called a *cyclic hemiacetal,* forms a ring structure that is the most stable form of aldopentoses and aldohexoses. In the following general diagram, the hydroxyl group on carbon 5 bonds with the carbonyl carbon 1 to produce a heterocyclic six-atom ring containing an oxygen atom and a new —OH group on carbon 1.

Open chain   Heterocyclic hemiacetal

Drawing Haworth Structures for Cyclic Forms

While the carbonyl group in an aldohexose could react with several of the —OH groups, the equilibrium for aldohexoses favors the formation of six-atom rings. Let's look at how we draw the cyclic hemiacetal for D-glucose starting with the Fischer projection.

Step 1   Think of turning the open chain of glucose clockwise to the right. Then the —OH groups written on the right, other than the one on carbon 5, are drawn down, and the —OH groups on the left are up.

D-Glucose (open chain)

Step 2   Rotate the groups around carbon 5 placing the —CH₂OH up, and the —OH group close to the carbonyl carbon 1. Form the cyclic hemiacetal by bonding the oxygen in the —OH group to the carbonyl carbon. This way of depicting a cyclic hemiacetal structure is known as a **Haworth structure**.

Carbon-5 oxygen bonds to carbonyl   Cyclic hemiacetal

Step 3   In the cyclic hemiacetal, carbon 1 is now a chiral carbon bonded to a new —OH group. There are two ways to place the —OH, either up or down, which gives two stereoisomers called **anomers**. The —OH group on carbon 1 is down in the α (alpha) anomer and up in the β (beta) anomer.

Such differences in structural forms may seem trivial. However, we can digest starch products such as pasta to obtain glucose because the polysaccharide contains the α isomers of glucose. We cannot digest paper or wood because cellulose consists of only β-D-glucose units. Humans have an α-amylase, an enzyme needed for the digestion of starches, but not a β-amylase for the digestion of cellulose.

Sometimes, the cyclic structure is simplified to show only the position of the hydroxyl groups on the six-atom ring structure.

α-D-Glucose (simplified structure) β-D-Glucose

Mutarotation

In solution, the α-D-glucose is in equilibrium with β-D-glucose. In a process called **mutarotation,** each isomer converts from the closed ring to the open chain and back again. As the ring opens and closes, the bond between carbons 1 and 2 can rotate, which allows the hydroxyl (——OH) group on carbon 1 to shift between the α and the β position. Although the open chain is an essential part of mutarotation, only a small amount of open chain is present at any given time.

Haworth Structures for α- and β-D-Glucose

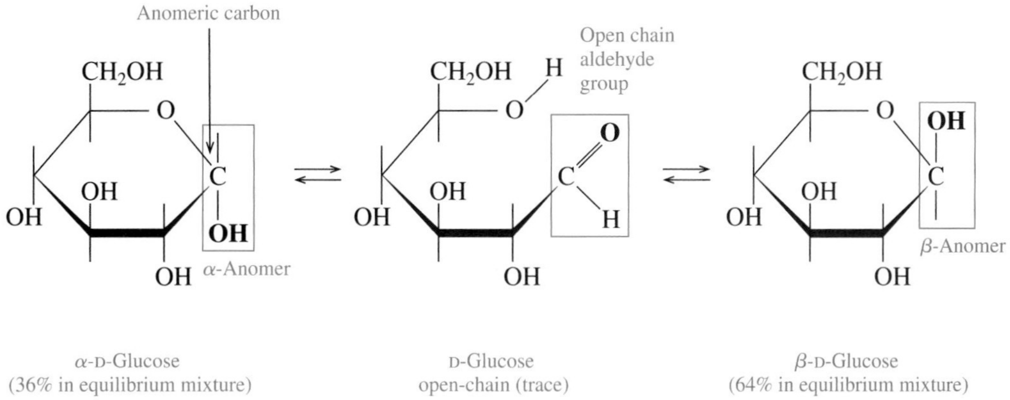

α-D-Glucose
(36% in equilibrium mixture)

D-Glucose
open-chain (trace)

β-D-Glucose
(64% in equilibrium mixture)

Cyclic Structures of Galactose

Galactose is an aldohexose like glucose, differing only in the arrangement of the ——OH group on carbon 4. Thus, its cyclic structure is also similar to glucose, except that in galactose the ——OH on carbon 4 is up. With the formation of a new hydroxyl group on carbon 1, galactose also exists as α and β anomers and undergoes mutarotation via the open-chain form in solution.

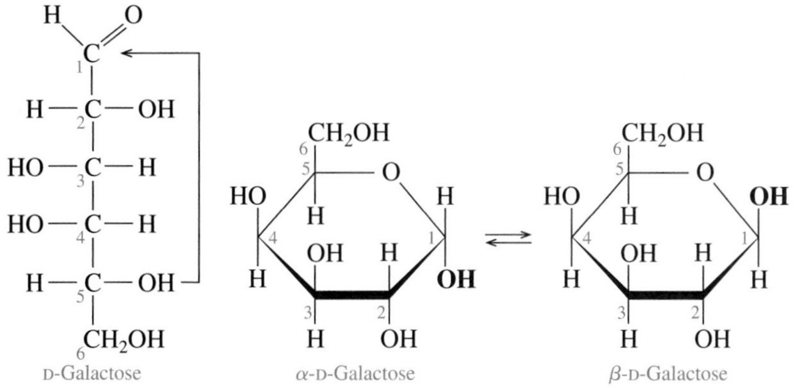

D-Galactose

α-D-Galactose

β-D-Galactose

Cyclic Structures of Fructose

In contrast to glucose and galactose, fructose is a ketohexose. It forms a hemiacetal when a hydroxyl group on carbon 5 reacts with the ketone group. The cyclic structure for fructose is a five-atom ring with carbon 2 at the right corner. A new hydroxyl group is on carbon 2 in addition to the carbon 1 from the CH_2OH group. There are also α and β anomers of fructose that undergo mutarotation in solution.

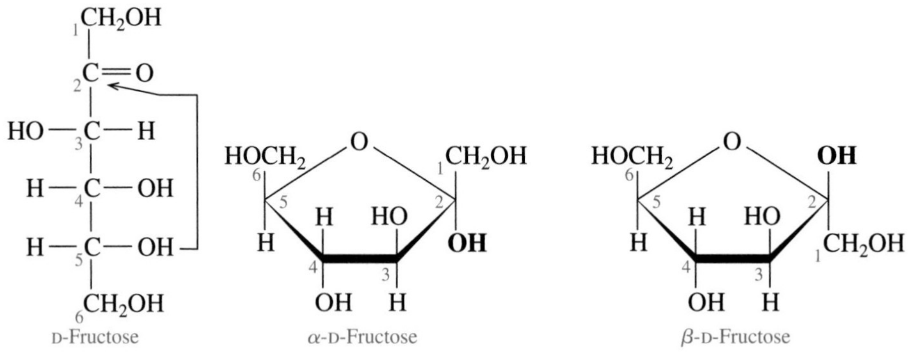
D-Fructose
α-D-Fructose
β-D-Fructose

SAMPLE PROBLEM 16.5

Drawing Cyclic Structures for Sugars

D-Mannose, a carbohydrate found in immunoglobulins, has the following open-chain structure. Draw the cyclic structure for β-D-mannose anomer.

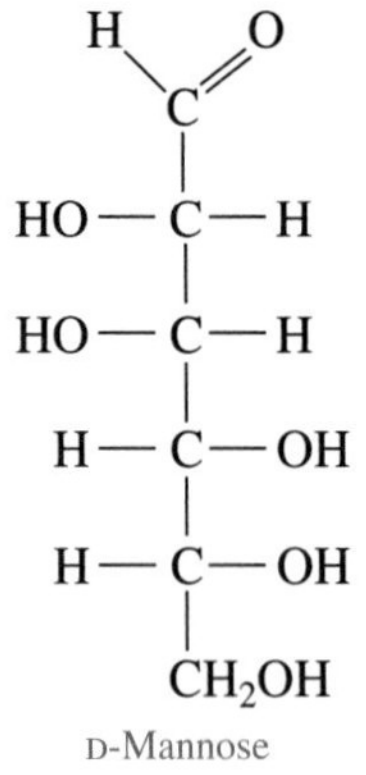
D-Mannose

Solution

Number the carbon atoms in the open chain starting at the aldehyde group. Turn the chain on its side and bend it into a hexagon so that the —OH group on carbon 5 is close to the carbon 1 carbonyl group. Draw the other —OH groups on the left of the open chain above the ring, and the —OH groups on the right below.

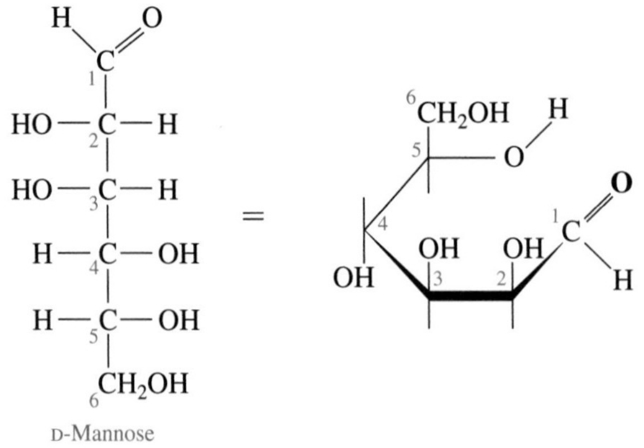
D-Mannose

Form the cyclic hemiacetal by bonding the oxygen in —OH to carbon 1. Write the new hydroxyl group upward to make the β-D-mannose anomer.

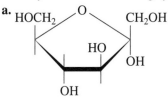

β-D-Mannose

Study Check

Draw the cyclic structure for α-D-glucose.

QUESTIONS AND PROBLEMS

Cyclic Structures of Monosaccharides

16.23 What are the kind and number of atoms in the ring portion of the cyclic structure of glucose?

16.24 What are the kind and number of atoms in the ring portion of the cyclic structure of fructose?

16.25 Draw the cyclic structures for the α and β anomers of D-glucose.

16.26 Draw the cyclic structures for the α and β anomers of D-fructose.

16.27 Identify each of the following cyclic structures as the α or β anomer:

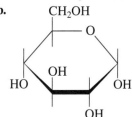

16.28 Identify each of the following cyclic structures as the α or β anomer.

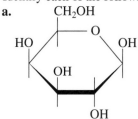
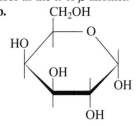

WEB TUTORIAL
Forms of Carbohydrates

16.6 Chemical Properties of Monosaccharides

Monosaccharides contain several functional groups that can undergo chemical reactions. In an aldose, the aldehyde group can be oxidized to a carboxylic acid. The carbonyl group in both an aldose and ketose can be reduced to give a hydroxyl group. The hydroxyl groups can react with other compounds to form a variety of derivatives. Several of these derivatives have important roles in metabolism and biological structures.

Oxidation of Monosaccharides

In Chapter 15, we looked at the Benedict's test for aldehydes with adjacent hydroxyl groups, which we find in monosaccharides. Although monosaccharides exist mostly in their cyclic forms, the aldehyde group of the open-chain structure in the equilibrium mixture does oxidize easily. When the monosaccharide oxidizes, a carboxylic group is produced. At the same time the Cu^{2+} in Benedict's reagent is reduced to Cu^{+}, forming a brick-red precipitate of Cu_2O. Monosaccharides that reduce another substance such as Benedict's reagent are called **reducing sugars.**

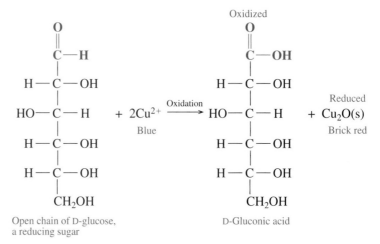

Open chain of D-glucose, a reducing sugar

D-Gluconic acid

Fructose is also a reducing sugar. In the open-chain form, a rearrangement between the hydroxyl group on carbon 1 and the ketone group provides an aldehyde group that can be oxidized.

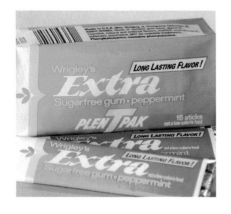

D-Fructose
(ketose)

D-Glucose
(aldose)

Reduction of Monosaccharides

The reduction of the carbonyl group in monosaccharides produces sugar alcohols, which are also called *alditols*. D-glucose is reduced to D-glucitol, better known as sorbitol. D-Mannose is reduced to give D-mannitol.

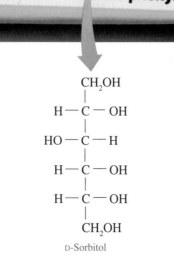

D-Sorbitol

D-Glucose

D-Glucitol or D-Sorbitol

Testing for Glucose in Urine

Normally, blood glucose flows through the kidneys and is reabsorbed into the bloodstream. However, if the blood level exceeds about 160 mg of glucose/dL of blood, the kidneys cannot reabsorb it all, and glucose spills over into the urine, a condition known as glucosuria. A symptom of diabetes mellitus is a high level of glucose in the urine.

Benedict's test can be used to determine the presence of glucose in urine. The amount of cuprous oxide (Cu_2O) formed is

proportional to the amount of reducing sugar present in the urine. Low to moderate levels of reducing sugar turn the solution green; solutions with high glucose levels turn Benedict's yellow or brick-red. Table 16.1 lists some colors associated with the concentration of glucose in the urine.

In another clinical test that is more specific for glucose, the enzyme glucose oxidase is used. The oxidase enzyme converts glucose to gluconic acid and hydrogen peroxide, H_2O_2. The peroxide produced reacts with a dye in the test strip to give different colors. The level of glucose present in the urine is found by matching the color produced to a color chart on the container.

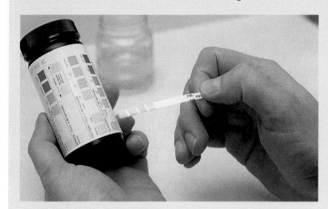

Table 16.1   Glucose Test Results

	Glucose Present in Urine	
Color	%	mg/dL
Blue	0	0
Blue-green	0.25	250
Green	0.50	500
Yellow	1.00	1000
Brick-red	2.00	2000

Sugar alcohols such as sorbitol, xylitol from xylose, and mannitol from mannose are used as sweeteners in many sugar-free products such as diet drinks and sugarless gum as well as products for people with diabetes. However, there are some side effects of these sugar substitutes. Some people experience some discomfort such as gas and diarrhea from the ingestion of sugar alcohols. The development of cataracts in diabetics is attributed to the accumulation of sorbitol in the lens of the eye.

CASE STUDY
Diabetes and Blood Glucose

Formation of Acetals (Glycosides)

In Chapter 15, we saw that hemiacetals react with other alcohol molecules to form acetals.

$$CH_3{-}\overset{O}{\overset{\|}{C}}{-}H \;+\; CH_3OH \;\overset{H^+}{\rightleftarrows}\; CH_3{-}\underset{OH}{\overset{OCH_3}{\underset{|}{\overset{|}{C}}}}{-}H \;+\; CH_3OH \;\overset{H^+}{\rightleftarrows}\; CH_3{-}\underset{OCH_3}{\overset{OCH_3}{\underset{|}{\overset{|}{C}}}}{-}H \;+\; H_2O$$

Aldehyde   Alcohol   Hemiacetal   Alcohol   Acetal

A similar reaction occurs when an alcohol reacts with a cyclic hemiacetal of a monosaccharide to give an acetal. These acetal products are called **glycosides**, and the acetal linkage in sugars is called a **glycosidic bond.** In cyclic acetals, the glycosidic bond cannot open to give the open-chain form. Thus, the acetals of monosaccharides cannot undergo mutarotation or oxidation, which means that acetals are nonreducing sugars. This property is indicated by the ending *–oside.* The glycosidic bond will remain in the α or β position, depending on the anomer that reacted. This is an important reaction in carbohydrate chemistry because it is the way that monosaccharides bond to form disaccharides and polysaccharides.

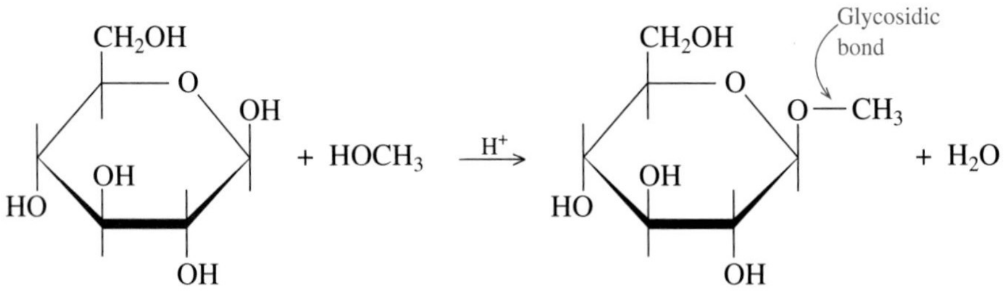

β-D-Glucose   Methanol   Methyl β-D-glucoside

SAMPLE PROBLEM 16.6

Reducing Sugars

Why is D-glucose called a *reducing sugar*?

Solution

D-glucose is easily oxidized by Benedict's reagent. A carbohydrate that reduces Cu^{2+} to Cu^+ is called a reducing sugar.

Study Check

A test using Benedict's reagent turns brick-red with a urine sample. According to Table 16.1, what might this result indicate?

QUESTIONS AND PROBLEMS

Chemical Properties of Monosaccharides

16.29 Draw the product xylitol produced from the reduction of D-xylose.

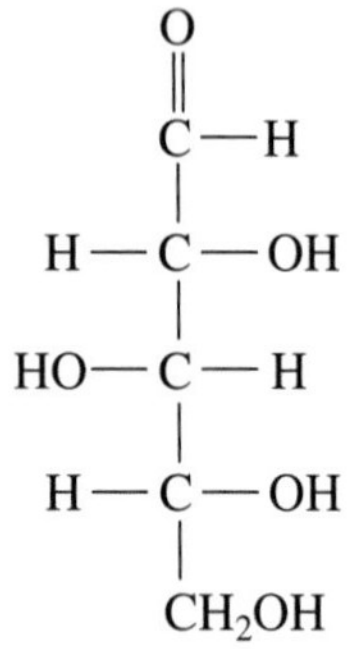

D-Xylose

16.30 Draw the product mannitol produced from the reduction of D-mannose.

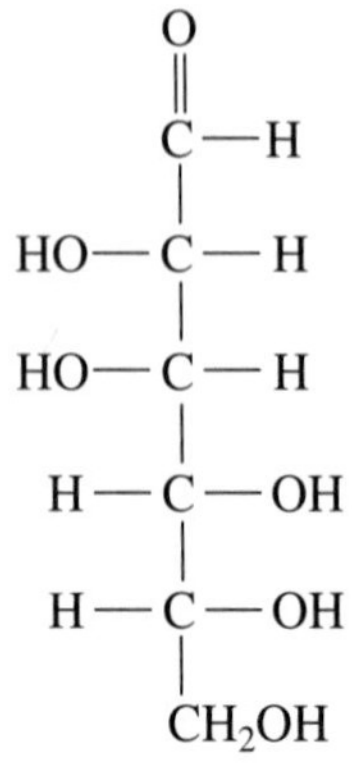

D-Mannose

16.31 Write the oxidation and reduction products of D-arabinose. What is the name of the sugar alcohol produced?

$$
\begin{array}{c}
O \\
\parallel \\
C-H \\
| \\
HO-C-H \\
| \\
H-C-OH \\
| \\
H-C-OH \\
| \\
CH_2OH
\end{array}
$$

D-Arabinose

16.32 Write the oxidation and reduction products of D-ribose. What is the name of the sugar alcohol produced?

$$
\begin{array}{c}
O \\
\parallel \\
C-H \\
| \\
H-C-OH \\
| \\
H-C-OH \\
| \\
H-C-OH \\
| \\
CH_2OH
\end{array}
$$

D-Ribose

16.33 Write the acetal anomers (α and β) of D-galactose and methyl alcohol.

16.34 Write the acetal anomers (α and β) of D-glucose and ethyl alcohol.

16.7 Disaccharides

A disaccharide is composed of two monosaccharides linked together. The most common disaccharides are maltose, lactose, and sucrose. Their hydrolysis, by an acid or an enzyme, gives the following monosaccharides.

$$\text{Maltose} + H_2O \xrightarrow{H^+} \text{glucose} + \text{glucose}$$

$$\text{Lactose} + H_2O \xrightarrow{H^+} \text{glucose} + \text{galactose}$$

$$\text{Sucrose} + H_2O \xrightarrow{H^+} \text{glucose} + \text{fructose}$$

Maltose

Maltose, or malt sugar, is a disaccharide obtained from starch. When maltose in barley and other grains is hydrolyzed by yeast enzymes, glucose is obtained that can undergo fermentation to give ethanol. Maltose is used in cereals, candies, and the brewing of beverages.

HEALTH NOTE

How Sweet Is My Sweetener?

Although many of the monosaccharides and disaccharides taste sweet, they differ considerably in their degree of sweetness. Dietetic foods contain sweeteners that are noncarbohydrate or carbohydrates that are sweeter. Some examples of sweeteners compared with sucrose are shown in Table 16.2.

Sucralose is made from sucrose by replacing some of the hydroxyl groups with chlorine atoms.

Sucralose

Aspartame, which is marketed as Nutra-Sweet, is used in a large number of sugar-free products. It is a noncarbohydrate sweetener made of aspartic acid and a methyl ester of phenylalanine. It does have some caloric value, but it is so sweet that a very small quantity is needed. However, one of the breakdown products phenylalanine, poses a danger to anyone who cannot metabolize it properly, a condition called phenylketonuria (PKU).

From aspartic acid　　From phenylalanine

Aspartame (Nutra-Sweet)

Table 16.2 Relative Sweetness of Sugars and Artificial Sweeteners

	Sweetness Relative to Sucrose (= 100)
Monosaccharides	
Galactose	30
Sorbitol	36
Glucose	75
Fructose	175
Disaccharides	
Lactose	16
Maltose	33
Sucrose	100 ⟵ reference standard
Artificial Sweeteners (Noncarbohydrate)	
Sucralose	600
Aspartame	18,000
Saccharin	45,000

Saccharin has been used as a noncarbohydrate artificial sweetener for the past 25 years. The use of saccharin has been banned in Canada because studies indicate that it may cause bladder tumors. However, it is still approved for use by the FDA in the United States.

Saccharin

As we saw in the previous section, the hemiacetal carbon of a monosaccharide reacts with the hydroxyl group of an alcohol. If the hydroxyl group is another monosaccharide, the glycoside product is a disaccharide.

To make maltose, the hydroxyl group on carbon 1 of one glucose molecule bonds with the hydroxyl group on carbon 4 of the second glucose molecule. A glycosidic bond joins the two glucose molecules with a loss of a molecule of water. The glycosidic bond is designated as an α-1,4 linkage to show that —OH on carbon 1 of the α anomer is joined to carbon 4 of the second glucose. Because the second glucose molecule has a free —OH on the anomeric carbon, there are α and β anomers of maltose. This anomeric carbon also opens up to give a free aldehyde group to oxidize, which makes maltose a reducing sugar.

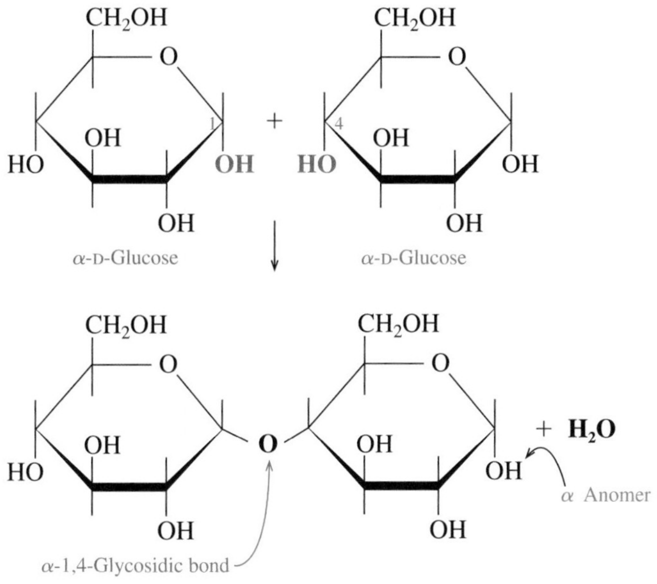

α-Maltose, a disaccharide

Lactose

Lactose, milk sugar, is a disaccharide found in milk and milk products. (See Figure 16.3.) It makes up 6–8% of human milk and about 4–5% of cow's milk and is used in products that attempt to duplicate mother's milk. Some people do not produce sufficient quantities of the enzyme needed to hydrolyze lactose, and the sugar remains undigested, causing abdominal cramps and diarrhea. In some commercial milk products, an enzyme called lactase is added to break down lactose.

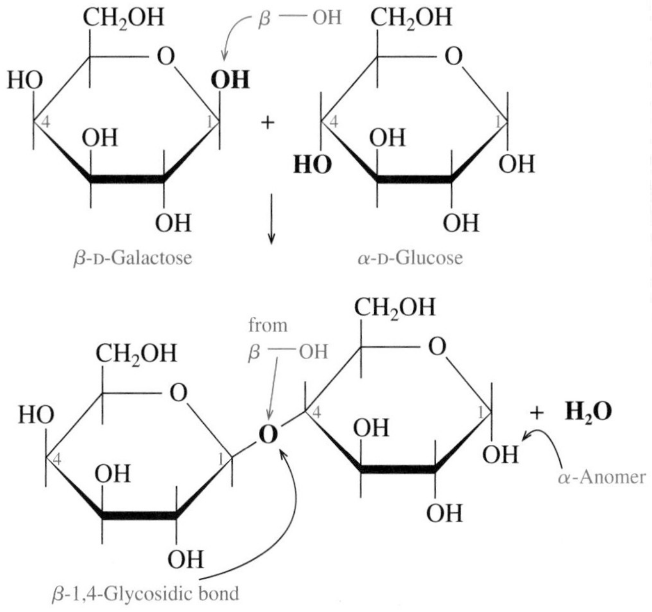

α-Lactose, a disaccharide

The bond in lactose is a β-1,4-glycosidic bond because the β anomer of galactose forms an acetal with a hydroxyl group on carbon 4 of glucose. In the lactose molecule, the acetal of galactose cannot open, however the hemiacetal carbon in glucose

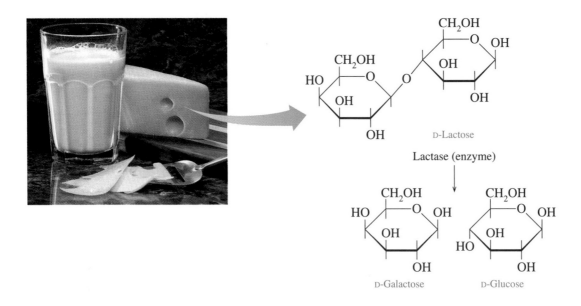

Figure 16.3 Lactose, a disaccharide found in milk and milk products undergoes digestion to give galactose and glucose.

Q What type of glycosidic bond links galactose and glucose in lactose?

undergoes mutarotation to give both α- and β-lactose. Because the open chain has an aldehyde group that can be oxidized, lactose is a reducing sugar.

Sucrose

You already know that ordinary table sugar is sucrose, a disaccharide that is the most abundant carbohydrate in the world. Most of the sucrose for table sugar comes from sugar cane (20% by mass) or sugar beets (15% by mass). (See Figure 16.4.) Both the raw and refined forms of sugar are sucrose. Some estimates indicate that each person in the United States consumes an average of 45 kg (100 lb) of sucrose every year either by itself or in a variety of food products.

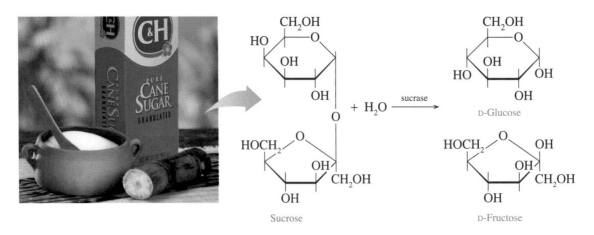

Figure 16.4 Sucrose, a disaccharide obtained from sugar beets and sugar cane, undergoes digestion to give glucose and fructose.

Q Why is sucrose a nonreducing sugar?

Sucrose consists of α-D-glucose and β-D-fructose molecules joined by an α, β-1,2-glycosidic bond. The structure of sucrose differs from the other disaccharides because the glycosidic bond ties up the anomeric carbons of both monosaccharide units. There are no isomers for sucrose. Without a free aldehyde group, there is no mutarotation. Sucrose cannot react with Benedict's reagent; sucrose is not a reducing sugar.

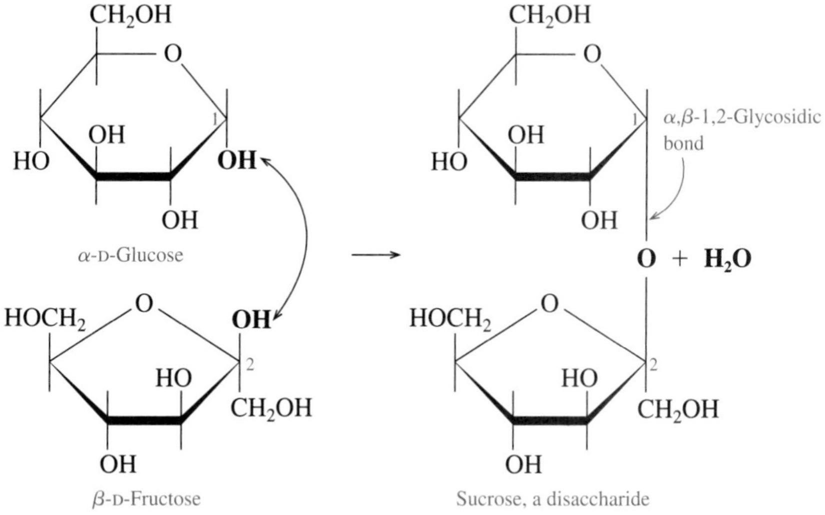

α-D-Glucose

β-D-Fructose

Sucrose, a disaccharide

α,β-1,2-Glycosidic bond

Fermentation

When yeast is added, the monosaccharides glucose and fructose (but not galactose) will undergo **fermentation** to produce ethanol and carbon dioxide gas.

$$C_6H_{12}O_6 \xrightarrow{\text{Yeast enzymes}} 2C_2H_5OH + 2CO_2(g)$$

Glucose or fructose   Ethanol

The disaccharides maltose and sucrose can also undergo fermentation because yeast contains enzymes for their hydrolysis to give glucose and fructose. However, lactose will not ferment because the enzyme lactase required for its hydrolysis is not present in yeast. Fermentation does not occur with polysaccharides.

SAMPLE PROBLEM 16.7

Glycosidic Bonds in Disaccharides

Melebiose is a disaccharide that has a sweetness of about 30 compared with sucrose (= 100).

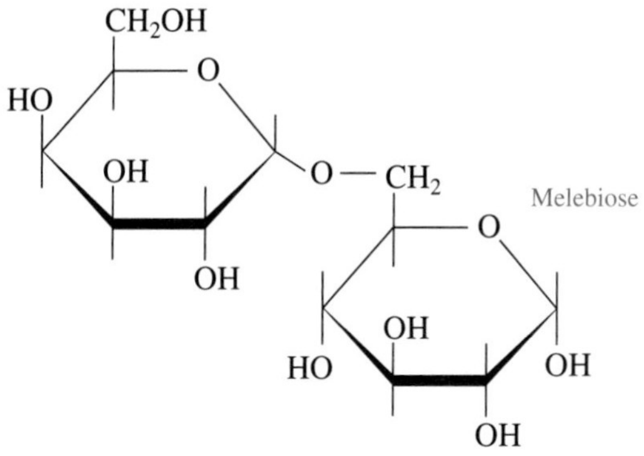

Melebiose

a. What are the monosaccharide units in melebiose?

b. What type of glycosidic bond links the monosaccharides?

c. Is the compound drawn as α- or β-melebiose?

Solution

a. The monosaccharide on the left side is α-D-galactose; on the right is α-D-glucose.

b. The monosaccharide units are linked by an α-1,6-glycosidic bond.

c. The downward position of the anomeric OH makes it α-melebiose.

Study Check

Cellobiose is a disaccharide composed of two β-D-glucose molecules linked by a β-1,4-glycosidic linkage. Draw a structural formula for β-cellobiose.

QUESTIONS AND PROBLEMS

Disaccharides

16.35 For each of the following disaccharides, give the monosaccharide units produced by hydrolysis, the type of glycosidic bond, and the identity of the disaccharide including the α or β anomer:

a.

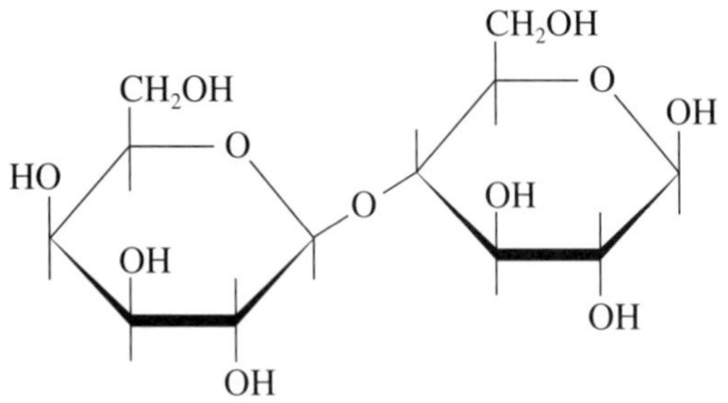

b.

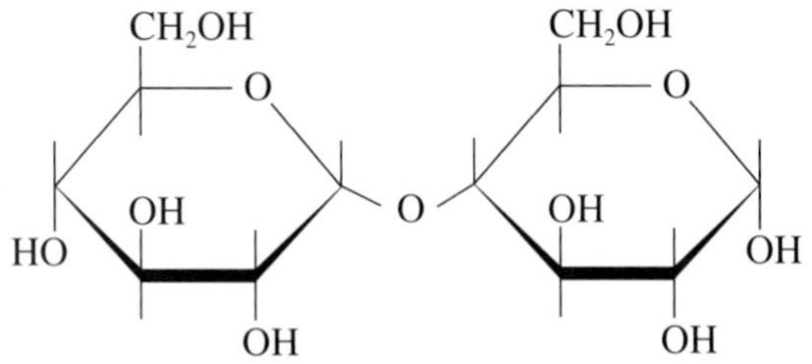

16.36 For each of the following disaccharides, give the monosaccharide units produced by hydrolysis, the type of glycosidic bond, and the identity of the disaccharide including the α or β anomer:

a.

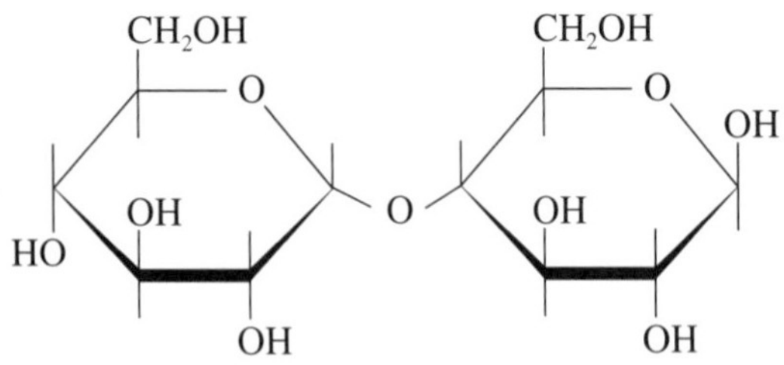

b.

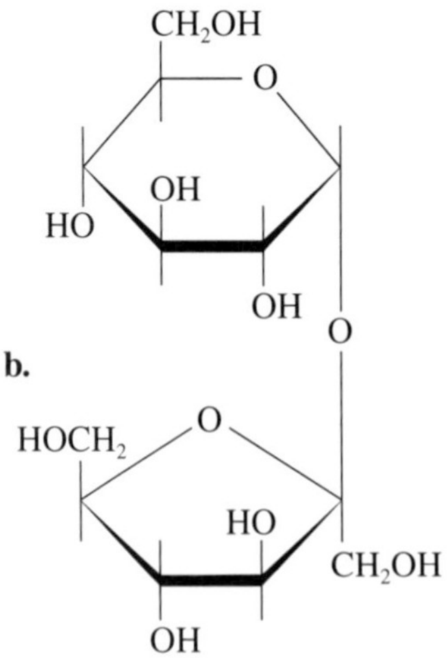

16.37 Indicate whether the sugars in problem 16.35 will undergo mutarotation and oxidation.

16.38 Indicate whether the sugars in problem 16.36 will undergo mutarotation and oxidation.

16.39 Identify disaccharides that fit each of the following descriptions:
  a. ordinary table sugar   **b.** found in milk and milk products
  c. also called *malt* sugar   **d.** hydrolysis gives galactose and glucose

16.40 Identify disaccharides that fit each of the following descriptions:
  a. not a reducing sugar   **b.** composed of two glucose units
  c. also called *milk* sugar   **d.** hydrolysis gives glucose and fructose

16.8 Polysaccharides

LEARNING GOAL

Describe the structural features of amylose, amylopectin, glycogen, and cellulose.

A polysaccharide is a polymer of many monosaccharides joined together. Three biologically important polysaccharides—starch, cellulose, and glycogen—are all polymers of D-glucose, which differ only in the type of glycosidic bonds and the amount of branching in the molecule.

WEB TUTORIAL
Polymers

Plant Starch: Amylose and Amylopectin

Starch, a storage form of glucose in plants, is found as insoluble granules in rice, wheat, potatoes, beans, and cereals. Starch is composed of two kinds of polysaccharides, amylose and amylopectin. **Amylose,** which makes up about 20% of starch, consists of α-D-glucose molecules connected by α-1,4-glycosidic bonds in a continuous chain. A typical polymer of amylose may contain from 250 to 4000 glucose units. Sometimes called a straight-chain polymer, polymers of amylose are actually coiled in helical fashion.

Amylopectin, which makes up as much as 80% of plant starch, is a branched-chain polysaccharide. Like amylose, the glucose molecules are connected by α-1,4-glycosidic bonds. However, at about every 25 glucose units, there is a branch of glucose molecules attached by an α-1,6-glycosidic bond between carbon 1 of the branch and carbon 6 in the main chain. (See Figure 16.5.)

Starches hydrolyze easily in water and acid to give smaller saccharides called dextrins, which then hydrolyze to maltose and finally glucose. In our bodies, these complex carbohydrates are digested by the enzymes amylase (in saliva) and maltase. The glucose obtained provides about 50% of our nutritional calories.

$$\text{Amylose, amylopectin} \xrightarrow{\text{H}^+ \text{ or amylase}} \text{dextrins} \xrightarrow{\text{H}^+ \text{ or amylase}} \text{maltose} \xrightarrow{\text{H}^+ \text{ or maltase}} \text{many D-glucose units}$$

Animal Starch: Glycogen

Glycogen, or animal starch, is a polymer of glucose that is stored in the liver and muscle of animals. It is hydrolyzed in our cells at a rate that maintains the blood level of glucose and provides energy between meals. The structure of glycogen is very similar to that of amylopectin found in plants except that glycogen is more highly branched. In glycogen, the glucose units are joined by α-1,4-glycosidic bonds, and branches occurring about every 10–15 glucose units are attached by α-1,6-glycosidic bonds.

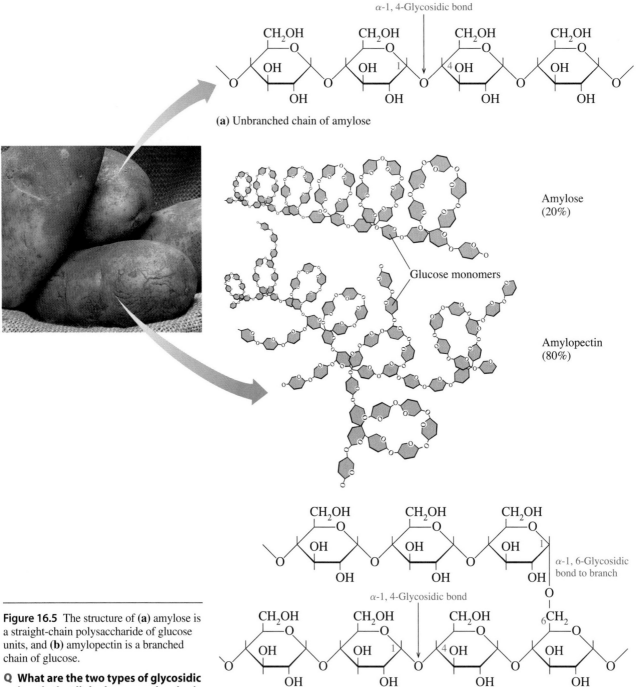

(a) Unbranched chain of amylose

Amylose (20%)

Glucose monomers

Amylopectin (80%)

(b) Branched-chain of amylopectin

Figure 16.5 The structure of **(a)** amylose is a straight-chain polysaccharide of glucose units, and **(b)** amylopectin is a branched chain of glucose.

Q **What are the two types of glycosidic bonds that link glucose molecules in amylopectin?**

Structural Polysaccharide: Cellulose

Cellulose is the major structural material of wood and plants. Cotton is almost pure cellulose. In **cellulose,** glucose molecules form a long unbranched chain similar to that of amylose. However, the glucose units in cellulose are linked by β-1,4-glycosidic bonds. The β isomers do not form coils like the α isomers but are aligned in parallel rows that are held in place by hydrogen bonds between hydroxyl groups in adjacent chains making cellulose insoluble in water. This gives a rigid structure to the cell walls in wood and fiber that is more resistant to hydrolysis than the starches. (See Figure 16.6.)

Cellulose

β-1, 4-glycosidic bond

Enzymes in our saliva and pancreatic juices hydrolyze the α-1,4-glycosidic bonds of the starches. However, there are no enzymes in humans that are able to hydrolyze the β-1,4-glycosidic bonds of cellulose; we cannot digest cellulose. Some animals such as goats and cows and insects like termites are able to obtain glucose from cellulose. Their digestive systems contain bacteria and protozoa with enzymes such as cellulase that can hydrolyze β-1,4-glycosidic bonds.

Figure 16.6 The polysaccharide cellulose is composed of β-1,4-glycosidic bonds.

Q Why are humans unable to digest cellulose?

Iodine Test

In the **iodine test**, iodine (I_2) is used to test for the presence of starch. The unbranched helical shape of the polysaccharide amylose in starch reacts strongly with iodine to form a deep blue-black complex. Amylopectin, cellulose, and glycogen produce reddish-purple and brown colors. Such colors do not develop when iodine is added to samples of mono- or disaccharides.

SAMPLE PROBLEM 16.8

Structures of Polysaccharides

Identify the polysaccharide described by each of the following statements:

a. A polysaccharide that is stored in the liver and muscle tissues
b. An unbranched polysaccharide containing β-1,4-glycosidic bonds
c. A starch containing α-1,4- and α-1,6-glycosidic bonds

Solution

a. glycogen   **b.** cellulose   **c.** amylopectin, glycogen

---

EXPLORE YOUR WORLD

Polysaccharides

Read the nutrition label on a box of crackers, cereal, bread, chips, or pasta. The major ingredient in crackers is flour, a starch. Chew on a single cracker for 4–5 minutes. An enzyme (amylase) in your saliva breaks apart the bonds in starch.

Questions

1. How are carbohydrates listed?
2. What other carbohydrates are listed?
3. How does the taste of the cracker change during the time that you chewed it?
4. What happened to the starches in the cracker as the amylase enzyme in your saliva reacted with the amylose or amylopectin?

Iodine Test for Starch

If you have some iodine solution for cuts, you can use it to test for starch. Collect samples of food such as crackers, bread, pasta, cereals, and candy. You may also test aspirin tablets, antacids, and the glue on an envelope. Place a few drops of iodine on each of the items. If starch is present, a deep blue-black color will develop. After recording your observations, throw all your samples away! Iodine is toxic.

Questions

1. Which of the foods test positive for starch?
2. What other items contain starch?
3. What type of carbohydrate is starch?
4. Why do you think starch is used in preparing tablets such as aspirin?
5. Why is starch present in the glue of the envelope flap?

Study Check

Cellulose and amylose are both unbranched glucose polymers. How do they differ?

QUESTIONS AND PROBLEMS

Polysaccharides

16.41 Describe the similarities and differences in the following polysaccharides:
  a. amylose and amylopectin   b. amylopectin and glycogen

16.42 Describe the similarities and differences in the following polysaccharides:
  a. amylose and cellulose   b. cellulose and glycogen

16.43 Give the name of one or more polysaccharides that matches each of the following descriptions:
  a. not digestible by humans
  b. the storage form of carbohydrates in plants
  c. contains only α-1,4-glycosidic bonds
  d. the most highly branched polysaccharide

16.44 Give the name of one or more polysaccharides that matches each of the following descriptions:
  a. the storage form of carbohydrates in animals
  b. contains only β-1,4-glycosidic bonds
  c. contains both α-1,4- and α-1,6-glycosidic bonds
  d. produces maltose during digestion

Chapter Review

16.1 Types of Carbohydrates

Carbohydrates are classified as monosaccharides (simple sugars), disaccharides (two monosaccharide units), and polysaccharides (many monosaccharide units).

16.2 Classification of Monosaccharides

Monosaccharides are polyhydroxy aldehydes (aldoses) or ketones (ketoses). Monosaccharides are also classified by their number of carbon atoms: *triose, tetrose, pentose,* or *hexose.*

16.3 D and L Notations from Fischer Projections

Chiral molecules can exist in two different forms which are mirror images of each other. These two different molecules are called enantiomers. In a Fischer projection (straight chain), the prefixes D- and L- are used to distinguish between the mirror images. In D-monosaccharides, the —OH is on the right of the chiral carbon farthest from the carbonyl carbon; it is on the left in L-monosaccharides.

16.4 Structures of Some Important Monosaccharides

Important monosaccharides are the aldohexoses glucose and galactose and the ketohexose fructose.

16.5 Cyclic Structures of Monosaccharides

The predominant form of monosaccharides is the cyclic form of five or six atoms. The cyclic structure forms by a reaction between an OH (usually the one on carbon 5 in hexoses) with the carbonyl group of the same molecule. The formation of a new hydroxyl group on carbon 1 (or 2 in fructose) gives α and β anomers of the cyclic monosaccharide. Because the molecule opens and closes continuously (mutarotation) while in solution, both anomers are present. Monosaccharides are reducing sugars because the open-chain aldehyde group (also available in ketoses) can be oxidized by a metal ion such as Cu^{2+}.

16.6 Chemical Properties of Monosaccharides

The aldehyde group in an aldose can be oxidized to a carboxylic acid, while the carbonyl group in an aldose or a ketose can be reduced to give a hydroxyl group. The anomeric hydroxyl groups can react with hydroxyl groups of an alcohol to form an acetal, which makes the compound a nonreducing sugar.

16.7 Disaccharides

Disaccharides are glycosides of two monosaccharide units joined together by a glycosidic bond. In the most common disaccharides, maltose, lactose, and sucrose, there is at least one glucose unit. In the glycoside, the acetal carbon cannot mutarotate or oxidize.

16.8 Polysaccharides

Polysaccharides are polymers of monosaccharide units. Starches consist of amylose, an unbranched chain of glucose, and amylopectin, a branched polymer of glucose. Glycogen, the storage form of glucose in animals, is similar to amylopectin with more branching. Cellulose is also a polymer of glucose, but in cellulose the glycosidic bonds are β bonds rather than α bonds as in the starches. Humans can digest starches, but not cellulose, to obtain energy. However, cellulose is important as a source of fiber in our diets.

Summary of Carbohydrates

Carbohydrate	Food Sources	Monosaccharides
Monosaccharides		
Glucose	Fruit juices, honey, corn syrup	Glucose
Galactose	Lactose hydrolysis	Galactose
Fructose	Fruit juices, honey, sucrose hydrolysis	Fructose
Disaccharides		
Maltose	Germinating grains, starch hydrolysis	Glucose + glucose
Lactose	Milk, yogurt, ice cream	Glucose + galactose
Sucrose	Sugar cane, sugar beets	Glucose + fructose
Polysaccharides		
Amylose	Rice, wheat, grains, cereals	Unbranched polymer of glucose joined by α-1,4-glycosidic bonds
Amylopectin	Rice, wheat, grains, cereals	Branched polymer of glucose joined by α-1,4- and α-1,6-glycosidic bonds
Glycogen	Liver, muscles	Highly branched polymer of glucose joined by α-1,4- and α-1,6-glycosidic bonds
Cellulose	Plant fiber, bran, beans, celery	Unbranched polymer of glucose joined by β-1,4-glycosidic bonds

Summary of Reactions

Glycoside (Acetal) Formation

Monosaccharide   Alcohol   Glycoside

Glycosidic bond

CH_2OH CH_2OH CH_2OH CH_2OH

O   O   O   O

OH   OH   →   OH   OH $+ H_2O$

HO   OH   HO   OH   HO   O   OH

OH   OH   OH   OH

Monosaccharide   Monosaccharide   Disaccharide, a glycoside

Oxidation and Reduction of Monosaccharides

CH_2OH CHO   COOH

H—C—OH   H—C—OH   H—C—OH

HO—C—H   reduction   HO—C—H   oxidation   HO—C—H
        ←   →
H—C—OH   H—C—OH   H—C—OH

H—C—OH   H—C—OH   H—C—OH

CH_2OH CH_2OH CH_2OH

D-Glucitol   D-Glucose   D-Gluconic acid

Hydrolysis of Disaccharides

Sucrose $+ H_2O$ ⟶ glucose + fructose

Lactose $+ H_2O$ ⟶ glucose + galactose

Maltose $+ H_2O$ ⟶ glucose + glucose

Key Terms

aldose Monosaccharides that contain an aldehyde group.

amylopectin A branched-chain polymer of starch composed of glucose units joined by α-1,4- and α-1,6-glycosidic bonds.

amylose An unbranched polymer of starch composed of glucose units joined by α-1,4-glycosidic bonds.

anomers The isomers of cyclic hemiacetals of monosaccharides that have a hydroxyl group on carbon 1 (or carbon 2). In the α anomer, the OH is drawn downward; in the β isomer the OH is up.

carbohydrate A simple or complex sugar composed of carbon, hydrogen, and oxygen.

cellulose An unbranched polysaccharide composed of glucose units linked by β-1,4-glycosidic bonds that cannot be hydrolyzed by the human digestive system.

disaccharides Carbohydrates composed of two monosaccharides joined by a glycosidic bond.

fermentation A reaction of glucose, fructose, maltose, or sucrose in which the sugar reacts with enzymes in yeast to give ethanol and carbon dioxide gas.

fructose A monosaccharide found in honey and fruit juices; it is combined with glucose in sucrose. Also called levulose and fruit sugar.

galactose A monosaccharide that occurs combined with glucose in lactose.

glucose The most prevalent monosaccharide in the diet. An aldohexose that is found in fruits, vegetables, corn syrup, and honey. Also known as blood sugar and dextrose. Combines in glycosidic bonds to form most of the polysaccharides.

glycogen A polysaccharide formed in the liver and muscles for the storage of glucose as an energy reserve. It is composed of glucose in a highly branched polymer joined by α-1,4- and α-1,6-glycosidic bonds.

glycosides Acetal products of a monosaccharide reacting with an alcohol or another sugar.

glycosidic bond The acetal bond that forms when an alcohol or a hydroxyl group of a monosaccharide adds to a hemiacetal. It is the type of bond that links monosaccharide units in di- or polysaccharides.

Haworth structure The cyclic structure that represents the closed chain of a monosaccharide.

iodine test A test for amylose that forms a blue-black color after iodine is added to the sample.

ketose A monosaccharide that contains a ketone group.

lactose A disaccharide consisting of glucose and galactose found in milk and milk products.

maltose A disaccharide consisting of two glucose units; it is obtained from the hydrolysis of starch and in germinating grains.

monosaccharide A polyhydroxy compound that contains an aldehyde or ketone group.

mutarotation The conversion between α and β anomers.

polysaccharides Polymers of many monosaccharide units, usually glucose. Polysaccharides differ in the types of glycosidic bonds and the amount of branching in the polymer.

reducing sugar A carbohydrate with a free aldehyde group capable of reducing the Cu^{2+} in Benedict's reagent.

saccharide A term from the Latin word *saccharum*, meaning "sugar"; it is used to describe the carbohydrate family.

sucrose A disaccharide composed of glucose and fructose; a nonreducing sugar, commonly called table sugar or "sugar."

Additional Problems

16.45 What are the structural differences in D-glucose and D-galactose?

16.46 What are the structural differences in D-glucose and D-fructose?

16.47 How do D-galactose and L-galactose differ?

16.48 How do α-D-glucose and β-D-glucose differ?

16.49 Consider the sugar D-gulose.

D-Gulose

a. What is the Fisher projection for L-gulose?
b. Draw the Haworth structure for α- and β-D-gulose.

16.50 Consider the structures for D-gulose in question 16.49.
a. What is the structure and name of the product formed by the reduction of D-gulose?
b. Write the structure and name of the product formed by the oxidation of D-gulose.

16.51 D-Sorbitol, a sweetener found in seaweed and berries, contains only hydroxyl functional groups. When D-sorbitol is oxidized, it forms D-glucose. What is the structural formula of D-sorbitol?

16.52 Raffinose is a trisaccharide found in Australian manna and in cottonseed meal. It is composed of three different monosaccharides. Identify the monosaccharides in raffinose.

16.53 If α-galactose is dissolved in water, β-galactose is eventually present. Explain how this occurs.

16.54 Why are lactose and maltose reducing sugars, but sucrose is not?

16.55 β-Cellobiose is a disaccharide obtained from the hydrolysis of cellulose. It is quite similar to maltose except it has a β-1,4-glycosidic bond. What is the structure of β-cellobiose?

16.56 The disaccharide trehalose found in mushrooms is composed of two α-D-glucose molecules joined by an α-1,1-glycosidic bond. Draw the structure of trehalose.

16.57 Gentiobiose is found in saffron.
a. Gentiobiose contains two glucose molecules linked by a β-1,6-glycoside bond. Draw the structure of α-gentiobiose.
b. Would gentiobiose be a reducing sugar? Why or why not?

16.58 From the compounds shown, select those that meet the following statements:

 a. is the L-enantiomer of mannose.

 b. a ketopentose.

 c. an aldopentose.

 d. a ketohexose.

17 Carboxylic Acids and Esters

"There are many carboxylic acids, including the alpha hydroxy acids, that are found today in skin products," says Dr. Ken Peterson, pharmacist and cosmetic chemist, Oakland. "When you take a carboxylic acid called a fatty acid and react it with a strong base, you get a salt called soap. Soap has a high pH because the weak fatty acid and the strong base won't have a neutral pH of 7. If you take soap and drop its pH down to 7, you will convert the soap to the fatty acid. When I create fragrances, I use my nose and my chemistry background to identify and break down the reactions that produce good scents. Many fragrances are esters, which form when an alcohol reacts with a carboxylic acid. For example, the ester that smells like pineapple is made from ethanol and butyric acid."

LOOKING AHEAD

the Chemistry place

www.chemplace.com/college

Visit the URL above or use the CD-ROM in the book for extra quizzing, interactive tutorials, career resources, and case studies.

Carboxylic acids are similar to the weak acids we studied in Chapter 10. They have a sour or tart taste, produce hydronium ions in water, and neutralize bases. You encounter carboxylic acids when you use a vinegar salad dressing, which is a solution of acetic acid and water, or experience the sour taste of citric acid in a grapefruit or lemon. When a carboxylic acid combines with an alcohol, an ester is produced. Aspirin is an ester as well as a carboxylic acid. Fats known as trigycerides are esters of glycerol and fatty acids, which are long-chain carboxylic acids. Many fruits and flavorings including bananas, oranges, and strawberries contain esters, which produce their pleasant aromas and flavors.

17.1 Carboxylic Acids

LEARNING GOAL

Give the common names, IUPAC names, and condensed structural formulas of carboxylic acids.

WEB TUTORIAL
Carboxylic Acids

In Chapter 15, we described the carbonyl group $(C{=}O)$ as the functional group in aldehydes and ketones. In a **carboxylic acid,** a hydroxyl group is attached to the carbonyl group, forming a **carboxyl group.** The carboxyl functional group may be attached to an alkyl (R) group or an aromatic (Ar) group.

Carbonyl group

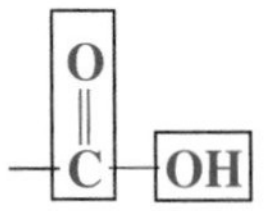

Hydroxyl group

Carboxyl group

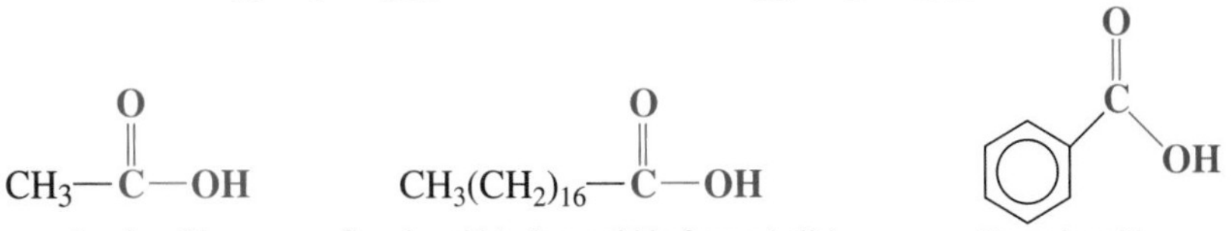

$$R{-}\overset{O}{\underset{\|}{C}}{-}OH \qquad Ar{-}\overset{O}{\underset{\|}{C}}{-}OH$$

$$CH_3{-}\overset{O}{\underset{\|}{C}}{-}OH \qquad CH_3(CH_2)_{16}{-}\overset{O}{\underset{\|}{C}}{-}OH$$

Acetic acid   Stearic acid (a fatty acid in fats and oils)   Benzoic acid (aromatic acid)

The carboxyl group can be written in several different ways. For example, the condensed structural formula for propanoic acid can be written as follows:

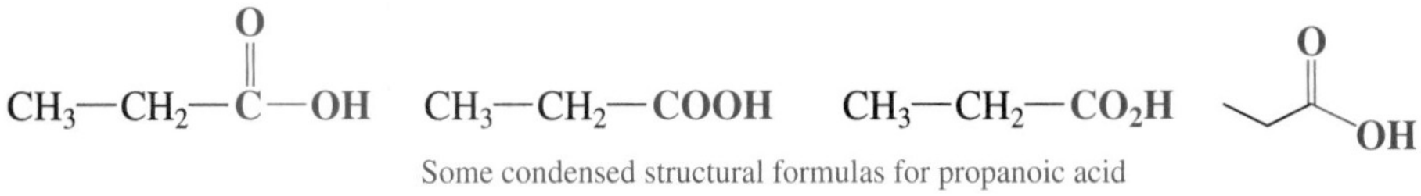

Figure 17.1 Red ants inject formic acid under the skin which causes burning and irritation.

Q What is the IUPAC name of formic acid?

$$CH_3{-}CH_2{-}\overset{O}{\underset{\|}{C}}{-}OH \quad CH_3{-}CH_2{-}COOH \quad CH_3{-}CH_2{-}CO_2H$$

Some condensed structural formulas for propanoic acid

Naming Carboxylic Acids

The IUPAC names of carboxylic acids use the alkane names of the corresponding carbon chains.

Step 1 Identify the longest carbon chain containing the carboxyl group and replace the *e* of the alkane name by *oic acid*.

Step 2 Number the carbon chain beginning with the carboxyl carbon as carbon 1.

Step 3 Give the location and names of substituents on the main chain. The carboxyl function group takes priority over all the functional groups we have discussed.

$$\text{H—C—OH}$$
Methanoic acid

$$\text{CH}_3\text{—CH—C—OH}$$
2-Methylpropanoic acid

$$\text{CH}_3\text{—CH—CH}_2\text{—C—OH}$$
3-Hydroxybutanoic acid

Methanoic acid
(formic acid)

Many carboxylic acids are still named by their common names, which are derived from their natural sources. In Chapter 15, we named aldehydes using the prefixes that represent the typical sources of carboxylic acids.

Formic acid is injected under the skin from bee or red ant stings and other insect bites. (See Figure 17.1.) Acetic acid is the oxidation product of the ethanol in wines and apple cider. The resulting solution of acetic acid and water is known as vinegar. Butyric acid gives the foul odor to rancid butter. (See Table 17.1.) Some ball-and-stick models of carboxylic acids are shown in Figure 17.2.

When using the common names, the Greek letters alpha (α), beta (β), and gamma (γ) are assigned to the carbons adjacent to the carboxyl carbon.

Ethanoic acid
(acetic acid)

$$\text{CH}_3\text{—CH—CH}_2\text{—COH}$$

IUPAC	4	3	2	1
Common	γ	β	α	

The aromatic carboxylic acid is called benzoic acid. With the carboxyl carbon as carbon 1, the ring is numbered in the direction that gives any substituents the smallest possible numbers. As we did with other aromatic compounds (Chapter 13), the prefixes *ortho, meta, and para* may be used to show the position of one other substituent.

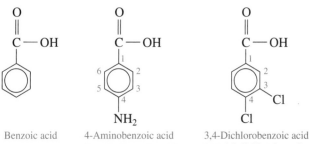

Benzoic acid

4-Aminobenzoic acid
(*p*-aminobenzoic acid)

3,4-Dichlorobenzoic acid
(*not* 4,5-dichlorobenzoic acid)

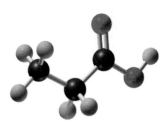

Propanoic acid
(propionic acid)

Figure 17.2 In carboxylic acids, a carbonyl group is bonded to a hydroxyl group.

Q What is the IUPAC and common name of a carboxylic acid with a chain of four carbons?

Table 17.1   Names and Natural Sources of Carboxylic Acids

Condensed Structural Formulas	IUPAC Name	Common Name	Occurs In
H—C—OH	Methanoic acid	Formic acid	Ant and bee stings (Latin *formica,* "ant")
$\text{CH}_3\text{—C—OH}$	Ethanoic acid	Acetic acid	Vinegar (Latin *acetum,* "sour")
$\text{CH}_3\text{—CH}_2\text{—C—OH}$	Propanoic acid	Propionic acid	Dairy products (Greek, *pro,* "first," *pion,* "fat")
$\text{CH}_3\text{—CH}_2\text{—CH}_2\text{—C—OH}$	Butanoic acid	Butyric acid	Rancid butter (Latin *butyrum,* "butter")

Naming Carboxylic Acids

Give the IUPAC and common name, if any, for each of the following carboxylic acids:

a. $CH_3—CH_2—\overset{\overset{\displaystyle O}{\|}}{C}—OH$ b. (structure shown)   c. (structure shown)

Solution

a. This carboxylic acid has 3 carbon atoms. In the IUPAC system, the *e* in propane is replaced by *oic acid,* to give the name, *propanoic acid.* Its common name is *propionic acid.*

b. This carboxylic acid has a methyl group on the second carbon. It has the IUPAC name *2-methylbutanoic acid.* In the common name, the Greek letter α specifies the carbon atom next to the carboxyl carbon, *α-methylbutyric acid.*

c. An aromatic carboxylic acid is named as benzoic acid. Counting from the carboxyl carbon places the Cl on carbon 3, or the *meta* carbon. The name is 3-chlorobenzoic acid or *meta*-chlorobenzoic acid.

Study Check

Write the condensed structural formula of 3-phenylpropanoic acid.

HEALTH NOTE

Alpha Hydroxy Acids

Alpha hydroxy acids (AHAs) are naturally occurring carboxylic acids found in fruits, milk, and sugarcane. Cleopatra reportedly bathed in sour milk to smooth her skin. Dermatologists have been using products with a high concentration of AHAs to remove acne scars and reduce irregular pigmentation and age spots. Now lower concentrations (8–10%) of AHAs have been added to skin care products for the purpose of smoothing fine lines, improving skin texture, and cleansing pores. Several alpha hydroxy acids may be found in skin care products singly or in combination. Glycolic acid and lactic acid are most frequently used.

Recent studies indicate that products with AHAs increase sensitivity of the skin to sun and UV radiation. It is recommended that a sunscreen with a sun protection factor (SPF) of at least 15 be used when treating the skin with products that include AHAs. Products containing AHAs at concentrations under 10% and pH values greater than 3.5 are generally considered safe. However, the Food and Drug Administration has reports of AHAs causing skin irritation including blisters, rashes, and discoloration of the skin. The FDA does not require product safety reports from cosmetic manufacturers, although they are responsible

for marketing safe products. The FDA advises that you test any product containing AHAs on a small area of skin before you use it on a large area.

Alpha Hydroxy Acid (Source)	Structure
Glycolic acid (Sugarcane, sugar beet)	$HO—CH_2—\overset{\overset{\displaystyle O}{\|}}{C}OH$
Lactic acid (Sour milk)	$CH_3—\overset{\overset{\displaystyle OH}{\|}}{C}H—\overset{\overset{\displaystyle O}{\|}}{C}OH$
Tartaric acid (Grapes)	$HO\overset{\overset{\displaystyle O}{\|}}{C}—\overset{\overset{\displaystyle OH}{\|}}{C}H—\overset{\overset{\displaystyle OH}{\|}}{C}H—\overset{\overset{\displaystyle O}{\|}}{C}OH$
Malic acid (Apples, grapes)	$HO\overset{\overset{\displaystyle O}{\|}}{C}—CH_2—\overset{\overset{\displaystyle OH}{\|}}{C}H—\overset{\overset{\displaystyle O}{\|}}{C}OH$
Citric acid (Citrus fruits: lemons, oranges, grapefruit)	$\begin{array}{c} CH_2—COOH \\ | \\ HO—C—COOH \\ | \\ CH_2—COOH \end{array}$

Preparation of Carboxylic Acids

Carboxylic acids can be prepared from primary alcohols or aldehydes. As we saw in Chapter 15, there is an increase in carbon–oxygen bonds as a primary alcohol is oxidized to an aldehyde. Oxidation continues easily as another oxygen is added to yield a carboxylic acid. For example, when ethyl alcohol in wine comes in contact with the oxygen in the air, vinegar is produced. The oxidation process converts the ethyl alcohol (primary alcohol) to acetaldehyde, and then to acetic acid, the carboxylic acid in vinegar. (See Figure 17.3.)

Figure 17.3 Vinegar is a 5% solution of acetic acid and water.

Q What is the IUPAC name for acetic acid?

$$CH_3-CH_2 \xrightarrow{[O]} CH_3-\overset{O}{\overset{\|}{C}}-H \xrightarrow{[O]} CH_3-\overset{O}{\overset{\|}{C}}-OH$$

Ethyl alcohol (OH)   Acetaldehyde   Acetic acid

SAMPLE PROBLEM 17.2

Preparation of Carboxylic Acids

Write an equation for the oxidation of 1-propanol and name each product.

Solution

A primary alcohol will oxidize to an aldehyde, which can oxidize further to a carboxylic acid.

$$CH_3-CH_2-CH_2-OH \xrightarrow{[O]} CH_3-CH_2-\overset{O}{\overset{\|}{C}}-H \xrightarrow{[O]} CH_3-CH_2-\overset{O}{\overset{\|}{C}}-OH$$

1-Propanol (propyl alcohol)   Propanal (propionaldehyde)   Propanoic acid (propionic acid)

Study Check

Write the condensed structural formula of the carboxylic acid produced by the oxidation of 1-butanol.

QUESTIONS AND PROBLEMS

Carboxylic Acids

17.1 What carboxylic acid is responsible for the pain of an ant sting?
17.2 What carboxylic acid is found in a solution of vinegar?
17.3 Explain the differences in the condensed structural formulas of propanal and propanoic acid.
17.4 Explain the differences in the condensed structural formulas of benzaldehyde and benzoic acid.
17.5 Give the IUPAC and common names (if any) for the following carboxylic acids:

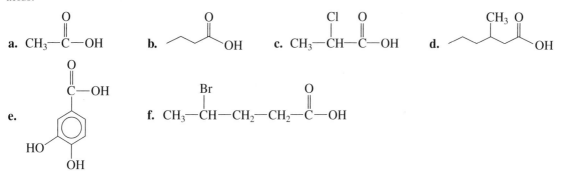

17.6 Give the IUPAC and common names (if any) for the following carboxylic acids:

a. $H-\overset{\overset{O}{\|}}{C}-OH$

b. (structure: butanoic chain with OH on carbonyl and Br on α-carbon)

c. (benzene ring with $\overset{\overset{O}{\|}}{C}-OH$)

d. (benzene ring with $\overset{\overset{O}{\|}}{C}-OH$ and Cl para)

e. $CH_3-\overset{\overset{CH_3}{|}}{CH}-CH_2-\overset{\overset{O}{\|}}{C}-OH$

f. $Cl-CH_2-\overset{\overset{O}{\|}}{C}-OH$

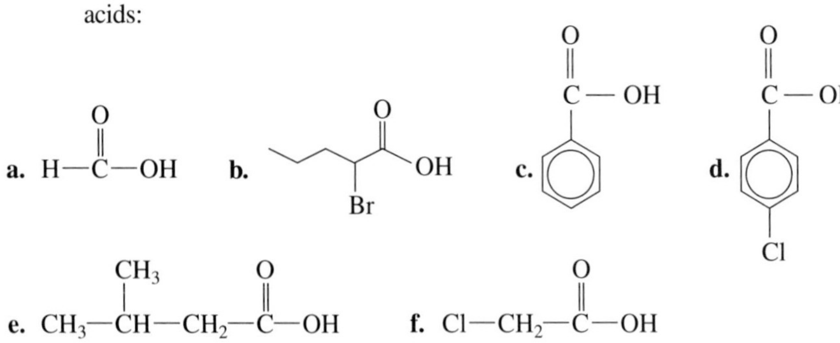

17.7 Draw the condensed structural formulas of each of the following carboxylic acids:

a. propionic acid
b. benzoic acid
c. 2-chloroethanoic acid
d. 3-hydroxypropanoic acid
e. α-methylbutyric acid
f. 3,5-dibromoheptanoic acid

17.8 Draw the condensed structural formulas of each of the following carboxylic acids:

a. butyric acid
b. 3-ethylbenzoic acid
c. α-hydroxyacetic acid
d. 2,4-dibromobutanoic acid
e. m-methylbenzoic acid
f. 4,4-dibromohexanoic acid

17.9 Draw the condensed structural formulas of the carboxylic acids formed by the oxidation of each of the following:

a. CH_3-OH

b. $CH_3-\overset{\overset{O}{\|}}{C}-H$

c. $CH_3-\overset{\overset{CH_3}{|}}{CH}-CH_2-CH_2-OH$

d. (cyclopentane ring with CH_2-CH_2-OH)

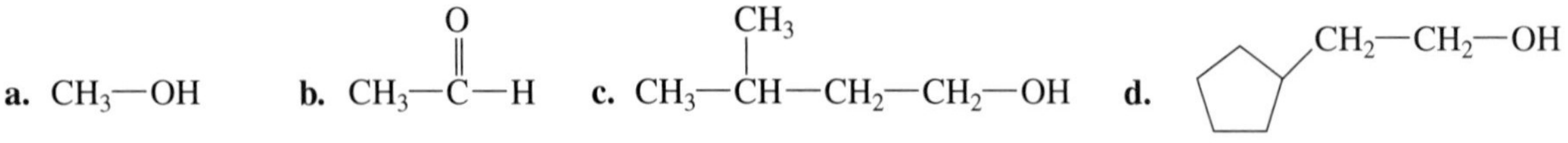

17.10 Draw the condensed structural formulas of the carboxylic acids formed by the oxidation of each of the following:

a. $CH_3-CH_2-CH_2-CH_2-CH_2-CH_2-OH$

b. $CH_3-CH_2-CH_2-CH_2-\overset{\overset{O}{\|}}{C}-H$

c. $CH_3-\overset{\overset{CH_3}{|}}{CH}-CH_2-\overset{\overset{O}{\|}}{C}-H$

d. (benzene ring with CH_2-CH_2-OH)

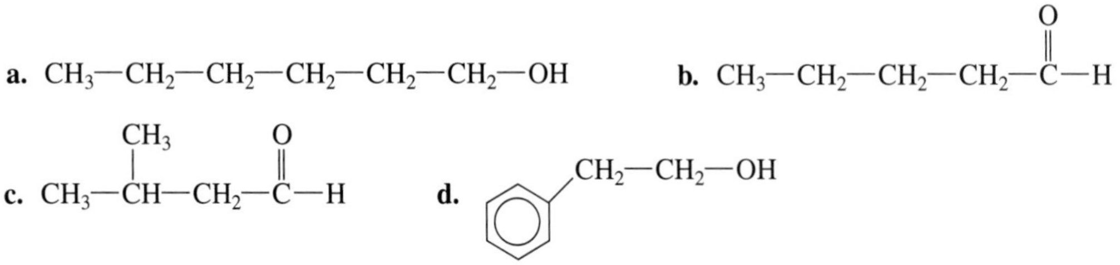

LEARNING GOAL
Describe the boiling points and solubility of carboxylic acids in water.

17.2 Physical Properties of Carboxylic Acids

Carboxylic acids are among the most polar organic compounds because the functional group consists of two polar groups: a hydroxyl (—OH) group and a carbonyl (C=O) group. This C=O double bond is similar to that of the aldehydes and ketones.

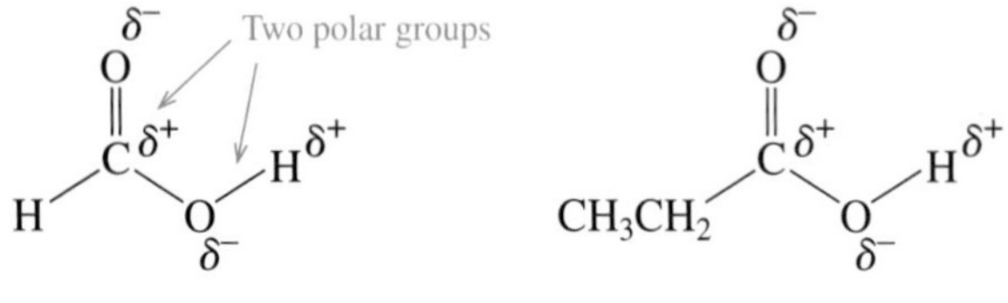

Therefore, carboxylic acids form hydrogen bonds with other carboxylic acid molecules and water. This ability to form hydrogen bonds has a major influence on both their boiling points and solubility in water.

WEB TUTORIAL
Carboxylic Acids

Boiling Points

Carboxylic acids have higher boiling points than alcohols, ketones, and aldehydes of similar mass.

$$CH_3-CH_2-\overset{\overset{\displaystyle O}{\|}}{C}-H \qquad CH_3-CH_2-CH_2OH \qquad CH_3-\overset{\overset{\displaystyle O}{\|}}{C}-OH$$

Compound	Propanal	1-Propanol	Acetic acid
Molar mass	58	60	60
bp	49°C	97°C	118°C

One reason for the high boiling points of carboxylic acids is the formation of hydrogen bonds between two carboxylic acids to give a dimer. Because the dimers are stable as gases, the mass of the carboxylic acid is effectively doubled, which means that higher temperatures are required to reach the boiling point and form gases.

Two hydrogen bonds

$$CH_3-C \overset{\displaystyle O\cdots H-O}{\underset{\displaystyle O-H\cdots O}{}} C-CH_3$$

A dimer of two acetic acid molecules

Solubility in Water

Carboxylic acids with one to four carbons are very soluble in water because the carboxyl group forms hydrogen bonds with several water molecules. (See Figure 17.4.) However, as the length of the carbon chain increases, the nonpolar portion reduces solubility. Carboxylic acids having five or more carbons are not very soluble in water. Table 17.2 lists boiling points and solubilities for some selected carboxylic acids.

Table 17.2 Physical Properties of Selected Carboxylic Acids

IUPAC name	bp (°C)	Solubility in Water
Methanoic acid	101	Very soluble
Ethanoic acid	118	Very soluble
Propanoic acid	141	Very soluble
Butanoic acid	164	Very soluble
Pentanoic acid	187	Slightly soluble
Hexanoic acid	205	Slightly soluble
Benzoic acid	250	Slightly soluble

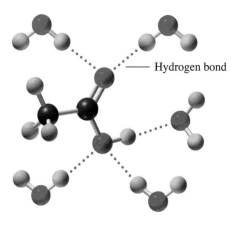

Hydrogen bond

Figure 17.4 Acetic acid forms hydrogen bonds with water molecules.

Q Why do the atoms in the carboxyl group hydrogen bond with water molecules?

SAMPLE PROBLEM 17.3

Physical Properties of Carboxylic Acids

Put the following organic compounds in order of increasing boiling points: butanoic acid, butanone, pentane, and 2-butanol.

Solution

The boiling point increases when the molecules of a compound can form hydrogen bonds or dipole-dipole interactions. The alkane has the lowest boiling point because alkanes cannot hydrogen bond. Ketones have low boiling points because they can only have dipole-dipole interactions via the $C=O$ group. Alcohols and carboxylic acids have higher boiling points because they hydrogen bond via the —OH. The higher boiling points of carboxylic acids are due to the formation of stable dimers, which increase the effective mass and therefore the boiling point.

Pentane < butanone < 2-butanol < butanoic acid

Study Check

At its boiling point, acetic acid forms gas molecules that have a mass of 120 rather than the molar mass of 60 g/mole for CH_3COOH. Explain.

QUESTIONS AND PROBLEMS

Physical Properties of Carboxylic Acids

17.11 Identify the compound in each pair that has the higher boiling point.
   a. acetic acid or butanoic acid   **b.** 1-propanol or propanoic acid
   c. butanone or butanoic acid

17.12 Identify the compound in each pair that has the higher boiling point.
   a. acetone or propanoic acid   **b.** propanoic acid or hexanoic acid
   c. ethanol or acetic acid

17.13 Place the compounds in each group in order of increasing solubility in water:
   a. propanoic acid, propanol, acetone
   b. butanoic acid, propanoic acid, acetic acid
   c. propane, ethanol, acetic acid

17.14 Place the compounds in each group in order of increasing solubility in water:
   a. butanone, butanoic acid, diethyl ether
   b. propanoic acid, acetic acid, formic acid
   c. butanone, 1-butanol, butanoic acid

WEB TUTORIAL
Carboxylic Acids

LEARNING GOAL

Describe the ionization of carboxylic acids as weak acids.

17.3 Acidity of Carboxylic Acids

One of the most important properties of carboxylic acids is their ionization in water, which makes them weak acids (Chapter 10). In the ionization, a carboxylic acid donates a proton to a water molecule to produce an anion called a **carboxylate ion** and a hydronium ion.

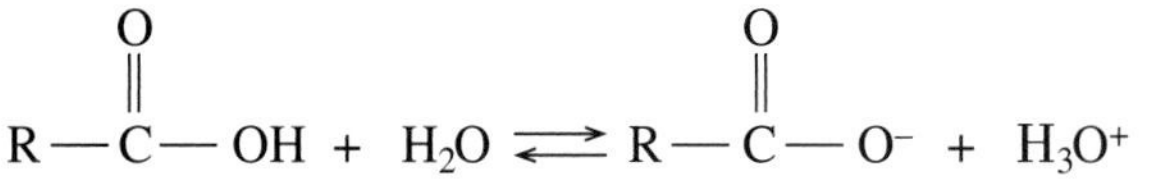

Carboxylic acid (weak acid) + H_2O ⇌ Carboxylate ion + Hydronium ion

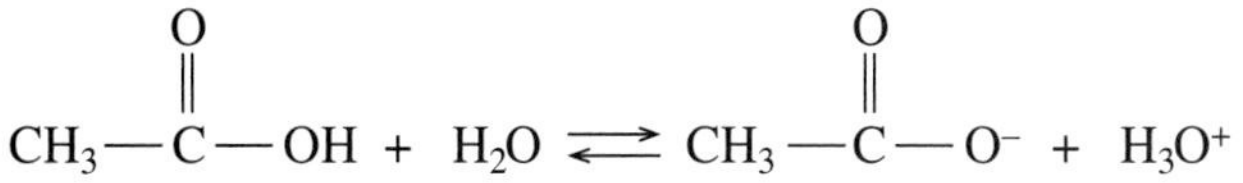

Ethanoic acid (acetic acid) + H_2O ⇌ Ethanoate ion (acetate ion) + Hydronium ion

---

SAMPLE PROBLEM 17.4

Ionization of Carboxylic Acids in Water

Write the equation for the ionization of propionic acid in water.

Solution

The ionization of propionic acid produces a carboxylate ion and a hydronium ion.

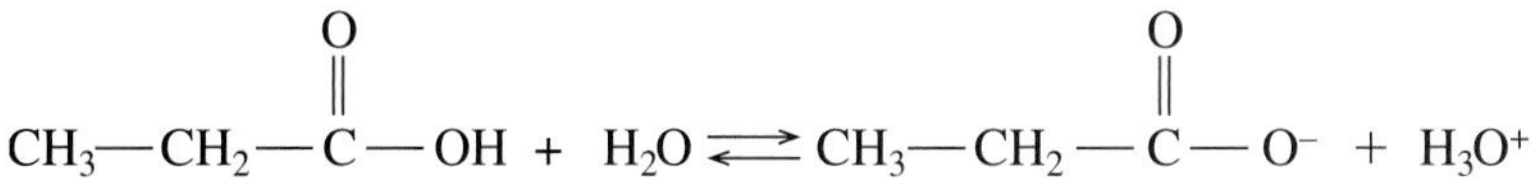

Study Check

Write an equation for the ionization of formic acid in water.

Carboxylic acids are more acidic than other organic compounds including phenols, but only a small percentage (~1%) of the carboxylic acid molecules in a dilute solution are ionized, which means that most of the acid is not ionized. The acid dissociation constants of carboxylic acid are between 10^{-4} to 10^{-5} as seen in Table 17.3.

Salts of Carboxylic Acids

Although carboxylic acids are weak acids, they are completely neutralized by strong bases such as NaOH and KOH. The products are water and a **carboxylic acid salt,** which is a carboxylate ion and the metal ion from the base. The carboxylate ion is named by replacing the *–oic acid* ending of the acid name with *ate*.

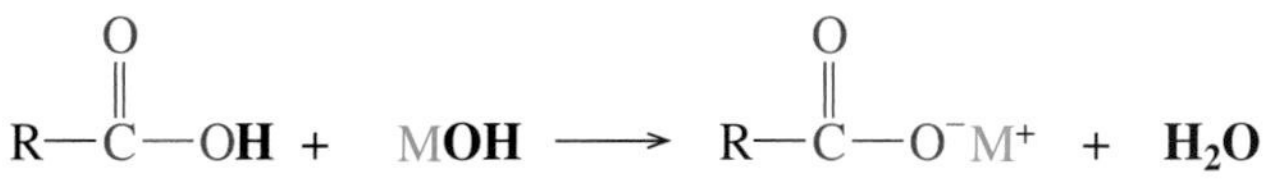

Table 17.3   Acid Dissociation Constants K_a for Selected Carboxylic Acids

Name	K_a (25°C)
Methanoic acid	1.8×10^{-4}
Ethanoic acid	1.8×10^{-5}
Propanoic acid	1.3×10^{-5}
Butanoic acid	1.5×10^{-5}
Pentanoic acid	1.5×10^{-5}
Benzoic acid	6.5×10^{-5}

Carboxylic acid + Strong base ⟶ Carboxylic acid salt + Water

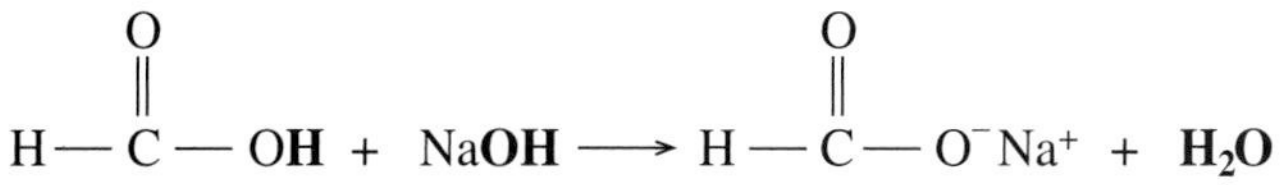

Formic acid + NaOH ⟶ Sodium formate + H_2O

Benzoic acid + KOH ⟶ Potassium benzoate + H_2O

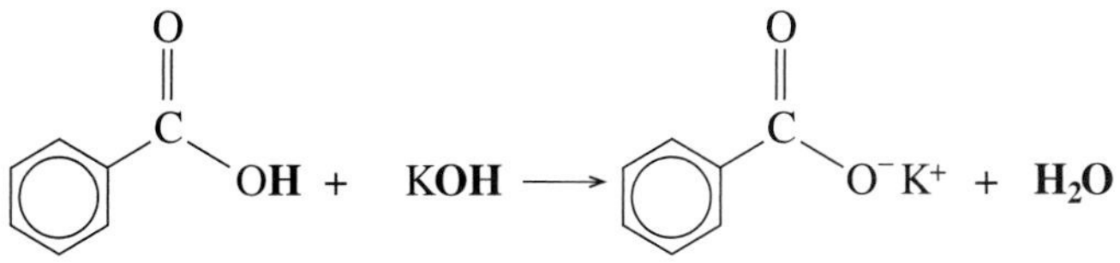

HEALTH NOTE

Carboxylic Acids in Metabolism

There are several carboxylic acids that are part of the metabolic processes within our cells. For example, during glycolysis, a molecule of glucose is broken down into two molecules of pyruvic acid or actually its carboxylate ion pyruvate. During strenuous exercise when oxygen levels are low (anaerobic), pyruvic acid is reduced to give lactic acid or the lactate ion. The buildup of lactate ion in the muscle leads to fatigue and pain.

$$CH_3-\overset{\overset{O}{\|}}{C}-\overset{\overset{O}{\|}}{C}-OH \quad +2H \xrightarrow{\text{Reduction}}$$

Pyruvic acid

$$CH_3-\overset{\overset{OH}{|}}{CH}-\overset{\overset{O}{\|}}{C}-OH$$

Lactic acid

In the citric acid cycle or Krebs cycle, several dicarboxylic acids are oxidized and decarboxylated (loss of CO_2) in order to produce energy for the cell. These carboxylic acids are normally referred to by their common names. At the start of the citric acid cycle, citric acid with six carbons is converted to five-carbon α-ketoglutaric acid. Citric acid is also the acid that gives the sour tastes to citrus fruits such as lemons and grapefruits.

$$
\begin{array}{ccc}
\text{COOH} & & \text{COOH} \\
| & & | \\
\text{CH}_2 & & \text{CH}_2 \\
| & \xrightarrow{[O]} & | \\
\text{HO}-\text{C}-\text{COOH} & & \text{CH}_2 \quad + \quad CO_2 \\
| & & | \\
\text{CH}_2 & & \text{C}=\text{O} \\
| & & | \\
\text{COOH} & & \text{COOH}
\end{array}
$$

Citric acid α-Ketoglutaric acid

The citric acid cycle continues as α-ketoglutaric acid loses CO_2 to give a four-carbon succinic acid. Then a series of reactions converts succinic acid to oxaloacetic acid. We see that some of the functional groups we have studied along with reactions such as hydration and oxidation are part of the metabolic processes that take place in our cells.

$$
\begin{array}{ccccc}
\text{COOH} & & \text{COOH} & & \\
| & & \| & & \\
\text{CH}_2 & \xrightarrow{[O]} & \text{C}-\text{H} & \xrightarrow{H_2O} & \\
| & & \| & & \\
\text{CH}_2 & & \text{H}-\text{C} & & \\
| & & | & & \\
\text{COOH} & & \text{COOH} & &
\end{array}
$$

Succinic acid   Fumaric acid

$$
\begin{array}{ccc}
\text{COOH} & & \text{COOH} \\
| & & | \\
\text{HO}-\text{C}-\text{H} & & \text{C}=\text{O} \\
| & \xrightarrow{[O]} & | \\
\text{CH}_2 & & \text{CH}_2 \\
| & & | \\
\text{COOH} & & \text{COOH}
\end{array}
$$

Malic acid   Oxaloacetic acid

At the pH of the aqueous environment in the cells, the carboxylic acids are ionized, which means it is actually the carboxylate ions that take part in the reactions of citric acid cycle. For example, in water, succinic acid is in equilibrium with its carboxylate ion succinate.

$$
\begin{array}{ccc}
\text{COOH} & & \text{COO}^- \\
| & & | \\
\text{CH}_2 & & \text{CH}_2 \\
| \quad + \quad 2H_2O & \rightleftharpoons & | \quad + \quad 2H_3O^+ \\
\text{CH}_2 & & \text{CH}_2 \\
| & & | \\
\text{COOH} & & \text{COO}^-
\end{array}
$$

Succinic acid   Succinate ion

In Chapters 23 and 24, we will study glycolysis and the citric acid cycle in more detail.

Sodium propionate, a preservative, is added to bread, cheeses, and bakery items to inhibit the spoilage of the food by microorganisms. Sodium benzoate, an inhibitor of mold and bacteria, is added to juices, margarine, relishes, salads, and jams. Monosodium glutamate (MSG) is added to meats, fish, vegetables, and bakery items to enhance flavor, although it causes headaches in some people. (See Figure 17.5.)

$$CH_3-CH_2-\overset{\overset{\displaystyle O}{\|}}{C}-O^-\,Na^+$$
Sodium propionate

$$\overset{\overset{\displaystyle O}{\|}}{C}-O^-\,Na^+$$
Sodium benzoate

$$HO-\overset{\overset{\displaystyle O}{\|}}{C}-\overset{\overset{\displaystyle NH_2}{|}}{CH}-CH_2-CH_2-\overset{\overset{\displaystyle O}{\|}}{C}-O^-Na^+$$
Monosodium glutamate

The carboxylic acid salts are solids at room temperature and have high melting points. Because they are ionic compounds, carboxylic acid salts of the alkali metals (Li^+, Na^+, and K^+) and NH_4^+ are usually soluble in water.

SAMPLE PROBLEM 17.5

Neutralization of a Carboxylic Acid

Write the equation for the neutralization of propionic acid with sodium hydroxide.

Solution

The neutralization of an acid with a base produces the salt of the acid and water.

$$CH_3-CH_2-\overset{\overset{\displaystyle O}{\|}}{C}-OH + NaOH \longrightarrow CH_3-CH_2-\overset{\overset{\displaystyle O}{\|}}{C}-O^-Na^+ + H_2O$$
Propionic acid   Sodium propionate

Study Check

What carboxylic acid will give potassium butyrate when it is neutralized by KOH?

Figure 17.5 Preservatives and flavor enhancers in soups and seasonings are often carboxylic acids or their salts.

Q What is the carboxylate salt produced by the neutralization of butanoic acid and lithium hydroxide?

QUESTIONS AND PROBLEMS

Acidity of Carboxylic Acids

17.15 Write equations for the ionization of each of the following carboxylic acids in water:

a. $H-\overset{\overset{\displaystyle O}{\|}}{C}-OH$ **b.** $CH_3-CH_2-\overset{\overset{\displaystyle O}{\|}}{C}-OH$ **c.** acetic acid

17.16 Write equations for the ionization of each of the following carboxylic acids in water:

a. $CH_3-\overset{\overset{\displaystyle CH_3}{|}}{CH}-\overset{\overset{\displaystyle O}{\|}}{C}-OH$ **b.** α-hydroxyacetic acid   **c.** butanoic acid

17.17 Write equations for the reaction of each of the following carboxylic acids with NaOH:
a. formic acid   **b.** propanoic acid   **c.** benzoic acid

17.18 Write equations for the reaction of each of the following carboxylic acids with KOH:

a. acetic acid   b. 2-methylbutanoic acid   c. *p*-chlorobenzoic acid

17.19 Give the IUPAC and common names, if any, of the carboxylic acid salts in problem 17.17.

17.20 Give the IUPAC and common names, if any, of the carboxylic acid salts in problem 17.18.

LEARNING GOAL

Write the products of the esterification reaction of an alcohol and carboxylic acid.

17.4 Esters of Carboxylic Acids

Esters are the derivatives of carboxylic acid in which an alkoxy group (—OR) replaces the hydroxyl (—OH) group in the carboxylic acid.

$$R-\overset{\overset{\textstyle O}{\|}}{C}-O-H \qquad R-\overset{\overset{\textstyle O}{\|}}{C}-O-R$$

Carboxylic acid   Ester

Esterification

In a reaction called **esterification,** a carboxylic acid reacts with an alcohol when heated in the presence of an acid catalyst (usually H_2SO_4). In the reaction, water is produced from the —OH removed from the carboxylic acid and an —H lost by the alcohol.

$$R-\overset{\overset{\textstyle O}{\|}}{C}-O-H \ + \ H-O-R \underset{\text{Heat}}{\overset{H^+}{\rightleftharpoons}} R-\overset{\overset{\textstyle O}{\|}}{C}-O-R \ + \ H-O-H$$

Acid   Alcohol   Ester

Example:

$$CH_3-\overset{\overset{\textstyle O}{\|}}{C}-O-H \ + \ H-O-CH_3 \underset{\text{Heat}}{\overset{H^+}{\rightleftharpoons}} CH_3-\overset{\overset{\textstyle O}{\|}}{C}-O-CH_3 \ + \ H-O-H$$

Acetic acid   Methyl alcohol   Methyl acetate

If we use acetic acid and 1-propanol, we can write an equation for the formation of the ester that is responsible for the flavor and odor of pears.

$$CH_3-\overset{\overset{\textstyle O}{\|}}{C}-OH \ + \ H-O-CH_2CH_2CH_3 \underset{\text{Heat}}{\overset{H^+}{\rightleftharpoons}} CH_3-\overset{\overset{\textstyle O}{\|}}{C}-O-CH_2CH_2CH_3 \ + \ \textbf{H}_2\textbf{O}$$

Acetic acid   1-Propanol   Propyl acetate
(pears)

SAMPLE PROBLEM 17.6

Writing Esterification Equations

The ester that gives the flavor and odor of apples can be synthesized from butyric acid and methyl alcohol. What is the equation for the formation of the ester in apples?

Solution

$$CH_3-CH_2-CH_2-\overset{\overset{\displaystyle O}{\|}}{C}-\textbf{OH} + \textbf{H}-O-CH_3 \underset{\text{Heat}}{\overset{H^+}{\rightleftharpoons}}$$

Butyric acid   Methyl alcohol

$$CH_3-CH_2-CH_2-\overset{\overset{\displaystyle O}{\|}}{C}-O-CH_3 + \textbf{H}_2\textbf{O}$$

Methyl butyrate

Study Check

What carboxylic acid and alcohol are needed to form the following ester, which gives the flavor and odor to apricots?

$$CH_3-CH_2-\overset{\overset{\displaystyle O}{\|}}{C}-O-CH_2-CH_2-CH_2-CH_2-CH_3$$

HEALTH NOTE

Salicylic Acid and Aspirin

Chewing on a piece of willow bark was a way of relieving pain for many centuries. By the 1800s, chemists discovered that salicylic acid was the agent in the bark responsible for the relief of pain. However, salicylic acid, which has both a carboxylic group and a hydroxyl group, irritates the stomach lining. An ester of salicylic acid and acetic acid called acetylsalicylic acid or "aspirin" that is less irritating was prepared in 1899 by the Bayer chemical company in Germany. In some aspirin preparations, a buffer is added to neutralize the carboxylic acid group and lessen its irritation of the stomach. Aspirin is used as an analgesic (pain reliever), antipyretic (fever reducer), and anti-inflammatory agent.

Salicylic acid   Acetic acid

Acetylsalicylic acid, "aspirin"

Salicylic acid   Methyl alcohol

Methyl salicylate
(oil of wintergreen)

Oil of wintergreen, or methyl salicylate, has a spearmint odor and flavor. Because it can pass through the skin, methyl salicylate is used in skin ointments where it acts as a counterirritant, producing heat to soothe sore muscles.

ENVIRONMENTAL NOTE

Plastics

Terephthalic acid (an acid with two carboxyl groups) is produced in large quantities for the manufacture of polyesters such as Dacron, and plastics. When terephthalic acid reacts with ethylene glycol, ester bonds can form on both ends of the molecules, allowing many molecules to combine until they have formed a long polymer known as a *polyester*.

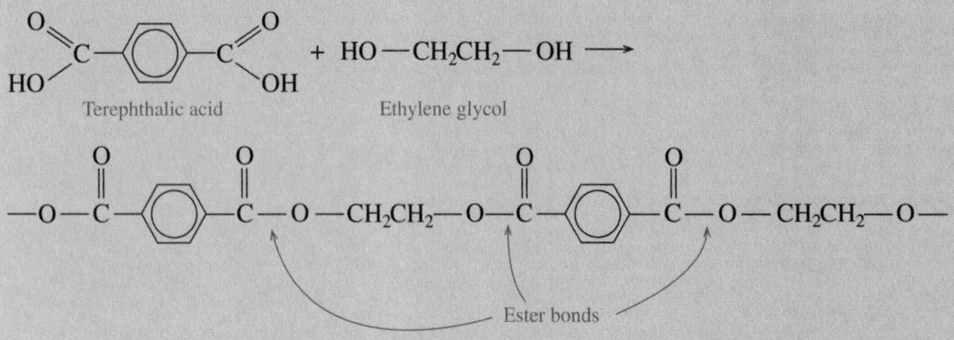

Terephthalic acid   Ethylene glycol

Ester bonds
A section of the polyester Dacron

Dacron polyester is used to make permanent press fabrics, carpets, and clothes. In medicine, artificial blood vessels and valves are made of Dacron, which is biologically inert and does not clot the blood. The polyester can also be made as a film called Mylar and as a plastic known as PETE (**poly**ethylene**ter**phthalate). PETE is used for plastic soft drink bottles as well as for containers of salad dressings, shampoos, and dishwashing liquids.

Today PETE is the most widely recycled of all the plastics. In 1992, 365 million pounds (166 million kilograms) of PETE were recycled. After it is separated from other plastics, PETE can be changed into other useful items including polyester fabric for T-shirts and coats, fill for sleeping bags, door mats, and tennis ball containers.

QUESTIONS AND PROBLEMS

Esters of Carboxylic Acids

17.21 Identify each of the following as an aldehyde, a ketone, a carboxylic acid, or an ester:

a. $CH_3-\overset{\overset{\displaystyle O}{\|}}{C}-H$ **b.** $CH_3-\overset{\overset{\displaystyle O}{\|}}{C}-O-CH_3$ **c.** $CH_3-CH_2-\overset{\overset{\displaystyle O}{\|}}{C}-CH_3$ **d.** $CH_3-CH_2-\overset{\overset{\displaystyle O}{\|}}{C}-O-H$

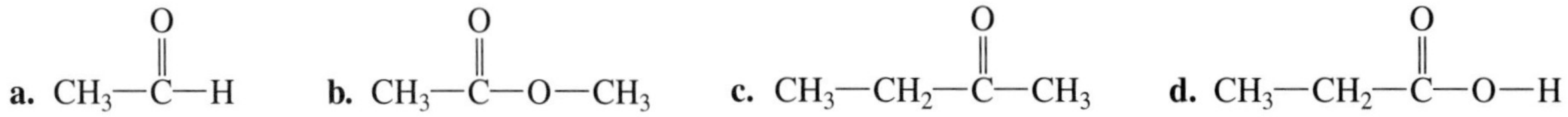

17.22 Identify each of the following as an aldehyde, a ketone, a carboxylic acid, or an ester:

a. $CH_3-\overset{\overset{\displaystyle O}{\|}}{C}-OH$ **b.** $CH_3-\overset{\overset{\displaystyle O}{\|}}{C}-O-CH_2-CH_3$ **c.** $CH_3-CH_2-\overset{\overset{\displaystyle O}{\|}}{C}-H$ **d.** $CH_3-\overset{\overset{\displaystyle CH_3}{|}}{CH}-\overset{\overset{\displaystyle O}{\|}}{C}-O-CH_2-CH_3$

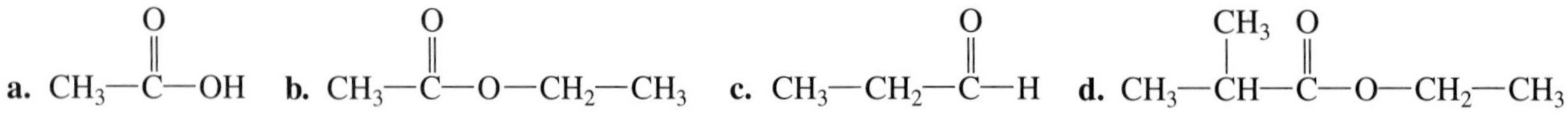

17.23 Write the condensed structural formula of the ester formed when each of the following react with methyl alcohol:
   a. acetic acid   **b.** butyric acid   **c.** benzoic acid

17.24 Write the condensed structural formula of the ester formed when each of the following react with methyl alcohol:
   a. formic acid   **b.** propionic acid   **c.** 2-methylpentanoic acid

17.25 Draw the condensed structural formulas of the ester formed when each of the following carboxylic acids and alcohols react:

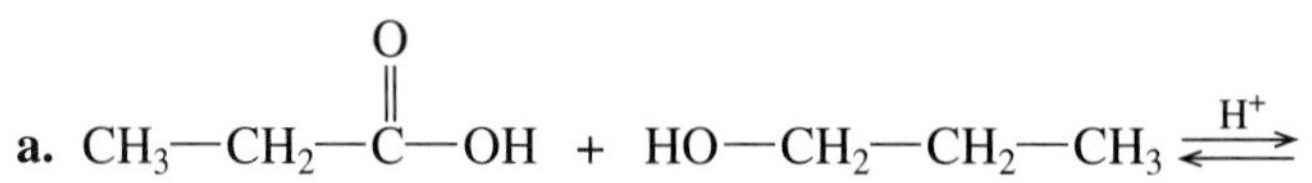

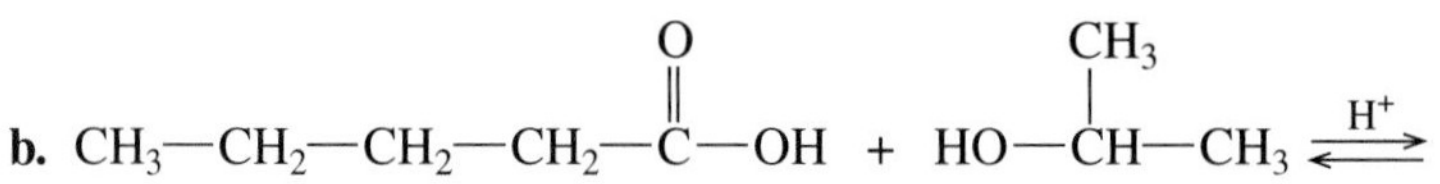

a. $CH_3-CH_2-\overset{\overset{\displaystyle O}{\|}}{C}-OH \ + \ HO-CH_2-CH_2-CH_3 \ \underset{\longleftarrow}{\overset{H^+}{\longrightarrow}}$

b. $CH_3-CH_2-CH_2-CH_2-\overset{\overset{\displaystyle O}{\|}}{C}-OH \ + \ HO-\overset{\overset{\displaystyle CH_3}{|}}{CH}-CH_3 \ \underset{\longleftarrow}{\overset{H^+}{\longrightarrow}}$

17.26 Draw the condensed structural formula of the ester formed when each of the following carboxylic acids and alcohols react:

a. $CH_3-CH_2-\overset{\displaystyle O}{\overset{\displaystyle \|}{C}}-OH \ + \ HO-CH_3 \ \overset{H^+}{\rightleftarrows}$

b. $\overset{\displaystyle O}{\overset{\displaystyle \|}{C}}-OH \ + \ HO-CH_2-CH_2-CH_2-CH_3 \ \overset{H^+}{\rightleftarrows}$

17.5 Naming Esters

The name of an ester consists of two words taken from the names of the alcohol and the acid. The first word indicates the *alkyl* part of the alcohol. The second word is the *carboxylate* name of the carboxylic acid. The IUPAC names of esters use the IUPAC names for the alkyl group and the carboxylate ion, while the common names of esters use the common names of each.

Ester

$$R-O\overset{\displaystyle}{\underset{}{\big|}}\overset{\displaystyle O}{\overset{\displaystyle \|}{C}}-R$$

From alcohol   From carboxylic acid
(alkyl)   (carboxylate)

Let's take a look at the following ester and break it into two parts, one from the alcohol and one from the acid. By writing and naming the alcohol and carboxylic acid that produced the ester, we can determine the name of the ester (Figure 17.6).

$$CH_3-O\overset{\displaystyle}{\underset{}{\big|}}\overset{\displaystyle O}{\overset{\displaystyle \|}{C}}-CH_3$$

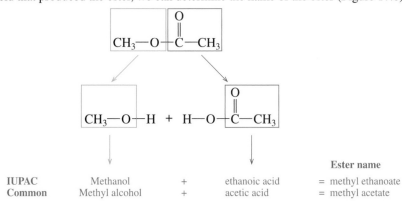

$$CH_3-O\big|H \ + \ H-O\overset{\displaystyle O}{\overset{\displaystyle \|}{C}}-CH_3$$

				Ester name
IUPAC	Methanol	+	ethanoic acid	= methyl ethanoate
Common	Methyl alcohol	+	acetic acid	= methyl acetate

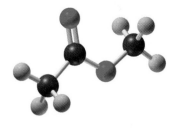

Figure 17.6 The ester methyl ethanoate (methyl acetate) is made from methyl alcohol and ethanoic acid (acetic acid).

Q What change is made in the name of the carboxylic acid used to make the ester?

The following examples of some typical esters show the IUPAC as well as the common names of esters.

$$CH_3-CH_3-O-\overset{\displaystyle O}{\overset{\displaystyle \|}{C}}-CH_3 \qquad CH_3-O-\overset{\displaystyle O}{\overset{\displaystyle \|}{C}}-CH_2-CH_3 \qquad CH_3-CH_2-O-\overset{\displaystyle O}{\overset{\displaystyle \|}{C}}$$

Ethyl ethanoate   Methyl propanoate   Ethyl benzoate
(ethyl acetate)   (methyl propionate)

Carboxylic Acids and Esters in Foods and Cosmetics

Smell some fruits such as a banana, strawberry, pear, and orange, and taste them if you like.

Look for the names of carboxylic acids and esters on the labels of food products, flavorings, and cosmetics you may use. List their names and write the formulas as you can. For complex names, use a *Merck Index* to look up the formulas of the substances whose names you read on the product labels.

Questions

1. For each fruit that you smell or taste, write the formula of the ester that is responsible for its characteristic smell and flavor.
2. What are the carboxylic acid or ester formulas found in some food products, flavorings, and cosmetics?
3. What are the formulas of some of the carboxylic acids and alcohols used in the esters?

SAMPLE PROBLEM 17.7

Naming Esters

Write the IUPAC and common names of the following ester:

$$CH_3-CH_2-\overset{\displaystyle O}{\overset{\|}{C}}-O-CH_2-CH_2-CH_3$$

Solution

The alcohol part of the ester is propyl, and the carboxylic acid part is propanoic (propionic) acid.

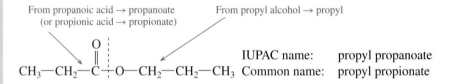

From propanoic acid → propanoate (or propionic acid → propionate)   From propyl alcohol → propyl

$$CH_3-CH_2-\overset{\displaystyle O}{\overset{\|}{C}}\overset{\vdots}{|}O-CH_2-CH_2-CH_3$$

IUPAC name: propyl propanoate
Common name: propyl propionate

Study Check

Draw the condensed structural formula of pentyl acetate.

Table 17.4   Some Naturally Occurring Esters in Fruits and Flavorings

Condensed Structural Formula	Name	Flavor/Odor
$HC\overset{\overset{\displaystyle O}{\|}}{}-O-CH_2CH_3$	Ethyl methanoate (ethyl formate)	Rum
$HC\overset{\overset{\displaystyle O}{\|}}{}-O-CH_2\overset{\overset{\displaystyle CH_3}{\|}}{C}HCH_3$	Isobutyl methanoate (isobutyl formate)	Raspberries
$CH_3\overset{\overset{\displaystyle O}{\|}}{C}-O-CH_2CH_2CH_3$	Propyl ethanoate (propyl acetate)	Pears
$CH_3\overset{\overset{\displaystyle O}{\|}}{C}-O-CH_2CH_2CH_2CH_2CH_3$	Pentyl ethanoate (pentyl acetate)	Bananas
$CH_3\overset{\overset{\displaystyle O}{\|}}{C}-O-CH_2CH_2CH_2CH_2CH_2CH_2CH_2CH_3$	Octyl ethanoate (octyl acetate)	Oranges
$CH_3CH_2CH_2\overset{\overset{\displaystyle O}{\|}}{C}-O-CH_2CH_3$	Ethyl butanoate (ethyl butyrate)	Pineapples
$CH_3CH_2CH_2\overset{\overset{\displaystyle O}{\|}}{C}-O-CH_2CH_2CH_2CH_2CH_3$	Pentyl butanoate (pentyl butyrate)	Apricots
$CH_3CH_2CH_2\overset{\overset{\displaystyle O}{\|}}{C}-SCH_3$	Methyl thiobutanoate (methyl thiobutyrate)	Strawberries

Esters in Plants

Many of the fragrances of perfumes and flowers and the flavors of fruits are due to esters. Small esters are volatile so we can smell them and soluble in water so we can taste them. (See Figure 17.7.) Several of these are listed in Table 17.4.

QUESTIONS AND PROBLEMS

Naming Esters

17.27 Give the names of the carboxylic acid and alcohol needed to produce each of the following esters:

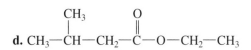

$$\textbf{a. } H—\overset{\overset{\textstyle O}{\|}}{C}—O—CH_3 \qquad \textbf{b. } CH_3—\overset{\overset{\textstyle O}{\|}}{C}—O—CH_3$$

$$\textbf{c. } CH_3—CH_2—CH_2—\overset{\overset{\textstyle O}{\|}}{C}—O—CH_3$$

$$\textbf{d. } CH_3—\overset{\overset{\textstyle CH_3}{|}}{CH}—CH_2—\overset{\overset{\textstyle O}{\|}}{C}—O—CH_2—CH_3$$

17.28 Give the names of the carboxylic acid and alcohol needed to produce each of the following esters:

$$\textbf{a. } CH_3—CH_2—\overset{\overset{\textstyle O}{\|}}{C}—O—CH_2—CH_3 \qquad \textbf{b. } CH_3—CH_2—CH_2—CH_2—CH_2—\overset{\overset{\textstyle O}{\|}}{C}—O—CH_3$$

$$\textbf{c. } CH_3—CH_2—\underset{\underset{\textstyle CH_3}{|}}{\overset{\overset{\textstyle O}{\|}}{CH}}—\overset{\overset{\textstyle O}{\|}}{C}—O—CH_3 \qquad \textbf{d. } CH_3—CH_2—\overset{\overset{\textstyle O}{\|}}{C}—O—CH_2—CH_2—CH_2—CH_3$$

17.29 Name each of the following esters:

$$\textbf{a. } CH_3—O—\overset{\overset{\textstyle O}{\|}}{C}—H \qquad \textbf{b. } CH_3—O—\overset{\overset{\textstyle O}{\|}}{C}—CH_3$$

$$\textbf{c. } CH_3—O—\overset{\overset{\textstyle O}{\|}}{C}—CH_2—CH_2—CH_3 \qquad \textbf{d. } CH_3—\overset{\overset{\textstyle CH_3}{|}}{CH}—CH_2—\overset{\overset{\textstyle O}{\|}}{C}—O—CH_2—CH_3$$

17.30 Name each of the following esters:

$$\textbf{a. } CH_3—CH_2—O—\overset{\overset{\textstyle O}{\|}}{C}—CH_2—CH_2—CH_3 \qquad \textbf{b. } CH_3—O—\overset{\overset{\textstyle O}{\|}}{C}—CH_2—CH_2—CH_2—CH_2—CH_3$$

$$\textbf{c. } CH_3—O—\overset{\overset{\textstyle O}{\|}}{C}—CH_2—\overset{\overset{\textstyle CH_3}{|}}{CH}—CH_3 \qquad \textbf{d. } CH_3—CH_2—\overset{\overset{\textstyle O}{\|}}{C}—O—CH_2—CH_2—CH_2—CH_3$$

17.31 Draw the condensed structural formulas of each of the following esters:
  a. methyl acetate   **b.** butyl formate
  c. ethyl pentanoate   **d.** 2-bromopropyl propanoate

17.32 Draw the condensed structural formulas of each of the following esters:
  a. hexyl acetate   **b.** propyl propionate
  c. ethyl-2-hydroxybutanoate   **d.** methyl benzoate

Figure 17.7 Esters are responsible for part of the odor and flavor of oranges, bananas, pears, pineapples, and strawberries.

Q **What is the ester found in pineapple?**

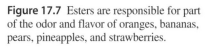

17.33 What is the ester responsible for the flavor and odor of the following fruit?
  a. banana   **b.** orange   **c.** apricot   **d.** raspberry

17.34 What flavor would you notice if you smelled or tasted the following?
  a. ethyl butanoate   **b.** propyl acetate
  c. methyl thiobutanoate   **d.** ethyl formate

LEARNING GOAL

Describe the boiling points and solubility of esters. Draw the condensed structural formulas of the hydrolysis products.

17.6 Properties of Esters

Esters have higher boiling points than alkanes, but lower than alcohols and carboxylic acids of similar mass. Because ester molecules do not have hydroxyl groups, they cannot hydrogen bond to each other.

CH_3—CH_2—CH_2—CH_3	CH_3—O—CH_2—CH_3	CH_3—O—$\overset{\overset{\displaystyle O}{\|\|}}{C}$—H	CH_3—CH_2—CH_2—OH	CH_3—$\overset{\overset{\displaystyle O}{\|\|}}{C}$—OH
Butane	Ethyl methyl ether	Methyl formate	1-Propanol	Acetic acid
Type Alkane	Ether	Ester	Alcohol	Carboxylic acid
Bp 0°C	11°C	32°C	97°C	118°C
Mass 58	60	60	60	60

Increasing boiling points →

Solubility in Water

It is possible for esters to use the lone pairs of electrons on the oxygen atom to hydrogen bond with water. Small esters having only a few carbon atoms are soluble in water. However, the solubility of esters decreases as the number of carbons increases.

Acid Hydrolysis of Esters

In **hydrolysis,** esters are split apart with water when heated in the presence of a strong acid, usually H_2SO_4 or HCl. The products of acid hydrolysis are the carboxylic acid and alcohol. Therefore, hydrolysis is the reverse of the esterification reaction. When hydrolysis of biological compounds occurs in the cells, an enzyme replaces the acid as the catalyst. In the hydrolysis reaction, the —OH from a water molecule replaces the —OR part of the ester to form the carboxylic acid.

$$R—\overset{\overset{\displaystyle O}{\|\|}}{C}—O—R \; + \; H—OH \; \underset{}{\overset{H^+}{\rightleftharpoons}} \; R—\overset{\overset{\displaystyle O}{\|\|}}{C}—O—H \; + \; R—OH$$

  Ester   Water   Carboxylic acid   Alcohol

Example

$$CH_3—\overset{\overset{\displaystyle O}{\|\|}}{C}—OCH_3 \; + \; H—OH \; \overset{H^+}{\rightleftharpoons} \; CH_3—\overset{\overset{\displaystyle O}{\|\|}}{C}—OH \; + \; H—OCH_3$$

  Methyl acetate   Acetic acid   Methyl alcohol

SAMPLE PROBLEM 17.8

Acid Hydrolysis of Esters

Aspirin that has been stored for a long time may undergo hydrolysis in the presence of water and heat. What are the hydrolysis products of aspirin? Why does a bottle of old aspirin smell like vinegar?

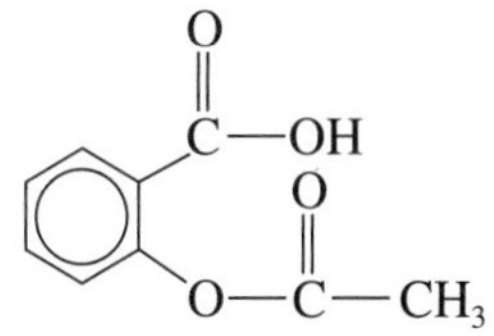

Aspirin, Acetylsalicylic acid

Solution

To write the hydrolysis products, separate the compound at the ester bond. Complete the formula of the carboxylic acid by adding —OH (from water) to the carbonyl group and the —H to complete the alcohol. The acetic acid in the products gives the vinegar odor to a sample of aspirin that has hydrolyzed.

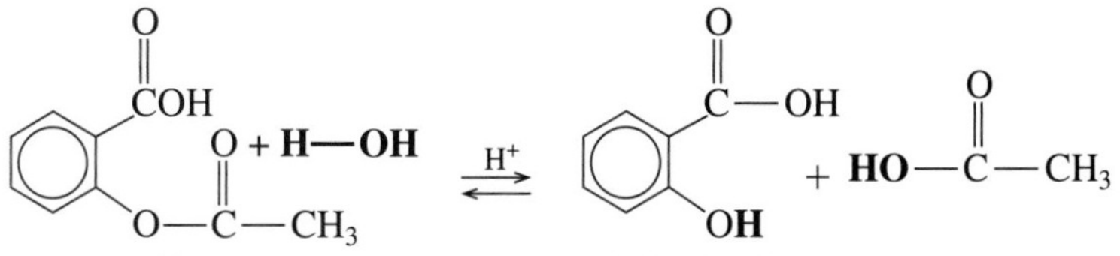

Aspirin Salicylic acid Acetic acid

Study Check

What are the names of the products from the acid hydrolysis of ethyl propionate?

Base Hydrolysis of Esters

When an ester undergoes hydrolysis with a strong base such as NaOH or KOH, the products are the carboxylic acid salt and the corresponding alcohol. The base hydrolysis reaction is also called **saponification,** which refers to the reaction of a long-chain fatty acid with NaOH to make soap. The carboxylic acid, which is produced in acid hydrolysis, is converted in an irreversible reaction to its carboxylate ion by the strong base.

$$\underset{\text{Ester}}{R-\overset{O}{\overset{\|}{C}}-O-R} + \underset{\text{Sodium hydroxide}}{NaOH} \longrightarrow \underset{\text{Carboxylic acid salt}}{R-\overset{O}{\overset{\|}{C}}-O^{-}Na^{+}} + \underset{\text{Alcohol}}{R-OH}$$

$$\underset{\substack{\text{Methyl ethanoate}\\ \text{(methyl acetate)}}}{CH_3-\overset{O}{\overset{\|}{C}}-O-CH_3} + \underset{\text{Sodium hydroxide}}{NaOH} \longrightarrow \underset{\substack{\text{Sodium acetate}\\ \text{(sodium ethanoate)}}}{CH_3-\overset{O}{\overset{\|}{C}}-O^{-}Na^{+}} + \underset{\substack{\text{Methanol}\\ \text{(methyl alcohol)}}}{CH_3-OH}$$

SAMPLE PROBLEM 17.9

Base Hydrolysis of Esters

Ethyl acetate is a solvent widely used for fingernail polish, plastics, and lacquers. Write the equation of the hydrolysis of ethyl acetate by NaOH.

Solution

The hydrolysis of ethyl acetate by NaOH gives the salt of acetic acid and ethyl alcohol.

$$CH_3-\overset{\overset{\displaystyle O}{\|}}{C}-OCH_2CH_3 \ + \ NaOH \ \xrightarrow{\text{Heat}} \ CH_3-\overset{\overset{\displaystyle O}{\|}}{C}-O^-Na^+ \ + \ HOCH_2CH_3$$

Ethyl acetate   Sodium acetate   Ethyl alcohol

Study Check

Write the condensed structural formulas of the products from the hydrolysis of methyl benzoate by KOH.

ENVIRONMENTAL NOTE

Cleaning Action of Soaps

For many centuries, soaps were made by heating a mixture of animal fats (tallow) with lye, a basic solution obtained from wood ashes. In the soap-making process, fatty acids, which are long-chain carboxylic acids, undergo saponification with the strong base in lye.

Fatty acid

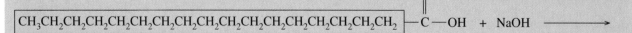

$$\boxed{CH_3CH_2CH_2CH_2CH_2CH_2CH_2CH_2CH_2CH_2CH_2CH_2CH_2CH_2CH_2CH_2CH_2}-\overset{\overset{\displaystyle O}{\|}}{C}-OH \ + \ NaOH \longrightarrow$$

Carboxylic acid salt, "soap"

$$\boxed{CH_3CH_2CH_2CH_2CH_2CH_2CH_2CH_2CH_2CH_2CH_2CH_2CH_2CH_2CH_2CH_2CH_2}-\overset{\overset{\displaystyle O}{\|}}{C}-O^-Na^+$$

Nonpolar tail   Polar head
(hydrophobic)   (hydrophilic)

Today soaps are also prepared from fats such as coconut oil. Perfumes are added to give a pleasant-smelling soap. Because a soap is the salt of a long-chain fatty acid, the two ends of a soap molecule have different polarities. The long carbon chain end is nonpolar and *hydrophobic* (water-fearing). It is soluble in nonpolar substances such as oil or grease; but it is not soluble in water. The carboxylate salt end is ionic and *hydrophilic* (water-loving). It is very soluble in water but not in oils or grease.

When a soap is used to clean grease or oil, the nonpolar ends of the soap molecules dissolve in the nonpolar fats and oils that accompany dirt. The water-loving salt ends of the soap molecules extend outside where they can dissolve in water. The soap molecules coat the oil or grease, forming clusters called *micelles*. The ionic ends of the soap molecules provide polarity to the micelles, which makes them soluble in water. As a result, small globules of oil and fat coated with soap molecules are pulled into the water and rinsed away.

One of the problems of using soaps is that the carboxylate end reacts with ions in water such as Ca^{2+} and Mg^{2+} and forms insoluble substances.

$$2CH_3(CH_2)_{16}COO^- \ + \ Mg^{2+} \longrightarrow \ [CH_3(CH_2)_{16}COO^-]_2Mg^{2+}$$

Stearate ion   Magnesium ion   Magnesium stearate
                                                          (insoluble)

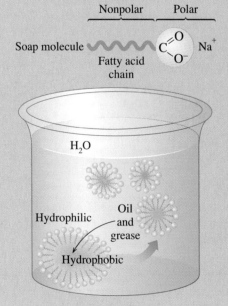

QUESTIONS AND PROBLEMS

Properties of Esters

17.35 For each of the following pairs of compounds, select the compound that has the higher boiling point:

a. 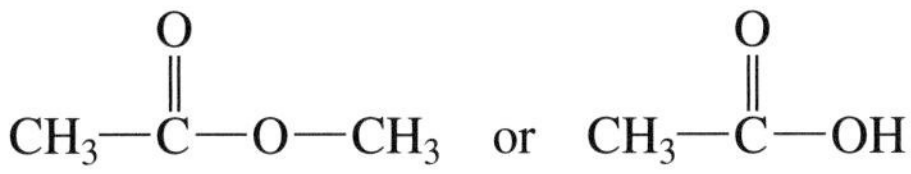 $CH_3-\overset{\overset{\displaystyle O}{\|}}{C}-O-CH_3$ or $CH_3-\overset{\overset{\displaystyle O}{\|}}{C}-OH$

b. $CH_3-\overset{\overset{\displaystyle O}{\|}}{C}-O-CH_3$ or $CH_3-CH_2-CH_2-CH_2-OH$

c. $CH_3-CH_2-CH_2-CH_3$ or $CH_3-O-\overset{\overset{\displaystyle O}{\|}}{C}-CH_3$

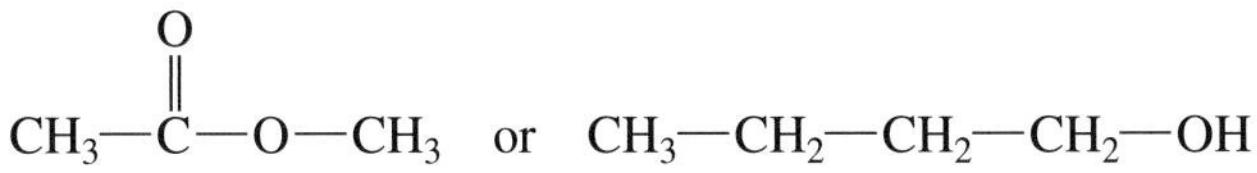

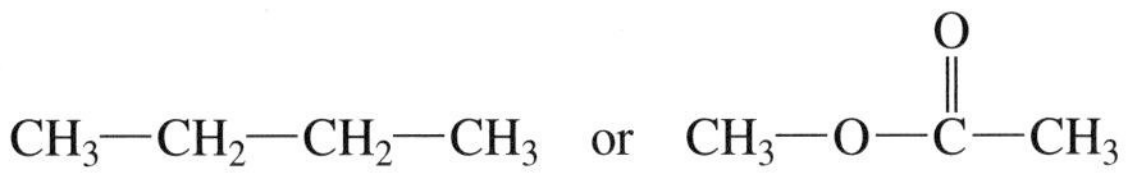

17.36 For each of the following pairs of compounds, select the compound that has the higher boiling point:

a. 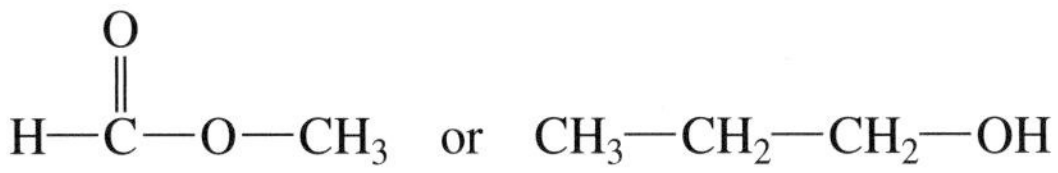 $H-\overset{\overset{\displaystyle O}{\|}}{C}-O-CH_3$ or $CH_3-CH_2-CH_2-OH$ **b.** $CH_3-\overset{\overset{\displaystyle O}{\|}}{C}-O-CH_3$ or $CH_3-\overset{\overset{\displaystyle O}{\|}}{C}-CH_2-CH_3$

c. $CH_3-O-CH_2-CH_3$ or $CH_3-O-\overset{\overset{\displaystyle O}{\|}}{C}-H$

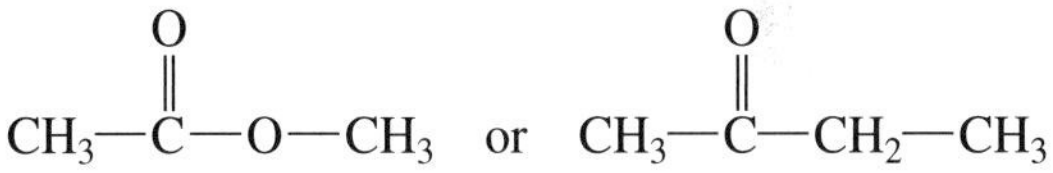

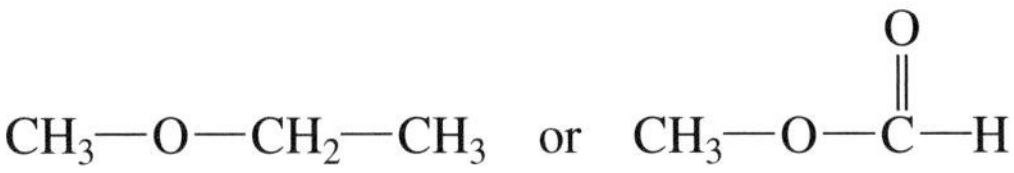

17.37 What are the products of the acid hydrolysis of an ester?

17.38 What are the products of the base hydrolysis of an ester?

17.39 Draw the condensed structural formulas of the products from the acid- or base-catalyzed hydrolysis of each of the following compounds:

a. 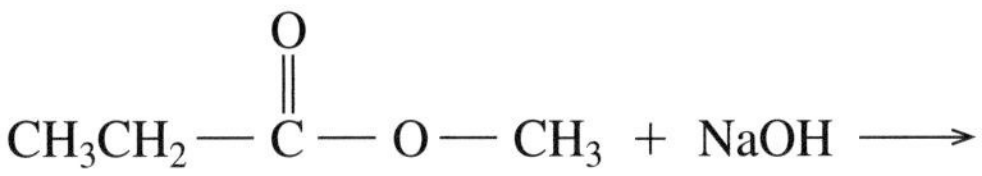 $CH_3CH_2-\overset{\overset{\displaystyle O}{\|}}{C}-O-CH_3 + NaOH \longrightarrow$ **b.** $CH_3-\overset{\overset{\displaystyle O}{\|}}{C}-O-CH_2CH_2CH_3 + H_2O \xrightarrow{H^+}$

c. $CH_3CH_2CH_2-\overset{\overset{\displaystyle O}{\|}}{C}-O-CH_2CH_3 + H_2O \xrightarrow{H^+}$ **d.** 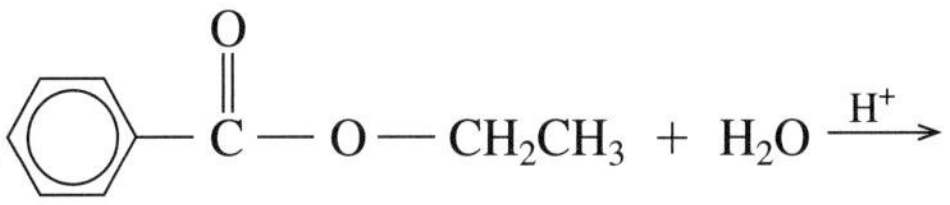 $\langle\bigcirc\rangle-\overset{\overset{\displaystyle O}{\|}}{C}-O-CH_2CH_3 + H_2O \xrightarrow{H^+}$

e. 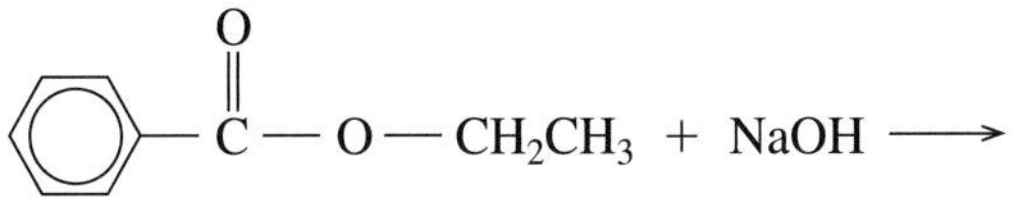 $\langle\bigcirc\rangle-\overset{\overset{\displaystyle O}{\|}}{C}-O-CH_2CH_3 + NaOH \longrightarrow$

17.40 Draw the condensed structural formulas of the products from the acid- or base-catalyzed hydrolysis of each of the following compounds:

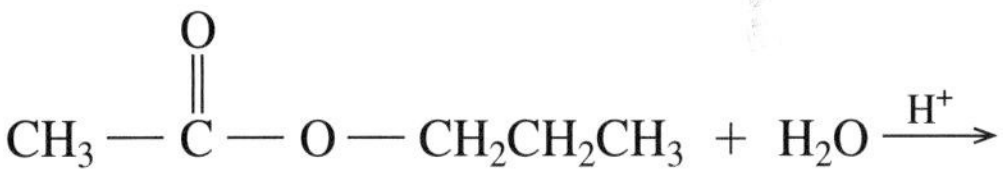

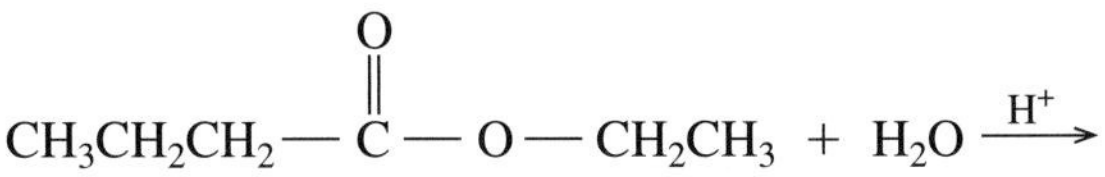

a. 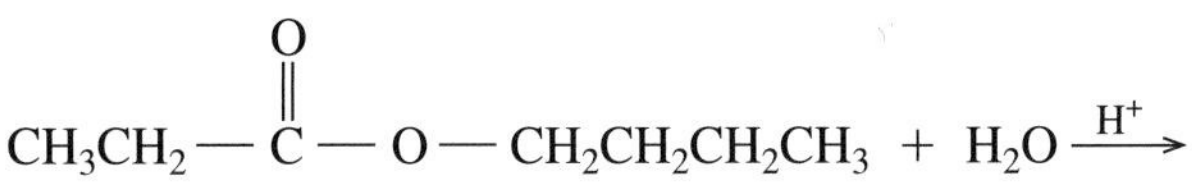 $CH_3CH_2-\overset{\overset{\displaystyle O}{\|}}{C}-O-CH_2CH_2CH_2CH_3 + H_2O \xrightarrow{H^+}$ **b.** $H-\overset{\overset{\displaystyle O}{\|}}{C}-O-CH_2CH_3 + NaOH \longrightarrow$

c. $CH_3CH_2-\overset{\overset{\displaystyle O}{\|}}{C}-O-CH_3 + H_2O \xrightarrow{H^+}$ **d.** 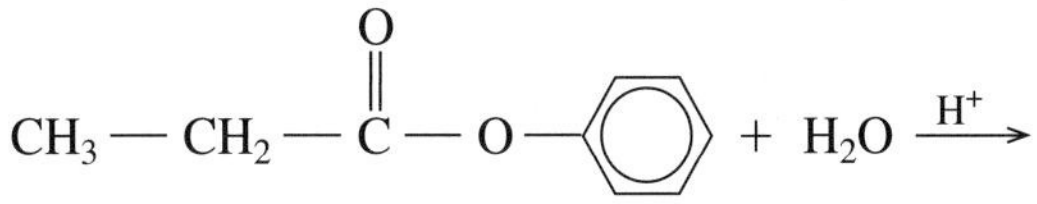 $CH_3-CH_2-\overset{\overset{\displaystyle O}{\|}}{C}-O-\langle\bigcirc\rangle + H_2O \xrightarrow{H^+}$

e. 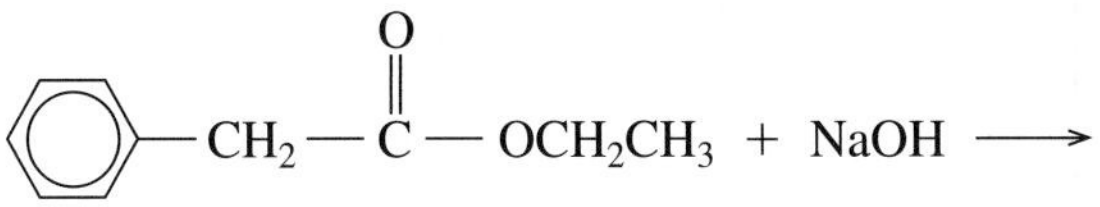 $\langle\bigcirc\rangle-CH_2-\overset{\overset{\displaystyle O}{\|}}{C}-OCH_2CH_3 + NaOH \longrightarrow$

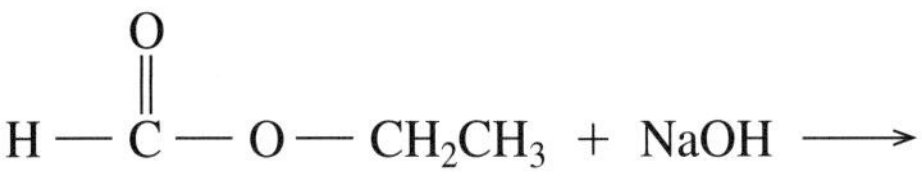

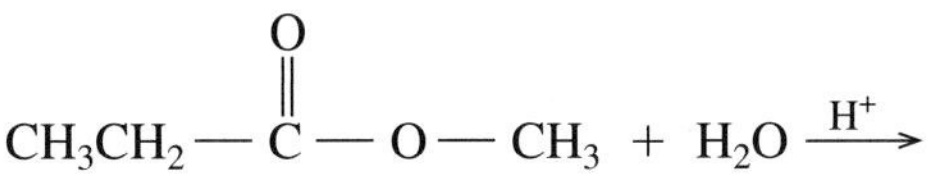

Chapter Review

17.1 Carboxylic Acids

A carboxylic acid contains the carboxyl functional group, which is a hydroxyl group connected to the carbonyl group.

17.2 Physical Properties of Carboxylic Acids

The carboxyl group contains polar bonds of $O-H$ and $C=O$, which makes a carboxylic acid with one to four carbon atoms very soluble in water.

17.3 Acidity of Carboxylic Acids

As weak acids, carboxylic acids ionize slightly by donating a proton to water to form carboxylate and hydronium ions. Carboxylic acids are neutralized by base, producing the carboxylate salt and water.

17.4 Esters of Carboxylic Acids

In an ester, an alkyl or aromatic group has replaced the H of the hydroxyl group of a carboxylic acid. In the presence of a strong acid, a carboxylic acid reacts with an alcohol to produce an ester. A molecule of water is removed: $-OH$ from the carboxylic acid and $-H$ from the alcohol molecule.

17.5 Naming Esters

The names of esters consist of two words, one from the alcohol and the other from the carboxylic acid with the *ic* ending replaced by *ate*.

17.6 Properties of Esters

Esters undergo acid hydrolysis by adding water to yield the carboxylic acid and alcohol (or phenol). Base hydrolysis or saponification of an ester produces the carboxylate salt and an alcohol.

Summary of Naming

Family	Condensed Structural Formula	IUPAC Name	Common Name
Carboxylic acid	$CH_3-\overset{\overset{\displaystyle O}{\|\|}}{C}-OH$	Ethanoic acid	Acetic acid
Carboxylic acid salt	$CH_3-\overset{\overset{\displaystyle O}{\|\|}}{C}-O^-Na^+$	Sodium ethanoate	Sodium acetate
Ester	$CH_3-\overset{\overset{\displaystyle O}{\|\|}}{C}-OCH_3$	Methyl ethanoate	Methyl acetate

Summary of Reactions

Ionization of a Carboxylic Acid in Water

$$R-\overset{\overset{\displaystyle O}{\|\|}}{C}-OH \; + \; H_2O \; \rightleftharpoons \; R-\overset{\overset{\displaystyle O}{\|\|}}{C}-O^- \; + \; H_3O^+$$

Carboxylic acid (weak acid) Water (weak base)   Carboxylate ion   Hydronium ion

$$CH_3-\overset{\overset{\displaystyle O}{\|\|}}{C}-OH \; + \; H_2O \; \rightleftharpoons \; CH_3-\overset{\overset{\displaystyle O}{\|\|}}{C}-O^- \; + \; H_3O^+$$

Ethanoic acid (acetic acid)   Ethanoate ion (acetate ion)   Hydronium ion

Neutralization of a Carboxylic Acid

$$R-\overset{\displaystyle O}{\underset{\displaystyle \|}{C}}-OH \ + \ NaOH \ \longrightarrow \ R-\overset{\displaystyle O}{\underset{\displaystyle \|}{C}}-O^-Na^+ \ + \ H_2O$$

Carboxylic acid   Base   Carboxylic acid salt

$$CH_3CH_2-\overset{\displaystyle O}{\underset{\displaystyle \|}{C}}-OH \ + \ NaOH \ \longrightarrow \ CH_3CH_2-\overset{\displaystyle O}{\underset{\displaystyle \|}{C}}-O^-Na^+ \ + \ H_2O$$

Propanoic acid   Sodium   Sodium propanoate
(propionic acid)   hydroxide   (sodium propionate)

Esterification: Carboxylic Acid and an Alcohol

$$R-\overset{\displaystyle O}{\underset{\displaystyle \|}{C}}-OH \ + \ HO-R \ \overset{H^+}{\rightleftharpoons} \ R-\overset{\displaystyle O}{\underset{\displaystyle \|}{C}}-O-R \ + \ H_2O$$

Carboxylic acid   Alcohol   Ester

$$CH_3-\overset{\displaystyle O}{\underset{\displaystyle \|}{C}}-OH \ + \ HO-CH_3 \ \overset{H^+}{\rightleftharpoons} \ CH_3-\overset{\displaystyle O}{\underset{\displaystyle \|}{C}}-O-CH_3 \ + \ H_2O$$

Ethanoic acid   Methanol   Methyl ethanoate
(acetic acid)   (methyl alcohol)   (methyl acetate)

Acid Hydrolysis of an Ester

$$R-\overset{\displaystyle O}{\underset{\displaystyle \|}{C}}-O-R \ + \ H-OH \ \overset{H^+}{\rightleftharpoons} \ R-\overset{\displaystyle O}{\underset{\displaystyle \|}{C}}-OH \ + \ R-OH$$

Ester   Carboxylic acid   Alcohol

$$CH_3-\overset{\displaystyle O}{\underset{\displaystyle \|}{C}}-OCH_3 \ + \ H-OH \ \overset{H^+}{\rightleftharpoons} \ CH_3-\overset{\displaystyle O}{\underset{\displaystyle \|}{C}}-OH \ + \ H-OCH_3$$

Methyl ethanoate   Ethanoic acid   Methanol
(methyl acetate)   (acetic acid)   (methyl alcohol)

Base Hydrolysis of an Ester

$$R-\overset{\displaystyle O}{\underset{\displaystyle \|}{C}}-O-R \ + \ NaOH \ \overset{Heat}{\longrightarrow} \ R-\overset{\displaystyle O}{\underset{\displaystyle \|}{C}}-O^-Na^+ \ + \ R-OH$$

Ester   Base   Carboxylate salt   Alcohol

$$CH_3CH_2-\overset{\displaystyle O}{\underset{\displaystyle \|}{C}}-OCH_3 \ + \ NaOH \ \overset{Heat}{\longrightarrow} \ CH_3CH_2C-O^-Na^+ \ + \ HOCH_3$$

Methyl propanoate   Sodium   Sodium propanoate   Methanol
(methyl propionate)   hydroxide   (sodium propionate)   (methyl
                                                                  alcohol)

Key Terms

carboxyl group A functional group found in carboxylic acids composed of carbonyl and hydroxyl groups.

$$-\overset{\displaystyle O}{\underset{\displaystyle \|}{C}}-OH \qquad \text{Carboxyl group}$$

carboxylate ion The anion produced when a carboxylic acid donates a proton to water.

carboxylic acids A family of organic compounds containing the carboxyl group.

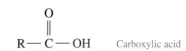

$$R-\overset{\displaystyle O}{\underset{\displaystyle \|}{C}}-OH \qquad \text{Carboxylic acid}$$

carboxylic acid salt The product of neutralization of a carboxylic acid; a carboxylate ion and the metal ion from the base.

esterification The formation of an ester from a carboxylic acid and an alcohol with the elimination of a molecule of water in the presence of an acid catalyst.

esters A family of organic compounds in which an alkyl group replaces the hydrogen atom in a carboxylic acid.

$$R-\overset{\overset{\displaystyle O}{\|}}{C}-O-R \quad \text{Ester}$$

hydrolysis The splitting of a molecule by the addition of water. Esters hydrolyze to produce a carboxylic acid and an alcohol.

saponification The hydrolysis of an ester with a strong base to produce a salt of the carboxylic acid and an alcohol.

Additional Problems

17.41 Give the IUPAC and common names (if any) for each of the following compounds:

a. $CH_3-\overset{\overset{\displaystyle CH_3}{|}}{CH}-CH_2-\overset{\overset{\displaystyle O}{\|}}{C}-OH$

b. (benzene ring)$-\overset{\overset{\displaystyle O}{\|}}{C}-O-CH_2-CH_3$

c. $CH_3-CH_2-O-\overset{\overset{\displaystyle O}{\|}}{C}-CH_2-CH_3$

d. (benzene ring with COOH and Cl)

e. $CH_3-\overset{\overset{\displaystyle OH}{|}}{CH}-CH_2-CH_2-\overset{\overset{\displaystyle O}{\|}}{C}-OH$

f. $CH_3-\overset{\overset{\displaystyle O}{\|}}{C}-O-\overset{\overset{\displaystyle CH_3}{|}}{CH}-CH_3$

17.42 Give the IUPAC and common names (if any) for each of the following compounds:

a. $CH_3-\overset{\overset{\displaystyle CH_3}{|}}{CH}-CH_2-CH_2-\overset{\overset{\displaystyle O}{\|}}{C}-OH$

b. (benzene ring with C—OH and two Cl)

c. (benzene ring)$-\overset{\overset{\displaystyle O}{\|}}{C}-O-CH_3$

d. $CH_3-CH_2-CH_2-\overset{\overset{\displaystyle O}{\|}}{C}-O-CH_3$

e. $CH_3-CH_2-O-\overset{\overset{\displaystyle O}{\|}}{C}-CH_2-\overset{\overset{\displaystyle CH_3}{|}}{CH}-CH_3$

f. $CH_3-\overset{\overset{\displaystyle CH_3}{|}}{CH}-CH_2-\overset{\overset{\displaystyle OH}{|}}{CH}-\overset{\overset{\displaystyle O}{\|}}{C}-OH$

17.43 Draw the structural formulas of at least three carboxylic acids with the molecular formula $C_5H_{10}O_2$.

17.44 Draw the structural formulas of at least three esters with the formula $C_4H_8O_2$.

17.45 Draw the condensed structural formulas of each of the following:
a. methyl acetate   **b.** p-chlorobenzoic acid
c. β-chloropropionic acid   **d.** ethyl butanoate
e. 3-methylpentanoic acid   **f.** ethyl benzoate

17.46 Draw the condensed structural formulas of each of the following:
a. α-bromobutyric acid
b. ethyl butyrate
c. 2-methylcyclohexanoic acid
d. 3,5-dimethylhexanoic acid
e. propyl acetate   **f.** 3,4-dibromobenzoic acid

17.47 In each of the following pairs of compounds, select the compound that would have the higher boiling point:

a. $CH_3-CH_2-CH_2-OH$ or $CH_3-\overset{\overset{\displaystyle O}{\|}}{C}-OH$

b. $CH_3-CH_2-CH_2-CH_3$ or $CH_3-CH_2-\overset{\overset{\displaystyle O}{\|}}{C}-OH$

c. $CH_3-\overset{O}{\underset{||}{C}}-OH$ or $CH_3-CH_2-CH_2-\overset{O}{\underset{||}{C}}-OH$

17.48 In each of the following pairs of compounds, select the compound that would have the higher boiling point:

a. $CH_3-CH_2-CH_2-OH$ or $CH_3-\overset{O}{\underset{||}{C}}-O-CH_3$

b. $CH_3-O-\overset{O}{\underset{||}{C}}-CH_3$ or $CH_3-CH_2-\overset{O}{\underset{||}{C}}-OH$

c. $CH_3-\overset{O}{\underset{||}{C}}-O-CH_3$ or $CH_3-CH_2-CH_2-CH_3$

17.49 Why does acetic acid have a higher boiling point than either 1-propanol or methyl formate when they all have the same molar mass?

17.50 Propionic acid, 1-butanol, and butanal all have the same molar mass. The possible boiling points are 76°C, 118°C, and 141°C. Match the compounds with the boiling points and explain your choice.

17.51 Which of the following compounds are soluble in water?
  a. $CH_3-CH_2-CH_2-CH_2-CH_3$
  b. $CH_3-CH_2-\overset{O}{\underset{||}{C}}-O^-Na^+$
  c. $CH_3-\overset{O}{\underset{||}{C}}-O-CH_3$
  d. $CH_3-CH_2-CH_2-OH$
  e. $CH_3-CH_2-\overset{O}{\underset{||}{C}}-OH$

17.52 Which of the following compounds are soluble in water?
  a. $CH_3-CH_2-CH_2-\overset{O}{\underset{||}{C}}-OH$
  b. $CH_3-CH_2-\overset{O}{\underset{||}{C}}-O-CH_2-CH_2-CH_3$
  c. $CH_3-CH_2-CH_2-CH_3$

d. $CH_3-(CH_2)_8-CH_2-OH$

e. $CH_3-CH_2-CH_2-O-CH_2-CH_2-CH_3$

17.53 Methyl benzoate is not soluble in water; but when it is heated with KOH, it dissolves. Write an equation for the reaction and explain what happens. When HCl is added to the solution, a white solid forms. What is the solid?

17.54 Hexanoic acid is soluble in NaOH solution, but hexanal is not. Explain.

17.55 Salicylic acid could be named *o*-hydroxybenzoic acid.
  a. What two reactive functional groups are present?
  b. Draw the structure of the ester product that forms when the hydroxyl group of salicylic acid reacts with acetic acid.
  c. Draw the structure of methyl salicylate, oil of wintergreen, formed when salicylic acid forms an ester with methyl alcohol.

17.56 What volume of 0.100 M NaOH is needed to neutralize 3.00 g of benzoic acid?

17.57 Write the products of the following reactions:
  a. $CH_3-CH_2-\overset{O}{\underset{||}{C}}-OH + H_2O \rightleftharpoons$
  b. $CH_3-CH_2-\overset{O}{\underset{||}{C}}-OH + KOH \longrightarrow$
  c. $CH_3-CH_2-\overset{O}{\underset{||}{C}}-OH + CH_3OH \overset{H^+}{\rightleftharpoons}$
  d. 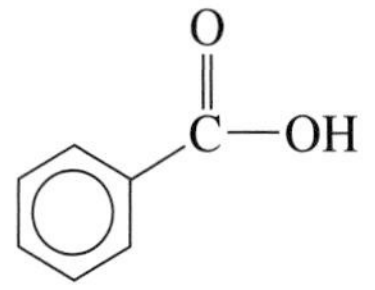 benzene ring with $\overset{O}{\underset{||}{C}}-OH$ + $CH_3-CH_2-OH \overset{H^+}{\rightleftharpoons}$

17.58 Write the products of the following reactions:
  a. $CH_3-\overset{O}{\underset{||}{C}}-OH + NaOH \longrightarrow$
  b. $CH_3-\overset{O}{\underset{||}{C}}-OH + H_2O \rightleftharpoons$

c. 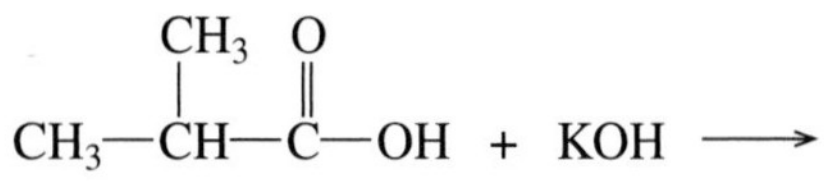 CH_3—$\overset{\overset{\displaystyle CH_3}{|}}{CH}$—$\overset{\overset{\displaystyle O}{\|}}{C}$—OH + KOH $\longrightarrow$

d. 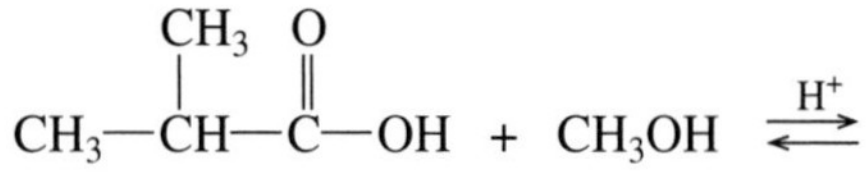 CH_3—$\overset{\overset{\displaystyle CH_3}{|}}{CH}$—$\overset{\overset{\displaystyle O}{\|}}{C}$—OH + CH_3OH $\overset{H^+}{\rightleftarrows}$

17.59 Give the IUPAC names of the carboxylic acid and alcohol needed to prepare each of the following esters:

a. 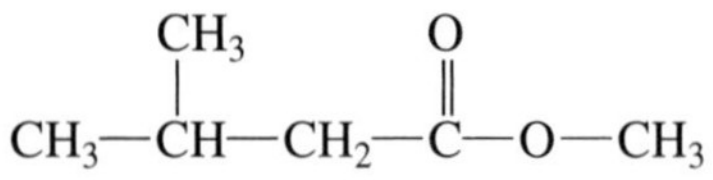 CH_3—$\overset{\overset{\displaystyle CH_3}{|}}{CH}$—$CH_2$—$\overset{\overset{\displaystyle O}{\|}}{C}$—O—$CH_3$

b.

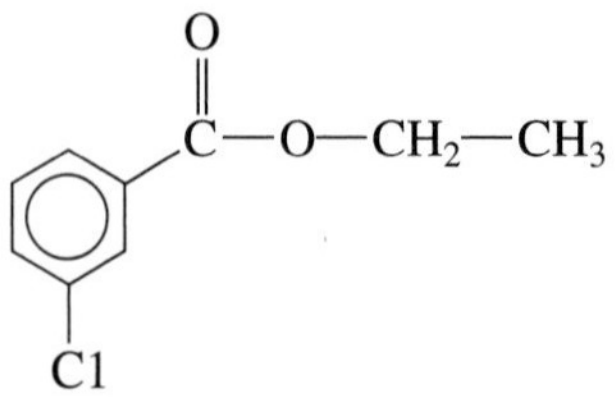

c. 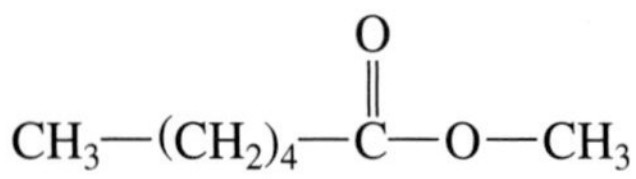 CH_3—$(CH_2)_4$—$\overset{\overset{\displaystyle O}{\|}}{C}$—O—$CH_3$

17.60 Give the IUPAC names of the carboxylic acid and alcohol needed to prepare each of the following esters:

a. 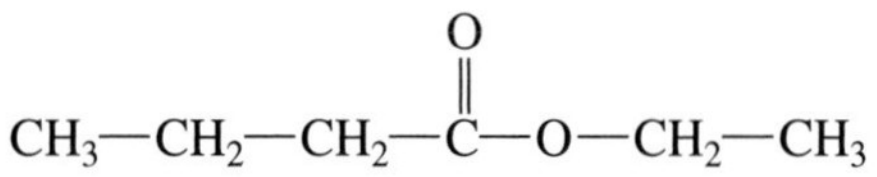 CH_3—CH_2—CH_2—$\overset{\overset{\displaystyle O}{\|}}{C}$—O—$CH_2$—$CH_3$

b.

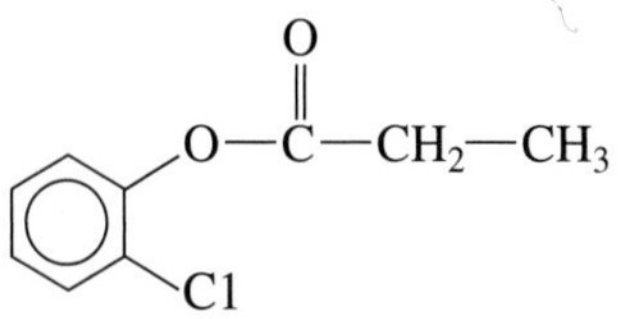

c. 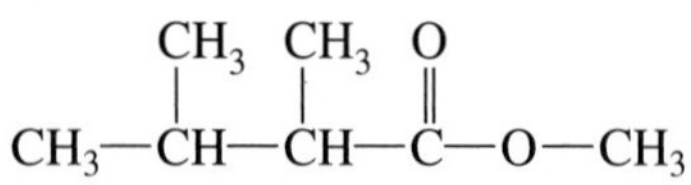 CH_3—$\overset{\overset{\displaystyle CH_3}{|}}{CH}$—$\overset{\overset{\displaystyle CH_3}{|}}{CH}$—$\overset{\overset{\displaystyle O}{\|}}{C}$—O—$CH_3$

17.61 Write the products of the following reactions:

a. 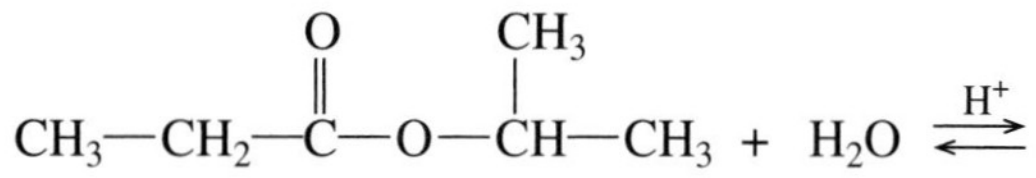 CH_3—CH_2—$\overset{\overset{\displaystyle O}{\|}}{C}$—O—$\overset{\overset{\displaystyle CH_3}{|}}{CH}$—$CH_3$ + H_2O $\overset{H^+}{\rightleftarrows}$

b. 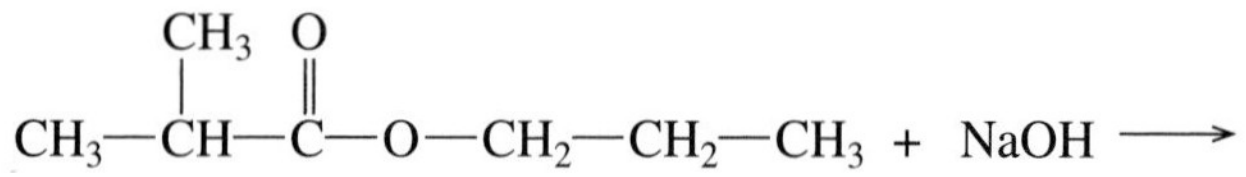 CH_3—$\overset{\overset{\displaystyle CH_3}{|}}{CH}$—$\overset{\overset{\displaystyle O}{\|}}{C}$—O—$CH_2$—$CH_2$—$CH_3$ + NaOH $\longrightarrow$

17.62 Write the products of the following reactions:

a. 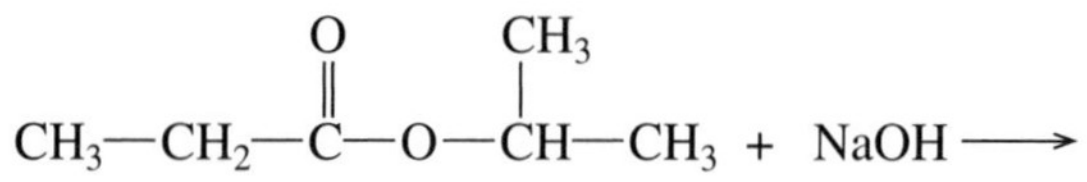 CH_3—CH_2—$\overset{\overset{\displaystyle O}{\|}}{C}$—O—$\overset{\overset{\displaystyle CH_3}{|}}{CH}$—$CH_3$ + NaOH $\longrightarrow$

b. 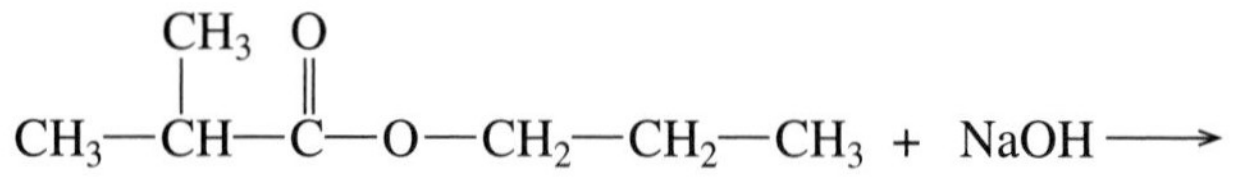 CH_3—$\overset{\overset{\displaystyle CH_3}{|}}{CH}$—$\overset{\overset{\displaystyle O}{\|}}{C}$—O—$CH_2$—$CH_2$—$CH_3$ + NaOH $\longrightarrow$

17.63 Using the reactions we have studied, indicate how you might prepare the following from the starting substance given:
   a. acetic acid from ethene
   b. butyric acid from 1-butanol

17.64 Using the reactions we have studied, indicate how you might prepare the following from the starting substance given:
   a. pentanoic acid from 1-pentanol
   b. ethyl acetate from 2 molecules of ethanol

18 Lipids

"In our toxicology lab, we measure the drugs in samples of urine or blood," says Penny Peng, assistant supervisor of chemistry, Toxicology Lab, Santa Clara Valley Medical Center. "But first we extract the drugs from the fluid and concentrate them so they can be detected in the machine we use. We extract the drugs by using different organic solvents such as methanol, ethyl acetate, or methylene chloride, and by changing the pH. We evaporate most of the organic solvent to concentrate any drugs it may contain. A small sample of the concentrate is placed into a machine called a gas chromatograph. As the gas moves over a column, the drugs in it are separated. From the results, we can identify as many as 10 to 15 different drugs from one urine sample."

LOOKING AHEAD

When we talk of fats and oils, waxes, steroids, cholesterol, and fat-soluble vitamins, we are discussing lipids. All the lipids are naturally occurring compounds that vary considerably in structure but share a common feature of being soluble in nonpolar solvents, but not in water. Fats, which are one family of lipids, have many functions in the body; they store energy and protect and insulate internal organs. Other types of lipids are found in nerve fibers and in hormones, which act as chemical messengers. Because they are not soluble in water, a major function of lipids is to build the cell membranes that separate the internal contents of cells from the surrounding aqueous environment.

Many people are concerned about the amounts of saturated fats and cholesterol in our diets. Researchers suggest that saturated fats and cholesterol are associated with diseases such as diabetes, cancers of the breast, pancreas, and colon, and artherosclerosis, a condition in which deposits of lipid materials accumulate in the coronary blood vessels. These plaques restrict the flow of blood to the tissue, causing necrosis (death) of the tissue. In the heart, this could result in a *myocardial infarction* (heart attack).

The American Institute for Cancer Research has recommended that our diet contain more fiber and starch by adding more vegetables, fruits, and whole grains with moderate amounts of foods with low levels of fat and cholesterol such as fish, poultry, lean meats, and low-fat dairy products. They also suggest that we limit our intake of foods high in fat and cholesterol such as eggs, nuts, fatty or organ meats, cheeses, butter, and coconut and palm oil.

18.1 Lipids

Lipids are a family of biomolecules that have the common property of being soluble in organic solvents but not in water. The word "lipid" comes from the Greek word *lipos,* meaning "fat" or "lard." Typically, the lipid content of a cell can be extracted using an organic solvent such as ether or chloroform. Lipids are an important feature in cell membranes, fat-soluble vitamins, and steroid hormones.

Types of Lipids

Within the lipid family, there are distinct structures that distinguish the different types of lipids. Lipids such as waxes, fats, oils, and glycerophospholipids are esters that can be hydrolyzed to give fatty acids along with other products including an alcohol. Sphingolipids contain an alcohol called sphingosine, and glycosphingolipids contain a carbohydrate. Steroids are characterized by the steroid nucleus of four fused carbon rings. They do not contain fatty acids and cannot be hydrolyzed. Figure 18.1 illustrates the general structure of lipids we will discuss in this chapter.

SAMPLE PROBLEM 18.1

Classes of Lipids

What type of lipid does not contain fatty acids?

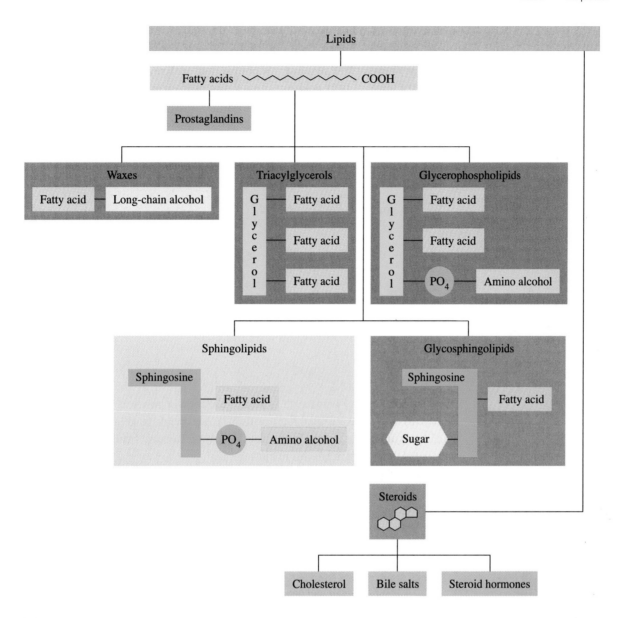

Figure 18.1 Structures for some classes of lipids that are naturally occurring compounds in cells and tissues.

Q **What chemical property do waxes, triacyclglycerols, and steroids have in common?**

Solution

The steroids are a group of lipids with no fatty acids.

Study Check

What type of lipid contains a carbohydrate?

QUESTIONS AND PROBLEMS

Lipids

18.1 What are some functions of lipids in the body?

18.2 What are some of the different kinds of lipids?

18.3 Lipids are not soluble in water. Are lipids polar or nonpolar molecules?

18.4 Which of the following solvents might be used to dissolve an oil stain?
    a. water   **b.** CCl_4 **c.** diethyl ether
    d. benzene   **e.** NaCl solution

18.2 Fatty Acids

The fatty acids are the simplest type of lipids and are found as components in more complex lipids. A **fatty acid** contains a long carbon chain attached to a carboxylic acid group at one end. Although the carboxylic acid part is hydrophilic, the long hydrophobic carbon chain makes long-chain fatty acids insoluble in water. Fatty acids have an even number of carbon atoms, usually between 10 and 20. An example of a fatty acid is lauric acid, a 12-carbon acid found in coconut oil. The structural formula of lauric acid can be written in several forms as follows.

Writing Formulas for Lauric Acid

$$CH_3-(CH_2)_{10}-\overset{\overset{\displaystyle O}{\|}}{C}-OH \qquad CH_3-(CH_2)_{10}-COOH$$

$$CH_3-CH_2-CH_2-CH_2-CH_2-CH_2-CH_2-CH_2-CH_2-CH_2-CH_2-\overset{\displaystyle O}{\underset{\displaystyle OH}{C}}$$
Condensed structural formula

Line-bond structural formula

Saturated fatty acids such as lauric acid contain only single bonds between carbons. **Monounsaturated fatty acids** have one double bond in the carbon chain, and **polyunsaturated fatty acids** have two or more double bonds. Table 18.1 lists some of the typical fatty acids in lipids.

In Chapter 13 we saw that compounds with double bonds can have cis and trans stereoisomers. This is also true of unsaturated fatty acids. For example, oleic acid, a monounsaturated fatty acid found in olives and corn, has one double bond at carbon 9. We can show its cis and trans structural formulas using the line-bond notation. The cis configuration is most prevalent in naturally occurring unsaturated fatty acids. In the cis isomer, the geometry of the carbon chain is not linear, but has a "kink" at the double bond site. As we will see, the cis bond has a major impact on the physical properties and uses of unsaturated fatty acids.

cis-Oleic acid
Cis double bond

trans-Oleic acid
Trans double bond

The human body is capable of synthesizing most fatty acids from carbohydrates or other fatty acids. However, humans do not synthesize sufficient amounts of fatty acids that have more than one double bond, such as linoleic acid, linolenic acid, and arachidonic acid. These fatty acids are called *essential* fatty acids because they must be provided by the diet. A deficiency of essential fatty acids can cause skin dermatitis in infants. However, the role of fatty acids in adult nutrition is not well understood. Adults do not usually have a deficiency of essential fatty acids.

Table 18.1 Structures and Melting Points of Common Fatty Acids

Name	Carbon Atoms	Structure	Melting Point (°C)	Source
Saturated Fatty Acids				
Capric acid	10		32	Saw palmetto
Lauric acid	12		43	Coconut
Myristic acid	14		54	Nutmeg
Palmitic acid	16		62	Palm
Stearic acid	18		69	Animal fat
Arachidic acid	20		76	Peanut oil, vegetable and fish oils
Monounsaturated Fatty Acids				
Palmitoleic acid	16		0	Butter
Oleic acid	18		13	Olives, corn
Polyunsaturated Fatty Acids				
Linoleic acid	18		−9	Soybean, safflower, sunflower
Linolenic acid	18		−17	Corn
Arachidonic acid	20		−50	Prostaglandins

Physical Properties of Fatty Acids

The uniform structure of the carbon chain in saturated fatty acids allows the molecules of saturated fatty acids to fit close together in a regular pattern. The close pattern of the saturated fatty acids allows strong attractions to occur between the car-

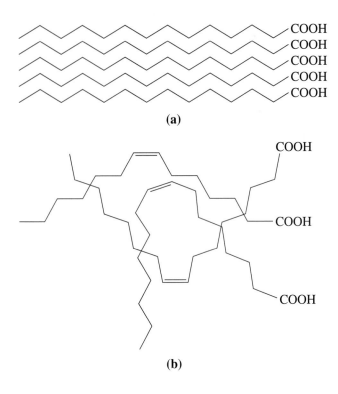

(a)

(b)

Figure 18.2 **(a)** In saturated fatty acids, the molecules fit closely together to give high melting points. **(b)** In unsaturated fatty acids, molecules cannot pack closely together resulting in lower melting points.

Q Why does the cis double bond affect the melting points of unsaturated fatty acids?

bon chains. (See Figure 18.2.) As a result, a significant amount of energy and high temperatures are required to separate the fatty acids and melt the fat. As the length of the carbon chain increases, more interactions occur between the carbon chains, requiring higher melting points. Saturated fatty acids are usually solids at room temperature.

In unsaturated fatty acids, the cis double bonds cause the carbon chain to bend, which gives the molecules an irregular shape. As a result, the molecules of an unsaturated fatty acid cannot stack together in a regular pattern, so fewer interactions occur between carbon chains. Consequently, the melting points of unsaturated fats are lower than those of saturated fats. (See Figure 18.3.) Most unsaturated fats are liquid oils at room temperature.

Prostaglandins

Prostaglandins are hormone-like substances produced in low amounts in most cells of the body. The variety of prostaglandins is formed from the unsaturated fatty acid arachidonic acid with 20 carbons. Prostaglandins are sometimes referred to as eicosanoids (eicos is the Greek word for 20). Most prostaglandins have a hydroxyl group on carbon 11 and carbon 15, and a trans double bond at carbon 13. Those with a ketone group on carbon 9 are designated as PGE, and as PGF, when there is an hydroxyl group on carbon 9. (See Figure 18.4.)

Although prostaglandins are broken down quickly, they have potent physiological effects. Some prostaglandins increase blood pressure, and others lower blood pressure. Other prostaglandins stimulate contraction and relaxation in the smooth muscle of the uterus. When tissues are injured, arachidonic acid is converted to prostaglandins such PGE and PGF that produce inflammation and pain in the area. (See Figure 18.5, p. 583.)

The treatment of pain, fever, and inflammation is based on inhibiting the enzymes that convert arachidonic acid to prostaglandins. Several nonsteroidal anti-inflammatory drugs (NSAIDs), such as aspirin, block the production of prostaglandins, and in doing so decrease pain and inflammation and reduce fever (antipyretics). Ibuprofen has similar anti-inflammatory and analgesic effects. Other NSAIDs include naproxen (Aleve and Naprosyn), ketoprofen (Actron), and nabumetone (Relafen). Long-term use of such products can result in liver, kidney, and gastrointestinal damage. Some forms of PGE are being tested as inhibitors of gastric secretion for use in the treatment of stomach ulcers.

Aspirin (acetylsalicyclic acid)   Ibuprofen (Advil, Motrin)   Naproxen (Aleve, Naprosyn)

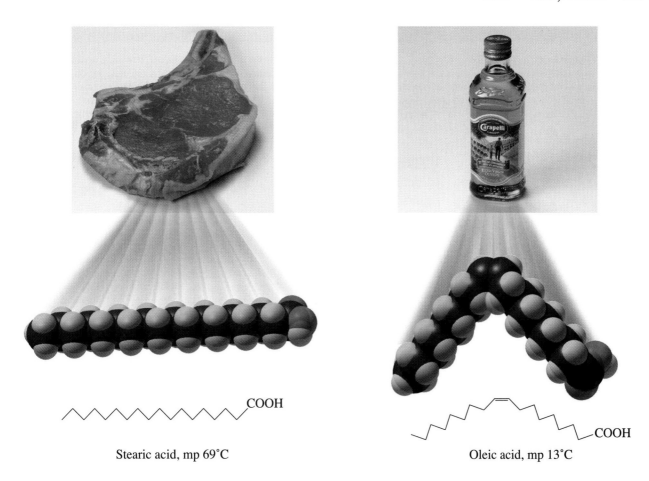

Stearic acid, mp 69°C   Oleic acid, mp 13°C

Figure 18.3 The cis double bond in oleic acid contributes to a lower melting point compared to stearic acid.

Q If both stearic acid and oleic acid have 18 carbons, why is there such a big difference in their respective melting points?

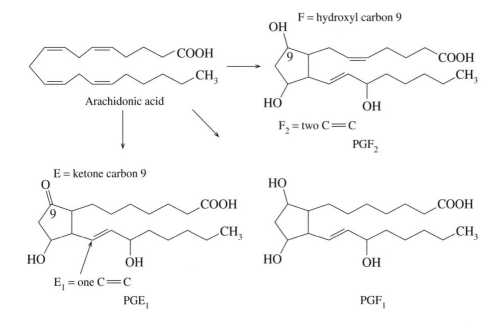

Figure 18.4 The structures of prostaglandins PGE have a ketone at carbon 9, while PGF has a hydroxyl group at carbon 9.

Q How is the number of carbon–carbon double bonds indicated in the abbreviation for prostaglandins?

Omega-3 Fatty Acids in Fish Oils

Over the past several decades, Americans have been changing their diets to include more polyunsaturated fats and fewer saturated fats. This change is a response to research that indicates that atherosclerosis and heart disease are associated with high levels of saturated fats in the diet. However, this association does not seem to be correct for the Inuit people of Alaska, who have a high-fat diet and high levels of blood cholesterol, but a very low occurrence of atherosclerosis and heart attacks. The fats in the Inuit diet are primarily from fish rather than from land animals, as in many other people's diets.

Both fish and vegetable oils have high levels of polyunsaturated fats. The fatty acids in vegetable oils are omega-6 acids, which means that the first double bond occurs at carbon 6 counting from the methyl group. Two common omega-6 acids are linoleic acid and arachidonic acid. However, the fatty acids in the fish oils are mostly the omega-3 type, in which the first double bond occurs at the third carbon counting from the methyl group. Three common omega-3 fatty acids in fish are linolenic acid, eicosapentaenoic acid (EPA), and docosahexaenoic acid (DHA).

In atherosclerosis and heart disease, cholesterol forms plaque that adhere to the walls of the blood vessels. Blood pressure rises as blood has to squeeze through a smaller opening in the blood vessel. As more plaque forms, there is also a possibility of blood clots blocking the blood vessels and causing a heart attack. Omega-3 fatty acids lower the tendency of blood platelets to stick together, thereby reducing the possibility of blood clots. However, high levels of omega-3 fatty acids can increase bleeding if the ability of the platelets to form blood clots is reduced too much. It does seem that a diet that includes fish such as salmon, tuna, and herring can provide higher amounts of the omega-3 fatty acids, which help lessen the possibility of developing heart disease.

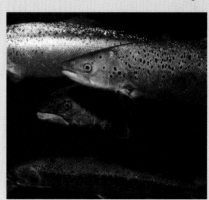

Omega-6 fatty acid

Linoleic acid

Omega-3 fatty acid

Linolenic acid

Eicosapentaenoic acid (EPA)

Docosahexaenoic acid (DHA)

Structures and Properties of Fatty Acids

Consider the structural formula of oleic acid.

$$CH_3(CH_2)_7CH{=}CH(CH_2)_7\overset{\displaystyle O}{\overset{\|}{C}}OH$$

a. Why is the substance called an acid?

b. How many carbon atoms are in oleic acid?

c. Is it a saturated or unsaturated fatty acid?

d. Is it most likely to be solid or liquid at room temperature?

e. Would it be soluble in water?

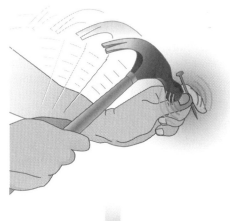

Figure 18.5 An injury to the body tissues releases arachidonic acid, which is converted to prostaglandins that cause pain, fever, and inflammation. Analgesics reduce the effects of prostaglandins by inhibiting the enzyme required for their synthesis.

Q How do analgesics reduce pain and inflammation of an injury?

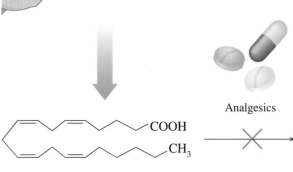

Analgesics

Arachidonic acid

PGF$_2$

Pain, fever, inflammation

Solution

a Oleic acid contains a carboxylic acid group.

b. It contains 18 carbon atoms.

c. It is an unsaturated fatty acid.

d. It is liquid at room temperature.

e. No, its long hydrocarbon chain makes it insoluble in water.

Study Check

Palmitoleic acid is a fatty acid with the following formula:

$$CH_3(CH_2)_5CH{=}CH(CH_2)_7\overset{\overset{\displaystyle O}{\|}}{C}OH$$

a. How many carbon atoms are in palmitoleic acid?

b. Is it a saturated or unsaturated fatty acid?

c. Is it most likely to be solid or liquid at room temperature?

WEB TUTORIAL
Triacylglycerols

QUESTIONS AND PROBLEMS

Fatty Acids

18.5 Describe some similarities and differences in the structures of a saturated fatty acid and an unsaturated fatty acid.

18.6 Stearic acid and linoleic acid both have 18 carbon atoms. Why does stearic acid melt at 69°C, but linoleic acid melts at −9°C?

18.7 Write the line-bond structure of the following fatty acids:
a. palmitic acid   **b.** oleic acid

18.8 Write the line-bond structure of the following fatty acids:
  a. stearic acid   **b.** linoleic acid

18.9 Which of the following fatty acids are saturated, and which are unsaturated?
  a. lauric acid   **b.** linolenic acid
  c. palmitoleic acid   **d.** stearic acid

18.10 Which of the following fatty acids are saturated, and which are unsaturated?
  a. linoleic acid   **b.** palmitic acid
  c. myristic acid   **d.** oleic acid

18.11 How does the structure of a fatty acid with a cis double bond differ from the structure of a fatty acid with a trans double bond?

18.12 In each pair, identify the fatty acid with the lower melting point. Explain.
  a. myristic acid and stearic acid
  b. stearic acid and linoleic acid
  c. oleic acid and linolenic acid

18.13 Describe the position of the first double bond in an omega-3 and an omega-6 fatty acid.

18.14 **a.** What are some sources of omega-3 and omega-6 fatty acids?
  b. How may omega-3 fatty acids help in lowering the risk of heart disease?

18.15 What are some structural and functional differences in arachidonic acid and prostaglandins such as PGE_2?

18.16 What is the structural difference in PGE and PGF?

18.17 What are some functions of prostaglandins in the body?

18.18 How does an anti-inflammatory drug reduce inflammation?

Figure 18.6 Honeycomb is made of beeswax, which is composed of esters of long-chain fatty acids and long-chain alcohols.

Q Write the structural formulas of the alcohol and carboxylic acid for beeswax.

18.3 Waxes, Fats, and Oils

Waxes are found in many plants and animals. Coatings of carnauba wax on fruits and the leaves and stems of plants help to prevent loss of water and damage from pests. Waxes on the skin, fur, and feathers of animals and birds provide a waterproof coating. A **wax** is an ester of a saturated fatty acid and a long-chain alcohol, each containing from 14 to 30 carbon atoms.

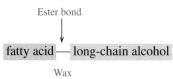

The formulas of some common waxes are given in Table 18.2. Beeswax obtained from honeycombs and carnauba wax obtained from palm trees are used to give a protective coating to furniture, cars, and floors. (See Figure 18.6.) Jojoba wax is used in making candles and cosmetics such as lipstick. Lanolin, a mixture of waxes obtained from wool, is used in hand and facial lotions to aid retention of water, which softens the skin.

Fats and Oils: Triacylglycerols

In the body, fatty acids are stored as fats and oils known as **triacylglycerols**. These substances, also called *triglycerides*, are triesters of glycerol (a trihydroxy alcohol) and fatty acids. The general formula of a triacylglycerol follows.

Table 18.2   Some Typical Waxes

Type	Structural Formula	Source	Uses
Beeswax	$CH_3(CH_2)_{14}$—$\overset{\overset{\displaystyle O}{\|\|}}{C}$—O—$(CH_2)_{29}CH_3$	Honeycomb	Candles, shoe polish, wax paper
Carnauba wax	$CH_3(CH_2)_{24}$—$\overset{\overset{\displaystyle O}{\|\|}}{C}$—O—$(CH_2)_{29}CH_3$	Brazilian palm tree	Waxes for furniture, cars, floors, shoes
Jojoba wax	$CH_3(CH_2)_{18}$—$\overset{\overset{\displaystyle O}{\|\|}}{C}$—O—$(CH_2)_{19}CH_3$	Jojoba	Candles, soaps, cosmetics

A triacylglycerol is produced by **esterification,** a reaction in which the hydroxyl groups of glycerol form ester bonds with the carboxyl groups of fatty acids. For example, glycerol and three molecules of stearic acid form tristearin (glyceryl tristearate).

Triacylglycerols are the major form of energy storage for animals. Animals that hibernate eat large quantities of plants, seeds, and nuts that contain large amounts of fats and oils. They gain as much as 14 kilograms a week. As the external temperature drops, the animal goes into hibernation. The body temperature drops to nearly freezing, and there is a dramatic reduction in cellular activity, respiration, and heart rate. Animals who live in extremely cold climates will hibernate for 4–7 months. During this time, stored fat is the only source of energy.

However, most fats and oils are mixed triacylglycerols that contain two or three different fatty acids. For example, a mixed triacylglycerol might be made from lauric acid, myristic acid, and palmitic acid. One possible structure for the mixed triacylglycerol follows:

$$
\begin{array}{l}
\quad\quad\quad\quad\quad O \\
\quad\quad\quad\quad\quad \| \\
CH_2\!-\!O\!-\!C(CH_2)_{10}CH_3 \quad \text{Lauric acid}\\
\,| \quad\quad\quad\quad O\\
\quad\quad\quad\quad \|\\
CH\!-\!O\!-\!C(CH_2)_{12}CH_3 \quad \text{Myristic acid}\\
\,| \quad\quad\quad\quad O\\
\quad\quad\quad\quad \|\\
CH_2\!-\!O\!-\!C(CH_2)_{14}CH_3 \quad \text{Palmitic acid}\\
\quad\quad \text{A mixed triacylglycerol}
\end{array}
$$

SAMPLE PROBLEM 18.3

Writing Structures for a Triacylglycerol

Draw the structural formula of triolein, a simple triacylglycerol that uses oleic acid.

Solution

Triolein is the triacylglycerol of glycerol and three oleic acid molecules. Each fatty acid is attached by an ester bond to one of the hydroxyl groups in glycerol.

$$
\begin{array}{l}
\quad\quad\quad\quad\quad O \\
\quad\quad\quad\quad\quad \| \\
CH_2\!-\!O\!-\!C(CH_2)_7CH\!=\!CH(CH_2)_7CH_3\\
\,| \quad\quad\quad\quad O\\
\quad\quad\quad\quad \|\\
CH\!-\!O\!-\!C(CH_2)_7CH\!=\!CH(CH_2)_7CH_3\\
\,| \quad\quad\quad\quad O\\
\quad\quad\quad\quad \|\\
CH_2\!-\!O\!-\!C(CH_2)_7CH\!=\!CH(CH_2)_7CH_3\\
\quad\quad \text{Glyceryl trioleate (triolein)}
\end{array}
$$

Study Check

Write the structure of the triacylglycerol containing 3 molecules of myristic acid.

Melting Points of Fats and Oils

A **fat** is a triacylglycerol that is solid at room temperature, such as fats in meat, whole milk, butter, and cheese. Most fats come from animal sources.

An **oil** is a triacylglycerol that is usually liquid at room temperature. The most commonly used oils come from plant sources. Olive oil and peanut oil are monounsaturated because they contain large amounts of oleic acid. Oils from corn, cottonseed, safflower, and sunflower are polyunsaturated because they contain large amounts of fatty acids with two or more double bonds. (See Figure 18.7.) A few oils such as palm oil and coconut oil are solid at room temperature because they consist mostly of saturated fatty acids.

The amounts of saturated, monounsaturated, and polyunsaturated fatty acids in some typical fats and oils are shown in Figure 18.8. Saturated fatty acids have higher melting points than unsaturated fatty acids because they pack together more tightly.

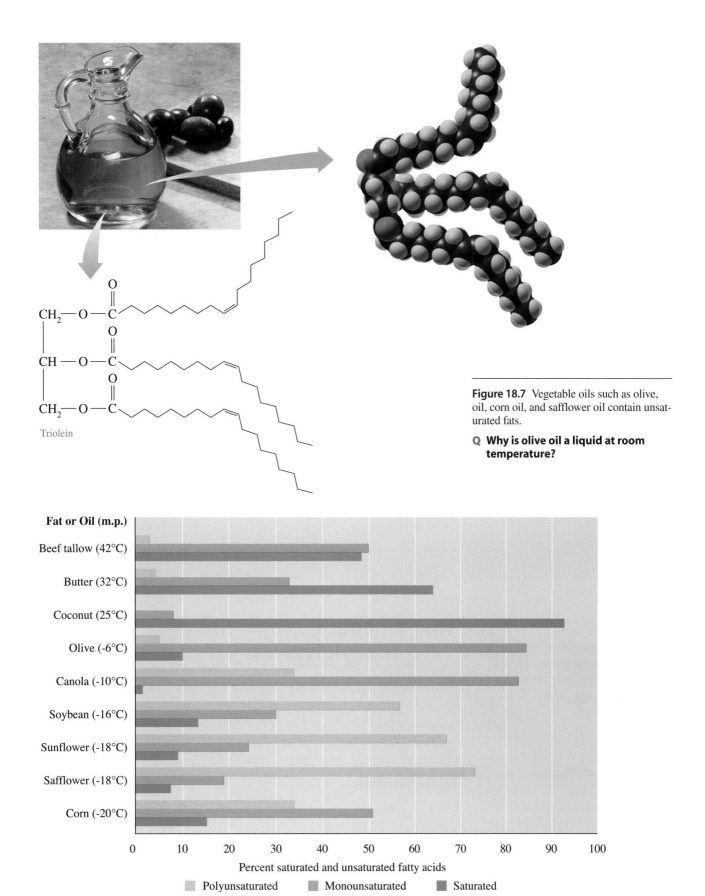

Figure 18.7 Vegetable oils such as olive, oil, corn oil, and safflower oil contain unsaturated fats.

Q **Why is olive oil a liquid at room temperature?**

Triolein

Figure 18.8 Fats and oils have low melting points because they have a higher percentage of unsaturated fatty acids than do animal fats.

Q **Why is the melting point of butter higher than olive or canola oil?**

Animal fats usually contain more saturated fatty acids than do vegetable oils. Therefore the melting points of animal fats are higher than those of vegetable oils.

QUESTIONS AND PROBLEMS

Waxes, Fats, and Oils

18.19 Draw the structure of a component of beeswax that contains myricyl alcohol, $CH_3(CH_2)_{29}OH$, and palmitic acid.

18.20 Draw the structure of a component of jojoba wax that contains arachidic acid, a 20-carbon saturated fatty acid, and 1-docosanol, $CH_3(CH_2)_{21}OH$.

18.21 Draw the structure of a triacylglycerol that contains stearic acid and glycerol.

18.22 A mixed triacylglycerol contains two palmitic acid molecules to every one oleic acid molecule. Write two possible structures (isomers) for the compound.

18.23 Draw the structure of tripalmitin.

18.24 Draw the structure of triolein.

18.25 Safflower oil is called a polyunsaturated oil, whereas olive oil is a monounsaturated oil. Explain.

18.26 Why does olive oil have a lower melting point than butter fat?

18.27 Why does coconut oil, a vegetable oil, have a melting point similar to fats from animal sources?

18.28 A label on a bottle of 100% sunflower seed oil states that it is lower in saturated fats than all the leading oils.
   a. How does the percentage of saturated fats in sunflower seed oil compare to that of safflower, corn, and canola oil? (See Figure 18.8.)
   b. Is the claim valid?

LEARNING GOAL

Draw the structure of the product from the reaction of a triacylglycerol with hydrogen, an acid or base, or an oxidizing agent.

WEB TUTORIAL
Triacylglycerols

18.4 Chemical Properties of Triacylglycerols

The chemical reactions of the triacylglycerols (fats and oils) are the same as we discussed for alkenes (Chapter 13), and carboxylic acids and esters (Chapter 17). We will look at the addition of hydrogen to the double bonds, the oxidation of double bonds, and the acid and base hydrolysis of the ester bonds of fats and oils.

Hydrogenation

The **hydrogenation** of unsaturated fats converts carbon–carbon double bonds to single bonds. The hydrogen gas is bubbled through the heated oil in the presence of a nickel catalyst.

$$-CH{=}CH- \;+\; H_2 \xrightarrow{Ni} \begin{array}{c} HH \\ | | \\ -C-C- \\ | | \\ HH \end{array}$$

For example, when hydrogen adds to all of the double bonds of triolein using a nickel catalyst, the product is the saturated fat tristearin.

HEALTH NOTE

Olestra: A Fat Substitute

In 1968, food scientists designed an artificial fat called *olestra* as a source of nutrition for premature babies. However, olestra could not be digested and was never used for that purpose. Then scientists realized that olestra had the flavor and texture of a fat without the calories.

Olestra is manufactured by obtaining the fatty acids from the fats in cottonseed or soybean oils and bonding the fatty acids with the hydroxyl groups on sucrose. Chemically, olestra is composed of six to eight long-chain fatty acids attached by ester links to a sugar (sucrose) rather than to a glycerol molecule found in fats. This makes olestra a very large molecule, which cannot be absorbed through the intestinal walls. The enzymes and bacteria in the intestinal tract are unable to break down the olestra molecule and it travels through the intestinal tract undigested.

In 1996 the Food and Drug Administration approved olestra for use in potato chips, tortilla chips, crackers and fried snacks. In 1996, olestra snack products were test marketed in parts of Iowa, Wisconsin, Indiana, and Ohio. By 1997, there were reports of some adverse reactions, including diarrhea, abdominal cramps, and anal leakage, indicating that olestra may act as a laxative in some people. However, the manufacturers contend there is no direct proof that olestra is the cause of those effects.

The large molecule of olestra also combines with fat-soluble vitamins (A, D, E, and K) as well as the carotenoids from the foods we eat before they can be absorbed through the intestinal wall. Carotenoids are plant pigments in fruits and vegetables that protect against cancer, heart disease, and macular degeneration, a form of blindness in the elderly. The FDA now requires manufacturers to add the four vitamins, but not the carotenoids, to olestra products. The label on an olestra product must state the following: "This product contains olestra. Olestra may cause abdominal cramping and loose stools. Olestra inhibits the absorption of some vitamins and other nutrients. Vitamins A, D, E, and K have been added." Some snack foods made with olestra are now in supermarkets nationwide. Since there are already low-fat snacks on the market, it remains to be seen whether olestra will have any significant effect on reducing the problem of obesity.

Fatty acids

$CH_3(CH_2)_6COOH$

$CH_3(CH_2)_8COOH$

Olestra

Glyceryl trioleate
(triolein)

Glyceryl tristearate
(tristearin)

In commercial hydrogenation, the addition of hydrogen is stopped before all the double bonds in an oil become completely saturated. Complete hydrogenation gives a very brittle product, whereas the partial hydrogenation of a liquid vegetable oil changes it to a soft, semisolid fat. As the oil becomes more saturated, the melting point increases and the fat becomes more solid at room temperature. Control of the degree of hydrogenation gives the various types of partially hydrogenated vegetable oil products on the market today—soft margarines, solid stick margarines, and solid shortenings. (See Figure 18.9.) Although these products now contain more saturated fatty acids than the original oils, they contain no cholesterol, unlike similar products from animal sources, such as butter and lard.

HEALTH NOTE

Trans Fatty Acids and Hydrogenation

In the early 1900s, margarine became a popular replacement for the highly saturated fats such as butter and lard. Margarine is produced by partially hydrogenating the unsaturated fats in vegetable oils such as safflower oil, corn oil, canola oil, cottonseed oil, and sunflower oil. Fats that are more saturated are more resistant to oxidation.

In vegetable oils, the unsaturated fats usually contain cis double bonds. As hydrogenation occurs, double bonds are converted to single bonds. However, some of the cis double bonds are converted to trans double bonds, which causes a change in the overall structure of the fatty acids. If the label on a product states that the oils have been "partially" or "fully hydrogenated," that product will also contain trans fatty acids. In the United States, it is estimated that 2–4% of our total calories come from trans fatty acids.

The concern about trans fatty acids is that their altered structure may make them behave like saturated fatty acids in the body. In the 1980s, research indicated that trans fatty acids have an effect on blood cholesterol similar to that of saturated fats, although study results vary. Several studies reported that trans fatty acids raise the levels of LDL-cholesterol, low-density lipoproteins containing cholesterol that can accumulate in the arteries. (LDLs and HDLs are described in the section on lipoproteins later in the chapter.) Some studies also report that trans fatty acids lower HDL-cholesterol, high-density lipoproteins that carry cholesterol to the liver to be excreted. But other studies did not report any decrease in HDL-cholesterol. In some American and European studies, an increased risk of breast cancer was associated with increased intake of trans fatty acids. However, these studies are not conclusive and not all studies have supported such findings. Current evidence does not yet indicate that the intake of trans fatty acids is a significant risk factor for heart disease. The trans fatty acids controversy will continue to be debated as more research is done.

Foods containing trans fatty acids include milk, bread, fried foods, ground beef, baked goods, stick margarine, butter, soft margarine, cookies, crackers, and vegetable shortening. The American Heart Association recommends that margarine should have no more than 2 grams of saturated fat per tablespoon and a liquid vegetable oil should be the first ingredient. They also recommend the use of soft margarine, which is lower in trans fatty acids because soft margarine is only slightly hydrogenated, and diet margarine because it has less fat and therefore fewer trans fatty acids.

Many health organizations agree that fat should account for less than 30% of daily calories (the current average for Americans is 34%) and saturated fat should be less than 10% of total calories. Lowering the overall fat intake would also decrease the amount of trans fatty acids. The Food and Drug Administration and the U.S. Department of Agriculture are encouraging the use of new food labels to inform consumers of the fat content of food. The best advice may be to reduce total fat in the diet by using fats and oils sparingly, cooking with little or no fat, substituting olive oil or canola oil for other oils, and limiting the use of coconut oil and palm oil, which are high in saturated fatty acids.

There are several products including peanut butter and butter-like spreads on the market that have 0% trans fatty acids. On the labels, they state that their products are nonhydrogenated, which avoids the production of the undesirable trans fatty acid. However, in the list of natural vegetable oils, such as soy and canola oil, there is also palm oil. Because palm oil has a melting point of 30°C, palm oil increases the overall melting point of the spread and gives a product that is solid at room temperature. However, palm oil contains high amounts of saturated fatty acids and has a similar effect in the body as does stearic acid (18 carbons) and fats derived from animal sources. Health experts recommend that we limit the amount of saturated fats, including palm oil, in our diets.

Cis-oleic acid

H₂/Ni

Double bond opens

Ni catalyst

Isomerization

Addition of H₂

Undesired side product (*trans*-oleic acid)

Desired saturated product (stearic acid)

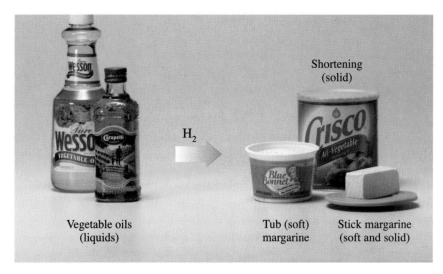

Figure 18.9 Many soft margarines, stick margarines, and solid shortenings are produced by the partial hydrogenation of vegetable oils.

Q How does hydrogenation change the structure of the fatty acids in the vegetable oils?

Oxidation of Unsaturated Fats

A fat or oil becomes rancid when its double bonds are oxidized in the presence of oxygen and microorganisms. The products are short-chain fatty acids and aldehydes that have disagreeable odors.

$$-CH=CH- \xrightarrow{[O]} -\overset{\overset{\displaystyle O}{\|}}{C}-H \ + \ H-\overset{\overset{\displaystyle O}{\|}}{C}- \xrightarrow{[O]} -\overset{\overset{\displaystyle O}{\|}}{C}-OH \ + \ HO-\overset{\overset{\displaystyle O}{\|}}{C}-$$

Unsaturated   Short-chain   Short-chain

fatty acids   aldehydes   carboxylic acids

If a vegetable oil does not contain an antioxidant, it will oxidize rather easily. You can detect an oil that has become rancid by its unpleasant odor. If an oil is covered tightly and stored in a refrigerator, the process of oxidation can be slowed and the oil will last longer.

Oxidation also occurs in the oils that accumulate on the surface of the skin during heavy exercise. (See Figure 18.10.) At body temperature, microorganisms on the skin promote rapid oxidation of the oils as they are exposed to oxygen and water. The resulting short-chain aldehydes and fatty acids account for the body odor associated with a workout and heavy perspiration.

Hydrolysis

In Chapter 17, we saw that esters in the presence of a strong acid undergo hydrolysis with water to give an alcohol and a carboxylic acid.

$$CH_3-O-\overset{\overset{\displaystyle O}{\|}}{C}-CH_3 \ + \ H_2O \xrightarrow{H^+} CH_3-OH \ + \ HO-\overset{\overset{\displaystyle O}{\|}}{C}-CH_3$$

Ester   + H_2O ⟶   alcohol   + carboxylic acid

Triacylglycerols are also hydrolyzed (split by water) in the presence of strong acids or digestive enzymes called *lipases*. The products of hydrolysis of the ester bonds are glycerol and three fatty acids. The polar glycerol is soluble in water, but the fatty acids with their long hydrocarbon chains are not.

Figure 18.10 After strenuous exercise, increased body temperature and perspiration promote the oxidation of oils that accumulate on the surface of the skin.

Q What types of organic products result from the oxidation of oils on the skin?

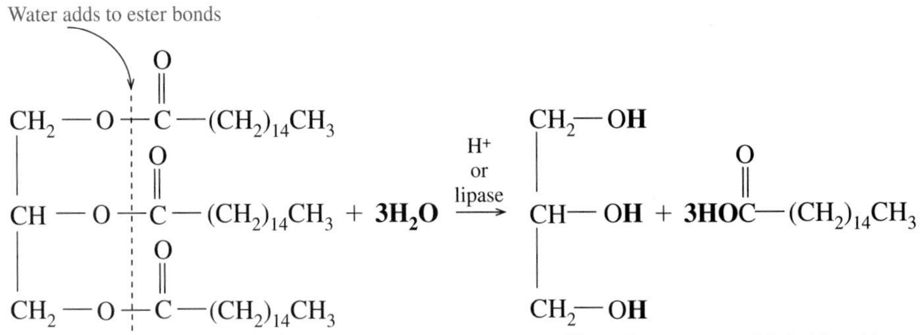

Water adds to ester bonds

$$CH_2-O \overset{\displaystyle O}{\underset{|}{\overset{||}{C}}}-(CH_2)_{14}CH_3$$

$$CH-O \overset{\displaystyle O}{\overset{||}{C}}-(CH_2)_{14}CH_3 + 3H_2O \xrightarrow[\text{lipase}]{H^+ \text{ or}} CH-OH + 3HOC-(CH_2)_{14}CH_3$$

$$CH_2-O \overset{\displaystyle O}{\overset{||}{C}}-(CH_2)_{14}CH_3$$

Glyceryl tripalmitate   Glycerol   3 Palmitic acid
(tripalmitin)   molecules

Saponification

When a strong base is used in the hydrolysis of an ester, the products are an alcohol and the carboxylic acid salt.

$$CH_3-O-\overset{\displaystyle O}{\overset{||}{C}}-CH_3 + NaOH \longrightarrow CH_3-OH + Na^+{}^-O-\overset{\displaystyle O}{\overset{||}{C}}-CH_3$$

Ester   + NaOH ⟶   alcohol   + carboxylic acid salt

When a fat is heated with a strong base such as sodium hydroxide, saponification of the fat gives glycerol and the sodium salts of the fatty acids, which are soaps. When NaOH is used, a solid soap is produced that can be molded into a desired shape; KOH produces a softer, liquid soap. Oils that are polyunsaturated produce softer soaps. Names like "coconut" or "avocado shampoo" tell you the sources of the oil used in the hydrolysis reaction.

Fat or oil + strong base ⟶ glycerol + salts of fatty acids (soaps)

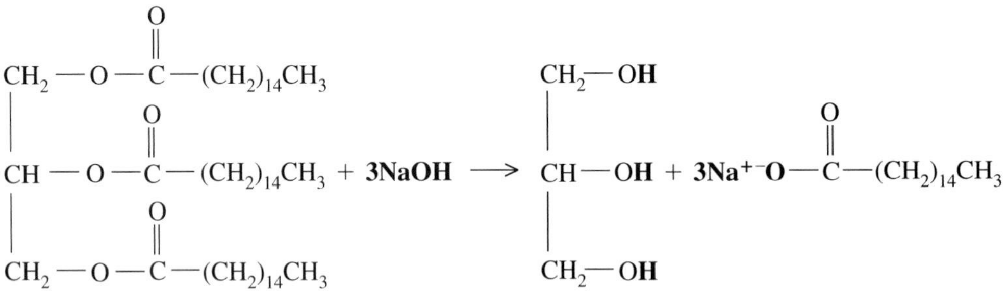

$$CH_2-O-\overset{\displaystyle O}{\overset{||}{C}}-(CH_2)_{14}CH_3$$

$$CH-O-\overset{\displaystyle O}{\overset{||}{C}}-(CH_2)_{14}CH_3 + 3NaOH \longrightarrow CH-OH + 3Na^+{}^-O-\overset{\displaystyle O}{\overset{||}{C}}-(CH_2)_{14}CH_3$$

$$CH_2-O-\overset{\displaystyle O}{\overset{||}{C}}-(CH_2)_{14}CH_3$$

Glyceryl tripalmitate   Glycerol   3 sodium palmitate
(tripalmitin)   (soap)

SAMPLE PROBLEM 18.4

Reactions of Lipids

Write the equation for the reaction catalyzed by the enzyme lipase that hydrolyzes trilaurin during the digestion process.

Solution

$$
\begin{array}{l}
CH_2\!-\!O\!-\!\overset{\displaystyle O}{\overset{\|}{C}}\!-\!(CH_2)_{10}CH_3 \\[2pt]
\ \ | \qquad\ \ \overset{\displaystyle O}{\overset{\|}{\ }} \\[2pt]
CH\!-\!O\!-\!\overset{\|}{C}\!-\!(CH_2)_{10}CH_3 \ + \ 3H_2O \xrightarrow{\ \text{Lipase}\ } \\[2pt]
\ \ | \qquad\ \ \overset{\displaystyle O}{\overset{\|}{\ }} \\[2pt]
CH_2\!-\!O\!-\!\overset{\|}{C}\!-\!(CH_2)_{10}CH_3
\end{array}
$$

Glyceryl trilaurate
(trilaurin)

$$
\begin{array}{l}
CH_2\!-\!OH \\[2pt]
\ \ | \qquad\qquad\quad \overset{\displaystyle O}{\overset{\|}{\ }} \\[2pt]
CH\!-\!OH \ + \ 3HOC\!-\!(CH_2)_{10}CH_3 \\[2pt]
\ \ | \\[2pt]
CH_2\!-\!OH
\end{array}
$$

Glycerol   3 Lauric acid
                  molecules

Study Check

What is the name of the product formed when a triacylglycerol containing oleic acid and linoleic acid is completely hydrogenated?

QUESTIONS AND PROBLEMS

Chemical Properties of Triacylglycerols

18.29 Write an equation for the hydrogenation of glyceryl trioleate, a fat containing glycerol and three oleic acid units.

18.30 Write an equation for the hydrogenation of glyceryl trilinolenate, a fat containing glycerol and three linolenic acid units.

18.31 A label on a container of margarine states that it contains partially hydrogenated corn oil.
 a. How has the liquid corn oil been changed?
 b. Why is the margarine product solid?

18.32 Why should a bottle of vegetable oil that has no preservatives be tightly covered and refrigerated?

18.33 a. Write an equation for the acid hydrolysis of glyceryl trimyristate (trimyristin).
 b. Write an equation for the NaOH saponification of glyceryl trimyristate (trimyristin).

18.34 a. Write an equation for the acid hydrolysis of glyceryl trioleate (triolein).
 b. Write an equation for the NaOH saponification of glyceryl trioleate (triolein).

18.35 Compare the structure of a triacylglycerol to the structure of olestra.

18.36 An oil is partially hydrogenated.
 a. Are all or just some of the double bonds converted to single bonds?
 b. What happens to many of the cis double bonds during hydrogenation?
 c. How can you reduce the amount of trans fatty acids in your diet?

18.37 Write the product of the hydrogenation of the following triacylglycerol.

$$CH_2-O-\overset{\overset{O}{\|}}{C}-(CH_2)_{16}CH_3$$
$$CH-O-\overset{\overset{O}{\|}}{C}-(CH_2)_7CH=CH(CH_2)_7CH_3$$
$$CH_2-O-\overset{\overset{O}{\|}}{C}-(CH_2)_{16}CH_3$$

18.38 Write all the products that would be obtained when the triacylglycerol in problem 18.37 undergoes complete hydrolysis.

LEARNING GOAL

Describe the characteristics of glycerophospholipids.

WEB TUTORIAL
Phospholipids

18.5 Glycerophospholipids

The **glycerophospholipids** are a family of lipids similar to triacylglycerols except that one hydroxyl group of glycerol is replaced by the ester of phosphoric acid and an amino alcohol, bonded through a phosphodiester bond. We can compare the general structures of a triacylglycerol and a glycerophospholipid as follows:

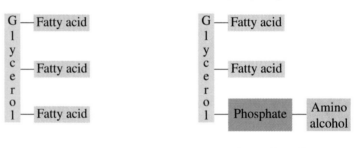

Triacylglycerol (triglyceride)   Glycerophospholipid

Phosphate Esters

In this group of compounds, ester bonds form between a hydroxyl group of an alcohol and phosphoric acid to give ester products similar to those formed with carboxylic acids.

$$HO-\overset{\overset{O}{\|}}{\underset{\underset{OH}{|}}{P}}-OH \; + \; HO-CH_3 \longrightarrow HO-\overset{\overset{O}{\|}}{\underset{\underset{OH}{|}}{P}}-O-CH_3 \; + \; H_2O$$

Phosphoric acid +   alcohol $\longrightarrow$ phosphate ester   + water

The phosphate ester forms a diester by reaction with another alcohol.

$$HO-\overset{\overset{O}{\|}}{\underset{\underset{OH}{|}}{P}}-O-CH_3 \; + \; HO-CH_3 \longrightarrow CH_3-O-\overset{\overset{O}{\|}}{\underset{\underset{OH}{|}}{P}}-O-CH_3 \; + \; H_2O$$

Phosphate ester   +   alcohol $\longrightarrow$ phosphate diester   + water

Three amino alcohols found in glycerophospholipids are choline, serine, and ethanolamine. Because these compounds exist in the body at physiological pH of 7.4, they are ionized. The nitrogen atom in the amino alcohol has a plus one charge while the —OH group from phosphoric acid gives up a proton.

$$HO-CH_2CH_2-\overset{+}{\underset{\underset{CH_3}{|}}{\overset{\overset{CH_3}{|}}{N}}}-CH_3 \qquad HO-CH_2-\overset{\overset{+}{NH_3}}{\underset{|}{CH}}-COO^- \qquad HO-CH_2CH_2-\overset{+}{N}H_3$$

Choline   Serine   Ethanolamine

Lecithins and **cephalins** are two types of glycerophospholipids that are particularly abundant in brain and nerve tissues as well as in egg yolks, wheat germ, and yeast. Lecithins contain choline, and cephalins contain ethanolamine and sometimes serine. In the following structural formulas, palmitic acid is used as an example of a fatty acid.

$$
\begin{array}{ll}
CH_2-O-\overset{\overset{O}{||}}{C}-(CH_2)_{14}CH_3 \\
CH-O-\overset{\overset{O}{||}}{C}-(CH_2)_{14}CH_3 \\
CH_2-O-\overset{\underset{O^-}{|}}{P}-O-CH_2CH_2\overset{+}{N}(CH_3)_3
\end{array}
$$

Nonpolar fatty acids — Polar — Choline
A lecithin   A cephalin

Glycerophospholipids contain both polar and nonpolar regions, which allow them to interact with both polar and nonpolar substances. The ionized alcohol and phosphate portion called "the head" is polar and can hydrogen bond with water. (See Figure 18.11.) The two fatty acids connected to the glycerol molecule repre-

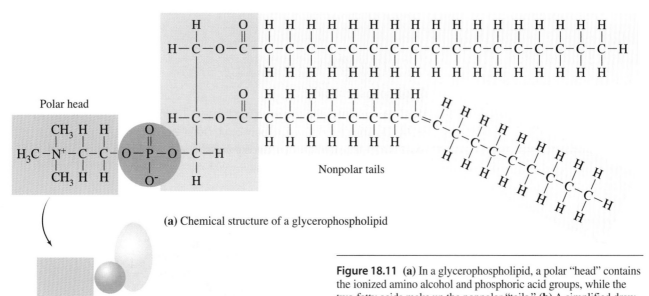

(a) Chemical structure of a glycerophospholipid

Polar head   Nonpolar tails

(b) Simplified way to draw a glycerophospholipid

Figure 18.11 (a) In a glycerophospholipid, a polar "head" contains the ionized amino alcohol and phosphoric acid groups, while the two fatty acids make up the nonpolar "tails." (b) A simplified drawing indicates the polar region and the nonpolar region.

Q Why are glycerophospholipids polar?

sent the nonpolar "tails" of the phospholipid. The hydrocarbon chains that make up the "tails" are only soluble in other nonpolar substances, mostly lipids.

Glycerophospholipids are the most abundant lipids in cell membranes, where they play an important role in cellular permeability. They make up much of the myelin sheath that protects nerve cells. In the body fluids, they combine with the less polar triglycerides and cholesterol to make them more soluble as they are transported in the body.

SAMPLE PROBLEM 18.5

Drawing Glycerophospholipid Structures

Draw the structure of cephalin, using stearic acid for the fatty acids and serine for the amino alcohol. Describe each of the components in the glycerophospholipid.

Solution

In general, glycerophospholipids are composed of a glycerol molecule in which two carbon atoms are attached to fatty acids such as stearic acid. The third carbon atom is attached in an ester bond to phosphate linked to an amino alcohol. In this example, the amino alcohol is serine.

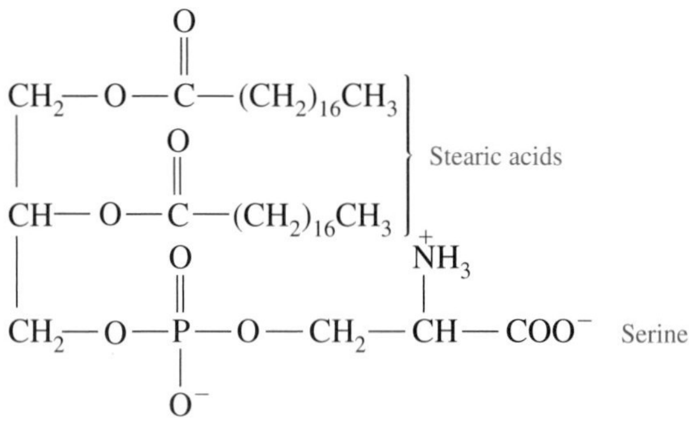

$$
\begin{array}{c}
\text{O} \\
\parallel \\
\text{CH}_2\text{—O—C—(CH}_2)_{16}\text{CH}_3 \\
\text{O} \\
\parallel \\
\text{CH—O—C—(CH}_2)_{16}\text{CH}_3 \\
\text{O} \quad \overset{+}{\text{N}}\text{H}_3 \\
\parallel \qquad | \\
\text{CH}_2\text{—O—P—O—CH}_2\text{—CH—COO}^- \\
| \\
\text{O}^-
\end{array}
$$

Stearic acids

Serine

Study Check

How do glycerophospholipids differ structurally from triacylglylcerols?

QUESTIONS AND PROBLEMS

Glycerophospholipids

18.39 Describe the differences between triacylglycerols and glycerophospholipids.

18.40 Describe the differences between lecithin and cephalin.

18.41 Draw the structure of a glycerophospholipid containing two molecules of palmitic acid, and ethanolamine. What is another name for this type of phospholipid?

18.42 Draw the structure of a glycerophospholipid that contains choline and palmitic acids.

18.43 Identify the following glycerophospholipid and list its components:

$$\begin{array}{l}
\text{CH}_2\!-\!\text{O}\!-\!\overset{\overset{\text{O}}{\|}}{\text{C}}(\text{CH}_2)_7\text{CH}\!=\!\text{CH}(\text{CH}_2)_7\text{CH}_3 \\[2pt]
\text{CH}\!-\!\text{O}\!-\!\overset{\overset{\text{O}}{\|}}{\text{C}}(\text{CH}_2)_{16}\text{CH}_3 \\[2pt]
\text{CH}_2\!-\!\text{O}\!-\!\overset{\overset{\text{O}}{\|}}{\underset{\underset{\text{O}^-}{|}}{\text{P}}}\!-\!\text{O}\!-\!\text{CH}_2\!-\!\text{CH}_2\!-\!\overset{+}{\text{N}}\text{H}_3
\end{array}$$

18.44 Identify the following glycerophospholipid and list its components:

$$\begin{array}{l}
\text{CH}_2\!-\!\text{O}\!-\!\overset{\overset{\text{O}}{\|}}{\text{C}}(\text{CH}_2)_{14}\text{CH}_3 \\[2pt]
\text{CH}\!-\!\text{O}\!-\!\overset{\overset{\text{O}}{\|}}{\text{C}}(\text{CH}_2)_{16}\text{CH}_3 \\[2pt]
\text{CH}_2\!-\!\text{O}\!-\!\overset{\overset{\text{O}}{\|}}{\underset{\underset{\text{O}^-}{|}}{\text{P}}}\!-\!\text{O}\!-\!\text{CH}_2\!-\!\text{CH}_2\!-\!\overset{\overset{\text{CH}_3}{|}}{\underset{\underset{\text{CH}_3}{|}}{\overset{+}{\text{N}}}}\!-\!\text{CH}_3
\end{array}$$

18.6 Sphingolipids

When the —NH_2 group in sphingosine bonds to a fatty acid by an *amide* link, a *ceramide* is obtained. The lipids known as sphingolipids form when an ester bond replaces the —OH group of the *ceramide* with a phosphate ester of an amino alcohol.

The *sphingolipids* are another group of phospholipids that are abundant in the biological membranes of the brain and nerve tissues. The **sphingolipids** are esters of an 18-carbon alcohol called sphingosine instead of glycerol.

$$\begin{array}{l}
\text{CH}_3(\text{CH}_2)_{12}\!-\!\text{CH}\!=\!\text{CH}\!-\!\text{CH}\!-\!\text{OH} \\[2pt]
\qquad\qquad\qquad\quad \text{CH}\!-\!\text{NH}_2 \\[2pt]
\qquad\qquad\qquad\quad \text{CH}_2\!-\!\text{OH}
\end{array}$$

Sphingosine

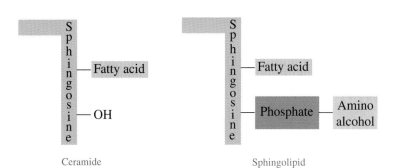

Ceramide   Sphingolipid

WEB TUTORIAL
Phospholipids

One of the most abundant sphingolipids is sphingomyelin. It is the white matter of the myelin sheath, a coating surrounding the nerve cells that increases the speed of nerve impulses and insulates and protects the nerve cells. In sphingomyelin, the sphingosine is linked by an amide bond to a fatty acid and to a phosphate ester of choline, an amino alcohol.

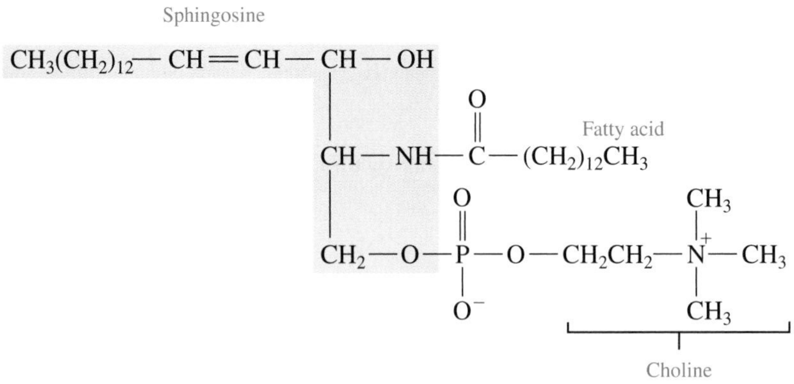

Sphingomyelin, a sphingolipid

Glycosphingolipids

Glycosphingolipids are sphingolipids that contain a carbohydrate such as galactose or glucose. They are important components of nerve membranes and muscle. In a glycosphingolipid, one or more monosaccharides, bonded by a glycosidic bond, replace the —OH of the ceramide.

Cerebrosides

Although they are only present in membranes in small amounts, certain types of glycosphingolipids on the cell surface are important to cellular recognition and tissue immunity. In glycosphingolipids such as **cerebrosides,** one monosaccharide (usually galactose, though sometimes glucose), replaces the —OH of the ceramide.

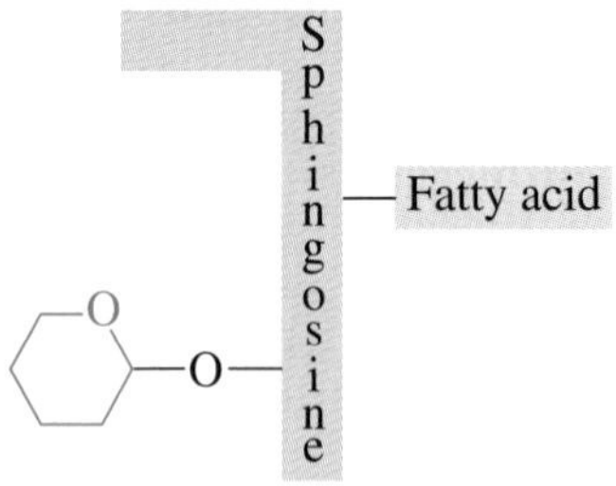

Cerebroside

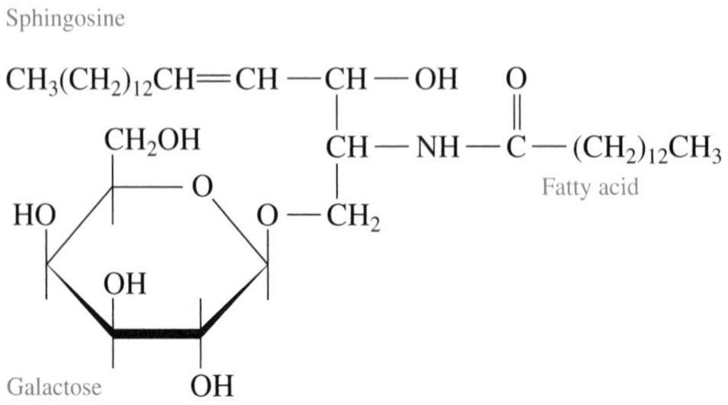

Galactocerebroside, a glycosphingolipid

Gangliosides

Gangliosides are similar to cerebrosides but contain two or more monosaccharides, such as glucose and galactose. Gangliosides are important in the membranes of neurons and act as receptors for hormones, viruses, and certain drugs. In Tay–Sachs disease, the ganglioside known as GM_2 accumulates because of a genetic defect in hexosaminidase A, an enzyme needed for the removal of the *N*-acetyl-D-galactosamine.

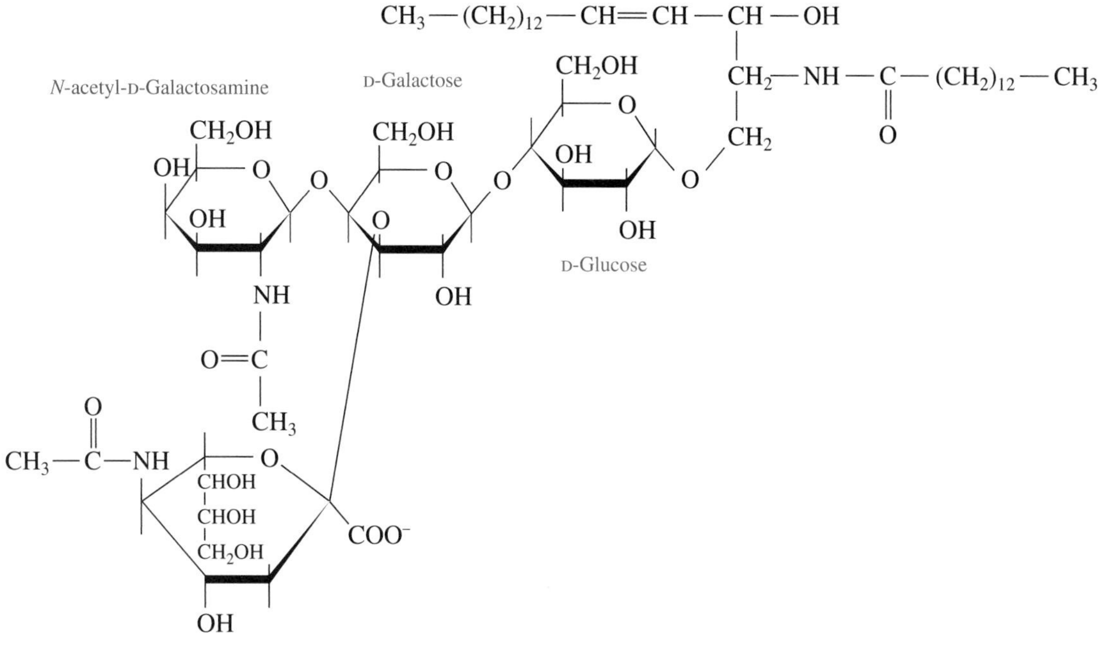

Glycosphingolipid GM_2 or Tay–Sachs ganglioside

Lipid Diseases

Many lipid diseases (**lipidoses**) involve the excessive accumulation of a sphingolipid or glycolipid because an enzyme needed for its breakdown is deficient or absent. The accumulation of these glycolipids may enlarge the spleen, liver, and bone marrow cells (Gaucher's disease) and cause mental retardation, seizures, blindness, and death in early infancy. Some lipid storage diseases are listed in Table 18.3.

In multiple sclerosis, sphingomyelins are lost from the myelin sheath, which is the protective membrane surrounding the neurons in the brain and spinal cord. As the disease progresses, the myelin sheath deteriorates. Scars form on the neurons and impair the transmission of nerve signals. The symptoms of multiple sclerosis include various levels of muscle weakness and loss of coordination and vision depending on the amount of damage. The cause of multiple sclerosis is not yet known, although some researchers suggest that a virus is involved.

Table 18.3   Lipid Diseases

Names of Disease	Lipid Stored	Type	Enzyme Absent
Fabry's	Gal-gal-glucosylceramide	Ganglioside	α-Galactosidase
Gaucher's	Glucosylceramide	Cerebroside	β-Glucosidase
Niemann–Pick	Sphingomyelin	Sphingolipid	Sphingomyelinase
Tay–Sachs	GM_2 ganglioside	Ganglioside	Hexosaminidase A

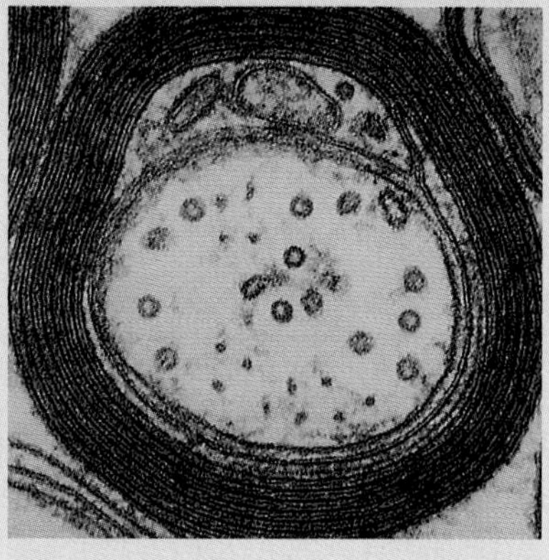

Myelin sheath

SAMPLE PROBLEM 18.6

Glycosphingolipid

In Fabry's disease, the ganglioside shown here accumulates due to a deficiency of α-galactosidase. Identify the components A–E in this glycolipid.

$$A$$

$$CH_3(CH_2)_{12}CH=CH-CH-OH \quad O$$

[Structure of the glycosphingolipid showing component A (sphingosine chain), component B (stearic acid chain $\|$ $-C-(CH_2)_{16}CH_3$), with $CH-NH$ linkage, and three sugar rings labeled C, D, and E bearing CH_2OH and OH groups, with HO group on ring C]

$$CH-NH-C-(CH_2)_{16}CH_3$$
$$B$$

Solution

In this glycolipid, the components are sphingosine (A); stearic acid (B), an 18-carbon fatty acid; two galactose units (C, D); and one glucose (E).

Study Check

How do we know that this glycolipid is a ganglioside rather than a cerebroside?

QUESTIONS AND PROBLEMS

Sphingolipids

18.45 Describe the difference between glycerophospholipids and sphingolipids.

18.46 Describe the differences between a cerebroside and a ganglioside.

18.47 Draw the structure of a cerebroside containing palmitic acid and galactose.

18.48 What amino alcohol is found in sphingomyelin? Draw the structure of a sphingomyelin containing palmitic acid.

LEARNING GOAL

Describe the structures of steroids.

WEB TUTORIAL
Phospholipids

18.7 Steroids: Cholesterol, Bile Salts, and Steroid Hormones

Steroids are compounds containing the steroid nucleus, which consists of three cyclohexane rings and one cyclopentane ring fused together. Although they are large molecules, steroids do not hydrolyze to give fatty acids and alcohols. The four rings in the steroid nucleus are designated A, B, C, and D. The carbon atoms are numbered beginning with the carbons in ring A and ending with the two methyl groups.

Steroid

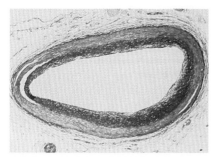

(a)

Cholesterol

Attaching other atoms and groups of atoms to the steroid structure forms a wide variety of steroid compounds. **Cholesterol**, which is one of the most important and abundant steroids in the body, is a *sterol* because it contains an oxygen atom as a hydroxyl (—OH) group on carbon 3. Like many steroids, cholesterol has methyl groups at carbon 10 and carbon 13 and a carbon chain at carbon 17 with a double bond between carbon 5 and carbon 6. In other steroids, the oxygen atom typically at carbon 3 forms a carbonyl (C=O) group.

Cholesterol

(b)

Figure 18.12 Excess cholesterol forms plaque that can block an artery, resulting in a heart attack. **(a)** A normal, open artery shows no buildup of plaque. **(b)** An artery that is almost completely clogged by atherosclerotic plaque.

Q What property of cholesterol would cause it to form deposits along the coronary arteries?

Cholesterol in the Body

Cholesterol is a component of cellular membranes, myelin sheath, and brain and nerve tissue. It is also found in the liver, bile salts, and skin, where it forms vitamin D. In the adrenal gland, it is used to synthesize steroid hormones. Cholesterol in the body is obtained from eating meats, milk, and eggs, and it is also synthesized by the liver from fats, carbohydrates, and proteins. There is no cholesterol in vegetable and plant products.

If a diet is high in cholesterol, the liver produces less. A typical daily American diet includes 400–500 mg of cholesterol, one of the highest in the world. The American Heart Association has recommended that we consume no more than 300 mg of cholesterol a day. The cholesterol contents of some typical foods are listed in Table 18.4.

When cholesterol exceeds its saturation level in the bile, gallstones may form. Gallstones are composed of almost 100% cholesterol with some calcium salts, fatty acids, and phospholipids. High levels of cholesterol are also associated with the accumulation of lipid deposits (plaque) that line and narrow the coronary arteries. (See Figure 18.12.) Clinically, cholesterol levels are considered elevated if the total plasma cholesterol level exceeds 200–220 mg/dL.

Table 18.4   Cholesterol Content of Some Foods

Food	Serving Size	Cholesterol (mg)
Liver (beef)	3 oz	370
Egg	1	250
Lobster	3 oz	175
Fried chicken	3½ oz	130
Hamburger	3 oz	85
Chicken (no skin)	3 oz	75
Fish (salmon)	3 oz	40
Butter	1 tablespoon	30
Whole milk	1 cup	35
Skim milk	1 cup	5
Margarine	1 tablespoon	0

Some research indicates that saturated fats in the diet may stimulate the production of cholesterol by the liver. A diet that is low in foods containing cholesterol and saturated fats appears to be helpful in reducing the serum cholesterol level. Other factors that may also increase the risk of heart disease are family history, lack of exercise, smoking, obesity, diabetes, gender, and age.

SAMPLE PROBLEM 18.7

Cholesterol

Observe the structure of cholesterol for the following questions:
a. What part of cholesterol is the steroid nucleus?
b. What features have been added to the steroid nucleus in cholesterol?
c. What classifies cholesterol as a sterol?

Solution

a. The four fused rings form the steroid nucleus.
b. The cholesterol molecule contains an alcohol group ($-$OH) on the first ring, one double bond in the second ring, and a branched carbon chain.
c. The alcohol group determines the sterol classification.

Study Check

Why is cholesterol in the lipid family?

Lipoproteins: Transporting Lipids

In the body, lipids must be transported through the bloodstream to tissues where they are stored, used for energy, or to make hormones. However, most lipids are nonpolar and insoluble in the aqueous environment of blood. They are made more soluble by combining them with phospholipids and proteins to form water-soluble complexes called **lipoproteins.** In general, lipoproteins are spherical particles with an outer surface of polar proteins and phospholipids that surround hundreds of nonpolar molecules of triacylglycerols and cholesteryl esters. (See Figure 18.13.) Cholesteryl esters are the prevalent form of cholesterol in the blood. They are formed by the esterification of the hydroxyl group in cholesterol with a fatty acid.

Cholesteryl ester

There are several types of lipoproteins that differ in density, lipid composition, and function. They include chylomicrons, very-low-density lipoprotein (VLDL), low-density lipoprotein (LDL), and high-density lipoprotein (HDL). (See Table 18.5.)

The chylomicrons formed in the mucosal cells of the small intestine and the VLDLs formed in the liver transport triacylglycerols, phospholipids, and choles-

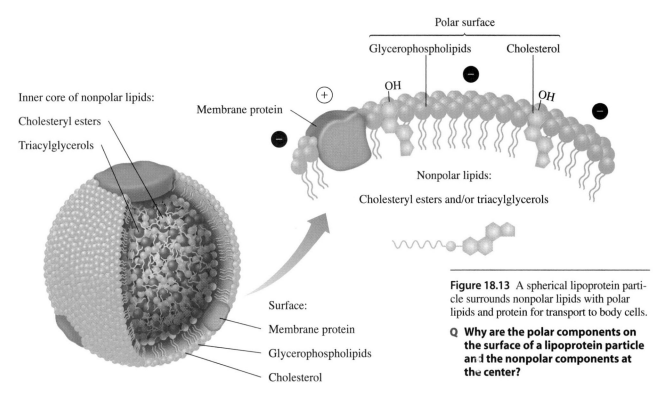

Inner core of nonpolar lipids:

Cholesteryl esters

Triacylglycerols

Polar surface

Glycerophospholipids

Cholesterol

OH

OH

Membrane protein

Nonpolar lipids:

Cholesteryl esters and/or triacylglycerols

Surface:

Membrane protein

Glycerophospholipids

Cholesterol

Figure 18.13 A spherical lipoprotein particle surrounds nonpolar lipids with polar lipids and protein for transport to body cells.

Q Why are the polar components on the surface of a lipoprotein particle and the nonpolar components at the center?

terol to the tissues for storage or to the muscles for energy. (See Figure 18.14.) The LDLs transport cholesterol to tissues to be used for the synthesis of cell membranes, steroid hormones, and bile salts. When the level of LDL exceeds the amount of cholesterol needed by the tissues, the LDLs deposit cholesterol in the arteries, which can restrict blood flow and increase the risk of developing heart disease and/or myocardial infarctions (heart attacks). This is why LDL cholesterol is called "bad" cholesterol.

Table 18.5   Composition and Properties of Plasma Lipoproteins

	Chylomicron	VLDL	LDL	HDL
Density (g/mL)	0.94	0.950–1.006	1.006–1.063	1.063–1.210
Composition (% by mass)				
Triacylglycerol	86	55	6	4
Phospholipids	7	18	22	24
Cholesterol	2	7	8	2
Cholesteryl esters	3	12	42	15
Protein	2	8	22	55

The HDLs remove excess cholesterol from the tissues and carry it to the liver where it is converted to bile salts and eliminated. When HDL levels are high, cholesterol that is not needed by the tissues is carried to the liver for elimination rather than deposited in the arteries, which gives the HDLs the name of "good" cholesterol. Most of the cholesterol in the body is synthesized in the liver, although some comes from the diet. However, a person on a high-fat diet reabsorbs cholesterol from the bile salts causing less cholesterol to be eliminated. In addition, higher levels of saturated fats stimulate the synthesis of cholesterol by the liver.

Because high cholesterol levels are associated with the onset of arteriosclerosis and heart disease, the serum levels of LDL and HDL are generally determined in a medical examination. For adults, recommended levels for total cholesterol are less than 200 mg/dL with LDL less than 130 mg/dL and HDL over 40 mg/dL. A lower level of serum cholesterol decreases the risk of heart disease. Higher HDL levels are found in people who exercise regularly and eat less saturated fat.

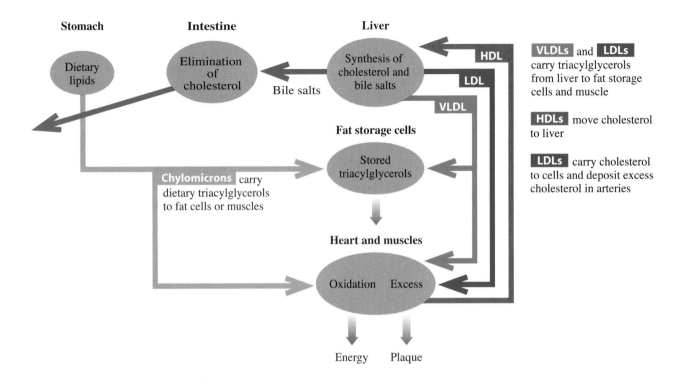

Figure 18.14 Lipoproteins such as HDLs and LDLs transport nonpolar lipids and cholesterol to cells and the liver.

Q What type of lipoprotein transports cholesterol to the liver?

Bile Salts

The *bile* salts are synthesized in the liver from cholesterol and stored in the gallbladder. When bile is secreted into the small intestine, the bile salts mix with the water-insoluble fats and oils in our diets. The bile salts with their nonpolar and polar regions act much like soaps, breaking apart and emulsifying large globules of fat. The emulsions that form have a larger surface area for the lipases, enzymes that digest fat. The bile salts also help in the absorption of cholesterol into the intestinal mucosa.

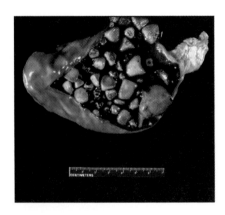

Figure 18.15 Gallstones form in the gallbladder when cholesterol levels are high.

Q What type of steroid is stored in the gallbladder?

If large amounts of cholesterol accumulate in the gallbladder, cholesterol can precipitate out and form gallstones. (See Figure 18.15.) If a gallstone passes into the bile duct, the pain can be severe. If the gallstone obstructs the duct, bile cannot be excreted. Then bile pigments known as bilirubin enter the blood where they cause jaundice, which gives a yellow color to the skin and eyes.

Steroid Hormones

The word *hormone* comes from the Greek "to arouse" or "to excite." Hormones are chemical messengers that serve as a kind of communication system from one part of the body to another. The *steroid* hormones, which include the sex hormones and the adrenocortical hormones, are closely related in structure to cholesterol and depend on cholesterol for their synthesis.

Two important male sex hormones, *testosterone* and *androsterone*, promote the growth of muscle and of facial hair and the maturation of the male sex organs and of sperm.

The *estrogens*, a group of female sex hormones, direct the development of female sexual characteristics: the uterus increases in size, fat is deposited in the breasts, and the pelvis broadens. *Progesterone* prepares the uterus for the implantation of a fertilized egg. If an egg is not fertilized, the levels of progesterone and estrogen drop sharply, and menstruation follows. Synthetic forms of the female sex hormones are used in birth-control pills. As with other kinds of steroids, side effects include weight gain and a greater risk of forming blood clots. The structures of some steroid hormones follow:

Hormone	Biological Effects
 Testosterone (androgen) (produced in testes)	Development of male organs; male sexual characteristics including muscles and facial hair; sperm formation
 Estradiol (estrogen) (produced in ovaries)	Development of female sexual characteristics; ovulation
 Progesterone (produced in ovaries)	Prepares uterus for fertilized egg
 Norethindrone (synthetic progestin)	Contraceptive (birth control) pill

Adrenal Corticosteroids

The adrenal glands, located on the top of each kidney, produce the corticosteroids. *Aldosterone*, a mineralocorticoid, is responsible for electrolyte and water balance by the kidneys. *Cortisone*, a glucocorticoid, increases the blood glucose level and stimulates the synthesis of glycogen in the liver from amino acids. Synthetic corticoids such as *prednisone* are derived from cortisone and used medically for reducing inflammation and treating asthma and rheumatoid arthritis, although precautions are given for long-term use.

Corticosteroids

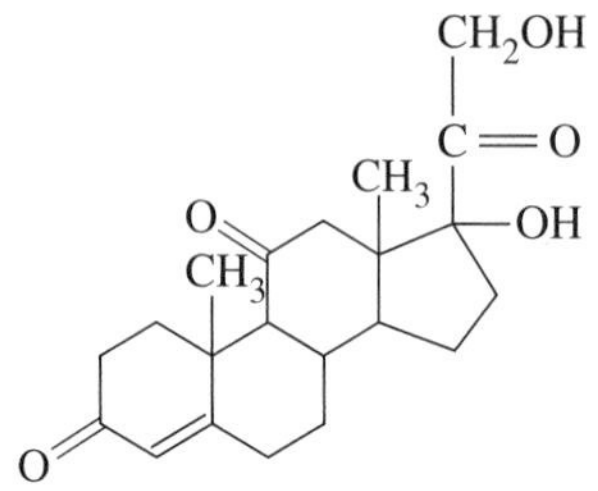

Cortisone
(produced in adrenal gland)

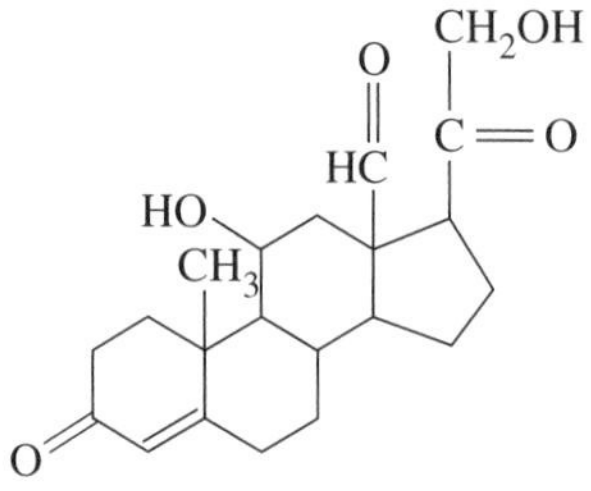

Aldosterone (mineralocorticoid)
(produced in adrenal gland)

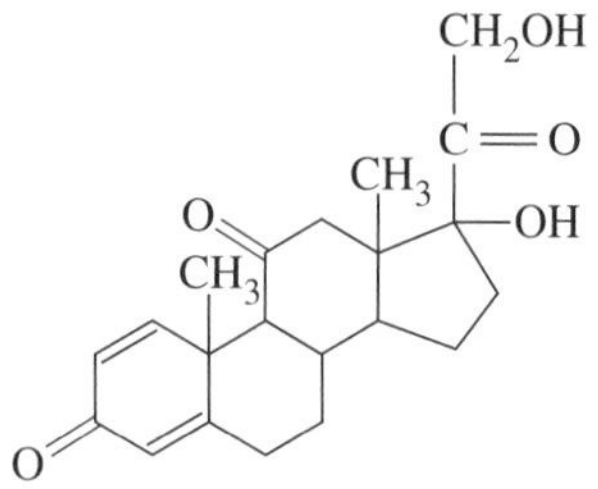

Prednisone
(synthetic corticoid)

Biological Effects

Increases the blood glucose and glycogen levels from fatty acids and amino acids

Increases the reabsorption of Na^+ in kidneys; retention of water

Reduces inflammation; treatment of asthma and rheumatoid arthritis

---

HEALTH NOTE

Anabolic Steroids

Some of the physiological effects of testosterone are to increase muscle mass and decrease body fat. Derivatives of testosterone called *anabolic steroids* that enhance these effects have been synthesized. Although they have some medical uses, anabolic steroids have been used in rather high dosages by some athletes in an effort to increase muscle mass. Such use is illegal.

Use of anabolic steroids in attempting to improve athletic strength can cause side effects including hypertension, fluid retention, increased hair growth, sleep disturbances, and acne. Over a long period of time, their use can be devastating and may cause irreversible liver damage and decreased sperm production.

Some Anabolic Steroids

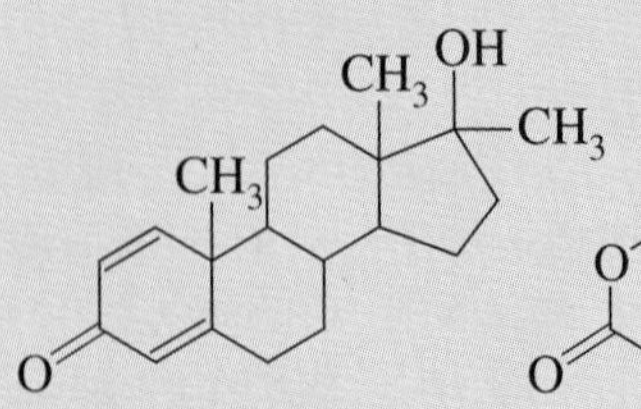

Methandienone

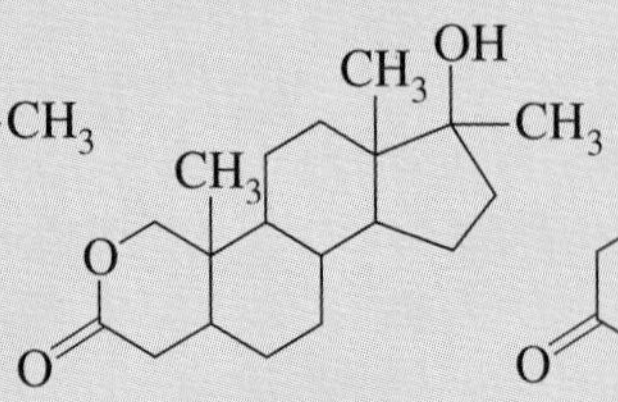

Oxandrolone

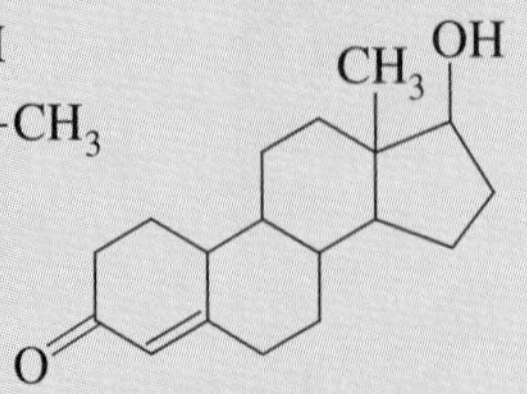

Nandrolone

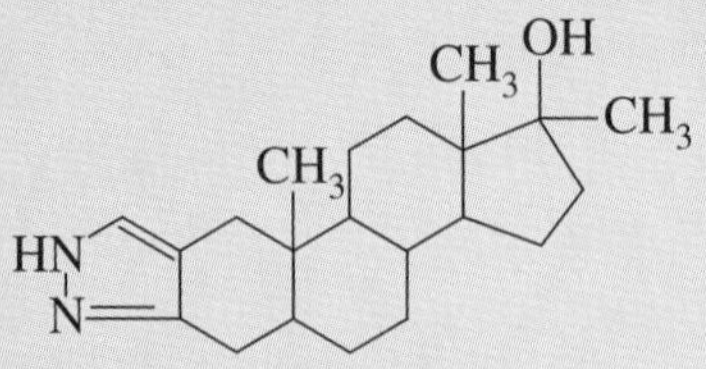

Stanozolol

Steroid Hormones

What are the functional groups on the steroid nucleus in the sex hormones estradiol and testosterone?

Solution

Estradiol contains a benzene ring and two alcohol groups. Testosterone contains a ketone group, a double bond, and an alcohol group.

Study Check

What are the similarities and differences in the structures of testosterone and the anabolic steroid nandrolone?

QUESTIONS AND PROBLEMS

Steroids: Cholesterol, Bile Salts, and Steroid Hormones

18.49 Draw the structure for the steroid nucleus.

18.50 Which of the following compounds are derived from cholesterol?
    a. glyceryl tristearate
    b. cortisone
    c. bile salts
    d. testosterone
    e. estradiol

18.51 What is the function of bile salts in digestion?

18.52 Why are gallstones composed of cholesterol?

18.53 What is the general structure of lipoproteins?

18.54 Why are lipoproteins needed to transport lipids in the bloodstream?

18.55 How do chylomicrons differ from very-low-density lipoproteins?

18.56 How do LDLs differ from HDLs?

18.57 Why are LDLs called "bad" cholesterol?

18.58 Why are HDLs called "good" cholesterol?

18.59 What are the similarities and differences between the sex hormones estradiol and testosterone?

18.60 What are the similarities and differences between the adrenal hormone cortisone and the synthetic corticoid prednisone?

18.61 Which of the following are male sex hormones?
    a. cholesterol
    b. aldosterone
    c. estrogen
    d. testosterone
    e. choline

18.62 Which of the following are adrenal steroids?
    a. cholesterol
    b. aldosterone
    c. estrogen
    d. testosterone
    e. choline

WEB TUTORIAL
Membrane Structure

Diffusion

Osmosis

Active Transport

18.8 Cell Membranes

The membrane of a cell separates the contents of a cell from the external fluids. It is semipermeable so that nutrients can enter the cell and waste products can leave. The main components of a cell membrane are the glycerophospholipids and sphingolipids. Earlier in this chapter we saw that the structures of such phospholipids consisted of a nonpolar region or tail with long-chain fatty acids and a polar region or head from phosphoric acid and amino alcohols that ionize at physiological pH. The lipid composition of the membranes of human red blood cells and bacterial cells is given in Table 18.6.

In a cell membrane, two rows of phospholipids are arranged like a sandwich. Their nonpolar tails, which are hydrophobic ("water-fearing"), move to the center, while their polar heads, which are hydrophilic ("water-loving"), align on the outer edges of the membrane. This double row arrangement of phospholipids is called a **lipid bilayer.** (See Figure 18.16.) One row of the phospholipids forms the outside surface of the membrane, which is in contact with the external fluids, and the other row forms the inside surface or edge of the membrane, which is in contact with the internal contents of the cell.

Most of the phospholipids in the lipid bilayer contain unsaturated fatty acids. Due to the kinks in the carbon chains at the cis double bonds, the phospholipids do not fit closely together. As a result, the lipid bilayer is not a rigid, fixed structure, but one that is dynamic and fluid-like. In this liquid-like bilayer, there are also proteins, carbohydrates, and cholesterol molecules. For this reason, the model of biological membranes is referred to as the **fluid mosaic model** of membranes.

In the fluid mosaic model, proteins known as peripheral proteins emerge on just one of the surfaces, outer or inner. The integral proteins extend through the entire lipid bilayer and appear on both surfaces of membrane. Some proteins and some lipids on the outer surface of the cell membrane are attached to carbohydrates. The saccharide chains of these glycoproteins and glycolipids project into the surrounding fluid environment where they are responsible for cell recognition and communication with chemical messengers such as hormones and neurotransmitters. In animals, cholesterol makes up 20–25% of the lipid bilayer. Embedded among the nonpolar tails of the fatty acids, cholesterol molecules add strength and rigidity to the cell membrane.

Table 18.6 Lipid Composition of Cell Membranes		
Type of Lipid	**Human Red Blood Cells**	**Bacterial Cells**
Glycerophospholipids		
–choline	19	0
–ethanolamine	18	65
–serine	8	0
Triacylglycerol	0	18
Sphinogmylelin	18	0
Glycosphingolipids	10	0
Cholesterol	25	0
Others	2	17

Data adapted from Mathews, C. K.; Van Holde, K. K.; Ahern, K. G. *Biochemistry*; Addison Wesley/Longman/Benjamin Cummings: New York, 2000, p. 322.

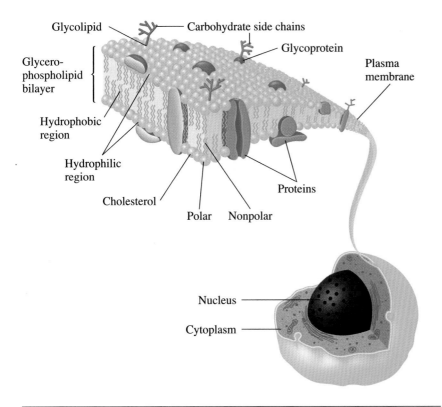

Figure 18.16 The fluid mosaic model of a cell membrane. Proteins and cholesterol are embedded in a lipid bilayer of phospholipids. The bilayer forms a membrane-type barrier with polar heads at the membrane surfaces and the nonpolar tails in the center away from the water.

Q What types of fatty acids are found in the phospholipids of the lipid bilayer?

Transport Through Cell Membranes

Ions and molecules flow in and out of the cell in several ways. In the simplest transport mechanism called diffusion or passive transport, ions and small molecules migrate from a higher concentration to a lower concentration. For example, some ions as well as small molecules such as O_2, urea, and water diffuse through cell membranes. If their concentration is greater outside the cell than inside, they diffuse into the cell. If water has a high concentration in the cell, it diffuses out of the cell.

Another type of transport called facilitated transport increases the rate of diffusion for substances that diffuse too slowly by passive diffusion to meet cell needs. This process utilizes the integral proteins that extend from one edge of the cell membrane to the other. Groups of integral proteins provide channels to transport chloride ion (Cl^-), bicarbonate ion (HCO_3^-), and glucose molecules in and out of the cell more rapidly.

Certain ions such as K^+, Na^+, and Ca^{2+}, move across a cell membrane against a concentration gradient. For example, the K^+ concentration is greater inside a cell, and the Na^+ concentration is greater outside. However, in the conduction of nerve impulses and contraction of muscles, K^+ moves into the cell, and Na^+ moves out. To move an ion from a lower to a higher concentration requires energy, which is accomplished by a process known as active transport. In active transport, a protein complex called a Na^+/K^+ pump breaks down ATP to ADP (Chapter 23), which

Figure 18.17 Substances are transported across a cell membrane by either simple diffusion, facilitated transport, or by active transport.

Q **What is the difference between simple diffusion and facilitated transport?**

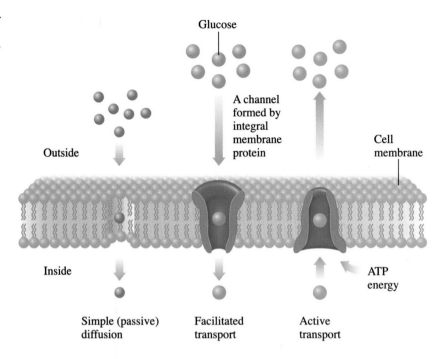

Glucose

Outside

A channel formed by integral membrane protein

Cell membrane

Inside

ATP energy

Simple (passive) diffusion

Facilitated transport

Active transport

releases energy to move Na^+ and K^+ against their concentration gradients. (See Figure 18.17.)

SAMPLE PROBLEM 18.9

Lipid Bilayer in the Cell Membranes

Describe the role of phospholipids in the lipid bilayer.

Solution

Phospholipids consist of polar and nonpolar parts. In a cell membrane, an alignment of the nonpolar sections toward the center with the polar sections on the outside produces a barrier that prevents the contents of a cell from mixing with the fluids on the outside of the cell.

Study Check

Why are protein channels needed in the lipid bilayer?

QUESTIONS AND PROBLEMS

Cell Membranes

18.63 What types of lipids are found in cell membranes?

18.64 Describe the structure of a lipid bilayer.

18.65 What is the function of the lipid bilayer in a cell membrane?

18.66 How do the unsaturated fatty acids in the phospholipids affect the structure of cell membranes?

18.67 Where are proteins located in cell membranes?

18.68 How are carbohydrates attached to the lipid bilayer?

18.69 What is the function of the carbohydrates on a cell membrane?

18.70 Why is a cell membrane semipermeable?

18.71 What are some ways that substances move in and out of cells?

18.72 Identify the type of transport described by each of the following:

  a. A molecule moves through a protein channel.

  b. O_2 moves into the cell from a higher concentration outside the cell.

  c. An ion moves from low to high concentration in the cell.

Chapter Review

18.1 Lipids

Lipids are nonpolar compounds that are not soluble in water. Classes of lipids include waxes, fats and oils, glycerophospholipids, and steroids.

18.2 Fatty Acids

Fatty acids are unbranched carboxylic acids that typically contain an even number (12–18) of carbon atoms. Fatty acids may be saturated, monounsaturated with one double bond, or polyunsaturated with two or more double bonds. The double bonds in unsaturated fatty acids are almost always cis.

18.3 Waxes, Fats, and Oils

A wax is an ester of a long-chain fatty acid and a long-chain alcohol. The triacylglycerols of fats and oils are esters of glycerol with three long-chain fatty acids. Fats contain more saturated fatty acids and have higher melting points than most vegetable oils.

18.4 Chemical Properties of Triacylglycerols

The hydrogenation of unsaturated fatty acids converts double bonds to single bonds. The oxidation of unsaturated fatty acids produces short-chain fatty acids with disagreeable odors. The hydrolysis of the ester bonds in fats or oils produces glycerol and fatty acids. In saponification, a fat heated with a strong base produces glycerol and the salts of the fatty acids or soaps.

18.5 Glycerophospholipids

Glycerophospholipids are esters of glycerol with two fatty acids and a phosphate group attached to an amino alcohol.

18.6 Sphingolipids

In sphingolipids, the alcohol sphingosine forms a bond with a fatty acid and a phosphate–amino alcohol group. In glycosphingolipids, sphingosine is bonded to a fatty acid and one or more monosaccharides.

18.7 Steroids: Cholesterol, Bile Salts, and Steroid Hormones

Steroids are lipids containing the steroid nucleus, which is a fused structure of four rings. Steroids include cholesterol, bile salts, and vitamin D. Lipids, which are nonpolar, are transported through the aqueous environment of the blood by forming lipoproteins. Lipoproteins such as chylomicrons and LDL transport triacylglycerols from the intestines and the liver to fat cells for storage and muscles for energy. HDLs transport cholesterol from the tissues to the liver for elimination. The steroid hormones are closely related in structure to cholesterol and depend on cholesterol for their synthesis. The sex hormones, such as estrogen and testosterone, are responsible for sexual characteristics and reproduction. The adrenal corticosteroids such as aldosterone and cortisone regulate water balance and glucose levels in the cells.

18.8 Cell Membranes

All animal cells are surrounded by a semipermeable membrane that separates the cellular contents from the external fluids. The membrane is composed of two rows of glycerophospholipids in a lipid bilayer. Nutrients and waste products move through the cell membrane using simple transport (diffusion), facilitated transport, or active transport.

Key Terms

cephalins Phospholipids found in brain and nerve tissues that incorporate the amino alcohol serine or ethanolamine.

cerebroside A glycolipid consisting of sphingosine, a fatty acid, and a monosaccharide (usually galactose).

cholesterol The most prevalent of the steroid compounds found in cellular membranes; needed for the synthesis of vitamin D, hormones, and bile acids.

esterification The reaction of an alcohol such as glycerol with acids to form ester bonds.

fat Another term for solid triacylglycerols.

fatty acids Long-chain carboxylic acids found in fats.

fluid mosaic model The concept that cell membranes are lipid bilayer structures that contain an assortment of polar lipids and proteins in a dynamic, fluid arrangement.

ganglioside A glycolipid consisting of sphingosine, a fatty acid, and two or more monosaccharides.

glycerophospholipids Polar lipids of glycerol attached to two fatty acids and a phosphate group connected to an amino group such as choline, serine, or ethanolamine.

glycosphingolipids The phospholipid that combines sphingosine with a fatty acid bonded to the nitrogen group and one or more monosaccharides bonded by a glycosidic link, which replaces the — OH group of sphingosine.

hydrogenation The addition of hydrogen to unsaturated fats.

lecithins Glycerophospholipids containing choline as the amino alcohol.

lipid bilayer A model of a cell membrane in which phospholipids are arranged in two rows interspersed with proteins arranged at different depths.

lipidoses Genetic diseases in which a deficiency of an enzyme for the hydrolysis of a lipid causes the accumulation of that lipid to toxic levels.

lipids A family of compounds that is nonpolar in nature and not soluble in water; includes fats, waxes, phospholipids, and steroids.

lipoprotein A combination of nonpolar lipids with glycerophospholipids and proteins to form a polar complex that can be transported through body fluids.

monounsaturated fatty acids Fatty acids with one double bond.

oil Another term for liquid triacylglycerols.

polyunsaturated fatty acids Fatty acids that contain two or more double bonds.

prostaglandins A number of compounds derived from arachidonic acid that regulate several physiological processes.

saturated fatty acids Fatty acids that have no double bonds; they have higher melting points than unsaturated lipids and are usually solid at room temperatures.

sphingolipids Phospholipids in which sphingosine has replaced glycerol.

steroids Types of lipid composed of a multicyclic ring system.

triacylglycerols A family of lipids composed of three fatty acids bonded through ester bonds to glycerol, a trihydroxy alcohol.

wax The ester of a long-chain alcohol and a long-chain saturated fatty acid.

Additional Problems

18.73 Among the ingredients of a lipstick are beeswax, carnauba wax, hydrogenated vegetable oils, and capric triglyceride. What types of lipids have been used? Draw the structure of capric triglyceride (capric acid is the saturated 10-carbon fatty acid).

18.74 Because peanut oil floats on the top of peanut butter, many brands of peanut butter are hydrogenated. A solid product then forms that is mixed into the peanut butter and does not separate. If a triacylglycerol in peanut oil that contains one palmitic acid, one oleic acid, and one linoleic acid is completely hydrogenated, what is the product?

18.75 Trans fats are produced during the hydrogenation of polyunsaturated oils.
  a. What is the typical configuration of a monounsaturated fatty acid?
  b. How does a trans fat compare with a cis fat?
  c. Draw the structure of *trans*-oleic acid.

18.76 One mole of triolein is completely hydrogenated. What is the product? How many moles of hydrogen are required? How many grams of hydrogen? How many liters of hydrogen are needed if the reaction is run at STP?

18.77 On the list of ingredients of a cosmetic product are glyceryl stearate and lecithin. What are the structures of these compounds?

18.78 Some typical meals at fast-food restaurants are listed here. Calculate the number of kilocalories from fat and the percentage of total kilocalories due to fat (1 gram of fat = 9 kcal). Would you expect the fats to be mostly saturated or unsaturated? Why?
  a. a chicken dinner, 830 kcal, 46 g of fat
  b. a quarter-pound cheeseburger, 518 kcal, 29 g of fat
  c. pepperoni pizza (three slices), 560 kcal, 18 g of fat
  d. beef burrito, 470 kcal, 21 g of fat
  e. deep-fried fish (three pieces), 480 kcal, 28 g of fat

18.79 Identify the following as fatty acids, soaps, triacyl-glycerols, wax, glycerophospholipid, sphingolipid, or steroid:

a. beeswax   **b.** cholesterol

c. lecithin

d. glyceryl tripalmitate (tripalmitin)

e. sodium stearate   **f.** safflower oil

g. sphingomyelin   **h.** whale blubber

i. adipose tissue   **j.** progesterone

k. cortisone   **l.** stearic acid

18.80 Why would an animal that lives in a cold climate have more unsaturated triacylglycerols in its body fat than an animal that lives in a warm climate?

18.81 Identify the components in each of the following lipids as:

1. glycerol   2. fatty acid   3. phosphate

4. amino alcohol   5. steroid nucleus

6. sphingosine

a. estrogen   **b.** cephalin

c. wax   **d.** triacylglycerol

e. glycerophospholipid

f. sphingomyelin

18.82 Which of the following are found in cell membranes?

a. cholesterol   **b.** triacylglycerols

c. carbohydrates   **d.** proteins

e. waxes   **f.** glycerophospholipids

g. sphingolipids   **h.** prostaglandins

18.83 Identify the lipoprotein in each description as:

1. chylomicrons   2. VLDL   3. LDL   4. HDL

a. "good" cholesterol

b. transports most of the cholesterol to the cells

c. carries triacylglycerols from the intestine to the fat cells

d. transports cholesterol to the liver

e. has the greatest abundance of protein

f. "bad" cholesterol

g. carries triacylglycerols synthesized in the liver to the muscles

h. has the lowest density

19 Amines and Amides

"The pharmacy is one of the many factors in the final integration of chemistry and medicine in patient care," says Dorothea Lorimer, pharmacist, Kaiser Hospital. "If someone is allergic to a medication, I have to find out if a new medication has similar structural features. For instance, some people are allergic to sulfur. If there is sulfur in the new medication, there is a chance it will cause a reaction."

A prescription indicates a specific amount of a medication. At the pharmacy, the chemical name, formula, and quantity in milligrams or micrograms are checked. Then the prescribed number of capsules is prepared and placed in a container. If it is a liquid medication, a specific volume is measured and poured into a bottle for liquid prescriptions.

LOOKING AHEAD

the
**Chemistry
place**

www.chemplace.com/college

Visit the URL above or use the CD-ROM in the book for extra quizzing, interactive tutorials, career resources, and case studies.

In earlier chapters, we looked at organic compounds that contain carbon, hydrogen, and oxygen. Now we are ready to discuss amines and amides, which are organic compounds that contain nitrogen. Many nitrogen-containing compounds are important to life as components of amino acids, proteins, and nucleic acids: DNA and RNA. Many amines that exhibit strong physiological activity are used in medicine as decongestants, anesthetics, and sedatives. Examples include dopamine, histamine, epinephrine, and amphetamine.

Alkaloids such as caffeine, nicotine, cocaine, and digitalis, which demonstrate powerful physiological activity, are naturally occurring amines obtained from plants. Amides, which are derived from carboxylic acids, are important in biology as the types of bonds that link amino acids in proteins. Some amides used medically include acetaminophen (Tylenol) used to reduce fever; phenobarbital, a sedative and anticonvulsant medication; and penicillin, an antibiotic.

19.1 Amines

Amines are considered as derivatives of ammonia (NH_3), in which one or more hydrogen atoms is replaced with alkyl or aromatic groups. For example, in methylamine, a methyl group replaces one hydrogen atom in ammonia. The bonding of two methyl groups gives dimethylamine, and the three methyl groups in trimethylamine replace all the hydrogen atoms in ammonia. (See Figure 19.1.)

Classification of Amines

When we classified alcohols in Chapter 14, we looked at the number of carbon atoms bonded to the alcohol carbon. Amines are classified in a similar way by counting the number of carbon atoms directly bonded to a nitrogen atom. In a *primary (1°) amine,* one carbon is bonded to a nitrogen atom. In a *secondary (2°) amine,* two carbons are bonded to the nitrogen atom, and a *tertiary (3°) amine* has three carbon atoms bonded to the nitrogen.

Ammonia	Primary (1°) amine	Secondary (2°) amine	Tertiary (3°) amine

Alkyl or Aromatic group

$$H-\overset{\overset{\displaystyle ..}{}}{N}-H \qquad R-\overset{\overset{\displaystyle ..}{}}{N}-H \qquad R-\overset{\overset{\displaystyle ..}{}}{N}-R \qquad R-\overset{\overset{\displaystyle ..}{}}{N}-R$$
$$\quad\underset{H}{|} \qquad\qquad\quad \underset{H}{|} \qquad\qquad\quad \underset{H}{|} \qquad\qquad\quad \underset{R}{|}$$

Examples:

$$CH_3-\overset{\overset{\displaystyle ..}{}}{N}-H \qquad CH_3-\overset{\overset{\displaystyle ..}{}}{N}-CH_3 \qquad CH_3-\overset{\overset{\displaystyle ..}{}}{N}-CH_3$$
$$\qquad\underset{H}{|} \qquad\qquad\qquad \underset{H}{|} \qquad\qquad\qquad \underset{CH_3}{|}$$

Line-Bond Formulas for Amines

We can draw line-bond formulas for amines just as we did for other organic compounds. For example, we can write the following line-bond formulas and classify each of the amines.

Ammonia

Methylamine (primary amine 1°)

Dimethylamine (secondary amine 2°)

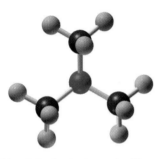

Trimethylamine (tertiary amine 3°)

Primary amine (1°)   Primary amine (1°)   Secondary amine (2°)   Tertiary amine (3°)

SAMPLE PROBLEM 19.1

Classifying Amines

Classify the following amines as primary, secondary, or tertiary:

a. NH_2

b.
$$CH_3{-}\overset{H}{\underset{}{N}}{-}CH_2CH_3$$

c. $\overset{}{\underset{H}{N}}{-}CH_3$

d.

Solution

a. This is a primary amine because there is one alkyl group (cyclohexyl) attached to a nitrogen atom.

b. This is a secondary amine. There are two alkyl groups (methyl and ethyl) attached to the nitrogen atom.

c. This is a secondary amine with two carbon groups, methyl and phenyl, bonded to the nitrogen atom.

d. The nitrogen atom in this line-bond formula is bonded to two carbon atoms, which makes it a secondary amine.

Study Check

Classify the following amine as primary, secondary, or tertiary:

$$CH_3CH_2{-}\overset{}{\underset{CH_3}{N}}{-}CH_2CH_3$$

Figure 19.1 Amines are classified according to the number of carbon atoms bonded to the N atom.

Q How many carbon atoms are bonded to the nitrogen atom in dimethylamine?

QUESTIONS AND PROBLEMS

Amines

19.1 What is a primary amine?

19.2 What is a tertiary amine?

19.3 Classify each of the following amines as primary, secondary, or tertiary:

a. $CH_3{-}CH_2{-}CH_2{-}NH_2$

b. $CH_3{-}\overset{H}{\underset{}{N}}{-}CH_2{-}CH_3$

c. NH_2

d. $\overset{CH_3}{\underset{CH_3}{N}}$

e. $CH_3{-}\overset{CH_3}{\underset{}{CH}}{-}\overset{CH_3}{\underset{}{N}}{-}CH_2{-}CH_3$

HEALTH NOTE

Amines in Health and Medicine

In response to allergic reactions or injury to cells, the body increases the production of histamine, which causes blood vessels to dilate and increases the permeability of the cells. Redness and swelling occur in the area. Administering an antihistamine such as diphenylhydramine helps block the effects of histamine.

Histamine

Diphenylhydramine

In the body, hormones called biogenic amines carry messages between the central nervous system and nerve cells. Epinephrine (adrenaline) and norepinephrine (noradrenaline) are released by the adrenal medulla in "fight or flight" situations to raise the blood glucose level and move the blood to the muscles. Used in remedies for colds, hay fever, and asthma, the norepinephrine contracts the capillaries in the mucous membranes of the respiratory passages. The prefix *nor–* in a drug name means there is one less CH_3— group on the nitrogen atom. Parkinson's disease is a result of a deficiency in another biogenic amine called dopamine.

Epinephrine (adrenaline)

Norepinephrine (noradrenaline)

Dopamine

Produced synthetically, amphetamines (known as "uppers") are stimulants of the central nervous system much like epinephrine, but they also increase cardiovascular activity and depress the appetite. They are sometimes used to bring about weight loss, but they can cause chemical dependency. Benzedrine and Neo-Synephrine (phenylephrine) are used in medications to reduce respiratory congestion from colds, hay fever, and asthma. Sometimes, Benzedrine is taken internally to combat the desire to sleep, but it has side effects. Methedrine is used to treat depression and in the illegal form is known as "speed" or "crank." The prefix *meth–* means that there is one more methyl group on the nitrogen atom.

Benzedrine (amphetamine)

Neo-Synephrine (phenylephrine)

Methamphetamine (methedrine)

19.4 Classify each of the following amines as primary, secondary, or tertiary:

a. $CH_3—CH_2—\overset{\overset{\displaystyle NH_2}{|}}{CH}—CH_3$

b. $CH_3—CH_2—\overset{\overset{\displaystyle CH_3}{|}}{N}—CH_2—CH_3$

c.

d.

e. $CH_3—\overset{\overset{\displaystyle H}{|}}{N}—\overset{\overset{\displaystyle CH_3}{|}}{\underset{\underset{\displaystyle CH_3}{|}}{C}}—CH_3$

the
Chemistry
place

WEB TUTORIAL
Amine and Amide
Functional Groups

19.2 Naming Amines

There are several systems in use for naming amines. For simple amines, the common names are often used. In the common name, the alkyl groups bonded to the nitrogen atom are listed in alphabetical order. The prefixes *di-* and *tri-* are used to indicate two and three identical substituents.

$$CH_3—NH_2 \qquad CH_3—NH—CH_3 \qquad CH_3—CH_2—CH_2—\overset{\overset{\displaystyle CH_3}{|}}{N}—CH_2—CH_3$$

Methylamine   Dimethylamine   Ethylmethylpropylamine

The IUPAC names for amines are similar to the names we used for alcohols, except that the *–e* in the parent alkane name is replaced with *amine.*

$$CH_4 \qquad CH_3—OH \qquad CH_3—NH_2$$

Methane   Methanol   Methan**amine**

Step 1   Identify the longest carbon chain bonded to the nitrogen atom. Replace the *–e* in the corresponding alkane name with *amine.*

Step 2   Number the carbon chain to show the position of the amine group and any other substituents.

$$CH_3—CH_2—NH_2 \qquad CH_3—CH_2—CH_2—NH_2 \qquad CH_3—\overset{\overset{\displaystyle NH_2}{|}}{CH}—CH_3$$

Ethan**amine**   1-Propan**amine**   2-Propan**amine**

$$CH_3—\overset{\overset{\displaystyle NH_2}{|}}{CH}—CH_2—CH_3 \qquad CH_3—\overset{\overset{\displaystyle CH_3}{|}}{CH}—CH_2—CH_2—NH_2$$

2-Butan**amine**   3-Methyl-1-butan**amine**

Step 3   In secondary and tertiary amines, the largest alkyl group attached to the nitrogen is named as the parent amine. The smaller alkyl groups are named with the prefix *N-* followed by the alkyl name, and listed alphabetically.

$$CH_3—CH_2—\overset{\overset{\displaystyle CH_3}{|}}{N}—CH_3$$
N, N-Dimethylethanamine

$$CH_3—CH_2—CH_2—\overset{\overset{\displaystyle CH_3}{|}}{N}—CH_3$$
N, N-Dimethyl-1-propanamine

$$CH_3—CH_2—CH_2—\overset{\overset{\displaystyle CH_3}{|}}{N}—CH_2—CH_3$$
N-Ethyl-*N*-methyl-1-propanamine

Heterocycles are cyclic amines that contain a nitrogen atom in the ring. For example, some of the pungent aroma and taste we associate with black pepper is due to a compound called piperidine, a six-atom ring containing a nitrogen atom. The fruit from the black pepper plant is dried and ground to give the black pepper we use to season our foods.

Piperidine

An amine with two amine functional groups is named as a *diamine.* For example, the amines 1,4-butanediamine and 1,5-pentanediamine contribute to the odors of decaying flesh.

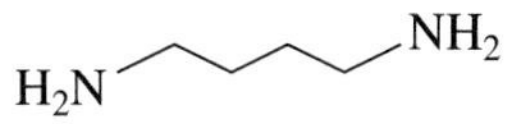

H_2N —— NH_2
1,4-Butanediamine
(putrescine)

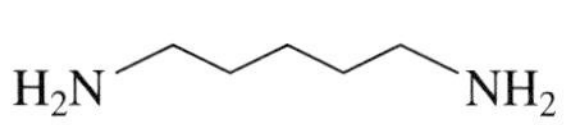

H_2N —— NH_2
1,5-Pentanediamine
(cadaverine)

In amines where another functional group takes priority, the —NH_2 group is named as a substituent *amino* group and numbered to show its location. For the major functional groups we have studied, the increasing priority follows the increase in oxidation:

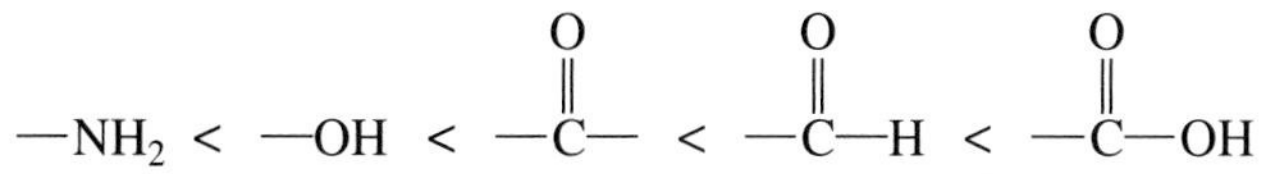

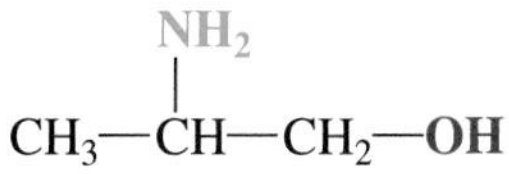

$$\text{low priority} \quad -NH_2 \; < \; -OH \; < \; -\overset{O}{\underset{}{\overset{\|}{C}}}- \; < \; -\overset{O}{\underset{}{\overset{\|}{C}}}-H \; < \; -\overset{O}{\underset{}{\overset{\|}{C}}}-OH \quad \text{high priority}$$

$$CH_3-\overset{NH_2}{\underset{}{\overset{|}{CH}}}-CH_2-OH$$
2-Amino-1-propanol

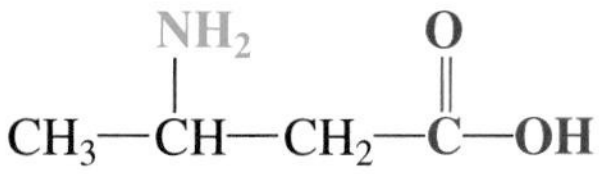

$$CH_3-\overset{NH_2}{\underset{}{\overset{|}{CH}}}-CH_2-\overset{O}{\underset{}{\overset{\|}{C}}}-OH$$
3-Aminobutanoic acid

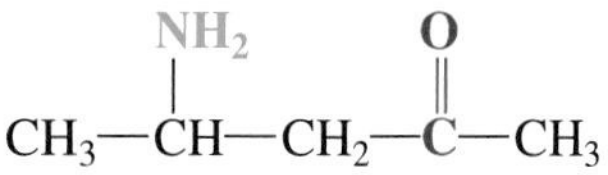

$$CH_3-\overset{NH_2}{\underset{}{\overset{|}{CH}}}-CH_2-\overset{O}{\underset{}{\overset{\|}{C}}}-CH_3$$
4-Amino-2-pentanone

Aromatic Amines

The aromatic amines use the name *aniline*, which is approved by IUPAC.

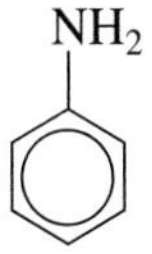

Aniline

4-Bromoaniline
(*p*-bromoaniline)

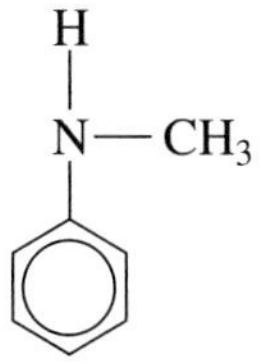

N-Methylaniline

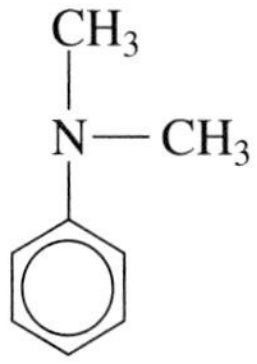

N,N-Dimethylaniline

SAMPLE PROBLEM 19.2

Naming Amines

Give the common name for each of the following amines:

a. CH_3CH_2—NH_2

b. 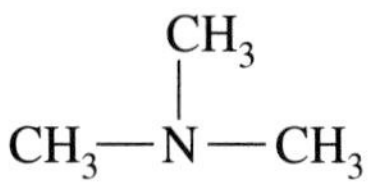 CH_3—$\overset{CH_3}{\underset{}{\overset{|}{N}}}$—$CH_3$

c. 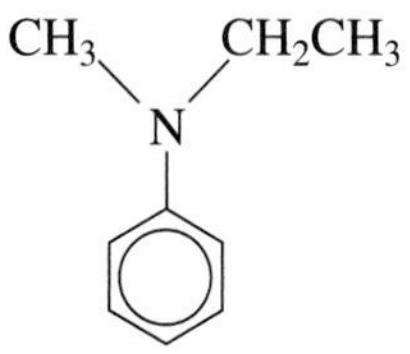 CH_3 —— CH_2CH_3 on N (attached to benzene ring)

Solution

a. This amine has one ethyl group attached to the nitrogen atom; its name is *ethylamine*.

b. This amine has three methyl groups attached to the nitrogen atom; its name is *trimethylamine*.

c. This amine is a derivative of aniline with a methyl and an ethyl group attached to the nitrogen atom; its name is *N-ethyl-N-methylaniline*.

Study Check

Draw the structure of ethylpropylamine.

QUESTIONS AND PROBLEMS

Naming Amines

19.5 Write the common and IUPAC names for each of the following:

a. $CH_3-CH_2-NH_2$

b. $CH_3-NH-CH_2-CH_2-CH_3$

c. 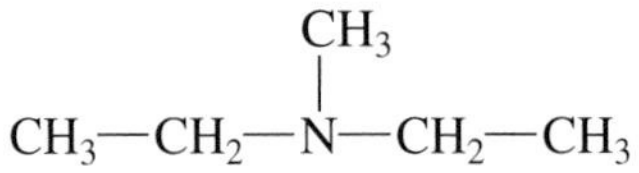 $CH_3-CH_2-\overset{\overset{\displaystyle CH_3}{|}}{N}-CH_2-CH_3$

d. 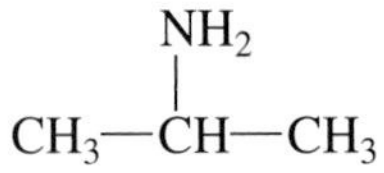 $CH_3-\overset{\overset{\displaystyle NH_2}{|}}{CH}-CH_3$

19.6 Write the common and IUPAC names for each of the following:

a. $CH_3-CH_2-CH_2-NH_2$

b. $CH_3-NH-CH_2-CH_3$

c. $CH_3-CH_2-CH_2-CH_2-NH_2$

d. 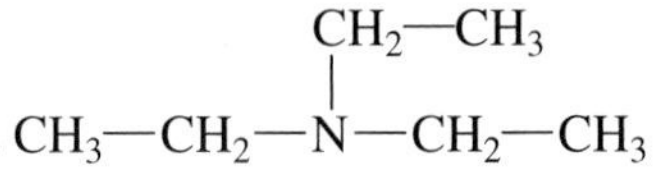 $CH_3-CH_2-\overset{\overset{\displaystyle CH_2-CH_3}{|}}{N}-CH_2-CH_3$

19.7 Write the IUPAC names for each of the following:

a. 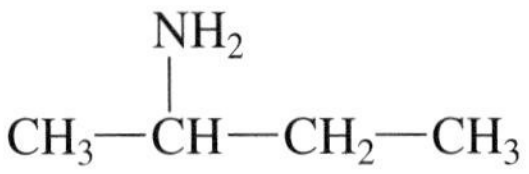 $CH_3-\overset{\overset{\displaystyle NH_2}{|}}{CH}-CH_2-CH_3$

b. 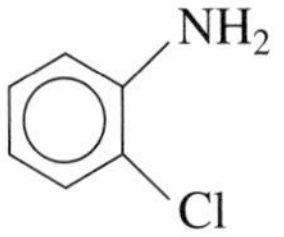 benzene ring with NH_2 and Cl substituents

c. 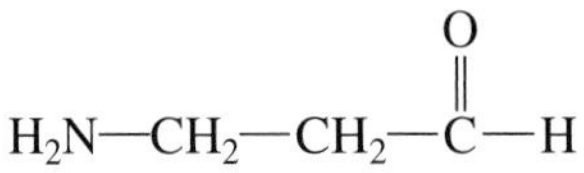 $H_2N-CH_2-CH_2-\overset{\overset{\displaystyle O}{||}}{C}-H$

d. 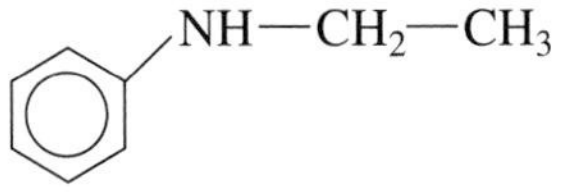 benzene ring with $NH-CH_2-CH_3$

19.8 Write the IUPAC names for each of the following:

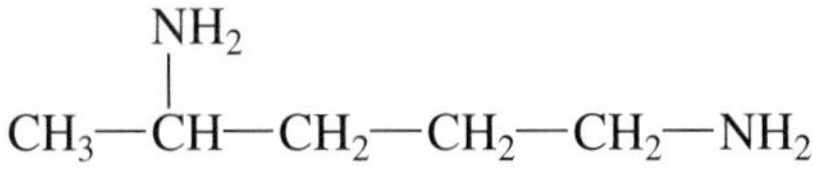

a. 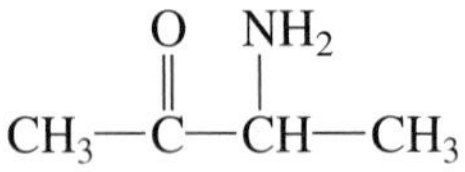 $CH_3-\overset{\overset{\displaystyle O}{||}}{C}-\overset{\overset{\displaystyle NH_2}{|}}{CH}-CH_3$

b. $CH_3-\overset{\overset{\displaystyle NH_2}{|}}{CH}-CH_2-CH_2-CH_2-NH_2$

c. benzene ring with $NH-CH_3$ and Br substituents

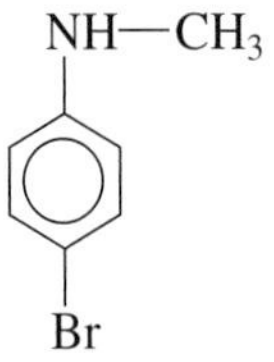

d. 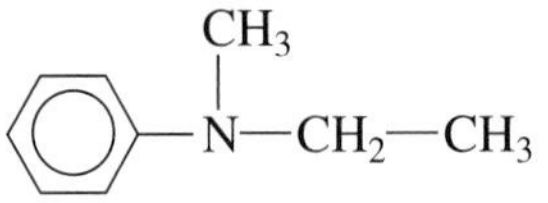 benzene ring with $\overset{\overset{\displaystyle CH_3}{|}}{N}-CH_2-CH_3$

19.9 Draw the condensed structural formulas for each of the following amines:
a. ethylamine
b. *N*-methylaniline
c. butylpropylamine
d. 2-pentanamine

19.10 Draw the condensed structural formulas for each of the following amines:
a. dimethylamine
b. *p*-chloroaniline
c. *N*,*N*-diethylaniline
d. 1-amino-3-pentanone

19.3 Physical Properties of Amines

Amines have higher boiling points than hydrocarbons of similar mass, but lower than the alcohols, as seen in Table 19.1.

Table 19.1   Comparison of Boiling Points (°C) of Amines, Alcohols, and Alkanes

NH$_3$			
	−33		

1 Carbon Atom		3 Carbon Atoms	
CH$_4$	−162	CH$_3$—CH$_2$—CH$_3$	−42
CH$_3$—NH$_2$	−7	CH$_3$—N(CH$_3$)—CH$_3$	3
CH$_3$—OH	65	CH$_3$—CH$_2$—NH—CH$_3$	36
2 Carbon Atoms		CH$_3$—CH$_2$—CH$_2$—NH$_2$	48
CH$_3$—CH$_3$	−89	CH$_3$—CH$_2$—CH$_2$—OH	97
CH$_3$—NH—CH$_3$	7		
CH$_3$—CH$_2$—NH$_2$	17		
CH$_3$—CH$_2$—OH	79		

Because amines contain a polar N—H bond, they form hydrogen bonds. (See Figure 19.2.) However, the nitrogen atom in amines is not as electronegative as the oxygen in alcohols, which makes the hydrogen bonds weaker in amines. The —NH$_2$ in primary amines can form more hydrogen bonds, which gives them higher boiling points than the secondary amines of the same mass. It is not possible for tertiary amines to hydrogen bond with each other (no N—H bonds), which makes their boiling points much lower and similar to those of alkanes and ethers.

CH$_3$—CH$_2$—CH$_2$—NH$_2$
Propylamine (1°)
bp 48°C

CH$_3$—CH$_2$—NH—CH$_3$
Ethylmethylamine (2°)
bp 36°C

CH$_3$—N(CH$_3$)—CH$_3$
Trimethylamine (3°)
bp 3°C

Solubility in Water

Like alcohols, the smaller amines, including tertiary ones, are soluble in water because they form hydrogen bonds with water. (See Figure 19.3.) However, in amines with more than six carbon atoms, the effect of hydrogen bonding is diminished, and their solubility in water decreases. Large amines are not soluble in water.

SAMPLE PROBLEM 19.3

Physical Properties of Amines

If the compounds trimethylamine and ethylmethylamine have the same molar mass, why is the boiling point of trimethylamine (3°C) lower than that of ethylmethylamine (37°C)?

Solution

With a polar N—H bond, hydrogen bonds form between ethylmethylamine molecules. Thus, a higher temperature is required to break the hydrogen bonds and form a gas. However, trimethylamine, which is a tertiary amine, has no N—H bond and cannot form hydrogen bonds between the amine molecules.

LEARNING GOAL

Describe the boiling points and solubility of amines.

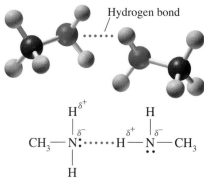

Figure 19.2 Hydrogen bonding occurs between 1° or 2° amines.

Q Why don't tertiary (3°) amines form hydrogen bonds?

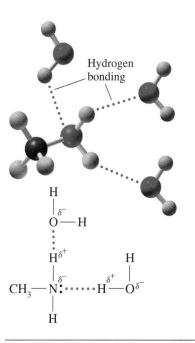

Figure 19.3 Hydrogen bonding occurs between amines and water molecules.

Q Why are tertiary (3°) amines soluble in water?

Study Check

Why is $CH_3CH_2NH_2$ more soluble in water than
$CH_3CH_2CH_2CH_2NHCH_2CH_2CH_3$?

QUESTIONS AND PROBLEMS

Physical Properties of Amines

19.11 Indicate the compound in each pair that would have the higher boiling point:

a. $CH_3-CH_2-NH_2$ or CH_3-CH_2-OH **b.** CH_3-NH_2 or $CH_3-CH_2-CH_2-NH_2$

c. $CH_3-\overset{\overset{\displaystyle CH_3}{|}}{N}-CH_3$ or $CH_3-CH_2-CH_2-NH_2$

19.12 Indicate the compound in each pair that would have the higher boiling point:

a. $CH_3-CH_2-CH_2-CH_3$ or $CH_3-CH_2-CH_2-NH_2$ **b.** CH_3-NH_2 or $CH_3-CH_2-NH_2$

c. $CH_3-CH_2-CH_2-OH$ or $CH_3-\overset{\overset{\displaystyle NH_2}{|}}{CH}-CH_3$

19.13 Propylamine (molar mass 59) has a boiling point of 48°C and ethylmethylamine (molar mass 59) has a boiling point of 37°C. Butane (molar mass 58) has a much lower boiling point -1°C. Explain.

19.14 Assign the boiling point of 3°C, 48°C, or 97°C to the appropriate compound: 1-propanol, propylamine, and trimethylamine.

19.15 Indicate if each of the following is soluble in water:
  a. $CH_3-CH_2-NH_2$

  b. $CH_3-NH-CH_3$

  c. $CH_3-CH_2-CH_2-\overset{\overset{\displaystyle CH_2-CH_2-CH_3}{|}}{N}-CH_2-CH_2-CH_3$

  d. $CH_3-\overset{\overset{\displaystyle NH_2}{|}}{CH}-CH_2-CH_3$

19.16 Indicate if each of the following is soluble in water:

  a. $CH_3-CH_2-CH_2-NH_2$ **b.** $CH_3-CH_2-CH_2-NH-CH_2-CH_3$

  c. $CH_3-\overset{\overset{\displaystyle CH_3}{|}}{N}-CH_3$ **d.**

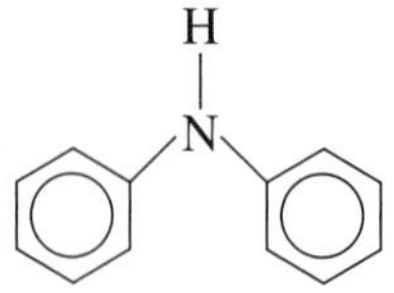

19.4 Amines React as Bases

In Chapter 10, we saw that ammonia (NH_3) acts as a Brønsted–Lowry base because it accepts a proton (H^+) from water to produce an ammonium ion (NH_4^+) and a hydroxide ion (OH^-). Let's review that equation:

$$\ddot{N}H_3 + H_2O \; \rightleftharpoons \; NH_4^+ \; + \; OH^-$$

Ammonia   Ammonium ion   Hydroxide ion

LEARNING GOAL

Write equations for the ionization and neutralization of amines.

the Chemistry place

WEB TUTORIAL
Amines as Bases

Ionization of an Amine in Water

In water, amines also act as Brønsted–Lowry bases because the lone electron pair on the nitrogen atom accepts a proton. The products are an alkyl ammonium ion and hydroxide ion. The name of the alkyl ammonium ion is similar to the common amine name, but *amine* is replaced by *ammonium ion*.

$$CH_3—\ddot{N}H_2 + H_2O \; \rightleftharpoons \; CH_3—\overset{+}{N}H_3 \; + \; OH^-$$

Methylamine   Methylammonium ion   Hydroxide ion

Secondary amines also accept a proton to form dialkyl ammonium ions.

$$\begin{array}{c} CH_3—\ddot{N}H + H_2O \\ | \\ CH_3 \end{array} \; \rightleftharpoons \; \begin{array}{c} CH_3—\overset{+}{N}H_2 \\ | \\ CH_3 \end{array} \; + \; OH^-$$

Dimethylamine   Dimethylammonium ion   Hydroxide ion

Basicity of Amines

Because amines act as weak bases by accepting protons from water and producing hydroxide ions, their aqueous solutions are basic. We can write the base dissociation constant K_b for methylamine as follows:

$$K_b = \frac{[CH_3—NH_3^+] \, [OH^-]}{[CH_3—NH_2]} = 4.4 \times 10^{-4}$$

Most of the K_b values for amines are less than 10^{-3}, which means that the equilibrium favors the undissociated amine molecules. Aqueous solutions of amines have basic pH values and turn red litmus paper blue. We can compare the strengths of some amines by looking at their K_b values as follows:

Ammonia	*1° amine*		*2° amine*	*3° amine*
NH_3	$CH_3—NH_2$	$CH_3—CH_2—NH_2$	$CH_3—NH—CH_3$	$\begin{array}{c} CH_3 \\ | \\ CH_3—N—CH_3 \end{array}$
K_b 1.8×10^{-5}	4.4×10^{-4}	5.6×10^{-4}	5.1×10^{-4}	5.3×10^{-5}

Amine Salts

When you squeeze lemon juice on fish, the "fishy odor" of the amines is removed by converting them to amine salts. In a neutralization reaction, an amine acts as a base and reacts with an acid to form an **amine salt**. The lone pair of electrons on the nitrogen atom accepts a proton H^+ from an acid to give an amine salt; no water

is formed. An amine salt is named by replacing the *-amine* part of the amine name with *-ammonium* followed by the name of the negative ion.

Neutralization of an Amine

$$CH_3-\ddot{N}H_2 + HCl \rightleftharpoons CH_3-\overset{+}{N}H_3\ Cl^-$$

Amine   **Acid**   **Amine salt**

Methyl**amine**   Methyl**ammonium chloride**

$$CH_3-\underset{\underset{CH_3}{|}}{\ddot{N}H} + HCl \rightleftharpoons CH_3-\underset{\underset{CH_3}{|}}{\overset{+}{N}H_2}\ Cl^-$$

Dimethyl**amine**   Dimethyl**ammonium chloride**

The ammonium ions are classified as primary, secondary, and tertiary.

$$CH_3-\underset{\underset{H}{|}}{\overset{\overset{H}{|}}{\overset{+}{N}}}-H \qquad CH_3-\underset{\underset{H}{|}}{\overset{\overset{CH_3}{|}}{\overset{+}{N}}}-H \qquad CH_3-\underset{\underset{CH_3}{|}}{\overset{\overset{CH_3}{|}}{\overset{+}{N}}}-H$$

Primary   Secondary   Tertiary

In a **quaternary ammonium ion**, a nitrogen atom bonds to four carbon groups. In the quaternary ion, the nitrogen atom has a positive charge just as it does in other amine salts. Choline, an amino alcohol present in glycerophospholipids is a quaternary ammonium ion.

$$CH_3-\underset{\underset{CH_3}{|}}{\overset{\overset{CH_3}{|}}{\overset{+}{N}}}-CH_3Cl^- \qquad\qquad HO-CH_2-CH_2-\underset{\underset{CH_3}{|}}{\overset{\overset{CH_3}{|}}{\overset{+}{N}}}-CH_3$$

Tetramethylammonium chloride   Choline

The quaternary salts differ from other amine salts because the nitrogen atom is not bonded to an H atom. Thus, quaternary salts do not react with bases.

Properties of Amine Salts

As ionic compounds, amine salts are solids at room temperature, odorless, and soluble in water and body fluids. For this reason, amines used as drugs are converted to their amine salts. The amine salt of ephedrine is used as a bronchodilator and in decongestant products such as Sudafed. The amine salt of diphenhydramine is used in products such as Benadryl for relief of itching and pain from skin irritations and rashes. (See Figure 19.4.) In pharmaceuticals, the naming of the amine salt follows an older method of giving the amine name followed by the name of the acid.

When an amine salt reacts with a strong base such as NaOH, it is converted back to the amine, which is also called the free amine or free base.

$$CH_3-NH_3^+\ Cl^- + NaOH \longrightarrow CH_3-NH_2 + NaCl + H_2O$$

The narcotic cocaine is typically extracted from coca leaves using an acidic HCl solution to give a white, solid amine salt, which is cocaine hydrochloride. This is the form in which cocaine is smuggled and sold illegally on the street to be snorted or injected. "Crack cocaine" is the free amine or free base of the amine obtained by treating the cocaine hydrochloride with NaOH and ether, a process known as "free-basing." The solid product is known as "crack cocaine" because it makes a crackling noise when heated. The free amine is rapidly absorbed when smoked and gives

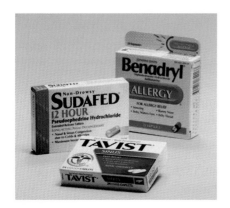

Figure 19.4 Decongestants and products that relieve itch and skin irritations can contain ammonium salts.

Q Why are the ammonium salts used rather than the biologically active amines?

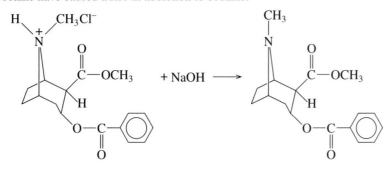

Ephedrine hydrochloride
Ephedrine HCl
Sudafed

Diphenhydramine hydrochloride
Diphenhydramine HCl
Benadryl

stronger highs than the cocaine hydrochloride. Unfortunately, these effects of crack cocaine have caused a rise in addiction to cocaine.

Cocaine hydrochloride

Cocaine ("free base")

the
Chemistry
place

CASE STUDY
Death by Chocolate?

SAMPLE PROBLEM 19.4

Reactions of Amines

Write an equation that shows ethylamine
a. ionizing as a weak base in water
b. neutralized by HCl

Solution

a. In water, ethylamine acts as a weak base by accepting a proton from water to produce ethylammonium hydroxide.

$$CH_3-CH_2-NH_2 + H-OH \rightleftharpoons CH_3-CH_2-NH_3^+ + OH^-$$

b. $CH_3-CH_2-NH_2 + HCl \longrightarrow CH_3-CH_2-NH_3^+Cl^-$

Study Check

What is the condensed structural formula of the salt formed by the reaction of trimethylamine and HCl?

QUESTIONS AND PROBLEMS

Amines React as Bases

19.17 Write an equation for the ionization of each of the following amines in water:
a. methylamine   **b.** dimethylamine   **c.** aniline

19.18 Write an equation for the ionization of each of the following amines in water:
a. ethylamine   **b.** propylamine   **c.** N-methylaniline

19.19 Write the condensed structural formula of the amine salt obtained when each of the amines in problem 19.17 reacts with HCl.

19.20 Write the condensed structural formula of the amine salt obtained when each of the amines in problem 19.18 reacts with HCl.

19.21 Novocain, a local anesthetic, is the hydrochloride salt of procaine.

Procaine

a. What is the formula of the amine salt formed when procaine reacts with HCl?
b. Why is Novocain, the amine salt, used rather than the amine procaine?

19.22 Lidocaine (xylocaine) is used as a local anesthetic and cardiac depressant.

Lidocaine (xylocaine)

a. What is the formula of the amine salt formed when lidocaine reacts with HCl?
b. Why is the amine salt of lidocaine used rather than the amine?

LEARNING GOAL

Identify heterocyclic amines; distinguish between the types of heterocyclic amines.

19.5 Heterocyclic Amines and Alkaloids

A **heterocyclic amine** is a cyclic organic compound that contains one or more nitrogen atoms in the ring. The heterocyclic amine rings typically consist of 5 or 6 atoms and one or more nitrogen atoms. Of the five-atom rings, the simplest one is pyrrolidine, which is a ring of four carbon atoms and a nitrogen atom, all with single bonds. Pyrrole is a five-atom ring with one nitrogen atom and two double bonds. Imidazole is a five-atom ring that contains two nitrogen atoms.

Pyrrolidine
(nicotine, etc.)

Pyrrole
(hemoglobin)

Imidazole
(histidine)

Many of the six-atom heterocyclic amines are aromatic. Pyridine is similar to benzene except that it has a nitrogen atom in place of a carbon atom. Pyrimidine, which is found in nucleic acids, is also similar to benzene except that it has two nitrogen atoms. In purine, another component of nucleic acids, a pyrimidine ring is fused with imidazole.

Piperidine
(quinine, drugs)

Pyridine
(nicotine, vitamins)

Pyrimidine
(nucleic acid: DNA, RNA)

Purine
(nucleic acid: DNA, RNA)

SAMPLE PROBLEM 19.5

Identify each of the following heterocyclic amines:

a.   **b.**

Solution

a. This five-atom ring with one nitrogen atom is pyrrole.
b. This aromatic ring with one nitrogen atom is pyridine.

Study Check

Identify the following heterocyclic amine.

Alkaloids: Amines in Plants

Alkaloids are physiologically active nitrogen-containing compounds produced by plants. The term *alkaloid* refers to the "alkali-like" or basic characteristics we have seen for amines. Certain alkaloids are used in anesthetics, in antidepressants, and as stimulants, although many are habit forming.

As a stimulant, nicotine increases the level of adrenaline in the blood, which increases the heart rate and blood pressure. It is well known that smoking cigarettes can damage the lungs and that exposure to tars and other carcinogens in cigarette smoke can lead to lung cancer. However, nicotine is responsible for the addiction of smoking. Nicotine has a simple alkaloid structure that includes a pyrrolidine ring. Coniine, which is obtained from hemlock, is an extremely toxic alkaloid that contains a piperidine ring.

Nicotine

Coniine

Figure 19.5 Coffee beans contain caffeine, which is an alkaloid that is a stimulant of the central nervous system.

Q Why is caffeine considered an alkaloid?

Caffeine

Caffeine contains an imidazole ring and is a stimulant of the central nervous system (CNS). Present in coffee, tea, soft drinks, chocolate, and cocoa, caffeine increases alertness, but may cause nervousness and insomnia. Caffeine is also used in certain pain relievers to counteract the drowsiness caused by an antihistamine. (See Figure 19.5.)

Several alkaloids are used in medicine. Quinine obtained from the bark of the cinchona tree has been used in the treatment of malaria since the 1600s. Atropine from belladonna is used in low concentrations to accelerate slow heart rates and as an anesthetic for eye examinations.

Quinine   Atropine

For many centuries morphine and codeine, alkaloids found in the oriental poppy plant, have been used as effective painkillers. (See Figure 19.6.) Codeine, which is structurally similar to morphine, is used in some prescription painkillers and cough syrups. Heroin, obtained by a chemical modification of morphine, is strongly addicting and is not used medically.

Heroin

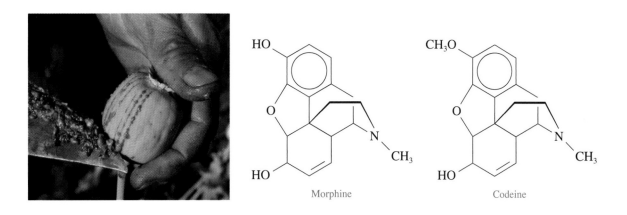

Figure 19.6 The green, unripe poppy seed capsule contains a milky sap (opium) that is the source of the alkaloids morphine and codeine.

Q Where is the piperidine ring in the structures of morphine and codeine?

HEALTH NOTE

Synthesizing Drugs

One area of research in pharmacology is the synthesis of compounds that retain the anesthetic characteristic of naturally occurring ring alkaloids such as cocaine and morphine without the addictive side effects. For example, cocaine is an effective anesthetic, but addictive. Research chemists modified the structure of cocaine, but kept the benzene group and nitrogen atom. The synthetic products procaine and lidocaine retain the anesthetic qualities of the natural alkaloid without the addictive side effects.

The structure of morphine was also modified to make a synthetic alkaloid, meperidine, or Demerol, which acts as an effective painkiller.

Cocaine

Procaine (Novocaine)

Lidocaine (Xylocaine)

Meperidine (Demerol)

SAMPLE PROBLEM 19.6

Heterocyclic Amines

Identify the heterocyclic amines in the alkaloids nicotine and caffeine.

Solution

In nicotine, the heterocyclic amine is the five-atom ring of pyrrolidine. Caffeine contains a purine, which is pyrimidine and imidazole fused together.

Study Check

What is the heterocyclic amine in meperidine (Demerol)?

QUESTIONS AND PROBLEMS

Heterocyclic Amines

19.23 Identify the following as amines or heterocyclic amines:

a. 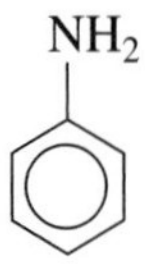 NH_2 (attached to benzene ring)   b. 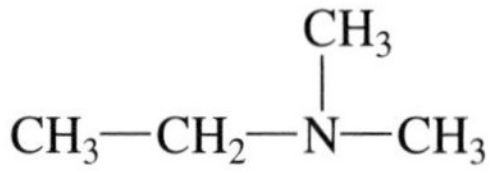 $CH_3-CH_2-\underset{\underset{CH_3}{|}}{N}-CH_3$ c.  (pyrimidine ring)   d. 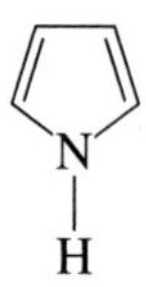 (pyrrole ring with N—H)

19.24 Identify the following as amines or heterocyclic amines:

a. 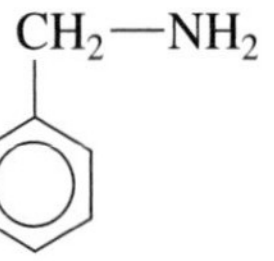 CH_2-NH_2 (attached to benzene ring)   b. 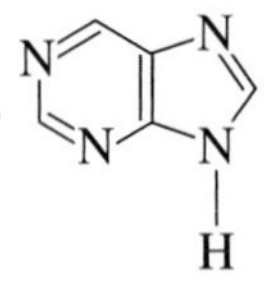 (purine ring with N—H)   c. 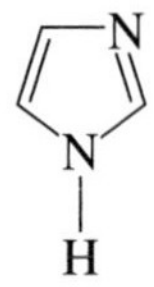 (imidazole ring with N—H)   d. 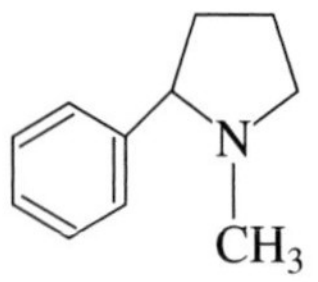 (phenyl attached to pyrrolidine ring with N—CH_3)

19.25 Identify the type of heterocyclic amines in problem 19.23.

19.26 Identify the type of heterocyclic amines in problem 19.24.

19.27 Low levels of serotonin in the brain appear to be associated with depressed states. What type of heterocyclic amine is serotonin?

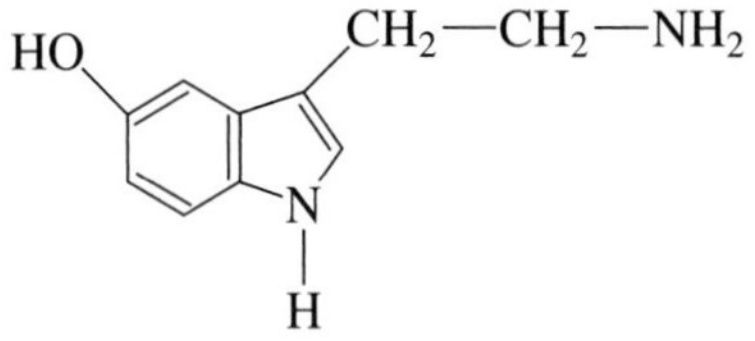

$$HO \qquad CH_2-CH_2-NH_2$$

(indole ring structure with N—H)

Serotonin

19.28 LSD is made from lysergic acid, which is produced by a fungus that grows on rye. What types of heterocyclic amines are in lysergic acid?

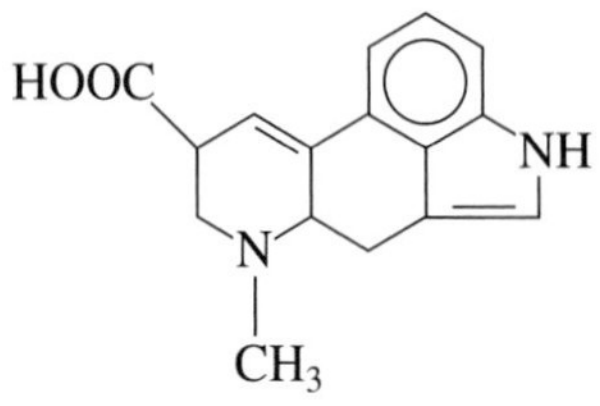

$$HOOC$$

(lysergic acid ring structure with NH and N—CH_3)

Lysergic acid

19.6 Structures and Names of Amides

The **amides** are derivatives of carboxylic acids in which an amino group replaces the hydroxyl group. (See Figure 19.7.) The amides are classified by the number of carbon atoms bonded to the nitrogen atom.

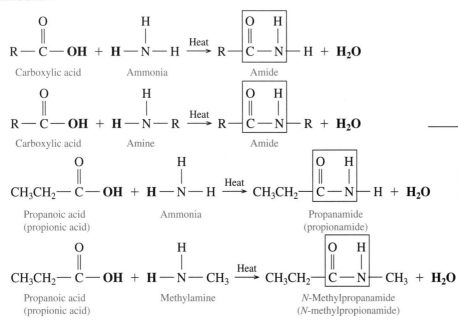

Carboxylic acid   Primary (1°) amide   Secondary (2°) amide   Tertiary (3°) amide

Preparation of Amides

An amide is produced in a reaction called **amidation,** in which a carboxylic acid reacts with ammonia or a primary or secondary amine. A molecule of water is eliminated, and the fragments of the carboxylic acid and amine molecules join to form the amide, much like the formation of ester. Because a hydrogen atom must be lost from the amines, only primary and secondary amines undergo amidation.

Amidation

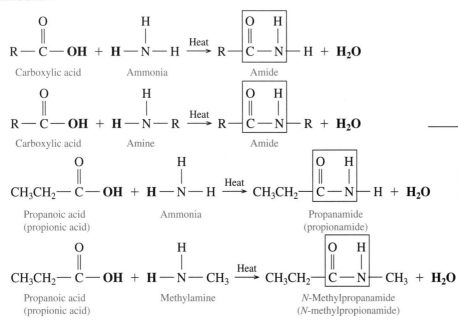

Carboxylic acid   Ammonia   Amide

Carboxylic acid   Amine   Amide

Propanoic acid (propionic acid)   Ammonia   Propanamide (propionamide)

Propanoic acid (propionic acid)   Methylamine   *N*-Methylpropanamide (*N*-methylpropionamide)

LEARNING GOAL

Write the amide products of amidation and give their common and IUPAC names.

Carboxylic acid

Ethanoic acid (Acetic acid)

Amide

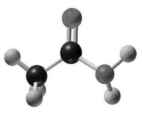

Ethanamide (Acetamide)

Figure 19.7 Amides are derivatives of carboxylic acids in which an amino group ($-NH_2$, $-NHR$, or $-NR_2$) replaces the hydroxyl group ($-OH$).

Q Why are amides with one to five carbon atoms soluble in water?

SAMPLE PROBLEM 19.7

Amidation Products

Give the structural formula of the amide product in each of the following reactions:

a.

$$\text{C}_6\text{H}_5\text{C}(=\text{O})\text{OH} + \text{NH}_3 \xrightarrow{\text{Heat}}$$

b.

$$\text{CH}_3\text{COH} + \text{NH}_2\text{CH}_2\text{CH}_3 \xrightarrow{\text{Heat}}$$

Solution

a. The structural formula of the amide product can be written by attaching the carbonyl group from the acid to the nitrogen atom of the amine. —OH is removed from the acid and —H from the amine to form water.

b.

$$CH_3\overset{\overset{O}{\|}}{C} - \overset{\overset{H}{|}}{N}CH_2CH_3$$

Study Check

What are the condensed structural formulas of the carboxylic acid and amine needed to prepare the following amide?

$$H - \overset{\overset{O}{\|}}{C} - \overset{\overset{CH_3}{|}}{N} - CH_3$$

Naming Amides

In both the common and IUPAC names, amides are named by dropping the *ic acid* or *oic acid* from the carboxylic acid names (IUPAC or common) and adding the suffix *amide*.

$$H - \overset{\overset{O}{\|}}{C} - NH_2 \qquad CH_3\overset{\overset{O}{\|}}{C} - NH_2 \qquad CH_3CH_2CH_2\overset{\overset{O}{\|}}{C} - NH_2$$

Methanamide   Ethanamide   Butanamide   Benzamide
(formamide)   (acetamide)   (butyramide)

When alkyl groups are attached to the nitrogen atom, the name of an amide is preceded by *N*- or *N,N*- depending on whether there are one or two groups.

$$CH_3\overset{\overset{O}{\|}}{C} - \overset{\overset{H}{|}}{N} - CH_3 \qquad CH_3CH_2\overset{\overset{O}{\|}}{C} - \overset{\overset{CH_3}{|}}{N} - CH_3$$

N-Methylethanamide   *N,N*-Dimethylpropanamide   *N*-Methylbenzamide
(*N*-methylacetamide)   (*N,N*-dimethylpropionamide)

$$CH_3\overset{\overset{CH_3}{|}}{C}HCH_2CH_2\overset{\overset{O}{\|}}{C} - NH_2 \qquad CH_3CH_2\overset{\overset{Br}{|}}{C}H - \overset{\overset{O}{\|}}{C} - \overset{\overset{CH_3}{|}}{N} - CH_3$$

4-Methylpentanamide   *N,N*-Dimethyl-2-bromobutanamide
                            (*N,N*-dimethyl-α-bromobutyramide)

SAMPLE PROBLEM 19.8

Naming Amides

Give the common and IUPAC names for each of the following amides:

a. $CH_3-CH_2-\overset{\overset{\displaystyle O}{\|}}{C}-NH_2$ b. $CH_3-\overset{\overset{\displaystyle Cl}{|}}{CH}-\overset{\overset{\displaystyle O}{\|}}{C}-NH-CH_2CH_3$

Solution

a. The IUPAC name of the carboxylic acid is propanoic acid; the common name is propionic acid. Replacing the *oic acid* or *ic acid* ending with *amide* gives the IUPAC name of *propanamide* and common name of *propionamide*.
b. Using the same carboxylic acid as in part a, the ethyl group attached to the nitrogen atom is named *N-ethyl*. The amide is named *N-ethyl-2-chloro-propanamide* (IUPAC), and *N-ethyl-α-chloropropionamide* (common).

Study Check

Draw the condensed structural formula of *N,N*-dimethyl-*p*-chlorobenzamide.

Physical Properties of Amides

The amides do not have the properties of bases that we saw for the amines. Only foramide is a liquid at room temperature, while the other amides are solids. For primary amides, the $-NH_2$ group can form several hydrogen bonds, which gives primary amides high melting points. The melting points of the secondary amides are lower because the number of hydrogen bonds decreases. Tertiary amides have even lower melting points because they cannot form hydrogen bonds with other tertiary amides.

Hydrogen bonding between amide molecules

The amides with one to five carbon atoms are soluble in water because they can hydrogen bond with water molecules.

Hydrogen bonding of amides with water

HEALTH NOTE

Amides in Health and Medicine

The simplest natural amide is urea, an end product of protein metabolism in the body. The kidneys remove urea from the blood and provide for its excretion in urine. If the kidneys malfunction, urea is not removed and builds to a toxic level, a condition called uremia. Urea is also used as a component of fertilizer, to increase nitrogen in the soil.

$$NH_2-\overset{\overset{\displaystyle O}{\|}}{C}-NH_2 \quad \text{Urea}$$

Synthetic amides are used as substitutes for sugar and aspirin. Saccharin is a very powerful sweetener and is used as a sugar substitute. The sweetener Aspartame is made from two amino acids, aspartic acid and phenylalanine.

Aspirin substitutes contain phenacetin or acetaminophen, which is used in Tylenol. Like aspirin, acetaminophen reduces fever and pain, but it has little anti-inflammatory effect.

Aspartic acid Phenylalanine Methyl ester
Aspartame

Saccharin

Phenacetin

Acetaminophen

Many barbiturates are cyclic amides of barbituric acid that act as sedatives in small dosages or sleep inducers in larger dosages. They are often habit forming. Barbiturate drugs include phenobarbital (Luminal), pentobarbital (Nembutal), and secobarbital (Seconal).

Luminal (phenobarbital)

Nembutal (pentobarbital)

Valium (diazepam)

Seconal (secobarbital)

Equanil (meprobamate)

QUESTIONS AND PROBLEMS

Structure and Names of Amides

19.29 Draw the condensed structural formulas of the amides formed in each of the following reactions:

$$\textbf{a.}\ CH_3\overset{\displaystyle O}{\overset{\|}{C}}OH\ +\ NH_3\ \xrightarrow{\text{Heat}}\qquad \textbf{b.}\ CH_3\overset{\displaystyle O}{\overset{\|}{C}}OH\ +\ NH_2CH_2CH_3\ \xrightarrow{\text{Heat}}\qquad \textbf{c.}\ \bigcirc\!\!-\!\!\overset{\displaystyle O}{\overset{\|}{C}}OH\ +\ NH_2CH_2CH_2CH_3\ \xrightarrow{\text{Heat}}$$

19.30 Draw the condensed structural formulas of the amides formed in each of the following reactions:

$$\textbf{a.}\ CH_3CH_2CH_2CH_2\overset{\displaystyle O}{\overset{\|}{C}}OH\ +\ NH_3\ \xrightarrow{\text{Heat}}\qquad \textbf{b.}\ CH_3\overset{\displaystyle CH_3}{\overset{|}{C}}HCH_2\overset{\displaystyle O}{\overset{\|}{C}}OH\ +\ NH_2CH_2CH_2CH_3\ \xrightarrow{\text{Heat}}$$

$$\textbf{c.}\ CH_3CH_2\overset{\displaystyle O}{\overset{\|}{C}}OH\ +\ \bigcirc\!\!-\!\!NH_2\ \xrightarrow{\text{Heat}}$$

19.31 Give the IUPAC and common names (if any) for each of the following amides:

$$\textbf{a.}\ CH_3\overset{\displaystyle O}{\overset{\|}{-C-}}NH-CH_3\qquad \textbf{b.}\ CH_3CH_2CH_2\overset{\displaystyle O}{\overset{\|}{-C-}}NH_2\qquad \textbf{c.}\ H\overset{\displaystyle O}{\overset{\|}{-C-}}NH_2\qquad \textbf{d.}\ \bigcirc\!\!-\!\!\overset{\displaystyle O}{\overset{\|}{C}}\overset{\displaystyle H}{\overset{|}{-N-}}CH_3$$

19.32 Give the IUPAC and common names (if any) for each of the following amides:

$$\textbf{a.}\ CH_3CH_2\overset{\displaystyle O}{\overset{\|}{C}}\overset{\displaystyle H}{\overset{|}{-N-}}CH_2CH_3\qquad \textbf{b.}\ CH_3CH_2CH_2CH_2CH_2\overset{\displaystyle O}{\overset{\|}{C}}-NH_2$$

$$\textbf{c.}\ CH_3\overset{\displaystyle O}{\overset{\|}{C}}\overset{\displaystyle CH_3}{\overset{|}{-N-}}CH_2CH_2CH_3\qquad \textbf{d.}\ \bigcirc\!\!-\!\!\overset{\displaystyle O}{\overset{\|}{C}}\overset{\displaystyle CH_2CH_3}{\overset{|}{-N-}}CH_2CH_3$$

19.33 Draw the condensed structural formulas for each of the following amides:
a. propionamide   **b.** 2-methylpentanamide   **c.** methanamide
d. N-ethylbenzamide   **e.** N-ethylbutyramide

19.34 Draw the condensed structural formulas for each of the following amides:
a. formamide   **b.** N,N-dimethylbenzamide
c. 3-methylbutyramide   **d.** 2,2-dichlorohexanamide
e. N-propyl-3-chloropentanamide

19.35 In each pair, identify the compound that has the higher melting point:
a. acetamide or N-methylacetamide
b. butane or propionamide
c. N,N-dimethylpropanamide or N-methylpropanamide

19.36 In each pair, identify the compound that has the higher melting point:
a. propane or acetamide
b. N-methylacetamide or propanamide
c. N,N-dimethylpropanamide or N-methylpropanamide

19.7 Hydrolysis of Amides

As we have seen, amide bonds are formed by the elimination of water. The reverse reaction called **hydrolysis** occurs when water is added back to the amide bond to split the molecule. When an acid is used, the hydrolysis products of an amide are the carboxylic acid and the ammonium salt. In base hydrolysis, the amide produces the salt of the carboxylic acid and ammonia or amine.

Acid Hydrolysis of Amides

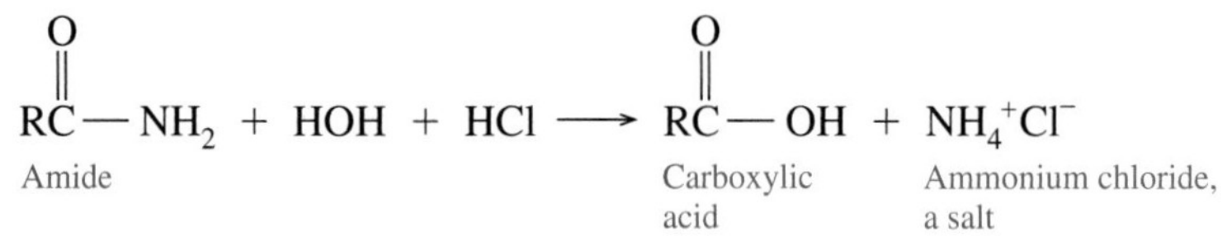

$$\underset{\text{Amide}}{RC\overset{\displaystyle O}{\overset{\|}{-}}NH_2} + HOH + HCl \longrightarrow \underset{\substack{\text{Carboxylic}\\\text{acid}}}{RC\overset{\displaystyle O}{\overset{\|}{-}}OH} + \underset{\substack{\text{Ammonium chloride,}\\\text{a salt}}}{NH_4^+Cl^-}$$

Example:

$$\underset{\substack{\text{Ethanamide}\\\text{(acetamide)}}}{CH_3\overset{\displaystyle O}{\overset{\|}{C}}-NH_2} + HOH + HCl \longrightarrow \underset{\substack{\text{Ethanoic acid}\\\text{(acetic acid)}}}{CH_3\overset{\displaystyle O}{\overset{\|}{C}}-OH} + \underset{\substack{\text{Ammonium}\\\text{chloride}}}{NH_4^+Cl^-}$$

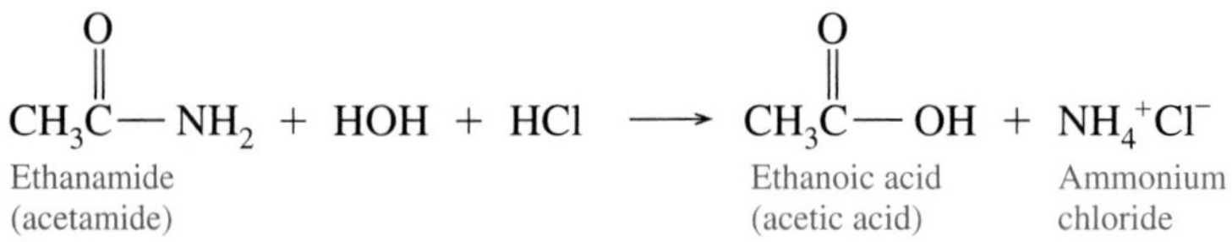

Base Hydrolysis of Amides

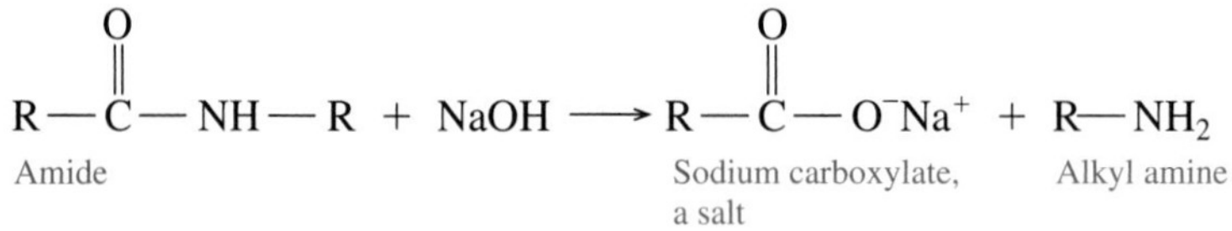

$$\underset{\text{Amide}}{R-\overset{\displaystyle O}{\overset{\|}{C}}-NH-R} + NaOH \longrightarrow \underset{\substack{\text{Sodium carboxylate,}\\\text{a salt}}}{R-\overset{\displaystyle O}{\overset{\|}{C}}-O^-Na^+} + \underset{\text{Alkyl amine}}{R-NH_2}$$

Example:

$$\underset{\substack{N\text{-Methylpropanamide}\\(N\text{-methylpropionamide})}}{CH_3-CH_2-\overset{\displaystyle O}{\overset{\|}{C}}-NH-CH_3} + NaOH \longrightarrow \underset{\substack{\text{Sodium propanoate, a salt}\\(\text{sodium propionate})}}{CH_3-CH_2-\overset{\displaystyle O}{\overset{\|}{C}}-O^-Na^+} + \underset{\substack{\text{Methanamine}\\(\text{methylamine})}}{NH_2CH_3}$$

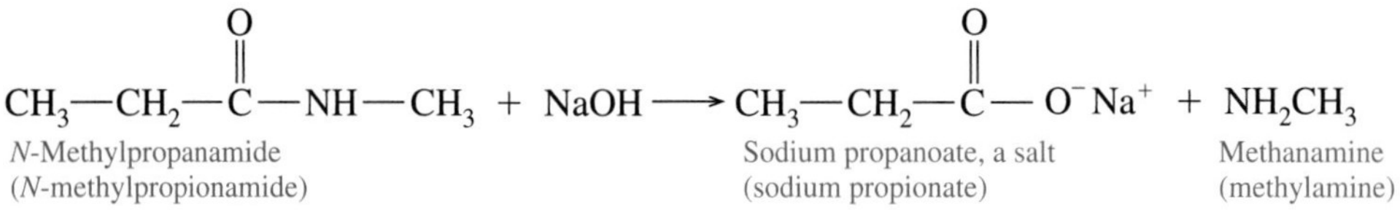

SAMPLE PROBLEM 19.9

Hydrolysis of Amides

Write the structural formulas for the products for the hydrolysis of *N*-methylpentanamide with NaOH.

Solution

Hydrolysis of the amide with a base produces a carboxylate salt (sodium pentanoate) and the corresponding amine (methylamine).

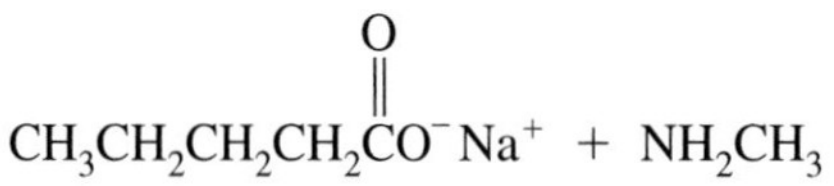

$$CH_3CH_2CH_2CH_2\overset{\displaystyle O}{\overset{\|}{C}}O^-Na^+ + NH_2CH_3$$

Study Check

What are the structures of the products from the hydrolysis of *N*-methylbutyramide with HBr?

QUESTIONS AND PROBLEMS

Hydrolysis of Amides

19.37 Write the condensed structural formulas for the products of the acid hydrolysis of each of the following amides with HCl:

a. 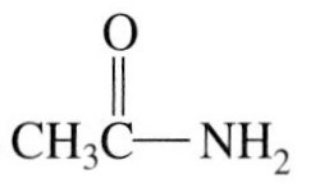 $CH_3\overset{\displaystyle O}{\overset{\displaystyle \|}{C}}-NH_2$

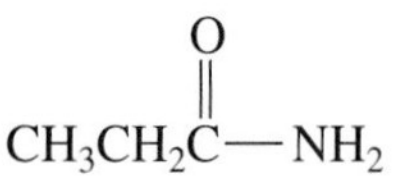

b. $CH_3CH_2\overset{\displaystyle O}{\overset{\displaystyle \|}{C}}-NH_2$

c. $CH_3CH_2CH_2\overset{\displaystyle O}{\overset{\displaystyle \|}{C}}-NH-CH_3$

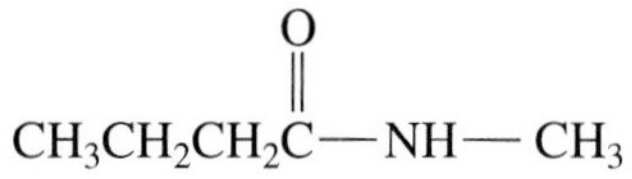

d. 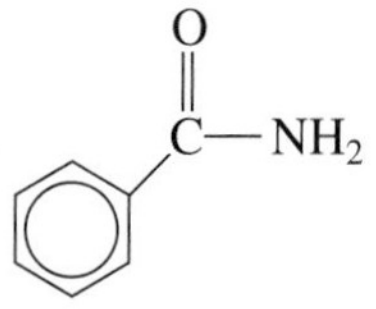 $\overset{\displaystyle O}{\overset{\displaystyle \|}{C}}-NH_2$ (benzene ring)

e. *N*-ethylpentanamide

19.38 Write the condensed structural formulas for the products of the base hydrolysis of each of the following amides with NaOH:

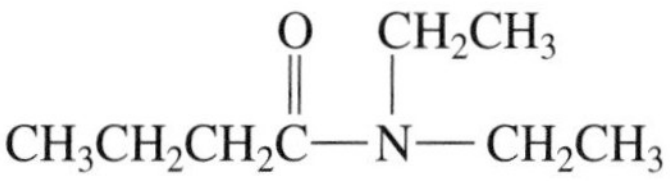

a. 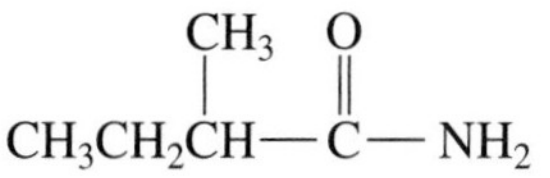 $CH_3CH_2\overset{\displaystyle CH_3}{\underset{\displaystyle |}{CH}}-\overset{\displaystyle O}{\overset{\displaystyle \|}{C}}-NH_2$

b. $CH_3CH_2CH_2\overset{\displaystyle O}{\overset{\displaystyle \|}{C}}-\overset{\displaystyle CH_2CH_3}{\underset{\displaystyle |}{N}}-CH_2CH_3$

c. $\overset{\displaystyle O}{\overset{\displaystyle \|}{C}}-\overset{\displaystyle CH_3}{\underset{\displaystyle |}{N}}-CH_2CH_2CH_2CH_3$ (benzene ring)

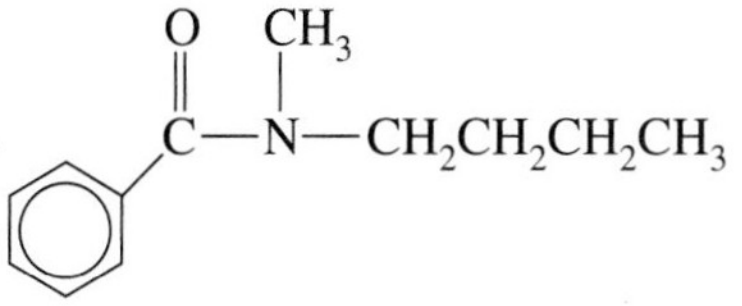

d. 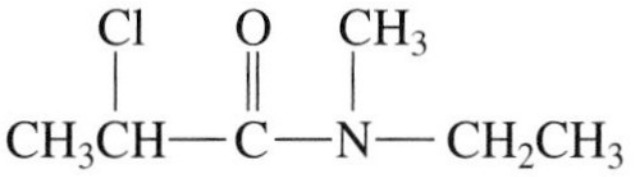 $CH_3\overset{\displaystyle Cl}{\underset{\displaystyle |}{CH}}-\overset{\displaystyle O}{\overset{\displaystyle \|}{C}}-\overset{\displaystyle CH_3}{\underset{\displaystyle |}{N}}-CH_2CH_3$

e. *N*-propyl benzamide

Chapter Review

19.1 Amines

A nitrogen atom attached to one, two, or three alkyl or aromatic groups forms a primary, secondary, or tertiary amine. Many amines, synthetic or naturally occurring, have physiological activity.

19.2 Naming Amines

In the common names of simple amines, the alkyl groups are listed alphabetically followed by the suffix *–amine*. In the IUPAC system, the *–amine* suffix is added to the alkane name of the longer carbon chain. Groups attached to the nitrogen atom use a *N*- prefix. When other functional groups are present, the $-NH_2$ is named as an amino group.

19.3 Physical Properties of Amines

Primary and secondary amines form hydrogen bonds, which make their boiling points higher than alkanes of similar mass, but lower than alcohols. Amines with up to six carbon atoms are soluble in water.

19.4 Amines React as Bases

In water, amines act as weak bases because the nitrogen atom accepts protons from water to produce ammonium and hydroxide ions. When amines react with acids, they form amine salts, which are named as ammonium salts. As ionic compounds, amine salts are solids, soluble in water, and odorless compared to the amines. Quaternary ammonium salts contain four carbon groups bonded to the nitrogen atom.

19.5 Heterocyclic Amines and Alkaloids

Heterocyclic amines are cyclic organic compounds that contain one or more nitrogen atoms in the ring. The amine rings typically consist of 5 or 6 atoms and one or more nitrogen atoms. Alkaloids such as caffeine and nicotine are naturally occurring amines derived from plants. Many are known for their physiological activity.

19.6 Structure and Names of Amides

Amides are derivatives of carboxylic acids in which the hydroxyl group is replaced by $-NH_2$, a primary, or secondary amine group. Amides are formed when carboxylic acids react with ammonia or primary or secondary amines in the presence of heat. Amides are named by replacing the *–ic acid* or *–oic acid* with *–amide*. Any carbon group attached to the nitrogen atom is named using the *N*- prefix.

19.7 Hydrolysis of Amides

Hydrolysis of an amide by an acid produces an amine salt. Hydrolysis by a base produces the salt of the carboxylic acid.

Summary of Naming

Family	Condensed Structural Formula	IUPAC name	Common Name
Amine	$CH_3CH_2—NH_2$	Ethanamine	Ethylamine
Amine salt	$CH_3CH_2—NH_3^+Cl^-$	Ethylammonium chloride	Ethylammonium chloride
Amide	$CH_3\overset{\overset{\displaystyle O}{\|}}{C}—NH_2$	Ethanamide	Acetamide

Summary of Reactions

Ionization of Amines in Water

$$R—NH_2 + HOH \rightleftharpoons R—NH_3^+ + OH^-$$

Amine   Alkyl ammonium hydroxide

Methylamine   Methylammonium hydroxide

Formation of Amine Salts

$$R—NH_2 + HCl \longrightarrow R—NH_3^+ \ Cl^-$$

Amine   Ammonium salt

Methylamine   Methylammonium chloride

Formation of Amides

Carboxylic acid   Ammonia   Amide

Carboxylic acid   Amine   Amide

Propanoic acid   Ammonia   Propanamide
(propionic acid)   (propionamide)

Propanoic acid   Methanamine   N-Methylpropanamide
(propionic acid)   (methylamine)   (N-methylpropionamide)

Hydrolysis of Amides

$$RC\overset{O}{\underset{||}{}}-NH_2 + H_2O + HCl \longrightarrow RC\overset{O}{\underset{||}{}}-OH + NH_4^+Cl^-$$

Amide   Carboxylic acid   Ammonium chloride

$$CH_3-\overset{O}{\underset{||}{C}}-NH_2 + HOH + HCl \longrightarrow CH_3-\overset{O}{\underset{||}{C}}-OH + NH_4^+Cl^-$$

Ethanamide (acetamide)   Ethanoic acid (acetic acid)   Ammonium chloride

$$CH_3-CH_2-\overset{O}{\underset{||}{C}}-NH-CH_3 + NaOH \longrightarrow CH_3-CH_2-\overset{O}{\underset{||}{C}}-O^-Na^+ + NH_2CH_3$$

N-Methylpropanamide (*N*-methylpropionamide)   Sodium propanoate (sodium propionate)   Methanamine (methylamine)

Key Terms

alkaloids Amines having physiological activity that are produced in plants.

amidation The formation of an amide from a carboxylic acid and ammonia or an amine.

amides Organic compounds containing the carbonyl group attached to an amino group or a substituted nitrogen atom:

$$R-\overset{O}{\underset{||}{C}}-NH_2 \qquad R-\overset{O}{\underset{||}{C}}-\overset{H}{\underset{|}{N}}-R$$

amines Organic compounds containing a nitrogen atom attached to one, two, or three hydrocarbon groups.

amine salt An ionic compound produced from an amine and an acid.

heterocyclic amine A cyclic organic compound that contains one or more nitrogen atoms in the ring.

hydrolysis The splitting of a molecule by the addition of water. Amides yield the corresponding carboxylic acid and amine or their salts.

quaternary ammonium ion An amine ion in which the nitrogen atom is bonded to four carbon groups.

Additional Problems

19.39 There are four amine isomers with the molecular formula C_3H_9N. Draw their condensed structural formulas. Name and classify each as a primary, secondary, or tertiary amine.

19.40 Name and classify each of the following compounds:

a. $CH_3-\overset{CH_2-CH_3}{\underset{|}{N}}-CH_2-CH_3$

b. $CH_3-CH_2-CH_2-CH_2-NH_2$

c. $CH_3-CH_2-CH_2-NH-CH_2-CH_3$

d.

e. NHCH$_3$

f. $CH_3-\overset{CH_3}{\underset{|}{CH}}-CH_2-\overset{CH_3}{\underset{|}{N}}-CH_2-CH_3$

g. $CH_3-\overset{CH_2-CH_3}{\underset{\underset{CH_3}{|}}{\overset{|+}{N}}}-CH_2-CH_3$ Cl^-

19.41 Draw the structure of each of the following compounds:
a. 3-pentanamine   **b.** cyclohexylamine
c. dimethylammonium chloride

d. triethylamine

e. 3-amino-2-hexanol

f. tetramethylammonium bromide

g. *N,N*-dimethylaniline

19.42 In each pair, indicate the compound that has the higher boiling point:

a. 1-butanol or butanamine

b. trimethylamine or propylamine

c. butylamine or diethylamine

d. butane or propylamine

19.43 In each pair, indicate the compound that is more soluble in water:

a. ethylamine or butylamine

b. trimethylamine or *N*-ethylcyclohexylamine

c. butylamine or pentane

d. $NH_2-CH_2-CH_2-CH_2-CH_2-CH_2-NH_2$

or

$CH_3-CH_2-CH_2-CH_2-CH_2-NH_2$

19.44 Give the IUPAC name for each of the following amides:

a. H—C—NH$_2$ (with =O on carbon)

b. $CH_3-CH_2-C(=O)-NH_2$

c. $CH_3-C(=O)-N(H)-CH_3$

d. $CH_3-CH_2-CH_2-C(=O)-NH-CH_2-CH_3$

e. $CH_3-C(=O)-N(CH_3)-CH_2-CH_2-CH_2-CH_3$

19.45 Indicate the name of the alkaloid in each of the following:

a. malaria treatment   **b.** tobacco

c. coffee and tea

d. a painkiller in oriental poppy plant

19.46 Identify the heterocyclic amines in each of the following:

a. caffeine   **b.** Demerol

c. nicotine   **d.** quinine

19.47 Write the structure of each product of the following reactions:

a. $CH_3-CH_2-NH_2 + H_2O \longrightarrow$

b. $CH_3-CH_2-NH_2 + HCl \longrightarrow$

c. $CH_3-CH_2-NH-CH_3 + H_2O \longrightarrow$

d. $CH_3-CH_2-NH-CH_3 + HCl \longrightarrow$

e. $CH_3-CH_2-CH_2-NH_3^+ Cl^- + NaOH \longrightarrow$

f. $CH_3-CH_2-NH_2^+(CH_3) Cl^- + NaOH \longrightarrow$

19.48 Toradol is used in dentistry to relieve pain. Name the functional groups in this molecule.

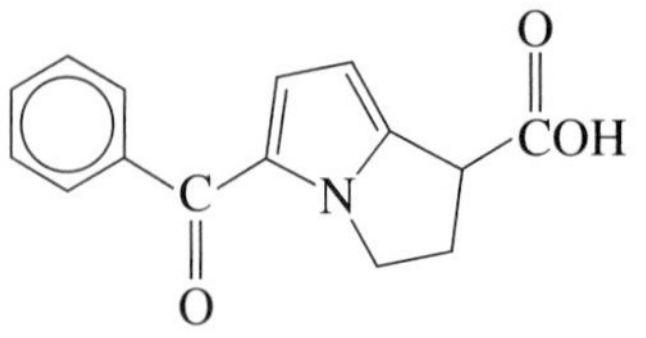

19.49 Voltaren is indicated for acute and chronic treatment of the symptoms of rheumatoid arthritis. Name the functional groups in this molecule.

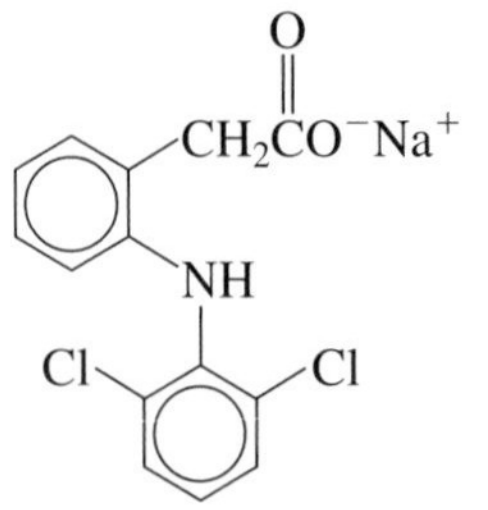

19.50 Many amine-containing drugs are given to patients in their salt form, such as hydrochloride or sulfate. What might be the reason?

19.51 Using a reference book such as the *Merck Index* or *Physicians' Desk Reference*, look up the structural formula of the following medicinal drugs and list the functional groups in the compounds. You may need to refer to the cross-index of names in the back of the reference book.

a. Keflex, an antibiotic

b. Inderal, a β-channel blocker used to treat heart irregularities

c. ibuprofen, an anti-inflammatory agent

d. Aldomet (methyldopa)

e. Percodan, a narcotic pain reliever

f. triamterene, a diuretic

20 Amino Acids and Proteins

"This lamb is fed with Lamb Lac, which is a chemically formulated replacement for ewe's milk," says part-time farmer Dennis Samuelson. "Its mother had triplets and didn't have enough milk to feed them all, so they weren't thriving the way the other lambs were. The Lamb Lac includes dried skim milk, dried whey, milk proteins, egg albumin, the amino acids methionine and lysine, vitamins, and minerals."

A veterinary technician diagnoses and treats diseases of animals, takes blood and tissue samples, and administers drugs and vaccines. Agricultural technologists assist in the study of farm crops to increase productivity and ensure a safe food supply. They look for ways to improve crop yields, develop safer methods of weed and pest control, and design methods to conserve soil and water.

LOOKING AHEAD

the Chemistry place

www.chemplace.com/college

Visit the URL above or use the CD-ROM in the book for extra quizzing, interactive tutorials, career resources, and case studies.

The word "protein" is derived from the Greek word *proteios*, meaning "first." Made of amino acids, proteins provide structure in membranes, build cartilage and connective tissue, transport oxygen in blood and muscle, direct biological reactions as enzymes, defend the body against infection, and control metabolic processes as hormones. They can even be a source of energy.

Compared with many of the compounds we have studied, protein molecules can be gigantic. Insulin has a molar mass of 5700, and hemoglobin has a molar mass of about 64,000. Some virus proteins are still larger, having molar masses of more than 40 million. Yet all proteins in humans are polymers made up of 20 different amino acids. Each kind of protein is composed of amino acids arranged in a specific order that determines the characteristics of the protein and its biological action.

Proteins perform many functions in the body: making up skin and hair, moving muscles, carrying oxygen, and regulating metabolism. All of these different functions of proteins depend on the structures and chemical behavior of amino acids, the building blocks of proteins. We will see how peptide bonds link amino acids and how the order of the amino acids in these protein polymers directs the formation of unique three-dimensional structures.

LEARNING GOAL

Classify proteins by their functions in the cells.

WEB TUTORIAL
Functions of Proteins

20.1 Functions of Proteins

The many kinds of proteins perform many different functions in the body. There are proteins that form structural components such as cartilage, muscles, hair, and nails. Wool, silk, feathers, and horns are some other proteins made by animals. Proteins called enzymes regulate biological reactions such as digestion and cellular metabolism. Still other proteins, hemoglobin and myoglobin, carry oxygen in the blood and muscle. (See Figure 20.1). Table 20.1 gives examples of proteins that are classified by their functions in biological systems.

SAMPLE PROBLEM 20.1

Classifying Proteins by Function

Give the class of protein that would perform each of the following functions.
a. catalyzes metabolic reactions of lipids
b. carries oxygen in the bloodstream
c. stores amino acids in milk

Solution

a. Enzymes catalyze metabolic reactions.
b. Transport proteins carry substances such as oxygen through the bloodstream.
c. A storage protein stores nutrients such as amino acids in milk.

Study Check

What proteins help regulate metabolism?

Table 20.1 Classification of Some Proteins and Their Functions

Class of Protein	Function in the Body	Examples
Structural	Provide structural components	*Collagen* is in tendons and cartilage. *Keratin* is in hair, skin, wool, and nails.
Contractile	Movement of muscles	*Myosin* and *actin* contract muscle fibers.
Transport	Carry essential substances throughout the body	*Hemoglobin* transports oxygen. *Lipoproteins* transport lipids.
Storage	Store nutrients	*Casein* stores protein in milk. *Ferritin* stores iron in the spleen and liver.
Hormone	Regulate body metabolism and nervous system	*Insulin* regulates blood glucose level. *Growth hormone* regulates body growth.
Enzyme	Catalyze biochemical reactions in the cells	*Sucrase* catalyzes the hydrolysis of sucrose. *Trypsin* catalyzes the hydrolysis of proteins.
Protection	Recognize and destroy foreign substances	*Immunoglobulins* stimulate immune responses.

Figure 20.1 The horns, feathers, and wool of animals are made of proteins.

Q What class of protein are wool, feathers, and horns?

QUESTIONS AND PROBLEMS

Functions of Proteins

20.1 Classify each of the following proteins according to its function:
a. hemoglobin, oxygen carrier in the blood
b. collagen, a major component of tendons and cartilage

c. keratin, a protein found in hair

d. amylase, an enzyme that hydrolyzes starch

20.2 Classify each of the following proteins according to its function:

a. insulin, a hormone needed for glucose utilization

b. antibodies, proteins that disable foreign proteins

c. casein, milk protein

d. lipases that hydrolyze lipids

LEARNING GOAL

Draw the structure for an amino acid.

20.2 Amino Acids

Proteins are composed of molecular building blocks called amino acids. An **amino acid** contains two functional groups, an amino group ($-NH_2$) and a carboxylic acid group ($-COOH$). In all of the 20 amino acids found in proteins, the amino group, the carboxylic group, and a hydrogen atom are bonded to a central carbon atom. Amino acids with this structure are called 2-amino acids, or α (alpha) amino acids. Although there are many amino acids, only 20 different amino acids are present in the proteins in humans. The unique characteristics of the 20 amino acids are due to a side chain (R), which can be an alkyl, hydroxyl, thiol, amino, sulfide, aromatic, or heterocyclic group.

General Structure of an α-Amino Acid

Classification of Amino Acids

A **nonpolar amino acid** contains an alkyl or aromatic side chain. Hydrocarbons are nonpolar and not soluble in water, which makes the nonpolar amino acids **hydrophobic** ("water-fearing"). **Polar amino acids** have side chains that contain polar groups such as hydroxyl ($-OH$), thiol ($-SH$), amide ($-CONH_2$), and heterocyclic amines. The electronegative atoms in the side chains make these amino acids **hydrophilic** ("water-attracting") by forming hydrogen bonds with water. The **acidic amino acids** have side chains that contain a carboxylic acid group ($-COOH$) and can ionize as a weak acid. The side chains of the **basic amino acids** contain an amino group that can ionize as a weak base. With the exception of proline, all the amino acids are primary amines. The structures of the side chains (R), common names, three-letter abbreviations, and isoelectric points, pI (see Section 20.3) of the 20 amino acids in proteins are listed in Table 20.2.

Table 20.2   The 20 Amino Acids in Proteins

Nonpolar Amino Acids

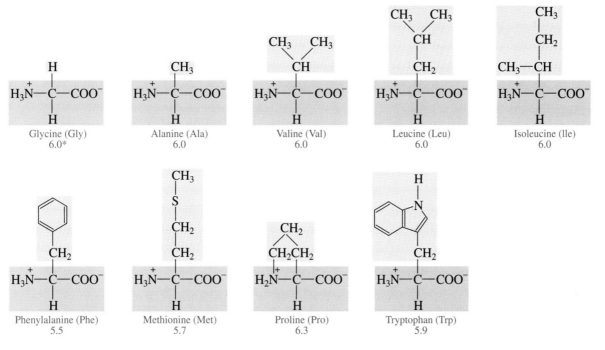

Glycine (Gly)
6.0*

Alanine (Ala)
6.0

Valine (Val)
6.0

Leucine (Leu)
6.0

Isoleucine (Ile)
6.0

Phenylalanine (Phe)
5.5

Methionine (Met)
5.7

Proline (Pro)
6.3

Tryptophan (Trp)
5.9

Polar Amino Acids (Neutral)

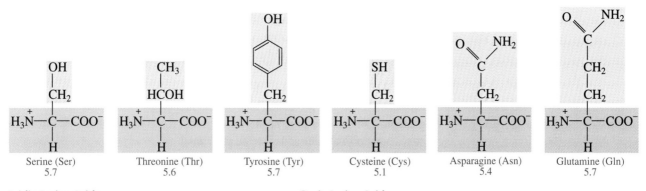

Serine (Ser)
5.7

Threonine (Thr)
5.6

Tyrosine (Tyr)
5.7

Cysteine (Cys)
5.1

Asparagine (Asn)
5.4

Glutamine (Gln)
5.7

Acidic Amino Acids   **Basic Amino Acids**

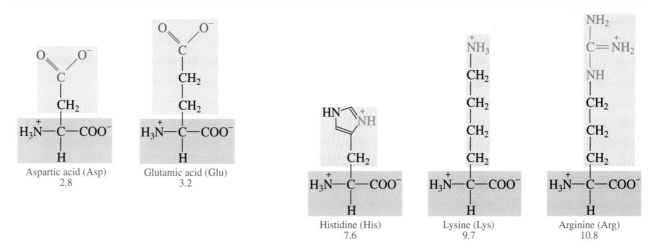

Aspartic acid (Asp)
2.8

Glutamic acid (Glu)
3.2

Histidine (His)
7.6

Lysine (Lys)
9.7

Arginine (Arg)
10.8

*Isoelectric points (pI)

Structural Formulas of Amino Acids

Write the structural formulas and abbreviations for the following amino acids:

a. alanine (R = $-CH_3$)   **b.** serine (R = $-CH_2OH$)

Solution

a. The structure of the amino acids is written by attaching the side group (R) to the central carbon atom of the general structure of an amino acid.

alanine (Ala)

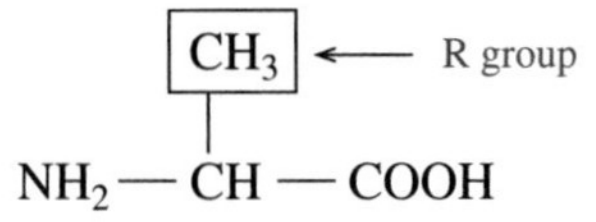

$$CH_3 \longleftarrow \text{R group}$$
$$|$$
$$NH_2 - CH - COOH$$

b. serine (Ser)

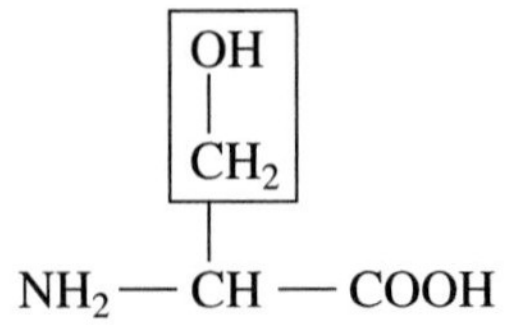

$$OH$$
$$|$$
$$CH_2$$
$$|$$
$$NH_2 - CH - COOH$$

Study Check

Classify the amino acids in the sample problem as polar or nonpolar.

Amino Acid Stereoisomers

All of the α-amino acids except for glycine are chiral because the α carbon is attached to four different atoms. Thus amino acids can exist as D and L enantiomers. We can write Fischer projections for amino acids as we did in Chapter 15 for aldehydes with the carboxylic acid group (the most highly oxidized carbon) at the top and the R group at the bottom. For the L isomer, the amino group, NH_2, is on the left, and in the D isomer, it is on the right. In biological systems, only L amino acids are incorporated into proteins. There are D amino acids found in nature, but not in proteins. Let's take a look at the enantiomers for glyceraldehyde and how the stereoisomers of alanine and cysteine are similar.

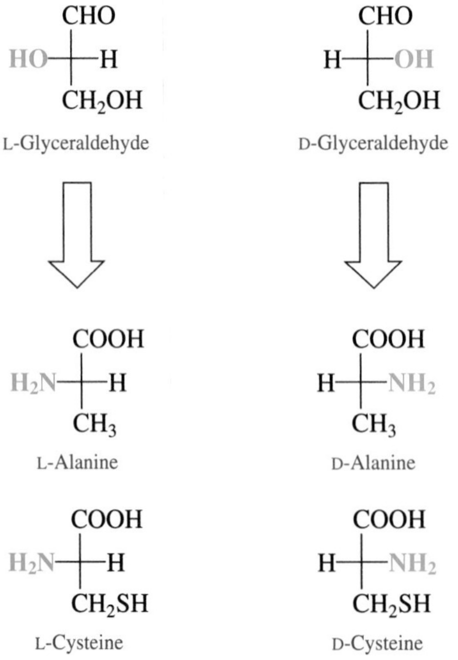

CHO	CHO
HO┼H	H┼OH
CH_2OH	CH_2OH
L-Glyceraldehyde	D-Glyceraldehyde

COOH	COOH
H_2N┼H	H┼NH_2
CH_3	CH_3
L-Alanine	D-Alanine

COOH	COOH
H_2N┼H	H┼NH_2
CH_2SH	CH_2SH
L-Cysteine	D-Cysteine

SAMPLE PROBLEM 20.3

Chiral Amino Acids

Write the Fischer projection for L-serine.

Solution

In L-serine, the —COOH is at the top and the R group —CH$_2$OH is at the bottom. The L-isomer has the —NH$_2$ on the left.

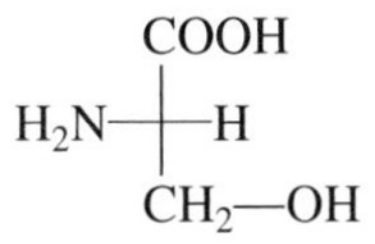

COOH

H$_2$N——H

CH$_2$—OH

L-Serine

Study Check

How does the Fischer projection for D-serine differ from L-serine?

QUESTIONS AND PROBLEMS

Amino Acids

20.3 Describe the functional groups found in all α-amino acids.

20.4 How does the polarity of the side chain in leucine compare to the side chain in serine?

20.5 Draw the structural formula for each of the following amino acids:
  a. alanine
  b. threonine
  c. glutamic acid
  d. phenylalanine

20.6 Draw the structural formula for each of the following amino acids:
  a. lysine
  b. aspartic acid
  c. leucine
  d. tyrosine

20.7 Classify the amino acids in problem 20.5 as hydrophobic (nonpolar), hydrophilic (polar, neutral), acidic, or basic.

20.8 Classify the amino acids in problem 20.6 as hydrophobic (nonpolar), hydrophilic (polar, neutral), acidic, or basic.

20.9 Give the name of the amino acid represented by each of the following three-letter abbreviations:
  a. Ala **b.** Val **c.** Lys **d.** Cys

20.10 Give the name of the amino acid represented by each of the following three-letter abbreviations:
  a. Trp **b.** Met **c.** Pro **d.** Gly

20.11 Draw the Fischer projections for the following amino acids.
  a. L-valine
  b. D-cysteine

20.12 Draw the Fischer projections for the following amino acids.
  a. L-threonine
  b. D-valine

LEARNING GOAL

Write the ionic form of an amino acid at pH values above, below, and equal to its isoelectric point.

20.3 Amino Acids as Acids and Bases

Although it is convenient to write amino acids with carboxyl ($-COOH$) group and amine ($-NH_2$) group, they are usually ionized. Depending on the pH, the carboxyl group loses a H^+, giving $-COO^-$, and the amino group accepts a H^+ to give an ammonium ion, $-NH_3^+$. The *dipolar* form of an amino acid called a **zwitterion** has a net charge of zero.

Amino Acid   Zwitterion (dipolar ion)

Solid amino acids have very high melting points because the zwitterion has the properties of a salt. For example, glycine melts at 260°C. However, most amino acids decompose when they are heated rather than melt. The ionic charges of the amino acids make them more soluble in water, but not in organic solvents.

At a certain pH known as the **isoelectric point (pI)**, the positive and negative charges are equal, which gives an overall charge of zero. However, when the pH is different from the pI, the zwitterion accepts or donates H^+. In a solution that is more acidic than the pI, the $-COO^-$ group acts as a base and accepts an H^+, which gives an overall positive charge to the amino acid.

Zwitterion ion accepts H^+ Positively charged ion

In a solution more basic than the pI, the $-NH_3^+$ group acts as an acid and loses an H^+, which gives the amino acid an overall negative charge.

Zwitterion donates H^+ Negatively charged ion

Let's take a look at the changes in all the ionic forms of alanine from its zwitterion (pI = 6.0), to the positive ion in a more acidic solution, and the negative ion in basic solution.

Alanine ion at acidic pH < 6 (charge = 1+)

Zwitterion of alanine pH = 6.0 (charge = 0)

Alanine ion at a pH > 6 (charge = 1−)

The zwitterions for polar and nonpolar amino acids typically exist at pH values of 5.0 to 6.0. However, the zwitterions of the acidic amino acids form at pH values of about 3 because the carboxyl group in their side chain must pick up H^+. The basic amino acids with amino groups in their side chains form zwitterions at high pI values from 7.6 to 10.8. The pI values are included in the list of the amino acids in Table 20.2.

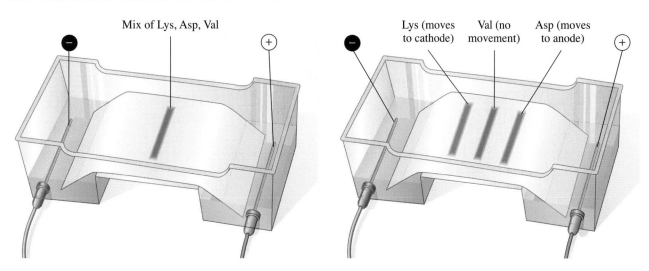

Aspartic acid
pH < 2
(charge = 1+)

Zwitterion
pH = 2.8
(charge = 0)

Aspartic acid
at pH 7
(charge = 1–)

Aspartic acid
basic pH
(charge = 2–)

Electrophoresis

It is possible to separate a mixture of amino acids by their isoelectric points using a laboratory method called **electrophoresis.** A buffered amino acid mixture is applied to a gel on a thin plate or piece of filter paper, which is connected to two electrodes. A voltage applied to the electrodes causes the positively charged amino acids to move toward the negative electrode and the negatively charged amino acids to move toward the positive electrode. Any amino acid at its isoelectric point with zero net charge would not move. After several hours, the sample is removed. It can be sprayed with a dye such as ninhydrin to make the amino acids visible. They are identified by their direction and rate of migration toward the electrodes. They are recovered separately by cutting up the filter paper or removing the amino acids from the gel.

Suppose we have a mixture of valine (pI 6.0), aspartic acid (pI 2.8), and lysine (pI 9.7) in a buffer of pH 6.0. When the mixture is placed between two electrodes at a high voltage, the aspartic acid, which would have a negative charge at pH 6.0, would move to the positive electrode (anode). (See Figure 20.2.) The lysine, which would be positively charged at a pH of 6.0, would move toward the negative electrode (cathode). Valine, which is a zwitterion at pH 6.0, would be neutral and not move in the presence of an electric field. Electrophoresis is a method used in medicine to screen for the sickle cell trait in newborn infants.

Figure 20.2 A positively charged amino acid (pH < pI) moves toward the negative electrode, a negatively charged amino acid (pH > pI) moves toward the positive electrode: An amino acid with no net charge (pH = pI) does not migrate.

Q How would the three amino acids migrate if the mixture were buffered to pH 9.7, the pI of lysine?

SAMPLE PROBLEM 20.4

Amino Acids in Acid or Base

Serine exists in its zwitterion form at a pH of 5.7. Draw the structural formula for the zwitterion of serine.

Solution

As a zwitterion, both the carboxylic acid group and the amino group are ionized.

$$
\overset{+}{N}H_3 - CH - \overset{\displaystyle O}{\overset{\|}{C}} - O^- \quad \text{Zwitterion of serine}
$$

with CH₂ and OH branches on the CH.

Study Check

Draw the structure of serine at a pH of 2.0.

QUESTIONS AND PROBLEMS

Amino Acids as Acids and Bases

20.13 Write the zwitterion of each of the following amino acids:
  a. glycine   **b.** cysteine   **c.** serine   **d.** alanine

20.14 Write the zwitterion of each of the following amino acids:
  a. phenylalanine   **b.** methionine   **c.** leucine   **d.** valine

20.15 Write the positive ion (acidic ion) of each of the amino acids in problem 20.13 at a pH below 1.0.

20.16 Write the negative ion (basic ion) of each of the amino acids in problem 20.13 at a pH above 12.0.

20.17 Would the following ions of valine exist at a pH above, below, or at pI?

a. $H_2N - CH - COO^-$ with CH(CH₃)(CH₃) branch

b. $H_3\overset{+}{N} - CH - COOH$ with CH(CH₃)(CH₃) branch

c. $H_3\overset{+}{N} - CH - COO^-$ with CH(CH₃)(CH₃) branch

20.18 Would the following ions of serine exist at a pH above, below, or at pI?

a. $H_3\overset{+}{N} - CH - COO^-$ with CH₂OH branch

b. $H_3\overset{+}{N} - CH - COOH$ with CH₂OH branch

c. $H_2N - CH - COO^-$ with CH₂OH branch

LEARNING GOAL

Draw the structure of a dipeptide from the zwitterions of two amino acids.

20.4 Formation of Peptides

The important reaction of amino acids is the formation of peptide bonds. This is the same amidation reaction we looked at in Chapter 19 in which a carboxylic acid reacts with an amine to form an amide. Let's review this reaction.

$$
\underset{\text{Carboxylic acid}}{R - \overset{\displaystyle O}{\overset{\|}{C}} - OH} + \underset{\text{Amine}}{H_2N - R} \xrightarrow{\text{Heat}} \underset{\text{Amide}}{R - \overset{\displaystyle O}{\overset{\|}{C}} - NH - R} + H_2O
$$

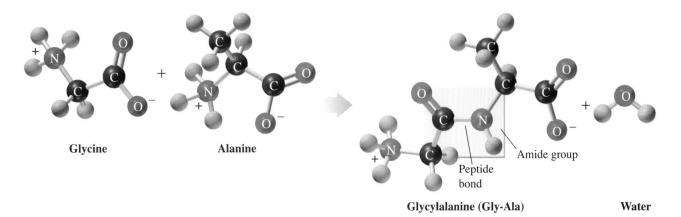

Glycine + **Alanine** → **Glycylalanine (Gly-Ala)** + **Water**

Amide group
Peptide bond

Figure 20.3 A peptide bond between glycine and alanine as zwitterions form the dipeptide glycylalanine.

Q **What functional groups in glycine and alanine form the peptide bond?**

WEB TUTORIAL
Structure of Proteins

The linking of two or more amino acids forms a **peptide**. A **peptide bond** is an amide bond that forms when the $-COO^-$ group of one amino acid reacts with the $-NH_3^+$ group of the next amino acid. We can write the amidation reaction for the zwitterion forms of two amino acids.

$$H_3\overset{+}{N}-CH-\boxed{\overset{O}{\overset{\|}{C}}-O^-}+\boxed{H_3\overset{+}{N}}-CH-\overset{O}{\overset{\|}{C}}-O^- \longrightarrow H_3\overset{+}{N}-CH-\boxed{\overset{O}{\overset{\|}{C}}-\overset{H}{\underset{}{N}}}-CH-\overset{O}{\overset{\|}{C}}-O^- + H_2O$$

R   R   R   R

Amino acid 1   Amino acid 2   Dipeptide

Two amino acids linked together by a peptide bond form a *dipeptide*. We can write the formation of the dipeptide glycylalanine between glycine and alanine as follows. (See Figure 20.3.) In a peptide, the amino acid written on the left with the unreacted or free amino group ($-NH_3^+$) is called the **N terminal** amino acid. The **C terminal** amino acid is the last amino acid in the chain with the unreacted or free carboxyl group ($-COO^-$).

N terminal   C terminal

$$H_3\overset{+}{N}-CH_2-\overset{O}{\overset{\|}{C}}-O^- + H_3\overset{+}{N}-CH-\overset{O}{\overset{\|}{C}}-O^- \longrightarrow H_3\overset{+}{N}-CH_2-\boxed{\overset{O}{\overset{\|}{C}}-\overset{H}{\underset{}{N}}}-CH-\overset{O}{\overset{\|}{C}}-O^- + H_2O$$

                    CH_3 CH_3

Glycine   Alanine   Glycylalanine (Gly-Ala)

Naming Peptides

In naming a peptide, each amino acid beginning from the N terminal is named with a–*yl* ending followed by the full name of the amino acid at the C terminal. Tryptophan becomes tryptophyl, aspartic acid aspartyl, glutamic acid glutamyl, and glutamine is glutaminyl. For convenience, the order of amino acids in the peptide is often written as the sequence of three-letter abbreviations. For example, a tripeptide consisting of alanine, glycine, and serine is named as ala**nyl**glyc**yl**serine.

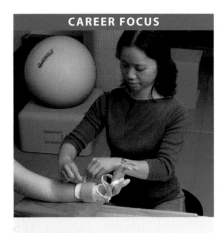

Rehabilitation Specialist

"I am interested in the biomechanical part of rehabilitation, which involves strengthening activities that help people return to the activities of daily living," says Minna Robles, rehabilitation specialist. "Here I am fitting a patient with a wrist extension splint that allows her to lift her hand. This exercise will also help the muscles and soft tissues in that wrist area to heal. An understanding of the chemicals of the body, how they interact, and how we can affect the body on a chemical level is important in understanding our work. One technique we use is called myofacial release. We apply pressure to a part of the body, which helps to increase circulation. By increasing circulation, we can move the soft tissues better, which improves movement and range of motion."

From alanine
alanyl

From glycine
glycyl

From serine
serine

Alanylglycylserine
(Ala-Gly-Ser)

SAMPLE PROBLEM 20.5

Writing Dipeptide Structures

Write a structural formula and three-letter abbreviation for the dipeptide valylserine.

Solution

Valine is joined to serine by a peptide bond; valine is the N terminal and serine is the C terminal.

Valylserine
Val-Ser

From valine   From serine

Study Check

Aspartame, an artificial sweetener 200 times sweeter than sucrose, contains the dipeptide Asp-Phe. Give the structure of the dipeptide in aspartame.

SAMPLE PROBLEM 20.6

Identifying a Tripeptide

Consider the following tripeptide.

a. What amino acid is the N terminal? What amino acid is the C terminal?

b. What is the three-letter abbreviation name for the order of amino acids in the tripeptide?

Solution

a. Threonine is the N terminal; phenylalanine is the C terminal.

b. Thr-Leu-Phe

Study Check

What is the full name of the tripeptide in Sample Problem 20.6?

QUESTIONS AND PROBLEMS

Formation of Peptides

20.19 Draw the structural formula of each of the following peptides and give the abbreviation for their names:
  a. alanylcysteine
  b. serylphenylalanine
  c. glycylalanylvaline
  d. valylisoleucyltryptophan

20.20 Draw the structural formula of each of the following peptides and give the abbreviation for their names:
  a. methionylaspartic acid
  b. alanyltryptophan
  c. methionylglutaminyllysine
  d. histidylglycylglutamylalanine

20.5 Protein Structure: Primary and Secondary Levels

LEARNING GOAL

Identify the primary and secondary structures of a protein.

The particular sequence of amino acids in a peptide or protein is referred to as the **primary structure.** For example, a hormone that stimulates the thyroid to release thyroxine consists of a tripeptide Glu-His-Pro.

the
Chemistry
place

WEB TUTORIAL
Primary and Secondary Structure

Although other sequences are possible for these three amino acids, only the tripeptide with the Glu-His-Pro sequence of amino acids has hormonal activity. Sequences such as His-Pro-Glu or Pro-His-Glu do not produce hormonal activity. Thus the biological function of peptides as well as proteins depends on the order of the amino acids.

When cells are damaged, a polypeptide called bradykinin is released at the site, which stimulates the release of prostaglandins. The presence of bradykinin, which contains nine amino acids, regulates blood pressure.

Arg — Pro — Pro — Gly — Phe — Ser — Pro — Phe — Arg
Bradykinin

Two hormones produced by the pituitary gland are the nonapeptides (nine-amino-acid peptide) oxytocin and vasopressin. Oxytocin stimulates uterine contractions in labor and vasopressin is an antidiuretic hormone that regulates blood pressure by adjusting the amount of water reabsorbed by the kidneys. The struc-

HEALTH NOTE

Natural Opiates in the Body

Painkillers known as enkephalins and endorphins are produced naturally in the body. They are polypeptides that bind to receptors in the brain to give relief from pain. This effect appears to be responsible for the runner's high, for the temporary loss of pain when severe injury occurs, and for the analgesic effects of acupuncture.

The *enkephalins*, found in the thalamus and the spinal cord, are pentapeptides, the smallest molecules with opiate activity.

The amino acid sequence of an enkephalin is found in the longer amino acid sequence of the endorphins.

Four groups of *endorphins* have been identified: α-endorphin contains 16 amino acids, β-endorphin contains 31 amino acids, γ-endorphin has 17 amino acids, and δ-endorphin has 27 amino acids. Endorphins may produce their sedating effects by preventing the release of substance P, a polypeptide with 11 amino acids, which has been found to transmit pain impulses to the brain.

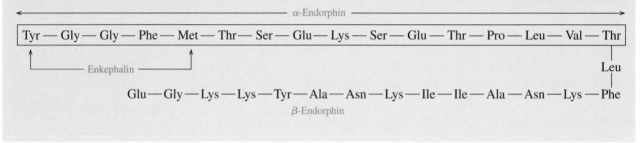

Figure 20.4 The structures of the nonapeptides oxytocin and vasopressin are very similar except for two different amino acids at positions 3 and 8, shown in red for oxytocin and blue for vasopressin. In both structures an amide group replaces the C terminal oxygen.

Q What amino acid has a side chain that can form a disulfide bond in oxytocin and vasopressin?

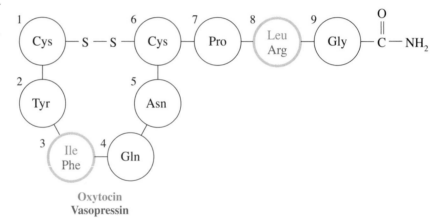

tures of these nonapeptides are very similar with nine amino acids in the same order and a cyclic structure with a disulfide bond. (See Figure 20.4.) Only the amino acids in positions 3 and 8 are different. However, the difference of two amino acids greatly affects how the two hormones function in the body.

Primary Structure of Proteins

When there are more than 50 amino acids in a chain, the polypeptide is usually called a **protein.** Each protein in our cells has a unique sequence of amino acids that determines its biological function. The primary structure of a protein is similar to the peptides we just looked at except that there are many more amino acids in the chain. In a protein, the primary structure is the order of the many amino acids held together by peptide bonds.

The first protein to have its primary structure determined was insulin, which is a hormone that regulates the glucose level in the blood. In the primary structure of human insulin, there are two polypeptide chains. In chain A, there are 21 amino acids, and chain B has 30 amino acids. The polypeptide chains are held together by disulfide bonds formed by the side chains of the cysteine amino acids in each of the chains. (See Figure 20.5.) This primary structure of insulin in humans is very similar to the primary structure of insulin in pigs and cows (bovine). Only the three amino acids at positions 8, 9, and 10 in chain A and position 30 in chain B vary from one species to another. For many years, bovine insulin obtained from the pancreas of cows was used to treat diabetics who lacked insulin. Today human insulin produced through genetic engineering is used in the treatment of diabetes.

Secondary Structure

The **secondary structure** of a protein describes the way the amino acids next to or near to each other along the polypeptide are arranged in space. The three most common types of secondary structure are the *alpha helix*, the *beta-pleated sheet*, and the *triple helix* found in collagen. In each type of secondary structure, we will look at the hydrogen bonding between the hydrogen atom of an amino group in the polypeptide chain and the oxygen atom of the carboxyl group in another part of the chain.

Hydrogen bond

$$-N-H\cdots\cdots O=C-$$

The Alpha Helix

The corkscrew shape of an **alpha helix** (α helix) is held in place by hydrogen bonds between each N—H group and the oxygen of a C=O group in the next turn of the helix, four amino acids down the chain. (See Figure 20.6.) Because many hydrogen bonds form along the peptide backbone, this portion of the protein takes the shape of a strong, tight coil that looks like a telephone cord or a Slinky toy. All the side chains (R groups) of the amino acids are located on the outside of the helix.

Beta-Pleated Sheet

Another type of secondary structure is known as the **beta-pleated sheet** (*β-pleated sheet*). In a β-pleated sheet, polypeptide chains are held together side by side by hydrogen bonds between the peptide chains. In a β-pleated sheet of silk fibroin, the small R groups of the prevalent amino acids, glycine, alanine, and serine, extend above and below the sheet. This results in a series of β-pleated sheets that can be stacked close together. The hydrogen bonds holding the β-pleated sheets tightly in place account for the strength and durability of fibrous proteins such as silk. (See Figure 20.7.)

In some proteins, the polypeptide chain consists of mostly the α-helix secondary structure, whereas other proteins consist of mostly the β-pleated sheet structure. Another group of proteins have a mixture with some sections of the polypeptide

Change in Amino Acids of Insulin in Pigs and Cows			
		Pig	Cow
Chain A	8	Thr	Ala
	9	Ser	Ser
	10	Ile	Val
Chain B	30	Ala	Ala

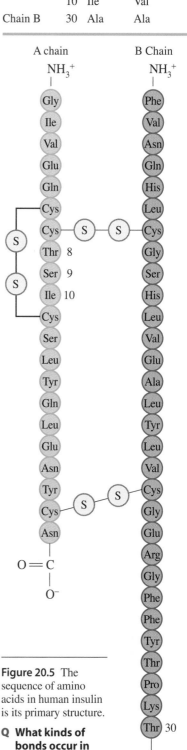

Figure 20.5 The sequence of amino acids in human insulin is its primary structure.

Q **What kinds of bonds occur in the primary structure of a protein?**

Figure 20.6 The α (alpha) helix acquires a coiled shape from hydrogen bonds between the N—H of the peptide bond in one loop and the C=O of the peptide bond in the next loop.

Q **What are partial charges of the H in N—H and the O in C=O that permits hydrogen bonds to form?**

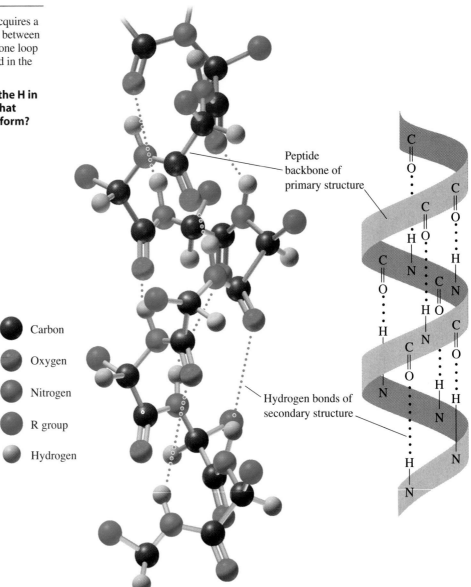

Carbon

Oxygen

Nitrogen

R group

Hydrogen

Peptide backbone of primary structure

Hydrogen bonds of secondary structure

chain in α-helixes and other sections in the β-pleated sheet structure. The tendency to form a certain type of secondary structure depends on the amino acids in a particular segment of the polypeptide chain. The amino acids that tend to form an α-helix or β-pleated sheet follow.

Amino acids found in *α-helix regions*		*Amino acids found in* *β-pleated sheets*	
Alanine	Glutamic acid	Valine	Proline
Cysteine	Glutamine	Serine	Arginine
Leucine	Histidine	Aspartic acid	
Methionine	Lysine	Asparagine	

Triple Helix

Collagen is the most abundant protein; it makes up as much as one-third of all the protein in vertebrates. It is found in connective tissue, blood vessels, skin, tendons,

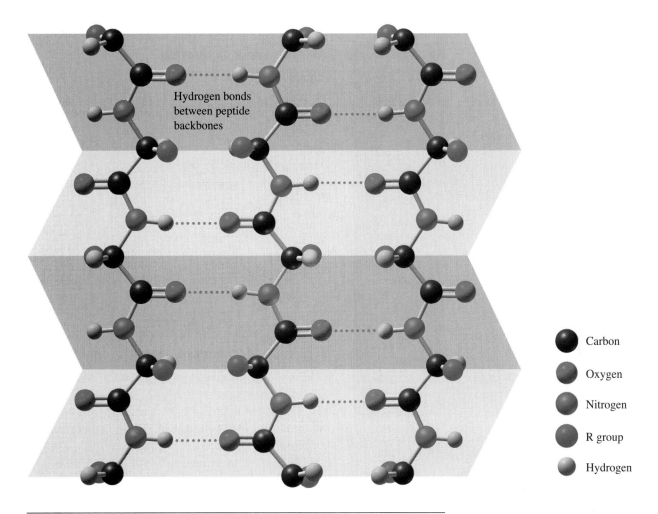

Figure 20.7 In a β (beta)-pleated sheet secondary structure, hydrogen bonds form between the peptide chains.

Q How do the hydrogen bonds differ in a β (beta)-pleated sheet from the alpha helix?

ligaments, the cornea of the eye, and cartilage. The strong structure of collagen is a result of three polypeptides woven together like a braid to form a **triple helix**, as seen in Figure 20.8. Collagen has a high content of glycine (33%), proline (22%), alanine (12%), and smaller amounts of hydroxyproline, and hydroxylysine. The hydroxy forms of proline and lysine contain —OH groups that form hydrogen bonds across the peptide chains and give strength to the collagen triple helix. When several triple helixes wrap together, they form the fibrils that make up connective tissues and tendons.

When a diet is deficient in vitamin C, collagen fibrils are weakened because the enzymes needed to form hydroxyproline and hydroxylysine require vitamin C. Without the —OH groups of hydroxyproline and hydroxylysine, there is less hydrogen bonding between collagen fibrils. Sores appear on the skin and gums, and blood vessels may be weakened, which can lead to aneurisms and rupture. In a young person, collagen is elastic. As a person ages, additional cross links form between the fibrils, which make collagen less elastic. Bones, cartilage, and tendons become more brittle, and wrinkles are seen as the skin loses elasticity. In connective tissue disorders such as lupus and rheumatoid arthritis, an over-active immune system produces increased amounts of collagen. Organs in the body containing large amounts of connective tissues such as joints, skin, kidneys, lungs, and heart are affected.

Triple helix   3 α-Helix peptide chains

Figure 20.8 Hydrogen bonds between polar R groups in three polypeptide chains form the triple helixes that combine to make fibers of collagen.

Q **What are some of the amino acids in collagen that form hydrogen bonds between the polypeptide chains?**

Identifying Secondary Structures

Indicate the secondary structure (α helix, β-pleated sheet, or triple helix) described in each of the following statements:

a. a coiled peptide chain held in place by hydrogen bonding between peptide bonds in the same chain

b. a structure that has hydrogen bonds between polypeptide chains arranged side by side.

Solution

a. α helix   **b.** β-pleated sheet

Study Check

What is the secondary structure in collagen?

Protein Structure: Primary and Secondary Levels

20.21 What type of bonding occurs in the primary structure of a protein?

20.22 How can two proteins with exactly the same number and type of amino acids have different primary structures?

20.23 Two peptides each contain one molecule of valine and two molecules of serine. What are their possible primary structures?

20.24 What are three different types of secondary protein structure?

20.25 What happens to the primary structure of a protein when a protein forms a secondary structure?

20.26 In an α helix, how does bonding occur between the amino acids in the polypeptide chain?

20.27 What is the difference in bonding between an α helix and a β-pleated sheet?

20.28 How is the secondary structure of a β-pleated sheet different from that of a triple helix?

Distinguish between the tertiary and quaternary structures of a protein.

20.6 Protein Structure: Tertiary and Quaternary Levels

The **tertiary structure** of a protein involves attractions and repulsions between the side chain groups of the amino acids in the polypeptide chain. As interactions occur between different parts of the peptide chain, segments of the chain twist and bend until the protein acquires a specific three-dimensional shape.

HEALTH NOTE

Essential Amino Acids

Of the 20 amino acids used to build the proteins in the body, only 10 can be synthesized in the body. The other 10 amino acids, listed in Table 20.3, are **essential amino acids** that cannot be synthesized and must be obtained from the proteins in the diet.

Table 20.3   Essential Amino Acids

Arginine (Arg)	Methionine (Met)
Histidine (His)	Phenylalanine (Phe)
Isoleucine (Ile)	Threonine (Thr)
Leucine (Leu)	Tryptophan (Trp)
Lysine (Lys)	Valine (Val)

Complete proteins, which contain all of the essential amino acids are found in most animal products such as eggs, milk, meat, fish, and poultry. However, gelatin and plant proteins such as grains, beans, and nuts are *incomplete proteins* because they are deficient in one or more of the essential amino acids. Diets that rely on plant foods for protein must contain a variety of protein sources to obtain all the essential amino acids. For example, a diet of rice and beans contains all the essential amino acids because they are complementary proteins. Rice contains the methionine and tryptophan deficient in beans, while beans contain the lysine that is lacking in rice. (See Table 20.4.)

Table 20.4   Amino Acid Deficiency in Selected Vegetables and Grains

Food Source	Amino Acids Missing
Eggs, milk, meat, fish, poultry	None
Wheat, rice, oats	Lysine
Corn	Lysine, tryptophan
Beans	Methionine, tryptophan
Peas	Methionine
Almonds, walnuts	Lysine, tryptophan
Soy	Low in methionine

Cross-Links in Tertiary Structures

The tertiary structure of a protein is stabilized by interactions between the R groups of the amino acids in one region of the polypeptide chain with R groups of amino acids in other regions of the protein. (See Figure 20.9.) Table 20.5 lists the stabilizing interactions of tertiary structures.

WEB TUTORIAL
Tertiary and Quaternary Structure

1. **Hydrophobic interactions** are interactions between two nonpolar R groups. For example, hydrophobic interactions would occur between the aromatic group in phenylalanine and the alky group of valine or leucine. Within the compact shape of a globular protein, the amino acids with nonpolar side chains push as far away from the aqueous environment as possible, which forms a hydrophobic center at the interior of the protein molecule.

2. **Hydrophilic interactions** are attractions between the external aqueous environment and amino acids that have polar or ionized side chains. The polar side chains pull toward the outer surface of globular proteins to hydrogen bond with water. The presence of the hydrophilic side chains on the exterior surface makes globular proteins soluble in water.

3. **Salt bridges** are ionic bonds between side groups of basic and acidic amino acids, which have positive and negative charges. For example, at a pH of 7.4, the side chain of lysine has a positive charge, and the side chain of glutamic acid has a negative charge. The attraction of the oppositely charged side chains forms a strong bond called a salt bridge. If the pH changes, the basic and acidic side chains lose their ionic charges and cannot form salt bridges, which causes a change in the shape of the protein.

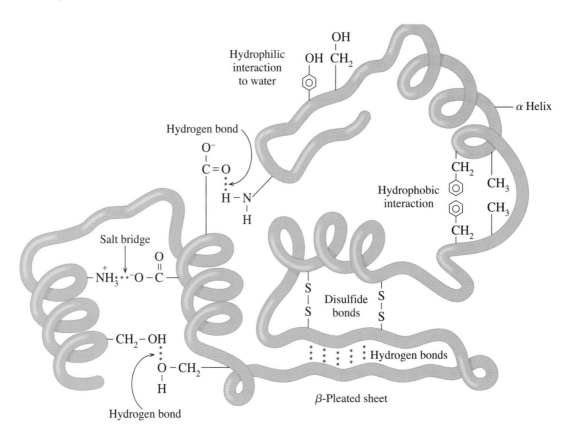

Figure 20.9 Interactions between amino acid R groups fold a protein into a specific three-dimensional shape called its tertiary structure.

Q Why would one section of the protein chain move to the center, while another section remains on the surface of the tertiary structure?

4. **Hydrogen bonds** form between polar amino acids. For example, a hydrogen bond can occur between the —OH of serine and the —NH_2 of glutamine.
5. **Disulfide bonds** (—S—S—) are covalent bonds that form between the —SH groups of cysteines in the polypeptide chain. In some proteins, there are several disulfide bonds between the R groups of cysteine in the polypeptide chain.

SAMPLE PROBLEM 20.8

Cross-Links in Tertiary Structures

What type of interaction would you expect between the R groups of the following amino acids?

a. cysteine and cysteine   **b.** glutamic acid and lysine

Solution

a. Because cysteine has an R group containing —SH, a disulfide bond will form.
b. An ionic bond (salt bridge) can form by the interaction of the —COO^- of glutamic acid and the —NH_3^+ of lysine.

Study Check

Would you expect to find valine and leucine in a globular protein on the outside or the inside of the tertiary structure? Why?

Table 20.5   Some Cross-Links in Tertiary Structures

	Nature of Bonding	Example
Hydrophobic interactions	Attractions between nonpolar alkyl and aromatic groups form a nonpolar center that is repelled by water	$-CH_3$ CH_3-
Hydrophilic Interactions	Attractions between polar or ionized R groups and water on the surface of the tertiary structure	$-CH_2OH \cdots O-H$ $\|$ H
Salt bridges	Ionic interactions between ionized R groups of acidic and basic amino acids	$\begin{matrix} O & & H \\ \| & & \|+ \\ -CO^- \cdots H-N- \\ & & \| \end{matrix}$
Hydrogen bonds	Occur between polar side groups of amino acids	$C=O \cdots HO-$ H $\|$ $C=O \cdots H-N-$
Disulfide bonds	Strong covalent links between sulfur atoms of two cysteine amino acids	$-SH + HS- \longrightarrow -S-S-$

HEALTH NOTE

Protein Structure and Mad Cow Disease

Up until recently, researchers thought that only virus or bacteria were responsible for transmitting diseases. Now a group of diseases have been found in which the infectious agents are proteins called *prions*. Bovine spongiform encephalopathy (BSE), or "mad cow disease," is a fatal brain disease of cattle in which the brain fills with cavities resembling a sponge. In the noninfectious form of the prion PrP[c], the N-terminal portion is a random coil. Although the noninfectious form may be ingested from meat products, its structure can change to what is known as PrP[sc] or *prion-related protein scrapie.* In this infectious form, the end of the peptide chain folds into a β-pleated sheet, which has disastrous effects on the brain and spinal cord. The conditions that cause this structural change are not yet known.

The human variant is called Creutzfeldt–Jakob (CJD) disease. Around 1955, Dr. Carleton Gajdusek was studying a disease known as Kuru, a neurological disease that was killing members of a tribe in Papua New Guinea. Because their diets were low in protein, it was a ritual to eat members of the tribe who died. As a result, the infectious agent Kuru was transmitted from one member to another. After Gajdusek identified the infectious agent in Kuru as similar to the prions that cause BSE, he received the Nobel prize.

BSE was diagnosed in Great Britain in 1986. The protein is present in nerve tissue, but is not found in meat. Control measures that exclude brain and spinal cord from animal feed are now in place to reduce the incidence of BSE. No cases of BSE have been found in the United States.

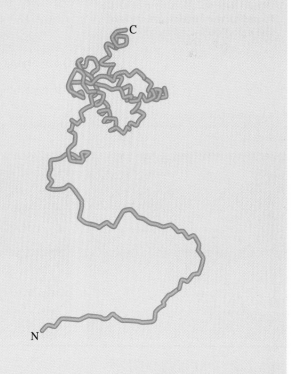

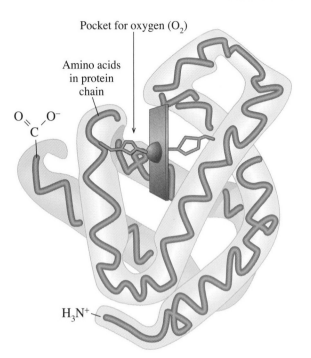

Pocket for oxygen (O₂)

Amino acids in protein chain

$O \underset{\parallel}{\overset{}{=}} \underset{C}{} \underset{}{O^-}$

H_3N^+

Figure 20.10 Myoglobin is a globular protein with a heme pocket in its tertiary structure that binds oxygen to be carried to the tissues.

Q **Would hydrophilic amino acids be found on the outside or inside of the myoglobin structure?**

Globular and Fibrous Proteins

A group of proteins known as **globular proteins** have compact, spherical shapes because their secondary structures of the polypeptide chain fold over on top of each other. It is the globular proteins that carry out the work of the cells: functions such as synthesis, transport, and metabolism.

Myoglobin is a globular protein that stores oxygen in skeletal muscle. High concentrations of myoglobin have been found in the muscles of sea mammals, such as seals and whales, that stay under the water for long periods. Myoglobin contains 153 amino acids in a single polypeptide chain with about three-fourths of the chain in the α-helix secondary structure. The polypeptide chain, including its helical regions, forms a compact tertiary structure by folding upon itself. (See Figure 20.10.) Within the tertiary structure, a pocket of amino acids and a heme group binds and stores oxygen (O_2).

The **fibrous proteins** are proteins that consist of long, thin, fiber-like shapes. They are typically involved in the structure of cells and tissues. Two types of fibrous protein are the α- and β-keratins. The **α-keratins** are the proteins that make up hair, wool, skin, and nails. In hair, three α-helixes coil together like a braid to form a fibril. Within the fibril, the α-helices are held together by disulfide ($-S-S-$) linkages between the R groups of the many cysteine amino acids in hair. Several fibrils bind together to form a strand of hair. (Figure 20.11.) The β-keratins are the type of proteins found in the feathers of birds and scales of reptiles. In β-keratins, the proteins consist of large amounts of β-pleated sheet structure.

Quaternary Structure: Hemoglobin

When a biologically active protein consists of two or more polypeptide subunits, the structural level is referred to as a **quaternary structure.** Hemoglobin, a globular protein that transports oxygen in blood, consists of four polypeptide chains or

α helix

Alpha keratin

Figure 20.11 The fibrous proteins of α-keratin wrap together to form fibrils that make up hair and wool. The proteins called β-keratins are found in the feathers of birds and scales of reptiles.

Q **Why does hair have a large amount of cysteine amino acids?**

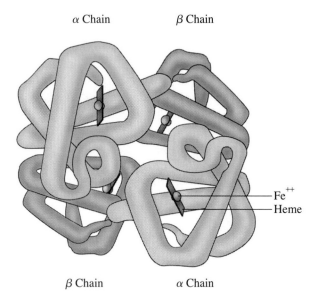

α Chain   β Chain

Fe^{++}
Heme

β Chain   α Chain

Figure 20.12 The quaternary structure of hemoglobin consists of four polypeptide subunits, each containing a heme group that binds an oxygen molecule.

Q What is the difference between a tertiary structure and a quaternary structure?

subunits, two α chains and two β chains. The subunits are held together in the quaternary structure by the same interactions that stabilize the tertiary structure, such as hydrogen bonds and salt bridges between side groups, disulfide links, and hydrophobic attractions. (See Figure 20.12.) Each subunit of the hemoglobin contains a heme group that binds oxygen. In the adult hemoglobin molecule, all four subunits ($\alpha_2\beta_2$) must be combined for the hemoglobin to properly function as an oxygen carrier. Therefore, the complete quaternary structure of hemoglobin can bind and transport four molecules of oxygen.

Hemoglobin and myoglobin have similar biological functions. Hemoglobin carries oxygen in the blood, whereas myoglobin carries oxygen in muscle. Myoglobin, a single polypeptide chain with a molar mass of 17,000, has about one-fourth the molar mass of hemoglobin (64,000). The tertiary structure of the single polypeptide myoglobin is almost identical to the tertiary structure of each of the subunits of hemoglobin. Myoglobin stores just one molecule of oxygen, just as each subunit of hemoglobin carries one oxygen molecule. The similarity in tertiary structures allows each protein to bind and release oxygen in a similar manner. Table 20.6 and Figure 20.13 summarize the structural levels of proteins.

SAMPLE PROBLEM 20.9

Identifying Protein Structure

Indicate whether the following conditions are responsible for primary, secondary, tertiary, or quaternary protein structures:

a. Disulfide bonds form between portions of a protein chain.

b. Peptide bonds form a chain of amino acids.

Solution

a. Disulfide bonds help to stabilize the tertiary structure of a protein.

b. The sequence of amino acids in a polypeptide is a primary structure.

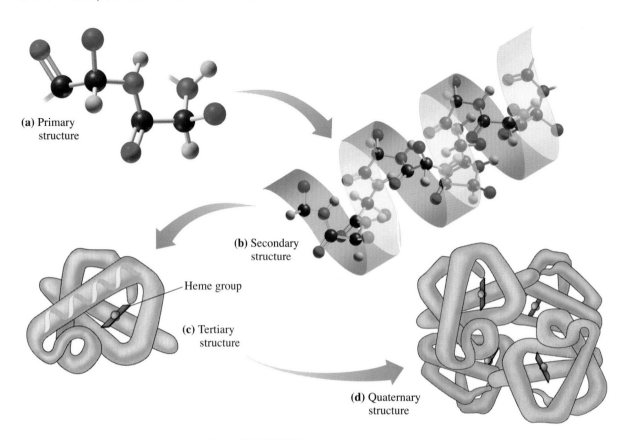

(a) Primary structure

(b) Secondary structure

Heme group

(c) Tertiary structure

(d) Quaternary structure

Figure 20.13 The quaternary structure of hemoglobin consists of four polypeptide subunits, each containing a heme group that binds an oxygen molecule.

Q What is the difference between a tertiary structure and a quaternary structure?

Study Check

What structural level is represented by the grouping of two subunits in insulin?

Table 20.6   Summary of Structural Levels in Proteins

Structural Level	Characteristics
Primary	The sequence of amino acids
Secondary	The coiled α-helix, β-pleated sheet, or a triple helix formed by hydrogen bonding between peptide bonds along the chain
Tertiary	A folding of the protein into a compact, three-dimensional shape stabilized by interactions between side R groups of amino acids
Quaternary	A combination of two or more protein subunits to form a larger, biologically active protein

QUESTIONS AND PROBLEMS

Protein Structure: Tertiary and Quaternary Levels

20.29 What type of interaction would you expect from the following R groups in a tertiary structure?
  a. two cysteine residues   **b.** glutamic acid and lysine
  c. serine and aspartic acid   **d.** two leucine residues

20.30 In myoglobin, about one-half of the 153 amino acids have nonpolar side chains.
  a. Where would you expect those amino acids to be located in the tertiary structure?

HEALTH NOTE

Sickle-Cell Anemia

Sickle-cell anemia is a disease caused by an abnormality in the shape of one of the subunits of the hemoglobin protein. In the β chain, the sixth amino acid, glutamic acid, which is polar, is replaced by valine, a nonpolar amino acid.

Because valine has a hydrophobic side chain, it draws the hydrophobic pocket that binds to oxygen to the surface of the hemoglobin. The affected red blood cells change from a rounded shape to a crescent shape, like a sickle, which interferes with their ability to transport adequate quantities of oxygen. Hydrophobic attractions cause several sickle-cell hemoglobin molecules to stick together, which forms long fibers of sickle-cell hemoglobin. The clumps of insoluble fibers clog capillaries, where they cause inflammation, pain, and organ damage. Critically low oxygen levels may occur in the affected tissues.

In sickle-cell anemia, both genes for the altered hemoglobin must be inherited. However, a few sickled cells are found in persons who carry one gene for sickle-cell hemoglobin, a condition that is also known to provide protection from malaria.

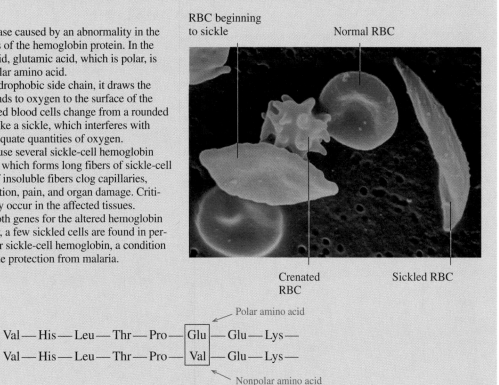

RBC beginning to sickle   Normal RBC

Crenated RBC   Sickled RBC

Polar amino acid

Normal β chain: Val—His—Leu—Thr—Pro—|Glu|—Glu—Lys—

Sickled β chain: Val—His—Leu—Thr—Pro—|Val|—Glu—Lys—

Nonpolar amino acid

   b. Where would you expect the polar side chains to be?

   c. Why is myoglobin more soluble in water than silk or wool?

20.31 A portion of a polypeptide chain contains the following sequence of amino acid residues:

   -Leu-Val-Cys-Asp-

   a. Which R groups can form a disulfide cross-link?

   b. Which amino acid residues are likely to be found on the inside of the protein structure? Why?

   c. Which amino acid residues would be found on the outside of the protein? Why?

   d. How does the primary structure of a protein affect its tertiary structure?

20.32 State whether the following statements apply to primary, secondary, tertiary, or quaternary protein structure:

   a. Side groups interact to form disulfide bonds or ionic bonds.

   b. Peptide bonds join amino acids in a polypeptide chain.

   c. Several polypeptides are held together by hydrogen bonds between adjacent chains.

   d. Hydrogen bonding between carbonyl oxygen atoms and nitrogen atoms of amide groups causes a polypeptide to coil.

   e. Hydrophobic side chains seeking a nonpolar environment move toward the inside of the folded protein.

   f. Protein chains of collagen form a triple helix.

   g. An active protein contains four tertiary subunits.

20.7 Protein Hydrolysis and Denaturation

We saw in Chapter 19 that amides can be hydrolyzed or split apart in the presence of an acid or base. Peptide bonds can be hydrolyzed too to give the individual amino acids. This is the process that occurs in the stomach when enzymes such as pepsin or trypsin catalyze the hydrolysis of proteins to give amino acids. This disrupts the primary structure by breaking the covalent amide bonds that link the amino acids. In the digestion of proteins, the amino acids are absorbed through the intestinal walls and carried to the cells where they can be used to synthesize proteins.

Alanylglycylserine (Ala-Gly-Ser)

H_2O | Enzyme

Alanine (Ala)   Glycine (Gly)   Serine (Ser)

Denaturation of Proteins

Denaturation of a protein occurs when there is a disruption of any of the bonds that stabilize the secondary, tertiary, or quaternary structure. However, the covalent amide bonds of the primary structure are not affected.

When the interactions between the R groups are undone or altered, a globular protein unfolds like a loose piece of spaghetti. With the loss of its overall shape, the protein is no longer biologically active. (See Figure 20.14.)

Denaturing agents include heat, acids and bases, organic compounds, heavy metal ions, and mechanical agitation.

Heat

Heat denatures proteins by breaking apart hydrogen bonds and the hydrophobic attraction between nonpolar side groups. Few proteins can remain biologically active above 50°C. Whenever you cook food, you are using heat to denature protein. The nutritional value of the proteins in food is not changed, but they are made more digestible. High temperatures are also used to disinfect surgical instruments and gowns by denaturing the proteins of any bacteria present.

Acids and Bases

Placing a protein in an acid or base affects the hydrogen bonding between polar R groups and disrupts the ionic bonds (salt bridges). In the preparation of yogurt and cheese, a bacteria that produces lactic acid is added to denature the milk protein and produce solid casein. Tannic acid, a weak acid used in burn ointments, coagulates proteins at the site of the burn, forming a protective cover and preventing further loss of fluid from the burn.

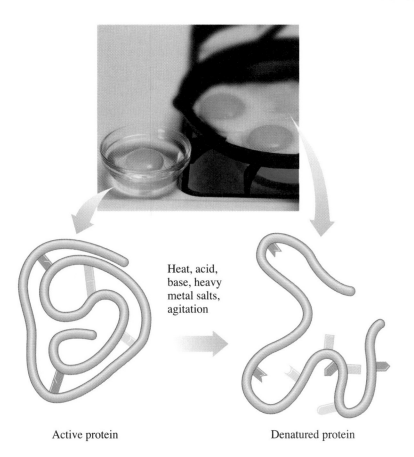

Active protein   Denatured protein

Figure 20.14 Denaturation of a protein occurs when the bonds of the tertiary structure are disrupted, which destroys the shape and renders the protein biologically inactive.

Q **What are some ways in which proteins are denatured?**

EXPLORE YOUR WORLD

Denaturation of Milk Protein

Place some milk in five glasses. Add the following to the milk samples in glasses 1–4. The fifth glass of milk is a reference sample.
1. Vinegar, drop by drop. Stir.
2. One-half teaspoon of meat tenderizer. Stir.
3. One teaspoon of fresh pineapple juice. (Canned juice has been heated.)
4. One teaspoon of fresh pineapple juice that you have heated to boiling.

Questions

1. How did the appearance of the milk change in each of the samples?
2. What enzyme is listed on the package label of the tenderizer?
3. How does the effect of the heated pineapple juice compare with that of the fresh juice? Explain.
4. Why is cooked pineapple used when making gelatin (a protein) desserts?

Organic Compounds

Ethanol and isopropyl alcohol act as disinfectants by forming their own hydrogen bonds with a protein and disrupting the hydrophobic interactions. An alcohol swab is used to clean wounds or to prepare the skin for an injection because the alcohol passes through the cell walls and coagulates the proteins inside the bacteria.

Heavy Metal Ions

When heavy metal ions like Ag^+, Pb^{2+}, and Hg^{2+} form ionic bonds or react with the disulfide ($-S-S-$) bonds, the denatured protein solidifies. In hospitals, dilute (1%) solutions of $AgNO_3$ are placed in the eyes of newborn babies to destroy the bacteria that causes gonorrhea. These heavy metal ions can be toxic. If ingested, they act as poisons by severely disrupting body proteins, especially enzymes. An antidote is a high-protein food such as milk, eggs, or cheese that will tie up the heavy metal ions until the stomach can be pumped.

Agitation

The whipping of cream and the beating of egg whites are examples of using mechanical agitation to denature protein. The whipping action stretches the polypeptide chains until the stabilizing interactions are disrupted.

SAMPLE PROBLEM 20.10

Effects of Denaturation

What happens to the tertiary structure of a globular protein when it is placed in an acidic solution?

Solution

An acid causes denaturation by disrupting the hydrogen bonds and the ionic bonds between the R groups. A loss in interactions causes the tertiary structure to lose stability. As the protein unfolds, both the shape and biological function are lost.

Study Check

Why is a dilute solution of $AgNO_3$ used to disinfect the eyes of newborn infants?

QUESTIONS AND PROBLEMS

Protein Hydrolysis and Denaturation

20.33 What products would result from the complete hydrolysis of Gly-Ala-Ser?

20.34 Would the hydrolysis products of the tripeptide Ala-Ser-Gly be the same or different from the products in problem 20.33? Explain.

20.35 What dipeptides would be produced from the partial hydrolysis of His-Met-Gly-Val?

20.36 What tripeptides would be produced from the partial hydrolysis of Ser-Leu-Gly-Gly-Ala?

20.37 What structural level of a protein is affected by hydrolysis?

20.38 What structural level of a protein is affected by denaturation?

20.39 Indicate the changes in protein structure for each of the following:
  a. An egg placed in water at 100°C is soft boiled in about 3 min.
  b. Prior to giving an injection, the skin is wiped with an alcohol swab.
  c. Surgical instruments are placed in a 120°C autoclave.
  d. During surgery, a wound is closed by cauterization (heat).

20.40 Indicate the changes in protein structure for each of the following:
  a. Tannic acid is placed on a burn.
  b. Milk is heated to 60°C to make yogurt.
  c. To avoid spoilage, seeds are treated with a solution of $HgCl_2$.
  d. Hamburger is cooked at high temperatures to destroy *E. coli* bacteria that may cause intestinal illness.

Chapter Review

20.1 Functions of Proteins

Some proteins are enzymes or hormones, whereas others are important in structure, transport, protection, storage, and muscle contraction.

20.2 Amino Acids

A group of 20 amino acids provides the molecular building blocks of proteins. Attached to the central (alpha) carbon of each amino acid is an amino group, a carboxyl group, and a unique side group (R). The R group gives an amino acid the property of being nonpolar, polar, acidic, or basic.

20.3 Amino Acids as Acids and Bases

Amino acids exist as dipolar ions called zwitterions, as positive ions at low pH, and as negative ions at high pH levels. At the isoelectric point, zwitterions are neutral.

20.4 Formation of Peptides

Peptide bonds join amino acids to form peptides by forming an amide bond between the carboxyl group of one amino acid and the amino group of the second. Long chains of amino acids are called proteins.

20.5 Protein Structure: Primary and Secondary Levels

The primary structure of a protein is its sequence of amino acids. In the secondary structure, hydrogen bonds between peptide groups produce a characteristic shape such as an α-helix, β-pleated sheet, or a triple helix.

20.6 Protein Structure: Tertiary and Quaternary Levels

In globular proteins, the polypeptide chain including its α-helical and β-pleated sheet regions folds upon itself to form a tertiary structure. A tertiary structure is stabilized by interactions that move hydrophobic R groups to the inside and hydrophilic R groups to the outside surface, and by attractions between R groups that form hydrogen, disulfide, and salt bridges. In a quaternary structure, two or more tertiary subunits must combine for biological activity. They are held together by the same interactions found in tertiary structures.

20.7 Protein Hydrolysis and Denaturation

Denaturation of a protein occurs when heat or other denaturing agents destroy the structure of the protein (but not the primary structure) until biological activity is lost.

Key Terms

acidic amino acid An amino acid that has a carboxylic acid side chain ($-COOH$), which ionizes as a weak acid.

α (alpha) helix A secondary level of protein structure, in which hydrogen bonds connect the NH of one peptide bond with the $C=O$ of a peptide bond later in the chain to form a coiled or corkscrew structure.

α-keratins Fibrous proteins that contain mostly α-helixes found in hair, nails, and skin.

amino acid The building block of proteins, consisting of an amino group, a carboxylic acid group, and a unique side group attached to the alpha carbon.

basic amino acid An amino acid that contains an amino ($-NH_2$) group that can ionize as a weak base.

β (beta)-pleated sheet A secondary level of protein structure that consists of hydrogen bonds between peptide links in parallel polypeptide chains.

C terminal The amino acid that is the last amino acid in a peptide chain with an unreacted or free carboxyl group ($-COO^-$).

collagen The most abundant form of protein in the body, which is composed of fibrils of triple helixes with hydrogen bonding between $-OH$ groups of hydroxyproline and hydroxylysine.

denaturation The loss of secondary and tertiary protein structure caused by heat, acids, bases, organic compounds, heavy metals, and/or agitation.

disulfide bonds Covalent $-S-S-$ bonds that form between the $-SH$ group of cysteines in a protein to stabilize the tertiary structure.

electrophoresis The use of electrical current to separate proteins or other charged molecules with different isoelectric points.

essential amino acids Amino acids that must be supplied by the diet because they are not synthesized by the body.

fibrous protein A protein that is insoluble in water; consists of polypeptide chains with α-helixes or β-pleated sheets, that make up the fibers of hair, wool, skin, nails, and silk.

globular proteins Proteins that acquire a compact shape from attractions between the R group of the amino acid residues in the protein.

hydrogen bonds Attractions between the polar R groups such as $-OH$, $-NH_2$, and $-COOH$ of amino acids in a polypeptide chain.

hydrophilic amino acid An amino acid having polar, acidic, or basic R groups that are attracted to water; "water-loving."

hydrophilic interactions the attraction between polar R groups on the protein surface and water

hydrophobic amino acid A nonpolar amino acid with hydrocarbon R groups; "water-fearing."

hydrophobic interactions The attraction between nonpolar R groups in a tertiary structure of a globular protein.

isoelectric point (pI) The pH at which an amino acid is in the neutral zwitterion form.

N terminal The amino acid in a peptide written on the left with the unreacted or free amino group ($-NH_3^+$).

nonpolar amino acids Amino acids that are not soluble in water because they contain a nonpolar R group.

peptide The combination of two or more amino acids joined by peptide bonds; dipeptide, tripeptide, and so on.

peptide bond The amide bond that joins amino acids in polypeptides and proteins.

polar amino acids Amino acids that are soluble in water because their R group is polar: hydroxyl (OH), thiol (SH), carbonyl ($C=O$), amino (NH_2), or carboxyl (COOH).

primary structure The sequence of the amino acids in a protein.

protein A term used for biologically active polypeptides that have many amino acids linked together by peptide bonds.

quaternary structure A protein structure in which two or more protein subunits form an active protein.

salt bridge The ionic bond formed between side groups of basic and acidic amino acids in a protein.

secondary structure The formation of an α-helix, β-pleated sheet, or triple helix.

tertiary structure The folding of the secondary structure of a protein into a compact structure that is stabilized by the interactions of R groups such as ionic and disulfide bonds.

triple helix The protein structure found in collagen consisting of three polypeptide chains woven together like a braid.

zwitterion The dipolar form of an amino acid consisting of two oppositely charged ionic regions, $—NH_3^+$ and $—COO^-$.

Additional Problems

20.41 Seeds and vegetables are often deficient in one or more essential amino acids. Using the following table, state whether the following combinations would provide all the essential amino acids:

Source	Lysine	Tryptophan	Methionine
Oatmeal	No	Yes	Yes
Rice	No	Yes	Yes
Garbanzo beans	Yes	No	Yes
Lima beans	Yes	No	No
Cornmeal	No	Yes	Yes

    a. rice and garbanzo beans
    b. lima beans and cornmeal
    c. a salad of garbanzo beans and lima beans
    d. rice and lima beans
    e. rice and oatmeal
    f. oatmeal and lima beans

20.42 How does denaturation of a protein differ from its hydrolysis?

20.43 What are some differences between the following pairs?
    a. secondary and tertiary protein structures
    b. essential and nonessential amino acids
    c. polar and nonpolar amino acids
    d. di- and tripeptides
    e. an ionic bond (salt bridge) and a disulfide bond
    f. fibrous and globular proteins
    g. α-helix and β-pleated sheet
    h. tertiary and quaternary structures of proteins

20.44 The proteins placed on a gel for electrophoresis have the following isoelectric points: albumin, 4.9, hemoglobin, 6.8, and lysozyme, 11.0. A buffer of pH 6.8 is placed on the gel.
    a. Which protein will migrate toward the positive electrode?
    b. Which protein will migrate toward the negative electrode?
    c. Which protein will remain at the same place it was originally placed?

20.45 **a.** What are some functions of α-keratins?
    b. What amino acids give strength to the α-keratins?

20.46 **a.** Where is collagen found?
    b. What type of secondary structure is used to form collagen?

20.47 **a.** Draw the structure of Ser-Lys-Asp.
    b. Would you expect to find this segment at the center or at the surface of a globular protein? Why?

20.48 **a.** Draw the structure of Val-Ala-Leu.
    b. Would you expect to find this segment at the center or at the surface of a globular protein? Why?

20.49 Would you expect the following segments in a polypeptide to have an α-helix or β-pleated sheet secondary structure?
    a. a segment with a high content of Val, Pro, and Ser
    b. a segment with a high content of His, Met and Leu.

20.50 What type of interaction would you expect from the following R groups in a tertiary structure?
    a. threonine and asparagine
    b. valine and alanine
    c. arginine and aspartic acid

20.51 If serine were replaced by valine in a protein, how would the tertiary structure be affected?

20.52 If you eat rice, what other vegetable protein source(s) could you eat to ingest all essential amino acids?

20.53 Draw the structure of each of the following amino acids at pH 4.
    a. serine   **b.** alanine   **c.** lysine

20.54 Draw the structure of each of the following amino acids at pH 11.
    a. cysteine   **b.** aspartic acid   **c.** valine

21 Enzymes and Vitamins

"At a time when we have a shortage of health care professionals, I think of myself as a physician extender," says Pushpinder Beasley, orthopedic physician assistant, Kaiser Hospital. "We can put a significant amount of time into our patient care. Just today, I examined a child's knee. One of the most common injuries to children is disruption of either knee ligaments or the soft tissue around the knees. In this child's case, we were checking her anterior ligaments, also known as ACL. I think an important role of the health care professional is to earn the trust of young people."

As part of a health care team, physician assistants examine patients, order laboratory tests, make diagnoses, report patient progress, order therapeutic procedures and, in most states, prescribe medications.

LOOKING AHEAD

the Chemistry place

www.chemplace.com/college

Visit the URL above or use the CD-ROM in the book for extra quizzing, interactive tutorials, career resources, and case studies.

Every second, thousands of chemical reactions occur in the cells of the human body. For example, many reactions occur to digest the food we eat, convert the products to chemical energy, and synthesize proteins and other macromolecules in our cells. In the laboratory, we can carry out reactions that hydrolyze polysaccharides, fats, or proteins, but we must use a strong acid or base, high temperatures, and long reaction times. In the cells of our body, these reactions must take place at rates that meet our physiological and metabolic needs. To make this happen, enzymes catalyze the chemical reactions in our cells, with a different enzyme for every reaction. Digestive enzymes in the mouth, stomach, and small intestine catalyze the hydrolysis of carbohydrate, fats, and proteins. Enzymes in the mitochondria extract energy from biomolecules to give us energy.

Every enzyme responds to what comes into the cells and to what the cells need. Enzymes keep reactions going when our cells need certain products, and turn off reactions when they don't need those products.

Many enzymes require cofactors to function properly. A cofactor can be an inorganic metal ion or an organic compound such as a vitamin. We obtain minerals such as zinc (Zn^{2+}) and iron (Fe^{3+}) and vitamins from our diets. A lack of minerals and vitamins can lead to certain nutritional diseases. For example, rickets is a deficiency of vitamin D and scurvy occurs when a diet is low in vitamin C.

LEARNING GOAL

Describe how enzymes function as biological catalysts.

21.1 Biological Catalysts

As a **catalyst,** an enzyme increases the rate of a reaction by changing the way a reaction takes place, but is itself not changed at the end of the reaction. An uncatalyzed reaction in a cell may take place eventually, but not at a rate fast enough for survival. For example, the hydrolysis of proteins in our diet would eventually occur without a catalyst, but not fast enough to meet the body's requirements for amino acids. The chemical reactions in our cells must occur at incredibly fast rates under mild conditions of pH 7.4 and a body temperature of 37°C. To do this, biological catalysts known as **enzymes** catalyze nearly all the chemical reactions that take place in the body. Because reactions in the cells are catalyzed, they produce only those products that are useful in the cell. Enzymes permit cells to use energy and materials efficiently while responding to cellular needs.

As catalysts, enzymes lower the activation energy for the reaction. (See Figure 21.1.) As a result, less energy is required to convert reactant molecules to products, which allows more reacting molecules to form product. However, as a catalyst an enzyme does not affect the equilibrium position, which means that there is an increase in the rates of both the forward and reverse directions. The rates of enzyme-catalyzed reactions are much faster than the rates of the uncatalyzed reactions. Some enzymes can increase the rate of a biological reaction by a factor of a billion or trillion or even a hundred million trillion compared to the rate of the uncatalyzed reaction. For example, an enzyme in the blood called carbonic anhydrase converts carbon dioxide (CO_2) and water (H_2O) to carbonic acid (H_2CO_3). In one minute, one molecule of anhydrase catalyzes the reaction of about one million (10^6) molecules.

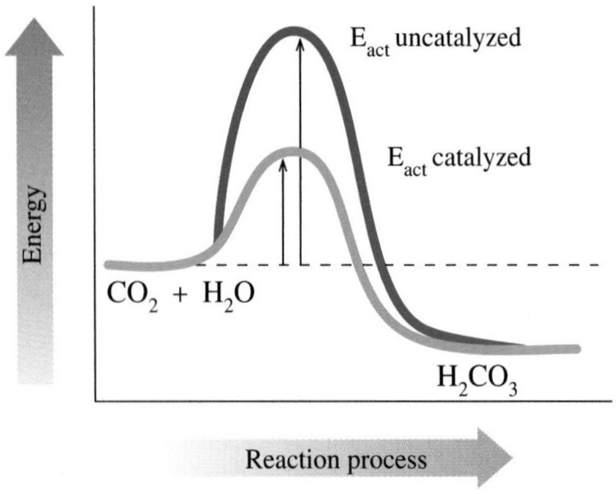

Figure 21.1 The activation energy needed for the reaction of CO_2 and H_2O is lowered by the enzyme carbonic anhydrase.

Q Why does an enzyme accelerate a reaction?

$$\text{CO}_2 + \text{H}_2\text{O} \xrightleftharpoons[\text{}]{\text{Carbonic anhydrase}} \text{H}_2\text{CO}_3$$

SAMPLE PROBLEM 21.1

Biological Catalysts

How do enzymes differ from catalysts used in the laboratory?

Solution

Enzymes function under mild physiological conditions of about pH 7 and 37°C. Catalysts in the laboratory are used at high temperatures and low or high pH.

Study Check

Why are enzymes called biological catalysts?

QUESTIONS AND PROBLEMS

Biological Catalysts

21.1 Why do cellular reactions require enzymes?

21.2 How do enzymes make cellular reactions proceed at faster rates?

21.2 Names and Classification of Enzymes

LEARNING GOAL

Classify enzymes according to the reaction they catalyze.

The names of enzymes describe the compound or the reaction that is catalyzed. The actual names of enzymes are derived by replacing the end of the name of the reaction or reacting compound with the suffix *ase*. For example, an *oxidase* catalyzes an oxidation reaction, and a *dehydrogenase* removes hydrogen atoms. The compound sucrose is hydrolyzed by the enzyme *sucrase*, and a lipid is hydrolyzed by a *lipase*. Some early known enzymes use names that end in the suffix *in*, such as *papain* found in papaya, *rennin* found in milk, and *pepsin* and *trypsin*, enzymes that catalyze the hydrolysis of proteins.

The International Commission on Enzymes has classified enzymes according to the six general types of reactions they catalyze. (See Table 21.1.)

SAMPLE PROBLEM 21.2

Naming Enzymes

What chemical reaction do the following enzymes catalyze?
a. amino transferase   **b.** lactate dehydrogenase

Solution

a. catalyzes the transfer of an amino group
b. catalyzes the removal of hydrogen from lactate

Study Check

What is the class of the enzyme lipase that catalyzes the hydrolysis of ester bonds in triglycerides?

QUESTIONS AND PROBLEMS

Names and Classification of Enzymes

21.3 What types of reaction are catalyzed by each of the following classes of enzymes?
  a. oxidoreductases   **b.** transferases
  c. hydrolases

21.4 What types of reaction are catalyzed by each of the following classes of enzymes?
  a. lyases   **b.** isomerases
  c. ligases

21.5 What class of enzyme catalyzes each of the following reactions?
  a. hydrolysis of sucrose   **b.** addition of oxygen
  c. converting glucose ($C_6H_{12}O_6$) to fructose ($C_6H_{12}O_6$)
  d. moving an amino group from one molecule to another

21.6 What class of enzyme catalyzes each of the following reactions?
  a. addition of water to a double bond
  b. removing hydrogen atoms
  c. splitting peptide bonds in proteins
  d. removing CO_2 from pyruvate

21.7 Identify the class of enzyme that catalyzes each of the following reactions:

a. $CH_3-\overset{\overset{O}{\|}}{C}-COO^- + H^+ \longrightarrow CH_3-\overset{\overset{O}{\|}}{C}-H + CO_2$

b. $CH_3-\overset{\overset{NH_3^+}{|}}{CH}-COO^- + {}^-OOC-\overset{\overset{O}{\|}}{C}-CH_2-CH_3 \longrightarrow$

$CH_3-\overset{\overset{O}{\|}}{C}-COO^- + {}^-OOC-\overset{\overset{NH_3^+}{|}}{CH}-CH_2-CH_3$

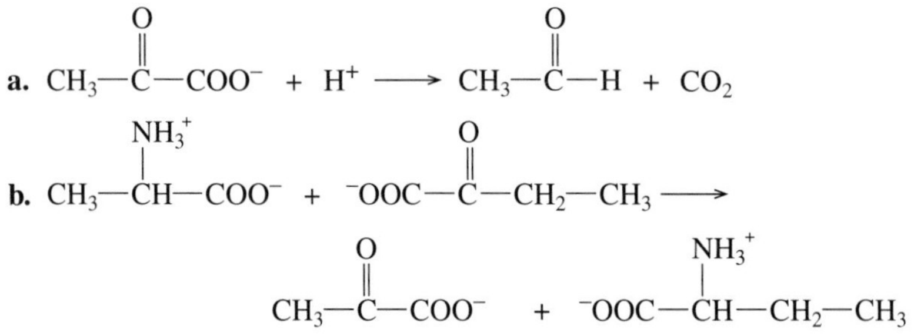

21.8 Identify the class of enzyme that would catalyze each of the following reactions:

a. $CH_3-\overset{\overset{O}{\|}}{C}-COO^- + CO_2 + ATP \longrightarrow {}^-OOC-CH_2-\overset{\overset{O}{\|}}{C}-COO^- + ADP + P_i$

b. $CH_3-CH_2-OH + NAD^+ \longrightarrow CH_3-\overset{\overset{O}{\|}}{C}-H + NADH + H^+$

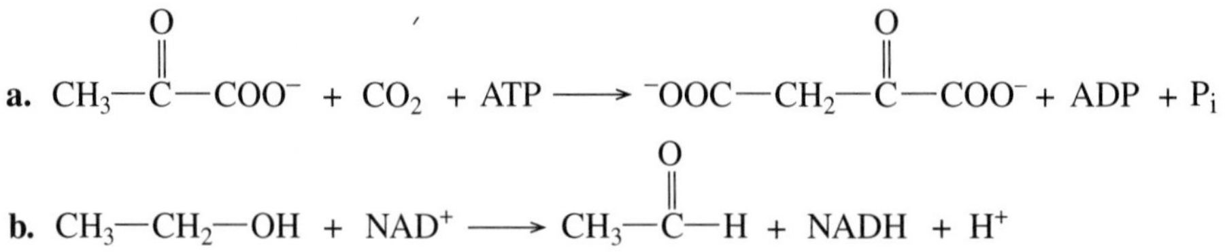

21.9 Assign a name to an enzyme that catalyzes each of the following reactions:
  a. oxidizes succinate   **b.** adds water to fumarate
  c. removes 2H from alcohol

21.10 Assign a name to an enzyme that catalyzes each of the following reactions:
  a. hydrolyzes sucrose
  b. transfers an amino group from aspartate
  c. removes a carboxylate group from pyruvate

Table 21.1   Classification of Enzymes

Class	General Reactions Catalyzed	Typical Subclasses	Function
1. Oxidoreductases	Oxidation–reduction reactions	Oxidases Reductases Dehydrogenases	Oxidation Reduction Remove 2H to form double bonds

$$CH_3-CH_2-OH \ + \ NAD^+ \ \xrightarrow{\text{Alcohol dehydrogenase}} \ CH_3-\overset{\overset{\textstyle O}{\|}}{C}-H \ + \ NADH \ + \ H^+$$

Ethanol   Coenzyme   Acetaldehyde   Coenzyme

Class	General Reactions Catalyzed	Typical Subclasses	Function
2. Transferases	Transfer of functional groups	Transaminases Kinases	Transfer amino groups Transfer phosphate groups

$$CH_3-\overset{\overset{\textstyle NH_3^+}{|}}{CH}-COO^- \ + \ {}^-OOC-\overset{\overset{\textstyle O}{\|}}{C}-CH_2CH_2-COO^- \ \rightleftharpoons \xrightarrow{\text{Alanine transaminase}} \ CH_3-\overset{\overset{\textstyle O}{\|}}{C}-COO^- \ + \ {}^-OOC-\overset{\overset{\textstyle NH_3^+}{|}}{CH}-CH_2CH_2-COO^-$$

Alanine   α-Ketoglutarate   Pyruvate   Glutamate

Class	General Reactions Catalyzed	Typical Subclasses	Function
3. Hydrolases	Hydrolysis reactions	Peptidases Lipases Amylases	Hydrolyze peptide bonds Hydrolyze ester bonds in lipids Hydrolyze 1,4-glycosidic bonds in amylose

$$-\underset{\underset{\textstyle H}{|}}{N}-\overset{\overset{\textstyle R}{|}}{CH}-\overset{\overset{\textstyle O}{\|}}{C}-\underset{\underset{\textstyle H}{|}}{N}-\overset{\overset{\textstyle R}{|}}{CH}-COO^- \ + \ H_2O \ \xrightarrow{\text{Peptidase}} \ -\underset{\underset{\textstyle H}{|}}{N}-\overset{\overset{\textstyle R}{|}}{CH}-\overset{\overset{\textstyle O}{\|}}{C}-O^- \ + \ H_3\overset{+}{N}-\overset{\overset{\textstyle R}{|}}{CH}-COO^-$$

Polypeptide C terminal   Shorter polypeptide   Amino acid from C terminal

Class	General Reactions Catalyzed	Typical Subclasses	Function
4. Lyases	Addition of a group to a double bond or removal of a group from a double bond without hydrolysis or oxidation	Decarboxylases Dehydrases Deaminases	Remove CO_2 Remove H_2O Remove NH_3

$$CH_3-\overset{\overset{\textstyle O}{\|}}{C}-COO^- \ + \ H^+ \ \xrightarrow{\text{Pyruvate decarboxylase}} \ CH_3-\overset{\overset{\textstyle O}{\|}}{C}-H \ + \ CO_2$$

Pyruvate   Acetaldehyde   Carbon dioxide

Class	General Reactions Catalyzed	Typical Subclasses	Function
5. Isomerases	Rearrangement of atoms to form isomers	Isomerases Epimerases	Convert cis and trans Convert D and L isomers

$$\underset{\underset{\textstyle H}{|}}{\overset{\overset{\textstyle {}^-OOC}{|}}{C}}=\underset{\underset{\textstyle H}{|}}{\overset{\overset{\textstyle COO^-}{|}}{C}} \ \xrightarrow{\text{Maleate isomerase}}\rightleftharpoons \ \underset{\underset{\textstyle H}{|}}{\overset{\overset{\textstyle {}^-OOC}{|}}{C}}=\underset{\underset{\textstyle COO^-}{|}}{\overset{\overset{\textstyle H}{|}}{C}}$$

Maleate   Fumarate

Class	General Reactions Catalyzed	Typical Subclasses	Function
6. Ligases	Bonding of molecules using ATP energy	Synthetases Carboxylases	Combine molecules Add CO_2

$${}^-OOC-\overset{\overset{\textstyle O}{\|}}{C}-CH_3 \ + \ CO_2 \ + \ ATP \ \xrightarrow{\text{Pyruvate carboxylase}} \ {}^-OOC-\overset{\overset{\textstyle O}{\|}}{C}-CH_2-COO^- \ + \ ADP \ + \ P_i \ + \ H^+$$

Pyruvate   Oxaloacetate

21.3 Enzymes as Catalysts

Nearly all enzymes are globular proteins. Each has a unique three-dimensional shape that recognizes and binds a small group of reacting molecules, which are called **substrates.** The tertiary structure of an enzyme plays an important role in how that enzyme catalyzes reactions.

Active Site

In a catalyzed reaction, an enzyme must first bind to a substrate in a way that favors catalysis. A typical enzyme is much larger than its substrate. However, within its large tertiary structure, there is a region called the **active site** where the enzyme binds a substrate or substrates and catalyzes the reaction. This active site is often a small pocket that closely fits the structure of the substrate. (See Figure 21.2) Within the active site, the side chains of amino acids bind the substrate with hydrogen bonds, salt bridges, or hydrophobic attractions. The active site of a particular enzyme fits the shape of only a few types of substrates, which makes enzymes very specific about the type of substrate they bind.

Some enzymes show absolute specificity by catalyzing one reaction of one specific substrate. Other enzymes catalyze a reaction for a group of substrates. Still other enzymes catalyze a reaction for a specific type of bond in a substrate. Types of enzyme specificity are listed in Table 21.2.

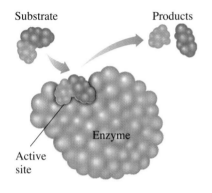

Substrate   Products

Enzyme

Active
site

Figure 21.2 On the surface of an enzyme, a small region called an active site binds a substrate and catalyzes a reaction of that substrate.

Q **Why does an enzyme catalyze a reaction of only certain substrates?**

the
**Chemistry
place**

WEB TUTORIAL
How Enzymes Work

Table 21.2   Types of Enzyme Specificity

Type	Reaction Type	Example
Absolute	Catalyze one type of reaction for a single substrate	Urease catalyzes only the hydrolysis of urea
Group	Catalyze one type of reaction for similar substrates	Hexokinase adds a phosphate group to hexoses
Linkage	Catalyze one type of reaction for a specific type of bond	Chymotrypsin catalyzes the hydrolysis of peptide bonds

Lock-and-Key and Induced Fit Models

In an early theory of enzyme action called the **lock-and-key model,** the active site is described as having a rigid, nonflexible shape. Thus only those substrates with shapes that fit exactly into the active site are able to bind with that enzyme. The shape of the active site is analogous to a lock, and the proper substrate is the key that fits into the lock. (See Figure 21.3a.)

While the lock-and-key model explains the binding of substrates for many enzymes, certain enzymes have a broader range of specificity than the lock and key model allows. In the **induced-fit model,** there is an interaction between both the enzyme and substrate. (See Figure 21.3b.) The active site adjusts to fit the shape of the substrate more closely. At the same time the substrate adjusts its shape to better adapt to the geometry of the active site. As a result, the reacting section of the substrate becomes aligned exactly with the groups in the active site that catalyze the reaction.

In the induced-fit model, substrate and enzyme work together to acquire a geometrical arrangement that lowers the activation energy. A different substrate could not induce these structural changes and no catalysis would occur. (See Figure 21.3c.)

What is the function of the active site in an enzyme?

Solution

The active site in an enzyme binds the substrate and contains the amino acid side chains that bind the substrate and catalyze the reaction.

Study Check

How do the lock-and-key and the induced-fit models differ in their description of the active site in an enzyme?

Enzyme Catalyzed Reaction

The proper alignment of a substrate within the active site forms an **enzyme–substrate (ES) complex.** This combination of enzyme and substrate provides an alternative pathway for the reaction that has a lower activation energy. Within the active site, the amino acid side chains take part in catalyzing the chemical reaction. For example, acidic and basic side chains remove protons from or provide protons for the substrate. As soon as the catalyzed reaction is complete, the products are quickly released from the enzyme so it can bind to a new substrate molecule. We can write the catalyzed reaction of an enzyme (E) with a substrate (S) to form product (P) as follows:

Step 1	E + S	$\rightleftharpoons$	ES		
Step 2			ES	$\longrightarrow$	E + P

	E + S	$\rightleftharpoons$	ES	$\longrightarrow$	E + P
	Enzyme + substrate		ES complex		Enzyme + product

Let's consider the hydrolysis of sucrose by sucrase. When sucrose binds to the active site of sucrase, the glycosidic bond of sucrose is placed into a geometry favorable for reaction. The amino acid side chains catalyze the cleavage of the sucrose to give the products glucose and fructose.

Sucrase + sucrose	$\rightleftharpoons$	sucrase-sucrose complex	$\longrightarrow$	sucrase + glucose + fructose
E + S		ES complex		E + P_1 + P_2

Because the structures of the products are no longer attracted to the active site, they are released and the sucrase binds another sucrose substrate. (See Figure 21.4).

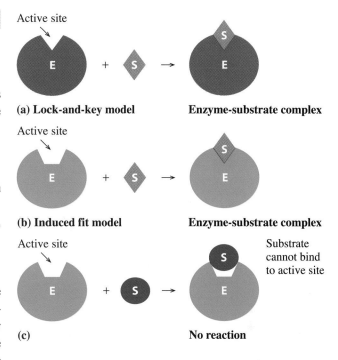

(a) Lock-and-key model   Enzyme-substrate complex

(b) Induced fit model   Enzyme-substrate complex

(c)   No reaction

Substrate cannot bind to active site

Figure 21.3 (a) In the lock-and-key model, a substrate fits the shape of the active site and forms an enzyme–substrate complex. **(b)** In the induced-fit model, a flexible active site and substrate adapt to provide a close fit to a substrate and proper orientation for reaction. **(c)** A substrate that does not fit or induce a fit in the active site cannot undergo catalysis by the enzyme.

Q How does the induced-fit model differ from the lock-and-key model?

QUESTIONS AND PROBLEMS

Enzymes as Catalysts

21.11 Match the following three terms, (1) enzyme–substrate complex, (2) enzyme, and (3) substrate, with these phrases:
  a. has a tertiary structure that recognizes the substrate
  b. the combination of an enzyme with the substrate
  c. has a structure that fits the active site of an enzyme

21.12 Match the following three terms, (1) active site, (2) lock-and-key model, and (3) induced-fit model, with these phrases:
  a. the portion of an enzyme where catalytic activity occurs

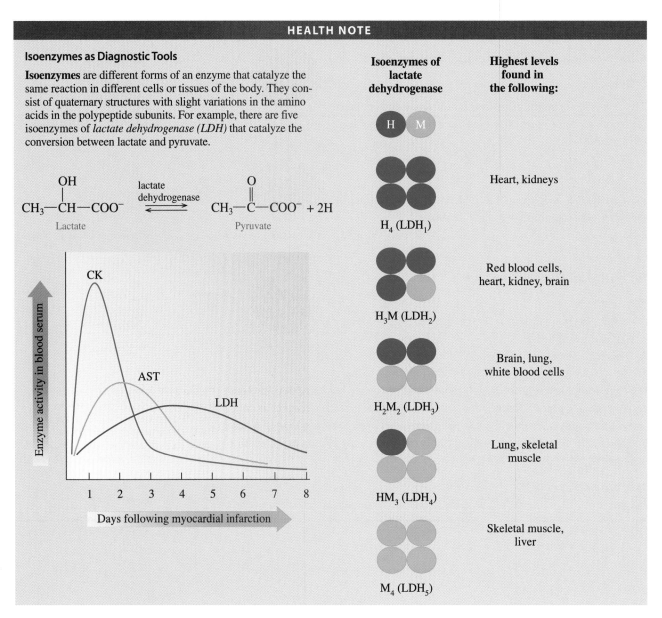

HEALTH NOTE

Isoenzymes as Diagnostic Tools

Isoenzymes are different forms of an enzyme that catalyze the same reaction in different cells or tissues of the body. They consist of quaternary structures with slight variations in the amino acids in the polypeptide subunits. For example, there are five isoenzymes of *lactate dehydrogenase (LDH)* that catalyze the conversion between lactate and pyruvate.

Isoenzymes of lactate dehydrogenase | **Highest levels found in the following:**

H_4 (LDH$_1$) — Heart, kidneys

H_3M (LDH$_2$) — Red blood cells, heart, kidney, brain

H_2M_2 (LDH$_3$) — Brain, lung, white blood cells

HM_3 (LDH$_4$) — Lung, skeletal muscle

M_4 (LDH$_5$) — Skeletal muscle, liver

Figure 21.4 After sucrose binds to sucrase at the active site, it is properly aligned for the hydrolysis reaction by the active site. The monosaccharide products dissociate from the active site and the enzyme is ready to bind to another sucrose molecule.

Q Why does the enzyme-catalyzed hydrolysis of sucrose go faster than the hydrolysis of sucrose in the chemistry laboratory?

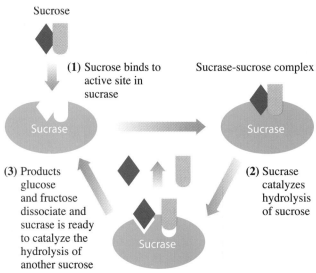

Sucrose

(1) Sucrose binds to active site in sucrase

Sucrase-sucrose complex

(2) Sucrase catalyzes hydrolysis of sucrose

(3) Products glucose and fructose dissociate and sucrase is ready to catalyze the hydrolysis of another sucrose

HEALTH NOTE (CONTINUED)

Each LDH isoenzyme contains a mix of polypeptide subunits, M and H. In the liver and muscle, lactate is converted to pyruvate by a LDH_5 isoenzyme with M subunits designated M_4. In the heart, the same reaction is catalyzed by a LDH_1 isoenzyme (H_4) containing four H subunits. Different combinations of the M and H subunits are found in the LDH isoenzymes of the brain, red blood cells, kidney, and white blood cells.

The different forms of an enzyme allow a medical diagnosis of damage or disease to a particular organ or tissue. In healthy tissues, isoenzymes function within the cells. However, when a disease damages a particular organ, cells die, which releases cell contents including the isoenzymes into the blood. Measurements of the elevated levels of specific isoenzymes in the blood serum help to identify the disease and its location in the body. For example, an elevation in the serum LDH_5, which is the M_4 isoenzyme of lactate dehydrogenase, indicates liver damage or disease. When a myocardial infarction (MI), or heart attack, damages the cells in heart muscle, an increase in the level of LDH_1 (H_4) isoenzyme is detected in the blood serum. (See Table 21.3.)

Another isoenzyme used diagnostically is creatine kinase (CK), which consists of two types of polypeptide subunits. One subunit (B) is prevalent in the brain and the other predominates in skeletal muscle (M). Normally only the CK_3 is present in low amounts in the blood serum. However, in a patient who has suffered an MI, the levels of CK_2 will be elevated soon after the heart attack. Table 21.4 lists some enzymes used to diagnose tissue damage and diseases of certain organs.

Table 21.3 Isoenzymes of Lactate Dehydrogenase and Creatinine Kinase

Isoenzyme	Abundant in	Subunits
Lactate Dehydrogenase (LDH)		
LDH_1	Heart, kidneys	H_4
LDH_2	Red blood cells, heart, kidney, brain	MH_3
LDH_3	Brain, lung, white blood cells	M_2H_2
LDH_4	Lung, skeletal muscle	M_3H
LDH_5	Skeletal muscle, liver	M_4
Creatinine Kinanse (CK)		
CK_1	Brain, lung	BB
CK_2	Heart	MB
CK_3	Skeletal muscle, red blood cells	MM

Table 21.4 Serum Enzymes Used in Diagnosis of Tissue Damage

Condition	Diagnostic Enzymes Elevated
Heart attack, or liver disease (cirrhosis, hepatitis)	Lactate dehydrogenase (LDH) Aspartate transaminase (AST)
Heart attack	Creatine kinase (CK)
Hepatitis	Alanine transaminase (ALT)
Liver (carcinoma) or bone disease (rickets)	Alkaline phosphatase (ALP)
Pancreatic disease	Amylase, cholinesterase, lipase (LPS)
Prostate carcinoma	Acid phosphatase (ACP)

    b. an active site that adapts to the shape of a substrate

    c. an active site that has a rigid shape

21.13 a. Write an equation that represents an enzyme-catalyzed reaction.

    b. How is the active site different from the whole enzyme structure?

21.14 a. Why does an enzyme speed up the reaction of a substrate?

    b. After the products have formed, what happens to the enzyme?

21.15 What are isoenzymes?

21.16 How is the LDH isoenzyme in the heart different from LDH isoenzyme in the liver?

21.17 A patient arrives in emergency complaining of chest pains. What enzymes would you test for in the blood serum?

21.18 A patient who is an alcoholic has elevated levels of LDH and AST. What condition might be indicated?

21.4 Factors Affecting Enzyme Activity

LEARNING GOAL

Describe the effect of temperature, pH, concentration of enzyme, and concentration of substrate on enzyme activity.

The **activity** of an enzyme describes how fast an enzyme catalyzes the reaction that converts a substrate to product. This activity is strongly affected by reaction conditions, which include the temperature, pH, concentration of the substrate, and concentration of the enzyme.

Temperature

Enzymes are very sensitive to temperature. At low temperatures, most enzymes show little activity because there is not a sufficient amount of energy for the catalyzed reaction to take place. At higher temperatures, enzyme activity increases as reacting molecules move faster to cause more collisions with enzymes. Enzymes are most active at **optimum temperature,** which is 37°C or body temperature for most enzymes. (See Figure 21.5.) At temperatures above 50°C, the tertiary structure and thus the shape of most proteins is destroyed causing a loss in enzyme activity. For this reason, equipment in hospitals and laboratories is sterilized in autoclaves where the high temperatures denature the enzymes in harmful bacteria.

pH

Enzymes are most active at their **optimum pH,** the pH that maintains the proper tertiary structure of the protein. (See Figure 21.6.) A pH value above or below the optimum pH causes a change in the three-dimensional structure of the enzyme that disrupts the active site. As a result the enzyme cannot bind substrate properly and no reaction occurs.

Enzymes in most cells have optimum pH values at physiological pH values around 7.4. However, enzymes in the stomach have a low optimum pH because they hydrolyze proteins at the acidic pH in the stomach. For example, pepsin, a digestive enzyme in the stomach has an optimum pH of 2. Between meals, the pH in the stomach is 4 or 5 and pepsin shows little or no digestive activity. When food enters the stomach, the secretion of HCl lowers the pH to about 2, which activates pepsin.

If small changes in pH are corrected, an enzyme can regain its structure and activity. However, large variations from optimum pH permanently destroy the structure of the enzyme. Table 21.5 lists the optimum pH values for selected enzymes.

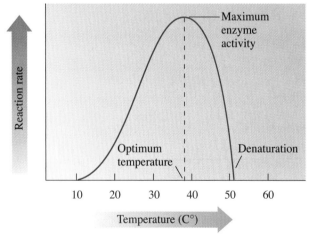

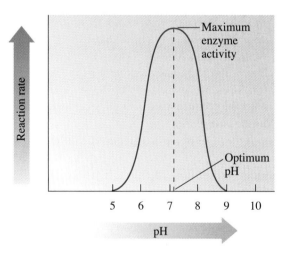

Figure 21.5 An enzyme attains maximum activity at its optimum temperature, usually 37°C. Lower temperatures slow the rate of reaction and temperatures above 50°C cause denaturation of the enzyme protein and loss of catalytic activity.

Q Why is 37°C the optimum temperature for many enzymes?

Figure 21.6 Enzymes are most active at their optimum pH. At a higher or lower pH, denaturation of the enzyme causes a loss of catalytic activity.

Q Why does the digestive enzyme pepsin have an optimum pH of 2?

Table 21.5 Optimum pH for Selected Enzymes

Enzyme	Location	Substrate	Optimum pH
Pepsin	Stomach	Peptide bonds	2
Urease	Liver	Urea	5
Sucrase	Small intestine	Sucrose	6.2
Pancreatic amylase	Pancreas	Amylose	7
Trypsin	Small intestine	Peptide bonds	8
Arginase	Liver	Arginine	9.7

Enzyme and Substrate Concentration

In any catalyzed reaction, the substrate must first bind with the enzyme to form the substrate–enzyme complex. Increasing the enzyme concentration when the substrate concentration remains constant increases the rate of the catalyzed reaction and thus enzyme activity. At higher concentrations more enzyme molecules are available to bind and catalyze the reaction of substrate molecules. When the enzyme concentration is increased to twice the initial concentration, the rate of the catalyzed reaction is twice as fast. If the enzyme concentration is increased to three times the initial enzyme concentration, the rate of reaction will increase also to three times as fast. There is a direct relationship between the enzyme concentration and enzyme activity. (See Figure 21.7a).

When enzyme concentration is kept constant, increasing the substrate concentration increases the rate of the catalyzed reaction as long as there are more enzyme molecules present than substrate molecules. At some point an increase in substrate concentration saturates the enzyme. With all the available enzyme molecules bonded to substrate, the rate of the catalyzed reaction reaches its maximum. Adding more substrate molecules cannot increase the rate further. (See Figure 21.7b.)

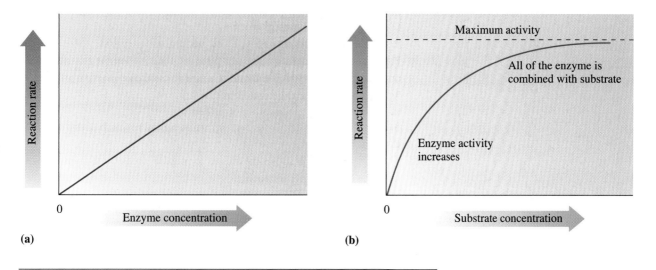

Figure 21.7 (a) The rate of reaction increases when the enzyme concentration increases at a fixed substrate concentration. **(b)** Increasing the substrate concentration increases the rate of reaction until the enzyme molecules are saturated with substrate.

Q What happens to the rate of reaction when substrate saturates enzyme?

SAMPLE PROBLEM 21.4

Factors Affecting Enzymatic Activity

Describe what effect the changes in parts a and b would have on the rate of the reaction catalyzed by urease.

$$H_2N-\overset{\overset{\displaystyle O}{\|}}{C}-NH_2 \;+\; H_2O \;\xrightarrow{\text{Urease}}\; 2NH_3 \;+\; CO_2$$
$$\text{Urea}$$

a. increasing the urea concentration
b. lowering the temperature to 10°C

Solution

a. An increase in urea concentration will increase the rate of reaction until all of the enzyme molecules are bound to the urea substrate. No further increase in rate occurs.
b. Because 10°C is lower than the optimum temperature of 37°C, the lower temperature will decrease the rate of the reaction.

Study Check

If urease has an optimum pH of 5, what is the effect of lowering the pH to 3?

QUESTIONS AND PROBLEMS

Factors Affecting Enzyme Action

21.19 Trypsin, a peptidase that hydrolyzes polypeptides, functions in the small intestine at an optimum pH of 8. How is the rate of a trypsin-catalyzed reaction affected by each of the following conditions?
　　a. lowering the concentration of polypeptides
　　b. changing the pH to 3
　　c. running the reaction at 75°C　　**d.** adding more trypsin

21.20 Pepsin, a peptidase that hydrolyzes proteins, functions in the stomach at an optimum pH of 2. How is the rate of a pepsin-catalyzed reaction affected by each of the following conditions?
　　a. increasing the concentration of proteins
　　b. changing the pH to 5　　　　**c.** running the reaction at 0°C
　　d. using less pepsin

21.21 The following graph shows the curves for pepsin, urease, and trypsin. Estimate the optimum pH for each.

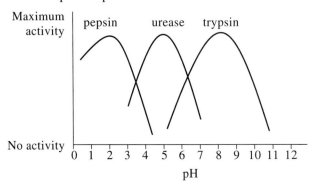

21.22 Refer to the graph in problem 21.21 to determine if the reaction rate in each condition will be at the optimum rate or not.
    a. trypsin, pH 5   **b.** urease, pH 5   **c.** pepsin, pH 4
    d. trypsin, pH 8   **e.** pepsin, pH 2

21.5 Enzyme Inhibition

Many kinds of molecules called **inhibitors** cause enzymes to lose catalytic activity. Although inhibitors act differently, they all prevent the active site from binding with a substrate. Some inhibitors cause a **reversible inhibition,** which means that the enzyme regains activity when the inhibitor dissociates from the enzyme. In an **irreversible inhibition,** an inhibitor bonds covalently with an enzyme and cannot be removed, which makes the loss of enzyme activity irreversible.

Reversible Competitive Inhibition

Reversible inhibition can be competitive or noncompetitive. In competitive inhibition, an inhibitor competes for the active site, whereas in noncompetitive inhibition, the inhibitor acts on a site that is not the active site.

A **competitive inhibitor** has a structure that is so similar to the substrate it can bond to the enzyme just like the substrate. Thus the competitive inhibitor competes with the substrate for the active site on the enzyme. As long as the inhibitor occupies the active site, the substrate cannot bind to the enzyme and no reaction takes place. (See Figure 21.8.)

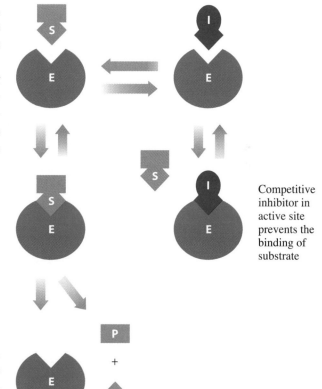

Competitive inhibitor in active site prevents the binding of substrate

$$S \rightleftharpoons ES \longrightarrow E + P \quad \text{Favored by increasing [S]}$$

$$+ \qquad \text{Enzyme–substrate complex}$$

$$E$$

$$+$$

$$I \rightleftharpoons EI \qquad \text{Favored when [S] is low}$$

$$\text{Enzyme–inhibitor complex}$$

As long as the concentration of the inhibitor is substantial, there is a loss of enzyme activity. However, increasing the substrate concentration displaces more of the inhibitor molecules. As more enzyme molecules bind to substrate (ES), enzyme activity is regained.

Malonate is a competitive inhibitor of the enzyme succinate dehydrogenase. Because malonate has a structure similar to the substrate, the two substances compete for the active site on the dehydrogenase. As long as the inhibitor occupies the active site on the enzyme, no reaction occurs. When more succinate is added, more active sites will fill with substrate, and there will be less inhibition.

Figure 21.8 With a structure similar to the substrate for an enzyme, a competitive inhibitor also fits the active site and competes with the substrate when both are present.

Q **Why does increasing the substrate concentration reverse the inhibition by a competitive inhibitor?**

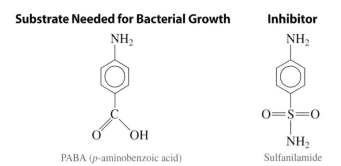

Enzyme + substrate ⟶ enzyme–substrate complex ⟶ product

Enzyme + inhibitor ⟶ enzyme–inhibitor complex ⟶ no product

Some bacterial infections are treated with competitive inhibitors called antimetabolites. Sulfanilamide, one of the first sulfa drugs, competes with PABA (*p*-aminobenzoic acid), which is an essential substance (metabolite) in the growth cycle of bacteria.

Substrate Needed for Bacterial Growth

PABA (*p*-aminobenzoic acid)

Inhibitor

Sulfanilamide

Reversible Noncompetitive Inhibition

The structure of a **noncompetitive inhibitor** does not resemble the substrate and does not compete for the active site. Instead a noncompetitive inhibitor binds to a site on the enzyme that is not the active site. When the noncompetitive inhibitor is bonded to the enzyme, the shape of the enzyme is distorted. Inhibition occurs because the substrate cannot fit in the active site, or it does not fit properly. Without the proper alignment of substrate with the amino acid side groups, no catalysis can take place. (See Figure 21.9.)

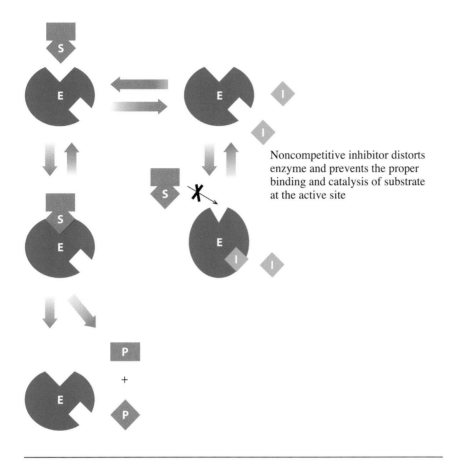

Noncompetitive inhibitor distorts enzyme and prevents the proper binding and catalysis of substrate at the active site

Figure 21.9 A noncompetitive inhibitor binds to an enzyme at a site other than the active site, which distorts the enzyme and prevents the proper binding and catalysis of the substrate at the active site.

Q **Why won't an increase in substrate concentration reverse the inhibition by a noncompetitive inhibitor?**

Because a noncompetitive inhibitor is not competing for the active site, the addition of more substrate does not reverse this type of inhibition. However, enzyme activity can be regained by lowering the concentration of the noncompetitive inhibitor making more enzyme molecules available. Examples of noncompetitive inhibitors are the heavy metal ions Pb^{2+}, Ag^+, and Hg^{2+} that bond with amino acid side groups such as $-COO^-$, or $-OH$. Catalytic activity is restored when chemical reagents remove the inhibitors.

WEB TUTORIAL
Enzyme Inhibition

Irreversible Inhibition

In irreversible inhibition, a molecule causes an enzyme to lose all enzymatic activity. Most irreversible inhibitors are toxic substances that destroy enzymes. Usually an irreversible inhibitor forms a covalent bond with an amino acid side group within the active site, which prevents the substrate from entering the active site or prevents catalytic activity.

Insecticides and nerve gases act as irreversible inhibitors of acetylcholinesterase, an enzyme needed for nerve conduction. The compound DFP (diisopropyl fluorophosphate) forms a covalent bond with the side chain $-CH_2OH$ of serine in the active site. When acetylcholinesterase is inhibited, the transmission of nerve impulses is blocked, and paralysis occurs.

DFP (Diisopropyl fluorophosphate)   Serine covalently bonded to DFP

Antibiotics produced by bacteria, mold, or yeast are irreversible inhibitors used to inhibit bacterial growth. For example, penicillin inhibits an enzyme needed for the formation of cell walls in bacteria, but not human cell membranes. With an incomplete cell wall, bacteria cannot survive, and the infection is stopped. However, some bacteria are resistant to penicillin because they produce penicillinase, an enzyme that breaks down penicillin. Over the years, derivatives of penicillin to which bacteria have not yet become resistant have been produced. Some irreversible enzyme inhibitors are listed in Table 21.6.

Table 21.6   Selected Irreversible Enzyme Inhibitors

Name	Structure	Natural/Synthetic Source	Inhibitory Action
Cyanide	CN^-	Bitter almonds	Bonds to metal ions in enzymes in the electron transport chain
Sarin	(structure)	Nerve gas	Similar to DFP
Parathion	(structure)	Insecticide	Similar to DFP
Penicillin	(structure)	*Penicillium* fungus	Inhibits enzymes that build cell walls in bacteria

R Groups for Penicillin Derivatives

SAMPLE PROBLEM 21.5

Enzyme Inhibition

State the type of reversible inhibition in the following:

a. The inhibitor has a structure that is similar to the substrate.

b. This inhibitor binds to the surface of the enzyme, changing its shape in such a way that it cannot bind to substrate.

Solution

a. competitive inhibition

b. noncompetitive inhibition

Study Check

Hydrogen cyanide (HCN) forms covalent bonds with catalase, an enzyme that contains iron (Fe^{3+}). What type of inhibitor is HCN?

QUESTIONS AND PROBLEMS

Enzyme Inhibition

21.23 Indicate whether the following describe a reversible competitive or a reversible noncompetitive enzyme inhibitor:
   a. The inhibitor has a structure similar to the substrate.
   b. The effect of the inhibitor cannot be reversed by adding more substrate.
   c. The inhibitor competes with the substrate for the active site.
   d. The structure of the inhibitor is not similar to the substrate.
   e. The addition of more substrate reverses the inhibition.

21.24 Oxaloacetate is an inhibitor of succinate dehydrogenase:

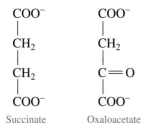

Succinate   Oxaloacetate

   a. Would you expect oxaloacetate to be a reversible competitive or a noncompetitive inhibitor? Why?
   b. Would oxaloacetate bind to the active site or elsewhere on the enzyme?
   c. How would you reverse the effect of the inhibitor?

21.25 Methanol and ethanol are oxidized by alcohol dehydrogenase. In methanol poisoning, a high concentration of ethanol is given intravenously as an antidote.
   a. Compare the structure of methanol and ethanol.
   b. Would ethanol compete for the active site or bind to a different site?
   c. Would ethanol be a competitive or noncompetitive inhibitor of methanol oxidation?

21.26 In humans, the antibiotic amoxycillin (a type of penicillin) is used to treat certain bacterial infections.
   a. Does the antibiotic inhibit enzymes in humans?
   b. Why does the antibiotic kill bacteria, but not humans?
   c. Are antibiotics reversible or irreversible inhibitors?

WEB TUTORIAL
Enzyme Inhibition

21.6 Regulation of Enzyme Activity

In any enzyme-catalyzed reaction, compounds are produced only in the amounts and at the times they are needed in the cell. This means that the rate of a catalyzed reaction must be controlled so it can speed up when more molecules of a compound are needed and slow down when an accumulation of that compound occurs. There are several factors in the cell that regulate enzyme activity.

Zymogens

Although most enzymes are active as soon as they are synthesized and acquire their tertiary structure, some are produced as inactive precursors called **zymogens** or *proenzymes*. Zymogens, which contain longer protein chains, are activated by the removal of peptide sections from the protein. Zymogens are often produced in an organ where they are stored and then transported to where they are needed, at which time they are activated.

Insulin

The protein hormone insulin is synthesized as inactive proinsulin. Recall that insulin contains two polypeptide chains (Chapter 20) linked by disulfide bonds. In proinsulin, the two chains are connected by a polypeptide of 33 amino acids. This peptide section is removed to form the active insulin hormone. (See Figure 21.10.)

Digestive Enzymes

Several of the digestive enzymes including trypsinogen, chymotrypsinogen, and procarboxypeptidase are produced as inactive forms and stored in the pancreas. After food is ingested, it reaches the small intestine. Hormones trigger the release of the zymogens from the pancreas. In the small intestine, the zymogens are converted into active forms by proteases that remove peptide sections from their protein chains. The change in tertiary structure activates the enzyme. For example, an enzyme called enteropeptidase removes a hexapeptide from trypsinogen to give active trypsin. Trypsin in turn cleaves peptide sections from chymotrypsinogen to

Figure 21.10 The removal of a peptide with 33 amino acids converts the zymogen proinsulin to active insulin.

Q Why is proinsulin called a zymogen?

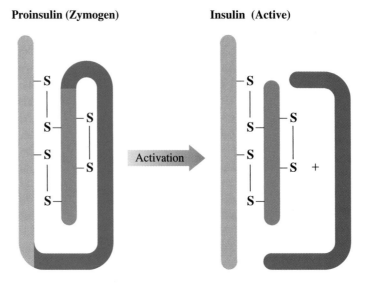

form active chymotrypsin and from procarboxypeptidase to yield active carboxypeptidase.

$$\text{Chymotrypsinogen} \xrightarrow{\text{trypsin}} \text{Chymotrypsin} + 2 \text{ dipeptides}$$

If the digestive zymogens were active in the pancreas, their catalytic action as proteases would digest the proteins of the pancreas. This can lead to conditions such as pancreatitis, which is an inflammation of the pancreas. Another zymogen, pepsinogen, is produced in the gastric mucosal cells that line the stomach. As food enters the stomach, HCl is secreted. Low pH levels cleave a peptide containing 42 amino acids from the pepsinogen protein to form pepsin, which digests proteins.

$$\text{Pepsinogen} \xrightarrow{\text{H}^+} \text{pepsin} + \text{peptides and amino acids}$$

Most protein hormones such as insulin, enzymes involved in digestion, and enzymes that catalyze blood clotting are initially synthesized as zymogens. (See Table 21.7.)

Table 21.7   Example of Zymogens and Their Active Forms

Zymogen	Produced in	Activated in	Active Form
Proinsulin	Pancreas	Blood	Insulin
Chymotrypsinogen	Pancreas	Small intestine	Chymotrypsin
Pepsinogen	Gastric mucosa	Stomach	Pepsin
Trypsinogen	Pancreas	Small intestine	Trypsin
Fibrinogen	Blood	Damaged tissues	Fibrin
Prothrombin	Blood	Damaged tissues	Thrombin

Allosteric Enzymes

Certain enzymes known as **allosteric enzymes** are capable of binding a regulator molecule that is different from the substrate. The binding of the regulator causes a change in the shape of the enzyme and therefore in the active site. There are both positive and negative regulators. A **positive regulator** speeds up a reaction by causing a change in the shape of the active site that permits the substrate to bind more effectively. A **negative regulator** slows down the rate of catalysis by preventing the proper binding of the substrate. In **feedback control,** the end product acts as a negative regulator. When the end product is produced in sufficient amounts for the cell, some binds to the first enzyme (E_1), which is an allosteric enzyme. (See Figure 21.11.) By inhibiting the reaction of the initial substrate, no intermediate compounds are produced for the other enzymes in the reaction pathway. The entire enzyme-catalyzed reaction sequence shuts down.

Eventually the concentration of end product becomes too low, which causes the end product inhibitor to dissociate from the allosteric enzyme (E_1). As its shape returns to its active form, the catalysis of initial substrate begins once again. Through feedback control, the catalysis of the initial substrate undergoes reaction only when the end product is needed somewhere in the cell. There is no accumulation of the end product, which conserves the materials needed in other reactions.

Let's look at the feedback control in a reaction pathway with five enzymes that converts the amino acid threonine to isoleucine, another amino acid.

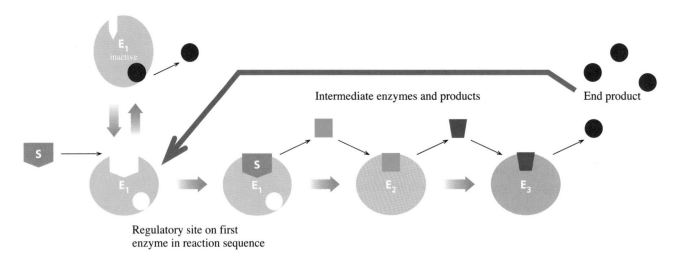

Figure 21.11 In feedback control, end product binds to a regulatory site on the first enzyme in the reaction pathway, which prevents the formation of all intermediate compounds needed in the synthesis of the end product.

Q **Why don't the intermediate enzymes in a reaction sequence have regulatory sites?**

When isoleucine begins to accumulate in the cell, it binds with the first enzyme (E_1) threonine deaminase in the pathway. The binding of isoleucine changes the shape of the deaminase, which prevents the substrate threonine from binding with the active site. The entire reaction pathway is turned off. None of the intermediate products from the other enzymes in the pathway can inhibit the first enzyme. As isoleucine is utilized in the cell, its concentration decreases, which causes the threonine deaminase to release the end product inhibitor. The tertiary shape of the deaminase returns to its active form and the reaction sequence once again converts threonine to isoleucine.

SAMPLE PROBLEM 21.6

Enzyme Regulation

How is the rate of a reaction sequence regulated in feedback control?

Solution

When the end product of a reaction sequence is produced at sufficient levels for the cell, some product molecules bind to the first enzyme in the sequence, which shuts down all the reactions that follow and stops the production of end product.

Study Check

Why is pepsin, a digestive enzyme, produced as a zymogen?

Regulation of Enzyme Activity

21.27 Why are many of the enzymes that act on proteins synthesized as zymogens?

21.28 The zymogen trypsinogen produced in the pancreas is activated in the small intestine where it catalyzes the digestion and hydrolysis of proteins. Explain how the activation of the zymogen while still in the pancreas can lead to an inflammation of the pancreas called pancreatitis.

21.29 In feedback inhibition, how does the end product of a reaction sequence regulate enzyme activity?

21.30 Why don't the second or third enzymes in a reaction sequence function as regulatory enzymes?

21.31 How does an allosteric enzyme function as a regulatory enzyme?

21.32 What is the difference between a negative regulator and a positive regulator?

21.33 Indicate if the following statements describe (1) a zymogen, (2) a positive regulator, (3) a negative regulator, or (4) allosteric enzyme.
   a. It slows down a reaction, but its shape is different from that of the substrate.
   b. It is the first enzyme in a sequence of reactions leading to an end product.
   c. It is produced as an inactive enzyme.

21.34 Indicate if the following statements describe (1) a zymogen, (2) a positive regulator, (3) a negative regulator, or (4) allosteric enzyme.
   a. It is activated when a peptide section is removed from its protein chain.
   b. It speeds up a reaction, but it is not the substrate.
   c. When it binds to end product, it stops the formation of more end product.

21.7 Enzyme Cofactors

LEARNING GOAL
Describe the types of cofactors found in enzymes.

Enzymes are known as **simple enzymes** when their functional forms consist only of proteins with tertiary structures. However, many enzymes require small molecules or metal ions called **cofactors** to catalyze reactions properly. When the cofactor is a small organic molecule, it is known as a **coenzyme.** If an enzyme requires a cofactor, neither the protein structure nor the cofactor alone has catalytic activity.

Forms of Active Enzymes

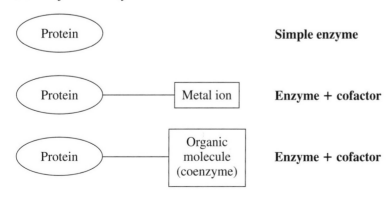

Protein — Simple enzyme

Protein — Metal ion — Enzyme + cofactor

Protein — Organic molecule (coenzyme) — Enzyme + cofactor

Metal Ions

Many enzymes must contain a metal ion to carry out their catalytic activity. The metal ions are bonded to one or more of the amino acid side chains. The metal ions from the minerals that we obtain from foods in our diet have various functions in catalysis. Ions such as Fe^{2+} and Cu^{2+} are used by oxidases where they lose or gain electrons in oxidation and reduction reactions. Other metals ions such as Zn^{2+} stabilize the amino acid side chains during hydrolysis reactions. Some metal cofactors required by enzymes are listed in Table 21.8.

Let's look at an example of a metal ion in an enzyme-catalyzed reaction. The enzyme carboxypeptidase A cleaves the C terminal amino acid of a protein when that amino acid has a bulky hydrophobic or aromatic side chain. (See Figure 21.12.) With the substrate in the active site, the Zn^{2+} helps to stabilize the negative charge on the oxygen atom of the carbonyl group and promotes the hydrolysis of the peptide bond.

Table 21.8   Enzymes and the Metal Ions Required as Cofactors

Metal Ion Cofactor	Function	Enzyme
Cu^{2+}	Oxidation-reduction	Cytochrome oxidase
Fe^{2+}/Fe^{3+}	Oxidation-reduction	Catalase
	Oxidation-reduction	Cytochrome oxidase
Zn^{2+}	Used with NAD^+	Alcohol dehydrogenase
		Carbonic anhydrase
		Carboxypeptidase A
Mg^{2+}	Hydrolyzes phosphate esters	Glucose-6-phosphatase
Mn^{2+}	Removes electrons	Arginase
Ni^{2+}	Hydrolyzes amides	Urease

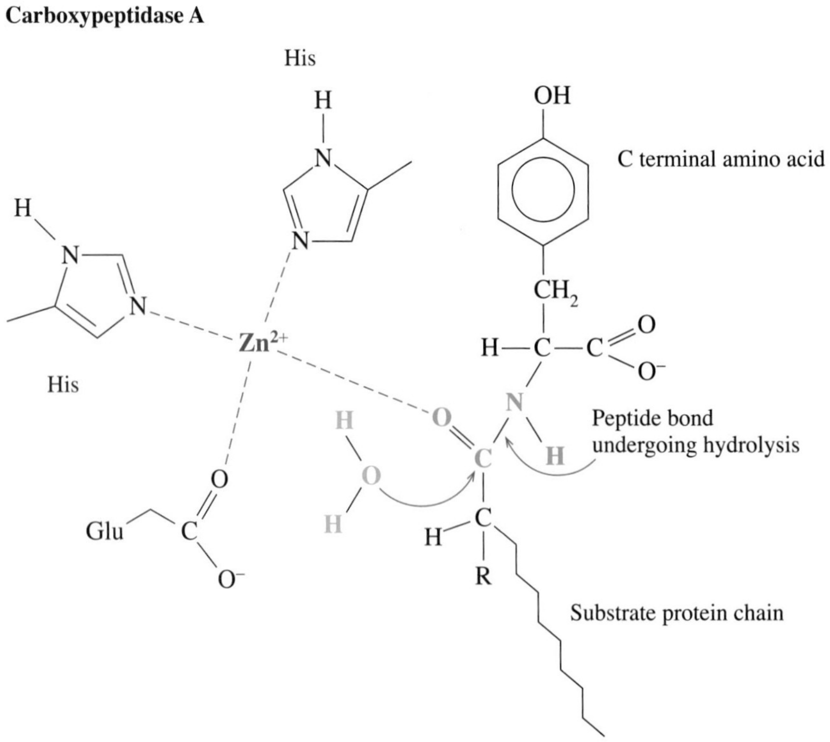

Figure 21.12 A Zn^{2+} cofactor aids in the hydrolysis of the peptide bond of a bulky C terminal amino acid by helping to stabilize the carbonyl oxygen.

Q When would the Zn^{2+} be utilized as a cofactor by other enzymes?

Carboxypeptidase A

Cofactors

Indicate whether each of the following enzymes are active as a simple enzyme or require a cofactor.

a. a polypeptide that needs Mg^{2+} for catalytic activity

b. an active enzyme composed only of a polypeptide chain

c. an enzyme that consists of a quaternary structure attached to vitamin B_6

Solution

a. The enzyme requires a cofactor.

b. An active enzyme that consists of only a polypeptide chain is a simple enzyme.

c. The enzyme requires a cofactor.

Study Check

Which of the nonprotein portions of the enzymes in Sample Problem 21.7 is a coenzyme?

Enzyme Cofactors

21.35 Is the enzyme described in each of the following statements a simple enzyme or one that requires a cofactor?
  a. requires vitamin B_1 (thiamine)
  b. needs Zn^{2+} for catalytic activity
  c. its active form consists of two polypeptide chains

21.36 Is the enzyme described in each of the following statements a simple enzyme or one that requires a cofactor?
  a. requires vitamin B_2 (riboflavin)
  b. its active form is composed of 155 amino acids
  c. uses Cu^{2+} during catalysis

21.8 Vitamins and Coenzymes

> **LEARNING GOAL**
>
> Describe the functions of the water-soluble and fat-soluble vitamins.

Vitamins are organic molecules that are essential for normal health and growth. They are required in trace amounts and must be obtained from the diet because they are not synthesized in the body. Before vitamins were discovered, it was known that lime juice prevented the disease scurvy in sailors and that cod liver oil could prevent rickets. In 1912, scientists found that, in addition to carbohydrates, fats, and proteins, certain other factors called vitamins must be obtained from the diet.

Classification of Vitamins

Vitamins are classified into two groups by solubility: water-soluble and fat-soluble. **Water-soluble vitamins** have polar groups such as —OH and —COOH, which make them soluble in the aqueous environment of the cells. The **fat-soluble vita-**

mins are nonpolar compounds, which are soluble in the fat (lipid) components of the body such as fat deposits and cell membranes.

Most water-soluble vitamins are not stored in the body and excess amounts are eliminated in the urine each day. Therefore, the water-soluble vitamins must be in the foods of our daily diets. Because many water-soluble vitamins are easily destroyed by heat, oxygen, and ultraviolet light, care must be taken in food preparation, processing, and storage. In the 1940s, a Committee on Food and Nutrition of the National Research Council began to recommend dietary enrichment of cereal grains. It was known that refining grains such as wheat caused a loss of vitamins. Thiamine (B_1), riboflavin (B_2), niacin, and iron were in the first group of added nutrients recommended. We now see the Recommended Daily Allowance (RDA) for many vitamins and minerals on food product labels such as cereals and bread.

The water-soluble vitamins are required by many enzymes as cofactors to carry out certain aspects of catalytic action. (See Table 21.9.) The coenzymes do not remain bonded to a particular enzyme, but are used over and over again by differ-

Table 21.9   Vitamins and Function

Water-Soluble Vitamins	Coenzyme	Function
Thiamine (vitamin B_1)	Thiamine pyrophosphate	Decarboxylation
Riboflavin (vitamin B_2)	Flavin adenine dinucleotide (FAD); Flavin mononucleotide (FMN)	Electron transfer
Niacin (vitamin B_3)	Nicotinamide adenine dinucleotide (NAD^+); Nicotinamide adenine dinucleotide phosphate ($NADP^+$)	Oxidation–reduction
Pantothenic acid (vitamin B_5)	Coenzyme A	Acetyl group transfer
Pyridoxine (vitamin B_6)	Pyridoxal phosphate	Transamination
Cobalamin (vitamin B_{12})	Methylcobalamin	Methyl group transfer
Ascorbic acid (vitamin C)	Vitamin C	Collagen synthesis, healing of wounds
Biotin	Biocytin	Carboxylation
Folic acid	Tetrahydrofolate	Methyl group transfer
Fat-Soluble Vitamins		
Vitamin A	Formation of visual pigments; development of epithelial cells	
Vitamin D	Absorption of calcium and phosphate; deposition of calcium and phosphate in bone	
Vitamin E	Antioxidant; prevents oxidation of vitamin A and unsaturated fatty acids	
Vitamin K	Synthesis of prothrombin for blood clotting	

Figure 21.13 The active forms of many enzymes require the combination of the protein with a coenzyme.

Q What is the function of water-soluble vitamins in enzymes?

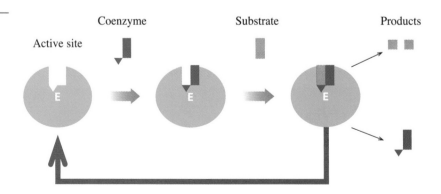

ent enzymes to facilitate an enzyme-catalyzed reaction. (See Figure 21.13.) Thus, only small amounts of coenzymes are required in the cells.

The fat-soluble vitamins A, D, E, and K are not involved as coenzymes, but they are important in processes such as vision, formation of bone, protection from oxidation, and proper blood clotting. Because the fat-soluble vitamins are stored in the body and not eliminated, it is possible to take too much, which could be toxic.

Water Soluble Vitamins

Thiamin (Vitamin B₁)

Thiamin (vitamin B₁) was the first B vitamin to be identified, thus the abbreviation B₁. The coenzyme thiamine pyrophosphate (TPP) is obtained when a synthetase adds two phosphate groups to the alcohol group of thiamine.

Thiamine (vitamin B₁)   Thiamine pyrophosphate (TPP)

The TPP coenzyme is involved in the decarboxylation reactions of α-keto carboxylic acids and reactions that cleave bonds to carbonyl carbons of α-hydroxyketones. A deficiency of thiamin may result in beriberi, which is characterized by fatigue, poor appetite, weight loss, nerve degeneration, and heart failure. The RDA (recommended daily allowance) for adults is about 2 mg. Dietary sources of thiamine include liver, yeast, whole grain bread, cereals, and milk.

Riboflavin (Vitamin B₂)

Riboflavin or vitamin B₂ is used to make the coenzymes flavin adenine dinucleotide (FAD) and flavin mononucleotide (FMN). The *ribo* part of the name comes from the sugar alcohol ritol in the *riboflavin* molecule. The coenzymes FAD and FMN are used by enzymes called *flavoenzymes* to catalyze oxidation–reduction reactions of carbohydrates, fats, and proteins. Riboflavin is needed for good vision and healthy skin and hair. The RDA for adults is 1.7 mg. Deficiency symptoms of riboflavin include dermatitis, dry skin, tongue inflammation, and cataracts. Riboflavin is provided in the diet by beef liver, chicken, eggs, green leafy vegetables, dairy foods, peanuts, and whole grains.

Riboflavin (vitamin B₂)   D-Ribitol

Niacin (Vitamin B₃)

Niacin or vitamin B_3 is a component of coenzymes nicotinamide adenine dinucleotide (NAD^+) and $NADP^+$, the phosphate form of NAD^+. The name niacin was assigned to the vitamin because its actual name, *nicotinic acid,* might be confused with nicotine. These coenzymes participate in oxidation–reduction, energy-production reactions in carbohydrate, fat, and protein metabolism. The RDA for niacin for adults is from 13–18 mg. A deficiency of niacin can result in *pellagra* characterized by dermatitis, muscle fatigue, loss of appetite, diarrhea, mouth sores, and mental disorders. In the diet, niacin is found in brewer's yeast, chicken, beef, fish, liver, brown rice, and whole grains.

Niacin (vitamin B_3)

Pantothenic Acid (Vitamin B₅)

Pantothenic acid or vitamin B_5 is part of a complex coenzyme known as coenzyme A. Coenzyme A transfers a two-carbon acetyl group from pyruvate to the citric acid cycle for the production of energy. The A in the name refers to the acetylation reaction. Coenzyme A is also involved in the conversion of amino acids and lipids to glucose as well as the synthesis of cholesterol and steroid hormones. The RDA of pantothenic acid for adults is 10 mg per day. A deficiency of pantothenic acid is characterized by fatigue, retarded growth, muscle cramps, and anemia. Dietary sources of pantothenic acid include salmon, beef, liver, eggs, brewer's yeast, whole grains, and fresh vegetables.

Pantothenic acid (vitamin B_5)

Pyridoxine (Vitamin B₆)

Pryidoxine and pyridoxal (an aldehyde) are forms of vitamin B_6 that are converted to the coenzymes pyridoxal phosphate (PLP). The PLP coenzyme participates in many enzyme-catalyzed reactions such as transaminations of amino acids and decarboxylations. The RDA for adults is 2 mg per day. Dietary sources include meat, liver, fish, nuts, whole grains, and spinach. Deficiency of pyridoxine may lead to dermatitis, fatigue, anemia, and retarded growth.

Pyridoxine (vitamin B_6)   Pyridoxal (vitamin B_6)   Pyridoxal phosphate (PLP)

Vitamin B₁₂ (cobalamin)

Cobalamin (Vitamin B₁₂)

Cobalamin or vitamin B₁₂ is a coenzyme consisting of four pyrrole rings with a cobalt ion (Co^{2+}) in the center. In its coenzyme form cobalamin participates in the transfer of methyl groups, molecular rearrangements, the formation of red blood cells, and the synthesis of acetylcholine for nerve cells. The RDA is about 3 μg per day for adults. A deficiency of cobalamin is found in pernicious anemia and results in malformed red blood cells, nerve damage, and some mental disorders. Dietary sources include liver, beef, kidney, chicken, fish such as salmon, halibut and tuna, yogurt, and milk. Because vitamin B₁₂ is not present in plants, strict vegetarians can experience symptoms of pernicious anemia.

Ascorbic Acid (Vitamin C)

Ascorbic acid or vitamin C has a simple chemical structure compared to most of the other vitamins. Its major function in the cells is its role in the synthesis of hydroxyproline and hydroxylysine, which are needed to form collagen. Collagen is the protein found in tendons, connective tissue, bone structure, and skin. The RDA for adults is 60 mg daily. A deficiency of ascorbic acid can lead to scurvy characterized by bleeding gums, weakened connective tissues, slow-healing wounds, and anemia. Dietary sources of ascorbic acid include blueberries, oranges, strawberries, cantaloupe, tomatoes, and vegetables especially red and green peppers, broccoli, cabbage, and spinach. (See Figure 21.14.)

Ascorbic acid (vitamin C)

Figure 21.14 Oranges, lemons, peppers, and tomatoes contain vitamin C or ascorbic acid.

Q What happens to excess vitamin C that may be consumed in a day?

Biotin

Biotin is a coenzyme for enzymes that transfer a carboxyl group in the reaction of pyruvate to oxaloacetate or acetyl-CoA to malonyl-CoA, which occurs in the synthesis of fatty acids. The RDA for adults is 0.3 mg per day. A deficiency of biotin can lead to dermatitis, loss of hair, fatigue, anemia, nausea, and depression. Dietary resources of biotin include liver, yeast, nuts, and eggs.

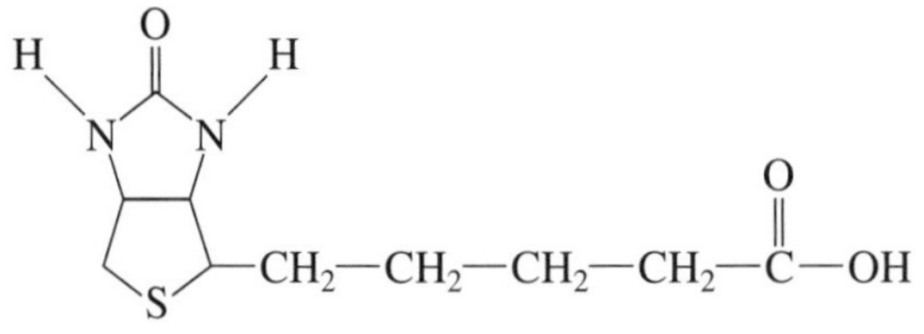

Biotin

Folic Acid (Folate)

Folic acid or folate is composed of a pyrimidine ring, *p*-aminobenzoic acid (PABA), and glutamate. The vitamin was discovered in the 1930s when people with a form of anemia were cured with extracts from liver or yeast. Folic acid is also found in spinach leaves, hence the name *folium,* Latin for leaf. In the cells, an enzyme called dihydrofolate reductase adds hydrogen atoms to the atoms in the heterocyclic ring of folate to yield the coenzyme tetrahydrofolate (THF). This coenzyme is used in reactions that transfer single-carbon groups and synthesize purines and pyrimidine to make DNA and RNA. It also plays a role with cobalamin in the production of red blood cells.

The RDA for folic acid is 0.4 mg for adults. A deficiency of folic acid can lead to abnormal red blood cells, anemia, intestinal-tract disturbances, loss of hair, growth impairment, depression, and spina bifida when there is a deficiency of folic acid during pregnancy. Dietary resources include green leafy vegetables, beans, meat, seafood, yeast, asparagus, and whole grain products, which are now enriched with folic acid.

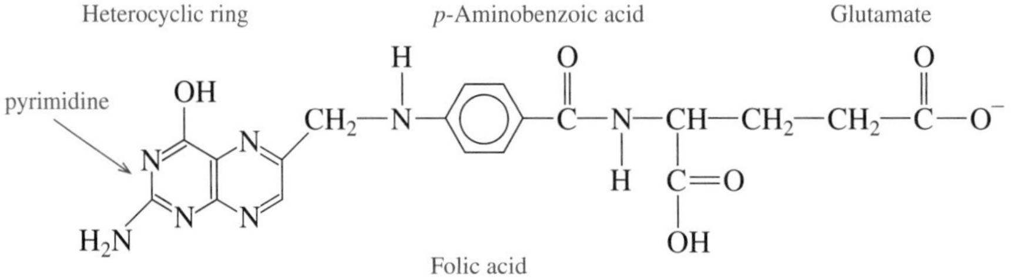

Folic acid

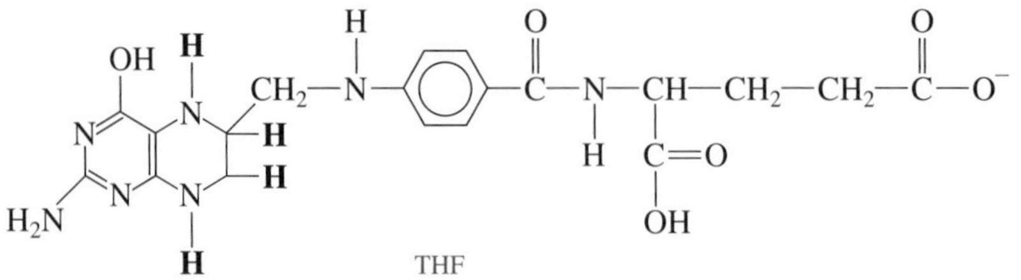

THF

Some compounds related to folate have been found that bring about remissions in people with leukemia. For example, 4-aminofolate referred to medically as *methotrexate* acts as a competitive inhibitor of the dihydrofolate reductase that forms THF. The growth of cells and tumor cells depends on THF to build purines and thymine. By inhibiting the reductase enzyme with methotrexate, THF cannot be produced, and the growth of tumor cells is blocked.

4- Aminofolate (methotrexate)

Fat-Soluble Vitamins

The fat-soluble vitamins are not used as coenzymes, but have important roles in vision, bone growth, and blood clotting.

Vitamin A

Vitamin A consists of three different forms depending on the oxidation of the functional group: retinol (alcohol), retinal (aldehyde), and retinoic acid (carboxylic acid). Vitamin A is obtained from animal sources in the diet or the β-carotenes of plants, which are converted to vitamin A in the liver. The retinol in the retinas of the eyes accumulates in the rod and cone cells where it plays a role in vision. Vitamin A is also involved in the synthesis of RNA and glycoproteins. The RDA for adults is 3 mg. A deficiency of vitamin A can cause night blindness, depress the immune response, and inhibit growth. Dietary resources of vitamin A include yellow and green fruits and vegetables such as apricots, broccoli, carrots, papaya, peaches, and spinach. (See Figure 21.15.)

Figure 21.15 Yellow and green fruits and vegetables contain vitamin A.

Q Why is vitamin A called a fat-soluble vitamin?

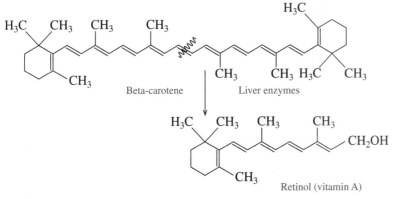

Beta-carotene   Liver enzymes

Retinol (vitamin A)

Vitamin D

The most prevalent form of vitamin D is vitamin D_3 or cholecalciferol. Technically, this is not a vitamin because it is not required in the diet. In skin, vitamin D_3 is synthesized from 7-dehydrocholesterol by the ultraviolet rays from sunlight. In regions of limited sunlight, vitamin D_3 is added to milk products to avoid vitamin D_3 deficiency. Its function in the body is to regulate the absorption of phosphorus and calcium during bone growth. The RDA for adults is 10 μg. A deficiency in vitamin D can result in weak bone structure, rickets in children, and osteomalacia in adults. The main source of vitamin D is sunlight with dietary sources including cod liver oil, egg yolk, and milk products enriched with vitamin D.

7-Dehydrocholesterol

UV (sunlight)

Vitamin D_3 (cholecalciferol)

Vitamin E

Vitamin E or tocopherol has a major role in cells as an antioxidant but not much is known about the mechanism of its activity. It appears to protect the cells in the body by removing damaging chemicals and by preventing the oxidation of unsaturated fatty acids. It has been used to reduce the damage to the retinas that can be caused by the high oxygen levels needed for respiration by premature infants. The RDA for adults is about 10 mg. A deficiency of vitamin E can cause hemolysis of red blood cells and anemia. Dietary resources of vitamin E include meat, whole grains, vegetables, and vegetable oils.

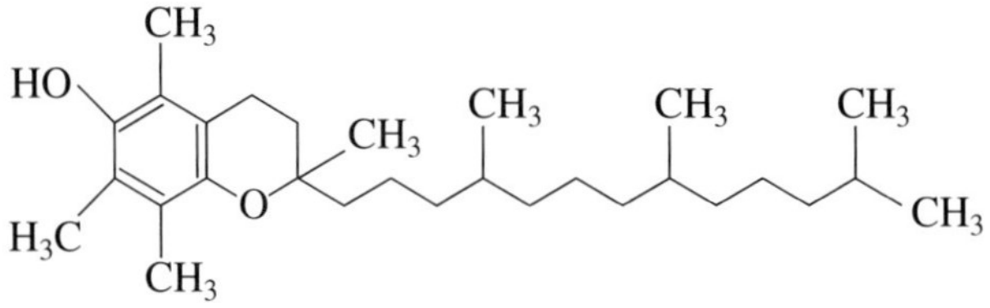

Vitamin E (tocopherol)

Vitamin K

Vitamin K_1, or phylloquinone, a substance found in plants, has a large saturated side chain. Vitamin K_2 or menaquinone, found in animals, has a very long unsaturated side chain. Vitamin K_2 takes part in the synthesis of zymogens needed for blood clotting. The RDA for vitamin K in adults is 80 μg. A deficiency of vitamin K can lead to bleeding from minor cuts, delayed clotting times, and bruising. Dietary resources of vitamin K include many foods such as liver, spinach, and cauliflower.

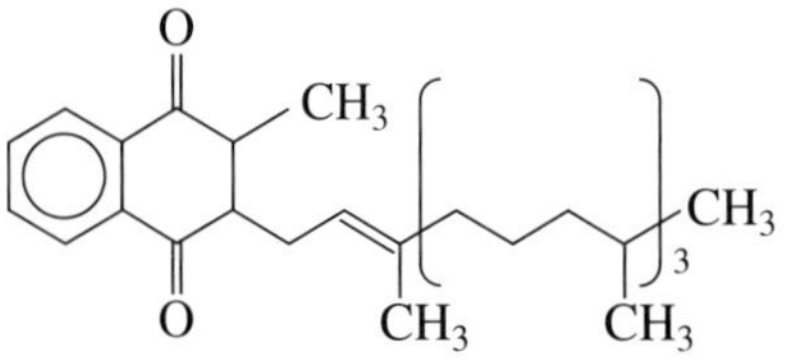

Vitamin K_1 (phylloquinone)

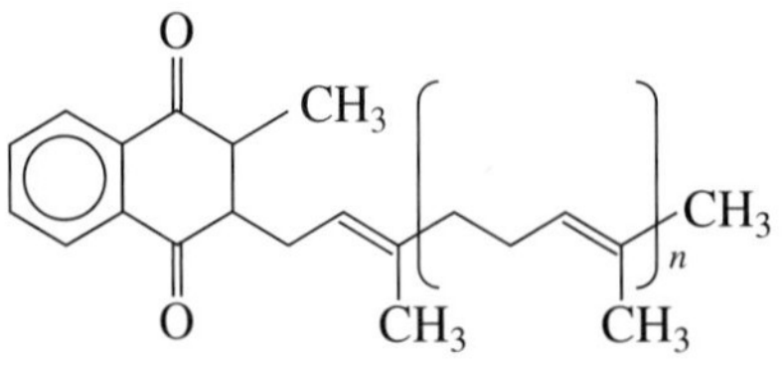

Vitamin K_2 (menaquinone)

SAMPLE PROBLEM 21.8

Vitamins

Why do you need a certain amount of thiamine and riboflavin in your diet every day, but not vitamin A or D?

Solution

Water-soluble vitamins like thiamine and riboflavin are not stored in the body whereas fat-soluble vitamins such as A and D are stored in the liver. Any excess of thiamine or riboflavin are eliminated in the urine and must be replenished each day from the diet.

Study Check

Why are fresh fruits rather than cooked fruits recommended as a source of vitamin C?

QUESTIONS AND PROBLEMS

Vitamins and Coenzymes

21.37 Give the abbreviation for each of the following coenzymes:
  a. tetrahydrofolate
  b. nicotinamide adenine dinucleotide

21.38 Give the abbreviation for each of the following coenzymes:
  a. flavin adenine dinucleotide **b.** thiamine pyrophosphate

21.39 Identify a vitamin that is a component of each of the following coenzymes:
  a. coenzyme A **b.** tetrahydrofolate (THF)
  c. NAD^+

21.40 Identify a vitamin that is a component of each of the following coenzymes:
  a. thiamine pyrophosphate **b.** FAD
  c. pyridoxal phosphate

21.41 What vitamin may be deficient in the following conditions?
  a. rickets **b.** scurvy **c.** pellagra

21.42 What vitamin may be deficient in the following conditions?
  a. poor night vision **b.** pernicious anemia **c.** beriberi

21.43 The RDA for pyridoxine (vitamin B_6) is 2 mg daily. Why will it not improve your nutrition to take 100 mg of pyridoxine daily?

21.44 The RDA for vitamin A is 3 mg daily. Why would an excess of vitamin A cause hypervitaminosis?

21.45 What is the change in the structure of pryidoxine (B_6) that yields the coenzyme PLP?

21.46 What is the change in the structure of folate that yields the coenzyme THF?

Chapter Review

21.1 Biological Catalysts

Enzymes are globular proteins that act as biological catalysts by lowering activation energy and accelerating the rate of cellular reactions.

21.2 Names and Classification of Enzymes

The names of most enzymes ending in *ase* describe the compound or reaction catalyzed by the enzyme. Enzymes are classified by the main type of reaction they catalyze, such as oxidoreductase, transferase, or isomerase.

21.3 Enzymes as Catalysts

Within the tertiary structure of an enzyme, a small pocket called the active site binds the substrates. In the lock-and-key model, a substrate precisely fits the shape of the active site. In the induced-fit model, substrates induce the active site to change structure to give an optimal fit by the substrate. In the enzyme–substrate complex, catalysis takes place when amino acid side chains react with a substrate. The products are released and the enzyme is available to bind another substrate molecule.

21.4 Factors Affecting Enzyme Action

Enzymes are most effective at optimum temperature and pH, usually 37°C and 7.4. The rate of an enzyme-catalyzed reaction decreases considerably at temperature and pH values above or below the optimum. An increase in substrate concentration increases the reaction rate of an enzyme-catalyzed reaction. If an enzyme is saturated, adding more substrate will not increase the rate further.

21.5 Enzyme Inhibition

An inhibitor reduces the activity of an enzyme or makes it inactive. A competitive inhibitor has a structure similar to the substrate and competes for the active site. When the active site is occupied, the enzyme cannot catalyze the reaction of the substrate. A noncompetitive inhibitor attaches elsewhere on the enzyme, changing the shape of both the enzyme and its active site. As long as the noncompetitive inhibitor is attached, the altered active site cannot bind with substrate.

21.6 Regulation of Enzyme Activity

Insulin and digestive enzymes are produced as inactive forms called zymogens. They are converted to active forms by removing a peptide portion from their protein chains. The rate of an enzyme-catalyzed reaction can be increased or decreased by regulator molecules that bind to a regulator site on an allosteric enzyme. The regulator molecule changes the shape of the enzyme and therefore the shape of the active site. A positive regulator increases the rate whereas a negative regulator decreases the rate. In feedback inhibition, the end product of a reaction sequence binds to a regulator site on the first enzyme, which is an allosteric enzyme, to decrease product formation.

21.7 Enzyme Cofactors

Simple enzymes are biologically active as a protein only, whereas other enzymes require small organic molecules or metals ions called cofactors. A cofactor may be a metal ion such as Cu^{2+} or Fe^{2+}, or an organic molecule called a coenzyme.

21.8 Vitamins and Coenzymes

A vitamin is a small organic molecule needed for health and normal growth. Vitamins are obtained in small amounts through the foods in the diet. The water-soluble vitamins B and C function as coenzymes. The fat-soluble vitamins are A, D, E, and K. Vitamin A is important in vision, vitamin D for proper bone growth, vitamin E is an antioxidant, and vitamin K is required for proper blood clotting.

Key Terms

active site A pocket in a part of the tertiary enzyme structure that binds substrate and catalyzes a reaction.

activity The rate at which an enzyme catalyzes the reaction that converts substrate to product.

allosteric enzyme An enzyme that regulates the rate of a reaction when a regulator molecule attaches to a site other than the active site.

antibiotic An irreversible inhibitor produced by bacteria, mold, or yeast that is toxic to bacteria.

catalyst A substance that takes part in a reaction to lower the activation energy, but is not changed.

coenzyme An organic molecule, usually a vitamin, required as a cofactor in enzyme action.

cofactor A nonprotein metal ion or an organic molecule that is necessary for a biologically functional enzyme.

competitive inhibitor A molecule with a structure similar to the substrate that inhibits enzyme action by competing for the active site.

enzymes Globular proteins that catalyze biological reactions.

enzyme–substrate (ES) complex An intermediate consisting of an enzyme that binds to a substrate in an enzyme-catalyzed reaction.

fat-soluble vitamins Vitamins that are not soluble in water and can be stored in the liver and body fat.

feedback control A type of inhibition in which an end product inhibits the first enzyme in a sequence of enzyme-catalyzed reactions.

induced-fit model A model of enzyme action in which a substrate induces an enzyme to modify its shape to give an optimal fit with the substrate structure.

inhibitors Substances that make an enzyme inactive by interfering with its ability to react with a substrate.

irreversible inhibition An inhibition caused by the covalent binding of an inhibitor to a part of the active site that cannot be reversed by adding more substrate.

isoenzymes Isoenzymes catalyze the same reaction, but have different combinations of polypeptide subunits.

lock-and-key model A model of an enzyme in which the substrate, like a key, exactly fits the shape of the lock, which is the specific shape of the active site.

negative regulator A molecule or end product that slows down or stops a catalytic reaction by binding to an allosteric enzyme.

noncompetitive inhibitor A substance that changes the shape of the enzyme, which prevents the active site from binding substrate properly.

optimum pH The pH at which an enzyme is most active.

optimum temperature The temperature at which an enzyme is most active.

positive regulator A molecule that increases the rate of an enzyme-catalyzed reaction by making the catalysis more favorable.

reversible inhibition Inhibition of an enzyme that is reversed by increasing the substrate concentration.

simple enzyme An enzyme that is active as a polypeptide only.

substrate The molecule that reacts in the active site in an enzyme-catalyzed reaction.

vitamins Organic molecules, which are essential for normal health and growth, obtained in small amounts from the diet.

water-soluble vitamins Vitamins that are soluble in water, cannot be stored in the body, are easily destroyed by heat, ultraviolet light, and oxygen, and function as coenzymes.

zymogen An inactive form of an enzyme that is activated by removing a peptide portion from one end of the protein.

Additional Problems

21.47 Why do the cells in the body have so many enzymes?

21.48 Are all the possible enzymes present at the same time in a cell?

21.49 How are enzymes different from the catalysts used in chemistry laboratories?

21.50 Why do enzymes function only under mild conditions?

21.51 Lactase is an enzyme that hydrolyzes lactose to glucose and galactose.
  a. What are the reactants and products of the reaction?
  b. Draw an energy diagram for the reaction with and without lactase.
  c. How does lactase make the reaction go faster?

21.52 Maltase is an enzyme that hydrolyzes maltose to two glucose molecules.
  a. What are the reactants and products of the reaction?
  b. Draw an energy diagram for the reaction with and without maltase.
  c. How does maltase make the reaction go faster?

21.53 Indicate whether each of the following would be a substrate (S) or an enzyme (E):
  a. lactose   **b.** lactase
  c. urease   **d.** trypsin
  e. pyruvate   **f.** transaminase

21.54 Indicate whether each of the following would be a substrate (S) or an enzyme (E):
  a. glucose   **b.** hydrolase
  c. maleate isomerase   **d.** alanine
  e. amylose   **f.** amylase

21.55 Give the substrate of each of the following enzymes:
  a. urease   **b.** lactase

  c. aspartate transaminase
  d. tyrosine synthetase

21.56 Give the substrate of each of the following enzymes:
  a. maltase
  b. fructose oxidase
  c. phenolase
  d. sucrase

21.57 Predict the major class of each of the following enzymes:
  a. acyltransferase
  b. oxidase
  c. lipase
  d. decarboxylase

21.58 Predict the major class for each of the following enzymes:
  a. cis–trans isomerase
  b. reductase
  c. carboxylase
  d. peptidase

21.59 What is the class of the enzyme that would catalyze each of the following reactions?

a. $CH_3\overset{\displaystyle O}{\overset{\displaystyle \|}{C}}H \longrightarrow CH_3\overset{\displaystyle O}{\overset{\displaystyle \|}{C}}OH$

b. $NH_2-CH_2-\overset{\displaystyle O}{\overset{\displaystyle \|}{C}}-NH-\overset{\displaystyle CH_3}{\overset{\displaystyle |}{C}}H-\overset{\displaystyle O}{\overset{\displaystyle \|}{C}}OH + H_2O \longrightarrow$

$NH_2-CH_2-\overset{\displaystyle O}{\overset{\displaystyle \|}{C}}OH + NH_2-\overset{\displaystyle CH_3}{\overset{\displaystyle |}{C}}H-\overset{\displaystyle O}{\overset{\displaystyle \|}{C}}OH$

c. $CH_3-CH=CH-CH_3 + H_2O \longrightarrow$

$CH_3-CH_2-\overset{\displaystyle OH}{\overset{\displaystyle |}{C}}H-CH_3$

21.60 What is the class of the enzyme that would catalyze each of the following reactions?

a. $CH_3-\overset{\overset{\displaystyle O}{\|}}{C}-\overset{\overset{\displaystyle O}{\|}}{C}OH \longrightarrow$

$CH_3-\overset{\overset{\displaystyle O}{\|}}{C}-OH + CO_2$

b. $CH_3-\overset{\overset{\displaystyle O}{\|}}{C}-\overset{\overset{\displaystyle O}{\|}}{C}OH + CO_2 + ATP \longrightarrow$

$HO-\overset{\overset{\displaystyle O}{\|}}{C}-CH_2-\overset{\overset{\displaystyle O}{\|}}{C}-\overset{\overset{\displaystyle O}{\|}}{C}-OH + ADP + P_i$

c. glucose-6-phosphate $\longrightarrow$ fructose-6-phosphate

21.61 How would the lock-and-key theory explain that sucrase hydrolyses sucrose, but not lactose?

21.62 How does the induced-fit model of enzyme action allow an enzyme to catalyze a reaction of a group of substrates?

21.63 If a blood test indicates a high level of LDH and CK, what could be the cause?

21.64 If a blood test indicates a high level of ALT, what could be the cause?

21.65 Indicate whether an enzyme is saturated or unsaturated in each of the following conditions:
  a. adding more substrate does not increase the rate of reaction
  b. doubling the substrate concentration doubles the rate of reaction

21.66 Indicate whether each of the following enzymes would be functional:
  a. pepsin, a digestive enzyme, at pH 2
  b. an enzyme at 37°C, if the enzyme is from a type of thermophilic bacteria that have an optimum temperature of 100°C

21.67 How does a reversible inhibition differ from an irreversible inhibition?

21.68 How does a competitive reversible inhibition differ from a noncompetitive reversible inhibition?

21.69 Ethylene glycol ($HO-CH_2-CH_2-OH$) is a major component of antifreeze. In the body, it is first converted to $HOOC-CHO$ (oxoethanoic acid) and then to $HOOC-COOH$ (oxalic acid), which is toxic.

  a. What class of enzyme catalyzes both of the reactions of ethylene glycol?
  b. The treatment for the ingestion of ethylene glycol is an intravenous solution of ethanol. How might this help prevent toxic levels of oxalic acid in the body?

21.70 Adults who are lactose intolerant cannot break down the disaccharide in milk products. To help digest dairy food, a product known as Lactaid can be added to milk and refrigerated for 24 hours.
  a. What enzyme is present in Lactaid and what is the major class?
  b. What might happen to the enzyme if the digestion product were stored in a warm area?

21.71 a. What type of an inhibitor is the antibiotic amoxicillin?
  b. Why can antibiotics be used to treat bacterial infections?

21.72 a. A gardener using Parathion develops a headache, dizziness, nausea, blurred vision, excessive salivation, and muscle twitching. What might be happening to the gardener?
  b. Why must humans be careful when using insecticides?

21.73 The enzyme pepsin is produced in the pancreas as the zymogen pepsinogen.
  a. How and where does pepsinogen become the active form pepsin?
  b. Why are proteases such as pepsin produced in inactive forms?

21.74 Thrombin is an enzyme that helps produce blood clotting when an injury and bleeding occurs.
  a. What would be the name of the zymogen of thrombin?
  b. Why would the active form of thrombin be produced only when an injury occurs to tissue?

21.75 What is an allosteric enzyme?

21.76 Why can some regulator molecules speed up a reaction, while others slow it down?

21.77 In feedback control, what type of regulator modifies the catalytic activity of the reaction pathway?

21.78 Why aren't the intermediate products in a reaction sequence used in feedback control?

21.79 Are each of the following statements describing a simple enzyme or one that requires a cofactor?

a. contains Mg^{2+} in the active site

b. has catalytic activity as a tertiary protein structure

c. requires folic acid for catalytic activity

21.80 Are each of the following statements describing a simple enzyme or one that contains a cofactor?

a. contains riboflavin or vitamin B_2

b. has four subunits of polypeptide chains

c. requires Fe^{3+} in the active site for catalytic activity

21.81 Match the following vitamins with their coenzymes:

a. pantothenic acid (B_5) NAD^+

b. niacin (B_3)   biocytin

c. biotin   coenzyme A

21.82 Match the following vitamins with their coenzymes:

a. folate   pyridoxal phosphate

b. riboflavin (B_2)   THF

c. pyridoxine   FAD

21.83 Why are only small amounts of vitamins needed in the cells when there are several enzymes that require coenzymes?

21.84 Why is there a daily requirement for vitamins?

21.85 Match each of the following vitamins with their deficiency symptoms or conditions:

a. niacin   night blindness

b. vitamin A   weak bone structure

c. vitamin D   pellagra

21.86 Match each of the following vitamins with their deficiency symptoms or conditions:

a. cobalamin   bleeding

b. vitamin C   anemia

c. vitamin K   scurvy

22 Nucleic Acids and Protein Synthesis

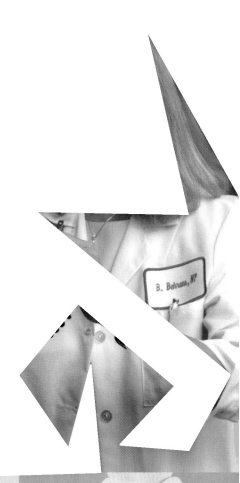

"I run the Hepatitis C Clinic, where patients are often anxious when diagnosed," says Barbara Behrens, nurse practitioner, Hepatitis C Clinic, Kaiser Hospital. "The treatment for hepatitis C can produce significant reactions such as a radical drop in blood count. When this happens, I get help within 24 hours. I monitor our patients very closely and many call me whenever they need to."

Hepatitis C is an RNA virus or retrovirus that causes liver inflammation, often resulting in chronic liver disease. Unlike many viruses to which we eventually develop immunity, the hepatitis C virus undergoes mutations so rapidly that scientists have not been able to produce vaccines. People who carry the virus are contagious throughout their lives and are able to pass the virus to other people.

the Chemistry place

www.chemplace.com/college

Visit the URL above or use the CD-ROM in the book for extra quizzing, interactive tutorials, career resources, and case studies.

Nucleic acids are the molecules in our cells that store and direct information for cellular growth and reproduction. Deoxyribonucleic acid (DNA), the genetic material in the nucleus of a cell, contains all the information needed for the development of a complete living system. The way you grow, your hair, your eyes, your physical appearance, the activities of the cells in your body are all determined by a set of directions contained in the DNA of your cells.

The nucleic acids are large molecules found in the nuclei of cells that contain all the information needed to direct the activities of a cell and its reproduction. All of the genetic information in the cell is called the *genome*. Every time a cell divides, the information in the genome is copied and passed on to the new cells. This replication process must duplicate the genetic instructions exactly. Some sections of DNA called *genes* contain the information to make a particular protein.

As a cell requires protein, another type of nucleic acid, RNA, translates the genetic information in DNA, and carries that information to the ribosomes where the synthesis of protein takes place. However, mistakes sometimes occur that lead to mutations that affect the synthesis of a certain protein.

22.1 Components of Nucleic Acids

Describe the nitrogen bases and ribose sugars that make up the nucleic acids DNA and RNA.

There are two closely related types of nucleic acids: *deoxyribonucleic acid* (**DNA**), and *ribonucleic acid* (**RNA**). Both are linear polymers of repeating monomer units known as *nucleotides*. A DNA molecule may contain several million nucleotides; smaller RNA molecules may contain up to several thousand. Each nucleotide has three components: a nitrogenous base, a five-carbon sugar, and a phosphate group. The sugar is bonded to a nitrogen base at carbon 1' (C1'), and to a phosphate group at carbon 5' (C5'). (See Figure 22.1.)

Nitrogen Bases

The **nitrogen-containing bases** in nucleic acids are derivatives of *pyrimidine* or *purine*.

Pyrimidine   Purine

In DNA, there are two purines, adenine (A) and guanine (G), and two pyrimidines, cytosine (C) and thymine (T). RNA contains the same bases, except thymine (5-methyluracil) is replaced by uracil (U). (See Figure 22.2.)

Ribose and Deoxyribose Sugars

The nucleotides of RNA and DNA contain five-carbon pentose sugars. In RNA, the five-carbon sugar is *ribose*, which gives the letter R in the abbreviation RNA. In DNA, the five-carbon sugar is *deoxyribose*, which is similar to ribose except that

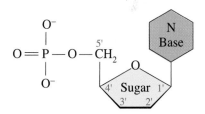

Figure 22.1 A diagram of the general structure of a nucleotide found in nucleic acids.

Q In a nucleotide, what types of groups are bonded to a five-carbon sugar?

WEB TUTORIAL
DNA and RNA Structure

Pyrimidines

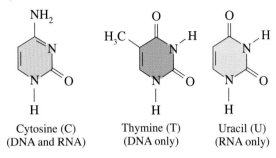

Cytosine (C)
(DNA and RNA)

Thymine (T)
(DNA only)

Uracil (U)
(RNA only)

Purines

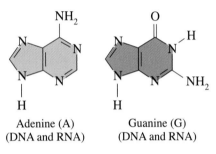

Adenine (A)
(DNA and RNA)

Guanine (G)
(DNA and RNA)

Figure 22.2 DNA contains the nitrogen bases A, G, C, and T; RNA contains A, G, C, and U.

Q Which nitrogen bases are found in DNA?

Pentose sugars in RNA and DNA

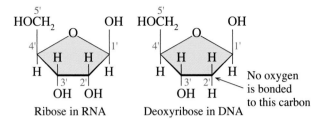

Ribose in RNA   Deoxyribose in DNA

Figure 22.3 The five-carbon pentose sugar found in RNA is ribose and deoxyribose in DNA.

Q What is the difference between ribose and deoxyribose?

there is no hydroxyl group (—OH) on C2' of ribose. The *deoxy–* prefix means "without oxygen" and provides the D in DNA.

The atoms in the pentose sugars are numbered with primes (1', 2', 3', 4', and 5') to differentiate them from the atoms in the nitrogen bases. (See Figure 22.3.) In the following sections, we will see that the presence or absence of a hydroxyl group (—OH) on the sugar strongly influences the structures and properties of RNA and DNA.

SAMPLE PROBLEM 22.1

Components of Nucleic Acids

Identify each of the following bases as a purine or pyrimidine.

a.

b.

Solution

a. Guanine is a purine.
b. Uracil is a pyrimidine.

Study Check

Indicate if the bases in problem 22.1 are found in RNA, DNA, or both.

QUESTIONS AND PROBLEMS

Components of Nucleic Acids

22.1 Identify each of the following bases as a purine or pyrimidine:
   a. thymine
   b.

22.2 Identify each of the following bases as a purine or pyrimidine:
  a. guanine
  b.

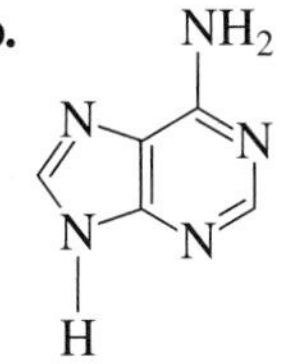

22.3 Identify the bases in problem 22.1 as present in RNA, DNA, or both.
22.4 Identify the bases in problem 22.2 as present in RNA, DNA, or both.

22.2 Nucleosides and Nucleotides

LEARNING GOAL

Identify the nucleosides and nucleotides in RNA and DNA.

the
Chemistry place

WEB TUTORIAL
DNA and RNA Structure

A **nucleoside** is produced when a nitrogen base of pyrimidine or a purine forms a glycosidic bond to C1' of a sugar, either ribose or deoxyribose. The name of a nucleoside is obtained from the name of the base and the ending *–osine* for purines and *–idine* for pyrimidines. For example, adenine, a purine, and ribose form a nucleoside called aden**osine**. When ribose combines with the base cytosine, a pyrimidine, the nucleoside is named cyt**idine.**

Sugar + base $\longrightarrow$ nucleoside

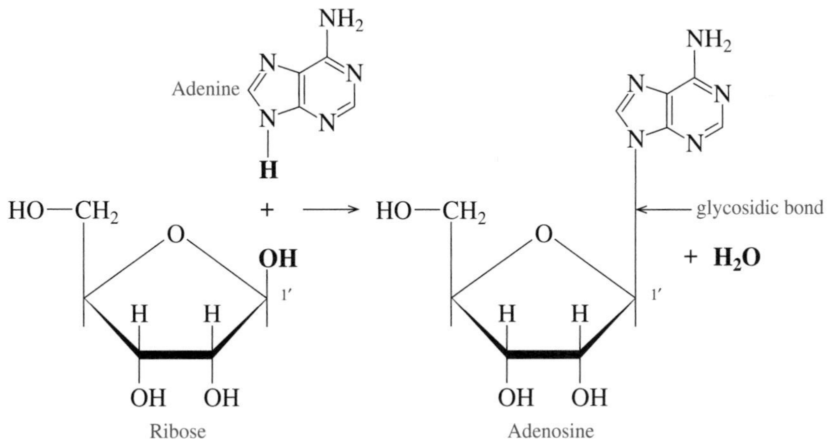

Formation of Nucleotides

In Chapter 18, we saw that phosphate esters form between a hydroxyl group and phosphoric acid to give ester products.

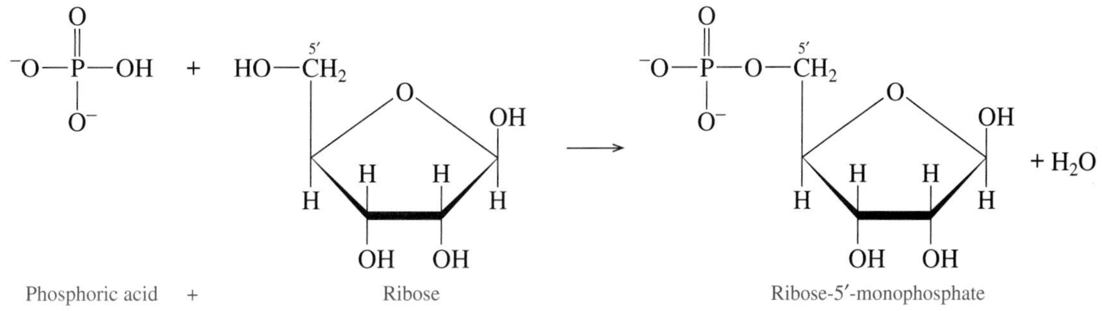

Nucleotides are formed when the C5' — OH group of ribose or deoxyribose in a nucleoside forms phosphate esters with phosphoric acid. Other hydroxyl groups on ribose can form phosphate esters too, but only the 5'-monophosphate nucleotides are found in RNA and DNA. All the nucleotides in RNA and DNA are shown in Figure 22.4.

Phosphoric acid + nucleoside $\longrightarrow$ nucleotide

Phosphoric acid   Deoxycytidine   Deoxycytidine 5'-monophosphate (dCMP)

Naming Nucleotides

The name of a nucleotide is obtained from the name of the nucleoside followed by 5'-monophosphate. A nucleotide in RNA is named as a *nucleoside 5'-monophosphate* or NMP. Nucleotides of DNA have the prefix *deoxy* added to the beginning of the nucleoside name, *deoxynucleoside 5'-monophosphate* or dNMP. For example, the nucleotide of adenosine in RNA is adenosine 5'-monophosphate or AMP. In DNA, it is named deoxyadenosine 5'-monophosphate or dAMP. Although the letters A, G, C, U, and T represent the bases, they are often used in the abbreviations of the respective nucleotides as well. The names and abbreviations of the bases, nucleosides, and nucleotides in DNA and RNA are listed in Table 22.1.

Table 22.1   Names of Nucleosides and Nucleotides in DNA and RNA

Base	Nucleosides	Nucleotides
RNA		
Adenine (A)	Adenosine (A)	Adenosine 5'-monophosphate (AMP)
Guanine (G)	Guanosine (G)	Guanosine 5'-monophosphate (GMP)
Cytosine (C)	Cytidine (C)	Cytidine 5'-monophosphate (CMP)
Uracil (U)	Uridine (U)	Uridine 5'-monophosphate (UMP)
DNA		
Adenine (A)	Deoxyadenosine (A)	Deoxyadenosine 5'-monophosphate (dAMP)
Guanine (G)	Deoxyguanosine (G)	Deoxyguanosine 5'-monophosphate (dGMP)
Cytosine (C)	Deoxycytidine (C)	Deoxycytidine 5'-monophosphate (dCMP)
Thymine (T)	Deoxythymidine (T)	Deoxythymidine 5'-monophosphate (dTMP)

Nucleosides   **Nucleotides**

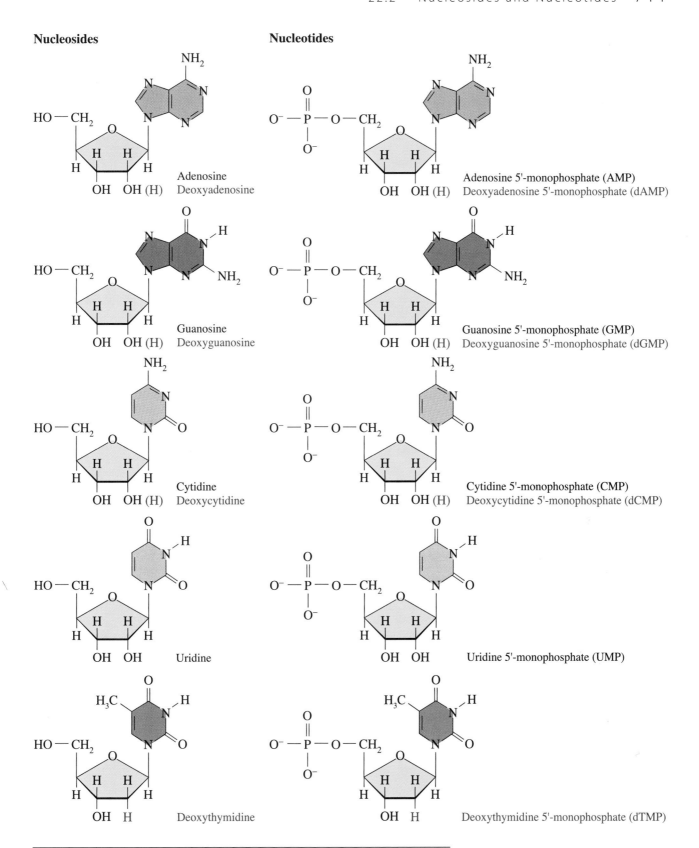

Figure 22.4 The nucleosides and nucleotides of RNA are identical to those of DNA except in DNA the sugar is deoxyribose and deoxythymidine replaces uridine.

Q What are two differences in the nucleotides of RNA and DNA?

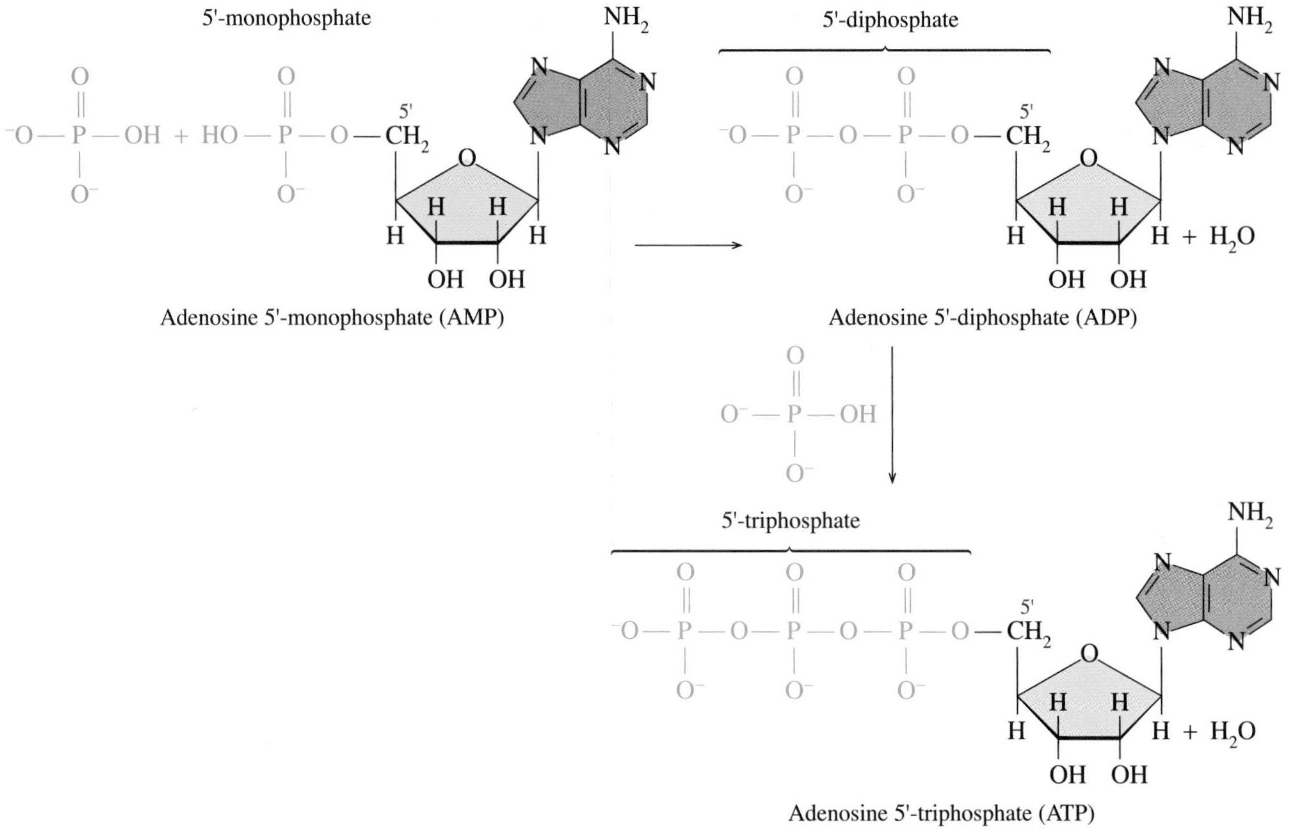

Figure 22.5 The addition of more phosphate groups forms adenosine 5'-diphosphate (ADP) and adenosine 5'-triphosphate (ATP).

Q How does the structure of deoxyguanidine triphosphate (dGTP) differ from ATP?

Formation of Nucleoside Diphosphates and Triphosphates

Any of the nucleoside 5'-monophosphates such as AMP can bond to additional phosphate groups. For example, adding another phosphate to AMP gives *ADP* (*adenosine 5'-diphosphate*) and ATP (*adenosine 5'-triphosphate*) when there are a total of three phosphates. (See Figure 22.5.) The other nucleotides also form di- and triphosphates. For example, GMP adds phosphate to yield GDP and GTP, dCMP forms dCDP and dCTP, and so on. Of the triphosphates, ATP is of particular interest because it is the major source of energy for most energy-requiring activities in the cell. GTP is an energy source for protein synthesis, and CTP is an intermediate in phospholipid synthesis.

SAMPLE PROBLEM 22.2

Nucleotides

Identify which nucleic acid (DNA or RNA) contains each of the following nucleotides; state the components of each nucleotide:

a. deoxyguanosine 5'-monophosphate (dGMP)

b. adenosine 5'-monophosphate (AMP)

Solution

a. This DNA nucleotide consists of deoxyribose, guanine, and phosphate.

b. This RNA nucleotide contains ribose, adenine, and phosphate.

What is the name and abbreviation of the DNA nucleotide of cytosine?

QUESTIONS AND PROBLEMS

Nucleosides and Nucleotides

22.5 What are the names and abbreviations of the four nucleotides in DNA?

22.6 What are the names and abbreviations of the four nucleotides in RNA?

22.7 Identify each of the following as a nucleoside or nucleotide:
  a. adenosine   **b.** deoxycytidine
  c. uridine   **d.** cytidine 5'-monophosphate

22.8 Identify each of the following as a nucleoside or nucleotide in DNA or RNA:
  a. deoxythymidine   **b.** guanosine
  c. adenosine   **d.** uridine 5'-monophosphate

22.9 Draw the structure of deoxyadenosine 5'-monophosphate (dAMP).

20.10 Draw the structure of uridine 5'-monophosphate (UMP).

22.3 Primary Structure of Nucleic Acids

The **nucleic acids** consist of polymers of many nucleotides. In a polymer of nucleotides, the 3'—OH group of the sugar in one nucleotide bonds to the phosphate group on the 5'—carbon atom in the sugar of the next nucleotide. This phosphate link between the sugars in adjacent nucleotides is referred to as a **phosphodiester bond.** As more nucleotides are added through phosphodiester bonds, a backbone forms that consists of alternating sugar and phosphate groups.

WEB TUTORIAL
DNA and RNA Structure

Free 5' end

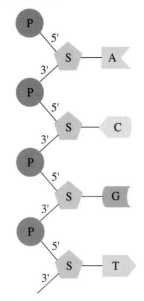

Free 3' end

Each nucleic acid has its own unique sequence of bases, which is known as its **primary structure.** It is this sequence of bases that carries the genetic information from one cell to the next. Along a polynucleotide of a DNA or RNA chain, the bases attached to each of the sugars extend out from the nucleic acid backbone. In RNA, ribose is the sugar used in the polynucleotide, whereas in DNA, the sugar is deoxyribose. In any nucleic acid, the sugar at the one end has an unreacted or free 5'-phosphate terminal end, and the 3'—terminal sugar at the other end has a free 3'-hydroxyl group.

A nucleic acid sequence is read from the sugar with free 5'-phosphate to the sugar with the free 3'-hydroxyl group. The order of bases is often written using only the letters of the bases. Thus, the nucleotides in a section of RNA shown in Figure 22.6 is read as 5'-ACGU-3'. This abbreviation starts with the base attached to the ribose with a free 5' phosphate, and ends with uracil, the base attached to the sugar with a free 3' hydroxyl group. In the nucleotide sequence in DNA, 5'-ACGT-3' (see Figure 22.7), adenine is the base attached to the deoxyribose with the free 5' phosphate and thymine is the base attached to the sugar with the free 3' hydroxyl group.

RNA (ribonucleic acid)

Adenine (A)

Cytosine (C)

3'–5' Phosphodiester bond

Guanine (G)

Uracil (U)

Free 3'

Figure 22.6 In the primary structure of an RNA, A, C, G, and U are linked by 3'–5' phosphodiester bonds.

Q Where are the free 5' phosphate and 3' hydroxyl groups of ribose?

DNA (deoxyribonucleic acid)

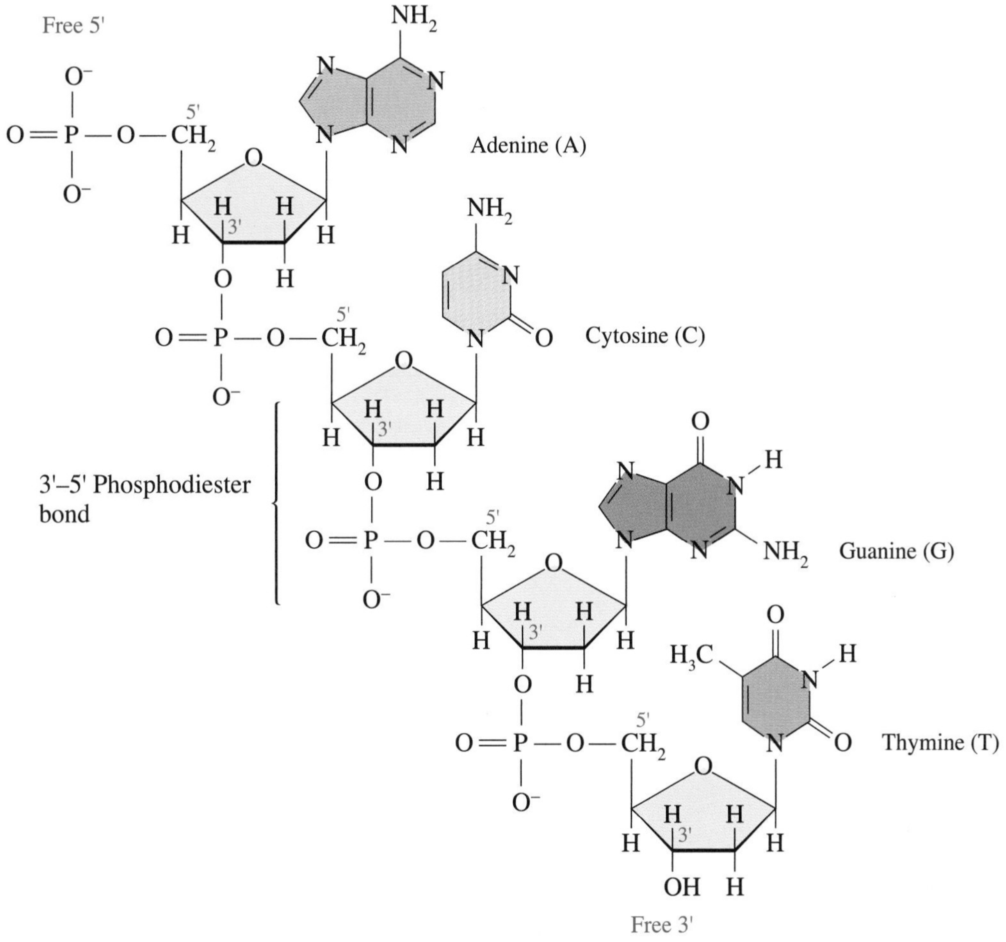

Figure 22.7 In the primary structure of a single DNA strand, A, C, G, and T are linked by 3'–5' phosphodiester bonds.

Bonding of Nucleotides

Draw the structure of an RNA dinucleotide formed by two cytidine monophosphates.

Solution

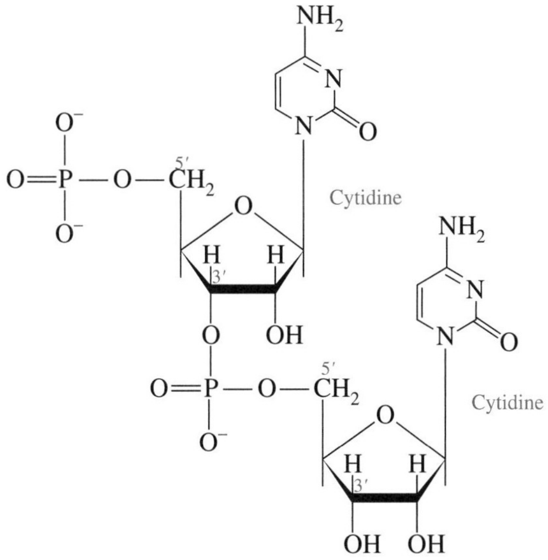

Figure 22.7 In the primary structure of a single DNA strand, A, C, G, and T are linked by 3'–5' phosphodiester bonds.

Q What hydroxyl groups does deoxyribose use to form a polymer of DNA?

Study Check

In the dinucleotide of cytidine shown in the solution, identify the free 5' phosphate group and the free 3' hydroxyl (—OH) group.

QUESTIONS AND PROBLEMS

Primary Structure of Nucleic Acids

22.11 How are the nucleotides held together in a nucleic acid chain?

22.12 How do the ends of a nucleic acid polymer differ?

22.13 Write the structure of the dinucleotide 5'-GC-3' that would be in RNA.

22.14 Write the structure of the dinucleotide 5'-AT-3' that would be in DNA.

LEARNING GOAL

Describe the double helix of DNA.

Figure 22.8 This space-filling model shows the double helix that is the characteristic shape of DNA molecules.

Q What is meant by the term double-helix?

22.4 DNA Double Helix: A Secondary Structure

In the 1940s, biologists determined that DNA in a variety of organisms had a specific relationship between bases: the percent of adenine (A) was equal to the percent of thymine (T), and the percent of guanine (G) was equal to cytosine (C). Although different organisms vary in the amounts of bases, adenine is paired only with thymine, and guanine is paired only with cytosine in 1:1 ratios. (See Table 22.2.) This relationship known as *Chargaff's rules* can be summarized as follows:

Amount of purines = Amount of pyrimidines

$$A = T$$

$$G = C$$

In 1953, James Watson and Francis Crick proposed that DNA was a **double helix** that consists of two polynucleotide strands winding about each other like a spiral staircase. (See Figure 22.8.) The hydrophilic sugar-phosphate backbones are analogous to the outside railings with the hydrophobic bases arranged like steps along the inside. The two strands run in opposite directions. One strand goes in the 5'-3' direction, and the opposite strand goes in the 3'-5' direction.

Table 22.2   Percent Bases in the DNAs of Selected Organisms

Organism	%A	%T	%G	%C
Humans	30	30	20	20
Chicken	28	28	22	22
Salmon	28	28	22	22
Corn (maize)	27	27	23	23
Neurospora	23	23	27	27

Source: Handbook of Biochemistry, 2nd ed. Sober, H.E., Ed.; CRC Press: Cleveland, OH, 1970.

Complementary Base Pairs

Each of the bases along one polynucleotide strand forms hydrogen bonds to a specific base on the opposite DNA strand. Adenine only bonds to thymine, and guanine only bonds to cytosine. (See Figure 22.9.) The pairs A—T and G—C are called **complementary base pairs.** The specificity of the base pairing is due to the fact that adenine and thymine form two hydrogen bonds, while cytosine and guanine form three hydrogen bonds. This explains why DNA has equal amounts of A and T bases and equal amounts of G and C.

Since each base pair contains a purine and a pyrimidine, the total width of the two pairs of bases A—T and G—C is the same. Thus the two polynucleotide strands of a DNA are the same distance apart all along the DNA polymer. There are no C—T, C—A, G—T, or G—A pairs in DNA because the other base-pair combinations cannot form as many hydrogen bonds or maintain a constant width

WEB TUTORIAL
DNA and RNA Structure

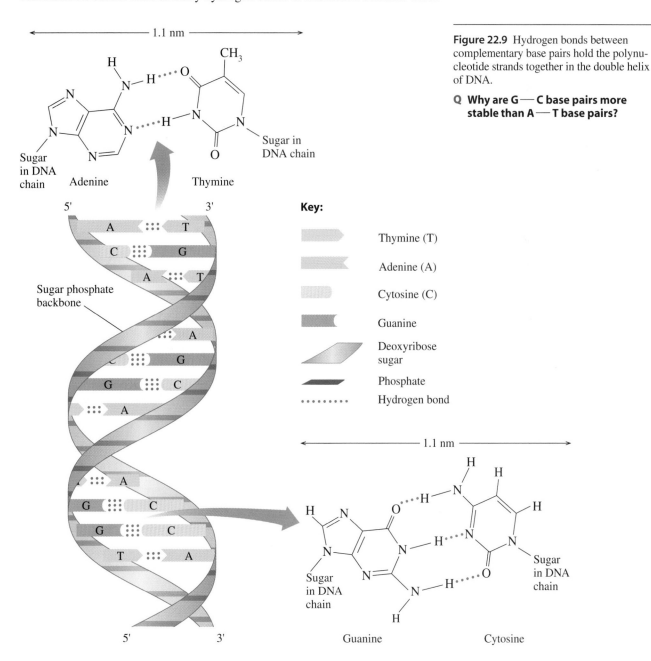

Figure 22.9 Hydrogen bonds between complementary base pairs hold the polynucleotide strands together in the double helix of DNA.

Q **Why are G—C base pairs more stable than A—T base pairs?**

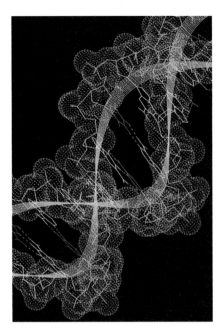

Figure 22.10 A computer-generated model of a DNA molecule.

Q **What is the complementary strand of a DNA section of 5'— GGCCTT— 3'?**

between the two DNA backbones. As we shall see, complementary base pairing plays a crucial role in cell replication and the transfer of hereditary information.

X-ray diffraction patterns of DNA indicate that DNA is a right-handed or alpha (α) helix. In one complete turn, there are about 10 pairs of nucleotides. (See Figure 22.10.) In mitochondria, bacteria, and viruses, DNA molecules are compact, highly coiled molecules. In the chromosomes, DNA strands are wrapped around proteins called *histones*, a structure that provides the most stable and orderly arrangement for the long DNA molecules.

SAMPLE PROBLEM 22.4

Complementary Base Pairs

Write the base sequence of the complementary segment for the following segment of a strand of DNA.

5'— A—C—G—A—T—C—T—3'

Solution

In the complementary strand of DNA, the base A pairs with T, and G pairs with C.

Given segment of DNA: 5'— A—C—G—A—T—C—T—3'
                                   :   :   :   :   :   :   :
Complementary segment: 3'— T—G—C—T—A—G—A—5'

Study Check

What is the sequence of bases that is complementary to a portion of DNA with a base sequence of 5'— G—G—T—T—A—A—C—C—3'?

QUESTIONS AND PROBLEMS

DNA Double Helix: A Secondary Structure

22.15 How are the two strands of nucleic acid in DNA held together?

22.16 What is meant by complementary base pairing?

22.17 Complete the base sequence in a second DNA strand if a portion of one strand has the following base sequence:
   a. 5'— AAAAAA— 3'   **b.** 5'— GGGGGG— 3'
   c. 5'— AGTCCAGGT— 3'   **d.** 5'— CTGTATACGTTA— 3'

22.18 Complete the base sequence in a second DNA strand if a portion of one strand has the following base sequence:
   a. 5'— TTTTTT— 3'   **b.** 5'— CCCCCCCCC— 3'
   c. 5'— ATGGCA— 3'   **d.** 5'— ATATGCGCTAAA— 3'

22.5 DNA Replication

DNA is found in *eukaryotic cells* of animals and plants including algae as well as in *prokaryotic cells* of bacteria. DNA from both types of cells is chemically similar and has the same function, which is to preserve genetic information. However, there are several differences in the cell types. Eukaryotic cells contain DNA in a nucleus with a nuclear membrane and combined with proteins called histones, have several organelles within membranes such as mitochondria and lysosomes, and divide by mitosis. Prokaryotic cells contain DNA as a circular chromosome with no membrane and not associated with histones, contain organelles without membranes, and divide by a simpler process called binary fission. As cells divide, copies of DNA must be produced in order to transfer the genetic information to the new cells. This is the process of DNA replication.

Replication and Energy

In DNA **replication**, the strands in the parent DNA separate, which allows each of the original strands to makes copies by synthesizing complementary strands. The

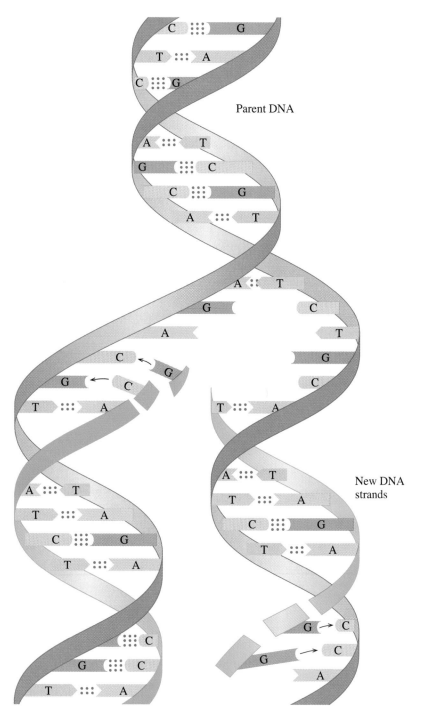

Figure 22.11 In DNA replication, the separate strands of the parent DNA are the templates for the synthesis of complementary strands, which produces two exact copies of DNA.

Q How many strands of the parent DNA are in each of the new double-stranded copies of DNA?

WEB TUTORIAL
DNA Replication

replication process begins when an enzyme called *helicase* catalyzes the unwinding of a portion of the double helix by breaking the hydrogen bonds between the complementary bases. These single strands now act as templates for the synthesis of new complementary strands. (See Figure 22.11.) Within the nucleus, nucleoside triphosphates for each base are available so that each exposed base on the template strand can form hydrogen bonds with its complementary base in the nucleoside triphosphate.

When a nucleoside triphosphate bonds to a sugar at the end of a growing strand, two phosphate groups are cleaved, which provides the energy for the reaction. For example, a T in the template strand forms a hydrogen bond with an A in ATP, and a G on the template strand forms hydrogen bonds with a CTP. After the base pairs are formed, *DNA polymerase* catalyzes the formation of phosphodiester bonds between the nucleotides. The hydrolysis of two phosphate groups releases energy for the new bonds. In this way, energy is provided to join each new nucleotide to the backbone of a growing DNA daughter strand. (See Figure 22.12.)

Eventually the entire double helix of the parent DNA is copied. In each new DNA molecule, one strand of the double helix is from the original DNA and one is a newly synthesized strand. This process, called *semi-conservative replication,* produces two new DNAs called *daughter DNA* that are identical to each other and exact copies of the original parent DNA. In the process of DNA replication, complementary base pairing ensures the correct placements of bases in the new DNA strands.

Direction of Replication

Now that we have seen the overall process, we can take a look at some of the details that are important in understanding DNA replication. The unwinding of DNA by *helicase* occurs simultaneously in several sections along the parent DNA molecule. As a result *DNA polymerase* can catalyze the replication process at each of these open DNA sections called **replication forks.** However, DNA polymerase catalyzes only phosphodiester bonds between the 5' phosphate of one nucleotide and the 3' hydroxyl of the next. That means that DNA polymerases have to move in opposite

Figure 22.12 As thymidine triphosphate bonds to the 3' —OH of the adjacent sugar, two phosphates are removed, which provides energy for the synthesis reaction.

Q Why are nucleoside triphosphates used to provide the complementary bases instead of nucleoside monophosphates?

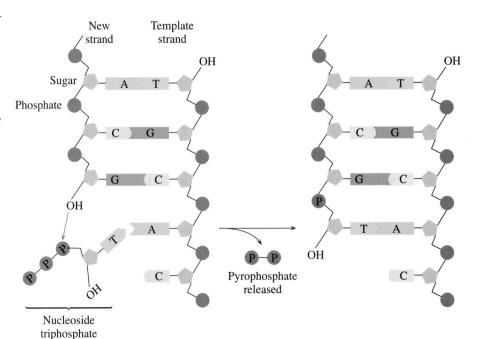

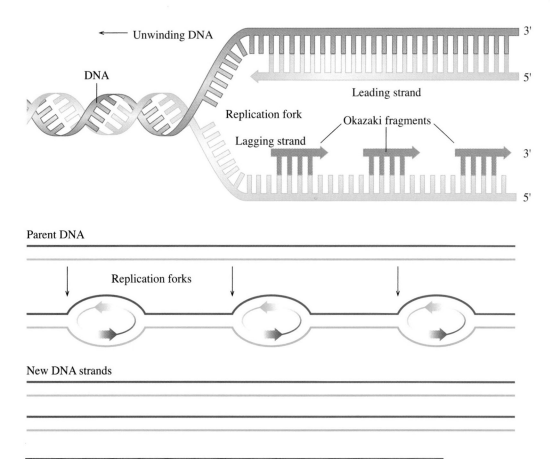

Figure 22.13 At each replication fork, DNA polymerase synthesizes a continuous DNA strand in the 5' to 3' direction. In the new 3' to 5' DNA strand, small Okazaki fragments are produced that are joined by DNA ligase.

Q Why is only one of the new DNA strands synthesized in a continuous direction?

directions along the separated strands of DNA. The new DNA strand that grows in the 5' to 3' direction, the *leading strand,* is synthesized continuously. The other new DNA, the *lagging strand,* is synthesized in the opposite direction, which is in the reverse 3' to 5' direction. In this lagging strand, short sections called **Okazaki fragments** are synthesized at the same time by several DNA polymerases and connected by DNA *ligase* to give a single 3' to 5' DNA strand. (See Figure 22.13.)

SAMPLE PROBLEM 22.5

DNA Replication

In an original DNA strand, there is a portion with the bases 5'—A—G—T—3'. What nucleotides are placed in a growing DNA daughter strand for this sequence?

Solution

Complementary base pairing allows only one possible nucleotide to pair with each base on the original strand. Thymine will pair only with adenine, cytosine with guanine, and adenine with thymine to give the sequence 3'—T—C—A—5'.

Study Check

Why would the new DNA strand you wrote for problem 22.5 be synthesized as Okazaki fragments that require a DNA ligase?

QUESTIONS AND PROBLEMS

DNA Replication

22.19 What is the function of the enzyme helicase in DNA replication?

22.20 What is the function of the enzyme DNA polymerase in DNA replication?

22.21 What process ensures that the replication of DNA produces identical copies?

22.22 Why are Okazaki fragments needed in the synthesis of the lagging strand?

LEARNING GOAL

Describe the structures and functions of the different RNAs.

WEB TUTORIAL
Overview of Protein Synthesis

22.6 Types of RNA

Ribonucleic acid, RNA, which makes up most of the nucleic acid found in the cell, is involved with transmitting the genetic information needed to operate the cell. Similar to DNA, RNA molecules are linear polymers of nucleotides. However, there are several important differences.

1. The sugar in RNA is ribose rather than the deoxyribose found in DNA.
2. The nitrogen base uracil replaces thymine.
3. RNA molecules are single, not double stranded.
4. RNA molecules are much smaller than DNA molecules.

There are three major types of RNA in the cells: *messenger RNA, ribosomal RNA,* and *transfer RNA.* Messenger RNA (**mRNA**) is a group of RNA molecules that carry instructions from the genes in the DNA to the ribosomes in the cell, where that information is used to synthesize proteins. Ribosomal RNA (**rRNA**), which makes up the largest percentage (about 75%) of RNA in the cell, is the RNA contained in the ribosomes. Transfer RNA (**tRNA**), which makes up about 15% of the RNA in a cell, is a group of small RNA molecules that bond to and carry specific amino acids to the ribosomes to be placed in peptide chains. The different types of RNA molecules found in a cell are classified according to their location and function in Table 22.3.

Table 22.3   Types of RNA Molecules

Type	Abbreviation	Percentage of Total RNA	Function in the Cell
Ribosomal RNA	rRNA	75	Major component of the ribosomes
Messenger RNA	mRNA	5–10	Carries information for protein synthesis from the DNA in the nucleus to the ribosomes
Transfer RNA	tRNA	10–15	Brings amino acids to the ribosomes for protein synthesis

Ribosomal RNA

Ribosomal RNA (rRNA), the most abundant type of RNA, makes up 65% of the structural material of the ribosomes; the other 35% is protein. Ribosomes, which are the sites for protein synthesis, consist of two subunits, a large subunit and a small subunit, which differ in their sedimentation rates (S values) when cen-

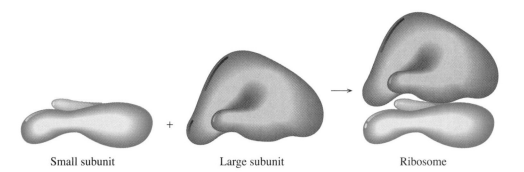

Small subunit   +   Large subunit   Ribosome

trifuged. In *E. coli* the small subunit is 30S and the large subunit is 50S. (See Figure 22.14.) Eukaryotic cells have larger ribosomes with 40S and 60S subunits. Cells that synthesize large numbers of proteins have thousands of ribosomes in their cells.

Figure 22.14 A typical prokaryotic ribosome consists of a small subunit and a large subunit.

Q Why would there be many thousands of ribosomes in a cell?

Messenger RNA

Messenger RNA (**mRNA**) carries genetic information from the DNA in the nucleus to the ribosomes in the cytoplasm for protein synthesis. Each gene, a segment of DNA, produces a separate mRNA molecule when a certain protein is needed in the cell, but then the mRNA is broken down quickly. The size of an mRNA depends on the number of nucleotides in that particular gene.

Transfer RNA

Transfer RNA (**tRNA**), the smallest of the RNA molecules, interprets the genetic information in mRNA and brings specific amino acids to the ribosome for protein synthesis. Only the tRNAs can translate the genetic information into amino acids for proteins. There are one or more different tRNAs for each of the 20 amino acids. The structures of the transfer RNAs are similar, consisting of 70–90 nucleotides. Hydrogen bonds between some of the complementary bases in the chain produce loops that give some double-stranded regions.

The actual structure of a tRNA is a three-dimensional L shape, but we often draw tRNA as a cloverleaf to illustrate its features. All tRNA molecules have a 3' end with the nucleotide sequence 3'— ACC, which is known as the *acceptor stem*. An enzyme attaches an amino acid by forming an ester bond with the free — OH at the end of the acceptor stem. Each tRNA contains an **anticodon**, which is a series of three bases that complements three bases on a mRNA. (See Figure 22.15.)

SAMPLE PROBLEM 22.6

Types of RNA

What is the function of mRNA in a cell?

Solution

mRNA carries the instructions for the synthesis of a protein from the DNA to the ribosomes.

Study Check

What is the function of tRNA in a cell?

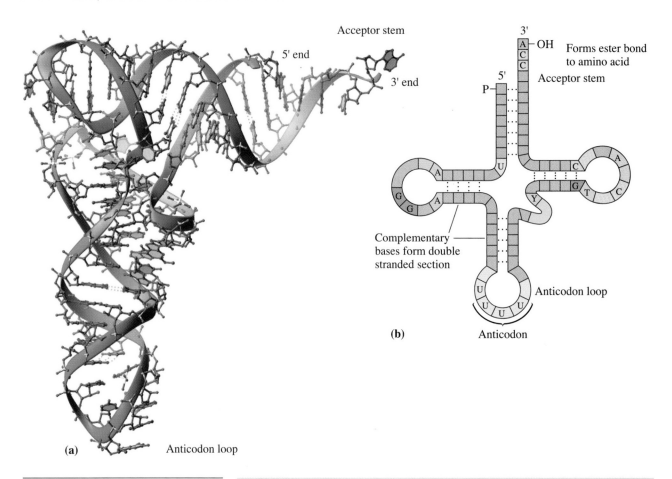

(a) Anticodon loop

(b) Anticodon

Figure 22.15 **(a)** In the ⌐ shape of a transfer RNA, some sections of the ribose–phosphate backbone form regions of complementary base bonding. **(b)** A typical tRNA molecule has an acceptor stem at the 3' end of the nucleic acid where an amino acid attaches, and an anticodon loop that complements three bases on mRNA.

Q Why will different tRNAs have different bases in the anticodon loop?

LEARNING GOAL

Describe the synthesis of mRNA from DNA.

the
Chemistry place

WEB TUTORIAL
Transcription

QUESTIONS AND PROBLEMS

Types of RNA

22.23 What are the three different types of RNA?

22.24 What are the functions of each type of RNA?

22.25 What is the composition of a ribosome?

22.26 What is the smallest RNA?

22.7 Transcription: Synthesis of mRNA

We now look at the processes involved in transferring genetic information encoded in the DNA to the production of proteins. First, the genetic information is copied from a gene in DNA to make a messenger RNA (mRNA), a process called **transcription**. As a carrier molecule, the mRNA moves out of the nucleus and goes to the ribosomes. During **translation,** tRNA molecules convert the information in the mRNA into amino acids, which are placed in the proper sequence to synthesize a protein. (See Figure 22.16.) The language that relates the series of nucleotides in mRNA with the amino acids specified is the *genetic code*.

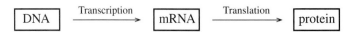

DNA $\xrightarrow{\text{Transcription}}$ mRNA $\xrightarrow{\text{Translation}}$ protein

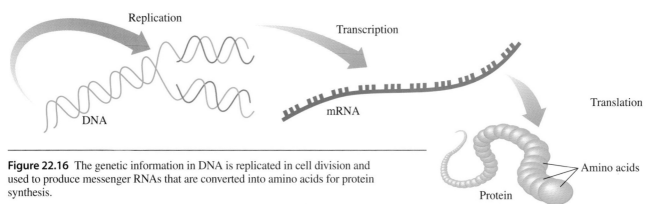

Figure 22.16 The genetic information in DNA is replicated in cell division and used to produce messenger RNAs that are converted into amino acids for protein synthesis.

Q What is the difference between transcription and translation?

Transcription begins when the section of a DNA that contains the gene to be copied unwinds. Within this unwound DNA, a polymerase enzyme identifies a nucleotide sequence of TATAAA, which is the starting or initiation point that is the signal to begin mRNA synthesis. Just as in DNA synthesis, C is paired with G, T pairs with A, and A pairs with U (not T). The polymerase enzyme moves along the unwound DNA in a 3' to 5' direction, forming bonds between the complementary bases in nucleoside triphosphates. Eventually the RNA polymerase reaches the termination point that signals the end of transcription, the new mRNA is released, and the DNA returns to its double helix structure. (See Figure 22.17.)

RNA polymerase

Section of bases on DNA template: 3'—G—A—A—C—T—5'

Complementary base sequence in mRNA: 5'—C—U—U—G—A—3'

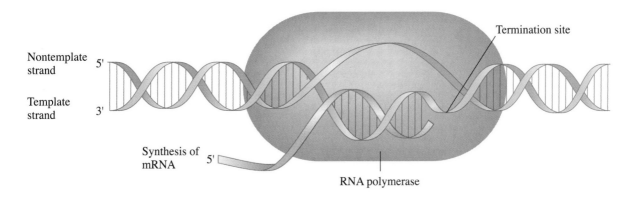

Nontemplate strand 5'

Template strand 3'

Synthesis of mRNA 5'

RNA polymerase

Termination site

Figure 22.17 DNA undergoes transcription when RNA polymerase makes a complementary copy of a gene using the 3' to 5' strand as the template.

Q Why is the mRNA connected in a 5' to 3' direction?

SAMPLE PROBLEM 22.7

RNA Synthesis

The sequence of bases in a part of the DNA template for mRNA is 3'—CGATCA—5'. What is the corresponding mRNA produced?

Solution

The nucleotides in DNA pair up with the ribonucleotides as follows: G ⟶ C, C ⟶ G, T ⟶ A, and A ⟶ U.

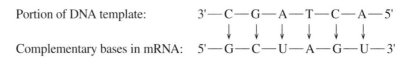

Portion of DNA template: 3'—C—G—A—T—C—A—5'

Complementary bases in mRNA: 5'—G—C—U—A—G—U—3'

Study Check

What is the DNA template that codes for the mRNA having the ribonucleotide sequence 5'—GGGUUUAAA—3'?

Processing of mRNA

In eukaryotes, the genes contain sections known as **exons** that code for proteins that are mixed in with sections called **introns** that do not code for protein. A newly formed mRNA is called a pre-mRNA because it is a copy of the entire DNA template including the noncoding introns. Before the newly synthesized pre-mRNA leaves the nucleus, it undergoes processing to remove the intron sections. The splicing of the pre-RNA produces a mature, functional mRNA that leaves the nucleus to deliver the genetic information to the ribosomes for the synthesis of protein. (See Figure 22.18.)

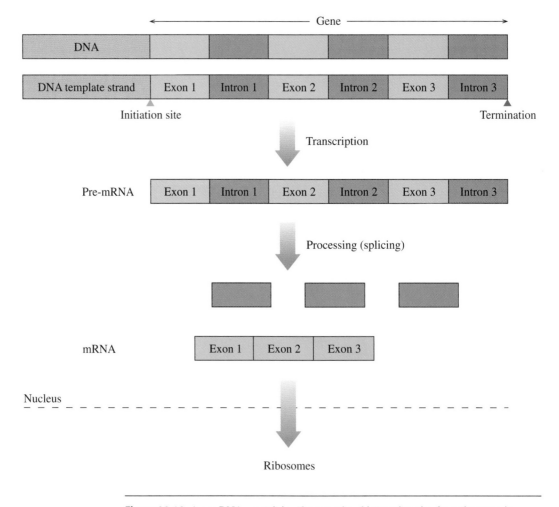

Figure 22.18 A pre-RNA, containing the exons' and introns' copies from the gene, is processed to remove the introns and form a mature mRNA that codes for a protein.

Q What is the difference between exons and introns?

Regulation of Transcription

The synthesis of mRNA does not occur randomly; rather, mRNA is synthesized in response to cellular needs for a particular protein. This regulation takes place at the transcription level where the absence or presence of end products speeds up or slows down the synthesis of mRNAs for specific enzymes. For example, *E. coli* that grow on lactose need β-galactosidase to hydrolyze lactose to glucose and galactose. However, when lactose is low or not present in the cell, there is little or no β-galactosidase present. The transcription of the mRNA for the enzyme is turned off. When lactose enters the cell, β-galactosidase levels rise because lactose induces the synthesis of its enzymes. In this process known as **enzyme induction:** lactose activates the transcription of the genes that produce the mRNAs that code for β-galactosidase.

In prokaryotes, each group of related proteins is regulated by an **operon,** which is a section of DNA containing a control site preceding the **structural genes** that code for proteins. The **control site** consists of a *promoter* and an *operator*. Preceding the operon is a regulatory gene. (See Figure 22.19.) A **regulatory gene** produces a mRNA for the synthesis of a **repressor** protein that binds to the operator and blocks the synthesis of β-galactosidase by RNA polymerase. When lactose enters the cell, lactose inactivates the repressor and it dissociates from the operator.

WEB TUTORIAL
The *lac* Operon in *E. coli*

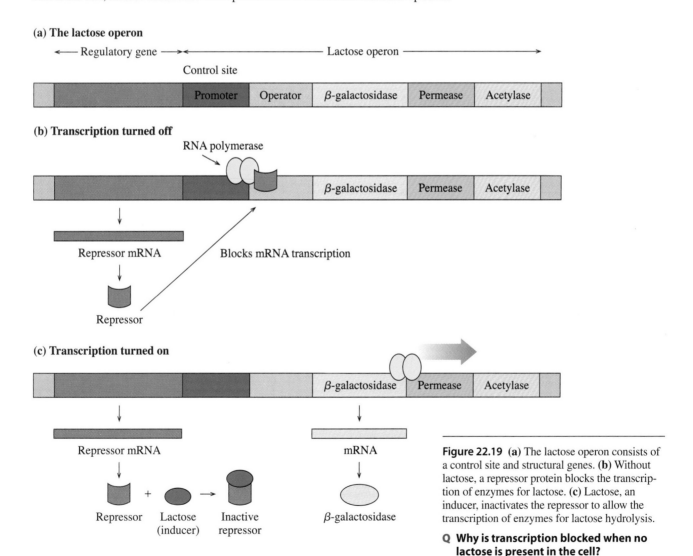

(a) The lactose operon

(b) Transcription turned off

(c) Transcription turned on

Figure 22.19 **(a)** The lactose operon consists of a control site and structural genes. **(b)** Without lactose, a repressor protein blocks the transcription of enzymes for lactose. **(c)** Lactose, an inducer, inactivates the repressor to allow the transcription of enzymes for lactose hydrolysis.

Q Why is transcription blocked when no lactose is present in the cell?

Without a repressor, RNA polymerase transcribes the genes that codes for the lactose enzymes.

Transcription

Indicate whether transcription takes place in each of the following conditions:

a. A repressor binds to the operator in the control site.

b. An inducer binds to the repressor protein.

Solution

a. No. A repressor blocks the synthesis of mRNA by RNA polymerase.

b. Yes. The lactose inactivates the repressor, which activates the transcription of the genes.

Study Check

How does a regulatory gene control the use of materials in a cell?

QUESTIONS AND PROBLEMS

Transcription: Synthesis of mRNA

22.27 What is meant by the term "transcription"?

22.28 What bases in mRNA are used to complement the bases A, T, G, and C in DNA?

22.29 Write the corresponding section of mRNA produced from the following section of DNA template:
3'—CCGAAGGTTCAC—5'

22.30 Write the corresponding section of mRNA produced from the following section of DNA template:
3'—TACGGCAAGCTA—5'

22.31 What are introns and exons?

22.32 What kind of processing do mRNA molecules undergo before they leave the nucleus?

22.33 What is an operon?

22.34 Why does the operon model control protein synthesis at the transcription level?

22.35 How is the operon turned off in *E. coli* that grow on lactose?

22.36 How is the operon activated in *E. coli* that grow on lactose?

22.8 The Genetic Code

The overall function of the RNAs in the cell is to facilitate the task of synthesizing protein. After the genetic information encoded in DNA is transcribed into mRNA molecules, the mRNAs move out of the nucleus to the ribosomes in the cytoplasm. At the ribosomes, the genetic information in the mRNAs is converted into a sequence of amino acids in protein.

$$\text{mRNA} \xrightarrow[\text{ribosomes in the cytoplasm}]{\text{Translation at the}} \text{protein}$$

Codons

Genetic information from DNA is encoded in the mRNA as a sequence of nucleotides. In the **genetic code,** a sequence of three bases (triplet), called a **codon,** specifies each amino acid in the protein. Early work on protein synthesis showed that repeating triplets of uracil (UUU) produced a polypeptide that contained only phenylalanine. Therefore, a sequence of 5'—UUU—UUU—UUU—3' codes is for three phenylalanines.

Codons in mRNA	5'—UUU—UUU—UUU—3'
Translation	↓ ↓ ↓
Amino acid sequence	Phe — Phe — Phe

The codons have now been determined for all 20 amino acids. A total of 64 codons are possible from the triplet combinations of A, G, C, and U. Three of these, UGA, UAA, and UAG, are stop signals that code for the termination of protein synthesis. All the other three-base codons shown in Table 22.4 specify amino acids, which means that one amino acid can have several codons. For example, glycine is the amino acid coded by the codons GGU, GGC, GGA, and GGG. The triplet AUG has two roles in protein synthesis. At the beginning of an mRNA, the codon AUG signals the start of protein synthesis. In the middle of a series of codons, the AUG codon specifies the amino acid methionine.

Table 22.4   mRNA Codons: The Genetic Code for Amino Acids

First Letter	Second Letter				Third Letter
	U	C	A	G	
U	UUU ⎱ Phe UUC ⎰ UUA ⎱ Leu UUG ⎰	UCU ⎱ UCC ⎰ Ser UCA UCG	UAU ⎱ Tyr UAC ⎰ UAA STOP UAG STOP	UGU ⎱ Cys UGC ⎰ UGA STOP UGG Trp	U C A G
C	CUU ⎱ CUC ⎰ Leu CUA CUG	CCU ⎱ CCC ⎰ Pro CCA CCG	CAU ⎱ His CAC ⎰ CAA ⎱ Gln CAG ⎰	CGU ⎱ CGC ⎰ Arg CGA CGG	U C A G
A	AUU ⎱ AUC ⎰ Ile AUA *AUG Met/start	ACU ⎱ ACC ⎰ Thr ACA ACG	AAU ⎱ Asn AAC ⎰ AAA ⎱ Lys AAG ⎰	AGU ⎱ Ser AGC ⎰ AGA ⎱ Arg AGG ⎰	U C A G
G	GUU ⎱ GUC ⎰ Val GUA GUG	GCU ⎱ GCC ⎰ Ala GCA GCG	GAU ⎱ Asp GAC ⎰ GAA ⎱ Glu GAG ⎰	GGU ⎱ GGC ⎰ Gly GGA GGG	U C A G

*Codon that signals the start of a peptide chain.
STOP codons signal the end of a peptide chain.

SAMPLE PROBLEM 22.9

Codons

What is the sequence of amino acids coded by the following codons in mRNA?
5'—GUC—AGC—CCA—3'

Solution

According to Table 22.4, GUC codes for valine, AGC for serine, and CCA for proline. The sequence of amino acids is Val-Ser-Pro.

Study Check

The codon UGA does not code for an amino acid. What is its function in mRNA?

QUESTIONS AND PROBLEMS

The Genetic Code

22.37 What is a codon?

22.38 What is the genetic code?

22.39 What amino acid is coded for by each codon?

    a. CUU   **b.** UCA

    c. GGU   **d.** AGG

22.40 What amino acid is coded for by each codon?

    a. AAA   **b.** GUC

    c. CGG   **d.** GCA

22.41 When does the codon AUG signal the start of a protein, and when does it code for the amino acid methionine?

22.42 The codons UGA, UAA, and UAG do not code for amino acids. What is their role as codons in mRNA?

22.9 Protein Synthesis: Translation

Once the mRNA is synthesized, it migrates out of the nucleus into the cytoplasm to the ribosomes. At the ribosomes, the **translation** process involves tRNA molecules, amino acids, and enzymes, all which convert the codons on mRNA into amino acids to make a protein.

Activation of tRNA

We have seen that the genetic code carried by the mRNA consists of triplets of nucleotides that correspond to one of 20 different amino acids. Now the tRNA are utilized to translate the codons into specific amino acids. This adaptor role of tRNA is possible because it reads both the triplet language of mRNA and picks up the amino acid that corresponds to that codon. The anticodon in the loop at the bottom of a tRNA contains a triplet of bases that complement a codon on mRNA. An amino acid is attached to the acceptor stem of the tRNA by enzymes called *aminoacyl-tRNA synthetases*. There is a different synthetase for each amino acid. The synthetase enzymes use energy from ATP to catalyze the formation of ester bonds between the carboxylic acid groups on the amino acids and the hydroxyl groups of the acceptor stem. (See Figure 22.20.) The matching of a tRNA to the correct amino acid is essential. If the wrong amino acid is attached, it would be placed into the protein and make an incorrect protein. However, each synthetase checks the attachment of an amino acid and tRNA and hydrolyzes any incorrect combinations.

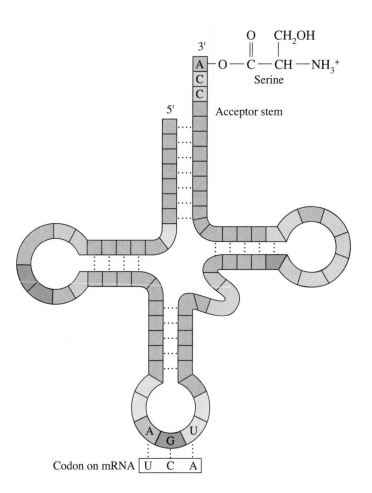

O   CH₂OH

3'
A — O — C — CH — NH_3^+
C Serine
C

5'   Acceptor stem

Codon on mRNA | U   C   A |

Figure 22.20 An activated tRNA with anti-codon AGU bonds to serine at the acceptor stem.

Q What is the codon for serine for this tRNA?

Initiation

Protein synthesis begins when a mRNA combines with the smaller subunit of a ribosome. Because the first codon in an mRNA is a *start* codon, AUG, a tRNA with anticodon UAC and carrying the amino acid methionine forms hydrogen bonds with the AUG codon. The larger ribosome completes the ribosomal unit and translation is ready to begin. (See Figure 22.21.) Protein synthesis takes place on the ribosome at two adjacent sites on the larger subunit. A tRNA that has the anticodon for the second codon with the second amino acid bonds to mRNA. A peptide bond forms between the amino acids and the first tRNA dissociates from the ribosome and returns to the cytoplasm. After the first tRNA detaches from the ribosome, the ribosome shifts to the next codon on the mRNA, a process called **translocation.** A new tRNA can now attach to the open binding site so that its amino acid forms a peptide bond with the previous amino acid. Again the earlier tRNA detaches and the ribosome shifts down the mRNA to read the next codon. Each time a peptide bond joins the new amino acid to the growing polypeptide chain. Sometimes several ribosomes, called a polysome, translate the same strand of mRNA at the same time to produce several copies of the peptide chain.

Termination

After all the amino acids for a particular protein have been linked together by peptide bonds, the ribosome encounters a stop codon. Because there are no tRNAs to complement the termination codon, protein synthesis ends. An enzyme releases the completed polypeptide chain from the ribosome. The initiating amino acid methio-

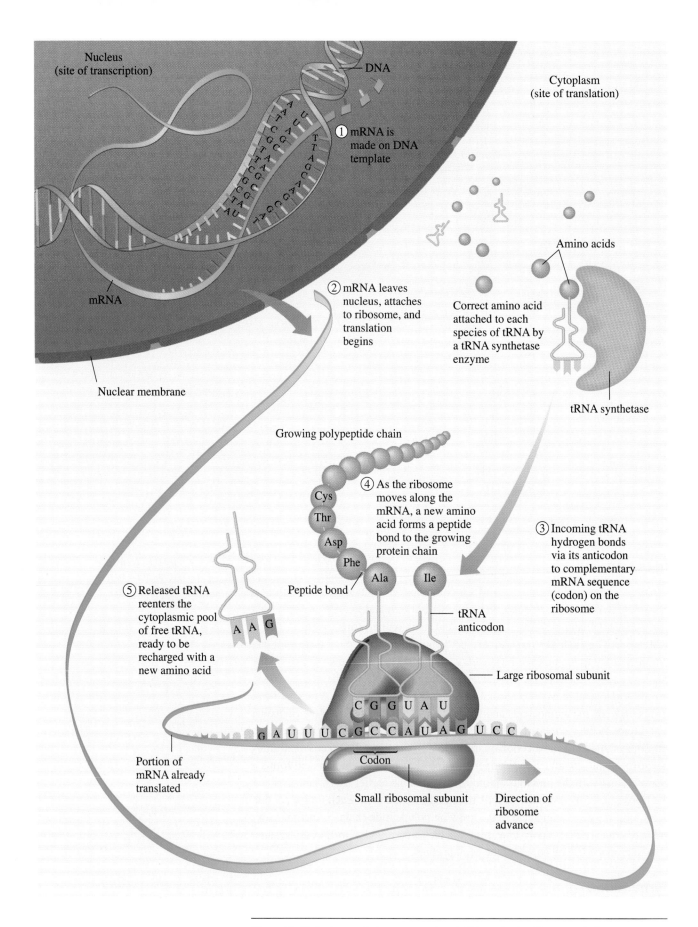

Figure 22.21 In the translation process, the mRNA synthesized by transcription attaches to a ribosome and tRNAs pick up their amino acids and place them in a growing peptide chain.

Q How is the correct amino acid placed in the peptide chain?

nine is often removed from the beginning of the peptide chain. Now the amino acids in the chain form the three-dimensional structure that makes the polypeptide into a biologically active protein.

SAMPLE PROBLEM 22.10

Protein Synthesis: Translation

What order of amino acids would you expect in a peptide for the mRNA sequence of 5'—UCA—AAA—GCC—CUU—3'?

Solution

Each of the codons specifies a particular amino acid. Using Table 22.4, we write a peptide with the following amino acid sequence:

RNA codons: 5'—UCA—AAA—GCC—CUU—3'
                                  ↓   ↓   ↓   ↓
Amino acid sequence: Ser — Lys — Ala— Leu

Study Check

Where would protein synthesis stop in the following series of bases in an mRNA?
5'—GGG—AGC—AGU—UAG—GUU—3'

QUESTIONS AND PROBLEMS

Protein Synthesis: Translation

22.43 What is the difference between a *codon* and an *anticodon*?

22.44 Why are there at least 20 different tRNAs?

22.45 What are the three steps of translation?

22.46 Where does protein synthesis take place?

22.47 What amino acid sequence would you expect from each of the following mRNA segments?
  a. 5'—AAA—AAA—AAA—3'
  b. 5'—UUU—CCC—UUU—CCC—3'
  c. 5'—UAC—GGG—AGA—UGU—3'

22.48 What amino acid sequence would you expect from each of the following mRNA segments?
  a. 5'—AAA—CCC—UUG—GCC—3'
  b. 5'—CCU—CGA—AGC—CCA—UGA—3'
  c. 5'—AUG—CAC—AAA—GAA—GUA—CUU—3'

22.49 How is a peptide chain extended?

22.50 What is meant by "translocation"?

22.51 The following portion of DNA is in the template DNA strand:

  3'—GCT—TTT—CAA—AAA—5'

  a. What is the corresponding mRNA section?
  b. What are the anticodons of the tRNAs?
  c. What amino acids will be placed in the peptide chain?

22.52 The following portion of DNA is in the template DNA strand:

  3'—TGT—GGG—GTT—ATT—5'

  a. What is the corresponding mRNA section?
  b. What are the anticodons of the tRNAs?
  c. What amino acids will be placed in the peptide chain?

HEALTH NOTE

Many Antibiotics Inhibit Protein Synthesis

Several antibiotics stop bacterial infections by interfering with the synthesis of proteins needed by the bacteria. Some antibiotics act only on bacterial cells by binding to the ribosomes in bacteria but do not act on human cells. A description of some of these antibiotics is given in Table 22.5.

Table 22.5   Antibiotics That Inhibit Protein Synthesis in Bacterial Cells

Antibiotic	Effect on Ribosomes to Inhibit Protein Synthesis
Chloramphenicol	Inhibits peptide bond formation and prevents the binding of tRNA
Erythromycin	Inhibits peptide chain growth by preventing the translocation of the ribosome along the mRNA
Puromycin	Causes release of an incomplete protein by ending the growth of the polypeptide early
Streptomycin	Prevents the proper attachment of tRNAs
Tetracycline	Prevents the binding of tRNAs

22.10 Genetic Mutations

A **mutation** is a change in the DNA nucleotide sequence that alters the sequence of amino acids, which may alter the structure and function of a protein in a cell. Some mutations are known to result from X rays, overexposure to sun (ultraviolet or UV light), chemicals called mutagens, and possibly some viruses. If a change in DNA occurs in a somatic cell (a cell other than a reproductive cell) the altered DNA will be limited to that cell and its daughter cells. If there is uncontrolled growth, the mutation could lead to cancer. If the mutation occurs in germ cell DNA (egg or sperm), then all the DNA produced in a new individual will contain the same genetic change. If the genetic change greatly affects the catalysis of metabolic reactions or the formation of important structural proteins, the new cells may not survive or the person may exhibit a genetic disease.

Consider a triplet of bases CCG in the coding strand of DNA, which produces the codon GGC in mRNA. At the ribosome, tRNA would place the amino acid glycine in the peptide chain. (See Figure 22.22a.) Now, suppose that T replaces the first C in the DNA triplet, which gives TCG as the triplet. Then the codon produced in the mRNA is AGC, which brings the amino acid serine to the peptide chain. The replacement of one base in the coding strand of DNA with another is called a **substitution** mutation. The change in the codon can lead to the insertion of a different amino acid at that point in the polypeptide. Substitution is the most common way in which mutations occur. (See Figure 22.22b.)

In a **frame shift mutation,** a base is added to or deleted from the normal order of bases in the coding strand of DNA. Suppose now an A is deleted from the triplet AAA, which gives a new triplet of AAC. The next triplet becomes CGA rather than CCG and so on. All the triplets shift over by one base, which changes all the codons that follow and leads to a different sequence of amino acids from that point. Figure 22.22c illustrates a frame shift mutation by deletion.

Effect of Mutations

When a mutation causes a change in the amino acid sequence, the structure of the resulting protein can be altered severely and it may lose biological activity. If the protein is an enzyme, it may no longer bind to its substrate or react with the substrate at the active site. When an altered enzyme cannot catalyze a reaction, certain substances may accumulate until they act as poisons in the cell, or substances vital to survival may not be synthesized. If a defective enzyme occurs in a major metabolic pathway or in the building of a cell membrane, the mutation can be lethal. When a protein deficiency is genetic, the condition is called a **genetic disease.**

X rays, UV sunlight, mutagens, viruses

DNA $\longrightarrow$ alteration of DNA $\longrightarrow$ defective protein $\longrightarrow$ genetic disease (germ cells) or cancer (somatic cells)

SAMPLE PROBLEM 22.11

Mutations

An mRNA has the sequence of codons 5'—CCC—AGA—GGG—3'. If a base substitution in the DNA changes the mRNA codon of AGA to ACA, how is the amino acid sequence affected in the resulting protein?

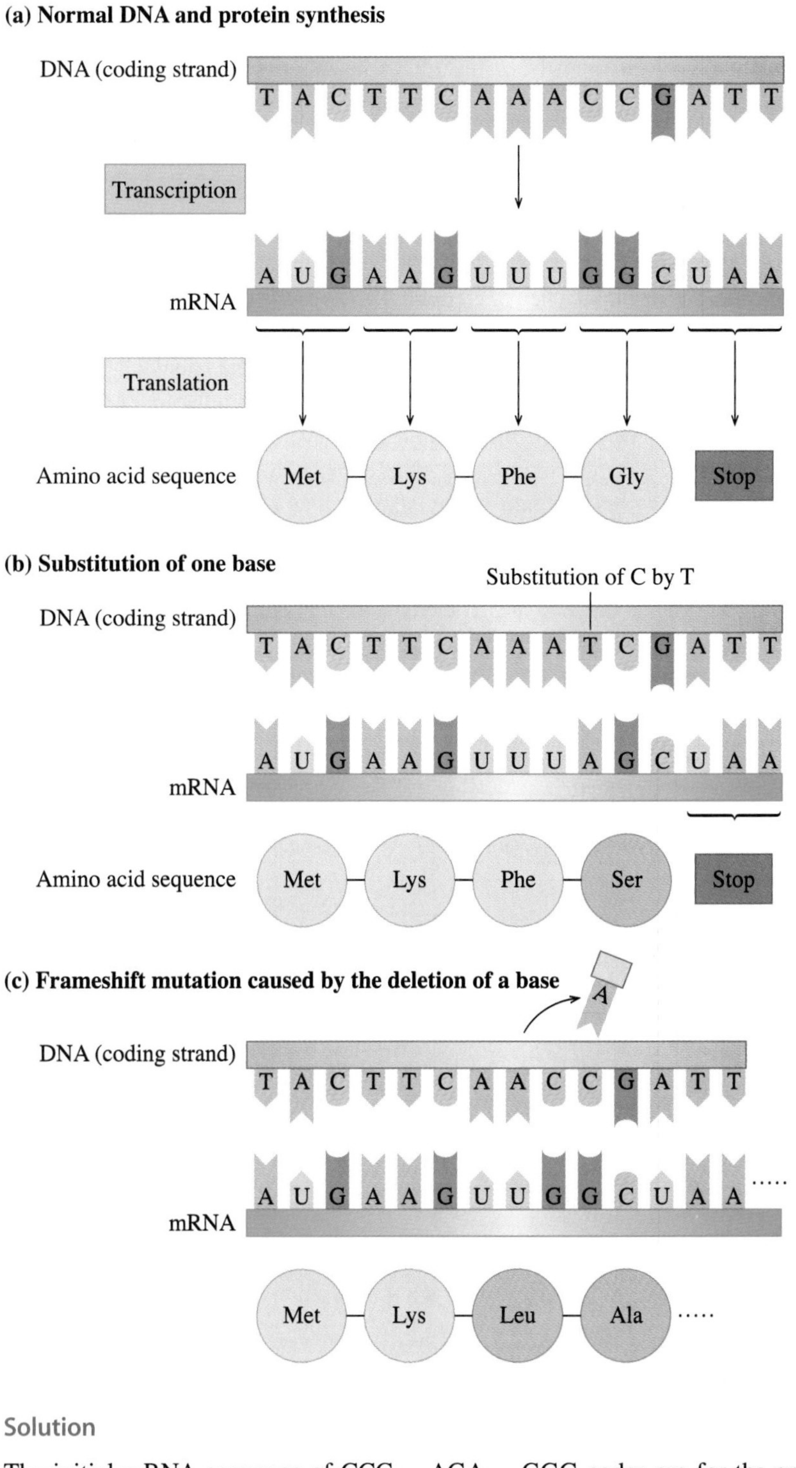

(a) Normal DNA and protein synthesis

DNA (coding strand)

T A C T T C A A A C C G A T T

Transcription

A U G A A G U U U G G C U A A

mRNA

Translation

Amino acid sequence   Met — Lys — Phe — Gly — Stop

(b) Substitution of one base

Substitution of C by T

DNA (coding strand)

T A C T T C A A A T C G A T T

A U G A A G U U U A G C U A A

mRNA

Amino acid sequence   Met — Lys — Phe — Ser — Stop

(c) Frameshift mutation caused by the deletion of a base

A

DNA (coding strand)

T A C T T C A A C C G A T T

A U G A A G U U G G C U A A .....

mRNA

Met — Lys — Leu — Ala .....

Figure 22.22 An alteration in the DNA coding strand (template) produces a change in the sequence of amino acids in the protein, which may lead to a mutation. **(a)** A normal DNA leads to the correct amino order in a protein. **(b)** The substitution of a base in DNA leads to a change in the mRNA codon and a change in the amino acid. **(c)** The deletion of a base causes a frame shift mutation, which changes the amino acid order.

Q When would a substitution mutation cause protein synthesis to stop?

Solution

The initial mRNA sequence of CCC—AGA—GGG codes are for the amino acids proline, arginine, and glycine. When the mutation occurs, the new sequence of mRNA codons CCC—ACA—GGG are for proline, threonine, and glycine. The amino acid arginine is replaced by threonine.

	Normal	*After Mutation*
mRNA codons	CCC—[AGA]—GGG	CCC—[ACA]—GGG
Amino acids	Pro — Arg — Gly	Pro — Thr — Gly

How might the protein made from this mRNA be affected by this mutation?

Genetic Diseases

Figure 22.23 This peacock with albinism does not produce the melanin needed to make bright colors of its feathers.

Q Why are traits such as albinism related to the gene?

A genetic disease is the result of a defective enzyme caused by a mutation in its genetic code. For example, phenylketonuria, PKU, results when DNA cannot direct the synthesis of the enzyme phenylalanine hydroxylase, required for the conversion of phenylalanine to tyrosine. In an attempt to break down the phenylalanine, other enzymes in the cells convert it to phenylpyruvate. The accumulation of phenylalanine and phenylpyruvate in the blood can lead to severe brain damage and mental retardation. If PKU is detected in a newborn baby, a diet is prescribed that eliminates all the foods that contain phenylalanine. Preventing the buildup of the phenylpyruvate ensures normal growth and development.

The amino acid tyrosine is needed in the formation of melanin, the pigment that gives the color to our skin and hair. If the enzyme that converts tyrosine to melanin is defective, no melanin is produced, a genetic disease known as albinism. Persons and animals with no melanin have no skin or hair pigment. (See Figure 22.23.) Table 22.6 lists some other common genetic diseases and the type of metabolism or area affected.

Phenylalanine $\longrightarrow$ Phenylpyruvate $\longrightarrow$ Phenylketonuria (PKU)

Phenylalanine hydroxylase

Tyrosine $\longrightarrow\!\!\times$ Melanin (pigments) $\longrightarrow$ Albinism

QUESTIONS AND PROBLEMS

Genetic Mutations

22.53 What is a substitution mutation?

22.54 How does a substitution mutation in the genetic code for an enzyme affect the order of amino acids in that protein?

22.55 What is the effect of a mutation on the amino acids in the polypeptide?

22.56 How can a mutation decrease the activity of a protein?

22.57 How is protein synthesis affected if the normal base sequence TTT in the DNA template is changed to TTC?

22.58 How is protein synthesis affected if the normal base sequence CCC is changed to ACC?

22.59 Consider the following portion of mRNA produced by the normal order of DNA nucleotides:

Table 22.6   Some Genetic Diseases

Genetic Disease	Result
Galactosemia	The transferase enzyme required for the metabolism of galactose-1-phosphate is absent. Accumulation of Gal-1-P leads to cataracts and mental retardation.
Cystic fibrosis	The most common inherited disease. Thick mucus secretions make breathing difficult and block pancreatic function.
Down syndrome	The leading cause of mental retardation, occurring in about 1 of every 800 live births. Mental and physical problems including heart and eye defects are the result of the formation of three chromosomes, usually chromosome 21, instead of a pair.
Familial hypercholesterolemia	A mutation of a gene on chromosome 19 results in high cholesterol levels that lead to early coronary heart disease in people 30–40 years old.
Muscular dystrophy (Duchenne)	One of 10 forms of MD. A mutation in the X chromosome results in the low or abnormal production of *dystrophin* by the X gene. This muscle-destroying disease appears at about age 5 with death by age 20 and occurs in about 1 of 10,000 males.
Huntington's disease (HD)	Appearing in middle age, HD affects the nervous system, leading to total physical impairment. It is the result of a mutation in a gene on chromosome 4, which can now be mapped to test people in families with HD.
Sickle-cell anemia	A defective hemoglobin from a mutation in a gene on chromosome 11 decreases the oxygen-carrying ability of red blood cells, which take on a sickled shape, causing anemia and plugged capillaries from red blood cell aggregation.
Hemophilia	One or more defective blood-clotting factors lead to poor coagulation, excessive bleeding, and internal hemorrhages.
Tay-Sachs disease	Hexosaminidase A is defective, causing an accumulation of gangliosides resulting in mental retardation, loss of motor control, and early death.

5'—ACA—UCA—CGG—GUA—3'

a. What is the amino acid order produced for normal DNA?
b. What is the amino acid order if a mutation changes UCA to ACA?
c. What is the amino acid order if a mutation changes CGG to GGG?
d. What happens to protein synthesis if a mutation changes UCA to UAA?
e. What happens if a G is added to the beginning of a chain?
f. What happens if the A is removed from the beginning of a chain?

22.60 Consider the following portion of mRNA produced by the normal order of DNA nucleotides:

5'—CUU—AAA—CGA—GUU—3'

a. What is the amino acid order produced for normal DNA?
b. What is the amino acid order if a mutation changes CUU to CCU?
c. What is the amino acid order if a mutation changes CGA to AGA?
d. What happens to protein synthesis if a mutation changes AAA to UAA?

22.61 a. A base substitution changes a codon for an enzyme from GCC to GCA. Why is there no change in the amino acid order in the protein?
b. In sickle-cell anemia, a base substitution in hemoglobin replaces glutamine (a polar amino acid) with valine. Why does the replacement of one amino acid cause such a drastic change in biological function?

22.62 a. A base substitution for an enzyme replaces leucine (a nonpolar amino acid) with alanine. Why does this change in amino acids have little effect on the biological activity of the enzyme?
b. A base substitution replaces cytosine in the codon UCA with adenine. How would this substitution affect the amino acids in the protein?

Describe the preparation and uses of recombinant DNA.

WEB TUTORIAL
Applications of DNA Technology

Restriction Enzymes

22.11 Recombinant DNA

Over the past two decades, new techniques called genetic engineering have permitted scientists to cut, splice, and recombine DNA from different kinds of cells. The new synthetic forms of DNA, which contain a DNA fragment from another organism, are known as **recombinant DNA.** Recombinant DNA technology is now used to produce human insulin for diabetics, the antiviral substance interferon, blood-clotting factor VIII, and human growth hormone.

Restriction Enzymes

Most of the work in recombinant DNA is done with *Escherichia coli (E. coli)* bacteria. The DNA in prokaryotes is present in the bacterial cells as several small circular molecules called *plasmids.* These plasmids, which are easy to isolate, are also capable of replication. The ability to insert foreign DNA into cells developed after scientists discovered that enzymes called **restriction enzymes** can cut DNA between a specific pair of nucleotides along the chain. The resulting DNA fragments have "sticky" ends that can form phosphodiester bonds when they are mixed with plasmids that are also cut open using restriction enzymes. After the foreign DNA fragment joins with the plasmid DNA, the recombined DNA in the plasmids is able to synthesize the proteins coded for by the new DNA. In one day, one *E. coli* bacterium is capable of producing a million copies of itself including the foreign DNA, a process known as gene cloning.

Preparing Recombinant DNA

Initially, *E. coli* cells are soaked in a detergent solution to dissolve the plasma membrane. The contents of the cells including the plasmids are released and collected.

Figure 22.24 Recombinant DNA is formed by placing a gene from another organism in a plasmid DNA of the bacterium, which causes the bacterium to produce a non-bacterial protein such as insulin or growth hormone.

Q How can recombinant DNA help a person with a genetic disease?

WEB TUTORIAL
Cloning a Gene in Bacteria

Gel Electrophoresis of DNA

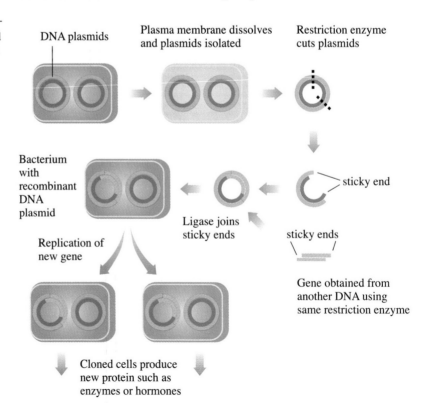

A *restriction enzyme,* which breaks phosphodiester bonds in DNA between specific nucleotides, is used to cut open the circular DNA strands in the plasmids. (See Figure 22.24.) The same enzymes are also used to cut out a piece of DNA from the chromosome of a different organism, such as the gene that produces insulin or growth hormone. The cut-out genes are then mixed with the plasmids that were cut open. The ends of the foreign DNA piece and the ends of the opened plasmids are joined by a DNA ligase. Then the altered plasmids containing the recombined DNA are placed in a fresh culture of *E. coli* bacteria where they can be reabsorbed into the bacterial cells. The new gene that was inserted in the plasmids is copied as the genetically engineered *E. coli* cells start to replicate. If the inserted DNA coded for the human insulin protein, the altered plasmids would begin to synthesize human insulin. Eventually, a large number of cells with the new DNA produce the insulin protein. Table 22.7 lists some of the products developed through recombinant DNA technology that are now used therapeutically.

WEB TUTORIAL
Analyzing DNA Fragments Using Gel Electrophoresis

Table 22.7   Therapeutic Products of Recombinant DNA

Product	Therapeutic Use
Human insulin	Treat diabetes
Erythropoietin (EPO)	Treat anemia; stimulate production of erythrocytes
Human growth hormone (HGH)	Stimulate growth
Interferon	Treat cancer and viral disease
Tumor necrosis factor (TNF)	Destroy tumor cells
Monoclonal antibodies	Transport drugs needed to treat cancer and transplant rejection
Epidermal growth factor (EGF)	Stimulate healing of wounds and burns
Interleukins	Stimulate immune system; treat cancer
Prourokinase	Destroy blood clots; treat myocardial infarctions

WEB TUTORIAL
DNA Fingerprinting

DNA Fingerprinting

In a process referred to as DNA fingerprinting or Southern transfer, restriction enzymes are used to cut fragments from DNA. The resulting DNA fragments called RFLPs (restriction fragment length polymorphisms) are then sorted by size using gel electrophoresis. The gel is treated with a radioactive isotope that adheres to specific base sequences in the RFLPs. A piece of x-ray film placed over the gel is exposed by the radioactivity emitted from RFLPs, which gives a pattern of dark and light bands known as a DNA fingerprint. (See Figure 22.25.) It has been estimated that the odds of two persons (who are not identical twins) producing the same DNA fingerprint is less than one in a billion.

One application of Southern transfer is genetic screening in which an individual can be screened for genes that are responsible for genetic diseases. It is being used to screen for genes responsible for sickle-cell disease, cystic fibrosis, breast cancer, colon cancer, Huntington's disease, and amyotropic lateral sclerosis also called Lou Gehrig's disease.

Human Genome Project

In the 1970s, scientists began to use restriction enzymes to map the location of genomes in viruses and plasmids. By 1987, the genome of *E. coli* was determined.

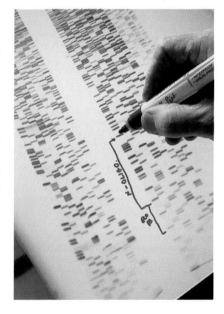

Figure 22.25 A scientist analyzes a nucleotide sequence in DNA from a human gene. Dark and light bands on the film represent the order of nucleotides. The marked sequences are involved in the growth of melanoma cancer cells.

Q   What causes DNA fragments to appear on x-ray film?

WEB TUTORIAL
The Human Genome Project:
Human Chromosome 17

Recently, these techniques combined with new computer programs have helped scientists compile the map of the human genome. In some recent reports, the total number of human genes is about 30,000. It now appears that most of the genome is not functional and perhaps carried over for millions of years. Large blocks of genes are copied from one human chromosome to another even though they no longer code for needed proteins. Thus, the coding portions of the genes seems to be even less than previously thought making up only about 1% of the total genome. The results of the genome project will help us identify defective genes that lead to genetic disease.

Polymerase Chain Reaction

The process of gene cloning using recombinant DNA requires living cells such as *E. coli*. In 1987 a process called **polymerase chain reaction (PCR)** made it possible to produce multiple copies of (amplify) the DNA in a short time. In the PCR technique, a sequence of a DNA is selected to copy, and the DNA is heated to separate the strands. Primers that are complementary to a small group of nucleotides on each side of the sequence to be copied are added to the ends of the templates. The DNA strands with their primers are mixed with DNA polymerase and a mixture of deoxyribonucleotides and complementary strands for the DNA section are produced. Then the process is repeated with the new batch of DNA. After several cycles of the PCR process, millions of copies of the initial DNA section are produced. (See Figure 22.26.)

SAMPLE PROBLEM 22.12

Recombinant DNA

What is the function of restriction enzymes in recombinant DNA?

Solution

Restriction enzymes are used to cut out a particular piece of DNA from a gene and to cut open the circular plasmids in a bacterium where the foreign DNA attaches.

Study Check

What is gene cloning?

QUESTIONS AND PROBLEMS

Recombinant DNA

22.63 Why are *E. coli* bacteria used in recombinant DNA procedures?

22.64 What is a plasmid?

22.65 How are plasmids obtained from *E. coli*?

22.66 Why are restriction enzymes mixed with the plasmids?

22.67 How is a gene for a particular protein inserted into a plasmid?

22.68 Why is DNA polymerase useful in criminal investigations?

22.69 What is a DNA fingerprint?

22.70 What beneficial proteins are produced from recombinant DNA technology?

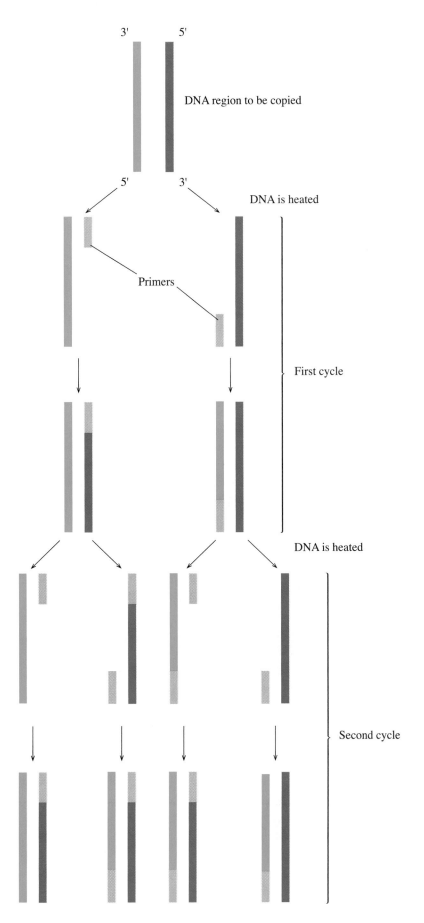

3'

5'

DNA region to be copied

5'

3'

DNA is heated

Primers

First cycle

DNA is heated

Second cycle

Figure 22.26 Each cycle of the polymerase chain reaction doubles the number of copies of the DNA section.

Q **Why are the DNA strands heated at the start of each cycle?**

Describe the methods by which a virus infects a cell.

22.12 Viruses

Viruses are small particles of 3 to 200 genes that cannot replicate without a host cell. A typical virus contains a nucleic acid, DNA or RNA, but not both, inside a protein coat. It does not have the necessary material such as nucleotides and enzymes to make proteins and grow. The only way a virus can replicate is to invade a host cell and take over the materials necessary for protein synthesis and growth. Some infections caused by viruses invading human cells are listed in Table 22.8. There are also viruses that attack bacteria, plants, and animals.

Table 22.8   Some Diseases Caused by Viral Infection

Disease	Virus
Common cold	Coronavirus (over 100 types)
Influenza	Orthomyxovirus
Warts	Papovavirus
Herpes	Herpesvirus
Leukemia, cancers, AIDS	Retrovirus
Hepatitis	Hepatitis A virus (HAV), hepatitis B virus (HBV)
Mumps	Paramyxovirus
Epstein-Barr	Epstein-Barr virus (EBV)

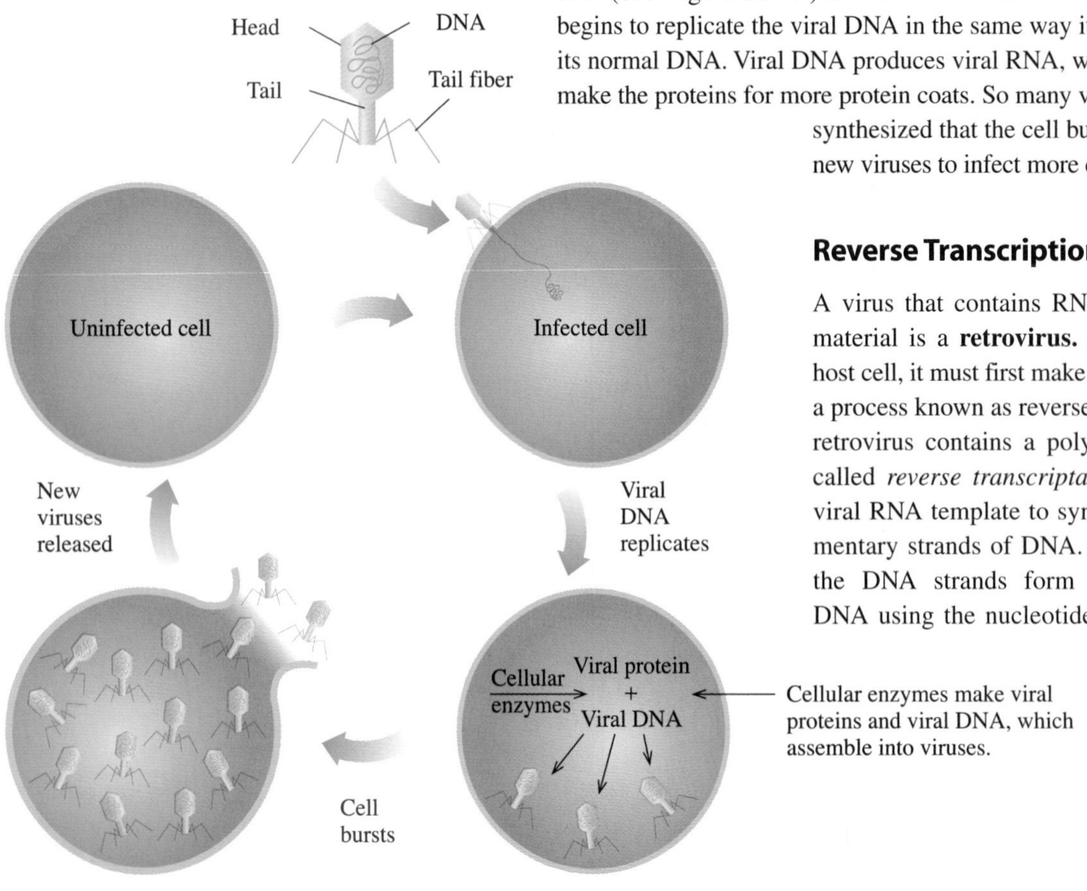

Figure 22.27 After a virus attaches to the host cell, it injects its viral DNA and uses the host cell's amino acids to synthesize viral protein and nucleic acids, enzymes, and ribosomes to make viral RNA. When the cell bursts, the new viruses are released to infect other cells.

Q **Why does a virus need a host cell for replication?**

A viral infection begins when an enzyme in the protein coat makes a hole in the host cell, allowing the nucleic acids to enter and mix with the materials in the host cell. (See Figure 22.27.) If the virus contains DNA, the host cell begins to replicate the viral DNA in the same way it would replicate its normal DNA. Viral DNA produces viral RNA, which proceeds to make the proteins for more protein coats. So many virus particles are synthesized that the cell bursts and releases new viruses to infect more cells.

Head — DNA
Tail — Tail fiber

Uninfected cell

Infected cell

New viruses released

Viral DNA replicates

Cellular enzymes — Viral protein + Viral DNA ← Cellular enzymes make viral proteins and viral DNA, which assemble into viruses.

Cell bursts

Reverse Transcription

A virus that contains RNA as its genetic material is a **retrovirus.** Once inside the host cell, it must first make viral DNA using a process known as reverse transcriptase. A retrovirus contains a polymerase enzyme called *reverse transcriptase* that uses the viral RNA template to synthesize complementary strands of DNA. Once produced, the DNA strands form double-stranded DNA using the nucleotides present in the

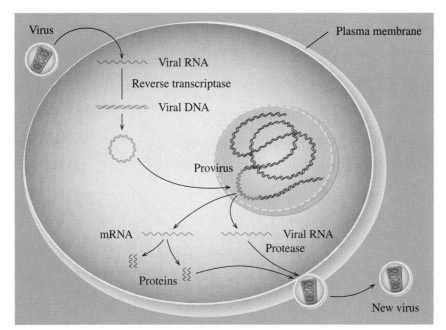

Figure 22.28 After a retrovirus injects its viral RNA into a cell, it forms a DNA strand by reverse transcription. The DNA forms a double-stranded DNA called a provirus, which joins the host cell DNA. When the cell replicates, the provirus produces the viral RNA needed to produce more virus particles.

Q What is reverse transcription?

host cell. This newly formed viral DNA called a provirus joins the DNA of the host cell. (See Figure 22.28.)

Vaccines are inactive forms of viruses that boost the immune response by causing the body to produce antibodies to the virus. Several childhood diseases such as polio, mumps, chicken pox, and measles can be prevented through the use of vaccines.

AIDS

In the early 1980s, a disease called AIDS (acquired immune deficiency syndrome) began to claim an alarming number of lives. An HIV-1 virus (human immunodeficiency virus type 1) is now known to be the AIDS-causing agent. (See Figure 22.29.) HIV is a retrovirus that infects and destroys T4 lymphocyte cells, which are involved in the immune response. After the HIV-1 virus binds to receptors on the surface of a T4 cell, the virus injects viral RNA into the host cell. As a retrovirus, the genes of the viral RNA direct the formation of viral DNA. The gradual depletion of T4 cells reduces the ability of the immune system to destroy harmful organisms. The AIDS syndrome is characterized by opportunistic infections such as *Pneumocystis carinii,* which causes pneumonia, and *Kaposi's sarcoma,* a skin cancer.

Treatment for AIDS is based on attacking the HIV-1 at different points in its life cycle including reverse transcription and protein synthesis. Nucleoside analogs mimic the structures of the nucleosides used for DNA synthesis. For example, the drug AZT (azidothymine) is similar to thymine, and ddI (dideoxyinosine) is similar to guanosine. Two other drugs include dideoxycytidine (ddC) and didehydro-3'-deoxythymidine (d4T). Such compounds are found in the "cocktails" that are providing extended remission of HIV infections. When a nucleoside analog is incorporated into viral DNA, the lack of a hydroxyl group on the 3' carbon in the sugar prevents the formation of the sugar–phosphate bonds, which prevents the replication of the virus.

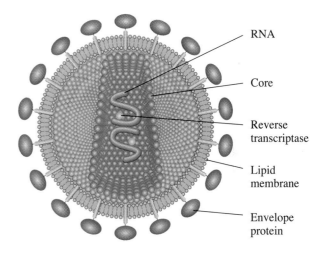

RNA

Core

Reverse transcriptase

Lipid membrane

Envelope protein

Figure 22.29 The HIV virus that causes AIDS syndrome destroys the immune system in the body.

Q Is the HIV virus a DNA virus or an RNA retrovirus?

WEB TUTORIAL
HIV Reproductive Cycle

AZT
Azidothymine

Dideoxyinosine (ddI)

Dideoxycytidine (ddC)

Didehydro-3'-deoxythymidine (d4T)

The newest and most powerful anti-HIV drugs are the protease inhibitors such as saquinavir (Invirase), indinavir, and ritonavir, which modify the three-dimensional structure of the protease active site. The inhibition of protease prevents the synthesis of proteins needed to make more copies of the virus. Researchers are not yet certain how long protease inhibitors will be beneficial for a person with AIDS, but they are encouraged by the current studies.

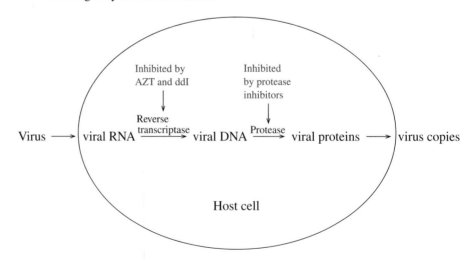

Cancer

In an adult body, most cells do not continue to reproduce. When cells in the body begin to grow and multiply without control, they invade neighboring cells and appear as a tumor or growth (neoplasm). When tumors interfere with normal functions of the body, they are cancerous. If they are limited, they are benign. Cancer can be caused by chemical and environmental substances, by radiation, or by oncogenic viruses.

Some reports estimate that 70–80% of all human cancers are initiated by chemical and environmental substances. A carcinogen is any substance that increases the chance of inducing a tumor. Known carcinogens include dyes, cigarette smoke, and asbestos. More than 90% of all persons with lung cancer are smokers. A carcinogen causes cancer by reacting with molecules in a cell, probably DNA, and altering the growth of that cell. Some known carcinogens are listed in Table 22.9.

Radiant energy from sunlight or medical radiation is another type of environmental factor. Skin cancer has become one of the

most prevalent forms of cancer. It appears that DNA damage in the exposed areas of the skin causes mutations. The cells lose their ability to control protein synthesis, and uncontrolled cell division leads to cancer. The incidence of malignant melanoma, one of the most serious skin cancers, has been rapidly increasing. Some possible factors for this increase may be the popularity of suntanning as well as the reduction of the ozone layer, which absorbs much of the harmful radiation from sunlight.

Oncogenic viruses cause cancer when cells are infected. Several viruses associated with human cancers are listed in Table 22.10. Some cancers such as retinoblastoma and breast cancer appear to occur more frequently in families. There is some indication that a missing or defective gene may be responsible.

Table 22.9   Some Chemical and Environmental Carcinogens

Carcinogen	Tumor Site
Asbestos	Lung, respiratory tract
Arsenic	Skin, lung
Cadmium	Prostate, kidneys
Chromium	Lung
Nickel	Lung, sinuses
Aflatoxin	Liver
Nitrites	Stomach
Aniline dyes	Bladder
Vinyl chloride	Liver

Table 22.10   Human Cancers Caused by Oncogenic Viruses

Virus	Disease
RNA viruses	
Human T-cell lymphotropic virus-type I (HTLV-I)	Leukemia
DNA viruses	
Epstein-Barr virus (EBV)	Burkitt's lymphoma (cancer of white blood B cells)
	Nasopharyngeal carcinoma
	Hodgkin's disease
Hepatitis B virus (HBV)	Liver cancer
Herpes simplex virus (type 2)	Cervical and uterine cancer
Papilloma virus	Cervical and colon cancer, genital warts

SAMPLE PROBLEM 22.13

Viruses

Why are viruses unable to replicate on their own?

Solution

Viruses only contain packets of DNA or RNA, but not the necessary replication machinery that includes enzymes and nucleosides.

Study Check

How do protease inhibitors affect the life cycle of the HIV-1 virus?

QUESTIONS AND PROBLEMS

Viruses

22.71 What type of genetic information is found in a virus?

22.72 Why do viruses need to invade a host cell?

22.73 A virus contains viral RNA.
   a. Why would reverse transcription be used in the life cycle of this type of virus?
   b. What is the name of this type of virus?
22.74 What is the purpose of a vaccine?
22.75 How do nucleoside analogs disrupt the life cycle of the HIV-1 virus?
22.76 How do protease inhibitors disrupt the life cycle of the HIV-1 virus?

Chapter Review

22.1 Components of Nucleic Acids
Nucleic acids, deoxyribonucleic acid (DNA), and ribonucleic acid (RNA) are polymers of nucleotides. A nucleoside is a combination of a pentose sugar and a nitrogen base.

22.2 Nucleosides and Nucleotides
A nucleotide is composed of three parts: a nitrogenous base, a sugar, and a phosphate group. In DNA, the sugar is deoxyribose and the nitrogen-containing base can be adenine, thymine, guanine, or cytosine. In RNA, the sugar is ribose and uracil replaces thymine.

22.3 Primary Structure of Nucleic Acids
Each nucleic acid has its own unique sequence of bases known as its primary structure. In a nucleic acid polymer, the 3'OH of each ribose sugar in RNA or deoxyribose sugar in DNA forms a phosphodiester bond to the phosphate group of the 5'-carbon atom group of the sugar in the next nucleotide to give a backbone of alternating sugar and phosphate groups. There is a free 5'-phosphate at one end of the polymer, and a free 3' OH group at the other end.

22.4 DNA Double Helix: A Secondary Structure
A DNA molecule consists of two strands of nucleotides that are wound around each other like a spiral staircase. The two strands are held together by hydrogen bonds between complementary base pairs, A with T, and G with C.

22.5 DNA Replication
During DNA replication, DNA polymerase makes new DNA strands along each of the original DNA strands that serve as templates. Complementary base pairing ensures the correct pairing of bases to give identical copies of the original DNA.

22.6 Types of RNA
The three types of RNA differ by function in the cell: ribosomal RNA makes up most of the structure of the ribosomes, messenger RNA carries genetic information from the DNA to the ribosomes, and transfer RNA places the correct amino acids in the protein.

22.7 Transcription: Synthesis of mRNA
Transcription is the process by which RNA polymerase produces mRNA from one strand of DNA. The bases in the mRNA are complementary to the DNA, except U is paired with A in RNA. The production of mRNA occurs when certain proteins are needed in the cell. In enzyme induction, the appearance of a substrate in a cell removes a repressor, which allows RNA polymerase to produce mRNA at the structural genes.

22.8 The Genetic Code
The genetic code consists of a sequence of three bases (triplet) that specifies the order for the amino acids in a protein. There are 64 codons for the 20 amino acids, which means there are several codons for some amino acids. The codon AUG signals the start of transcription and codons UAG, UGA, and UAA signal it to stop.

22.9 Protein Synthesis: Translation
Proteins are synthesized at the ribosomes in a translation process that includes three steps: initiation, translocation, and termination. During translation, tRNAs bring the appropriate amino acids to the ribosome and peptide bonds form. When the polypeptide is released, it takes on its secondary and tertiary structures and becomes a functional protein in the cell.

22.10 Genetic Mutations
A genetic mutation is a change of one or more bases in the DNA sequence that alters the structure and ability of the resulting protein to function properly. In a substitution, one base may be altered, and a frame shift mutation inserts or deletes a base and changes all the codons after the base change.

22.11 Recombinant DNA

A recombinant DNA is prepared by inserting a DNA segment, a gene, into the DNA present in plasmids of *E. coli* bacteria. As the altered bacterial cells replicate, the protein expressed by the foreign DNA segment is produced. In criminal investigation, large quantities of DNA are obtained from smaller amounts by DNA polymerase chain reactions.

22.12 Viruses

Viruses containing DNA or RNA must invade host cells to use the machinery within the cell for the synthesis of more viruses. For a retrovirus containing RNA, a viral DNA is synthesized by reverse transcription using the nucleotides and enzymes in the host cell. In the treatment of AIDS, nucleoside analogs inhibit the reverse transcriptase of the HIV-1 virus, and protease inhibitors disrupt the catalytic activity of protease needed to produce proteins for the synthesis of more viruses.

Key Terms

anticodon The triplet of bases in the center loop of tRNA that is complementary to a codon on mRNA.

codon A sequence of three bases in mRNA that specifies a certain amino acid to be placed in a protein. A few codons signal the start or stop of transcription.

complementary base pairs In DNA, adenine is always paired with thymine (A—T or T—A), and guanine is always paired with cytosine (G—C or C—G). In forming RNA, adenine is paired with uracil (A—U).

control site A section of DNA composed of a promoter and operator that regulates protein synthesis.

DNA Deoxyribonucleic acid; the genetic material of all cells containing nucleotides with deoxyribose sugar, phosphate, and the four nitrogenous bases adenine, thymine, guanine, and cytosine.

double helix The helical shape of the double chain of DNA that is like a spiral staircase with a sugar–phosphate backbone on the outside and base pairs like stair steps on the inside.

enzyme induction A model of cellular regulation in which protein synthesis is induced by a substrate.

exons The sections in a DNA template that code for proteins.

frame shift mutation A mutation that inserts or deletes a base in a DNA sequence.

genetic code The sequence of codons in mRNA that specifies the amino acid order for the synthesis of protein.

genetic disease A physical malformation or metabolic dysfunction caused by a mutation in the base sequence of DNA.

introns The sections in DNA that do not code for protein.

mRNA Messenger RNA; produced in the nucleus by DNA to carry the genetic information to the ribosomes for the construction of a protein.

mutation A change in the DNA base sequence that alters the formation of a protein in the cell.

nitrogen-containing base Nitrogen-containing compounds found in DNA and RNA: adenine (A), thymine (T), cytosine (C), guanine (G), and uracil (U).

nucleic acids Large molecules composed of nucleotides, found as a double helix in DNA, and as the single strands of RNA.

nucleoside The combination of a pentose sugar and a nitrogen-containing base.

nucleotides Building blocks of a nucleic acid consisting of a nitrogen-containing base, a pentose sugar (ribose or deoxyribose), and a phosphate group.

Okazaki fragments The short segments formed by DNA polymerase in the daughter DNA strand that runs in the 3' to 5' direction.

operon A group of genes, including a control site and structural genes, whose transcription is controlled by the same regulatory gene.

phosphodiester bond The phosphate link that joins the 3' hydroxyl group in one nucleotide to the phosphate group on the 5'-carbon atom in the next nucleotide.

polymerase chain reaction (PCR) A strand of DNA is copied many times by mixing it with DNA polymerase and a mixture of deoxyribonucleotides.

primary structure The sequences of nucleotides in nucleic acids.

recombinant DNA DNA spliced from different organisms to form new, synthetic DNA.

regulatory gene A gene in front of the control site that produces a repressor.

replication The process of duplicating DNA by pairing the bases on each parent strand with their complementary base.

replication forks The open sections in unwound DNA strands where DNA polymerase begins the replication process.

repressor A protein that interacts with the operator gene in an operon to prevent the transcription of mRNA.

restriction enzyme An enzyme that cuts a large DNA strand into smaller fragments to isolate a gene or to remove a portion of the DNA in the plasmids of *E. coli.*

retrovirus A virus that contains RNA as its genetic material and that synthesizes a complementary DNA strand inside a cell.

RNA Ribonucleic acid, a type of nucleic acid that is a single strand of nucleotides containing adenine, cytosine, guanine, and uracil.

rRNA Ribosomal RNA; the most prevalent type of RNA; a major component of the ribosomes.

structural genes The sections of DNA that code for the synthesis of proteins.

substitution A mutation that replaces one base in a DNA with a different base.

transcription The transfer of genetic information from DNA by the formation of mRNA.

translation The interpretation of the codons in mRNA as amino acids in a peptide.

translocation The shift of a ribosome along mRNA from one codon (three bases) to the next codon during translation.

tRNA Transfer RNA; an RNA that places a specific amino acid into a peptide chain at the ribosome. There is one or more tRNA for each of the 20 different amino acids.

virus Small particles containing DNA or RNA in a protein coat that require a host cell for replication.

Additional Problems

22.77 Identify each of the following nitrogen bases as a pyrimidine or a purine:
  a. cytosine   **b.** adenine   **c.** uracil
  d. thymine   **e.** guanine

22.78 Indicate if each of the nitrogen bases in problem 22.77 are found in DNA only, RNA only, or both DNA and RNA.

22.79 Identify the nitrogen base and sugar in each of the following nucleosides:
  a. deoxythymidine   **b.** adenosine
  c. cytidine   **d.** deoxyguanosine

22.80 Identify the nitrogen base and sugar in each of the following nucleotides:
  a. CMP   **b.** dAMP
  c. dGMP   **d.** UMP

22.81 How do the bases of thymine and uracil differ?

22.82 How do the bases of cytosine and uracil differ?

22.83 Draw the structure of CMP.

22.84 Draw the structure of dGMP.

22.85 What is similar about the primary structure of RNA and DNA?

22.86 What is different about the primary structure of RNA and DNA?

22.87 If the DNA double helix in salmon contains 28% adenine, what is the percent of thymine, guanine, and cytosine?

22.88 If the DNA double helix in humans contains 20% cytosine, what is the percent of guanine, adenine, and thymine?

22.89 In DNA, how many hydrogen bonds form between each of the following?
  a. adenine and thymine
  b. guanine and cytosine

22.90 Why are there no base pairs in DNA between adenine and guanine, or thymine and cytosine?

22.91 Write the complementary base sequence for each of the following DNA segments:
  a. 5'—GACTTAGGC—3'
  b. 3'—TGCAAACTAGCT—5'
  c. 5'—ATCGATCGATCG—3'

22.92 Write the complementary base sequence for each of the following DNA segments:
  a. 5'—TTACGGACCGC—3'
  b. 5'—ATAGCCCTTACTGG—3'
  c. 3'—GGCCTACCTTAACGACG—5'

22.93 In DNA replication, what is the difference in the synthesis of the leading strand and the lagging strand?

22.94 How are the Okazaki fragments joined to the growing DNA strand?

22.95 Where are the DNA strands of the original DNA found in the double helix of each of the daughter DNA molecules?

22.96 How can replication occur at several places along a DNA double helix?

22.97 Match the following statements with rRNA, mRNA, or tRNA.
   a. is the smallest type of RNA
   b. makes up the highest percent of RNA in the cell
   c. carries genetic information from the nucleus to the ribosomes

22.98 Match the following statements with rRNA, mRNA, or tRNA.
   a. have two subunits
   b. brings amino acids to the ribosomes for protein synthesis
   c. acts as a template for protein synthesis

22.99 What are the possible codons for each of the following amino acids?
   a. threonine   **b.** serine   **c.** cysteine

22.100 What are the possible codons for each of the following amino acids?
   a. valine   **b.** proline   **c.** histidine

22.101 What is the amino acid for each of the following codons?
   a. AAG   **b.** AUU   **c.** CGG

22.102 What is the amino acid for each of the following codons?
   a. CAA   **b.** GGC   **c.** AAC

22.103 Endorphins are polypeptides that reduce pain. What is the amino acid order for the following mRNA that codes for a pentapeptide that is an endorphin called leucine enkephalin?

5'—AUG—UAC—GGU—GGA—UUU—CUA—UAA—3'

22.104 Endorphins are polypeptides that reduce pain. What is the amino acid order for the following mRNA that codes for a pentapeptide that is an endorphin called methionine enkephalin?

5'—AUG—UAC—GGU—GGA—UUU—AUG—UAA—3'

22.105 What is the anticodon on tRNA for each of the following codons in an mRNA?
   a. AGC   **b.** UAU   **c.** CCA

22.106 What is the anticodon on tRNA for each of the following codons in an mRNA?
   a. GUG   **b.** CCC   **c.** GAA

22.107 Oxytocin is a nonapeptide with nine amino acids. How many nucleotides would be found in the mature mRNA for this protein?

22.108 A protein contains 35 amino acids. How many nucleotides would be found in the mature mRNA for this protein?

22.109 What is the difference between a DNA virus and a retrovirus?

22.110 How does polymerase chain reaction (PCR) produce many copies of a DNA section?

23 Metabolic Pathways for Carbohydrates

"I was checking this dog's ears for foxtails and her eyes for signs of conjunctivitis," says Joyce Rhodes, veterinary assistant at the Sonoma Animal Hospital. "We always check a dog's teeth for tartar, because dental care is very important to the well-being of the animal. When I do need to give a medication to an animal, I use my chemistry to prepare the proper dose that the pet should take. Dosages may be in milligrams, kilograms, or milli-liters."

As a member of the veterinary health care team, a veterinary techni-cian (VT) assists a veterinarian in the care and handling of animals. A VT takes medical histories, collects spec-imens, performs laboratory proce-dures, prepares an animal for surgery, assists in surgical procedures, takes X rays, talks with animal owners, and cleans teeth.

LOOKING AHEAD

the Chemistry place

www.chemplace.com/college

Visit the URL above or use the CD-ROM in the book for extra quizzing, interactive tutorials, career resources, and case studies.

When we eat food such as a tuna fish sandwich, the polysaccharides, lipids, and proteins are digested to smaller molecules that are absorbed into the cells of our body. As these molecules of glucose, fatty acids, and amino acids are broken down further, energy is released. This energy is used in the cells to synthesize high-energy compounds such as adenosine triphosphate (ATP). Our cells utilize ATP energy when they do work such as contracting muscles, synthesizing large molecules, sending nerve impulses, and moving substances across cell membranes.

All the chemical reactions that take place in living cells to break down or build molecules are known as *metabolism*. In a metabolic pathway, reactions are linked together in a series, each catalyzed by a specific enzyme to produce an end product. In this and the following chapters, we will look at these pathways and the way they produce energy and cellular compounds.

23.1 Metabolism and Cell Structure

LEARNING GOAL
Describe three stages of metabolism.

The term **metabolism** refers to all the chemical reactions that provide energy and the substances required for continued cell growth. There are two types of metabolic reactions: catabolic and anabolic. In **catabolic reactions,** complex molecules are broken down to simpler ones with an accompanying release of energy. **Anabolic reactions** utilize energy available in the cell to build large molecules from simple ones.

We can think of the catabolic processes in metabolism as consisting of three stages. (See Figure 23.1.) Let's use that tuna fish sandwich for our example. In stage 1 of metabolism, the processes of digestion break down the large macromolecules into small monomer units. For example, the polysaccharides in bread break down to monosaccharides, the lipids in the mayonnaise break down to glycerol and fatty acids, and the proteins from the tuna yield amino acids. These digestion products diffuse into the bloodstream for transport to cells. In stage 2, they are broken down to two- and three-carbon compounds such as pyruvate and acetyl CoA in the cells. Stage 3 begins with the oxidation of the two-carbon acetyl CoA in the citric acid cycle, which produces several reduced coenzymes. As long as the cells have oxygen, the hydrogen ions and electrons are transferred to the electron transport chain, where most of the energy in the cell is produced. This energy is used to synthesize adenosine triphosphate (ATP), which provides energy for the anabolic pathways in the cell.

Cell Structure for Metabolism

To understand the relationship of metabolic reactions, we need to look at where metabolic reactions take place in the cells. (See Figure 23.2.) In Chapter 22, we described two types of cells: prokaryotic and eukaryotic. *Prokaryotic* (before nucleus) cells have no nucleus and form single-celled organisms such as bacteria. The cells in plants and animals are *eukaryotic* cells, which have a nucleus.

In a eukaryotic cell, a *plasma membrane* is a lipid bilayer that separates the materials inside the cell from the aqueous environment surrounding the cell. In addition, the outer surface of the plasma membrane contains structures that allow

Stages of Metabolism

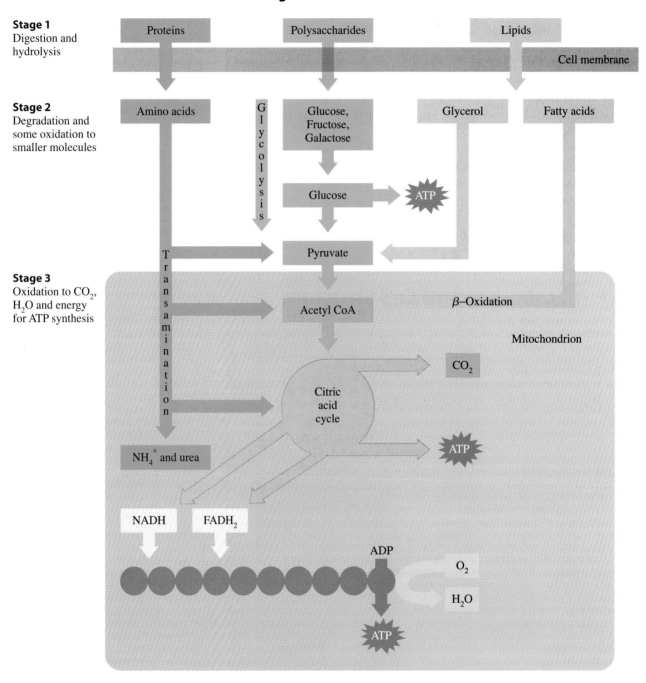

Stage 1
Digestion and
hydrolysis

Stage 2
Degradation and
some oxidation to
smaller molecules

Stage 3
Oxidation to CO_2,
H_2O and energy
for ATP synthesis

Proteins

Polysaccharides

Lipids

Cell membrane

Amino acids

Glycolysis

Glucose,
Fructose,
Galactose

Glycerol

Fatty acids

Glucose

ATP

Pyruvate

Transamination

Acetyl CoA

β–Oxidation

Mitochondrion

CO_2

Citric
acid
cycle

ATP

NH_4^+ and urea

NADH

$FADH_2$

ADP

O_2

H_2O

ATP

Figure 23.1 In the three stages of catabolic
metabolism, foods are digested and
degraded into smaller molecules, which are
oxidized to produce energy.

**Q Where is most of the ATP energy
produced in the cells?**

cells to communicate with each other. The *nucleus* contains the genes that control
DNA replication and protein synthesis of the cell. The **cytoplasm** consists of all the
materials between the nucleus and the plasma membrane. The **cytosol,** which is the
fluid part of the cytoplasm, is an aqueous solution of electrolytes and enzymes that
catalyze many of the cell's chemical reactions.

Within the cytoplasm are specialized structures called *organelles* that carry out
specific functions in the cell. We have already seen (Chapter 22) that the *ribosomes*
are the sites of protein synthesis using mRNA templates. The *endoplasmic reticulum* consists of two forms, a rough endoplasmic reticulum where proteins are

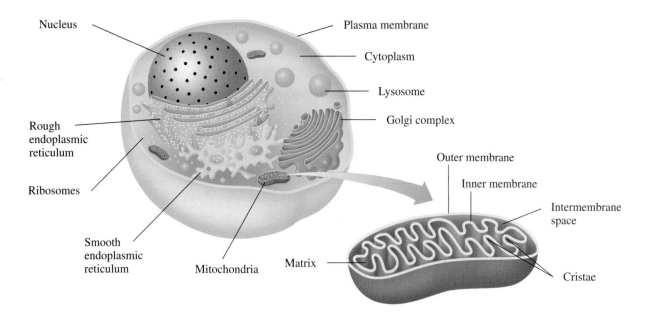

Nucleus

Plasma membrane

Cytoplasm

Lysosome

Golgi complex

Rough endoplasmic reticulum

Ribosomes

Smooth endoplasmic reticulum

Mitochondria

Matrix

Outer membrane

Inner membrane

Intermembrane space

Cristae

Figure 23.2 The diagram illustrates the major components of a typical eukaryotic cell.

Q What is the cytoplasm in a cell?

processed for secretion and phospholipids are synthesized, and a smooth endoplasmic reticulum where fats and steroids are synthesized. The *Golgi complex* modifies proteins it receives from the rough endoplasmic reticulum and secretes these modified proteins into the extracellular fluid as well as forming glycoproteins and cell membranes. *Lysosomes* contain many hydrolytic enzymes that break down recyclable cellular structures that are no longer needed by the cell. The **mitochondria** are the energy-producing factories of the cells. A mitochondrion consists of an outer membrane and an inner membrane with an intermembrane space between them. The fluid section surrounded by the inner membrane is called the *matrix.* Enzymes located in the matrix and along the inner membrane catalyze the oxidation of carbohydrates, fats, and amino acids. All of these oxidation pathways lead to CO_2, H_2O, and energy, which is used to form energy-rich compounds. Table 23.1 summarizes some of the functions of the cellular components in eukaryotic cells.

Table 23.1   Locations and Functions of Components in Eukaryotic Cells

Component	Description and Function
Plasma membrane	Separates the contents of a cell from the external environment and contains structures that communicate with other cells
Cytoplasm	Consists of all of the cellular contents between the plasma membrane and nucleus
Cytosol	Is the fluid part of the cytoplasm that contains enzymes for many of the cell's chemical reactions including glycolysis, glucose, and fatty acid synthesis
Endoplasmic reticulum	Rough type processes proteins for secretion and synthesizes phospholipids; smooth type synthesizes fats and steroids
Golgi complex	Modifies and secretes proteins from the endoplasmic reticulum and synthesizes glycoproteins and cell membranes
Lysosomes	Contain hydrolytic enzymes that digest and recycle old cell structures
Mitochondria	Contain the structures for the synthesis of ATP from energy-producing reactions
Nucleus	Contains genetic information for the replication of DNA and the synthesis of protein
Ribosomes	Are the sites of protein synthesis using mRNA templates

Metabolism and Cell Structure

Identify the following as catabolic or anabolic reactions.
a. Digestion of polysaccharides
b. Synthesis of proteins
c. Oxidation of glucose to CO_2 and H_2O

Solution

a. The breakdown of large molecules is a catabolic reaction.
b. The synthesis of large molecules requires energy and involves anabolic reactions.
c. The breakdown of monomers such as glucose involves catabolic reactions.

Study Check

What is the difference between the cytoplasm and cytosol?

QUESTIONS AND PROBLEMS

Metabolism and Cell Structure

23.1 What stage of metabolism involves the digestion of polysaccharides?

23.2 What stage of metabolism involves the conversion of small molecules to CO_2, H_2O, and energy for the synthesis of ATP?

23.3 What is meant by a catabolic reaction in metabolism?

23.4 What is meant by an anabolic reaction in metabolism?

23.5 Match each of the following organelles with their function:
    (1) lysosome   (2) Golgi complex   (3) smooth endoplasmic reticulum
    a. synthesis of fats and steroids   **b.** contains hydrolytic enzymes
    c. modifies products from rough endoplasmic reticulum

23.6 Match each of the following organelles with their function:
    (1) mitochondria   (2) rough endoplasmic reticulum
    (3) plasma membrane
    a. separates cell contents from external surroundings
    b. sites of energy production   **c.** synthesizes proteins for secretion

LEARNING GOAL
Describe the role of ATP in catabolic and anabolic reactions.

the
Chemistry
place

WEB TUTORIAL
ATP

23.2 ATP and Energy

In our cells, the energy released from the oxidation of the food we eat is used to form a compound called *adenosine triphosphate* (**ATP**). As we saw in Chapter 22, the ATP molecule is composed of the nitrogen base adenine, a ribose sugar, and three phosphate groups replacing the 5'—OH of ribose. (See Figure 23.3.)

Hydrolysis of ATP Yields Energy

In the cells, there are a variety of compounds known as "high-energy" compounds. The most important of these is ATP, which transfers a phosphate group to water when it undergoes hydrolysis. The cleavage of one phosphate group releases energy of 7.3 kcal per mole of ATP or 31 kJ per mole of ATP. The equilibrium strongly favors the products, which are adenosine diphosphate (**ADP**) and hydrogen phosphate ion (HPO_4^{2-}) abbreviated as P_i (inorganic phosphate).

Adenosine—O—P(=O)(O⁻)—O—P(=O)(O⁻)—O—P(=O)(O⁻)—O⁻ + H₂O ⟶

ATP (Adenosine triphosphate)

Adenosine—O—P(=O)(O⁻)—O—P(=O)(O⁻)—O⁻ + HO—P(=O)(O⁻)—O⁻ + H⁺ + 7.3 kcal (31 kJ)/mole

ADP (Adenosine diphosphate)   Inorganic phosphate P_i

We can write this reaction in its simplified form as

$$\text{ATP} + H_2O \longrightarrow \text{ADP} + P_i + 7.3 \text{ kcal (31 kJ)/mole}$$

Every time we contract muscles, move substances across cellular membranes, send nerve signals, or synthesize an enzyme, we use energy from ATP hydrolysis. In a cell that is doing work (anabolic processes), 1–2 million ATP molecules may be hydrolyzed in one second. The amount of ATP hydrolyzed in one day can be as much as our body mass, even though only about 1 gram of ATP is present in all our cells at any given time.

When we take in food, the resulting catabolic reactions provide energy to regenerate ATP in our cells. Then 7.3 kcal (31 kJ)/mole is used to make ATP from ADP and P_i. (See Figure 23.4.)

$$\text{ADP} + P_i + 7.3 \text{ (31 kJ) kcal/mole} \longrightarrow \text{ATP}$$

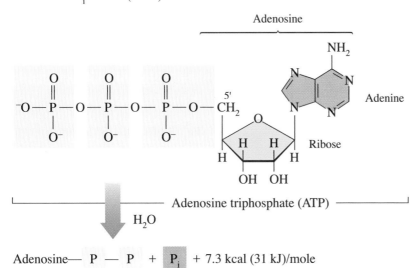

Adenosine

NH₂

Adenine

Ribose

Adenosine triphosphate (ATP)

H₂O

Adenosine— P — P + P_i + 7.3 kcal (31 kJ)/mole

Adenosine diphosphate (ADP)

H₂O

Adenosine— P + P_i + 7.3 kcal (31 kJ)/mole   P = Phosphate group

Adenosine monophosphate (AMP)

P_i = Inorganic phosphate

Figure 23.3 Adenosine triphosphate hydrolyzes to form ADP and AMP, along with a release of energy.

Q How much energy is released when a phosphate group is cleaved from one mole of ATP?

Figure 23.4 ATP, the energy-storage molecule, connects the energy-producing reactions with the energy-requiring reactions that do work in the cells.

Q What type of reaction provides energy for ATP synthesis?

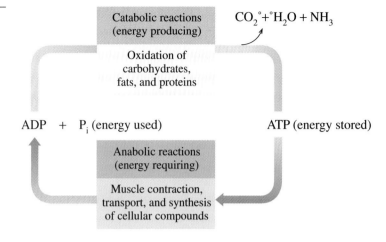

In some pathways, ATP is hydrolyzed to yield AMP and pyrophosphate (PP_i).

$$\text{Adenosine}-O-\overset{\overset{\displaystyle O}{\|}}{\underset{\underset{\displaystyle O^-}{|}}{P}}-O-\overset{\overset{\displaystyle O}{\|}}{\underset{\underset{\displaystyle O^-}{|}}{P}}-O-\overset{\overset{\displaystyle O}{\|}}{\underset{\underset{\displaystyle O^-}{|}}{P}}-O^- \; + \; H_2O \longrightarrow$$

ATP (Adenosine triphosphate)

$$\text{Adenosine}-O-\overset{\overset{\displaystyle O}{\|}}{\underset{\underset{\displaystyle O^-}{|}}{P}}-O^- \; + \; HO-\overset{\overset{\displaystyle O}{\|}}{\underset{\underset{\displaystyle O^-}{|}}{P}}-O-\overset{\overset{\displaystyle O}{\|}}{\underset{\underset{\displaystyle O^-}{|}}{P}}-O^- \; + \; H^+ \; + \; 7.3 \text{ kcal (31 kJ)/mole}$$

AMP (Adenosine monophosphate)   Pyrophosphate (PP_i)

Usually the PP_i, which is also an energy-rich compound, is immediately hydrolyzed to give 2 P_i. The combination of these two reactions releases energy that is similar to the hydrolysis of two ATP to ADP and P_i.

$$HO-\overset{\overset{\displaystyle O}{\|}}{\underset{\underset{\displaystyle O^-}{|}}{P}}-O-\overset{\overset{\displaystyle O}{\|}}{\underset{\underset{\displaystyle O^-}{|}}{P}}-O^- \; + \; H_2O \longrightarrow 2 \; HO-\overset{\overset{\displaystyle O}{\|}}{\underset{\underset{\displaystyle O^-}{|}}{P}}-O^- \; + \; H^+ \; + \; 7.8 \text{ kcal (33 kJ)/mole}$$

PP_i 2 P_i

ATP Is Coupled with Reactions

ATP and other energy-rich compounds are coupled with metabolic reactions and processes that require energy. ATP is very useful in the body because it can be used to drive reactions that require energy and do not occur on their own. For example, the glucose obtained from carbohydrates must add a phosphate group to start its breakdown in the cell. However, the cost of adding a phosphate to glucose is 3.3 kcal (14 kJ)/mole, which means that the reaction does not occur spontaneously in the cell. By coupling the reaction with the hydrolysis of an energy-rich compound such as ATP, the reaction takes place because the energy from hydrolysis pushes or "drives" the energy-requiring reaction.

Glucose + P$_i$ + 3.3 kcal (14 kJ)/mole $\longrightarrow$ glucose-6-phosphate   Requires energy

ATP $\longrightarrow$ ADP + P$_i$ + 7.3 kcal (31 kJ)/mole   Provides energy

Glucose + ATP $\longrightarrow$ ADP + glucose-6-phosphate + 4.0 kcal (17 kJ)/mole

The coupling of a reaction that requires energy with a reaction that supplies energy is a very important concept in biochemical pathways. Many of the reactions essential to a cell for survival cannot proceed by themselves, but can be made to proceed by coupling them with a reaction that releases energy. Similar kinds of coupled reactions are also used to transmit nerve impulses, transport substances across membranes to higher concentrations, and to contract muscles.

SAMPLE PROBLEM 23.2

Hydrolysis of ATP

Write an equation for the hydrolysis of ATP.

Solution

The hydrolysis of ATP produces ADP, P$_i$, and energy.

$$ATP + H_2O \longrightarrow ADP + P_i + energy$$

HEALTH NOTE

ATP Energy and Ca^{2+} Needed to Contract Muscles

Our muscles consist of thousands of parallel fibers. Within these muscle fibers are fibrils composed of two kinds of proteins called filaments. Arranged in alternating rows, the thick filaments of myosin overlap the thin filaments containing actin. During a muscle contraction, the thin filaments slide inward over the thick filaments causing a shortening of the muscle fibers.

Calcium ion, Ca^{2+}, and ATP play an important role in muscle contraction. An increase in the Ca^{2+} concentration in the muscle fibers causes the filaments to slide, while a decrease stops the process. In a relaxed muscle, the Ca^{2+} concentration is low. When a nerve impulse reaches the muscle, calcium channels in the membrane open and Ca^{2+} flows into the fluid surrounding the filaments. The Ca^{2+} combines with troponin, a protein that covers the binding site on actin. When the troponin detaches from actin, myosin can bind to actin, and start to pull the actin filaments inward. The energy used for the contraction is provided by splitting ATP that is attached to myosin to ADP + P$_i$ and energy.

Muscle contraction continues as long as both ATP and Ca^{2+} levels are high around the filaments. Relaxation of the muscle occurs as the nerve impulse ends and the Ca^{2+} channels close. The Ca^{2+} concentration is lowered as ATP energy is used to pump the remaining calcium ions out of the area. Troponin re-attaches to the actin, which causes the thin filaments to return to their relaxed state. In rigor mortis, Ca^{2+} concentration remains high within the muscle fibers causing a continued state of rigidity for approximately 24 hours. After that, cellular deterioration causes a decrease in calcium ions and muscles relax.

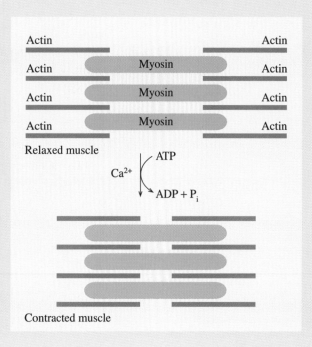

What are the components of the ATP molecule?

QUESTIONS AND PROBLEMS

ATP and Energy

23.7 Why is ATP considered an energy-rich compound?

23.8 What is meant when we say that the hydrolysis of ATP is used to "drive" a reaction?

23.9 Phosphoenolpyruvate (PEP) is a high-energy phosphate compound that releases 14.8 kcal/mole of energy when it hydrolyzes to pyruvate and P_i. This reaction can be coupled with the synthesis of ATP from ADP and P_i.
 a. Write an equation for the energy-releasing reaction of PEP.
 b. Write an equation for the energy-requiring reaction that forms ATP.
 c. Write the overall equation for the coupled reaction including the net energy change.

23.10 The phosphorylation of glycerol to glycerol-3-phosphate requires 2.2 kcal/mole and is driven by the hydrolysis of ATP.
 a. Write an equation for the energy-releasing reaction of ATP.
 b. Write an equation for the energy-requiring reaction that forms glycerol-3-phosphate.
 c. Write the overall equation for the coupled reaction including the net energy change.

LEARNING GOAL

Describe the components and functions of the coenzymes FAD, NAD$^+$, and coenzyme A.

23.3 Important Coenzymes in Metabolic Pathways

Before we look at the metabolic reactions that extract energy from our food, we need to review some ideas about oxidation and reduction reactions. (Chapters 6 and 14). An **oxidation** involves the loss of hydrogen and electrons by a substance or an increase in oxygen. **Reduction** is the gain of hydrogen and electrons or a decrease in oxygen.

Oxidation: Loss of H, loss of e⁻, or increase of O

$$CH_3-CH_3 \overset{[O]}{\rightleftharpoons} CH_3-CH_2\overset{OH}{|} \overset{[O]}{\rightleftharpoons} CH_3-\overset{O}{\overset{\|}{C}}-H \overset{[O]}{\rightleftharpoons} CH_3-\overset{O}{\overset{\|}{C}}-OH$$

Alkane   Alcohol (1°)   Aldehyde   Carboxylic acid

$$CH_3-\overset{OH}{\overset{|}{CH}}-CH_3 \overset{[O]}{\rightleftharpoons} CH_3-\overset{O}{\overset{\|}{C}}-CH_3$$

Alcohol (2°)   Ketone

Reduction: Gain of H, gain of e⁻, or decrease of O

When an enzyme catalyzes an oxidation, hydrogen atoms are removed from a substrate as hydrogen ions, $2H^+$, and electrons, $2e^-$.

$$\text{2 H atoms (removed in oxidation)} \longrightarrow 2H^+ + 2e^-$$

Those hydrogen ions and electrons are picked up by a coenzyme for the reaction, which is reduced. As we saw in Chapter 22, the structures of many coenzymes include the water-soluble B vitamins we obtain from the foods in our diets. Now we can look at the structure of several important coenzymes.

NAD^+

NAD^+ (nicotinamide adenine dinucleotide) is an important coenzyme in which the vitamin *niacin* provides the *nicotinamide* group, which is bonded to adenosine diphosphate (ADP). (See Figure 23.5.) In the oxidized form of NAD^+, the nitrogen atom in nicotinamide has a positive charge because it has four bonds rather than three. The NAD^+ coenzyme participates in reactions that produce a carbon–oxygen ($C=O$) double bond such as the oxidation of alcohols to aldehydes and ketones. The NAD^+ is reduced when the carbon in the pyridine ring of nicotinamide accepts a hydrogen ion and two electrons leaving one H^+. Let's look at the reactions that take place when ethanol is oxidized in the liver to acetaldehyde using NAD^+.

Oxidation

$$CH_3-CH_3-OH \underset{\text{dehydrogenase}}{\overset{\text{Alcohol}}{\rightleftharpoons}} CH_3-\overset{\overset{\displaystyle O}{\|}}{C}-H + 2H^+ + 2e^-$$

Ethanol   Acetaldehyde

Reduction

$$NAD^+ + 2H^+ + 2e^- \rightleftharpoons NADH + H^+$$

Overall oxidation–reduction reaction

$$CH_3-CH_2-OH + NAD^+ \underset{\text{dehydrogenase}}{\overset{\text{Alcohol}}{\rightleftharpoons}} CH_3-\overset{\overset{\displaystyle O}{\|}}{C}-H + NADH + H^+$$

Ethanol   Acetaldehyde

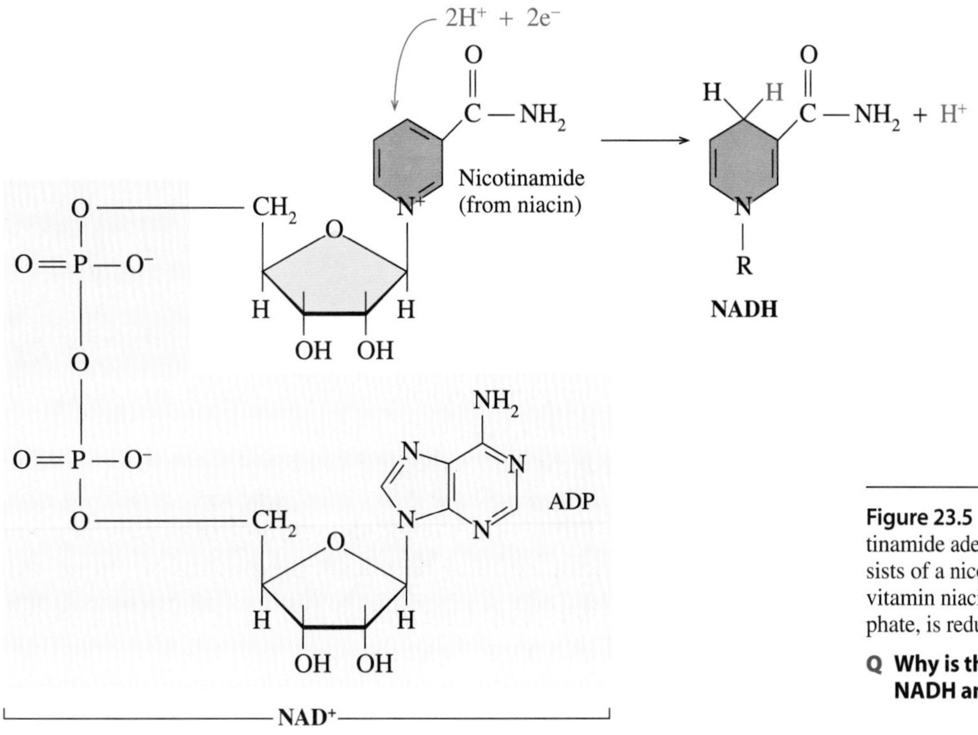

Figure 23.5 The coenzyme NAD^+ (nicotinamide adenine dinucleotide), which consists of a nicotinamide portion from the vitamin niacin, ribose, and adenine diphosphate, is reduced to $NADH + H^+$.

Q Why is the conversion of NAD^+ to $NADH$ and H^+ called a reduction?

FAD

FAD (flavin adenine dinucleotide) is a coenzyme derived from adenosine diphosphate (ADP) and riboflavin. Riboflavin, also known as vitamin B_2, consists of ribitol, a sugar alcohol, and flavin, which is a ring structure found in flavoenzymes. (See Figure 23.6.) As a coenzyme, two nitrogen atoms in the flavin part of the FAD coenzyme accept the hydrogen, which reduces the FAD to $FADH_2$.

$$FAD + 2H^+ + 2e^- \rightleftharpoons FADH_2$$

FAD typically participates in oxidation reactions that produce a carbon–carbon ($C=C$) double bond.

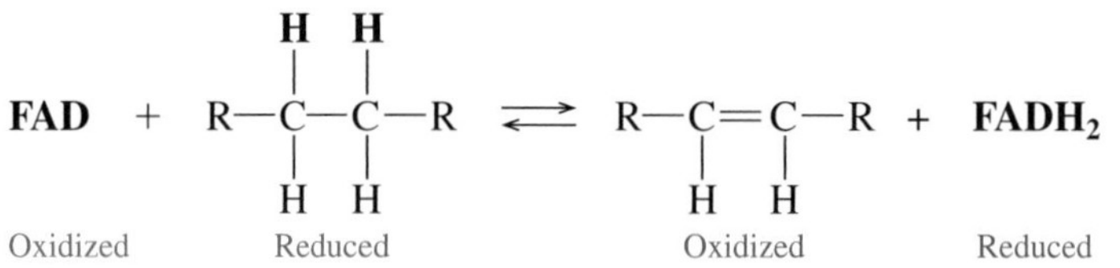

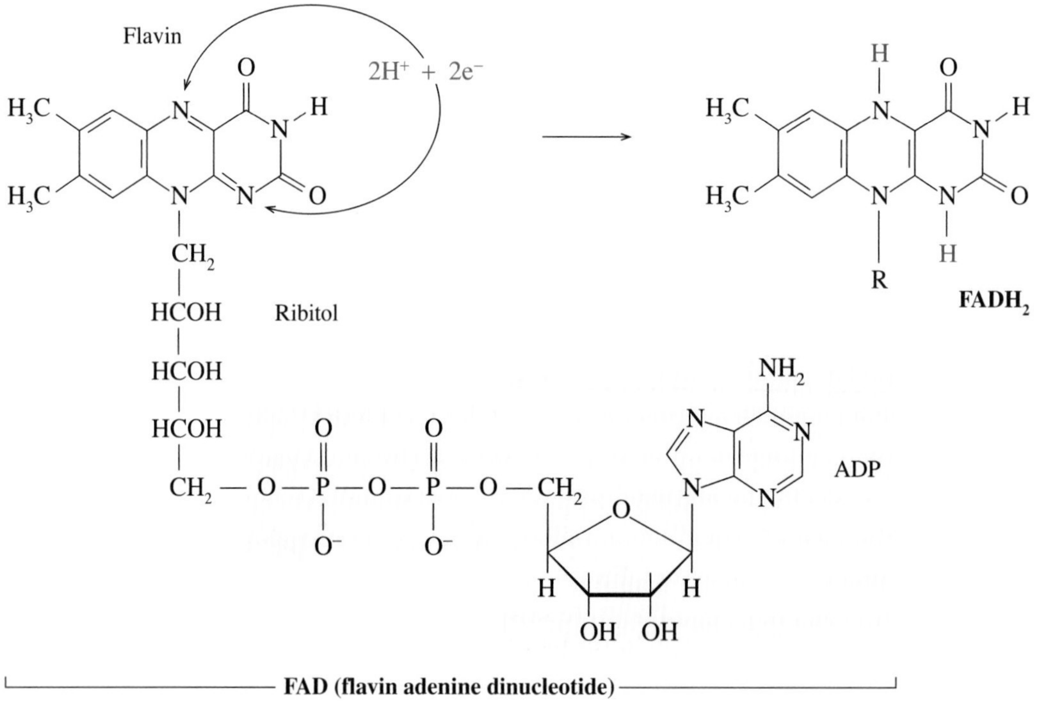

Figure 23.6 The coenzyme FAD (flavin adenine dinucleotide) made from riboflavin (vitamin B_2) and adenine diphosphate is reduced to $FADH_2$.

Q What is the type of reaction in which FAD accepts hydrogen?

Coenzyme A

Coenzyme A (CoA) is made up of several components: pantothenic acid (vitamin B_5), adenosine diphosphate (ADP) with a phosphate group on C3′, and an aminoethanethiol. (See Figure 23.7.) The main function of coenzyme A is to activate acyl groups (indicated by the letter A in CoA), particularly the acetyl group. When the free thiol (—SH) group of CoA bonds to the two-carbon acetyl group, or to longer chain acyl groups such as fatty acids, the products are energy-rich thioesters. Then the acetyl or acyl groups can be easily transferred to other substrates.

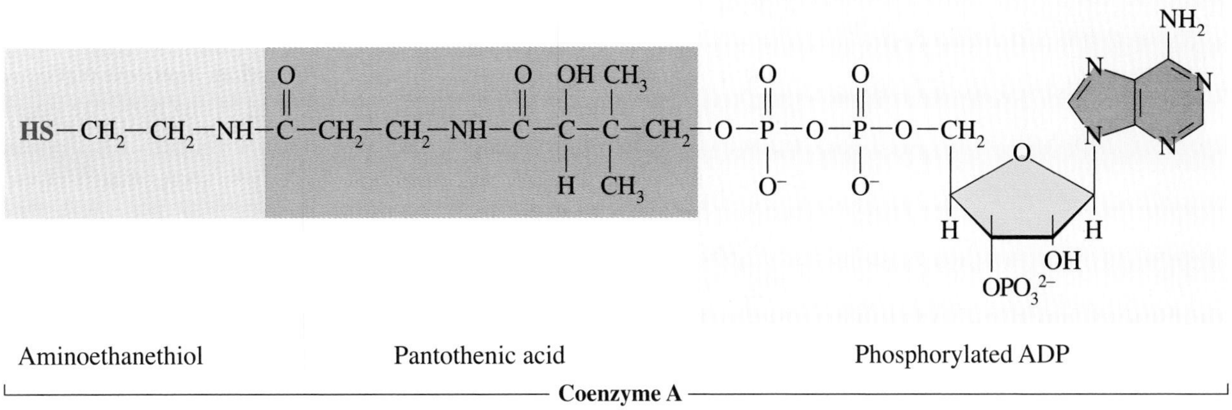

Aminoethanethiol   Pantothenic acid   Phosphorylated ADP

Coenzyme A

Thioester bond

Figure 23.7 Coenzyme A is derived from a phosphorylated (3') adenine diphosphate (ADP) and pantothenic acid bonded by an amide linked to aminoethanethiol, which contains the —SH reactive part of the molecule.

Q What part of coenzyme A reacts with a two-carbon acetyl group?

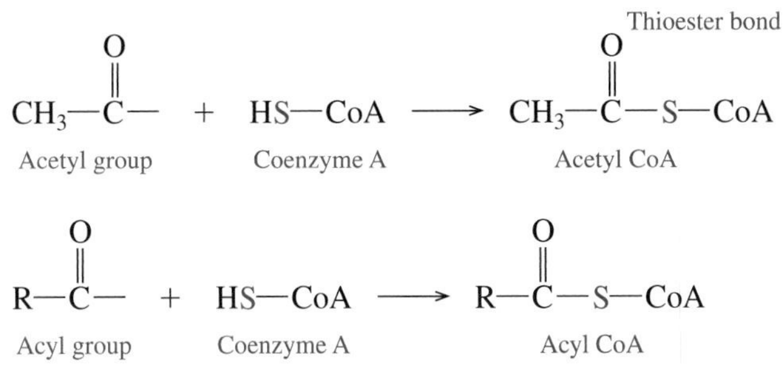

SAMPLE PROBLEM 23.3

Coenzymes in Metabolic Pathways

What vitamin is part of the coenzyme FAD?

Solution

FAD, flavin adenine dinucleotide, is made from the vitamin riboflavin.

Study Check

What is the abbreviation of the reduced form of FAD?

QUESTIONS AND PROBLEMS

Important Coenzymes in Metabolic Pathways

23.11 Identify one or more coenzymes with each of the following components:
   a. pantothenic acid   **b.** niacin   **c.** ribitol

23.12 Identify one or more coenzymes with each of the following components:
   a. riboflavin   **b.** adenine   **c.** aminoethanethiol

23.13 Give the abbreviation for the following:
   a. the reduced form of NAD^+ **b.** the oxidized form of $FADH_2$

23.14 Give the abbreviation for the following:
   a. the reduced form of FAD   **b.** the oxidized form of NADH

23.15 What coenzyme picks up hydrogen when a carbon–carbon double bond is formed?

23.16 What coenzyme picks up hydrogen when a carbon–oxygen double bond is formed?

LEARNING GOAL

Give the sites and products of digestion for carbohydrates.

23.4 Digestion of Carbohydrates

In the first stage of catabolism, foods undergo **digestion,** a process that converts large molecules to smaller ones that can be absorbed by the body. We begin the digestion of carbohydrates as soon as we chew food. (See Figure 23.8.) An enzyme produced in the salivary glands called **amylase** hydrolyzes some of the α-glycosidic bonds in the starches amylose and amylopectin, producing smaller polysaccharides called dextrins, which contain three to eight glucose units, maltose, and some glucose. However, amylase does not cleave the branched chains of amylopectin. After swallowing, the partially digested starches enter the acid environment of the stomach, where the low pH soon stops further carbohydrate digestion.

Digestion of Disaccharides

In the small intestine, which has a pH of about 8, an α-amylase produced in the pancreas hydrolyzes the remaining polysaccharides to maltose and glucose. A branching enzyme hydrolyzes the glycosidic bonds to the branches in amylopectin. Then enzymes produced in the mucosal cells that line the small intestine hydrolyze maltose as well as lactose and sucrose from milk products and natural sugar. The hydrolysis reactions for the three common dietary disaccharides are written as follows.

$$\text{Lactose} + H_2O \xrightarrow{\text{Lactase}} \text{galactose} + \text{glucose}$$

$$\text{Sucrose} + H_2O \xrightarrow{\text{Sucrase}} \text{fructose} + \text{glucose}$$

$$\text{Maltose} + H_2O \xrightarrow{\text{Maltase}} \text{glucose} + \text{glucose}$$

The monosaccharides are absorbed through the intestinal wall into the bloodstream, which carries them to the liver where fructose and galactose are converted to glucose.

SAMPLE PROBLEM 23.4

Digestion of Carbohydrates

Indicate the carbohydrate that undergoes digestion in each of the following sites:
a. mouth   **b.** stomach   **c.** small intestine

Solution

a. Starches amylose and amylopectin (α-1,4-glycosidic bonds only).
b. Essentially no digestion occurs in the stomach.
c. Dextrins, maltose, sucrose, and lactose.

Study Check

Describe the digestion of amylose, a polymer of glucose molecules joined by α-glycosidic bonds.

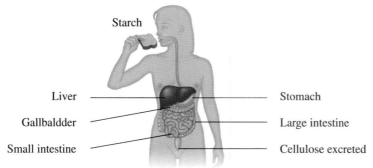

Figure 23.8 In stage 1 of metabolism, the digestion of carbohydrates begins in the mouth with the action of salivary amylase to yield dextrins (small polysaccharides), maltose, and some glucose and is completed in the small intestine by pancreatic amylase and enzymes for the disaccharides.

Q **Why is there little or no digestion of carbohydrates in the stomach?**

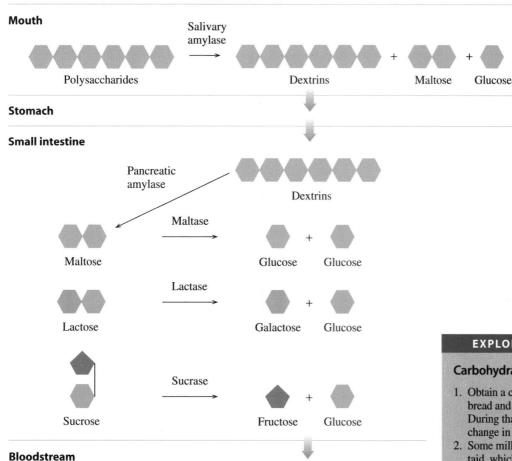

Mouth

Stomach

Small intestine

Bloodstream

QUESTIONS AND PROBLEMS

Digestion of Foods

23.17 What is the general type of reaction that occurs during the digestion of carbohydrates?

23.18 Why is α-amylase produced in the salivary glands and in the pancreas?

23.19 Complete the following equation by filling in the missing words:
   a. _____ + H_2O ⟶ galactose + glucose
   b. Sucrose + H_2O ⟶ _____ + _____
   c. Maltose + H_2O ⟶ glucose + _____

23.20 Give the site and the enzyme for each of the reactions in problem 23.19.

EXPLORE YOUR WORLD

Carbohydrate Digestion

1. Obtain a cracker or small piece of bread and chew it for 4–5 minutes. During that time observe any change in the taste.
2. Some milk products contain Lactaid, which is the lactase that digests lactose. Look for the brands of milk and ice cream that contain Lactaid or lactase enzyme.

Questions

1a. How does the taste of the cracker or bread change after you have chewed it for 4–5 minutes? What could be an explanation?
1b. What part of carbohydrate digestion occurs in the mouth?
2a. Write an equation for the digestion of lactose.
2b. Where does lactose undergo digestion?

HEALTH NOTE

Lactose Intolerance

The disaccharide in milk is lactose, which is broken down by *lactase* in the intestinal tract to monosaccharides that are a source of energy. Infants and small children produce lactase to break down the lactose in milk. It is rare for an infant to lack the ability to produce lactase. As they mature, many people experience a decrease in the production of the lactase enzyme. By the time they are adults, many people have little or no lactase production, causing lactose intolerance. This condition may affect 25% of the people in the United States. A deficiency of lactase occurs in adults throughout the world, but in the United States it is more prevalent among the African–American, Hispanic, and Asian populations.

When lactose is not broken down into glucose and galactose, it cannot be absorbed through the intestinal wall and remains in the intestinal tract. In the intestines, the lactose undergoes fermentation to products that include lactic acid and gases such as methane (CH_4) and CO_2. Symptoms of lactose intolerance, which appear approximately $\frac{1}{2}$ to 1 hour after ingesting milk or milk products, include nausea, abdominal cramps, and diarrhea. The severity of the symptoms depends on how much lactose is present in the food and how much lactase a person produces.

Treatment of Lactose Intolerance

One way to reduce the reaction to lactose is to avoid products that contain lactose. However, it is important to consume foods that provide the body with calcium. Many people with lactose

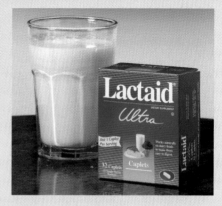

intolerance seem to tolerate yogurt, which is a good source of calcium. Although there is lactose in yogurt, the bacteria in yogurt may produce some lactase, which helps to digest the lactose. A person who is lactose intolerant should also know that some foods that may not seem to be dairy products contain lactose. For example, baked goods, cereals, breakfast drinks, salad dressings, and even lunchmeat can contain lactose in their ingredients. Labels must be read carefully to see if the ingredients include "milk" or "lactose."

The enzyme lactase is now available in many forms such as tablets that are taken with meals, drops that are added to milk, or as additives in many dairy products such as milk. When lactase is added to milk that is left in the refrigerator for 24 hours, the lactose level is reduced by 70–90%. Lactase pills or chewable tablets are taken when a person begins to eat a meal that contains dairy foods. If taken too far ahead of the meal, too much of the lactase will be degraded by stomach acid. If taken following a meal, the lactose will have entered the lower intestine.

LEARNING GOAL

Describe the conversion of glucose to pyruvate in glycolysis.

the Chemistry place

WEB TUTORIAL
Glycolysis

23.5 Glycolysis: Oxidation of Glucose

Our major source of energy is the glucose produced when we digest the carbohydrates in our food, or from glycogen, a polysaccharide stored in the liver and skeletal muscle. Glucose in the bloodstream enters our cells for further degradation in a pathway called glycolysis. Early organisms used this pathway to produce energy from simple nutrients long before there was any oxygen in Earth's atmosphere. Glycolysis is an **anaerobic** process; no oxygen is required.

All the reactions in glycolysis take place in the cytoplasm of the cell where all the enzymes for glycolysis are located. We can now look at the details of the individual steps in glycolysis, which is one of the metabolic pathways in stage 2. In **glycolysis,** a six-carbon glucose molecule is broken down to yield two three-carbon pyruvate molecules. (See Figure 23.9.) In glycolysis glucose is converted to pyruvate in a sequence of 10 reactions. In the first five reactions (1–5), the energy of 2 ATPs is invested to add phosphate groups to form sugar phosphates. Then the six-carbon sugar phosphate is cleaved to yield two three-carbon sugar phosphate molecules. In the last five reactions (6–10), the phosphate groups in these energy-rich trioses are hydrolyzed, which generates energy in the form of 4 ATPs. The final products are 2 pyruvate molecules and 2 reduced NADH molecules. (See Figure 23.10.)

Stage 1

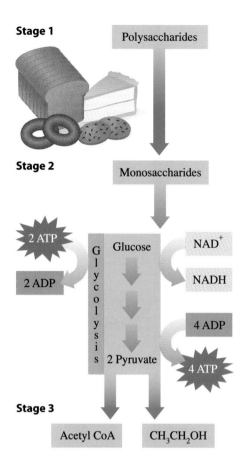

Stage 2

Stage 3

Figure 23.9 Glucose obtained from the digestion (stage 1) of polysaccharides is degraded in a metabolic pathway called glycolysis to give pyruvate.

Q What is the end product of glycolysis?

Figure 23.10 In glycolysis, the six-carbon glucose molecule is degraded to yield two three-carbon pyruvate molecules. A net of two ATPs are produced along with two NADH.

Q Where in the glycolysis pathway is glucose cleaved to yield two three-carbon compounds?

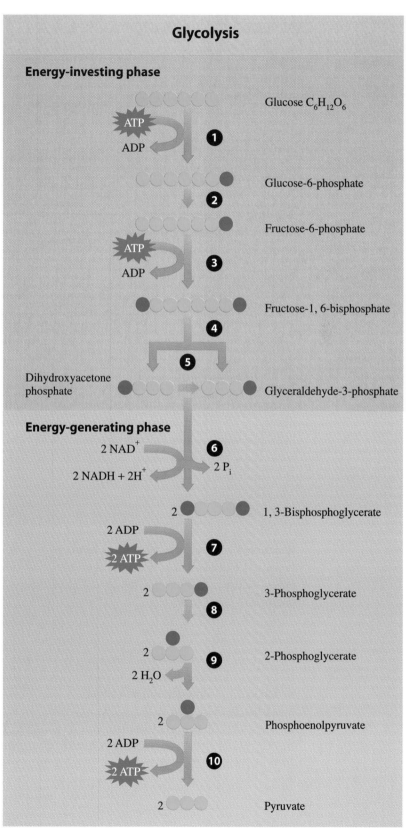

Energy-Investing Reactions: Steps 1–5

Reaction 1   Phosphorylation: first ATP invested

Glucose is converted to glucose-6-phosphate by a reaction with ATP catalyzed by *hexokinase*.

$$P = -O-\overset{\displaystyle O}{\underset{\displaystyle O^-}{\overset{\|}{\underset{|}{P}}}}-O^-$$

Reaction 2   Isomerization

The enzyme *phosphoglucoisomerase* converts glucose-6-phosphate, an aldose, to fructose-6-phosphate, a ketose.

Reaction 3   Phosphorylation: second ATP invested

A second ATP reacts with fructose-6-phosphate to give fructose-1,6-bisphosphate. The word *bisphosphate* is used to show that the phosphates are on different carbons in fructose and not connected to each other.

Reaction 4   Cleavage: two trioses form

Fructose-1,6-bisphosphate is "split" into two triose phosphates—dihydroxyacetone phosphate and glyceraldehyde-3-phosphate—catalyzed by *aldolase*.

Reaction 5   Isomerization of a triose

In reaction 5, *triose phosphate isomerase* converts one of the triose products, dihydroxyacetone phophate, to the other, glyceraldehyde-3-phosphate. Now all 6 carbon atoms from glucose are in two identical triose phosphates.

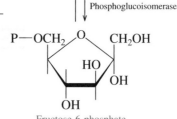

Glucose

ATP
Hexokinase
ADP

Glucose-6-phosphate

Phosphoglucoisomerase

Fructose-6-phosphate
ATP
Phosphofructokinase
ADP

Fructose-1,6-bisphosphate

Fructose-1,6-bisphosphate aldolase

Dihydroxyacetone phosphate

Glyceraldehyde-3-phosphate

Triosephosphate isomerase

Glyceraldehyde-3-phosphate

Energy-Generating Reactions: Steps 6–10

Reaction 6   First energy-rich compound

The aldehyde group of glyceraldehyde-3-phosphate is oxidized and phosphorylated by *glyceraldehyde-3-phosphate dehydrogenase*. The coenzyme NAD^+ is reduced to NADH and H^+. The 1,3-bisphosphoglycerate is an extremely high-energy compound that releases 12 kcal (49 kJ) per mole when the phosphate bond is broken.

Reaction 7   Formation of first ATP

The energy-rich 1,3-bisphosphoglycerate now drives the formation of ATP when *phosphoglycerate kinase* transfers a phosphate group to ADP. This process is called a **substrate-level phosphorylation.** At this point, glycolysis pays back the debt of two ATP invested in the early reactions.

Reaction 8   Formation of 2-phosphoglycerate

A *phosphoglycerate mutase* transfers the phosphate group from carbon 3 to carbon 2 to yield 2-phosphoglycerate.

Reaction 9   Second energy-rich compound

An *enolase* catalyzes the removal of water to yield phosphoenolpyruvate, a high-energy compound that releases 15 kcal or 62 kJ per mole when the phosphate bond is hydrolyzed.

Reaction 10   Formation of second ATP

This last reaction catalyzed by *pyruvate kinase* transfers the phosphate to ADP in a second direct substrate phosphorylation to yield pyruvate and ATP.

Summary of Glycolysis

In the glycolysis pathway, glucose is converted to two pyruvates. Initially, two ATP molecules were required to form sugar phosphate. Later, four ATP were generated, which gives a net gain of two ATP. Overall, glycolysis yields two ATP and two NADH for each glucose that is converted to two pyruvates.

$$C_6H_{12}O_6 + 2\ NAD^+ \xrightarrow{\quad 2\ ADP\ +\ 2\ P_i \quad 2\ ATP \quad} 2\ CH_3{-}\overset{\overset{\displaystyle O}{\|}}{C}{-}COO^- + 2\ NADH + 4H^+$$

Glucose

Pyruvate

It appears right now that glycolysis does a lot of work to produce only 2 ATP, 2 NADH, and pyruvate. However, under aerobic conditions, stage 3 operates to reoxidize NADH to produce more ATP, and pyruvate enters the citric acid cycle where it generates considerably more energy. We will look at the oxidative pathways of stage 3 in Chapter 24.

Other Hexoses Enter Glycolysis

Other monosaccharides can enter glycolysis if they are first converted to intermediates of the pathway. Digestion of carbohydrates produces galactose from lactose in milk products and fructose from fruits and sucrose. Galactose reacts with ATP to yield galactose-1-phosphate, which is converted to glucose-6-phosphate, an intermediate of glycolysis in reaction 2. In the liver, fructose is reacted with ATP to yield fructose-1-phosphate, which is cleaved in a reaction similar to reaction 4 to give dihydroxyacetone phosphate and glyceraldehyde. Dihydroxyacetone phosphate isomerizes to glyceraldehyde-3- phosphate, and glyceraldehyde is phosphorylated to glyceraldehyde-3-phosphate, an intermediate in reaction 6. In muscle and kidney, fructose is phosphorylated to fructose-6-phosphate, which enters glycolysis in reaction 3.

$$\text{Galactose} \xrightarrow[\text{ADP}\ +\ P_i]{\text{ATP}} \text{galactose-1-phosphate} \longrightarrow \text{glucose-6-phosphate}$$

Muscle and Kidney

$$\text{Fructose} \longrightarrow \text{fructose-6-phosphate} \xrightarrow{\quad 3 \quad} \boxed{\text{Gycolysis}}$$

glyceraldehyde-3-phosphate

Liver   ATP   ADP + P_i

$$\text{Fructose} \xrightarrow{} \text{fructose-1-phosphate} \longrightarrow \text{glyceraldehyde} + \text{dihydroxyacetone phosphate}$$

ADP + P_i ATP

Regulation of Glycolysis

Metabolic pathways such as glycolysis do not run at the same rates all the time. The amount of glucose that is broken down is controlled by the requirements in the cells for pyruvate, ATP, and other intermediates of glycolysis. Within the glycolysis sequence, three enzymes respond to the levels of ATP and other products continually speed up or slow down the flow of glucose into the pathway.

Reaction 1   Hexokinase

The amount of glucose entering the glycolysis pathway decreases when high levels of glucose-6-phosphate are present in the cell. This phosphorylation product inhibits hexokinase, which prevents glucose from reacting with ATP. This is a feedback inhibition, which is a type of enzyme regulation we discussed in Chapter 21.

Reaction 3   Phosphofructokinase

The reaction catalyzed by phosphofructokinase is a very important control point for glycolysis. Once fructose-1,6-bisphosphate is formed, it must continue through the remaining reactions to pyruvate. As an allosteric enzyme, phosphofructokinase is inhibited by high levels of ATP and activated by high levels of ADP and AMP. High levels of ADP and AMP indicate that the cell has used up much of its ATP. As a regulator, phosphofructokinase increases the rate of pyruvate production for ATP synthesis when the cell needs to replenish ATP and slows or stops the reaction when ATP is plentiful.

Reaction 10   Pyruvate Kinase

In the last reaction of glycolysis, high levels of ATP as well as acetyl CoA inhibit pyruvate kinase, which is another allosteric enzyme.

Summary of Regulation

These are examples of how metabolic pathways shut off enzymes to stop the production of molecules that are not needed. Since pyruvate can be used to synthesize ATP, several enzymes in glycolysis respond to ATP levels in the cell. When ATP levels are high, enzymes in glycolysis slow or stop the synthesis of pyruvate. With phosphofructokinase and pyruvate kinase inhibited by ATP, glucose-6-phosphate accumulates and inhibits the first reaction and glucose does not enter the glycolysis pathway. The glycolysis pathway is shut down until ATP is once again needed in the cell. When ATP levels are low or AMP/ADP levels are high, these enzymes are activated and pyruvate production starts again.

SAMPLE PROBLEM 23.5

Glycolysis

What are the steps in glycolysis that generate ATP?

Solution

ATP is produced when phosphate groups are transferred directly to ADP from 1,3-bisphosphoglycerate (step 7) and from phosphoenolpyruvate (step 10).

Study Check

If four ATP molecules are produced in glycolysis, why is there a net yield of two ATP?

QUESTIONS AND PROBLEMS

Glycolysis: Oxidation of Glucose

23.21 What is the starting product of glycolysis?

23.22 What is the end product of glycolysis?

23.23 How is ATP used in the initial steps of glycolysis?

23.24 How many ATP molecules are used in the initial steps of glycolysis?

23.25 What trioses are obtained when fructose-1,6-diphosphate splits?

23.26 Why does one of the triose products undergo isomerization?

23.27 How does direct phosphorylation account for the production of ATP in glycolysis?

23.28 Why are there two ATP molecules formed for one molecule of glucose?

23.29 Indicate the enzyme that catalyzes the following reactions in glycolysis:
   a. phosphorylation
   b. direct transfer of a phosphate group

23.30 Indicate the enzyme that catalyzes the following reactions in glycolysis:
   a. isomerization
   b. formation of a ketotriose and an aldotriose

23.31 How many ATP or NADH are produced (or required) in each of the following steps in glycolysis?
   a. glucose to glucose-6-phosphate
   b. glyceraldehyde-3-phosphate to 1,3-bisphosphoglycerate
   c. glucose to pyruvate

23.32 How many ATP or NADH are produced (or required) in each of the following steps in glycolysis?
   a. 1,3-bisphosphoglycerate to 3-phosphoglycerate
   b. fructose-6-phosphate to fructose-1,6-bisphosphate
   c. phosphoenolpyruvate to pyruvate

23.33 Which step in glycolysis involves the following?
   a. the first ATP molecule is hydrolyzed
   b. direct substrate phosphorylation occurs
   c. six-carbon sugar splits into two three-carbon molecules

23.34 Which step in glycolysis involves the following?
   a. isomerization takes place   **b.** NAD$^+$ is reduced
   c. a second ATP molecule is synthesized

23.35 How do galactose and fructose, obtained from the digestion of carbohydrates, enter glycolysis?

23.36 What are three enzymes that regulate glycolysis?

23.37 Indicate whether each of the following would activate or inhibit phosphofructokinase.
   a. low levels of ATP   **b.** high levels of ATP

23.38 Indicate whether each of the following would activate or inhibit pyruvate kinase.
   a. low levels of ATP
   b. high levels of fructose-1,6-bisphosphate

LEARNING GOAL

Give the conditions for the conversion of pyruvate to lactate, ethanol, and acetyl coenzyme A.

23.6 Pathways for Pyruvate

The pyruvate produced from glucose can now enter one of three pathways that continue to extract energy. The available pathway depends on whether there is sufficient oxygen in the cell. During **aerobic** conditions, oxygen is available to convert pyruvate to acetyl coenzyme A (CoA). When oxygen levels are low, pyruvate is reduced to lactate. In yeast cells, which are anaerobic, pyruvate is converted to ethanol.

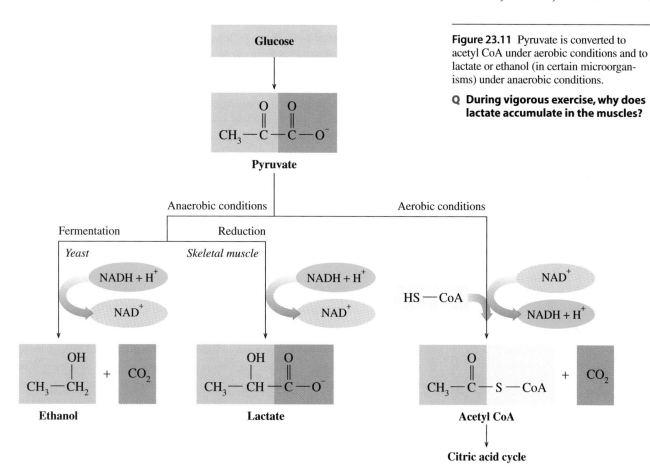

Figure 23.11 Pyruvate is converted to acetyl CoA under aerobic conditions and to lactate or ethanol (in certain microorganisms) under anaerobic conditions.

Q During vigorous exercise, why does lactate accumulate in the muscles?

Pyruvate to Acetyl CoA (Aerobic Conditions)

In glycolysis, two ATP molecules were generated when glucose was converted to pyruvate. However, much more energy is still available. The greatest amount of the energy is obtained from glucose when oxygen levels are high in the cells. Under aerobic conditions, pyruvate moves from the cytoplasm (where glycolysis took place) into the matrix of the mitochondria to be oxidized further. In a complex reaction, pyruvate is oxidized and a carbon atom is removed from pyruvate as CO_2. The coenzyme NAD^+ is required for the oxidation. The resulting two-carbon acetyl compound is attached to CoA, producing **acetyl CoA,** an important intermediate in many metabolic pathways. (See Figure 23.11.)

$$CH_3-\overset{O}{\underset{\|}{C}}-\overset{O}{\underset{\|}{C}}-O^- + HS-CoA + NAD^+ \xrightarrow{\overset{\text{Pyruvate}}{\text{dehydrogenase}}} CH_3-\overset{O}{\underset{\|}{C}}-S-CoA + CO_2 + NADH + H^+$$

Pyruvate   Acetyl CoA

Pyruvate to Lactate (Anaerobic Conditions)

When we engage in strenuous exercise, the oxygen stored in our muscle cells is quickly depleted. Under anaerobic conditions, pyruvate remains in the cytoplasm where it is reduced to lactate. NAD^+ is produced and used to oxidize more glyceraldehyde-3-phosphate in the glycolysis pathway, which produces a small but needed amount of ATP.

$$CH_3-\overset{\overset{O}{\|}}{C}-\overset{\overset{O}{\|}}{C}-O^- \underset{\text{Lactate dehydrogenase}}{\rightleftharpoons} CH_3-\overset{\overset{OH}{|}}{\underset{\overset{|}{H}}{C}}-\overset{\overset{O}{\|}}{C}-O^-$$

NADH + H⁺   NAD⁺

Pyruvate (oxidized)   Lactate (reduced)

The accumulation of lactate causes the muscles to tire rapidly and become sore. After exercise, a person continues to breathe rapidly to repay the oxygen debt incurred during exercise. Most of the lactate is transported to the liver where it is converted back into pyruvate. Under anaerobic conditions, the only ATP production in glycolysis occurs during the steps that phosphorylate ADP directly, giving a net gain of only two ATP molecules:

$$C_6H_{12}O_6 + 2\,ADP + 2\,P_i \longrightarrow 2\,CH_3-\overset{\overset{OH}{|}}{CH}-COO^- + 2\,ATP$$

Glucose   Lactate

Bacteria also convert pyruvate to lactate under anaerobic conditions. In the preparation of kimchee and sauerkraut, cabbage is covered with a salt brine. The glucose from the starches is converted to lactate. This acid environment acts as a preservative that prevents the growth of other bacteria. The pickling of olives and cucumbers gives similar products. When cultures of bacteria that produce lactate are added to milk, the acid denatures the milk proteins to give sour cream and yogurt.

Pyruvate to Ethanol (Anaerobic)

the Chemistry place

WEB TUTORIAL
Fermentation

Some microorganisms, particularly yeast, convert sugars to ethanol under anaerobic conditions by a process called **fermentation.** After pyruvate is formed in glycolysis, a carbon atom is removed in the form of CO_2 (**decarboxylation**). The NAD^+ for continued glycolysis is regenerated when the acetaldehyde is reduced to ethanol.

$$CH_3-\overset{\overset{O}{\|}}{C}-\overset{\overset{O}{\|}}{C}-O^- \xrightarrow[\text{decarboxylase}]{\text{Pyruvate}} CH_3-\overset{\overset{O}{\|}}{C}-H \xrightarrow[\text{Alcohol dehydrogenase}]{} CH_3-\overset{\overset{H}{|}}{\underset{\overset{|}{H}}{C}}-OH$$

CO₂   NADH + H⁺   NAD⁺

Pyruvate   Acetaldehyde   Ethanol

The process of fermentation by yeast is one of the oldest known chemical reactions. Enzymes in the yeast convert the sugars in a variety of carbohydrate sources to glucose and then to ethanol. The evolution of CO_2 gas produces the bubbles in beer, sparkling wines, and champagne. The type of carbohydrate used determines the taste associated with a particular alcoholic beverage. Beer is made from the fermentation of barley malt, wine from the sugars in grapes, vodka from potatoes, sake from rice, and whiskeys from corn or rye. Fermentation produces solutions up to about 15% alcohol by volume. At this concentration, the alcohol kills the yeast, and fermentation stops. Higher concentrations are obtained by distilling the alcohol.

SAMPLE PROBLEM 23.6

Fates of Pyruvate

Is each of the following products from pyruvate produced under anaerobic or aerobic conditions?

a. acetyl CoA   **b.** lactate

Solution

a. aerobic conditions   **b.** anaerobic conditions

Study Check

After strenuous exercise, some lactate is oxidized back to pyruvate by lactate dehydrogenase using NAD^+. Write an equation to show this reaction.

QUESTIONS AND PROBLEMS

Pathways for Pyruvate

23.39 What condition is needed in the cell to convert pyruvate to acetyl CoA?

23.40 What coenzymes are needed for the oxidation of pyruvate to acetyl CoA?

23.41 Write the overall equation for the conversion of pyruvate to acetyl CoA.

23.42 What are the possible products of pyruvate under anaerobic conditions?

23.43 How does the formation of lactate permit glycolysis to continue under anaerobic conditions?

23.44 After running a marathon, a runner has muscle pain and cramping. What might have occurred in the muscle cells to cause this?

23.45 In fermentation, a carbon atom is removed from pyruvate. What is the compound formed by that carbon atom?

23.46 Some students decided to make some wine by placing yeast and grape juice in a container with a tight lid. A few weeks later, the container explodes. What reaction could account for the explosion?

23.7 Glycogen Metabolism

LEARNING GOAL

Describe the breakdown and synthesis of glycogen.

We have just eaten a large meal that has supplied us with all the glucose we need to produce pyruvate and ATP by glycolysis. Then we use excess glucose to replenish our energy reserves by synthesizing glycogen that is stored in limited amounts in our skeletal muscle and liver. When glycogen stores are full, any remaining glucose is converted to triacylglycerols and stored as body fat as we will see in Chapter 25. When our diet does not supply sufficient glucose, or we have utilized our blood glucose, we degrade the stored glycogen and release glucose.

Glycogenesis

Glycogen is a polymer of glucose with α-1,4 glycosidic bonds and multiple branches attached by α-1,6 glycosidic bonds, as seen in Chapter 16. **Glycogenesis** is the synthesis of glycogen from glucose molecules, which occurs when the digestion of polysaccharides produces high levels of glucose. The synthesis of glycogen starts with glucose-6-phosphate, which can be obtained from the first reaction in glycolysis. (See Figure 23.12.) It is converted to an isomer glucose-1-phosphate, which is activated using high-energy UTP (uridine triphosphate) to yield UDP-glucose. The reaction is driven by the energy released from the hydrolysis of pyrophosphate (PP_i).

Glucose-6-phosphate $\rightleftharpoons$ glucose-1-phosphate

Glucose-1-phosphate $+$ UTP $\longrightarrow$ UDP-glucose $+$ PP_i

$$PP_i \longrightarrow 2\,P_i$$

UDP-Glucose (uridine diphosphate glucose)

The UDP-glucose attaches to the end glucose of a glycogen chain releasing UDP, which reacts with ATP to regenerate UTP.

$$\text{UDP-Glucose} + \text{glycogen} \xrightarrow{\text{Glycogen synthase}} \text{glucose—glycogen} + \text{UDP}$$

$$\text{UDP} + \text{ATP} \longrightarrow \text{UTP} + \text{ADP}$$

The overall reaction of glycogenesis starting with glucose is written simply as

$$\text{Glucose} \longrightarrow \text{glycogen}$$

Figure 23.12 In glycogenesis, glucose is used to synthesize glycogen.

Q What is the function of UTP in glycogen synthesis?

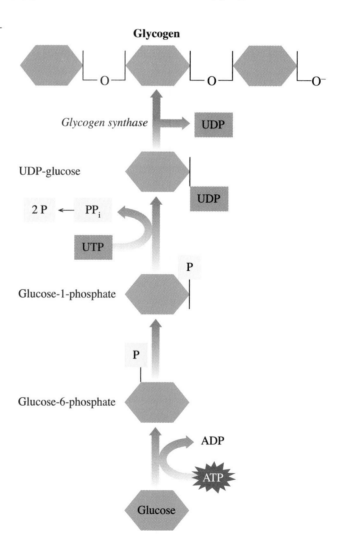

Glycogenolysis

Glucose is the primary energy source for muscle contraction and the brain. When blood glucose is depleted, glycogen breaks down to glucose in a process called **glycogenolysis.** Glucose molecules are removed one by one from the end of the glycogen chain and phosphorylated to yield glucose-1-phosphate.

$$\text{Glucose}-\text{glycogen} + P_i \xrightarrow{\substack{\text{Glycogen} \\ \text{phosphorylase}}} \text{glucose-1-phosphate} + \text{glycogen}$$

Glucose-1-phosphate may be converted to glucose-6-phosphate, which enters the glycolysis pathway to replenish ATP.

$$\text{Glucose-1-phosphate} \underset{\longleftarrow}{\overset{\text{Phosphoglucomutase}}{\longrightarrow}} \text{glucose-6-phosphate}$$

Free glucose is needed for energy by the brain and muscle. While glucose can diffuse across cell membranes, glucose phosphates cannot. Only cells in the liver and kidneys have a *gluco*se-6-*phosphatase* that hydrolyzes the phosphate to yield free glucose.

$$\text{Glucose-6-phosphate} \xrightarrow{\text{Glucose-6-phosphatase}} \text{Glucose} + P_i$$

The overall reaction of glycogenolysis, which converts glycogen to glucose, is written as

$$\text{Glycogen} \longrightarrow \text{glucose}$$

Let's summarize the events and compounds involved in the breakdown and synthesis of glycogen as follows.

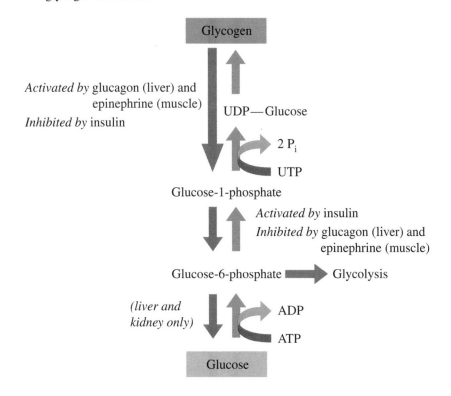

Regulation of Glycogen Metabolism

The brain, skeletal muscles, and red blood cells require large amounts of glucose every day to function properly. To protect the brain, hormones with opposing actions control blood glucose levels. When glucose is low, glucagon, a hormone produced in the pancreas, is secreted into the bloodstream. In the liver, glucagon accelerates the rate of glycogenolysis, which raises blood glucose levels. At the same time, glucagon inhibits the synthesis of glycogen.

Glycogen in skeletal muscle is broken down quickly when the body requires a "burst of energy" often referred to as "fight or flight." Epinephrine released from the adrenal glands converts *glycogen phosphorylase* from an inactive to an active form. The secretion of only a few molecules of epinephrine can break down a huge number of glycogen molecules.

Soon after we have eaten and digested a meal, our blood glucose levels rise. Rising glucose levels stimulate the pancreas to secrete the hormone insulin into our bloodstreams. Insulin promotes the use of glucose in the cells by accelerating glycogen synthesis as well as degradation reactions such as glycolysis. At the same time, insulin inhibits the synthesis of glucose, which we will discuss in the next section.

SAMPLE PROBLEM 23.7

Glycogen Metabolism

Identify each of the following as part of the reaction pathways of (1) glycolysis, (2) glycogenolysis, or (3) glycogenesis.

a. Glucose-1-phosphate is converted to glucose-6-phosphate.

b. Glucose-1-phosphate forms UDP-glucose.

c. An isomerase converts glucose-6-phosphate to fructose-6-phosphate.

Solution

a. (2) glycogenolysis

b. (3) glycogenesis

c. (1) glycolysis

Study Check

Why do cells in liver and kidney provide glucose to raise blood glucose levels, but cells in skeletal muscle do not?

QUESTIONS AND PROBLEMS

Glycogen Metabolism

23.47 What is meant by the term *glycogenesis*?

23.48 What is meant by the term *glycogenolysis*?

23.49 How do muscle cells use glycogen to provide energy?

23.50 How does the liver raise blood glucose levels?

23.51 What is the function of *glycogen phosphorylase*?

23.52 Why is the enzyme *phosphoglucomutase* used in both glycogenolysis and glycogenesis?

23.8 Gluconeogenesis: Glucose Synthesis

Glycogen stored in our liver and muscles can supply us with about one day's requirement of glucose. However, glycogen stores are quickly depleted if we fast for more than one day, run a marathon, or participate in other heavy exercise. Then glucose is synthesized from carbon atoms obtained from noncarbohydrate compounds in a process called **gluconeogenesis** ("new glucose"). Most glucose is synthesized in the cytosol of liver cells. (See Figure 23.13.)

Carbon atoms for glucose can be obtained from lactate and other food sources such as amino acids, and glycerol from fats. Each is converted to pyruvate or an intermediate for the synthesis of glucose. Most of the reactions in gluconeogenesis are the reverse of glycolysis and catalyzed by the same enzymes. However, three of the reactions are not reversible: the ones catalyzed by hexokinase, phosphofructokinase, and pyruvate kinase, glycolysis reactions 1, 3, and 10, respectively. Different enzymes are used to replace them, but all the other reactions simply reverse glycolysis and use the same enzymes. We will now look at these three reactions in gluconeogenesis that differ from the reverse of glycolysis.

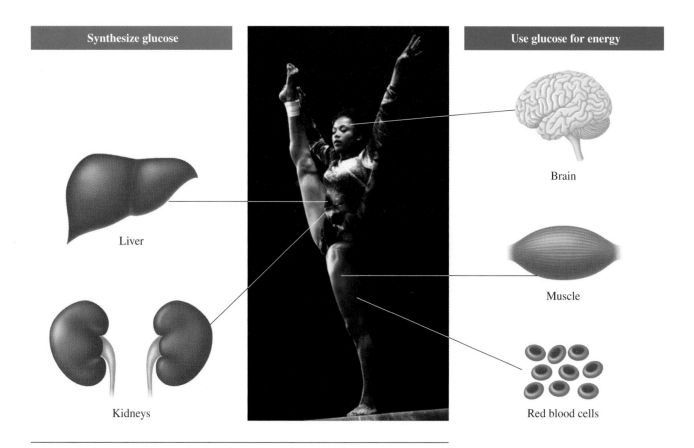

Figure 23.13 Glucose is synthesized in the tissues of the liver and kidneys. Tissues that use glucose as their main energy source are the brain, skeletal muscle, and red blood cells.

Q Why does the body need a pathway for the synthesis of glucose from noncarbohydrate sources?

Converting Pyruvate to Phosphoenolpyruvate

To start the synthesis of glucose, two steps are needed. The first step converts pyruvate to oxaloacetate, and the second step converts oxaloacetate to phosphoenolpyruvate. The hydrolysis of ATP and GTP are used to drive the reactions. (See Figure 23.14.)

$$\underset{\text{Pyruvate}}{CH_3-\overset{\overset{\textstyle O}{\|}}{C}-COO^-} + \mathbf{CO_2} + ATP + H_2O \xrightarrow{\overset{\text{Pyruvate}}{\text{carboxylase}}} \underset{\text{Oxaloacetate}}{{}^-OOC-CH_2-\overset{\overset{\textstyle O}{\|}}{C}-COO^-} + ADP + P_i$$

$$\underset{\text{Oxaloacetate}}{{}^-OOC-CH_2-\overset{\overset{\textstyle O}{\|}}{C}-COO^-} + GTP \xrightarrow{\overset{\text{Phosphoenolpyruvate}}{\text{carboxykinase}}} \underset{\text{Phosphoenolpyruvate}}{CH_2{=}\overset{\overset{\textstyle P}{|}}{C}-COO^-} + \mathbf{CO_2} + GDP$$

Molecules of phosphoenolpyruvate now enter the next five reverse reactions in glycolysis using the same enzymes to form fructose-1,6-bisphosphate.

Converting Fructose-1,6-bisphosphate to Fructose-6-phosphate

The second irreversible reaction in glycolysis is bypassed using *fructose-1,6-bisphosphatase* to cleave a phosphate from fructose-1,6-bisphosphate by hydrolysis with water and releasing the energy that drives the reaction.

$$\text{Fructose-1,6-bisphosphate} + H_2O \xrightarrow{\text{Fructose-1,6-bisphosphatase}} \text{fructose-6-phosphate} + P_i$$

Then fructose-6-phosphate undergoes a reversible reaction to yield glucose-6-phosphate.

Converting Glucose 6-phosphate to Glucose

In the final reaction, glucose-6-phosphate is converted to glucose by a different enzyme than used in glycolysis. *Glucose-6-phosphatase* catalyzes the hydrolysis of glucose-6-phosphate with water.

$$\text{Glucose-6-phosphate} + H_2O \xrightarrow{\text{Glucose-6-phosphatase}} \text{glucose} + P_i$$

Energy Cost of Gluconeogensis

The pathway of gluconeogenesis consists of seven reversible reactions of glycolysis and four new reactions that replace the three irreversible reactions. Overall, this synthesis of glucose requires 4 ATPs, 2GTPs, and 2 NADH. If all the reactions were simply the reverse of glycolysis, the synthesis of glucose would not be energetically favorable. By using the energy resources and bypassing the three irreversible and energy-requiring reactions, gluconeogenesis becomes favorable in terms of energy. The overall equation for gluconeogenesis is written as follows:

$$2\,\text{Pyruvate} + 4\,ATP + 2\,GTP + 2\,NADH + 2H^+ + 6H_2O \longrightarrow$$
$$\text{glucose} + 4\,ADP + 2\,GDP + 6\,P_i + 2\,NAD^+$$

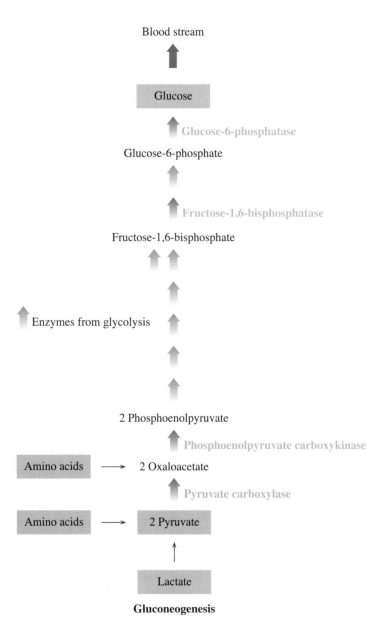

Gluconeogenesis

Figure 23.14 In gluconeogenesis, three irreversible reactions of glycolysis are bypassed using four different enzymes.

Q **Why are 11 enzymes required for gluconeogenesis, and only 10 for glycolysis?**

Lactate and the Cori Cycle

When a person exercises vigorously, the anaerobic conditions cause the reduction of pyruvate to lactate, which accumulates in the muscle. This reaction is necessary to oxidize NADH to NAD^+, which allows glycolysis to continue to produce a small amount of ATP. Lactate is an important source of carbon for gluconeogenesis. Lactate is transported to the liver where it is oxidized to pyruvate, which is used to synthesize glucose. Glucose enters the bloodstream and returns to the muscle to rebuild glycogen stores. This flow of lactate and glucose between the muscle and liver, known as the **Cori cycle,** is very active when a person has just completed a period of vigorous exercise. (See Figure 23.15.)

Regulation of Gluconeogenesis

Gluconeogenesis is a pathway that protects the brain and nervous system from experiencing a loss of glucose, which causes impairment of function. It is also a

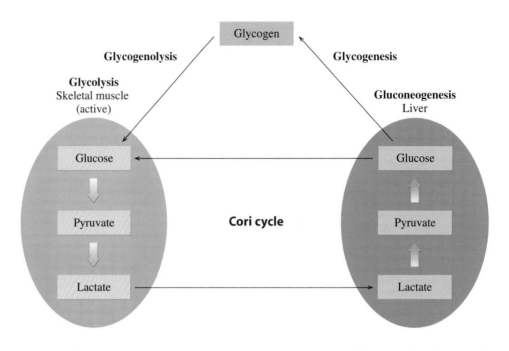

Figure 23.15 Different pathways connect the utilization and synthesis of glucose.

Q Why is lactate formed in the muscle converted to glucose in the liver?

pathway that is utilized when vigorous activity depletes blood glucose and glyco-gen stores. Thus, the level of carbohydrate available from the diet controls gluco-neogenesis. When a diet is high in carbohydrate, the gluconeogenesis pathway is not utilized. However, when a diet is low in carbohydrate, the pathway is very active.

As long as conditions in a cell favor glycolysis, there is no synthesis of glucose, but when the cell requires synthesis of glucose, glycolysis is turned off. The same three reactions that control glycolysis also control gluconeogenesis, but with differ-ent enzymes. Let's look at how high levels of certain compounds activate or inhibit the two processes. (See Table 23.2.)

Table 23.2 Regulation of Glycolysis and Gluconeogenesis

	Glycolysis	Gluconeogenesis
Enzyme	**Hexokinase**	**Glucose-6-phosphatase**
Activated by	High glucose levels Insulin, epinephrine	Low glucose levels Glucose-6-phosphate
Inhibited by	Glucose-6-phosphate	
Enzyme	**Phosphofructokinase**	**Fructose-1,6-bisphosphatase**
Activated by	AMP	Low glucose levels, glucagon
Inhibited by	ATP	AMP, insulin
Enzyme	**Pyruvate kinase**	**Pyruvate carboxylase**
Activated by	Fructose-1,6-bisphosphate	Low glucose level, glucagon
Inhibited by	ATP Acetyl CoA	Insulin

SAMPLE PROBLEM 23.8

Gluconeogenesis

The conversion of fructose-1,6-bisphosphate to fructose-6-phosphate is an irreversible reaction using the glycolytic enzyme. How does gluconeogenesis make this reaction happen?

Solution

This reverse reaction is catalyzed by a different enzyme *fructose-1,6-bisphosphatase,* which cleaves a phosphate using a hydrolysis reaction, a reaction that is energetically favorable.

Study Check

Why is hexokinase in glycolysis replaced by glucose-6-phosphatase in gluconeogenesis?

QUESTIONS AND PROBLEMS

Gluconeogenesis: Glucose Synthesis

23.53 What is the function of gluconeogenesis in the body?

23.54 What enzymes in glycolysis are not used in gluconeogenesis?

23.55 What enzymes in glycolysis are used in gluconeogenesis?

23.56 How is the lactate produced in skeletal muscle used for glucose synthesis?

23.57 Indicate whether each of the following activates or inhibits gluconeogenesis:
    a. low glucose levels   **b.** glucagon   **c.** insulin

23.58 Indicate whether each of the following activates or inhibits glycolysis?
    a. low glucose levels   **b.** insulin   **c.** glucagon

Chapter Review

23.1 Metabolism and Cell Structure

Metabolism includes all the catabolic and anabolic reactions that occur in the cells. Catabolic reactions degrade large molecules into smaller ones with an accompanying release of energy. Anabolic reactions require energy to synthesize larger molecules from smaller ones. The 3 stages of metabolism are digestion of food, degradation of monomers such as glucose to pyruvate, and the extraction of energy from the two- and three-carbon compounds from stage 2. Many of the metabolic enzymes are present in the cytosol of the cell where metabolic reactions take place.

23.2 ATP and Energy

Energy obtained from catabolic reactions is stored primarily in adenosine triphosphate (ATP), a high-energy compound that is hydrolyzed when energy is required by anabolic reactions.

23.3 Important Coenzymes in Metabolic Pathways

FAD and NAD^+ are the oxidized forms of coenzymes that participate in oxidation–reduction reactions. When they pick up hydrogen ions and electrons, they are reduced to $FADH_2$ and $NADH + H^+$. Coenzyme A contains a thiol group that bonds with a two-carbon acetyl group (acetyl CoA) or a longer chain acyl group (acyl CoA).

23.4 Digestion of Carbohydrates

The digestion of carbohydrates is a series of reactions that breaks down polysaccharides into hexose monomers such as glucose, galactose, and fructose. These monomers can be absorbed through the intestinal wall into the bloodstream to be carried to cells where they provide energy and carbon atoms for synthesis of new molecules.

23.5 Glycolysis: Oxidation of Glucose

Glycolysis, which occurs in the cytosol, consists of 10 reactions that degrade glucose (six carbons) to two pyruvate molecules (three carbons each). The overall series of reactions yields two molecules of the reduced coenzyme NADH and two ATP.

23.6 Pathways for Pyruvate

Under aerobic conditions, pyruvate is oxidized in the mitochondria to acetyl CoA. In the absence of oxygen, pyruvate is reduced to lactate and NAD^+ is regenerated for the continuation of glycolysis, while microorganisms such as yeast reduce pyruvate to ethanol, a process known as fermentation.

23.7 Glycogen Metabolism

Glycogenolysis breaks down glycogen to glucose when glucose and ATP levels are low. When blood glucose levels are high, glycogenesis converts glucose to glycogen, which is stored in the liver.

23.8 Gluconeogenesis: Glucose Synthesis

When blood glucose levels are low and glycogen stores in the liver are depleted, glucose is synthesized from compounds such as pyruvate and lactate.

Summary of Key Reactions

Hydrolysis of ATP

$$ATP + H_2O \longrightarrow ADP + P_i + 7.3 \text{ kcal (31 kJ)/mole}$$

$$ATP + H_2O \longrightarrow AMP + PP_i + 7.3 \text{ kcal (31 kJ)/mole}$$

Formation of ATP

$$ADP + P_i + 7.3 \text{ kcal/ (31 kJ)/mole} \longrightarrow ATP$$

Reduction of FAD and NAD^+

$$FAD + 2H^+ + 2e^- \longrightarrow FADH_2$$

$$NAD^+ + 2H^+ + 2e^- \longrightarrow NADH + H^+$$

Hydrolysis of Disaccharides

$$\text{Lactose} + H_2O \xrightarrow{\text{Lactase}} \text{galactose} + \text{glucose}$$

$$\text{Sucrose} + H_2O \xrightarrow{\text{Sucrase}} \text{fructose} + \text{glucose}$$

$$\text{Maltose} + H_2O \xrightarrow{\text{Maltase}} \text{glucose} + \text{glucose}$$

Glycolysis

$$C_6H_{12}O_6 + 2 \text{ ADP} + 2 P_i + 2 NAD^+ \rightarrow 2CH_3\underset{\text{Pyruvate}}{-\overset{\overset{\text{O}}{\|}}{C}-COO^-} + 2 \text{ ATP} + 2 \text{ NADH} + 4H^+$$

Glucose   Pyruvate

Oxidation of Pyruvate to Acetyl-CoA

$$CH_3\underset{\text{Pyruvate}}{-\overset{\overset{\text{O}}{\|}}{C}-COO^-} + NAD^+ + HS-CoA \longrightarrow CH_3\underset{\text{Acetyl CoA}}{-\overset{\overset{\text{O}}{\|}}{C}-S-CoA} + NADH + H^+ + CO_2$$

Reduction of Pyruvate to Lactate

$$\text{Pyruvate} + \text{NADH} + \text{H}^+ \longrightarrow \text{Lactate} + \text{NAD}^+$$

$$\underset{\text{Pyruvate}}{\text{CH}_3-\overset{\overset{\displaystyle O}{\|}}{\text{C}}-\text{COO}^-} + \text{NADH} + \text{H}^+ \longrightarrow \underset{\text{Lactate}}{\text{CH}_3-\overset{\overset{\displaystyle OH}{|}}{\text{CH}}-\text{COO}^-} + \text{NAD}^+$$

Oxidation of Glucose to Lactate

$$\text{Glucose} + 2\,\text{ADP} + 2\,\text{P}_i \longrightarrow 2\,\text{Lactate} + 2\,\text{ATP}$$

Reduction of Pyruvate to Ethanol

$$\underset{\text{Pyruvate}}{\text{CH}_3-\overset{\overset{\displaystyle O}{\|}}{\text{C}}-\text{COO}^-} + \text{NADH} + \text{H}^+ \longrightarrow \underset{\text{Ethanol}}{\text{CH}_3-\text{CH}_2-\text{OH}} + \text{NAD}^+ + \text{CO}_2$$

Glycogenesis

$$\text{Glucose} \longrightarrow \text{glycogen}$$

Glycogenolysis

$$\text{Glycogen} \longrightarrow \text{glucose}$$

Gluconeogensis

$$\text{Pyruvate (or lactate)} \longrightarrow \text{glucose}$$

$$2\,\text{Pyruvate} + 4\,\text{ATP} + 2\,\text{GTP} + 2\,\text{NADH} + 2\text{H}^+ + 6\text{H}_2\text{O} \longrightarrow$$
$$\text{Glucose} + 4\,\text{ADP} + 2\,\text{GDP} + 6\,\text{P}_i + 2\,\text{NAD}^+$$

Key Terms

acetyl CoA A two-carbon acetyl unit from oxidation of pyruvate that bonds to coenzyme A.

ADP Adenosine diphosphate, a compound of adenine, a ribose sugar, and two phosphate groups, it is formed by the hydrolysis of ATP.

aerobic An oxygen-containing environment in the cells.

amylase An enzyme that hydrolyzes the glycosidic bonds in polysaccharides during digestion.

anabolic reaction A metabolic reaction that requires energy.

anaerobic A condition in cells when there is no oxygen.

ATP Adenosine triphosphate, a high-energy compound that stores energy in the cells, consists of adenine, a ribose sugar, and three phosphate groups.

catabolic reaction A metabolic reaction that produces energy for the cell by the degradation and oxidation of glucose and other molecules.

coenzyme A (CoA) A coenzyme that transports acyl and acetyl groups.

Cori cycle A cyclic process in which lactate produced in muscle is transferred to the liver to be synthesized to glucose, which can be used again by muscle.

cytoplasm The material in eukaryotic cells between the nucleus and the plasma membrane.

cytosol The fluid of the cytoplasm, which is an aqueous solution of electrolytes and enzymes.

decarboxylation The loss of a carbon atom in the form of CO_2.

digestion The processes in the gastrointestinal tract that break down large food molecules to smaller ones that pass through the intestinal membrane into the blood stream.

FAD A coenzyme (flavin adenine dinucleotide) for dehydrogenase enzymes that form carbon–carbon double bonds.

fermentation The anaerobic conversion of glucose by enzymes in yeast to yield alcohol and CO_2.

gluconeogenesis The synthesis of glucose from noncarbohydrate compounds.

glycogenesis The synthesis of glycogen from glucose molecules.

glycogenolysis The breakdown of glycogen into glucose molecules.

glycolysis The ten oxidation reactions of glucose that yield two pyruvate molecules.

metabolism All the chemical reactions in living cells that carry out molecular and energy transformations.

mitochondria The organelles of the cells where energy-producing reactions take place.

NAD^+ The hydrogen acceptor used in oxidation reactions that form carbon–oxygen double bonds.

oxidation The loss of hydrogen as hydrogen ions and electrons or the gain of oxygen by a substrate that is degraded to smaller molecules or a coenzyme.

reduction The gain of hydrogen ions and electrons or the loss of oxygen by a substrate or a coenzyme.

Additional Problems

23.59 What is meant by the term metabolism?

23.60 How do catabolic reactions differ from anabolic reactions?

23.61 What stage of metabolism involves the digestion of large food polymers?

23.62 What state of metabolism degrades monomers such as glucose into smaller molecules?

23.63 What type of cell has a nucleus?

23.64 What are the organelles in a cell?

23.65 What is the full name of ATP?

23.66 What is the full name of ADP?

23.67 Write an equation for the hydrolysis of ATP to ADP.

23.68 At the gym, you expend 300 kcal riding the stationary bicycle for 1 hr. How many moles of ATP will this require?

23.69 What is the full name of FAD?

23.70 What type of reaction uses FAD as the coenzyme?

23.71 What is the full name of NAD^+?

23.72 What type of reaction uses NAD^+ as the coenzyme?

23.73 Write the letters for the reduced forms of
a. FAD   **b.** NAD^+

23.74 What is the name of the vitamin in the structure of each of the following?
a. FAD   **b.** NAD^+ **c.** coenzyme A

23.75 How and where does lactose undergo digestion in the body? What are the products?

23.76 How and where does sucrose undergo digestion in the body? What are the products?

23.77 How do galactose and fructose enter glycolysis?

23.78 What is the general type of reaction that takes place in the digestion of carbohydrates?

23.79 What are the reactant and product of glycolysis?

23.80 What is the coenzyme used in glycolysis?

23.81 In glycolysis, which reactions involve phosphorylation and which reactions involve a direct substrate phosphorylation to generate ATP?

23.82 How do ADP and ATP regulate the glycolysis pathway?

23.83 What reaction and enzyme in glycolysis convert a hexose bisphosphate into two triose phosphates?

23.84 How does the investment and generation of ATP give a net gain of ATP for glycolysis?

23.85 What compound is converted to fructose-6-phosphate by phosphoglucoisomerase?

23.86 What product forms when glyceraldehyde-3-phosphate adds a phosphate group?

23.87 When is pyruvate converted to lactate in the body?

23.88 When pyruvate is used to form acetyl CoA or ethanol in fermentation, the product has only two carbon atoms. What happened to the third carbon?

23.89 How does phosphofructokinase regulate the rate of glycolysis?

23.90 How does pyruvate kinase regulate the rate of glycolysis?

23.91 When does the rate of glycogenolysis increase in the cells?

23.92 If glucose-1-phosphate is the product from glycogen, how does it enter glycolysis?

23.93 What is the end product of glycogenolysis in the liver?

23.94 What is the end product of glycogenolysis in skeletal muscle?

23.95 Why is glucose for blood glucose provided by glycogenolysis in the liver, but not in skeletal muscle?

23.96 When does the rate of glycogenesis increase in the cells?

23.97 How do the hormones insulin and glucagon affect the rate of glycogenesis, glycogenolysis, and glycolysis?

23.98 What is the function of gluconeogenesis?

23.99 Where does the Cori cycle operate?

23.100 Identify each of the following as part of glycolysis, glycogenolysis, glycogenesis, or gluconeogenesis:
a. Glycogen is broken down to glucose in the liver.
b. Glucose is synthesized from noncarbohydrate sources.
c. Glucose is degraded to pyruvate.
d. Glycogen is synthesized from glucose.

23.101 Indicate whether each of the following conditions would increase or decrease the rate of glycogenolysis in the liver:
a. low blood glucose level
b. secretion of insulin
c. secretion of glucagon
d. high levels of ATP

23.102 Indicate whether each of the following conditions would increase or decrease the rate of glycogenesis in the liver:
a. low blood glucose level
b. secretion of insulin
c. secretion of glucagon
d. high levels of ATP

23.103 Indicate whether each of the following conditions would increase or decrease the rate of gluconeogenesis:
a. high blood-glucose level
b. secretion of insulin
c. secretion of glucagon
d. high levels of ATP

23.104 Indicate whether each of the following conditions would increase or decrease the rate of glycolysis:
a. high blood-glucose level
b. secretion of insulin
c. secretion of glucagon
d. high levels of ATP

24 Metabolism and Energy Production

"I am trained in basic life support. I work with the ER staff to assist in patient care," says Mandy Dornell, emergency medical technician at Seaton Medical Center. "In the ER, I take vital signs, patient assessment, and do CPR. If someone has a motor vehicle accident, I may suspect a neck or back injury. Then I may use a backboard or a cervical collar, which prevents the patient from moving and causing further damage. When people have difficulty breathing, I insert an airway—nasal or oral—to assist ventilation. I also set up and monitor IVs, and I am trained in childbirth."

When someone is critically ill or injured, the quick reactions of emergency medical technicians (EMTs) and paramedics provide immediate medical care and transport to an ER or trauma center.

LOOKING AHEAD

www.chemplace.com/college

Visit the URL above or use the CD-ROM in the book for extra quizzing, interactive tutorials, career resources, and case studies.

In Chapter 23 we described the digestion of carbohydrates to glucose and the degradation of glucose to pyruvate. We saw that pyruvate is converted to lactate when no oxygen is available in the cell or to two-carbon acetyl CoA when oxygen is plentiful. Although glycolysis produces a small amount of ATP, most of the ATP in the cells is produced during the conversion of pyruvate, when oxygen is available in the cell. In a process known as *respiration,* oxygen is required to complete the oxidation of glucose to CO_2 and H_2O.

In the *citric acid cycle,* a series of metabolic reactions in the mitochondria oxidize acetyl CoA to carbon dioxide, which releases energy to produce NADH and $FADH_2$. These reduced coenzymes enter the *electron transport chain* or *respiratory chain* where they provide hydrogen ions and electrons that combine with oxygen (O_2) to form H_2O. The energy released during electron transport is used to synthesize ATP from ADP.

24.1 The Citric Acid Cycle

The **citric acid cycle** is a series of reactions that degrades acetyl CoA to yield CO_2 and energy, which is used to produce $NADH + H^+$ and $FADH_2$. (See Figure 24.1.) The citric acid cycle connects the products from stages 1 and 2 with the electron transport chain and the synthesis of ATP of stage 3. As a central pathway in metabolism, the citric acid cycle uses acetyl CoA from the degradation of carbohydrates as well as lipids and proteins.

The citric acid cycle is also called the tricarboxlyic acid cycle because citric acid is the product that forms in the first reaction. Although citric acid is present as citrate at cellular pH, its acid name is retained. The citric acid cycle is also known as the Krebs cycle for H. A. Krebs, who recognized it as the major pathway for the oxidation of carbohydrate.

Overview of the Citric Acid cycle

The fuel for the citric acid cycle is acetyl CoA, which is primarily produced from the oxidative decarboxylation of pyruvate. Acetyl CoA is also obtained from fats and proteins when they are used to generate energy, as we will see in Chapter 25. There are a total of 8 reactions in the citric acid cycle, which we can separate into two parts. In part 1, a two-carbon acetyl group bonds with four-carbon oxaloacetate to yield citrate. Then decarboxylation reactions remove two carbon atoms as CO_2, which produces a four-carbon compound. Then the reactions in part 2 convert the four-carbon compound back to oxaloacetate, which bonds with another acetyl CoA and goes through the cycle again. (See Figure 24.2.) In one turn of the citric acid cycle, four oxidation reactions provide hydrogen ions and electrons, which are used to reduce FAD and NAD^+ coenzymes.

Stages of Metabolism

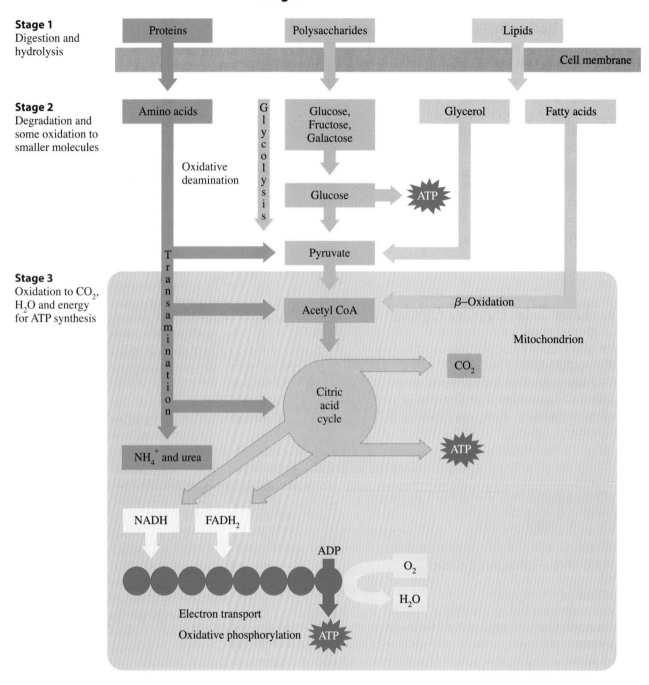

Figure 24.1 The citric acid cycle connects the catabolic pathways that begin with the digestion and degradation of foods in stages 1 and 2 with the oxidation of substrates in stage 3 that generates most of the energy for ATP synthesis.

Q Why is the citric acid cycle called a central metabolic pathway?

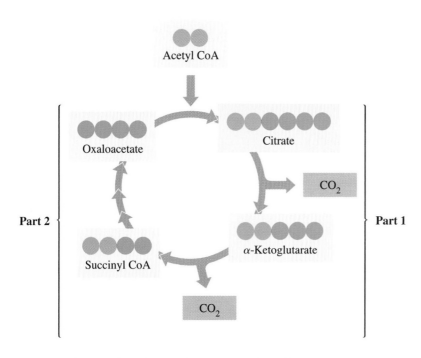

Figure 24.2 In part 1 of the citric acid cycle, two carbon atoms are removed as CO_2 from six-carbon citrate to give four-carbon succinyl CoA, which is converted in part 2 to four-carbon oxaloacetate.

Q What is the difference in part 1 and part 2 of the citric acid cycle?

Part 1   Decarboxylation Removes Two Carbon Atoms

Reaction 1   Formation of Citrate

In the first reaction, the two-carbon acetyl group in acetyl CoA bonds with the four-carbon oxaloacetate to yield citrate and CoA. The formation of citrate is an important regulation point in the cycle. (See Figure 24.3.)

$$CH_3{-}\overset{\overset{\textstyle O}{\|}}{C}{-}S{-}CoA \ + \ \underset{\underset{\textstyle COO^-}{\underset{\textstyle |}{CH_2}}}{\overset{\overset{\textstyle COO^-}{|}}{\underset{}{C}}}{=}O \ + \ H_2O \ \xrightarrow{\overset{\text{Citrate}}{\text{synthase}}} \ HO{-}\underset{\underset{\textstyle COO^-}{\underset{\textstyle |}{CH_2}}}{\overset{\overset{\textstyle COO^-}{\overset{\textstyle |}{CH_2}}}{C}}{-}COO^- \ + \ HS{-}CoA \ + \ H^+$$

Acetyl CoA   Oxaloacetate   Citrate

Reaction 2   Isomerization to Isocitrate

In order to continue oxidation, citrate undergoes isomerization to yield isocitrate. This is necessary because the secondary hydroxyl group (Chapter 14) in isocitrate can be oxidized in the next reaction, while the tertiary hydroxyl group in citrate cannot be oxidized. The isomerization actually consists of two steps. Citrate loses water (dehydration) to yield aconitate, which is rehydrated to form isocitrate. This reaction converts the tertiary hydroxyl ($-$OH) group to a secondary hydroxyl group, because a tertiary hydroxyl ($-$OH) group cannot be oxidized further.

$$
\begin{array}{ccccc}
COO^- & & COO^- & & COO^- \\
| & & | & & | \\
CH_2 & & CH_2 & & CH_2 \\
| & \xrightarrow[\text{Aconitase}]{H_2O} & \| & \xrightarrow[\text{Aconitase}]{H_2O} & | \\
HO{-}C{-}COO^- & & C{-}COO^- & & H{-}C{-}COO^- \\
| & & \| & & | \\
H{-}C{-}H & & C{-}H & & HO{-}C{-}H \\
| & & | & & | \\
COO^- & & COO^- & & COO^- \\
\text{Citrate} & & \text{Aconitate} & & \text{Isocitrate}
\end{array}
$$

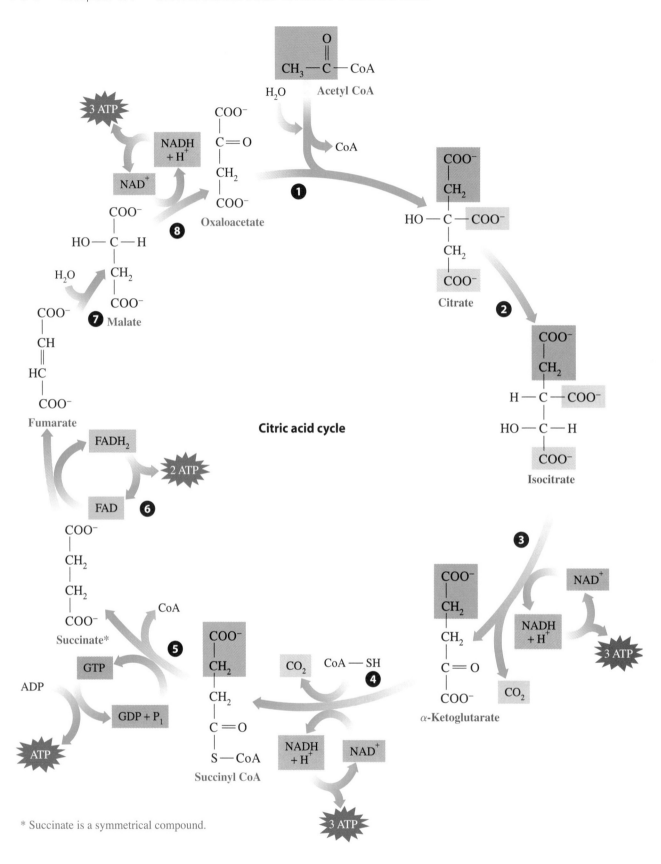

Figure 24.3 In the citric acid cycle, oxidation reactions produce two CO_2, reduced coenzymes NADH and $FADH_2$, and regenerate oxaloacetate.

Q How many reactions in the citric acid cycle produce a reduced coenzyme?

Reaction 3   First Oxidative Decarboxylation (CO₂)

This is the first time in the citric acid cycle that both an oxidation and a decarboxylation occur together. The oxidation converts the hydroxyl group to a ketone and the decarboxylation removes a carbon as a CO_2 molecule. That carbon is from a carboxylate group (COO^-) in the original oxaloacetate and not from the acetyl group. The loss of CO_2 shortens the carbon chain to yield a five-carbon α-ketoglutarate. The energy from the oxidation is used to transfer hydrogen ions and electrons to NAD^+. We can summarize the reactions as follows:

1. The hydroxyl group ($-OH$) is oxidized to a ketone ($C=O$).
2. NAD^+ is reduced to yield NADH.
3. A carboxylate group (COO^-) is removed as CO_2.

$$
\begin{array}{c}
COO^- \\
| \\
CH_2 \\
| \\
H-C-COO^- \;+\; NAD^+ \\
| \\
HO-C-H \\
| \\
COO^-
\end{array}
\xrightarrow[\text{dehydrogenase}]{\text{Isocitrate}}
\begin{array}{c}
COO^- \\
| \\
CH_2 \\
| \\
H-C-H \;+\; CO_2 \;+\; NADH \\
| \\
C=O \\
| \\
COO^-
\end{array}
$$

Isocitrate α-Ketoglutarate

Reaction 4   Second Oxidative Decarboxylation (CO₂)

In this reaction, a second CO_2 is removed as α-ketoglutarate undergoes oxidative decarboxylation. The resulting four-carbon group combines with coenzyme A to form succinyl CoA and hydrogen ions and electrons are transferred to NAD^+. The two reactions that occur are as follows:

1. A second carbon is removed as CO_2.
2. NAD^+ is reduced to yield NADH.

$$
\begin{array}{c}
COO^- \\
| \\
CH_2 \\
| \\
CH_2 \;+\; NAD^+ \;+\; CoA-SH \\
| \\
C=O \\
| \\
COO^-
\end{array}
\xrightarrow[\text{dehydrogenase}]{\alpha\text{-Ketoglutarate}}
\begin{array}{c}
COO^- \\
| \\
CH_2 \\
| \\
CH_2 \;+\; CO_2 \;+\; NADH \\
| \\
C=O \\
| \\
S-CoA
\end{array}
$$

α-Ketoglutarate   Succinyl CoA

Part 2

Reaction 5   Hydrolysis of Succinyl CoA

The energy released by the hydrolysis of the thioester bond in succinyl CoA is used to add a phosphate group (P_i) directly to GDP (guanosine diphosphate). The products are succinate and GTP, which is a high-energy compound similar to ATP.

$$
\begin{array}{c}
COO^- \\
| \\
CH_2 \\
| \\
CH_2 \;+\; GDP \;+\; P_i \;+\; H^+ \\
| \\
C=O \\
| \\
S-CoA
\end{array}
\xrightarrow[\text{synthetase}]{\text{Succinyl CoA}}
\begin{array}{c}
COO^- \\
| \\
CH_2 \\
| \\
CH_2 \;+\; GTP \;+\; CoA-SH \\
| \\
COO^-
\end{array}
$$

Succinyl CoA   Succinate

The hydrolysis of GTP is used to add a phosphate group to ADP, which regenerates GDP for the citric acid cycle. This is the only time in the citric acid cycle that a direct substrate phosphorylation is used to produce ATP.

$$GTP + ADP \longrightarrow GDP + ATP$$

Reaction 6   Dehydrogenation of Succinate

In this oxidation reaction, hydrogen is removed from two carbon atoms in succinate, which produces fumarate, a compound with a trans double bond. This is the only place in the citric acid cycle where FAD is reduced to $FADH_2$.

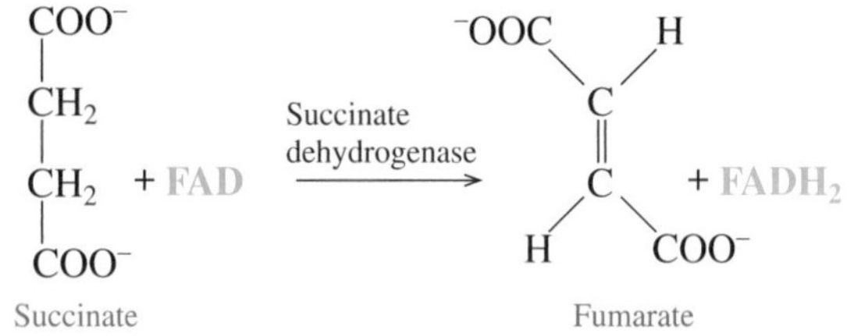

Succinate   Fumarate

Reaction 7   Hydration

In a hydration reaction, water adds to the double bond of fumarate to yield malate.

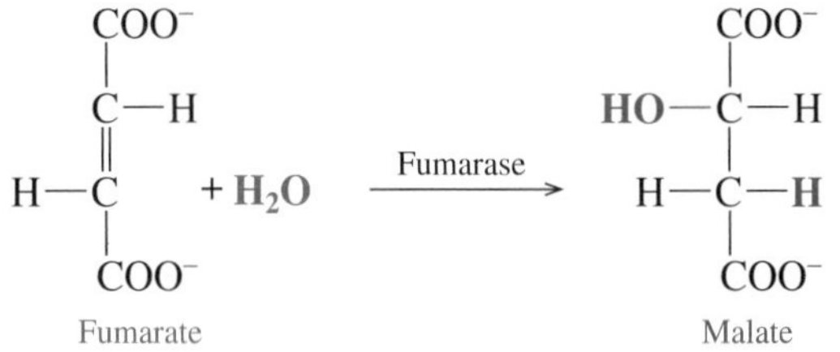

Fumarate   Malate

Reaction 8   Dehydrogenation Forms Oxaloacetate

In the last step of the citric acid cycle, the hydroxyl ($-$OH) group in malate is oxidized to yield oxaloacetate, which has a ketone group. The coenzyme NAD^+ is reduced to $NADH + H^+$.

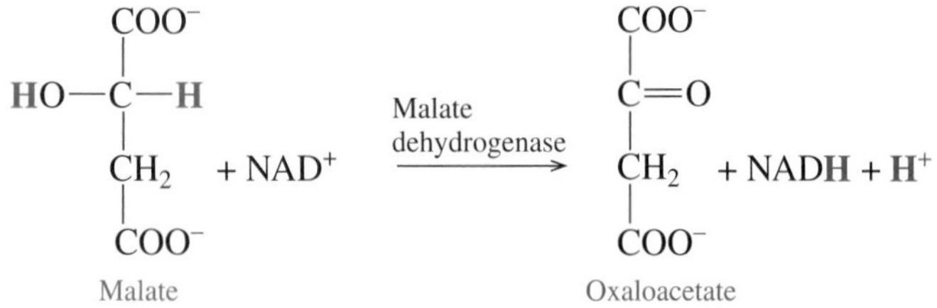

Malate   Oxaloacetate

Summary of Products from the Citric Acid Cycle

We have seen that the citric acid cycle begins when a two-carbon acetyl group from acetyl CoA combines with oxaloacetate. In part 1 of the cycle, two carbon atoms are removed to yield two CO_2 and a four-carbon compound that undergoes more reactions in part 2 to regenerate oxaloacetate. In four oxidation reactions, energy is released that reduces three NAD^+ and one FAD. In one reaction, GTP produced by a direct phosphorylation is used to form ATP from ADP. A summary of the products from one turn of the citric acid cycle is as follows:

2 CO_2 molecules

3 NADH molecules

1 $FADH_2$ molecule

1 GTP molecule used to form ATP

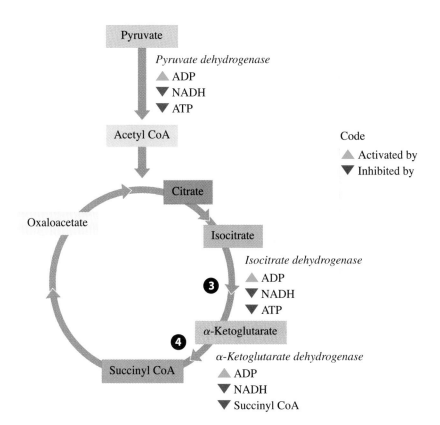

Figure 24.4 High levels of ADP activate enzymes for the production of acetyl CoA and the citric acid cycle, whereas high levels of ATP, NADH, and succinyl CoA inhibit enzymes in the citric acid cycle.

Q **How do high levels of ATP affect the rate of the citric acid cycle?**

We can write the overall chemical equation for one turn of the citric acid cycle as follows:

$$\text{Acetyl CoA} + 3\,\text{NAD}^+ + \text{FAD} + \text{GDP} + \text{P}_i + 2\text{H}_2\text{O} \longrightarrow$$

$$2\text{CO}_2 + 3\,\text{NADH} + 2\text{H}^+ + \text{FADH}_2 + \text{HS—CoA} + \text{GTP}$$

Regulation of Citric Acid Cycle

The primary function of the citric acid cycle is to produce high-energy compounds for ATP synthesis. When the cell needs energy, low levels of ATP stimulate the conversion of pyruvate to acetyl CoA, the fuel for the citric acid cycle. When ATP and NADH levels are high, there is a decrease in the production of acetyl CoA from pyruvate.

In the citric acid cycle, the enzymes that catalyze reactions 3 and 4 respond to allosteric activation and inhibition (Chapter 21). In reaction 3, isocitrate dehydrogenase is activated by high levels of ADP and inhibited by high levels of ATP and NADH. In reaction 4, α-ketoglutarate dehydrogenase is activated by high levels of ADP and inhibited by high levels of NADH and succinyl CoA. (See Figure 24.4.)

SAMPLE PROBLEM 24.1

Citric Acid Cycle

When one acetyl CoA completes the citric acid cycle, how many of each of the following are produced?

a. NADH **b.** ketone group **c.** CO_2

Solution

a. One turn of the citric acid cycle produces three molecules of NADH.

b. Two ketone groups form when the secondary alcohol groups in isocitrate and malate are oxidized by NAD^+.

c. Two molecules of CO_2 are produced by the decarboxylation of isocitrate and α-ketoglutarate.

Study Check

What substance is a substrate in the first reaction of the citric acid cycle and a product in the last reaction?

QUESTIONS AND PROBLEMS

The Citric Acid Cycle

24.1 What other names are used for the citric acid cycle?

24.2 What compounds are needed to start the citric acid cycle?

24.3 What are the products from one turn of the citric acid cycle?

24.4 What compound is regenerated in each turn of the citric acid cycle?

24.5 Which reactions of the citric acid cycle involve oxidative decarboxylation?

24.6 Which reactions of the citric acid cycle involve a dehydration reaction?

24.7 Which reactions of the citric acid cycle reduce NAD^+?

24.8 Which reactions of the citric acid cycle reduce FAD?

24.9 Where does a direct substrate phosphorylation occur?

24.10 What is the total NADH and total $FADH_2$ produced in one turn of the citric acid cycle?

24.11 Refer to the diagram of the citric acid cycle to answer each of the following:
a. What are the six-carbon compounds?
b. How is the number of carbon atoms decreased?
c. What are the five-carbon compounds?
d. Which reactions are oxidation reactions?
e. In which reactions are secondary alcohols oxidized?

24.12 Refer to the diagram of the citric acid cycle to answer each of the following:
a. What is the yield of CO_2 molecules?
b. What are the four-carbon compounds?
c. What is the yield of GTP molecules?
d. What are the decarboxylation reactions?
e. Where does a hydration occur?

24.13 Indicate the name of the enzyme for each of the following reactions in the citric acid cycle:
a. joins acetyl CoA to oxaloacetate
b. forms a carbon–carbon double bond
c. adds water to fumarate

24.14 Indicate the name of the enzyme for each of the following reactions in the citric acid cycle:
a. isomerizes citrate
b. oxidizes and decarboxylates α-ketoglutarate
c. adds P_i to GDP

24.15 State the acceptor for hydrogen or phosphate in each of the following reactions:
a. isocitrate $\longrightarrow$ α-ketoglutarate
b. succinyl CoA $\longrightarrow$ succinate   **c.** succinate $\longrightarrow$ fumarate

24.16 State the acceptor for hydrogen or phosphate in each of the following reactions:
a. malate $\longrightarrow$ oxaloacetate
b. α-ketoglutarate $\longrightarrow$ succinyl CoA
c. pyruvate $\longrightarrow$ acetyl CoA

24.17 What enzymes in the citric acid cycle are allosteric enzymes?

24.18 How does NADH affect the rate of the citric acid cycle?

24.19 How do high levels of ADP affect the rate of the citric acid cycle?

24.20 Why does the rate of the oxidation of pyruvate affect the rate of the citric acid cycle?

24.2 Electron Carriers

At this point, the metabolic cycles of glycolysis, oxidation of two pyruvate, and the citric acid cycle for two acetyl CoA would produce four ATP along with ten NADH and two $FADH_2$ from the degradation of glucose.

Glycolysis:	2 NADH and 2 ATP
Oxidation of 2 pyruvate:	2 NADH
Citric acid cycle (2 acetyl CoA):	6 NADH, 2 $FADH_2$, and 2 ATP

Now we will see how the oxidation of these reduced coenzymes provides the energy for the synthesis of considerably more ATP. In the **electron transport chain** or *respiratory chain,* hydrogen ions and electrons from NADH and $FADH_2$ are passed from one electron acceptor to the next until they combine with oxygen to form H_2O. The electron acceptors in this transport system are known as **electron carriers.** The energy released during electron transport is used to synthesize ATP from ADP and P_i, a process called *oxidative phosphorylation.* As long as oxygen is available for the mitochondria in the cell, electron transport and oxidative phosphorylation function to produce most of the ATP energy manufactured in the cell.

Oxidation and Reduction of Electron Carriers

The electron carriers in the electron transport include flavins, iron–sulfur proteins, coenzyme Q, and cytochromes. Each type of electron carrier contains a group or ion that is reduced and then oxidized as the electrons are accepted and then passed on. We can illustrate this process with two electron carriers A and B. As hydrogen ions and electrons are transferred from the reduced AH_2 to the oxidized carrier B, reduced BH_2 is formed along with the oxidized form of carrier A.

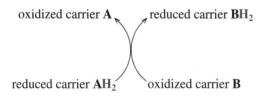

There are four types of electron carriers that make up the electron transport system.

1. **FMN (flavin mononucleotide)** is a coenzyme derived from riboflavin (vitamin B_2). FMN contains a flavin ring system that is also found in FAD. In riboflavin, the ring system is attached to ribitol, the sugar alcohol of ribose. (See Figure 24.5.) The reduced product is $FMNH_2$.

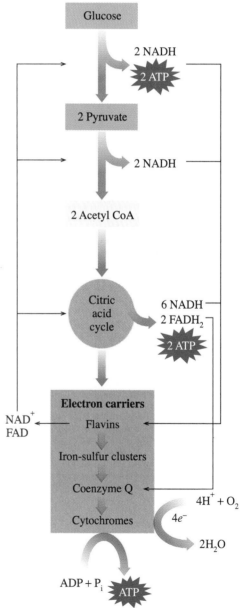

Figure 24.5 The electron carrier FMN consists of a flavin ring system containing the reactive center, ribitol, and a phosphate group.

Q What part of the FMN molecule is reduced when hydrogen ions and electrons are accepted?

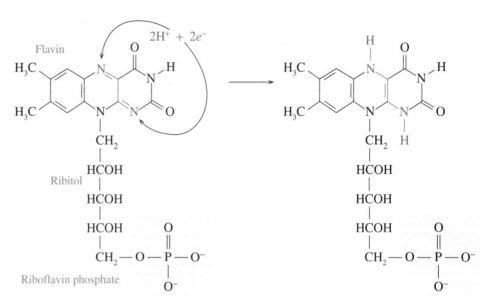

FMN (flavin mononucleotide)   FMNH$_2$

Figure 24.6 In a typical iron–sulfur cluster, an iron ion bonds to sulfur atoms in the thiol (— SH) groups of four cysteine groups in proteins.

Q In an iron–sulfur cluster, what are the ionic charges of the oxidized and reduced iron ions?

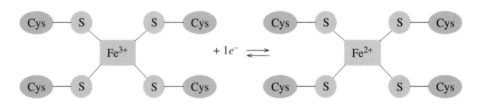

Typical Fe-S cluster

2. **Fe–S** clusters is the name given to a group of iron–sulfur proteins that contain iron–sulfur clusters embedded in the proteins of the electron transport chain. The clusters contain iron ions, inorganic sulfides, and several cysteine groups. The iron in the clusters is reduced to Fe^{2+} and oxidized to Fe^{3+} as electrons are accepted and lost. (See Figure 24.6.)

3. **Coenzyme Q (Q** or **CoQ)** is derived from quinone, which is a six-carbon cyclic compound with two double bonds and two keto groups attached to a long carbon chain. (See Figure 24.7.) Coenzyme Q is reduced when the keto groups of quinone accept hydrogen ions and electrons.

4. **Cytochromes (cyt)** are proteins that contain an iron ion in a heme group. The different cytochromes are indicated by the letters following the abbreviation for cytochrome (cyt): cyt *b*, cyt c_1, cyt *c*, cyt *a*, and cyt a_3. In each cytochrome, the Fe^{3+} accepts a single electron to form Fe^{2+}, which is oxidized back to Fe^{3+} when the electron is passed to the next cytochrome. (See Figure 24.8.)

$$Fe^{3+} + 1e^- \rightleftarrows Fe^{2+}$$

SAMPLE PROBLEM 24.2

Electron Carriers

Give the abbreviation for each of the following carriers:

a. the oxidized form of flavin mononucleotide

b. the reduced form of coenzyme Q

Solution

a. FMN   **b.** QH$_2$

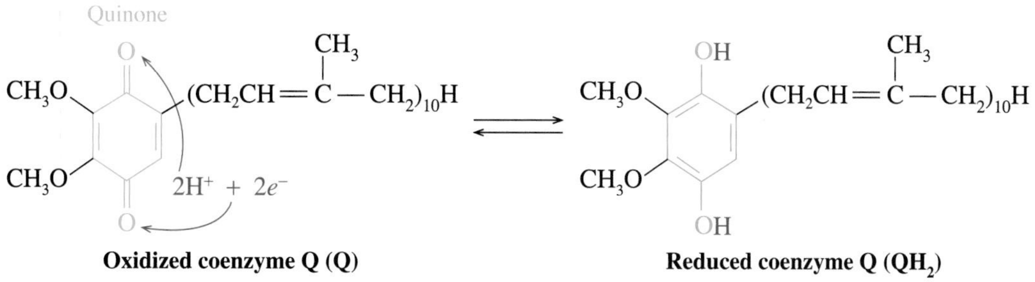

Quinone

Oxidized coenzyme Q (Q)   **Reduced coenzyme Q (QH₂)**

Figure 24.7 The electron carrier coenzyme Q accepts electrons from FADH₂ and FMNH₂ and passes them to the cytochromes.

Q **How does reduced coenzyme Q compare to the oxidized form?**

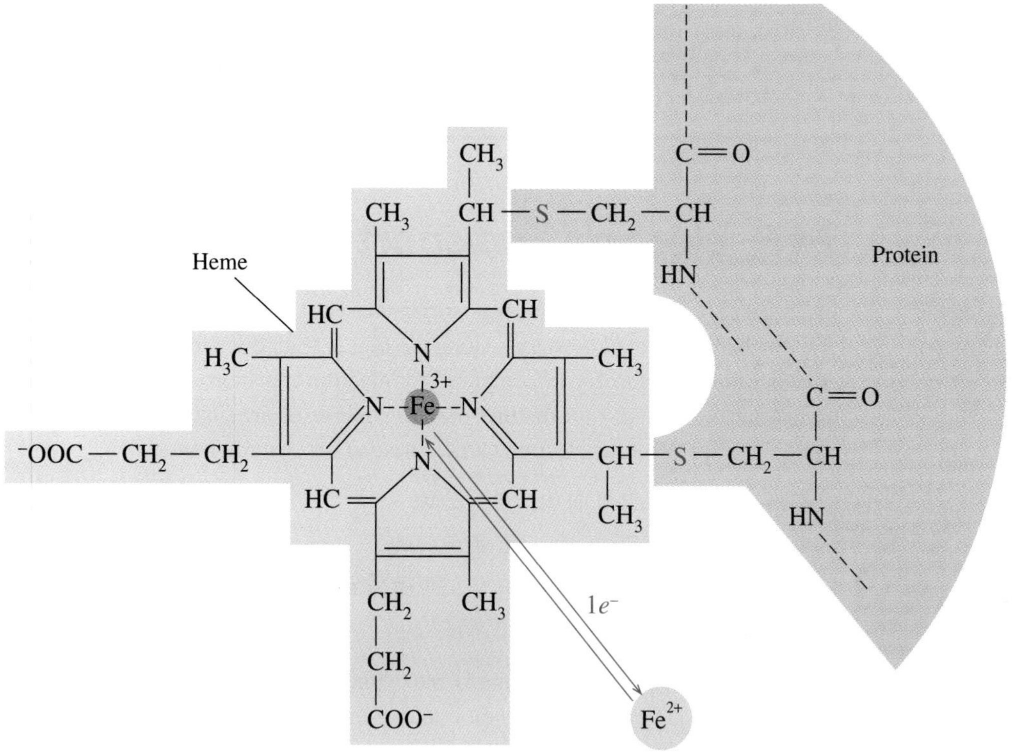

Figure 24.8 The iron-containing proteins known as cytochromes are identified as *b*, *c*, *c₁*, *a*, and *a₃*.

Q **What are the reduced and oxidized forms of the cytochromes?**

Study Check

What is the oxidized form of iron in cyt a_3?

SAMPLE PROBLEM 24.3

Oxidation and Reduction

Identify the following steps in the electron transport chain as oxidation or reduction:

a. $FMN + 2H^+ + 2e^- \longrightarrow FMNH_2$
b. Cyt c (Fe^{2+}) $\longrightarrow$ cyt c (Fe^{3+}) $+ e^-$

Solution

a. The gain of hydrogen is reduction.   **b.** The loss of electrons is oxidation.

Study Check

Identify each of the following as oxidation or reduction:

a. $QH_2 \longrightarrow Q$ **b.** Cyt b (Fe^{3+}) $\longrightarrow$ cyt b (Fe^{2+})

QUESTIONS AND PROBLEMS

Electron Carriers

24.21 Is cyt b (Fe^{3+}) the abbreviation for the oxidized or reduced form of cytochrome b?

24.22 Is $FMNH_2$ the abbreviation for the oxidized or reduced form of flavin mononucleotide?

24.23 Identify the following as oxidation or reduction:

a. $FMNH_2 \longrightarrow FMN + 2H^+ + 2e^-$

b. $Q + 2H^+ + 2e^- \longrightarrow QH_2$

24.24 Identify the following as oxidation or reduction:

a. Cyt c (Fe^{3+}) + $e^- \longrightarrow$ cyt c (Fe^{2+})

b. $Fe^{2+}S$ cluster $\longrightarrow Fe^{3+}S$ cluster + e^-

LEARNING GOAL

Describe the role of the electron carriers in electron transport.

WEB TUTORIAL
Electron Transport

24.3 Electron Transport

In Chapter 23, we saw that a mitochondrion consists of inner and outer membranes with the matrix located between. Along the highly folded inner membrane are the enzymes and electron carriers required for electron transport. When these membranes are broken up, four distinct protein complexes are obtained. Within each complex are some of the electron carriers needed for electron transport.

Complex I	NADH dehydrogenase
Complex II	Succinate dehydrogenase
Complex III	Coenzyme Q-cytochrome c reductase
Complex IV	Cytochrome c oxidase

Two electron carriers, coenzyme Q and cytochrome c, are not firmly attached to the membrane. They function as mobile carriers shuttling electrons between the protein complexes that are tightly bound to the membrane. The mobile carrier coenzyme Q moves electrons from complexes I and II to complex III. The other mobile carrier, cytochrome c, transfers electrons from complex III to complex IV. (See Figure 24.9.)

Complex I   NADH Dehydrogenase

At complex I, all of the NADH generated in the cell transfers hydrogen ions and electrons to the electron carrier FMN. The reduced $FMNH_2$ forms, while NADH is reoxidized to NAD^+, which returns to oxidative pathways such as the citric acid cycle to oxidize more substrates.

$$NADH + H^+ + FMN \longrightarrow NAD^+ + FMNH_2$$

Within complex I, the electrons are transferred to iron–sulfur (Fe–S) clusters and then to coenzyme (Q).

$$FMNH_2 + Q \longrightarrow QH_2 + FMN$$

The overall reaction sequence in complex I can be written as follows:

$$NADH + H^+ + Q \longrightarrow QH_2 + NAD^+$$

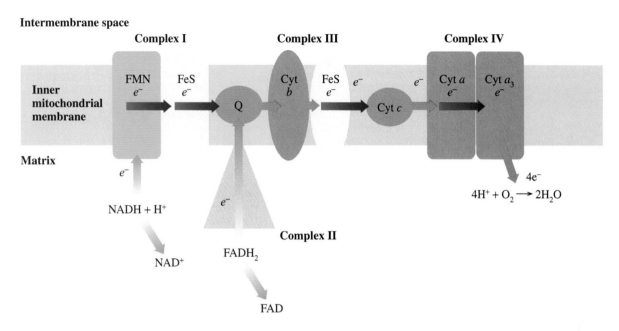

Complex II   Succinate Dehydrogenase

Complex II is specifically used when $FADH_2$ is generated by the conversion of succinate to fumarate in the citric acid cycle. The electrons from $FADH_2$ are transferred to coenzyme Q to yield QH_2. Because complex II is at a lower energy level than complex I, the electrons from $FADH_2$ enter electron transport at a lower energy level than those from NADH.

$$FADH_2 + Q \longrightarrow FAD + QH_2$$

Complex III   Coenzyme Q-Cytochrome c Reductase

The mobile carrier QH_2 transfers the electrons it has collected from NADH and $FADH_2$ to an iron–sulfur (Fe–S) cluster, and then to cytochrome *b*, the first cytochrome in complex III.

$$QH_2 + 2 \text{ cyt } b \text{ (Fe}^{3+}) \longrightarrow Q + 2 \text{ cyt } b \text{ (Fe}^{2+}) + 2H^+$$

From cyt *b*, the electron is transferred to an Fe–S cluster and then to cytochrome c_1 and then to cytochrome *c*. Each time an Fe^{3+} ion accepts an electron, it is reduced to Fe^{2+}, and then oxidized back to Fe^{3+} as the electron is passed on along the chain. Cytochrome *c* is another mobile carrier; it moves the electron from complex III to complex IV.

Complex IV   Cytochrome c Oxidase

At complex IV, electrons are transferred from cytochrome *c* to cytochrome *a*, and then to cytochrome a_3, the last cytochrome. In the final step of electron transport, electrons and hydrogen ions combine with oxygen (O_2) to form water.

$$4H^+ + 4e^- + O_2 \longrightarrow 2H_2O$$

Figure 24.9 Most of the electron carriers in the electron transport chain are found in protein complexes bound to the inner membrane of the mitochondria. Two are mobile carriers that carry electrons between the protein complexes.

Q What is the function of the electron carriers coenzyme Q and cytochrome c?

HEALTH NOTE

Toxins: Inhibitors of Electron Transport

There are several substances that inhibit the electron carriers in the electron transport chain. Rotenone, a product from a plant root used in South America to poison fish, blocks electron transport between the NADH dehydrogenase (complex I) and coenzyme Q. The barbiturates amytal and demerol also inhibit the NADH dehydrogenase complex. Another inhibitor is the antibiotic antimycin A, which blocks the flow of electrons between cytochrome b and cytochrome c_1 (complex III). Another group of compounds including cyanide (CN^-) and carbon monoxide inhibit cytochrome c oxidase (complex IV). The toxic nature of these compounds makes it clear that an organism relies heavily on the process of electron transport.

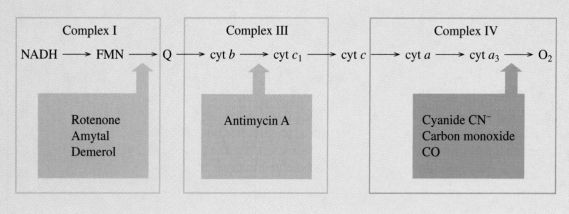

Rotenone

Amytal

Antimycin A

When an inhibitor blocks a step in the electron transport chain, the carriers preceding that step are unable to transfer electrons and remain in their reduced forms. All the carriers after the blocked step remain oxidized without a source of electrons. Thus, any of these inhibitors shut down the flow of electrons through the electron transport chain.

Complex I	Complex III	Complex IV
NADH → FMN → Q	→ cyt b → cyt c_1	→ cyt c → cyt a → cyt a_3 → O_2
Rotenone Amytal Demerol	Antimycin A	Cyanide CN^- Carbon monoxide CO

SAMPLE PROBLEM 24.4

Electron Transport

Identify the following electron carriers that are mobile carriers.

a. Cyt c **b.** FMN   **c.** Fe–S clusters   **d.** Q

Solution

a. and **d.** Cyt c and Q are mobile carriers.

What is the final substance that accepts electrons in the electron transport chain?

QUESTIONS AND PROBLEMS

Electron Transport

24.25 What reduced coenzymes provide the electrons for electron transport?

24.26 What happens to the energy level as electrons are passed along the electron transport chain?

24.27 Arrange the following in the order they appear in electron transport: cytochrome c, cytochrome b, FAD, and coenzyme Q.

24.28 Arrange the following in the order they appear in electron transport: O_2, NAD^+, cytochrome a_3, and FMN.

24.29 How are electrons carried from complex I to complex III?

24.30 How are electrons carried from complex III to complex IV?

24.31 How is NADH oxidized in electron transport?

24.32 How is $FADH_2$ oxidized in electron transport?

24.33 Complete the following reactions in electron transport:
a. $NADH + H^+ +$ _____ $\longrightarrow$ _____ $+ FMNH_2$
b. $QH_2 + 2\ cyt\ b\ (Fe^{3+}) \longrightarrow$ _____ $+$ _____ $+ 2H^+$

24.34 Complete the following reactions in electron transport:
a. $Q +$ _____ $\longrightarrow$ _____ $+ FAD$
b. $2\ cyt\ a\ (Fe^{3+}) + 2\ cyt\ a_3\ (Fe^{2+}) \longrightarrow$ _____ $+$ _____

24.4 Oxidative Phosphorylation and ATP

We have seen that energy is generated when electrons from the oxidation of substrates flow through the electron transport chain. Now we will look at how that energy is coupled with the production of ATP for the cell, which is a process called **oxidative phosphorylation.**

Chemiosmotic Model

In 1978, Peter Mitchell received the Nobel Prize for his theory called the **chemiosmotic model,** which links the energy from electron transport to a proton gradient that drives the synthesis of ATP. In this model, three of the complexes (I, III, and IV) extend through the inner membrane with one end of each complex in the matrix and the other end in the intermembrane space. In the chemiosmotic model, each of these complexes act as a **proton pump** by pushing protons (H^+) out of the matrix and into the intermembrane space. This increase in protons in the intermembrane space lowers the pH and creates a proton gradient. Because protons are positively charged, both the lower pH and the electrical charge of the proton gradient make it an electrochemical gradient. (See Figure 24.10.)

To equalize the pH between the intermembrane space and the matrix, there is a tendency by the protons to return the matrix. However, protons cannot diffuse

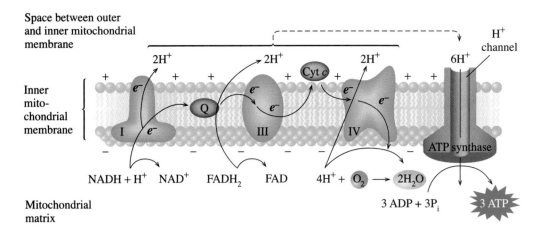

Figure 24.10 In the electron transport chain, protein complexes oxidize and reduce coenzymes to provide electrons and protons that move into the intermembrane space where they create a proton gradient that drives ATP synthesis.

Q What is the major source of NADH for the electron transport chain?

through the inner membrane. The only way protons can return to the matrix is to pass through a protein complex called **ATP synthase.** As the protons flow through ATP synthase, energy generated from the proton gradient is used to drive the ATP synthesis. Thus the process of oxidative phosphorylation couples the energy from electron transport to the synthesis of ATP from ADP and P_i.

$$\text{ADP} + P_i + \text{energy} \xrightarrow{\text{ATP synthase}} \text{ATP}$$

Details of ATP Synthase

ATP synthase consists of two enzyme complexes. (See Figure 24.11.) The F_0 complex contains the channel for the return of protons to the matrix. The F_1 section consists of a center subunit (γ) and three surrounding protein subunits, which have three active sites with different shapes or conformations known as loose (L), tight (T), and open (O). As the protons flow through the F_0 channel, the energy released turns the center subunit (γ). We might think of the flow of protons as a stream or river that turns a water wheel. As the center unit supplies energy to the three active sites, their shapes change. ATP synthesis begins when the substrates ADP and P_i enter a loose (L) active site. As the loose (L) site shape converts to a tight (T) shape, ATP is formed. However, the ATP is tightly bound to the active site (T). When energy turns the γ unit, the tight site (T) converts to an open (O) site, which releases the ATP. The open site will convert to an L site and accept new substrates ADP and P_i. According to Paul Boyer, who earned the 1997 Nobel Prize in chemistry for his work on ATP synthase, the formation of ATP is spontaneous, whereas its release from the synthase requires the energy supplied by the proton gradient. The following steps summarize the changes in the active site in the synthesis of ATP. (See Figure 24.12.)

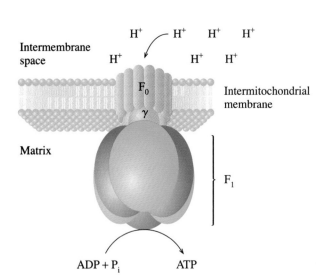

Figure 24.11 ATP synthase consists of two protein complexes. An F_0 section contains the channel for proton flow, and an F_1 section uses the energy from the proton gradient to drive the synthesis of ATP.

Q What are the functions of the F_0 and F_1 sections of ATP synthase?

Step 1   The synthesis of ATP begins when ADP and P_i bind to a loose (L) site.

Step 2   Energy from the γ center now converts all the sites: an L site becomes a tight (T) site, an O site changes to an L site, and a T site changes to an O site.

Step 3   Now the ADP and P_i—which are now in a tight (T) site—spontaneously form ATP, which is held tightly in this active site.

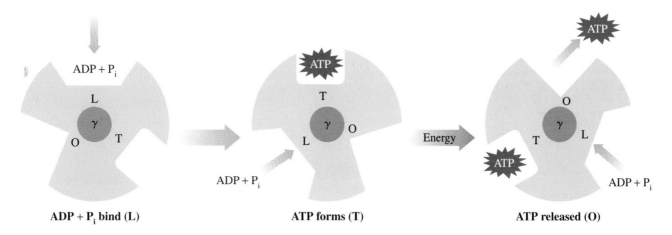

ADP + P$_i$ bind (L)   ATP forms (T)   ATP released (O)

Step 4   Another input of energy from proton flow changes all the active sites again. The ATP site changes to an open (O) site, which has little affinity for ATP, and ATP is released. This site is open once again and ready to accept more ADP and P$_i$.

In summary, the energy from protons flowing through F$_0$ turns the center γ unit in F$_1$, which causes a change in the shapes of the active sites from loose (L), where ADP, and P$_i$ bind, to tight (T) where ATP forms, and to open (O), which releases ATP. This process of oxidative phosphorylation continues as long as energy from the electron transport system is generated, which pumps the protons into the inner membrane space and produces the proton gradient to fuel ATP synthase.

Figure 24.12 In the F$_1$ ATP synthase, ATP is formed when an L active site containing ADP and P$_i$ converts to a T site. When energy from the proton flow in F$_0$ changes the site to an open (O) site, ATP is released.

Q What shape of an active site in F$_1$ ATP synthase accepts the substrates, and which shape releases the ATP?

Electron Transport and ATP Synthesis

We have seen that oxidative phosphorylation couples the energy from electron transport with the synthesis of ATP. Because NADH enters the electron transport chain at complex I, energy is released from the oxidation of NADH to synthesize three ATP. However, FADH$_2$, which enters the chain at a lower energy level at complex II, provides energy to drive the synthesis of only two ATP. (See Figure 24.13.) The overall equation for the oxidation of NADH and FADH$_2$ can be written as follows:

$$\boxed{\textbf{NADH + H}^+} + \tfrac{1}{2}O_2 + 3\,ADP + 3P_i \longrightarrow NAD^+ + H_2O + \boxed{\textbf{3 ATP}}$$

$$\boxed{\textbf{FADH}_2} + \tfrac{1}{2}O_2 + 2\,ADP + 2P_i \longrightarrow FAD + H_2O + \boxed{\textbf{2 ATP}}$$

Regulation of the Electron Transport Chain and Oxidative Phosphorylation

The electron transport chain is regulated by the availability of ADP, P$_i$, oxygen (O$_2$), and NADH. Low levels of any of these compounds will decrease the activity of the electron transport chain and formation of ATP. When a cell is active and ATP is consumed rapidly, the elevated levels of ADP will activate the synthesis of ATP. Therefore, the activity of the electron transport chain is strongly dependent on the levels of ADP for ATP synthesis.

High energy

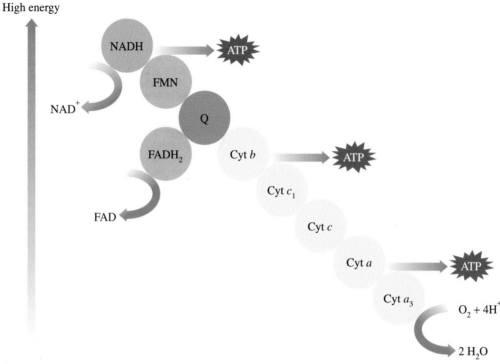

Low energy

Figure 24.13 As the energy levels decrease in the flow of electrons along the major electron carriers, three of the electron transfers release sufficient energy to drive ATP synthesis.

Q Why does electron transfer from FADH$_2$ provide less energy than from NADH and H$^+$?

ATP Synthesis

Why does the oxidation of NADH provide energy for the formation of three ATP whereas FADH$_2$ produces two ATP?

Solution

Electrons from the oxidation of NADH enter the electron chain at a higher energy level than FADH$_2$, providing energy to pump three pairs of protons to the intermembrane. However, FADH$_2$ transfers electrons to Q that go through two sites that pump protons and therefore provide energy for two ATP.

Study Check

What complexes act as proton pumps?

Oxidative Phosphorylation and ATP

24.35 What is meant by oxidative phosphorylation?

24.36 How is the proton gradient established?

24.37 According to the chemiosmotic theory, how does the proton gradient provide energy to synthesize ATP?

24.38 How does the phosphorylation of ADP occur?

24.39 How are glycolysis and the citric acid cycle linked to the production of ATP by the electron transport chain?

24.40 Why does FADH$_2$ have a yield of two ATP via the electron chain, but NADH yields three ATP?

24.41 What are the parts of ATP synthase?

ATP Synthase and Heating the Body

Some types of compounds called *uncouplers* separate the electron transport system from the ATP F_0F_1 synthase. They do this by disrupting the proton gradient needed for the synthesis of ATP. The electrons are transported to O_2 in electron transport, but ATP is not formed by ATP synthase.

Some uncouplers transport the protons through the inner membrane, which is normally impermeable to protons; others block the channel in the F_0 portion of ATP synthase. Compounds such as dicumarol and 2,4-dinitrophenol (DNP) are hydrophobic and bind with protons to carry them across the inner membrane. An antibiotic, oligomycin, binds to the F_0 complex, and blocks the channel, which does not allow any protons to return to the matrix. By removing protons or blocking the F_0 channel, there is no proton flow through the F_0 channel to generate energy for ATP synthesis.

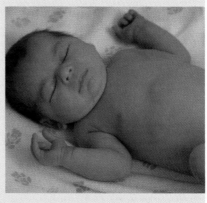

heat production. These animals have large amounts of a tissue called brown fat, which contains a high concentration of mitochondria. This tissue is brown because of the color of iron in the cytochromes of the mitochondria. The proton pumps still operate in brown tissue but a protein embedded in the cell wall allows the protons to bypass ATP synthase. The energy that would be used to synthesize ATP is released as heat. In newborn babies brown fat is used to generate heat because newborns have not stored much fat. The brown fat deposits are located near major blood vessels, which carry the warmed blood to the body. Infants have a small mass but a large surface area and need to produce more heat than the adult. Most adults have little or no brown fat, although someone who works outdoors in a cold climate will develop some brown fat deposits.

Plants also use uncouplers. Some plants use uncoupling agents to volatize fragrant compounds that attract insects to pollinate the plants. Skunk cabbage uses this system. In early spring, heat is used to warm early shoots of plants under the snow, which helps them melt the snow around the plants.

Dicumarol   2,4-Dinitrophenol (DNP)

When there is no mechanism for ATP synthesis, the energy of electron transport is released as heat. Certain animals that are adapted to cold climates have developed their own uncoupling system, which allows them to use electron transport energy for

24.42 What is the role of each part of ATP synthase in ATP synthesis?

24.43 What type of active site in ATP synthase binds ADP and P_i?

24.44 How is the ATP released from ATP synthase?

24.5 ATP Energy from Glucose

Under aerobic conditions, the oxidation of glucose through glycolysis, oxidation to pyruvate, the citric acid cycle, the electron transport chain produces ATP from many NADH molecules and some $FADH_2$. Let's see how much ATP is associated with each of these metabolic cycles.

ATP from Glycolysis

In glycolysis, the oxidation of glucose stores energy in two NADH molecules as well as two ATP molecules from direct substrate phosphorylation. However, glycolysis occurs in the cytoplasm, and the NADH produced cannot pass through the mitochondrial membrane to the electron transport chain.

Therefore the hydrogen ions and electrons from NADH in the cytoplasm are transferred to compounds that can enter the mitochondria. In this shuttle system, dihydroxyacetone phosphate, a compound in glycolysis, is reduced to glycerol-3-phosphate and NAD^+ is regenerated. After glycerol-3-phosphate crosses the mitochondrial membrane, the hydrogen ions and electrons are transferred to FAD. $FADH_2$ is produced along with glycerol-3-phosphate, which returns to the cytoplasm. The overall reaction for the glycerol-3-phosphate shuttle is:

$$\text{NADH} + \text{H}^+ + \text{FAD} \longrightarrow \text{NAD}^+ + \text{FADH}_2$$

Cytoplasm   Mitochondria

Therefore the transfer of electrons from NADH in the cytoplasm to $FADH_2$ produces only two ATP, rather than three. In glycolysis, four ATP from two NADH and two ATP from direct phosphorylation add up to six ATP from one glucose molecule.

$$\text{Glucose} \longrightarrow 2\text{ pyruvate} + 2\text{ ATP} + 2\text{ NADH } (\longrightarrow 2\text{ FADH}_2)$$

$$\text{Glucose} \longrightarrow 2\text{ pyruvate} + 6\text{ ATP}$$

ATP from the Oxidation of Two Pyruvate

Under aerobic conditions, pyruvate enters the mitochondria, where it is oxidized to give acetyl CoA, CO_2, and NADH. Because each glucose molecule yields two pyruvate, two NADH enter electron transport. The oxidation of two pyruvate molecules leads to the production of six ATP molecules.

$$2\text{ Pyruvate} \longrightarrow 2\text{ acetyl CoA} + 6\text{ ATP}$$

ATP from the Citric Acid Cycle

One turn of the citric acid cycle produces two CO_2, three NADH, one $FADH_2$, and one ATP by direct substrate phosphorylation. When the NADH and $FADH_2$ enter electron transport, three NADH produce a total of nine ATP molecules, and one $FADH_2$ produces two more ATP.

$$
\begin{aligned}
3\text{ NADH} \times 3\text{ ATP} &= 9\text{ ATP} \\
1\text{ FADH}_2 \times 2\text{ ATP} &= 2\text{ ATP} \\
\underline{1\text{ GTP} \times 1\text{ ATP}} \quad &= \underline{1\text{ ATP}} \\
\text{Total (one turn)} &= 12\text{ ATP}
\end{aligned}
$$

Thus, one turn of the citric acid cycle generates energy for the synthesis of a total of 12 ATP molecules. Because two acetyl CoA molecules are produced from each glucose, two turns of the citric acid cycle produces 24 ATP.

$$\text{Acetyl CoA} \longrightarrow 2\text{CO}_2 + 12\text{ ATP (one turn of citric acid cycle)}$$

$$2\text{ Acetyl CoA} \longrightarrow 4\text{CO}_2 + 24\text{ ATP (two turns of citric acid cycle)}$$

ATP from the Complete Oxidation of Glucose

We can now determine the total ATP for the complete oxidation of glucose by combining the equations from glycolysis, oxidation of pyruvate, and citric acid cycle. (See Figure 24.14.) The details of the reaction pathways are given in Table 24.1.

ATP and Complete Oxidation of Glucose

Metabolic Pathway	Substrate(s) Oxidized	Products	ATP
Glycolysis	1 Glucose	2 Pyruvate, 2 ATP, 2 NADH	6 ATP
Oxidation	2 Pyruvate	2 Acetyl CoA, 2 NADH, 2CO$_2$	6 ATP
Citric acid cycle	2 Acetyl CoA	6 NADH, 2 FADH$_2$, 2 ATP, 4CO$_2$	24 ATP
Complete oxidation	Glucose + 6O$_2$ $\longrightarrow$ 6CO$_2$ + 6H$_2$O + 36 ATP		36 ATP

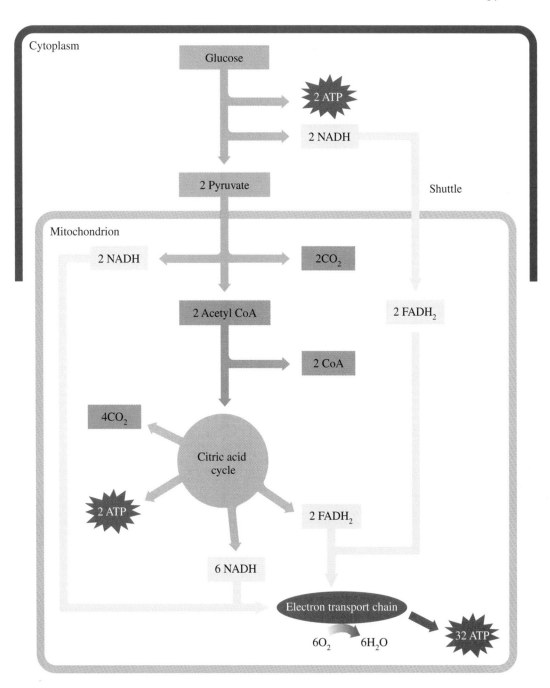

Figure 24.14 The complete oxidation of glucose to CO_2 and H_2O yields a total of 36 ATP.

ATP Production

Indicate the amount of ATP produced by each of the following oxidation reactions:

a. pyruvate to acetyl CoA   **b.** glucose to acetyl CoA

Solution

a. The oxidation of pyruvate to acetyl CoA produces one NADH, which yields three ATP.

b. Six ATP are produced from the oxidation of glucose to two pyruvate molecules.

Q What metabolic pathway produces most of the ATP from the oxidation of glucose?

Another six ATP result from the oxidation of two pyruvate molecules to two acetyl CoA molecules.

Study Check

What are the sources of ATP in the citric acid cycle?

Table 24.1　ATP Produced by the Complete Oxidation of Glucose

Reaction Pathway	ATP for One Glucose
ATP from Glycolysis	
Activation of glucose	−2 ATP
Oxidation of glyceraldehyde-3-phosphate (2 NADH)	6 ATP
Transport of 2 NADH across membrane	−2 ATP
Direct ADP phosphorylation (two triose phosphate)	4 ATP
Summary: $C_6H_{12}O_6 \longrightarrow$ 2 pyruvate + $2H_2O$ Glucose	6 ATP
ATP from Pyruvate	
2 Pyruvate $\longrightarrow$ 2 acetyl CoA (2 NADH)	6 ATP
ATP from Citric Acid Cycle	
Oxidation of 2 isocitrate (2 NADH)	6 ATP
Oxidation of 2 α-ketoglutarate (2 NADH)	6 ATP
2 Direct substrate phosphorylations (2 GTP)	2 ATP
Oxidation of 2 succinate (2 $FADH_2$)	4 ATP
Oxidation of 2 malate (2 NADH)	6 ATP
Summary: 2 Acetyl CoA $\longrightarrow$ $4CO_2$ + $2H_2O$	24 ATP
Overall ATP Production for One Glucose	
$C_6H_{12}O_6 + 6O_2 + 36\,ADP + 36\,P_i \longrightarrow 6CO_2 + 6H_2O + 36\,ATP$ Glucose	

When glucose is not immediately used by the cells for energy, it is stored as glycogen in the liver and muscles. When the levels of glucose in the brain or blood become low, the glycogen reserves are hydrolyzed and glucose is released into the blood. If glycogen stores are depleted, some glucose can be synthesized from non-carbohydrate sources. It is the balance of all these reactions that maintains the necessary blood glucose level available to our cells and provides the necessary amount of ATP for our energy needs. (See Figure 24.15.)

QUESTIONS AND PROBLEMS

ATP Energy from Glucose

24.45 Why does the NADH produced in glycolysis yield only two ATP?

24.46 Under anaerobic conditions, what is the maximum number of ATP molecules that can be produced from one glucose molecule?

24.47 What is the energy yield in ATP molecules associated with each of the following?
   a. NADH $\longrightarrow$ NAD$^+$
   b. glucose $\longrightarrow$ 2 pyruvate

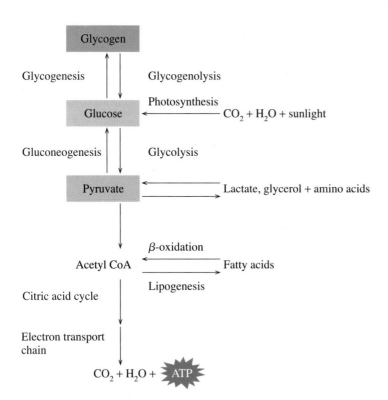

Figure 24.15 The ATP level is maintained by a balance of metabolic pathways that increase or decrease the glucose according to the energy requirements in the cell.

Q What metabolic pathways are stimulated by low ATP levels?

HEALTH NOTE

Efficiency of ATP Production

In a laboratory, a calorimeter is used to measure the heat energy from the combustion of glucose. In a calorimeter, one mole of glucose produces a total of 687 kcal (2870 kJ).

$$C_6H_{12}O_6 + 6O_2 \longrightarrow 6CO_2 + 6H_2O + 687 \text{ kcal (2870 kJ)/mole}$$

Let's compare the amount of energy produced from one mole of glucose in a calorimeter with the ATP energy produced in the mitochondria. We can use the energy of the hydrolysis of ATP, which is 7.3 kcal (31 kJ)/mole ATP. Because one mole of glucose generates energy for 36 moles of ATP, the total energy from the oxidation of one mole of glucose in the cells would be 263 kcal (1100 kJ)/per mole.

$$\frac{36 \text{ mole ATP}}{1 \text{ mole glucose}} \times \frac{7.3 \text{ kcal (31 kJ)}}{1 \text{ mole ATP}} = 263 \text{ kcal (1100 kJ) per 1 mole glucose}$$

Compared to the energy produced by burning glucose in a calorimeter, our cells are about 38% (263 kcal/687 kcal) efficient in converting the total available chemical energy in glucose to ATP.

$$\frac{263 \text{ kcal (cells)}}{687 \text{ kcal (calorimeter)}} \times 100 = 38\%$$

The rest of the energy from glucose produced during the oxidation of glucose in our cells is lost as heat.

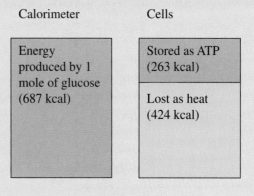

Calorimeter	Cells
Energy produced by 1 mole of glucose (687 kcal)	Stored as ATP (263 kcal)
	Lost as heat (424 kcal)

c. 2 pyruvate $\longrightarrow$ 2 acetyl CoA + 2CO$_2$

d. acetyl CoA $\longrightarrow$ 2CO$_2$

24.48 What is the energy yield in ATP molecules associated with each of the following?

    a. FADH$_2$ $\longrightarrow$ FAD

    b. glucose + 6O$_2$ $\longrightarrow$ 6CO$_2$ + 6H$_2$O

    c. glucose $\longrightarrow$ 2 lactate

    d. pyruvate $\longrightarrow$ lactate

Chapter Review

24.1 The Citric Acid Cycle

In a sequence of reactions called the citric acid cycle, an acetyl group is combined with oxaloacetate to yield citrate. Citrate undergoes oxidation and decarboxylation to yield two CO_2, GTP, three NADH, and $FADH_2$ with the regeneration of oxaloacetate. The direct phosphorylation of ADP by GTP yields ATP.

24.2 Electron Carriers

Electron carriers that transfer hydrogen ions and electrons include FMN, iron–sulfur proteins, coenzyme Q, and several cytochromes. Both iron–sulfur proteins and cytochromes contain iron ions that are reduced to Fe^{2+} and reoxidized to Fe^{3+} as electrons are accepted and then passed to the next electron carrier.

24.3 Electron Transport

The reduced coenzymes NADH and $FADH_2$ from various metabolic pathways are oxidized to NAD^+ and FAD when their protons and electrons are transferred to the electron transport chain. The final acceptor, O_2, combines with protons and electrons to yield H_2O. At three transfer points in electron transport, the energy decrease provides the necessary energy for ATP synthesis.

24.4 Oxidative Phosphorylation and ATP

The protein complexes in electron transport act as proton pumps to move protons into the inner membrane space, which produces a proton gradient. As the protons return to the matrix by way of ATP synthase, energy is generated. This energy is used to drive the synthesis of ATP in a process known as oxidative phosphorylation. The available ADP and ATP levels in the cells control the activity of the electron transport chain.

24.5 ATP Energy from Glucose

With the exception of the NADH produced from glycolysis, the oxidation of NADH yields three ATP molecules, and $FADH_2$ yields two ATP. The energy from the NADH produced in the cytoplasm is used to form $FADH_2$. Under aerobic conditions, the complete oxidation of glucose yields a total of 36 ATP from the oxidation of the reduced coenzymes NADH and $FADH_2$ by electron transport, oxidative phosphorylation, and from some direct substrate phosphorylation.

Summary of Key Reactions

Citric Acid Cycle

$$\text{Acetyl CoA} + 3\,NAD^+ + FAD + GDP + P_i + 2H_2O \longrightarrow$$
$$2CO_2 + 3\,NADH + 3H^+ + FADH_2 + CoA + GTP$$

Electron Transport Chain

$$NADH + H^+ + 3\,ADP + 3\,P_i + \tfrac{1}{2}O_2 \longrightarrow NAD^+ + 3\,ATP + H_2O$$
$$FADH_2 + 2\,ADP + 2\,P_i + \tfrac{1}{2}O_2 \longrightarrow FAD + 2\,ATP + H_2O$$

Phosphorylation of ADP

$$ADP + P_i \longrightarrow ATP + H_2O$$

Complete Oxidation of Glucose

$$C_6H_{12}O_6 + 6O_2 + 36\,ADP + 36\,P_i \longrightarrow 6\,CO_2 + 6\,H_2O + 36\,ATP$$

Key Terms

ATP synthase (F_0F_1)   An enzyme complex that links the energy released by protons returning to the matrix with the synthesis of ATP from ADP and P_i. The F_0 section contains the channel for proton flow, and the F_1 section uses the energy from the proton flow to drive the synthesis of ATP.

chemiosmotic model   The conservation of energy from transfer of electrons in the electron transport chain by pumping protons into the intermembrane space to produce a proton gradient that provides the energy to synthesize ATP.

citric acid cycle   A series of oxidation reactions in the mitochondria that convert acetyl CoA to CO_2 and yield NADH and $FADH_2$. It is also called the tricarboxylic acid cycle and the Krebs cycle.

coenzyme Q (CoQ, Q)   A mobile carrier that transfers electrons from NADH and $FADH_2$ to cytochrome *b* in complex III.

cytochromes (cyt)   Iron-containing proteins that transfer electrons from QH_2 to oxygen.

electron carriers   A group of proteins that accept and pass on electrons as they are reduced and oxidized. Most of the carriers are tightly attached to the inner mito-

chondrial membrane, but two are mobile carriers, which move electrons between the complexes containing the other carriers.

electron transport chain   A series of reactions in the mitochondria that transfer electrons from NADH and $FADH_2$ to electron carriers, which are arranged from higher to lower energy levels, and finally to O_2, which produces H_2O. Energy changes during three of these transfers provide energy for ATP synthesis.

Fe–S (iron–sulfur) clusters   Proteins containing iron and sulfur in which the iron ions accept electrons from $FMNH_2$ and cytochrome *b*.

FMN (flavin mononucleotide)   An electron carrier derived from riboflavin (vitamin B_2) that transfers hydrogen ions and electrons from NADH entering in the electron transport chain.

oxidative phosphorylation   The synthesis of ATP from ADP and P_i using energy generated by the oxidation reactions in the electron transport chain.

proton pumps   The enzyme complexes I, III, and IV that move protons from the matrix into the intermembrane space, creating a proton gradient.

Additional Problems

24.49   What is the main function of the citric acid cycle in energy production?

24.50   Most metabolic pathways are not considered cycles. Why is the citric acid cycle considered to be a metabolic cycle?

24.51   If there are no reactions in the citric acid cycle that used oxygen, O_2, why does the cycle operate only in aerobic conditions?

24.52   What products of the citric acid cycle are needed for the electron transport chain?

24.53   Identify the compounds in the citric acid cycle that have the following:
   a. six carbon atoms
   b. five carbon atoms
   c. a keto group

24.54   Identify the compounds in the citric acid cycle that have the following:
   a. four carbon atoms
   b. a hydroxyl group
   c. a double bond

24.55   In which reaction of the citric acid cycle does each of the following occur?
   a. a five-carbon keto acid is decarboxylated
   b. a double bond is hydrated
   c. NAD^+ is reduced
   d. a secondary hydroxyl group is oxidized

24.56   In which reaction of the citric acid cycle does each of the following occur?
   a. FAD is reduced
   b. a six-carbon keto acid is decarboxylated
   c. a carbon–carbon double bond is formed
   d. GDP undergoes direct phosphorylation

24.57 Indicate the coenzyme for each of the following reactions:

    a. isocitrate $\longrightarrow$ α-ketoglutarate

    b. α-ketoglutarate $\longrightarrow$ succinyl CoA

24.58 Indicate the coenzyme for each of the following reactions:

    a. succinate $\longrightarrow$ fumarate

    b. malate $\longrightarrow$ oxaloacetate

24.59 How do each of the following regulate the citric acid cycle?

    a. high levels of NADH   **b.** high levels of ATP

24.60 How do each of the following regulate the citric acid cycle?

    a. high levels of ADP   **b.** low levels of NADH

24.61 Identify each as part of the structure of one of the following components in the electron transport chain as (1) FMN, (2) Fe–S cluster (3) CoQ, or (4) cytochrome

    a. a heme group   **b.** contains a ribitol group

24.62 Identify each as part of the structure of one of the following components in the electron transport chain as (1) FMN, (2) Fe–S cluster (3) CoQ, or (4) cytochrome

    a. contains the three-ring system of flavins

    b. a six-atom ring attached to a long-carbon chain

24.63 Identify each of the following electron carriers as part of a complex or as a mobile carrier? If part of a complex, indicate which one.

    a. CoQ   **b.** Fe–S clusters   **c.** cyt a_3

24.64 Identify each of the following electron carriers as part of a complex or as a mobile carrier. If part of a complex, indicate which one.

    a. cyt b **b.** cyt c **c.** FMN

24.65 Identify the complex where each of the following are oxidized or reduced and complete the equation:

    a. $FADH_2 + Q \longrightarrow$

    b. cyt a (Fe^{2+}) + cyt a_3 (Fe^{3+}) $\longrightarrow$

24.66 Identify the complex where each of the following are oxidized or reduced and complete the equation:

    a. cyt c (Fe^{2+}) + cyt a (Fe^{3+}) $\longrightarrow$

    b. NADH + FMN $\longrightarrow$

24.67 Complete the following by adding the substances that are missing:

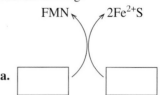

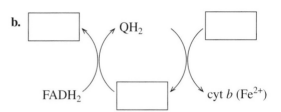

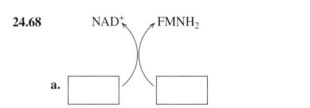

24.68

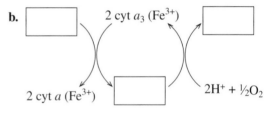

24.69 At which parts of the electron transport system are protons pumped into the intermembrane space?

24.70 What is the effect of proton accumulation in the intermembrane space?

24.71 In the chemiosmotic model, how is energy provided to synthesize ATP?

24.72 In what part of the electron transport chain does the synthesis of ATP take place?

24.73 Why do protons tend to leave the intermembrane space and return to the matrix?

24.74 Why do the enzymes that pump protons extend across the membrane from the matrix to the intermembrane space?

24.75 How many ATP molecules are produced by energy generated when electrons flow from $FADH_2$ to oxygen (O_2)?

24.76 How many ATP molecules are produced by energy generated when electrons flow from NADH to oxygen (O_2)?

24.77 What part of the electron transport chain is inhibited by each of the following?
   a. amytal and rotenone
   b. antimycin A
   c. cyanide and carbon monoxide

24.78 **a.** When an inhibitor blocks the electron transport chain, how are the coenzymes that precede the blocked site affected?
   b. When an inhibitor blocks the electron transport chain, how are the coenzymes that follow the blocked site affected?

24.79 How many ATP are produced when glucose is oxidized to pyruvate compared to when glucose is oxidized to CO_2 and H_2O?

24.80 Why do the two NADH produced in glycolysis provide a net of two ATP and not three?

24.81 Using the value of 7.3 kcal/mole for ATP, how many kcal are conserved as ATP from one mole of glucose in each of the following?
   a. glycolysis
   b. oxidation of pyruvate to acetyl CoA

   c. citric acid cycle
   d. complete oxidation of glucose to CO_2 and H_2O

24.82 What percent of ATP energy is conserved from one mole of glucose in problem 24.81a–c?

24.83 What does it mean to say that the cell is 38% efficient in storing the energy from the complete combustion of glucose?

24.84 Considering the efficiency of ATP synthesis, how many kcal of energy would be conserved from the complete oxidation of four moles of glucose?

24.85 Where is ATP synthase for oxidative phosphorylation located in the cell?

24.86 How is a proton gradient developed in a cell?

24.87 How is the energy from the proton gradient utilized by ATP synthase?

24.88 How do the active sites on F_1 ATP synthase change during ATP production?

24.89 Why would a bear that is hibernating have more brown fat than one that is active?

24.90 A student is considering using a 2,4-dinitrophenol, which is an uncoupler, to lose weight. Explain how the uncoupler will affect the body temperature of the student. How would you advise the student if the dosage of DNP needed to lose weight is close to toxic levels?

25 Metabolic Pathways for Lipids and Amino Acids

"Occupational therapists teach children and adults the skills they need for the job of living," says occupational therapist Leslie Wakasa. "When working with the pediatric population, we are crucial in educating children with disabilities, their families, caregivers, and school staff in ways to help them be as independent as they can be in all aspects of their daily lives. It's rewarding when you can show children how to feed themselves, which is a huge self-esteem issue for them. The opportunity to help people become more independent is very rewarding."

A combination of technology and occupational therapy helps children who are nonverbal to communicate and interact with their environment. By leaning on a red switch, Alex is learning to use a computer.

LOOKING AHEAD

the Chemistry place

www.chemplace.com/college

Visit the URL above or use the CD-ROM in the book for extra quizzing, interactive tutorials, career resources, and case studies.

In previous chapters, we focused on carbohydrates because glucose is the primary fuel for the synthesis of ATP. However, lipids and proteins also play an important role in metabolism and energy production. In this chapter we will look at how the digestion of lipids produces fatty acids and glycerol and digestion of proteins gives amino acids. Almost all our energy is stored in the form of triacylglycerols in fat cells of **adipose tissue.** Many people go on diets after they discover that adipose tissue can store unlimited quantities of fat. This fact has become quite apparent in the large number of people in the U.S. that are considered obese. When our caloric intake exceeds the nutritional and metabolic needs of the body, excess carbohydrates and fatty acids are converted to triacylglycerols and added to our fat cells.

The digestion and degradation of dietary proteins as well as body proteins provides amino acids, which are needed to synthesize nitrogen-containing compounds in our cells such as new proteins and nucleic acids. Although amino acids are not considered a primary source of fuel, energy can be extracted from amino acids if glycogen and fat reserves have been depleted. However, when a person who is fasting or starving utilizes amino acids as the only source of energy, the breakdown of the body's own proteins eventually destroys essential body tissues, particularly the heart muscle.

25.1 Digestion of Triacylglycerols

LEARNING GOAL

Describe the sites and products obtained from the digestion of triacylglycerols.

Our adipose tissue is made of fat cells called *adipocytes*, which store triacylglycerols. (See Figure 25.1.) Let's compare the amount of energy stored in the fat cells to the energy from glucose, glycogen, and protein. A typical 70 kg (150 lb) person has about 135,000 kcal of energy stored as fat, 24,000 kcal as protein, 720 kcal as glycogen reserves, and 80 kcal as blood glucose. Therefore, the energy available from stored fats is about 85% of the total energy available in the body. Thus body fat is our major source of stored energy.

Digestion of Dietary Fats

The digestion of dietary fats begins in the small intestine, when the hydrophobic fat globules mix with bile salts released from the gallbladder. In a process called *emulsification,* the bile salts break the fat globules into smaller droplets called micelles. Then *pancreatic lipases* released from the pancreas hydrolyze the triacylglycerols

Figure 25.1 The fat cells (adipocytes) that make up adipose tissue are capable of storing unlimited quantities of triacylglycerols.

Q What are some sources of fats in our diet?

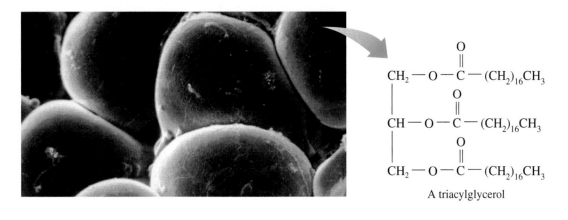

A triacylglycerol

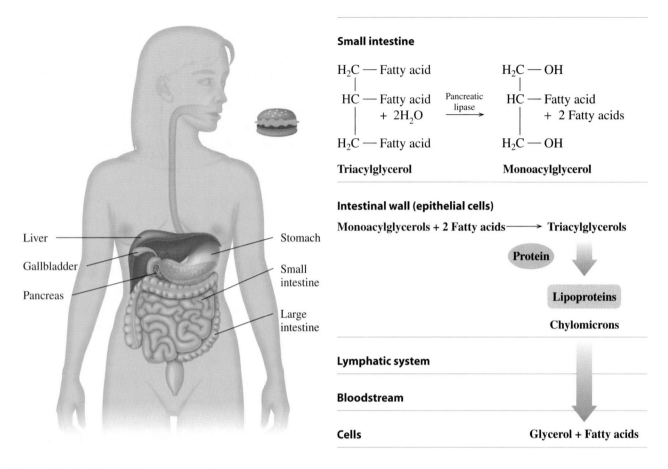

Figure 25.2 The triacylglycerols that re-form in the intestinal wall from the digestion products monoacylglycerols and fatty acids bind to proteins for transport through the lymphatic system and bloodstream to the cells.

Q What kinds of enzymes are secreted from the pancreas into the small intestine to hydrolyze triacylglycerols?

in the micelles to yield monoacylglycerols and free fatty acids. These digestion products are absorbed into the intestinal lining where they recombine to form triacylglycerols, which are coated with proteins to form lipoproteins called **chylomicrons.** The chylomicrons transport the triacylglycerols through the lymph system and into the bloodstream to be carried to cells of the heart, muscle, and adipose tissues. (See Figure 25.2.) We can write the overall equation for the digestion of triacylglycerols as follows:

$$\text{Triacylglycerols} + 2H_2O \xrightarrow{\text{Pancreatic lipase}} \text{Monoacylglycerols} + 2\text{ Fatty acids}$$

In the cells, enzymes hydrolyze the triacylglycerols to yield glycerol and free fatty acids, which can be used for energy production. The preferred fuel of the heart is fatty acids that are oxidized to acetyl CoA units for ATP synthesis. However, the brain and red blood cells cannot utilize fatty acids. Fatty acids cannot diffuse across the blood-brain barrier, and red blood cells have no mitochondria, which is where fatty acids are oxidized. Therefore, glucose and glycogen are the only source of energy for the brain and red blood cells.

Mobilization of Fat Stores

When blood glucose is depleted and glycogen stores are low, the process of **fat mobilization** breaks down triacylglycerols in the adipose tissue to fatty acids and glycerol. The mobilization occurs when the hormones glucagon or epinephrine are secreted into the bloodstream and bind to receptors on the membrane of adipose cells. This activates enzymes within the fat cells that begin the hydrolysis of a triacylglycerol. A fatty acid is hydrolyzed from carbon 1 or carbon 3 followed by the hydrolysis of the second and third fatty acids.

We can write the overall reaction for the mobilization of fats in fat cells as follows:

$$\text{Triacylglycerols} + 3H_2O \xrightarrow{\text{Lipases}} \text{Glycerol} + 3 \text{ Fatty acids}$$

The products of fat mobilization, glycerol and fatty acids, diffuse into the bloodstream and bind with plasma proteins (albumin) to be transported to the tissues. Most of the glycerol goes into the liver where it is converted to glucose.

Metabolism of Glycerol

Using two steps, enzymes in the liver convert glycerol to dihydroxyacetone phosphate. In the first step, glycerol is phosphorylated using ATP to yield glycerol-3-phosphate. In the second step, the hydroxyl group is oxidized to yield dihydroxyacetone phosphate, which is an intermediate in several metabolic pathways including glycolysis and gluconeogenesis. (See Chapter 23.)

The overall reaction for the metabolism of glycerol is written as follows:

$$\text{Glycerol} + \text{ATP} + \text{NAD}^+ \longrightarrow \text{Dihydroxyacetone phosphate} + \text{ADP} + \text{NADH} + H^+$$

SAMPLE PROBLEM 25.1

Fats and Digestion

What are the sites, enzymes, and products for the digestion of triacylglycerols?

Solution

The digestion of triacylglycerols takes place in the small intestine where pancreatic lipase catalyzes their hydrolysis to monoacylglycerols and fatty acids.

Study Check

What happens to the products from the digestion of triacylglycerols in the membrane of the small intestine?

QUESTIONS AND PROBLEMS

Digestion of Triacylglycerols

25.1 What is the role of bile salts in lipid digestion?

25.2 How are insoluble triacylglycerols transported to the tissues?

25.3 When are fats released from fat stores?

25.4 What happens to the glycerol produced from the hydrolysis of triacylglycerols in adipose tissues?

25.5 How is glycerol converted to an intermediate of glycolysis?

25.6 How can glycerol be used to synthesze glucose?

25.2 Oxidation of Fatty Acids

A large amount of energy is obtained when fatty acids undergo oxidation in the mitochondria to yield acetyl CoA. In stage 2 of fat metabolism, fatty acids undergo **beta oxidation (β oxidation),** which removes two-carbon segments, one at a time, from a fatty acid.

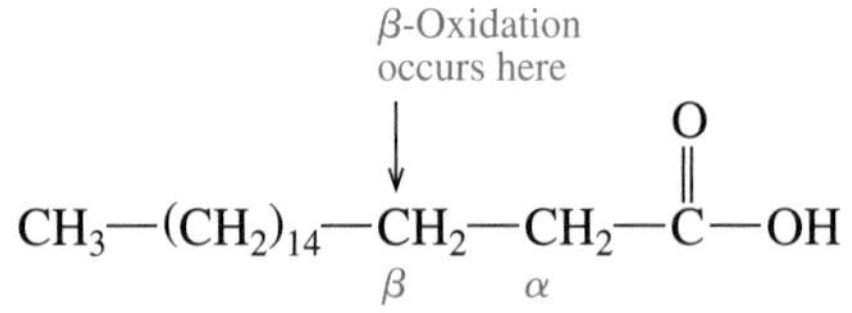

$$\underset{\text{Stearic acid}}{CH_3-(CH_2)_{14}-\underset{\beta}{CH_2}-\underset{\alpha}{CH_2}-\overset{\overset{\textstyle O}{\|}}{C}-OH}$$

β-Oxidation occurs here

Each cycle in β oxidation produces acetyl CoA and a fatty acid that is shorter by two carbons. The cycle repeats until the original fatty acid is completely degraded to two-carbon acetyl CoA units. Each acetyl CoA can then enter the citric acid cycle in the same way as the acetyl CoA units derived from glucose.

Fatty Acid Activation

Before a fatty acid can enter the mitochondria, it undergoes activation in the cytosol. The activation process combines a fatty acid with coenzyme A to yield fatty acyl CoA. The energy released by the hydrolysis of two phosphate groups from ATP is used to drive the reaction. The products are AMP and pyrophosphate (PP_i), which hydrolyzes to yield two inorganic phosphates ($2\ P_i$).

$$R-CH_2-\overset{\overset{\displaystyle O}{\|}}{C}-O^- \ + \ ATP \ + \ HS-CoA \ \xrightarrow[synthetase]{Acyl\ CoA} \ R-CH_2-\overset{\overset{\displaystyle O}{\|}}{C}-S-CoA \ + \ AMP \ + \ 2\ P_i$$

Fatty acid   Fatty acyl CoA

Transport of Fatty Acyl CoA

The long hydrocarbon chain prevents the fatty acyl CoA molecule from crossing the inner mitochondrial membrane. However, it is transported into the matrix when it binds with a charged carrier called *carnitine*. (See Figure 25.3.)

$$R-CH_2-\overset{\overset{\displaystyle O}{\|}}{C}-S-CoA + H-\overset{\overset{\displaystyle \overset{+}{N}(CH_3)_3}{\underset{\displaystyle \underset{\displaystyle COO^-}{CH_2}}{\overset{\displaystyle CH_2}{|}}}}{C}-OH \ \rightleftharpoons \ H-\overset{\overset{\displaystyle \overset{+}{N}(CH_3)_3}{\underset{\displaystyle \underset{\displaystyle COO^-}{CH_2}}{\overset{\displaystyle CH_2}{|}}}}{C}-O-\overset{\overset{\displaystyle O}{\|}}{C}-CH_2-R + HS-CoA$$

Fatty acyl CoA   Carnitine   Fatty acyl carnitine

After the fatty acyl group is released in the matrix, it recombines with coenzyme A, and the carnitine returns to the inner membrane. While it may seem like a complicated way to move fatty acyl CoA into the matrix, this transport system provides a way to regulate degradation (oxidation) and synthesis of fatty acids. When fatty acids are being synthesized in the cytosol, the transport of fatty acyl CoA into the matrix is blocked, which prevents fatty acid degradation.

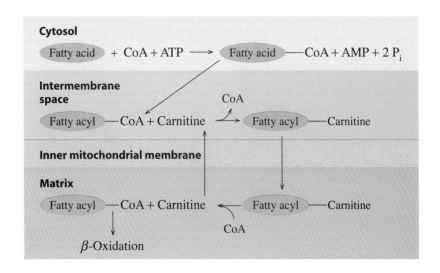

Figure 25.3 Fatty acids are activated and transported by carnitine through the inner mitochondrial membrane into the matrix.

Q Why is carnitine used to transport a fatty acid into the matrix?

Reactions of β-Oxidation Cycle

In the matrix, fatty acyl CoA molecules undergo β oxidation, which is a cycle of four reactions that convert the $-CH_2-$ of the β carbon to a β-keto group. Once the β-keto group is formed, a two-carbon acetyl group can be split from the chain, which shortens the fatty acyl group.

β-Oxidation Pathway

Reaction 1   Oxidation (Dehydrogenation)

In the first reaction of β oxidation, the FAD coenzyme removes hydrogen atoms from the α and β carbons of the activated fatty acid to form a trans carbon–carbon double bond, and $FADH_2$.

Reaction 2   Hydration

A water molecule now adds across the trans double bond, which places a hydroxyl group (—OH) on the β carbon.

Reaction 3   Oxidation (Dehydrogenation)

The hydroxy group on the β carbon is oxidized to yield a ketone. The hydrogen atoms removed in the dehydrogenation reduce coenzyme NAD^+ to $NADH + H^+$. At this point, the β carbon has been oxidized to a keto group.

Reaction 4   Cleavage of Acetyl CoA

In the final step of β oxidation, the C_α—C_β bond splits to yield free acetyl CoA and a fatty acyl CoA molecule that is shorter by two carbon atoms. This shorter fatty acyl CoA is ready to go through the β-oxidation cycle again.

The reaction for one cycle of β oxidation is written as follows:

$$R-CH_2-CH_2-\overset{\overset{\displaystyle O}{\|}}{C}-S-CoA + NAD^+ + FAD + H_2O + HS-CoA \longrightarrow$$

$$R'-CH_2-CH_2-\overset{\overset{\displaystyle O}{\|}}{C}-S-CoA \quad + \quad CH_3-\overset{\overset{\displaystyle O}{\|}}{C}-S-CoA + NADH + H^+ + FADH_2$$

New fatty acyl (–2 C) CoA   Acetyl CoA

Fatty Acyl Length Determines Cycle Repeats

The number of carbon atoms in a fatty acid determines the number of times the cycle repeats and the number of acetyl CoA units it produces. For example, the complete β oxidation of myristic acid (C_{14}) produces seven acetyl CoA groups, which is equal to one-half the number of carbon atoms in the chain. Because the final turn of the cycle produces two acetyl CoA groups, the total number of times the cycle repeats is one less than the total number of acetyl groups it produces. Therefore, the C_{14} fatty acid goes through the cycle six times. (See Figure 25.4.)

Fatty acid	Number of Acetyl CoA	β-Oxidation cycles
Myristic acid C_{14}	7 acetyl CoA	6
Palmitic acid C_{16}	8 acetyl CoA	7
Stearic acid C_{18}	9 acetyl CoA	8

We can write an overall equation for the complete oxidation of myristyl-CoA as follows:

$$\text{Myristyl-CoA} + 6\,\text{CoA} + 6\,\text{FAD} + 6\,\text{NAD}^+ + 6\,\text{H}_2\text{O} \longrightarrow$$
$$7\,\text{Acetyl CoA} + 6\,\text{FADH}_2 + 6\,\text{NADH} + 6\text{H}^+$$

Oxidation of Unsaturated Fatty Acids

The β-oxidation sequence we have described applies to the most typical type of fatty acids, which are saturated with an even number of carbon atoms. However the fats in our diets, particularly the oils, contain unsaturated fatty acids, which have one or more cis double bonds. The hydration reaction adds water to trans double bonds, not cis. When the double bond in an unsaturated fatty acid is ready for hydration, an isomerase forms a trans double bond between the α and β carbons, which is the arrangement needed for the hydration reaction.

Because the isomerization provides the trans double bond for the hydration in reaction 2, it bypasses the first reaction. Therefore, the energy released by the β oxidation of an unsaturated fatty acid is slightly less since no $FADH_2$ is produced in that cycle.

SAMPLE PROBLEM 25.2

β Oxidation

Match each of the following with reactions in the β-oxidation cycle:

(1) first oxidation (2) hydration (3) second oxidation (4) cleavage

Figure 25.4 Myristic acid (C_{14}) undergoes six oxidation cycles that repeat reactions 1–4 to yield 7 acetyl CoA molecules.

Q **How many NADH and $FADH_2$ are produced in one turn of the fatty acid cycle of β oxidation?**

Matrix

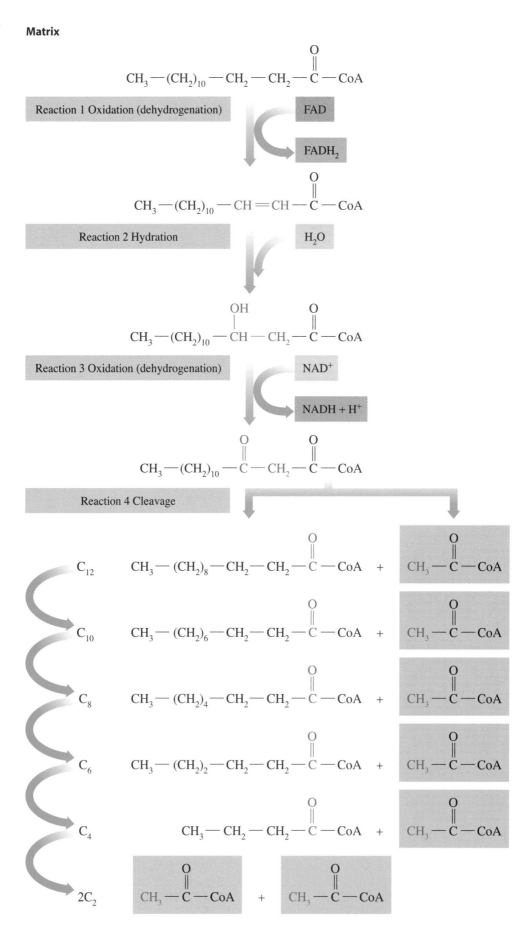

a. water is added to a trans double bond

b. an acetyl CoA is removed

c. FAD is reduced to $FADH_2$

d. bypassed during the oxidation of unsaturated fatty acids

Solution

a. (2) hydration   **b.** (4) cleavage

c. (1) first oxidation   **d.** (1) first oxidation

Study Check

Which coenzyme is needed in reaction 3 when a β-hydroxyl group is converted to a β-keto group?

QUESTIONS AND PROBLEMS

Oxidation of Fatty Acids

25.7 Where in the cell is a fatty acid activated?

25.8 What is the function of carnitine in the degradation of fatty acids?

25.9 What coenzymes are required for β oxidation?

25.10 When does an isomerization occur during the β oxidation of a fatty acid?

25.11 In each of the following acyl CoA molecules, identify the β carbon.

a.
$$CH_3-CH_2-CH_2-CH_2-CH_2-CH_2-CH_2-\overset{\overset{\displaystyle O}{\|}}{C}-S-CoA$$

b.
$$CH_3-(CH_2)_{14}-CH_2-CH_2-\overset{\overset{\displaystyle O}{\|}}{C}-S-CoA$$

c.
$$CH_3-CH_2-CH=CH-CH_2-CH_2-CH_2-CH_2-CH_2-\overset{\overset{\displaystyle O}{\|}}{C}-S-CoA$$

25.12 Write the product when each of the following undergoes the indicated reaction in β oxidation:

a.
$$CH_3-(CH_2)_{12}-CH=CH-\overset{\overset{\displaystyle O}{\|}}{C}-S-CoA + H_2O \xrightarrow{\text{Enoyl CoA hydratase}}$$

b.
$$CH_3-(CH_2)_6-CH_2-CH_2-\overset{\overset{\displaystyle O}{\|}}{C}-S-CoA \xrightarrow{\text{Acyl CoA dehydrogenase}}$$

c.
$$CH_3-(CH_2)_4-\overset{\overset{\displaystyle O}{\|}}{C}-CH_2-\overset{\overset{\displaystyle O}{\|}}{C}-S-CoA + HS-CoA \xrightarrow{\text{Thiolase}}$$

25.13 Capric acid, $CH_3(CH_2)_8COOH$, is a C_{10} fatty acid.

    a. Write the formula of the activated form of capric acid.

    b. Indicate the α and β carbon atoms in the fatty acid.

    c. Write the overall equation for the first cycle of β oxidation for capric acid.

    d. Write the overall equation for the complete β oxidation of capric acid.

25.14 Arachidic acid, CH_3—$(CH_2)_{18}$—$COOH$, is a C_{20} fatty acid.
  a. Write the formula of the activated form of arachidic acid.
  b. Indicate the α and β carbon atoms in the fatty acid.
  c. Write the overall equation for the first cycle of β oxidation for arachidic acid.
  d. Write the overall equation for the complete β oxidation of arachidic acid.

LEARNING GOAL

Calculate the total ATP produced by the complete oxidation of a fatty acid.

25.3 ATP and Fatty Acid Oxidation

We can now determine the total energy yield from the oxidation of a particular fatty acid. The total ATP is the sum of the ATP produced from the $FADH_2$, NADH, and acetyl CoA units. In each β-oxidation cycle, one NADH, one $FADH_2$, and one acetyl CoA are produced. From Chapter 24, we know that hydrogen ions and electrons transferred from NADH to coenzyme Q in the electron transport chain generate sufficient energy to synthesize three ATP, whereas $FADH_2$ leads to the synthesis of two ATP. However, the greatest amount of energy produced from a fatty acid is generated by the production of the acetyl CoA units that enter the citric acid cycle. We saw in Chapter 24 that one acetyl CoA leads to the synthesis of a total of 12 ATP.

Let's continue with our example of myristic acid and calculate the total ATP it produces. So far we know that the C_{14} acid produces seven acetyl CoA units and goes through six turns of the cycle. We also need to remember that activation of the myristic acid requires the equivalent of two ATP. We can set up the calculation as follows.

ATP Production for Myristic Acid

Activation	-2 ATP
7 acetyl CoA	
7 acetyl CoA $\times$ 12 ATP/acetyl CoA	84 ATP
6 β-oxidation cycles	
6 $FADH_2$ $\times$ 2 ATP/$FADH_2$	12 ATP
6 NADH $\times$ 3 ATP/NADH	18 ATP
Total	112 ATP

The Energy Yield from Fats

Myristic acid, $C_{14}H_{28}O_2$, has a molar mass of 228 g/mole. We can calculate the ATP produced per gram of the fatty acid as follows:

$$\frac{112 \text{ moles ATP}}{1 \text{ mole myristic acid}} \times \frac{1 \text{ mole myristic acid}}{228 \text{ g myristic acid}} = 0.49 \text{ mole ATP per 1 g myristic acid (fat)}$$

In Chapter 24 we saw that the complete oxidation of glucose generated a total of 36 ATP. Glucose has the formula $C_6H_{12}O_6$ and a molar mass 180 g. We can calculate the ATP produced per gram of glucose as follows:

$$\frac{36 \text{ moles ATP}}{1 \text{ mole glucose}} \times \frac{1 \text{ mole glucose}}{180 \text{ g glucose}} = 0.20 \text{ mole ATP per 1 g glucose}$$

HEALTH NOTE

Stored Fat and Obesity

The storage of fat is an important survival feature in the lives of many animals. In hibernating animals, large amounts of stored fat provide the energy for the entire hibernation period, which could be several months. In camels, such as dromedary camels, large amounts of food are stored in the camel's hump, which is actually a huge fat deposit. When food resources are low, the camel can survive months without food or water by utilizing the fat reserves in the hump. Migratory birds preparing to fly long distances also store large amounts of fat. Whales are kept warm by a layer of body fat called "blubber" under their skin, which can be as thick as 2 feet. Blubber also provides energy when whales must survive long periods of starvation. Penguins also have blubber, which protects them from the cold and provides energy when they are sitting on a nest of eggs.

Humans also have the capability to store large amounts of fat, although they do not hibernate or usually have to survive for long periods of time without food. When humans survived on sparse diets that were mostly vegetarian, the fat content was about 20%. Today, a typical diet includes more dairy products and foods with high fat levels, which increases the daily fat intake to as much as 60% of the diet. The U.S. Public Health Service now estimates that in the United States, more than one-third of adults are obese. Obesity is defined as a body weight that is more than 20 percent over an ideal weight. Obesity is a major factor in health problems such as diabetes, heart disease, high blood pressure, stroke, and gallstones as well as some cancers and forms of arthritis.

At one time we thought that obesity was simply a problem of eating too much. However research now indicates that certain pathways in lipid and carbo-

hydrate metabolism may cause excessive weight gain in some people. In 1995, scientists discovered that a hormone called *leptin* is produced in fat cells. When fat cells are full, high levels of leptin signal the brain to limit the intake of food. When fat stores are low, leptin production decreases, which signals the brain to increase food intake. Some obese persons have high levels of leptin, which means that leptin did not cause them to decrease how much they ate.

Research on obesity has become a major research field. Scientists are studying differences in the rate of leptin production, degrees of resistance to leptin, and possible combinations of these factors. After a person has dieted and lost weight, the leptin level drops. This decrease in leptin may cause an increase in hunger, slow metabolism, and increased food intake, which starts the weight-gain cycle all over again. Currently, studies are being made to assess the safety of leptin therapy following a weight loss.

From these calculations, we see that one gram of fat produces more than twice the ATP energy as 1 gram of glucose. This also means that we obtain more than double the number of nutritional calories from 1 g of fat (9 kcal/g fat) than we do from 1 g of carbohydrate (4 kcal/g). This is one reason why a low-fat diet is recommended when we are trying to lose weight.

SAMPLE PROBLEM 25.3

ATP Production from β Oxidation

How much ATP will be produced from the β oxidation of palmitic acid, a C_{16} saturated fatty acid?

Solution

A 16-carbon fatty acid will produce 8 acetyl CoA units and go through 7 β oxidation cycles. Each acetyl CoA can produce 12 ATP by way of the citric acid cycle. In the electron transport chain, each $FADH_2$ produces 2 ATP, and each NADH produces 3 ATP.

ATP Production for Palmitic Acid ($C_{16}H_{32}O_2$)

Activation of palmitic acid to palmitoyl-CoA	−2 ATP
8 acetyl CoA × 12 ATP (citric acid cycle)	96 ATP
7 FADH$_2$ × 2 ATP (electron transport chain)	14 ATP
7 NADH × 3 ATP (electron transport chain)	21 ATP
Total	129 ATP

Study Check

Compare the percent of the total ATP from reduced coenzymes and from acetyl CoA in the β oxidation of palmitic acid.

QUESTIONS AND PROBLEMS

ATP and Fatty Acid Oxidation

25.15 Why is the energy of fatty acid activation from ATP to AMP considered the same as the hydrolysis of 2 ATP $\longrightarrow$ 2 ADP?

25.16 What is the number of ATP obtained from one acetyl CoA in the citric acid cycle?

25.17 Consider the complete oxidation of capric acid CH_3—$(CH_2)_8$—COOH, a C_{10} fatty acid.
  a. How many acetyl CoA units are produced?
  b. How many cycles of β oxidation are needed?
  c. How many ATPs are generated from the oxidation of capric acid?

25.18 Consider the complete oxidation of arachidic acid, CH_3—$(CH_2)_{18}$—COOH, a C_{20} fatty acid.
  a. How many acetyl CoA units are produced?
  b. How many cycles of β oxidation are needed?
  c. How many ATPs are generated from the oxidation of arachidic acid?

25.4 Ketogenesis and Ketone Bodies

When carbohydrates are not available to meet energy needs, the body breaks down body fat. However, the oxidation of large amounts of fatty acids can cause acetyl CoA molecules to accumulate in the liver. Then acetyl CoA molecules combine to form keto compounds called **ketone bodies** in a pathway known as **ketogenesis.** (See Figure 25.5.)

1. In ketogenesis, two molecules of acetyl CoA combine to form acetoacetyl CoA, which reverses the last reaction in β oxidation.
2. The hydrolysis of acetoacetyl CoA forms acetoacetate, a ketone body, which reacts further to produce two other ketone bodies.
3. Acetoacetate can be reduced to yield β-hydroxybutyrate, which is considered a ketone body even though it does not contain a keto group.
4. Acetone forms when acetoacetate undergoes decarboxylation and loses CO_2.

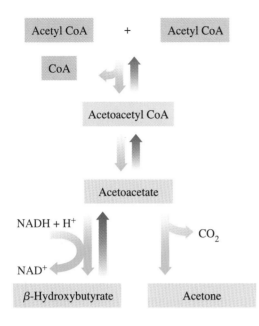

Figure 25.5 In ketogenesis, acetyl CoA molecules combine to produce ketone bodies: acetoacetate, β-hydroxybutyrate, and acetone.

Q What condition in the body leads to the formation of ketone bodies?

Ketogenesis

$$CH_3-\overset{\displaystyle O}{\overset{\|}{C}}-S-CoA + CH_3-\overset{\displaystyle O}{\overset{\|}{C}}-S-CoA$$

$$HS-CoA \overset{}{\underset{\mathbf{1}}{\leftarrow}} \quad \text{Thiolase}$$

$$CH_3-\overset{\displaystyle O}{\overset{\|}{C}}-CH_2-\overset{\displaystyle O}{\overset{\|}{C}}-S-CoA +$$
Acetoacetyl CoA

2

$$CH_3-\overset{\displaystyle O}{\overset{\|}{C}}-CH_2-\overset{\displaystyle O}{\overset{\|}{C}}-O^- + HS-CoA$$
Acetoacetate

β-Hydroxybutyrate
dehydrogenase

3   **4**

$$NADH + H^+$$
$$NAD^+$$
$$CO_2$$

$$CH_3-\underset{\underset{\displaystyle \text{β-Hydroxybutyrate}}{}}{\overset{\displaystyle OH}{\overset{|}{C}H}}-CH_2-\overset{\displaystyle O}{\overset{\|}{C}}-O^- \quad \underset{\displaystyle \text{Acetone}}{CH_3-\overset{\displaystyle O}{\overset{\|}{C}}-CH_3}$$

Ketone bodies are produced mostly in the liver and transported to cells in the heart, brain, and skeletal muscle, where small amounts of energy can be obtained by converting acetoacetate or β-hydroxybutyrate back to acetyl CoA.

β-Hydroxybutyrate ⟶ acetoacetate ⟶ acetoacetyl-CoA ⟶ 2 acetyl CoA

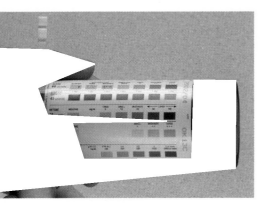

Ketosis

When ketone bodies accumulate, they may not be completely metabolized by the body. This may lead to a condition called **ketosis,** which is found in severe diabetes, diets high in fat and low in carbohydrates, and starvation. Because two of the ketone bodies are acids, they can lower the blood pH below 7.4, which is **acidosis,** a condition that often accompanies ketosis. A drop in blood pH can interfere with the ability of the blood to carry oxygen and cause breathing difficulties.

SAMPLE PROBLEM 25.4

Ketogenesis

When does ketogenesis take place in the liver?

Solution

When excess acetyl CoA cannot be processed by the citric acid cycle, acetyl CoA units enter the ketogenesis pathway where they form ketone bodies.

Study Check

What are the names of the three compounds that are called ketone bodies?

HEALTH NOTE

Ketone Bodies and Diabetes

The blood glucose is elevated within 30 minutes following a meal containing carbohydrates. The elevated level of glucose stimulates the secretion of the hormone insulin from the pancreas, which increases the flow of glucose into muscle and adipose tissue for the synthesis of glycogen. As blood glucose levels drop, the secretion of insulin decreases. When blood glucose is low, another hormone, glucagon, is secreted by the pancreas, which stimulates the breakdown of glycogen in the liver to yield glucose.

In *diabetes mellitus,* glucose cannot be utilized or stored as glycogen because insulin is not secreted or does not function properly. In Type I, *insulin-dependent diabetes,* which often occurs in childhood, the pancreas produces inadequate levels of insulin. This type of diabetes can result from damage to the pancreas by viral infections or from genetic mutations. In Type II, *insulin-resistant diabetes,* which usually occurs in adults, insulin is produced, but insulin receptors are not responsive. Thus a person with Type II diabetes does not respond to insulin therapy.

Gestational diabetes can occur during pregnancy, but blood glucose levels usually return to normal after the baby is born. Mothers with diabetes tend to gain weight and have large babies.

In all types of diabetes, insufficient amounts of glucose are available in the muscle, liver, and adipose tissue. As a result liver cells synthesize glucose from noncarbohydrate sources (gluconeogenesis) and break down fat, which elevates the acetyl CoA level. Excess acetyl CoA undergoes ketogenesis and ketone bodies accumulate in the blood. As the level of acetone increases, its odor can be detected on the breath of a person with uncontrolled diabetes who is in ketosis.

In uncontrolled diabetes, the concentration of blood glucose exceeds the ability of the kidney to reabsorb glucose, and glucose appears in the urine. High levels of glucose increase the osmotic pressure in the blood, which leads to an increase in urine output. Symptoms of diabetes include frequent urination and excessive thirst. Treatment for diabetes includes a change to a diet to limit carbohydrate intake, and may require medication such as a daily injection of insulin.

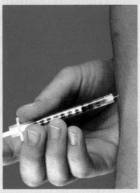
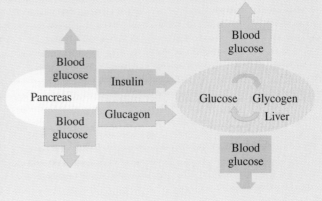

QUESTIONS AND PROBLEMS

Ketogenesis and Ketone Bodies

25.19 What is ketogenesis?

25.20 If a person is fasting, why would they have high levels of acetyl CoA?

25.21 What type of reaction converts acetoacetate to β-hydroxybutyrate?

25.22 How is acetone formed from acetoacetate?

25.23 What is ketosis?

25.24 Why do diabetics produce high levels of ketone bodies?

25.5 Fatty Acid Synthesis

LEARNING GOAL
Describe the biosynthesis of fatty acids from acetyl CoA.

When the body has met all its energy needs and the glycogen stores are full, acetyl CoA from the breakdown of carbohydrates and fatty acids is used to form new fatty acids. In the pathway called **lipogenesis,** two-carbon acetyl units are linked together to give a 16-carbon fatty acid, palmitic acid. Although the reactions appear much like the reverse of the reactions we discussed in fatty acid oxidation, the synthesis of fatty acids proceeds in a separate pathway with different enzymes. Fatty acid oxidation occurs in the mitochondria and uses FAD and NAD^+, whereas fatty acid synthesis occurs in the cytosol and uses the reduced coenzyme NADPH. NADPH is similar to NADH, except it has a phosphate group.

Acyl Carrier Protein (ACP)

In β oxidation, acetyl and acyl molecules are activated using coenzyme A (CoA). In fatty acid synthesis, acyl compounds also are activated, but by an acyl carrier protein (ACP—SH). In the ACP—SH molecule, the thiol and pantothenic acid found in CoA are attached to a protein.

ACP—SH

$$HS-CH_2-CH_2-NH-\overset{O}{\underset{\|}{C}}-CH_2-CH_2-NH-\overset{O}{\underset{\|}{C}}-\underset{H}{\overset{OH}{\underset{|}{CH}}}-\overset{CH_3}{\underset{|}{C}}-CH_2-O-\overset{O}{\underset{O^-}{\overset{\|}{\underset{|}{P}}}}-O-CH_2-Protein$$

2-Aminoethanethiol   Pantothenic acid

For fatty acid synthesis, the activated forms of malonyl-ACP and acetyl-ACP are produced by transferring the acetyl or acyl group from CoA—SH to ACP—SH.

$$CH_3-\overset{O}{\underset{\|}{C}}-S-CoA + HS-ACP \xrightarrow{\text{Acetyl CoA transacylase}} CH_3-\overset{O}{\underset{\|}{C}}-S-ACP + HS-CoA$$
Acetyl CoA   Acetyl ACP

$$^-O-\overset{O}{\underset{\|}{C}}-CH_2-\overset{O}{\underset{\|}{C}}-S-CoA + HS-ACP \xrightarrow{\text{Malonyl CoA transacylase}} {}^-OOC-CH_2-\overset{O}{\underset{\|}{C}}-S-ACP + HS-CoA$$
Malonyl CoA   Malonyl ACP

Synthesis of Malonyl CoA

Fatty acid synthesis begins when acetyl CoA combines with bicarbonate to form a three-carbon compound malonyl CoA. The hydrolysis of ATP provides the energy for the reaction.

$$
\underset{\text{Acetyl CoA}}{CH_3-\overset{\overset{\displaystyle O}{\|}}{C}-CoA} + HCO_3^- + ATP \xrightarrow{\substack{\text{Acetyl CoA}\\\text{carboxylase}}} \underset{\text{Malonyl CoA}}{{}^-O-\overset{\overset{\displaystyle O}{\|}}{C}-CH_2-\overset{\overset{\displaystyle O}{\|}}{C}-CoA} + ADP + P_i + H^+
$$

Synthesis of Palmitate

The next four reactions occur in a cycle that adds two-carbon acetyl groups to a carbon chain. (See Figure 25.6.)

Reaction 1   Condensation
Acetyl ACP and malonyl ACP condense to yield acetoacetyl-ACP and CO_2.

Reaction 2   Reduction
The keto group on the β carbon is reduced to a hydroxyl group using hydrogen from the reduced coenzyme NADPH.

Reaction 3   Dehydration
The alcohol is dehydrated to form a trans double bond in *trans*-enoyl-ACP.

Reaction 4   Reduction
NADPH reduces the double bond to a single bond, which forms butyryl-ACP, a saturated four-carbon compound.

Cycle Repeats

The cycle repeats as the longer four-carbon butyryl-ACP reacts with another malonyl-ACP to produce hexanoyl-ACP. After seven cycles, the product, C_{16} palmitoyl-ACP, is hydrolyzed to yield palmitate and free ACP.

The overall reaction starts with acetyl CoA and requires ATP and NADPH. We can write the overall equation for the synthesis of palmitate as follows:

$$8\,\text{Acetyl CoA} + 14\,\text{NADPH} + 14H^+ + 7\,\text{ATP} \longrightarrow$$

$$\text{palmitate} + 8\,\text{CoA} + 14\,\text{NADP}^+ + 7H_2O + 7\,\text{ADP} + 7\,P_i$$

Longer and Shorter Fatty Acids

Although we have looked at the synthesis of the fatty acid palmitate, shorter and longer fatty acids are also produced in cells. Shorter fatty acids are released before there are 16 carbon atoms in the chain. Longer fatty acids are produced with special enzymes that add two-carbon units to the carboxyl end of the fatty acid chain. An unsaturated cis bond can also be incorporated into a 10-carbon fatty acid followed by the same elongation reactions we have seen.

Regulation of Fatty Acid Synthesis

Fatty acid synthesis takes place primarily in the adipose tissue, where triacylglycerols are formed and stored. The hormone insulin stimulates the formation of fatty

Cycle 1

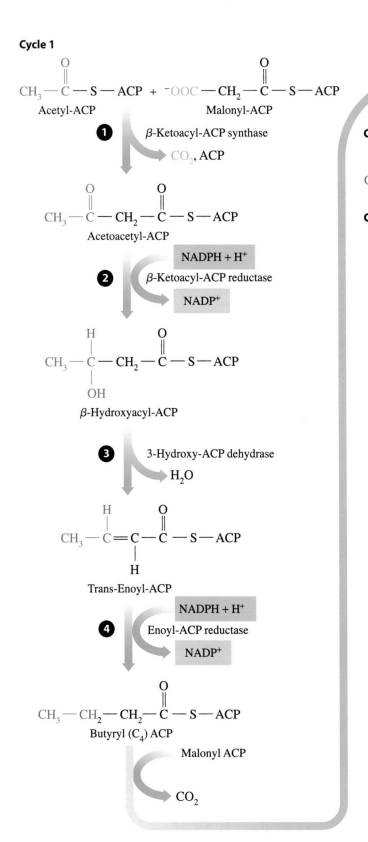

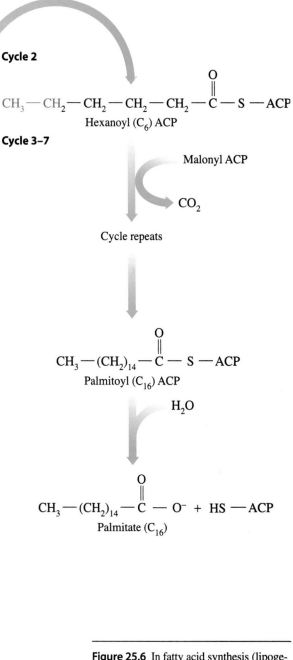

Figure 25.6 In fatty acid synthesis (lipogenesis), two-carbon units from acetyl CoA are added together to form palmitate.

Q Identify the types of reactions in 1–4 as reduction, dehydration, or condensation.

acids. When blood glucose is high, insulin moves glucose into the cells. In the cell, insulin stimulates glycolysis and the oxidation of pyruvate, thereby producing acetyl CoA for fatty acid synthesis. During lipogenesis, the production of malonyl CoA blocks the transport of fatty acyl groups into the matrix of the mitochondria, which prevents their oxidation.

Comparison of β Oxidation and Fatty Acid Synthesis

We have seen that many of the steps in the synthesis of palmitate are similar to those that occur in the β oxidation of palmitate. Synthesis combines two-carbon units, whereas β oxidation removes two-carbon units. Synthesis involves reduction and dehydration, whereas β oxidation involves oxidation and hydration. We can distinguish between the two pathways by comparing some of their features in Table 25.1.

Table 25.1   A Comparison of β Oxidation and Fatty Acid Synthesis

	β Oxidation	Fatty Acid Synthesis (lipogenesis)
Site	Mitochondrial matrix	Cytosol
Activated by	Glucagon Low blood glucose	Insulin High blood glucose
Activator	Coenzyme A (CoA)	Acyl carrier protein (ACP)
Initial substrate	Fatty acid	Acetyl CoA $\longrightarrow$ Malonyl CoA
Coenzymes	FAD, NAD$^+$	NADPH, NADP$^+$
Types of Reaction	Oxidation Hydration Cleavage	Reduction Dehydration Condensation
Function	Cleaves two-carbon acyl group	Adds two-carbon acyl group
Final product	Acetyl CoA units	Palmitate (C_{16}) and other fatty acids

SAMPLE PROBLEM 25.5

Fatty Acid Synthesis

Malonyl ACP is required for the elongation of fatty acid chains. How is malonyl ACP formed from the starting material acetyl CoA?

Solution

Acetyl CoA combines with bicarbonate to form malonyl CoA, which reacts with ACP to form malonyl ACP.

Study Check

If malonyl ACP is a three-carbon acyl group, why are only two carbon atoms added each time malonyl ACP is combined with a fatty acid chain?

QUESTIONS AND PROBLEMS

Fatty Acid Synthesis

25.25 Where does fatty acid synthesis occur in the cell?

25.26 What compound is involved in the activation of acyl compounds in fatty acid synthesis?

25.27 What is the starting material for fatty acid synthesis?

25.28 What is the function of malonyl ACP in fatty acid synthesis?

25.29 Identify the reaction catalyzed by each of the following enzymes:

    (1) acetyl CoA carboxylase   (2) acetyl CoA transacylase

    (3) malonyl CoA transacylase

  a. converts malonyl CoA to malonyl ACP

  b. combines acetyl CoA with bicarbonate to give malonyl CoA

      c. converts acetyl CoA to acetyl ACP

25.30 Identify the reaction catalyzed by each of the following enzymes:
      (1) β-ketoacyl-ACP synthase   (2) β-ketoacyl-ACP reductase
      (3) 3-hydroxy-ACP dehydrase   (4) enoyl-ACP reductase
      a. catalyzes the dehydration of an alcohol
      b. converts a carbon–carbon double bond to a carbon–carbon single bond
      c. combines a two-carbon acyl group with a three-carbon acyl group accompanied by the loss of CO_2
      d. reduces a keto group to a hydroxyl group

25.31 Determine the number of each of the following involved in the synthesis of one molecule of capric acid, a fatty acid with 10 carbon atoms, $C_{10}H_{20}O_2$.
      a. HCO_3^- **b.** ATP   **c.** acetyl CoA
      d. malonyl ACP   **e.** NADPH   **f.** CO_2 removed

25.32 Determine the number of each of the following needed in the synthesis of one molecule of myristic acid, a fatty acid with 14 carbon atoms $C_{14}H_{28}O_2$.
      a. HCO_3^- **b.** ATP   **c.** acetyl ACP
      d. malonyl ACP   **e.** NADPH   **f.** CO_2 removed

25.6 Digestion of Proteins

LEARNING GOAL

Describe the hydrolysis of dietary protein and absorption of amino acids.

The major role of proteins is to provide amino acids for the synthesis of new proteins for the body and nitrogen atoms for the synthesis of compounds such as nucleotides. We have seen that carbohydrates and lipids are the major sources of energy, but when they are not available, amino acids are degraded to substrates that enter energy-producing pathways.

In stage 1, the digestion of proteins begins in the stomach, where hydrochloric acid (HCl) at pH 2 denatures proteins and activates enzymes such as pepsin that begin to hydrolyze peptide bonds. Polypeptides from the stomach move into the small intestine where trypsin and chymotrypsin complete the hydrolysis of the peptides to amino acids. The amino acids are absorbed through the intestinal walls into the blood stream for transport to the cells. (See Figure 25.7.)

SAMPLE PROBLEM 25.6

Digestion of Proteins

What are the sites and end products for the digestion of proteins?

Solution

Proteins begin digestion in the stomach and complete digestion in the small intestine to yield amino acids.

Study Check

What is the function of HCl in the stomach?

Protein Turnover

Our bodies are constantly replacing old proteins with new ones. The process of synthesizing proteins and breaking them down is called **protein turnover.** Many

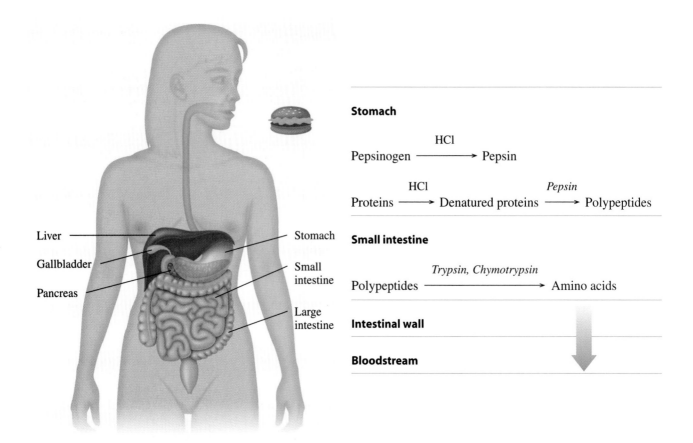

Stomach

$$\text{Pepsinogen} \xrightarrow{\text{HCl}} \text{Pepsin}$$

$$\text{Proteins} \xrightarrow{\text{HCl}} \text{Denatured proteins} \xrightarrow{\textit{Pepsin}} \text{Polypeptides}$$

Small intestine

$$\text{Polypeptides} \xrightarrow{\textit{Trypsin, Chymotrypsin}} \text{Amino acids}$$

Intestinal wall

Bloodstream

Liver — Stomach
Gallbladder — Small intestine
Pancreas — Large intestine

Figure 25.7 Proteins are hydrolyzed to polypeptides in the stomach and to amino acids in the small intestine.

Q What enzyme, secreted into the small intestine, hydrolyzes peptides?

types of proteins including enzymes, hormones, and hemoglobin are synthesized in the cells and then degraded. For example, the hormone insulin has a half-life of 10 minutes, whereas the half-lives of lactate dehydrogenase is about 2 days and hemoglobin is 120 days. Damaged and ineffective proteins are also degraded and replaced. While most amino acids are used to build proteins, other compounds also require nitrogen for their synthesis as seen in Table 25.2. (See Figure 25.8.)

Usually we maintain a nitrogen balance in the cells so that the amount of protein we break down is equal to the amount that is reused. A diet that is high in protein has a positive nitrogen balance because it supplies more nitrogen than we need. Because the body cannot store nitrogen, the excess is excreted as urea. A diet that does not provide sufficient nitrogen has a negative nitrogen balance, which is a condition that occurs during starvation and fasting.

Table 25.2   Nitrogen-Containing Compounds

Type of Compound	Example
Nonessential amino acids	Alanine, aspartate, cysteine, glycine
Proteins	Muscle protein, enzymes
Neurotransmitters	Acetylcholine, dopamine, serotonin
Amino alcohols	Choline, ethanolamine
Heme	Hemoglobin
Hormones	Thyroxine, epinephrine, insulin
Nucleotides (nucleic acids)	Purines, pyrimidines

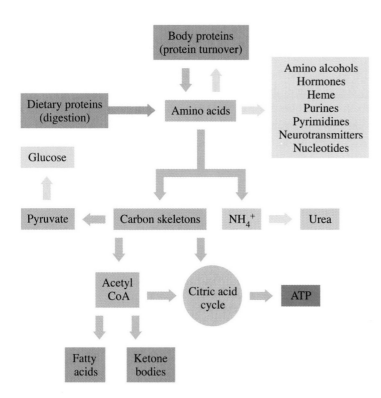

Figure 25.8 Proteins are used in the synthesis of nitrogen-containing compounds or degraded to carbon skeletons that enter other metabolic pathways and urea.

Q **What are some compounds that require nitrogen for their synthesis?**

Energy from Amino Acids

Normally, only a small amount (about 10%) of our energy needs is supplied by amino acids. However, more energy is extracted from amino acids in conditions such as fasting or starvation, when carbohydrate and fat stores are exhausted. If amino acids remain the only source of energy for a long period of time, the breakdown of body proteins eventually leads to a destruction of essential body tissues.

SAMPLE PROBLEM 25.7

In positive nitrogen balance, why are excess amino acids excreted?

Solution

Because the body cannot store nitrogen, amino acids that are not needed for the synthesis of proteins are excreted.

Study Check

Under what condition does the body have a negative nitrogen balance?

QUESTIONS AND PROBLEMS

Digestion of Proteins

25.33 Where do dietary proteins undergo digestion in the body?

25.34 What is meant by protein turnover?

25.35 What are some nitrogen-containing compounds that need amino acids for their synthesis?

25.36 What is the fate of the amino acids obtained from a high protein diet?

LEARNING GOAL

Describe the reactions of transamination and oxidative deamination in the degradation of amino acids.

25.7 Degradation of Amino Acids

When dietary protein exceeds the nitrogen needed for protein synthesis, the excess amino acids are degraded in a similar way. The α-amino group is removed to yield a keto acid, which can be converted to an intermediate of other metabolic pathways. The carbon atoms from amino acids are used in the citric acid cycle as well as the synthesis of fatty acids, ketone bodies, and glucose. Most of the amino groups are converted to urea.

Transamination

The degradation of amino acids occurs primarily in the liver. In a **transamination** reaction, an α-amino group is transferred from an amino acid to an α-keto acid, usually α-ketoglutarate. A new amino acid and a new α-keto acid are produced. The enzymes for the transfer of amino groups are known as transaminases or aminotransferases.

$$R_1-\overset{\overset{+}{N}H_3}{\underset{|}{C}}H-COO^- \;+\; R_2-\overset{O}{\overset{\|}{C}}-COO^- \xrightarrow{\text{Aminotransferase}} R_1-\overset{O}{\overset{\|}{C}}-COO^- \;+\; R_2-\overset{\overset{+}{N}H_3}{\underset{|}{C}}H-COO^-$$

Amino acid α-Keto acid   New α-Keto acid   New amino acid

We can write an equation to show the transfer of the amino group from alanine to α-ketoglutarate to yield glutamate, the new amino acid, and the α-keto acid pyruvate. The α-keto acid often used in transamination reactions is α-ketoglutarate, which is converted to glutamate.

$$CH_3-\overset{\overset{+}{N}H_3}{\underset{|}{C}}H-COO^- \;+\; {}^-OOC-\overset{O}{\overset{\|}{C}}-CH_2-CH_2-COO^- \underset{\longleftarrow}{\overset{\text{Alanine aminotransferase}}{\longrightarrow}}$$

Alanine α-Ketoglutarate

$$CH_3-\overset{O}{\overset{\|}{C}}-COO^- \;+\; {}^-OOC-\overset{\overset{+}{N}H_3}{\underset{|}{C}}H-CH_2-CH_2-COO^-$$

Pyruvate   Glutamate

SAMPLE PROBLEM 25.8

Transamination

Write the formula of the amino acid and α-keto acid produced from the transamination of oxaloacetate by alanine.

$$CH_3-\overset{\overset{+}{N}H_3}{\underset{|}{C}}H-COO^- \;+\; {}^-OOC-\overset{O}{\overset{\|}{C}}-CH_2-COO^- \longrightarrow$$

Alanine   Oxaloacetate

Solution

$$CH_3-\overset{O}{\overset{\|}{C}}-COO^- \;+\; {}^-OOC-\overset{\overset{+}{N}H_3}{\underset{|}{C}}H-CH_2-COO^-$$

Pyruvate   Aspartate

Study Check

What is a possible name for the enzyme that catalyzes this reaction?

Oxidative Deamination

In a process called **oxidative deamination,** the amino group in glutamate is removed as an ammonium ion NH_4^+. This reaction is catalyzed by *glutamate dehydrogenase,* which uses either NAD^+ or $NADP^+$ as a coenzyme.

$$^-OOC-\underset{\text{Glutamate}}{\overset{\overset{+}{N}H_3}{\overset{|}{C}H}}-CH_2-CH_2-COO^- + H_2O + NAD^+ \text{ (or } NADP^+) \xrightarrow[\text{dehydrogenase}]{\text{Glutamate}}$$

$$^-OOC-\underset{\alpha\text{-Ketoglutarate}}{\overset{O}{\overset{\|}{C}}}-CH_2-CH_2-COO^- + NH_4^+ + NADH \text{ (or NADPH)} + H^+$$

Therefore the amino group from any amino acid can be used to form glutamate, which undergoes oxidative deamination converting the amino group to an ammonium ion. Then the ammonium ion is converted to urea, which we will discuss in the next section.

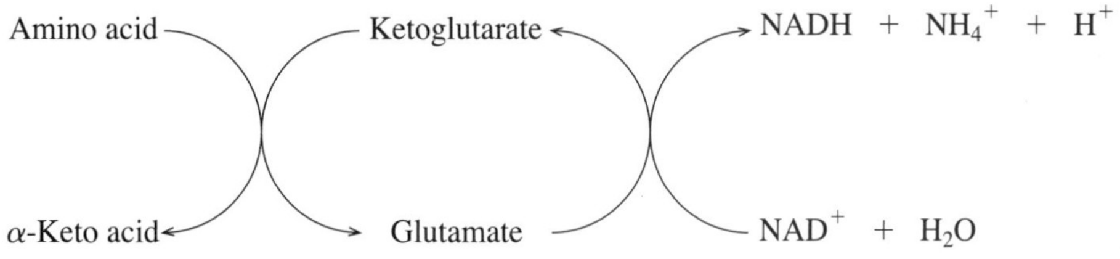

SAMPLE PROBLEM 25.9

Transamination and Oxidative Deamination

Indicate whether each of the following represents a transamination or an oxidative deamination:

a. Glutamate is converted to α-ketoglutarate and NH_4^+.
b. Alanine and α-ketoglutarate react to form pyruvate and glutamate.
c. A reaction is catalyzed by glutamate dehydrogenase, which requires NAD^+.

Solution

a. oxidative deamination
b. transamination
c. oxidative deamination

Study Check

How is α-ketoglutarate regenerated to participate in more transamination reactions?

QUESTIONS AND PROBLEMS

Degradation of Amino Acids

25.37 What are the reactants and products in transamination reactions?

25.38 What types of enzymes catalyze transamination reactions?

25.39 Write the structure of the α–keto acid produced from each of the following in transamination:

$$
\begin{array}{c} \overset{+}{N}H_3 \\ | \\ \textbf{a. } H-CH-COO^- \quad glycine \end{array}
\qquad
\begin{array}{c} \overset{+}{N}H_3 \\ | \\ \textbf{b. } CH_3-CH-COO^- \quad alanine \end{array}
$$

$$
\begin{array}{c} CH_3 \quad \overset{+}{N}H_3 \\ | \qquad | \\ \textbf{c. } CH_3-CH-CH-COO^- \quad valine \end{array}
$$

25.40 Write the structure of the α–keto acid produced from each of the following in transamination:

$$
\begin{array}{c} \overset{+}{N}H_3 \\ | \\ \textbf{a. } {}^-OOC-CH_2-CH-COO^- \quad aspartate \end{array}
\qquad
\begin{array}{c} CH_3 \quad \overset{+}{N}H_3 \\ | \qquad | \\ \textbf{b. } CH_3-CH_2-CH-CH-COO^- \quad isoleucine \end{array}
$$

$$
\begin{array}{c} CH_3 \quad \overset{+}{N}H_3 \\ | \qquad | \\ \textbf{c. } HO-CH_2-CH-COO^- \quad serine \end{array}
$$

25.41 Write the reaction for the oxidative deamination of glutamate.

25.42 How do the amino groups from all 20 amino acids produce ammonium ions when glutamate undergoes oxidative deamination?

LEARNING GOAL

Describe the formation of urea from ammonium ion.

25.8 Urea Cycle

The ammonium ion, which is the end product of amino acid degradation, is toxic if it is allowed to accumulate. Therefore, a pathway called the **urea cycle** detoxifies ammonium ions by converting them to urea, which is transported to the kidneys to form urine.

$$
\begin{array}{c} O \\ \| \\ H_2N-C-NH_2 \end{array}
$$
Urea

In one day, a typical adult may excrete about 25–30 g of urea in the urine. This amount increases when a diet is high in protein. If urea is not properly excreted, it builds up quickly to a toxic level. To detect renal disease, the blood urea nitrogen (BUN) level is measured. If the BUN is high, protein intake must be reduced, and hemodialysis may be needed to remove toxic nitrogen waste from the blood.

Urea Cycle

The urea cycle in the liver cells consists of reactions that take place in both the mitochondria and cytosol. (See Figure 25.9.) In preparation for the urea cycle, the ammonium ions react with carbon dioxide from the citric acid cycle and two ATP to yield carbamoyl phosphate.

$$NH_4^+ + CO_2 + 2\,ATP + H_2O \xrightarrow{\substack{\text{Carbamoyl} \\ \text{phosphate} \\ \text{synthetase}}} \underset{\text{Carbamoyl phosphate}}{H_2N-\overset{\displaystyle O}{\overset{\|}{C}}-O-\overset{\displaystyle O}{\underset{\displaystyle O^-}{\overset{\|}{P}}}-O^-} + 2\,ADP + P_i$$

Reaction 1   Transfer of Carbamoyl Group

In the mitochondria, the carbamoyl group is transferred from carbamoyl phosphate to ornithine to yield citrulline, which is transported across the mitochondrial membrane into the cytosol. The hydrolysis of the phosphate bond provides the energy to drive the reaction.

Reaction 2   Condensation with Aspartate

In the cytosol, citrulline condenses with the amino acid aspartate to form argininosuccinate. The hydrolysis of ATP to AMP and two inorganic phosphates provides the energy for the reaction. The nitrogen atom in aspartate becomes the other nitrogen atom in the urea that is produced in the final reaction.

Reaction 3   Cleavage of Fumarate

The argininosuccinate undergoes a cleavage to yield fumarate, a citric acid cycle intermediate, and arginine.

Reaction 4   Hydrolysis to Form Urea

The hydrolysis of arginine yields urea and ornithine, which returns to the mitochondria to repeat the cycle.

Summary of Urea Formation

The formation of urea starts with two nitrogen atoms, one from NH_4^+ and one from aspartate, and a carbon atom from CO_2. In the urea cycle, a total of four phosphate bonds are hydrolyzed to provide the energy for the reactions. We can write an overall reaction starting with ammonium ion as follows:

$$NH_4^+ + CO_2 + 3\,ATP + aspartate + 2H_2O \longrightarrow Urea + 2\,ADP + AMP + 4P_i + fumarate$$

SAMPLE PROBLEM 25.10

Urea Cycle

Indicate the reaction in the urea cycle where each of the following compounds is a reactant.

a. aspartate   **b.** ornithine   **c.** arginine

Solution

a. Aspartate condenses with citrulline in reaction 2.
b. Ornithine accepts the carbamoyl group in reaction 1.
c. Arginine is cleaved in reaction 4.

Study Check

Name the products of each reaction in problem 25.10.

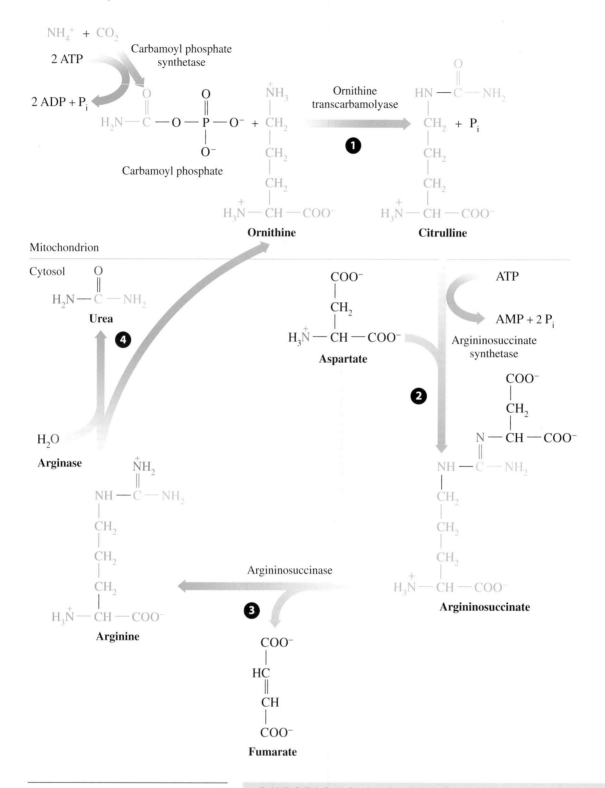

Figure 25.9 In the urea cycle, urea is formed from a carbon and nitrogen (blue) from carbamoyl phosphate (initially an ammonium ion from oxidative deamination) and a nitrogen from aspartate (pink).

Q Where in the cell is urea formed?

QUESTIONS AND PROBLEMS

Urea Cycle

25.43 Why does the body convert NH_4^+ to urea?

25.44 Where is the energy source for the formation of urea?

25.45 What is the structure of urea?

25.46 What is the structure of carbamoyl phosphate?

25.47 What is the source of carbon in urea?

25.48 How much ATP energy is required to drive one turn of the urea cycle?

25.9 Fates of the Carbon Atoms from Amino Acids

LEARNING GOAL

Describe where carbon atoms from amino acids enter the citric acid cycle or other pathways.

The carbon skeletons from the transamination of amino acids are used as intermediates of the citric acid cycle or other metabolic pathways. We can classify the amino acids according to the number of carbon atoms in those intermediates. (See Figure 25.10.) The amino acids with three carbons or those that are converted into carbon skeletons with three carbons are converted to pyruvate. The four-carbon group consists of amino acids that are converted to oxaloacetate, and the five-carbon group provides α-ketoglutarate. Some amino acids are listed twice because they can enter different pathways to form citric acid cycle intermediates.

A **glucogenic amino acid** (orange) generates pyruvate or oxaloacetate, which can be converted to glucose by gluconeogenesis. A **ketogenic amino acid** (green) produces acetoacetyl CoA or acetyl CoA, which can enter the ketogenesis pathway to form ketone bodies or fatty acids.

Carbon Skeletons from the Three-Carbon Amino Acids

The carbon atoms from alanine, serine, and cysteine are converted to pyruvate in one step or several steps. Alanine is converted by a simple transamination.

Alanine + α-ketoglutarate $\longrightarrow$ pyruvate + glutamate

Glycine is converted to serine and then to pyruvate. When tryptophan is degraded, the three carbon atoms that are not part of the ring system form alanine, which goes to pyruvate. Although it has more than three carbon atoms, threonine can be degraded to glycine and acetaldehyde. The oxidation of acetaldehyde gives acetyl-CoA.

Carbon Skeletons from the Four-Carbon Amino Acids

Oxaloacetate is the α-keto acid that forms when the four-carbon aspartate undergoes transamination.

Aspartate + α-ketoglutarate $\longrightarrow$ oxaloacetate + glutamate

Asparagine hydrolyzes to give NH_4^+ and aspartate, which goes to oxaloacetate. The degradation pathways of other four-carbon amino acids, isoleucine, methionine, and valine, produce succinyl CoA, another intermediate of the citric acid cycle.

Carbon Skeletons from the Five-Carbon Amino Acids

The five-carbon amino acids glutamine, glutamate, proline, arginine, and histidine are converted to glutamate, which undergoes oxidation–deamination to yield α-ketoglutarate and ammonium ion.

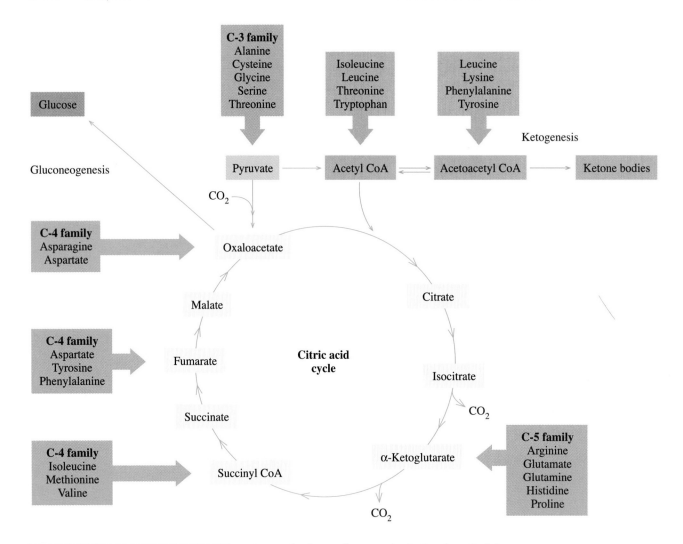

Figure 25.10 Carbon atoms from degraded amino acids are converted to the intermediates of the citric acid cycle or other pathways. Glucogenic amino acids (orange boxes) produce carbon skeletons that can form glucose, and ketogenic amino acid (green boxes) can produce ketone bodies.

Q Why is aspartate glucogenic, but leucine is ketogenic?

Degradation of Ketogenic Amino Acids

Two ketogenic amino acids, leucine and lysine, require a series of reactions that degrade them to acetoacetate, which is an intermediate in the production of ketone bodies. Lysine also produces acetyl CoA. Some of the carbon atoms of the aromatic amino acids phenylalanine and tyrosine are converted to acetoacetate as well as fumarate.

SAMPLE PROBLEM 25.11

What degradation products of amino acids are citric acid intermediates?

Solution

Carbon atoms from amino acids enter the citric acid cycle as acetyl CoA, α-ketoglutarate, succinyl CoA, fumarate, or oxaloacetate.

Study Check

Which amino acids provide carbon atoms that enter the citric acid as α-ketoglutarate?

Fates of the Carbon Atoms from Amino Acids

25.49 What is a glucogenic amino acid?

25.50 What is a ketogenic amino acid?

25.51 What metabolic substrate(s) can be produced from the carbon atoms of each of the following amino acids?

    a. alanine   **b.** aspartate

    c. valine   **d.** glutamine

25.52 What metabolic substrate(s) can be produced from the carbon atoms of each of the following amino acids?

    a. leucine   **b.** asparagine

    c. cysteine   **d.** arginine

25.10 Synthesis of Amino Acids

LEARNING GOAL

Illustrate how some nonessential amino acids are synthesized from intermediates in the citric acid cycle and other metabolic pathways.

Plants and bacteria such as *E. coli* produce all of their amino acids using NH_4^+ and NO_3^-. However, humans can synthesize only 10 of the 20 amino acids found in their proteins. The **nonessential amino acids** are synthesized in the body, whereas the **essential amino acids** must be obtained from the diet. (See Table 25.3.) Two amino acids, arginine and histidine, are essential in diets for children due to their rapid growth requirements, but not adults.

Some Pathways for Amino Acid Synthesis

There are a variety of pathways involved in the synthesis of nonessential amino acids. When the body synthesizes nonessential amino acids, the α-keto acid skeletons are obtained from the citric acid cycle or glycolysis and converted to amino acids by transamination. (See Figure 25.11.)

Some of the amino acids are formed from a simple transamination. For example, the transfer of the amino group from glutamate to pyruvate, a three-carbon α-keto acid, produces alanine, an amino acid with three carbons.

Table 25.3 Essential and Nonessential Amino Acids in Humans

Essential		Nonessential	
Arginine*	Methionine	Alanine	Glutamine
Histidine*	Phenylalanine	Asparagine	Glycine
Isoleucine	Threonine	Aspartate	Proline
Leucine	Tryptophan	Cysteine	Serine
Lysine	Valine	Glutamate	Tyrosine

*Essential for children only

In another transamination using glutamate, the four-carbon oxaloacetate from the citric acid cycle is converted to aspartate.

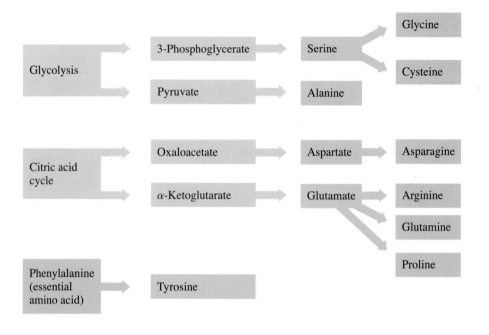

Figure 25.11 Nonessential amino acids are synthesized from intermediates of glycolysis and the citric acid cycle.

Q How is alanine formed from pyruvate formed in glycolysis?

$$^-OOC-CH_2-\overset{\overset{\displaystyle O}{\|}}{C}-COO^- \; + \; ^-OOC-\overset{\overset{\displaystyle \overset{+}{N}H_3}{|}}{CH}-CH_2-CH_2-COO^- \quad \xrightarrow{\text{Glutamate oxaloacetate transaminase}}$$

Oxaloacetate   Glutamate

$$^-OOC-CH_2-\overset{\overset{\displaystyle \overset{+}{N}H_3}{|}}{CH}-COO^- \; + \; ^-OOC-\overset{\overset{\displaystyle O}{\|}}{C}-CH_2-CH_2-COO^-$$

Aspartate   α-Ketoglutarate

These two transaminases are abundant in the cells of the liver and heart, but are present only in low levels in the bloodstream. When an injury or disease occurs, they are released from the damaged cells into the bloodstream. The elevated levels of *serum glutamate-pyruvate transaminase* (SGPT) and *serum glutamate-oxaloacetate transaminase* (SGOT) provide a means to diagnose the extent of damage to the liver or heart.

Synthesis of Glutamine

The synthesis of the other nonessential amino acids requires several other reactions in addition to transamination. For example, glutamine is synthesized when a second amino group is added to glutamate using the energy from the hydrolysis of ATP.

$$^-OOC-\overset{\overset{\displaystyle \overset{+}{N}H_3}{|}}{CH}-CH_2-CH_2-COO^- \; + \; NH_3 \quad \xrightarrow[\text{ATP} \quad \text{ADP} + P_i]{\text{Glutamine synthetase}} \quad ^-OOC-\overset{\overset{\displaystyle \overset{+}{N}H_3}{|}}{CH}-CH_2-CH_2-\overset{\overset{\displaystyle O}{\|}}{C}-NH_2$$

Glutamate   Glutamine

Synthesis of Serine and Cysteine

In the synthesis of serine, three steps are required starting with 3-phosphoglycerate from glycolysis. In this pathway, the —OH group of glycerate is oxidized to give an α-keto acid, which undergoes transamination by glutamate accompanied by the loss of the phosphate group.

$$\underset{\text{3-Phosphoglycerate}}{{}^-OOC-\overset{\overset{\displaystyle OH}{|}}{C}H-CH_2-O-\overset{\overset{\displaystyle O}{\|}}{\underset{\underset{\displaystyle O^-}{|}}{P}}-O^-} + NAD^+ + Glutamate + H_2O \longrightarrow$$

$$\underset{\text{Serine}}{{}^-OOC-\overset{\overset{\displaystyle \overset{+}{N}H_3}{|}}{C}H-CH_2-OH} + NADH + H^+ + \alpha\text{-Ketoglutarate} + P_i$$

Once serine is formed, its —OH group is replaced by —SH from a reaction with homocysteine.

$$\underset{\text{Serine}}{{}^-OOC-\overset{\overset{\displaystyle \overset{+}{N}H_3}{|}}{C}H-CH_2-OH} + \underset{\text{Homocysteine}}{{}^-OOC-\overset{\overset{\displaystyle \overset{+}{N}H_3}{|}}{C}H-CH_2-CH_2-SH} \xrightarrow[\text{H}_2\text{O}]{\overset{\text{Cysteine}}{\text{synthase}}}$$

$$\underset{\text{Cysteine}}{{}^-OOC-\overset{\overset{\displaystyle \overset{+}{N}H_3}{|}}{C}H-CH_2-SH} + \underset{\alpha\text{-Ketobutyrate}}{{}^-OOC-\overset{\overset{\displaystyle O}{\|}}{C}-CH_2-CH_3}$$

Synthesis of Tyrosine

Tyrosine, an aromatic amino acid with a hydroxyl group, is formed from phenylalanine, an essential amino acid.

$$\underset{\text{Phenylalanine}}{\bigcirc\!\!\!-CH_2-\overset{\overset{\displaystyle \overset{+}{N}H_3}{|}}{C}H-COO^-} + O_2 \xrightarrow{\overset{\text{Phenylalanine}}{\text{hydroxylase}}} HO-\bigcirc\!\!\!-CH_2-\underset{\text{Tyrosine}}{\overset{\overset{\displaystyle \overset{+}{N}H_3}{|}}{C}H-COO^-} + H_2O$$

HEALTH NOTE

Homocysteine and Coronary Heart Disease

In the body, the amino acid methionine is degraded to homocysteine. We obtain most of our methionine, an essential amino acid, from the proteins in meat. In turn, homocysteine can be used to synthesize methionine in a process that requires folic acid and vitamin B_{12} (cobalamin). In a study at Harvard in the 1960s, children suffering from a genetic disorder, *homocystinuria,* were found to have high homocysteine levels. They were also found to have advanced atherosclerosis, which led to strokes and heart attacks early in life. This was one of the first indications of a link between elevated homocysteine levels and heart disease. In the past few years, additional clinical research has indicated that elevated blood levels of homocysteine are associated with increased risk of coronary heart disease, which can also lead to stroke and myocardial infarction (heart attack). It was also found that low levels of folic acid accompanied the elevated levels of homocysteine. This finding suggests that inadequate levels of folic acid limit the synthesis of methionine from homocysteine causing an accumulation of homocysteine.

When the body has adequate amounts of vitamin B_6, B_{12}, and folic acid, the synthesis of methionine maintains proper levels of homocysteine, which we need for healthy tissues. However, a deficiency of any one of these vitamins such as folic acid can lead to increased homocysteine levels, which may be damaging to the heart. Folic acid is recommended as supplement for pregnant women to avoid folic acid deficiencies during the growth of a fetus.

HEALTH NOTE

Phenylketonurea (PKU)

In the genetic disease phenylketonurea (PKU), a person cannot convert phenylalanine to tyrosine because the gene for an enzyme in the conversion is defective. As a result, large amounts of phenylalanine accumulate. In a different pathway, phenylalanine undergoes transamination to form phenylpyruvate, which is decarboxylated to phenylacetate. Large amounts of these compounds are excreted in the urine.

In infants, high levels of phenylpyruvate and phenylacetate cause severe mental retardation. However, the defect can be identified at birth, and all newborns are now tested for PKU. By detecting PKU early, retardation is avoided by using a diet with proteins that are low in phenylalanine and high in tyrosine. It is also important to avoid the use of sweeteners and soft drinks containing aspartame, which contains phenylalanine as one of two amino acids in its structure. In adulthood, persons with PKU can eat a nearly normal diet as long as they are checked for phenylpyruvate periodically.

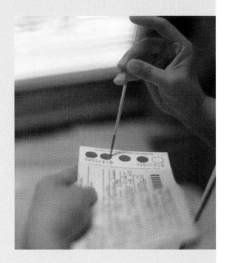

$$\underset{\text{Phenylalanine}}{\bigcirc-CH_2-\overset{\overset{+}{N}H_3}{\underset{|}{CH}}-COO^-} \xrightarrow{\text{Transanimation}} \underset{\text{Phenylpyruvate}}{\bigcirc-CH_2-\overset{\overset{O}{\|}}{C}-COO^-}$$

Phenylalanine hydroxylase

CO_2

$$\underset{\text{Tyrosine}}{HO-\bigcirc-CH_2-\overset{\overset{+}{N}H_3}{\underset{|}{CH}}-COO^-} \qquad \underset{\text{Phenylacetate}}{\bigcirc-CH_2-COO^-}$$

SAMPLE PROBLEM 25.12

Synthesis of Amino Acids

What compound is most often the source of amino groups when transamination is used to synthesize nonessential amino acids?

Solution

Glutamate is the usual source of amino groups in the synthesis of nonessential amino acids.

Study Check

What are the sources of the substrates used to synthesize nonessential amino acids?

Overview of Metabolism

In these chapters, we have seen that catabolic pathways degrade large molecules to small molecules that are used for energy production via the citric acid cycle and the electron transport chain. We have also looked at the anabolic pathways that lead to the synthesis of larger molecules in the cell. In the overall view of metabolism, there are several branch points from which compounds may be degraded for energy or used to synthesize larger molecules. For example, glucose can be degraded to acetyl CoA for the citric acid cycle to produce energy or converted to glycogen for storage. When glycogen stores are depleted, fatty acids are degraded for energy. Amino acids normally used to synthesize nitrogen-containing compounds in the

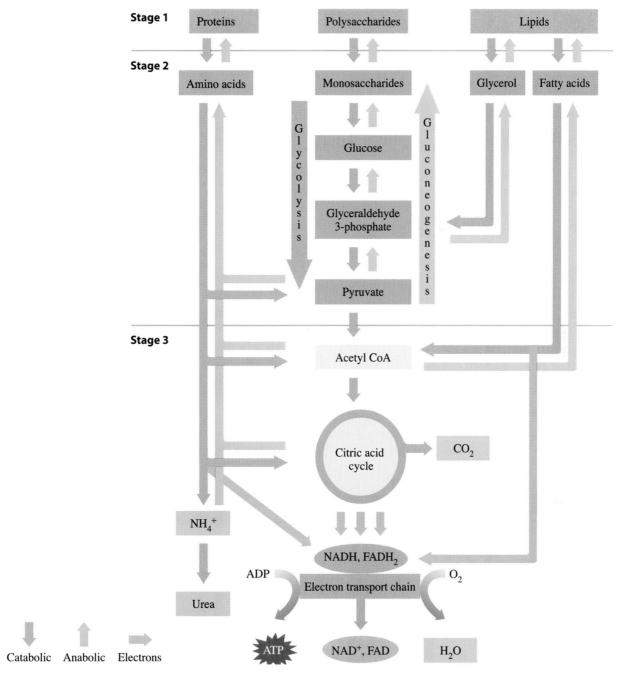

Stage 1

| Proteins | Polysaccharides | Lipids |

Stage 2

| Amino acids | Monosaccharides | Glycerol | Fatty acids |

Glycolysis

Glucose

Gluconeogenesis

Glyceraldehyde 3-phosphate

Pyruvate

Stage 3

Acetyl CoA

Citric acid cycle → CO_2

NH_4^+

NADH, $FADH_2$

ADP   Electron transport chain O_2

Urea

Catabolic   Anabolic   Electrons

ATP NAD^+, FAD H_2O

cells can also be used for energy after they are degraded to intermediates of the citric acid cycle. In the synthesis of nonessential amino acids, α-keto acids of the citric acid cycle enter a variety of reactions that convert them to amino acids through transamination by glutamate. (See Figure 25.12.)

Figure 25.12 Catabolic and anabolic pathways in the cells provide the energy and necessary compounds for the cells.

Q Under what conditions in the cell are amino acids degraded for energy?

QUESTIONS AND PROBLEMS

Synthesis of Amino Acids

25.53 What do we call the amino acids that humans can synthesize?

25.54 How do humans obtain the amino acids that cannot be synthesized in the body?

25.55 How is glutamate converted to glutamine?

25.56 What amino acid is needed for the synthesis of tyrosine?

25.57 What do the letters PKU mean?

25.58 How is PKU treated?

Chapter Review

25.1 Digestion of Triacylglycerols

Triacylglycerols are hydrolyzed in the small intestine to monoacylglycerol and fatty acids, which enter the intestinal wall and form new triacylglycerols. They bind with proteins to form chylomicrons, which transport them through the lymphatic system and bloodstream to the tissues.

25.2 Oxidation of Fatty Acids

When needed as an energy source, fatty acids are linked to coenzyme A and transported into the mitochondria where they undergo β oxidation. The fatty acyl chain is oxidized to yield a shorter fatty acid, acetyl CoA, and the reduced coenzymes NADH and $FADH_2$.

25.3 ATP and Fatty Acid Oxidation

Although the energy from a particular fatty acid depends on its length, each oxidation cycle yields 5 ATP with another 12 ATP from the acetyl CoA that enter the citric acid cycle.

25.4 Ketogenesis and Ketone Bodies

When high levels of acetyl CoA are present in the cell, they enter the ketogenesis pathway forming ketone bodies such as acetoacetate, which cause ketosis and acidosis.

25.5 Fatty Acid Synthesis

When there is an excess of acetyl CoA in the cell, the two-carbon acetyl CoA units link together to synthesize palmitate, which is converted to triacylglycerols and stored in the adipose tissue.

25.6 Digestion of Proteins

The digestion of proteins, which begins in the stomach and continues in the small intestine, involves the hydrolysis of peptide bonds by proteases to yield amino acids that are absorbed through the intestinal wall and transported to the cells.

25.7 Degradation of Amino Acids

When the amount of amino acids in the cells exceeds that needed for synthesis of nitrogen compounds, the process of transamination converts them to α-keto acids and glutamate. Oxidative deamination of glutamate produces ammonium ions and α-ketoglutarate.

25.8 Urea Cycle

Ammonium ions from oxidative deamination combine with bicarbonate and ATP to form carbamoyl phosphate, which is converted to urea.

25.9 Fates of the Carbon Atoms from Amino Acids

The carbon atoms from the degradation of glucogenic amino acids enter the citric acid cycle or glucose synthesis, whereas ketogenic amino acids provide acetyl CoA or acetoacetate for ketogenesis.

25.10 Synthesis of Amino Acids

Nonessential amino acids are synthesized when amino groups from glutamate are transferred to an α-keto acid obtained from glycolysis or the citric acid cycle.

Summary of Key Reactions

Digestion of Triacylglycerols

$$\text{Triacylglycerols} + 2H_2O \xrightarrow{\text{Pancreatic lipase}} \text{monoacylglycerols} + 2 \text{ fatty acids}$$

Mobilization of Fats

$$\text{Triacylglycerols} + 3H_2O \xrightarrow{\text{Lipases}} \text{glycerol} + 3 \text{ fatty acids}$$

Metabolism of Glycerol

Glycerol $+$ ATP $+$ NAD$^+$ $\longrightarrow$ dihydroxyacetone phosphate $+$ ADP $+$ NADH $+$ H$^+$

β Oxidation of Fatty Acid

Myristyl-CoA $+$ 6 CoA $+$ 6 FAD $+$ 6 NAD$^+$ $+$ 6 H$_2$O $\longrightarrow$

7 acetyl CoA $+$ 6 FADH$_2$ $+$ 6 NADH $+$ 6H$^+$

Fatty Acid Synthesis

8 Acetyl CoA $+$ 14 NADPH $+$ 14H$^+$ $+$ 7 ATP $\longrightarrow$

Palmitate (C$_{16}$) $+$ 8 CoA $+$ 14 NADP$^+$ $+$ 7 H$_2$O $+$ 7 ADP $+$ 7 P$_i$

Transamination

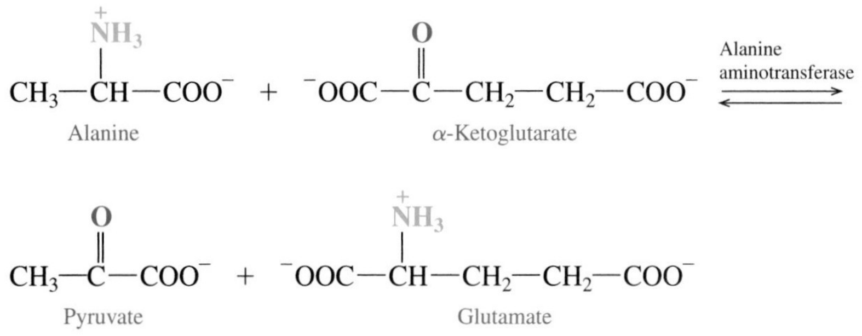

Oxidative Deamination

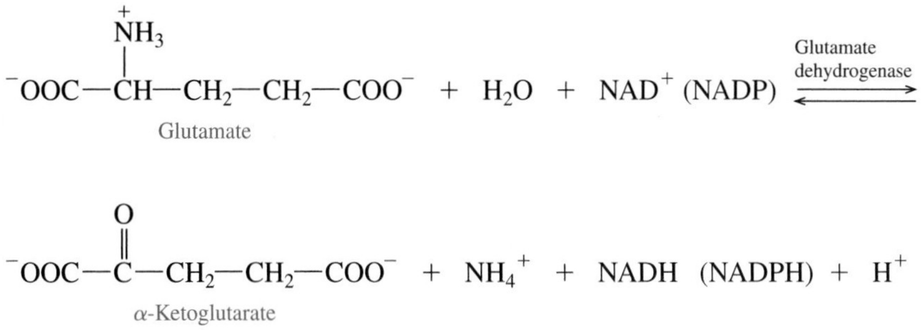

Urea Cycle

NH$_4^+$ $+$ CO$_2$ $+$ 3 ATP $+$ aspartate $+$ 2H$_2$O $\longrightarrow$ Urea $+$ 2 ADP $+$ AMP $+$ 4 P$_i$ $+$ fumarate

Key Terms

acidosis   Low blood pH resulting from the formation of acidic ketone bodies.

adipose tissue   Tissues containing cells in which triacylglycerols are stored.

beta (β) oxidation   The degradation of fatty acids that removes two-carbon segments from an oxidized β carbon in the chain.

chylomicron   Lipoproteins formed by coating triacylglycerols with proteins for transport in the lymph and bloodstream.

essential amino acid   An amino acid that must be obtained from the diet because it cannot be synthesized in the body.

fat mobilization   The hydrolysis of triacylglycerols in the adipose tissue to yield fatty acids and glycerol for energy production.

glucogenic amino acid   An amino acid that provides carbon atoms for the synthesis of glucose.

ketogenesis   The pathway that converts acetyl CoA to four-carbon acetoacetate and other ketone bodies.

ketogenic amino acid   An amino acid that provides carbon atoms for the fatty acid synthesis or ketone bodies.

ketone bodies   The products of ketogenesis: acetoacetate, β-hydroxybutyrate, and acetone.

ketosis   A condition in which high levels of ketone bodies cannot be metabolized, leading to lower blood pH.

lipogenesis   The synthesis of fatty acid in which two-carbon acetyl units link together to yield palmitic acid.

nonessential amino acid   An amino acid that can be synthesized by reactions including transamination of α-keto acids in the body.

oxidative deamination   The loss of ammonium ion when glutamate is degraded to α-ketoglutarate.

protein turnover   The amount of protein that we break down from our diet and utilize for synthesis of proteins and nitrogen-containing compounds.

transamination   The transfer of an amino group from an amino acid to an α-keto acid.

urea cycle   Ammonium ions from the degradation of amino acids and CO_2 form carbamoyl phosphate, which is converted to urea.

Additional Problems

25.59 How are dietary triacylglycerols digested?

25.60 What is a chylomicron?

25.61 Why are the fats in the adipose tissues of the body considered as the major form of stored energy?

25.62 How are fatty acids obtained from stored fats?

25.63 Why doesn't the brain utilize fatty acids for energy?

25.64 Why don't red blood cells utilize fatty acids for energy?

25.65 A triacylglycerol is hydrolyzed in fat cells of adipose tissues and the fatty acid is transported to the liver.
  a. What happens to the glycerol?
  b. Where in the liver cells is the fatty acid activated for β oxidation?
  c. What is the energy cost for activation of the fatty acid?
  d. What is the purpose of activating fatty acids?

25.66 Consider the β oxidation of a saturated fatty acid.
  a. What is the activated form of the fatty acid?
  b. Why is the oxidation called β oxidation?
  c. What reactions in the fatty acid cycle require coenzymes?
  d. What is the yield in ATP for one cycle of β oxidation?

25.67 Lauric acid, $CH_3—(CH_2)_{10}—COOH$, is a saturated fatty acid.
  a. Write the formula of the activated form of lauric acid.

  b. Indicate the α and β carbon atoms in the fatty acyl molecule.
  c. Write the overall equation for the complete β oxidation for lauric acid.
  d. How many acetyl CoA units are produced?
  e. How many cycles of β oxidation are needed?
  f. Account for the total ATP yield from β oxidation of lauric acid (C_{12} fatty acid) by completing the following calculation:

activation	⟶	-2 ATP
___acetyl CoA	⟶	___ATP
___FADH$_2$	⟶	___ATP
___NADH	⟶	___ATP
Total		___ATP

25.68 Calculate the total ATP produced in the complex oxidation of caproic acid, $C_6H_{12}O_2$, with the total ATP produced from the oxidation of glucose, $C_6H_{12}O_6$.

25.69 Identify each of the following as involved in β oxidation or in fatty acid synthesis:
  a. NAD^+
  b. occurs in the mitochondrial matrix
  c. malonyl ACP
  d. cleavage of two-carbon acyl group
  e. acyl protein carrier
  f. acetyl CoA carboxylase

25.70 Identify each of the following as involved in β oxidation or in fatty acid synthesis:
  a. NADPH   **b.** takes place in the cytosol

c. FAD   **d.** oxidation of a hydroxyl group

e. coenzyme A   **f.** hydration of a double bond

25.71 The metabolism of triacylglycerols and carbohydrates is influenced by the hormones insulin and glucagon. Indicate the results of each of the following as stimulating (1) fatty acid oxidation or (2) the synthesis of fatty acids.

     a. low blood glucose   **b.** glucagon secreted

25.72 The metabolism of triacylglycerols and carbohydrates is influenced by the hormones insulin and glucagon. Indicate the results of each of the following as stimulating (1) fatty acid oxidation or (2) the synthesis of fatty acids.

     a. high blood glucose   **b.** insulin secreted

25.73 Why is ammonium ion produced in the liver converted immediately to urea?

25.74 What compound is regenerated in the urea cycle?

25.75 Indicate the corresponding reactant in the urea cycle for each of the following compounds:

     a. aspartate   **b.** ornithine

25.76 Indicate the products in the urea cycle for each of the following compounds:

     a. arginine   **b.** argininosuccinate

25.77 What metabolic substrate(s) can be produced from the carbon atoms of each of the following amino acids?

     a. serine   **b.** lysine

     c. methionine   **d.** glutamate

25.78 What metabolic substrate(s) can be produced from the carbon atoms of each of the following amino acids?

     a. leucine   **b.** isoleucine

     c. cysteine   **d.** phenylalanine

25.79 How much ATP can the degradation product of serine provide?

25.80 A camel hump contains 14 kg of triacylglycerols. Using the value of 0.49 moles of ATP per gram of fat, how many mole of ATP could be produced by the fat in the camel's hump?

Answers

Chapter 1

Answers to Study Checks

1.1 length

1.2 a. 4.25×10^5 m   **b.** 8×10^{-7} g

1.3 exact

1.4 14 inches

1.5 a. two   **b.** one   **c.** five

1.6 a. 36 m   **b.** 0.0026 L   **c.** 3800 g   **d.** 1.3 kg

1.7 a. 0.4924   **b.** 0.0080 or 8.0×10^{-3} **c.** 2.0

1.8 a. 83.70 mg   **b.** 0.5 L

1.9 a. mega   **b.** centi

1.10 a. 1000   **b.** 0.001 or $\frac{1}{1000}$

1.11 Equality: 1 in. = 2.54 cm

Conversion Factors: $\dfrac{1 \text{ in.}}{2.54 \text{ cm}}$ and $\dfrac{2.54 \text{ cm}}{1 \text{ in.}}$

1.12 1.888 L

1.13 0.44 oz

1.14 10 mL

1.15 5.0 mg/dL is hypocalcemia.

1.16 17 students got As

1.17 8.9 g/cm^3

1.18 2.2 g/mL

1.19 Air has a density of 1.29 g/L. Because helium has a density of 0.179 g/L, a helium-filled balloon is less dense than air and floats in air.

1.20 3.4 g of mercury

1.21 0.916

1.22 207 mL

1.23 12°F

1.24 39.8°C

1.25 464 K

1.26 night, −260.°C; day, 410.°C

Answers to Selected Questions and Problems

1.1 In the U.S., weight is measured in pounds (lb), height in feet and inches, gasoline in gallons, and temperature in Fahrenheit (°F). In France, weight is measured in kilograms, height in meters, gasoline in liters, and temperature in Celsius (°C).

1.3 In the metric system, the meter (m) is used in length measurements. In the U.S. system, several units are used including inch (in.), feet (ft), and miles (mi).

1.5 a. meter; length
   b. gram; mass
   c. milliliter; volume
   d. meter; length
   e. Celsius; temperature

1.7 a. 5.5×10^4 m
   b. 4.8×10^2 g
   c. 5×10^{-6} cm
   d. 1.4×10^{-4} s

1.9 a. 7.2×10^3
   b. 3.2×10^{-2}
   c. 1×10^4
   d. 6.8×10^{-2}

1.11 a. 12000
   b. 0.0825
   c. 4 000 000

1.13 a. first decimal place (0.6)
   b. second decimal place (0.05)
   c. first decimal place (0.0)

1.15 The number four was determined by counting the number of chairs. No measuring device was used.

1.17 a. measured   **b.** exact
   c. exact   **d.** measured

1.19 a. not significant   **b.** significant   **c.** significant
   d. significant   **e.** not significant

1.21 a. 5   **b.** 2   **c.** 2
   d. 3   **e.** 4   **f.** 3

1.23 a. 5.0×10^3 L   **b.** 3.0×10^4 g
   c. 1.0×10^5 m   **d.** 2.5×10^{-4} cm

1.25 The precision of the answer is limited by the precision of the measurements used in the calculation.

1.27 An answer from multiplication or division can have only as many significant figures as the measurement that has the smallest number of significant figures.

1.29 a. 1.85   **b.** 184   **c.** 0.00474
   d. 8810   **e.** 1.83

1.31 a. 1.6   **b.** 0.01
   c. 27.6   **d.** 3.5

1.33 a. 53.54 cm   **b.** 127.6 g
   c. 121.5 mL   **d.** 0.50 L

1.35 Changing to the metric system would give a uniform decimal system of measurement that would be used throughout the world. However, the change may be expensive and frustrating to many U.S. citizens.

1.37 km/h is kilometers per hour; mi/h is miles per hour

1.39 The prefix *kilo* means to multiply by 1000. One kg is the same mass as 1000 g.

1.41 a. mg   **b.** dL   **c.** km
   d. kg   **e.** μL

1.43 a. 0.01   **b.** 1000   **c.** 0.001
   d. 0.1   **e.** 1 000 000

1.45 a. 100 cm   **b.** 1000 m
   c. 0.001 m   **d.** 1000 mL

1.47 a. kilogram   **b.** milliliter
   c. km   **d.** kL

1.49 A conversion factor can be inverted to give a second conversion factor.

1.51 1 kg = 1000 g

1.53 a. 3 ft = 1 yd; $\dfrac{3 \text{ ft}}{1 \text{ yd}}$ and $\dfrac{1 \text{ yd}}{3 \text{ ft}}$

   b. 1 mile = 5280 feet; $\dfrac{5280 \text{ ft}}{1 \text{ mi}}$ and $\dfrac{1 \text{ mi}}{5280 \text{ ft}}$

   c. 1 min = 60 sec; $\dfrac{60 \text{ sec}}{1 \text{ min}}$ and $\dfrac{1 \text{ min}}{60 \text{ sec}}$

   d. 1 gal = 27 mi; $\dfrac{1 \text{ gal}}{27 \text{ mi}}$ and $\dfrac{27 \text{ mi}}{1 \text{ gal}}$

1.55 a. 100 cm = 1 m; $\dfrac{100 \text{ cm}}{1 \text{ m}}$ and $\dfrac{1 \text{ m}}{100 \text{ cm}}$

   b. 1000 mg = 1 g; $\dfrac{1000 \text{ mg}}{1 \text{ g}}$ and $\dfrac{1 \text{ g}}{1000 \text{ mg}}$

   c. 1 L = 1000 mL; $\dfrac{1000 \text{ mL}}{1 \text{L}}$ and $\dfrac{1 \text{ L}}{1000 \text{ mL}}$

   d. 1 dL = 100 mL; $\dfrac{100 \text{ mL}}{1 \text{ dL}}$ and $\dfrac{1 \text{ dL}}{100 \text{ mL}}$

1.57 The unit in the denominator must cancel with the preceding unit.

1.59 a. 8.0 yd   **b.** 9.0×10^2 s   **c.** 0.88 gal

1.61 a. 1.75 m   **b.** 5.5 L   **c.** 5.5 g

1.63 a. 710 mL   **b.** 75.0 kg

   c. 495 mm   **d.** 9.5 kg

1.65 a. 66 gal   **b.** 3 tablets   **c.** 1800 mg

1.67 a. 152 g of oxygen   **b.** 0.026 g magnesium

1.69 lead; 11.3 g/mL

1.71 a. 1.20 g/mL   **b.** 4.4 g/mL   **c.** 3.10 g/mL

1.73 a. 210 g   **b.** 575 g   **c.** 62 oz

1.75 a. 1.030   **b.** 1.13   **c.** 0.85 g/mL

1.77 In the U.S. we still use the Fahrenheit temperature scale. In °F, normal body temperature is 98.6. On the Celsius scale, her temperature would be 37.7°C, a mild fever.

1.79 a. 98.6°F   **b.** 18.5°C   **c.** 246 K
   d. 335 K   **e.** 46°C   **f.** 295 K

1.81 a. 41°C
   b. No. The temperature is equivalent to 39°C.

1.83 42 min

1.85 $0.58 per lb

1.87 16 onions

1.89 a. 96 crackers
   b. 0.2 oz of fat
   c. 110 g of sodium

1.91 0.35 oz

1.93 185 mg/dL

1.95 790 g

1.97 720 cm^3

1.99 44 kg water

1.101 a. 1.4 kg (1400 g) body fat
b. 6.2 lb

1.103 a. 1.012 g/mL
b. 5.11 g

Chapter 2

Answers to Study Checks

2.1 Si, S, and Ag

2.2 magnesium, aluminum, and fluorine

2.3 Shiny, liquid, gray color, forms spherical drops if the thermometer breaks.

2.4 a. Period 2: Li, Be, B, C, N, O, F, Ne
b. Group 2A: Be, Mg, Ca, Sr, Ba, Ra

2.5 a. P   **b.** Rn

2.6 nitrogen

2.7 A proton, symbol p or p^+ with a mass of 1 amu and a charge of $1+$, is found in the nucleus of an atom.

2.8 a. 26   **b.** 26   **c.** iron, Fe

2.9 Because the atomic number of silver is 47, it has 47 protons. The mass number is 107, which is the sum of 47 protons and 60 neutrons.

2.10 45 neutrons

2.11 a. $^{15}_{7}N$ **b.** $^{42}_{20}Ca$ **c.** $^{27}_{13}Al$

2.12 10.8 amu

2.13 magnesium

2.14 5A;15; phosphorus

2.15 a. 8 electrons
b. 6 electrons
c. 2 electrons

2.16 a. $1s^2\,2s^2\,2p^6\,3s^2\,3p^6\,4s^2\,3d^7$
or $[Ar]4s^2\,3d^7$

Answers to Selected Questions and Problems

2.1 a. Cu   **b.** Si   **c.** K   **d.** N
e. Fe   **f.** Ba   **g.** Pb   **h.** Sr

2.3 a. carbon   **b.** chlorine   **c.** iodine
d. mercury   **e.** fluorine   **f.** argon
g. zinc   **h.** nickel

2.5 a. sodium, chlorine
b. calcium, sulfur, oxygen
c. carbon, hydrogen, chlorine, nitrogen, oxygen
d. calcium, carbon, oxygen

2.7 a. Period 2   **b.** Group 8A
c. Group 1A   **d.** Period 2

2.9 a. alkaline earth metal   **b.** transition element
c. noble gas   **d.** alkali metal
e. halogen

2.11 a. C   **b.** He   **c.** Na   **d.** Ca   **e.** Al

2.13 a. metal   **b.** nonmetal   **c.** metal
d. nonmetal   **e.** nonmetal   **f.** nonmetal
g. metal   **h.** metal

2.15 a. electron   **b.** proton   **c.** electron   **d.** neutron

2.17 Rutherford determined that an atom contains a small, compact nucleus that is positively charged.

2.19 b.

2.21 In the process of brushing hair, strands of hair become charged with like charges that repel each other.

2.23 a. atomic number   **b.** both
c. mass number   **d.** atomic number

2.25 a. lithium, Li   **b.** fluorine, F   **c.** calcium, Ca
d. zinc, Zn   **e.** neon, Ne   **f.** silicon, Si
g. iodine, I   **h.** oxygen, O

2.27 a. 12   **b.** 30   **c.** 53   **d.** 19

2.29 *See Table 2.17*

Table 2.17

Name of Element	Symbol	Atomic Number	Mass Number	Number of Protons	Number of Neutrons	Number of Electrons
Aluminum	Al	13	27	13	14	13
Magnesium	Mg	12	24	12	12	12
Potassium	K	19	39	19	20	19
Sulfur	S	16	31	16	15	16
Iron	Fe	26	56	26	30	26

2.31 a. 13 protons, 14 neutrons, 13 electrons
b. 24 protons, 28 neutrons, 24 electrons
c. 16 protons, 18 neutrons, 16 electrons
d. 26 protons, 30 neutrons, 26 electrons

2.33 a. $^{31}_{15}P$ **b.** $^{80}_{35}Br$ **c.** $^{27}_{13}Al$ **d.** $^{35}_{17}Cl$

2.35 a. $^{32}_{16}S$ $^{33}_{16}S$ $^{34}_{16}S$ $^{36}_{16}S$
b. They all have the same number of protons and electrons.
c. They have different numbers of neutrons, which gives them different mass numbers.
d. The atomic mass of S listed on the periodic table is the average atomic mass of all the isotopes.

2.37 You need to know the percentage and the mass for each of the isotopes that make up a particular element.

2.39 69.8 amu

2.41 The energy of electrons is of a specific quantity for each energy level. Electrons cannot have energies that are between the energy levels.

2.43 a. 8 **b.** 5 **c.** 8 **d.** 0 **e.** 8

2.45 a. 2, 4 **b.** 2, 8, 8 **c.** 2, 8, 6
d. 2, 8, 4 **e.** 2, 8, 3 **f.** 2, 5

2.47 a. Li **b.** Mg **c.** H
d. Cl **e.** O

2.49 a. gain **b.** emit

2.51 a. B 2,3 Al 2, 8, 3
b. 3
c. Group 3A

2.53 a. $2e^-$, Group 2A **b.** $7e^-$, Group 7A
c. $6e^-$, Group 6A **d.** $5e^-$, Group 5A
e. $2e^-$, Group 2A **f.** $7e^-$, Group 7A

2.55 They each have two electrons in their outer shells.

2.57 a. $2e^-$ **b.** $6e^-$ **c.** $8e^-$ **d.** $2e^-$

2.59 a. $1s^2\ 2s^2\ 2p^6\ 3s^2$ **b.** $1s^2\ 2s^2\ 2p^6\ 3s^2\ 3p^3$
c. $1s^2\ 2s^2\ 2p^6\ 3s^2\ 3p^6$ **d.** $1s^2\ 2s^2\ 2p^6\ 3s^2\ 3p^4$
e. $1s^2\ 2s^2\ 2p^6\ 3s^2\ 3p^5$
f. $1s^2\ 2s^2\ 2p^6\ 3s^2\ 3s^6\ 4s^2\ 3d^{10}$ or [Ar] $4s^2\ 3d^{10}$
g. $1s^2\ 2s^2\ 2p^6\ 3s^2\ 3p^6\ 4s^2\ 3d^{10}\ 4p^6\ 5s^2$ or [Kr] $5s^2$
h. $1s^2\ 2s^2\ 2p^6\ 3s^2\ 3p^6\ 4s^2\ 3d^{10}\ 4p^6\ 5s^2\ 4d^{10}\ 5p^5$ or [Kr] $5s^2\ 4d^{10}\ 5p^5$

2.61 a. H hydrogen **b.** N nitrogen **c.** Na sodium
d. Ne neon **e.** Si silicon **f.** Br bromine

2.63 The first letter of a symbol is a capital, but a second letter is lowercase. The symbol Co is for cobalt, but the symbols in CO are for carbon and oxygen.

2.65 a. Mg, magnesium **b.** Br, bromine
c. Al, aluminum **d.** O, oxygen

2.67 a. The proton is a <u>positive</u> particle.
b. Electrons are found <u>in the outer part of the atom</u>.
c. The nucleus is a <u>very small</u> part of the atom.
d. The <u>electron</u> has a negative charge.
e. Most of the mass of the atom is due to its <u>protons and neutrons</u>.

2.69 a. 26 protons, 30 neutrons, 26 electrons
b. $^{51}_{26}Fe$ **c.** $^{51}_{24}Cr$

2.71 a. An isotope is one kind of atom that makes up a particular element. Most elements are mixtures of two or more isotopes.
b. Atomic number gives the number of protons in an atom; mass number gives the total number of protons and neutrons.

2.73 a. $^{16}_{8}X$, $^{17}_{8}X$, $^{18}_{8}X$ all have 8 protons
b. $^{16}_{8}X$, $^{17}_{8}X$, $^{18}_{8}X$ all isotopes of O
c. $^{16}_{8}X$ and $^{16}_{9}X$ have mass number 16, and $^{18}_{10}X$ and $^{18}_{8}X$ have a mass number of 18.
d. $^{16}_{8}X$ and $^{18}_{10}X$ both have 8 neutrons.

2.75 63.6 amu

2.77 8.09×10^7 sodium atoms

2.79 a. any two elements in Group 2A such as magnesium, calcium, and barium
b. any two elements in Group 6A such as oxygen and sulfur
c. any two elements in Group 8A except helium: neon, argon, krypton, xenon, or radon
d. any two elements in Group 3A such as boron and aluminum

2.81 a. any two elements in Group 7A such as fluorine, chlorine, bromine, or iodine
b. any two elements in Group 8A such as helium, neon, argon, krypton, xenon, or radon
c. any two elements in Group 1A such as lithium, sodium, and potassium, except hydrogen.
d. any two elements in Group 2A such as magnesium, calcium, and barium

2.83 a. 3 **b.** 3 **c.** 10 **d.** 2

2.85 a. Ga **b.** C **c.** Rb
d. Zr **e.** Kr

2.87 a. $1s^2\ 2s^2\ 2p^6\ 3s^2\ 3p^6\ 4s^2\ 3d^{10}\ 4p^3$ or [Ar] $4s^2\ 3d^{10}\ 4p^3$
b. $1s^2\ 2s^2\ 2p^6\ 3s^2\ 3p^6\ 4s^2\ 3d^{10}\ 4p^6\ 5s^1$ or [Kr] $5s^1$
c. $1s^2\ 2s^2\ 2p^6\ 3s^2\ 3p^6\ 4s^2\ 3d^{10}\ 4p^6\ 5s^2\ 4d^{10}\ 5p^6\ 6s^2$ or [Xe] $6s^2$

d. $1s^2\,2s^2\,2p^6\,3s^2\,3p^2$ or [Ne] $3s^2\,3p^2$

e. $1s^2\,2s^2\,2p^6\,3s^2\,3p^6\,4s^2\,3d^{10}\,4p^5$ or [Ar] $4s^2\,3d^{10}\,4p^5$

Chapter 3

Answers to Study Checks

3.1 β or $_{-1}^{0}e$

3.2 Limiting the time one spends near a radioactive source and staying as far away as possible will reduce exposure to radiation.

3.3 $_{84}^{214}\text{Po} \longrightarrow {}_{82}^{210}\text{Pb} + {}_{2}^{4}\text{He}$

3.4 $_{53}^{131}\text{I} \longrightarrow {}_{54}^{131}\text{Xe} + {}_{-1}^{0}e$

3.5 $_{13}^{27}\text{Al} + {}_{2}^{4}\text{He} \longrightarrow {}_{15}^{30}\text{P} + {}_{0}^{1}\text{n}$

3.6 $_{7}^{14}\text{N}$

3.7 5.6×10^{11} iodine-131 atoms

3.8 0.50 g

3.9 17,000 years

3.10 An iodine-131 sample with a higher activity is used in radiation therapy when radiation is needed to destroy some of the cells in the thyroid gland.

3.11 fusion

Answers to Selected Questions and Problems

3.1 a. Both an alpha particle and a helium nucleus have 2 protons and 2 neutrons. However, an α-particle is emitted from a nucleus during radioactive decay.
b. α, $_{2}^{4}\text{He}$

3.3 a. $_{19}^{39}\text{K}$, $_{19}^{40}\text{K}$, $_{19}^{41}\text{K}$
b. They all have 19 protons and 19 electrons, but they differ in the number of neutrons.

3.5

Medical Use	Isotope Symbol	Mass Number	Number of Protons	Number of Neutrons
Heart imaging	$_{81}^{201}\text{Tl}$	201	81	120
Radiation therapy	$_{27}^{60}\text{Co}$	60	27	33
Abdominal scan	$_{31}^{67}\text{Ga}$	67	31	36
Hyperthyroidism	$_{53}^{131}\text{I}$	131	53	78
Leukemia treatment	$_{15}^{32}\text{P}$	32	15	17

3.7 a. α, $_{2}^{4}\text{He}$ **b.** $_{0}^{1}\text{n}$ **c.** β, $_{-1}^{0}e$
d. $_{7}^{15}\text{N}$ **e.** $_{53}^{125}\text{I}$

3.9 a. β or e^{-}
b. α or He
c. n
d. Na
e. C

3.11 a. Because β particles move faster than α particles, they can penetrate further into tissue.
b. Ionizing radiation breaks bonds and forms reactive species that cause undesirable reactions in the cells.
c. X-ray technicians leave the room to increase the distance between them and the radiation. Also a wall that contains lead shields them.
d. Wearing gloves shields the skin from α and β radiation.

3.13 a. $_{84}^{208}\text{Po} \longrightarrow {}_{82}^{204}\text{Pb} + {}_{2}^{4}\text{He}$
b. $_{90}^{232}\text{Th} \longrightarrow {}_{88}^{228}\text{Ra} + {}_{2}^{4}\text{He}$
c. $_{102}^{251}\text{No} \longrightarrow {}_{100}^{247}\text{Fm} + {}_{2}^{4}\text{He}$
d. $_{86}^{220}\text{Rn} \longrightarrow {}_{84}^{216}\text{Po} + {}_{2}^{4}\text{He}$

3.15 a. $_{11}^{25}\text{Na} \longrightarrow {}_{12}^{25}\text{Mg} + {}_{-1}^{0}e$
b. $_{8}^{20}\text{O} \longrightarrow {}_{9}^{20}\text{F} + {}_{-1}^{0}e$
c. $_{38}^{92}\text{Sr} \longrightarrow {}_{39}^{92}\text{Y} + {}_{-1}^{0}e$
d. $_{19}^{42}\text{K} \longrightarrow {}_{20}^{42}\text{Ca} + {}_{-1}^{0}e$

3.17 a. $_{14}^{28}\text{Si}$
b. $_{36}^{87}\text{Kr}$
c. $_{-1}^{0}e$
d. $_{92}^{238}\text{U}$

3.19 a. $_{4}^{10}\text{Be}$ **b.** $_{-1}^{0}e$
c. $_{13}^{27}\text{Al}$ **d.** $_{1}^{1}\text{H}$

3.21 a. When radiation enters the Geiger counter, it ionizes a gas in the detection tube, which produces a burst of current that is detected by the instrument.
b. becquerel (Bq), curie (Ci)
c. gray (Gy), rad
d. 1000 Gy

3.23 a. 2.2×10^{12} disintegrations
b. 294 μCi

3.25 When pilots are flying at high altitudes, there is less atmosphere to protect them from cosmic radiation.

3.27 A half-life is the time it takes for one-half of a radioactive sample to decay.

3.29 a. 40.0 mg   **b.** 20.0 mg
c. 10.0 mg   **d.** 5.00 mg

3.31 128 days, 192 days

3.33 a. Since the elements Ca and P are part of bone, their radioactive isotopes will also become part of the bony structures of the body where their radiation can be used to diagnose or treat bone diseases.

  b. Strontium (Sr) acts much like calcium (Ca) because both are Group 2A elements. The body will accumulate radioactive strontium in bones in the same way that it incorporates calcium. Radioactive strontium is harmful to children because the radiation it produces causes more damage in cells that are dividing rapidly.

3.35 180 μCi

3.37 Nuclear fission is the splitting of a large atom into smaller fragments with the release of large amounts of energy.

3.39 $^{103}_{42}\text{Mo}$

3.41 a. fission **b.** fusion **c.** fission **d.** fusion

3.43 Each has 6 protons and 6 electrons, but carbon-12 has 6 neutrons and carbon-14 has 8 neutrons. Carbon-12 is a stable isotope, but carbon-14 is radioactive and will emit radiation.

3.45 a. α radiation consists of a helium nucleus emitted from the nucleus of a radioisotope. β radiation is an electron and γ radiation is high-energy radiation emitted from the nucleus of a radioisotope.

  b. α particle: $^{4}_{2}\text{He}$
  β particle: $^{0}_{-1}e$
  γ radiation: γ

  c. α particles penetrate 0.05 mm into tissue, β particles 4–5 mm, and γ rays 50 cm or more.

  d. Paper or clothing will shield you from α particles; heavy clothing, lab coats, and gloves will shield you from β particles; and lead and concrete are needed for shielding from γ radiation.

3.47 a. $^{225}_{90}\text{Th} \longrightarrow {}^{221}_{88}\text{Ra} + {}^{4}_{2}\text{He}$
  b. $^{210}_{83}\text{Bi} \longrightarrow {}^{206}_{81}\text{Tl} + {}^{4}_{2}\text{He}$
  c. $^{137}_{55}\text{Cs} \longrightarrow {}^{137}_{56}\text{Ba} + {}^{0}_{-1}e$
  d. $^{126}_{50}\text{Sn} \longrightarrow {}^{126}_{51}\text{Sb} + {}^{0}_{-1}e$

3.49 a. $^{17}_{8}\text{O}$ **b.** $^{1}_{1}\text{H}$ **c.** $^{143}_{54}\text{Xe}$

3.51 a. $^{16}_{8}\text{O} + {}^{16}_{8}\text{O} \longrightarrow {}^{4}_{2}\text{He} + {}^{28}_{14}\text{Si}$
  b. $^{249}_{98}\text{Cf} + {}^{18}_{8}\text{O} \longrightarrow {}^{263}_{106}\text{Sg} + 4\,{}^{1}_{0}\text{n}$
  c. $^{222}_{86}\text{Rn} \longrightarrow {}^{218}_{84}\text{Po} + {}^{4}_{2}\text{He}$
  $^{218}_{84}\text{Po} \longrightarrow {}^{214}_{82}\text{Pb} + {}^{4}_{2}\text{He}$

3.53 14 days

3.55 a. $^{131}_{53}\text{I} \longrightarrow {}^{0}_{-1}e + {}^{131}_{54}\text{Xe}$
  b. 0.375 g
  c. 32 days

3.57 16 μCi

3.59 7.5 mg

3.61 A film badge detects the amount of radiation exposure. The badge is checked periodically to make sure that a technician has not received more than the maximum permissible radiation dose.

3.63 The irradiation of meats, fruits, and vegetables kills bacteria such as *E. coli* that can cause foodborne illnesses. In addition, spoilage is deterred, and shelf life is extended.

3.65 In the fission process, an atom splits into smaller nuclei. In fusion, small nuclei combine (fuse) to form a larger nucleus.

3.67 Fusion occurs naturally in the sun and other stars.

Chapter 4

Answers to Study Checks

4.1 six electrons

4.2 $\cdot\text{Ca}\cdot$

4.3 c. krypton (Kr)

4.4 $35p^+$ and $36e^-$

4.5 K^+ and S^{2-}

4.6 Li $\overset{\cdot\cdot}{\cdot\text{S}\cdot} \longrightarrow$ Li$^+$ $\left[\overset{\cdot\cdot}{:\text{S}:}\right]^{2-} \longrightarrow$ Li$_2$S
  Li $\qquad$ Li$^+$

4.7 CaCl_2

4.8 a. calcium ion **b.** nitride ion
  c. bromide ion

4.9 calcium chloride

4.10 copper(I) ion; cuprous ion

4.11 gold(III) chloride; auric chloride

4.12 Al_2S_3

4.13 H $\overset{\cdot\cdot}{:\text{N}:}$ H
  $\quad\;\;$ H

4.14 Total number of valence electrons: $1e^- + 4e^- + 5e^-$
$= 10\,e^-$

A triple bond gives octets to C and N.   H $:$ C $::\!:$ N $:$
or H—C≡N$:$

4.15 xenon hexafluoride

4.16 IF$_5$

4.17 a. nonpolar covalent
  b. polar covalent H$^{\delta+}$—O$^{\delta-}$

4.18 a. polar covalent   **b.** ionic
  c. nonpolar covalent

4.19 (NH$_4$)$_3$PO$_4$

4.20 calcium phosphate

4.21 iron(II) nitrate

4.22 With four bonded atoms and no lone pairs, the shape would be tetrahedral.

4.23 polar

Answers to Selected Questions and Problems

4.1 The valence electrons are the electrons in the highest energy level of an atom.

4.3 a. $1s^2 2s^2 2p^3$ 5 valence electrons
  b. $1s^2 2s^2 2p^4$ 6 valence electrons
  c. $1s^2 2s^2 2p^6 3s^2 3p^6$ 8 valence electrons
  d. $1s^2 2s^2 2p^6 3s^2 3p^6 4s^1$ 1 valence electron
  e. $1s^2 2s^2 2p^6 3s^2 3p^4$ 6 valence electrons

4.5 a. Group 6A $:\overset{\cdot\cdot}{\underset{\cdot}{S}}\cdot$
  b. Group 5A $\cdot\overset{\cdot\cdot}{\underset{\cdot}{N}}\cdot$
  c. Group 2A $\cdot$Ca$\cdot$
  d. Group 1A   Na$\cdot$
  e. Group 1A   K$\cdot$

4.7 a. M$\cdot$ **b.** $\cdot$M$\cdot$

4.9 They are all in Group 1A; each has 1 valence electron.

4.11 a. By losing 1 valence electron from the third energy level, sodium achieves an octet in the second shell.
  b. The sodium ion Na$^+$ has the same electron arrangement as Ne (2, 8).
  c. Group 1A and 2A elements become stable by losing electrons to form compounds. Group 8A elements are stable.

4.13 a. $:\overset{\cdot\cdot}{\underset{\cdot\cdot}{Ne}}:$ Stable, has a complete valence shell

  b. $:\overset{\cdot\cdot}{O}\cdot$ Needs 2 electrons to complete valence shell
  c. Li$\cdot$ Needs to lose 1 electron to obtain a complete valence shell
  d. $:\overset{\cdot\cdot}{\underset{\cdot\cdot}{Ar}}:$ Stable, has an octet

4.15 a. 1   **b.** 2   **c.** 3   **d.** 1   **e.** 2

4.17 a. neon   **b.** neon   **c.** argon
  d. neon   **e.** neon

4.19 a. lose $2e^-$ **b.** gain $3e^-$ **c.** gain $1e^-$
  d. lose $1e^-$ **e.** lose $3e^-$

4.21 a. Li$^+$ **b.** F$^-$ **c.** Mg^{2+}
  d. Fe^{3+} **e.** Zn^{2+}

4.23 a. Cl$^-$ **b.** K$^+$ **c.** O^{2-} **d.** Al^{3+}

4.25 a. potassium   **b.** sulfide
  c. calcium   **d.** nitride

4.27 a, c

4.29 a. K$\cdot$ $\cdot\overset{\cdot\cdot}{\underset{\cdot\cdot}{Cl}}:$ $\longrightarrow$ K$^+$ $\left[:\overset{\cdot\cdot}{\underset{\cdot\cdot}{Cl}}:\right]^-$ $\longrightarrow$ KCl

  b. $\cdot$Ca$\cdot$ $\cdot\overset{\cdot\cdot}{\underset{\cdot\cdot}{Cl}}:$ $\longrightarrow$ Ca^{2+} $\left[:\overset{\cdot\cdot}{\underset{\cdot\cdot}{Cl}}:\right]^-$ $\longrightarrow$ CaCl$_2$

  $\cdot\overset{\cdot\cdot}{\underset{\cdot\cdot}{Cl}}:$ $\left[:\overset{\cdot\cdot}{\underset{\cdot\cdot}{Cl}}:\right]^-$

  c. Na$\cdot$ Na$^+$
  Na$\cdot$ $\cdot\overset{\cdot}{\underset{\cdot}{N}}\cdot$ $\longrightarrow$ Na$^+$ $\left[:\overset{\cdot\cdot}{\underset{\cdot\cdot}{N}}:\right]^{3-}$ $\longrightarrow$ Na$_3$N
  Na$\cdot$ Na$^+$

4.31 a. Na$_2$O   **b.** AlBr$_3$ **c.** BaO
  d. MgCl$_2$ **e.** Al$_2$S$_3$

4.33 a. Na$_2$S   **b.** K$_3$N   **c.** AlI$_3$ **d.** Li$_2$O

4.35 a. aluminum oxide   **b.** calcium chloride
  c. sodium oxide   **d.** magnesium nitride
  e. potassium iodide

4.37 Most of the transition metals form more than one positive ion. The specific ion is indicated in a name by writing a Roman numeral that is the same as the ionic charge. For example, iron forms Fe^{2+} and Fe^{3+} ions, which are named iron(II) and iron(III).

4.39 a. iron(II) or ferrous   **b.** copper(II) or cupric
  c. zinc   **d.** lead (IV) or plumbic

4.41 a. tin(II) chloride, stannous chloride
  b. iron(II) oxide, ferrous oxide
  c. copper(I) sulfide, cuprous sulfide
  d. copper(II) sulfide, cupric sulfide

4.43 a. Au^{3+} **b.** Fe^{3+} **c.** Pb^{4+} **d.** Sn^{2+}

4.45 a. $MgCl_2$ **b.** Na_2S **c.** Cu_2O
d. Zn_3P_2 **e.** AuN

4.47 a. :Br:Br:

b. H:H

c. H:F:

d. :O:F:
        :F:

4.49 a. H:C:::C:H

b. H:C:H
        with :O: above and below

c. :O:S::O: with :O: above

4.51 a. phosphorus tribromide   **b.** carbon tetrabromide
c. silicon dioxide   **d.** hydrogen fluoride
e. nitrogen triiodide

4.53 a. dinitrogen trioxide
b. nitrogen trichloride
c. silicon tetrabromide
d. phosphorus pentachloride
e. sulfur trioxide

4.55 a. CCl_4 **b.** CO **c.** PCl_3 **d.** N_2O_4

4.57 a. OF_2 **b.** BF_3 **c.** N_2O_3 **d.** SF_6

4.59 a. nonpolar covalent   **b.** none
c. nonpolar covalent   **d.** ionic
e. polar covalent

4.61 a. polar covalent   **b.** ionic
c. polar covalent   **d.** nonpolar covalent
e. polar covalent

4.63 a. F   **b.** F   **c.** Cl   **d.** Br   **e.** Cl

4.65 a. $\overset{\delta^+\ \ \delta^-}{H-F}$

b. $\overset{\delta^+\ \ \delta^-}{C-Cl}$

c. $\overset{\delta^+\ \ \delta^-}{N-O}$

d. $\overset{\delta^+\ \ \delta^-}{N-F}$

4.67 a. HCO_3^- **b.** NH_4^+
c. PO_4^{3-} **d.** HSO_4^-

4.69 a. sulfate   **b.** carbonate
c. phosphate   **d.** nitrate

4.71

	OH^-	NO_2^-	CO_3^{2-}	HSO_4^-	PO_4^{3-}
Li^+	LiOH	$LiNO_2$	Li_2CO_3	$LiHSO_4$	Li_3PO_4
Cu^{2+}	$Cu(OH)_2$	$Cu(NO_2)_2$	$CuCO_3$	$Cu(HSO_4)_2$	$Cu_3(PO_4)_2$
Ba^{2+}	$Ba(OH)_2$	$Ba(NO_2)_2$	$BaCO_3$	$Ba(HSO_4)_2$	$Ba_3(PO_4)_2$

4.73 a. CO_3^{2-}, sodium carbonate
b. NH_4^+, ammonium chloride
c. PO_4^{3-}, lithium phosphate
d. NO_2^-, copper(II) nitrite
e. SO_3^{2-}, iron(II) sulfite

4.75 a. $Ba(OH)_2$ **b.** Na_2SO_4 **c.** $Fe(NO_3)_2$
d. $Zn_3(PO_4)_2$ **e.** $Fe_2(CO_3)_3$

4.77 a. aluminum sulfate   **b.** calcium carbonate
c. dinitrogen oxide   **d.** sodium phosphate
e. ammonium sulfate   **f.** iron(III) oxide

4.79 a. linear   **b.** pyramidal

4.81 The four electron groups in PCl_3 have a tetrahedral arrangement, but three bonded atoms around a central atom give a pyramidal shape.

4.83 The electron-dot structure of BH_3 has three bonded atoms and no lone pairs, which gives BH_3 the shape trigonal planar. In the molecule NH_3, there are four electrons pairs of which three are bonded to atoms and one is a lone pair. The shape of NH_3 is pyramidal.

4.85 a. linear   **b.** bent   **c.** linear   **d.** tetrahedral

4.87 a. $\left[:\overset{..}{O}: \ \ :\overset{..}{O}:\overset{..}{C}::\overset{..}{O}: \right]^{2-}$ trigonal planer

b. $\left[\begin{matrix} :\overset{..}{O}: \\ :\overset{..}{O}:\overset{..}{S}:\overset{..}{O}: \\ :\overset{..}{O}: \end{matrix} \right]^{2-}$

tetrahedral

4.89 Cl_2 is a nonpolar molecule because there is a nonpolar covalent bond between Cl atoms, which have identical electronegativity values. In HCl, the bond is a polar bond, which makes HCl a polar molecule.

4.91
a. $\overset{\delta^+\ \ \delta^-}{Cl-Cl}$ **b.** $\overset{\delta^+\ \ \delta^-}{C-Cl}$ **c.** $\overset{\delta^++\delta^-}{N-O}$

d. $\overset{\delta^+\ \ \delta^-}{H-O}$ **e.** $\overset{\delta^+\ \delta^-}{P-F}$ **f.** $\overset{\delta^-\ \delta^+}{S-H}$

4.93 a. nonpolar   **b.** polar
     c. nonpolar   **d.** nonpolar

4.95 a. P   **b.** Na   **c.** Al   **d.** Si

4.97 a. $1s^2 2s^2 2p^6$
     b. $1s^2 2s^2 2p^6$
     c. $1s^2 2s^2 2p^6$
     d. $1s^2 2s^2 2p^6$
     e. $1s^2$

4.99 a. Group 2A
     b. $\cdot X \cdot$
     c. Be

4.101 a. Sn^{4+} **b.** 50 protons, 46 electrons
     c. SnO_2 **d.** $Sn_3(PO_4)_4$

4.103 a. ionic   **b.** covalent   **c.** covalent
     d. covalent   **e.** ionic   **f.** covalent
     g. ionic   **h.** ionic

4.105 a. iron(III) chloride   **b.** dichlorine heptoxide
     c. bromine   **d.** calcium phosphate
     e. phosphorus trichloride   **f.** aluminum carbonate
     g. lead(IV) chloride   **h.** magnesium carbonate
     i. nitrogen dioxide   **j.** tin(II) sulfate
     k. barium nitrate   **l.** copper(II) sulfide

4.107 a. trigonal planar, nonpolar   **b.** bent, polar
     c. linear, nonpolar

4.109 a. bent, polar   **b.** pyramidal, polar
     c. trigonal planar, nonpolar

4.111 a. trigonal planar
     b. tetrahedral
     c. tetrahedral

Chapter 5

Answers to Study Checks

5.1 potential energy

5.2 140 cal

5.3 350 J

5.4 180 g

5.5 4.0 kcal/g

5.6 380 kcal

5.7 liquid

5.8 H_2O forms hydrogen bonds.

5.9 40. kJ

5.10 Gas changes to liquid.

5.11 14 kcal

5.12 18,000 cal

Answers to Selected Questions and Problems

5.1 When the car is at the top of the ramp, it has its maximum potential energy. As it descends, potential energy changes to kinetic energy. At the bottom, all the energy is kinetic.

5.3 a. potential   **b.** kinetic
     c. potential   **d.** potential

5.5 a. Electrical energy changes to heat.
     b. Electrical energy changes to kinetic energy.
     c. Chemical energy changes to kinetic energy.
     d. Radiant energy changes to heat.

5.7 Copper has the lowest specific heat of the samples and will reach the highest temperature.

5.9 a. 3.5 kcal   **b.** 120 J
     c. 102 cal   **d.** 1100 cal

5.11 a. 250 cal   **b.** 47,000 J
      c. 980 J   **d.** 39 kJ
      e. 530 J

5.13 a. 1200 g   **b.** 89 g
      c. 0.13 cal/g °C   **d.** 0.19 cal/g °C

5.15 a. 5. Cal   **b.** 210 Cal   **c.** 25 Cal

5.17 a. 110 Cal   **b.** 18 g
      c. 130 Cal   **d.** 280 Cal

5.19 210 kcal; 880 kJ

5.21 a. gas   **b.** gas   **c.** solid

5.23 a. dipole–dipole   **b.** ionic   **c.** dispersion
      d. hydrogen bond   **e.** dispersion

5.25 a. HF; hydrogen bonds are stronger than dipole–dipole interactions of HBr.
      b. NaF; ionic bonds are stronger than the hydrogen bonds in HF.
      c. $MgBr_2$; ionic bonds are stronger than the dipole–dipole interactions in PBr_3.
      d. C_4H_{10} has more electrons and therefore more dispersion forces than CH_4.

5.27 b. freezing   **d.** melting

5.29 a. melting   **b.** melting
      c. sublimation   **d.** freezing

5.31 a. 5200 cal absorbed   **b.** 5700 J absorbed
      c. 18 kcal released   **d.** 17 kJ released

5.33 a. condensation   **b.** evaporation
      c. boiling   **d.** condensation

5.35 a. The heat from the skin is used to evaporate the water (perspiration). Therefore the skin is cooled.
      b. On a hot day, there are more molecules with sufficient energy to become water vapor.
      c. In a closed bag, some molecules evaporate but they cannot escape and will condense back to liquid; the clothes will not dry.

5.37 a. 5400 cal absorbed   **b.** 110,000 J absorbed
      c. 4300 kcal released   **d.** 380 kJ released

5.39

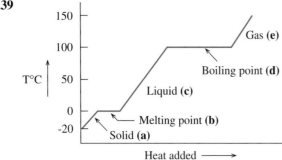

5.41 a. 1100 cal   **b.** 7300 cal
      c. 40. kJ   **d.** 17 kcal

5.43 26 kcal; 110 kJ

5.45 In Table 5.1, sand has a lower specific heat than water. When both substances absorb the same amount of heat, the final temperature of the sand will be higher than that of water.

5.47 When water vapor condenses or liquid water freezes, heat is released, which warms the air.

5.49 60°C

5.51 a. 45 g protein, 140 g carbohydrate, 53 g fat
      b. 71 g protein, 210 g carbohydrate, 84 g fat
      c. 98 g protein, 290 g carbohydrate, 120 g fat

5.53 3500 kcal, 15,000 kJ

5.55 85 kJ

5.57 a. HF   **b.** H_2O
      c. KCl   **d.** NH_3

5.59 In nonpolar compounds, there are no dipoles. However, they do interact by dispersion forces from temporary dipoles that result from a momentary shift of electrons. Because octane C_8H_{18} has more electrons than ethane C_2H_6, octane can form larger temporary dipoles, which results in a higher melting point.

5.61 a. 3   **b.** 3
      c. 4   **d.** 2
      e. 3   **f.** 1

5.63 320 g of ice melts

5.65 68 kcal

Chapter 6

Answers to Study Checks

6.1 a. physical   **b.** chemical   **c.** physical

6.2 There are 4 carbon atoms, 12 hydrogen atoms, and 14 oxygen atoms on the reactant side of the equation, and on the product side.

6.3 a. $3Fe + 2O_2 \longrightarrow Fe_3O_4$
      b. $C_3H_8 + 5O_2 \longrightarrow 3CO_2 + 4H_2O$

6.4 $2NO + O_2 \longrightarrow 2NO_2$ Combination reaction

6.5 Lithium is oxidized: $2Li \longrightarrow 2Li^+ + 2e^-$
     Fluorine is reduced: $F_2 + 2e^- \longrightarrow 2F^-$

6.6 Iron is oxidized: $2Fe \longrightarrow 2Fe^{3+} + 6e^-$
     Bromine is reduced: $3Br_2 + 6e^- \longrightarrow 6Br^-$

6.7 $2C_2H_5OH + 6O_2 \longrightarrow 4CO_2 + 6H_2O + 326\,kcal$

6.8 The rate of reaction will decrease because the number of collisions between reacting particles will be less, and a smaller number of the collisions that do occur will have sufficient activation energy.

6.9 Removing H_2 favors the reactants; the equilibrium shifts to the reactants.

Answers to Selected Questions and Problems

6.1 a. physical   **b.** chemical   **c.** physical
      d. chemical   **e.** physical   **f.** chemical

6.3 a. Reactant side: 2 N atoms, 4 O atoms
           Product side: 2 N atoms, 4 O atoms
      b. Reactant side: 5 C atoms, 2 S atoms, 4 O atoms
           Product side: 5 C atoms, 2 S atoms, 4 O atoms
      c. Reactant side: 4 C atoms, 4 H atoms, 10 O atoms
           Product side: 4 C atoms, 4 H atoms, 10 O atoms

d. Reactant side: 2 N atoms, 8 H atoms, 4 O atoms
  Product side: 2 N atoms, 8 H atoms, 4 O atoms

6.5 **a.** not balanced   **b.** balanced
  c. not balanced   **d.** balanced

6.7 **a.** 2 Na atoms, 2 Cl atoms
  b. 1 P atom, 3 Cl atoms, 6 H atoms
  c. 4 P atoms, 16 O atoms, 12 H atoms

6.9 **a.** $N_2 + O_2 \longrightarrow 2NO$
  b. $2HgO \longrightarrow 2Hg + O_2$
  c. $4Fe + 3O_2 \longrightarrow 2Fe_2O_3$
  d. $2Na + Cl_2 \longrightarrow 2NaCl$
  e. $2Cu_2O + O_2 \longrightarrow 4CuO$

6.11 **a.** $Mg + 2AgNO_3 \longrightarrow Mg(NO_3)_2 + 2Ag$
  b. $CuCO_3 \longrightarrow CuO + CO_2$
  c. $2Al + 3CuSO_4 \longrightarrow 3Cu + Al_2(SO_4)_3$
  d. $Pb(NO_3)_2 + 2NaCl \longrightarrow PbCl_2 + 2NaNO_3$
  e. $2Al + 6HCl \longrightarrow 2AlCl_3 + 3H_2$

6.13 **a.** A single reactant splits into two simpler substances (elements).
  b. One element in the reacting compound is replaced by the other reactant.

6.15 **a.** combination; combustion
  b. single replacement
  c. decomposition
  d. double replacement
  e. combustion
  f. decomposition
  g. double replacement
  h. combination; combustion

6.17 **a.** $Mg + Cl_2 \longrightarrow MgCl_2$
  b. $2HBr \longrightarrow H_2 + Br_2$
  c. $Mg + Zn(NO_3)_2 \longrightarrow Mg(NO_3)_2 + Zn$
  d. $K_2S + Pb(NO_3)_2 \longrightarrow PbS + 2KNO_3$
  e. $2C_2H_6 + 7O_2 \longrightarrow 4CO_2 + 6H_2O$

6.19 **a.** oxidation   **b.** reduction
  c. oxidation   **d.** reduction

6.21 **a.** Zn is oxidized; Cl_2 is reduced.
  b. Br^- in NaBr is oxidized; Cl_2 is reduced.
  c. The O^{2-} in PbO is oxidized; the Pb^{2+} is reduced.
  d. Sn^{2+} is oxidized; Fe^{3+} is reduced.

6.23 **a.** reduction   **b.** oxidation

6.25 Linoleic acid gains hydrogen atoms and is reduced.

6.27 **a.** The energy of activation is the energy required to break the bonds of the reacting molecules.
  b. A catalyst provides a pathway that lowers the activation energy and speeds up a reaction.

c. In exothermic reactions, the energy of the products is lower than the reactants.

d.

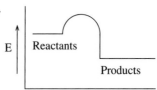

6.29 **a.** exothermic   **b.** endothermic   **c.** exothermic

6.31 **a.** exothermic   **b.** endothermic   **c.** exothermic

6.33 **a.** The rate of a reaction tells how fast the products are formed.
  b. Reactions go faster at higher temperatures.

6.35 **a.** increase   **b.** increase
  c. increase   **d.** decrease

6.37 Reversible reactions are reactions that proceed in both the forward and reverse directions.

6.39 **a.** Products are favored.
  b. Reactants are favored.
  c. Products are favored.
  d. Reactants are favored.

6.41 **a.** Products are favored.
  b. Reactants are favored.
  c. Products are favored.
  d. Reactants are favored.

6.43 The escape of CO_2 gas removes CO_2; equilibrium favors the formation of products. The soda no longer bubbles.

6.45 **a.** $2Al + 3Cl_2 \longrightarrow 2AlCl_3$ Combination
  b. $2Fe + 3H_2SO_4 \longrightarrow Fe_2(SO_4)_3 + 3H_2$
      Single replacement
  c. $2AgNO_3 + H_2S \longrightarrow Ag_2S + 2HNO_3$
      Double replacement
  d. $Cl_2 + 2KI \longrightarrow 2KCl + I_2$
      Single replacement

6.47 **a.** $C_4H_8 + 6O_2 \longrightarrow 4CO_2 + 4H_2O$
  b. $C_5H_{12} + 8O_2 \longrightarrow 5CO_2 + 6H_2O$
  c. $C_6H_{12}O_6 + 6O_2 \longrightarrow 6CO_2 + 6H_2O$

6.49 **a.** reduction   **b.** oxidation
  c. oxidation   **d.** reduction

6.51 **a.** S is oxidized: $S \longrightarrow S^{2+} + 2e^-$
      Cl is reduced: $Cl_2 + 2e^- \longrightarrow 2Cl^-$
  b. Br^- is oxidized; $2Br^- \longrightarrow Br_2 + 2e^-$
      Cl is reduced: $Cl_2 + 2e^- \longrightarrow 2Cl^-$

c. Al is oxidized: $Al \longrightarrow Al^{3+} + 3e^-$
Fe^{3+} is reduced: $Fe^{3+} + 3e^- \longrightarrow Fe$
d. Cu^{2+} is reduced: $Cu^{2+} + 2e^- \longrightarrow Cu$
C is oxidized: $C \longrightarrow C^{4+} + 4e^-$
(in CO_2, $2 \times O^{2-}$ means that C is 4+)

6.53 $N_2(g) + 2O_2(g) + heat \rightleftharpoons 2NO_2(g)$
a. Products are favored.
b. Reactants are favored.
c. Products are favored.
d. Products are favored.
e. Reactants are favored.

6.55 An increase in hemoglobin molecules favors the products, which provides more oxygenated hemoglobin in the blood.

Chapter 7

Answers to Study Check

7.1 3.0×10^{23} molecules CO_2

7.2 6 moles of C, 8 moles of H, 6 moles of O

7.3 138.0 g

7.4 24.4 g Au

7.5 62 g $(NH_4)_2CO_3$

7.6 0.00621 mole of $CaCO_3$, 0.00550 mole of $MgCO_3$

7.7 56.6% K, 8.7% C, 34.7% O

7.8 Na_2SO_4

7.9 K_2SO_4

7.10 Fe_2O_3

7.11 The total mass of the reactants is 164.0 g (68.0 g NH_3 + 96.0 g O_2), which equals the total mass of the products of 164.0 g (56.0 g N_2 + 108.0 g H_2O).

7.12 $\dfrac{2 \text{ moles } Na_2O}{1 \text{ mole } O_2}$ and $\dfrac{1 \text{ mole } O_2}{2 \text{ moles } Na_2O}$

7.13 1.8 moles of Fe

7.14 2.5 moles of O_2

7.15 15 g NO

7.16 27.5 g CO_2

7.17 9.17 g C_3H_8

7.18 77.7%

7.19 $2NO(g) + O_2(g) \rightleftharpoons 2NO_2(g)$

7.20 A K_{eq} of about 1 indicates that the reactants and products are about equally favored.

Answers to Selected Questions and Problems

7.1 1.00 mole contains 6.02×10^{23} atoms of an element, molecules of a covalent substance, or formula units of an ionic substance.

7.3 **a.** There are 6.02×10^{23} C atoms in 1.00 mole of C.
b. 1.2×10^{24} Fe atoms
c. 6.0×10^{22} CO_2 molecules

7.5 **a.** 24 moles of H
b. 6.0×10^{25} atoms of C
c. 2.4×10^{22} atoms of N

7.7 **a.** 1.5×10^{24} atoms of C
b. 3.6×10^{24} atoms of Cl
c. 2.8 moles of SO_3

7.9 **a.** 58.5 g   **b.** 159.8 g   **c.** 73.8 g
d. 342.3 g   **e.** 58.3 g   **f.** 365.1 g

7.11 **a.** 46.0 g   **b.** 112 g   **c.** 14.8 g

7.13 **a.** 29.3 g   **b.** 109 g   **c.** 4.05 g

7.15 **a.** 5.87 g   **b.** 10.7 g   **c.** 15.9 g

7.17 **a.** 602 g   **b.** 11g

7.19 **a.** 0.463 mole   **b.** 0.0167 mole
c. 0.882 mole   **d.** 1.17 moles

7.21 **a.** 0.568 mole   **b.** 0.321 mole   **c.** 0.262 mole

7.23 12 moles

7.25 **a.** 0.78 mole S   **b.** 1.95 moles S   **c.** 5.49 moles S

7.27 **a.** 1.72×10^{24} atoms of N
b. 1.8×10^{24} atoms of N
c. 2.4×10^{24} atoms of N

7.29 **a.** 39.0% Mg and 61.0% F
b. 54.1% Ca, 43.2% O, and 2.7% H
c. 40.0% C, 6.7% H, and 53.3% O

7.31 **a.** N_2O **b.** CH_3 **c.** HNO_3

7.33 **a.** K_2S **b.** GaF_3 **c.** B_2O_3

7.35 **a. (1)** Two molecules of sulfur dioxide react with 1 molecule of oxygen to produce 2 molecules of sulfur trioxide.
a. (2) Two moles of sulfur dioxide react with 1 mole of oxygen to produce 2 moles of sulfur trioxide.

b. (1) Four atoms of phosphorus react with 5 molecules of oxygen to produce 2 molecules of diphosphorus pentoxide.

b. (2) Four moles of phosphorus react with 5 moles of oxygen to produce 2 moles of diphosphorus pentoxide.

7.37 a. 160.2 g reactants = 160.2 g products
b. 284 g reactants = 284 g products

7.39 a. $\dfrac{2 \text{ moles } SO_2}{1 \text{ mole } O_2}$ and $\dfrac{1 \text{ mole } O_2}{2 \text{ moles } SO_2}$

$\dfrac{2 \text{ moles } SO_2}{2 \text{ moles } SO_3}$ and $\dfrac{2 \text{ moles } SO_3}{2 \text{ moles } SO_2}$

$\dfrac{2 \text{ moles } SO_3}{1 \text{ mole } O_2}$ and $\dfrac{1 \text{ mole } O_2}{2 \text{ moles } SO_3}$

b. $\dfrac{4 \text{ moles } P}{5 \text{ moles } O_2}$ and $\dfrac{5 \text{ moles } O_2}{4 \text{ moles } P}$

$\dfrac{4 \text{ moles } P}{2 \text{ moles } P_2O_5}$ and $\dfrac{2 \text{ moles } P_2O_5}{4 \text{ moles } P}$

$\dfrac{5 \text{ moles } O_2}{2 \text{ moles } P_2O_5}$ and $\dfrac{2 \text{ moles } P_2O_5}{5 \text{ moles } O_2}$

7.41 a. 1.0 mole O_2 **b.** 10. moles of H_2
c. 5.0 moles H_2O

7.43 a. 1.25 moles C   **b.** 0.96 mole CO
c. 1.0 mole SO_2 **d.** 0.50 mole CS_2

7.45 a. 78 g Na_2O **b.** 6.26 g O_2 **c.** 19.4 g O_2

7.47 a. 190 g O_2 **b.** 3.79 g N_2 **c.** 54 g H_2O

7.49 a. 70.9%   **b.** 63.2%

7.51 70.8g Al_2O_3

7.53 a. $K_{eq} = \dfrac{[CS_2][H_2]^4}{[CH_4][H_2S]^2}$

b. $K_{eq} = \dfrac{[N_2][O_2]}{[NO]^2}$

c. $K_{eq} = \dfrac{[CS_2][O_2]^4}{[SO_3]^2[CO_2]}$

7.55 a. Products   **b.** Reactants

7.57 a. 40.3% K, 26.8% Cr, and 33.0% O
b. 12.9% Al, 1.4% H, 17.1% C, and 68.6% O
c. 40.0% C, 6.7% H, and 53.3% O

7.59 a. SF_6 **b.** $AgNO_3$ **c.** Au_2O_3

7.61 2330 moles of water

7.63 a. 1.35 moles of glucose
b. 123 g of ethanol

7.65 $2NH_3 + 5F_2 \longrightarrow N_2F_4 + 6HF$
a. 1.33 moles of NH_3 and 3.33 moles of F_2
b. 143 g F_2
c. 10.4 g of N_2F_4

7.67 81.4%

Chapter 8

Answers to Study Checks

8.1 The gas molecules that carry the odor of the food move throughout the house until they reach the room that you are in.

8.2 The mass in grams gives the amount of gas.

8.3 0.862 atm

8.4 245 torr

8.5 50.0 mL

8.6 560 mL

8.7 16°C

8.8 241 mm Hg

8.9 7.50 L

8.10 7.0 g N_2

8.11 20. g Cl_2

8.12 1.24 atm

8.13 752 torr

Answers to Selected Questions and Problems

8.1 a. At a higher temperature, gas particles have greater kinetic energy, which makes them move faster.
b. Because there are great distances between the particles of a gas, they can be pushed closer together and still remain a gas.
c. Gas particles are very far apart, which means that the mass of a gas in a certain volume is very small, resulting in a low density.

8.3 a. temperature   **b.** volume
c. amount   **d.** pressure

8.5 atmospheres (atm), mm Hg, torr, lb/in.², pascals

8.7 a. 1520 torr   **b.** 29.4 lb/in.²   **c.** 1520 mm Hg

8.9 As a diver ascends to the surface, external pressure decreases. If the air in the lungs were not exhaled, its volume would expand and severely damage the lungs. The pressure in the lungs must adjust to changes in the external pressure.

8.11 In order to exhale (expiration) air containing CO_2, the pressure in the lungs must increase, which is done when the volume of the lungs is decreased as the diaphragm relaxes.

8.13 a. The pressure is greater in cylinder A. According to Boyle's law, a decrease in volume pushes the gas particles closer together, which will cause an increase in the pressure.
b.

Property	Initial	Final
Pressure (P)	650 mm Hg	1.2 atm
Volume (V)	220 mL	160 mL

8.15 a. The pressure doubles.
b. The pressure falls to one-third the initial pressure.
c. The pressure increases to ten times the original pressure.

8.17 a. 328 mm Hg   **b.** 2620 mm Hg
c. 4400 mm Hg   **d.** 55,000 mm Hg

8.19 1700 mm Hg

8.21 a. 25 L   **b.** 25 L   **c.** 100. L   **d.** 45 L

8.23 25 L

8.25 a. inspiration   **b.** expiration   **c.** inspiration

8.27 a. C   **b.** A   **c.** B

8.29 When a gas is heated at constant pressure, the volume of the gas increases to fill the hot air balloon.

8.31 a. 2400 mL   **b.** 4900 mL
c. 1800 mL   **d.** 1700 mL

8.33 An increase in temperature increases the pressure inside the can. When the pressure exceeds the pressure limit of the can, it explodes.

8.35 When the temperature increases, the pressure of the oxygen inside the filled tanks increases causing the tanks to explode.

8.37 a. 770 torr   **b.** 1.51 atm

8.39 a. boiling point   **b.** vapor pressure
c. atmospheric pressure   **d.** boiling point

8.41 a. On top of a mountain, water boils below 100°C because the atmospheric (external) pressure is less than 1 atm.
b. Because the pressure inside a pressure cooker is greater than 1 atm, water boils above 100°C. At a higher temperature, food cooks faster.

8.43 $\dfrac{P_1 V_1}{T_1} = \dfrac{P_2 V_2}{T_2}$
Boyle's, Charles', and Gay-Lussac's laws are combined to make this law.

8.45 a. 4.25 atm   **b.** 3.07 atm
c. 0.605 atm

8.47 110 mL

8.49 The volume increases because the amount of gas particles is increased.

8.51 a. 4.00 L
b. 14.6 L
c. 26.7 L

8.53 a. 2.00 moles of O_2 **b.** 0.179 mole of CO_2
c. 4.48 L   **d.** 55 400 mL

8.55 4.93 atm

8.57 29.4 g O_2

8.59 565 K (292°C)

8.61 In a gas mixture, the pressure that each gas exerts as part of the total pressure is called the partial pressure of that gas. Because the air sample is a mixture of gases, the total pressure is the sum of the partial pressures of each gas in the sample.

8.63 765 torr

8.65 425 torr

8.67 a. The partial pressure of oxygen will be lower than normal.
b. Breathing a higher concentration of oxygen will help to increase the supply of oxygen in the lungs and blood and raise the partial pressure of oxygen in the blood.

8.69 The increase in temperature increases the motion of the gas particles. They hit the walls with more force, which creates more pressure.

8.71 a. The volume of the chest and lungs is decreased.
b. The decrease in volume increases the pressure, which can dislodge the food in the trachea.

8.73 12 atm

8.75 $-223°C$

8.77 He 600 torr, O_2 1800 torr

8.79 370 torr

8.81 1.43 g/L

8.83 **a.** The P_{O_2} is high in the lungs and low in the blood coming to the lungs.
  b. The P_{O_2} is high in arterial blood and low in venous blood.
  c. The P_{CO_2} is high in the tissues and low in arterial blood.
  d. The P_{CO_2} is high in the venous blood and low in the lungs.

8.85 44.8 L CO_2

8.87 3.7 L O_2

8.89 1.1×10^{24} molecules of CO_2

8.91 4.4 g

8.93 2170 mL

8.95 42.1 g/mole

8.97 8.56 L of H_2 (g)

8.99 **a.** 16 L of $O_2(g)$ **b.** 1.02 g of $NH_3(g)$

8.101 3.4 L of $O_2(g)$

Chapter 9

Answers to Study Checks

9.1 Within liquid water, the water molecules form hydrogen bonds in all directions with surrounding water molecules. However, the water molecules on the surface can only hydrogen bond in a downward direction, which pulls them more tightly together.

9.2 Iodine is the solute, and ethyl alcohol is the solvent.

9.3 Yes. Both the solute and solvent are nonpolar substances; "like dissolves like."

9.4 $AlCl_3$ • $6H_2O$ is 44.6% by mass water.

9.5 A solution of a weak electrolyte will contain both molecules and ions.

9.6 71 g Cl^-

9.7 6.88 g Cl^-

9.8 57 g $NaNO_3$

9.9 The value of 65 g/100 g H_2O is more likely because the solubility of most solids increases when the temperature increases.

9.10 **a.** No; $PbCl_2$ is an insoluble chloride
  b. Yes; a salt containing K^+ ion is soluble
  c. No; $FeCO_3$ is insoluble

9.11 3.4% (m/m) NaCl solution

9.12 4.8% (m/v) Br_2 in CCl_4

9.13 30 g NaCl

9.14 2.1 M KNO_3

9.15 4.5 moles of HCl

9.16 123 g $NaHCO_3$

9.17 250 mL

9.18 colloids

9.19 5% glucose

9.20 The red blood cell will shrink (crenate).

Answers to Selected Questions and Problems

9.1 **a.**

The oxygen in the water molecule has a partial negative charge and the hydrogen atoms have partial positive charges.

  b.

9.3 **a.** NaCl, solute; water, solvent
  b. water, solute; ethanol, solvent
  c. oxygen, solute; nitrogen, solvent

9.5 **a.** water   **b.** CCl_4
  c. water   **d.** CCl_4

9.7 The polar water molecules pull the K^+ and I^- ions away from the solid and into solution, where they are hydrated.

9.9 Na_2SO_4 • 10 H_2O; 55.9% by mass water.

9.11 In a solution of KF, only the ions of K^+ and F^- are present in the solvent. In an HF solution, there are a few ions of H^+ and F^- present but mostly dissolved HF molecules.

9.13 a. $KCl \xrightarrow{H_2O} K^+ + Cl^-$

b. $CaCl_2 \xrightarrow{H_2O} Ca^{2+} + 2Cl^-$

c. $K_3PO_4 \xrightarrow{H_2O} 3K^+ + PO_4^{3-}$

d. $Fe(NO_3)_3 \xrightarrow{H_2O} Fe^{3+} + 3NO_3^-$

9.15 a. mostly molecules and a few ions
b. ions only
c. molecules only

9.17 a. strong electrolyte
b. weak electrolyte
c. nonelectrolyte

9.19 a. 1 Eq **b.** 2 Eq **c.** 2 Eq **d.** 6 Eq

9.21 a. 0.704 Eq **b.** 0.805 Eq
c. 0.20 Eq **d.** 1.0 Eq

9.23 0.18 g Mg^{2+}

9.25 3.5 g Na^+, 5.5 g Cl^-

9.27 55 mEq/L

9.29 a. saturated **b.** unsaturated

9.31 a. unsaturated **b.** unsaturated
c. saturated

9.33 a. 68.0 g KCl
b. 12.0 g KCl

9.35 a. The solubility of solid solutes typically increases as temperature increases.
b. The solubility of a gas is less at a higher temperature.
c. Gas solubility is less at a higher temperature and the CO_2 pressure in the can is increased.

9.37 a. soluble **b.** insoluble **c.** insoluble
d. soluble **e.** soluble

9.39 a. no solid forms
b. Ag_2S is insoluble; $2AgNO_3 + K_2S \longrightarrow Ag_2S(s) + 2KNO_3$
c. $CaSO_4$ is insoluble; $CaCl_2 + Na_2SO_4 \longrightarrow CaSO_4(s) + 2NaCl$

9.41 5% (m/m) is 5 g of glucose in 100 g of solution, whereas 5% (m/v) is 5 g of glucose in 100 mL of solution.

9.43 a. 17%
b. 10.%
c. 5.3%

9.45 a. 30.%
b. 3.3%
c. 11%

9.47 a. 2.5 g KCl **b.** 50. g NH_4Cl
c. 25 mL acetic acid

9.49 79.9 mL of alcohol

9.51 a. 20 g mannitol **b.** 300 g mannitol

9.53 a. 20. mL
b. 400. mL **c.** 20. mL

9.55 2 L

9.57 a. 0.50 M glucose **b.** 2.5 M NaCl
c. 0.036 M KOH

9.59 a. 3.0 moles NaCl **b.** 0.40 mole KBr
c. 1.0 mole NaCl

9.61 a. 120 g NaOH
b. 60. g KCl
c. 1.5 g NaCl

9.63 a. 1.5 L
b. 10. L
c. 0.067 L

9.65 a. solution **b.** colloid **c.** suspension

9.67 An emulsion is a colloid in which a liquid is dispersed in another liquid or a solid.

9.69 a. Water flows from wet soil through plant membranes toward the higher solute concentration in the plants.
b. Because a brine solution (salt water) has a higher solute concentration, water flows from the cucumber to the brine and the cucumber becomes a pickle.

9.71 a. starch
b. from pure water into the starch
c. starch

9.73 a. B 10% glucose **b.** B 8% albumin
c. B 10% NaCl

9.75 a. hypotonic **b.** hypotonic **c.** isotonic
d. isotonic

9.77 a. NaCl **b.** alanine **c.** NaCl **d.** urea

9.79 a. 0.40 Osm
b. 1.5 Osm
c. 0.40 Osm

9.81 Because iodine is a nonpolar molecule, it will dissolve in hexane, a nonpolar solvent. Iodine does not dissolve in water because water is a polar solvent.

9.83 222 g water

9.85 When solutions of $NaNO_3$ and KCl are mixed, no insoluble products are formed. All the combinations of salts are soluble. When KCl and $Pb(NO_3)_2$ solutions are mixed, the insoluble salt $PbCl_2$ forms.

9.87 17.0% m/m

9.89 **a.** 60 g of amino acids, 380 g of glucose and 100 g of lipids
b. 2600 kcal

9.91 38 mL of solution

9.93 To make a 2.0 M KCl solution, weigh out 37.3 g KCl (0.500 mole) and place into a volumetric flask. Add water to dissolve the KCl and give a final volume of 0.250 liter.

9.95 **a.** 35.0% HNO_3 **b.** 165 mL
c. 42.4%   **d.** 6.73 M

9.97 **a.** 1600 g $Al(NO_3)_3$
b. 6.8 g of $C_6H_{12}O_6$

9.99 The solution will dehydrate the flowers because water will flow out of the cells of the flowers into the more concentrated salt solution.

9.101 Drinking seawater will cause water to flow out of the body cells and further dehydrate a person.

9.103 **a.** 0.40 Osm   **b.** 0.15 Osm
c. 0.6 Osm   **d.** 0.3 Osm

9.105 **a.** B
b. B
c. stay the same
d. stay the same
e. A

9.107 **a.** 10. g of $PbCl_2$
b. 19 mL

Chapter 10

Answers to Study Checks

10.1 Sulfuric acid; potassium hydroxide

10.2 $Ca(OH)_2 \xrightarrow{H_2O} Ca^{2+} + 2OH^-$

10.3 $HNO_3 + H_2O \longrightarrow H_3O^+ + NO_3^-$

10.4 A base accepts a proton to form its conjugate acid.
a. H_2S **b.** HCl   **c.** HNO_2

10.5 The conjugate acid/base pairs are HCN/CN^- and SO_4^{2-}/HSO_4^-.

10.6 NH_3

10.7 $HNO_3 + H_2O \rightleftharpoons H_3O^+ + NO_3^-$

The products are favored because HNO_3 is a stronger acid than H_3O^+.

10.8 Because nitrous acid has a larger K_a than carbonic acid, it forms more $[H_3O^+]$ and is a stronger acid compared to carbonic acid.

10.9 $[H_3O^+] = 2.5 \times 10^{-11}$ M; basic

10.10 11.38

10.11 $[H_3O^+] = 1.0 \times 10^{-10}$ M; pH = 10.00

10.12 $[H_3O^+] = 1 \times 10^{-5}$ M; $[OH^-] = 1 \times 10^{-9}$ M

10.13 pOH = 3.30; pH = 10.70

10.14 $H_2SO_4 + 2NaHCO_3 \longrightarrow Na_2SO_4 + 2CO_2 + 2H_2O$

10.15 The anion PO_4^{3-} reacts with H_2O forming the weak acid HPO_4^{2-} and OH^-, which makes the solution basic.

10.16 No. Both substances are salts; the mixture has no weak acid present.

10.17 pH = 7.21

10.18 40 mL

10.19 63 mL

10.20 0.200 M HCl

10.21 30.0 mL

Answers to Selected Questions and Problems

10.1 **a.** acid   **b.** acid   **c.** acid
d. base   **e.** base

10.3 **a.** hydrochloric acid   **b.** calcium hydroxide
c. carbonic acid   **d.** nitric acid
e. sulfurous acid

10.5 **a.** $Mg(OH)_2$ **b.** HF   **c.** H_3PO_4
d. LiOH   **e.** $Cu(OH)_2$

10.7 **a.** HI is the acid (proton donor) and H_2O is the base (proton acceptor).
b. H_2O is the acid (proton donor) and F^- is the base (proton acceptor).

10.9 **a.** F^-, fluoride ion   **b.** OH^-, hydroxide ion
c. HCO_3^-, bicarbonate ion   **d.** SO_4^{2-}, sulfate ion

10.11 a. HCO_3^-, bicarbonate ion
b. H_3O^+, hydronium ion
c. H_3PO_4, phosphoric acid
d. HBr, hydrobromic acid

10.13 a. acid H_2CO_3; conjugate base HCO_3^-
base H_2O; conjugate acid H_3O^+
b. acid NH_4^+; conjugate base NH_3
base H_2O; conjugate acid H_3O^+
c. acid HCN; conjugate base CN^-
base NO_2^-; conjugate acid HNO_2

10.15 $NH_4^+ + H_2O \rightleftharpoons NH_3 + H_3O^+$

10.17 A strong acid is a good proton donor, whereas its conjugate base is a poor proton acceptor.

10.19 a. HBr   **b.** HSO_4^- **c.** H_2CO_3

10.21 a. HSO_4^- **b.** HF   **c.** HCO_3^-

10.23 a. reactants   **b.** reactants   **c.** products

10.25 The reactants are favored because NH_4^+ is a weaker acid than HSO_4^-.
$NH_4^+ + SO_4^{2-} \rightleftharpoons NH_3 + HSO_4^-$

10.27 a. H_2SO_3 **b.** HSO_3^- **c.** H_2SO_3
d. HS^- **e.** H_2SO_3

10.29 $H_3PO_4 + H_2O \rightleftharpoons H_3O^+ + H_2PO_4^-$

$$K_a = \frac{[H_3O^+][H_2PO_4^-]}{[H_3PO_4]}$$

10.31 In pure water, $[H_3O^+] = [OH^-]$ because one of each is produced every time a proton transfers from one water to another.

10.33 In an acid solution, the $[H_3O^+]$ is greater than the $[OH^-]$.

10.35 a. acidic   **b.** basic   **c.** basic   **d.** acidic

10.37 a. 1.0×10^{-9} M   **b.** 1.0×10^{-6} M
c. 2.0×10^{-5} M   **d.** 4.0×10^{-13} M

10.39 a. 1.0×10^{-11} M   **b.** 2.0×10^{-9} M
c. 5.6×10^{-3} M   **d.** 2.5×10^{-2} M

10.41 In a neutral solution, the $[H_3O^+]$ is 1.0×10^{-7} M and the pH is 7.00, which is the negative value of the power of 10.

10.43 a. basic   **b.** acidic   **c.** basic
d. acidic

10.45 An increase or decrease of 1 pH unit changes the $[H_3O^+]$ by a factor of 10. Thus a pH of 3 (10^{-3} M) is 10 times more acid than a pH of 4 (10^{-4} M).

10.47 a. 4.00   **b.** 8.52   **c.** 9.00   **d.** 3.40

10.49

$[H_3O^+]$	$[OH^-]$	pH	pOH	Acidic, Basic, or Neutral?
1.0×10^{-8} M	1.0×10^{-6} M	8.00	6.00	Basic
1.0×10^{-3} M	1.0×10^{-11} M	3.00	11.00	Acidic
2.8×10^{-5} M	3.6×10^{-10} M	4.55	9.45	Acidic
1.0×10^{-12} M	1.0×10^{-2} M	12.00	2.00	Basic

10.51 a. $ZnCO_3 + 2HCl \longrightarrow ZnCl_2 + CO_2 + H_2O$
b. $Zn + 2HCl \longrightarrow ZnCl_2 + H_2$
c. $HCl + NaHCO_3 \longrightarrow NaCl + H_2O + CO_2$
d. $2HNO_3 + Mg(OH)_2 \longrightarrow Mg(NO_3)_2 + 2H_2O$

10.53 a. $2HCl + Mg(OH)_2 \longrightarrow MgCl_2 + 2H_2O$
b. $H_3PO_4 + 3LiOH \longrightarrow Li_3PO_4 + 3H_2O$

10.55 a. $H_2SO_4 + 2NaOH \longrightarrow Na_2SO_4 + 2H_2O$
b. $3HCl + Fe(OH)_3 \longrightarrow FeCl_3 + 3H_2O$
c. $H_2CO_3 + Mg(OH)_2 \longrightarrow MgCO_3 + 2H_2O$

10.57 The anion from the weak acid removes a proton from H_2O to make a basic solution.

10.59 a. neutral
b. acidic $NH_4^+ + H_2O \rightleftharpoons NH_3 + H_3O^+$
c. basic $CO_3^{2-} + H_2O \rightleftharpoons HCO_3^- + OH^-$
d. basic $S^{2-} + H_2O \rightleftharpoons HS^- + OH^-$

10.61 (b) and (c) are buffer systems. (b) contains the weak acid H_2CO_3 and its salt $NaHCO_3$. (c) contains HF, a weak acid, and its salt KF.

10.63 a. A buffer system keeps the pH constant.
b. To neutralize any H^+ added.
c. The added H^+ reacts with F^- from NaF.
d. The added OH^- is neutralized by the HF.

10.65 $[H_3O^+] = 4.5 \times 10^{-4} \times \dfrac{[0.10 \text{ M}]}{[0.10 \text{ M}]} = 4.5 \times 10^{-4}$

$pH = -\log[4.5 \times 10^{-4}] = 3.35$

10.67 The pH of the 0.10 M HF/0.10 M NaF buffer is 3.14. The pH of the 0.060 M HF/0.120 M NaF buffer is 3.44.

10.69 Adding water to the soup concentrate makes a larger volume of the soup, which is called dilution.

10.71 a. $\dfrac{25.0 \text{ mL} \times 6.0 \text{ M}}{200 \text{ mL}} = 0.75$ M
b. $\dfrac{25\% \times 10.0 \text{ mL}}{500 \text{ mL}} = 0.50\%$

10.73 a. $\dfrac{250 \text{ mL} \times 0.200 \text{ M}}{4.00 \text{ M}} = 13 \text{ mL}$

b. $\dfrac{750 \text{ mL} \times 0.100 \text{ M}}{6.0 \text{ M}} = 13 \text{ mL}$

10.75 $\dfrac{25.0 \text{ mL} \times 3.0 \text{ M}}{0.15 \text{ M}} = 500 \text{ mL}$

Add 475 mL water to the 25 mL of 3.0 M solution to make 500 mL of solution.

10.77 To a known volume of formic acid, add a few drops of indicator. Place a NaOH solution of known molarity in a buret. Add base to acid until one drop changes the color of the solution. Use the volume and molarity of NaOH and the volume of formic acid to calculate the concentration of the formic acid in the sample.

10.79 a. 8.8 M HCl
b. 0.75 M H_2SO_4
c. 0.53 M H_3PO_4

10.81 a. sulfuric acid
b. potassium hydroxide
c. calcium hydroxide
d. hydrochloric acid
e. nitrous acid

10.83 Both strong and weak acids produce H_3O^+ in water. Weak acids are only slightly ionized, whereas a strong acid exists as ions in solution.

10.85 a. The $Mg(OH)_2$ that dissolves is completely ionized, making it a strong base.
b. $Mg(OH)_2 + 2HCl \longrightarrow MgCl_2 + 2H_2O$

10.87 a. HF **b.** H_3O^+
c. HNO_2 **d.** HCO_3^-

10.89 a. pH 7.70, pOH 6.30 **b.** pH 1.30, pOH 12.70
c. pH 10.54, pOH 3.46 **d.** pH 11.7, pOH 2.3

10.91 a. $[H_3O^+] = 1 \times 10^{-3}$ M; $[OH^-] = 1 \times 10^{-11}$ M
b. $[H_3O^+] = 1 \times 10^{-6}$ M; $[OH^-] = 1 \times 10^{-8}$ M
c. $[H_3O^+] = 1 \times 10^{-8}$ M; $[OH^-] = 1 \times 10^{-6}$ M
d. $[H_3O^+] = 1 \times 10^{-11}$ M; $[OH^-] = 1 \times 10^{-3}$ M

10.93 $[OH^-] = 0.020$ M

10.95 0.16 M HCl; pH = 0.79 and pOH = 13.21

10.97 a. basic **b.** basic
c. acidic **d.** neutral

10.99 a. acid: $H_2PO_4^- + H_3O^+ \longrightarrow H_3PO_4 + H_2O$
b. base: $H_3PO_4 + OH^- \longrightarrow H_2PO_4^- + H_2O$
c. pH = 2.12

10.101 a. Hyperventilation will lower the CO_2 level in the blood and shift the equilibrium of the carbonate buffer toward bicarbonate, lowering the H_2CO_3 concentration, which decreases the H_3O^+.
b. Treat the diabetes condition to reduce the production of acidic compounds; perhaps an infusion of bicarbonate.

10.103 a. 5.0 mL **b.** 80. mL

10.105 a. 5.3 mL **b.** 10. mL

Chapter 11

Answers to Study Checks

11.1 The structure needs more hydrogen atoms to complete four bonds to carbon and three bonds to nitrogen.

$$H-\overset{\overset{\displaystyle H}{|}}{\underset{\underset{\displaystyle H}{|}}{C}}-\overset{\overset{\displaystyle H}{|}}{\underset{\underset{\displaystyle H}{|}}{C}}-\overset{\overset{\displaystyle H}{|}}{N}-H$$

11.2 In CBr_4, the shape that allows the bonds to be as far apart as possible is a tetrahedron.

11.3 The CH_3F molecule is polar because the stronger C—F dipole is not canceled by the C—H bonds.

11.4 Octane is not soluble in water; it is an organic compound.

11.5 CH_3CH_2—O—CH_3 contains the functional group C—O—C; it is an ether.

$$CH_3-\overset{\overset{\displaystyle OH}{|}}{C}H-CH_3$$

contains the —OH functional group; it is an alcohol.

11.6 A primary amine has one carbon atom attached to nitrogen. A secondary amine has two carbon atoms attached to nitrogen.

11.7 Because the two compounds are constitutional isomers, they probably have different boiling points.

11.8 $CH_3-CH_2-NH_2$ and $CH_3-\overset{\overset{\displaystyle H}{|}}{N}-CH_3$

Answers to Selected Questions and Problems

11.1 a.

$$\begin{array}{ccc} H & H & H \\ | & | & | \\ H-C-&C-&C-H \\ | & | & | \\ H & H & H \end{array}$$

b.

$$\begin{array}{cc} H & H \\ | & | \\ H-C-&C-O-H \\ | & | \\ H & H \end{array}$$

c.

$$\begin{array}{ccc} H & H & H \\ | & | & | \\ H-C=&C-&N-H \end{array}$$

11.3 a. incorrect   **b.** incorrect
c. correct   **d.** correct

11.5 VSEPR theory predicts that the four bonds in CH_4 will be as far apart as possible, which means that the hydrogen atoms are at the corners of a tetrahedron.

11.7 a. nonpolar   **b.** nonpolar
c. polar   **d.** polar

11.9 a. inorganic   **b.** organic
c. organic   **d.** inorganic
e. inorganic   **f.** organic

11.11 a. inorganic   **b.** organic
c. organic   **d.** inorganic

11.13 a. alcohol   **b.** alkene
c. aldehyde   **d.** ester

11.15 a. ether   **b.** alcohol   **c.** ketone
d. carboxylic acid   **e.** amine

11.17 a. constitutional isomers
b. identical compounds
c. constitutional isomers
d. constitutional isomers
e. constitutional isomers
f. different compounds

11.19

$$CH_3-CH_2-\overset{\displaystyle O}{\overset{\displaystyle \|}{C}}-O-H \qquad CH_3-\overset{\displaystyle O}{\overset{\displaystyle \|}{C}}-O-CH_3$$

$$H-\overset{\displaystyle O}{\overset{\displaystyle \|}{C}}-O-CH_2-CH_3$$

11.21 a. Organic compounds have covalent bonds; inorganic compounds have ionic as well as polar covalent and a few have nonpolar covalent bonds.
b. Most organic compounds are insoluble in water; many inorganic compounds are soluble in water.
c. Most organic compounds have low melting points; inorganic compounds have high melting points.
d. Most organic compounds are flammable; inorganic compounds are not flammable.

11.23 a. butane   **b.** butane
c. potassium chloride   **d.** potassium chloride
e. butane

11.25

a.

$$\begin{array}{cc} H & H \\ | & | \\ H-C-&C-H \\ | & | \\ H & H \end{array}$$

b.

$$\begin{array}{cc} H & H \\ | & | \\ H-C-&C-O-H \\ | & | \\ H & H \end{array}$$

c.

$$\begin{array}{ccc} H & H & O \\ | & | & \| \\ H-C-&C-&C-H \\ | & | & \\ H & H & \end{array}$$

d.

$$\begin{array}{cccc} H & & H & H \\ | & & | & | \\ H-C-&O-&C-&C-H \\ | & & | & | \\ H & & H & H \end{array}$$

11.27 a. A hydroxyl group is —OH; a carbonyl group is C=O.
b. An alcohol contains the hydroxyl (—OH) functional group; an ether contains a C—O—C functional group.
c. A carboxylic acid contains the COOH functional group; an ester contains a —COOC— functional group.

11.29 a. constitutional isomers
b. different compounds
c. constitutional isomers
d. different compounds
e. constitutional isomers
f. identical compounds
g. different compounds

11.31 a. polar   **b.** polar
c. nonpolar   **d.** polar
e. polar   **f.** polar

11.33 **a.** alcohol **b.** unsaturated hydrocarbon
 c. aldehyde **d.** saturated hydrocarbon
 e. carboxylic acid **f.** amine
 g. tetrahedral **h.** constitutional isomers
 i. ester **j.** hydrocarbon
 k. ether **l.** unsaturated hydrocarbon
 m. functional group **n.** ketone

Chapter 12

Answers to Study Checks

12.1 $CH_3-CH_2-CH_2-CH_2-CH_2-CH_2-CH_3$

12.2 This is a constitutional isomer. There is a five-carbon chain with a methyl group bonded to the middle (third) carbon.

12.3 3-ethyl-5-methylheptane

12.4

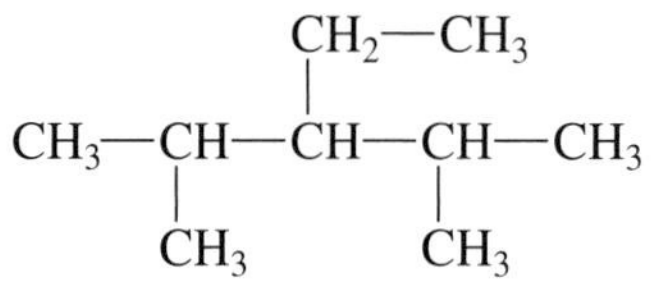

$$CH_3-CH-CH-CH-CH_3$$
with CH_2-CH_3 on second carbon, CH_3 and CH_3 below

12.5 2,2-dimethylpentane; 3,3-dimethylpentane
 2,3-dimethylpentane; 2,4-dimethylpentane

12.6 1-bromo-1-chloro-2,2,2-trifluoroethane

12.7 1-bromo-2-chloro-3-methylcyclobutane

12.8

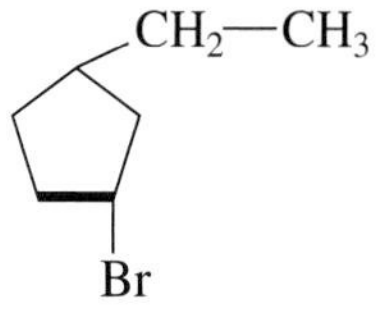

CH_2-CH_3 and Br on cyclopentane ring

12.9

H_3C and CH_3 on cyclopropane ring

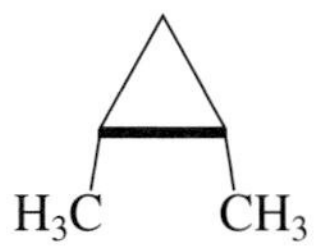

12.10 Cyclohexane

12.11

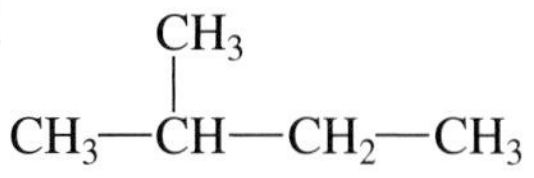

$$CH_3-CH-CH_2-CH_3 = C_5H_{12}$$
with CH_3 above second carbon

$$C_5H_{12} + 8O_2 \longrightarrow 5CO_2 + 6H_2O$$

12.12

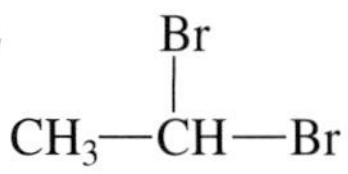

$$CH_3-CH-Br \qquad Br-CH_2-CH_2-Br$$
with Br above

Answers to Selected Questions and Problems

12.1 **a.** C_7H_{16} **b.** 12 **c.** 4

12.3

a.

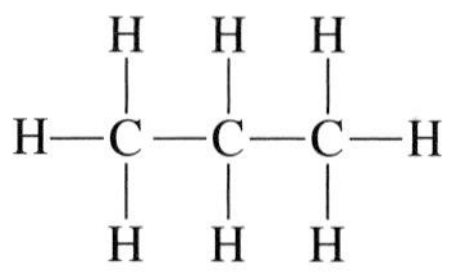

$$H-C-C-C-H$$
with H H H above and H H H below

b. $CH_3-CH_2-CH_2-CH_2-CH_2-CH_3$

c.  ∿ (zigzag structure)

12.5 **a.** pentane **b.** heptane **c.** hexane

12.7 **a.** two different conformations
 b. constitutional isomers
 c. constitutional isomers

12.9 **a.** propyl **b.** isopropyl
 c. butyl **d.** methyl

12.11 **a.** 2-methylpropane
 b. 2-methylpentane
 c. 4-ethyl-2-methylhexane
 d. 2,3,4-trimethylheptane
 e. 5-isopropyl-3-methyloctane

12.13 **a.** 2,3-dimethylpentane
 b. 3-ethyl-5-methylheptane
 c. 3-isopropyl-2,4-dimethylpentane

12.15

a. 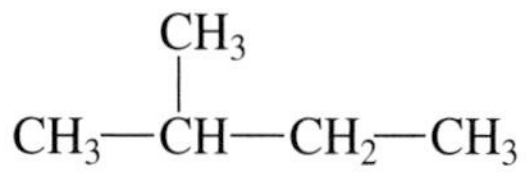 $CH_3-CH-CH_2-CH_3$ with CH_3 above second carbon

b. 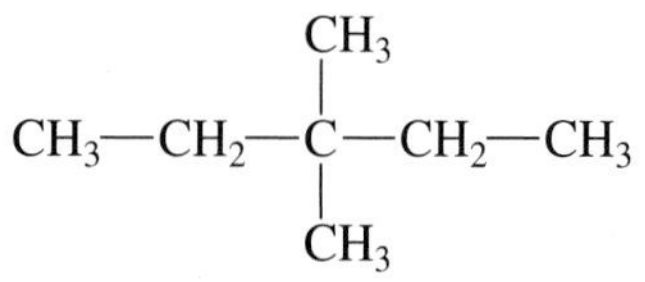 $CH_3-CH_2-C-CH_2-CH_3$ with CH_3 above and CH_3 below middle carbon

c. $CH_3-CH-CH-CH_2-CH-CH_3$ with CH_3, CH_3, CH_3 above

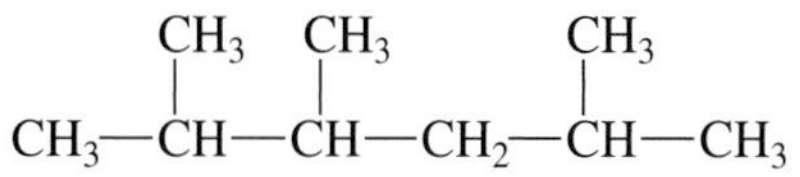

d. 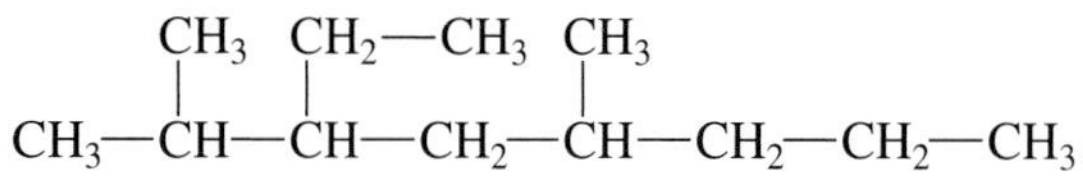 $CH_3-CH-CH-CH_2-CH-CH_2-CH_2-CH_3$ with CH_3, CH_2-CH_3, CH_3 above

e. 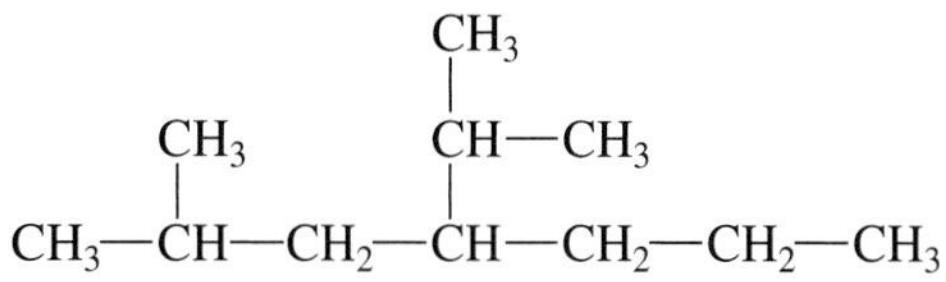 $CH_3-CH-CH_2-CH-CH_2-CH_2-CH_3$ with CH_3 above, and $CH-CH_3$ with CH_3 above

f. 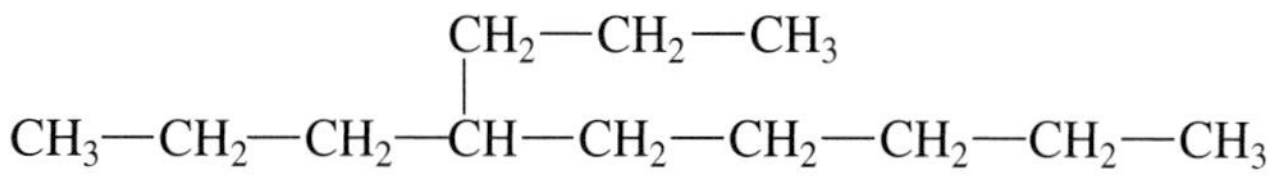 $CH_3-CH_2-CH_2-CH-CH_2-CH_2-CH_2-CH_2-CH_3$ with $CH_2-CH_2-CH_3$ above

12.17 a.2-methylheptane; 3-methylheptane;
4-methylheptane
   b. 2,2-dimethylpentane; 3,3-dimethylpentane;
2,3-dimethylpentane; 2,4-dimethylpentane;
3-ethylpentane

12.19 a.bromoethane; ethyl bromide
   b. 1-fluoropropane; propyl fluoride
   c. 2-chloropropane; isopropyl chloride
   d. trichloromethane; chloroform

12.21 a.2-bromo-3-methylbutane
   b. 3-bromo-2-chloropentane
   c. 2-fluoro-2-methylbutane

12.23

a. CH₃—CH—CH₃ (Cl on center carbon)

b. CH₃—CH—CH—CH₃ (Br and Cl)

c. CH₃Br

d. CH₃—CH₂—CH₂—CH₂—Cl

e. Br—CH—CH—CH₂—CH—CH₃ (Br, Cl, F)

f. CBr₄

12.25 methyl chloride is CH₃Cl; ethyl chloride is
CH₃CH₂Cl.

12.27

Cl—CH₂—CH₂—CH₂—CH₃
1-chlorobutane

CH₃—CH—CH₂—CH₃ (Cl)
2-chlorobutane

Cl—CH₂—CH—CH₃ (CH₃)
1-chloro-2-methylpropane

CH₃—C—CH₃ (CH₃ top, Cl bottom)
2-chloro-2-methylpropane

12.29 a. C₅H₁₀   **b.** 8   **c.** 6

12.31 a. cyclobutane
   b. chlorocyclopentane
   c. methylcyclohexane
   d. 1-bromo-3-methylcyclobutane
   e. 1-bromo-2-chlorocyclopentane
   f. 1,3-dibromo-5-methylcyclohexane

12.33

a. (cyclopentane with CH₃)   **b.** (cyclohexane with CH—CH₃ and CH₃)

c. (cyclobutane with two CH₃)   **d.** (cyclopentane with Br, CH₃, CH₃)

12.35 Choice of three: 1,1-dimethylcyclopropane, 1,2-
dimethylcyclopropane, ethylcyclopropane, methyl-
cyclobutane

12.37 a. Does not have cis-trans isomers
   b. *trans*-1,2-dimethylcyclopropane
   c. *cis*-1,2-dichlorocyclobutane
   d. *trans*-1,3-dimethylcyclopentane

12.39 a.

(cyclopentane with H₃C and CH₃)

b.

c.

12.41 a. CH₃—CH₂—CH₂—CH₂—CH₂—CH₂—CH₃
   b. liquid
   c. insoluble in water
   d. float
   e. lower

12.43 a. heptane   **b.** cyclopropane   **c.** hexane

12.45 a.2C₂H₆ + 7O₂ ⟶ 4CO₂ + 6H₂O
   b. 2C₃H₆ + 9O₂ ⟶ 6CO₂ + 6H₂O
   c. 2C₈H₁₈ + 25O₂ ⟶ 16CO₂ + 18H₂O
   d. C₆H₁₂ + 9O₂ ⟶ 6CO₂ + 6H₂O

12.47 **a.** CH_3-CH_2-Cl

b.

c. $CH_3-\overset{\overset{\displaystyle CH_3}{|}}{CH}-CH_2-Cl$ $CH_3-\overset{\overset{\displaystyle CH_3}{|}}{\underset{\underset{\displaystyle Cl}{|}}{C}}-CH_3$

12.49 a. constitutional isomers   **b.** constitutional isomers
c. same molecule   **d.** constitutional isomers

12.51 a. methyl   **b.** propyl   **c.** isopropyl

12.53 a. 2,2-dimethylbutane
b. chloroethane
c. 2-bromo-4-ethylhexane
d. 1,1-dibromocyclohexane

12.55

a. $CH_3-CH_2-\overset{\overset{\displaystyle CH_2-CH_3}{|}}{CH}-CH_2-CH_2-CH_3$

b. $CH_3-\overset{\overset{\displaystyle CH_3}{|}}{CH}-\overset{\overset{\displaystyle CH_3}{|}}{CH}-CH_2-CH_3$

c. $Cl-CH_2-CH_2-\overset{\overset{\displaystyle Cl}{|}}{\underset{\underset{\displaystyle CH_3}{|}}{C}}-CH_2-CH_2-CH_2-CH_3$

d.

e.

f. $CH_3-CH_2-\overset{\overset{\displaystyle CH_3}{|}}{CH}-Cl$

12.57 a.   **b.**

c.

12.59 **a.** You should have two of the following four possibilities.

$CH_3-\overset{\overset{\displaystyle CH_3}{|}}{CH}-CH_2-CH_2-CH_3$ 2-methylpentane

$CH_3-CH_2-\overset{\overset{\displaystyle CH_3}{|}}{CH}-CH_2-CH_3$ 3-methylpentane

$CH_3-\overset{\overset{\displaystyle CH_3}{|}}{CH}-\overset{\overset{\displaystyle CH_3}{|}}{CH}-CH_3$ 2,3-dimethylbutane

$CH_3-\overset{\overset{\displaystyle CH_3}{|}}{\underset{\underset{\displaystyle CH_3}{|}}{C}}-CH_2-CH_3$ 2,2-dimethylbutane

b. The molecular formula C_7H_{16} would be the formula for the following constitutional isomers:

$CH_3-CH_2-CH_2-CH_2-CH_2-CH_2-CH_3$ heptane

$CH_3-\overset{\overset{\displaystyle CH_3}{|}}{CH}-CH_2-CH_2-CH_2-CH_3$ 2-methylhexane

$CH_3-CH_2-\overset{\overset{\displaystyle CH_3}{|}}{CH}-CH_2-CH_2-CH_3$ 3-methylhexane

$CH_3-\overset{\overset{\displaystyle CH_3}{|}}{\underset{\underset{\displaystyle CH_3}{|}}{C}}-CH_2-CH_2-CH_3$ 2,2-dimethylpentane

$CH_3-CH_2-\overset{\overset{\displaystyle CH_3}{|}}{\underset{\underset{\displaystyle CH_3}{|}}{C}}-CH_2-CH_3$ 3,3-dimethylpentane

$CH_3-\overset{\overset{\displaystyle CH_3}{|}}{CH}-\overset{\overset{\displaystyle CH_3}{|}}{CH}-CH_2-CH_3$ 2,3-dimethylpentane

$CH_3-\overset{\overset{\displaystyle CH_3}{|}}{CH}-CH_2-\overset{\overset{\displaystyle CH_3}{|}}{CH}-CH_3$ 2,4-dimethylpentane

$CH_3-\overset{\overset{\displaystyle CH_3}{|}}{\underset{\underset{\displaystyle CH_3}{|}}{C}}-\overset{\overset{\displaystyle CH_3}{|}}{CH}-CH_3$ 2,2,3-trimethylbutane

c. CH_3—CH_2—CH_2—CH_2—CH_2—CH_3 hexane

$$CH_3-\underset{\underset{\textstyle CH_3}{|}}{CH}-CH_2-CH_2-CH_3 \quad \text{2-methylpentane}$$

$$CH_3-CH_2-\underset{\underset{\textstyle CH_3}{|}}{CH}-CH_2-CH_3 \quad \text{3-methylpentane}$$

$$CH_3-\underset{\underset{\textstyle CH_3}{|}}{CH}-\underset{\underset{\textstyle CH_3}{|}}{CH}-CH_3 \quad \text{2,3-dimethylbutane}$$

d.

1,2-dibromo-cyclohexane   1,3-dibromo-cyclohexane   1,4-dibromo-cyclohexane

12.61 a. 2-methylbutane
b. 2,3-dimethylpentane
c. 1,3-dibromocyclohexane
d. hexane

12.63 CH_3—CH_2—CH_2—CH_2—CH_2—Cl

$$CH_3-\underset{\underset{\textstyle Cl}{|}}{CH}-CH_2-CH_2-CH_3$$

$$CH_3-CH_2-\underset{\underset{\textstyle Cl}{|}}{CH}-CH_2-CH_3$$

$$CH_3-\underset{\underset{\textstyle CH_3}{|}}{CH}-CH_2-CH_2-Cl$$

$$CH_3-\underset{\underset{\textstyle CH_3}{|}}{CH}-\underset{\underset{\textstyle Cl}{|}}{CH}-CH_3$$

$$CH_3-\overset{\overset{\textstyle CH_3}{|}}{\underset{\underset{\textstyle Cl}{|}}{C}}-CH_2-CH_3$$

$$Cl-CH_2-\underset{\underset{\textstyle CH_3}{|}}{CH}-CH_2-CH_3$$

$$CH_3-\overset{\overset{\textstyle CH_3}{|}}{\underset{\underset{\textstyle CH_3}{|}}{C}}-CH_2-Cl$$

12.65

a. $CH_3-CH_2-\underset{\underset{\textstyle Br}{|}}{\overset{\overset{\textstyle CH_3}{|}}{CH}}$

b.
Br   Cl

c.

CH₃

CH₃

d.

Br Br

12.67 Condensed structural formula

$$CH_3-\overset{\overset{\textstyle CH_3}{|}}{\underset{\underset{\textstyle CH_3}{|}}{C}}-CH_2-\underset{\underset{\textstyle CH_3}{|}}{CH}-CH_3$$

molecular formula of C_8H_{18}

The combustion reaction: $2C_8H_{18} + 25O_2 \longrightarrow 16CO_2 + 18H_2O$

12.69 a. heptane   **b.** cyclopentane
c. hexane   **d.** cyclohexane

12.71 a. $C_3H_8 + 5O_2 \longrightarrow 3CO_2 + 4H_2O$
b. $C_5H_{12} + 8O_2 \longrightarrow 5CO_2 + 6H_2O$
c. $C_4H_8 + 6O_2 \longrightarrow 4CO_2 + 4H_2O$
d. $2C_8H_{18} + 25O_2 \longrightarrow 16CO_2 + 18H_2O$

12.73 a. CH_3—CH_2—Cl

b. $CH_3-CH_2-CH_2-Cl$ and $CH_3-\underset{\underset{\textstyle Cl}{|}}{CH}-CH_3$

c.
Cl

12.75 a. $C_5H_{12} + 8O_2 \longrightarrow 5CO_2 + 6H_2O$
b. 72.0 g/mole
c. 2.8×10^4 kcal
d. 3700 L

Chapter 13

Answers to Study Checks

13.1 This is cylcobutene, a cycloalkene.

13.2. In an alkene, the three bonded atoms are as far apart as possible at 120°. In an alkyne, the two bonded atoms are as far apart as possible at 180°.

13.3 a. $CH_3C{\equiv}CCH_2CH_3$ **b.**   Cl

13.4 CH_3CH_2 $CH_2CH_2CH_3$

                        C=C

              H   H

           cis-3-Heptene

13.5 CH_3

           $CH_3CHCH_2CH_3$

13.6 1,2-dichlorobutane

13.7 CH_3 Br   1-bromo-1-methylcyclopentane

13.8 $CH_2{=}CHCH_2CH_2CH_3$ + HCl $\longrightarrow$

              Cl

           $CH_3CHCH_2CH_2CH_3$

13.9 CH_3

           $CH_3CCH_2CH_3$

              OH

13.10 The monomer of PVC, polyvinyl chloride, is chloroethene:

           H   Cl

           C = C

           H   H

13.11 1, 3-diethylbenzene; *m*-diethylbenzene

13.12 Chlorobenzene can be prepared from benzene and chlorine, using $FeCl_3$ as a catalyst.

                        Cl

           ⬡ + Cl_2 $\xrightarrow{FeCl_3}$ ⬡ + HCl

Answers to Selected Questions and Problems

13.1 a. An alkane has only sigma bonds.
   b. An alkyne has a sigma bond and two pi bonds.
   c. An alkene has the groups on the carbon atom arranged at 120°.

13.3 a. An alkane has the general formula C_nH_{2n+2}.
   b. An alkene has a double bond.
   c. An alkyne has a triple bond.
   d. An alkene has a double bond.
   e. A cycloalkene has a double bond in a ring.

13.5. $CH_2{=}CH{-}CH_3$ Propene

   Cyclopropane

13.7 a. Propene contains a double bond, and propyne has a triple bond.
   b. Cyclohexane is a saturated cycloalkane, and cyclohexene contains a double bond and is unsaturated.

13.9 a. ethene   **b.** 2-methyl-1-propene
   c. 4-bromo-2-pentyne   **d.** cyclobutene
   e. 4-ethylcyclopentene   **f.** 4-ethyl-2-hexene

13.11 a. $CH_3CH{=}CH_2$ **b.** $CH_2{=}CHCH_2CH_2CH_3$

                 CH_3

   c. $CH_2{=}CCH_2CH_3$ **d.** ⬡ CH_3

                 Cl

   e. $CH_3{-}CH{-}C{\equiv}C{-}CH_2{-}CH_3$

13.13 Constitutional isomers have the same molecular formula, but different arrangements of the atoms. Geometric isomers in alkenes have different geometric relationship of the groups attached to a double bond.

13.15 There are four constitutional isomers.

                                        Cl

   $Cl{-}CH_2{-}CH{=}CH_2$ $CH_3{-}C{=}CH_2$

                    Cl   Cl

   $CH_3{-}CH{=}CH$ △

13.17 a. There are no cis–trans isomers.
   b. This alkene has cis–trans isomers.
   c. There are no cis–trans isomers.

13.19 a. *cis*-2-butene   **b.** *trans*-3-octene
   c. *cis*-3-heptene

13.21 a.

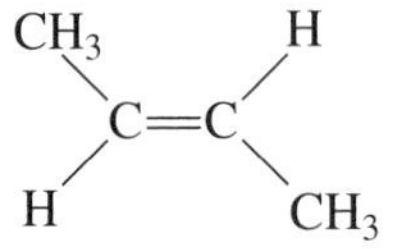

b.

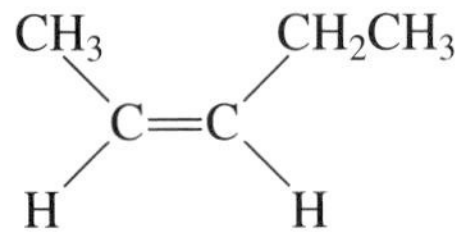

c.

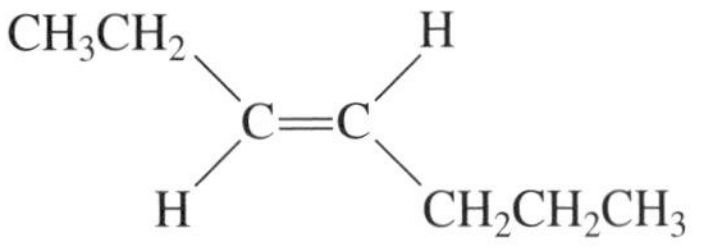

13.23 a. $CH_3-CH_2-CH_2-CH_2-CH_3$ Pentane

b.

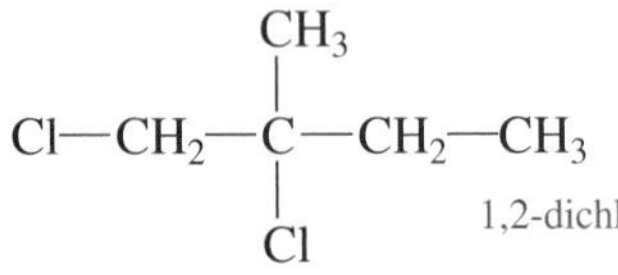

1,2-dichloro-2-methylbutane

c.

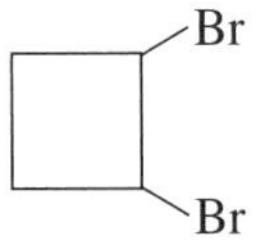

1.2-dibromocyclobutane

d.

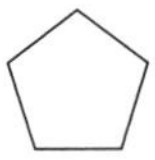

Cyclopentane

e.

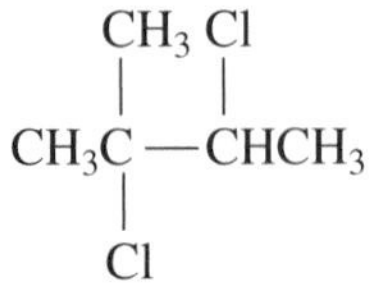

2.3-dichloro-2-methylbutane

f. $CH_3-CH_2-CH_2-CH_2-CH_3$ Pentane

13.25 a.

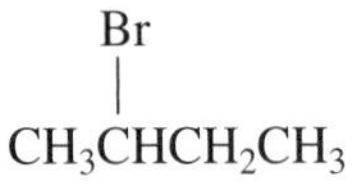

b.

c.

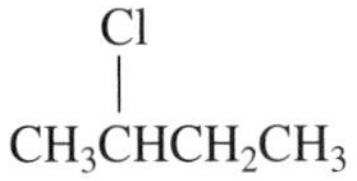

d.

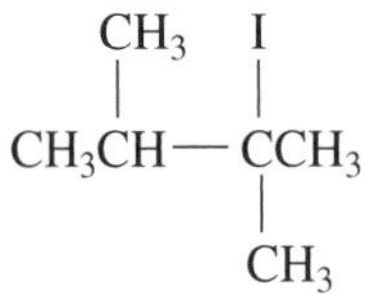

e.

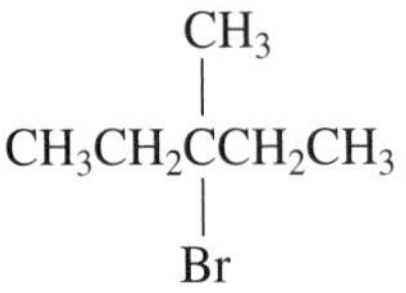

f.

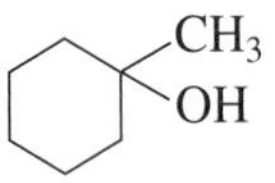

13.27

a.

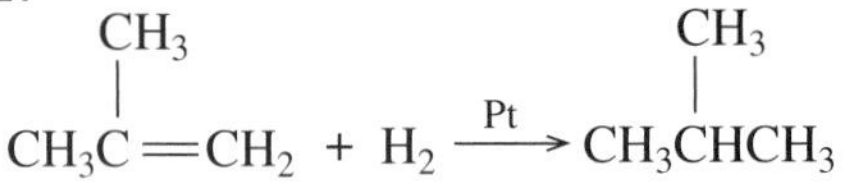

b.

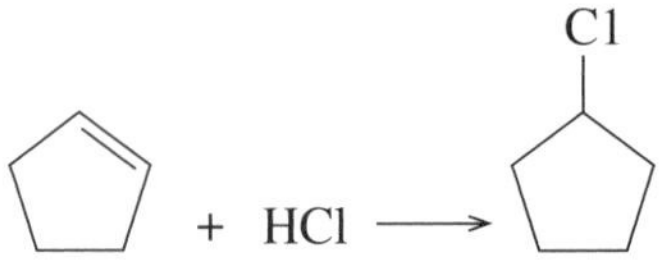

c.

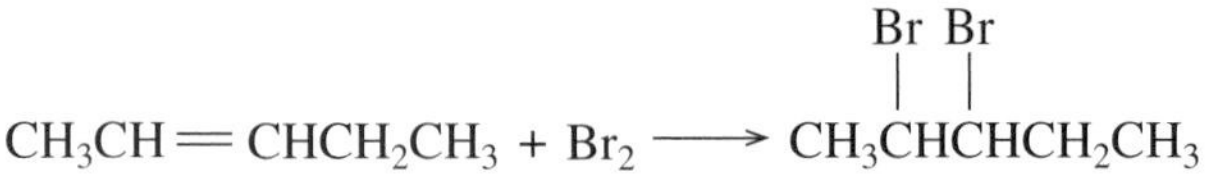

d.

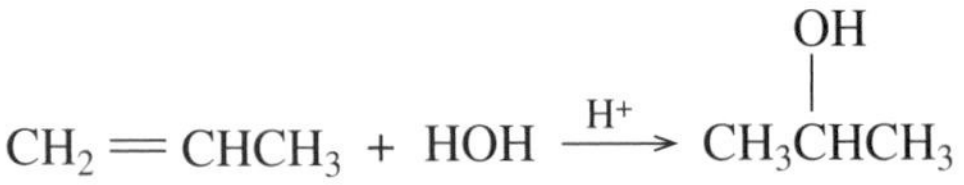

e.

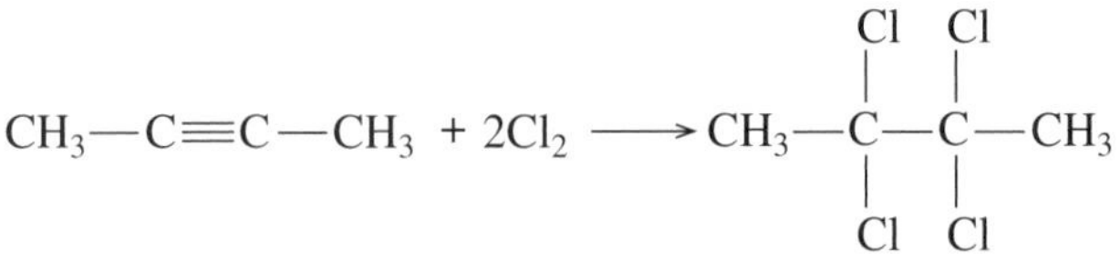

13.29 A polymer is a very large molecule composed of small units that are repeated many times.

13.31

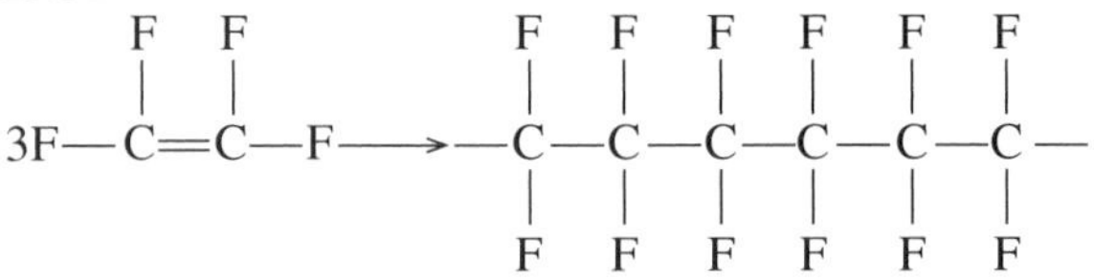

13.33

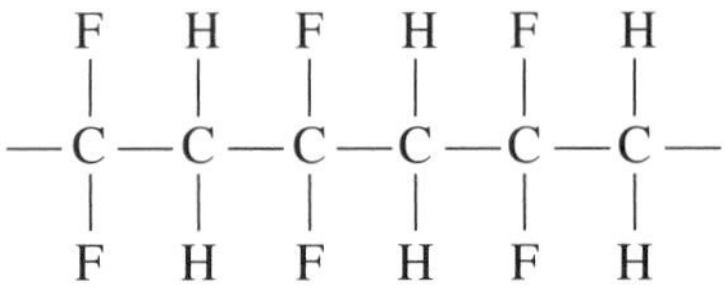

13.35 Cyclohexane, C_6H_{12}, is a cycloalkane in which six carbon atoms in a ring are linked by single bonds. In benzene, C_6H_6, alternating single and double bonds link the six carbon atoms.

13.37 a. 1-chloro-2-methylbenzene; *o*-chloromethylbenzene; *o*-chlorotoluene; 2-chlorotoluene
   b. ethylbenzene
   c. 1,3,5-trichlorobenzene
   d. *m*-xylene; *m*-methyltoluene;1,3-dimethylbenzene; 3-methyltoluene
   e. 1-bromo-3-chloro-5-methylbenzene; 3-bromo-5-chlorotoluene
   f. isopropyl benzene

13.39

a.

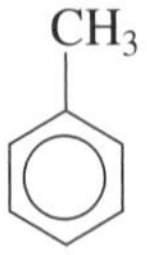

b.

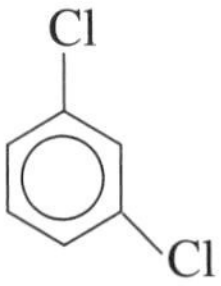

c.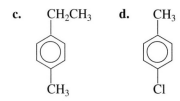

13.41 Benzene undergoes substitution reactions because a substitution reaction allows benzene to retain the stability of the aromatic system.

13.43 a. no reaction

b. Cl 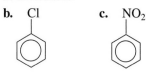 **c.** NO₂

13.45 All the compounds have three carbon atoms. Propane is a saturated alkane, and cyclopropane is a saturated cyclic hydrocarbon. Both propene and propyne are unsaturated hydrocarbons, but propene has a double bond and propyne has a triple bond.

13.47 a. chlorocyclopentane
b. 2-chloro-4-methylpentane
c. 2-methyl-1-pentene
d. 2-pentyne
e. 1-chlorocyclopentene
f. *trans*-2-pentene
g. 1,3-dichlorocyclohexene

13.49 a. constitutional isomers **b.** cis-trans isomers
c. identical   **d.** constitutional isomers

13.51

13.53
a.

CH₃ ... H   *trans*-2-pentene
H ... CH₂CH₃

CH₃ ... CH₂CH₃   *cis*-2-pentene
H ... H

b.

CH₃CH₂ ... H   *trans*-3-hexene
H ... CH₂CH₃

CH₃CH₂ ... CH₂CH₃   *cis*-3-hexene
H ... H

c.

CH₃ ... H   *trans*-2-butene
H ... CH₃

CH₃ ... CH₃   *cis*-2-butene
H ... H

d.

CH₃ ... H   *trans*-2-hexene
H ... CH₂CH₂CH₃

CH₃ ... CH₂CH₂CH₃   *cis*-2-hexene
H ... H

13.55

a.   **b.** CH₃CH=CHCH₂CH₃

c. CH₂=CHCH₃ **d .**

13.57

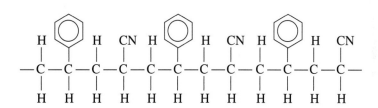

13.59 a. methylbenzene; toluene
b. 1-chloro-2-methylbenzene; *o*-chlorotoluene; 2-chlorotoluene
c. 1-ethyl-4-methylbenzene; *p*-ethylmethylbenzene; *p*-ethyltoluene
d. 1,3-diethylbenzene; *m*-diethylbenzene

13.61 a. chlorobenzene
b. *o*-bromotoluene, *m*-bromotoluene, *p*-bromotoluene
c. benzenesulfonic acid
d. no products

Chapter 14

Answers to Study Checks

14.1 3°; The carbon attached to the hydroxyl group is bonded to three carbon atoms.

14.2 3-bromo-4-methylcyclopentanol

14.3

CH₃CH₂CH—CH—CHCH₂CH₃
with OH, CH₃, Cl substituents

14.4 CH₃CH₂—SH

14.5 ethanol

14.6 ethyl phenyl ether

14.7 1-butanol, 2-butanol, methoxy-1-propane, ethoxy ethane.

14.8 Both are unsaturated cyclic ethers, but furan has five atoms in the ring and pyran has six atoms.

14.9 Ethanol molecules can hydrogen bond with each other, but ether molecules cannot. Thus, a higher temperature is required to break the hydrogen bonds between ethanol molecules.

14.10 cyclopentene

14.11 2-methyl-1-propanol, 2-methyl-2-propanol

14.12 cyclohexyl alcohol

14.13 Propene adds water in the presence of an acid to form 2-propanol, which loses two hydrogen atoms to form propanone (acetone).

CH₂=CHCH₃ + HOH —H⁺→ CH₃CHCH₃ (with OH) —[O]→

Propene   2-Propanol

CH₃CCH₃ (with =O)

Propanone

Answers to Selected Questions and Problems

14.1 a. 1°   **b.** 1°   **c.** 3°
       d. 2°   **e.** 3°

14.3 a. ethanol   **b.** 2-butanol
       c. 3-hexanol   **d.** 3-methyl-1-butanol
       e. 3,4-dimethylcyclohexanol
       f. 3,5,5-trimethyl-1-heptanol

14.5

a. CH₃CH₂CH₂OH   **b.** CH₃OH

c. CH₃CH₂CHCH₂CH₃ (with OH)   **d.** CH₃CCH₂CH₃ (with OH and CH₃)

e. cyclohexane with OH

f. HOCH₂CH₂CH₂CH₂OH

14.7 a. phenol
       b. 2-bromophenol, *o*-bromophenol
       c. 3,5-dichlorophenol
       d. 3-bromophenol, *m*-bromophenol

14.9 a. 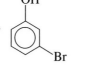   **b.**

c. phenol with two Cl (2,5)   **d.** phenol with attached phenyl ring

e.
       4-ethylphenol (OH top, CH₂CH₃ bottom)

14.11 a. methanethiol   **b.** 2-propanethiol
        c. 2,3-dimethyl-1-butanethiol
        d. cyclobutanethiol

14.13 a. ethanol   **b.** menthol   **c.** *ortho*-phenylphenol

14.15 a. methoxyethane, ethyl methyl ether
        b. methoxycyclohexane, cyclohexyl methyl ether
        c. ethoxycyclobutane, cyclobutyl ethyl ether
        d. 1-methoxypropane, methyl propyl ether

14.17 a. CH₃CH₂—O—CH₂CH₂CH₃

b. CH₃CH₂—O—△ (cyclopropyl)

c. OCH₃ on cyclopentane ring

d. CH₃CH₂—O—CH₂CHCH₂CH₃ (with CH₃)

e. CH₃CHCHCH₂CH₃ (with OCH₃ and OCH₃)

14.19 a. Constitutional isomers (C₅H₁₂O)
        b. Different compounds
        c. Constitutional isomers (C₅H₁₂O)

14.21 **a.** tetrahydrofuran
b. 3-methylfuran
c. 5-methyl-1,3-dioxane

14.23 **a.** methanol
b. 1-butanol
c. 1-butanol

14.25 **a.** yes, hydrogen bonding
b. yes; hydrogen bonding
c. no; long carbon chain diminishes effect of —OH group
d. no; alkanes are nonpolar
e. yes: —OH ionizes

14.27 **a.** CH_3CH_2CH=CH_2 **b.**

c.   **d.** CH_3CH_2CH=$CHCH_3$

14.29 **a.** CH_3OCH_3
b. $CH_3CH_2CH_2OCH_2CH_2CH_3$

14.31 **a.** CH_3CH_2OH **b.** $CH_3OH + CH_3CH_2OH$
c.

![cyclohexanol structure with OH]

14.33 **a.** $CH_3CH_2CH_2CH_2\overset{O}{\overset{\|}{C}}H$ **b.** $CH_3CH_2\overset{O}{\overset{\|}{C}}CH_3$
c. ![cyclohexanone]
d. $CH_3\overset{O}{\overset{\|}{C}}CH_2\overset{CH_3}{\underset{}{CH}}CHCH_3$
e. $CH_3\overset{CH_3}{\underset{}{CH}}CH_2\overset{O}{\overset{\|}{C}}H$

14.35 **a.** CH_3OH
b. ![cyclopentanol with OH]
c. $CH_3CH_2\overset{OH}{\underset{}{CH}}CH_3$
d. ![benzyl alcohol CH2OH]
e. ![3-methylcyclohexanol OH CH3]

14.37 **a.** 2°   **b.** 1°   **c.** 1°
d. 2°   **e.** 1°   **f.** 3°

14.39 **a.** alcohol   **b.** ether   **c.** thiol
d. alcohol   **e.** ether   **f.** cyclic ether
g. alcohol   **h.** phenol

14.41 **a.** 2-chloro-4-methylcyclohexanol
b. methyl phenyl ether
c. 2-propanethiol
d. 2,4-dimethyl-2-pentanol
e. methyl propyl ether   **f.** 3-methyl furan
g. 4-bromo-2-pentanol   **h.** *meta*-cresol

14.43

a. ![cyclopentane ring with OH and CH3]

b. ![4-chlorophenol with OH and Cl]

c. $H_3C-\overset{CH_3}{\underset{}{CH}}-\overset{OH}{\underset{}{CH}}-CH_2-CH_3$

d. ![phenyl ether with O—CH2—CH3]

e. $CH_3-CH_2-\overset{SH}{\underset{}{CH}}-CH_2-CH_3$ **f.** ![methyl phenol with CH3 and OH]

g. ![phenol with OH, two Br]

14.45 **a.** glycerol
b. 1,2-ethanediol; ethylene glycol
c. ethanol

14.47 $CH_3-CH_2-CH_2-CH_2-OH$

$CH_3-\overset{CH_3}{\underset{}{CH}}-CH_2-OH$

$CH_3-\overset{OH}{\underset{}{CH}}-CH_2-CH_3$

$CH_2-\overset{OH}{\underset{CH_3}{C}}-CH_3$

14.49 **a.** 1-propanol; hydrogen bonding
b. 1-propanol; hydrogen bonding
c. 1-butanol; higher molar mass

14.51 **a.** soluble; hydrogen bonding
b. soluble; hydrogen bonding
c. insoluble; long carbon chain diminishes effect of polar —OH on hydrogen bonding

14.53 **a.** CH$_3$—CH=CH$_2$

b. CH$_3$—CH$_2$—$\overset{\displaystyle O}{\overset{\|}{C}}$—H

c. CH$_3$—CH=CH—CH$_3$

d. CH$_3$—CH$_2$—$\overset{\displaystyle O}{\overset{\|}{C}}$—CH$_3$

e. CH$_3$—CH$_2$—CH$_2$—O—CH$_2$—CH$_2$—CH$_3$

f.    **g.**

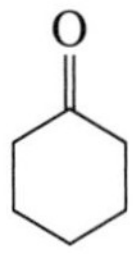

14.55

a. CH$_3$—CH$_2$—CH$_2$—OH $\xrightarrow{\text{H}^+,\ \text{heat}}$ CH$_3$—CH=CH$_2$ + HCl $\longrightarrow$ CH$_3$—$\overset{\displaystyle Cl}{\underset{}{CH}}$—CH$_3$

b. CH$_3$—$\overset{\displaystyle OH}{\underset{\displaystyle CH_3}{C}}$—CH$_3$ $\xrightarrow{\text{H}^+,\ \text{heat}}$ CH$_3$—$\overset{}{\underset{\displaystyle CH_3}{C}}$=CH$_2$ + H$_2$ $\xrightarrow{\text{Pt}}$ CH$_3$—$\overset{}{\underset{\displaystyle CH_3}{CH}}$—CH$_3$

c. CH$_3$—CH$_2$—CH$_2$—OH $\xrightarrow{\text{H}^+,\ \text{heat}}$ CH$_3$—CH=CH$_2$ + H$_2$O $\xrightarrow{\text{H}^+,\ \text{heat}}$ CH$_3$—$\overset{\displaystyle OH}{\underset{}{CH}}$—CH$_3$

$\xrightarrow{\text{[O]}}$ CH$_3$—$\overset{\displaystyle O}{\overset{\|}{C}}$—CH$_3$

14.57 **a.** cycloalkene, ketone, alcohol

14.59

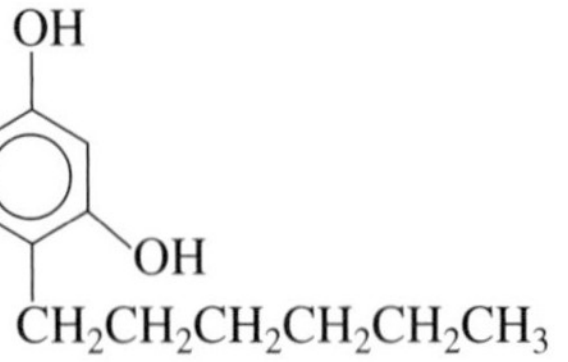

14.61 **a.**

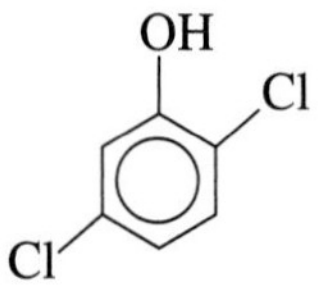

b. CH$_3$—$\overset{\displaystyle CH_3}{\underset{}{CH}}$—CH$_2$—CH$_2$—SH

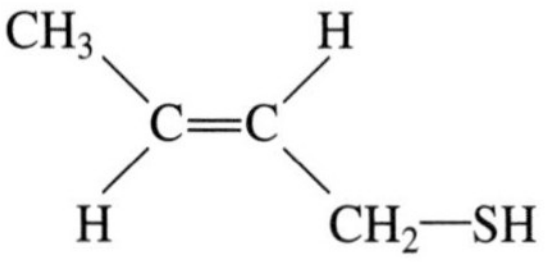

c.

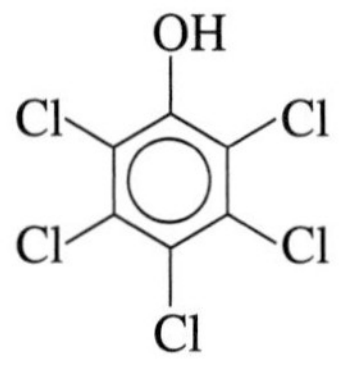

Chapter 15

Answers to Study Checks

15.1 CH$_3$—CH$_2$—$\overset{\displaystyle O}{\overset{\|}{C}}$—CH$_2$—CH$_3$

15.2 The oxygen atom is much more electronegative than the carbon atom in the carbonyl group.

15.3 ethyl propyl ketone

15.4 propanal (IUPAC), propionaldehyde (common)

15.5   HO—CH₂—C(=O)—CH₂—OH

15.6 The oxygen atom in the carbonyl group of acetone hydrogen bonds with water molecules.

15.7 achiral. A bowling pin has a symmetrical shape.

15.8   HO—CH₂—CH(OH)—CH(OH)—C(=O)—H

15.9

CHO	CHO
H——OH	HO——H
CH₃	CH₃
D-2-hydroxypropanal	L-2-hydroxypropanal

15.10 cyclohexanol

15.11 The oxidation of benzaldehyde reduces Ag^+ to metallic silver, which forms a silvery coating on the walls of the test tube.

15.12 1-propanol

15.13 hemiacetal

15.14   CH₃—C(OCH₃)(OCH₃)—CH₃

Answers to Selected Problems

15.1   a. ketone   **b.** aldehyde
       c. ketone   **d.** aldehyde

15.3   a. 1   **b.** 1   **c.** 2

15.5   a. propanal   **b.** 2-methyl-3-pentanone
       c. 3-hydroxybutanal   **d.** 2-pentanone
       e. 3-methylcyclohexanone   **f.** 4-chlorobenzaldehyde

15.7   a. acetaldehyde
       b. methyl propyl ketone
       c. formaldehyde

15.9   a. CH₃—C(=O)—H

b. CH₃—C(=O)—CH₂—CH(OH)—CH₃

c. CH₃—CH(Br)—CH(Br)—C(=O)—H

d. CH₃—C(=O)—CH₂—CH₂—CH₂—CH₃

e. CH₃—CH₂—CH(CH₃)—CH₂—C(=O)—H

f. CH₃—CH₂—C(=O)—CH₂—C(=O)—H

15.11   CHO — (benzene ring) — OCH₃

15.13   a. benzaldehyde
       b. acetone; propanone
       c. formaldehyde

15.15   a. CH₃—CH₂—C(=O)—H has a polar carbonyl group.
       b. pentanal has a higher molar mass and thus a higher boiling point
       c. 1-butanol because it can hydrogen bond with other 1-butanol molecules

15.17   a. CH₃—C(=O)—C(=O)—CH₃; more hydrogen bonding
       b. acetaldehyde; more hydrogen bonding
       c. acetone; lower number of carbon atoms

15.19 No. The long carbon chain diminishes the effect of the carbonyl group.

15.21   a. achiral

b. chiral   CH₃—CH(Br)—CH₂CH₃
                     Br chiral carbon

c. chiral   CH₃—CH(Br)—C(=O)—H
                     chiral carbon

d. achiral

15.23   a. CH₃—C(CH₃)=CH—CH₂—CH₂—CH(CH₃)—CH₂—CH₂—OH
          chiral carbon

b. H₂N—CH(CH₃)—C(=O)—OH
       chiral carbon

15.25 a.

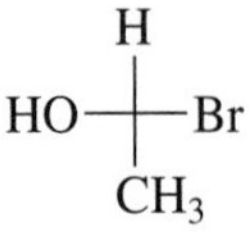

HO—C—Br with H on top and CH₃ on bottom

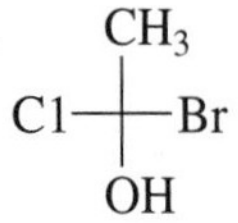

b.

Cl—C—Br with CH₃ on top and OH on bottom

c.

HO—C—H with CHO on top and CH₂CH₃ on bottom

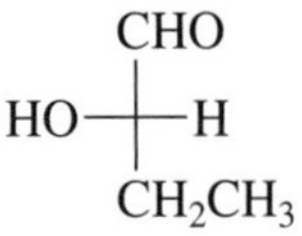

15.27 a. identical   **b.** enantiomers
c. enantiomers   **d.** enantiomers

15.29 a. CH_3OH

b. 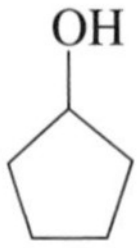 cyclopentane with OH

c. 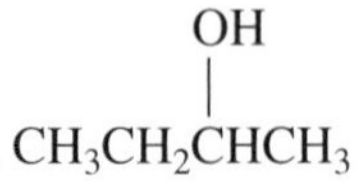 $CH_3CH_2CHCH_3$ with OH

d. 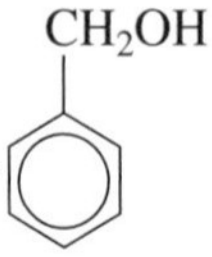 benzene ring with CH₂OH

e. 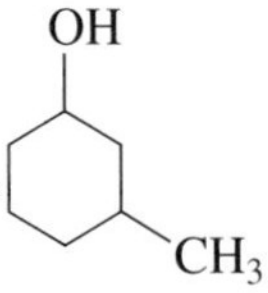 cyclohexane with OH and CH₃

15.31 a. 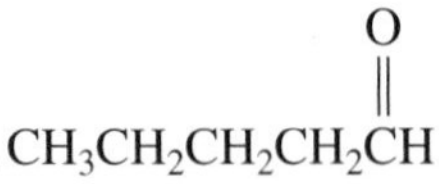 $CH_3CH_2CH_2CH_2CH$ with =O

b. 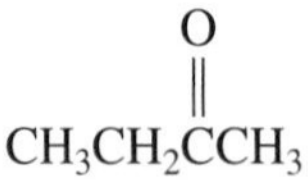 $CH_3CH_2CCH_3$ with =O

c. 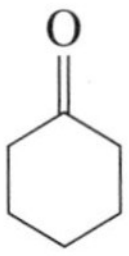 cyclohexanone

d. 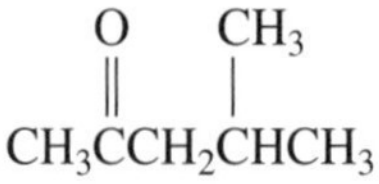 $CH_3CCH_2CHCH_3$ with =O and CH₃

e. 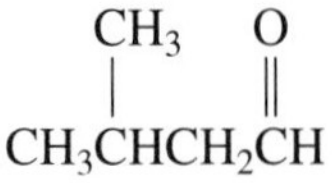 CH_3CHCH_2CH with CH₃ and =O

15.33 a. $CH_3CH_2CH_2CH_2OH$

b. 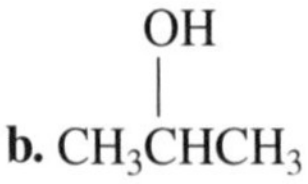 CH_3CHCH_3 with OH

c. 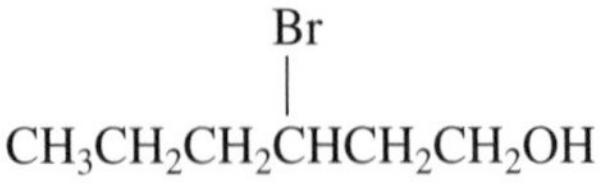 $CH_3CH_2CH_2CHCH_2CH_2OH$ with Br

d. 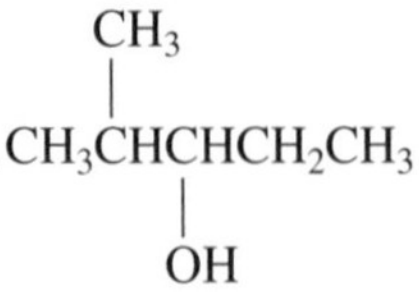 $CH_3CHCHCH_2CH_3$ with CH₃ and OH

15.35 a. 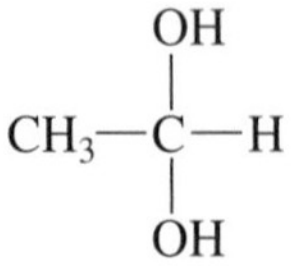 $CH_3—C—H$ with OH and OH

b. 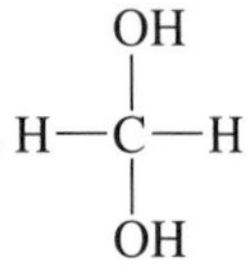 $H—C—H$ with OH and OH

15.37 a. hemiacetal   **b.** hemiacetal   **c.** acetal
d. hemiacetal   **e.** acetal

15.39 a. 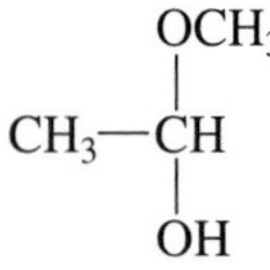 $CH_3—CH$ with OCH₃ and OH

b. 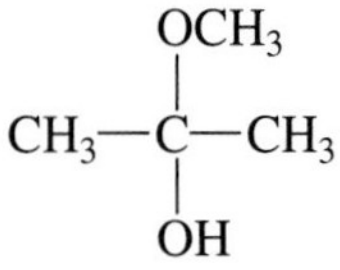 $CH_3—C—CH_3$ with OCH₃ and OH

c. cyclopentane with HO and OCH₃

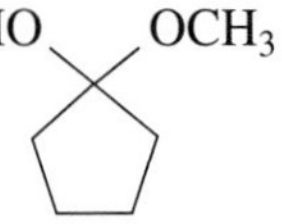

d. 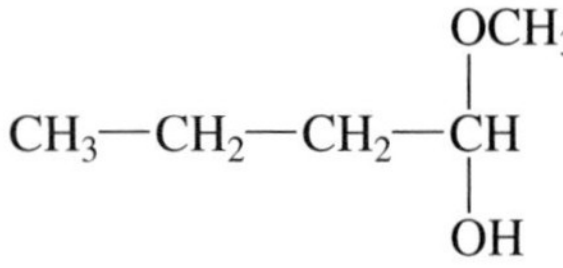 $CH_3—CH_2—CH_2—CH$ with OCH₃ and OH

15.41 a. 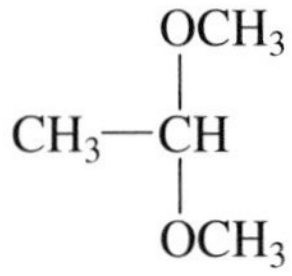 $CH_3—CH$ with OCH₃ and OCH₃

b. 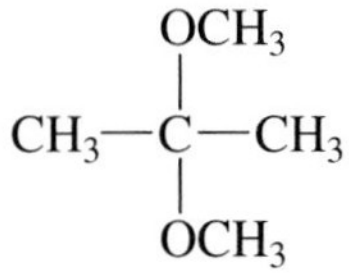 $CH_3—C—CH_3$ with OCH₃ and OCH₃

c. cyclopentane with CH₃O and OCH₃

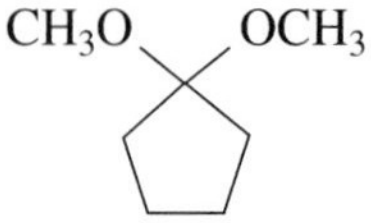

d. 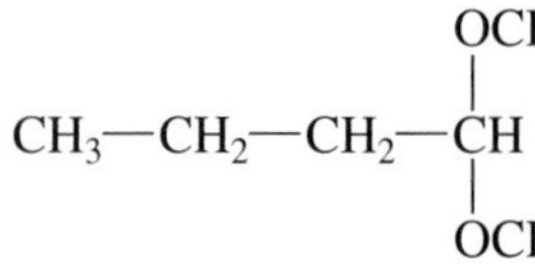 $CH_3—CH_2—CH_2—CH$ with OCH₃ and OCH₃

15.43 The carbonyl group consists of a sigma bond and a pi bond, which is an overlapping of the *p* orbitals of the carbon and oxygen atom.

15.45 $CH_3—CH_2—CH_2—C—H$ with =O; $CH_3—CH—C—H$ with CH₃ and =O;

$CH_3—CH_2—C—CH_3$ with =O

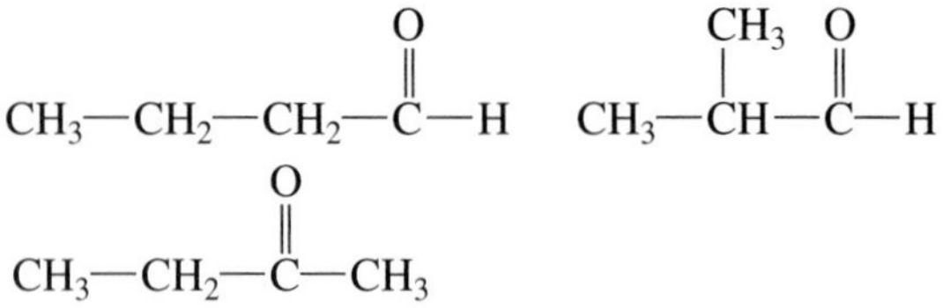

15.47 a. 2-bromo-4-chlorocyclopentanone
b. 4-chloro-3-hydroxybenzaldehyde
c. 3-chloropropanal
d. 5-hydroxy-3-hexanone
e. 2-chloro-3-pentanone
f. 3-methylcyclohexanone

15.49

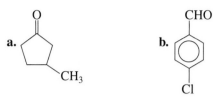

c. Cl—CH₂—CH₂—$\overset{\overset{\textstyle O}{\|}}{C}$—H

d. CH₃—CH₂—$\overset{\overset{\textstyle O}{\|}}{C}$—CH₃

e. CH₃—CH₂—CH₂—$\overset{\overset{\textstyle CH_3}{|}}{CH}$—CH₂—$\overset{\overset{\textstyle O}{\|}}{C}$—H

f. CH₃—CH₂—CH₂—CH₂—CH₂—$\overset{\overset{\textstyle O}{\|}}{C}$—CH₃

15.51 b, c, and d

15.53 a. CH₃—CH₂—OH   **b.** CH₃—CH₂—$\overset{\overset{\textstyle O}{\|}}{C}$—H

c. CH₃—CH₂—CH₂—OH

15.55

a. H—$\overset{\overset{\textstyle Cl}{|}}{\underset{\underset{\textstyle Cl}{|}}{C}}$—$\overset{\overset{\textstyle Cl}{|}}{\underset{\underset{\textstyle Cl}{|}}{Ⓒ}}$—O—H   **b.** none   **c.** none

d. CH₃—$\overset{\overset{\textstyle NH_2}{|}}{Ⓒ H}$—$\overset{\overset{\textstyle O}{\|}}{C}$—H

e. CH₃—CH₂—$\overset{\overset{\textstyle Br}{|}}{Ⓒ H}$—CH₂—CH₂—CH₃

f. none

15.57 a. identical   **b.** enantiomers
c. enantiomers (turn 180°)   **d.** identical

15.59 a. CH₃—CH₂—$\overset{\overset{\textstyle O}{\|}}{C}$—H $\xrightarrow[\text{oxidation}]{\text{Further}}$ $\overset{\overset{\textstyle O}{\|}}{\underset{\underset{\textstyle CH_3}{|}}{CH_2}}$—C—OH

b. CH₃—$\overset{\overset{\textstyle O}{\|}}{C}$—CH₂—CH₂—CH₃

c. CH₃—CH₂—CH₂—$\overset{\overset{\textstyle O}{\|}}{C}$—OH

d.
(cyclohexanone)

15.61

a. CH₃—$\overset{\overset{\textstyle OH}{|}}{CH}$—CH₃

b. (benzyl) CH₂—CH₂—OH

c. CH₃—$\overset{\overset{\textstyle CH_3}{|}}{CH}$—CH₂—$\overset{\overset{\textstyle OH}{|}}{CH}$—CH₃

15.63

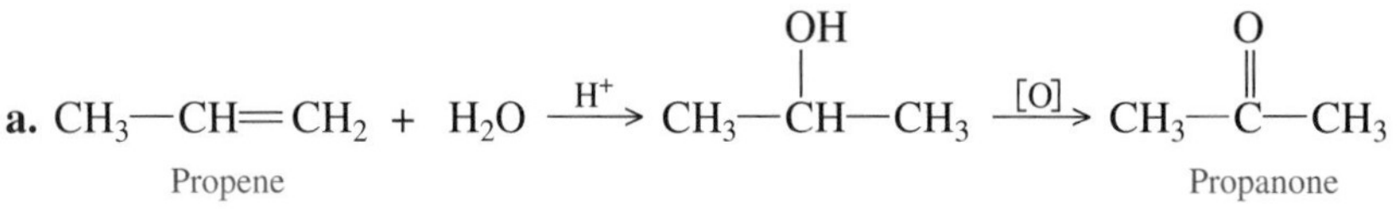

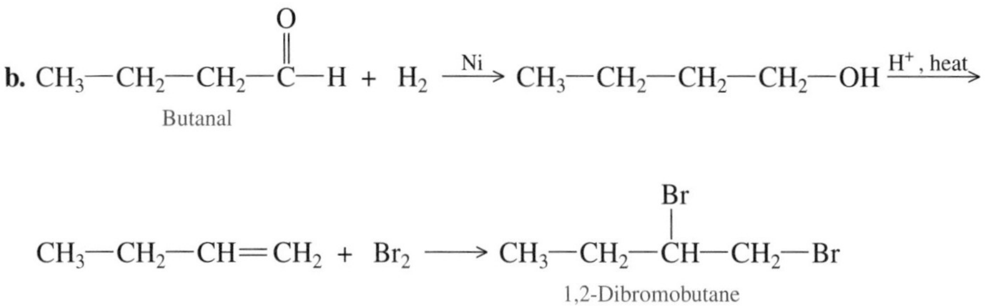

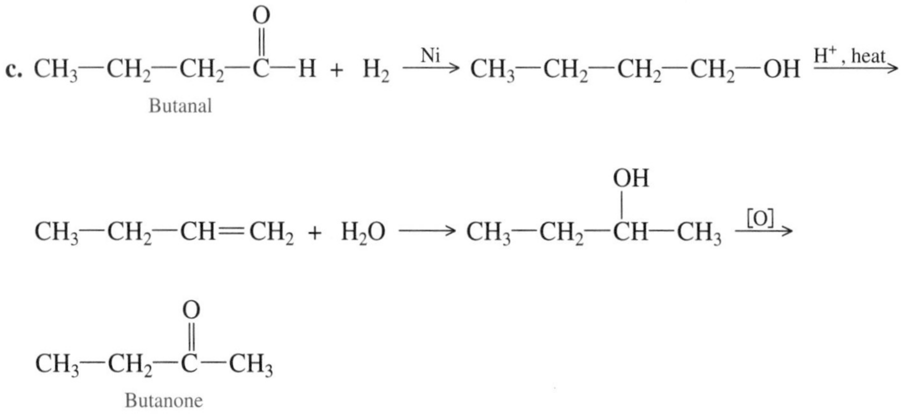

a. $CH_3—CH{=}CH_2$ + H_2O $\xrightarrow{H^+}$ $CH_3—\overset{\overset{\displaystyle OH}{|}}{CH}—CH_3$ $\xrightarrow{[O]}$ $CH_3—\overset{\overset{\displaystyle O}{\|}}{C}—CH_3$

Propene   Propanone

b. $CH_3—CH_2—CH_2—\overset{\overset{\displaystyle O}{\|}}{C}—H$ + H_2 $\xrightarrow{Ni}$ $CH_3—CH_2—CH_2—CH_2—OH$ $\xrightarrow{H^+,\ heat}$

Butanal

$CH_3—CH_2—CH{=}CH_2$ + Br_2 $\longrightarrow$ $CH_3—CH_2—\overset{\overset{\displaystyle Br}{|}}{CH}—CH_2—Br$

1,2-Dibromobutane

c. $CH_3—CH_2—CH_2—\overset{\overset{\displaystyle O}{\|}}{C}—H$ + H_2 $\xrightarrow{Ni}$ $CH_3—CH_2—CH_2—CH_2—OH$ $\xrightarrow{H^+,\ heat}$

Butanal

$CH_3—CH_2—CH{=}CH_2$ + H_2O $\longrightarrow$ $CH_3—CH_2—\overset{\overset{\displaystyle OH}{|}}{CH}—CH_3$ $\xrightarrow{[O]}$

$CH_3—CH_2—\overset{\overset{\displaystyle O}{\|}}{C}—CH_3$

Butanone

15.65 a. acetal; propanal and methanol
   b. hemiacetal; butanone and ethanol
   c. acetal; cyclohexanone and ethanol

Chapter 16

Answers to Study Checks

16.1 a monosaccharide

16.2 $\begin{array}{c} CH_2OH \\ | \\ C{=}O \\ | \\ CH_2OH \end{array}$

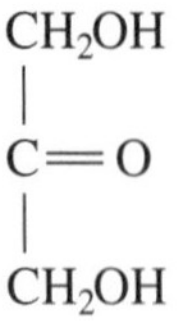

16.3

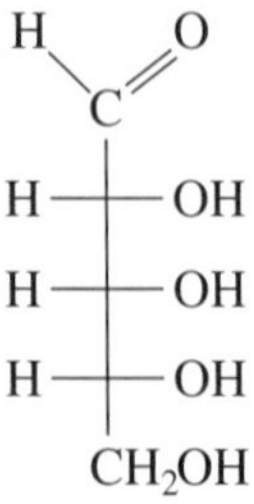

16.4 Ribulose is a ketopentose.

16.5

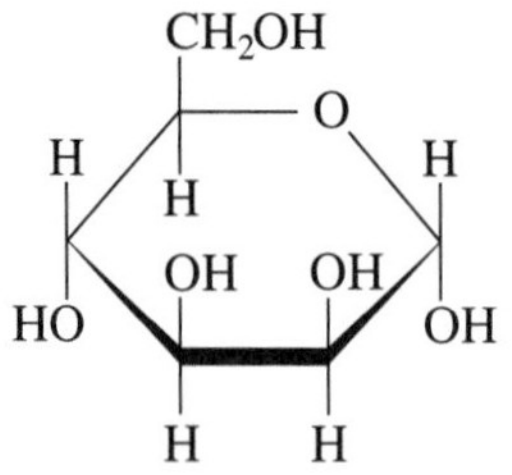

16.6 This indicates a high level of reducing sugar (probably glucose) in the urine. One common cause of this condition is diabetes mellitus.

16.7

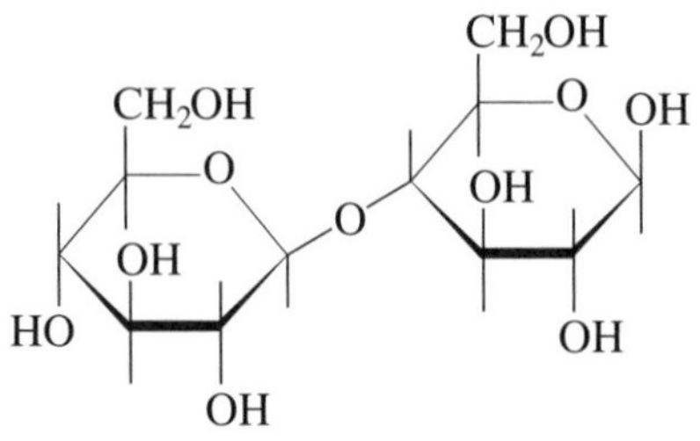

16.8 Cellulose contains glucose units connected by β-1,4-glycosidic bonds, whereas the glucose units in amylose are connected by α-1,4-glycosidic bonds.

Answers to Selected Questions and Problems

16.1 Photosynthesis requires CO_2, H_2O, and the energy from the sun. Respiration requires O_2 from the air and glucose from our foods.

16.3 A monosaccharide is a simple sugar composed of three to six carbon atoms. A disaccharide is composed of two monosaccharide units.

16.5 Hydroxyl groups are found in all monosaccharides along with a carbonyl on the first or second carbon.

16.7 A ketopentose contains hydroxyl and ketone functional groups and has five carbon atoms.

16.9 **a.** ketose **b.** aldose **c.** ketose
d. aldose **e.** aldose

16.11 A Fischer projection is a two-dimensional representation of the three-dimensional structure of a molecule.

16.13 **a.** D **b.** D **c.** L **d.** D

16.15 **a.** CHO / H—OH / HO—H / CH₂OH **b.** CH₂OH / C=O / H—OH / HO—H / CH₂OH

c. CHO / HO—H / HO—H / H—OH / H—OH / CH₂OH **d.** CHO / HO—H / HO—H / HO—H / HO—H / CH₂OH

16.17 D-Glucose and L-Glucose Fischer projections.

16.19 In D-galactose the hydroxyl on carbon four extends to the left. In glucose this hydroxyl goes to the right.

16.21 **a.** glucose **b.** galactose **c.** fructose

16.23 In the cyclic structure of glucose, there are five carbon atoms and an oxygen.

16.25 α-D-Glucose and β-D-Glucose cyclic structures.

16.27 **a.** α-anomer **b.** α-anomer

16.29 Xylitol Fischer projection.

16.31 Oxidation product and Reduction product (sugar alcohol): D-Arabitol.

16.33

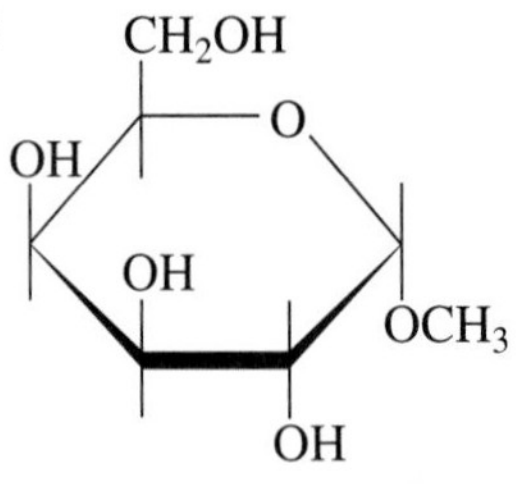

α-Anomer   β-Anomer

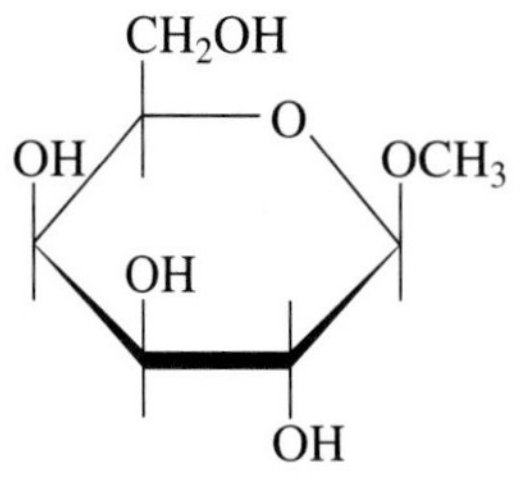

16.35 a. galactose and glucose; β-1,4 bond; β-lactose
  b. glucose and glucose; α-1,4 bond; α-maltose

16.37 a. Will undergo mutarotation; can be oxidized
  b. Will undergo mutarotation; can be oxidized

16.39 a. sucrose   **b.** lactose
  c. maltose   **d.** lactose

16.41 a. Amylose is an unbranched polymer of glucose units joined by α-1,4 bonds; amylopectin is a branched polymer of glucose joined by α-1,4 and α-1,6 bonds.
  b. Amylopectin, which is produced in plants, is a branched polymer of glucose, joined by α-1,4 and α-1,6 bonds. Glycogen, which is produced in animals, is a highly branched polymer of glucose, joined by α-1,4 and α-1,6 bonds.

16.43 a. cellulose   **b.** amylose, amylopectin
  c. amylose   **d.** glycogen

16.45 They differ only at carbon 4 where the —OH in D-glucose is on the right side and in D-galactose it is on the left side.

16.47 D-galactose is the mirror image of L-galactose. In D-galactose, the —OH group on carbon 5 is on the right side whereas in L-galactose, the —OH group on carbon 5 is on the left side.

16.49 a.

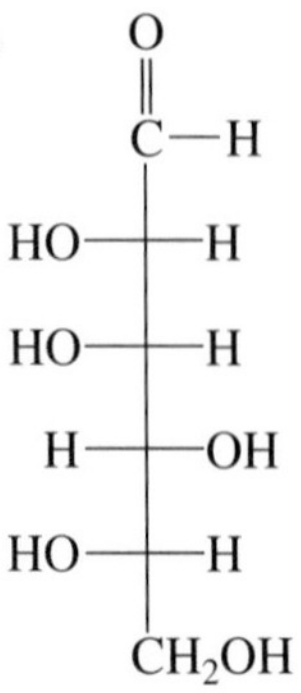

L-Gulose

b.

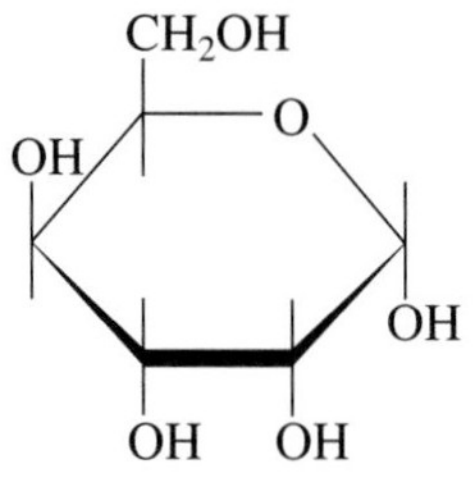

α-D-Gulose   β-D-Gulose

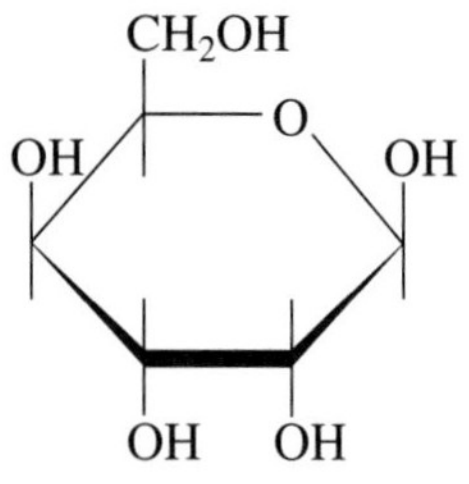

16.51

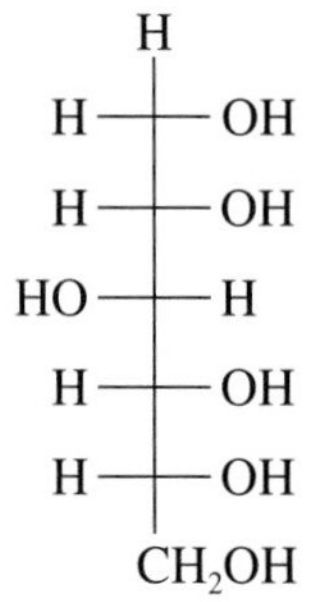

16.53 The α-galactose forms an open-chain structure, and when the chain closes, it can form both α- and β-galactose.

16.55

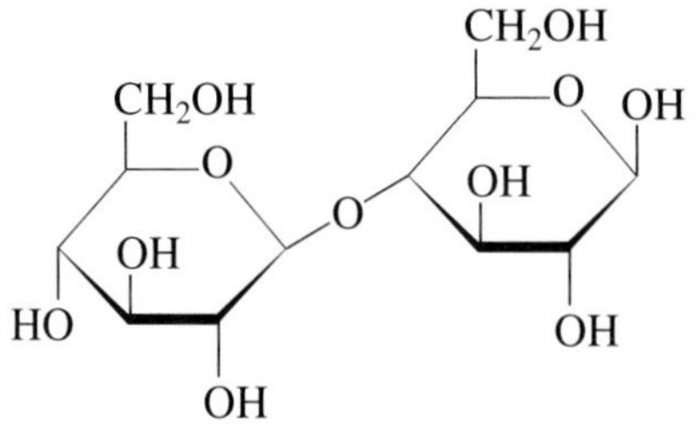

16.57 a.

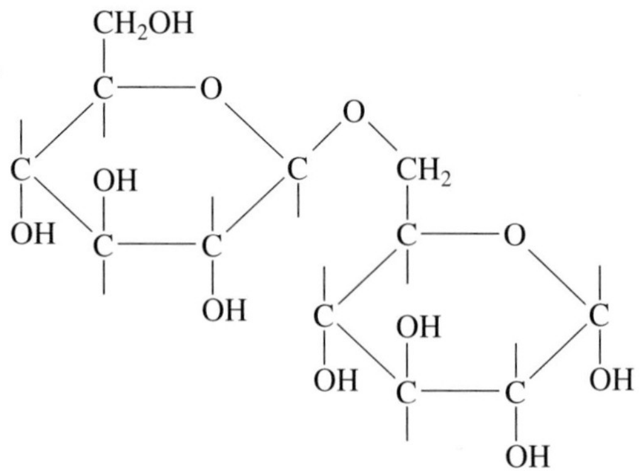

  b. Yes. The hemiacetal on the right side can open up to form the open chain with the aldehyde.

Chapter 17

Answers to Study Checks

17.1

17.2

$CH_3CH_2CH_2\overset{\overset{\displaystyle O}{\|}}{C}OH$

17.3 Two carboxylic acid molecules hydrogen bond, forming a dimer, which has a mass twice that of the single acid molecule.

17.4 $H\overset{\overset{\displaystyle O}{\|}}{C}OH + H_2O \rightleftharpoons H\overset{\overset{\displaystyle O}{\|}}{C}O^- + H_3O^+$

17.5 butanoic acid, butyric acid

17.6 propanoic (propionic) acid and 1-pentanol

17.7 $CH_3\overset{\overset{\displaystyle O}{\|}}{C}\!-\!OCH_2CH_2CH_2CH_2CH_3$

17.8 propanoic (propionic) acid and ethanol

17.9

$+ CH_3OH$

Answers to Selected Problems

17.1 methanoic acid (formic acid)

17.3 Each compound contains three carbon atoms. They differ because propanal, an aldehyde, contains a carbonyl group bonded to a hydrogen. In propanoic acid, the carbonyl group connects to a hydroxyl group, forming a carboxyl group.

17.5 **a.** ethanoic acid (acetic acid)
b. butanoic acid (butyric acid)
c. 2-chloropropanoic acid (α-chloropropionic acid)
d. 3-methylhexanoic acid
e. 3,4-dihydroxybenzoic acid
f. 4-bromopentanoic acid

17.7 **a.** $CH_3\!-\!CH_2\!-\!\overset{\overset{\displaystyle O}{\|}}{C}\!-\!OH$ **b.**

c. $Cl\!-\!CH_2\!-\!\overset{\overset{\displaystyle O}{\|}}{C}\!-\!OH$

d. $HO\!-\!CH_2\!-\!CH_2\!-\!\overset{\overset{\displaystyle O}{\|}}{C}\!-\!OH$

e. $CH_3\!-\!CH_2\!-\!\underset{\underset{\displaystyle CH_3}{|}}{CH}\!-\!\overset{\overset{\displaystyle O}{\|}}{C}\!-\!OH$

f. $CH_3\!-\!CH_2\!-\!\underset{\underset{\displaystyle Br}{|}}{CH}\!-\!CH_2\!-\!\underset{\underset{\displaystyle Br}{|}}{CH}\!-\!CH_2\!-\!\overset{\overset{\displaystyle O}{\|}}{C}\!-\!OH$

17.9 a. $H\overset{\overset{\displaystyle O}{\|}}{C}OH$ **b.** $CH_3\!-\!\overset{\overset{\displaystyle O}{\|}}{C}\!-\!OH$

c. $CH_3\!-\!\underset{\underset{\displaystyle CH_3}{|}}{CH}\!-\!CH_2\!-\!\overset{\overset{\displaystyle O}{\|}}{C}\!-\!OH$

d.

17.11 a. Butanoic acid has a higher molar mass and would have a higher boiling point.
b. Propanoic acid can form more hydrogen bonds and would have a higher boiling point.
c. Butanoic acid can form hydrogen bonds and would have a higher boiling point.

17.13 a. acetone, propanol, propanoic acid because of the greater number of hydrogen bonds.
b. butanoic acid, propanoic acid, acetic acid because of the decrease in molar mass
c. propane, ethanol, acetic acid because of the greater number of hydrogen bonds.

17.15

a. $H\!-\!\overset{\overset{\displaystyle O}{\|}}{C}\!-\!OH + H_2O \rightleftharpoons H\!-\!\overset{\overset{\displaystyle O}{\|}}{C}\!-\!O^- + H_3O^+$

b. $CH_3\!-\!CH_2\!-\!\overset{\overset{\displaystyle O}{\|}}{C}\!-\!OH + H_2O \rightleftharpoons CH_3\!-\!CH_2\!-\!\overset{\overset{\displaystyle O}{\|}}{C}\!-\!O^- + H_3O^+$

c. $CH_3\!-\!\overset{\overset{\displaystyle O}{\|}}{C}\!-\!OH + H_2O \rightleftharpoons CH_3\!-\!\overset{\overset{\displaystyle O}{\|}}{C}\!-\!O^- + H_3O^+$

17.17 a.
$$H-\overset{\overset{\displaystyle O}{\|}}{C}-OH \ + \ NaOH \longrightarrow \ H-\overset{\overset{\displaystyle O}{\|}}{C}-O^-Na^+ \ + \ H_2O$$

b.
$$CH_3-CH_2-\overset{\overset{\displaystyle O}{\|}}{C}-OH \ + \ NaOH \longrightarrow CH_3-CH_2-\overset{\overset{\displaystyle O}{\|}}{C}-O^-Na^+ \ + \ H_2O$$

c.
$$\underset{\bigcirc}{\overset{\overset{\displaystyle O}{\|}}{C}-OH} \ + \ NaOH \longrightarrow \underset{\bigcirc}{\overset{\overset{\displaystyle O}{\|}}{C}-O^-Na^+} \ + \ H_2O$$

17.19 a. sodium methanoate, sodium formate
   b. sodium propanoate, sodium propionate
   c. sodium benzoate

17.21 a. aldehyde   **b.** ester
   c. ketone   **d.** carboxylic acid

17.23 a. $CH_3-\overset{\overset{\displaystyle O}{\|}}{C}-O-CH_3$

b. $CH_3-CH_2-CH_2-\overset{\overset{\displaystyle O}{\|}}{C}-O-CH_3$

c. $\underset{\bigcirc}{\overset{\overset{\displaystyle O}{\|}}{C}-O-CH_3}$

17.25 a. $CH_3-CH_2-\overset{\overset{\displaystyle O}{\|}}{C}-O-CH_2-CH_2-CH_3$

b. $CH_3-CH_2-CH_2-CH_2-\overset{\overset{\displaystyle O}{\|}}{C}-O-\overset{\overset{\displaystyle CH_3}{|}}{CH}-CH_3$

17.27 a. formic acid (methanoic acid) and methyl alcohol (methanol)
   b. acetic acid (ethanoic acid) and methyl alcohol (methanol)
   c. butyric acid (butanoic acid) and methyl alcohol (methanol)
   d. β-methylbutyric acid (3-methylbutanoic acid) and ethyl alcohol (ethanol)

17.29 a. methyl formate (methyl methanoate)
   b. methyl acetate (methyl ethanoate)
   c. methyl butyrate (methyl butanoate)
   d. ethyl-β-methyl butyrate (ethyl-3-methyl butanoate)

17.31 a. $CH_3-\overset{\overset{\displaystyle O}{\|}}{C}-O-CH_3$

b. $H-\overset{\overset{\displaystyle O}{\|}}{C}-O-CH_2-CH_2-CH_2-CH_3$

c. $CH_3-CH_2-CH_2-CH_2-\overset{\overset{\displaystyle O}{\|}}{C}-O-CH_2-CH_3$

d. $CH_3-CH_2-\overset{\overset{\displaystyle O}{\|}}{C}-O-CH_2-\overset{\overset{\displaystyle Br}{|}}{CH}-CH_3$

17.33 a. pentyl ethanoate (pentyl acetate)
   b. octyl ethanoate (octyl acetate)
   c. pentyl butanoate (pentyl butyrate)
   d. isobutyl methanoate (isobutyl formate)

17.35 a. $CH_3-\overset{\overset{\displaystyle O}{\|}}{C}-OH$

b. $CH_3-CH_2-CH_2-CH_2-OH$

c. $CH_3-O-\overset{\overset{\displaystyle O}{\|}}{C}-CH_3$

17.37 The products of the acid hydrolysis of an ester are an alcohol and a carboxylic acid.

17.39

a. CH$_3$CH$_2$CO$^-$Na$^+$ and   CH$_3$OH (with O double bond on carbonyl)

b. CH$_3$COH   and   CH$_3$CH$_2$CH$_2$OH

c. CH$_3$CH$_2$CH$_2$COH   and   CH$_3$CH$_2$OH

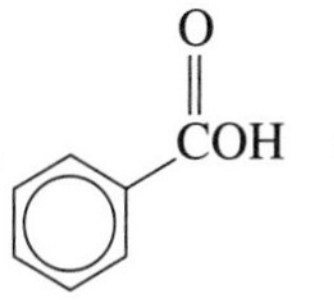

d. COH   and   CH$_3$CH$_2$OH

e. O$^-$Na$^+$ and   CH$_3$CH$_2$OH

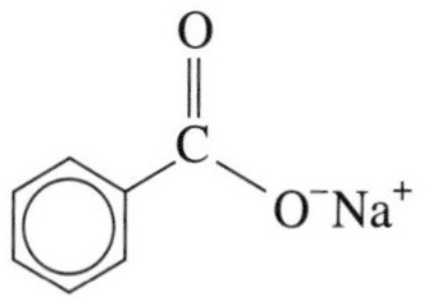

17.41 a. 3-methylbutanoic acid; β-methylbuyric acid
b. ethyl benzoate
c. ethyl propanoate; ethyl propionate
d. 2-chlorobenzoic acid; *ortho*-chlorobenzoic acid
e. 4-hydroxypentanoic acid
f. 2-propyl ethanoate; isopropyl acetate

17.43

CH$_3$—CH$_2$—CH$_2$—CH$_2$—C—OH   CH$_3$—CH$_2$—CH—C—OH (CH$_3$)

CH$_3$—CH—CH$_2$—C—OH (CH$_3$)   CH$_3$—C—C—OH (CH$_3$, CH$_3$)

17.45 a. CH$_3$—O—C—CH$_3$ **b.** 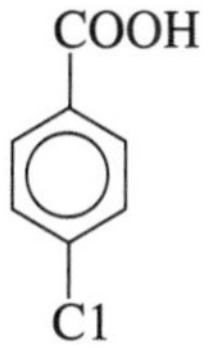 (ring COOH, Cl)

c. Cl—CH$_2$—CH$_2$—C—OH

d. CH$_3$—CH$_2$—O—C—CH$_2$—CH$_2$—CH$_3$

e. CH$_3$—CH$_2$—CH(CH$_3$)—CH$_2$—C—OH

f. 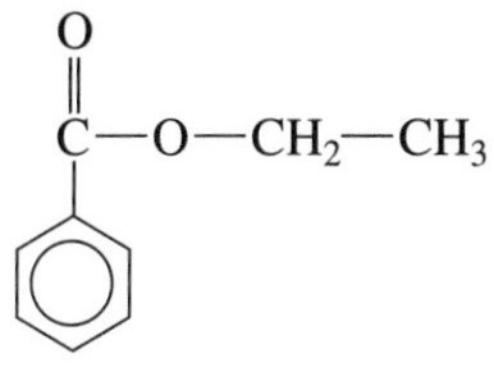 C—O—CH$_2$—CH$_3$ (benzene ring)

17.47 a. CH$_3$—C—OH

b. CH$_3$—CH$_2$—C—OH

c. CH$_3$—CH$_2$—CH$_2$—C—OH

17.49 The presence of two polar groups in the carboxyl group allows hydrogen bonding and the formation of a dimer that doubles the effective molar mass.

17.51 b, c, d, and e are all soluble in water

17.53

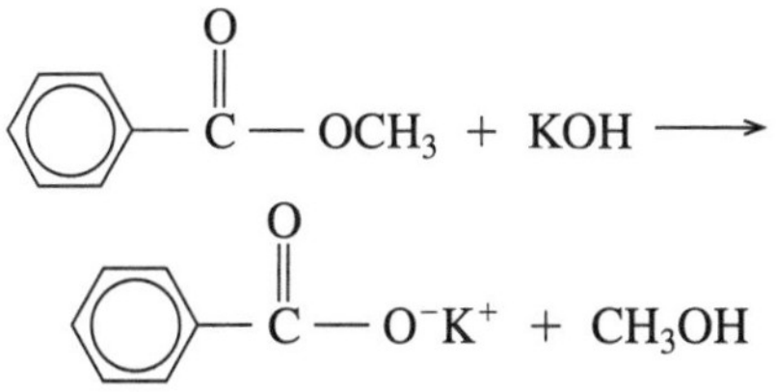

C—OCH$_3$ + KOH ⟶ C—O$^-$K$^+$ + CH$_3$OH

A soluble salt, potassium benzoate, is formed. When acid is added, the salt is converted to insoluble benzoic acid.

17.55 a. hydroxyl and carboxylic acid

b. 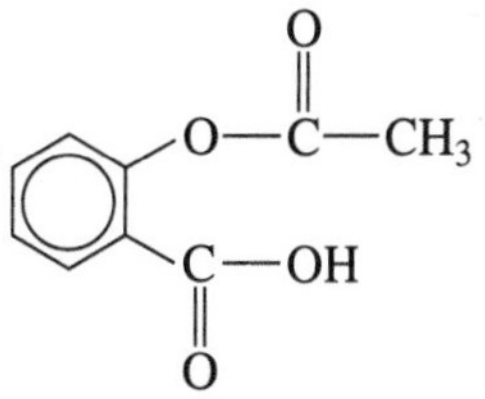 O—C—CH$_3$, C—OH

c. OH, C—OCH$_3$

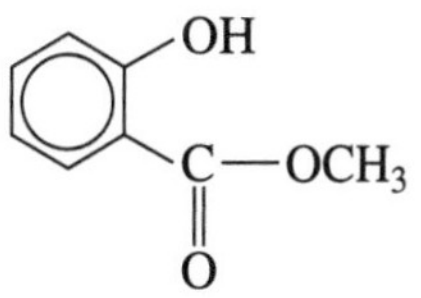

17.57

a. CH$_3$—CH$_2$—C—O$^-$ + H$_3$O$^+$

b. CH$_3$—CH$_2$—C—O$^-$K$^+$ + H$_2$O

c. CH$_3$—CH$_2$—C—O—CH$_3$ + H$_2$O

d.

$$\underset{\underset{}{}}{C}-O-CH_2-CH_3 \ + \ H_2O$$

17.59 a. 3-methylbutanoic acid and methanol

b. 3-chlorobenzoic acid and ethanol

c. hexanoic acid and methanol

17.61

a. $CH_3-CH_2-\underset{\underset{}{}}{C}-OH$ and $HO-\underset{\underset{}{CH_3}}{CH}-CH_3$

b. $CH_3-\underset{\underset{CH_3}{}}{CH}-\underset{\underset{}{}}{C}-O^-Na^+$ and $HO-CH_2-CH_2-CH_3$

17.63

a. $CH_2{=\!=}CH_2 \ + \ H_2O \xrightarrow{H^+} CH_3CH_2OH \xrightarrow{[O]} CH_3\underset{\underset{}{}}{C}OH$

b. $CH_3-CH_2-CH_2-CH_2-OH \xrightarrow{[O]} CH_3-CH_2-CH_2-\underset{\underset{}{}}{C}-OH$

Chapter 18

Answers to Study Checks

18.1 a glycolipid

18.2 a. 16   **b.** unsaturated   **c.** liquid

18.3

$$CH_2-O-\underset{\underset{}{}}{C}-(CH_2)_{12}-CH_3$$
$$CH-O-\underset{\underset{}{}}{C}-(CH_2)_{12}-CH_3$$
$$CH_2-O-\underset{\underset{}{}}{C}-(CH_2)_{12}-CH_3$$

18.4 tristearin

18.5 Triacylglycerols contain three fatty acids connected by ester bonds to glycerol. Glycerophospholipids con-tain glycerol esterified to two fatty acids and a phos-phate connected to an amino alcohol.

18.6 Cerebrosides contain only one monosaccharide, and gangliosides contain two or more monosaccharide units.

18.7 Cholesterol is not soluble in water; it is classified with the lipid family.

18.8 Testosterone and nandrolone both contain a steroid nucleus with one double bond and a ketone group in the first ring, and a methyl and alcohol group on the five-carbon ring. Nandrolone does not have the sec-ond methyl group at the first and second ring fusion that is seen in the structure of testosterone.

18.9 Protein channels allow electrolytes to flow in and out of the cell through the lipid bilayer.

Answers to Selected Questions and Problems

18.1 Lipids provide energy and protection and insulation for the organs in the body.

18.3 Because lipids are not soluble in water, a polar sol-vent, they are nonpolar molecules.

18.5 All fatty acids contain a long chain of carbon atoms with a carboxylic acid group. Saturated fats contain only carbon-to-carbon single bonds; unsaturated fats contain one or more double bonds. More saturated fats are found in animal fats, whereas vegetable oils contain more unsaturated fats.

18.7 a. palmitic acid

COOH

b. oleic acid

COOH

18.9 a. saturated   **b.** unsaturated
  c. unsaturated   **d.** saturated

18.11 In a cis fatty acid, the hydrogen atoms are on the same side of the double bond, which produces a bend in the carbon chain. In a trans fatty acid, the hydrogen atoms are on opposite sides of the double bond, which gives a carbon chain without any bend.

18.13 In an omega-3 fatty acid, there is a double bond on carbon 3 counting from the methyl group, whereas in an omega-6 fatty acid, there is a double bond begin-ning at carbon 6 counting from the methyl group.

18.15 Arachidonic acid contains four double bonds and no side groups. In PGE_2, a part of the chain forms cyclopentane and there are hydroxyl and ketone functional groups.

18.17 Prostaglandins affect blood pressure and stimulate contraction and relaxation of smooth muscle.

18.19

$$CH_3(CH_2)_{14}\overset{\overset{\displaystyle O}{\|}}{C}O(CH_2)_{29}CH_3$$

18.21

$$\begin{array}{l} CH_2O\overset{\overset{\displaystyle O}{\|}}{C}(CH_2)_{16}CH_3 \\ | \quad\;\; O \\ CHO\overset{\|}{C}(CH_2)_{16}CH_3 \\ | \quad\;\; O \\ CH_2O\overset{\|}{C}(CH_2)_{16}CH_3 \end{array}$$

18.23

$$\begin{array}{l} CH_2O\overset{\overset{\displaystyle O}{\|}}{C}(CH_2)_{14}CH_3 \\ | \quad\;\; O \\ CHO\overset{\|}{C}(CH_2)_{14}CH_3 \\ | \quad\;\; O \\ CH_2O\overset{\|}{C}(CH_2)_{14}CH_3 \end{array}$$

18.25 Safflower oil contains fatty acids with two or three double bonds; olive oil contains a large amount of oleic acid, which has only one (monounsaturated) double bond.

18.27 Although coconut oil comes from a vegetable, it has large amounts of saturated fatty acids and small amounts of unsaturated fatty acids.

18.29

$$\begin{array}{l} CH_2O\overset{\overset{\displaystyle O}{\|}}{C}(CH_2)_7CH{=}CH(CH_2)_7CH_3 \\ | \quad\;\; O \\ CHO\overset{\|}{C}(CH_2)_7CH{=}CH(CH_2)_7CH_3 \;+\; 3H_2 \xrightarrow{\text{Pt}} \\ | \quad\;\; O \\ CH_2O\overset{\|}{C}(CH_2)_7CH{=}CH(CH_2)_7CH_3 \end{array}$$

$$\begin{array}{l} CH_2O\overset{\overset{\displaystyle O}{\|}}{C}(CH_2)_{16}CH_3 \\ | \quad\;\; O \\ CHO\overset{\|}{C}(CH_2)_{16}CH_3 \\ | \quad\;\; O \\ CH_2O\overset{\|}{C}(CH_2)_{16}CH_3 \end{array}$$

18.31 a. Some of the double bonds in the unsaturated fatty acids have been converted to single bonds.
 b. It is mostly saturated fatty acids.

18.33 a.

$$\begin{array}{l} CH_2O\overset{\overset{\displaystyle O}{\|}}{C}(CH_2)_{12}CH_3 \\ | \quad\;\; O \\ CHO\overset{\|}{C}(CH_2)_{12}CH_3 \;+\; 3H_2O \xrightarrow{H^+} \\ | \quad\;\; O \\ CH_2O\overset{\|}{C}(CH_2)_{12}CH_3 \end{array}$$

$$\begin{array}{l} CH_2OH \\ | \qquad\qquad\qquad\quad O \\ CHOH \;+\; 3CH_3(CH_2)_{12}\overset{\|}{C}OH \\ | \\ CH_2OH \end{array}$$

 b.

$$\begin{array}{l} CH_2O\overset{\overset{\displaystyle O}{\|}}{C}(CH_2)_{12}CH_3 \\ | \quad\;\; O \\ CHO\overset{\|}{C}(CH_2)_{12}CH_3 \;+\; 3\,NaOH \longrightarrow \\ | \quad\;\; O \\ CH_2O\overset{\|}{C}(CH_2)_{12}CH_3 \end{array}$$

$$\begin{array}{l} CH_2OH \\ | \qquad\qquad\qquad\quad O \\ CHOH \;+\; 3CH_3(CH_2)_{12}\overset{\|}{C}O^-Na^+ \\ | \\ CH_2OH \end{array}$$

18.35 A triacylglycerol is composed of glycerol with three hydroxyl groups that form ester links with three long-chain fatty acids. In olestra, six to eight long-chain fatty acids form ester links with the hydroxyl groups on sucrose, a sugar. The olestra cannot be digested because our enzymes cannot break down the large olestra molecule.

18.37

$$\begin{array}{l} CH_2O{-}\overset{\overset{\displaystyle O}{\|}}{C}{-}(CH_2)_{16}CH_3 \\ | \qquad\;\; O \\ CHO{-}\overset{\|}{C}{-}(CH_2)_{16}CH_3 \\ | \qquad\;\; O \\ CH_2O{-}\overset{\|}{C}{-}(CH_2)_{16}CH_3 \end{array}$$

18.39 A triacylglycerol consists of glycerol and three fatty acids. A glycerophospholipid consists of glycerol, two fatty acids, a phosphate group, and an amino alcohol.

18.41

$$CH_2OC(CH_2)_{14}CH_3$$
(with C=O above)
$$CHOC(CH_2)_{14}CH_3$$
(with C=O above)
$$CH_2OPOCH_2CH_2\overset{+}{N}H_3$$
(with P=O above, O^- below)

This is a cephalin

18.43 This phospholipid is a cephalin. It contains glycerol, oleic acid, stearic acid, a phosphate, and ethanolamine.

18.45 A sphingolipid contains the amino alcohol sphingosine (instead of glycerol) and only one fatty acid. A glycerophospholipid consists of glycerol, two fatty acids, a phosphate group, and an amino alcohol.

18.47

$$CH_3(CH_2)_{12}CH=CHOH$$
$$CHNHC(CH_2)_{14}CH_3$$

HOCH₂ ... HO ... OH ... O—CH₂ ... OH (sugar ring structure)

18.49

(steroid nucleus structure)

18.51 Bile salts act to emulsify fat globules, allowing the fat to be more easily digested.

18.53 Lipoproteins are large, spherically shaped molecules that transport lipids in the bloodstream. They consist of an outside layer of phospholipids and proteins surrounding an inner core of hundreds of nonpolar lipids and cholesterol esters.

18.55 Chylomicrons have a lower density than VLDLs. They pick up triacylglycerols from the intestine, whereas VLDLs transport triacylglycerols synthesized in the liver.

18.57 "Bad" cholesterol is the cholesterol carried by LDLs that can form deposits in the arteries called plaque, which narrow the arteries.

18.59 Both estradiol and testosterone contain the steroid nucleus and a hydroxyl group. Testosterone has a

ketone group, a double bond, and two methyl groups. Estradiol has a benzene ring, a hydroxyl group in place of the ketone, and a methyl group.

18.61 Testosterone is a male sex hormone.

18.63 Phospholipids with smaller amounts of glycolipids and cholesterol.

18.65 The lipid bilayer in a cell membrane surrounds the cell and separates the contents of the cell from the external fluids.

18.67 The peripheral proteins in the membrane emerge on the inner or outer surface only, whereas the integral proteins extend through the membrane to both surfaces.

18.69 The carbohydrates as glycoproteins and glycolipids on the surface of cells act as receptors for cell recognition and chemical messengers such as neurotransmitters.

18.71 Substances move through cell membrane by simple transport, facilitated transport, and active transport.

18.73 Beeswax and carnauba are waxes. Vegetable oil and capric triglyceride are triacylglycerols.

$$CH_2OC(CH_2)_8CH_3$$
$$CHOC(CH_2)_8CH_3$$
$$CH_2OC(CH_2)_8CH_3$$
Capric triacylglycerol

18.75 **a.** A typical fatty acid has a cis double bond.
b. A trans fatty acid has a trans double bond.
c.

$$CH_3(CH_2)_6CH_2 \quad H$$
$$C=C$$
$$H \quad CH_2(CH_2)_6COH$$

18.77

$$CH_2OC(CH_2)_{16}CH_3$$
$$CHOC(CH_2)_{16}CH_3$$
$$CH_2OC(CH_2)_{16}CH_3$$
Glyceryl stearate

$$CH_2OC(CH_2)_{14}CH_3$$

(structure, O double bonds shown)

CH$_2$OC(CH$_2$)$_{14}$CH$_3$

CHOC(CH$_2$)$_{14}$CH$_3$

CH$_2$OPOCH$_2$CH$_2$$\overset{+}{N}$(CH$_3$)$_3$

O$^-$

Lecithin

(Note: This lecithin structure is shown with choline, but ethanolamine or serine are also possible amino alcohols in lecithin.)

18.79 Stearic acid is a fatty acid. Sodium stearate is a soap. Glyceryl tripalmitate, safflower oil, whale blubber, and adipose tissue are triacylglycerols. Beeswax is a wax. Lecithin is a glycerophospholipid. Sphingomyelin is a sphingolipid. Cholesterol, progesterone, and cortisone are steroids.

18.81 a. 5   **b.** 1, 2, 3, 4   **c.** 2
      d. 1, 2   **e.** 1, 2, 3, 4   **f.** 2, 3, 4, 6

18.83 a. 4   **b.** 3   **c.** 1
      d. 4   **e.** 4   **f.** 3
      g. 2   **h.** 1

Chapter 19

Answers to Study Checks

19.1 tertiary (3°)

19.2 CH$_3$CH$_2$NHCH$_2$CH$_2$CH$_3$

19.3 Hydrogen bonding makes amines with six or fewer carbon atoms soluble in water.

19.4

CH$_3$

CH$_3$—$\overset{+}{N}$—H Cl$^-$

CH$_3$

19.5 pyrmidine

19.6 piperidine

19.7

O

HCOH   and   HNCH$_3$ (CH$_3$)

19.8

Cl—(benzene ring)—C—N—CH$_3$ (with O double bond, CH$_3$ and N—CH$_3$)

19.9 CH$_3$CH$_2$CH$_2$COH   and   CH$_3$NH$_3^+$Br$^-$

Answers to Selected Problems

19.1 In a primary amine, there is one alkyl group (and two hydrogens) attached to a nitrogen atom.

19.3 a. primary   **b.** secondary
      c. primary   **d.** tertiary   **e.** tertiary

19.5 a. ethylamine; ethanamine
      b. methylpropylamine; N-methyl-1-propanamine
      c. diethylmethylamine; N-methyl-N-ethylethanamine
      d. isopropylamine; 2-propanamine

19.7 a. 2-butanamine
      b. 2-chloroaniline
      c. 3-aminopropanal
      d. N-ethylaniline

19.9 a. CH$_3$CH$_2$NH$_2$ **b.** (benzene ring with NHCH$_3$)

      c. CH$_3$CH$_2$CH$_2$CH$_2$—N—CH$_2$CH$_2$CH$_3$ (H on N)

      d. CH$_3$—CH—CH$_2$—CH$_2$—CH$_3$ (NH$_2$ on CH)

19.11 a. CH$_3$—CH$_2$—OH
      b. CH$_3$—CH$_2$—CH$_2$—NH$_2$
      c. CH$_3$—CH$_2$—CH$_2$—NH$_2$

19.13 As a primary amine, propylamine can form two hydrogen bonds, which gives it the highest boiling point. Ethylmethylamine, a secondary amine, can form one hydrogen bond, and butane cannot form hydrogen bonds. Thus butane has the lowest boiling point of the three compounds.

19.15 a. yes   **b.** yes   **c.** no   **d.** yes

19.17

a. $CH_3NH_2 + H_2O \rightleftarrows CH_3-NH_3^+ + OH^-$

b. $CH_3-NH-CH_3 + H_2O \rightleftarrows CH_3-\overset{+}{N}H_2-CH_3 + OH^-$

c.

NH₂ benzene + H₂O ⇌ NH₃⁺ benzene + OH⁻

19.19 **a.** $CH_3-NH_3^+\ Cl^-$

b. $CH_3-\overset{+}{N}H_2-CH_3\ Cl^-$

c. $NH_3^+\ Cl^-$ benzene ring

19.21

a. H_2N- benzene $-\overset{O}{\overset{\|}{C}}-O-CH_2-CH_2-\overset{CH_2CH_3}{\underset{CH_2CH_3}{\overset{|}{\underset{|}{N^+}}}}-H\ Cl^-$

b. Amine salts are soluble in body fluids.

19.23 **a.** amine   **b.** amine
c. heterocyclic amine   **d.** heterocyclic amine

19.25 **c.** pyrimidine   **d.** pyrrole

19.27 pyrrole

19.29 **a.** $CH_3\overset{O}{\overset{\|}{C}}-NH_2$

b. $CH_3\overset{O}{\overset{\|}{C}}-NHCH_2CH_3$

c. benzene $-\overset{O}{\overset{\|}{C}}-\overset{H}{\overset{|}{N}}-CH_2CH_2CH_3$

19.31 **a.** *N*-methylethanamide (*N*-methylacetamide)
b. butanamide (butyramide)
c. methanamide (formamide)
d. *N*-methylbenzamide

19.33

a. $CH_3CH_2\overset{O}{\overset{\|}{C}}-NH_2$

b. $CH_3CH_2CH_2\overset{CH_3}{\underset{|}{CH}}-\overset{O}{\overset{\|}{C}}-NH_2$

c. $H\overset{O}{\overset{\|}{C}}NH_2$

d. benzene $-\overset{O}{\overset{\|}{C}}-\overset{H}{\overset{|}{N}}-CH_2CH_3$

e. $CH_3CH_2CH_2\overset{O}{\overset{\|}{C}}-\overset{H}{\overset{|}{N}}-CH_2CH_3$

19.35 **a.** acetamide   **b.** propionamide
c. *N*-methylpropanamide

19.37

a. $CH_3\overset{O}{\overset{\|}{C}}OH\ +\ NH_4^+Cl^-$

b. $CH_3CH_2\overset{O}{\overset{\|}{C}}OH\ +\ NH_4^+Cl^-$

c. $CH_3CH_2CH_2\overset{O}{\overset{\|}{C}}OH\ +\ CH_3NH_3^+Cl^-$

d. benzene $-\overset{O}{\overset{\|}{C}}OH\ +\ NH_4^+Cl^-$

e. $CH_3CH_2CH_2CH_2\overset{O}{\overset{\|}{C}}OH\ +\ CH_3CH_2-NH_3^+Cl^-$

19.39 $CH_3-CH_2-CH_2-NH_2$
Propylamine 1°

$CH_3-CH_2-NH-CH_3$
Ethymethylamine 2°

$CH_3-\overset{CH_3}{\underset{}{\overset{|}{N}}}-CH_3$
Trimethylamine 3°

$CH_3-\overset{CH_3}{\underset{}{\overset{|}{CH}}}-NH_2$
Isopropylamine 1°

19.41 a.

$$\underset{\underset{NH_2}{|}}{CH_3-CH_2-CH-CH_2-CH_3}$$

b.

cyclohexane with NH_2

c.

$$\underset{\underset{CH_3}{|}}{CH_3-NH_2{}^+ \ Cl^-}$$

d.

$$\underset{\underset{CH_2-CH_3}{|}}{CH_3-CH_2-N-CH_2-CH_3}$$

e.

$$\underset{\underset{OH}{|}\ \underset{NH_2}{|}}{CH_3-CH-CH-CH_2-CH_2-CH_3}$$

f.

$$\underset{\underset{CH_3}{|}}{\overset{\overset{CH_3}{|}}{CH_3-N^+-CH_3} \ Br^-}$$

g.

CH_3, CH_3 on N attached to phenyl

19.43 a. ethylamine   **b.** trimethylamine
     c. butylamine
     d. $NH_2-CH_2CH_2CH_2CH_2CH_2-NH_2$

19.45 a. quinine   **b.** nicotine
     c. caffeine   **d.** morphine, codeine

19.47 a. $CH_3-CH_2-NH_3{}^+ OH^-$

     b. $CH_3-CH_2-NH_3{}^+ Cl^-$

     c. $CH_3-CH_2-\overset{+}{N}H_2-CH_3 \ OH^-$

     d. $CH_3-CH_2-\overset{+}{N}H_2-CH_3 \ Cl^-$

     e. $CH_3-CH_2-CH_2-NH_2 + NaCl + H_2O$

     f. $\underset{\underset{}{}}{\overset{\overset{CH_3}{|}}{CH_3-CH_2-NH}} + NaCl + H_2O$

19.49 carboxylate salt, aromatic, amine, haloaromatic

19.51 a. aromatic, amine, amide, carboxylic acid, cycloalkene
     b. aromatic, ether, alcohol, amine
     c. aromatic, carboxylic acid
     d. phenol, amine, carboxylic acid
     e. aromatic, ether, alcohol, amine, ketone
     f. aromatic, amine

Chapter 20

Answers to Study Checks

20.1 hormones

20.2 a. nonpolar   **b.** polar

20.3 In the Fischer projection of D-serine, the —NH_2 group is on the right side.

20.4

$$\underset{\overset{+}{N}H_3-CH-\overset{O}{\overset{||}{C}}-OH}{\overset{\overset{OH}{|}}{\overset{CH_2}{|}}}$$

20.5

structure of dipeptide

20.6 threonylleucylphenylalanine

20.7 a triple helix

20.8 Both are nonpolar and would be found on the inside of the tertiary structure.

20.9 quaternary

20.10 The heavy metal Ag^+ denatures the protein in bacteria that cause gonorrhea.

Answers to Selected Questions and Problems

20.1 a. transport   **b.** structural
     c. structural   **d.** enzyme

20.3 All amino acids contain a carboxylic acid group and an amino group on the α carbon.

20.5

a. 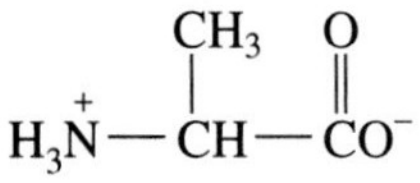
$$H_3\overset{+}{N}-CH-CO^-$$
with CH_3 and O (CO⁻)

b. 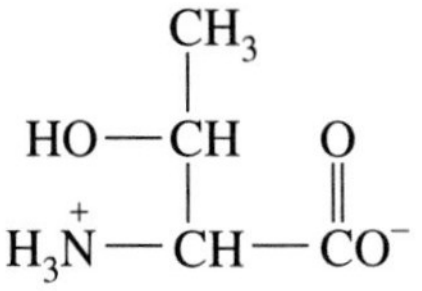
$$H_3\overset{+}{N}-CH-CO^-$$
with $HO-CH$, CH_3, and O

c. 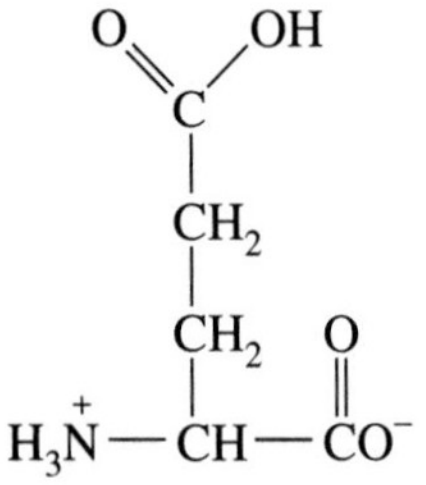
$$H_3\overset{+}{N}-CH-CO^-$$
with CH_2, CH_2, C bearing O and OH

d. 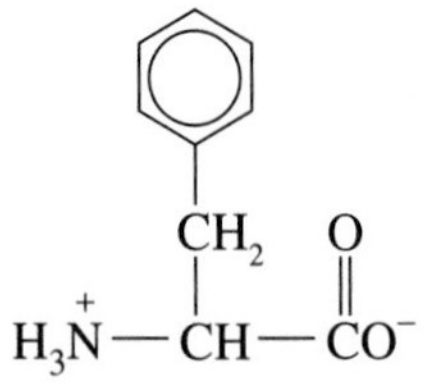
$$H_3\overset{+}{N}-CH-CO^-$$
with CH_2 and a benzene ring

20.7 **a.** hydrophobic nonpolar **b.** hydrophilic polar
c. acidic **d.** hydrophobic nonpolar

20.9 **a.** alanine **b.** valine
c. lysine **d.** cysteine

20.11

a. 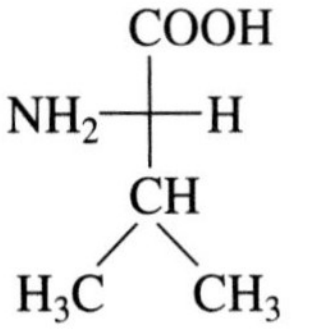
$$\text{COOH}$$
$$NH_2 - | - H$$
$$CH$$
$$H_3C \quad CH_3$$

b. 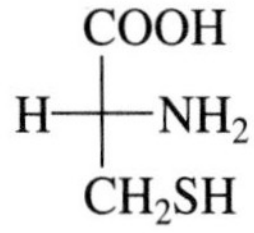
$$\text{COOH}$$
$$H - | - NH_2$$
$$CH_2SH$$

20.13 **a.** 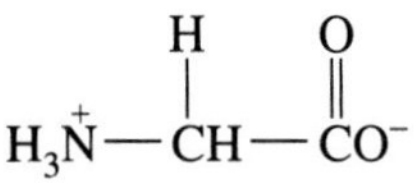 $H_3\overset{+}{N}-CH-CO^-$ (with H and O)
b. 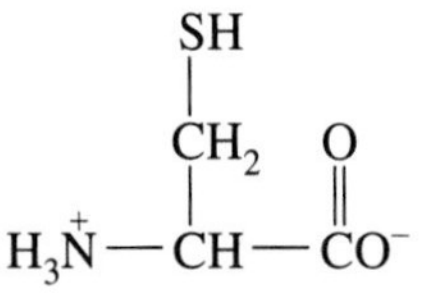 $H_3\overset{+}{N}-CH-CO^-$ (with SH, CH_2, O)
c. 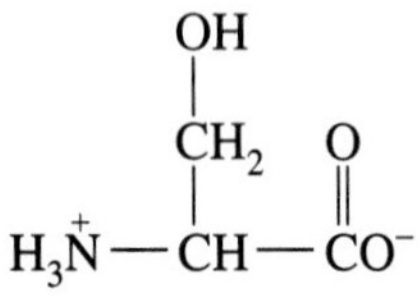 $H_3\overset{+}{N}-CH-CO^-$ (with OH, CH_2, O)
d. 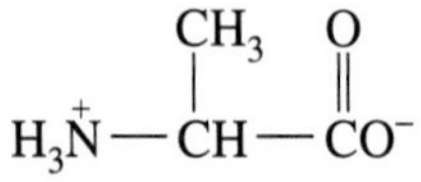 $H_3\overset{+}{N}-CH-CO^-$ (with CH_3, O)

20.15 **a.** 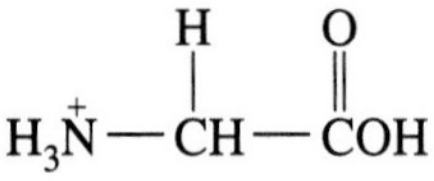 $H_3\overset{+}{N}-CH-COH$ (with H, O)
b. 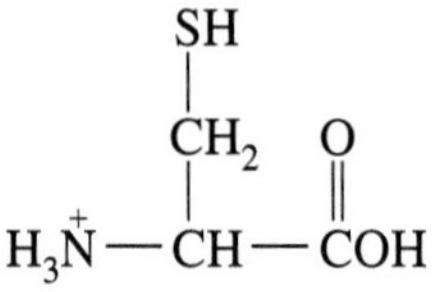 $H_3\overset{+}{N}-CH-COH$ (with SH, CH_2, O)
c. 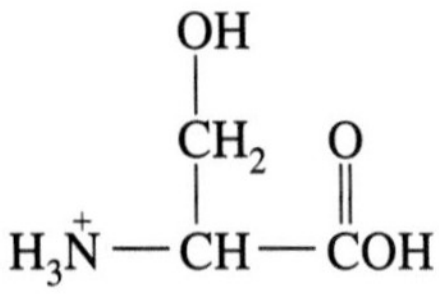 $H_3\overset{+}{N}-CH-COH$ (with OH, CH_2, O)
d. 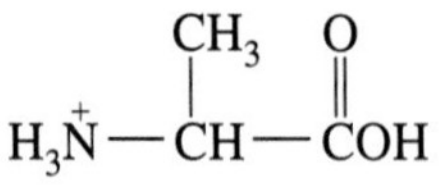 $H_3\overset{+}{N}-CH-COH$ (with CH_3, O)

20.17 **a.** above pI **b.** below pI **c.** at pI

20.19

a. 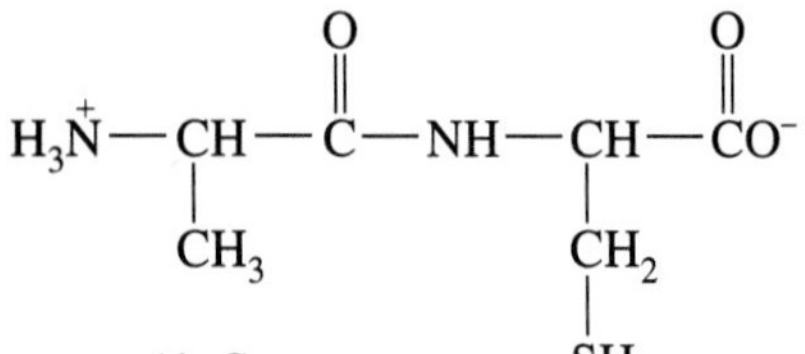
$$H_3\overset{+}{N}-CH-\overset{O}{C}-NH-CH-CO^-$$
with CH_3 and CH_2, SH
Ala-Cys

b. 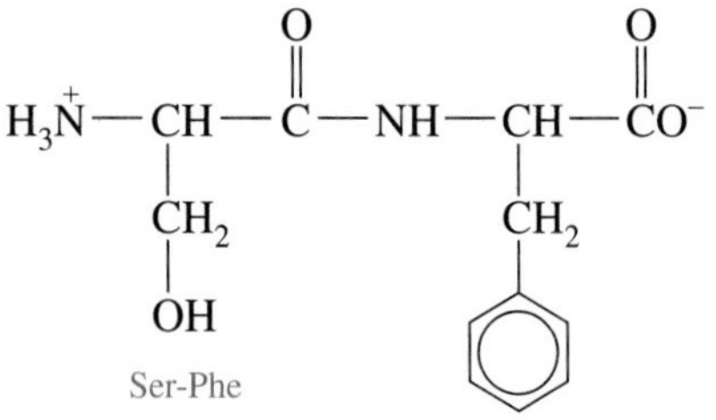
$$H_3\overset{+}{N}-CH-\overset{O}{C}-NH-CH-CO^-$$
with CH_2, OH and CH_2, benzene ring
Ser-Phe

c. 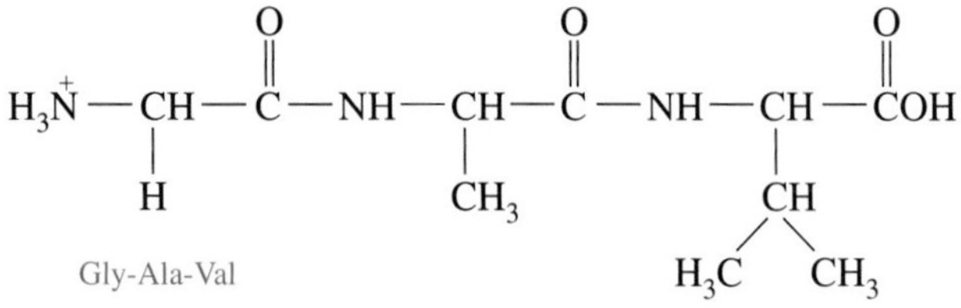
$$H_3\overset{+}{N}-CH-\overset{O}{C}-NH-CH-\overset{O}{C}-NH-CH-COH$$
with H, CH_3, CH (H_3C CH_3)
Gly-Ala-Val

d. 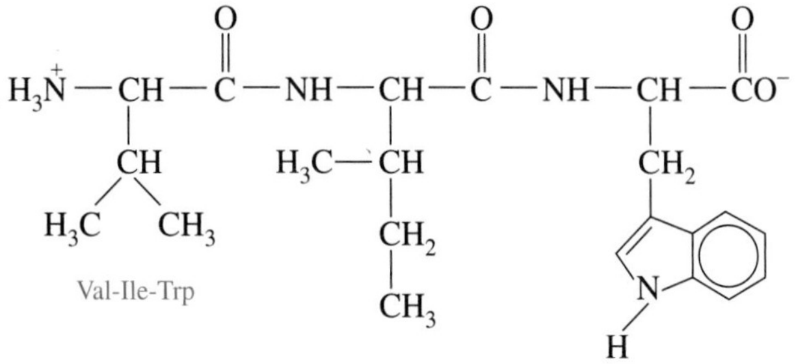
$$H_3\overset{+}{N}-CH-\overset{O}{C}-NH-CH-\overset{O}{C}-NH-CH-CO^-$$
with CH (H_3C CH_3), H_3C-CH / CH_2 / CH_3, CH_2 indole ring
Val-Ile-Trp

20.21 Amide bonds form to connect the amino acids that make up the protein.

20.23 Val-Ser-Ser, Ser-Val-Ser, or Ser-Ser-Val

20.25 The primary structure remains unchanged and intact as hydrogen bonds form between carbonyl oxygen atoms and amino hydrogen atoms.

20.27 In the α-helix, hydrogen bonds form between the carbonyl oxygen atom and the amino hydrogen atom of the fourth amino acid in the sequence. In the β-pleated sheet, hydrogen bonds occur between parallel peptides or across sections of a long polypeptide chain.

20.29 **a.** a disulfide bond
b. salt bridge
c. hydrogen bond
d. hydrophobic interaction

20.31 **a.** cysteine
b. Leucine and valine will be found on the inside of the protein because they are hydrophobic.
c. The cysteine and aspartic acid would be on the outside of the protein because they are polar.
d. The order of the amino acids (the primary structure) provides the R groups, whose interactions determine the tertiary structure of the protein.

20.33 The products would be the amino acids glycine, alanine, and serine.

20.35 His-Met; Met-Gly; Gly-Val

20.37 Hydrolysis splits the amide linkages in the primary structure.

20.39 a. Placing an egg in boiling water coagulates the protein of the egg.
  b. Using an alcohol swab coagulates the protein of any bacteria present.
  c. The heat from an autoclave will coagulate the protein of any bacteria on the surgical instruments.
  d. Heat will coagulate the surrounding protein to close the wound.

20.41 a. yes   **b.** yes   **c.** no
  d. yes   **e.** no   **f.** yes

20.43 a. The secondary structure of a protein depends on hydrogen bonds to form a helix or a pleated sheet; the tertiary structure is determined by the interaction of R groups and determines the three-dimensional structure of the protein.
  b. Nonessential amino acids can be synthesized by the body; essential amino acids must be supplied by the diet.
  c. Polar amino acids have hydrophilic side groups, whereas nonpolar amino acids have hydrophobic side groups.
  d. Dipeptides contain two amino acids, whereas tripeptides contain three.
  e. An ionic bond is an interaction between a basic and acidic side group; a disulfide bond links the sulfides of two cysteines.
  f. Fibrous proteins consist of three to seven α helixes coiled like a rope. Globular proteins form a compact spherical shape.
  g. The α-helix is the secondary shape like a staircase or corkscrew. The β-pleated sheet is a secondary structure that is formed by many protein chains side by side like a pleated sheet.
  h. The tertiary structure of a protein is its three-dimensional structure. The quaternary structure involves the grouping of two or more peptide units for the protein to be active.

20.45 a. α-keratins are fibrous proteins that provide structure to hair, wool, skin, and nails.
  b. α-keratins have a high content of cysteine.

20.47

b. This segment contains polar R groups, which would be found on the surface of a globular protein where they can hydrogen bond with water.

20.49 a. β-pleated sheet   **b.** α-helix

20.51 Serine is a polar amino acid, whereas valine is nonpolar. Valine would be in the center of the tertiary structure. However, serine would pull that part of the chain to the outside surface of the protein where it forms hydrophilic bonds with water.

20.53

Chapter 21

Answers to Study Checks

21.1 Enzymes act as catalysts in biological systems because they lower the activation energy of chemical reactions in the cells.

21.2 hydrolase

21.3 In the lock-and-key model, the shape of a substrate fits the shape of the active site exactly. In the induced-fit model, the substrate induces the active site to adjust its shape to fit the substrate.

21.4 At a pH lower than the optimum pH, denaturation will decrease the activity of urease.

21.5 HCN is a noncompetitive inhibitor.

21.6 Pepsin hydrolyzes proteins in the foods we ingest. It is synthesized as a zymogen, pepsinogen, to prevent its digestion of the proteins that make up the organs in the body

21.7 vitamin B_6

21.8 Water-soluble vitamins are easily destroyed by heat.

Answers to Selected Questions and Problems

21.1 The chemical reactions can occur without enzymes, but the rates are too slow. Catalyzed reactions, which are many times faster, provide the amounts of products needed by the cell at a particular time.

21.3 **a.** oxidation–reduction
b. transfer of a group from one substance to another
c. hydrolysis (splitting) of molecules with the addition of water

21.5 **a.** hydrolase   **b.** oxidoreductase
c. isomerase   **d.** transferase

21.7 **a.** lyase   **b.** transferase

21.9 **a.** succinate oxidase
b. fumarate hydrase
c. alcohol dehydrogenase

21.11 **a.** enzyme   **b.** enzyme–substrate complex
c. substrate

21.13 **a.** $E + S \rightleftharpoons ES \longrightarrow E + P$
b. The active site is a region or pocket within the tertiary structure of an enzyme that accepts the substrate, aligns the substrate for reaction, and catalyzes the reaction.

21.15 Isoenzymes are slightly different forms of an enzyme that catalyze the same reaction in different organs and tissues of the body.

21.17 A doctor might run tests for the enzymes CK, LDH, and AST to determine if the patient had a heart attack.

21.19 **a.** The rate would decrease.
b. The rate would decrease.
c. The rate would decrease.
d. The rate would increase if $[S] < [E]$.

21.21 pepsin, pH 2; urease, pH 5; trypsin, pH 8

21.23 **a.** competitive   **b.** noncompetitive
c. competitive   **d.** noncompetitive
e. competitive

21.25 **a.** methanol, CH_3OH; ethanol, CH_3CH_2OH
b. Ethanol has a similar structure to methanol and could compete for the active site.
c. Ethanol is a competitive inhibitor of methanol oxidation.

21.27 Digestive enzymes are proteases and would digest the proteins of the organ where they are produced if they were active immediately upon synthesis.

21.29 In feedback inhibition, the product binds to the first enzyme in a series and changes the shape of the active site. If the active site can no longer bind the substrate effectively, the reaction will stop.

21.31 When a regulator molecule binds to an allosteric site, the shape of the enzyme is altered, which makes the active site more reactive or less reactive and thereby increases or decreases the rate of the reaction.

21.33 **a.** 3; negative regulator
b. 4; allosteric enzyme
c. 1; zymogen

21.35 **a.** an enzyme that requires a cofactor
b. an enzyme that requires a cofactor
c. a simple enzyme

21.37 **a.** THF   **b.** NAD^+

21.39 **a.** pantothenic acid (vitamin B_5)   **b.** folic acid
c. niacin (vitamin B_3)

21.41 **a.** vitamin D or cholecaliferol
b. ascorbic acid or vitamin C
c. niacin or vitamin B_3

21.43 Vitamin B_6 is a water-soluble vitamin, which means that each day any excess of vitamin B_6 is eliminated from the body.

21.45 The side chain —CH_2OH on the ring is oxidized to —CHO, and the other —CH_2OH forms a phosphate ester.

21.47 The many different reactions that take place in cells require different enzymes because enzymes react with only a certain type of substrate.

21.49 When exposed to conditions of strong acids or bases, or high temperatures, the proteins of enzymes are denatured rapidly causing a loss of tertiary structure and catalytic activity.

21.51 **a.** The reactant is lactose and the products are glucose and galactose.
b.

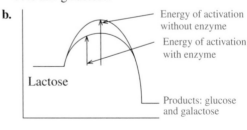

c. By lowering the energy of activation, the enzyme furnishes a lower energy pathway by which the reaction can take place.

21.53 a. S b. E
c. E d. E
e. S f. E

21.55 a. urea b. lactose
c. aspartate d. tyrosine

21.57 a. transferase b. oxidoreductase
c. hydrolase d. lyase

21.59 a. oxidoreductase b. hydrolase c. lyase

21.61 Sucrose fits the shape of the active site in sucrase, but lactose does not.

21.63 A heart attack may be the cause.

21.65 a. saturated b. unsaturated

21.67 In a reversible inhibition, the inhibitor can dissociate from the enzyme, whereas in irreversible inhibition, the inhibitor forms a strong covalent bond with the enzyme and does not dissociate. Irreversible inhibitors act as poisons to enzymes.

21.69 a. oxidoreductase
b. At high concentration, ethanol, which acts as a competitive inhibitor of ethylene glycol, would saturate the alcohol dehydrogenase enzyme to allow ethylene glycol to be removed from the body without producing oxalic acid.

21.71 a. Antibiotics such as amoxicillin are irreversible inhibitors.
b. Antibiotics inhibit enzymes needed to form cell walls in bacteria, not humans.

21.73 a. When pepsinogen enters the stomach, the low pH cleaves a peptide from its protein chain to form pepsin.
b. An active protease would digest the proteins of the pancreas rather than the proteins in the foods entering the stomach.

21.75 An allosteric enzyme contains sites for regulators that alter the enzyme and speed up or slow down the rate of the catalyzed reaction.

21.77 This would be a negative regulator because the end product of the reaction pathway binds to the enzyme to decrease or stop the first reaction in the reaction pathway.

21.79 a. requires a cofactor
b. simple enzyme
c. requires a cofactor (coenzyme)

21.81 a. pantothenic acid (B_5); coenzyme A
b. niacin (B_3); NAD^+
c. biotin; biocytin

21.83 A vitamin combines with an enzyme only when the enzyme and coenzyme are needed to catalyze a reaction. When the enzyme is not need, the vitamin dissociates for use by other enzymes in the cell.

21.85 a. niacin; pellagra
b. vitamin A; night blindness
c. vitamin D; weak bone structure

Chapter 22

Answers to Study Checks

22.1 a. Guanine is found in both RNA and DNA.
b. Uracil is found only in RNA.

22.2 deoxycytidine 5'—monophosphate (dCMP)

22.3

22.4 5'—C—C—A—A—T—T—G—G—3'

22.5 In DNA, the lagging strand is synthesized in the 3'–5' direction in short sections that are joined by DNA ligase.

22.6 Each type of tRNA matches a specific codon to a specific amino acid.

22.7 3'—CCCAAATTT—5'

22.8 A regulatory gene produces mRNA for the production of a protein repressor, which binds to the operator and blocks protein synthesis.

22.9 UGA is a stop codon that signals the termination of translation.

22.10 at UAG

22.11 If the substitution of an amino acid in the polypeptide affects an interaction essential to functional structure on the binding of a substrate, the resulting protein could be less effective or nonfunctional.

22.12 Gene cloning is the process by which recombinant DNA technology inserts the DNA of a gene into the plasmid of *E. coli* bacteria that multiply rapidly to make many copies of the gene.

22.13 A protease inhibitor modifies the three-dimensional structure of protease, which prevents the enzymes from synthesizing the proteins needed to produce more viruses.

Answers to Selected Questions and Problems

22.1 **a.** pyrimidine   **b.** pyrimidine

22.3 **a.** DNA   **b.** both DNA and RNA

22.5 deoxyadenosine 5'—monophosphate (dAMP), deoxythymidine 5'—monophosphate (dTMP), deoxycytidine 5'—monophosphate (dCMP), and deoxyguanosine 5'—monophosphate (dGMP)

22.7 **a.** nucleoside   **b.** nucleoside
       c. nucleoside   **d.** nucleotide

22.9

22.11 The nucleotides in nucleic acids are held together by phosphodiester bonds between the 3'—OH of a sugar (ribose or deoxyribose) and a phosphate group on the 5'—carbon of another sugar.

22.13

22.15 The two DNA strands are held together by hydrogen bonds between the bases in each strand.

22.17 **a.** 3'—TTTTTT—5'
       b. 3'—CCCCCC—5'
       c. 3'—TCAGGTCCA—5'
       d. 3'—GACATATGCAAT—5'

22.19 The enzyme helicase unwinds the DNA helix to prepare the parent DNA strand for the synthesis of daughter DNA strands.

22.21 The DNA strands separate and the DNA polymerase pairs each of the bases with its complementary base and produces two exact copies of the original DNA.

22.23 Ribosomal RNA, messenger RNA, and transfer RNA

22.25 A ribosome, which is about 65% rRNA and 35% protein, consists of a small subunit and a large subunit.

22.27 In transcription, the sequence of nucleotides on a DNA template (one strand) is used to produce the base sequences of a messenger RNA.

22.29 5'—GGCUUCCAAGUG—3'

22.31 In eukaryotic cells, genes contain sections called exons that code for protein and sections called introns that do not code for protein.

22.33 An operon is a section of DNA that regulates the synthesis of one or more proteins.

22.35 When the lactose level is low in *E. coli*, a repressor produced by the mRNA from the regulatory gene binds to the operator blocking the synthesis of mRNA from the genes and preventing the synthesis of protein.

22.37 A three-base sequence in mRNA that codes for a specific amino acid in a protein.

22.39 **a.** leucine   **b.** serine
c. glycine   **d.** arginine

22.41 When AUG is the first codon, it signals the start of protein synthesis. Thereafter, AUG codes for methionine.

22.43 A codon is a base triplet in the mRNA. An anticodon is the complementary triplet on a tRNA for a specific amino acid.

22.45 Initiation, translocation, and termination.

22.47 **a.** —Lys—Lys—Lys—
b. —Phe—Pro—Phe—Pro—
c. —Tyr—Gly—Arg—Cys—

22.49 The new amino acid is joined by a peptide bond to the peptide chain. The ribosome moves to the next codon, which attaches to a tRNA carrying the next amino acid.

22.51 **a.** 5'—CGA—AAA—GUU—UUU—3'
b. GCU, UUU, CAA, AAA
c. Using codons in mRNA: Arg—Lys—Val—Phe

22.53 A base in DNA is replaced by a different base.

22.55 If the resulting codon still codes for the same amino acid, there is no effect. If the new codon codes for a different amino acid, there is a change in the order of amino acids in the polypeptide.

22.57 The normal triplet TTT forms a codon AAA, which codes for lysine. The mutation TTC forms a codon AAG, which also codes for lysine. There is no effect on the amino acid sequence.

22.59 **a.** —Thr—Ser—Arg—Val—
b. —Thr—Thr—Arg—Val—
c. —Thr—Ser—Gly—Val
d. —Thr—STOP   Protein synthesis would terminate early. If this occurs early in the formation of the polypeptide, the resulting protein will probably be nonfunctional.
e. The new protein will contain the sequence —Asp—Ile—Thr—Gly—.
f. The new protein will contain the sequence —His—His—Gly—.

22.61 **a.** GCC and GCA both code for alanine.

b. A vital ionic cross-link in the tertiary structure of hemoglobin cannot be formed when the polar glutamine is replaced by valine, which is nonpolar. The resulting hemoglobin is malformed and less capable of carrying oxygen.

22.63 *E. coli* bacterial cells contain several small circular plasmids of DNA that can be isolated easily. After the recombinant DNA is formed, *E. coli* multiply rapidly, producing many copies of the recombinant DNA in a relatively short time.

22.65 *E. coli* are soaked in a detergent solution that dissolves the cell membrane and releases the cell contents including the plasmids, which are collected.

22.67 When a gene has been obtained using restriction enzymes, it is mixed with the plasmids that have been opened by the same enzymes. When mixed together in a fresh *E. coli* culture, the sticky ends of the DNA fragments bond with the sticky ends of the plasmid DNA to form a recombinant DNA.

22.69 In DNA fingerprinting, restriction enzymes cut a sample DNA into fragments, which are sorted by size by gel electrophoresis. After tagging the DNA fragments with a radioactive isotope, a piece of x-ray film placed over the gel is exposed by the radioactivity to give a pattern of dark and light bands known as a DNA fingerprint.

22.71 DNA or RNA, but not both.

22.73 **a.** A viral RNA is used to synthesize a viral DNA to produce the proteins for the protein coat, which allows the virus to replicate and leave the cell.
b. retrovirus

22.75 Nucleoside analogs such as AZT and ddI are similar to the nucleosides required to make viral DNA in reverse transcription. However, they interfere with the ability of the DNA to form and thereby disrupt the life cycle of the HIV-1 virus.

22.77 **a.** pyrimidine   **b.** purine   **c.** pyrimidine
d. pyrimidine   **e.** purine

22.79 **a.** thymine and deoxyribose
b. adenine and ribose
c. cytosine and ribose
d. guanine and deoxyribose

22.81 They are both pyrimidines, but thymine has a methyl group.

22.83

22.85 They are both polymers of nucleotides connected through phosphodiester bonds between alternating sugar and phosphate groups with bases extending out from each sugar.

22.87 28% T, 22% G, and 22% C

22.89 **a.** two   **b.** three

22.91 **a.** 3'—CTGAATCCG—5'
b. 5'—ACGTTTGATCGT—3'
c. 3'—TAGCTAGCTAGC—5'

22.93 DNA polymerase synthesizes the leading strand continuously in the 5' to 3' direction. The lagging strand is synthesized in small segments called Okazaki fragments because it must grow in the 3' to 5' direction.

22.95 One strand of the parent DNA is found in each of the two copies of the daughter DNA molecule.

22.97 **a.** tRNA.   **b.** rRNA   **c.** mRNA

22.99 **a.** ACU, ACC, ACA, and ACG
b. UCU, UCC, UCA, UCG, AGU, and AGC
c. UGU and UGC

22.101 What is the amino acid for each of the following codons?
a. lysine   **b.** isoleucine   **c.** arginine

22.103 start—Tyr—Gly—Gly—Phe—Leu—stop

22.105 **a.** UCG   **b.** AUA   **c.** GGU

22.107 Three nucleotides are needed for each amino acid plus a start and stop triplet, which makes a minimum total of 33 nucleotides.

22.109 A DNA virus attaches to a cell and injects viral DNA that uses the host cell to produce copies of DNA to make viral RNA. A retrovirus injects viral RNA from which complementary DNA is produced by reverse transcription.

Chapter 23

Answers to Study Checks

23.1 The cytoplasm is all the cellular material including cytosol and organelles between the plasma membrane and the nucleus. The cytosol, the aqueous part of the cytoplasm, is a solution of electrolytes and enzymes.

23.2 Adenine, ribose, and three phosphate groups.

23.3 $FADH_2$

23.4 The digestion of amylose begins in the mouth when salivary amylase hydrolyzes some of the glycosidic bonds. In the small intestine, pancreatic amylase hydrolyzes more glycosidic bonds, and finally maltose is hydrolyzed by maltase to yield glucose.

23.5 In the initial reactions of glycolysis, energy in the form of two ATP is invested to activate glucose and convert it to fructose-1,6-bisphosphate.

23.6

23.7 Only liver and kidney cells contain the phosphatase enzyme that converts glucose-6-phosphate to free glucose.

23.8 The reaction catalyzed by hexokinase in glycolysis is irreversible

23.9 When a cell needs a supply of ribose-5-phophsate for nucleotide synthesis.

Answers to Selected Questions and Problems

23.1 The digestion of polysaccharides takes place in stage 1.

23.3 In metabolism, a catabolic reaction breaks apart large molecules, releasing energy.

23.5 **a.** (3) smooth endoplasmic reticulum
b. (1) lysosome
c. (2) Golgi complex

23.7 The phosphoric anhydride bonds (P—O—P) in ATP release energy that is sufficient for energy-requiring processes in the cell.

23.9 **a.** PEP + H_2O $\longrightarrow$ pyruvate + P_i + 14.8 kcal /mole
b. ADP + P_i + 7.3 kcal/mole $\longrightarrow$ ATP

c. Coupled: PEP + ADP $\longrightarrow$ ATP + pyruvate + 7.5 kcal /mole

23.11 a. coenzyme A **b.** NAD^+ **c.** FAD

23.13 a. NADH **b.** FAD

23.15 FAD

23.17 Hydrolysis is the main reaction involved in the digestion of carbohydrates.

23.19 a. lactose
 b. glucose and fructose
 c. glucose

23.21 glucose

23.23 ATP is required in phosphorylation reactions.

23.25 glyceraldehyde-3-phosphate and dihydroxyacetone phosphate

23.27 ATP is produced in glycolysis by transferring a phosphate from 1,3-bisphosphoglycerate and from phosphoenolpyruvate directly to ADP.

23.29 a. hexokinase **b.** phosphokinase

23.31 a. 1 ATP required
 b. 1 NADH is produced for each triose.
 c. 2 ATP and 2 NADH

23.33 a. In reaction 1, a hexokinase uses ATP to phosphorylate glucose.
 b. In reactions 7 and 10, phosphate groups are transferred from 1,3-bisphosphoglycerate and phosphoenolpyruvate directly to ADP to produce ATP.
 c. In reaction 4, the six-carbon molecule fructose-1,6-bisphosphate is split into two three-carbon molecules, glyceraldehyde-3-phosphate and dihydroxyacetone phosphate.

23.35 Galactose reacts with ATP to yield galactose-1-phosphate, which is converted to glucose-6-phosphate, an intermediate in glycolysis. Fructose reacts with ATP to yield fructose-1-phosphate, which is cleaved to give dihydroxyacetone phosphate and glyceraldehyde. Dihydroxyacetone phosphate isomerizes to glyceraldehyde-3-phosphate, and glyceraldehyde is phosphorylated to glyceraldehyde-3-phosphate, which is an intermediate in glycolysis.

23.37 a. activate **b.** inhibit

23.39 Aerobic (oxygen) conditions are needed.

23.41 The oxidation of pyruvate converts NAD^+ to NADH and produces acetyl CoA and CO_2.

Pyruvate + NAD^+ + CoA $\longrightarrow$ Acetyl CoA + CO_2 + NADH + H^+

23.43 When pyruvate is reduced to lactate, NAD^+ is produced, which can be used for the oxidation of more glyceraldehyde-3-phosphate in glycolysis. This recycles NADH.

23.45 Carbon dioxide, CO_2

23.47 Glycogenesis is the synthesis of glycogen from glucose molecules.

23.49 Muscle cells break down glycogen to glucose-6-phosphate, which enters glycolysis.

23.51 Glycogen phosphorylase cleaves the glycosidic bonds at the ends of glycogen chains to remove glucose monomers as glucose-1-phosphate.

23.53 When there are no glycogen stores remaining in the liver, gluconeogenesis synthesizes glucose from noncarbohydrate compounds such as pyruvate and lactate.

23.55 Phosphoglucoisomerase, aldolase, triosephosphate isomerase, glyceraldehyde-3-phoshpate dehydrogenase, phosphoglycerokinase, phosphoglyceromutase, and enolase.

23.57 a. activates **b.** activates **c.** inhibits

23.59 Metabolism includes all the reactions in cells provide energy and material for cell growth.

23.61 Stage 1

23.63 Eukaryotic cell

23.65 Adenosine triphosphate

23.67 ATP + H_2O $\longrightarrow$ ADP + P_i + 7.3 kcal (31kJ)/mole

23.69 Flavin adenine dinucleotide

23.71 Nicotinamide adenine dinucleotide

23.73 a. $FADH_2$ **b.** NADH + H^+

23.75 Lactose undergoes digestion in the mucosal cells of the small intestine to yield galactose and glucose.

23.77 Galactose and fructose are converted in the liver to glucose phosphate compounds that can enter the glycolysis pathway.

23.79 Glucose is the reactant and pyruvate is the product of glycolysis.

23.81 Reaction 1 and 3 involve phosphorylation of hexoses with ATP, and reactions 7 and 10 involve direct substrate phosphorylation that generates ATP.

23.83 Reaction 4, which converts fructose-1,6-bisphosphate into two triose phosphates, is catalyzed by aldolase.

23.85 Glucose-6-phosphate

23.87 Pyruvate is converted to lactate when oxygen is not present in the cell (anaerobic) to regenerate NAD^+ for glycolysis.

23.89 Phosphofructokinase is an allosteric enzyme that is activated by high levels of AMP and ADP because the cell needs to produce more ATP. When ATP levels are high due to a decrease in energy needs, ATP inhibits phosphofructokinase, which reduces its catalysis of fructose-6-phosphate.

23.91 The rate of glycogenolysis increases when blood glucose levels are low and glucagon has been secreted, which accelerate the breakdown of glycogen.

23.93 Glucose

23.95 The cells in the liver, but not skeletal muscle, contain a phosphatase enzyme needed to convert glucose-6-phosphate to free glucose that can diffuse through cell membranes into the blood stream. Glucose-6-phosphate, which is the end product of glycogenolysis in muscle cells, cannot diffuse easily across cell membranes.

23.97 Insulin increases the rate of glycogenolysis and glycolysis and decreases the rate of glycogenesis. Glucagon decreases the rate glycogenolysis and glycolysis and increases the rate of glycogenesis.

23.99 The Cori cycles is a cyclic process that involves the transfer of lactate from muscle to the liver where glucose is synthesized, which can be used again by the muscle.

23.101 **a.** increase   **b.** decrease
      c. increase   **d.** decrease

23.103 **a.** decrease   **b.** decrease
      c. increase   **d.** decrease

Chapter 24

Answers to Study Checks

24.1 oxaloacetate

24.2 Fe^{3+}

24.3 **a.** oxidation   **b.** reduction

24.4 Oxygen (O_2) is the last substance that accepts electrons.

24.5 The protein complexes I, III, and IV pump protons from the matrix to the intermembrane space.

24.6 Three NADH provide nine ATP, one $FADH_2$ provides two ATP, and one direct phosphorylation provides one ATP.

Answers to Selected Questions and Problems

24.1 Krebs cycle and tricarboxlyic acid cycle

24.3 $2CO_2$, 3 NADH + $3H^+$, $FADH_2$, GTP (ATP), and HS—CoA.

24.5 Two reactions, 3 and 4, involve oxidative decarboxylation.

24.7 NAD^+ is reduced in reactions 3, 4, and 8 of the citric acid cycle.

24.9 In reaction 5, GDP undergoes a direct substrate phosphorylation.

24.11 **a.** citrate and isocitrate
b. In decarboxylation, a carbon atom is lost as CO_2.
c. α-ketoglutarate
d. isocitrate $\longrightarrow$ α-ketoglutrate; α-ketoglutarate $\longrightarrow$ succinyl CoA; succinate $\longrightarrow$ fumarate; malate $\longrightarrow$ oxaloacetate
e. steps 3, 8

24.13 **a.** citrate synthase
b. succinate dehydrogenase and aconitase
c. fumarase

24.15 **a.** NAD^+ **b.** GDP   **c.** FAD

24.17 Isocitrate dehydrogenase and α-ketoglutarate dehydrogenase are allosteric enzymes

24.19 High levels of ADP increase the rate of the citric acid cycle.

24.21 oxidized

24.23 **a.** oxidation   **b.** reduction

24.25 NADH and $FADH_2$

24.27 FAD, coenzyme Q, cytochrome b, cytochrome c

24.29 The mobile carrier Q transfers electrons from complex I to III.

24.31 NADH transfers electrons to FMN in complex I to give NAD^+.

24.33 **a.** $NADH + H^+ + FMN \longrightarrow NAD^1 + FMNH_2$
b. $QH_2 + 2 \text{ cyt } b \text{ Fe}^{3+} \longrightarrow Q + 2 \text{ cyt } b \text{ (Fe}^{2+}) + 2H^+$

24.35 In oxidation phosphorylation, the energy from the oxidation reactions in the electron transport chain is used to drive ATP synthesis.

24.37 As protons return to the lower energy environment in the matrix, they pass through ATP synthase where they release energy to drive the synthesis of ATP.

24.39 The oxidation of the reduced coenzymes NADH and $FADH_2$ by the electron transport chain generates energy to drive the synthesis of ATP.

24.41 ATP synthase consists of two protein complexes known as F_0 and F_1.

24.43 The loose (L) site in ATP synthase binds ADP and P_i.

24.45 Glycolysis takes place in the cytoplasm, not in the mitochondria. Because NADH cannot cross the mitochondrial membrane, one ATP is hydrolyzed to transport the electrons from NADH to FAD. The resulting $FADH_2$ produces only 2 ATP for each NADH produced in glycolysis.

24.47 **a.** 3 ATP   **b.** 2 ATP   **c.** 6 ATP   **d.** 12 ATP

24.49 The oxidation reactions of the citric acid cycle produce a source of reduced coenzymes for the electron transport chain and ATP synthesis.

24.51 The oxidized coenzymes NAD^+ and FAD needed for the citric acid cycle are regenerated by the electron transport chain.

24.53 **a.** citrate, isocitrate
b. α-ketoglutarate
c. α-ketoglutarate, succinyl CoA, oxaloacetate

24.55 **a.** In reaction 4, α-ketoglutarate, a five-carbon keto acid, is decarboxylated.
b. In reaction 1 and reaction 7, double bonds in aconitate and fumarate are hydrated.
c. NAD^+ is reduced in reactions 3, 4, and 8.
d. In reactions 3 and 8, a secondary hydroxyl group in isocitrate and malate is oxidized.

24.57 **a.** NAD^+ **b.** NAD^+ and CoA

24.59 **a.** High levels of NADH inhibit isocitrate dehydrogenase and α-ketoglutarate dehydrogenase to slow the rate of the citric acid cycle.
b. High levels of ATP inhibit the citric acid cycle.

24.61 **a.** (4) cytochrome   **b.** (1) FMN

24.63 **a.** CoQ is a mobile carrier
b. Fe–S clusters are found in complex I, III, and IV
c. cyt a_3 is part of complex IV

24.65 **a.** complex II; $FADH_2 + Q \longrightarrow FAD + QH_2$
b. complex IV; cyt a (Fe^{2+}) + cyt a_3 (Fe^{3+}) $\longrightarrow$ cyt a (Fe^{3+}) + cyt a_3 (Fe^{2+})

24.67

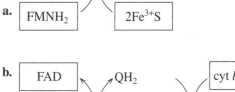

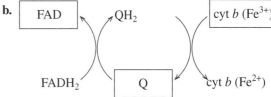

24.69 complex I, III, and IV

24.71 Energy released as protons flow through ATP synthase back to the matrix is utilized for the synthesis ATP.

24.73 Protons flow into the matrix where H^+ concentration is lower.

24.75 Two ATP molecules are produced from $FADH_2$

24.77 **a.** NADH dehydrogenase (complex I)
b. electron flow from cyt b to cyt c_1 (complex III)
c. cytochrome c oxidase (complex IV)

24.79 The oxidation of glucose to pyruvate produces 6 ATP whereas the oxidation of glucose to CO_2 and H_2O produces 36 ATP.

24.81 **a.** 6 ATP $\times$ 7.3 kcal/mole = 44 kcal
b. 6 ATP $\times$ 7.3 kcal/mole = 44 kcal (2 pyruvate to 2 acetyl CoA)
c. 24 ATP $\times$ 7.3 kcal/ mole = 175 kcal (2 acetyl CoA citric acid cycle)
d. 36 $\times$ 7.3 kcal/mol = 263 (complete oxidation of glucose to CO_2 and H_2O)

24.83 In a calorimeter, the complete combustion of glucose gives 687 kcal. Using the value for glucose in problem 24.81, 283 /687 × 100 = 38.3%

24.85 The ATP synthase extends through the inner mitochondrial membrane with the F_0 part in contact with the proton gradient in the intermembrane space, while the F_1 complex is in the matrix.

24.87 As protons from the proton gradient move through the ATP synthase to return to the matrix, energy is released and used to drive ATP synthesis at F_1 ATP synthase.

24.89 A hibernating bear has stored fat as brown fat, which can be used during the winter for heat rather than ATP energy.

Chapter 25

Answers to Study Checks

25.1 In the membrane of the small intestine, monoacylglycerols and fatty acids recombine to form new triacylglycerols that bind with proteins to form chylomicrons for transport to the lymphatic system and the bloodstream.

25.2 NAD^+

25.3

8 acetyl CoA – 2 ATP $\dfrac{94 \text{ ATP}}{129 \text{ ATP}} \times 100 = 73\%$

35 ATP from NADH and $FADH_2$ $\dfrac{35 \text{ ATP}}{129 \text{ ATP}} \times 100 = 27\%$

25.4 The ketone bodies are acetoacetate, β-hydroxybutyrate, and acetone.

25.5 In each cycle of fatty acid synthesis, a two-carbon atom acyl group from malonyl ACP adds to a fatty acid chain and one carbon atom forms CO_2.

25.6 HCl denatures proteins and activates enzymes such as pepsin.

25.7 In conditions such as fasting or starvation, a diet insufficient in protein leads to a negative nitrogen balance.

25.8 Alanine oxaloacetate transaminase or transaminotransferase.

25.9 The process of oxidative deamination regenerates α-ketoglutarate from glutamate.

25.10 **a.** argininosuccinate
b. citrulline
c. urea and ornithine

25.11 Arginine, glutamate, glutamine, histidine, and proline provide carbon atoms for α-ketoglutarate.

25.12 Substrates for the synthesis of nonessential amino acids are obtained from glycolysis and the citric acid cycle.

Answers to Selected Questions and Problems

25.1 The bile salts emulsify fat to give small fat globules for lipase hydrolysis.

25.3 When blood glucose and glycogen stores are depleted

25.5 Glycerol is converted to glycerol-3-phosphate, which is converted to dihydroxyacetone phosphate, an intermediate of glycolysis.

25.7 In the cytosol at the outer mitochondrial membrane

25.9 FAD and NAD^+

25.11

a. $CH_3-CH_2-CH_2-CH_2-CH_2-\overset{\beta}{C}H_2-CH_2-\overset{\displaystyle O}{\overset{\|}{C}}-S-CoA$

b. $CH_3-(CH_2)_{14}-\overset{\beta}{C}H_2-CH_2-\overset{\displaystyle O}{\overset{\|}{C}}-S-CoA$

c. $CH_3-CH_2-CH=CH-CH_2-CH_2-CH_2-\overset{\beta}{C}H_2-CH_2-\overset{\displaystyle O}{\overset{\|}{C}}-S-CoA$

25.13

a. and b. $CH_3-(CH_2)_6-\underset{\beta}{CH_2}-\underset{\alpha}{CH_2}-\overset{\overset{O}{\parallel}}{C}-S-CoA$

c. $CH_3-(CH_2)_8-\overset{\overset{O}{\parallel}}{C}-S-CoA + NAD^+ + FAD + H_2O + SH-CoA \longrightarrow$

$CH_3-(CH_2)_6-\overset{\overset{O}{\parallel}}{C}-S-CoA + CH_3-\overset{\overset{O}{\parallel}}{C}-S-CoA + NADH + H^+ + FADH_2$

d. $CH_3-(CH_2)_8-COOH + 4CoA + 4FAD + 4NAD^+ + 4H_2O \longrightarrow$

5Acetyl CoA + 4FADH$_2$ + 4NADH + 4H$^+$

25.15 The hydrolysis of ATP to AMP hydrolyzes ATP to ADP, and ADP to AMP, which provides the same amount of energy as the hydrolysis of 2 ATP to 2 ADP.

25.17 a. 5 acetyl CoA units
b. 4 cycles of β oxidation
c. 60 ATP from 5 acetyl CoA (citric acid cycle) + 12 ATP from 4 NADH + 8 ATP from 4 FADH$_2$ −2 ATP (activation) = 80 −2 = 78 ATP

25.19 Ketogenesis is the synthesis of ketone bodies from excess acetyl CoA from fatty acid oxidation, which occurs when glucose is not available for energy, particularly in starvation, fasting, and diabetes.

25.21 Acetoacetate undergoes reduction using NADH + H$^+$ to yield β-hydroxybutyrate.

25.23 High levels of ketone bodies lead to ketosis, a condition characterized by acidosis (a drop in blood pH values), excessive urination, and strong thirst.

25.25 in the cytosol of cells in liver and adipose tissue

25.27 acetyl CoA, HCO$_3^-$, and ATP

25.29 a. (3) malonyl CoA transacylase
b. (1) acetyl CoA carboxylase
c. (2) acetyl CoA transacylase

25.31 a. 4 HCO$_3^-$ **b.** 4 ATP
c. 5 acetyl CoA   **d.** 4 malonyl ACP
e. 8 NADPH   **f.** 4 CO$_2$ removed

25.33 The digestion of proteins begins in the stomach and is completed in the small intestine.

25.35 Hormones, heme, purines and pyrimidines for nucleotides, proteins, nonessential amino acids, amino alcohols, and neurotransmitters require nitrogen obtained from amino acids.

25.37 The reactants are an amino acid and an α-keto acid, and the products are a new amino acid and a new α-keto acid.

25.39

a. $H-\overset{\overset{O}{\parallel}}{C}-COO^-$ **b.** $CH_3-\overset{\overset{O}{\parallel}}{C}-COO^-$

c. $CH_3-\overset{\overset{CH_3}{|}}{CH}-\overset{\overset{O}{\parallel}}{C}-COO^-$

25.41

$^-OOC-\overset{\overset{\overset{+}{N}H_3}{|}}{CH}-CH_2-CH_2-COO^- + H_2O + NAD^+ (NADP^+) \xrightarrow{\text{Glutamate dehydrogenase}}$

Glutamate

$^-OOC-\overset{\overset{O}{\parallel}}{C}-CH_2-CH_2-COO^- + NH_4^+ + NADH (NADPH) + H^+$

α-Ketoglutarate

25.43 NH_4^+ is toxic if allowed to accumulate in the liver.

25.45

$$H_2N\overset{\overset{\displaystyle O}{\|}}{—C}—NH_2$$

25.47 CO_2 from the citric acid cycle

25.49 Glucogenic amino acids produce compounds used to synthesis glucose.

25.51 **a.** pyruvate   **b.** oxaloacetate, fumarate
          c. succinyl CoA   **d.** α-ketoglutarate

25.53 nonessential amino acid

25.55 Glutamine synthetase catalyzes the addition of an amino group to glutamate using energy from the hydrolysis of ATP.

25.57 phenylketournia

25.59 Triacylglycerols are hydrolyzed to monoacylglycerols and fatty acids in the small intestine, which reform triacylglycerols in the intestinal lining for transport as lipoproteins to the tissues.

25.61 Fats can be stored in unlimited amounts in adipose tissue compared to the limited storage of carbohydrates as glycogen.

25.63 The fatty acids cannot diffuse across the blood brain barrier.

25.65 **a.** Glycerol is converted to glycerol-3-phosphate and to dihydroxyacetone phosphate, which enters glycolysis or gluconeogensis.
          b. Activation of fatty acids occurs on the outer mitochondrial membrane.
          c. The energy cost is equal to 2 ATP.
          d. Only fatty acyl CoA can move into the intermembrane space for transport by carnitine into the matrix.

25.67 **a.** and **b.**

$$CH_3—(CH_2)_8—CH_2—\underset{\beta}{CH_2}—\underset{\alpha}{\overset{\overset{\displaystyle O}{\|}}{C}}—CoA$$

   c. Lauryl-CoA + 5 CoA + 5 FAD + 5 NAD^+ + 5 H_2O $\longrightarrow$
   6 Acetyl CoA + 5 $FADH_2$ + 5 NADH + $5H^+$
   d. Six acetyl CoA units are produced
   e. Five cycles of β oxidation are needed
   f.

activation		$\longrightarrow$	-2 ATP
6 acetyl CoA x 12		$\longrightarrow$	72 ATP
5 $FADH_2$	$\times 2$	$\longrightarrow$	10 ATP
5 NADH	$\times 3$	$\longrightarrow$	15 ATP
		Total	95 ATP

25.69 **a.** β oxidation   **b.** β oxidation
          c. Fatty acid synthesis   **d.** β oxidation
          e. Fatty acid synthesis   **f.** Fatty acid synthesis

25.71 **a.** (1) fatty acid oxidation
          b. (2) the synthesis of fatty acids.

25.73 Ammonium ion is toxic if allowed to accumulate in the liver.

25.75 **a.** citrulline   **b.** carbamoyl phosphate

25.77 **a.** pyruvate   **b.** acetoacetyl-CoA, acetyl CoA
          c. succinyl-CoA   **d.** α-ketoglutarate

25.79 Serine is degraded to pyruvate, which is oxidized to acetyl CoA. The oxidation produces NADH + H^+, which provides 3 ATP. In one turn of the citric acid cycle, the acetyl CoA provides 12 ATP. Thus, serine can provide a total of 15 ATP.

Credits

Unless otherwise acknowledged, all photographs are the property of Benjamin Cummings.

Chapter 1

p. 30   Stephen Frink/Stone.
p. 32   David R. Frazier Photolibrary.
p. 33   © Tom Pantages.

Chapter 2

p. 46   © Daniel Aubry/The Stock Market.
p. 53   Courtesy of Lawrence Berkeley Laboratory, University of California.
p. 67   top: Courtesy of the National Museum of American History/Smithsonian Institution.
         bottom: Courtesy of the Anderson Cancer Center University of Texas, Austin.

Chapter 3

p. 84   bottom: © John Coletti/Stock Boston.
p. 91   © David Parker/SPL/Photo Researchers, Inc.
p. 92   Don Farrall/PhotoDisc.
p. 99   Zuckerman, Bruce E./Corbis.
p. 101   © SIU/Visuals Unlimited.
p. 102   © 1997 PhotoDisc/Lawrence Berkeley National Library.
p. 103   © 1997 PhotoDisc/Lawrence Berkeley Laboratory. © GJLP/CNRI/Phototake, NYC.
p. 106   Roger Ressmeyer/Corbis.

Chapter 4

p. 115   © Robert E. Lyons/Visuals Unlimited.
p. 143   © Professor P. Motta, Department of Anatomy, University "La Sapienza," Rome/SPL/Photo Researchers, Inc.

Chapter 5

p. 164   © Custom Medical Stock Photo.
p. 168   Siede Preis/PhotoDisc.

Chapter 6

p. 198   Patrick Clark/ PhotoDisc.
p. 212   top: © 1997 PhotoDisc/Jack Star/Photolink.
         bottom: © 1997 PhotoDisc/Sean Thompson.

Chapter 7

p. 239   Steve Mason/PhotoDisc.
p. 241   Robert Holmes/Corbis.

Chapter 8

p. 249   © 1997 PhotoDisc/T. O'Keefe/Photolink.
p. 256   © 1997 PhotoDisc/Larry Brownstein.
p. 272   © SIU/Visuals Unlimited.

Chapter 9

p. 278   top: Layne Kennedy/Corbis.
         bottom: © George I. Bernard/Animals Animals.
p. 295   top: © CNRI/SPL/Photo Researchers, Inc.
p. 297   left: NMSB/Custom Medical Stock Photo.
         right: M. Kalab/Custom Medical Stock Photo.
p. 311   Dennis Kunkel/Dennis Kunkel Microscopy, Inc.
p. 314   Jeff Greenberg/Visuals Unlimited.

Chapter 10

p. 342   left: © NYC Parks Photo Archive/Fundamentals Photographs.
         right: © Kristen Brochmann/Fundamental Photographs.

Chapter 11

p. 376   Paul Hanna/Reuters NewMedia Inc./Corbis.

Chapter 12

p. 396   © Stephen Kline/Bruce Coleman, Inc.
p. 397   Total Ozone Mapping Spectrometer/NASA.
p. 405   Natalie Fobes/Corbis
p. 407   © 1997 PhotoDisc/K. Knudson/Photolink.

Chapter 13

p. 417   © 1997 PhotoDisc/Don Tremain.
p. 420   © 1997 PhotoDisc/Tim Hall.
p. 422   Ian O'Leary/Stone.
p. 425   bottom: Alastair Shay/Papilio/Corbis.
p. 435   top: Courtesy of NASA.
p. 441   © Dianora Niccolini/Medical Images, Inc.

Chapter 14

p. 463   © Tom Pantages.
p. 468   © Al Assid/The Stock Market.
p. 473   Owen Franken/Corbis.
p. 475   Shelley Gazin/Corbis.

Chapter 16

p. 525   John Wilson White/Addison Wesley Longman.
p. 533   © Custom Medical Stock Photo.
p. 543   left: Richard Hamilton Smith/Corbis.
         right: David Toase/PhotoDisc.

Chapter 17

p. 550   Robert and Linda Mitchell/Robert and Linda
         Mitchell Photography.
p. 558   left: Warren Morgan/Corbis

Chapter 18

p. 582   Lawson Wood/Corbis.
p. 584   Ralph A. Clevenger/Corbis.
p. 586   George D. Lepp/Corbis.
p. 591   Lori Adamski Peek/Corbis.
p. 599   © C. Raines/Visuals Unlimited.
p. 601   Courtesy of the National Heart, Lung, and Blood
         Institute/National Institutes of Health.
p. 604   Custom Medical Stock Photo.
p. 606   AFP Photo/Paul Jones/Corbis.

Chapter 19

p. 624   bottom: Dr. Morley Read/Science Photo
         Library/Photo Researchers, Inc.
p. 628   top left: © Ed Drews/Photo Researchers, Inc.
p. 629   Michael S. Yamashita/Corbis.

Chapter 20

p. 643   left: Long Horn Cow/Corbis.
         top right: D. Robert & Lorri Franz/Corbis.
         bottom right: Richard Hamilton Smith/Corbis.
p. 658   Jilly Wendell/Stone.
p. 662   left: Richard Hamilton Smith/Corbis.
         middle: Eric and David Hosking/Corbis.
         right: Bruce Wilson/Stone.
p. 665   © Lewin/Royal Free Hospital/Photo Researchers,
         Inc.

Chapter 22

p. 716   © 1997 PhotoDisc/M. Freeman/Photolink.
p. 736   Morton Beebe, S.F./Corbis.

Chapter 23

p. 757   Lori Adamski Peek/Stone.
p. 772   top: Joel W. Rogers/Corbis.
p. 777   Neal Preston/Corbis.

Chapter 24

p. 805   Philip James Corwin/Corbis.

Chapter 25

p. 815   Hossler/Custom Medical Stock Photo.
p. 825   left: Gunter Marx Photography/Corbis.
         right: Chase Swift/Corbis.
p. 828   bottom left: Custom Medical Stock Photo.
         bottom right: SIU/Custom Medical Stock Photo.
p. 846   Custom Medical Stock Photo.

Glossary/Index

Some Useful Conversion Factors

Length
1 meter (m) = 100 centimers (cm)
1 meter = 1000 millimeters (mm)
1 meter = 39.4 inches (in.)
1 inch = 2.54 cm

Volume
1 liter (L) = 1000 mililiters (mL) = 1000 cm^3
1 L = 1.06 quarts (qt)
1 qt = 946 mL

Mass
1 kilogram (kg) = 1000 grams (g)
1 kg = 2.20 pounds (lb)
1 lb = 454 g

Temperature
°F = 1.8°C + 32
$°C = \dfrac{(°F - 32)}{1.8}$
K = °C + 273

Gases
1 atm = 760 mm Hg
1 mole (STP) = 22.4 L

H_2O
density = 1.00 g/1 mL
melt: 80. cal/g
boil: 540 cal/g
SH = 1 cal/g°C

Energy
1 Cal = 1 kcal = 1000 cal
1 calorie (cal) = 4.18 joules(J)

Important Common Ions

1+		2+		3+		3−		2−		1−	
H^+	hydrogen	Mg^{2+}	magnesium	Al^{3+}	aluminum	N^{3-}	nitride	O^{2-}	oxide	F^-	fluoride
Li^+	lithium	Ca^{2+}	calcium			P^{3-}	phosphide	S^{2-}	sulfide	Cl^-	Chloride
Na^+	sodium	Sr^{2+}	strontium							Br^-	bromide
K^+	potassium	Ba^{2+}	barium							I^-	iodide
Ag^+	silver	Zn^{2+}	zinc								

Important Ions with Variable Valence

1+ or 2+		2+ or 3+		1+ or 3+		2+ or 4+	
Cu^+	copper (I)	Fe^{2+}	iron (II)	Au^+	gold (I)	Sn^{2+}	tin (II)
Cu^{2+}	copper (II)	Fe^{3+}	iron (III)	Au^{3+}	gold (III)	Sn^{4+}	tin (IV)
						Pb^{2+}	lead (II)
						Pb^{4+}	lead (IV)

Important Polyatomic Ions

Cation		Anions					
1+		1−		2−		3−	
NH_4^+	ammonium ion	HCO_3^-	hydrogen carbonate (or bicarbonate)	CO_3^{2-}	carbonate	PO_4^{3-}	phosphate
		HSO_4^-	hydrogen sulfate (or bisulfate)	SO_4^{2-}	sulfate	PO_3^{3-}	phosphite
		HSO_3^-	hydrogen sulfite (or bisulfite)	SO_3^{2-}	sulfite		
		NO_3^-	nitrate				
		NO_2^-	nitrite				
		ClO_3^-	chlorate				
		ClO_2^-	chlorite				
		OH^-	hydroxide				
		HPO_4^-	dihydrogen phosphate				
		CN^-	cyanide				